W9-BUL-478

Preventive Nutrition

Fourth Edition

Nutrition and Health

Adrianne Bendich, PhD, FACN, Series Editor

For further volumes:
http://www.springer.com/series/7659

PREVENTIVE NUTRITION

The Comprehensive Guide for Health Professionals

Fourth Edition

Edited by

ADRIANNE BENDICH, Ph.D., FACN

GlaxoSmithKline Consumer Healthcare, Parsippany, NJ

and

RICHARD J. DECKELBAUM, MD, FRCP(C)

Columbia University, Institute of Human Nutrition, New York, NY

Foreword by

ALFRED SOMMER, MD, MHS

Johns Hopkins Bloomberg School of Public Health, Johns Hopkins University, Baltimore, MD

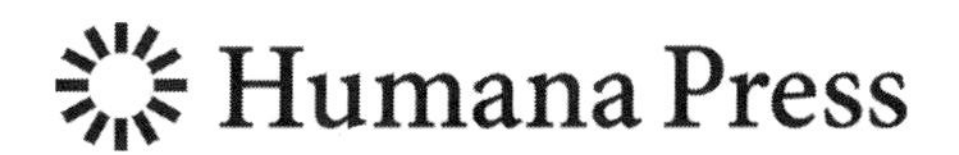

Editors
Adrianne Bendich
GlaxoSmithKline Consumer Healthcare
1500 Littleton Road
Parsippany, NJ 07054
USA
adrianne.4.bendich@gsk.com

Richard J. Deckelbaum
Institute of Human Nutrition
Columbia University
630 West 168th St.
New York, NY 10032
BHN7-702
USA
rjd20@columbia.edu

ISBN 978-1-60327-541-5 e-ISBN 978-1-60327-542-2
DOI 10.1007/978-1-60327-542-2

Library of Congress Control Number: 2009934689

Printed on acid-free paper

springer.com

Dedication

AB dedicates this volume to her core support team with deepest gratitude: David, Debbie, Debra, Elaine, Harriet, Jacob, John, Jorden, Oscar, Rebecca, and Tyler.

RJD dedicates this volume to all his students, fellows, and colleagues who have helped him understand the interdisciplinary requirements for translating basic nutrition science into policy, actions, and programs.

Series Editor Introduction

The Nutrition and Health Series of books has, as an overriding mission, to provide health professionals with texts that are considered essential because each includes (1) a synthesis of the state of the science; (2) timely, in-depth reviews by the leading researchers in their respective fields; (3) extensive, up-to-date fully annotated reference lists; (4) a detailed index; (5) relevant tables and figures; (6) identification of paradigm shifts and the consequences; (7) virtually no overlap of information between chapters, but targeted, inter-chapter referrals; (8) suggestions of areas for future research; and (9) balanced, data-driven answers to patient/health professionals' questions that are based upon the totality of evidence rather than the findings of any single study.

The goal of the Series is to develop volumes that are adopted as the standard text in each area of nutritional sciences that the volume reviews. Evidence of the success of the Series is the publication of second, third, and even fourth editions of more than half of the volumes published since the Nutrition and Health Series was initiated in 1997. The series volumes that are considered for subsequent editions have clearly demonstrated their value to health professionals. New editions provide readers with updated information as well as new chapters that contain relevant up-to-date information. Each editor of new and updated volumes has the potential to examine a chosen area with a broad perspective, both in subject matter and in the choice of chapter authors. The international perspective, especially with regard to public health initiatives, is emphasized where appropriate. The editors, whose trainings are both research and practice oriented, have the opportunity to develop a primary objective for their book, define the scope and focus, and then invite the leading authorities from around the world to be part of their initiative. The authors are encouraged to provide an overview of the field, discuss their own research, and relate the research findings to potential human health consequences. Because each book is developed de novo, the chapters are coordinated so that the resulting volume imparts greater knowledge than the sum of the information contained in the individual chapters.

Preventive Nutrition: Fourth edition is very special for me and my coeditor, Dr. Richard Deckelbaum. Each of the volumes' Table of Contents is included in the Appendix as there have been many contributors to the prior volumes, and we have added new topics as these emerge with findings that are relevant to health providers and their patients, clients, and/or family members. The overarching goal of the volumes is to provide readers with the most up-to-date and comprehensive review of the state of the science in each chapter and then to integrate the information so that the synergies between chapters are visible. The overriding driver for the timing of the fourth edition was the publishing of the results from the Women's Health Initiative (WHI), the largest placebo-controlled intervention study in postmenopausal women ever undertaken. Two chapters in this volume review the findings. The dietary modification component of WHI examined outcomes for coronary heart disease and breast and colorectal cancer and the calcium and vitamin D arm looked at osteoporotic fracture risk and colorectal cancer risk. In addition to the WHI results, there are several new chapters that expand the scope of the volume, including a chapter on human deficiency virus (HIV) infection and how this affects nutritional status; a chapter on econutrition, two on cancers of the female and male reproductive systems; two new chapters examine the role of micronutrients and brain functions such as cognitive functions and dementia and another on psychiatric conditions. There are four new chapters in the areas of obesity and diabetes; a new chapter on gastric acid levels, treatments, and nutritional consequences; a chapter describing the role of the food industry in preventive nutrition, and a final chapter for practicing health providers concerning ways to support patients in making permanent, beneficial changes in their diets.

The first chapter in the section on Global Issues begins with the thoughtful synthesis by Dr. Walter Willett of well over 100 epidemiological and intervention studies that provide the basis for a positive public health perspective on the importance of preventive nutrition strategies. The following chapter attempts to estimate the value, in dollars and cents, of currently accepted preventive nutrition strategies. Econutrition, a new chapter, is relevant to determining how food intakes can be improved in developing nations where food insecurity and undernutrition remain critical issues. Although it is difficult to develop and conduct studies on the effects of agrodiversity programs, the authors have tabulated the results of interventions around the globe and the totality of the evidence points to the value of these programs for the health of the communities. Econutrition is equally relevant in developed nations as fuel costs increase, food security issues become more prevalent, and the concept of consuming locally grown produce and locally manufactured foods is being embraced. The final chapter in this section provides an in-depth examination of the potential effects of drugs on nutritional status, including absorption, metabolism, and excretion of both macro- and micronutrients. This encyclopedic chapter, with more than 300 references, contains numerous tables with practical listings of interactions between drugs and food intake and disposition, drugs that affect body weight, drug-induced changes in blood glucose levels, drugs that alter specific vitamin levels and those that alter macro and trace mineral status.

The second section examines the importance of *Preventive Nutrition* to primary and secondary cancer prevention. The chapter on diet and childhood cancers updates the findings from the first edition and concludes that this area remains one where the data from survey studies are suggestive, but not as yet conclusive concerning the benefit of multivitamin use during pregnancy and reduced risk of two of the major childhood cancers, leukemia, and brain tumors. There is an insightful evaluation of the differences between the timing of cancer development in childhood, especially young childhood cancers versus adult cancer development. Virtually all the epidemiological studies of maternal diet, supplement use, and major childhood cancers have been tabulated for the reader. Gastric and esophageal cancers are closely related to lifestyle factors, including alcohol and tobacco use; chronic consumption of very hot liquids; salted, pickled, and smoked foods; and obesity. The link between *Helicobacter pylori* infection, lifestyle factors and diet, and development of gastric and esophageal cancers is discussed in depth; treatment guidelines for *H. pylori* are included. Even though gastric cancer incidence has fallen in the last 30 years, esophageal cancer and precancerous lesions have increased, and the studies of dietary factors that may reduce the risk of these cancers are outlined in informative tables, and the chapter includes over 200 references. The next two chapters review the major cancers of the female and male reproductive organs. Breast cancer is the most common cancer in women worldwide. As with all adult cancers, the incidence of breast and ovarian cancer increases with age. The chapter reviews the genetics, lifestyle factors, obesity and importance of exercise, female hormone status, other hormonal considerations including insulin levels, and dietary components including those found in fruits and vegetables and phytoestrogens. Prostate cancer is among the most prevalent cancers in men, and the incidence appears to be directly related to environmental factors as there is a 60-fold difference in prostate cancer rates between black men in the United States and Japanese and Chinese men in their native countries. The incidence increases with age and prostate cancer is a common cause of death in developed nations. Many dietary components have been associated with decreased risk of prostate cancer, including tomatoes and carotenoids in tomatoes, vitamin E, and selenium, yet intervention trials have not shown decreased risk when these nutrients have been provided as supplements. The chapter includes in-depth discussion of these nutrients as well as dietary fats, soy and phytoestrogens, dairy products, vitamin D and calcium, cruciferous vegetables, and zinc with respect to both survey data and intervention study findings. The last chapter in this section reviews the use of dietary supplements and their association with reduction in certain cancer risks as well as the increased risk seen with beta-carotene supplementation in cigarette smokers. Supplements reviewed include beta-carotene, vitamins C and E, selenium, calcium and folic acid, as well as multivitamins. The chapter provides the reader with critical information about the difficulties in collecting information about use of dietary supplements, the timing of their use, and the motivation for use before or after diagnosis of cancer.

Cardiovascular and cerebrovascular diseases are the focus of the third section. Long-chain omega-3 fatty acids (LC PUFA) have been associated with reduced risk of CVD in survey studies for the past 50 years, and the potential mechanisms for reduction in risk continue to be elucidated, but include anti-inflammatory, antiarrhythmic and antiproliferative actions at the cellular level and beneficial alterations in lipid profiles, especially triglycerides. This chapter contains a detailed description of the effects of LC PUFA on lipids, lipoproteins, triglycerides, apolipoprotein levels, turnover, and metabolism. New data on the role of LC PUFA in diabetics are also included. LC PUFA are examples of "good fats." The next chapter describes the adverse effects to the cardiovascular system caused by trans fats, examples of "bad fats." The majority of trans fats are found in foods that contained hydrogenated oils or contained fats that went through other production processes that resulted in their formation. The concentration of trans fats in processed foods in the United States and Europe has significantly decreased over the past few years. Nevertheless, trans fats are still part of the diets, as these are formed when foods are fried or when hydrogenated margarines are used in baked goods or in other foods. The adverse effects may last over many years. The effects of trans fats on lipid metabolism, lipoprotein oxidation, platelet aggregation, inflammatory actions, endothelial function are discussed, and excellent figures are included to help the reader understand the complexities of trans fat metabolism. The next chapter comprehensively reviews the areas of antioxidants and B vitamins and atherosclerosis. Atherosclerosis is the major cause of cardiovascular disease (CVD) and oxidative damage and inflammation of the arteries are the primary mechanisms of action. Both the survey studies and the intervention data from primary as well as secondary prevention studies are reviewed for dietary as well as supplemental intakes of vitamins E and C, beta-carotene, and B vitamins. Arterial imaging trials using angiography and ultrasound in supplemented patients are included, and the new area of nutragenomics is discussed. There is also a complete listing of the ongoing trials with antioxidants and/or B vitamins, and four comprehensive tables of all the past trials are provided for the reader. The final chapter examines the roles of the B vitamins in the prevention of cognitive decline and vascular dementia. Of note, stroke mortality has declined significantly in the United States and Canada post mandatory folate fortification. B vitamins are required for DNA synthesis, neurotransmitter metabolism and maintenance of normal homocysteine levels. Higher than normal homocysteine levels may damage brain tissues including the vasculature. The chapter includes an extensive review of the functions of folate, vitamins B6 and B12 and discusses the clinical study results that suggest that B vitamin status may affect age-related cognitive declines.

Obesity is the major risk factor for type 2 diabetes and increases the risk of CVD, some cancers, arthritis, and many other chronic conditions. One of the critical factors that led to the development of the fourth edition of *Preventive Nutrition* was the conclusion of the Women's Health Initiative (WHI) that included the major clinical study of the potential for reduction in fat intake to reduce CVD risk and risk of cancer of the colon and breast. Although this was not planned as an obesity study, 74% of women in the study were overweight (36%) or obese (38%). About 34% of the women self-reported that they had metabolic syndrome yet only about 4% were using pills or other medications to treat diabetes. All five chapters in this section on obesity and diabetes are new. The chapter that describes the WHI study and its findings was written by investigators who were part of the study and thus this important chapter provides insights into the difficulties in enrolling subjects and maintaining their enthusiasm for about 9 years. The figures and tables included in the chapter are excellent resources for anyone who wants to know more about this landmark study, and the chapter serves as an important introduction to the next four chapters that deal in depth with diabetes and obesity. The next chapter includes a detailed description of the development of diabetes and provides the health professional with an overview of carbohydrate metabolism, insulin resistance, the guidelines for the diagnosis of prediabetes, metabolic syndrome and type 2 diabetes as well as nutritional strategies for the treatment of prediabetic states and prevention of diabetes.

There is a critical review of the findings from high fat versus low fat, high carbohydrate diet studies; pharmacological studies; studies where the types of fats are altered; and studies where the types of carbohydrates are altered. Genetics is also discussed and an important discussion of gene–environmental interactions is included. Type 2 diabetes incidence increases with age as indicated by the fact that more than 40% of those with diabetes are over 65 years and over 15% are over age 75. Moreover, the risk of co-morbidities associated with diabetes including diabetic retinopathy, peripheral vascular disease and CVD, and renal failure increases with age. The chapter provides guidelines for dietary intakes to help control insulin levels as well as information about medication use, special nutrition interventions, hypoglycemia, and depression. The next two chapters are new to the volume and include the latest data on the biology of obesity and a chapter that describes the growing problem of obesity in developing nations and potential ways to address the problem. Obesity is now considered a disease that includes a chronic state of inflammation, and adipose tissue is the site of synthesis of numerous cytokines (adipokines) that are pro-inflammatory. There is an in-depth chapter that thoroughly describes the significant interactions between white adipose tissue adipokines (including leptin, adiponectin, resistin, visfatin, and adipsin) and the immune cells and cytokines (including interleukin-6 and tumor necrosis factor) involved in the inflammatory process. Consequences of this interaction include an inflammation-mediated dysregulation of insulin synthesis, insulin resistance, and type 2 diabetes ultimately. The next chapter contains an in-depth discussion of the interactions between transitions in diets and food availability, work environments, political priorities, and the effects on health outcomes in developing countries. The chapter reviews the consequences of national initiatives that have worked in the past to reduce obesity, and consequent cardiovascular disease risk by altering the types of fats that are consumed through taxation and other government programs. Focus is placed upon the potential for the educational and the medical communities' guidance for the political leadership in the transition to more healthy diets and lifestyles.

The fifth section examines strategies for prevention of major disabilities, including osteoporosis, gastric acid dysfunction, age-related eye diseases, and nutritionally related immune depression in the elderly and in other at-risk groups. The effects of calcium and vitamin D supplementation on fracture risk in the WHI trial are reviewed as are new clinical and epidemiological studies linking vitamin D to bone and other functional outcomes. Nutrients examined include calcium, vitamin D, protein, phosphorous, magnesium, vitamin K, trace minerals, as well as overall diet. Requirements for the nutrients throughout the life cycle are included. The limited positive findings from the WHI are placed in perspective, and valuable discussions of drug-induced risk factors for fracture that impact nutrient status are also included. The next chapter looks at the role of the stomach, gastric acid secretions, and the effects of reduced acid secretion on absorption of essential nutrients. This is an important new chapter as millions of adults and children are treated for acid reflux problems, and the condition as well as the treatments can alter bioavailability of certain nutrients. Recent epidemiological studies have shown an association between treatment of gastric acid reflux disease (GERD) with gastric acid suppressive drugs and altered nutritional status that may be related to increased risk of fractures and infections.

Age-related blindness is another serious disability that affects millions of seniors globally and the numbers affected will increase as populations age. Both cataract and age-related macular degeneration (AMD) risks have been reduced in individuals whose diets are rich in antioxidants such as vitamins C and E, zinc and carotenoids including beta-carotene, lutein, and zeaxanthin. The development of these diseases, the role of nutrients, including carbohydrates, in prevention of progression, and the lifestyle factors that can increase as well as decrease risk are examined in depth in the next chapter. The elderly are also at risk for loss of certain immune functions such as those linked to fighting infections and responding to vaccinations. The importance of essential micronutrients in maintaining and even enhancing immune responses in the elderly is outlined in the following chapter. The three chapters contain well-organized reviews of the functions of the immune system, the organs and the cells, their receptors and interleukins, and other molecules involved in the responses. Immunosuppression seen with micronutrient deficiencies in the elderly is also seen in the most at-risk populations of children and women of reproductive age in developing countries. The next chapter reviews the specific effects of vitamin A, zinc, iron, vitamin C, and other essential micronutrient deficiencies; the infections that are most prevalent in the undernourished populations; and it cites over 300 references, including many that document the reduction in infections when dietary intakes are improved. This chapter provides a compelling picture of the importance of public health initiatives that deal with provision of clean water, sanitation, reduction in poverty, universal education, and other health areas in addition to improvements in diets. This chapter and the one on econutrition confirm the importance of agronutrition to the health of developing nations.

Of great concern in developing countries especially in southern Africa is the epidemic of human immunodeficiency virus (HIV) infection. The volume contains a new chapter on HIV and nutrition and concentrates on the consequences of the infection on nutritional status in the adequately nourished as well as the malnourished. Most of the cited data are from studies in Africa as 2/3 of all HIV-infected persons live in Africa. The chapter examines the role of undernutrition followed by HIV infection and vice versa; the effects of antiviral therapy on nutritional status; and the pathophysiology of this disease, including the specific effects on the gastrointestinal (GI) tract. For instance, patients with HIV have increased rates of protein turnover and increased losses of nutrients through the GI tract. The synergistic adverse effects of HIV, insufficient food intake compounded with inefficient absorption of nutrients further exacerbates the reduced immune function that defines this disease. This chapter sensitively explores the effects of HIV in women of childbearing potential who become pregnant and breast feed their infants especially in Africa where infant food choices may be limited. There is also an important, balanced review of the nutritional benefits and potential risks of antiretroviral therapies. Also, there is an in-depth examination of studies in women and children, including a review of ongoing programs that include a comprehensive team approach that addresses food insecurity, disease prevention, as well as care of the HIV-infected patient and treatment for this disease and the related medical issues.

Some of the most remarkable findings in preventive nutrition have been in the understanding of the value of micronutrients in improving pregnancy outcomes, including reductions in major, serious birth defects as well as preterm births. The updated chapter continues to document the reduction in not only neural tube birth defect (NTD) prevention but also decreases in cardiovascular birth defects, reductions in cleft lip, limb reductions, and renal deformities. New data concerning the reduction in preterm births to women who have taken the folic acid-containing periconceptional multivitamin supplement are included. The Hungarian women were relatively well nourished, not teens and most were married, in contrast to most of the US cohorts who are mainly unmarried, pregnant teens. This updated chapter is of great relevance as the rate of preterm births in the United States continues to increase. Maternal nutritional status, especially undernutrition or obesity, iron, zinc, and/or folate deficiency in the first trimester, low calcium and omega-3 fatty acid intakes, lack of periconceptional multivitamin supplementation, and adverse lifestyle habits are reviewed, and their impact on risk of preterm and/or low birth weight birth outcomes is tabulated. The next and new chapter concerns prenatal nutrition and its association with adult mental health and further documents the critical importance of periconceptional and maternal nutrition including the time periods between pregnancies. Epidemiological studies have linked increased risk of adult schizophrenia in offspring when pregnant women were exposed to severe undernutrition. The chapter reviews the importance of three essential micronutrients, folate, selenium, and zinc in fetal neurodevelopment. Analysis of the data from the Dutch and Chinese famines points to increases in neurodevelopmental disorders in children, adolescents, and adults, including increased risk of addictions and psychiatric disorders. Potential mechanisms of action are reviewed including genetic and environmental effects.

A unique section that has been part of all of the *Preventive Nutrition* volumes has included informative chapters about nutritional transitions around the world. This last section contains five chapters with an in-depth review of Latin America, including South America, Asia, and developing countries. Obesity is a growing concern in all of these countries and there are significant increases in obesity in childhood. Urbanization and the increase in fast-food restaurants and prepared food choices have been associated with the increases in daily caloric intakes. Strategies to combat the rise in obesity prevalence include community physical activity programs and increased nutrition education at all grade levels. The least developed countries have the double burden of under- and overnutrition and have the least resources to deal with both problems. This chapter reviews the progress made in the lowest to highest income countries that are still considered as developing nations. The two final chapters, also new to the volume, examine the role of food companies with global reach in shaping the nutritional status of developing countries, populations in economic transition and in poverty. The global food companies support research that has resulted in safe, long-lasting foods that have incorporated essential nutrients to help assure adequate intakes that locally available food sources may not provide at sufficient levels. Collaborative efforts with the WHO and other nongovernmental organizations and governments continue to result in enhanced products for specific population groups. The chapter enumerates ongoing efforts where global food and beverage companies are contributing to improving global nutrition by developing innovative products and also by working together to develop nutritionally important advertising geared to children. The chapters point to the critical importance of reaching the Millennium Development Goals of eradicating extreme poverty and hunger and provide readers with the strategies that have been put in place to try to achieve these and other Millennium Goals. The last chapter reviews the importance of educating health providers about preventive nutrition strategies so that countries in transition can take full advantage of the consequent beneficial health effects. The chapter contains informative tables that outline the different dietary guidelines as well as guidance for assessment of patient motivation and readiness to make behavioral changes in diet/nutrient intake and access to screening tools. Finally, the chapter includes strategies for integrating preventive nutrition into overall clinical preventive services provided by the health practitioner.

Preventive nutrition initiatives and potential implementation strategies are complex, yet we, as the editors, and our authors have provided chapters that balance the most technical information with practical discussions of the importance for clients and patients as well as graduate and medical students, health professionals, and academicians. Hallmarks of the chapters include Key Points and Key Words at the beginning of each chapter, recommendations at the conclusion of each chapter, complete definitions of terms with the abbreviation fully defined, and consistent use of terms between chapters. There are over 120 relevant tables, graphs, and figures as well as more than 4,000 up-to-date references; all of the 32 chapters include highlights of major findings. The volume contains a highly annotated index and within chapters, readers are referred to relevant information in other chapters. We have chosen 68 of the most well-recognized and respected authors who are internationally distinguished researchers, clinicians, and epidemiologists who provide a broad foundation for understanding the role nutritional status, dietary intakes, life stages, disease states, and genetic factors in the implementation of the most promising preventive nutrition strategies.

In conclusion, "*Preventive Nutrition: The Comprehensive Guide for Health Professionals*, Fourth Edition" edited by Adrianne Bendich and Richard J. Deckelbaum provides health professionals in many areas of research and practice with the most up-to-date, well-referenced volume on the importance of preventive nutrition strategies for determining the potential for optimal health in populations around the world. This volume will serve the reader as the benchmark in this complex area of interrelationships between nutritional status, physiological functioning of organ systems, disease status, age, sex, public health initiatives, and critical drug–nutrient interactions. Moreover, the interactions between genetic and environmental factors and the numerous co-morbidities seen especially in the aging population are clearly delineated so that students as well as practitioners can better understand the complexities that can impact the success of preventive nutrition strategies. Dr. Deckelbaum and I thank Dr. Alfred Sommer for his eloquent Foreword, and we applaud our outstanding authors for their efforts to develop the most authoritative resource in the field to date, and this excellent text is a very welcome addition to the Nutrition and Health Series.

Adrianne Bendich, PhD, FACN

Foreword

The title says it all. This volume has become the go-to, comprehensive review of our understanding of the relationships between nutrition, health, and disease. It is only 4 years since the last edition, but relevant data, and changes in the state of global health and nutrition, have changed greatly in the interim. True knowledge, on the other hand, has not progressed nearly as much; all the more reason why nutritionists and health professionals will benefit from this thoughtfully updated tome.

As I can't avoid an opportunity to be provocative, and must avoid plagiarizing my foreword from previous editions, I ask that the reader reflects on some aspects of the nature of nutrition and health, as I have come to view them.

Several serious relationships are obvious: some people are obese (almost always because they eat too much) and some are thin (and sometimes starving). These two macronutritional states, and their relationship to diet, are pretty obvious. So, too, are the classic nutritional deficiency states (iron deficiency anemia, rickets, scurvy, beri–beri, pellagra, goiter and associated iodine deficiency states, xerophthalmia). After these, the ice gets a little thin. Yes, we now know the importance of insuring that women receive an "adequate" supply of folic acid preconceptionally to reduce the risk of a host of congenital anomalies in their offspring; in most societies, including the United States, this requires supplementation (or, as in the United States, fortification of common dietary staples). We've also learned in recent years that most children (and their mothers!) in the developing world are deficient in vitamin A, and improving their vitamin A status reduces their risk of both blindness and death. But we also know that the reason most children (and their mothers) in the West are vitamin A sufficient is that they receive supplemental, nondietary sources of this important micronutrient.

So, here is the conundrum. If, as many would have us believe, everyone would achieve maximum health if only they consumed a "balanced" diet, why must the West use supplements to achieve physiologically adequate intakes of vitamin A, folic acid, and in many areas, iodine? How did my perfectly healthy, well-nourished daughter become anemic from iron deficiency within 8 months of moving to Eritrea as a Peace Corps Volunteer? Indeed, what constitutes a "balanced" diet, and has it ever occurred in nature on its own?

Many of this compendium's chapters attempt to tease apart the bits of evidence that relate specific nutrients (sometimes in specific amounts) to specific clinical and "health" outcomes. As the authors and the book's editors acknowledge, this is all heady stuff, and we must be grateful that the authors have taken on the challenge. But, as they regularly make clear, most of the existing evidence is far from conclusive and rarely provides a solid basis upon which to provide prescriptive advice, or even any but the vaguest of guidelines, for "healthy living." Thanks to studies like the Women's Health Initiative, which this volume appropriately deals with in considerable depth, we are beginning to obtain the objective evidence for action that only randomized, controlled trials can provide.

The most immediate nutrition–health dyads stare us in the face daily, whether it is the explosion in obesity and the related epidemic of "metabolic syndrome" or insufficiencies of macro- and micronutrition that accompany food insecurity in the developing world (or the worst paradox of them all: the "double burden" of "malnutrition" suffered by poor countries, whose elites have grown obese on calorie-dense fast foods while their masses starve). But there are, in addition, tantalizing new syndromes few of us had previously even dreamed of. The "Barker Hypothesis" is a "hypothesis" no longer; instead it now undergirds the swelling number of investigators attempting to unravel the seemingly implacable impact that fetal nutrition has many decades later on adult, chronic diseases—an impact that may well be passed along from generation to generation.

The "Holy Grail" of "preventive nutrition" remains what Jim Fries first termed "the compression of morbidity," in his classic paper of 1980: healthier living (and eating) that minimizes morbidity until that brief interval preceding death. In Fries' prescient vision, nothing we do will extend the natural lifespan (which he estimated to average roughly 85 years), but it can help to ensure we live a longer period of disability-free existence.

People seek (and deserve) the best advice available from their health professionals. Given the growing numbers of chronic diseases and the direct and indirect costs associated with them, relatively inexpensive nutritional interventions need to be discovered, promulgated, and actually employed. This latest edition of *Preventive Nutrition* provides a comprehensive integration of the best available data on the role of nutrition in health. It is an excellent place to discover what still needs scientific inquiry and what health professionals might reasonably advise their patients and the public.

Alfred Sommer, MD, MHS

Preface

Preventive Nutrition was first published in 1997; the second edition appeared in 2001. Also in 2001, we edited a volume entitled *Primary and Secondary Preventive Nutrition*. In 2005, we published the third edition that was greatly expanded to include topics in *Primary and Secondary Preventive Nutrition*. The volumes were well accepted as critical resources for medical school libraries and health care professionals, including physicians, nutritionists, dieticians, nurses, pharmacists. The volumes were also used by academic researchers as the basis for developing graduate courses in preventive nutrition; *Preventive Nutrition* has been adopted as a text in many of the most highly rated graduate nutrition programs in the United States, Europe, South America, and Asia. Copies of chapters from the earlier editions are available at the Springer website (springer.com), and this is important as these remain valuable resources and can help to fill any gaps in subsequent volumes since each edition contains unique chapters, and new chapters covering emerging issues are added to each edition.

One overriding factor that necessitated the development of the fourth edition was the inclusion of the results from the Women's Health Initiative (WHI), the largest intervention study ever undertaken and completed in postmenopausal women. Additionally, we have introduced many new, timely chapters that review the totality of the evidence concerning econutrition, micronutrients and major cancers, cognitive function, and other mental health areas, the importance of gastric acid secretions and the nutritional effects of current therapies, and the latest information on the biology of obesity and its relationship to type 2 diabetes. Two other new chapters are included: one examines the potential for implementing preventive nutrition strategies with the development of new foods and the last chapter provides practical information to health providers about behavioral strategies to help assure compliance and therefore the health benefits of preventive nutrition. Development of the fourth edition also provided our world-renown authors the ability to update their excellent chapters. In toto, there are 32 chapters with more than 4,000 references and over 100 informative tables and figures.

Our mission for the volume continues to be the resource for health professionals with up-to-date, comprehensive reviews that evaluate the dietary practices and interventions that have been shown to reduce disease risk and/or to improve health outcomes. Evidence-based nutritional interventions are a critical component of preventive medicine approaches to prevent an increasing number of diseases and diminish their consequences. Within the past 5 years, many studies have demonstrated the potential for optimizing macro- and micronutrient intakes not only to improve individual outcomes throughout the lifecycle, from fetal development through aging, but also to reduce individual and national health care costs.

The fourth edition is comprehensive and well organized. The volume is divided into seven sections beginning with an overview of (1) global issues affecting preventive nutrition; (2) cancer prevention; (3) cardiovascular disease prevention; (4) diabetes and obesity; (5) prevention of major disabilities and improvement in health outcomes; (6) optimal pregnancy and infancy outcomes; (7) nutrition transitions around the world. Key Points are included at the beginning of each chapter to help define the chapter's critical learning points. There is a list of Key Words and the final section provides Recommendations. The fourth edition of *Preventive Nutrition* continues to include a comprehensive index as well as helpful listings of relevant books, journals, and websites.

In acknowledgement of the growing commonality of nutrition issues around the world, we begin this volume with an examination of Global Issues that affect public health and disease risks. There is an important comprehensive chapter that examines the effects of drug–nutrient interactions. In the section on cancer, there is a comprehensive updating of the components in the diet that affect cancer risk, including an overview on information relating to cancers of the male and female reproductive tracts. In the cardiovascular disease section, there are comprehensive revisions to the chapters on antioxidants, trans fats, and the long chain omega-3 fatty acids, and a new chapter on B vitamins and vascular dementia has been included.

Diabetes and obesity are major health concerns that are significantly impacted by diet and dietary components. We have included the comprehensive review of the findings from the WHI study, including a discussion of the colon cancer data. Other new chapters in this section provide a review of the effects type 2 diabetes and obesity in the elderly and impacts on cardiovascular health. There is a new chapter that provides insights into the inflammatory processes associated with obesity and another that looks at obesity's health care costs in developing countries.

In the section on the prevention of major disabilities, there is a new chapter on cataract and age-related macular degeneration, antioxidants, and carbohydrate intakes. There are updated chapters on the importance of calcium, vitamin D, protein, and other nutrients in protecting bones from fractures—this chapter also reviews the results for the calcium and vitamin D intervention in the WHI study. Recent studies on the importance of micronutrients in maintaining the immune function of healthy older adults are included in the updated chapter. There is a revised chapter from the second edition that examines the effects of micronutrient deficits on the immune function of the very young in developing countries and the efforts to restore micronutrient status. A new chapter reports on recent nutritional studies in developing countries where a high proportion of children and their mothers are infected with HIV.

Another major objective of the volume is to provide the latest compelling data on the critical importance of periconceptional nutritional status for the prevention of birth defects and optimization of gestation. This area is gaining in importance as low birth weight and size, and preterm delivery are being linked to risk of chronic diseases in adulthood. One of the critical nutrient-based discoveries of the last century is the ability to prevent neural tube birth defects by folic acid supplementation. The updated chapter contains the latest data that confirm and strengthen the original findings that periconceptional supplementation with a folic acid-containing prenatal supplement significantly reduced the risk of first time neural tube birth defects as well as cardiovascular and other serious birth defects and also new data on reduced risk of preterm delivery. The next comprehensive chapter reviews the significant association between micronutrient status and risk of preterm delivery and low birth weight outcomes. The final chapter in this section deals with the findings of increased risk of mental illness in children of mothers who lived through famines during their pregnancies. The psychiatric illnesses in the exposed children may not begin until adulthood.

Preventive Nutrition contains important chapters for investigators who are interested in the impact of nutrition on world health in the unique section on Nutrition Transitions Around the World. This section examines the successes and consequent public health implications of national preventive nutrition strategies, not only in the United States and Europe but also in "Westernizing" nations and less developed countries. There are two insightful chapters on the growing presence of obesity in Latin and South America as well as across Asia. The chapters confirm that westernization of eating habits has negatively impacted the health of these nations.

The last two chapters are new to the volume and these serve as very useful resources for many of our readers. There is now a chapter concerning the role of global food companies, their development of safe, nutritious foods, and the implementation of preventive nutrition around the world. The last chapter is complementary in that it reviews the behavioral changes needed for individuals to choose the diets and nutrients that can best improve their health and reduce their risks of chronic disease. Many of our readers are involved in the development of curricula for graduate and medical schools, and these unique chapters can help by providing new perspectives to students and faculty.

The fourth edition provides access to practicing health professionals, including physicians, nutritionists, dentists, pharmacists, dieticians, health educators, policy makers, and research investigators to the recent original research demonstrating that the risk of major diseases can be prevented, or at least delayed, with simple nutritional approaches. Many health professionals are frequently asked about recent findings on vitamins or other nutrients mentioned in their local newspaper or on the evening news. Patients are looking for credible information from their health care providers on ß-carotene, lycopene, antioxidants, soy, fiber, folate, calcium, and the myriad of bioactive phytochemicals, such as those found in garlic and other foods, and *Preventive Nutrition* is a balanced, data-driven resource for health professionals.

As the demographics of US and European populations change and become more multicultural, it is increasingly important for health professionals to understand the nutritional backgrounds and diversities of their patients and clients. As important, there may be significant national dietary initiatives that provide roadmaps for effective implementation of preventive nutrition within an overall strategy of health improvement, especially for vulnerable members of the population. As editors, we have endeavored to include the most compelling areas of nutrition research that can have immediate impact on the health and education of our readers. The fourth edition of *Preventive Nutrition* can serve as a valued resource to health professionals and students who are interested in reducing disease risk and enhancing health by implementing the well-documented preventive nutrition strategies outlined in this volume.

We thank Richard Hruska, Paul Dolgert, and the staff at Springer/Humana Press for their confidence in our vision and their assistance in bringing this volume to fruition. We remain grateful to our authors for their insightful chapters and to Dr. Alfred Sommer for his eloquent Foreword.

Adrianne Bendich, PhD, FACN
Richard J. Deckelbaum, MD, FRCP(C)

Contents

Dedication v

Series Editor Introduction vii

Foreword xv

Preface xvii

Contributors xxv

PART I: GLOBAL ISSUES

1 Public Health Benefits of Preventive Nutrition 3
Walter C. Willett

2 Health Economics of Preventive Nutrition 23
Adrianne Bendich and Richard J. Deckelbaum

3 Econutrition: Preventing Malnutrition with Agrodiversity Interventions 51
Bram P. Wispelwey and Richard J. Deckelbaum

4 Nutrition in the Age of Polypharmacy 79
Lindsey R. Lombardi, Eugene Kreys, Susan Gerry, and Joseph I. Boullata

PART II: CANCER PREVENTION

5 Diet and Childhood Cancer: Preliminary Evidence 127
Greta R. Bunin and Jaclyn L.F. Bosco

6 Prevention of Upper Gastrointestinal Tract Cancers 145
Elizabeth T.H. Fontham and L. Joseph Su

7 Factors in the Causation of Female Cancers and Prevention 175
Donato F. Romagnolo and Ivana Vucenik

8 The Role of Nutrition and Diet in Prostate Cancer 195
Lorelei Mucci and Edward Giovannucci

9 Dietary Supplements and Cancer Risk: Epidemiologic Research and Recommendations 219
Marian L. Neuhouser and Cheryl L. Rock

Part III: Cardiovascular Disease Prevention

10 *N*–3 Fatty Acids from Fish and Plants: Primary and Secondary Prevention of Cardiovascular Disease 249
William E. Connor and Sonja L. Connor

11 Cardiovascular Effects of *Trans* Fatty Acids 273
Ronald P. Mensink

12 Antioxidant and B-vitamins and Atherosclerosis 285
Juliana Hwang-Levine, Wendy J. Mack, and Howard N. Hodis

13 B Vitamins in the Prevention of Cognitive Decline and Vascular Dementia 325
Kristen E. D'Anci and Irwin H. Rosenberg

Part IV: Diabetes and Obesity

14 The Women's Health Initiative: Lessons for Preventive Nutrition 337
Cynthia A. Thomson, Shirley A.A. Beresford, and Cheryl Ritenbaugh

15 Role of Nutrition in the Pathophysiology, Prevention, and Treatment of Type 2 Diabetes and the Spectrum of Cardiometabolic Disease 371
Cristina Lara-Castro and W. Timothy Garvey

16 Nutrition, Metabolic Syndrome, and Diabetes in the Senior Years 389
Barbara Stetson and Sri Prakash L. Mokshagundam

17 Adipokines, Nutrition, and Obesity 419
Melissa E. Gove and Giamila Fantuzzi

18 Diet, Obesity, and Lipids: Cultural and Political Barriers to Their Control in Developing Economies 433
Henry Greenberg, Anne Marie Thow, Susan R. Raymond, and Stephen R. Leeder

Part V: Prevention of Major Disabilities; Improvement in Health Outcomes

19 Diet, Osteoporosis, and Fracture Prevention: The Totality of the Evidence 443
Robert P. Heaney

20 Gastric Acid Secretions, Treatments, and Nutritional Consequences 471
Ronit Zilberboim and Adrianne Bendich

21 Nutritional Antioxidants, Dietary Carbohydrate, and Age-Related Maculopathy and Cataract 501
Chung-Jung Chiu and Allen Taylor

22 Micronutrients and Immunity in Older People 545
John D. Bogden and Donald B. Louria

23 Micronutrients: Immunological and Infection Effects on Nutritional Status and Impact on Health in Developing Countries 567
Ian Darnton-Hill and Faruk Ahmed

24 HIV and Nutrition 611
Kevin A. Sztam and Murugi Ndirangu

Part VI: Optimal Pregnancy/Infancy Outcomes

25 Folic Acid/Folic Acid-Containing Multivitamins and Primary Prevention of Birth Defects and Preterm Birth 643
Andrew E. Czeizel

26 Maternal Nutrition and Preterm Delivery 673
Theresa O. Scholl

27 Linking Prenatal Nutrition to Adult Mental Health 705
Kristin Harper, Ezra Susser, David St. Clair, and Lin He

Part VII: Nutrition Transitions Around the World

28 Nutritional Habits and Obesity in Latin America: An Analysis of the Region 723
Jaime Rozowski, Oscar Castillo, Yéssica Liberona, and Manuel Moreno

29 Effects of Western Diet on Risk Factors of Chronic Disease in Asia 743
Koji Takemoto, Supawadee Likitmaskul, and Kaichi Kida

30 Goals for Preventive Nutrition in Developing Countries 757
Osman M. Galal and Gail G. Harrison

31 Preventive Nutrition and the Food Industry: Perspectives on History, Present, and Future Directions 769
Derek Yach, Zoë Feldman, Dondeena Bradley, and Robert Brown

32 The Role of Preventive Nutrition in Clinical Practice 793
A. Julian Munoz, Jamy D. Ard, and Douglas C. Heimburger

Appendix A: Table of Contents from Volumes 1 to 3 and Primary and Secondary Preventive Nutrition 823

Table of Contents of Second Edition 827

Table of Contents of 3rd Edition 829

Appendix B: Nutrition and Health Series 833

Appendix C: Books Related to Preventive Nutrition 835

Appendix D: Web Sites of Interest 839

Subject Index 843

Contributors

FARUK AHMED, BSC (HONORS), MSC, PHD • *Department of Family Sciences, College for Women, Kuwait University, Safat, Kuwait; Senior Lecturer (Adjunct), School of Population Health,University of Queensland, Brisbane, Australia*
JAMY D. ARD, MD • *Departments of Nutrition Sciences and Medicine, University of Alabama at Birmingham, UAB Department of Nutrition Sciences, 1675 University Blvd, Webb 439, Birmingham, AL 35294-3360, USA*
ADRIANNE BENDICH, PHD, FACN • *GSK Consumer Healthcare, Parsippany, NJ 07054, USA, adrianne.4.bendich@gsk.com*
SHIRLEY A. A. BERESFORD, PHD • *Department of Epidemiology, University of Washington, Seattle, Washington, DC 98199, USA*
JOHN D. BOGDEN, PHD • *Department of Preventive Medicine and Community Health, UMDNJ-New Jersey Medical School, Newark, NJ 07101-1709, USA, bogden@umdnj.edu*
JACLYN L. F. BOSCO, MPH • *Department of Epidemiology, Boston University School of Public Health, Boston, MA 02118, USA*
JOSEPH I. BOULLATA, PHARM.D, RPH, BCNSP • *Associate Professor of Pharmacology & Therapeutics, University of Pennsylvania and Clinical Specialist in Nutrition Pharmacotherapy, Hospital of the University of Pennsylvania, 418 Curie Boulevard, Philadelphia, PA 19104, USA, boullata@nursing.upenn.edu*
DONDEENA BRADLEY, PHD • *VP Nutrition, PepsiCo Headquarters, 700 Anderson Hill Road, Purchase, New York 10577, USA*
BOB BROWN, PHD MPH • *PepsiCo Headquarters, 700 Anderson Hill Road, Purchase, New York 10577, USA*
GRETA R. BUNIN, PHD • *Research Associate Professor, Children's Hospital of Philadelphia, Philadelphia, PA 19104, USA, bunin@email.chop.edu*
OSCAR CASTILLO, MSC • *Department of Nutrition, Nutrition Intervention Unit, Diabetes and Metabolism, Pontificia Universidad Católica de Chile, Santiago, Chile*
CHUNG-JUNG CHIU, PHD • *Laboratory for Nutrition and Vision Research, USDA Human Nutrition Research Center on Aging, Tufts University, Boston, MA 02111, USA*
SONJA L. CONNOR, MS, RD, LD • *Division of Endocrinology, Diabetes and Clinical Nutrition, Oregon Health Sciences University, Portland, OR 97239-3098, USA*
WILLIAM E. CONNOR, MD • *Division of Endocrinology, Diabetes and Clinical Nutrition, Oregon Health Sciences University, Portland, OR 97239-3098, USA*
ANDREW E. CZEIZEL • *Foundation for the Community Control of Hereditary Diseases, Törökvész lejtő 32, Budapest H-1026, Hungary, czeizel@interware.hu*
KRISTEN E. D'ANCI, PHD • *Department of Psychology , Jean Mayer USDA Human Nutrition Research Center on Aging, Tufts University, Medford, MA 02155, USA*
IAN DARNTON-HILL, MD, MPH, MSC • *Friedman School of Nutrition Science and Policy, Tufts University, Boston, USA; Special Adviser to the Executive Director on Ending Child Hunger & Undernutrition UNICEF, 3 United Nations Plaza, New York 10017, USA, idarntonhill@unicef.org*

RICHARD J. DECKELBAUM, MD • *Department of Pediatrics & Institute of Human Nutrition, College of Physicians and Surgeons, Columbia University, New York 10032, USA, rjd20@columbia.edu*

GIAMILA FANTUZZI, PHD • *Department of Kinesiology and Nutrition, University of Illinois at Chicago, Chicago, IL 60612, USA, giamila@uic.edu*

ZOË FELDMAN, MPH • *Consultant for Global Nutrition and Public Policy, Pepsico Headquarters, 700 Anderson Hill Road, Purchase, New York 10577, USA*

ELIZABETH T. H. FONTHAM, MPH, DRPH • *Louisiana State University Health Sciences Center, School of Public Health, New Orleans, LA 70112, USA, efonth@lsuhsc.edu*

OSMAN M. GALAL, MD, PHD • *Department of Community Health Sciences, School of Public Health, University of California, Los Angeles, CA 90095, USA, ogalal@ucla.edu*

W. TIMOTHY GARVEY, MD • *Department of Nutrition Sciences, University of Alabama at Birmingham; UAB Diabetes Research and Training Center, GRECC Investigator and Staff Physician, Birmingham VA Medical Center, 1675 University Boulevard, Birmingham, AL 35294-3360, USA, garveyt@uab.edu*

SUSAN GERRY, PHARM.D • *Clinical Pharmacist, Hospital of the University of Pennsylvania, Philadelphia, PA, 19104, USA*

EDWARD GIOVANNUCCI, MD, SCD • *Departments of Epidemiology and Nutrition, Harvard School of Public Health, Boston, MA, Boston, MA 02111, USA; Channing Laboratory, Brigham and Women's Hospital, Harvard Medical School, Boston, MA 02111, USA, egiovann@hsph.harvard.edu*

MELISSA E. GOVE • *Department of Kinesiology and Nutrition, University of Illinois at Chicago, Chicago, IL 60612, USA*

HENRY GREENBERG, MD • *St. Luke's Roosevelt Hospital Center Roosevelt Division, 1000 Tenth Ave, 3B-30, New York 10019, USA, hmg1@columbia.edu*

KRISTIN HARPER, PHD, MPH • *Robert Wood Johnson Health & Society Scholar Program, Columbia University, New York 10019, USA*

GAIL G. HARRISON, PHD • *Department of Community Health Sciences, UCLA Center for Health Policy Research, School of Public Health, University of California, Los Angeles, CA 90024, USA, gailh@ucla.edu*

JULIANA HWANG-LEVINE, PHARM.D • *Department of Pharmacology and Pharmaceutical Sciences, Atherosclerosis Research Unit, School of Pharmacy, University of Southern California, Los Angeles, CA 90033, USA, julianah@usc.edu*

LIN HE, PHD • *Institute for Nutritional Sciences, Shanghai Institutes for Biological Sciences, Shanghai, China; Shanghai JiaoTong University, Shanghai, China*

ROBERT P. HEANEY, MD • *Department of Medicine, Creighton University, Omaha, Nebraska 68178, USA, rheaney@creighton.edu*

DOUGLAS C. HEIMBURGER, MD, MS • *Departments of Nutrition Sciences and Medicine, University of Alabama at Birmingham, UAB Department of Nutrition Sciences, 1675 University Boulevard, Birmingham, AL 35294-3360, USA, dheimbur@uab.edu*

HOWARD N. HODIS, MD • *Division of Cardiovascular Medicine, Keck School of Medicine, University of Southern California, Los Angeles, CA 90033, USA, athero@usc.edu*

KAICHI KIDA • *Department of Pediatrics, Ehime University School of Medicine, Shigenobu, Japan*

EUGENE KREYS, PHARM.D • *Clinical Pharmacist, University of Pittsburgh Medical Center, Pittsburgh, PA, USA*

CRISTINA LARA-CASTRO, MD, PHD • *Department of Nutrition Sciences, University of Alabama at Birmingham, Birmingham, AL 35294-3360, USA; The Birmingham Veterans Affairs Medical Center, Birmingham, AL 35233, USA*
STEPHEN R. LEEDER, MD, PHD • *Menzies Centre for Health Policy, Australian Health Policy Institute; University of Sydney, Sydney, Australia*
YESSICA LIBERONA, MSC • *Department of Nutrition, Nutrition Intervention Unit, Diabetes and Metabolism, Pontificia Universidad Católica de Chile, Santiago, Chile*
SUPAWADEE LIKITMASKUL, MD • *Department of Pediatrics, Siriraji Hospital, Bangkok, Thailand*
LINDSEY R. LOMBARDI, PHARM.D • *Resident in Oncology Pharmacotherapy, Hospital of the University of Pennsylvania, Philadelphia, PA, 19014, USA*
DONALD B. LOURIA, MD • *Department of Preventive Medicine and Community Health, UMDNJ-New Jersey Medical School, Newark, NJ 07101-1709, USA*
WENDY J. MACK, PHD • *Department of Preventive Medicine, Atherosclerosis Research Unit, Keck School of Medicine, University of Southern California, Los Angeles, CA 90033, USA, wmack@usc.edu*
RONALD P. MENSINK, MD • *Department of Human Biology, Maastricht University, Postbus 616, 6200 MD, Maastricht, The Netherlands, r.mensink@hb.unimaas.nl*
SRI PRAKASH L. MOKSHAGUNDAM, MD • *Division of Endocrinology and Metabolism, University of Louisville, 530 S. Jackson Street, Louisville, KY 40202, USA*
MANUEL MORENO, MD • *Department of Nutrition, Diabetes and Metabolism, Faculty of Medicine, Pontificia Universidad Católica de Chile, Santiago, Chile*
LORELEI MUCCI, SCD • *Department of Epidemiology, Channing Laboratory, Brigham and Women's Hospital, Harvard Medical School, Boston, MA 02111, USA, lmucci@hsph.harvard.edu*
A. JULIAN MUNOZ, MD, MSPH • *Departments of Nutrition Sciences and Medicine, UAB Department of Nutrition Sciences, University of Alabama at Birmingham, 1675 University Boulevard, Birmingham, AL 35294-3360, USA*
MURUGI NDIRANGU, PHD • *Division of Nutrition, School of Health Professions, College of Health and Human Sciences, Georgia State University, Atlanta, GA 30302-3995, USA*
MARIAN L. NEUHOUSER, PHD, RD • *Cancer Prevention Program, Fred Hutchinson Cancer Research Center, 1100 Fairview Avenue North, M4B402 Seattle, WA 98109-1024, USA, mneuhous@fhcrc.org*
SUSAN R. RAYMOND, PHD • *Institute of Human Nutrition, Columbia University College of Physicians and Surgeons, Changing our World Inc, New York 10019, USA*
CHERYL RITENBAUGH, PHD, MPH • *Department of Family and Community Medicine and Nutritional Sciences, University of Arizona, Tucson, AZ 85721-0038, USA*
CHERYL L. ROCK, PHD, RD • *Department of Family and Preventive Medicine, School of Medicine, University of California, San Diego, La Jolla, CA 92093-0901, USA, clrock@ucsd.edu*
DONATO F. ROMAGNOLO, PHD • *Department of Nutritional Sciences and BIO5 Research Institute, The University of Arizona, Tucson, AZ 85721-0038, USA, donato@u.arizona.edu*
IRWIN H. ROSENBERG, MD • *Jean Mayer USDA Human Nutrition Research Center on Aging, Gerald J. and Dorothy R. Friedman, School of Nutrition Science and Policy, Tufts University, Boston, MA 02111, USA, irwin.rosenberg@tufts.edu*
JAIME ROZOWSKI, PHD • *Department of Nutrition, Diabetes and Metabolism, Faculty of Medicine, Pontificia Universidad Católica de Chile, Santiago, Chile, jrozowski@puc.cl*
THERESA O. SCHOLL, PHD, MPH • *Department of OB/GYN, University of Medicine and Dentistry of New Jersey-SOM, Science Center, Stratford, NJ 08104, USA*

DAVID ST. CLAIR, MD, PHD • *Department of Mental Health, University of Aberdeen, Foresterhill, Aberdeen, Scotland, UK*

BARBARA STETSON, PHD • *Department of Psychological and Brain Sciences, 317 Life Sciences Building, University of Louisville, Louisville, KY 40292, USA, barbara.stetson@louisville.edu*

L. JOSEPH SU, PHD, MPH • *School of Public Health, Louisiana State University Health Sciences Center, New Orleans, LA 70112, USA*

EZRA SUSSER, MD, DRPH • *Department of Epidemiology, Mailman School of Public Health, Columbia University, New York 10032, USA, ess8@columbia.edu*

KEVIN A. SZTAM, MD, MPH • *Division of Gastroenterology and Nutrition, Harvard Medical School, Children's Hospital Boston, Hunnewell Ground, Boston, MA 02111, USA, Kevin.Sztam@childrens.harvard.edu*

KOJI TAKEMOTO, MD • *Department of Pediatrics, Ehime University School of Medicine, Shigenobu, Japan, takemoto@m.ehime-u.ac.jp*

ALLEN TAYLOR, PHD • *Laboratory for Nutrition and Vision Research, USDA Human Nutrition Research Center on Aging, Tufts University, Boston, MA 02111, USA, allen.taylor@tufts.edu*

CYNTHIA A. THOMSON, PHD, RD • *Nutritional Sciences, Public Health and Medicine, University of Arizona, Tucson, AZ 85721-0038, USA, cthomson@email.arizona.edu*

ANNE MARIE THOW, MPP • *Australian Health Policy Institute, University of Sydney, Sydney, Australia*

IVANA VUCENIK, PHD • *Department of Medical and Research Technology, University of Maryland School of Medicine, Baltimore, MD 21201, USA, ivucenik@som.umaryland.edu*

WALTER C. WILLETT, MD, MPH • *Department of Nutrition, Harvard University School of Public Health, Boston, MA 02115, USA, wwillett@sph.harvard.edu*

BRAM WISPELWEY, MS • *Institute of Human Nutrition, College of Physicians and Surgeons, Columbia University, New York 10032, USA*

DEREK YACH, MBCHC, MPH • *VP Global Health Policy, Pepsico Headquarters, 700 Anderson Hill Road, Purchase, New York 10577, USA, derek.yach@pepsico.com*

RONIT ZILBERBOIM, PHD • *Principle Clinical Scientist, Medical Affairs, GSK Consumer Healthcare, Parsippany, NJ 07054, USA*

I Global Issues

1 Public Health Benefits of Preventive Nutrition

Walter C. Willett

Key Points

- Staying lean and physically active throughout adult life has major health benefits.
- Diets low in the percentage of energy from fat have not been associated with lower risks of heart disease, cancer, or better long-term weight control.
- Avoiding industrially produced trans fat, keeping saturated fat low, and emphasizing unsaturated fats will minimize risks of heart disease and type 2 diabetes.
- Consuming grains in their original high fiber/whole grain form is likely to reduce risk of type 2 diabetes and heart disease. Consumption of sugary beverages increases risk of type 2 diabetes and probably heart disease. High intake of fruits and vegetables will help prevent risks of cardiovascular disease, but the benefits for cancer reduction appear modest.
- High consumption of alcohol and alcoholism have many adverse health and social consequences, and intakes as low as one drink per day or less are associated with greater risks of breast cancer. In contrast, moderate consumption of alcohol reduces risks of coronary heart disease and type 2 diabetes.

Key Words: Diet; nutrition; health; preventation; cancer; heart disease

1.1. INTRODUCTION

Until very recently, most populations had no choice but to consume foods that were produced locally, and availability was often extremely seasonal. This resulted in diets that were highly variable across the globe; for example, in some Arctic climates, almost no carbohydrates, fruits, or vegetables were consumed, and diets consisted mainly of fat and protein from animal sources. In other regions, populations subsisted on primarily vegetarian diets with the large majority of calories from carbohydrate sources. The fact that humans could survive and reproduce with such varied dietary patterns is a testimony to the adaptability of human biology. Yet, disease rates and overall mortality varied dramatically among these various populations, and formal studies of these relationships provided early clues about the importance of diet in human health and disease; these "ecological" studies are described in more detail below.

In the last few decades, enormous changes have occurred in the diets of most populations. These changes were due to combination of increased wealth of some groups, new processing and preservation technologies, and greatly expanded transportation infrastructures. Collectively, these changes have allowed foods to be transported across and among continents and to be available virtually the whole year. At first, these changes globally were described as the "westernization" of diets because of increases in meat, dairy products, and processed foods. However, many of the more recent changes are not necessarily toward the diets of western countries, but instead emphasize refined starches, sugar

A. Bendich, R.J. Deckelbaum (eds.), *Preventive Nutrition*, Nutrition and Health, DOI 10.1007/978-1-60327-542-2_1,

and sugary beverages, and partially hydrogenated vegetable fats. These patterns, which have been described as "industrial diets," are usually the cheapest source of calories, and they have permeated poor populations of both rural and urban countries around the world.

The recent changes in diets, along with changes in physical activity and tobacco use, have profoundly affected rates of disease, sometimes positively but often adversely. On the positive side, we have seen dramatic declines in rates of coronary heart disease (CHD) in many western populations, where this has been the leading cause of death. For example, in Finland, which at one time had the highest rates of CHD, mortality from this cause has declined by more than 80% (*see* Fig. 1.1) *(1)*. On the other hand, in Japan, formerly a country with very low rates of colon cancer, rates of this malignancy have increased greatly and now have surpassed those of the United States *(2)*. Most importantly, at present an epidemic of obesity and diabetes has affected almost all the world's populations, rich and poor. This epidemic, which could reverse important gains in life expectancy *(3)*, is likely to be the greatest challenge to public health in this century, unless an unforeseen problem emerges.

In this brief overview, I will address the components of diet and nutrition that have well-documented relationships to human health and disease. The focus is on the prevention of major illness in adults, and most of the evidence is based on studies of diet during midlife and later. A fundamental conclusion is that the vast majority of deaths due to coronary heart disease, stroke, diabetes, and some important

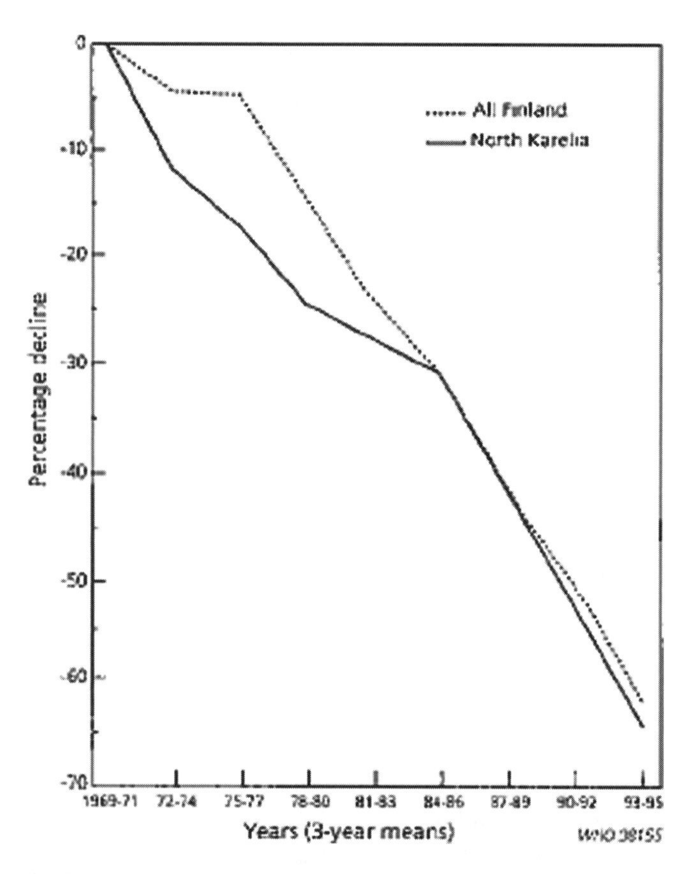

Fig. 1.1. Percentage decline in age-adjusted coronary mortality of 35–64-year-old males in Finland *(1)*.

cancers are preventable by healthy diets in combination with regular physical activity and avoidance of tobacco *(4)*. The relation of diet during pregnancy, infancy, and early childhood to childhood mortality, unfortunately still a major issue in many poor countries, has been extensively studied and is addressed in other chapters of this volume. This study builds on earlier reviews *(5)*, emphasizing newer evidence and understanding.

1.2. SOURCES OF EVIDENCE

Traditionally, animal experiments and small human metabolic studies formed the basis of dietary recommendations. Inevitably, the study of chronic disease in humans has required epidemiologic approaches. Initially, investigations compared dietary intakes and disease rates among populations in various countries, which were termed ecological studies. These analyses highlighted the large differences in disease rates worldwide and provided many hypotheses; however, such studies are limited as many other factors besides diet vary across cultures and the data are inherently aggregated. The next generation of studies was primarily case–control investigations, which mainly examined dietary factors retrospectively in relation to risk of cancer and other diseases. Now, large prospective studies of many thousands of persons are beginning to provide data based on both biochemical indicators of diet and dietary questionnaires that have been rigorously validated *(6)*. Prospective studies are less subject to biases resulting from the retrospective reporting of dietary intakes or the effects of disease on biochemical indicators. Micronutrient supplements can potentially be evaluated in randomized trials; however, trials of dietary interventions may often be infeasible due to difficulties in maintaining compliance for the necessary long periods, which could be decades. Recent advances in molecular biology have yet to contribute substantially to dietary recommendations, but in the future these approaches may provide useful intermediary endpoints, allow the study of gene–diet interactions, and enhance our understanding of the mechanisms by which dietary factors influence disease. Ultimately, our knowledge is best based on a synthesis of epidemiologic, metabolic, animal, and mechanistic studies.

1.3. SPECIFIC DIETARY COMPONENTS

1.3.1. Dietary Fat

Until recently, reviews and dietary guidelines have consistently emphasized reducing total fat intake, usually to 30% of energy or less *(7–9)*, to prevent coronary heart disease (CHD). The classical diet–heart hypothesis has rested heavily on the repeated observation that serum total cholesterol levels predict CHD risk; serum cholesterol has thus functioned as a surrogate marker of risk in hundreds of metabolic studies. These studies, summarized as equations by Keys *(10)* and Hegsted *(11)*, indicated that compared to carbohydrates, saturated fats and dietary cholesterol increase, and polyunsaturated fat decreases, serum cholesterol, whereas monounsaturated fat has no influence. These widely used equations, while valid for total cholesterol, have become less relevant as surrogate variables for CHD risk, with the recognition that the high-density lipoprotein cholesterol fraction (HDL) is strongly and inversely related to CHD risk and that the ratio of total cholesterol to HDL is a better predictor *(12,13)*.

Substitution of carbohydrate for saturated fat (the basis of the American Heart Association diets) tends to reduce HDL as well as total and low-density lipoprotein (LDL) cholesterol; thus, the ratio does not change appreciably *(14)*. In contrast, substituting poly- or monounsaturated fat for saturated fat reduces LDL without affecting HDL, thus providing an improved ratio *(14)*. In addition, monounsaturated fats, compared to carbohydrate, reduce blood sugar and triglycerides in adult onset diabetics *(15)*. Questions have been raised as to whether the reductions in HDL resulting from a high-carbohydrate diet have the same adverse effect as reductions caused by other factors *(16)*. Although this is difficult to address directly, other factors that influence HDL levels, including alcohol, estrogens, obesity, smoking, exercise, and some medications, usually affect CHD risk as predicted *(17,18)*.

The use of the usual cholesterol prediction equations has been further complicated by the recognition that different saturated fats vary in their influence on LDL levels: 18:0, stearic acid (the main fat in chocolate and a major saturated fat in beef fat) has little effect; 16:0, palmitic acid (the main fat in palm oil also found in beef fat) modestly increases LDL; and 14:0, myristic acid (the main saturated fat in butter and other dairy fats) most strongly increases LDL *(19,20)*. However, this usually does not have practical importance in usual diets because intakes of the various saturated fats are strongly correlated with each other. However, stearic acid produced by the complete hydrogenation of vegetable oils is currently being considered as a replacement for trans-fatty acids (*see* below). The assumption that stearic acid is "neutral" is not warranted as long-term studies are limited; the available evidence suggests that it may be more strongly related to risk of cardiovascular disease than other saturated fats *(21,22)*, and in a controlled feeding study, high amounts of stearic acid (as in interesterified fat) had adverse effects on glucose regulation *(23)*.

The optimal amount of polyunsaturated fat intake in the diet remains uncertain. The earlier metabolic studies predicting total serum cholesterol *(10,11)* suggested that intakes should be maximized, and the American Heart Association recommended intakes of 10% of energy (compared to United States averages of about 3% in the 1950 s and 6% at present). Concerns have arisen from animal studies in which omega-6 polyunsaturated fat (typically as corn oil) has promoted tumor growth *(24)* and the possibility that high intakes of omega-6 relative to omega-3 fatty acids might promote coronary thrombosis *(25,26)*. However, as described below, available evidence from human studies has not supported these concerns at levels of omega-6 fatty acid intake up to about 10% of calories.

The relation to CHD incidence has been examined in many epidemiologic studies. In Keys' pioneering ecological study of diets and CHD in seven countries *(27,28)*, total fat intake had little association with population rates of CHD; indeed, the lowest rate was in Crete, which had the highest fat intake due to the large consumption of olive oil. Saturated fat intake, however, was positively related to CHD in Keys' study. In contrast to international comparisons, little relationship has been seen with saturated fat intake in many prospective studies of individuals *(6,29,30)*. Some studies, however, tend to support a modest association between dietary cholesterol and CHD risk *(31)*, and inverse associations have been seen with polyunsaturated fat *(29,30)*. Similarly, dietary intervention trials have generally shown little effect on CHD incidence when carbohydrate replaces saturated fat, but replacing saturated fat with polyunsaturated fat has been associated with lower incidence of CHD *(32–35)*. In the Women's Health Initiative (WHI), by far the largest trial to examine fat intake and incidence of CHD *(36)*, 48,000 women were randomized to a low-fat diet or their normal diet. No effect was seen, which is consistent with epidemiologic studies because the type of fat was not modified, but any conclusion from this study must be tempered because the compliance with the low-fat diet was poor.

Trans-fatty acids are formed by the partial hydrogenation of liquid vegetable oils in the production of margarine and vegetable shortening and can account for as much as 40% of these products. Even higher levels of trans fats are found in "vegetable ghee," which is widely used in the Middle East and south Asia *(37)*. In the United States, intake of trans-fatty acids from partially hydrogenated vegetable fats (which increased from nothing in 1900 to a peak of about 5.5% of total fat by about the 1960s) has closely paralleled the epidemic of CHD during this century, in contrast to intake of animal fat, which has steadily declined over this period *(38)*. Trans-fatty acids increase LDL and decrease HDL *(38–44)*, raise Lp(a), another lipid fraction implicated in CHD etiology *(42,45)*, and increase C-reactive protein and other inflammatory markers *(46)*. Positive associations between intake of trans-fatty acids and CHD have been seen among regions in the Seven Countries Study *(47)*. In the most detailed prospective study, trans-fatty acid intake was strongly associated with risk of CHD *(30)* and, as predicted by metabolic studies, this association was stronger than for saturated fat. The association between trans-fatty acid intake and risk of CHD has been confirmed in other prospective studies; in a meta-analysis, a 2% energy increase in trans fat intake was associated with a 23% increase in risk of CHD *(44)*. Higher intake of trans fat has also been associated with risk of type 2 diabetes *(48)*, gall stones *(49)*, dementia *(50)*, weight gain *(51)*, and an adverse effect on insulin resistance has been shown in a long-term study in monkeys *(52)*.

Since 2005 the US Food and Drug Administration has required that food labels include the trans fat content (*see* chapter 11). Denmark has banned the sale of industrially produced trans fat, which was hardly noticed by consumers. In the United States, the use of trans fats in restaurants has been banned by many cities, Puerto Rico, and the state of California. This has caused manufacturers to reformulate their products, and intake is now rapidly declining. High intake of omega-3 fatty acids from fish reduces platelet aggregability and prolongs bleeding time *(26)*, slightly reduces blood pressure *(53)*, decreases serum triglycerides, but increases LDL cholesterol *(54)*. Fish consumption was associated with a greatly reduced risk of myocardial infarction (MI) in one prospective study *(55)* and in a randomized trial among postinfarction patients *(56)*. Subsequent data have been less supportive of a major effect of fish consumption on overall risk of CHD *(57–59)*, but the benefits of omega-3 fatty acids appear to be primarily in prevention of fatal arrhythmias that can complicate CHD, rather than in prevention of infarction *(60–62)*. The amount of omega-3 fatty acids needed to prevent arrhythmia is remarkably small—on the order of 1 g/day or perhaps even less *(62)*. Some plant oils, including soybean, rapeseed (canola), and flaxseed, also contain substantial amounts of the 18-carbon omega-3 fatty acid, alpha-linolenic acid (ALA). Because vast regions of the world consume little omega-3 fatty acids from any source, and the global supply of fish is limited, whether ALA can provide similar benefits as the longer-chain fish oils is a crucial public health issue. More data are needed, but available evidence suggests that higher intakes of ALA can prevent both fatal and nonfatal CHD *(60,63)*; in eastern Europe, increases in rapeseed oil have been associated with rapid declines in CHD mortality *(64)*.

1.3.2. Dietary Fat and Cancer

Another major justification for reduction of dietary fat has been anticipated decreases in the risk of cancers of the breast, colon and rectum, and prostate *(65,66)*. The primary evidence has been that countries with low fat intake, also the less affluent areas, have had low rates of these cancers *(66,67)*. These correlations have been primarily with animal fat and meat intake, rather than with vegetable fat consumption.

The hypothesis that fat intake increases breast cancer risk has been supported by most animal models *(68,69)*, although no association was seen in a large study that did not use an inducing agent *(70)*. Moreover, much of the effect of dietary fat in the animal studies appears to be owing to an increase in total energy intake, and energy restriction profoundly decreases incidence *(24,68,70)*. Many large prospective studies have been published *(71–77)*, and little or no association has been seen in all. In a pooled analysis of prospective studies including 351,821 women and 7,329 cases of breast cancer, the relative risk for a 5% of energy increment in total fat was 1.00 (95% CI 0.98–1.03) *(75)*. In the Nurses' Health Study, no decrease in risk was seen with less than 20% of energy from fat *(78)*, and with 20 years of follow-up and multiple measures of diet, there was no hint of any positive association with intake of total or specific types of fat *(76)*. In the AARP cohort, a weak positive association (RR = 1.11) was found for women with the highest compared to the lowest intake of fat that was statistically significant due to the large number of cases *(77)*. In the WHI trial of dietary fat reduction, only a slight and not statistically significant reduction in risk was seen *(79)*, and even this slight difference could be due to the transient loss of weight in the intervention group (*see* Chapter 14). Thus, over a wide range of intake, dietary fat consumed by middle-aged women appears to have little or no influence on breast cancer risk. However, higher intake of animal fat, but not vegetable fat, by young adult women has been associated with a greater risk of breast cancer before menopause, suggesting that some components of animal foods rather than fat per se may increase risk (*80*). The relation of fat intake during childhood to risk of breast cancer has been minimally studied.

As with breast cancer, prospective studies have not supported the hypothesized associations between dietary fat and risks of colorectal or prostate cancer (*81*). However, positive association between consumption of red meat, particularly processed meat, and risk of colorectal cancer has been seen in many prospective studies (*81–84*). This suggests that other components of red meat such as heat-induced carcinogens, the high content of heme iron, or nitroso compounds might be responsible for the elevated risk.

Although dietary fat does not appear to explain the high rates of breast, colon, and prostate cancer in Western countries, a massive body of evidence indicates that excessive body fat, the result of excessive energy intake in relation to physical activity, is an important risk factor for cancers of the endometrium, breast (after menopause), pancreas, colon, kidney, esophagus (adenocarcinoma), and some hematologic malignancies *(85,81)*. Excess body fat is now second only to smoking as a cause of cancer in the United States. This appears to be mediated through multiple mechanisms, including increases in circulating estrogen levels (breast and endometrial cancers), gastric reflux (esophageal cancer), insulin resistance (colon and pancreatic cancer), and possibly other pathways.

1.3.3. Dietary Fat and Body Fatness

In addition to being a major risk factor for cancer, overweight is an important cause of diabetes, cardiovascular disease, and other important diseases (*see* below), and short-term studies have suggested that reducing the fat content of the diet induces weight loss. However, population differences in weight do not appear to be due primarily to fat intake; in Europe, southern countries with relatively low fat intake have higher rates of obesity than Northern European countries *(86)*. Also, among 65 counties in China, no correlation was seen between body weight and fat intake, which varied from approximately 6–30% of energy *(87)*. Inconsistent associations have been observed in cross-sectional and prospective studies within countries, but such observations are particularly prone to distortion because subjects may alter their diets to modify their weight. In randomized trials of fat reduction, the optimal way to study this relationship, modest weight reductions are typically seen in the short term. However, in randomized studies lasting a year or longer, reductions in fat from greater than 30% of energy to 18–25% of energy had minimal effects on overall long-term body weight *(88)*. Several recent randomized trials have compared very low fat, moderate fat, and low carbohydrate diets; weight loss over 1–2 years has been similar in all groups *(89)* or *(90)* on low fat/high carbohydrate diets. As predicted by shorter studies, cardiovascular risk factors have tended to be least desirable on low fat diets *(90)*. Very low fat intakes, less than 10% of energy, in conjunction with a high volume of bulky food as consumed by some traditional societies, may induce weight loss *(91)*, but long-term studies are needed. However, available evidence suggests that reductions in dietary fat over the ranges currently recommended will not have sustained benefits on body fatness and that this is likely to have adverse metabolic effects (*see* Chapter 15).

What can we now say about dietary fat and health? In 1989, a major review concluded that dietary fat per se is not associated with risk of CHD *(65)*. This was generally ignored but subsequent studies have added further support for this conclusion and have also failed to support suggested major reductions in cancer and other risks. Both metabolic and epidemiological data strongly indicate that intake of partially hydrogenated vegetable fats should be minimized. Metabolic data and epidemiologic data suggest that saturated fat intake should be as low as reasonably feasible, but these data also suggest that the benefits will be minimal if carbohydrate rather than unsaturated fats replace the saturated fat. Definitive data are not available on the optimal intake of polyunsaturated fat, but intakes of up to at least 10% of energy from linoleic acid (omega-6) have positive health benefits and no evidence of harm has been documented. Consumption of omega-3 fatty acids is essential and several servings of fish per week appears to provide adequate amounts for most healthy people. Whether ALA from plant sources can provide the same benefits as longer-chain fish oils is not fully resolved; this is a major global nutritional issue. Metabolic data as well as the experience of Southern European populations suggest that consuming a substantial proportion of energy as monounsaturated fat would be desirable. Although available evidence suggests that low total fat intakes have little benefit, consuming low amounts of red meat intake may decrease the incidence of colorectal cancer.

1.3.4. Vegetables and Fruits

Recommendations to eat a generous amount of vegetables and fruits *(65)* are supported by epidemiologic studies of cardiovascular risk *(92)*. Many early studies also suggested that high intake of these foods would greatly reduce the risk of a wide range of cancers *(93,94)*. However, most of these studies were case–control investigations, and more recent cohort studies have tended to show much weaker—or no—relation between overall fruit and vegetable consumption and risks of common cancers, include those of the breast, lung, and large bowel *(81,95,96)*. In case–control studies apparent protective effects with greater fruit and vegetable consumption have been seen for cancers of the stomach, oral cavity, larynx, pancreas, bladder, and cervix *(93,94,97)*, but, given the experience with other cancers, these findings require prospective confirmation.

Plants contain numerous components that have potential anticancer activity *(94)*. Considerable evidence suggests that folic acid reduces risk of colorectal cancer *(83,98)*, but vitamin supplements and fortification are now greater sources in the United States than fruits and vegetables. Other chemical constituents of plants could reduce the formation of carcinogens, induce detoxifying enzymes, and block the effects of endogenous estrogens. Further details about the amounts of these substances in foods could permit more informative investigations as lumping fruits and vegetables all together has little biological rationale.

In contrast to the weakened evidence that high intake of fruits and vegetables reduces cancer incidence, evidence has been strengthened that greater consumption will reduce risk of cardiovascular disease *(92)*. High intake of fruits and vegetables reduces blood pressure *(99)*, a major risk factor for cardiovascular disease, and potassium appears to be the primary explanation *(100)*. Evidence that elevated blood homocysteine is an independent risk factor for coronary heart and cerebrovascular disease *(101–103)* and that levels can be reduced by supplements of folic acid and vitamin B6 *(104,105)* suggests one mechanism.

Randomized trials of folate supplementation show a reduction in risk of stroke *(106)*. The evidence from randomized trials of folic acid in reduction of myocardial infarction has generally not supported the apparent benefit seen in epidemiologic studies *(107)*. However, this may be due to the existence of advanced coronary disease in most studies, the use of many drugs in these studies of ill patients, and the relatively short-term nature of these studies. An association between a polymorphism in the methylenetetrahydrofolate reductase gene and risk of CHD supports a benefit *(107)*, as does the evidence of benefit for stroke, which would presumably involve similar mechanisms.

Suboptimal dietary folic acid, which is mainly obtained from fortified breakfast cereals, vegetables, and fruits, definitively increases risk of neural tube defects, the most common severe birth defect

(108,109) and may account for more than half of these cases. The effect of low folate intake may be particularly adverse among the approximately 10% of the population who are genetically less efficient in utilizing the ingested form of this vitamin *(110)*.

In both case–control *(111)* and prospective studies *(112,113)*, intake of dietary antioxidants, including the carotenoids lutein and zeaxanthin, and vitamin C has been inversely related to risk of cataracts. As cataract formation, which is increased by sunlight and cigarette smoking *(114)*, involves the accumulation of oxidized and denatured proteins, this lesion may represent a convenient marker of long-term oxidative damage. High intake of lutein and zeaxanthin in the form of spinach has been associated with a decreased risk of advanced macular degeneration *(115)*. This is particularly notable because lutein and zeaxanthin are the carotenoids specifically concentrated in the macula, where they apparently play a protective role against photodamage *(116)*.

1.3.5. Starches and Complex Carbohydrates

As protein varies only modestly across a wide range of human diets, a higher carbohydrate consumption is, in practice, the reciprocal of a low-fat diet. For reasons discussed under the topic of fat, a high-carbohydrate diet may have adverse metabolic consequences. In particular, such diets are associated with an increase in triglycerides and a reduction in HDL cholesterol *(19)*. These adverse responses are aggravated in the context of insulin resistance *(117,118)*, which to some degree is highly prevalent in western populations. Although Asian populations had been thought to be at lower risk for insulin resistance and type 2 diabetes, much evidence now indicates that these populations, and also Hispanic and African populations, have a higher risk of type 2 diabetes, probably due to genetic susceptibility, compared to European populations, given the same diet, activity level, and BMI *(119)*. This has enormous implications because many of these populations have traditionally consumed large amounts of carbohydrate, which was well tolerated as long they were lean and active, which may become deleterious in the background of lower activity and even modest amounts of weight gain.

Several reasons exist to emphasize whole grains and other less refined complex carbohydrates as opposed to the highly refined products and sugar generally consumed in the United States. Adverse consequences of highly refined grains appear to result from both the rapid digestion and absorption of these foods, as well as from the loss of fiber and micronutrients in the milling process. The glycemic response after carbohydrate intake, which has been characterized by the glycemic index, is greater with highly refined foods as compared to less-refined, whole grains *(120)*. The greater glycemic response owing to highly refined carbohydrates is accompanied by increased plasma insulin levels and appears to augment the other adverse metabolic changes due to carbohydrate consumption noted above *(120)* to a greater degree than with less refined foods. Diets with a high-glycemic index or glycemic load (the product of dietary glycemic index and total carbohydrate intake) appear to increase the risk of noninsulin-dependent diabetes *(121,122)* and possibly risk of CHD, particularly among women with greater insulin resistance *(123)*.

Fiber intake, particularly from grain sources, has consistently been inversely related to risk of coronary heart disease and type 2 diabetes *(6,124–126)*. Risk of MI appears to be reduced by higher intake of dietary fiber from grains to a greater degree than can be explained by the effect of fiber on blood lipids alone *(127)*. Anticipated reductions in colon cancer risk by diets high in grain fiber have been difficult to document epidemiologically *(128–130)*. However, reduced constipation and risk of colonic diverticular disease *(114)* are clear benefits of such diets. The role of soluble fiber, found in oat bran and some other plant foods, in lowering blood lipids has been hotly debated; current evidence suggests that a small effect may exist with large intakes *(131,132)*.

The importance of micronutrients in the prevention of many chronic conditions, discussed below, has reemphasized the problem of "empty calories" associated with diets high in sugar and highly refined carbohydrates. In the standard milling of white flour, as much as 60–90% of vitamins B6 and E, folate, and other nutrients are lost *(133)*; this may be nutritionally critical for persons with otherwise marginal intakes. In the United States, thiamin, riboflavin, folate, and niacin are presently replaced by fortification, but other nutrients remain substantially reduced. Fortification of grains with folic acid has not been implemented in many countries despite clear benefits for reduction of neural tube defects and stroke (*see* above). One reason expressed for not doing so is the potential promotion of existing neoplasias, especially those of the colon *(134)*. This concern was heightened by an apparent transient pause in the decline in incidence of colon cancer in the United States and Canada, but this may also have been due to increased diagnosis due to screening by colonoscopy. Importantly, in the United States, there has been no suggestion of any increase in colon cancer mortality after folic acid fortification (http://progressreport.cancer.gov/).

Sugar in the form of soda and other beverages is of special concern because of the large amounts consumed by many populations, and because this appears to result in excess energy intake due to failure to suppress satiety *(135)*. Not surprisingly, daily consumption of sugary beverages has been associated with increased risks of type 2 diabetes *(136)*.

1.3.6. Protein

Average protein consumption in the United States and other affluent countries substantially exceeds conventional requirements *(65)*, and adequate intake can be maintained on most reasonable diets, including those without animal products. High intake of animal protein can increase urinary calcium loss *(137)*, contribute to homocysteinemia *(138)*, and has been hypothesized to increase risk of various cancers *(139)*; however, there is little evidence for the latter effect. Substituting protein for carbohydrate improves blood lipids and blood pressure. Also, because protein from foods is not consumed in isolation, the effects of these foods will depend mainly on the quality of fat and carbohydrate that they contain *(140)*.

1.3.7. Calcium, Vitamin D, and Dairy Products

Recommendations to maintain adequate calcium intake *(65)* and to consume dairy products on a daily basis *(141)* derive primarily from the role of calcium in maintaining bone strength. Calcium supplements in conjunction with vitamin D have reduced fracture incidence in older adults *(142,143)*, but in such studies benefits of calcium cannot be distinguished from those of vitamin D. In a meta-analysis of randomized trials, no reduction in overall fracture risk was seen with supplemental calcium alone *(144)*, and in a meta-analysis of prospective studies, calcium intake over about 500 mg/day was not associated with lower risk of fractures. Uncertainty remains regarding the optimal intake. In the United States, intakes as high as 1,500 mg/d have been recommended for postmenopausal women at risk of fractures *(145)*, which are difficult to achieve without supplements, but in the United Kingdom 700 mg/d is considered adequate for those over 19 years of age (http://www.foodstandards.gov.uk/news/newsarchive/foodpromotionplans). However, many populations have low-fracture rates despite minimal dairy product consumption and low overall calcium intake by adults *(146)*, and for this reason the WHO considers 500 mg/day to be an adequate intake *(5)*.

Milk and other dairy products may not be directly equivalent to calcium from supplements, as these foods contain a substantial amount of protein, which can enhance renal calcium losses *(137)*. Several prospective studies have directly addressed the relation of dairy product consumption to fracture incidence, with the exception of one small study *(147)*; higher consumption of calcium or dairy products as an adult has not been associated with lower fracture incidence *(148,149)*. At best, the benefits of high calcium intake are minor compared with those from regular physical activity *(150–153)*. Low-calcium intake has been associated with risk of colon cancer in large prospective studies *(154)*; evidence from a randomized trial that calcium supplementation modestly reduces colon adenoma recurrence adds important evidence of causality to the epidemiologic findings *(155)*.

Although calcium intakes can be increased by a high consumption of greens and certain other vegetables, greatly increased intakes would be required for most women to achieve the high calcium recommended levels by diet without regular use of milk and other dairy products. Calcium supplements are an inexpensive form of calcium without accompanying calories or saturated fat. Thus, dairy product consumption can be considered an optional rather than a necessary dietary component. Enthusiasm regarding high dairy consumption should also be tempered by the suggestion in many studies that this is associated with increased risks of prostate cancer *(81,156–158)* and possibly ovarian cancer *(159)*. Whether an increased risk is due to the calcium, lactose, or endogenous hormones in milk remains uncertain.

Until recently, the consequences of low vitamin D status were thought to be limited to rickets, osteoporosis, and fractures. However, almost every organ has been found to have vitamin D receptors, and inadequate vitamin D status has also been associated with greater risks of infections *(160)*, many cancers *(161)*, multiple sclerosis *(162,163)* , muscle weakness *(164)*, coronary heart disease *(165)*, and other conditions. The optimal blood levels of vitamin D (25 OH vitamin D) appear to be in the range of 70–100 nmol/mL *(160)*; this implies that the large majority of US residents have suboptimal status, and among persons with dark skin, this may be as high as 90%. The alternatives for increasing blood levels are primarily to increase sun exposure, which if not done carefully will increase risks of skin cancer, or to take supplements; the levels of vitamin D naturally present in fish or fortified milk can prevent rickets, but it is difficult to reach optimal levels from these sources (*see* Chapter 19).

1.3.8. Salt and Processed Meats

Reduction of salt (sodium chloride) intake will decrease blood pressure to a small degree. Law et al. *(166)* have concluded that a 3-g/d decrease would reduce the incidence of stroke by 22% and of CHD by 16%. Although the decrease in risk of cardiovascular disease achieved by reducing salt consumption is small for most individuals, the overall number of deaths potentially avoided is large, supporting policies to reduce consumption, particularly in processed foods and by institutions. In several case–control studies, the consumption of salty and pickled foods has been associated with stomach cancer *(167,81)*.

1.4. BODY WEIGHT

Until recently, the issue of optimal body weight was controversial due to analyses that did not account for confounding influences of factors such as smoking (which is a strong cause of premature death and is also associated with low body weight) or the fact that many individuals, particularly at older ages, have low body weights because of chronic illness *(168)*. More detailed analyses indicate that middle-aged persons with a body mass index (BMI) even close to 25 kg/m^2 have a high prevalence of abnormal blood glucose, lipids, and blood pressure *(169)* and experience substantial increases in MI *(170,171)*, diabetes *(172)*, hypertension *(173)*, many cancers *(85,174)*, gallstones *(175)*, and total mortality rates *(176)* compared to their leaner counterparts. Thus, the current guidelines based on a BMI range of 18–25 kg/m^2 are generally considered optimal, and the best health experience is achieved by avoiding increases in weight during adulthood *(168)*. As noted earlier, dietary fat composition over a wide range appears to have little relationship with weight maintenance; in contrast, low consumption of sugary beverages *(135)*, trans fat *(51)*, and higher intake of dietary fiber *(177)* appear to be helpful for weight control, and regular exercise and avoidance of extreme inactivity such as excessive television watching are crucial *(178)* (*see* Chapters 16 and 18).

1.5. ALCOHOL

Many adverse influences of heavy alcohol consumption are well recognized, but moderate consumption has both beneficial and harmful effects, greatly complicating decisions for individuals. Overwhelming epidemiologic data indicate that moderate consumption reduces risk of MI *(179–181)*, one to two drinks a day decrease risk by approximately 30–40%. Although it has been suggested that this effect may be a result of antioxidants in red wine *(182)*, similar protective effects for equivalent amounts of alcohol have been seen for all types of alcoholic beverages *(183,184)*. On the other hand, modest positive associations with risk of breast cancer incidence have been observed in dozens of studies with even one alcoholic drink per day *(185)*, possibly because alcohol increases endogenous estrogen levels *(186,187)* and interferes with folate metabolism *(188)*. The overall effect of alcohol, as represented by total mortality, appears beneficial up to about two drinks per day in men *(189)*. Overall, a similar relation with total mortality is seen among women, but no net benefit was observed among those at low risk of coronary heart disease because of age less than 50 years or lack of coronary risk factors *(190)*. Furthermore, the risk of transition from moderate alcohol consumption to addiction and uncontrolled drinking has not been well quantified.

1.6. VITAMIN SUPPLEMENTS

The most firmly established benefit of vitamin supplements, based on case–control, cohort, and randomized studies, is that folic acid supplements in the amounts contained in multiple vitamins can reduce the risks of neural tube defects by approximately 70% *(108,191)*. As noted above, correction on low folate levels can reduce the risk of stroke *(106)* and probably also the risks of coronary heart disease *(107)* and several cancers *(192)*. The cardiovascular benefits are likely mediated through reductions of homocysteine, and in some populations correction of low levels of vitamin B6 and B12 as well as low folate will probably have similar benefits. Vitamin B12 absorption declines with age, and supplements can prevent deficiency in older persons.

In prospective epidemiologic studies, healthy men and women who consumed the highest amounts of vitamin E (mostly from supplements) had an approximately 40% lower risk of MI compared to those having low vitamin E intakes *(193,194)*. However, in randomized trials, mainly among patients with existing coronary heart disease, little benefit has been seen *(195)*. The apparent difference may relate to the study populations because persons with existing coronary disease were excluded from the epidemiologic studies, and they were typically on many drugs that could overlap in mechanisms with vitamin E. Vitamin C supplementation was associated with lower risk of CHD in one international comparison *(196)*, but the data available did not distinguish vitamin C from other supplements. The association between vitamin C and CHD risk has been inconsistent in other prospective studies *(194,197)*. Apart from a possible reduction in risk of cataracts *(114)*, only limited evidence exists at present that high doses of vitamin C have substantial benefits.

Intake of preformed vitamin A (retinol) just above the RDA has been associated with excess risk of hip fracture in prospective studies *(198,199)*, possibly by competing with vitamin D at the receptor level, and elevated risks were seen for both use of multiple vitamins and specific supplements of vitamin A. In a more recent study, a modest positive association between vitamin A intake and risk of fractures was limited to those with low vitamin D intake, adding further evidence for an interaction with these vitamins *(200)*. The weaker association seen in this recent study may have resulted from reductions during the follow-up in the retinol content of breakfast cereals and multiple vitamins made in response to the evidence on fracture incidence. Serum levels of retinyl esters have not been associated with bone mineral density *(201)*, but these findings are difficult to interpret because retinyl esters are highly variable and the degree to which a single measure represents long-term vitamin A intake is unclear.

In a randomized trial conducted in a region of China with low consumption of fruits and vegetables, a supplement containing beta-carotene, vitamin E, and selenium reduced incidence of stomach cancer *(202)*.

Current evidence, although far from complete, suggests that supplements of folate and probably other vitamins, at the RDA level, contained in most nonprescription multivitamin preparations, have substantial benefits for at least an important, but unidentified, population subgroup, perhaps characterized by increased requirements or suboptimal diets. As intakes of folate as well as other micronutrients appear marginal for many Americans *(93,97)*, if the amount of retinol is kept low (as in the United States at present) the risks of using multivitamins appear low, the cost of supplements is low (especially compared to that of fresh fruits and vegetables), and the use of a daily or several-times-a-week multiple vitamin appears rational for the majority of Americans, given current knowledge. Multiple vitamins may have little benefit in someone consuming an optimal diet, but such persons are not common in the United States *(203)* and probably rare in poor populations. Further, inclusion of vitamin D, at doses of at least 1,000 IU per day, will provide a critical nutrient that cannot be obtained in sufficient amounts by diet, although many people may require additional amounts as a separate supplement to reach adequate levels.

Vitamin E supplements do not benefit persons with established coronary heart disease, but for others at risk of CHD it can be rational to use these while waiting for further data. For other vitamins and minerals, there is presently limited evidence of benefit of supplements over the RDA levels. Intake of vitamin A at levels above the RDA can potentially be harmful. In one study, intake of supplements containing more than 10,000 IU/d of preformed vitamin A was associated with risk of specific birth defects *(204)*.

1.7. RECOMMENDATIONS

Any set of dietary or nutritional recommendations must be made with the qualification that information is currently incomplete, and some conclusions may be modified with new data. Most importantly, the major causes of morbidity and mortality in the United States develop over many decades, and large-scale nutritional epidemiologic studies have only begun in the last 25 years; a full picture of the relation between diet and disease will require additional decades of careful investigation. Nevertheless, combining metabolic, clinical, and epidemiologic evidence, several general recommendations that are unlikely to change substantially can be made to those who are interested in consuming a healthy diet.

1. Stay lean and active throughout life. For most individuals, body weight should not increase by more than 5–10 pounds after age 21. Because most of us work at sedentary jobs, weight control will usually require conscious regular daily exercise as well as some effort to avoid overconsumption of calories.
2. Trans-fatty acids from partially hydrogenated vegetable oils should be avoided completely. Unless explicitly stated otherwise, it is safest to assume that deep-fried fast foods and most commercially prepared foods contain trans-fatty acids. These unhealthy fats can be replaced with a combination of vegetable oils that include a mix of monounsaturated and polyunsaturated fats. These should include good sources of omega-3 fatty acids, such as canola (rapeseed) or soybean oil.

3. Grains should be consumed primarily in a minimally refined, whole grain form and intake of simple sugars, especially as beverages, should be low.
4. Vegetables and fruits should be consumed in abundance (five servings/d is minimal) and include green leafy and orange vegetables daily.
5. Red meat should be consumed only occasionally and in low amounts if at all; nuts and legumes as well as poultry and fish in moderation are healthy alternatives.
6. The optimal consumption of dairy products and calcium intake is not known, and dairy products should be considered as optional. High consumption of milk (e.g., more than two servings per day) is not likely to be beneficial for middle-aged and older adults and may increase risk of prostate and ovarian cancer. Adequate calcium intake may be particularly important for growing children, adolescents, and lactating women; supplements should be considered if dietary sources are low.
7. Unless one is extremely careful about a healthy food selection at every meal, consuming a daily RDA-level (DV) multiple vitamin containing folic acid and at least 1,000 IU of vitamin D provides a sensible nutritional safety net. Because menstrual losses of iron may not be adequately replaced by iron intake on the low-energy diets of women in a sedentary society, it makes sense for most premenopausal women to use a multiple vitamin/multimineral that also contains iron. Pending further data, the use of a vitamin E supplement at 400–800 IU/d is reasonable for most middle-aged and older healthy persons as available evidence suggests that this may reduce risk of MI. Personal physicians should be made aware of any nutritional supplements that are being consumed in the event of possible interactions with medications or diagnostic tests. Further, use of supplements should not be considered as an alternative to eating a healthy diet because foods contain a wide variety of additional factors that are likely to contribute to good health.
8. Finally, be adventuresome in eating! Unfortunately, most of us in the United States are heirs to the rather monotonous Northern European dietary tradition centered on the consumption of meat, dairy products, and potatoes. Contemporary food processing has added to the deleterious effects of this diet by the removal of dietary fiber and micronutrients through overrefining of foods and has profoundly and adversely altered the biological effects of vegetable oils through the process of partial hydrogenation. To further aggravate matters, the worst aspects of diet tend to be the most heavily marketed and promoted. Fortunately, healthy diets do not have to be invented or discovered through new technological advances. Existing foods together with the lessons of various cultural models of eating based primarily around minimally processed foods from plant sources provide a means of achieving a diet that is healthy as well as interesting and enjoyable.

REFERENCES

1. Puska P, Vartiainen E, Tuomilehto J, Salomaa V, Nissinen A. Changes in premature deaths in Finland: successful long-term prevention of cardiovascular diseases. Bull World Health Organ 1998;76:419–425.
2. Bosetti C, Malvezzi M, Chatenoud L, Negri E, Levi F, La Vecchia C. Trends in colorectal cancer mortality in Japan, 1970–2000. Int J Cancer 2005;113:339–341.
3. Ludwig DS. Childhood obesity—the shape of things to come. N Engl J Med 2007;357:2325–2327.
4. Willett WC. Balancing life-style and genomics research for disease prevention. Science. 2002;296:695–698.
5. World Health Organization, FAO. Diet, nutrition and the prevention of chronic diseases: report of a joint WHO/FAO expert consultation (Report 916). World Health Organization, Geneva, 2003.
6. Willett WC. Nutritional Epidemiology, 2nd Ed. Oxford University Press, New York, 1998.
7. Department of Health and Human Services. The Surgeon General's Report on Nutrition and Health, Government Printing Office, Washington, DC (DHHS publication [PHS] 50210), 1988.
8. U.S. Department of Agriculture, U.S. Department of Health and Human Services. Nutrition and Your Health: Dietary Guidelines for Americans, 5th Ed. U.S. Government Printing Office, Washington, DC, 2000.
9. World Health Organization. Obesity: preventing and managing the global epidemic. WHO Technical Report Series no 894. ISBN no. 92 4 120894 5Geneva, 3–5 June 1997. World Health Organization, Geneva 2000.
10. Keys A. Serum-cholesterol response to dietary cholesterol. Am J Clin Nutr 1984;40:351–359.

11. Hegsted DM. Serum-cholesterol response to dietary cholesterol: a re-evaluation. Am J Clin Nutr 1986;44:299–305.
12. Castelli WP, Abbott RD, McNamara PM. Summary estimates of cholesterol used to predict coronary heart disease. Circulation 1983;67:730–734.
13. Ginsberg HN, Barr SL, Gilbert A, et al. Reduction of plasma cholesterol levels in normal men on an American Heart Association Step 1 diet or a Step 1 diet with added monounsaturated fat. N Engl J Med 1990;322:574–579.
14. Mensink RP, Zock PL, Kester AD, Katan MB. Effects of dietary fatty acids and carbohydrates on the ratio of serum total to HDL cholesterol and on serum lipids and apolipoproteins: a meta-analysis of 60 controlled trials. Am J Clin Nutr 2003;77:1146–1155.
15. Garg A, Grundy SM, Koffler M. Effect of high carbohydrate intake on hyperglycemia, islet cell function, and plasma lipoproteins in NIDDM. Diabetes Care 1992;15:1572–1580.
16. Brinton EA, Eisenberg S, Breslow JL. Increased apo A-I and apo A-II fractional catabolic rate in patients with low high density lipoprotein-cholesterol levels with or without hypertriglyceridemia. J Clin Invest 1991;87:536–544.
17. Sacks FM, Willett WC. More on chewing the fat—the good fat and the good cholesterol. N Engl J Med 1991;325: 1740–1742.
18. Mänttäri M, Huttunen JK, Koskinen P, et al. Lipoproteins and coronary heart disease in the Helsinki Heart Study. Eur Heart J 1990;11:26 h–31 h.
19. Mensink RP, Katan MB. Effect of dietary fatty acids on serum lipids and lipoproteins: a meta-analysis of 27 trials. Arterioscler Thromb 1992;12:911–919.
20. Denke MA, Grundy SM. Effects of fats high in stearic acid on lipid and lipoprotein concentrations in men. Am J Clin Nutr 1991;54:1036–1040.
21. Hu FB, Stampfer MJ, Manson JE, et al. Dietary saturated fats and their food sources in relation to the risk of coronary heart disease in women. Am J Clin Nutr 1999;70:1001–1008.
22. Kabagambe EK, Baylin A, Siles X, Campos H. Individual saturated fatty acids and nonfatal acute myocardial infarction in Costa Rica. Eur J Clin Nutr 2003;57:1447–1457.
23. Sundram K, Karupaiah T, Hayes KC. Stearic acid-rich interesterified fat and *trans*-rich fat raise the LDL/HDL ratio and plasma glucose relative to palm olein in humans. Nutr Metab (Lond) 2007;4:3.
24. Welsch CW. Relationship between dietary fat and experimental mammary tumorigenesis: a review and critique. Cancer Res 1992;52(Suppl. 7):2040S–2048S.
25. Renaud S, Kuba K, Goulet C, Lemire Y, Allard C. Relationship between fatty-acid composition of platelets and platelet aggregation in rat and man. Relation to thrombosis. Circ Res 1970;26:553–564.
26. Leaf A, Weber PC. Cardiovascular effects of n-3 fatty acids. N Engl J Med 1988;318:549–557.
27. Keys A. Seven Countries: A Multivariate Analysis of Death and Coronary Heart Disease. Harvard University Press, Cambridge, MA, 1980.
28. Verschuren WM, Jacobs DR, Bloemberg BP, et al. Serum total cholesterol and long-term coronary heart disease mortality in different cultures. Twenty-five-year follow-up of the Seven Countries Study. J Am Med Assoc 1995;274: 131–136.
29. Shekelle RB, Shryock AM, Paul O, et al. Diet, serum cholesterol, and death from coronary heart disease: The Western Electric Study. N Engl J Med 1981;304:65–70.
30. Hu F, Stampfer MJ, Manson JE, et al. Dietary fat intake and the risk of coronary heart disease in women. N Engl J Med 1997;337:1491–1499.
31. Shekelle RB, Stamler J. Dietary cholesterol and ischemic heart disease. Lancet 1989;1:1177–1179.
32. Multiple Risk Factor Intervention Trial Research Group. Multiple Risk Factor Intervention Trial: risk factor changes and mortality results. J Am Med Assoc 1982;248:1465–1477.
33. Stamler J, Wentworth D, Neaton JD. Is the relationship between serum cholesterol and risk of premature death from coronary heart disease continuous and graded? Findings in 356,222 primary screenees of the Multiple Risk Factor Intervention Trial (MRFIT). JAMA 1986;256:2823–2828.
34. Frantz IDJ, Dawson EA, Ashman PL, et al. Test of effect of lipid lowering by diet on cardiovascular risk: the Minnesota Coronary Survey. Arteriosclerosis 1989;9:129–135.
35. Sacks F. Dietary fats and coronary heart disease. Overview. J Cardiovascular Risk 1994;1:3–8.
36. Howard BV, Van Horn L, Hsia J, et al. Low-fat dietary pattern and risk of cardiovascular disease: the Women's Health Initiative Randomized Controlled Dietary Modification Trial. JAMA 2006;295:655–666.
37. Mozaffarian D, Abdollahi M, Campos H, Houshiarrad A, Willett WC. Consumption of *trans* fats and estimated effects on coronary heart disease in Iran. Eur J Clin Nutr 2007;61:1004–1010.
38. Booyens J, Louwrens CC. The Eskimo diet. Prophylactic effects ascribed to the balanced presence of natural *cis* unsaturated fatty acids and to the absence of unnatural *trans* and *cis* isomers of unsaturated fatty acids. Med Hypotheses 1986;21:387–408.

39. Mensink RPM, Katan MB. Effect of dietary *trans*-fatty acids on high-density and low-density lipoprotein cholesterol levels in healthy subjects. N Engl J Med 1990;323:439–445.
40. Zock PL, Katan MB. Hydrogenation alternatives: effects of *trans* fatty acids and stearic acid versus linoleic acid on serum lipids and lipoproteins in humans. J Lipid Res 1992;33:399–410.
41. Judd JT, Clevidence BA, Muesing RA, Wittes J, Sunkin ME, Podczasy JJ. Dietary *trans* fatty acids: effects of plasma lipids and lipoproteins on healthy men and women. Am J Clin Nutr 1994;59:861–868.
42. Nestel P, Noakes M, Belling Bea. Plasma lipoprotein and Lp[a] changes with substitution of elaidic acid for oleic acid in the diet. J Lipid Res 1992;33:1029–1036.
43. Sundram K, Ismail A, Hayes KC, Jeyamalar R, Pathmanathan R. *Trans* (elaidic) fatty acids adversely affect the lipoprotein profile relative to specific saturated fatty acids in humans. J Nutr 1997;127:514S–520S.
44. Mozaffarian D, Katan MB, Ascherio A, Stampfer MJ, Willett WC. *Trans* fatty acids and cardiovascular disease. N Engl J Med 2006;354:1601–1613.
45. Mensink RP, Zock PL, Katan MB, Hornstra G. Effect of dietary*cis* and *trans* fatty acids on serum lipoprotein [a] levels in humans. J Lipid Res 1992;33:1493–1501.
46. Mozaffarian D, Willett WC. *Trans* fatty acids and cardiovascular risk: a unique cardiometabolic imprint? Curr Atheroscler Rep 2007;9:486–493.
47. Kromhout D, Menotti A, Bloemberg B, et al. Dietary saturated and *trans* fatty acids and cholesterol and 25-year mortality from coronary heart disease: the Seven Countries Study. Prev Med 1995;24:308–315.
48. Hu FB, Manson JE, Stampfer MJ, et al. Diet, lifestyle, and the risk of type 2 diabetes mellitus in women. N Engl J Med 2001;345:790–797.
49. Tsai CJ, Leitzmann MF, Willett WC, Giovannucci EL. Long-term intake of *trans*-fatty acids and risk of gallstone disease in men. Arch Intern Med 2005;165:1011–1015.
50. Morris MC, Evans DA, Bienias JL, Tangney CC, Wilson RS. Dietary fat intake and 6-year cognitive change in an older biracial community population. Neurology 2004;62:1573–1579.
51. Field AE, Willett WC, Lissner L, Colditz GA. Dietary fat and weight gain among women in the Nurses' Health Study. Obesity (Silver Spring) 2007;15:967–976.
52. Kavanagh K, Jones KL, Sawyer J, et al. *Trans* fat diet induces abdominal obesity and changes in insulin sensitivity in monkeys. Obesity (Silver Spring) 2007;15:1675–1684.
53. Bonaa KH, Bzerve KS, Staume B, Gram IT, Thelle D. Effect of eicosapentaenoic and docosahexaenoic acids on blood pressure in hypertension: a population-based intervention trial from the Tromso study. N Engl J Med 1990;322: 795–801.
54. Kestin M, Clifton P, Belling GB, Nestel PJ. N-3 fatty acids of marine origin lower systolic blood pressure and triglycerides but raise LDL cholesterol compared with n-3 and n-6 fatty acids from plants. Am J Clin Nutr 1990;51: 1028–1034.
55. Kromhout D, Bosscheiter EB, de Lezenne Coulander C. The inverse relation between fish consumption and 20-year mortality from coronary heart disease. N Engl J Med 1985;312:1205–1209.
56. Burr ML, Fehily AM, Gilbert JF, et al. Effects of changes in fat, fish, and fibre intakes on death and myocardial reinfarction: diet and reinfarction trial (DART). Lancet 1989;2:757–761.
57. Vollset SE, Heuch I, Bjelke E. Fish consumption and mortality from coronary heart disease (letter). N Engl J Med 1985;313:820–821.
58. Ascherio A, Rimm EB, Stampfer MJ, Giovannucci E, Willett WC. Dietary intake of marine n-3 fatty acids, fish intake and the risk of coronary disease among men. N Engl J Med 1995;332:977–982.
59. Morris MC, Manson JE, Rosner B, Buring JE, Willett WC, Hennekens CH. Fish consumption and cardiovascular disease in the Physicians' Health Study: a prospective study. Am J Epidemiol 1995;142:166–175.
60. de Lorgeril M, Renaud S, Mamelle N, et al. Mediterranean alpha-linolenic acid-rich diet in secondary prevention of coronary heart disease [Erratum in: Lancet 1995;345:738]. Lancet 1994;343:1454–1459.
61. Leaf A. Omega-3 fatty acids and prevention of ventricular fibrillation. Prostaglandins Leukot Essent Fatty Acids 1995;52:197–198.
62. GISSI-Prevention Investigators. Dietary supplementation with n-3 polyunsaturated fatty acids and vitamin E after myocardial infarction: results of the GISSI-Prevenzione trial. Lancet 1999;354:447–455.
63. Campos H, Baylin A, Willett WC. Alpha-linolenic acid and risk of nonfatal acute myocardial infarction. Circulation 2008;118:339–345.
64. Zatonski W, Campos H, Willett W. Rapid declines in coronary heart disease mortality in Eastern Europe are associated with increased consumption of oils rich in alpha-linolenic acid. Eur J Epidemiol 2008;23:3–10.
65. National Research Council—Committee on Diet and Health. Diet and Health: Implications for Reducing Chronic Disease Risk, National Academy Press, Washington, DC. 1989.

66. Prentice RL, Sheppard L. Dietary fat and cancer. Consistency of the epidemiologic data, and disease prevention that may follow from a practical reduction in fat consumption. Cancer Causes Control 1990;1:81–97.
67. Armstrong B, Doll R. Environmental factors and cancer incidence and mortality in different countries, with special reference to dietary practices. Int J Cancer 1975;15:617–631.
68. Ip C. Quantitative assessment of fat and calorie as risk factors in mammary carcinogenesis in an experimental model. In: Mettlin CJ, Aoki K, eds. Recent Progress in Research on Nutrition and Cancer: Proceedings of a Workshop Sponsored by the International Union Against Cancer, Held in Nagoya, Japan, November 1–3, 1989. Wiley-Liss, Inc., New York, 1990, pp. 107–117.
69. Freedman LS, Clifford C, Messina M. Analysis of dietary fat, calories, body weight, and the development of mammary tumors in rats and mice: a review. Cancer Res 1990;50:5710–5719.
70. Appleton BS, Landers RE. Oil gavage effects on tumor incidence in the National Toxicology Program's 2-year carcinogenesis bioassay. Adv Exp Med Biol 1986;206:99–104.
71. Kushi LH, Sellers TA, Potter JD, et al. Dietary fat and postmenopausal breast cancer. J Natl Cancer Inst 1992;84: 1092–1099.
72. Howe GR, Friedenreich CM, Jain M, Miller AB. A cohort study of fat intake and risk of breast cancer. J Natl Cancer Inst 1991;83:336–340.
73. Graham S, Zielezny M, Marshall J, et al. Diet in the epidemiology of postmenopausal breast cancer in the New York State cohort. Am J Epidemiol 1992;136:1327–1337.
74. Van den Brandt PA, Van't Veer P, Goldbohm RA, et al. A prospective cohort study on dietary fat and the risk of postmenopausal breast cancer. Cancer Res 1993;53:75–82.
75. Smith-Warner SA, Spiegelman D, Adami HO, et al. Types of dietary fat and breast cancer: a pooled analysis of cohort studies. Int J Cancer 2001;92:767–774.
76. Kim EH, Willett WC, Colditz GA, et al. Dietary fat and risk of postmenopausal breast cancer in a 20-year follow-up. Am J Epidemiol 2006;164:990–997.
77. Thiebaut A, Kipnis V, Chang S-C, et al. Dietary fat and postmenopausal invasive breast cancer in the National Institutes of Health—AARP Diet and Health Study Cohort. J Natl Cancer Inst 2007;99:451–462.
78. Holmes MD, Hunter DJ, Colditz GA, et al. Association of dietary intake of fat and fatty acids with risk of breast cancer. JAMA 1999;281:914–920.
79. Prentice RL, Thomson CA, Caan B, et al. Low-fat dietary pattern and cancer incidence in the Women's Health Initiative Dietary Modification Randomized Controlled Trial. J Natl Cancer Inst 2007;99:1534–1543.
80. Cho E, Spiegelman D, Hunter DJ, et al. Premenopausal fat intake and risk of breast cancer. J Natl Cancer Inst 2003;95:1079–1085.
81. W.C.R.F./A.I.C.R. Second Expert Report: Food, Nutrition, Physical Activity, and the Prevention of Cancer: A Global Perspective, 2007.
82. Willett WC, Stampfer MJ, Colditz GA, Rosner BA, Speizer FE. Relation of meat, fat, and fiber intake to the risk of colon cancer in a prospective study among women. N Engl J Med 1990;323:1664–1672.
83. Giovannucci E, Rimm EB, Ascherio A, Stampfer MJ, Colditz GA, Willett WC. Alcohol, low-methionine-low-folate diets, and risk of colon cancer in men. J Natl Cancer Inst 1995;87:265–273.
84. Norat T, Lukanova A, Ferrari P, Riboli E. Meat consumption and colorectal cancer risk: dose-response meta-analysis of epidemiological studies. Int J Cancer 2002;98:241–256.
85. Calle EE, Rodriguez C, Walker-Thurmond K, Thun MJ. Overweight, obesity, and mortality from cancer in a prospectively studied cohort of U.S. adults. N Engl J Med 2003;348:1625–1638.
86. Seidell JC, Derenberg I. Obesity in Europe—prevalences and consequences for use of medical care. Pharmacoeconomics 1994;5(Suppl. 1):38–44.
87. Chen X, Yang GQ, Chen J, Chen X, Wen Z, Ge K. Studies on the relations of selenium and Keshan disease. Biol Trace Elem Res 1980;2:91–107.
88. Willett WC, Leibel RL. Dietary fat is not a major determinant of body fat. Am J Med 2002;113(Suppl. 9B):47S–59S.
89. Dansinger ML, Gleason JA, Griffith JL, Selker HP, Schaefer EJ. Comparison of the Atkins, Ornish, Weight Watchers, and Zone diets for weight loss and heart disease risk reduction: a randomized trial. JAMA 2005;293:43–53.
90. Shai I, Schwarzfuchs D, Henkin Y, et al. Weight loss with a low-carbohydrate, Mediterranean, or low-fat diet. N Engl J Med 2008;359:229–241.
91. Shintani TT, Hughes CK, Beckman S, O'Connor HK. Obesity and cardiovascular risk intervention through the ad libitum feeding of a traditional Hawaiian diet. Am J Clin Nutr 1991;6:1647 s–1651 s.
92. Hung HC, Joshipura KJ, Jiang R, et al. Fruit and vegetable intake and risk of major chronic disease. J Natl Cancer Inst 2004;96:1577–584.

93. Block G, Patterson B, Subar A. Fruit, vegetables, and cancer prevention: a review of the epidemiological evidence. Nutr Cancer 1992;18:1–29.
94. Steinmetz KA, Potter JD. Vegetables, fruit and cancer. I. Epidemiology. Cancer Causes Control 1991;2:325–357.
95. Michels KB, Giovannucci E, Joshipura KJ, et al. Fruit and vegetable consumption and colorectal cancer incidence. IARC Sci Publ 2002;156:139–140.
96. Smith-Warner SA, Spiegelman D, Yaun SS, et al. Intake of fruits and vegetables and risk of breast cancer: a pooled analysis of cohort studies. JAMA 2001;285:769–776.
97. Block G, Abrams B. Vitamin and mineral status of women of childbearing potential. Ann N Y Acad Sci 1993;678: 244–254.
98. Giovannucci E, Stampfer MJ, Colditz GA, et al. Folate, methionine, and alcohol intake and risk of colorectal adenoma. J Natl Cancer Inst 1993;85:875–884.
99. Sacks FM, Obarzanek E, Windhauser MM, et al. Rationale and design of the Dietary Approaches to Stop Hypertension Trial (DASH)—A multicenter controlled-feeding study of dietary patterns to lower blood pressure. Annu Epidemiol 1995;5:108–118.
100. Sacks FM, Willett WC, Smith A, Brown LE, Rosner B, Moore TJ. Effect on blood pressure of potassium, calcium, and magnesium in women with low habitual intake. Hypertension 1998;31:131–138.
101. Stampfer MJ, Malinow MR, Willett WC, et al. A prospective study of plasma homocyst(e)ine and risk of myocardial infarction in US physicians. J Am Med Assoc 1992;268:877–881.
102. Kang SS, Wong PWK, Norusis M. Homocysteinemia due to folate deficiency. Metabolism 1987;36:458–462.
103. Selhub J, Jacques PF, Bostom AG, et al. Association between plasma homocysteine concentrations and extracranial carotid-artery stenosis. N Engl J Med 1995;332:286–291.
104. Kang SS, Wong PWK, Cook HY, Norusis M, Messer JV. Protein bound homocyst(e)ine—a possible risk factor for coronary artery disease. J Clin Invest 1986;77:1482–1486.
105. Wilcken DEL, Dudman NPB, Tyrrell PA. Homocystinuria due to cystathionine B-synthase deficiency—the effects of betaine treatment in pyridoxine-responsive patients. Metabolism 1985;34:1115–1121.
106. Wang X, Qin X, Demirtas H, et al. Efficacy of folic acid supplmentation in stroke prevention: a meta-analysis. Lancet 2007;369:1876–1882.
107. Wald DS, Wald NJ, Morris JK, Law M. Folic acid, homocysteine, and cardiovascular disease: judging causality in the face of inconclusive trial evidence. BMJ 2006;333:1114–1117.
108. MRC Vitamin Study Research Group. Prevention of neural tube defects: results of the Medical Research Council Vitamin Study. Lancet 1991;338:131–137.
109. Werler MM, Shapiro S, Mitchell AA. Periconceptional folic acid exposure and risk of occurrent neural tube defects. JAMA 1993;269:1257–1261.
110. van der Put NM, Steegers-Theunissen RP, Frosst P, et al. Mutated methylenetetrahydrofolate reductase as a risk factor for spina bifida. Lancet 1995;346:1070–1071.
111. Jacques PF, Hartz SC, Chylack LT, McGandy RB, Sadowski JA. Nutritional status in persons with and without senile cataract: blood vitamin and mineral levels. Am J Clin Nutr 1988;48:152–158.
112. Hankinson SE, Stampfer MJ, Seddon JM, et al. Nutrient intake and cataract extraction in women: a prospective study. Br Med J 1992;305:335–339.
113. Chasan-Taber L, Willett WC, Seddon JM, et al. A propective study of carotenoid and vitamin A intakes and risk of cataract extraction in US women. Am J Clin Nutr 1999;70:509–516.
114. Hankinson SE, Willett WC, Colditz GA, et al. A prospective study of smoking and risk of cataract surgery in women. J Am Med Assoc 1992;268:994–998.
115. Seddon JM, Ajani UA, Sperduto RD, et al. Dietary carotenoids, vitamins A, C, and E, and advanced age-related macular degeneration. J Am Med Assoc 1994;272:1413–1420.
116. Schalch W. Carotenoids in the retina: a review of their possible role in preventing or limiting damage caused by light and oxygen. In: Emerit I, Chance B, eds. Free Radicals and Aging. Birkhauser Verlag, Basel, Switzerland, 1992, pp. 280–298.
117. Jeppesen J, Hollenbeck CB, Zhou MY, et al. Relation between insulin resistance, hyperinsulemia, postheparin plasma lipoprotein lipase activity, and postprandial lipemia. Arterioscler Thromb Vasc Biol 1995;15:320–324.
118. Jeppesen J, Chen YDI, Zhou MY, Schaaf P, Coulston A, Reaven GM. Postprandial triglyceride and retinyl ester responses to oral fats effects of fructose. Am J Clin Nutr 1995;61:787–791.
119. Shai I, Jiang R, Manson JE, et al. Ethnicity, obesity, and risk of type 2 diabetes in women: a 20-year follow-up study. Diabetes Care 2006;29:1585–1590.

120. Jenkins DJ, Wolever TM, Taylor RH, et al. Glycemic index of foods: a physiological basis for carbohydrate exchange. Am J Clin Nutr 1981;34:362–366.
121. Salmeron J, Manson JE, Stampfer MJ, Colditz GA, Wing AL, Willett WC. Dietary fiber, glycemic load, and risk of non-insulin-dependent diabetes mellitus in women. JAMA 1997;277:472–477.
122. Salmeron J, Ascherio A, Rimm EB, et al. Dietary fiber, glycemic load, and risk of NIDDM in Men. Diabetes Care 1997;20:545–550.
123. Liu S, Willett WC, Stampfer MJ, et al. A prospective study of dietary glycemic load, carbohydrate intake, and risk of coronary heart disease in US women. Am J Clin Nutr 2000;71:1455–1461.
124. Morris JN, Marr JW, Clayton DG. Diet and heart: a postscript. Br Med J 1977;2:1307–1314.
125. Khaw KT, Barrett-Connor E. Dietary fiber and reduced ischemic heart disease mortality rates in men and women: a 12-year prospective study. Am J Epidemiol 1987;126:1093–1102.
126. Hu FB, Willett WC. Optimal diets for prevention of coronary heart disease. JAMA 2002;288:2569–2578.
127. Rimm EB, Ascherio A, Giovannucci E, Spiegelman D, Stampfer MJ, Willett WC. Vegetable, fruit, and cereal fiber intake and risk of coronary heart disease among men. J Am Med Assoc 1996;275:447–451.
128. Willett W. The search for the causes of breast and colon cancer. Nature 1989;338:389–394.
129. Fuchs CS, Colditz GA, Stampfer MJ, et al. Dietary fiber and the risk of colorectal cancer and adenoma in women. N Engl J Med 1999;340:169–176.
130. Terry P, Giovannucci E, Michels KB, et al. Fruit, vegetables, dietary fiber, and risk of colorectal cancer. J Natl Cancer Inst 2001;93:525–533.
131. Jenkins DJ, Wolever TM, Rao AV, et al. Effect of blood lipids of very high intakes of fiber in diets low in saturated fat and cholesterol. N Engl J Med 1993;329:21–26.
132. Brown L, Rosner B, Willett WC, Sacks FM. Cholesterol-lowering effects of dietary fiber: a meta-analysis. Am J Clin Nutr 1999;69:30–42.
133. Schroeder HA. Losses of vitamins and trace minerals resulting from processing and preservation of foods. Am J Clin Nutr 1971;24:562–573.
134. Mason JB. Folate and colonic carcinogenesis: searching for a mechanistic understanding. J Nutr Biochem 1994;5: 170–175.
135. Popkin BM, Armstrong LE, Bray GM, Caballero B, Frei B, Willett WC. A new proposed guidance system for beverage consumption in the United States. Am J Clin Nutr 2006;83:529–542.
136. Schulze MB, Manson JE, Ludwig DS, et al. Sugar-sweetened beverages, weight gain, and incidence of type 2 diabetes in young and middle-aged women. JAMA 2004;292:927–934.
137. Lutz J, Linkswiler HM. Calcium metabolism in postmenopausal women and osteoporotic women consuming two levels of dietary protein. Am J Clin Nutr 1981;34:2178–2186.
138. Gruberg ER, Raymond SA. Beyond Cholesterol. St. Martin's, New York, 1981.
139. Youngman LD, Campbell TC. The sustained development of preneoplastic lesions depends on high protein intake. Nutr Cancer 1992;18:131–142.
140. Halton TL, Willett WC, Liu S, et al. Low-carbohydrate-diet score and the risk of coronary heart disease in women. N Engl J Med 2006;355:1991–2002.
141. Welsh S, Davis C, Shaw A. Development of the food guide pyramid. Nutr Today 1992;27:12–23.
142. Chapuy MC, Arlof ME, Duboeuf F, et al. Vitamin D3 and calcium to prevent hip fractures in elderly women. N Engl J Med 1992;327:1637–1642.
143. Heaney RP. Thinking straight about calcium. N Engl J Med 1993;328:503–504.
144. Bischoff-Ferrari HA, Dawson-Hughes B, Baron JA, et al. Calcium intake and hip fracture risk in men and women: a meta-analysis of prospective cohort studies and randomized controlled trials. Am J Clin Nutr 2007;86:1780–1790.
145. Institute of Medicine. Dietary Reference Intakes: Calcium, Phosphorus, Magnesium, Vitamin D, Fluoride. National Academy Press, Washington, DC, 1997.
146. Hegsted DM. Calcium and osteoporosis. J Nutr 1986;116:2316–2319.
147. Holbrook TL, Barrett-Conner E, Wingard DL. Dietary calcium and risk of hip fracture: 14-year prospective population study. Lancet 1988;2:1046–1049.
148. Feskanich D, Willett WC, Colditz GA. Calcium, vitamin D, milk consumption, and hip fractures: a prospective study among postmenopausal women. Am J Clin Nutr 2003;77:504–11.
149. Michaelsson K, Melhus H, Bellocco R, Wolk A. Dietary calcium and vitamin D intake in relation to osteoporotic fracture risk. Bone 2003;32:694–703.
150. Feskanich D, Willett WC, Colditz GA. Walking and leisure-time activity and risk of hip fracture in postmenopausal women. JAMA 2002;288:2300–2306.

151. Wickham CAC, Walsh K, Cooper C, et al. Dietary calcium, physical activity, and risk of hip fracture: a prospective study. Br Med J 1989;299:889–892.
152. Michaelsson K, Holmberg L, Mallmin H, et al. Diet and hip fracture risk: a case-control study. Intl J Epidemiol 1995;24:771–782.
153. Feskanich D, Willett WC, Stampfer MJ, Colditz GA. Milk, dietary calcium, bone fractures in women: a 12-year prospective study. Am J Public Health 1997;87:992–997.
154. Cho E, Smith-Warner S, Spiegelman D, et al. Dairy foods and calcium and colorectal cancer: a pooled analysis of 10 cohort studies. J Natl Cancer Inst 2004:96: 1015–1022.
155. Baron JA, Beach M, Mandel JS, et al. Calcium supplements for the prevention of colorectal adenomas. The Calcium Polyp Prevention Study Group. N Engl J Med 1999;340:101–107.
156. Giovannucci E, Rimm EB, Wolk A, et al. Calcium and fructose intake in relation to risk of prostate cancer. Cancer Res 1998;58:442–447.
157. Giovannucci E. Nutritional and environmental epidemiology of prostate cancer. In: Kantoff PW, Carroll PR, D'Amico AV, eds. Prostate Cancer: Principles and Practice. Lippincott Williams & Wilkins, Philadelphia, PA, 2002, pp. 117–139.
158. Giovannucci E, Liu Y, Stampfer MJ, Willett WC. A prospective study of calcium intake and incident and fatal prostate cancer. Cancer Epidemiol Biomarkers Prev 2006;15:203–210.
159. Genkinger JM, Hunter DJ, Spiegelman D, et al. Dairy products and ovarian cancer: a pooled analysis of 12 cohort studies. Cancer Epidemiol Biomarkers Prev 2006;15:364–372.
160. Bischoff-Ferrari HA, Giovannucci E, Willett WC, Dietrich T, Dawson-Hughes B. Estimation of optimal serum concentrations of 25-hydroxyvitamin D for multiple health outcomes. Am J Clin Nutr 2006;84:18–28.
161. Giovannucci E. The epidemiology of vitamin D and cancer incidence and mortality: a review. Cancer Causes Control 2005:16: 83–95.
162. Munger KL, Levin LI, Hollis BW, Howard NS, Ascherio A. Serum 25-hydroxyvitamin D levels and risk of multiple sclerosis. JAMA 2006;296:2832–2838.
163. Munger KL, Zhang SM, O'Reilly E, et al. Vitamin D intake and incidence of multiple sclerosis. Neurology 2004;62: 60–65.
164. Broe KE, Chen TC, Weinberg J, Bischoff-Ferrari HA, Holick MF, Kiel DP. A higher dose of vitamin D reduces the risk of falls in nursing home residents: a randomized, multiple-dose study. J Am Geriatr Soc 2007;55:234–239.
165. Giovannucci E, Liu Y, Hollis BW, Rimm EB. 25-hydroxyvitamin D and risk of myocardial infarction in men: a prospective study. Arch Intern Med 2008;168:1174–1180.
166. Law MR, Frost CD, Wald NJ. By how much does dietary salt reduction lower blood pressure? III-Analysis of data from trials of salt reduction. Br Med J 1991;302:819–824.
167. Correa P, Fontham E, Pickle LW, Chen V, Lin Y, Haenszel W. Dietary determinants of gastric cancer in south Louisiana inhabitants. J Natl Cancer Inst 1985;75:645–654.
168. Willett WC, Dietz WH, Colditz GA. Guidelines for healthy weight. N Engl J Med 1999;341:427–434.
169. Garrison RJ, Kannel WB. A new approach for estimating healthy body weights. Int J Obesity 1993;17:417–423.
170. Lew EA, Garfinkel L. Variations in mortality by weight among 750,000 men and women. J Chronic Dis 1979;32: 563–576.
171. Willett WC, Manson JE, Stampfer MJ, et al. Weight, weight change, and coronary heart disease in women: risk within the "normal" weight range. J Am Med Assoc 1995;273:461–465.
172. Colditz GA, Willett WC, Stampfer MJ, et al. Relative weight and increased risk of diabetes in a cohort of US women (abstract). Am J Epidemiol 1987;126:750–751.
173. Witteman JC, Willett WC, Stampfer MJ, et al. Relation of moderate alcohol consumption and risk of systemic hypertension in women. Am J Cardiol 1990;65:633–637.
174. International Agency for Research on Cancer. Weight Control and Physical Activity. IARCPress, Lyon, 2002.
175. Maclure KM, Hayes KC, Colditz GA, Stampfer MJ, Speizer FE, Willett WC. Weight, diet and risk of symptomatic gallstones in middle-aged women. N Engl J Med 1989;321:563–569.
176. Manson JE, Willett WC, Stampfer MJ, et al. Body weight and mortality among women. N Engl J Med 1995;333: 677–685.
177. van Dam RM, Seidell JC. Carbohydrate intake and obesity. Eur J Clin Nutr 2007;61 Suppl 1:S75–S99.
178. Gortmaker SL, Dietz WH, Cheung LW. Inactivity, diet, and the fattening of America. J Am Diet Assoc 1990;90: 1247–1252.
179. Klatsky AL, Armstrong MA, Friedman GD. Risk of cardiovascular mortality in alcohol drinkers, ex-drinkers, and nondrinkers. Am J Cardiol 1990;66:1237–1242.

180. Rimm EB, Giovannucci EL, Willett WC, et al. Prospective study of alcohol consumption and risk of coronary disease in men. Lancet 1991;338:464–468.
181. Rimm EB, Klatsky A, Grobbee D, Stampfer MJ. Review of moderate alcohol consumpton and reduced risk of coronary heart disease: is the effect due to beer, wine, or spirits? Br Med J 1996;312:731–736.
182. Renaud S, de Lorgeril M. Wine, alcohol, platelets, and the French paradox for coronary disease. Lancet 1992;339:1523–1526.
183. Maclure M. Demonstration of deductive meta-analysis: ethanol intake and risk of myocardial infarction. Epidemiol Rev 1993;15:328–351.
184. Mukamal KJ, Conigrave KM, Mittleman MA, et al. Roles of drinking pattern and type of alcohol consumed in coronary heart disease in men. N Engl J Med 2003;348:109–118.
185. Smith-Warner SA, Spiegelman D, Yaun S-S, et al. Alcohol and breast cancer in women: a pooled analysis of cohort studies. J Am Med Assoc 1998;279:535–540.
186. Reichman ME, Judd JT, Longcope C, et al. Effects of alcohol consumption on plasma and urinary hormone concentrations in premenopausal women. J Natl Cancer Inst 1993;85:722–727.
187. Hankinson SE, Willett WC, Manson JE, et al. Alcohol, height, and adiposity in relation to estrogen and prolactin levels in postmenopausal women. J Natl Cancer Inst 1995;87:1297–1302.
188. Zhang S, Hunter DJ, Hankinson SE, et al. A prospective study of folate intake and the risk of breast cancer. JAMA 1999;281:1632–1637.
189. Boffetta P, Garfinkel L. Alcohol drinking and mortality among men enrolled in a American Cancer Society prospective study. Epidemiology 1990;1:342–348.
190. Fuchs CS, Stampfer MJ, Colditz GA, et al. Alcohol consumption and mortality among women. N Engl J Med 1995;332:1245–1250.
191. Willett WC. Folic acid and neural tube defect: can't we come to closure? Am J Publ Hlth 1992;82:666–668.
192. Giovannucci E. Epidemiologic studies of folate and colorectal neoplasia: a review. Journal of Nutrition 2002;132:2350S–2355S.
193. Stampfer MJ, Hennekens CH, Manson JE, Colditz GA, Rosner B, Willett WC. Vitamin E consumption and the risk of coronary disease in women. N Engl J Med 1993;328:1444–1449.
194. Rimm EB, Stampfer MJ, Ascherio A, Giovannucci E, Colditz GA, Willett WC. Vitamin E consumption and the risk of coronary heart disease in men. N Engl J Med 1993;328:1450–1456.
195. Yusuf S, Dagenais G, Pogue J, Bosch J, Sleight P. Vitamin E supplementation and cardiovascular events in high-risk patients. The Heart Outcomes Prevention Evaluation Study Investigators. N Engl J Med 2000;342:154–160.
196. Enstrom JE, Kanim LE, Klein MA. Vitamin C intake and mortality among a sample of the United States population. Epidemiology 1992;3:194–202.
197. Osganian SK, Stampfer MJ, Rimm E, et al. Vitamin C and risk of coronary heart disease in women. J Am Coll Cardiol 2003;42:246–252.
198. Feskanich D, Singh V, Willett WC, Colditz GA. Vitamin A intake and hip fractures among postmenopausal women. Jama 2002;287:47–54.
199. Melhus H, Michaelsson K, Kindmark A, et al. Excessive dietary intake of vitamin A is associated with reduced bone mineral density and increased risk for hip fracture. Ann Intern Med 1998;129:770–778.
200. Caire-Juvera G, Ritenbaugh C, Wactawski-Wende J, Snetselaar LG, Chen Z. Vitamin A and retinol intakes and the risk of fractures among participants of the Women's Health Initiative Observational Study. Am J Clin Nutr 2009;89: 323–330.
201. Ballew C, Galuska D, Gillespie C. High serum retinyl esters are not associated with reduced bone mineral density in the Third National Health And Nutrition Examination Survey, 1988–1994. J Bone Miner Res 2001;16:2306–2312.
202. Blot WJ, Li JY, Taylor PR, et al. Nutrition intervention trials in Linxian, China: supplementation with specific vitamin/mineral combinations, cancer incidence, and disease-specific mortality in the general population. J Natl Cancer Inst 1993;85:1483–1492.
203. Stampfer MJ, Hu FB, Manson JE, Rimm EB, Willett WC. Primary prevention of coronary heart disease in women through diet and lifestyle. N Engl J Med 2000;343:16–22.
204. Rothman KJ, Moore LL, Singer MR, et al. Teratogenicity of high vitamin A intake. N Engl J Med 1995;333: 1369–1373.

2 Health Economics of Preventive Nutrition

Adrianne Bendich and Richard J. Deckelbaum

Key Points

- The economic burden of obesity, nutritionally related adverse pregnancy outcomes, cardiovascular and cerebrovascular disease, cancer, age-related blindness, and osteoporosis is a global problem and costs will continue to rise as the developed nations' population age and the developing nations continue to have high birth rates.
- Major discoveries have been made in preventive nutrition within the past 20 years including utility of folic acid in neural tube birth-defect prevention, high-dose antioxidants plus high doses of zinc for reduction of age-related macular degeneration progression, and calcium and vitamin D in hip fracture prevention.
- Implementation of preventive nutrition strategies has been shown to be cost-effective in developing nations and in more economically advantaged populations.
- Daily use of Certain nutritional supplements, such as multi vitamins, reduce the risk of several adverse pregnancy outcomes, is associated with reduced risk of age-related visual and neurological deterioration, and several types of cancers.
- Improvements in diets, cessation of lifestyle habits that increase the risk of chronic disease and adverse pregnancy outcomes, daily exercise, and the consumption of dietary supplements where indicated will decrease health-care costs.

Key Words: Adverse pregnancy outcomes; age-related blindness; cancer; cardiovascular disease; cost-effectiveness; economic burden; obesity; osteoporosis

2.1. INTRODUCTION

Health economics examines both the costs of care (economic burden) as well as the potential costs savings of disease/treatment avoidance. The objectives of this chapter are to first examine the current epidemiological data that describe the health status of adults and the economic burden of disease; second, to review the potential for disease/adverse event prevention associated with consumption of healthy diets/nutritional interventions in both developed and developing nations, and finally, to look at the cost-effectiveness of preventive nutrition strategies in relevant populations. The major diseases/conditions examined in this chapter include obesity, nutritionally related adverse pregnancy outcomes, cardiovascular and cerebrovascular diseases, cancer, age-related blindness, and osteoporosis. Our goal is to document the totality and the consistency of the data and then attempt to quantify the savings that could be seen once preventive nutrition strategies are put in place.

2.2. HEALTH STATUS: INTERACTIONS WITH DIET AND OTHER LIFESTYLE FACTORS

In the United States, life expectancy has increased annually since 1975. As of 2006, life expectancy at birth for males is 75.4 years and for females 80.7 years *(1)*. Heart disease, cancer, and cerebrovascular disease remain the number 1–3 leading causes of death.

A. Bendich, R.J. Deckelbaum (eds.), *Preventive Nutrition*, Nutrition and Health, DOI 10.1007/978-1-60327-542-2_2,

2.2.1. Nutrition and General Health Status

The US Centers for Disease Control's (CDC) National Health Interview Survey, 2006, monitors the health of the civilian, non-institutionalized US population and was published in early 2008 *(2)*. The questionnaire is administered in person to over 29,000 selected households by a trained interviewer and is designed to be representative of national health status. In 2006, 66% of adults 18 years of age and over self-reported excellent or very good health. However, when the CDC conducted the National Health Interview Survey in over 58,000 individuals in 2000–2003 in a sample that was nationally representative of individuals 55 years of age and older, the responses to health status questions were reflective of the age of the respondent *(3)*. Individuals were asked if they would describe their health in general as excellent, very good, good, fair, or poor. About 20% of those 55–64 years responded with fair or poor health and the percentage increased to about 34% in those 85 years and older. Of interest, in the eldest group, 50% reported high blood pressure, about 39% reported heart disease, and 11% reported diabetes. The survey collected self-reported information on physical functioning and found that from 55 to 64 years of age, between 10 and 20% had some difficulty walking one-quarter of a mile, walking up 10 stairs, or shopping. By age 85 or older, these percentages increased to 56, 46, and 36. Relevant to nutritional status, this survey also collected information on total loss of teeth. The percentage of individuals 55–64 years of age who were edentulous (toothless) was about 14% and steadily increased to 40% in adults of 85 years and older.

Limitations in physical functioning and consequently the capability of completing the activities for daily living including meal preparation can adversely affect nutritional status. Of interest are new data that suggest that food choices may actually impact upon mobility. There are few studies that have examined the association between physical functioning and dietary habits. Recently, the Study of Women's Health Across the Nation (SWAN) reported significant associations between baseline dietary intake of fats, fruits, vegetables, and fiber and decreases in physical functioning over 4 years in over 3,000 healthy women who were 42–52 years of age at baseline. Women in the highest quartiles of cholesterol and total fat intakes had 40 or 50%, respectively, greater odds of being more limited in activity than women in the lowest quartiles. With regard to fruit, vegetable, and fiber intakes, women with the lowest intakes had a 60, 50, or 80% increased risk of more physical limitations compared to the women who met or exceeded the recommended intake of these dietary components *(4)*.

The CDC's National Health and Nutrition Examination Survey (NHANES) includes detailed in-person home interviews as well as mobile health examination centers where biological samples are collected and body measurements are recorded. Weiss et al. *(5)*, using NHANES data from over 4,000 adults aged 65 years or older, determined the percentage of the cohort who had one or more of the five following conditions: arthritis, cerebrovascular accident, chronic lower respiratory tract disease (CLRT), coronary heart disease (CHD), or diabetes. They found that the majority of participants who had one of these conditions had at least one other coincident disease. Among women, less than 20% had only CHD or CLRT or diabetes; in men, about 30% had only one disease. Arthritis was reported in almost half the men and women in this cohort as their only disease of the five enumerated in this study. Thus, the other 50% with arthritis had another one or more of these serious conditions. Comorbidities increase the risk of drug–drug interactions as well as drug–nutrient interactions, which can increase costs associated with disease treatment. Still, there are very few analyses that look at the health economics of multiple diseases, and none look at primary or secondary prevention with nutritional interventions.

Perceptions about risks for chronic disease can impact decisions about diet and health. For many postmenopausal women, breast cancer is considered to be a much greater risk than heart disease, and risk of hip fracture is often underestimated. However, recent data from the Women's Health Initiative (WHI) Observational Study of more than 83,000 women document that more postmenopausal women experienced a fracture in any year than the combined prevalence of new-onset breast cancer and serious heart disease *(6)*. Although cardiovascular disease (CVD), cancer, and stroke are the major causes of death in older women, and fractures are less prevalent in black compared to white women, mortality following a hip fracture is higher among black women than white women. There remain consistent findings of a lack of physician recommendation of adequate calcium and vitamin D intake even in older women who have suffered hip fractures *(7)*.

2.2.2. Diet, Smoking, and Health

National surveys also examine the diet and health-related habits in adolescents and have documented that cigarette smoking, physical inactivity, alcohol consumption, and poor diets are frequently seen. Almost one-quarter of high school students smoked cigarettes on one or more of the past 30 days, had less-than-recommended intakes of nutrients, and were not exercising. A recent publication from a survey study of over 4,500 adolescents (aged 11–18; 1/3 in middle school and 2/3 in high school) from Minneapolis, MN, documented that about 17% smoked cigarettes. Those who smoked daily were significantly less likely to eat breakfast and lunch and significantly more likely to consume more than three fast food meals/week than non-smokers. Daily smokers consumed significantly more calories, and less fruits and vegetables, fiber, calcium, vitamin C, and folate/day than non-smokers. Daily smokers also engaged in significantly less vigorous physical activity/week than non-smokers and also had less moderate exercise/week *(8)*. These findings are similar to those reported by Wilson et al. from an even larger survey study of teens in Virginia *(9)*. The Youth Risk Behavior Surveillance Survey in 2007 documented that in students in grades 9–12 about 13% were obese; about 79% had not eaten five or more fruits and vegetables in the past 7 days; about 20% had smoked cigarettes in the past 30 days; about 30% had driven in a vehicle that was driven by someone who had been drinking alcohol; 75% had consumed alcohol in the past 12 months—almost 45% had at least one alcoholic drink in the past 30 days and 26% had five or more drinks within a few hours at least one time during the past 30 days; about 20% of students used marijuana one or more times in the past 30 days; over 4% had used methamphetamines; and about 65% had not met recommended levels of physical activity. It should also be noted that there are approximately 757,000 pregnancies among young women aged 15–19 years annually and more than 9 million cases of sexually transmitted diseases reported annually in persons 15–24 years old *(10)*.

The habits developed during adolescence often continue into adulthood and are likely to influence the risk of chronic diseases later in life. NHANES data from 1999 to 2002 reported that 24% of the US adults were current cigarette smokers and 25% were former smokers. Of current smokers, 58% started smoking before the age of 18 and currently smoked an average of 16 cigarettes/day. The survey also found that 70% of adults were current users of alcohol and that 28% had consumed five or more drinks in 1 day in the prior year; about twice as many men as women were moderate to heavy drinkers *(11)*.

With regard to obesity, discussed in detail below, in US adults 20 years old and older with body mass index >25, 64% were overweight or obese and 30% were frankly obese. Twelve percent of adults had ever been told by a doctor or other health professional that they had heart disease and 21% had been told on two or more visits that they had hypertension *(12)*. Of interest as well are the results from another US survey of non-institutionalized adults that found that in any given week, 81% of adults are taking at least one medication and 27% take at least five. In this survey medication includes both prescription and non-prescription drugs as well as vitamin/mineral/herbal or natural supplement. Even so, 50% of the individuals were taking prescription drugs, and 95% of the women 65 years or older took at least one prescription drug *(13)*.

2.3. GUIDELINES FOR DIET/NUTRIENTS AND REDUCED DISEASE RISK

The major chronic diseases share common cellular and biochemical mechanisms in their pathogenesis *(14–16)*. For example, oxidative damage, inflammation, and cell proliferation are common in both atherosclerosis and cancer and are of importance to some of the complications associated with diabetes and age-related eye diseases. Changes in signal transduction and gene expression relate to cancer, atherosclerosis, obesity, and diabetes. As well, DNA modifications likely contribute mechanistically to each of these disease classes. Similarly, there are consistencies in the requirements for the essential nutrients needed to prevent the pathogenic changes at the cellular level. Thus, it is not surprising that the sets of nutrition recommendations aimed at reducing the risk of a number of chronic diseases and developed by a number of different private and government organizations show far more commonalities than differences. The recommendations of most nutrition and public health groups are very similar to the 2000 United States Dietary Guidelines of the US Department of Health and Human Services and US Department of Agriculture, which are outlined in Table 2.1 *(17)*.

An important question is how to implement the current dietary recommendations for all population groups regardless of their economic and educational status. Where will food fortification and micronutrient supplementation be of most benefit? The reality is that the majority of populations, both in the United States and elsewhere, do not follow dietary recommendations *(18–21)*. Given the available evidence that higher fruit and vegetable as well as whole grain intakes can have marked effects on reduction of chronic disease risk, it is quite likely that improved diets alone would be a major step

Table 2.1
2000 US Dietary Guidelines

Aim, Build, Choose—For Good Health
Aim for fitness
• Aim for healthy weight
• By physically active each day
Build a healthy base
• Let the pyramid guide your food choices
• Choose a variety of grains daily, especially whole grains
• Choose a variety of fruits and vegetables daily
• Keep food safe to eat
Choose sensibly
• Choose a diet that is low in saturated fat and cholesterol and moderate in total fat
• Choose beverages and foods that limit your intake of sugars
• Choose and prepare foods with less salt
• If you drink alcoholic beverages, do so in moderation

forward in chronic disease risk reduction and economic cost. However, it is also clear that fortification and supplementation of certain nutrients have been associated with reduced risk of chronic diseases in populations that have not routinely consumed the highest intakes of recommended foods *(22–24)*. An inclusive approach of combining the best of many nutritional delivery systems and including physical activity and smoking reduction may provide a greater potential for cost-effectiveness that diet recommendations alone.

Relevant to the developing countries, a report from UNICEF and the Micronutrient Initiative *(25)* finds that lack of basic vitamins and minerals in the diet is holding back their economic development. A lack of key vitamins and minerals is responsible for impairing intellectual development, compromising immune systems, and increasing the risk of serious birth defects. The report found that fortification of wheat flour with iron and folic acid in the 75 neediest countries would be expected to reduce iron deficiency by 10%, and birth defects could be lowered by a third. Such fortification would cost a total of about $85 million, which is about 4 cents per person. The estimated benefit is that these countries would gain $275 million from increased productivity and $200 million from enhanced earning potential.

2.3.1. Guidelines for Macronutrient Intakes and Implementations

The United States Department of Agriculture (USDA) has developed a Healthy Eating Index (HEI) that includes the five food groups (meat, dairy, fruits, vegetables, and grains), four nutrients (total fat, saturated fat, cholesterol, and sodium), and dietary variety. The index is used to provide an overall picture of the type and quantity of foods people eat, their compliance with dietary recommendations, and the variety of their diets *(26)*. Each of the dietary components is weighted from 5 to a maximum value of 20 and a perfect score on the HEI would equal 100. An HEI score of over 80 is labeled as a "good" diet; between 51 and 80 indicates a diet that "needs improvement"; and less than 51 is considered a "poor" diet. The criteria for the index are included in Table 2.2. The index was applied to the US adults 60 years of age and over who participated in the NHANES 1999–2002. Overall, only 17% of the US adults over 60 had a diet that met the criteria of "good"; 68% had a diet that "needs improvement"; and 14% had a diet considered as "poor." Current smokers had the lowest index (60.4) of any group including those who had no teeth (edentulous) and those with less than a high school education, both of which had scores of 63.3 or 63.4. With regard to the individual nutrients, only 18% met the dietary recommendations for grains; 23% for dairy; 26% for meat; 27% for fruit; and 32% for vegetables. These data suggest that calcium intakes from dietary sources are low for more than 75% of the population at greatest risk for osteoporosis. Animal protein also appears to be low for more than 70% of this population. The dietary sources of the nutrients in fruits, vegetables, and grains would also be low for more than 70% of this sector of the population. A recent analysis of data from a cohort of 738 US women found that the average HEI score was about 60%. Consumption of whole grains and dark green and orange vegetables and legumes was again found to be less than half of the recommended intake levels *(27)*. As low intakes of many of these nutrients are consistently shown to be associated with increased risk of cancer, cardiovascular disease, diabetes, age-related blindness, and dementia, these data raise concerns about implementation of preventive nutrition strategies in the aging population. Perhaps, if there were economic incentives to encourage healthy eating habits, there would be an improvement in the HEI in the future and an improvement in health outcomes as well.

It is not surprising that major national efforts have been undertaken to provide practical advice and evaluation of diet programs that might reduce risk factors for chronic disease, especially obesity and hypertension. The US Institute of Medicine developed guidelines for macronutrient intakes called acceptable macronutrient distribution ranges (AMDRs) for carbohydrate, fat, *n*–3 and *n*–6

Table 2.2
USDA Healthy Eating Index and Standards for Scoring Criteria—2005

Component	Maximum Points	Standard for Maximum Score	Standard for Minimum Score of 0
Total fruit (includes 100% juice)	5	≥0.8 cup equiv./1,000 kcal	No fruit
Whole fruit (no juice)	5	≥0.4 cup equiv./1,000 kcal	No whole fruit
Total vegetables	5	≥1.1 cup equiv./1,000 kcal	No vegetables
Dark green and orange vegetables and legumes[a]	5	≥0.4 cup equiv./1,000 kcal	No dark green or orange vegetables or legumes
Total grains	5	≥3.0 oz equiv./1,000 kcal	No grains
Whole grains	5	≥1.5 oz equiv./1,000 kcal	No whole grains
Milk[b]	10	≥1.3 cup equiv./1,000 kcal	No milk
Meat and beans	10	≥2.5 oz equiv./1,000 kcal	No meat or beans
Oils[c]	10	≥1.2 g equiv./1,000 kcal	No oil
Saturated fat	10	≤7% of energy[d]	≥15% of energy
Sodium	10	≤0.7 gram/1,000 kcal[d]	≥2.0 g/1,000 kcal
Calories from solid fats, alcoholic beverages, and added sugars (SoFAAS)	20	≤20% of energy	≥50% of energy

Intakes between the minimal and maximal levels are scored proportionately except saturated fat and sodium.
[a]Legumes counted as vegetables only after meat and beans standard is met. [b]Includes all milk products such as fluid milk, yogurt, cheese, and soy beverages. [c]Includes non-hydrogenated vegetable oils and oils in fish, nuts, and seeds. [d]Saturated fat and sodium get a score of 8 for the intake levels that reflect the 2005 Dietary Guidelines, <10% of calories from saturated fat and 1.1 g of sodium/1,000 kcal, respectively.

polyunsaturated fatty acids, and protein based on evidence from completed intervention trials as well as epidemiological evidence that suggested a role of macronutrient intakes in chronic disease prevention. The AMDR for adults for fat is 20–35% of calories (5–10% of energy should be from *n*–6 fatty acids and about 10% can be consumed from long chain *n*–3 fatty acids); for carbohydrate, it is 45–65% of calories (with no more than 25% as added sugar); and for protein, it is 10–35% of calories *(28)*.

Since the publication of the AMDRs, a number of intervention studies with different diet plans have been studied. During the same time period, several professional societies also developed their own dietary guidelines for primary and/or secondary prevention of heart disease, diabetes, or cancer. It is of interest to note that even though the organizations developed their guidelines independently, these were more alike than different so that the dietary recommendations from the American Heart Association, the American Diabetes Association, and the American Cancer Society were very similar. De Souza et al. compared the macronutrient content of five popular diet plans as well as the Dietary Approaches to Stop Hypertension (DASH) diet and the three OmniHeart diets with the ranges recommended in the AMDR *(28,29)*. Of relevance, the DASH diet had been shown to significantly reduce blood pressure and serum lipid risk factors in populations with and without hypertension and the OmniHeart diets had improved cardiovascular disease (CVD) risk.

2.4. OBESITY

Obesity is an increasing global problem. The World Health Organization (WHO) estimates that over 300,000,000 are obese and that over a billion are overweight.

Obesity has been designated a disease recently by a consensus of prominent researchers in 2008 *(30)*. Obesity increases the risk of other diseases including diabetes and cardiovascular disease as well as certain cancers. The prevalence of obesity in the US adults doubled between 1980 and 2004 *(31)*. The magnitude of the obesity epidemic is clearly obvious: more than 1/3 of all adults or 72 million people were obese in the United States in 2005–2006. Even more disturbing are the data that indicate that 40% of adults 40–59 years of age in the United States are obese and in this age group, more than 50% of Black and Mexican-American women are obese. The population with the highest prevalence of obesity is Black women 60 years of age or older: 61% are obese. Obesity is linked to between 280,000 and 325,000 deaths/year in the United States, and is more costly than smoking-related deaths, the number one killer. It is estimated that obesity-related expenditures reached $75 billion in 2003. Currently, about 27% of the US adults and children are classified as obese *(32)*. A recent economic analysis of the costs associated with obesity in Australia found that there were about 3.1 million obese adults in 2005 and that the cost of obesity in 2005 was over $1.7 billion that included direct health costs and indirect costs for lost work productivity, absenteeism, and unemployment *(33)*. In the United States, more than 25% of health-care costs are associated with obesity and inactivity *(34)*. Direct health-care costs of obesity exceed $61 billion annually and $76 billion are spent on health care for the inactive. Wang et al. *(35)* have projected the future prevalence of obesity and consequent health-care costs in the United States. They predict that by 2030, over 85% of adults will be overweight and 51% will be obese. Health-care costs attributable to obesity/overweight are predicted to double every decade and reach close to $1 trillion by 2030 (*see* Chapters 14–18).

Decreased fat intake, particularly saturated fat intake, is clearly linked to decreased serum cholesterol levels and decreased prevalence of cardiovascular complications such as coronary artery disease *(36)*. While recent studies place some doubt on the contribution of fat intake toward increasing risk of breast cancer, high levels of fat intake are still associated with increased production of inflammatory adipokines that are associated with increased risk of some cancers and type 2 diabetes (*see* Chapter 17) *(2,37–39)*. Foods rich in other dietary components including fiber, complex carbohydrates, and micronutrients appear to decrease the risk of certain forms of cancer as well as coronary heart disease and manifestations of diabetes *(40–43)*. With regard to obesity-related diabetes, a recent analysis of the direct medical costs of care, based upon actual cases, showed that the lowest costs were associated with those diabetics who were diet controlled (*see* Chapters 15 and 16). Obesity added significantly to health costs as did use of insulin and more advanced disease resulting in kidney failure and need for dialysis. Dialysis alone increased costs 11-fold *(44)*.

The increase in obesity is not confined to developed nations. The status of obesity in China was reviewed in the March 2008 issue of *Obesity Reviews*. Chen *(45)* documented that in 2002, 22.8% of adult Chinese were overweight and 7.1% were obese using BMI criteria of >28 for obesity (a total of 300 million adults). From 1992 to 2002, there was a 40% increase in the prevalence of overweight and a remarkable 97% increase in obesity. The associated direct medical costs attributable to overweight and obesity is estimated to be about $2.74 billion accounting for about 25% of the medical costs associated with hypertension, type 2 diabetes, coronary heart disease, and stroke. The consequences of medical costs associated with obesity-related diseases to the economy in developing countries can be overwhelming and are reviewed in detail in Chapter 18.

The prevalence of childhood obesity in the United States is growing faster than that of adult obesity and has grown from 6.5% in 1980 to 16.3% in 2004. Childhood obesity has resulted in the development of adult diseases in children such as type 2 diabetes and hypertension *(34)*. The unexpected large increase in childhood obesity and overweight, reviewed extensively by the leading researchers *(46–48)*, documents that the unfortunate consequences are not only seen in childhood but also in adulthood even if the obese or overweight child develops into a normal weight adult. China also reports a significant increase in childhood obesity from 1985 to 1995 that has continued in 2000 where the prevalence of overweight/obesity in 7–12-year-old boys from urban environments approached 29% and was about 16% for girls *(45)*.

The critical issue with obesity and overweight is how can one lose the excess weight and then maintain the weight loss. A secondary question is whether one can reduce risk factors for chronic disease, and consequently the costs of treatment, even without losing weight (*see* Chapter 14). In fact, the second question has been answered positively in several (but not all) well-controlled studies in overweight adults including the Dietary Approaches to Stop Hypertension (DASH) study and the OmniHeart study as well as in smaller clinical trials that have shown that both systolic and diastolic blood pressure, serum cholesterol, serum low-density lipoproteins (LDLs), and other markers of cardiovascular disease can be significantly reduced with dietary modifications including reduction in salt intake and increased potassium intake, even without weight loss *(29,49,50)*. DASH unfortunately did not improve all risk factors as levels of protective high-density lipoprotein (HDL) cholesterol were reduced and triglycerides were not reduced. The dietary refinements tested in the OmniHeart study suggested that lowering carbohydrate intake and increasing protein or unsaturated fat intake (from vegetables mainly) resulted in reduction of blood pressure, LDL cholesterol, and triglycerides, with modest or no reductions in HDLs even when weight remained constant. In the largest intervention study on dietary fat restriction without weight loss in postmenopausal overweight women, small reduction of total fat intake did not have beneficial effects on these risk indices in the Women's Health Initiative (WHI) dietary modification intervention trial *(51,52)*.

Numerous weight loss diets are available to consumers and their efficacy for weight loss may be dependent on compliance more so than the ratios of fat to carbohydrate to protein or even the types of these macronutrients *(29)*. However, it appears that different weight loss diets can have differing effects on risk factors especially for cardiovascular disease. Also, there are few data from well-controlled studies that point to the best diets for maintaining weight loss although there are data in pre-menopausal overweight women using diets that result in weight loss, blood pressure, and triglyceride reduction but not LDL reduction *(53)*. Some suggest that physical activity may play a key role in weight maintenance; however, there are few studies with consistent findings. Nevertheless, it is obvious that blood pressure reduction as well as LDL reduction reduce the risk of CVD and may reduce the need for drug therapy to control these risk factors. The costs associated with such dietary changes appear to be relatively modest *(49)* and the reduction in health-care costs may outweigh these costs. Kahn et al. using a model system based on the US NHANES data suggest that reduction in BMI would reduce the risk of myocardial infarcts by about 30% over 30 years and is one of the 11 preventive strategies that could greatly reduce the risk of CVD and significantly lower health-care costs *(54)*. Unfortunately, this model system relied upon prescription drugs and other expensive interventions to reduce blood pressure and lower LDL independent of weight loss in the obese and thus reported an overall increase in health-care costs to prevent between 36–63% of all heart attacks and 20–31% of strokes.

A major factor in the increase in health-care costs is the morbidity associated with obesity. Annual health-care costs attributed to obesity from all causes were estimated at over $99 billion in the United States *(40)*. The dissection of obesity costs from other factors that are associated with increased cardiovascular disease and cancer remains difficult. The association of obesity with hyperlipidemia, hypertension, and type 2 diabetes contributes to increased risk (and costs) for cardiovascular disease and stroke *(55)*. Despite the difficulty in assigning exact dollar costs to each of these risk factors and diseases it seems obvious that combining nutritional approaches with a healthy lifestyle can have a marked impact on reducing health-care costs. Of note, only 0.25% of health-care cost dollars go to prevention strategies that would be very cost-effective *(19)*.

2.5. OPTIMIZING PREGNANCY OUTCOMES

The "Barker hypothesis" suggests that nutritional exposure in utero has effects that last a lifetime, and birth weight and gestation length are predictive of heart disease decades post exposure. Other chronic diseases of aging have also been linked to maternal nutrition *(56–59)*. In addition to these potential long-term effects, maternal diet and nutritional status can have more immediate effects on maternal health as well as the health of the fetus and newborn. Over the past 20 years, the importance of preventive nutrition in enhancing pregnancy outcomes has been a very valuable area of research. The key advances have been in reduction of major birth defects, prolonging pregnancy to term and avoiding maternal complications *(15,60,61)*. Each of these conditions that are prevented results in cost savings and also reductions in attendant emotional and psychological costs. Optimizing pregnancy outcomes is also a high priority in developing nations and certain essential micronutrient deficiencies have profound effects on the physical as well as the mental capabilities of the offspring and can significantly increase the risk of maternal morbidity and mortality.

2.5.1. Birth Defects

A recent UNICEF report indicates that developing countries are examining the economic burden of adverse pregnancy outcomes *(25)*. Iodine deficiency, prevalent in many African and Asian nations, has serious and irreparable consequences in pregnancy. As many as 20 million babies a year are born mentally impaired as a result of maternal iodine deficiency. Severe iron deficiency anemia contributes to the deaths of an estimated 50,000 women a year during childbirth. Folate deficiency is linked to approximately 200,000 severe birth defects every year, and is associated with roughly 1 in 10 adult deaths from heart disease. Food fortification with essential vitamins and minerals in foods that are regularly consumed by most people (such as flour, salt, sugar, cooking oil, and margarine) is the preferred method to increase consumption in the developing countries and is predicted to cost only a few cents per person per year. Supplementation is considered only for the most vulnerable groups (particularly children and women of childbearing age) with vitamin and mineral supplements in the form of tablets, capsules, and syrups. Although the actual costs of the supplements are low, implementation costs are higher than for fortification. However, in some settings, fortification and supplementation programs cannot be successful without a parallel program to control diseases such as malaria, diarrhea, and parasitic infections which reduce the ability of the body to absorb and retain essential vitamins and minerals. Nevertheless, it is estimated that in the 75 neediest countries, if wheat flour was fortified with iron and folic acid, iron deficiency would be reduced by 10% and birth defects could be lowered by a third.

Birth defects remain the leading cause of infant mortality and morbidity in the United States. Cardiovascular birth defects are the most common birth defect; neural tube birth defects (NTD) are the second leading defect. One major preventive nutrition finding of the 20th century was that periconceptional use (use of the supplement prior to conception and during the first trimester) of a folic acid-containing multivitamin/mineral supplement significantly reduced the occurrence of NTD *(62)*. Another key finding was that supplementation with folic acid alone could more than halve the risk of having a second child affected by NTD *(63)*.

The only intervention trial that examined the potential to reduce the risk of first occurrence of NTD was done in Hungary using a prenatal multivitamin supplement containing 800 μg of folic acid and several other essential micronutrients. The Hungarian trial *(64)* found a significant reduction in two major types of birth defects in addition to more than a 50% reduction in NTD. Defects of the urinary tract, especially obstructive anomalies of the urinary tract were reduced by more than 50% in the supplemented group. Cardiovascular defects, which are 10 times more common than NTD, were halved in the multivitamin group. In addition to showing that periconceptional use of multivitamins significantly reduced NTDs, Czeizel and Dudas found that women who took the multivitamin containing 12 vitamins (including 800 μg of folic acid, 4 μg of vitamin B_{12}, 2.6 mg of vitamin B_6 as well as 100 mg of vitamin C and 15 IU of vitamin E) also had less morning sickness than the women taking the matched placebo *(64)*. In the United Kingdom's Medical Research Council *(63)* trial for prevention of neural tube birth defect recurrence, the multivitamin used in that study contained only eight of the vitamins and did not include vitamins E and B_{12}; it only contained 40 mg of vitamin C and 1 mg of vitamin B_6. Even though it contained 4 mg of folic acid (five times the level in the Czeizel and Dudas trial) there was no decrease in cardiovascular birth defects in either the group taking folic acid alone or the group taking folic acid plus their multivitamin *(63)*. The supplement used in the Czeizel and Dudas trial is more comparable to the typical one-a-day type multivitamin/mineral supplement sold in the United States than the supplement used in the MRC trial. Thus, it is not surprising that survey data from the United States also report lowered cardiovascular birth-defect risk with periconceptional use of multivitamin/mineral supplements (*see* Chapter 25).

Because cardiovascular birth defects are about 10-fold more prevalent than NTD, the total medical costs associated with these types of defects are high (about $50,000/patient; $3 billion/year). The risk reduction associated with multivitamin use was high (about 50%) *(64,65)* and thus the potential cost savings approached $1.5 billion/year for hospitalization costs alone *(22)*. The cost savings are particularly great because the costs of the multivitamins are relatively low. The cost savings would probably be much greater than $1.5 billion/year because this calculation did not include the costs associated with renal birth defects, cleft lip and palate as well as limb reduction birth defects, all of which were also found to be reduced with multivitamin supplementation periconceptionally *(66)*. Maternal supplementation with multivitamins has also been associated with decreased risk of major childhood brain tumors and leukemia *(67)*, and as presented in Chapters 5 and 25, with a decreased risk of preterm delivery. Still, there are few women (about 20%), even today, who take a multivitamin during the periconceptional period, before they have confirmed their pregnancy *(44)*.

Oakley, in an editorial in the *New England Journal of Medicine*, has suggested that the right advice to American women is to eat the best diets possible and also take a multivitamin containing folic acid to assure the birth-defect preventive level of folic acid is consumed daily *(68)*. This advice was given after the FDA initiated its policy for fortification of enriched grain products and flour with folic acid. It appears that this level, which raises folic acid intakes on an average by 100 μg/day, may not be sufficient to prevent folic acid-responsive birth defects. Further verification of the need for intake of the 400 μg level of folic acid is seen in the recent study from China *(69)*. There was a fourfold reduction in NTD in the high-risk area and a 40% reduction in a lower risk area when women anticipating pregnancy took 400 μg of supplemental folic acid during the periconceptional period.

Obesity also has been associated with an increased risk of birth defects including NTD and cardiovascular birth defects and multiple birth defects. Possible mechanisms to explain the association include metabolic disturbances, such as undiagnosed diabetes, hyperglycemia, or elevated insulin levels, nutritional deficiencies, and increased requirements for certain nutrients. Beyond birth defects, the Barker hypothesis suggests that maternal obesity may increase the risk of chronic diseases during childhood, adolescence, and/or adulthood *(70)*.

2.5.2. Low Birth Weight and Premature Birth Prevention

Low birth weight and preterm delivery often occur together. Preterm delivery is defined as birth following less than 37 weeks of gestation; very preterm delivery is defined as less than 33 weeks gestation. Low birth weight is defined as less than 2,500 grams (g) and very low birth weight is less than 1,500 g. Of all low birth weight infants born in the United States, 60–70% are also preterm. Preterm delivery associated with low birth weight is the second leading cause of infant hospitalization and is also the second leading cause of infant mortality, following birth defects *(71)*. Low birth weight is second only to cardiovascular birth defects in contributing to annual hospitalization costs, exceeding $2.5 billion/year *(22)*. The vast majority of these costs are borne by public funds in the United States and the affected mothers and children are often the least equipped to deal with this event. Preterm births are more prevalent in teens, those with less than a high school education and those with the lowest incomes. The most recent data on infants born in the United States with low birth weight was published in 2002 and documents that about 6.5% of infants born to white women have LBW; for black women, the rate was considerably higher, 13% *(72,73)*.

Scholl et al. *(74)* found in a prospective, case–control study that pregnant women in Camden, NJ, who took prenatal multivitamins during the first trimester had a fourfold reduction in very preterm births and a twofold reduction in preterm births. Even if supplementation began in the second trimester, there was a significant twofold reduction in very preterm as well as preterm births. The risk of very low birth outcomes (which was highly correlated with preterm delivery) was dramatically reduced by six- to sevenfold when prenatal multivitamins were taken during the first two trimesters. Low birth weight was also reduced significantly with supplementation. These results were found in women who were at high risk for preterm/low birth weight outcomes: they were poor, teens, and many had low weight gain during pregnancy. Prenatal supplements increased iron and folate status significantly, but did not alter serum zinc level (*see* Chapter 26). Previously, low iron and/or folate status has been associated with increased risk of preterm and low birth weight *(61)*. Zinc-containing multivitamins have also been shown to reduce preterm births in an intervention study *(75)*. Because there was such a dramatic reduction in preterm births in the women who took zinc and (presumably) folic acid-containing prenatals, the cost savings were predicted to reach over $1.5 billion/year *(22)*. Calcium, antioxidants, and long-chain polyunsaturated fatty acids have also reduced the risk of preterm births *(76)*. It would appear that there is significant overlap between the optimal intakes of the nutrients linked to reducing preterm births and those nutrients that reduce the risk of major birth defects. Thus, such an intervention would be highly cost-effective, saving billions and perhaps trillions of dollars/year in health-care costs globally.

Recently, there have been links made between maternal diet during pregnancy, preterm birth, and cardiovascular disease in the offspring 50 or more years after their birth *(56,57)*. Whether women who used supplements before and during pregnancy have children who, at middle age, have less chronic diseases such as cardiovascular disease is an unanswered question. Thus, it appears certain that the cost savings associated with a program to provide folic acid-containing multivitamins to all women of childbearing potential before as well as during the entire pregnancy would be far greater than that based on either cardiovascular and other birth defect or preterm birth reduction alone. Moreover, the reduction in suffering for the infant, parents, and family is substantial.

2.5.2.1. Studies of Birth Defect and LBW Prevention from Developing Countries

In developing countries, premature births are a serious problem and an additional problem is intrauterine growth retardation that is often coincident with premature birth, but may also occur with term delivery. De Onis et al. *(77)* reviewed the 12 nutritionally based interventions that have been examined for their potential to reduce the incidence of intrauterine growth retardation and often preterm birth and low birth weight (LBW). Only one of the interventions, balanced protein/energy supplementation during pregnancy significantly reduced the risk of low birth weight. Many of the micronutrient supplementation strategies in the other studies are likely to prove beneficial but larger cohorts would be required to prove efficacy. The micronutrients expected to be effective include vitamin D, folic acid, zinc, calcium, magnesium, and iron. Ramakrishnan *(78)* has also extensively reviewed both observational and intervention studies from developing and developed countries that examined the role of micronutrients in optimizing pregnancy outcomes. The authors remind us that in developing countries, one out of every five infants is LBW (20%) compared to a rate of 6% in developed countries. Moreover, in developing countries, the majority of LBW infants are carried to term. Thus, intrauterine growth retardation is a significant problem in developing countries and impacts the potential physical and most likely, the mental growth of the child throughout their lifetime. Low maternal micronutrient intake of zinc, calcium, magnesium, vitamin A, vitamin C and possibly B vitamins, copper, and selenium is associated with premature birth and LBW. Deficiency of iodine, folate, and/or iron is also linked to adverse pregnancy outcomes.

There have been recent data published from two large, well-controlled intervention studies in developing countries. West et al. *(79)* examined, in a placebo-controlled trial, the effects of weekly supplementation with the recommended dietary allowance of vitamin A provided as either preformed vitamin A (retinol) or beta-carotene in over 20,000 pregnant women in Nepal. They found a significant 40% reduction in maternal mortality with both vitamin A interventions compared to placebo. It is important to note that women were provided with the supplements before conception and that this relatively modest dose of vitamin A in the form of beta-carotene appeared more effective than the preformed supplement of retinol. Low beta-carotene status has been reported in cases of pre-eclampsia in women from developing countries, and as discussed below, the antioxidant potential of beta-carotene (compared to the much lower antioxidant potential of retinol) may have been involved in reducing maternal mortality in the study in Nepal.

The second major study involved over 1,000 HIV-infected pregnant women from Tanzania *(80)*. In this placebo-controlled trial, the pregnant women received either placebo, beta-carotene, and preformed vitamin A, a multivitamin-containing vitamins B_1, B_2, B_6, B_{12}, niacin, folic acid, and vitamins C and E; or the multivitamin plus beta-carotene and retinol from 12 to 27 weeks gestation to birth. There was a 40% reduction in fetal death, 44% reduction in low birth weight, 39% reduction in very preterm birth, and a 43% reduction in small for gestational age outcomes in the groups supplemented with the multivitamin independent of the vitamin A. Additionally, mothers taking the multivitamin had significantly heavier babies than those not taking the multivitamin ($p = 0.01$). Even though the supplement did not contain iron, the women in the multivitamin group had a significant increase in their hemoglobin levels compared to those not taking the multivitamin. Because this study involved HIV-positive women, the investigators also measured the concentration of total T cells (CD3), T helper cells (CD4), and T suppressor cells (CD8). HIV-positive pregnant women who took the multivitamin supplement had significant increases in total T cells, which was mainly due to increases in CD3 cells; CD8 cells also increased. Although vitamin A did not show any effect in this study, it is important to note that the HIV-infected women had low vitamin A status at the onset and the level provided may not have been sufficient and/or may have not been absorbed sufficiently to show an effect. It may also be that the vitamin A was administered too late in these pregnancies to see its effect. In the West et al. *(79)* study, vitamin A supplementation was started before conception. Vitamin A is critical for early embryonic growth. The critical finding in the Fawzi et al. *(80)* study was that a multivitamin supplement containing modest doses of micronutrients significantly improved birth outcomes at levels similar to that seen when poor non-HIV-positive women in the United States used multivitamins during their pregnancies *(74)*.

In both developed and underdeveloped countries, there is a direct relationship between the level of poverty and the risk of low birth weight outcomes. Ramakrishnan *(81)* has rightly argued that improving the nutritional status of poor women prior to and during pregnancy has the potential to significantly reduce the risk of preterm and low birth weight births. It is expected that such programs would still be cost-effective given the many adverse outcomes associated with premature birth and/or low birth weight outcomes.

2.5.3. Reduction in Pre-eclampsia and Other Pregnancy-Associated Adverse Effects

Pre-eclampsia is defined as hypertension and proteinuria beginning during the second half of gestation. Approximately 5% of US pregnant women develop this condition during pregnancy. Pre-eclampsia is the leading cause of maternal death and accounts for more than 40% of premature births worldwide *(82–84)*. The costs associated with pre-eclampsia, therefore, would not only include 40% of the costs associated with the infant of a preterm delivery, but would also include the costs associated with maternal care.

Several mechanisms have been proposed for the initiation and progression of pre-eclampsia including oxidative damage to the placenta and/or imbalance in blood pressure regulation. Because of the importance of micronutrients such as vitamins C and E and beta-carotene as antioxidants and the association of low calcium status and high blood pressure, these micronutrients have been studied in well-designed protocols to determine their effects on pre-eclampsia.

Fourteen randomized clinical trials that involved the use of calcium supplements during pregnancy to determine effects on maternal blood pressure and pre-eclampsia were reviewed by Bucher et al. *(85)*. They concluded that calcium supplementation (usually 1,500–2,000 mg/day from week 20 to birth) led to a significant reduction in maternal systolic and diastolic blood pressure and 62% reduction in the risk of pre-eclampsia. The authors recommended that calcium supplementation be provided for all women at risk for pre-eclampsia. It should be noted that the majority of the populations included in this analysis were not from the United States and had intakes of calcium of about 300–500 mg/day, well below their nation's recommended intake levels *(86,87)*.

Several important studies have been published following this review. Levine et al. *(84)* randomized over 4,500 US pregnant women to either 2,000 mg/day of calcium or placebo and found that there was no difference in the rate of pre-eclampsia between the groups (about 7%). It must be noted, however, that these women were not at an increased risk for pre-eclampsia and that they had dietary calcium intakes of over 1,100 mg/day—very close to the recommended intake level during pregnancy, yet well above the national average. In a recent re-evaluation of the data concerning calcium and risk of pre-eclampsia, DerSimonian and Levine *(83)* again found that calcium supplementation did not appear to further lower the risk of pre-eclampsia, but for high-risk pregnancies, calcium supplementation appeared to significantly lower the risk. High-risk pregnancies include teens, women with pre-existing hypertension, and women carrying multiple fetuses. However, in the most recent trial of 1,800 mg/day of elemental calcium supplementation in over 400 pregnant Australian women with apparently adequate dietary calcium intakes (median 1,100–1,200 mg/day) and other characteristics very similar to those in the Levine et al. *(84)*, they reported a greater than twofold reduction in the risk of pre-eclampsia and a reduction in the risk of preterm delivery from 10% (placebo) to 4.4% (calcium supplemented). There were also reductions in hospital admissions for threatened preterm labor ($p = 0.03$), and preterm premature rupture of membranes ($p = 0.08$), which would reduce health-care costs above and beyond those reduced by the benefits seen with calcium supplementation. In fact, the authors calculated that 73 low-risk women or 42 high-risk women would need to be supplemented to prevent one case of pre-eclampsia. Given that supplementation was only for at most 20 weeks, and the approximate cost of the calcium supplements are about $20.00/woman, it would cost about $1,500 to provide the supplements to 73 women, well below the expected medical costs associated with one case of pre-eclampsia and the potential pre-term and/or low birth weight infant *(88)*.

One other study of interest examined the blood pressure of the children of mothers who took calcium during pregnancy to prevent pre-eclampsia. Belizan et al. *(89)* showed that at 7 years of age, there was a significantly lower systolic blood pressure in the group of children whose mothers had taken 2,000 mg/day of calcium during pregnancy. The greatest antihypertensive effect was seen in the overweight children. As mentioned previously, it is not possible to account for all of the savings that might accrue from calcium supplementation if one limits the cost savings to those associated with pre-eclampsia alone. The greatest health-care savings may not be seen until decades after the intervention.

As mentioned above, oxidative stress is another mechanism that has been examined in pre-eclampsia. Vitamin E is the major lipid-soluble antioxidant in serum, and vitamin C is the major water-soluble antioxidant in the blood. Wang et al. *(90)* found and Jain and Wise *(16)* confirmed that serum lipid peroxide levels were significantly higher and serum vitamin E levels were significantly lower in women with pre-eclampsia compared to women with normal pregnancies.

Chappell et al. *(91)* enrolled 283 pregnant women at risk for pre-eclampsia in a placebo-controlled trial using daily supplements of vitamin C (1,000 mg) and vitamin E (400 IU). There was a significant 61% reduction in risk of pre-eclampsia in the antioxidant-supplemented group. These investigators hypothesized that antioxidants might stabilize the maternal endothelium and placenta and thus reduce pre-eclampsia risk. They found that both the plasma marker for endothelial activation and the index for placental dysfunction significantly decreased in the supplemented group. The levels of the antioxidants used in this study were well above national recommended intake levels. Following the publication of these results, four other intervention studies using these doses of vitamins E and C have been conducted and none of these have replicated the Chappell findings *(92,93)*. Of concern is the recent publication of a secondary analysis of the latest pre-eclampsia study that reports a significant increased risk of premature rupture of membranes in the supplemented women *(94)*. Given the inconsistencies in the findings, that may be due to different population groups, different pre-existing conditions and other factors independent of the basic research data linking oxidative damage to increased risk of pre-eclampsia, the supplementation with high doses of vitamin E and vitamin C for prevention of pre-eclampsia is not being recommended.

2.5.4. Preventive Nutrition in Childhood

Preventive nutrition strategies in childhood center on two major areas of concern: the effects of severe and marginal deficiencies of essential nutrients in developing countries and the health effects of obesity in the children in developed as well as developing countries. With regard to the former, the UNICEF report in 2004 *(25)* indicates that iron deficiency impairs mental development in young children and is lowering national IQs. It also undermines adult productivity, with estimated losses of 2% of GDP in the worst-affected countries. Vitamin A deficiency compromises the immune systems of approximately 40% of children under 5 years in the developing world, leading to the deaths of 1 million youngsters each year. Undernutrition of essential vitamins and minerals has its most devastating and durable effects when it occurs early in life and continues through early childhood.

Undernutrition and essential micronutrient deficiencies are not exclusively seen in the underdeveloped countries. In the United States, for instance, the Pediatric Nutrition Surveillance System that monitors the nutritional status of low-income children reports that about 18% of the children under 2 years old and 17% of children 2–5 years of age had iron deficiency anemia *(72)*. Unfortunately, the percentage of African-American children with anemia actually rose from 21% in 1989 to about 25% in 1997.

In addition to iron deficiency's adverse effects on cognitive function, deficiencies in iodine, polyunsaturated fatty acids, and vitamin B_{12} have also been linked to decreased cognitive abilities. Other studies have shown that supplementation of well-fed children with zinc, folic acid, vitamins B_6 and B_{12}, iron, or vitamin A improved cognitive function *(95)*.

Childhood cancers are relatively rare events, but are the number 2 killers of children after accidents. Bunin and Bosco examined all of the epidemiological studies associating diet and supplement use and risk of leukemia and brain tumors and found the strongest association between maternal use of multivitamins during pregnancy and decreases in cancer risk (Chapter 5).

2.6. THE ADULT CHRONIC DISEASES

There is a critical convergence of the micronutrients discussed as preventive nutrition strategies related to pregnancy and those associated with reduction in risk of chronic diseases throughout adulthood. In fact, the origins of adult chronic disease may well be at conception or very soon thereafter so that the similarity in the list of essential nutrients may well be because these are required throughout the lifespan. It is only now, when there is a critical mass of data that we are able to see the connecting nutritional threads that run throughout life. Therefore, we now examine the parallel data from studies associating certain nutritional factors with reducing the risk of adult chronic diseases.

Five of the 10 leading causes of death in the United States are conditions that can be modified by diet *(96)*. The major broad categories of nutrition-related chronic diseases include cardiovascular disease, stroke, cancer, diabetes, and osteoporosis. Deterioration of visual and mental/neurological function could also be added to this list *(14)*. In trying to assign health dollar costs to each of these individual categories, it is evident that these conditions overlap in terms of the cost of morbidities and mortalities. For example, a definable co-morbidity of obesity is often type 2 diabetes, and the presence of insulin resistance or type 2 diabetes markedly increases risk for cardiovascular disease *(31)*. As mentioned earlier, more than 50% of adults that have one chronic disease, have another.

The costs of care for those with chronic diseases depends upon the disease, the extent of disease, and the length of the illness *(19)*. Examining the three leading causes of death in the United States, the annual economic costs of cardiovascular disease, cancer, and stroke have been estimated to be $138 billion, $104 billion, and $30 billion, respectively, a total exceeding $250 billion annually. Nutritional interventions that reasonably could delay the onset of cardiovascular disease and stroke for 5 years have been calculated to potentially result in annual health cost savings of $84 billion.

2.6.1. Cardiovascular Disease and Cancer

As indicated above, national survey data indicate that the foods that provide essential nutrients are often not consumed in the quantities recommended. Other sources of essential vitamins and minerals include fortified foods and dietary supplements. For cardiovascular disease prevention, a major preventive nutrition strategy that has been implemented first in Europe and now in the United States, is the removal of trans fats from widely consumed packaged foods (*see* Chapter 11). In contrast to the removal of trans fat from the diet is the inclusion of long-chain polyunsaturated fatty acids (LC PUFA) in many more foods than in prior years as well as the use of fish oil supplements or supplements of LC PUFA from algal sources. Survey data have consistently associated fish consumption and diets containing higher levels of *n*–3 PUFA with decreased risk of CVD. A possible mechanism is the substitution of *n*–3 LC PUFA for more pro-inflammatory *n*–6 fatty acids. The *n*–3 LC PUFA also have been shown to beneficially affect the electrical conductance in the heart at relatively modest intake levels (*see* Chapter 10). Survey data have consistently found a lowered risk of CVD with the highest intakes of antioxidants, especially vitamins E and C. However, as reviewed in Chapter 12 and previously *(97)* not all recent intervention studies have confirmed secondary cardiovascular risk reduction with supplemental vitamin E. Ongoing studies will help to evaluate the potential for this essential vitamin with antioxidant and other functions to affect the primary prevention of cardiovascular disease.

Damage to the vessels in the brain can result in stroke, and inflammation in the vessels has been linked to the development of vascular dementia and Alzheimer's disease. B vitamins that are involved in the maintenance of normal homocysteine levels may be a part of the protective responses to inflammation in the brain and help prevent age-related cognitive declines. Costs of care for elderly with cognitive impairments are increasing annually and any nutritional intervention that can slow the progression of dementia would be of great value (*see* Chapter 13). In addition to antioxidants, such as vitamin E, there has also been a continued association of low B vitamin status (especially folic acid and vitamins $B_6$and B_{12}) with increased risk of CVD as well as stroke and vascular dementia *(21,23,98–100)*. Multivitamin use has been associated with improved B vitamin status and reduction in homocysteine levels, an important risk factor in CVD and related morbidities *(101)*. Although the cost-effectiveness of multivitamin supplementation has not been reported in the literature, this intervention could also prove to be a cost-saving health strategy for cardiovascular disease prevention and other health areas discussed below.

Toole et al. *(102)* examined the potential for supplements of vitamins—folic acid, vitamin B_6, and vitamin B_{12}—to reduce the risk of recurrent strokes in a 2-year double-blind, randomized study of 3,680 patients who had suffered an ischemic stroke and had higher-than-average serum homocysteine levels. The comparison was between a multivitamin that included a relatively high level of the three B vitamins versus one with lower levels. Serum homocysteine levels were lowered moderately in a dose-response manner, and the lower the baseline homocysteine, the lower the risk of recurrent stroke. There was no difference in the rate of recurrent stroke, coronary heart disease, or deaths between the groups. The authors suggest that one potential confounder was the mandatory folic acid fortification that took place during their study. Another possibility is that the study was too short in duration although there are indications in other studies that some improvements can be seen with other nutritional interventions in 2 years (*see* Chapter 10).

Cancer is the number 2 killer in developed nations and recent progress in treatment has assured that cancer patients are surviving for more years than ever before. Survey data have consistently found that individuals who consume the greatest number of servings of fruits and vegetables and those who take multivitamins have the lowest risk for the major cancers as well as many of the less common malignancies (*see* Chapter 9). As with cardiovascular disease prevention, intervention studies for cancer prevention, especially in individuals at high risk for certain cancers have not consistently proven that supplements can reduce risk. Similar micronutrients have been investigated including the antioxidants with mixed results including increased risk for lung cancer in heavy smokers taking high-dose beta-carotene supplements.

There is greater appreciation that specific genetic characteristics may affect the response to preventive strategies, and the area of nutragenomics has helped to clarify why certain interventions may work best under specific genetic backgrounds. Targeted interventions have the greatest chance for cost savings and this strategy is being implemented for breast cancer and prostate cancer as described in Chapters 7 and 8.

In addition to genetic factors, infections can affect cancer risk, especially with regard to gastric cancers. Again, inflammation and oxidative damage are implicated in the initiation of malignancy, and certain antioxidants have been shown to lower gastric cancer in a number of, but not all, studies (*see* Chapter 6). One of the most consistent associations in survey studies is between calcium (with or without vitamin D) from food and/or supplements and reduced risk of colon cancer. Intervention studies have also shown reduction in precancerous lesions of the colon in the calcium-supplemented cohorts. For instance, the Calcium Polyp Prevention Study, a double-blind, placebo-controlled study, found that at risk subjects had fewer precancerous lesions with supplementation *(103)*. The subjects had previously had one colon polyp removed prior to entering this study. Those that took 1,200 mg of calcium/day for up to 4 years had significantly fewer recurrences of polyps than the placebo group. The mechanism of action of calcium in reducing the risk of colon cancer by reducing the number of precancerous lesions is thought to be via the reduction in bile acids that are thought to be carcinogenic. Women were not included in the WHI study because of a prior colon polyp. However, colon cancer was an end point that was examined in the WHI as the women took 1,000 mg of elemental calcium and 400 IU of vitamin D/day for many years. There was no effect of the calcium and vitamin D supplementation on risk of colon cancer in this cohort *(104)*.

McCarron and Heaney *(36)* looked at the estimated costs associated with diseases or conditions that can be beneficially affected by a diet rich in calcium/dairy products. Direct medical costs associated with obesity and diabetes alone reach $100 billion annually. Osteoporosis costs are about $17 billion and hypertension adds another $34 billion. Data from intervention studies and recent epidemiological data consistently find a significant reduced risk of colon cancer associated with higher-than-average calcium intakes; costs associated with colon cancer alone are estimated at $5 billion annually. Using the findings of the Dietary Approaches to Stop Hypertension (DASH) intervention study, the epidemiological data from the Coronary Artery Risk Development in Young Adults (CARDIA) study as well as several intervention studies using calcium supplements, the authors estimate that $26 billion in health-care costs could be saved annually with the inclusion of two additional servings of dairy foods/day; savings would also be seen with increased calcium intake *(36)*.

2.6.2. Age-Related Eye Diseases

The two major age-related eye diseases are cataracts and age-related macular degeneration (AMD), both of which result in blindness. Both of these diseases share a common pathophysiological etiology of oxidative damage. Oxidative damage from cigarette smoke and/or sun exposure is a modifiable risk factor for these diseases. Both conditions develop over many years and there are consistent survey data that suggest that loss of sight can be delayed with diets rich in fruits and vegetables, and for cataracts, higher-than-recommended intakes of vitamin C. The macula's yellow color is derived from carotenoids, lutein, and zeaxanthin, and individuals with higher-than-average intakes of diets rich in these antioxidant carotenoids have a decreased risk of developing AMD. Even though these carotenoids and others are not found in the lens, higher-than-average carotenoid intakes have been associated with reduced cataract risk. Some, but not all, survey and intervention studies have found that antioxidant supplements and multivitamin supplements can reduce the risk of cataract progress and even reduce the need for cataract surgery (*see* Chapter 21). Since cataract surgery is the most common type of surgery in the elderly, a reduction in this medical expense would help to balance the costs of the supplements. With AMD, there are currently no effective treatments to reverse the loss of sight so the strategy is to delay the full loss of sight. The Age-Related Eye Disease intervention study proved that progress of AMD could be slowed with supplementation *(105)*. Higher-than-recommended intake levels of vitamin C, E, beta-carotene, and zinc taken for several years slowed the progression of AMD in this well-controlled study. The economic impact of prevention of blindness is difficult to assess as it would likely include the costs for nursing home or other care; however, the cost for the supplement is about $500/year and the charge for minimal nursing home care is about $3,000/month.

2.6.3. Osteoporosis

Although it is difficult to determine the exact cost savings involved with a preventive nutrition strategy that could impact more than one disease condition, it is, nevertheless, worthwhile to examine the "minimal" health economic benefit of implementing such a strategy. We have chosen to examine in detail the data from well-controlled intervention studies that tested the hypothesis that nutritional intervention could reduce the risk of osteoporosis-related hip fracture. We chose this example because there appears to be less overlap between osteoporosis and the other chronic degenerative diseases discussed. We are cognizant, however, that the intervention which was common among the studies reviewed, calcium and vitamin D, has been associated with reducing the risk of colon cancer *(103)*, hypertension *(106)*, and several other disease states *(107)* so that even in this example, the economic benefits may well be underestimated.

More than 6 million US adults have osteoporosis *(108)*, which is defined as having bones that are two standard deviations below the peak bone mineral density (BMD) seen in young adults *(109,110)*. Osteoporosis is a known risk factor for hip fracture *(111)*. Osteoporotic hip fractures pose enormous human and economic costs. The nearly 300,000 annual hip fractures in the United States have been estimated to cost the nation around $5.6 billion based on hospital and other costs. Those with hip fractures experience increased risk of institutionalization and death *(112)*.

Supplemental calcium with/without vitamin D has been shown to reduce the risk of hip fractures, especially in compliant study participants *(7,113–116)*. Four placebo-controlled, double-blind studies have shown that calcium supplementation (with or without vitamin D) significantly reduced the risk of hip fracture in individuals over the age of 50. Earlier attempts to determine the health-care savings prior to the publication of the WHI results *(117)* found a combined 47% reduction in the risk of hip fracture in those individuals who took supplemental calcium at levels that ranged from 500 to 1,200 mg/day for up to 3.4 years. The WHI study reported a 29% reduction in hip fractures in the compliant women who took 1,000 mg/day of supplemental calcium for about 7 years. It is not surprising that there was a lower percentage of hip fracture reduction in the WHI study as the women were not at an increased risk for hip fracture at baseline whereas the earlier three studies only enrolled osteoporotic women.

Using the risk reductions seen in the studies of osteoporotic women and US National Data on the hospital and certain other costs associated with hip fracture, the potential reduction in cost of care for hip fracture patients was calculated. Over 134,000 hip fractures and $2.6 billion could have been avoided if all US adults over 50 years of age took 1,200 mg of supplemental calcium/day. Universal calcium supplementation was cost-effective for women over the age of 75 based on the need to take the supplemental calcium for 34 months before the hip fracture reduction was seen. If, however, the assumption is made that the benefit would be seen after 12–14 months of supplementation as was seen in the Chapuy study *(113)*, then universal calcium supplementation becomes cost-effective for all US men and women 65 years and over. Another key finding from this study was that calcium supplementation that commences in women over the age of 85 is very cost-effective as the intervention data suggest that bone mineral density can be lost less quickly even at this age if calcium supplementation is instituted.

Even though the cost analysis did not include vertebral fractures, it would appear that fractures of the spine would also be reduced in many older women with osteoporosis who supplemented with 1,200 mg/day of calcium. Recker et al. *(118)*, in a placebo-controlled, double-blind study in at-risk women over 60 years old who consumed less than 1,000 mg/day of calcium, found that supplementation significantly reduced the risk of spine fractures, especially in women with a history of bone fractures. The WHI study did not find a significant decrease in spine fracture for the entire cohort and has yet to report these data for the compliant women *(7)*.

The National Institutes of Health and the National Academy of Sciences recommend that postmenopausal women should consume daily about 1,200 mg of elemental calcium *(77,86)*. The benefits of antiresorptive therapies for osteoporosis prevention are predicated on the daily consumption of 1,000 mg of calcium *(119)*. However, data from a telephone survey of a representative sample of US households revealed that the average daily intake of dietary calcium falls far short of the minimum recommended daily amount of 1,000 mg. The telephone survey found that only half of the adults 60–94 years of age drank one glass of milk, which provides 300 mg of calcium, everyday *(120)*. LeBoff et al. *(121)* measured the vitamin D and calcium status of postmenopausal women with hip fractures and found that 50% had deficient vitamin D levels and over 80% had low calcium levels. Since vitamin D is required for calcium absorption, the authors suggest that the low calcium status was linked to the low vitamin D status. Thus, these data suggest that individuals at risk for hip and other fractures should increase their calcium as well as vitamin D intakes (*see* Chapter 19).

Cost-savings analyses are based on databases that capture the costs associated with certain events. Thus, when health economists are limited to hospitalization costs, it is not possible to determine the cost savings associated with reducing the occurrence of osteoporosis as there are no hospital expenses associated with osteoporosis itself. It is well recognized that increasing bone mass during the years of bone growth and early adulthood is the best osteoporosis preventive strategy. Yet, because there are so few hospitalizations of young people who fracture bones, universal calcium and vitamin D supplementation may not seem cost-effective. Recently, however, Singer et al. *(122)* documented the incidence of fractures in individuals 15–94 years of age in Edinburgh, Scotland. They reported that between the ages of 15 and 49 years, men had 2.9 times the fractures as age-matched women; fractures of the wrist began to increase in women at 40 years of age, before menopause and that over the age of 60, women had 2.3 times the risk of fractures as men. Although this study did not examine nutritional factors, other studies have linked increased risk of fractures in young adults with low intakes of calcium and other micronutrients and low sun exposure as found in Scotland. It may be that the establishment of newer databases that capture the total costs (loss of work as well as medical costs) associated with fractures will permit the evaluation of the cost-effectiveness of earlier nutritional interventions for improvement of bone health.

One other potential benefit of calcium supplementation may be a reduction in blood pressure in both men and women *(106)*. As described above, calcium supplementation lowered blood pressure during pregnancy and also reduced blood pressure in the male and female offspring of women who took calcium supplements during pregnancy *(85,88)*. Since high blood pressure is a strong risk factor in stroke and aggravates other cardiovascular diseases, calcium supplementation may be very cost-effective if all of its biological functions are included in the calculations.

It is also important to note that, in addition to calcium, antioxidant status has also been associated with hip fracture risk. Lifestyle factors, such as smoking, also increase the risk of hip fracture. Melhus et al. *(123)* found a threefold increased risk of hip fracture in women who were current smokers and had the lowest intakes of either vitamin E or vitamin C compared to non-smoking women with the highest antioxidant intakes. If the smokers had the lowest intakes of both vitamins, the odds ratio increased to 4.9.

Smoking, independent of antioxidant status, has been shown to increase the risk of hip fracture, which may be due to its association with decreased calcium absorption *(124)*. To complete the circle of life events, it is most interesting to note that Jones et al. *(125)* found that maternal smoking during pregnancy resulted in their children having shorter stature that was linked to lower bone mass. It may well be that the children of smoking mothers are at greater risk for osteoporosis because their bones never accumulate the bone mass needed to prevent this disease in later life. Importantly, infant bone mass has been shown to be increased if mothers are supplemented with calcium during pregnancy.

Koo et al. *(126)* showed that total bone mineral content was significantly greater in infant children born to mothers supplemented with 2,000 mg/day of calcium during pregnancy compared to women in the placebo group who consumed less than 600 mg/day of calcium.

2.7. KEY MICRONUTRIENTS FOR DISEASE PREVENTION

A number of vitamins, minerals, and long-chain fatty acids have been consistently associated with primary prevention of age-related chronic diseases including cataracts and age-related macular degeneration *(127)*. Oxidative damage to sugars has recently also been associated with increased risk of age-related eye diseases and is extensively reviewed in Chapter 21. Coincidentally, the vast majority of these essential micronutrients also appear to reduce the risk of adverse pregnancy outcomes. The list includes the antioxidant vitamins, C and E; B vitamins, folic acid, B_{12} and B_6; and vitamin D; minerals associated with improved health outcomes for young and old include calcium, magnesium, iron, zinc, and selenium. It should be noted that iron, zinc, and selenium are also required for the functioning of antioxidant enzymes (Table 2.3).

The majority of data associating micronutrients with primary disease prevention come from epidemiological studies. There appear to be certain chronic diseases that are good candidates for preventive nutrition strategies (Table 2.4). Findings suggest that individuals with the highest intakes of foods containing these micronutrients over the longest time spans have the greatest reduced risk of disease.

Table 2.3
Association of Essential Nutrient Status with Reduced Risk of Adverse Pregnancy Outcomes

Vitamin/Mineral	*Low Status Associated with Increased Risk*	*Supplementation Reduces Risk*
Folic acid	Neural tube birth defects Damage to sperm DNA	Neural tube birth defects
Iron	Low birth weight Premature delivery Maternal morbidity and mortality Cognitive deficits	Cognitive deficits
Calcium	Pre-eclampsia	Pre-eclampsia
Iodine	Cognitive deficits	
Vitamins C and E		Pre-eclampsia
Long-chain polyunsaturated fatty acids	Premature delivery	
Vitamin A as retinol and as beta-carotene		Low birth weight and premature delivery
Magnesium, copper, selenium, niacin	Premature delivery and low birth weight	
Multivitamin/mineral supplement		Neural tube birth defects, cleft lip/cleft palate, cardiovascular birth defects, renal birth defects, low birth weight, premature delivery
Multivitamin-containing zinc		Premature delivery

Table 2.4
Diseases/Conditions That Are Good Candidates for Preventive Nutrition Strategies

Disease/Conditions—in Alphabetical Order	
Age-related macular degeneration	Hypertension
Birth defects	Kidney stones
Cataracts	Low birth weight
Childhood cancers (major)	Obesity
Colon cancer	Osteoporosis
Coronary artery disease	Pre-eclampsia
Diabetes type 2	Stroke

Some of the survey studies have included questions about dietary supplement use, and in many cases, those with the highest intakes and longest use of the micronutrients from supplements also have lower risks of the major degenerative diseases. Of interest, often those with the highest intakes of vitamins from naturally occurring sources or from fortified foods also have the highest total intakes because they are often users of vitamin and/or mineral supplements.

It is important to note that for cardiovascular disease prevention, there appears to be a link between antioxidant nutrients and the B vitamins *(128)*. Although it is now well accepted that recommended intake levels or higher of folic acid, vitamins B_6, and vitamin B_{12} are required to lower homocysteine levels, new data suggest that high levels of vitamins E and C may also lower several of the cardiovascular risk factors associated with elevated homocysteine levels. Antioxidant nutrients and the B vitamins affect different aspects of the cardiovascular system, so that both classes of nutrients are needed for protection *(129)*. It appears that in nature, evolution has required man's need for 13 essential vitamins and at least an equal number of minerals. It is only reasonable to find that these dietary components work best when all of their levels are optimized, and not just a "single magic bullet" is at an "optimal" level. This may be one of many reasons why the intervention studies with one or more, but not all, of the essential nutrients are not able to replicate the associations seen in the survey studies.

2.8. CONCLUSIONS AND RECOMMENDATIONS

The goal of preventive nutrition is to maximize the years of HEALTHY life expectancy so that individuals can enjoy their lives and be competent in performing the activities for daily living independently. Within the past 5 years, there has been a growing appreciation of the value of nutritional interventions as cost-effective and safe means of maintaining health as well as decreasing risk of major chronic diseases. Moreover, an increasing body of evidence suggests that there are more commonalities than differences between the nutrition recommendations to reduce risk factors for diseases: for instance, both cancer and cardiovascular disease prevention recommendations include increased consumption of fruits and vegetables *(98,130–137)*. Likewise, increased whole grain intake (as compared to refined grains) has been associated with decreased prevalence of cardiovascular disease, cancer, and also type 2 diabetes *(17,18,138–141)*. A simple dietary change—increasing the number of servings of low fat, calcium-rich dairy foods/day by two has significant potential for an annual $26 billion in health-care savings in the costs associated with hypertension, obesity, type 2 diabetes, coronary artery disease, stroke, osteoporosis, kidney stones, colorectal cancer, and pregnancy-associated pre-eclampsia *(36)*. Some recommendations have already been adopted by many nations including the recommendation that all women of childbearing potential consume 400 μg of folic acid/day. Compelling data point to choosing a multivitamin supplement as a source of the recommended level of folic acid. Calcium and vitamin D supplementation has been shown to significantly lower the risk of hip fracture in both osteoporotic women and in women with less risk of fracture who actually take the supplements daily. Both of these interventions would appear to be highly cost-effective, especially if these could be targeted to the appropriate populations.

The examples in this chapter provide ample evidence that there are important, simple, safe, and economically sound interventions that can be implemented to reduce disease risk throughout the life span. Quoting from Blumberg's conclusion in his chapter *(19)* on "Public Health Implications of Preventive Nutrition" his counsel seems particularly appropriate at this time.

> Shifting the health care system from its current emphasis on treatment to prevention will take time. Even if such changes are implemented, more time will be required before its impact on chronic disease mortality will become apparent because of long latency periods, although a delay in the onset of clinical symptoms will be detected earlier. The dividends of prevention in reducing the population illness burden and enhancing of the quality of life can be substantial. Efforts must be strengthened to encourage all segments of the population to adopt preventive nutrition strategies, not just those who are high risk. Food habits develop early in life, and this is a useful time to develop preventive nutrition behaviors, although an emphasis on older adults appears more critical at this juncture. . . . Together with an increase in physical activity and the cessation of tobacco use, dietary modification and improvements in nutritional status present us with the greatest potential for reducing the incidence of chronic disease, improving public health, and limiting the growth of health care expenditures.

REFERENCES

1. Heron MP, Hoyert DL, Xu J, Scott C, Tejada-Vera B. Deaths: preliminary data for 2006. National Vital Statistics Reports 2008;56(16).
2. Adams PF, Lucas JW, Barnes PM. Summary Health Statistics for U.S. Adults, 2008 (Ref Type: Electronic Citation). Available at: http://www.cdc.gov/nchs/data/series/sr_10/sr10_236.pdf.
3. Schoenborn CA, Vickerie JL, Powell-Griner E. Health characteristics of adults 55 years of age and over: United States, 2000–2003. Adv Data 2006;370:1–31.
4. Tomey KM, Sowers MR, Crandall C, Johnston J, Jannausch M, Yosef M. Dietary intake related to prevalent functional limitations in midlife women. Am J Epidemiol 2008;167(8):935–943.
5. Weiss CO, Boyd CM, Yu Q, Wolff JL, Leff B. Patterns of prevalent major chronic disease among older adults in the United States. JAMA 2007;298(10):1160–1162.
6. Cauley JA, Wampler NS, Barnhart JM, Wu L, Allison M, Chen Z, et al. Incidence of fractures compared to cardiovascular disease and breast cancer: the Women's Health Initiative Observational Study. Osteoporos Int 2008;19(12): 1717–1723.
7. Jackson RD, LaCroix AZ, Gass M, Wallace RB, Robbins J, Lewis CE, et al. Calcium plus vitamin D supplementation and the risk of fractures. N Engl J Med 2006;354(7):669–683.
8. Larson NI, Story M, Perry CL, Neumark-Sztainer D, Hannan PJ. Are diet and physical activity patterns related to cigarette smoking in adolescents? Findings from Project EAT. Prev Chronic Dis 2007;4(3):A51.
9. Wilson DB, Smith BN, Speizer IS, Bean MK, Mitchell KS, Uguy LS, et al. Differences in food intake and exercise by smoking status in adolescents. Prev Med 2005;40(6):872–879.
10. Eaton DK, Kann L, Kinchen S, Shanklin S, Ross J, Hawkins J, et al. Youth risk behavior surveillance—United States, 2007. MMWR Surveill Summ 2008;57(4):1–131.
11. Fryar CD, Hirsch R, Porter KS, Kottiri B, Brody DJ, Louis T. Smoking and alcohol behaviors reported by adults: United States, 1999–2002. Adv Data 2007;378:1–28.
12. Qato DM, Alexander GC, Conti RM, Johnson M, Schumm P, Lindau ST. Use of prescription and over-the-counter medications and dietary supplements among older adults in the United States. JAMA 2008;300(24):2867–2878.
13. Patterns of Medication Use in the United States. Slone Epidemiology Center, 2005 (Ref Type: Electronic Citation). http://www.bu.edu/Slone/SloneSurvey/AnnualRpt/SloneSurveyWebReport 2006.pdf.

14. Anderson R. Antioxidant nutrients and prevention of oxidant-mediated diseases. In: Bendich A, Deckelbaum RJ, eds. Preventive Nutrition: The Comprehensive Guide for Health Professionals. Humana Press, Totowa, NJ, 2005, pp. 505–520.
15. Woodall AA, Ames BN. Nutritional prevention of DNA damage to sperm and consequent risk reduction in birth defects and cancer in offspring. In: Bendich A, Deckelbaum RJ, eds. Preventive Nutrition: The Comprehensive Guide for Health Professionals. Humana Press, Totowa, NJ, 1997, pp. 373–385.
16. Jain SK, Wise R. Relationship between elevated lipid peroxides, vitamin E deficiency and hypertension in preeclampsia. Mol Cell Biochem 1995;151(1):33–38.
17. Dietary Guidelines Advisory Committee. The Report of the Dietary Guidelines Advisory Committee on Dietary Guidelines for Americans, 2000 (Ref Type: Report).
18. Willet WC. Potential benefits of preventive nutrition strategies: lessons for the United States. In: Bendich A, Deckelbaum RJ, eds. Preventive Nutrition: The Comprehensive Guide for Health Professionals. Humana Press, Totowa, NJ, 2005, pp. 713–734.
19. Blumberg J. Public health implications of preventative nutrition. In: Bendich A, Deckelbaum RJ, eds. Preventive Nutrition: The Comprehensive Guide for Health Professionals. Humana Press, Totowa, NJ, 1997, pp. 1–17.
20. Krebs-Smith SM, Smiciklas-Wright H, Guthrie HA, Krebs-Smith J. The effects of variety in food choices on dietary quality. J Am Diet Assoc 1987;87(7):897–903.
21. Selhub J, Jacques PF, Rosenberg IH, Rogers G, Bowman BA, Gunter EW, et al. Serum total homocysteine concentrations in the third National Health and Nutrition Examination Survey (1991–1994): population reference ranges and contribution of vitamin status to high serum concentrations. Ann Intern Med 1999;131(5):331–339.
22. Bendich A, Mallick R, Leader S. Potential health economic benefits of vitamin supplementation. West J Med 1997;166(5):306–312.
23. Ross GW, Petrovitch H, White LR, Masaki KH, Li CY, Curb JD, et al. Characterization of risk factors for vascular dementia: the Honolulu-Asia Aging Study. Neurology 1999;53(2):337–343.
24. Spencer AP, Carson DS, Crouch MA. Vitamin E and coronary artery disease. Arch Intern Med 1999;159(12): 1313–1320.
25. Vitamin and Mineral Deficiency. 2004 (Ref Type: Electronic Citation). Available at: http://www.unicef.org/media/files/davos_micronutrient.pdf, Accessed 9/14/2004.
26. Ervin RB. Healthy Eating Index scores among adults, 60 years of age and over, by sociodemographic and health characteristics: United States, 1999–2002. Adv Data 2008;395:1–16.
27. Freedman LS, Guenther PM, Krebs-Smith SM, Kott PS. A population's mean Healthy Eating Index-2005 scores are best estimated by the score of the population ratio when one 24-hour recall is available. J Nutr 2008;138(9):1725–1729.
28. Institute of Medicine. Dietary Reference Intakes for Energy, Carbohydrate, Fiber, Fat, Fatty Acids, Cholesterol, Protein, and Amino Acids. The National Academies Press, Washington, DC, 2005.
29. de Souza RJ, Swain JF, Appel LJ, Sacks FM. Alternatives for macronutrient intake and chronic disease: a comparison of the OmniHeart diets with popular diets and with dietary recommendations. Am J Clin Nutr 2008;88(1):1–11.
30. Allison DB, Downey M, Atkinson RL, Billington CJ, Bray GA, Eckel RH, et al. Obesity as a disease: a white paper on evidence and arguments commissioned by the Council of the Obesity Society. Obesity (Silver Spring) 2008;16(6):1161–1177.
31. Ogden CL, Carroll MD, Curtin LR, McDowell MA, Tabak CJ, Flegal KM. Prevalence of overweight and obesity in the United States, 1999–2004. JAMA 2006;295(13):1549–1555.
32. Ogden CL, Carroll MD, McDowell MA, Flegal KM. Obesity among adults in the United States—no statistically significant changes since 2003–2004. NCHS Data Brief [1]. 2007 (Ref Type: Generic).
33. Kouris-Blazos A, Wahlqvist ML. Health economics of weight management: evidence and cost. Asia Pac J Clin Nutr 2007;16(Suppl. 1):329–338.
34. Trust for America's Health. F as in fat: how obesity policies are failing in America, 2008 (Ref Type: Generic).
35. Wang Y, Beydoun MA, Liang L, Caballero B, Kumanyika SK. Will all Americans become overweight or obese? estimating the progression and cost of the US obesity epidemic. Obesity (Silver Spring) 2008;16(10):2323–2330.
36. McCarron DA, Heaney RP. Estimated healthcare savings associated with adequate dairy food intake. Am J Hypertens 2004;17(1):88–97.
37. Holmes MD, Hunter DJ, Colditz GA, Stampfer MJ, Hankinson SE, Speizer FE, et al. Association of dietary intake of fat and fatty acids with risk of breast cancer. JAMA 1999;281(10):914–920.
38. Rozowski J, Moreno M. Effect of Westernization of Nutritional Habits on Obesity Prevalence in Latin America: Analysis and Recommendations. In: Bendich A, Deckelbaum RJ, eds. Preventive Nutrition: The Comprehensive Guide for Health Professionals. Humana Press, Totowa, NJ, 1997 pp. 487–504.

39. Zhang S, Hunter DJ, Rosner BA, Colditz GA, Fuchs CS, Speizer FE, et al. Dietary fat and protein in relation to risk of non-Hodgkin's lymphoma among women. J Natl Cancer Inst 1999;91(20):1751–1758.
40. Frier HI, Greene HL. Obesity and Chronic Disease: Impact of Weight Reduction. In: Bendich A, Deckelbaum RJ, eds. Preventive Nutrition: The Comprehensive Guide for Health Professionals. Humana Press, Totowa, NJ, 2005, pp. 383–401.
41. Zhang S, Hunter DJ, Hankinson SE, Giovannucci EL, Rosner BA, Colditz GA, et al. A prospective study of folate intake and the risk of breast cancer. JAMA 1999;281(17):1632–1637.
42. Williams CL. Can Childhood Obesity Be Prevented?: Preschool Nutrition and Obesity. In: Bendich A, Deckelbaum RJ, eds. Preventive Nutrition: The Comprehensive Guide for Health Professionals.: Humana Press, 2005: 345–381.
43. Calle EE, Thun MJ, Petrelli JM, Rodriguez C, Heath CW, Jr. Body-mass index and mortality in a prospective cohort of U.S. adults. N Engl J Med 1999;341(15):1097–1105.
44. Brandle M, Zhou H, Smith BR, Marriott D, Burke R, Tabaei BP, et al. The direct medical cost of type 2 diabetes. Diabetes Care 2003;26(8):2300–2304.
45. Chen CM. Overview of obesity in Mainland China. Obes Rev 2008;9(Suppl. 1):14–21.
46. Bergstorm E, Hernell O. Obesity and insulin resistance in childhood and adolescence. In: Bendich A, Deckelbaum RJ, eds. Preventive Nutrition: The Comprehensive Guide for Health Professionals. Humana Press, Totowa, NJ, 2005, pp. 293–320.
47. Faith MS, Calamaro CJ, Pietrobelli A, Dolan MS, Allison DB, Heymsfield SB. Prevention of pediatric obesity: examining the issues and forecasting research directions. In: Bendich A, Deckelbaum RJ, eds. Preventive Nutrition: The Comprehensive Guide for Health Professionals. Humana Press, Totowa, NJ, 2005, pp. 321–343.
48. Kiess W, Bottner A, Bluher S, Raile K, Seidel B, Kapellen T, et al. Pharmacoeconomics of obesity management in childhood and adolescence. Expert Opin Pharmacother 2003;4(9):1471–1477.
49. Appel LJ, Sacks FM, Carey VJ, Obarzanek E, Swain JF, Miller ER, III, et al. Effects of protein, monounsaturated fat, and carbohydrate intake on blood pressure and serum lipids: results of the OmniHeart randomized trial. JAMA 2005;294(19):2455–2464.
50. Appel LJ, Moore TJ, Obarzanek E, Vollmer WM, Svetkey LP, Sacks FM, et al. A clinical trial of the effects of dietary patterns on blood pressure. DASH Collaborative Research Group. N Engl J Med 1997;336(16):1117–1124.
51. Howard BV, Van Horn L, Hsia J, Manson JE, Stefanick ML, Wassertheil-Smoller S, et al. Low-fat dietary pattern and risk of cardiovascular disease: the Women's Health Initiative Randomized Controlled Dietary Modification Trial. JAMA 2006;295(6):655–666.
52. Anderson CA, Appel LJ. Dietary modification and CVD prevention: a matter of fat. JAMA 2006;295(6):693–695.
53. Gardner CD, Kiazand A, Alhassan S, Kim S, Stafford RS, Balise RR, et al. Comparison of the Atkins, Zone, Ornish, and LEARN diets for change in weight and related risk factors among overweight premenopausal women: the A TO Z Weight Loss Study: a randomized trial. JAMA 2007;297(9):969–977.
54. Kahn R, Robertson RM, Smith R, Eddy D. The impact of prevention on reducing the burden of cardiovascular disease. Diabetes Care 2008;31(8):1686–1696.
55. Goldstain DJ. The Management of Eating Disorders and Obesity, 1st ed. Humana Press, Totowa, NJ, 1999.
56. Barker DJ. Fetal origins of cardiovascular disease. Ann Med 1999;31(Suppl. 1):3–6.
57. Klebanoff MA, Secher NJ, Mednick BR, Schulsinger C. Maternal size at birth and the development of hypertension during pregnancy: a test of the Barker hypothesis. Arch Intern Med 1999;159(14):1607–1612.
58. Moore SE. Nutrition, immunity and the fetal and infant origins of disease hypothesis in developing countries. Proc Nutr Soc 1998;57(2):241–247.
59. Yarbrough DE, Barrett-Connor E, Kritz-Silverstein D, Wingard DL. Birth weight, adult weight, and girth as predictors of the metabolic syndrome in postmenopausal women: the Rancho Bernardo Study. Diabetes Care 1998;21(10): 1652–1658.
60. Czeizel AE. Folic Acid-Containing Multivitamins and Primary Prevention of Birth Defects. In: Bendich A, Deckelbaum RJ, eds. Preventive Nutrition The Comprehensive Guide for Health Professionals. Humana Press, Totowa, NJ, 1997, pp. 351–371.
61. Scholl TO. Maternal nutrition and preterm delivery. In: Bendich A, Deckelbaum RJ, eds. Preventive Nutrition: The Comprehensive Guide for Health Professionals. Humana Press, Totowa, NJ, 1997, pp. 405–421.
62. Czeizel AE, Dudas I. Prevention of the first occurrence of neural-tube defects by periconceptional vitamin supplementation. N Engl J Med 1992;327(26):1832–1835.
63. Prevention of neural tube defects: results of the Medical Research Council Vitamin Study. MRC Vitamin Study Research Group. Lancet 1991; 338 (8760): 131–137.

64. Czeizel AE. Folic Acid-Containing Multivitamins and Primary Prevention of Birth Defects. In: Bendich A, Deckelbaum RJ, eds. Preventive Nutrition the Comprehensive Guide for Health Professionals. Humana Press, Totowa, NJ, 2005: 603–628.
65. Botto LD, Khoury MJ, Mulinare J, Erickson JD. Periconceptional multivitamin use and the occurrence of conotruncal heart defects: results from a population-based, case-control study. Pediatrics 1996;98(5):911–917.
66. Butterworth CE, Jr., Bendich A. Folic acid and the prevention of birth defects. Annu Rev Nutr 1996;16:73–97.
67. Bunin GR, Cary JM. Diet and Childhood Cancer. In: Bendich A, Deckelbaum RJ, eds. Preventive Nutrition the Comprehensive Guide for Health Professionals. Humana Press, Totowa, NJ, 1997, pp. 17–32.
68. Oakley GP, Jr. Eat right and take a multivitamin. N Engl J Med 1998;338(15):1060–1061.
69. Berry RJ, Li Z, Erickson JD, Li S, Moore CA, Wang H, et al. Prevention of neural-tube defects with folic acid in China. China-U.S. Collaborative Project for Neural Tube Defect Prevention. N Engl J Med 1999;341(20):1485–1490.
70. Obesity and reproduction: an educational bulletin. Fertil Steril 2008;90(5 Suppl.):S21–S29.
71. Preterm singleton births—United States, 1989–1996. MMWR Morb Mortal Wkly Rep 1999; 48 (9): 185–189.
72. Polhamus B, Dalenius K, Thompson D, Scanlon K, Borland E, Smith B, et al. Pediatric nutrition surveillance. Nutr Clin Care 2003;6(3):132–134.
73. Infant Mortality and Low Birth Weight Among Black and White Infants—United States, 1980–2000, 2002 (Ref Type: Electronic Citation). Available at: http://www.cdc.gov/mmwr/preview/mmwrhtml/mm5127a1.htm.
74. Scholl TO, Hediger ML, Bendich A, Schall JI, Smith WK, Krueger PM. Use of multivitamin/mineral prenatal supplements: influence on the outcome of pregnancy. Am J Epidemiol 1997;146(2):134–141.
75. Goldenberg RL, Tamura T, Neggers Y, Copper RL, Johnston KE, DuBard MB, et al. The effect of zinc supplementation on pregnancy outcome. JAMA 1995;274(6):463–468.
76. Scholl TO. Maternal Nutrition and Preterm Delivery. In: Bendich A, Deckelbaum RJ, eds. Preventive Nutrition: The Comprehensive Guide for Health Professionals. Humana Press, Totowa, NJ 2005: 629–663.
77. de Onis M, Villar J, Gulmezoglu M. Nutritional interventions to prevent intrauterine growth retardation: evidence from randomized controlled trials. Eur J Clin Nutr 1998;52(Suppl. 1):S83–S93.
78. Ramakrishnan U, Manjrekar R, Rivera J, Gonzβles-Cossøo T, Martorell R. Micronutrients and pregnancy outcome: a review of the literature. Nutr Res 1999;19(1):103–159.
79. West KP, Jr., Katz J, Khatry SK, LeClerq SC, Pradhan EK, Shrestha SR, et al. Double blind, cluster randomised trial of low dose supplementation with vitamin A or beta carotene on mortality related to pregnancy in Nepal. The NNIPS-2 Study Group. BMJ 1999;318(7183):570–575.
80. Fawzi WW, Msamanga GI, Spiegelman D, Urassa EJ, McGrath N, Mwakagile D, et al. Randomised trial of effects of vitamin supplements on pregnancy outcomes and T cell counts in HIV-1-infected women in Tanzania. Lancet 1998;351(9114):1477–1482.
81. Ramakrishnan U. Nutrition and low birth weight: from research to practice. Am J Clin Nutr 2004;79(1):17–21.
82. Chappell LC, Seed PT, Briley AL, Kelly FJ, Lee R, Hunt BJ, et al. Effect of antioxidants on the occurrence of pre-eclampsia in women at increased risk: a randomised trial. Lancet 1999;354(9181):810–816.
83. DerSimonian R, Levine RJ. Resolving discrepancies between a meta-analysis and a subsequent large controlled trial. JAMA 1999;282(7):664–670.
84. Levine RJ, Hauth JC, Curet LB, Sibai BM, Catalano PM, Morris CD, et al. Trial of calcium to prevent preeclampsia. N Engl J Med 1997;337(2):69–76.
85. Bucher HC, Guyatt GH, Cook RJ, Hatala R, Cook DJ, Lang JD, et al. Effect of calcium supplementation on pregnancy-induced hypertension and preeclampsia: a meta-analysis of randomized controlled trials. JAMA 1996;275(14): 1113–1117.
86. NIH Consensus conference. Optimal calcium intake. NIH Consensus Development Panel on Optimal Calcium Intake. JAMA 1994; 272 (24): 1942–1948.
87. Institute of Medicine. Calcium. Dietary reference intakes for calcium, phosphorus, magnesium, vitamin D, and fluoride. National Academy Press, Washington, DC, 1997, pp. 71–145.
88. Crowther CA, Hiller JE, Pridmore B, Bryce R, Duggan P, Hague WM, et al. Calcium supplementation in nulliparous women for the prevention of pregnancy-induced hypertension, preeclampsia and preterm birth: an Australian randomized trial. FRACOG and the ACT Study Group. Aust N Z J Obstet Gynaecol 1999;39(1):12–18.
89. Belizan JM, Villar J, Bergel E, del Pino A, Di Fulvio S, Galliano SV, et al. Long-term effect of calcium supplementation during pregnancy on the blood pressure of offspring: follow up of a randomised controlled trial. BMJ 1997;315(7103):281–285.
90. Wang YP, Walsh SW, Guo JD, Zhang JY. Maternal levels of prostacyclin, thromboxane, vitamin E, and lipid peroxides throughout normal pregnancy. Am J Obstet Gynecol 1991;165(6 Pt 1):1690–1694.

91. Blair SN, Brodney S. Effects of physical inactivity and obesity on morbidity and mortality: current evidence and research issues. Med Sci Sports Exerc 1999;31(11 Suppl.):S646–S662.
92. Polyzos NP, Mauri D, Tsappi M, Tzioras S, Kamposioras K, Cortinovis I, et al. Combined vitamin C and E supplementation during pregnancy for preeclampsia prevention: a systematic review. Obstet Gynecol Surv 2007;62(3):202–206.
93. Spinnato JA, Freire S, Pinto E Silva JL, Cunha Rudge MV, Martins-Costa S, Koch MA, et al. Antioxidant therapy to prevent preeclampsia: a randomized controlled trial. Obstet Gynecol 2007;110(6):1311–1318.
94. Spinnato JA, Freire S, Pinto E Silva JL, Rudge MV, Martins-Costa S, Koch MA, et al. Antioxidant supplementation and premature rupture of the membranes: a planned secondary analysis. Am J Obstet Gynecol 2008;199(4):433–438.
95. Hughes D, Bryan J. The assessment of cognitive performance in children: considerations for detecting nutritional influences. Nutr Rev 2003;61(12):413–422.
96. Ernst ND, McGinnis JM. Preventive Nutrition: A Historic Perspective and Future Economic Outlook. In: Bendich A, Deckelbaum RJ, eds. Preventive Nutrition: The Comprehensive Guide for Health Professionals. Humana Press, Totowa, NJ, 2005, pp. 3–22.
97. Hodis HN, Mack WJ, Sevanian A. Antioxidant vitamin supplementation and cardiovascular disease. In: Bendich A, Deckelbaum RJ, eds. Preventive Nutrition: The Comprehensive Guide for Health Professionals. Humana Press, Totowa, NJ, 2005, pp. 245–277.
98. Deckelbaum RJ, Fisher EA, Winston M, Kumanyika S, Lauer RM, Pi-Sunyer FX, et al. Summary of a scientific conference on preventive nutrition: pediatrics to geriatrics. Circulation 1999;100(4):450–456.
99. Fassbender K, Mielke O, Bertsch T, Nafe B, Froschen S, Hennerici M. Homocysteine in cerebral macroangiography and microangiopathy. Lancet 1999;353(9164):1586–1587.
100. Ridker PM, Manson JE, Buring JE, Shih J, Matias M, Hennekens CH. Homocysteine and risk of cardiovascular disease among postmenopausal women. JAMA 1999;281(19):1817–1821.
101. Jacques PF, Selhub J, Bostom AG, Wilson PW, Rosenberg IH. The effect of folic acid fortification on plasma folate and total homocysteine concentrations. N Engl J Med 1999;340(19):1449–1454.
102. Toole JF, Malinow MR, Chambless LE, Spence JD, Pettigrew LC, Howard VJ, et al. Lowering homocysteine in patients with ischemic stroke to prevent recurrent stroke, myocardial infarction, and death: the Vitamin Intervention for Stroke Prevention (VISP) randomized controlled trial. JAMA 2004;291(5):565–575.
103. Baron JA, Beach M, Mandel JS, van Stolk RU, Haile RW, Sandler RS, et al. Calcium supplements for the prevention of colorectal adenomas. Calcium Polyp Prevention Study Group. N Engl J Med 1999;340(2):101–107.
104. Wactawski-Wende J, Kotchen JM, Anderson GL, Assaf AR, Brunner RL, O'Sullivan MJ, et al. Calcium plus vitamin D supplementation and the risk of colorectal cancer. N Engl J Med 2006;354(7):684–696.
105. AREDS report no.8. A randomized, placebo-controlled, clinical trial of high-dose supplementation with vitamins C and E, beta carotene, and zinc for age-related macular degeneration and vision loss: AREDS report no. 8. Arch Ophthalmol 2001; 119 (10): 1417–1436.
106. McCarron DA, Reusser ME. Finding consensus in the dietary calcium-blood pressure debate. J Am Coll Nutr 1999;18(5 Suppl.):398S–405S.
107. Holick MF. Vitamin D deficiency. N Engl J Med 2007;357(3):266–281.
108. Looker AC, Orwoll ES, Johnston CC, Jr., Lindsay RL, Wahner HW, Dunn WL, et al. Prevalence of low femoral bone density in older U.S. adults from NHANES III. J Bone Miner Res 1997;12(11):1761–1768.
109. Heaney RP. Osteoporosis: Minerals, Vitamins, and other Micronutrients. Preventive Nutrition The Comprehensive Guide for Health Professionals. Humana Press, Totowa, NJ, 1997, pp. 285–302.
110. Holick MF. Vitamin D physiology, molecular biology and clinical applications. Humana Press, Totowa, NJ, 1999.
111. Marshall D, Johnell O, Wedel H. Meta-analysis of how well measures of bone mineral density predict occurrence of osteoporotic fractures. BMJ 1996;312(7041):1254–1259.
112. Kleerekoper A. Evaluation and treatment of postmenopausal osteoporosis. In: Fauvus M, ed. Primer on the Metabolic Bone Disease and Disorders of Mineral Metabolism. Lippincott-Raven, Philadelphia, PA, 1996, pp. 264–271.
113. Chapuy MC, Arlot ME, Duboeuf F, Brun J, Crouzet B, Arnaud S, et al. Vitamin D3 and calcium to prevent hip fractures in the elderly women. N Engl J Med 1992;327(23):1637–1642.
114. Dawson-Hughes B, Harris SS, Krall EA, Dallal GE. Effect of calcium and vitamin D supplementation on bone density in men and women 65 years of age or older. N Engl J Med 1997;337(10):670–676.
115. Reid IR, Ames RW, Evans MC, Gamble GD, Sharpe SJ. Long-term effects of calcium supplementation on bone loss and fractures in postmenopausal women: a randomized controlled trial. Am J Med 1995;98(4):331–335.
116. Prentice RL, Anderson GL. The women's health initiative: lessons learned. Annu Rev Public Health 2008;29:131–150.
117. Bendich A, Leader S, Muhuri P. Supplemental calcium for the prevention of hip fracture: potential health-economic benefits. Clin Ther 1999;21(6):1058–1072.

118. Recker RR, Hinders S, Davies KM, Heaney RP, Stegman MR, Lappe JM, et al. Correcting calcium nutritional deficiency prevents spine fractures in elderly women. J Bone Miner Res 1996;11(12):1961–1966.
119. Eastell R. Calcium requirements during treatment of osteoporosis in women. In: Bendich A, Deckelbaum RJ, eds. Preventive Nutrition: The Comprehensive Guide for Health Professionals. Humana Press Inc., Totowa, NJ, 2006, pp. 425–431.
120. Elbon SM, Johnson MA, Fischer JG. Milk consumption in older Americans. Am J Public Health 1998;88(8): 1221–1224.
121. LeBoff MS, Kohlmeier L, Hurwitz S, Franklin J, Wright J, Glowacki J. Occult vitamin D deficiency in postmenopausal US women with acute hip fracture. JAMA 1999;281(16):1505–1511.
122. Singer BR, McLauchlan GJ, Robinson CM, Christie J. Epidemiology of fractures in 15,000 adults: the influence of age and gender. J Bone Joint Surg Br 1998;80(2):243–248.
123. Melhus H, Michaelsson K, Holmberg L, Wolk A, Ljunghall S. Smoking, antioxidant vitamins, and the risk of hip fracture. J Bone Miner Res 1999;14(1):129–135.
124. Krall EA, Dawson-Hughes B. Smoking and bone loss among postmenopausal women. J Bone Miner Res 1991;6(4):331–338.
125. Jones G, Riley M, Dwyer T. Maternal smoking during pregnancy, growth, and bone mass in prepubertal children. J Bone Miner Res 1999;14(1):146–151.
126. Koo WW, Walters JC, Esterlitz J, Levine RJ, Bush AJ, Sibai B. Maternal calcium supplementation and fetal bone mineralization. Obstet Gynecol 1999;94(4):577–582.
127. Siegal M, Chiu C, Taylor A. Antioxidant Status and Risk for Cataract. In: Bendich A, Deckelbaum RJ, eds. Preventive Nutrition: The Comprehensive Guide for Health Professionals. Humana Press, Totowa, NJ, 2005, pp. 463–503.
128. Nappo F, De Rosa N, Marfella R, De Lucia D, Ingrosso D, Perna AF, et al. Impairment of endothelial functions by acute hyperhomocysteinemia and reversal by antioxidant vitamins. JAMA 1999;281(22):2113–2118.
129. Woodside JV, Young IS, Yarnell JW, Roxborough HE, McMaster D, McCrum EE, et al. Antioxidants, but not B-group vitamins increase the resistance of low-density lipoprotein to oxidation: a randomized, factorial design, placebo-controlled trial. Atherosclerosis 1999;144(2):419–427.
130. Beresford SAA, Motulsky AG. Homocysteine, Folic Acid, and Cardiovascular Disease Risk. Preventive Nutrition: The Comprehensive Guide for Health Professionals. Humana Press, Totowa, NJ, 2005, pp. 191–220.
131. Bostick RM. Diet and Nutrition in the Etiology and Primary Prevention of Colon Cancer. In: Bendich A, Deckelbaum RJ, eds. Preventive Nutrition The Comprehensive Guide for Health Professionals. Humana Press, Totowa, NJ, 2001, pp. 47–96.
132. Buring JE, Gaziano JM. Antioxidant vitamins and cardiovascular disease. In: Bendich A, Deckelbaum RJ, eds. Preventive Nutrition: The Comprehensive Guide for Health Professionals. Humana Press, Totowa, NJ, 1997, pp. 171–180.
133. Fontham ETH. Prevention of upper gastrointestinal tract cancers. In: Bendich A, Deckelbaum RJ, eds. Preventive Nutrition: The Comprehensive Guide for Health Professionals. Humana Press, Totowa, NJ, 2005: 25–54.
134. Hertog MG, Bueno-de-Mesquita HB, Fehily AM, Sweetnam PM, Elwood PC, Kromhout D. Fruit and vegetable consumption and cancer mortality in the Caerphilly Study. Cancer Epidemiol Biomarkers Prev 1996;5(9):673–677.
135. Howe GR. Nutrition and breast cancer. In: Bendich A, Deckelbaum RJ, eds. Preventive Nutrition: The Comprehensive Guide for Health Professionals. Humana Press, Totowa, NJ, 1997, pp. 97–108.
136. Joshipura KJ, Ascherio A, Manson JE, Stampfer MJ, Rimm EB, Speizer FE, et al. Fruit and vegetable intake in relation to risk of ischemic stroke. JAMA 1999;282(13):1233–1239.
137. Comstock GW, Helzlsouer KJ. Preventive nutrition and lung cancer. In: Bendich A, Deckelbaum RJ, eds. Preventive Nutrition: The Comprehensive Guide for Health Professionals. Humana Press, Totowa, NJ, 1997, pp. 109–134.
138. Kushi LH, Folsom AR, Prineas RJ, Mink PJ, Wu Y, Bostick RM. Dietary antioxidant vitamins and death from coronary heart disease in postmenopausal women. N Engl J Med 1996;334(18):1156–1162.
139. Jacobs DR, Jr., Meyer KA, Kushi LH, Folsom AR. Whole-grain intake may reduce the risk of ischemic heart disease death in postmenopausal women: the Iowa Women's Health Study. Am J Clin Nutr 1998;68(2):248–257.
140. Jacobs DR, Jr., Marquart L, Slavin J, Kushi LH. Whole-grain intake and cancer: an expanded review and meta-analysis. Nutr Cancer 1998;30(2):85–96.
141. Jarvi AE, Karlstrom BE, Granfeldt YE, Bjorck IM, Vessby BO, Asp NG. The influence of food structure on postprandial metabolism in patients with non-insulin-dependent diabetes mellitus. Am J Clin Nutr 1995;61(4):837–842.

3 Econutrition: Preventing Malnutrition with Agrodiversity Interventions

Bram P. Wispelwey and Richard J. Deckelbaum

Key Points

- Macronutrient and micronutrient deficiencies continue to be serious problems in lower income countries.
- Based on available published studies, agrodiversity interventions can significantly improve nutritional outcomes, and these results improve further with well-planned designs as well as with additional inputs above and beyond physical inputs (e.g., nutrition education and gender considerations).
- There exists a strong micronutrient bias in the agrodiversity literature, favoring vitamin A over iron in the vast majority of projects undertaken.
- Home gardening: there is a strong rationale for its status as the preferred agrodiversity intervention, yet more investments in scale-up projects are essential.
- There are many incentives for—and challenges to—investing in agrodiversity interventions.
- There is great need to infuse studies with uniform, quality methodology and to utilize indigenous expertise and crops in the design and implementation stages.
- Interventions should be tailored to the specific needs and realities of environments and cultures, and interdisciplinary cooperation among researchers is essential.
- Study quality, capital investments, education, new market opportunities, and gender considerations are all important when designing an ideal agrodiversity intervention.

Key Words: Agrodiversity; dietary diversity; econutrition; home gardening; malnutrition

3.1. INTRODUCTION

While it may seem intuitive that agricultural biodiversity (agrodiversity) interventions have an impact on the nutritional status of participating households and communities in lower income countries, this very premise has been debated consistently over the last quarter century *(1–3)*. A goal of this chapter is to explore this somewhat controversial hypothesis and to proceed further in establishing the conditions and inputs that determine the success or the failure of an intervention's outcome on health. Despite the scholarly debate on the above-stated premise, little research has been conducted to test its validity. Many reasons likely account for this paucity of published data, not least of which are the expense, the perceived obviousness of correlation, the selection of adequate controls, and the difficulty in controlling for confounders and biologically assessing the nutritional outcomes. Research constraints notwithstanding, the available literature will be carefully assessed below, followed by an accounting of key inputs for success and open questions not yet adequately addressed in the literature.

Given that the "growing consensus is that the union between agriculture and nutrition requires cultural, economic and social conditioning factors" *(4)*, it seems that there has never been a better time to avoid academic and research "silos" or "stovepipes." An important idea behind agrodiversity

A. Bendich, R.J. Deckelbaum (eds.), *Preventive Nutrition*, Nutrition and Health, DOI 10.1007/978-1-60327-542-2_3,

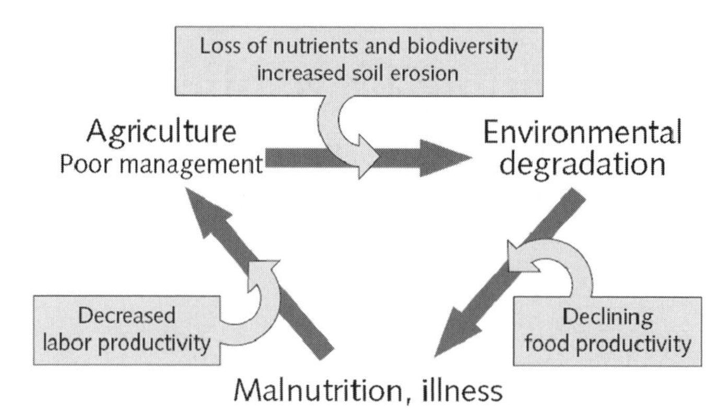

Fig. 3.1. The cycles of agriculture mismanagement, agricultural degradation, and malnutrition, demonstrating the interdisciplinary nature of econutrition. For development interventions to be successful, they must target and understand the interactions among the three disciplines. Reprinted with permission from the *Food and Nutrition Bulletin (6)*.

interventions is that since many nutrient deficiencies are concurrent and interrelated, the integrated solutions must be as well. For clarification and simplification, the new comprehensive term *econutrition (5)* provides an essential understanding of the "interrelationships among nutrition and human health, agriculture and food production, environmental health, and economic development" (*see* Fig. 3.1) *(6)*. Careful consideration of these interrelationships is of fundamental necessity for adequately addressing the ecological and nutritional problems. Vertical, narrow interventions have often failed to acknowledge this reality, which has in part prompted the multisectoral concomitant interventions of the Millennium Villages Project (MVP) in Africa *(7)*. In 2001, the Director General of the International Food Policy Research Institute Per Pinstrup-Anderson stated that "Coordination between sectors is crucial. Multidisciplinary teams must join together to design and implement global strategies to attack the problem in a sustainable way" *(8)*.

Despite the experiences over the last few decades with agriculture interventions, the question of their efficacy has remained. Marie Ruel points out, "In the end, the same question posed in reviews published decades ago remains: what can food-based interventions ... really achieve? Food-based approaches are an essential part of the long-term strategy to alleviate micronutrient deficiencies, but their real potential has not been explored adequately" *(9)*. Since 2001, a few projects have been undertaken and some additional studies have been published with encouraging results. Still, the primary intervention methods addressed in this chapter are the intervention types for which there is significant published evidence.

3.2. AGRODIVERSITY INTERVENTIONS: A BRIEF HISTORY

While "theory" behind agrodiversity interventions and home gardening efficacy may appear sound, investments in pilot projects and full-scale experiments have been few and relatively recent. In the earliest recorded projects dating from the 1970s and the 1980s, study design frequently became a barrier to achieving positive nutritional outcomes. This was true of home gardening interventions where the primary focus was on promoting food production and a particular agricultural strategy. Little effort was put into the promotion of nutrition education and communication, and most of these studies failed to demonstrate any impact on either food intake or nutritional status *(10)*, (*(11)* cited in *(9)*).

Results from nonhome gardening, larger scale agricultural interventions from that period proved to be even more discouraging. A broad 1970s pilot project in sub-Saharan Africa looked to determine the nutritional impact after providing for livestock development, soil conservation, dry-land agriculture, small-scale irrigation, afforestation, rural industries, and transportation infrastructures. While certain aspects of the project were successful (creating a market for local products like honey and inland fisheries), the promoted shift to crop farming from livestock-only created some problems. The irrigation schemes were set up to benefit those closest to the water and with most land—i.e., the wealthier. The poorer farmers did not benefit, a problem worsened by the fact that the seeds provided were not as drought- or pest-resistant as promised. The ultimate nutritional outcome was one of heightened inequality—malnutrition decreased for the better-off and increased for the poorest in the region *(1)*.

A 5-year integrated development program in another African country was set up to specifically address the needs of small farmers: better training, physical inputs, loans/credit, extension services, and marketing facilities. Due to unfortunate planning, however, a certain amount of land had to be owned to qualify for a loan and credit in the form of seeds and fertilizer. Cash crops were more heavily rewarded, leading to increased income for some, but less local food production overall. This caused a decrease in area food security, particularly for the landless and small-scale landowners *(1)*.

Other projects led to even greater exacerbation of malnutrition by creating situations where increased availability did not translate into better nutrition, food supplies decreased, or food prices increased (or some combination of these), leading to a call from the Food and Agriculture Organization (FAO) to have nutritionists involved in the early stages of intervention planning *(1)*. In another set of studies, the availability of seed, the access to credit, and the provision of training services led to a sharp rise in yields and productivity, but the link could not be established between this productivity and local nutrition outcomes *(2)*. Despite the discouraging nature of many of the early agrodiversity projects, many of the daunting challenges were elucidated:

> It is clear that it is not enough just to "hand out the hoes." We have to make sure that we are giving the hoe to the right people, that there is enough land to hoe, that the hands we are trying to put the hoe into are not too busy and overworked to use it, and that the people with the hoes know enough about how to feed themselves and their families that they can actually do so, given changes in agriculture and increases in income *(1)*.

The above list that Lunven declared as necessary for success 25 years ago is surprisingly close to the proposed protocols for the most successful agrodiversity interventions to be described below.

While the smaller home garden interventions also had their share of failures initially, there are still many reports of success from the 1970s and the 1980s, particularly in their impact on production and income (cited in *(9,12)*). As far as vitamin A-focused and food-based agrodiversity interventions are concerned, the express purpose is generally to increase the diversity and quantity of what is usually a primarily grain-based diet. Other less-frequent motivations include increasing food supplies during the lean season, increasing micronutrient-rich fruits and vegetables throughout the entire year, increasing household income through sale of products, and increasing women's control over income *(9)*.

After a flurry of education-only and mass-media nutritional interventions in Indonesia and Thailand in the early 1990s, home gardening interventions began to change and increasingly resemble their modern structure *(9)*. Nutrition education was shown to impact intake of vitamin A-rich foods even without any physical inputs or agricultural intervention (cited in *(9)*). Since this point in the mid-1990s, education has become a standard component of home gardening projects, increasing their efficacy significantly (although this is often difficult to test explicitly). While many studies have documented the impact of interventions on intake, however, evidence is still scarce on other indicators of nutritional status. In some perhaps suboptimally designed early studies, home gardening interventions were associated with a decreased risk of vitamin A deficiency *(13)* and less signs of clinical deficiency in the eyes *(14)*. A 1994 review analyzing the effect of food-based approaches on vitamin A deficiency also found that agriculture interventions could be effective in impacting human nutritional status *(15)*. The evolution of agrodiversity interventions up to this point has been a long, troubled process filled with setbacks and disappointments, but it is ultimately one that has led to a newer, improved generation of interventions that are marked for their quality.

3.3. HOME GARDENING: RATIONALE FOR ITS STATUS AS A PREFERRED AGRODIVERSITY INTERVENTION

From the raw data and figures, it appears that higher quality studies and those with the greatest number of different capital inputs (*see* Table 3.1) achieve better nutritional outcomes. An analysis of the studies included in Table 3.2 leads one to conclude that agrodiversity interventions can be successful approaches toward addressing micronutrient deficiencies in at-risk populations in lower income countries. Given the limited number of studies and their diversity in many respects, these relationships cannot yet be adequately assessed in a quantitative manner. It can be noted, however, that intervention studies were found to be mixed with regard to nutritional outcomes. Some studies had no impact on primary outcomes, while others actually describe poorer outcomes after the intervention. In this current analysis, only studies that were specifically designed to impact nutritional status via agrodiversity were included, and the outcomes were more frequently positive—particularly when additional capital investments, better design quality, and nutrition education components were included.

The utilization of home gardening as the primary and/or preferred means of improving human nutritional status via an agrodiversity approach becomes evident when searching the relevant literature. While the lack of large-scale, field-based agrodiversity interventions is disconcerting, this reality simultaneously speaks to the many merits of home gardens. Among the numerous qualities of home gardening that are associated with its popularity is its compatibility with ecological considerations. Given the logistical problems in lower income countries, this is especially true in the poorest parts of

Table 3.1
Explanations of the Types of Capital Available for Agrodiversity Interventions

Natural Capital	*Physical Capital*	*Social Capital*	*Human Capital*	*Financial Capital*
Use of sustainable agriculture practices; intensification of existing systems; diversification by adding new systems	Support for the increase in land, tools, seeds, fertilizer, livestock, etc.	Use of social and participatory processes	Use of agriculture training programs; nutrition education programs; other training programs; gender considerations	Provide access to credit, grants, subsidies; value-added products; value-added marketing; other financial benefits

Summaries of capital types (above) from Berti et al. *(4)*, used with permission from Cambridge University Press.

Table 3.2
Summary of Agrodiversity Intervention Results

Part A: All studies, data, and the table categories are from Berti et al. *(4)*, used with permission from Cambridge University Press.
Abbreviations: NA—not applicable; hh(s)—household(s); veg—vegetable(s); XN—night blindness; ARI—acute respiratory infection; UTRI—upper respiratory tract infection; VA—vitamin A; RF—riboflavin; B_6—vitamin B_6; VC—vitamin C; P/L—pregnant/lactating; Hb—hemoglobin; OFSP—orange-fleshed sweet potato; sig—significant(ly)
Types of capital input that projects invested in: N—natural; P—physical; S—social; H—human (italics indicate nutrition education included); F—financial

Country (references)	*Type of Study*	*Improvement in Agricultural Indicators?*	*Improvement in Dietary Intake Indicators?*	*Improvement in Anthropometric Indicators?*	*Improvement in Biochemical/Clinical Indicators?*	*Improvement in Morbidity/Mortality Indicators?*	*Capital Inputs*	*Weighting*
North Bangladesh *(16), (63)*	Pre–post with control group	Yes. Intervention hhs with gardens increased to 100%; average size of garden increased to 130%; number of varieties increased 5×	Yes. Veg intake increased in hhs and specifically in infants and children	Yes. Improvement in stunting and underweight	Yes. Children: anemia 30% less than control; XN decreased by 50%, no change in control. Women: XN less in intervention than control	No difference in diarrhea prevalence. Intervention children had less severe ARI and less URTI than controls	N P S *H* F	High
Nepal *(64)*	Pre–post, no control	Yes. Number of gardens	NA	No (indicators worsened drastically)	NA	NA	NA	Low

(Continued)

Table 3.2
(Continued)

Country (references)	*Type of Study*	*Improvement in Agricultural Indicators?*	*Improvement in Dietary Intake Indicators?*	*Improvement in Anthropometric Indicators?*	*Improvement in Biochemical/Clinical Indicators?*	*Improvement in Morbidity/Mortality Indicators?*	*Capital Inputs*	*Weighting*
Vietnam *(50,49)*	Intervention vs. control, some pre–post	Yes. Production and sale of veg, fruit, fish, and meat	Yes. Intervention children ~50% higher intake of veg, fruit, energy, protein, VA, and iron	Yes. Stunting decreased from 50 to 42% in intervention children	NA	Yes. Intervention children: incidence of respiratory infections decreased from 50 to 11% (no change in controls); diarrhea decreased from 18 to 5%	N P S *H* F	High
Bangladesh *(65)*	Pre–post with control (but 'pre' is after second year of intervention)	Yes. Small increase in hhs growing VA-rich crops (intervention and control)	Yes. 10–20% increased intake of VA-rich veg (also in control) and other veg	NA	No difference. No change in XN	NA	N P S *H*	Low
Kenya *(66)*	Intervention (training in marketing and nutrition) vs. control (promotion, no training), pre–post	Yes. Yield of sweet potato was ~0 pre and 5–19 tons per hectare post	Yes. Where VA intake initially low, it improved to almost adequate (no increase in controls)	NA	NA	NA	N P S *H* F	High

Tanzania *(67)*	Intervention vs. control 5 years postintervention	Yes. More gardens with guava and papaw	Yes. Intake of VA-rich foods ~50% greater than control	NA	No	No. Helminth infection:intervention 79% vs. control 49%	N P S *H*	Mid
Vietnam *(68)*	Pre–post	Yes. Per capita home veg production increased 5×	Yes. Increase in intake of energy, protein, fat, and veg of 17, 23, 75, 250%, respectively	NA	Yes. Xerophthalmia decreased almost to zero	NA	N S *H*	Low
Guatemala *(69)*	Pre–post, with control	No	Yes. Control children (without garden with leaf veg) 3.5× more VA deficiency	NA	NA	NA	P *H*	Mid
Philippines *(70)*	Pre–post, with control	Yes. Production of five types of veg increased 37–700%	Yes. Increased veg consumption; VA intake increased by 12%, control decreased by 48%	NA	NA	NA	N P S *H* F	High
NE Thailand *(53,52)*	Cross-sectional pre–post, with control	NA	Yes. Increased VA and iron intakes in children, schoolgirls and P/L women; in some cases also in controls	NA	Yes. Schoolgirls' serum retinol increased from moderately deficient to nondeficient (no increase in control)	NA	N S *H* F	High

(Continued)

Table 3.2
(Continued)

Country (references)	*Type of Study*	*Improvement in Agricultural Indicators?*	*Improvement in Dietary Intake Indicators?*	*Improvement in Anthropometric Indicators?*	*Improvement in Biochemical/Clinical Indicators?*	*Improvement in Morbidity/Mortality Indicators?*	*Capital Inputs*	*Weighting*
NE Thailand *(53,71)*	Pre–post, with control	NA	Yes. VA intake increased in intervention: preschoolers (3×), schoolchildren (2×), lactating women (2×). No change in control women	NA	NA	NA	N S *H*	High
Philippines *(72,14,18)*	Paired pre–post	NA	Yes. Increased children's VA intake	Yes. Improved weight for height and reduced severe wasting	No difference. No change in serum VA levels or prevalence of xerophthalmia	NA	P *H*	Mid
NE Senegal *(11)*	Survey of those with and without gardens at baseline and 10–12 years later	NA	Yes/no. Some nutrients increased, some decreased	NA	NA	NA	N P *H* F	Low
Egypt *(73)*	Pre–post with control	Increased yields of maize, peanut, wheat (41–74%)	Yes. Hh protein and iron increased (10%, 20%)	No difference	NA	No difference in mortality	P S *H* F	Low

Part B: Studies located by Ruel *(9)*; table categories from Berti et al. *(4)*; summaries from Wispelwey *(40)*

Abbreviations: NA—not applicable; hh(s)—household(s); veg—vegetable(s); XN—night blindness; ARI—acute respiratory infection; UTRI—upper respiratory tract infection; VA—vitamin A; RF—riboflavin; B_6—vitamin B_6; VC—vitamin C; P/L—pregnant/lactating; Hb—hemoglobin; OFSP—orange-fleshed sweet potato; sig—significant(ly)

Types of capital input that projects invested in: N—natural; P—physical; S—social; H—human (italics indicate nutrition education included); F—financial

Country (references)	*Type of Study*	*Improvement in Agricultural Indicators?*	*Improvement in Dietary Intake Indicators?*	*Improvement in Anthropometric Indicators?*	*Improvement in Biochemical/Clinical Indicators?*	*Improvement in Morbidity/Mortality Indicators?*	*Capital Inputs*	*Weighting*
Ethiopia *(74)*	Intervention vs. control after 9 months of intervention	NA	Yes. More diversified diets and VA-containing foods in intervention hhs	NA	Yes. Less XN and less Bitot's spots (1% intervention vs. 4% in control)	NA	N P S *H* F	Low
India *(75)*	Pre–post	Yes. Increase in percentage of households growing leafy veg; increase in production of all veg types	Yes. Weekly leafy veg intake increased by >100%	NA	Yes. Reduction in conjunctival xerosis, Bitot's spots, and XN	NA	N P S *H*	Low
Bangladesh *(51)*	Pre–post, with control	Yes. Increased production of vegetables and fish	Yes/no. Increase in vegetable consumption but not fish consumption	NA	No difference. No effect on hemoglobin from vegetable or fishpond production	NA	N P S *H* F	Mid

(*Continued*)

Table 3.2
(Continued)

Part C: Table categories by Berti et al. *(4)*; studies located and summarized by Wispelwey *(40)*

Abbreviations: NA—not applicable; hh(s)—household(s); veg—vegetable(s); XN—night blindness; ARI— acute respiratory infection; UTRI—upper respiratory tract infection; VA— vitamin A; RF—riboflavin; B_6—vitamin B_6; VC—vitamin C; P/L—pregnant/lactating; Hb—hemoglobin; OFSP—orange-fleshed sweet potato; sig—significant(ly)

Types of capital input that projects invested in: N—natural; P—physical; S—social; H—human (italics indicate nutrition education included); F—financial

[a]This intervention is currently ongoing with the eventual goal of publishing data on the nutritional impact on the local children.

Country (references)	*Type of Study*	*Improvement in Agricultural Indicators?*	*Improvement in Dietary Intake Indicators?*	*Improvement in Anthropometric Indicators?*	*Improvement in Biochemical/Clinical Indicators?*	*Improvement in Morbidity/Mortality Indicators?*	*Capital Inputs*	*Weighting*
South Africa *(23,76)*	Pre–post, with control; 1-year follow-up	Yes. More butternuts, carrots, spinach, sweet potato, pawpaw in intervention	Yes. Increase in both, but sig higher in intervention group for VA, RF, B_6,VC	NA	NA	NA	N P *H* S	Mid
South Africa *(22)*	Pre–post, with control; 20-month follow-up	Yes. More VA-containing fruits and veg in intervention	Yes. Intervention group consumed more VA-containing fruits and veg	No difference between groups at baseline or follow-up	Yes. Intervention group has sig higher serum retinol status	NA	N P *H* S	Mid
Malawi *(24,25)*	Pre–post, with control; complete study (with all nutrition indicators) not yet concluded	Yes. Average legume plot increased from 100 m^2 to 862 m^2 in 5 years	Yes. Greater consumption of soybeans and groundnuts in intervention hhs	NA[a]	NA[a]	NA[a]	N P *H* S	High

Burkina Faso *(42)*	Pre–post, no control; one group added financial assistance	NA	Yes/no. Both groups ate more liver but less mango after intervention	NA	Yes. Serum retinol increased in both, but slightly less in group with financial assistance	NA	S *H* F	Low
Mozambique *(21)*	Pre–post, with control; 2-year follow-up	Yes. Intervention plot areas of OFSP increased 10×	Yes. Intervention children more likely to eat adequate OFSP serving (55% vs. 8%); 7× higher VA intake overall; higher intake of iron, zinc, VC, energy, protein, and B vitamins	Yes; less wasting and less low weight for age among intervention children	Yes. Serum retinol increased in intervention children, but not control	Yes; less morbidity in both groups	N P *H* S F	Very high

the world. In the absence of a thriving market, particularly in rural areas, achieving a diverse diet is very difficult without home gardens *(16)*. Home gardens are the oldest known production systems in the world, and their global ubiquity and longevity speak to their vast economic and nutritional benefits *(16)*. Usually home gardens complement field agriculture—while field crops provide for most energy needs, gardens supplement the diet with the essential vitamins and minerals *(11)*.

Despite the many merits and popularity of home gardening, this intervention has faced its share of critics in the past. Detractors have criticized study designs, management, monitoring, unrealized expectations, and the failure to definitively prove sustainability *(16)*. The foremost failure, which has been acknowledged for many years, is the lack of understanding and adaptation to local conditions and communities by project planners. Planners often failed to recognize the need for formulating a design based on the specific local environment as well as on social and resource supply conditions (*(17)* cited in *(16)*). Such a critique, however, speaks only to the need for better gardening projects rather than their inherent inferiority. Some have suggested that home gardening is not as cost effective as fortification, supplementation, and target subsidies (*(17,18)* cited in *(16)*). As noted below, however, more recent economic analyses support the adoption of community agriculture-based interventions as highly cost effective *(19)*. In any case, cost-effective analyses must be received with caution due to the fact that they generally focus on a narrow outcome (such as vitamin A deficiency) rather than accounting for all of the benefits that home gardening may provide to a household or a community. And while some argue that gardening leaves out many of the food insecure, particularly those without land and water *(16)*, this critique does not speak to the *rural* realities most frequently encountered in these types of interventions. Due to the lack of cultivable land and the presence of inexpensive markets in urban areas, other types of intervention may be necessary to improve the dietary diversity of the food insecure in urban settings.

Home gardens have many advantages. While animal foods generally supply a more abundant and more readily absorbable conglomeration of macro- and micronutrients, fruits and vegetables are often the only reliable nutrition source for many poor families in lower income countries. Home production may fulfill many nutritional needs by providing access to foods that cannot be afforded or are not available in local markets. Furthermore, home gardening can serve purposes beyond its role as a more sustainable micronutrient supplement. While the broad range of social benefits from home gardening may be difficult to measure and may not fit well into traditional vertical planning and assessment, it is this aspect of the intervention that fuses well with the concept of econutrition.

Home gardens may increase diet diversity and improve household food security. Fruits and vegetables have the added benefit of making other foods more palatable by limiting redundancy, further increasing overall energy intake. Home gardens may also lead to improved soil quality by the planting of nitrogen-fixing species, providing tangible results that may lead local landowners to invest in similar techniques in local staple farming or large-scale agriculture. They can be adapted to hostile environments (i.e., very dry and very wet) and are a low-risk, low-cost technology. Of note, the poor can actively partake in this intervention, while large-scale field projects by their nature must involve state or wealthy private ownership. Furthermore, community and school gardening can still benefit landless households *(16)*. Where large-scale interventions may impact only a small number of landowners, home gardening specifically allows for widespread ownerships as well as the significant empowerment of women who typically form the link between food access and consumption. For long-term economic progress and sustainability, home gardens can be a source of additional income for households. This allows for further diet diversification via greater purchasing power, as well as local economic stimulation and even some possibility for monetary savings (*see* Table 3.1 for list of potential capital inputs and benefits in addition to nutritional outcomes).

The benefits of home gardening can be felt in a variety of food insecurity situations. In areas dominated by a staple crop that grows only part of the year, home gardens can provide year-round food security by offering seed options to fit a variety of soil states *(20)*. This is especially important in areas that are rife with famine from soil erosion, war, drought, and other natural disasters, as is the (currently worsening) case in many sub-Saharan locales. The harsh geographic conditions would seem to demand that this type of agrodiversity be practiced on a larger scale, or at least be the subject of more research. Still, Talukder et al.'s statement that "there remains limited experience and success in moving from small-scale pilot programmes [sic] to large-scale programmes, and there are only a few opportunities to share experiences across countries and development sectors *(20)*" remains true today.

While this gap in knowledge and experience needs to be better-addressed, a few larger scale projects have been successfully completed in recent years in Bangladesh *(20)* and Mozambique *(21)*. In order to scale-up to larger garden projects, Marsh suggests that it may be advisable to invest in an experimental garden site for varietal trials *(16)*. This suggestion has been successfully implemented by some of the most recent and ongoing trials in South Africa *(22,23)*, Malawi *(24,25)*, and Mozambique *(21)* (*see* Table 3.2, Part C for brief summaries). Additional requirements for large-scale projects are sustainable seed production and availability, as well as the prioritizing of traditional plants and building on indigenous knowledge *(16)*. Due to the increasingly out-of-reach expense of fertilizers, other low-cost methods such as composting and leguminous planting should be explored *(16)*.

Historically, large-scale farming system interventions dealing with income generation and conservation have failed to include a human nutritional evaluation component *(26)*. Thus, for assessment there are mostly small-scale projects that are nutrition focused and large-scale projects that are agriculture and income focused. To rectify this, it is likely easier to expand and scale up, little by little, small home gardening pilot projects than to invest in expensive large-scale projects with only theory to back up the design protocol. The former is the route taken by Helen Keller International, which formed a nationwide Bangladeshi project that has been a highly successful and still under-replicated home gardening scale-up intervention. Despite the overall modest amount of data on agrodiversity interventions currently available, large-scale projects may be justified, given the large-scale examples and recent robust successes in Bangladesh *(20)*, Mozambique *(21)*, and elsewhere (*see* Table 3.2). While some argue that the lack of scientific data is hampering project development *(9)*, there are others who cite a dearth of political and organizational commitment as the principal impediments *(26)*. In any case, there are many reasons to believe that investing in agrodiversity interventions will lead to positive outcomes—and not only in terms of human nutrition.

3.4. INCENTIVES FOR—AND CHALLENGES TO—INVESTING IN AGRODIVERSITY INTERVENTIONS

It has recently been estimated that "implementing interventions that improve child nutrition and provide clean water and sanitation and clean household fuels to all children younger than 5 years would result in an estimated annual reduction in child deaths of 49,700 in Latin America and the Caribbean, 0.8 million (24%) in South Asia, and 1.47 million (31%) in sub-Saharan Africa" *(27)*. Likely these interventions would decrease regional micronutrient deficiency and nutritional morbidity among adults as well.

While investments in nutrition are often couched in ethical or human rights terms, there are also strong economic arguments for investing in nutritional status improvements. Returns for programs designed to improve nutrition are predicted to significantly outweigh their costs: in China, for example, preventing micronutrient deficiencies is estimated to lead to \$2.5–\$5 billion annually in increased GDP *(19)*. Lack of adequate policies and programs to address anemia in women will lead to agricultural productivity losses in Sierra Leone, reaching nearly $100 million over the next 5 years (cited in *(19)*). One reason for the "favoritism" shown to vitamin A interventions (discussed in depth below) is that eliminating vitamin A deficiency alone will save 16% of the global burden of disease in children, inevitably leading to a more productive generation (cited in *(19)*).

Agrodiversity interventions have garnered interest from the international community as a sustainable solution, in conjunction with other nutrition interventions, for addressing the epidemic of severe undernutrition worldwide. Perhaps most encouragingly, community-based crop growth promotion (without supplementary food) has been estimated to be up to 100 times as cost effective as direct food supplementation *(19)*. Nutrition education is also 14–69 times less expensive than direct food supplementation and has been shown to lead to better outcomes when compared to other forms of intervention *(9,19)*. Furthermore, community-focused agriculture interventions are very similar in the combined cost to iron supplementation, vitamin A fortification of sugar, and salt iodization *(19)*, but still, adequacy of these micronutrients needs to be assured.

Based on the assumption that the financial benefits of improving nutrition outweigh the investment, market-focused approaches are often proposed as solutions to diminishing poverty and concomitant undernutrition. However, market availability may fail to adequately address nutritional problems for two major reasons—access and affordability. This may not be obvious as, for example, healthy growth rates cannot be detected with the naked eye; parents cannot often tell when their children are stunted or becoming malnourished *(19)*. Micronutrient deficiencies often cannot be detected without clinical tests unless severe. Also, adequate nutrition is not intuitive, a reality that is increasingly evident in wealthier nations *(19)*. While countries may list the importance of eating a variety of foods in their nutrition guidelines, the word "variety" carries different meanings to different people *(28,29)*. How food variety can achieve healthy micronutrient intake levels is depicted in Fig. 3.2.

Free markets may affect nutrition negatively, as food advertising and marketing often change preferences in unhealthy ways. These realities partially explain why nutrition does not always improve when per capita income increases in lower income countries *(9)*. Paradoxically, the fact "that good economic growth is not a prerequisite for hunger reduction is exemplified by some of the most successful countries in hunger reduction, which have not necessarily achieved better economic growth rates" *(30)*. Of note, better nutrition may reduce the spread of contagious diseases and provide other health benefits as well as increase national economic productivity *(19)*.

While often a difficult task, achieving a highly varied diet helps achieve optimal health. Ironically perhaps, nearly two-thirds of hungry people worldwide are farmers and pastoralists living in Asia and Africa *(31)*. While it may seem a formidable task to alleviate the suffering of the 800 million who are energy-deficient and the 2 billion with micronutrient deficiency, most of these deficiencies can be resolved with relatively small increases in the variety of food items consumed (*see* Fig. 3.2) *(6,32,33)*. It was recently determined that "for every 1% increase in dietary diversity, there is a 1% increase in per capita food consumption for poor and middle-income countries *(34)*," with the implication that increased dietary diversity may improve macronutrient deficiencies in addition to micronutrient deficiencies. Evidence from a 10-country analysis suggests that dietary diversity could be a useful indicator of food security *(33)*, which, if true, would provide a simple method for identifying households and communities at risk for malnutrition. An unpublished study in Senegal showed that dietary diversity is positively correlated with intakes of many key nutrients including calcium, iron, zinc, vitamin A, vitamin C, thiamin, riboflavin, and vitamin B_6 *(32)*. Another study revealed that dietary diversity negatively correlates with the incidence of respiratory and gastrointestinal infection. In this pilot study, inclusion of wild-gathered foods and animal-source food provided the strongest correlation between diversity indices and the disease outcomes *(32)*. The Millennium Villages Project in Sauri, Kenya, has reported a significant positive correlation between crop species diversity and crop functional diversity

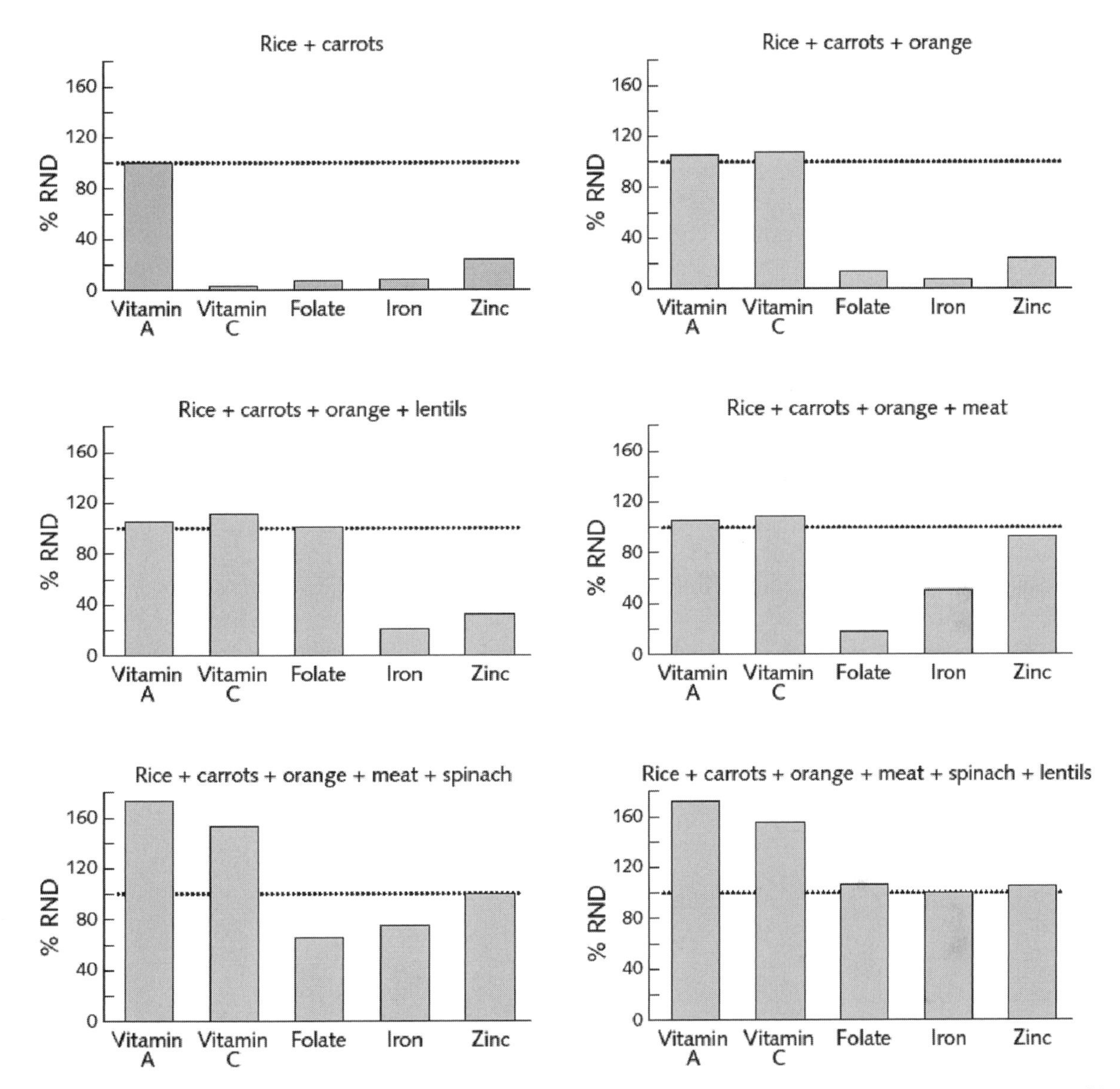

Fig. 3.2. The effect on micronutrient adequacy of sequentially adding small amounts of nutrient-dense foods, with rice as an example of a staple food. Dietary adequacy is expressed as percentage of required nutrient density (RND). Reprinted with permission from the *Food and Nutrition Bulletin (6)*.

when considering seven different vitamins and minerals (Econutrition Task Force personal communication). With the link between crop diversity and crop functional diversity established in these data, the remaining challenge (explored in this chapter) is to connect functional diversity to nutrition, health, and more productive human populations. Indeed, the International Union of Nutritional Sciences (IUNS) has set up an Econutrition Task Force to begin addressing these very issues *(34)*. As stated in the Introduction, however, establishing the link between functional agrodiversity and human nutrition and health is easier theorized than demonstrated.

While it has been slow to change its focus, nutrition science has come a long way since its assumption a few decades ago that high staple production and consumption is very desirable *(5)*. No longer are staples alone considered the ideal basis of a healthy human diet *(34,26)*. In reality, this notion has made sustainable food supplies and health more precarious in many areas of the world: when fluctuations in climate and pestilence compromise production, communities relying on a single or few crops may face serious food insecurity *(5)*. Furthermore, the act of producing enough food to feed an entire population may not adequately address undernutrition without additional, thorough interventions. South Africa, for example, produces enough food to feed its entire nation, yet 35% of its people are food insecure due to poverty and other forms of inaccessibility *(35)*.

The Green Revolution has greatly increased the amount of energy available from food staples per capita—population in lower income countries doubled from 1965 to 1999, but cereal production nearly tripled (Econutrition Task Force personal communication). However, this large cereal production may also be partially responsible (along with subsidies, transportation, and other factors) for higher prices of vegetables, fruits, and meats, effectively diminishing the ability of the poor and landless to achieve adequate diets *(26)*. At low incomes, the poor are often forced to choose between adequate energy in staple form and other food products. To keep from going hungry, they choose mainly staples, a direction which often leads to micronutrient deficiencies since major staples are generally poor in vitamins and minerals (*see* Fig. 3.3). For this reason, micronutrient intakes increase more rapidly with an increase in income than do energy intakes. Since there has been no similar Green Revolution for nonstaple food sectors, relative prices of many nonstaple foods have increased over time (personal communication). The needs and benefits of home gardens and other agrodiversity interventions become more obvious as they represent a potentially cheap and effective way to access nonstaple foods on a regular basis.

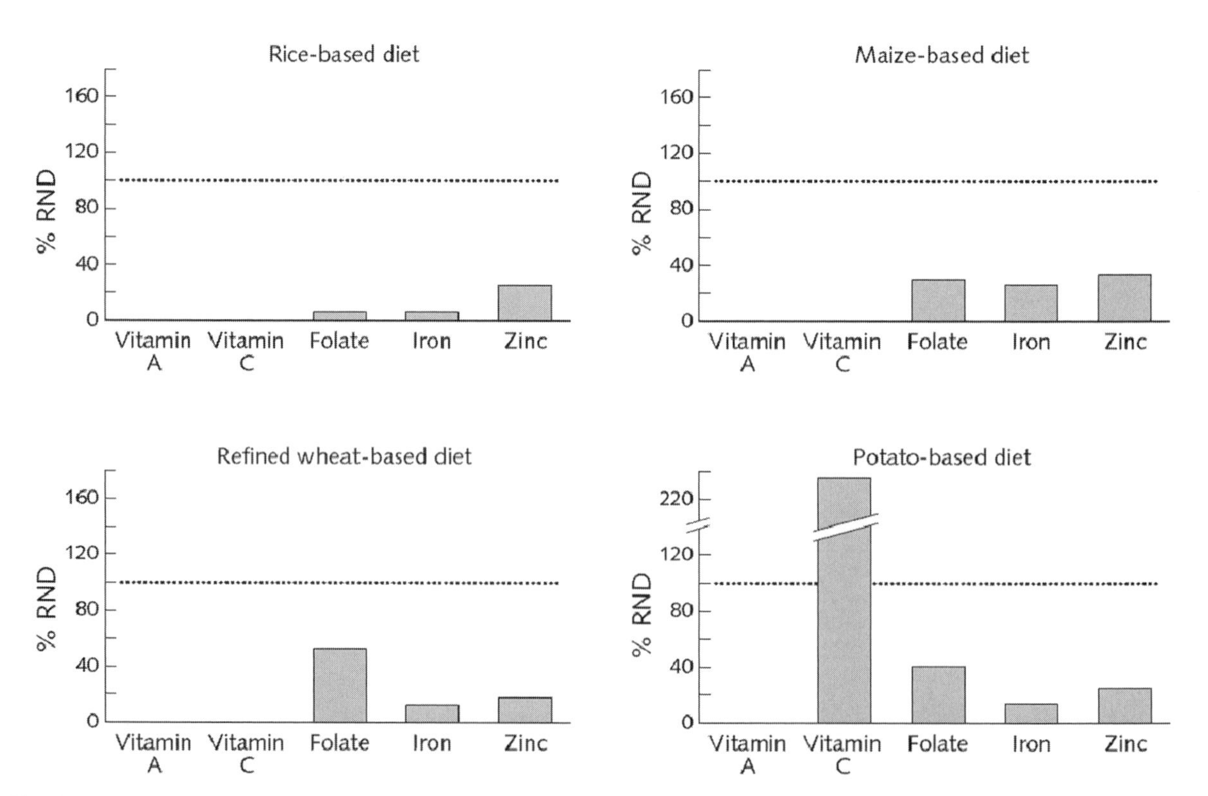

Fig. 3.3. The insufficiency of common staple foods to meet critical micronutrient needs, expressed as percentages of required nutrient density (RND). Reprinted with permission from the *Food and Nutrition Bulletin (6)*.

Changes in food habits in lower income countries have been reported to lead to decreased dietary variety, which is being linked more strongly to levels of household food security *(36)* and to micronutrient deficiencies—particularly vitamin A, iron, and zinc *(26)*. Since 1970, cereal production has quadrupled in South Asia, while micronutrient-rich pulse production has declined by 20% *(37)*. While per capita energy intake has increased in the last three decades, iron intake has decreased, increasing the prevalence of anemia *(26)*. Traditional cereals that generally take more time to prepare are being replaced by maize, wheat, rice, and potatoes that are less nutrient-dense and significantly lower in iron. In Kenya, a country increasingly hampered by overnutrition and overweight in addition to undernutrition, the drop in per capita pulse and legume intake mirrors the energy increase supplied by fats and oils almost exactly *(26)*.

Arguments for utilizing agrodiversity interventions go beyond the economical and nutritional impacts. It has been predicted that fossil fuel availability will decline significantly over the next 50 years and alternative energy sources may not be able to make up for this loss of energy *(33)*. In this case, communities around the globe would stand to benefit from food produced locally and in ways that are both sustainable and ecologically sound *(38)*. It is important to pursue interventions that improve ecological systems in addition to human nutrition precisely because the two are mutually reinforcing.

In summary, there are multiple ways in which agrodiversity interventions may improve human well-being:

- Agrodiversity allows people, particularly those who are poor or without access to a diverse local market, to meet their significant energy, vitamin, and mineral needs.
- Agrodiversity makes harvests more stable by improving disease and pest resistance, as well as providing the means for withstanding environmental fluctuations *(39)*. These outcomes lead to improved security in the local food supply.
- Local agrodiversity diminishes excessive carbon fuel use and improves soil quality.

The aspects of agrodiversity interventions that are most strongly associated with positive nutrition outcomes need to be defined. In determining the make-up of ideal agrodiversity interventions, agrodiversity intervention history and the most recent results will be reviewed.

3.5. AGRODIVERSITY INTERVENTIONS AND STUDY QUALITY

3.5.1. *Study Quality*

Among studies included in Table 3.2, those that were determined to be higher quality *(4,40)* were significantly ($p = 0.0075$) more likely to achieve positive outcomes in nutritional indicators than those deemed "low" or "mid." This result is seen most clearly in terms of the dietary intake and anthropometric indicators, which displayed a trend of increasing positive outcomes from "low" to "high/very high" weighting. Biochemical and clinical indicators, while achieving all positive outcomes for the "high/very high" weighted studies, did not follow the same predictable track through "low" and "mid." Indeed, "low"-weighted studies were more than twice as likely to achieve positive biochemical or clinical indicator outcomes than "mid" studies based on the data collected.

3.5.2. *Approaches and Endpoints*

The difficulty of comparing studies with different objectives and inputs has previously been noted *(4)*. Figures 3.4–3.6 summarize varying approaches and endpoints. Figure 3.4, comparing subjective study quality to outcome success, illustrates that the highest quality studies were successful in improving dietary intake, anthropometrics, and biomedical or clinical indicators. In terms of dietary intake

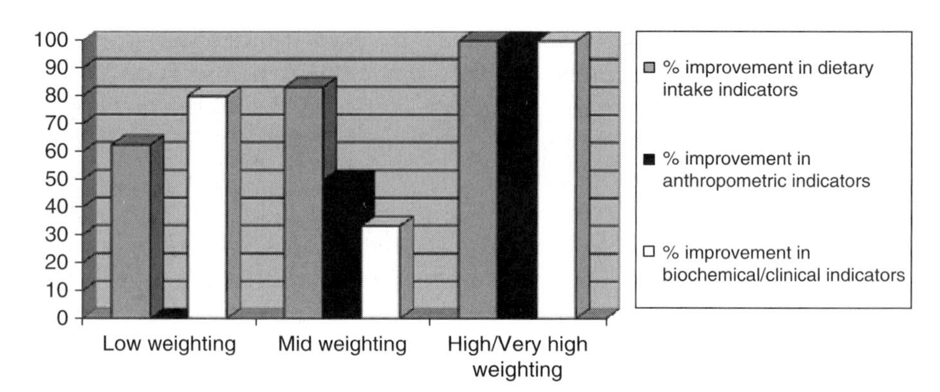

Fig. 3.4. Percentage of studies that found improvements in three types of nutritional indicators, organized by the quality weighting of the study. Overall, higher quality studies had a greater percentage of positive outcomes. The results from a systematic search of agriculture, nutrition, and health-care literature for agrodiversity interventions designed to improve nutritional outcomes were analyzed (for complete methods and results, contact authors).

and anthropometrics, the results were "dose-dependent" in that as study quality increased so too did the percentage achieving successful nutrition outcomes. Still for biochemical/clinical indicators, the "mid"-quality studies performed worse than the "low"-quality studies. This finding may be due in part to the necessity of reaching a certain threshold of design quality in order to achieve significant results with more sensitive (biochemical indicators) and potentially confounder-tainted (clinical indicators) outcomes. Studies in the "mid"-quality range perhaps were not adequately powered to detect biochemical differences between control and intervention groups.

3.5.3. Agrodiversity Interventions and Capital Inputs

Studies that invested in all five types of capital inputs (*see* Table 3.1) achieved overall better outcomes than those that invested in four or less, although this result was not significant. The benefit of studies with all five types of capital input was relatively slight, as dietary intake indicators were the same between these two groups and the difference between biochemical and clinical indicators between these subsets was marginal. Figure 3.5 shows the trend in added benefit from investing broadly in five types of capital (natural, physical, social, human, and financial). While the results were fairly similar between these two groups when examining dietary intake (the easiest outcome to achieve successfully) and biochemical/clinical indicators, anthropometric outcomes were positive far more frequently in the 5-capital group. This is particularly encouraging, given that maintaining healthy growth patterns may be a crucial and difficult outcome to achieve via agrodiversity or other types of interventions. The results support Berti et al.'s statement that "investing broadly in the target population—and not just in the agriculture intervention—does seem to improve prospects for positively impacting on the health of the people" *(4)*.

3.5.4. Agrodiversity Interventions and Education

Studies with a nutrition education component as part of their design achieved greater percentages of positive outcomes (in dietary intake indicators, anthropometric indicators, biochemical/clinical indicators, and morbidity/mortality indicators) than those without any nutrition education component (Fig. 3.6). Only 17% of the outcomes in the three studies that did not have a nutrition education component were positive. Out of the 19 studies supplementing the agrodiversity intervention with nutrition

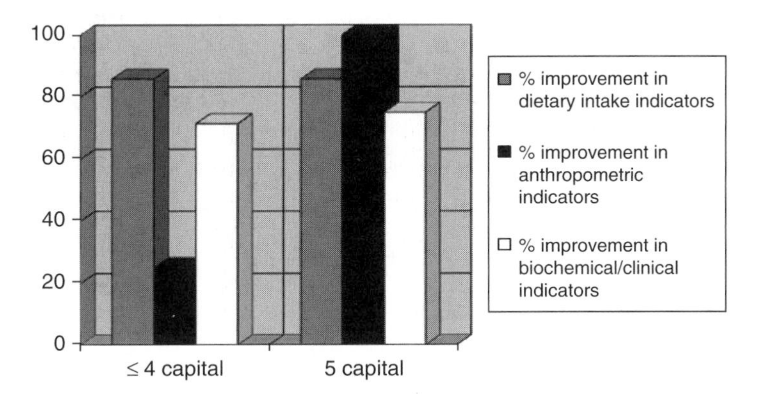

Fig. 3.5. Percentage of studies displaying nutritional improvement by capital inputs listed in Table 3.1. Percentage of studies that found improvements in three types of nutritional indicators, organized by the number of capital inputs the intervention invested in. Overall, studies that invested in all five types of capital had a greater percentage of positive outcomes. The results from a systematic search of agriculture, nutrition, and health-care literature for agrodiversity interventions designed to improve nutritional outcomes were analyzed (for complete methods and results, contact authors).

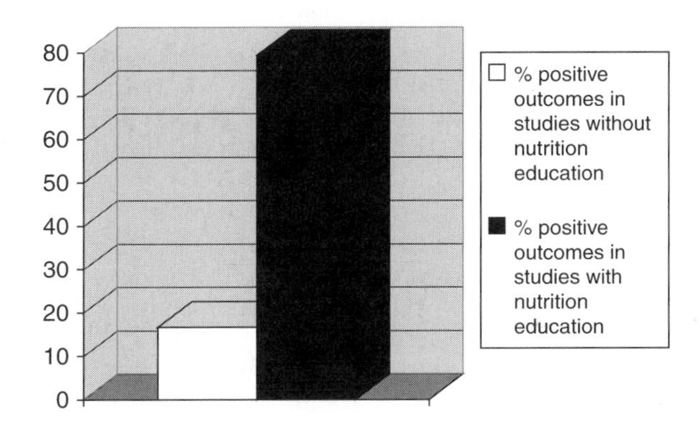

Fig. 3.6. Percentage of studies that found improvements in four types of nutritional and health indicators, organized by the inclusion or the lack of inclusion of a nutrition education component. Overall, studies with a nutrition education component had a greater percentage of positive outcomes. The results from a systematic search of agriculture, nutrition, and health-care literature for agrodiversity interventions designed to improve nutritional outcomes were analyzed (for complete methods and results, contact authors).

education, 79% of outcomes were successful in positively impacting nutritional indicators. Clearly, studies with a nutrition education component were significantly ($p = 0.005$) more likely to achieve these positive nutritional outcomes.

Home gardening projects report higher success rates than other types of agrodiversity interventions, as has previously been found *(4,9)*, but this finding is difficult to interpret, given the small number of nonhome gardening interventions available for review with this chapter's inclusion criteria. Often nutrition education, strong community ownership, and gender considerations are also a part of these home gardening interventions, obscuring whether home gardening itself is responsible for successful outcomes. Home gardening almost always implies a strong investment in human capital *(4)*, making it inherently difficult to determine what aspect of the intervention (or what combination of aspects) is actually leading to the successful outcomes.

It has previously been found that nutrition education alone can lead to improvements in nutritional outcomes without any agriculture or food inputs *(9,41)*. Still, for many rural poor it is necessary to have the means and opportunities to act on any gained knowledge of nutrition *(35)*. While the vast majority of the studies included in this analysis had a nutrition education component (19 of the 22), the results still imply that education may be an essential part of ensuring successful interventions. It is possible that a failure to include a nutrition education component is a marker for an overall poorly designed study, although the sample size does not allow for much insight on this point. Regardless, these data solidly support a continuation of the already widespread and popular nutrition education components in agriculture interventions.

3.5.5. Agrodiversity Interventions and Household Income Generation

While some interventions resulted in increasing household income, in the recent studies in South Africa *(22,23)*, Malawi *(24,25)*, and Mozambique *(21)*, this was not a universal intended outcome of the projects. The Mozambique project is the exception, however, as it set out to establish new markets and increase household incomes in addition to improving nutritional outcomes *(21)*. In this respect, as well as in terms of nutritional outcomes, the Mozambique project was successful. Other studies also noted the unexpected arrival of new markets sometimes just months after implementing an intervention *(22,23)*. In a vitamin A study in Burkina Faso that investigated the impact of adding a stipend to the physical inputs and education component, however, the added funding did not additionally impact serum retinol status beyond what was achieved by the education and agriculture aspects of the program *(42)*. More studies specifically looking to impact or supplement income through agrodiversity interventions are needed.

3.5.6. Micronutrient Bias in the Literature

Studies primarily or significantly concerned with vitamin A intake and status related to 19 of the 22 interventions are listed in Table 3.2. After some years of controversy, several recent feeding trials have shown that serum retinol or improved vitamin A status was documented after feeding beta-carotene-rich foods for short periods (between just 3 weeks up to 4 months), sometimes with an addition of fat and/or deworming (*(43–48)* cited in *(21)*). It is a given that among nutritional deficits vitamin A and iron deficiencies have major public health impacts *(9)*, but while several recent vitamin A agrodiversity studies have been conducted, it is somewhat surprising that there is a paucity of conducted studies utilizing iron-related endpoints. Only 4 of the 22 studies from Table 3.2 examined iron intake.

Almost one-third of children in lower income countries suffer from some level of vitamin A deficiency, impairing growth, development, vision, immune function, and occasionally even leading to blindness and death *(9)*. However, iron is also recognized as a common deficiency in the world—one-half of pregnant women are anemic, along with one-third of school-age children *(9)*. Iron deficiency may also impair physical growth, cognitive development, and immunity in children, and it may lead to fatigue and reduced work capacity in adults *(9)*. One proposed reason for the lack of agrodiversity interventions focusing on iron intake is that while heme iron (animal sources) is highly bioavailable, nonheme iron (plant sources) is less bioavailable *(9)*. Nonheme iron is also very sensitive to inhibitors such as phytic acid, tannins, and some dietary fibers, which can be very common in human diets. Among low-income populations that consume little animal foods, staples generally make up the only significant portion of iron intake, but these staples are also generally high in phytic acid.

"Home gardening and promotional and education interventions clearly have the potential to improve vitamin A nutrition, especially when they are combined" *(9)*–the same cannot be said for iron interventions. Most researchers believe that increasing iron via food-based means requires animal foods. In a study done in Vietnam, increased intake of iron among children in intervention households was noted as compared with those from control households, but the source of the increased iron was not specified *(49,50)*. This study failed to utilize blood tests to determine iron status either before or after the intervention. In another study in Bangladesh, intervention households increased vegetable but not fish intake, and iron status did not improve *(51)*. In a pair of Thai studies, social marketing increased iron intake in lactating women, and ferritin levels among school-age girls also increased *(52,53)*. These girls were also given iron supplements, however, making it impossible to determine the impact of the food-based component on the intervention. Limited experience with quality iron studies does not allow for significant conclusions regarding optimal intervention approaches.

Still, recent data indicate impressive iron density in many traditional cereals (such as millets) as well as other indigenous leafy vegetables in Africa *(54)*. These traditional crops have been largely replaced by low-iron crops such as maize and wheat, and studies reincorporating the indigenous plants could provide some very interesting data. With the recognition of the importance and success of nutrition education components in this review and previous studies, perhaps donors and researchers will have more confidence in implementing nonheme iron agrodiversity interventions in the future.

3.5.7. Gender Considerations

Agricultural interventions have often failed to address the production-oriented constraints of women, including their lack of control over resources and household decision making *(55)*. Additionally, women and girls often receive less nonstaple food than males, increasing their risk for micronutrient deficiencies (Econutrition Task Force personal communication). Studies from West Africa have shown that women will not use yield-increasing technologies if they require a greater time commitment, due to already-full days *(55)*. One study described that if women have access to the same agricultural inputs as men, agricultural outputs would increase by somewhere between 7 and 24% (*(56)* cited in *(55)*). As more and more interventions also aim to increase household income to help alleviate nutritional deficiencies, these projects should specifically target and appeal to women since they are the primary providers of food, childcare, and health services *(3)*. Home gardening may be the only source of income for many poor women *(16)*. Giving women a voice in design and implementation planning will also increase their desire to participate.

Focusing on women is especially important in studies looking to impact nutrition outcomes. Both men and women are often engaged in agricultural production and marketing, but women and girls are generally responsible for family nutrition *(35)*. In a study undertaken in Uganda, men preferred bean varieties that held especially high market value, while women preferred the varieties that were easier to process and prepare for meals *(55)*. This allowed extension agents to tailor the intervention to suit the needs of both men and women, ultimately leading to increased bean consumption and protein intake. There has also been some evidence in parts of Africa that as gardens become more profitable, men will take over management and marketing tasks *(16)*. Agrodiversity interventions must take these obstacles into account and assist women in preserving their control over production and income. An earlier review also found that gender issue considerations lead to better interventions *(4)*, a finding recently confirmed *(40)*. High quality studies with the largest number of inputs almost always accounted for gender considerations, and these studies achieved the most successful outcomes.

Gender considerations are particularly important when households have young children. If a project is established that requires a mother to spend more time away from her young children, the likelihood of poor nutritional outcomes may actually increase even if food security improves. Indeed, the likelihood of wasting or stunting increases if young children are away from their mothers and cared for by older siblings *(57)*. However, when women control the sources of food entering the household, improvements are seen in children's weight for age *(58)*. Therefore, some balance is needed where women have more control over household agriculture but still do not compromise their children's well-being by being frequently absent. Technologies that reduce preparation and field time for women are important because they allow more time to be spent with children and on household nutrition *(35)* (*see* Chapter 18).

In a follow-up study to determine some of the long-term effects on the women who participated in Helen Keller International's study in Bangladesh, significantly more intervention than control women "perceived that they had increased their economic contribution to their households since the time the program was launched in their subdistricts (>85% vs. 52%) *(59)*." Additionally, the women in the intervention groups (formerly and currently active in the project) were more likely to have year-round gardens, greater vegetable intake, and more gardening income. The level of influence in household decision making also increased for intervention women in a "dose-dependent" fashion based on the length of time they had been involved in home gardening *(59)*, further highlighting the empowering benefits (especially for poor women) that rarely show up in cost-effective or nutrition analyses. This is particularly encouraging in light of the fact that female seclusion and subordination are common for women in Bangladesh, and they are often relegated to a restricted role involving mainly domestic work *(59)*. Seldom are these women consulted in household decisions, and they are generally forced to rely on their husbands. It is argued that gains are made in self-esteem and empowerment when women believe they are significantly contributing to household economic well-being, as confirmed in a recent long-term social follow-up study *(59)* (*see* Chapter 23).

The benefits of taking gender considerations into account when designing an intervention seem clear, and while the connection to improved nutrition may be indirect, it is still critical. The recently posited Agriculture–Nutrition Advantage Framework (*see* Fig. 3.7) describes this as an intrinsic aspect of a successfully linked agriculture–nutrition intervention *(55)*. Johnson-Welch et al. address some concrete ways of adding value to interventions via gender sensitivity:

- Include gender awareness training sessions among senior staff; address gender-related issues in terms of labor differences, decision-making power, etc. and add this to any nutrition education component.
- Integrate gender into government and NGO policies; consult women on intervention planning/make women project leaders in the community.
- Facilitate women's access to labor-enhancing technologies, extension services, and learning opportunities *(55)*.

3.6. RECOMMENDATIONS

3.6.1. Open Questions, Suggestions, and Future Research

One aspect of agrodiversity and econutrition in need of greater exploration is the role of incorporating indigenous crops and indigenous knowledge into the intervention designs. There is a gap of scientific data on the nutritional value of many plants, especially in the more remote parts of Africa and Asia *(26)*. Without adequate empirical data on local biodiversity, "wild" food plants cannot be fully implemented into agrodiversity and food security interventions. Data from some areas, particularly Mauritius, Micronesia, and parts of Africa, are starting to become available *(26)*. Encouragingly,

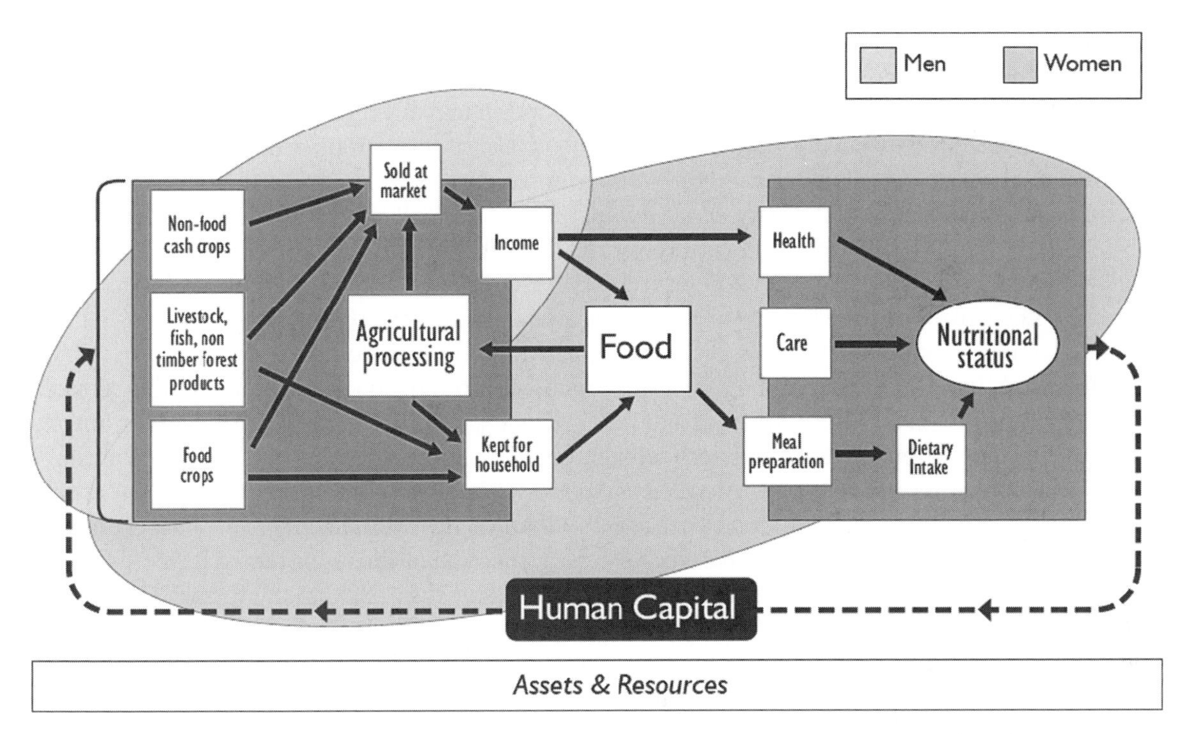

Fig. 3.7. The Agriculture–Nutrition Advantage Conceptual Framework. The overlapping domains of men and women demonstrate that a gender-informed approach is essential for linking successful agriculture to successful nutrition. Permission to reproduce from the International Center for Research on Women (ICRW) *(55)*.

many of the traditional foods long thought to be low in nutrients are being found to have immense nutritional value *(60)*. African leafy vegetables (ALVs) have been found to constitute a sizeable percentage of vitamin A consumption among the rural poor *(61)*, and these plants are also proving to be high in antioxidants *(26)*.

Currently, the International Plant Genetic Resources Institute (IPGRI) is involved in some major projects to promote dietary diversity within traditional food systems, one of which focuses specifically on ALVs as a means of increasing dietary diversity *(54)*. While sub-Saharan Africa has the most severe nutrition and micronutrient problems, it also has many exceptionally nutrient-dense native fruits and vegetables that are simply underutilized *(54)*. The specific goal of this project is to reintroduce indigenous ALVs into the diet and to increase income via agriculture (specifically for women). Unfortunately, no measurements are being used to ascertain the nutrition outcomes of the program *(26)*. Regardless, it is important to continue these attempts at increasing consumption, processing, and suitable management of these valuable ALVs.

A question that is only beginning to be explored (notably in an ongoing project in Malawi *(24,25)*) is that of holistically combining ecological outcomes with nutrition outcomes in an agrodiversity intervention. Many sub-Saharan African soils are very low in nitrogen, and fertilizers have become prohibitively expensive in that part of the world. Incorporating legumes into agriculture systems can add significant amounts of nitrogen to the soil, however, while concurrently adding to agricultural production and nutritional availability *(6)*. While the study in Malawi is not yet complete, the results thus far are promising *(24,25)*.

An area within agrodiversity interventions that requires more clarification is the issue of how best to assess dietary diversity and nutritional outcomes. In studies performed in developed countries, dietary diversity is correlated with dietary quality *(33)*. More research needs to be done, however, to elucidate the mechanisms of this association. Two common methods of assessing dietary diversity include the "food variety score," which estimates the number of food items consumed during a recording period, and the "dietary diversity score," which estimates the number of food groups consumed *(26)*. There has been some controversy about which method is superior (or if either is adequate). One study in Burkina Faso found that the "dietary diversity score" was a better method because it had a stronger link to anthropometric indices than the "food variety score," but other studies have found both methods to be equally acceptable *(26)*. Overall, the "dietary diversity score" is favored, particularly in sub-Saharan Africa and other lower income countries where it has been better tested *(26,33)*.

A shortcoming of available literature is the failure to address long-term sustainability issues, especially since sustainability is one of the major arguments put forth in favor of implementing agrodiversity interventions. Even some of the better and more recent studies do not appear to plan for long-term follow-up. One exception is the aforementioned Helen Keller International project in Bangladesh. This study found that active and former participants (those who had been out of the program for at least 3 years) were better-off economically and nutritionally than control households *(59)*. Interestingly, former participants were less likely to have year-round gardens than those currently in the project, but they also had a greater income from gardening *(59)*. This result suggests that the former project participants had shifted their garden production to be more economically productive and efficient. Despite these encouraging results from Bangladesh, it is important that agrodiversity interventions be evaluated over 5–10-year periods to adequately determine behavior change, growth, and health effects *(8)*.

Until very recently, nearly all agrodiversity interventions suffered from at least one of the following three design, evaluation, and analysis flaws as discussed by Ruel *(9,12)*:

- a lack of replicate units of intervention and analysis;
- inappropriate selection of control and comparison groups;
- inappropriate control for confounding factors and intermediary outcomes.

The "lack of replicate units of intervention" relates to a failure to set up adequate means to identify or to account for changes that may occur in either the intervention or control areas that could impact outcome (such as weather or additional interventions). To avoid this flaw, replicates of intervention and control areas should be included so that any fundamental differences that occur during the study can be averaged out *(9)*. In the Mozambique study discussed above *(22)*, this problem was addressed by randomly selecting a large number of villages to be part of either the intervention or the control groups.

Inappropriate control group selection is common in agrodiversity interventions. Often the intervention group consists of volunteers while the control group simply consists of the nonparticipators. The volunteers may be inherently different from those who do not wish to participate (perhaps in terms of poverty), so this design must be avoided at all costs. While randomization may be difficult on ethical and logistical grounds, researchers in Mozambique were able to successfully achieve this type of cohort selection by randomly selecting villages as well as the participants within these villages.

Another problem is the failure to appropriately control for confounding factors and intermediate outcomes. Often studies simply fail to gather data on potential confounders, or they fail to adjust for these factors with a final multivariate analysis *(9)*. While confounder data are frequently overlooked, it is especially important to include in studies that cannot achieve ideal, randomized, double-blind formats. Measuring intermediate outcomes throughout (as accounted for, again, in the Mozambique study) adds plausibility to any nutrition outcomes uncovered at the study's endpoint.

There have been failures in the past in managing tropic and nontemperate agroecosystems due to a failure of adequately adapting project protocols to "fit" a given area *(62)*. Interventions need to be tailored to each area based on local macronutrient and micronutrient needs, economic conditions, seasonal climate issues, social customs, and other factors. For example, many foods are designated a "status" level by a community, and training, education, and communication may be necessary to break down stereotypes against low-status but nutrient-dense foods *(3)*. Local crop portfolios should be assessed to determine which key nutrients are missing or are low. Then substitutes can be found and adopted through intervention programs.

Different sectors need to work together to address agriculture, nutrition, and health problems that are often closely linked. Overreliance on vertical thinking must also be avoided, as it leads to sector isolation. Sustainable changes must involve communities as well as organizations—farmer-to-farmer (horizontal) communication is a key part of integrating a successful intervention, in addition to organization or researcher-to-farmer (vertical) communication *(55)*. This type of communication must be two-way to be effective because understanding traditional diet patterns is essential to successfully presenting new foods and ideas *(16)*. While home production and consumption may be the primary goals of the researchers, households will often be more interested in participating if additional income is also one of the project's goals. In many cases, gardening can actually have higher returns relative to land and labor than large-scale field agriculture, and this information should be passed on to community members to pique their interest *(16)*.

3.6.2. Conclusions and Overall Recommendations

A number of important questions relating to agrodiversity interventions have recently been addressed; many still remain. Current research suggests that fruit and vegetable sources of vitamin A are adequately bioavailable and capable of improving serum retinol status. Nutrition education is a vitally important component of agrodiversity interventions, and adequately addressing gender considerations further improves outcomes. Higher quality study designs and broader capital inputs appear to improve nutrition outcomes. However, design protocols and statistical analyses need to improve. Dietary diversity measurements should be formalized and applied for better comparisons, and inputs and outcomes should be more carefully realized.

Agrodiversity interventions can address multiple nutrient adequacies simultaneously. While cereal-based interventions and aid may be more "controversial" now due to the increasing prevalence of the "double burden of disease" in even the poorest countries, micronutrient deficiencies are generally prevalent among the under- *and* overfed. Agrodiversity interventions now have the potential to provide a sustainable source of micronutrients to needy populations, and they should be scaled up and utilized in conjunction with supplements and other methods of addressing malnutrition. As Wenhold et al. state, "action-oriented strategies should go beyond the question of 'why' an agriculture–nutrition-linked, gender-informed approach should be used, to 'how' it could be implemented *(35)*." With the recent advances in agrodiversity intervention quality, that "how" may perhaps be a little clearer.

REFERENCES

1. Lunven, P. (1982) The nutritional consequences of agricultural and rural development projects. *Food Nutr Bull* **4**, 17–22.
2. Kirsten, J., Townsend, R., Gibson, C. (1998) Determining the contribution of agricultural production to household nutritional status in KwaZulu-Natal, South Africa. *Dev South Afr* **15**, 573–87.
3. Bonnard, P. (1999) Increasing the Nutritional Impacts of Agricultural Interventions. In: Food and Nutrition Technical Assistance Project. Washington, DC, 1–20.
4. Berti, P. R., Krasevec, J., FitzGerald, S. (2004) A review of the effectiveness of agriculture interventions in improving nutrition outcomes. *Public Health Nutr* **7**, 599–609.
5. Wahlqvist, M. L., Specht, R. L. (1998) Food variety and biodiversity: econutrition. *Asia Pacific J Clin Nutr* **7**, 314–9.
6. Deckelbaum, R. J., Palm, C., Mutuo, P., DeClerck, F. (2006) Econutrition: implementation models from the Millennium Villages Project in Africa. *Food Nutr Bull* **27**, 335–42.

7. Sachs, J. D. (2005) The End of Poverty: Economic Possibilities for Our Time. New York: Penguin Books, 1–416.
8. Pinstrup-Anderson, P. (2001) Forward to: Ruel, M. T. Can Food-Based Strategies Help Reduce Vitamin A and Iron Deficiencies? Washington, DC: International Food Policy Research Institute, 1–54.
9. Ruel, M. T. (2001) Can Food-Based Strategies Help Reduce Vitamin A and Iron Deficiencies?: A Review of Recent Evidence. Washington, DC: International Food Policy Research Institute, 1–54.
10. Ensing, B., Sangers, S. (1986) Home Gardening in a Sri Lankan Wet Zone Village: Can it Contribute to Improved Nutrition? Wageningen, The Netherlands: Wageningen Agricultural University.
11. Brun, T., Reynaud, J., Chevassus-Agnes, S. (1989) Food and nutritional impact of one home garden project in Senegal. *Ecol Food Nutr* **23**, 91–108.
12. Ruel, M. T., Levin, C. E. (2000) Assessing the potential for food-based strategies to reduce vitamin A and iron deficiencies: a review of recent evidence. In: International Food Policy Research Institute: Food Consumption and Nutrition Division, Discussion Paper No 92, 1–62.
13. Cohen, N., Jalil, M., Rahman, H., et al. (1985) Landholding, wealth and risk of blinding malnutrition in rural Bangladesh households. *Soc Sci Med* **21**, 1269–72.
14. Solon, F., Fernandez, T. L., Latham, M. C., Popkin, B. M. (1979) An evaluation of strategies to control vitamin A deficiency in the Philippines. *Am J Clin Nutr* **32**, 1445–53.
15. Gillespie, S., Mason, J. (1994) Controlling vitamin A deficiency. In: Nutrition Policy Discussion Paper. United Nations Administrative Committee on Coordination. Subcommittee on Nutrition state-of-the-art series, No 14. Geneva, 1–87.
16. Marsh, R. (1998) Building on traditional gardening to improve household food security. *Food Nutr Agric* **22**, 4–21.
17. Brownrigg, L. (1985) Home Gardening in International Development: What the Literature Shows. Washington, DC: League for International Food Education, 1–330.
18. Popkin, B. M., Solon, F. S., Fernandez, T., Latham, M. C. (1980) Benefit-cost analysis in the nutrition area: a project in the Philippines. *Soc Sci Med* **14C**, 207–16.
19. WorldBank (2005) Why invest in nutrition? In: Shekar M, Heaver R, Lee Y, eds. Repositioning Nutrition as Central to Development: A Strategy for Large Scale Action. Washington, DC. World Bank Publications, 29–41.
20. Talukder, A., Kiess, L., Huq, N., de Pee, S., Darnton-Hill, I., Bloem, M. W. (2000) Increasing the production and consumption of vitamin A-rich fruits and vegetables: lessons learned in taking the Bangladesh homestead gardening programme to a national scale. *Food Nutr Bull* **21**, 165–72.
21. Low, J. W., Arimond, M., Osman, N., Cunguara, B., Zano, F., Tschirley, D. (2007) A food-based approach introducing orange-fleshed sweet potatoes increased vitamin A intake and serum retinol concentrations in young children in rural Mozambique. *J Nutr* **137**, 1320–7.
22. Faber, M., Phungula, M. A. S., Venter, S. L., Dhansay, M. A., Benade, A. J. S. (2002) Home gardens focusing on the production of yellow and dark-green leafy vegetables increase the serum retinol concentrations of 2–5-years old children in South Africa. *Am J Clin Nutr* **76**, 1048–54.
23. Faber, M., Venter, S. L., Benade, A. J. S. (2002) Increased vitamin A intake in children aged 2–5 years through targeted home-gardens in a rural South African community. *Public Health Nutr* **5**, 11–6.
24. Bezner Kerr, R., Chirwa, M. (2004) Participatory research approaches and social dynamics that influence agricultural practices to improve child nutrition in Malawi. *EcoHealth* **1**, 109–19.
25. Bezner Kerr, R., Snapp, S., Chirwa, M., Shumba, L., Msachi, R. (2007) Participatory research on legume diversification with Malawian smallholder farmers for improved human nutrition and soil fertility. *Exp Agric* **43**, 437–53.
26. Frison, E. A., Smith, I. F., Johns, T., Cherfas, J., Eyzagirre, P. B. (2006) Agricultural biodiversity, nutrition, and health: making a difference to hunger and nutrition in the developing world. *Food Nutr Bull* **27**, 167–79.
27. Gakidou, E., Oza, S., Vidal Fuertes, C., et al. (2007) Improving child survival through environmental and nutritional interventions: the importance of targeting interventions toward the poor. *JAMA* **298**, 1876–87.
28. Maunder, E. M. W., Matji, J., Hlatshwayo-Molea, T. (2001) Enjoy a variety of foods-difficult but necessary in developing countries. *S Afr J Clin Nutr* **14**, S7–11.
29. Labadarios, D., Steyn, N. P. (2001) South African food-based dietary guidelines–guidelines for whom? *S Afr J Clin Nutr* **14**, 1–4.
30. Shetty, P. (2005) The Boyd Orr Lecture: achieving the goal of halving global hunger by 2015. In: The Summer Meeting of the Nutrition Society. Norwich: Proceedings of the Nutrition Society, 7–18.
31. Borlaug, N. (2007) Feeding a Hungry World. *Science* **318**, 359.
32. Johns, T., Eyzagirre, P. B. (2005) Symposium on "Wild-gathered plants: basic nutrition, health and survival": linking biodiversity, diet and health in policy and practice. In: The Summer Meeting of the Nutrition Society. Norwich: Proceedings of the Nutrition Society, 182–9.

33. Ruel, M. T. (2003) Operationalizing dietary diversity: a review of measurement issues and research priorities. *Am Soc Nutr Sci J Nutr* **133**, 3911S–26S.
34. Wahlqvist, M. L. (2004) Requirements for healthy nutrition: Integrating food sustainability, food variety, and health. *J Food Sci* **69**, CRH16–8.
35. Wenhold, F. A. M., Faber, M., van Averbeke, W., et al. (2007) Linking smallholder agriculture and water to household food security and nutrition. *Water SA* **33**, 327–36.
36. Yeudall, F., Sebastian, R., Cole, D. C., Ibrahim, S., Lubowa, A., Kikafunda, J. (2007) Food and nutritional security of children of urban farmers in Kampala, Uganda. *Food Nutr Bull* **28**, S237–46.
37. Graham, R. D., Welch, R. M., Saunders, D. A., et al. (2007) Nutritious subsistence food systems. *Adv Agron* **92**, 1–74.
38. Provenza, F. D. (2008) What does it mean to be locally adapted and who cares anyway? *J Anim Sci* **86**, E271–84.
39. Frison, E. (2006) Agricultural biodiversity and human well-being. In: International Workshop on Tropical Agricultural Development. Brasilia, Brazil, 1–2.
40. Wispelwey, B. P. (2008) Preventing malnutrition: a systematic review assessing the impact of agrodiversity interventions on nutritional status in lower income countries. In: Master's Thesis. New York: Available upon request from Columbia University Institute of Human Nutrition, 1–78.
41. Walsh, C. M., Dannhauser, A., Joubert, G. (2002) The impact of a nutrition education programme on the anthropometric nutritional status of low-income children in South Africa. *Public Health Nutr* **5**, 3–9.
42. Nana, C. P., Brouwer, I. D., Zagre, N. M., Kok, F. J., Traore, A. S. (2006) Impact of promotion of mango and liver as sources of vitamin A for young children: a pilot study in Burkina Faso. *Public Health Nutr* **9**, 808–13.
43. Jalal, F., Nesheim, M. C., Agus, Z., Sanjur, D., P, H. J. (1998) Serum retinol concentrations in children are affected by food sources of beta-carotene, fat intake, and antihelmintic drug treatment. *Am J Clin Nutr* **68**, 623–9.
44. Tang, G., Gu, X., Hu, S., et al. (1999) Green and yellow vegetables can maintain stores of vitamin A in Chinese children. *Am J Clin Nutr* **70**, 1069–76.
45. Takyi, E. E. (1999) Children's consumption of dark green, leafy vegetables with added fat enhances serum retinol. *J Nutr* **129**, 1549–54.
46. Drammeh, B. S., Marquis, G. S., Funkhouser, E., Bates, C., Eto, I., Stephensen, C. B. (2002) A randomized, 4-month mango and fat supplementation trial improved vitamin A status among young Gambian children. *J Nutr* **132**, 3693–9.
47. van Jaarsveld, P. J., Faber, M., Tanumihardjo, S. A., Nestel, P., Lombard, C. J., Benade, A. J. (2005) Beta-carotene-rich orange-fleshed sweet potato improves the vitamin A status of primary school children assessed with the modified-relative-dose-response test. *Am J Clin Nutr* **81**, 1080–7.
48. Haskell, M. J., Jamil, K. M., Hassan, F., et al. (2004) Daily consumption of Indian spinach (*Basella alba*) or sweet potatoes has a positive effect on total-body vitamin A stores in Bangladeshi men. *Am J Clin Nutr* **80**, 705–14.
49. English, R., Badcock, J. (1998) A community nutrition project in Vietnam: effects on child morbidity. *Food Nutr Agric* **22**, 15–21.
50. English, R. M., Badcock, J. C., Giay, T., Ngu, T., Waters, A. M., Bennett, S. A. (1997) Effect of nutrition improvement project on morbidity from infectious diseases in preschool children in Vietnam: comparison with control commune. *BMedJ* **315**, 1122–5.
51. International Food Policy Research Institute. (2000) Commercial Vegetable and Polyculture Fish Production in Bangladesh: Their Impacts on Income, Household Resource Allocation, and Nutrition. Washington, DC. Project Brief No. 2, 1–2.
52. Smitasiri, S., Sa-ngobwarchar, K., Kongpunya, P., et al. (1999) Sustaining behavioural change to enhance micronutrient status through community and women-based interventions in northeast Thailand: vitamin A. *Food Nutr Bull* **20**, 243–51.
53. Smitasiri, S., Dhanamitta, S. (1999) Sustaining Behavior Change to Enhance Micronutrient Status: Community and Women-Based Interventions in Thailand. Washington, DC: Research Report No. 2. International Center for Research on Women (ICRW)/Opportunities for Micronutrient Interventions (OMNI), 1–28.
54. Oniang'o, R. K., Shiundu, K., Maundu, P., Johns, T. (2005) Diversity, nutrition and food security: the case of African leafy vegetables. In: Bala Ravi S, Hoeschle-Zeledon I, Swaminathan MS, Frison E, eds. Hunger and Poverty: The Role of Biodiversity Report of an International Consultation on the Role of Biodiversity in Achieving the UN Millennium Development Goal of Freedom from Hunger and Poverty. Chennai, India: M.S. Swaminathan Research Foundation, 83–100.
55. Johnson-Welch, C., MacQuarrie, K., Bunch, S. (2005) A Leadership Strategy for Reducing Hunger and Malnutrition in Africa: The Agriculture–Nutrition Advantage. Washington, DC: International Center for Research on Women, 1–28.
56. Quisumbing, A., Haddad, L. (1998) Gender issues for food security in developing countries: Implications for project design and implementation. *Can J Dev Stud* **19**, 185–208.

57. Begin, F., Frongillo Jr, E. A., Delisle, H. (1999) Caregiver behaviors and resources influence child height-for-age in rural Chad. *J Nutr* **129**, 680–6.
58. von Braun, J., Johm, K. B., Puetz, D. (1994) Nutritional effects of commercialization of a woman's crop: Irrigated rice in The Gambia. In: von Braun J, Kennedy E, eds. Agricultural Commercialization, Economic Development and Nutrition. Baltimore: Johns Hopkins University Press, 343–62.
59. Bushamuka, V. N., de Pee, S., Talukder, A., et al. (2005) Impact of a homestead gardening program on household food security and empowerment of women in Bangladesh. *Food Nutr Bull* **26**, 17–25.
60. Maunder, E. M. W., Meaker, J. L. (2007) The current and potential contribution of home-grown vegetables to diets in South Africa. *Water SA* **33**, 401–6.
61. Mulokozi, G., Mselle, L., Mgoba, C., Mugyabuso, J. K. L., Ndossi, G. D. (2000) Improved Solar Drying of Vitamin A-Rich Foods by Women's Groups in the Singida District of Tanzania. Washington, DC: Research Report No. 5. International Center for Research on Women (ICRW)/Opportunities for Micronutrient Interventions (OMNI), 1–28.
62. Janzen, D. H. (1973) Tropical agroecosystems: these habitats are misunderstood by the temperate zones, mismanaged by the tropics. *Science* **182**, 1212–9.
63. Helen Keller International, Asian Vegetable and Research Development Centre (1993) Home Gardening in Bangladesh: Evaluation Report. Dhaka, Bangladesh.
64. CARE-Nepal. (1995) A study of the Evaluation of Home Gardening Program in Bajura and Mahottari Districts. Nepal Project Report: CARE Nepal.
65. Greiner, T., Mitra, S. N. (1995) Evaluation of the impact of a food-based approach to solving vitamin A deficiency in Bangladesh. *Food Nutr Bull* **16**, 193–205.
66. Hagenimana, V., Anyango Oyunga, M., Low, J., Njoroge, S. M., Gichuki, S. T., Kabira, J. (1999) The Effects of Women Farmers' Adoption of Orange-Fleshed Sweet Potatoes: Raising Vitamin A Intake in Kenya. Washington, DC: Research Report No. 3. International Center for Research on Women (ICRW)/Opportunities for Micronutrient Interventions (OMNI), 1–24.
67. Kidala, D., Greiner, T., Gebre-Medhin, M. (2000) Five-year follow-up of a food-based vitamin A intervention in Tanzania. *Publich Health Nutr* **3**, 425–31.
68. Ngu, T., Quang, N. D., Ha, P. H., Giay, T., Badcock, J. C., Fitzgerald, S. (1994) A food based approach to nutrition improvement through household food security in Vietnam, with special reference to vitamin A deficiency. In: 16th International Vitamin A Consultative Group Meeting. Chang Rai, Thailand: International Life Sciences Institute, 77.
69. Phillips, M., Sanghvi, T., Suarez, R., McKigney, J., Fiedler, J. (1996) The costs and effectiveness of three vitamin A interventions in Guatemala. *Soc Sci Med* **42**, 1661–8.
70. Solon, F., Briones, H., Fernandez, J. R., Shafritz, L. B. (1996) Moving to a long-term strategy: increasing vegetable gardening and consumption in the Philippines. In: Seidel R. E., ed. Strategies for Promoting Vitamin A Production, Consumption and Supplementation Four Case Studies. Washington, DC: Academy for Educational Development.
71. Attig, G. A., Smitasiri, S., Ittikom, K., Dhanamitta, S. (1993) Promoting home gardening to control vitamin A deficiency in northeastern Thailand. *Food Nutr Agric* **7**, 18–25.
72. Brun, T. A., Geissler, C., Kennedy, E. (1991) The impact of agricultural projects on food, nutrition, and health. *World Review of Nutrition Diatetics* **65**, 99–123.
73. Galal, O. M., Harrison, G. G., Abdou, A. I., Zein el Abedin, A. (1987) The impact of a small-scale agricultural intervention on socioeconomic and health status. *Food Nutr* **13**, 35–43.
74. Ayalew, W., Gebriel, Z. W., Kassa, H. (1999) Reducing Vitamin A Deficiency in Ethiopia: Linkages with a Women-Focused Dairy Goat Farming Project. Washington, DC: Research Report No. 4. International Center for Research on Women (ICRW)/Opportunities for Micronutrient Interventions (OMNI), 1–28.
75. Chakravarty, I. (2000) Food-based strategies to control vitamin A deficiency. *Food Nutr Bull* **21**, 135–43.
76. Faber, M., Jogessar, V. B., Benade, A. J. S. (2001) Nutritional status and dietary intakes of children aged 2–5 years and their caregivers in a rural South African community. *Int J Food Sci Nutr* **52**, 401–11.

4 Nutrition in the Age of Polypharmacy

Lindsey R. Lombardi, Eugene Kreys, Susan Gerry, and Joseph I. Boullata

Key Points

- Drug-induced changes to nutritional status may be considered a subclass of adverse drug effects.
- Medication may alter food intake, as well as cause weight loss or weight gain.
- Medication may alter macronutrient digestion, absorption, metabolism, and excretion.
- Medication may alter vitamin absorption, distribution, metabolism, and excretion.
- Medication may alter mineral absorption, distribution, metabolism, and excretion.
- Significant drug-induced alterations in nutritional status can be prevented or managed by selecting a therapeutic alternative without the unwanted nutritional effect or by providing supplementation in the case of nutrient losses.

Key Words: Absorption; digestion; drug interaction; excretion; food intake; medication; metabolism; nutrient; nutritional status

4.1. INTRODUCTION

Preventive medicine includes nutritional approaches for promoting health and for preventing, delaying, or modifying disease processes. Avoiding poor nutritional status is an aspect that has wide ranging benefits on health outcomes. Recognition of risk factors for poor nutritional status is vital to this goal. Alterations to nutritional status may occur for any number of reasons, including the influence of medication. This chapter will provide a review on the potential changes in nutritional status resulting from medication use.

4.1.1. Medication Use

Medication use continues to be a part of daily life for many people. It is estimated that about 80% of American adults use medication *(1)*. This widespread use occurs in other countries as well, although differences exist in usage patterns of specific drugs *(2)*. Nearly 290 billion dollars per year are spent on medication in the United States, with just over 3.8 trillion prescriptions dispensed annually *(3)*. At almost 23 billion dollars annually, dietary supplement sales have risen fivefold since 1994 *(4)*. Prescription drug use and expenditures on medication are expected to accelerate *(5)*. More complex and specialty medications account for ongoing increases in drug expenditures in all care settings *(6)*.

Aside from the promise of benefit, these thousands of prescription, non-prescription, and natural health products each have their own set of potential adverse effects. Medication use is associated with adverse drug effects due to predictable side effects, less predictable idiosyncratic and allergic reactions, and interactions. The prevalence of medication use in the elderly, in particular, is widely recognized and associated with greater adverse drug effects *(7)*. Adverse drug reactions may account

A. Bendich, R.J. Deckelbaum (eds.), *Preventive Nutrition*, Nutrition and Health, DOI 10.1007/978-1-60327-542-2_4,

for up to 13% of hospital admissions *(8)*. Drugs and their adverse effects can impact on nutritional status. Interestingly, some of the well-recognized adverse effects of individual drugs occur as a direct result of alterations in nutritional status. In fact, drug-induced changes in nutritional status may be considered as a subclass of adverse drug effects. Given the enormous economic impact of drug use and the untoward consequences of this use, it is concerning that more attention is not regularly paid to drug influences on nutritional status. Of the many issues that influence patient safety within health care, drug–nutrient interactions are not routinely identified and as yet do not parallel the systematic surveillance seen with drug–drug interactions.

4.1.2. Interactions Between Drugs and Nutrients

A drug–nutrient interaction can be broadly defined as the result of a physical, chemical, physiologic, or pathophysiologic relationship between a drug and a nutrient, multiple nutrients, food in general, or nutritional status *(9)*. A sundry of potential drug–nutrient interactions exist that occur through various mechanisms, some of which can be clinically significant *(10)*. An interaction is considered significant from a clinical perspective if it alters therapeutic response or if nutritional status is compromised. The clinical consequences of an interaction are related to alterations in the disposition and effect of the drug or nutrient. The term *disposition* refers to the absorption, distribution, and elimination of a drug or nutrient which can involve physiologic transporters and metabolizing enzymes. And the term *effect* refers to the physiologic action of a drug or nutrient at the level of cellular or subcellular targets.

Interactions requiring drug avoidance are rare; instead, close patient monitoring with modification based on findings is the usual approach. Mechanisms of interactions may be physicochemical or physiologic—involving transporters, enzymes, and other physiologic functions *(10)*. Physiologic manifestations of a drug–nutrient interaction may differ based on gene polymorphism (e.g., methotrexate and folic acid) *(11,12)*. Based on the above definition, drug–nutrient interactions can be classified into one of five categories (Table 4.1). The many drug–nutrient interactions can be categorized with each having an identified precipitating factor and an object of the interaction. In some cases the drug is the precipitating factor (i.e., causing a change to nutritional status). This is the focus of the chapter (i.e., the first two categories of interactions listed in Table 4.1).

The risk for developing a clinically significant drug–nutrient interaction is greater in patients with chronic diseases who use multiple medications. Additional risk factors include either end of the age spectrum, the presence of genetic variants in transporters, enzymes, or receptors, impaired organ

Table 4.1
Classification of Drug–Nutrient Interactions

Precipitating Factor	*Object of the Interaction*	*Potential Consequence*
Drug	Nutritional status	Altered global nutritional status
Drug	Nutrient status	Altered status of an individual nutrient
Nutritional status	Drug	Treatment failure or drug toxicity
Food or food component	Drug	Treatment failure or drug toxicity
Specific nutrient or other dietary supplement ingredient	Drug	Treatment failure or drug toxicity

function, and poor underlying nutritional status. Given the number of individuals with poor nutritional status (i.e., protein-calorie malnutrition, obesity, specific nutrient deficits), each risk factor may impact the potential safety and efficacy of commonly used medications.

The breadth of this topic is not only overlooked by many clinicians but also by regulatory agencies *(10)*. This is particularly true in terms of identifying the nutritional consequences of each newly approved drug beyond simply the effect on appetite or body weight. A number of less well-acknowledged alterations in nutritional status do exist. In fact some listed side effects of medication may be the result of an imbalance of one or more nutrients. Such a clinical effect may be just as likely to be identified initially by a nutrition specialist as a nutritional disorder as it would a drug side effect. The interference of drugs with nutritional status or the status of a specific nutrient is often less readily evident clinically and more insidious in developing. Although not all drug-induced changes to nutritional status are manifest overtly, particularly with chronic medication use, even marginal deficits or tissue-specific deficits may have profound clinical effects *(13)*. Non-nutrient supplements (e.g., herbal medicines) have the same potential to alter nutritional status and nutrient disposition, as do other medicines *(14,15)*.

4.1.3. Clinical Implications

Adverse drug effects may not be attributed by the clinician to a nutrient deficit but instead to the drug itself or the disease states being managed. An impact on the status of a nutrient may be clinically subtle and not typically resulting in a classic deficiency syndrome. Subclinical states of nutrient deficits are not as well appreciated as classic deficiencies. Sensitive and specific parameters for nutrient status would, therefore, need to be assessed regularly. Alterations in status should be identified long before a deficit is evident, as the goal is to maintain or improve a patient's nutritional status. Ideally, every clinician should make an attempt to identify or predict drug–nutrient interactions to maximize patient outcome.

Other findings may increase the risk of drug-induced malnutrition. The clinical influence of a drug on nutritional status is not necessarily a fixed property of the drug itself but involves patient-related variables as well. The latter would include physiologically altered nutrient requirements, pharmacogenetic variability, a marginal diet, malabsorption, a chronic or catabolic disease, altered organ function, and concurrently ingested substances or environmental exposures. In other words, clinically significant outcomes of an interaction are patient specific, not necessarily just drug specific. All this presumes that nutritional status of patients receiving medication is regularly assessed in all health-care settings. A complete drug history is also required to identify all pharmacologically active substances administered whether prescription drugs, over-the-counter medicine, or natural health products. The latter group includes micronutrient and protein supplements, herbals and other botanicals, and non-nutrient/non-herbal supplements. A complete medical history and physical examination would need to be performed to identify diseases also associated with reduced nutritional status.

Recommendations for improving or altering dietary patterns need to take into account any influence that may have on medication use. Evidence-based nutritional recommendations should help prevent adverse drug effects and maximize drug benefit. Significant drug-induced alterations in nutritional status can be prevented or managed by selecting a therapeutic alternative without the unwanted nutritional effect or providing supplementation in the case of nutrient losses. Data supporting the use of nutrient supplementation are still limited in most cases, however. Individuals using pharmaceutical products need to make sure that their dietary habits and nutrient intakes reduce the risk of adverse drug effects.

Extensive lists of drug-induced nutrient depletions and in-depth discussions are available *(16–19)*. Special mention needs to be made of examples not discussed later in the chapter. The many potential metabolic derangements (macronutrient and micronutrient) associated with the use of parenteral nutrition, a pharmaceutically complex prescription drug preparation, are well recognized *(20,21)*. Ethanol is another drug responsible for significant alterations in nutritional status. It should be kept in mind that the use of ethanol will increase the risk of those drug-induced nutritional deficits discussed in this chapter. Thorough reviews describing the effect of alcohol consumption on nutritional status can be found elsewhere *(22–24)*. Furthermore, toxic ingestions beyond the therapeutic use of medications and related substances can also cause nutrient derangements. For example, acetaminophen overdose depletes glutathione and its rate-limiting amino acid cysteine, isopropyl alcohol ingestion causes hypoglycemia, and ethylene glycol causes hypocalcemia and depletes several B vitamins as well as *(25)*. Chemical exposures may interact with poor nutritional status increasing the costs to health, particularly in resource-poor individuals and communities *(26)*.

4.2. MECHANISMS OF ALTERED NUTRITIONAL STATUS

The impact of drugs on nutritional status has been recognized for many years, particularly from the time that synthetic drug production developed *(27,28)*. A few circumstances account for most clinically common drug-induced changes to nutritional status. These include drugs that cause significant gastrointestinal (GI) effects including anorexia and malabsorption, or drugs that are nutrient antagonists by virtue of similar chemical structure and function, and nutrients that are involved in multiple metabolic pathways. Antivitamin attributes have been ascribed to a number of drugs based on influences on transporters and/or enzymes. Yet, even this may be the result of one or more different factors including decreased absorption, reduced conversion to active form, interference with vitamin-dependent pathways, or increased vitamin clearance. An interaction is more likely with nutrients not stored in great amounts (e.g., thiamin), or those with complex absorption (e.g., vitamin B_{12}), distribution (e.g., zinc), and metabolism (e.g., vitamin B_6) characteristics, or required for rapidly dividing tissues (e.g., folic acid, iron). Nutritional effects involving a few drug classes have been more frequently described. This is in part due to their chronic use in individual patients, as a result of being in clinical use for decades, or their obvious impact on the GI tract. These include antimicrobial agents especially those used to manage tuberculosis, antiepileptics, antineoplastics, and cardiovascular agents. Many other drug classes are likely implicated as well, but the drug approval process does not require studies to evaluate short- or long-term influences on nutritional status.

Several mechanisms exist to explain drug-induced changes to nutritional status. These may involve influences on food intake, on digestion, and on absorption. Additional mechanisms involve influences on nutrient distribution, metabolism, and excretion. More than one mechanism can be in play for a given drug and nutrient. For example, the antiepileptic agent, carbamazepine, decreases biotin status both by inhibiting intestinal absorption and by accelerating metabolism to inactive metabolites *(29,30)*. A single nutrient may be influenced by more than one drug through a combination of mechanisms. For example, folate status may be influenced by drugs that impair its absorption, alter protein binding in the circulation, block release from tissue sites, or enhance hepatic metabolism *(31)*. A single drug can have an influence on more than one nutrient through one or more mechanisms. For example, by inducing a general malabsorption, colchicine increases loss of fat, nitrogen, sodium, and potassium in the feces with decreases in absorption of xylose, vitamin B_{12}, and possibly some carotenoids *(32)*.

Corticosteroids, as used in a wide variety of disorders, offer a further example. They have been noted to reduce levels of several nutrients (e.g., folate, vitamin B_{12}, calcium, and selenium) *(33,34)*. Consideration needs to be given, however, to the role played by the pathophysiology that prompted the use of the drug (e.g., corticosteroids) in the first place. This is not always evaluated in published reports, which makes it less clear whether the mechanism of altered nutritional status is exclusively drug induced. In some cases the exact mechanism of altered nutritional status is not fully understood. For example, the use of kava (*Piper methysticum*) may cause a yellowing of the skin unrelated to the hepatotoxic potential of this herbal medicine. It has been suggested that this may occur due to a niacin deficiency although the mechanism remains unclear *(35)*. Of course, the influence of some drugs on nutritional status is the primary therapeutic action sought clinically. For example, orlistat decreases fat absorption, pamidronate lowers serum calcium, and warfarin reduces vitamin K availability.

Aside from appreciating the mechanism of an interaction, the clinical significance of many drug-induced alterations to nutritional status is described less frequently than many clinicians would desire. This alone limits the ability to make recommendations such as supplementation of the affected nutrient(s). Clinical judgment is needed in combination with available evidence to assess individual risk from an interaction.

The impact of drugs on nutritional status is not always predicted from animal studies and is not routinely assessed during the drug approval process. Given this, clinicians should operate on the assumption that any variability in nutritional status is the result of a drug-induced change unless proven otherwise. Manifestations are typically acute for a drug that interferes with a nutrient's metabolic activities, compared to interactions with nutrient intake, absorption, or clearance which are expected to take longer to manifest. The biochemical, functional, or clinical manifestations will depend on the degree of nutrient deficit (or excess) and the tissue compartment affected. Of course gender, age, and genetic factors all need to be taken into account. There is increased risk if the effects on nutritional status are additive, and with chronic medication use, or if the patient has marginal nutritional status to begin with. Risk to a patient from acute use of a single drug would be expectedly less, especially, in the face of adequate nutritional status, compared to an elderly patient with poor nutritional status requiring long-term use of one or more medications impacting on the status of a nutrient. Indeed for many nutrients, subclinical deficits (i.e., non-classic deficiency) may still be significant even when not recognized or identified.

There is no single best way to present information on drug-induced changes to nutritional status—for example, it could be described by drug class (*If I take this drug what might happen?*) or by nutrient class (*If I have this deficit, what drug might have caused it?*). Varying degrees of drug-induced changes in nutritional status will be described in the following sections on food intake and digestion. This will be followed by sections on the influence of drugs on the absorption, distribution, metabolism, and excretion of macronutrients, vitamins, and minerals. Of note, the apparent contradiction of seeing the same drug on countervailing lists (Tables 4.2–4.8) reflects the limitations of the data on drug-induced influences on nutritional status.

4.3. FOOD INTAKE

A wide number of medications have the potential to alter food intake, thereby increasing the risk for poor nutritional status. Table 4.2 provides a list of select drugs that influence food intake through an impact on appetite. In addition to drugs with an obvious impact on appetite control and neural circuits of feeding, there are those that influence GI tract structure and function, and some with an even less apparent role through alterations in sensation and cognition *(36)*.

Table 4.2
Select Drugs That Alter Food Intake

Appetite Stimulants	*Appetite Suppressants*	*Appetite Suppressants* (Contd.)
Antihistamines	Abacavir	Methylphenidate
Benzodiazepines	Amantadine	Metronidazole
Cannabinoids	Amiodarone	Naloxone
Clozapine	Aldesleukin	Nicotine
Corticosteroids	Amphetamines	Paroxetine
Insulin	Antineoplastics	Pentamidine
Lithium	Atomexetine	Perhixiline
Mirtazapine	Bethionol	Phentermine
Oral hypoglycemics	Bromocriptine	Pimozide
Oxandrolone	Bupropion	Pramipexole
Phenothiazines	Caffeine	Procarbazine
Tricyclic antidepressants	Cisplatin	Ropinirole
Valproic acid	Dacarbazine	Selegiline
	Dextroamphetamine	Sertraline
	Digoxin	Sibutramine
	Epirubicin	Tacrolimus
	Etoposide	Temozolomide
	Felbamate	Thiazides
	Fenfluramine	Topiramate
	Fluorouracil	Trazodone
	Fluoxetine	Zonisamide
	Fluvoxamine	
	Interferon	
	Irinotecan	
	Levodopa	

Food intake may be reduced by drug-induced drowsiness (e.g., antidepressants, antihistamines, antipsychotics, benzodiazepines, clonidine, methyldopa, phenytoin, skeletal muscle relaxants) and drug-induced depression (e.g., β-blockers, digoxin, efavirenz, goserelin, interferon, leuprolide, levodopa, methyldopa, phenobarbital, reserpine, and the triptans). Drugs that alter vision or gait may limit the ability to gather food and prepare meals. Excessive hypotension and dizziness caused by medication may also restrict this ability. Cognitive disturbances caused by medication (e.g., amiodarone, anticholinergics, antidepressants, digoxin, opioid analgesics, and sedative-hypnotics) may also lead to reduced food intake and subsequent weight loss. Consideration should even be given to drug-induced tremor as a cause for reducing food intake when severe or uncontrolled. Although unrelated to pharmacologic effect, the impact, particularly for chronic medications, that out-of-pocket drug expense has on food choice compromise, cannot be overlooked especially in the elderly *(37)*.

Given the wide range of innervating neural pathways that affect the function of the GI tract, it is easy to appreciate that medication with pharmacologic effect on cholinergic, histaminic, dopaminergic, opioid, or serotonergic receptors can influence GI function. Any drug that causes altered taste, dry mouth, oral pain, anorexia, nausea, vomiting, altered gastric emptying, diarrhea, or constipation can lead to impaired oral intake and nutrient absorption, increasing the risk for malnutrition.

4.3.1. *Appetite and Body Weight*

4.3.1.1. Decreased Appetite and Body Weight

Weight loss has been associated with a number of medications beyond those causing GI dysfunction (Table 4.3). Decreased food intake can occur as a result of primary appetite suppression. Intake may also be reduced by secondary effects of medication. Most of these are drugs that stimulate the nervous system centrally and peripherally and include the amphetamines, caffeine, methylphenidate, and theophylline. Other agents that can produce an anorexic effect include amantadine, the antihistamines, bethionol, dacarbazine, epirubicin, etoposide, fenfluramine, fluoxetine, fluvoxamine, levodopa, lithium, metformin, naloxone, nicotine, perhixiline, phentermine, pimozide, quinine, sibutramine, temozalomide, topiramate, trazodone, and zonisamide. Other causes of weight loss still need to be ruled out in patients using any of these medications.

4.3.1.2. Increased Appetite and Body Weight

Numerous drugs are associated with increased appetite and weight gain (Table 4.3). This is most commonly reported with psychotropic medications, which may reinforce negative self-images held by some patients *(38)*. These drugs include amitriptyline, chlorpromazine, clozapine, imipramine, lithium, mirtazapine, and valproic acid *(39)*. Additional medications include insulin, oxandrolone and other steroids, and β-blockers. Drug-induced weight gain can be accompanied by increased risk of morbidity associated with increased body weight from other causes. The weight gain from β-blockers is predominantly from accumulation of abdominal fat, and may be more likely in younger patients and those with β-adrenergic receptor polymorphisms *(40)*. Weight gain induced by medication may be difficult to reverse given that much of it is adipose tissue gain. A number of adverse metabolic effects are associated with the use of second generation antipsychotic agents (i.e., weight gain, hyperglycemia, dyslipidemia) *(41)*. Weight gain is more likely with risperidone, quetiapine, and olanzapine than with aripiprazole, ziprasidone, or clozapine *(42)*.

4.3.2. *GI Tract*

4.3.2.1. Oral Cavity

Change in taste perception can have a significant influence on subsequent dietary intake. Medications can cause loss of taste (aguesia), distortion of taste (dysguesia), decreased sense of taste (hypoguesia), and even gustatory hallucination (phantoguesia) *(19,43,44)*. Xerostomia as a result of inhibited saliva production, most commonly a result of drugs with anticholinergic properties, can also alter taste sensation. The drugs that cause dry mouth and altered taste sensation include the antihistamines, diuretics, antipsychotics, and tricylic antidepressants (Table 4.4) *(19,44)*. Metered-dose inhalers that deliver medication allow some drug to precipitate in the oropharynx. This may also directly alter taste and reduce food intake. In the case of corticosteroid inhalers, the potential exists for fungal overgrowth in the region that also can reduce food intake. Oropharyngeal candidiasis may occur with chronic use of broad-spectrum antibiotics as well. If the causative agent cannot be discontinued or reduced in dose, use of breath mints, lozenges, and sugarless gum may offer relief. Oral care regimens and spacers should be encouraged for those using metered-dose inhalers, and appropriate treatment implemented for candidiasis. If taste disturbances are the direct result of zinc deficiency then supplementation of this mineral may be of benefit *(43,45)*. Cytotoxic drugs can create inflammation of the oropharyngeal mucosal surface. This stomatitis severely curtails food intake because of the pain associated with eating. Additionally, several agents from this group of medications can cause a profound

Table 4.3
Drug-Induced Changes to Body Weight

Increase	*Decrease*
Amitriptyline	Amphetamines
Aripiprazole	Caffeine
Atenolol	Diethylpropion
Carbamazepine	Ephedrine
Chlorpromazine	Fluoxetine
Clomipramine	Mazindol
Clozapine	Orlistat
Cyclophosphamide	Phendimetrazine
Desipramine	Phentermine
Doxepin	Selegiline
Fluorouracil	Sibutramine
Fluphenazine	
Gabapentin	
Glucocorticoids	
Haloperidol	
Imipramine	
Insulin	
Lithium	
Maprotiline	
Methotrexate	
Mirtazapine	
Nortriptyline	
Olanzapine	
Perphenazine	
Phenelzine	
Progestins	
Propranolol	
Protease inhibitors	
Quetiapine	
Risperidone	
Tamoxifen	
Thioridazine	
Thiothixene	
Trifluoroperazine	
Valproic acid	
Ziprasidone	

degree of nausea and vomiting. The mechanism for the sensitivity of GI function to medication has not received adequate attention *(46)*.

4.3.2.2. Nausea and Vomiting

Significant nausea may be centrally mediated (e.g., cytotoxic agents, opioid analgesics), or the result of local GI irritation. Vomiting may be a reported side effect of many medications; however, in most patients continued intolerance to the medication only occurs with a select number of these drugs.

Table 4.4
Drug-Induced Changes to Gastrointestinal Function

Altered Taste (and Smell)	*Dry Mouth*	*Dysphagia*	*Increased Gastric Emptying*
ACE inhibitors	Amantadine	Alendronate	Bethanachol
Acetazolamide	Amitriptyline	Anticholinergic drugs	Erythromycin
Albuterol	Antihistamines	Antineoplastics	Laxatives
Allopurinol	Antipsychotics	Antipsychotics	Metoclopramide
Amphotericin B	Atropine	CNS depressants	Misoprostol
Antihistamines	Benzodiazepines	Immunosuppressants	
Aspirin	Brompheniramine	Corticosteroids	
Azathioprine	Bumetanide	Iron	
β-Lactam antibiotics	Captopril	Lidocaine (local)	
Baclofen	Cetirizine	NSAIDs	
Bismuth salts	Clonidine	Neuromuscular blockers	
Bromocriptine	Cyclopentolate	Potassium	
Carbamazepine	Cyproheptadine	Quinidine	
Chloral hydrate	Decongestants	Salicylates	
Chlorpromazine	Didanosine		
Clarithromycin	Diphenhydramine	*Delayed Gastric Emptying*	
Clofibrate	Diuretics	Anticholinergics	
Digoxin	Flecainide	Caffeine	
Diltiazem	Flunitrazepam	Calcium channel blockers	
Diuretics	Granisetron	Clonidine	
Ethambutol	Imipramine	Dicyclomine	
Flurazepam	Isoniazid	Iron	
Gold	Loratidine	Meperidine	
Griseofulvin	Mesalamine	Nitrates	
Iron	Molindone	Opiates	
Levodopa	Nizatidine	Oxybutynin	
Linezolid	Nortriptyline	Theophylline	
Lithium	Ondansetron	Tricyclic antidepressants	
Maribavir	Oxybutynin	Verapamil	
Metformin	Pentoxifylline		
Metronidazole	Procainamide		
Nifedipine	Propantheline		
Opioid analgesics	Rimantadine		
Penicillamine	Selegiline		
Pentamidine	Sertraline		
Phenytoin	Trazodone		
Propranolol	Tricyclic antidepressants		
Rifabutin	Trimethobenzamide		
Quinidine			
Selegilene			
Sulfasalazine			
Thioridazine			

(*Continued*)

Table 4.4
(Continued)

Nausea/Vomiting	*Nausea/Vomiting* (Contd.)	*Diarrhea*
Aldesleukin	Irinotecan	Acarbose
Alfentanil	Isoflurane	Aminosalicylic acid
Altretamine	Ketamine	Auranofin
Amantadine	Ketorolac	Azathioprine
Amifostine	Lithium	Busulfan
Amiodarone	Lomustine	Calcitonin
Atracurium	Lorazepam	Carboplatin
Azacitidine	Mechlorethamine	Carmustine
Azathioprine	Metformin	Chlorambucil
Bromocriptine	Methadone	Cisapride
Buprenorphine	Methohexital	Clindamycin
Butorphanol	Methotrexate	Colchicine
Busulfan	Metronidazole	Cyclophosphamide
Carbamazepine	Midazolam	Cytarabine
Carboplatin	Mitoxantrone	Dactinomycin
Carmustine	Mivacurium	Darbepoetin
Cisatracurium	Morphine	Daunorubicin
Cisplatin	Nalbuphine	Didanosine
Clofarabine	Nitrous oxide	Digoxin
Codeine	Oxaliplatin	Docetaxel
Cyclophosphamide	Oxycodone	Doxorubicin
Cytarabine	Oxymorphone	Epirubicin
Dacarbazine	Pancuronium	Erythromycin
Dactinomycin	Pentostatin	Erythropoietin
Daunorubicin	Pipecuronium	Etoposide
Desflurane	Potassium	Exemestane
Diazepam	Procarbazine	Fluoroquinolones
Digoxin	Propofol	Fluorouracil
Doxacurium	Pyridostigmine	Fluoxetine
Doxorubicin	Quinidine	Hydralazine
Enflurane	Remifentanil	Hydroxyurea
Erythromycin	Repaglinide	Idarubicin
Estrogens	Rocuronium	Ifosfamide
Etomidate	Sevoflurane	Irinotecan
Epirubicin	Sirolimus	Lactulose
Fentanyl	Streptozocin	Lansoprazole
Fluconazole	Sufentanil	Lomustine
Fluoxetine	Tacrolimus	Magnesium
Halothane	Thiopental	Mechlorethamine
Hydrocodone	Topiramate	Melphalan
Hydromorphone	Tramadol	Mercaptopurine
Ibuprofen	Valacyclovir	Metformin
Idarubicin	Valproic acid	Methotrexate
Ifosfamide	Vecuronium	Metoclopramide
Ipecac	Zidovudine	Misoprostol

Diarrhea (Contd.)	*Constipation*
Mitomycin	Alosetron
Mitoxantrone	Aluminum
Neostigmine	Amitriptyline
Omeprazole	Anticholinergics
Orlistat	β-Blockers
Paclitaxel	Calcium
Pantoprazole	Cilansetron
Paroxetine	Clonidine
Procarbazine	Dicyclomine
Quinidine	Diltiazem
Serotonin antagonists	Imipramine
Sertraline	Iron
Sorbitol	Loperamide
Tegaserod	Ondansetron
Temozolomide	Opioid analgesics
Tetracycline	Phenothiazines
Theophylline	Verapamil
Thioguanine	
Topotecan	
Tretinoin	
Vinblastine	
Vincristine	

Interference with nutritional status is observed with severe and prolonged emesis as seen with cytotoxic chemotherapy. Poorly controlled nausea and vomiting significantly impact not only nutritional status but also quality of life. Among the antineoplastic agents, some are much more emetogenic than others. The most likely to cause emesis—immediate or delayed—include aldesleukin, altretamine, carboplatin, carmustine, cisplatin, cyclophosphamide, dacarbazine, dactinomycin, daunorubicin, doxorubicin, epirubicin, idarubicin, ifosphamide, irinotecan, lomustine, mechlorethamine, mitoxantrone, pentostatin, and streptozocin *(47,48)*. However, drugs used to combat chemotherapy-induced emesis (e.g., aprepitant, ondansetron) may also cause fatigue, taste disturbances, constipation, and diarrhea, thereby compounding a patient's drug-induced anorexia *(49)*.

Drugs that increase GI motility are most likely to cause abdominal pain and diarrhea that, if severe or prolonged, will impair nutritional status by reducing intake, limiting absorption, and increasing nutrient losses. Drugs with direct effects on the GI tract mucosa include the non-steroidal anti-inflammatory drugs and iron, while metoclopramide, erythromycin, and cisapride increase motility.

4.3.2.3. Diarrhea

Although many drugs are known to cause diarrhea by one or more mechanisms, a recalcitrant few cause the most severe form. Diarrhea may be secretory or osmotic in nature, but in either case, if the diarrhea is severe, it will decrease dietary intake and create fluid, electrolyte, and other mineral losses. Drugs commonly associated with diarrhea include magnesium-containing antacids and cathartics, antibiotics, antineoplastics, cholinergics, colchicine, digoxin, ezetimibe, gold salts, mycophenolate mofetil, orlistat, prostaglandins, quinidine, and the laxatives *(48,50)*. Some infrequently used antihypertensive agents (reserpine, guanethidine, methyldopa, guanabenz, guanadrel) can also cause diarrhea.

4.3.2.4. Constipation

Constipation can be caused most commonly by agents that induce an ileus (e.g., opioid analgesics, anticholinergics, calcium or aluminum-containing antacids, barium sulfate, calcium channel blockers, clonidine, diuretics, ganglionic blockers, iron preparations, nonsteroidal anti-inflammatory drugs, polystyrene sodium sulfonate, and bile acid sequestrants) *(48,50)*. Drugs with anticholinergic properties—including amitriptyline, atropine, belladonna, benztropine, diphenhydramine, hyoscyamine, imipramine, ipratropium, isopropamide, olanzapine, oxybutynin, procainamide, scopolamine, trihexyphenidyl, and zotepine—can reduce GI motility.

4.4. DIGESTION

The impact of medication on the digestive process is not well documented. Drugs that are associated with pancreatitis or that cause damage to specialized cells of the GI tract can play a role in altering digestion. Many drugs have been reported to cause pancreatitis but the association is strongest for only a few. These include asparaginase, azathioprine/mercaptopurine, bumetanide, cisplatin, didanosine, the estrogens, furosemide, metronidazole, pentamidine, the sulfonamides, thiazide diuretics, and valproic acid *(48,51)*. The effect on digestion is expected to be most significant if chronic or recurrent bouts of pancreatitis are experienced. Natural products used for their medical benefits may reduce carbohydrate digestion by interfering with α-glucosidase and α-amylase *(52)*. This is similar to the effect of the prescription drug acarbose used in the management of diabetes. Any drug that creates a barrier between food particles and digestive components can reduce digestion. Maldigestion of fat is most likely to occur before impaired digestion and absorption of other macronutrients. Micronutrient malabsorption associated with changes in the digestive process includes the diminished availability of dietary vitamin B_{12} in the presence of chronic acid-suppressive drug therapy *(53,54)*. It can be expected that rapid intestinal transit induced by drugs that increase GI motility would also increase the risk for maldigestion and subsequent malabsorption. This has the potential to influence both macronutrients and micronutrients.

4.5. MACRONUTRIENTS

4.5.1. Absorption and Distribution

Medication may cause malabsorption of carbohydrate, fat, and protein. Malabsorption of carbohydrate and fat is an expected therapeutic effect from the use of acarbose and orlistat, respectively. Both oral neomycin and colchicine are associated with carbohydrate and fat malabsorption. Neomycin produces the classic example of dose-related drug-induced malabsorption *(55)*. By causing brush border damage, an enteropathy arises in part the result of disaccharide intolerance and bile acid precipitation. Ezetimibe and the bile acid sequestrants (e.g., cholestyramine, colestipol, colesevelam) inhibit the absorption of cholesterol in the management of hyperlipidemia. Cytotoxic drugs can damage the absorptive surface of the GI mucosa. For example, methotrexate can cause macronutrient malabsorption following epithelial cell desquamation. The effects that antimicrobials, which alter the gut flora, have on the recovery of carbohydrate in the colon are not well investigated. Lipodystrophy, including fat redistribution syndrome, associated with the highly active antiretroviral therapy regimens used in patients with HIV infection is likely multifactorial *(56)*. Cholestasis and steatohepatitis caused by drugs Commonly reduces intake and may contribute to altered digestion or macronutrient disposition in these patients *(57,58)*.

4.5.2. Metabolism

Medication may also be associated with metabolic interference of the macronutrients which may result in altered circulating levels of glucose, cholesterol, triglycerides, or amino acids. Medication with a primary therapeutic effect of reducing blood glucose, triglycerides, or cholesterol are well known and not mentioned further. Metabolic changes are in many cases transient but can be life-threatening or chronic in nature.

4.5.2.1. Glucose

Quite a number of drugs are associated with hyperglycemia or hypoglycemia (Table 4.5) *(59–61)*. Clinically hyperglycemia is most commonly seen with corticosteroids, β-blockers, thiazide diuretics, caffeine, calcium channel blockers, cyclosporine, growth hormone, morphine, nicotine, octreotide, phenytoin, protease inhibitors, sympathomimetics, tacrolimus, theophylline, and thyroid products *(59,61)*. The calcineurin inhibitors (i.e., cyclosporine, tacrolimus) decrease pancreatic insulin secretion and contribute to hyperglycemia and diabetes. Epidemiologic data suggest no predictable difference between first- and second-generation antipsychotic agents in resultant diabetes mellitus *(62)*. The atypical antipsychotic agents (aripiprazole, clozapine, olanzapine, paliperidone, quetiapine, risperidone, and ziprasidone) have the potential to cause hyperglycemia and may increase the risk of diabetes *(63,64)*. The conventional antispychotics may also cause glucose intolerance although more likely related to weight gain than to central glucose dysregulation *(64)*. As would be expected, hypoglycemia is most often associated with insulin and oral hypoglycemics, but also with anabolic steroids, angiotensin-converting enzyme inhibitors, insulin-like growth factor-1, octreotide, salicylates, tetracycline, and warfarin *(60,61)*. Gatifloxacin is associated with both hyper- and hypoglycemia in elderly hospitalized patients *(65)*. The antiepileptic drug topiramate has been shown to improve glycemic control (especially, fasting blood glucose) and reduce body weight in diabetics.

4.5.2.2. Lipid

Hyperlipemia is also an adverse effect of anabolic hormones, β-blockers, diuretics, progestins including those in combination oral contraceptives, danazol, immunosuppressive agents, propofol, and protease inhibitors *(66)*. Capecitabine may cause severe hypertriglyceridemia, especially in at-risk individuals *(67)*. Many other drugs are associated with lipid abnormalities. Postmenopausal estrogen and progesterone use has been associated with improved lipid profiles.

4.5.2.3. Protein

Corticosteroids increase endogenous protein catabolism, likely by increasing activity of the ubiquitin-proteasome proteolytic pathway *(68,69)*. Protein synthesis can be increased as a result of growth hormone, anabolic hormones, and insulin-like growth factor-1 without impacting protein degradation *(70)*. Drug-induced azotemia is more likely a result of impairing renal excretory function than interfering with protein metabolism. Very little information is available regarding the impact of medication on specific amino acid absorption, transport, or metabolism. Several drugs are absorbed by peptide transporters setting up the potential for competition. Some significant adverse effects attributed to valproic acid (e.g., hepatotoxicity, teratogenicity) may involve drug-induced alteration in the methionine cycle *(71)*, and oral contraceptives may increase tryptophan metabolism *(72)*. A drug used in the

Table 4.5
Drug-Induced Changes to Blood Glucose

Hyperglycemia	*Hypoglycemia*
Acetazolamide	Acetaminophen
Aripiprazole	β-Agonists
β_2-Agonists	Benzodiazepines
β-Blockers	Biguanides
Bumetanide	Captopril
Caffeine	Cotrimoxazole
Calcium channel blockers	Disopyramide
Clozapine	Enalapril
Corticosteroids	Ethanol
Cyclosporine	Fluoxetine
Cyproheptadine	Ganciclovir
Dapsone	Gatifloxacin
Diazoxide	Insulin
Furosemide	Lisinopril
Gatifloxacin	Lithium
Growth hormone	MAO inhibitors
Isoniazid	Mebendazole
Lithium	Nonselective β-blockers
Minoxidil	Octreotide
Morphine	Pentamidine
Nicotine	Quinidine
Octreotide	Quinine
Olanzapine	Salicylates
Oral contraceptives	Sertraline
Paliperidone	Sulfonylureas
Pentamidine	Tetracycline
Phenothiazines	Tricyclic antidepressants
Phenytoin	Warfarin
Protease inhibitors	
Quetiapine	
Rifampin	
Risperidone	
Sympathomimetics	
Tacrolimus	
Theophylline	
Thiazide diuretics	
Thyroid hormone	
Torsemide	
Ziprasidone	

management of tardive dyskinesia was found clinically to cause severe lactic acidosis, hypoglycemia, hyperammonemia, and acute hepatic encephalopathy in patients. This set of adverse drug effects was eventually traced not to altered macronutrient disposition but to the induction of pantothenic acid deficiency by the drug *(73)*. This illustrates the complexity of drug-induced changes in nutrient status and the difficulty in identifying them.

4.5.3. Excretion

Drug-induced diarrhea or impairment of renal function (reabsorption or concentrating ability) will lead to the unrecoverable loss of macronutrients possibly with significant consequences.

4.6. VITAMINS

An assortment of drugs can influence nutritional status with a particular effect on vitamins (Table 4.6). They may affect the way vitamins are absorbed, distributed, metabolized, and/or excreted. With

Table 4.6
Drugs That Alter Vitamin Status

Vitamin A	*Vitamin D*	*Vitamin E*	*Pyridoxine*
Cholestyramine	Allopurinol	Aspirin	Aminoglycosides
Colchicine	Aluminum	Cholestyramine	Aminophylline
Colesevelam	Cholestyramine	Colesevelam	Bumetanide
Colestipol	Cimetidine	Colestipol	Carbamazepine
Corticosteroids	Colesevelam	Corticosteroids	Cephalosporins
Ethanol	Colestipol	Gemfibrozil	Cycloserine
Mineral Oil	Corticosteroids	Loop diuretics	Dopamine
Neomycin	Isoniazid	Mineral oil	Estrogen
Orlistat	Ketoconazole	Oral contraceptives	Ethanol
Thiamin	Mineral oil	Orlistat	Fluoroquinolones
Aminoglycosides	Neomycin	*Niacin*	Furosemide
Antacids	Orlistat	Aminoglycosides	Gentamicin
Bumetanide	Phenobarbital	Azathioprine	Hydralazine
Cephalosporins	Phenytoin	Cephalosporins	Isoniazid
Digoxin	Rifampin	Fluoroquinolones	Oral contraceptives
Ethanol	Sunscreen	Isoniazid	Penicillamine
Fluoroquinolones	*Riboflavin*	Mercaptopurine	Phenelzine
Furosemide	Aminoglycosides	Sulfonamides	Phenobarbital
Oral contraceptives	Amitriptyline	Tetracyclines	Phenytoin
Penicillins	Cephalosporins	Thioguanine	Sulfonamides
Phenytoin	Chlorpromazine	Valproic acid	Tetracyclines
Sulfonamides	Doxorubicin	*Vitamin K*	Theophylline
Tetracyclines	Fluoroquinolones	Aminoglycosides	Torsemide
Theophylline	Imipramine	Cephalosporins	Valproic acid
Torsemide	Oral contraceptives	Cholestyramine	
	Phenobarbital	Colesevelam	
	Probenecid	Colestipol	
	Quinacrine	Corticosteroids	
	Sulfonamides	Fluoroquinolones	
	Tetracyclines	Mineral oil	
		Oral contraceptives	
		Orlistat	
		Phenobarbital	
		Phenytoin	
		Sulfonamides	
		Tetracyclines	

(*Continued*)

Table 4.6
(Continued)

Vitamin B_{12}	*Folic Acid*	*Choline*	*Vitamin C*
Amikacin	Aspirin	Cyclosporine	Aspirin
Antivirals	Azathioprine	Daunorubicin	Hydrocortisone
Cephalosporins	Carbamazepine	Nifedipine	Oral contraceptives
Chloramphenicol	Celecoxib	Verapamil	Salsalate
Cholestyramine	Cholestyramine	*Biotin*	Tetracycline
Clofibrate	Colchicine	Carbamazepine	
Colchicine	Colesevelam	Phenobarbital	
Colesevelam	Colestipol	Phenytoin	
Colestipol	Corticosteroids	Primidone	
Fluoroquinolones	H_2 antagonists		
Gentamicin	Indomethacin		
H_2 antagonists	Metformin		
Metformin	Methotrexate		
Neomycin	Neomycin		
Nitrous oxide	NSAIDs		
Oral contraceptives	Olsalazine		
Phenytoin	Oral contraceptives		
Proton pump inhibitors	Pancreatic enzymes		
Sulfonamides	Pemetrexed		
Tetracyclines	Phenobarbital		
Tobramycin	Phenytoin		
Zidovudine	Primidone		
	Pyridoxine		
	Pyrimethamine		
	Sulfasalazine		
	Triamterene		
	Valproic acid		

the relatively recent focus on drug–nutrient interactions involving vitamins, the realization of the impact that medication therapy can have on a patient's nutritional status comes to the forefront. Given that many transporters and enzymes are common to both drugs and vitamins, as well as the influence of drugs on the expression of transporters and enzymes, interactions may be expected in vitamin absorption, distribution, metabolism, and excretion.

4.6.1. Absorption

4.6.1.1. Water-Soluble Vitamins

4.6.1.1.1. Folic Acid. Nutrients that rely on a carrier-mediated intestinal transport system for absorption pose a potential risk for interaction with drugs that use the same or associated transport mechanisms. Several drugs that use the folic acid transporters (reduced-folate carrier and/or proton-coupled folate transporter), including sulfasalazine, olsalazine, and the antifolate drugs (e.g., methotrexate, pemetrexed), have the potential to decrease absorption of folic acid *(76)*. Methotrexate competes with folic acid for absorption and contributes to folate deficiency in patients treated with even low doses of this drug *(77,78)*. In addition, patients with rheumatoid arthritis concurrently receiving sulfasalazine and methotrexate, as compared to monotherapy with either agent, were found to have consistently elevated homocysteine levels which are associated with cellular folate depletion *(79)*. The folate deficit may account for some of the toxicity of the drug, and repletion of folic acid during methotrexate therapy likely does not interfere with drug efficacy *(80,81)*.

Triamterene has been reported to cause megaloblastosis, which may result from competitive inhibition of folic acid absorption *(82–84)*. Nonetheless, folate depletion does not occur in all patients treated with this diuretic *(85)*. Pancreatic extracts and cholestyramine have also been reported to impair folic acid absorption and reduce bioavailability *(86–88)*. Chronic ethanol use may induce a downregulation of the reduced folate transporter resulting in folate malabsorption *(89,90)*.

4.6.1.1.2. Vitamin B_{12}. A reversible vitamin B_{12} malabsorption has been described with both parenteral and oral administration of colchicine *(32,91)*. It is unclear whether this represents a specific inhibition of transport involving the intrinsic factor–vitamin B_{12} receptor or if it is a result of general mucosal damage *(92,93)*. Serum vitamin B_{12} concentrations remain normal, given the typically large body stores of vitamin B_{12}; however, normality depends on the patient's baseline vitamin B_{12} status, as well as the duration of drug treatment *(94)*.

Although vitamin B_{12} malabsorption is identified among individuals with advanced stages of HIV infection, subnormal serum concentrations of this vitamin may occur more frequently in those receiving zidovudine therapy *(95)*.

Biguanides (e.g., metformin) can induce malabsorption of vitamin B_{12} in as many as 30% of patients receiving the drug *(96,97)*; both a dose- and time-dependent relationship is implicated. In one retrospective analysis the risk for developing deficiency (based only on serum vitamin B_{12} values) increased for every 1 g of metformin administered per day (OR 2.88). The risk of vitamin B_{12} deficiency increases with duration of metformin use. Individuals receiving the drug for greater than 3 years had an OR of 2.39 for deficiency as compared to those on therapy for less than 3 years after adjustment for age, vegetarian status, and acid-suppressive therapy *(98)*. Although this may be reversible in some patients, it is likely that approximately half of patients exhibit depressed secretion of intrinsic factor that may not be reversible *(96)*. It is possible that the patient may experience a concomitant elevation in serum homocysteine, however, not all patients exhibit a clinically symptomatic deficiency *(97,99)*. Hematological and especially neurological complications of vitamin B_{12} deficiency may not become evident for several years.

Type-2 histamine receptor (H_2) antagonists and proton pump inhibitors significantly decrease dietary vitamin B_{12} absorption as a result of decreased cleavage from food in the presence of lower levels of gastric acid and pepsin *(53,54,100–102)*. H_2 antagonists may even reduce intrinsic factor secretion *(103)*. However, conflicting clinical evidence has been reported which may be related to heterogeneity in baseline vitamin B_{12} status and duration of therapy. In one study, 150 patients were receiving H_2 antagonists for up to 72 months; approximately 23% of those were also receiving vitamin B_{12} supplementation. An influence on serum vitamin B_{12} status by H_2 antagonists was not observed *(104)*. In the same study, 141 patients had been prescribed a proton pump inhibitor for 72 months, 24% of those were also receiving supplementation; a statistically significant decrease in serum vitamin B_{12} levels was seen with increasing duration of therapy *(104)*. With proton pump inhibitor use vitamin B_{12} status is statistically lower at year 4 of therapy compared to year 2, although still remaining in the normal range *(105)*. If this reduction is to become clinically evident, it is not likely to occur until at least 1–2 years of therapy *(53,106,107)*. Although the population's average diet yields more than sufficient amounts of cobalamin, it is possible that use of a proton pump inhibitor chronically would result in a clinically significant deficiency. In addition, it should be considered that patients with inadequate intake, the elderly or malnourished may be at increased risk *(105,107)*. A drawback of most reports has been the exclusive use of serum vitamin B_{12} values to evaluate status rather than methylmalonic acid. As expected, gastric acid-suppressive therapy does not interfere with the absorption of crystalline vitamin B_{12} preparations which can easily improve vitamin B_{12} status *(100,107)*. Consideration should be given to vitamin B_{12} monitoring and oral supplementation in patients using these medications chronically. Closer examination is needed to determine whether drug-induced increases in gastric pH influence the absorption of other vitamins. For example, β-carotene absorption may be reduced at lower levels of gastric acidity *(108)*.

4.6.1.1.3. Others. Several medications (e.g., cyclosporine, daunorubicin, nifedipine, verapamil) may inhibit choline transporter uptake directly or indirectly *(74)*. Additionally, intestinal absorption of biotin can be limited by competition with carbamazepine for the sodium-dependent multivitamin transporter *(75)*. Many gut microorganisms, including several strains of Bifidobacteria, can synthesize B vitamins—particularly thiamin, niacin, and folate *(109,110)*—and can be a source of several vitamins *(111)*. Although the clinical importance of bacterially derived B vitamins or vitamin K to human nutrition remains unclear, antimicrobials, which drastically reduce the number of bacteria that produce these nutrients in the GI lumen, may impact individuals with borderline status of these vitamins.

4.6.1.1.4. Carnitine. In the mitochondria, the vitamin-like substance carnitine is involved in long-chain fatty acid metabolism. To reach its site of effect carnitine relies on membrane transport. Some drugs may interact with carnitine transport at this level. For example, the antimicrobial levofloxacin (and grepafloxacin - no longer on the U.S. market) can inhibit carnitine absorption at the GI tract by interfering with the function of the organic cation transporter OCTN2 *(112)*. OCTN2 also transports verapamil, quinidine, cephaloridine, and other cationic compounds; competition for the transporter can reduce carnitine absorption and may cause deficiency *(113)*.

4.6.1.2. Lipid-Soluble Vitamins

Orally administered mineral oil can cause gross malabsorption of nutrients. Lipid-soluble nutrients, including vitamins A, D, E, and K as well as the carotenoids, solubilize in the oil and become trapped within the gut lumen. Given the rapid GI transit of mineral oil, digestion is incomplete and absorption is reduced, partly because of the swift transit but also as a result of the mechanical barrier formed around food particles *(114)*.

By binding bile acids, cholestyramine reduces the absorption of the fat-soluble vitamins *(87,115)*. Bile acid sequestrants, along with probucol, significantly reduce β-carotene and lycopene concentrations *(116)*. The potential for developing osteomalacia and bleeding disorders has been suggested following chronic treatment *(117)*. One study suggested no significant effect on vitamin D status in patients using cholestyramine chronically, although bone density measurements were not performed, and serum vitamin D samples were obtained during spring and summer *(118)*.

A patient with a history of gastric stapling and subclinical osteomalacia (T-score –1.7 at hip) received an intravenous bisphosphonate (pamidronate). Three weeks later, the patient presented with atrial fibrillation and tetany in the setting of profound hypocalcemia, hyperparathyroidism, and a decreased serum 25(OH)-vitamin D level *(119)*. The patient was aggressively replenished with calcium (intravenous and oral) and dihydrotachysterol for 7 days. Thereafter, for 2 months ergocalciferol at a dose of 50,000 units 3 times weekly replaced the twice daily tachysterol. Oral calcium at 4 g daily was continued. By 2 months the symptoms had resolved. Patients with a clinical indication for bisphosphonates should receive a baseline vitamin D assessment; those who are deficient should receive replacement prior to intravenous bisphosphonate administration *(119)*.

By inhibiting lipase action, orlistat not only induces lipid malabsorption but also decreases the absorption of fat-soluble vitamins *(120)*. Vitamin E absorption may be reduced by over 50%, whereas vitamin A is not as significantly influenced *(121)*. Vitamin supplementation may be required for some patients. However, in a study in which patients received a lipid-soluble vitamin-containing supplement during treatment with orlistat, serum concentrations of vitamin D and vitamin K both remained low *(122)*. In fact, the reduction in vitamin D, despite supplementation, warranted a recommendation for regular monitoring in patients receiving orlistat, even when supplementation is included as part of the regimen. Vitamin supplementation should be separated by at least 2 h from the administration of orlistat. The absorption of a pharmacological dose of β-carotene may also be reduced by about one-third during treatment with orlistat *(123)*.

Hepatic menaquinone (vitamin K_2) content may be reduced following broad-spectrum antibiotic use *(124)*, but it remains to be determined whether this is a cause–effect relationship or if it is clinically significant. In fact, alternative explanations to vitamin K deficits (i.e., antimicrobial structure-related) have been offered for changes in blood clotting in patients receiving antibiotics, which includes cases of bleeding following cephalosporin use *(125)*.

4.6.2. Distribution

4.6.2.1. Water-Soluble Vitamins

4.6.2.1.1. Folic Acid & Thiamin. Folate binding and serum folate may be reduced in patients who are using aspirin, with a possible rise in urinary folate excretion *(126,127)*. Red blood cell folate levels are lower in patients treated with the antiepileptics phenytoin and carbamazepine, but this is not correlated with dietary folate intake *(128)*. Phenytoin may reduce tissue uptake of folate *(129)*. Additionally, thiamin levels in the cerebrospinal fluid are lower in patients being treated with phenytoin compared to controls *(130)*. Uptake of thiamin and its phosphoesters may vary by nervous tissue region during phenytoin treatment *(131)*. Intracellular loss of thiamin from cardiac cells occurs in the presence of furosemide or digoxin, because of a reduced ability of the cells to take up the vitamin *(132)*.

4.6.2.1.2. Other Vitamins. Aspirin also inhibits the uptake of ascorbic acid into tissues including white blood cells which are used to assess vitamin C status. The saturable, high-affinity ascorbic acid transporter may be inhibited by hydrocortisone *(133)*. This can reduce tissue saturation, possibly leading to tissue depletion in patients chronically treated with aspirin or hydrocortisone and creating opportunity for greater renal excretion *(134–136)*.

4.6.2.1.3. Carnitine. Zidovudine can deplete carnitine in muscle, which is associated with drug-induced myopathy *(137)*. Tissue carnitine depletion during treatment with valproic acid may partly result from an inhibition of tissue uptake *(138)*. It is possible that some antiepileptic drugs are associated with decreased carnitine levels. One study examined the effect of gabapentin, lamotrigine, phenytoin, tiagabine, topiramate, vigabatrin, and valproic acid for effect on carnitine transport by the human placental carnitine transporter. At Varying concentrations the antiepileptic drugs inhibited the transport, with tiagabine having the greatest inhibitory effect and lamotrigine the least. The authors concluded that valproic acid and phenytoin, however, would likely have the greatest clinical impact due to unbound therapeutic drug concentrations *(139)*.

4.6.2.2. Lipid-Soluble Vitamins

The mechanism that leads to decreased levels of α-tocopherol, γ-tocopherol, and ubiquinone-10 (coenzyme Q_{10}) as described with gemfibrozil therapy remains unclear *(140)*. The dose-dependent reduction in tissue and circulating levels of ubiquinone-10 levels with hydroxymethylglutaryl-coenzyme A reductase inhibitors (e.g., lovastatin, pravastatin, simvastatin) likely result from inhibited synthesis *(141–143)*. The lower ubiquinone levels may contribute to impaired mitochondrial function, as determined by lactate-to-pyruvate ratio *(144)*. This may partly explain the drug-induced hepatic dysfunction seen with this class of agents. However, the decline in coenzyme Q_{10} may be the result of total cholesterol decline, which may not require further attention *(145)*. Supplementation with an appropriate coenzyme Q-containing product may prevent the depletion, although the clinical value of this intervention has yet to be determined prospectively *(146)*.

4.6.3. Metabolism

4.6.3.1. Water-Soluble Vitamins

4.6.3.1.1. Vitamin B_6. The metabolism of pyridoxine is impaired by several drugs, including isoniazid, cycloserine, theophylline, and hydralazine. The potential for a secondary deficiency of niacin also exists. Reduced circulating levels of pyridoxine and thiamin occur more frequently (and in a dose-dependent manner) in patients treated with theophylline than in similar patients not treated with this agent *(147–149)*. Plasma pyridoxal-5′-phosphate appears to be more affected than plasma pyridoxal or urinary 4-pyridoxic acid *(149)*. Theophylline is a pyridoxal kinase antagonist that, with chronic use, may require concurrent vitamin B_6 supplementation to limit nervous system side effects such as tremor *(150,151)*. This differs from the effects of dopamine, isoniazid, cycloserine, hydralazine, and penicillamine, which form covalent complexes with pyridoxal or pyridoxal-5′-phosphate and, thereby inhibit pyridoxal kinase *(150)*. The poor availability of this vitamin, which then reduces GABA levels in the brain, may play a role in the seizure risk of theophylline. An 81-year-old woman treated with theophylline presented with complex partial seizures despite serum drug concentrations within the normal range but with an undetectable serum vitamin B_6 level *(152)*. Discontinuation of theophylline and supplementation with intravenous vitamin B_6 resolved her seizures. The authors concluded that vitamin B_6 deficiency may be present in a patient with theophylline-induced seizure activity *(152)*.

Hydralazine-induced deficiency of vitamin B_6 has been reported to cause neuropathy *(153)*. Through its interaction, hydralazine appears to inhibit a number of enzymes that require pyridoxal as a cofactor and this might even contribute to the drug's blood pressure lowering effects *(154)*.

4.6.3.1.2. Folic Acid. A large portion of patients treated with sulfasalazine develop folate deficiency *(155)*. The deficiency may be responsible for severe megaloblastic anemia and pancytopenia, which are associated with the drug *(156,157)*. Although malabsorption may play a partial role, metabolic abnormalities likely contribute to the deficit *(158)*. Several nonsteroidal anti-inflammatory drugs can also interfere with the impact of folate metabolism on biosynthesis of other nutrients *(158)*. Although not all patients treated with these agents exhibit a significant decrease in serum folate levels, tissue levels of folate may be reduced in the absence of serum folate abnormalities.

Elevated homocysteine levels have been identified as a risk factor for venous thromboembolism. Patients with a history of venous thromboembolism had a high prevalence of hyperhomocysteinemia after completing a course of warfarin therapy (159). This retrospective analysis illustrated a positive association between homocysteine level and duration of warfarin therapy although this observation may be an indirect influence of the decreased green-leafy vegetable diet that many are inappropriately advised to follow while on therapy, rather than a direct influence on folate status *(159)*.

Nearly every antiepileptic drug has been linked to fetal malformations *(160)*. Of special concern is folate deficiency in women of childbearing age receiving these drugs; as insufficient folate levels are an independent risk factor for fetal malformations and pregnancy complications *(161)*. Although the benefit of folic acid supplementation for reducing the risk of neural tube defects in the general patient population is well recognized, it is less established for women receiving antiepileptic drugs. Numerous cases have been reported of women receiving antiepileptic drugs, often valproate, and giving birth to a child with a neural tube defect *(160,162)*. It is recommended that women, even if not planning for pregnancy, receive ongoing supplementation of folic acid if receiving antiepileptics. A population-based, case–control study evaluated the impact of folic acid supplementation in early pregnancy on the prevalence of congenital abnormalities in children of mothers using antiepileptic drugs (carbamazepine, phenobarbital, phenytoin, primidone) at termination of pregnancy, birth, and 3 months of age. The findings suggest that the risk of congenital abnormality of an in utero exposure to these drugs is decreased but not removed with folic acid supplementation *(163)*.

The enzyme inducing antiepileptic drugs (i.e., carbamazepine, phenobarbital, phenytoin) interfere with folate metabolism *(164,165)*. Disturbances of folate and thiamin status have been suggested as an explanation for adverse neuropsychiatric effects of these medications *(166,167)*. The associated elevation in homocysteine concentrations can be normalized with folic acid supplementation *(161)*. Serum folate levels decrease following initiation of phenytoin therapy. Conversely, folic acid supplementation alters the pharmacokinetics of phenytoin. Because of the two-way interaction between this drug–nutrient pair, administration of folic acid supplementation at the initiation of phenytoin therapy prevents the decrease in folate concentrations and allows quicker achievement of steady state phenytoin concentrations *(168)*. Negative studies may involve patients who have maintained adequate folate status since before initiation of the antiepileptic drug *(169)*. The effects of carbamazepine may be more severe than the others, are likely dose related, and supplementation with a folic acid analog may improve some cognitive measures in patients treated with the drug *(126,164)*. The effects from carbamazepine may take approximately 2 months to manifest *(170,171)*. In a recent study, 45 patients receiving chronic carbamazepine, oxcarbazapine, or valproate monotherapy were compared to 23 healthy controls in order to evaluate folic acid, erythrocyte folate, vitamin B_{12}, and homocysteine levels as well as, blood smears, and the risk of megaloblastic anemia. Although no statistical difference was detected in vitamin B_{12} or erythrocyte folate levels, the patients had statistically lower serum folic acid levels than did control subjects. Female patients had statistically higher homocysteine levels. Of note, 75% of those patients with a low serum vitamin B_{12} level were receiving carbamazepine. Hypochromia/microcytosis was noted in approximately 26% of patients and 17% of the control group *(172)*. Phenytoin, carbamazepine, phenobarbital, and primidone have been associated with decreased serum vitamin B_{12} concentration, as well as megaloblastic anemia. It is possible that the anemia is a result of reduced folic acid metabolism in these patients *(172)*.

4.6.3.1.3. Biotin. Biotin catabolism can occur with chronic use of the antiepileptics carbamazepine, phenytoin, phenobarbital, and primidone but this has not been demonstrated with valproic acid *(173–175)*. This catabolism can be associated with increased organic aciduria. The loss of biotin may contribute to the adverse effects of these anticonvulsants, including poor seizure control and teratogenicity. Biotin deficiency may be the most significant of the nutrient deficits that are caused by the anticonvulsants *(176)*. Supplementation of biotin to help control seizure activity awaits clinical intervention trials.

4.6.3.1.4. Other B Vitamins. Early descriptions of the influence of oral contraceptives on nutritional status described decreased levels of several B vitamins and ascorbic acid, whereas levels of vitamin K increased *(92)*. Compared to nonusers, oral contraceptive users exhibit lower levels of erythrocyte folate and riboflavin, erythrocyte transketolase activity, plasma pyridoxal phosphate, and serum vitamin B_{12} within a few cycles of initiating treatment *(177–179)*. Increases in urinary formiminoglutamic acid in women using oral contraceptives indicate alteration in folate metabolism *(178)*. The reduced folate status in these women of childbearing age should be addressed prior to discontinuing use of oral contraceptives. Oral contraceptives may increase tryptophan metabolism through induction of tryptophan oxygenase, and it has been theorized that this may lead to a pyridoxine deficiency *(72,179)*. The deficit has been associated with clinical symptoms, including depression and anxiety. In such cases of deficiency, vitamin B_6 repletion may require doses of at least 40 mg daily to correct levels and clinical manifestations *(180)*, although signs and symptoms are not universally responsive to vitamin supplementation at pharmacological doses *(181)*. Despite several reports of poor riboflavin status in users *(182,183)*, when diet is controlled, the need for supplementation vanishes *(184)*. Several drugs can impair the conversion of riboflavin to active coenzymes. These include chlorpromazine, amitriptyline, imipramine, quinacrine, and doxorubicin *(185,186)*.

4.6.3.1.5. Vitamin C. Oral contraceptive formulations may also have an influence on ascorbic acid status *(187–189)*. Reductions in plasma and leukocyte levels of this vitamin have been described, although the mechanism remains unclear. Oral contraceptive users appear to have a more rapid turnover of ascorbic acid *(190,191)*. Although increased utilization and metabolism is a possible mechanism for this rapid turnover, it may also be explained by alterations in tissue distribution *(188)*. It is possible that with adequate dietary intake, no significant change in ascorbic acid status occurs, and individuals using oral contraceptives would rarely be expected to require supplementation *(192,193)*.

4.6.3.1.6. Carnitine. As discussed, reductions in folate status have been noted for phenobarbital, phenytoin, and carbamazepine; treatment with another antiepileptic drug—valproic acid—does not appear to disturb this vitamin *(126)*. In contrast, carnitine deficiency can occur with valproic acid treatment *(194,195)*. Reductions are noted in both plasma free carnitine and plasma total carnitine concentrations *(194,195)*. A reduction in urinary total and free carnitine has also been described for chronic valproic acid treatment *(196)*. Valproic acid can inhibit the hepatic synthesis of carnitine and contribute to a deficiency state. Reductions occur in both plasma free carnitine and plasma total carnitine concentrations (*194,195*). This appears to occur at the level of butyrobetaine hydroxylase but is without direct inhibition, probably a result of reduced α-ketoglutarate levels, which are required as a cofactor *(197)*. This deficit may contribute to the drug's adverse effects, including hyperammonemia. Valproic acid treatment is associated with altered acylcarnitine subspecies that reflect impaired intermediary metabolism likely responsible for drug-induced toxicity *(198)*. Management of clinical deficiency required a significant dose of carnitine in children, in which case symptoms resolved within 1 week of the intervention *(195)*. An appropriate prophylactic dose has not been described. It has been suggested that oral L-carnitine supplementation be considered for patients with symptomatic valproic acid-associated hyperammonemia or for those with multiple risk factors for valproic acid hepatotoxicity as well as infants and children using valproic acid *(199)*. The recommended oral dose of L-carnitine is 100 mg/kg daily to a maximum of 2 g daily. Intravenous administration of L-carnitine is recommended for patients with valproic acid-induced hepatotoxicity or other acute metabolic crises associated with carnitine deficits *(199)*. Supplementation may not be needed in patients receiving valproic acid who are otherwise healthy and ingest a regular diet *(200)*.

4.6.3.2. Lipid-Soluble Vitamins

4.6.3.2.1. Vitamin D. Several drug classes are associated with osteomalacia, osteopenia, and osteoporosis. These include antiepileptic drugs, highly active antiretroviral therapy, corticosteroids, and cimetidine. Patients treated with antiepileptics may have reduced bone mass in the absence of obvious vitamin D deficiency *(201)*. Several mechanisms have been hypothesized to account for phenobarbital-induced osteomalacia. These include suppression of genes and enzymes involved in the hepatic 25-hydroxylation of vitamin D and the resultant increase in risk of bone loss and fracture with chronic therapy *(202)*. Phenobarbital, phenytoin, and carbamazepine can cause deficits by altering vitamin D metabolism through enzyme induction, thereby reducing serum 25(OH)-vitamin D_3 *(203–207)*. Determining bone mineral densities may be useful in detecting bone loss, especially in children receiving antiepileptic agents *(208)*; however, skeletal changes may be identified only by bone biopsy *(209)*. One pediatric study supplemented patients receiving chronic antiepileptics with vitamin D_3 (400 IU = 10 μg/day); a significant difference in bone mineral density was not detected between patients on monotherapy and polytherapy *(210)*. Elevated alkaline phosphatase in patients who had low serum 25(OH)-vitamin D_3 while receiving carbamazepine normalized following vitamin D supplementation *(211)*. However, the use of vitamin D supplementation may not be warranted in all patients receiving carbamazepine *(212)*. Nevertheless, monitoring vitamin D status is justified.

Interference with vitamin K metabolism may be a lesser factor in side effects of phenytoin and phenobarbital on bone *(213)*. Lastly, cimetidine inhibits 25-hydroxylase activity in the liver, thereby reducing vitamin D hydroxylation, but serum 25(OH)-vitamin D rises within 1 month of discontinuing therapy *(214,215)*. Effects of medication on vitamin D vary, which may partly result from the complexities of formation, activation, and metabolism, not to mention polymorphism of the vitamin D receptor and the vitamin D-binding protein.

A meta-analysis addressing the impact of HMG-CoA reductase inhibitors (e.g., atorvastatin, lovastatin, simvastatin) on bone health revealed that statin therapy was associated with a reduction in hip fracture, increased hip BMD, and minimal effects on bone biomarkers *(216)*. A group of 83 patients with acute coronary syndrome in whom 75% had vitamin D deficiency received atorvastatin for secondary prevention and their deficits improved after 12 months *(217)*. The role played by increasing availability of 7-dehydrocholesterol as the vitamin D precursor to 25(OH)-vitamin D is unclear.

Eliciting the majority of its effects through agonist activity at the vitamin D receptor, vitamin D metabolites are acted upon by CYP isoenzymes. The active metabolite, through binding to the vitamin D receptor, also induces the expression of CYP3A4, and to a lesser degree, CYP2B6 and CYP2C9 in human hepatocytes *(218)*. It is accepted that a number of drugs can induce osteomalacia, possibly by accelerating catabolism of the active form of vitamin D. Many of the drugs that result in osteomalacia (e.g., antiepileptics) may also activate the nuclear transcription factor pregnane X receptor (PXR). Activation of these receptors may increase the expression of the vitamin D receptor target gene, which would increase the breakdown of the active metabolite of vitamin D. A report identified this target to be CYP3A4, rather than CYP24, of which activation would increase degradation of active vitamin D and lead to drug-induced osteomalacia *(219)*.

Highly active antiretroviral therapy has been associated with loss of bone mass, although there are some conflicting data, and the infection itself may contribute to bone mineral loss *(220)*. In one study, for example, DEXA scans were performed and compared in antiretroviral-naïve patients, patients receiving protease inhibitors, and patients on non-protease inhibitor regimens. A high percentage of the patients, regardless of treatment allocation, were osteopenic. Osteoporosis was noted only in patients on antiretrovirals and a statistical difference was detected with regard to mean T-score and bone mineral density of the spine in patients receiving a protease inhibitor as compared to the other arms of the study. All three arms of the study, as a group, had significantly lower levels of 1,25$(OH)_2$-vitamin D_3 than did controls. The authors discuss that antiretroviral therapy, protease inhibitors, in particular, may effect the activity of enzymes essential in the synthesis of 1,25$(OH)_2$-vitamin D_3. Therefore, highly active antiretroviral therapy, and/or specifically protease inhibitors (e.g., indinavir, lopinavir, nelfinavir, ritonavir, saquinavir), may worsen the bone loss associated with HIV infection *(221)*. In patients treated with the nucleoside reverse transcriptase inhibitor tenofovir with a history of pathological bone fracture or at risk for osteopenia, monitoring 25(OH)-vitamin D and BMD should be considered. Tenofovir with lamivudine and efavirenz was compared to stavudine with lamuvidine and efavirenz in antiretroviral treatment-naïve patients *(222)*. No significant differences were noted between treatment with respect to incidence or progression of osteopenia or osteoporosis. From a meta-analysis, the prevalence of osteoporosis in patients with HIV was more than 3 times higher than in controls *(223)*, the prevalence of osteoporosis was 2 times greater in patients treated with antiretrovirals compared with antiretroviral-naïve patients, and protease inhibitor-containing regimens were 1.6 times more likely to be associated with osteoporosis compared to non-protease inhibitor regimens.

Steroid-induced osteoporotic fracture commonly manifests in the vertebrae with trabecular bone most often affected. Many mechanisms are thought to contribute to this adverse effect; such as inhibition of osteoblasts and decreased intestinal absorption of calcium resulting in secondary hyperparathyroidism. In a meta-analysis of five double-blind, placebo-controlled trials addressing calcium and vitamin D for corticosteroid-induced osteoporosis, the authors concluded that supplementation with both calcium and vitamin D is more effective than placebo or calcium alone in the slowing of bone loss in the forearm and lumbar spine *(224)*.

Although not often mentioned, most topically applied sunscreen drug products form a barrier to ultraviolet light that can reduce vitamin D formation in the skin *(225)*. Osteomalacia resulting from alterations in nutrient availability has been attributed to a number of medications beyond sunscreens. These include cholestyramine, phenytoin, phenobarbital, isoniazid, aluminum antacids, bisphosphonates, fluoride, isoniazid, and rifampin.

Isoniazid is noted to impair both hepatic and renal vitamin D metabolism *(226)*. However, the combination of isoniazid and rifampin did not appear to alter vitamin D metabolism over a 9-month treatment period *(227)*, despite a study in normal subjects in which a brief course of oral rifampin decreased circulating 25(OH)-vitamin D_3 levels by more than 50% *(228)*.

As vitamin K positively influences BMD and fracture risk, attention has focused on the influence of coumarin-based anticoagulants on bone health—specifically on osteocalcin. Adequate calcium and vitamin D status are necessary for vitamin K to be effective *(229)*. But it is still premature to make recommendations for routine vitamin K supplementation. Clearly further data are required to better evaluate the role of vitamin K in bone health and to clarify the influence of chronic warfarin anticoagulation *(229)*.

4.6.3.2.2. Others. The effect of vitamin A administration with BCG vaccine at birth on longer term vitamin A status was evaluated. Between 6 weeks and 4 months, it was noted that retinol-binding protein concentrations improved. It was observed that this increase was significantly and inversely correlated with the number of diphtheria–tetanus–pertussis vaccines during the interval; specifically in females and in those who received vitamin A as compared to placebo. In order to explain the results, the authors performed a subanalysis to correct for acute phase reactions leading to decreased retinol-binding protein; however, the result remained unaffected.

It is not currently known whether drugs that increase CYP activity alter the metabolism of vitamin E isomers or vitamin status given the relevance of this enzyme system to the initial steps in vitamin E metabolism *(230)*. The effect of vitamin E in reducing LDL peroxidation is attenuated when coadministered with atorvastatin but without influencing the drug's lipid lowering effect *(231)*.

4.6.4. Excretion of Vitamins

As discussed, there appears to be a depletion of folic acid and vitamin B_{12} with long-term use of metformin *(99)*. This depletion is also noted with long-term use of diuretics *(232)*. Diuretics may increase renal losses of thiamin. This reduction in thiamin may contribute to development of symptomatic heart failure, although these patients often have poor thiamin intake as well *(232)*. Patients with heart failure are likely to exhibit thiamin deficiency, which not only resolves with thiamin treatment but also improves left ventricular ejection fraction *(233,234)*.

Chlorpromazine and tricyclic antidepressants (e.g., amitriptyline) increase urinary excretion of riboflavin *(235,236)*. It is unclear whether this is the end result of decreased tissue distribution or the result of reduced metabolic incorporation. Flavin adenine dinucleotide production from riboflavin may be inhibited by these drugs, given the structural similarities *(237)*. Furthermore, tricyclic antidepressants may inhibit a number of riboflavin-dependent enzymes (e.g., nicotinamide adenine dinucleotide phosphate oxidase) involved in mitochondrial respiration in cardiac tissue. It has not been established whether this inhibition relates to the potential cardiotoxicity of these agents, which is otherwise attributed to their quinidine-like effect. However, the presence of an abundant amount of ubiquinone can overcome the difficulty posed by inhibition of those enzymes *(238)*. Riboflavin supplementation may be problematic because it may increase the autoinduction of amitriptyline.

Valproic acid, as well as pivalic acid—used to form salts that improve drug absorption, can inhibit renal OCTN2 in an animal model; hence a reduced ability of the kidney to reasbsorb L-carnitine may result. This could result in increased excretion of L-carnitine and changes in carnitine homeostasis *(239)*. As previously discussed, levofloxacin (and grepafloxacin - no longer on the U.S. market) can also inhibit OCTN2 *(112)*.

4.7. MINERALS

4.7.1. Absorption

The following discussion addresses the effect of medication on the status of electrolytes and trace elements (Tables 4.7 and 4.8). Alterations of mineral status may be more clinically significant in the presence of renal dysfunction. Osteomalacia may result from poor mineralization of the bone during deficits of calcium and phosphorus as well as vitamin D. Several drugs, including aluminum-containing antacids, cholestyramine, isoniazid, phenobarbital, phenytoin, and rifampin, may contribute to this adverse event by reducing absorption or altering distribution and metabolism *(240)*.

Aluminum-containing antacids bind phosphate and fluoride within the gut, thereby reducing the absorption of these minerals *(241)*. Formation of aluminum phosphates within the GI lumen accounts for the reduced phosphate absorption in the presence of aluminum-containing medication (e.g., antacids, sucralfate). In the past, this interaction was used as a therapeutic approach to manage hyperphosphatemia in patients with chronic kidney disease. Long-term use of these binders can exacerbate osteomalacia that results in part from aluminum accumulation as well as the loss of phosphate and calcium.

Through reduction of vitamin D, some of the antiepileptic drugs can decrease active intestinal calcium absorption, thereby placing patients receiving these medications at risk for osteomalacia in the absence of vitamin D and calcium supplementation *(242)*. Reduction in serum calcium, along with increased alkaline phosphatase, has been noted to occur following the first month of phenytoin therapy *(243)*. Hypocalcemia may be more severe with phenobarbital treatment than with phenytoin or carbamazepine *(206)*. These antiepileptics appear to induce hepatic vitamin D_3 metabolism and subsequently lead to a reduction in the intestinal absorption of calcium *(244)*. Valproic acid monotherapy has not been associated with hypocalcemia. Corticosteroids are also associated with loss of bone mass, in part resulting from reduced vitamin D-dependent calcium absorption *(245)*.

The H_2 antagonist cimetidine has the potential to lower serum calcium levels after several weeks of therapy without affecting parathyroid hormone (PTH) levels (perhaps through an influence on hepatic vitamin D 25-hydroxylase activity) *(214,215,246–248)*. As a result of decreasing gastric acid, this class of agents may also reduce the absorption of dietary iron and place patients at risk for iron deficiency *(101)*. However, this finding was not confirmed in patients with Zollinger–Ellison syndrome who were chronically treated with H_2 antagonists or proton pump inhibitors *(249)*. Managing this risk through iron supplementation may actually lead to decreased drug and iron absorption (at least in the case of cimetidine), because a drug–nutrient complex is formed; however, reduction in drug absorption may not be clinically relevant *(250,251)*. Absorption of zinc is also reduced in individuals who are treated with cimetidine *(252)*.

Table 4.7
Drugs That Alter Macro-mineral Status

Sodium

Hyponatremia

- Acetazolamide
- Amitriptyline
- Bromocriptine
- Carbamazepine
- Citalopram
- Chlorpropamide
- Clofibrate
- Cyclophosphamide
- Fluoxetine
- Fluvoxamine
- Hydrochlorothiazide
- Imipramine
- Metolazone
- Morphine
- Oxcarbazepine
- Oxytocin
- Paroxetine
- Pentamidine
- Sertraline
- Spironolactone
- Tolterodine
- Venlafaxine
- Vinblastine
- Vincristine

Hypernatremia

- Bumetanide
- Cidofoavir
- Colchicine
- Conivaptan
- Demeclocycline
- Ethacrynic acid
- Foscarnet
- Furosemide
- Hypertonic Na salts
- Lithium
- Methoxyflurane
- Phenytoin
- Torsemide
- Vinblastine

Potassium

Hypokalemia

- Acetazolamide
- Activated charcoal
- Albuterol
- Alvimopan
- Amiloride
- Ammonium chloride
- Amphotericin
- Ampicillin
- Aspirin
- β_2 Agonists
- Betamethasone
- Bisacodyl
- Bumetanide
- Carbenicillin
- Carboplatin
- Carmustine
- Chlorothiazide
- Chlorpropamide
- Chlorthalidone
- Cisplatin
- Colchicine
- Corticosteroids
- Corticotropin
- Cortisone
- Cyanocobalamin
- Cytarabine
- Desirudin
- Dexamethasone
- Dextrose
- Didanosine
- Digoxin immune fab
- Dobutamine
- Doxorubicin
- Ephedrine
- Epinephrine
- Ethacrynic acid
- Fluconazole
- Fluoxetine
- Foscarnet
- Furosemide
- Ganciclovir
- Gentamicin
- Hydrochlorothiazide
- Hydrocortisone
- Indapamide

Potassium

Hypokalemia (Contd.)

- Insulin
- Isoflurane
- Isoproterenol
- Isosorbide mononitrate
- Itraconazole
- Levalbuterol
- Levodopa/carbidopa
- Lithium
- Methylprednisolone
- Metolazone
- Mezlocillin
- Nafcillin
- Nifedipine
- Nylidrin
- Ondansetron
- Oxacillin
- Pamidronate
- Penicillin G
- Phosphates
- Piperacillin
- Polymyxin B
- Prednisolone
- Prednisone
- Pseudoephedrine
- Rifampin
- Risperidone
- Ritodrine
- Saline laxatives
- Salmeterol
- Sargramostim
- Sirolimus
- Sodium bicarbonate
- Sodium lactate
- Sodium polystyrene sulfonate
- Sorbitol
- Sotalol
- Tacrolimus
- Terbutaline
- Testosterone
- Theophylline
- Thiazides
- Ticarcillin
- Tobramycin
- Torsemide
- Triamcinolone
- Vincristine

Potassium
- *Hyperkalemia*
  - Aliskiren
  - Amiloride
  - Angiotensin blockers
  - Atenolol
  - β-Blockers
  - Benazepril
  - Candesartan
  - Captopril
  - Conivaptan
  - Cotrimoxazole
  - Cyclosporine
  - Digoxin
  - Enalapril
  - Eplerenone
  - Eprosartan
  - Fosinopril
  - Glucagon
  - Heparin
  - Ibuprofen
  - Indomethacin
  - Irbesartan
  - Lisinopril
  - Lithium
  - Losartan
  - Metoprolol
  - Naproxen
  - Olmesartan
  - Pentamidine
  - Perindopril
  - Potassium salts
  - Propranolol
  - Quinapril
  - Ramipril
  - Spironolactone
  - Succinylcholine
  - Tacrolimus
  - Telmisartan
  - Trandolapril
  - Trimethoprim
  - Triamterene
  - Valsartan

Magnesium
- *Hypomagnesemia*
  - Albuterol
  - Amikacin
  - Amphotericin B
  - Bumetanide
  - Carboplatin
  - Chlorothiazide
  - Cholestyramine
  - Cisplatin
  - Corticosteroids
  - Cyclosporine
  - Dextrose
  - Didanosine
  - Digoxin
  - Docusate
  - Estrogen
  - Ethacrynic acid
  - Ethanol
  - Foscarnet
  - Furosemide
  - Gentamicin
  - Hydrochlorothiazide
  - Insulin
  - Laxatives
  - Oral contraceptives
  - Pamidronate
  - Penicillamine
  - Pentamidine
  - Phosphates
  - Sargramostim
  - Sulfonamides
  - Tacrolimus
  - Tetracyclines
  - Torsemide
  - Zoledronic acid
- *Hypermagnesemia*
  - Lithium
  - Magnesium salts

Phosphorus
- *Hypophosphatemia*
  - Acetazolamide
  - Alendronate
  - Al–Mg antacids
  - Arginine
  - Calcitonin
  - Calcium salts
  - Carmustine
  - Cefotetan
  - Cholestyramine
  - Cisplatin
  - Demeclocycline
  - Dextrose
  - Digoxin
  - Erythropoietin
  - Ethanol
  - Felbamate
  - Foscarnet
  - Glucagon
  - Insulin
  - Magnesium
  - Osmotic diuretics
  - Pamidronate
  - Sevelamer
  - Sirolimus
  - Sucralfate
  - Tacrolimus
  - Zoledronic acid
- *Hyperphosphatemia*
  - Antileukemia regimens
  - Phosphate salts

(*Continued*)

Table 4.7
(Continued)

Calcium

Hypocalcemia

- Alendronate
- Amphotericin
- Antacids
- Bleomycin
- Bumetanide
- Calcitonin
- Carboplatin
- Cholestyramine
- Cimetidine
- Cisplatin
- Citrate salts
- Codeine
- Corticosteroids
- Corticotropin
- Cyclosporine
- Cytarabine
- Daunorubicin
- Didanosine
- Diethylstilbestrol
- Digoxin
- Doxorubicin
- Edetate disodium
- Estrogens
- Ethacrynic acid
- Etidronate
- Famotidine
- Fluocortolone
- Fluoride
- Fluorouracil
- Foscarnet
- Furosemide
- Gentamicin
- Interferon
- Isoniazid
- Ketoconazole
- Lansoprazole
- Leucovorin
- Magnesium
- Mineral oil
- Mithramycin
- Nizatadine
- Pamidronate
- Pentamidine
- Pentobarbital

Calcium

Hypocalcemia (Contd.)

- Phenobarbital
- Phenytoin
- Phosphates
- Polymyxin B
- Propylthiouracil
- Ranitidine
- Rituximab
- Saline laxatives
- Sargramostim
- Sodium polystyrene
- Sulfonamides
- Terbutaline
- Tetracyclines
- Tobramycin
- Torsemide
- Triamterene
- Zoledronic acid

Hypercalcemia

- Al–Mg antacids
- Calcium salts
- Ganciclovir
- Hydrochlorothiazide
- Lithium
- Tamoxifen
- Theophylline
- Vitamin A
- Vitamin D

Table 4.8
Drugs That Alter Trace Mineral Status

Chromium	*Iron*	*Zinc*
Corticosteroids	Aspirin	Amiloride
Copper	Bisphosphonates	Antivirals
Antacids	Calcium	Bumetanide
Antivirals	Carbidopa	Captopril
Cimetidine	Cefdinir	Cholestyramine
Ciprofloxacin	Cholestyramine	Cimetidine
Clofibrate	Ciprofloxacin	Ciprofloxacin
Corticosteroids	Colesevelam	Colesevelam
Ethambutol	Colestipol	Colestipol
Famotidine	Darbepoetin	Corticosteroids
NSAIDs	Deferasirox	Deferiprone
Oral contraceptives	Deferoxamine	Edetate calcium disodium
Penicillamine	EDTA	Enalapril
Trientine	Erythropoietin	Estrogen
Valproic acid	Ethanol	Ethambutol
Zidovudine	Indomethacin	Folic acid
Zinc salts	Levodopa	Furosemide
	Neomycin	Hydrochlorothiazide
	Penicillamine	NSAIDs
	Sulfonamides	Oral contraceptives
	Tetracyclines	Penicillamine
	Trientine	Tetracycline
	Selenium	Torsemide
	Clozapine	Triamterene
	Corticosteroids	Valproic acid
	Oral contraceptives	Zidovudine
	Valproic acid	

Cholestyramine may negatively influence the status of magnesium, calcium, iron, and zinc *(253)*. The absorption of multivalent cations (e.g., calcium, iron, magnesium) may be reduced through chelation with the tetracycline antibiotics *(254)*. Malabsorption of electrolytes can occur as a result of the GI effects of colchicine, and other medication that adversely increase gut motility including and laxatives.

4.7.2. Distribution

Disorders of serum sodium concentration must be evaluated in conjunction with volume status. For example, although a number of drugs can cause sodium retention (e.g., NSAIDs, corticosteroids, estrogens), there may not be an influence on serum sodium concentrations because of simultaneous increases in water retention; however, edema may be evident once retention is significant. Thiazide diuretics, selective serotonin reuptake inhibitors, and others are more likely to cause hyponatremia in the elderly as a result of impaired water excretion.

Several medications including tricyclic antidepressants, serotonin reuptake inhibitors, carbamazepine and oxcarbazepine, vincristine and vinblastine, may cause hyponatremia (especially in the elderly) through an influence on antidiuretic hormone *(255–265)*. Other medications, including lithium therapy, can cause hypernatremia as a consequence of decreased renal responsiveness to antidiuretic hormone (i.e., nephrogenic diabetes insipidus) *(266–268)*. In either case, the alteration in serum sodium concentration is related to a disorder of water regulation, which either dilutes (hyponatremia) or concentrates (hypernatremia) the sodium content of the serum. Therefore, management of these disorders includes water restriction (hyponatremia) or water replacement (hypernatremia). If the hypernatremia associated with lithium is managed with sodium restriction rather than water replacement, the risk of lithium toxicity is actually increased as the drug is reabsorbed renally in place of sodium.

Intracellular concentrations of potassium and magnesium may be reduced in patients who are treated with diuretics *(269)*. This tissue depletion can occur without a concomitant reduction in the serum concentration and despite supplementation *(270)*. Magnesium deficits are especially important in light of the poor dietary intake of magnesium among adults *(271)*. On the other hand, thiazide diuretics, in particular, allow calcium retention, which may even improve bone mineral content in chronic users *(272)*.

Patients with tumor-related hypercalcemia can be treated with pamidronate to lower serum calcium levels. However, hypocalcemia may occur in about one-fourth of patients who are treated with pamidronate, although the patients are usually asymptomatic *(273)*. By inhibiting bone resorption, bisphosphonates (e.g., pamidronate) may cause or exacerbate uncorrected hypocalcemia in the presence of vitamin D deficits or in the absence of adequate calcium and vitamin D intake *(119)*.

Estrogen use can cause a decrease in serum magnesium which has been suggested to result from intracellular shifting of the ion *(274,275)*. Low dietary magnesium intake in women using oral contraceptives or estrogen replacement, particularly in the presence of calcium supplementation, may increase the risk of adverse drug effects, including thrombosis *(276)*. The influence of oral contraceptives on copper and zinc are widely described, but the mechanism for this effect remains unclear *(92,193)*. Both whole blood and plasma levels of copper increase in patients using oral contraceptives, whereas the data regarding decrease of zinc concentrations is much less consistent *(277–279)*.

Corticosteroids can cause a dose-dependent, transient decrease in serum zinc concentrations following an initial rise *(280)*. No decrease was identified 48 h following an intravenous dose. A slower decline in serum copper levels that is attributed to corticosteroids may last as long as 96 h after a dose *(280)*. Serum selenium levels have been noted to increase during corticosteroid treatment *(281)*. The clinical significance of the redistribution of these minerals with corticosteroid administration is not clear but may play a role in the immunomodulating effects of corticosteroids.

Patients treated with zidovudine were noted to have reduced serum copper and zinc levels compared to similar patients who were not receiving this antiretroviral medication *(282)*. Although it has only been investigated in an animal model, the drug ethambutol may decrease distribution of copper to the liver and heart, and distribution of zinc to the liver and kidneys *(283)*.

Beta-2 adrenergic receptors have a profound stimulatory effect on Na^+/K^+-ATPase so that affecting these receptors may lead to transient transcellular shifts of potassium, thereby altering intracellular/extracellular potassium ratio. As a result, medications with β_2-agonist activity, like albuterol, pseudoephedrine, and theophylline, cause a shift of potassium into the cells that may lead to hypokalemia. Likewise, medications with β_2-antagonist activity, such as non-selective β-blockers, can bring about a shift of potassium out of the cell and may result in hyperkalemia *(284)*. Succinylcholine can also cause hyperkalemia through an extracellular shift.

4.7.3. *Excretion*

4.7.3.1. Electrolytes

Electrolyte loss may occur through the gastrointestinal tract or the kidneys. Sodium, potassium, and water depletion following chronic use of laxatives has been well recognized for years as the so-called "laxative abuse syndrome," but still continues to occur in patients *(285,286)*. Elevated levels of renin and aldosterone respond well to potassium, sodium, and fluid repletion *(287)*. Prophylactic potassium supplementation may be beneficial in patients requiring bowel regimens who may have a history of cardiovascular disease *(288)*. Of course, appropriate use of laxatives is the best management approach in this case. Besides increasing the risk for hyperphosphatemia, phosphate enemas can cause hypocalcemia and hypomagnesemia as well as hypokalemia *(289–291)*. The phosphate load of the concentrated product is excessive—particularly for infants, the elderly, and others with impaired renal function.

4.7.3.1.1. Cardiopulmonary. A number of drugs can increase renal losses of electrolytes (e.g., antimicrobials, chemotherapy, diuretics). The pharmacology of most diuretics involves the renal excretion of sodium and water. Loop diuretics increase loss of calcium in the urine as PTH levels increase *(292)*. Loop diuretics also increase renal wasting of potassium and magnesium *(293,294)*; the latter likely occurs through an inhibition of passive magnesium absorption *(295)*. Magnesium wasting occurs with a number of other drugs as well *(296)*. Several anticancer medications are inherently nephrotoxic and lead to electrolyte wasting that accompanies kidney damage. Cisplatin and carboplatin commonly result in hypomagnesemia related to their cumulative dose, while carmustine and azacitidine often cause hypophosphatemia and hypokalemia *(297)*. There is sufficient evidence that theophylline may produce clinically significant hypomagnesemia at toxic and possibly even at therapeutic levels. Total body magnesium deficits may exist in the face of serum magnesium concentrations within the normal range. In fact, unrecognized drug-induced hypomagnesemia may be responsible for refractory episodes of hypokalemia and hypocalcemia. Excessive loss of electrolytes through diuresis increases the risk for arrhythmia and sudden death in patients treated with these agents *(298,299)*.

Aldosterone also plays a major part in potassium homeostasis. There are many medications that affect the renin–angiotensin–aldosterone system and may lead to hyperkalemia. ACE inhibitors and angiotensin receptor blockers disrupt the effects of angiotensin II *(284)*. NSAIDs indirectly suppress renin and aldosterone by inhibiting prostaglandin synthesis *(300)*. Unfractionated heparin and possibly low molecular weight heparins inhibit aldosterone synthesis in the adrenal gland *(301)*. Spironolactone blocks aldosterone at the receptor site *(302)*. Finally, drospirenone, a novel progestin, demonstrates aldosterone antagonist properties *(303)*. Conversely, fludrocortisone, an oral agent with strong mineralocorticoid activity, directly causes potassium secretion *(304)*.

4.7.3.1.2. Antimicrobials. Patients with HIV infection often experience electrolyte abnormalities. However, with the introduction of HAART therapy the causes of these abnormalities have dramatically shifted from opportunistic infections to the antiretroviral therapy itself. Tenofovir has been associated with a Fanconi-like syndrome that leads to both hypophosphatemia and hypokalemia *(305,306)*. Foscarnet can result in hypokalemia, hypomagnesemia, hypocalcemia, and hypophosphatemia *(296,307,308)*. The effect of foscarnet on calcium and phosphorus may be minimized when the drug is administered as a liposome-encapsulated formulation *(309)*. Foscarnet binding to ionized calcium may explain some of the abnormalities *(310)*.

The penicillins (e.g., ampicillin, carbenicillin, nafcillin, oxacillin, ticarcillin) are reported to cause hypokalemia, probably because of increased potassium secretion resulting from the solute load (nonresorbable anion) presented to the distal renal tubules *(311–315)*. Hypocalcemia, hypokalemia, and hypomagnesemia have been noted in patients treated with aminoglycoside antibiotics (i.e., amikacin, gentamicin, tobramycin) *(316)*. Treatment with the antifungal agent amphotericin is known to result in renal depletion of potassium and magnesium. Hypokalemia, in turn, may potentiate the tubular toxicity of amphotericin *(317)*. This may be less likely in patients treated with liposomal formulations of this drug *(318)*.

4.7.3.1.3. Immunosuppressants. Corticosteroids—administered orally or parenterally—increase the risk of hypokalemia in patients receiving diuretics *(319)*. Urinary potassium excretion increases with dose and duration of corticosteroid therapy *(320)*. Mineralocorticoids stimulate renal tubular potassium secretion directly, whereas glucocorticoids do so indirectly through the sodium load presented to the distal tubule *(321)*. Corticosteroids may also increase renal magnesium losses, resulting in hypomagnesemia *(322)*. Corticosteroids can decrease intestinal calcium absorption, increase renal calcium excretion, and decrease osteoblast activity, all of which contribute to osteoporosis. This drug-induced osteoporosis may partly depend on the dose and duration of corticosteroid use as well as bone mineral density prior to initiation of the drug regimen. Supplementation with vitamin D and calcium can prevent some of the increases in PTH and loss of bone mineral density that occur with corticosteroid therapy *(323,324)*. The altered calcium absorption may vary among glucocorticoids through a mechanism that apparently does not involve 1,25$(OH)_2$-vitamin D_3 *(325)*. Immunosuppressants other than corticosteroids can cause electrolyte disturbances. Cyclosporine and tacrolimus can cause increased renal magnesium losses and possible potassium losses as well *(326,327)*. The proton pump inhibitor, lansoprazole, has been reported to cause severe, symptomatic hypocalcemia through an unknown mechanism *(328)*.

4.7.3.2. Trace Minerals

Loss of blood at the GI tract is higher than physiologic loss as a result of aspirin or other NSAID therapy and if chronic may increase the risk of developing iron deficiency *(329,330)*. Aspirin use has been associated with lower serum ferritin concentrations in some patients, although confounding factors could have played a role *(331)*. Sirolimus may interfere with iron homeostasis *(332)*.

Captopril and, to a lesser extent, enalapril can increase urinary zinc excretion after several months of treatment in patients with hypertension *(333,334)*. This is not reflected by serum values of zinc, which remain unchanged *(335)*. No documentation exists regarding whether this will lead to clinically apparent deficits, but it may play a role in the altered taste experienced by patients receiving this class of medication. Diuretics (including the thiazides and chlorthalidone) also increase urinary zinc losses *(336,337)*. Again, serum levels remain normal in most patients, although this is expected as a result of redistribution from the intracellular space.

Penicillamine is used to lower serum copper concentrations in patients with Wilson's disease by increasing renal elimination of a drug–mineral complex. Although penicillamine is useful in Wilson's disease, this nutrient loss may explain the drug's teratogenic effect *(338)*. Penicillamine also has the potential to lower levels of other nutrients. Loss of zinc may be increased, but iron, calcium, and magnesium losses are not readily apparent *(339)*. Circulating zinc levels improve within a few months of therapy; it is not clear whether this is a recovery of zinc status or merely redistribution from other compartments *(340)*. Supplementation of magnesium, zinc, and pyridoxine has been suggested for patients receiving penicillamine *(341)*. Zidovudine and valproic acid may also reduce serum copper concentrations *(342)*.

Corticosteroids may increase chromium losses and, therefore, may contribute to the hyperglycemia associated with these agents *(343)*. Corticosteroids may also increase urinary loss of selenium *(344)*.

Full appreciation of the clinical relevance of drug-induced losses of the trace minerals will require additional study.

4.8. SUMMARY AND RECOMMENDATIONS

The clinical relevance of any drug-induced changes to the status of a specific nutrient or overall nutritional status will be determined on an individual basis. It may be viewed as a finding of little value, or may help to explain debilitating adverse effects associated with chronic use of a medication. Aside from the few widely recognized interactions, most of these associations are infrequently made in clinical practice. Many of these medications may not cause clinically significant nutritional deficiencies when administered at therapeutic doses. Yet, when combined with certain comorbidities or concomitant medications, they may cause additive deleterious effects that lead to clinically significant nutritional deficiencies. Therefore, everyone requiring medication should have a thorough nutritional assessment performed at baseline and periodically during chronic treatment.

A nutritionally focused patient history, physical examination, along with appropriate laboratory marker evaluations is important to correctly identify nutrient deficits or excesses in patients using medication. All clinicians should be aware of nutritional aspects of drug use. When a change in nutritional status is recognized by a practitioner the work-up should include a drug-induced etiology. Any abnormalities in nutritional status or status of a specific nutrient identified by this diligence should also be documented and considered for publication to expand on the currently available literature.

When a drug-induced change poses a significant nutritional concern for a patient, a therapeutically equivalent or alternative agent may be selected. For those drugs recognized to cause altered nutritional status, specific interventions should take place at the time the medication is started in order to prevent adverse effects from occurring. Current evidence does not support indiscriminant use of nutrient supplements to manage losses that may result from drug use. Indeed the dose of a nutrient needed to offset the adverse effects of many medications is not known. And in some instances pharmacologic nutrient dosing may increase drug clearance with clinical consequences, as in the case of folic acid and phenytoin *(345)*.

Ideally, the approval process of new drugs should evaluate for effects on nutritional status. Clinicians should be sensitized to identify and report such otherwise unrecognized effects of drugs and develop guidelines to address altered nutritional status that take into account the most appropriate test(s) of status. Prospective study of methods to correct nutritional status should evaluate not just parameters of nutrient status, but influence on clinical manifestations. In this way, patients required to use medication can be managed optimally with regard to maintaining adequate nutritional status.

REFERENCES

1. Kaufman DW, Kelly JP, Rosenberg L, et al. Recent patterns of medication use in the ambulatory adult population of the U.S.: the Slone survey. JAMA 2002;287:337–344.
2. Lidell E, Luepker R, Baigi A, Lagiou A, Hildingh C. Medication usage among young adult women: a comparison between Sweden, the USA, and Greece. Nurs Health Sci 2008;10:4–10.
3. IMS Health, Inc. Global pharmaceutical sales by region, 2007. Available from: www.imshealth.com . Accessed July 2008.
4. Nutrition Business Journal. NBJ's supplement business report 2007: an analysis of markets, trends, competition and strategy in the U.S. dietary supplement industry. October 2007. Available from: www.nbj.stores.yahoo.net/nbjsubure20.html. Accessed July 2008.
5. Centers for Medicare & Medicaid Services, Office of the Actuary. National health expenditure projections. Available from: http://www.cms.hhs.gov/NationalHealthExpendData/Downloads/proj2007.pdf . Accessed July 2008.

6. Hoffman JM, Shah ND, Vermeulen LC, et al. Projecting future drug expenditures, 2008. Am J Health Syst Pharm 2008;65:234–253.
7. Salazar JA, Poon I, Nair M. Clinical consequences of polypharmacy in elderly: expect the unexpected, think the unthinkable. Expert Opin Drug Saf 2007;6:695–704.
8. Kongkaew C, Noyce PR, Ashcroft DM. Hospital admissions associated with adverse drug reactions: a systematic review of prospective observational studies. Ann Pharmacother 2008;42:1017–1025.
9. Santos C, Boullata J. An approach to evaluating drug-nutrient interactions. Pharmacotherapy 2005;25:1789–1800.
10. Boullata JI. An introduction to drug–nutrient interactions. In: Boullata JI, Armenti VT (eds). Handbook of drug-nutrient interactions, 2nd edition. Totowa, NJ: Humana Press, 2009.
11. Evans WE. Differing effects of methylenetetrahydrofolate reductase single nucleotide polymorphism on methotrexate efficacy and toxicity in rheumatoid arthritis. Pharmacogenetics 2002;12:181–182.
12. Drozdzik M, Rudas T, Pawlik A, Gornik W, Kurzawski M, Herczynska M. Reduced folate carrier-1 80G>A polymorphism affects methotrexate treatment outcome in rheumatoid arthritis. Pharmacogenomics J 2007;7:404–407.
13. Rathman SC, Blanchard RK, Badinga L, et al. Dietary carbamazepine administration decreases liver pyruvate carboxylase activity and biotinylation by decreasing protein and mRNA expression in rats. J Nutr 2003;133:2119–2124.
14. Boullata J. Natural health product interactions with medication. Nutr Clin Pract 2005;20:33–51.
15. Chan L-N. Interaction of natural products with medication and nutrients. In: Boullata JI, Armenti VA (eds). Handbook of drug–nutrient interactions, 2nd edition. Totowa, NJ: Humana Press, 2009.
16. Roe DA. Drug-induced nutritional deficiencies, 2nd edition. Westport, CT: AVI Publishing Co., Inc., 1985.
17. Pelton R, LaValle JB, Hawkins EB, et al. (eds). Drug-induced nutrient depletion handbook. Second edition. Hudson, OH: Lexi-Comp, Inc., 2001.
18. Stargrove MB, Treasure J, McKee DL (eds). Herb, nutrient, and drug interactions: clinical implications and therapeutic strategies. St. Louis, MO: Mosby Elsevier, 2008.
19. Gervasio JM. Drug-induced changes to nutritional status. In: Boullata JI, Armenti VA (eds). Handbook of drug-nutrient interactions, 2nd edition. Totowa, NJ: Humana Press, 2009.
20. Matarese LE. Metabolic complications of parenteral nutrition therapy. In: Gottschlich MM, Fuhrman MP, Hammond KA, et al. (eds). The science and practice of nutrition support. Dubuque, IA: Kendall/Hunt Publishing Company, 2001:269–286.
21. Kumpf VJ, Gervasio J. Complications of parenteral nutrition. In: Gottschlich MM (ed.). The A.S.P.E.N. nutrition support core curriculum: a case-based approach; the adult patient. Silver Spring, MD: American Society for Parenteral and Enteral Nutrition, 2007:323–339.
22. Lieber CS. Alcohol: its metabolism and interaction with nutrients. Annu Rev Nutr 2000;20:395–430.
23. Seitz HK, Suter PM. Ethanol toxicity and nutritional status. In: Kotsonis FN, Mackey MA (eds). Nutritional toxicology, 2nd edition. London, UK: Taylor & Francis, 2002:122–154.
24. Lands WEM. Alcohol: the balancing act. In: Bendich A, Deckelbaum RJ (eds). Preventive nutrition, 3rd edition. Totowa, NJ: Humana Press, 2005:807–831.
25. Mowry JB, Furbee RB, Chyka PA. Poisoning. In: Chernow B (ed.). The pharmacologic approach to the critically ill patient, 3rd edition. Baltimore, MD: Williams & Wilkins, 1994:975–1008.
26. Kordas K, Lönnerdal B, Stoltzfus RJ. Interactions between nutrition and environmental exposures: effects on health outcomes in women and children. J Nutr 2007;137:2794–2797.
27. Biehl JP, Vilter RW. Effect of isoniazid on vitamin B6 metabolism: its possible significance in producing isoniazid neuritis. Proc Soc Exp Biol Med 1954;85:389–392.
28. Levy L, Higgins LJ, Burbridge TN. Isoniazid-induced vitamin B6 deficiency. Am Rev Resp Dis 1967;96:910–917.
29. Said HM, Redha R, Nylander W. Biotin transport in the human intestine: inhibition by anticonvulsant drugs. Am J Clin Nutr 1989;49:127–131.
30. Mock DM, Dyken ME. Biotin catabolism is accelerated in adults receiving long-term therapy with anticonvulsants. Neurology 1997;49:1444–1447.
31. Lambie DG, Johnson RH. Drugs and folate metabolism. Drugs 1985;30:145–155.
32. Race TF, Paes IC, Faloon WW. Intestinal malabsorption induced by oral colchicines: comparison with neomycin and cathartic agents. Am J Med Sci 1970;259:32–41.
33. Frequin ST, Wevers RA, Braam M, et al. Decreased vitamin B12 and folate levels in cerebrospinal fluid and serum of multiple sclerosis patients after high-dose intravenous methylprednisolone. J Neurol 1993;240:305–308.
34. Peretz AM, Neve JD, Vertongen F, et al. Selenium status in relation to clinical variables and corticosteroid treatment in rheumatoid arthritis. J Rheumatol 1987;14:1104–1107.
35. Ruze P. Kava-induced dermopathy: a niacin deficiency? Lancet 1990;335:1442–1445.

36. Adan RAH, Vanderschuren LJMJ, la Fleur SE. Anti-obesity drugs and neural circuits of feeding. Trend Pharmacol Sci 2008;29:208–217.
37. Wolfe WS, Frongillo EA, Valois P. Understanding the experience of food insecurity by elders suggests ways to improve its measurement. J Nutr 2003;133:2762–2769.
38. Umbricht D, Kane J. Medical complications of new antipsychotic drugs. Schizophr Bull 1996;22:475–483.
39. Vanina Y, Podalskaya A, Sedky K, et al. Body weight changes associated with psychopharmacology. Psych Serv 2002;53:842–847.
40. Pischon T, Sharma AM. Use of beta-blockers in obesity hypertension: potential role of weight gain. Obes Rev 2001;2:275–280.
41. American Diabetes Association. Consensus development conference on antipsychotic drugs and obesity and diabetes. Diabetes Care 2004;27:596–601.
42. Brixner DI, Said Q, Corey-Lisle PK, et al. Naturalistic impact of second generation antispychotics on weight gain. Ann Pharmacother 2006;40:626–632.
43. Mott AE, Leopold DA. Disorders of taste and smell. Med Clin North Am 1991;75:13231–13253.
44. Ackerman BH, Kasbekar N. Disturbances of taste and smell induced by drugs. Pharmacotherapy 1997;17:482–496.
45. Henkin RI, Schecter PJ, Friedewald WT, et al. A double-blind study of the effects of zinc sulfate on taste and smell dysfunction. Am J Med Sci 1976;272:285–299.
46. Tack J. Chemosensitivity of the human gastrointestinal tract in health and disease. Neurogastroenterol Motil 2007;19:241–244.
47. DiPiro CV, Taylor AT. Nausea and vomiting. In: DiPiro JT, et al. (eds). Pharmacotherapy: a pathophysiologic approach, 6th edition. New York: McGraw-Hill Medical Publishing Division, 2005:665–676.
48. Chisholm-Burns MA, Wells BG, Schwinghammer TL, et al. (eds). Pharmacotherapy: principles and practice. New York, NY: McGraw-Hill Medical, 2008.
49. Genser D. Food and drug interaction: consequences for the nutrition/health status. Ann Nutr Metab 2008;52 (Suppl. 1):29–32.
50. Spruill WJ, Wade WE. Diarrhea, constipation, and irritable bowel syndrome. In: DiPiro JT, et al. (eds). Pharmacotherapy: a pathophysiologic approach, 6th edition. New York: McGraw-Hill Medical Publishing Division, 2005: 677–692.
51. Berardi RR, Montgomery PA. Pancreatitis. In: DiPiro JT, et al. (eds). Pharmacotherapy: a pathophysiologic approach, 6th edition. New York: McGraw-Hill Medical Publishing Division, 2005:721–736.
52. Thalapaneni NR, Chidambaram KA, Ellappan T, Sabapathi ML, Mandal SC. Inhibition of carbohydrate digestive enzymes by *Talinum portulacifolium* (Forssk) leaf extract. J Complement Integr Med 2008;5:11.
53. Saltzman JR, Kemp JA, Golner BB, et al. Effect of hypochlorhydria due to omeprazole treatment or atrophic gastritis on protein-bound vitamin B12 absorption. J Am Coll Nutr 1994;13:584–591.
54. Termanini B, Gibril F, Sutliff VE, et al. Effect of long-term gastric acid suppressive therapy on serum vitamin B12 levels in patients with Zollinger-Ellison syndrome. Am J Med 1998;104:422–430.
55. Jacobson ED. Depletion of vitamin B12, iron, beta-carotene, and fat malabsorptive effects of neomycin in commonly used doses. JAMA 1961;175:187–190.
56. Carr A, Samaras K, Burton S, et al. A syndrome of peripheral lipodystrophy, hyperlipidemia and insulin resistance in patients receiving HIV protease inhibitors. AIDS 1998;12:F51–F58.
57. Stravitz RT, Sanyal AJ. Drug induced steatohepatitis. Clin Liver Dis 2003;7(2):435–451.
58. Levy C, Lindor KD. Drug-induced cholestasis. Clin Liver Dis 2003;7:311–330.
59. Pandit MK, Burke J, Gustafson AB, et al. Drug-induced disorders of glucose tolerance. Ann Intern Med 1993;118: 529–539.
60. Marks V, Teale JD. Drug-induced hypoglycemia. Endocrinol Metab Clin North Am 1999;28:555–577.
61. Luna B, Feinglos MN. Drug-induced hyperglycemia. JAMA 2001;286:1945–1948.
62. Citrome LL, Holt RIG, Zachary WM, et al. Risk of treatment-emergent diabetes mellitus in patients receiving antipsychotics. Ann Pharmacother 2007;41:1593–1603.
63. Buse JB, Cavazonni P, Hornbuckle K, et al. A retrospective cohort study of diabetes mellitus and antipsychotic treatment in the United States. J Clin Epidemiol 2003;56:164–170.
64. Gianfrancesco F, White R, Wang RH, et al. Antispychotic-induced type 2 diabetes: evidence from a large health plan database. J Clin Psychopharmacol 2003;23:328–335.
65. Lodise T, Graves J, Miller C, Mohr JF, Lomaestro B, Smith RP. Effects of gatifloxacin and levofloxacin on rates of hypoglycemia and hyperglycemia among elderly hospitalized patients. Pharmacotherapy 2007;27:1498–1505.
66. Mantel-Teeuwisse AK, Kloosterman JM, Maitland-van der Zee AH, et al. Drug-induced lipid changes: a review of the unintended effects of some commonly used drugs on serum lipid levels. Drug Saf 2001;24:443–456.

67. Kurt M, Babaoglu MO, Yasar U, Shorbagi A, Guler N. Capecitabine-induced severe hypertriglyceridemia: report of two cases. Ann Pharmacother 2006;40:328–331.
68. Brillon DJ, Zheng B, Campbell RG, et al. Effect of cortisol on energy expenditure and amino acid metabolism in humans. Am J Physiol 1995;268:E501–E513.
69. Ferrando AA, Stuart CA, Sheffield-Moore M, et al. Inactivity amplifies the catabolic response of skeletal muscle to cortisol. J Clin Endocrinol Metab 1999;84:3515–3521.
70. Strobl JS, Thomas MJ. Human growth hormone. Pharmacol Rev 1994;46:1–34.
71. Úbeda N, Alonso-Aperte E, Varela-Moreiras G. Acute valproate administration impairs methionine metabolism in rats. J Nutr 2002;132:2737–2742.
72. Slap GB. Oral contraceptives and depression: impact, prevalence and cause. J Adolesc Health Care 1981;2:53–64.
73. Noda S, Haratake J, Sasaki A, et al. Acute encephalopathy with hepatic steatosis induced by pantothenic acid antagonist, calcium hopantenate, in dogs. Liver 1991;11:134–142.
74. Kamath AV, Darling IM, Morris ME. Choline uptake in human intestinal Caco-2 cells is carrier-mediated. J Nutr 2003;133:2607–2611.
75. Said HM. Recent advances in carrier-mediated intestinal absorption of water-soluble vitamins. Annu Rev Physiol 2004;66:419–446.
76. Zimmerman J. Drug interactions in intestinal transport of folic acid and methotrexate: further evidence for the heterogeneity of folate transport in the human small intestine. Biochem Pharmacol 1992;44:1839–1842.
77. Chungi VS, Bourne DW, Dittert LW. Competitive inhibition between folic acid and methotrexate for transport carrier in the rat small intestine. J Pharm Sci 1979;68:1552–1553.
78. Leeb BF, Witzmann G, Orgis E, et al. Folic acid and cyanocobalamin levels in serum and erythrocytes during low-dose methotrexate therapy for rheumatoid arthritis and psoriatic arthritis patients. Clin Exp Rheumatol 1995;13:459–463.
79. Jansen G, van der Heijden J, Oerlemans R, et al. Sulfasalazine is a potent inhibitor of the reduced folate carrier: implications for combination therapies with methotrexate in rheumatoid arthritis. Arthritis Rheum 2004;50: 2130–2139.
80. Dijkmans BA. Folate supplementation and methotrexate. Br J Rheumatol 1995;34:1172–1174.
81. Morgan SL, Baggott JE, Lee JY, et al. Folic acid supplementation prevents deficient blood folate levels and hyperhomocysteinemia during long term, low dose methotrexate therapy for rheumatoid arthritis: implications for cardiovascular disease prevention. J Rheumatol 1998;25:441–446.
82. Corcino J, Waxman S, Herbert V. Mechanisms of triamterene-induced megaloblastosis. Ann Intern Med 1970;73: 419–424.
83. Joosten E, Pelemans W. Megaloblastic anaemia in an elderly patient treated with triamterene. Meth J Med 1991;38:209–211.
84. Zimmerman J, Selhub J, Rosenberg IH. Competitive inhibition of folic acid absorption in rat jejunum by triamterene. J Lab Clin Med 1986;108:272–276.
85. Mason JB, Zimmerman J, Otradovec CL, et al. Chronic diuretic therapy with moderate doses of triamterene is not associated with folate deficiency. J Lab Clin Med 1991;117:365–369.
86. Russell DM, Dutta SK, Rosenberg IH, et al. Impairment of folic acid by oral pancreatic extracts. Dig Dis Sci 1980;25:369–373.
87. West RJ, Lloyd JK. The effect of cholestyramine on intestinal absorption. Gut 1975;16:93–98.
88. Hoppner K, Lampi B. Bioavailability of folate following ingestion of cholestyramine in the rat. Int J Vitam Nutr Res 1991;61:130–134.
89. Hamid A, Wani NA, Rana S, Vaiphei K, Mahmood A, Kaur J. Down-regulation of reduced folate carrier may result in folate malabsorption across intestinal brush border membrane during experimental alcoholism. FEBS J 2007;274:6317–6328.
90. Hamid A, Kaur J, Mahmood A. Evaluation of the kinetic properties of the folate transport system in intestinal absorptive epithelium during experimental ethanol ingestion. Mol Cell Biochem 2007;304:265–271.
91. Faloon WW, Chodos RB. Vitamin B12 absorption studies using colchicine, neomycin and continuous 57Co B12 administration. Gastroenterology 1969;56:1251.
92. Webb DI, Chodos RB, Mahar CQ, et al. Mechanism of vitamin B12 malabsorption in patients receiving colchicines. N Engl J Med 1968;279:845–850.
93. Stopa EG, O'Brien R, Katz M. Effect of colchicine on guinea pig intrinsic factor-vitamin B12 receptor. Gastroenterology 1979;76:309–314.
94. Ehrenfeld M, Levy M, Sharon P, et al. Gastrointestinal effects of long-term colchicine therapy in patients with recurrent polyserositis. Dig Dis Sci 1982;27:723–727.

95. Paltiel O, Falutz J, Veilleux M, et al. Clinical correlates of subnormal vitamin B12 levels in patients infected with the human immunodeficiency virus. Am J Hematol 1995;49:318–322.
96. Adams JF, Clark JS, Ireland JT, et al. Malabsorption of vitamin B12 and intrinsic factor secretion during biguanide therapy. Diabetologia 1983;24:16–18.
97. Berger W. Incidence of severe side effects during therapy with sulfonylureas and biguanides. Horm Metab Res Suppl 1985;15:111–115.
98. Ting RZW, Szeto CC, Chan MHM, Ma KK, Chow KM. Risk factors of vitamin B12 deficiency in patients receiving metformin. Arch Intern Med 2006;166:1975–1979.
99. Carlsen SM, Folling I, Grill V, et al. Metformin increases total serum homocysteine levels in nondiabetic male patients with coronary heart disease. Scan J Clin Lab Invest 1997;57:521–527.
100. Salom IL, Silvis SE, Doscherholmen A. Effect of cimetidine on the absorption of vitamin B12. Scan J Gastroenterol 1982;17:129–131.
101. Aymard JP, Aymard B, Netter P, et al. Haematological adverse effects of histamine H2-receptor antagonists. Med Toxicol Adverse Drug Exp 1988;3:430–448.
102. Force RW, Nahata MC. Effect of histamine H2-receptor antagonists on vitamin B12 absorption. Ann Pharmacother 1992;26:1283–1286.
103. Festen HP. Intrinsic factor secretion and cobalamin absorption: physiology and pathophysiology in the gastrointestinal tract. Scand J Gastroenterol 1991;188:1–7.
104. Dharmarajan TS, Kanagala MR, Murakonda P, Lebelt AS, Norkus EP. Do acid-lowering agents affect vitamin B12 status in older adults? J Am Med Dir Assoc 2008;9:162–167.
105. Howden CW. Vitamin B12 levels during prolonged treatment with proton pump inhibitors. J Clin Gastroenterol 2000;30:29–33.
106. Schenk BE, Festen HP, Kuipers EJ, et al. Effect of short- and long-term treatment with omeprazole on the absorption and serum levels of cobalamin. Aliment Pharmacol Ther 1996;10:541–545.
107. Ruscin JM, Page RL, Valuck RJ. Vitamin B12 deficiency associated with histamine2-receptor antagonists and a proton-pump inhibitor. Ann Pharmacother 2002;36(5):812–816.
108. Tang G, Serfaty-Lacrosniere C, Camilo ME, et al. Gastric acidity influences the blood response to a beta-carotene dose in humans. Am J Clin Nutr 1996;64:622–626.
109. Deguchi Y, et al. Comparative studies on synthesis of water-soluble vitamins among human species of Bifidobacteria. Agric Biol Chem 1985;19:13–19.
110. Stevens CE, Hume ID. Contributions of microbes in vertebrate gastrointestinal tract to production and conservation of nutrients. Physiol Rev 1998;78:393–427.
111. Hill MJ. Intestinal flora and endogenous vitamin synthesis. Eur J Cancer Prev 1997;6(Suppl. 1):S43–S45.
112. Hirano T, Yasuda S, Osaka Y, Kobayashi M, Itagaki S, Iseki K. Mechanism of the inhibitory effect of zwitterionic drugs (levofloxacin and grepafloxacin) on carnitine transporter (OCTN2) in Caco-2 cells. Biochim Biophys Acta 2006;1758:1743–1750.
113. Tune BM, Hsu CY. Toxicity of cephaloridine to carnitine transport and fatty acid metabolism in rabbit renal cortical mitochondria: structure-activity relationships. J Pharmacol Exp Ther 1994;270:873–878.
114. Becker GL. The case against mineral oil. Am J Dig Dis 1953;19:344–347.
115. Hathcock JN. Metabolic mechanisms of drug-nutrient interactions. Fed Proc 1985;44(1Pt1):124–129.
116. Elinder LS, Hadell K, Johansson J, et al. Probucol treatment decreases serum concentrations of diet-derived antioxidants. Arterioscler Thromb Vasc Biol 1995;15:1057–1063.
117. Knodel LC, Talbert RL. Adverse effects of hypolipidemic drugs. Med Toxicol 1987;2:10–32.
118. Hoogwerf BJ, Hibbard DM, Hunninghake DB. Effects of long-term cholestyramine administration on vitamin D and parathormone levels in middle-aged men with hypercholesterolemia. J Lab Clin Med 1992;119:407–411.
119. Rosen CJ, Brown S. Severe hypocalcemia after intravenous bisphosphonate therapy in occult vitamin D deficiency. N Engl J Med 2003;348:1503–1504.
120. Finer N, James WP, Kopelman PG, et al. One-year treatment of obesity: a randomized, double-blind, placebo-controlled, multicentre study of orlistat, a gastrointestinal lipase inhibitor. Int J Obes 2000;24:306–313.
121. Melia AT, Koss-Twardy SG, Zhi J. The effect of orlistat, an inhibitor of dietary fat absorption, on the absorption of vitamins A and E in healthy volunteers. J Clin Pharmacol 1996;36:647–653.
122. McDuffie JR, Calis KA, Booth SL, et al. Effects of orlistat on fat-soluble vitamins in obese adolescents. Pharmacotherapy 2002;22:814–822.
123. Zhi J, Melia AT, Koss-Twardy SG, et al.The effect of orlistat, an inhibitor of dietary fat absorption, on the pharmacokinetics of beta-carotene in healthy volunteers. J Clin Pharmacol 1996;36:152–159.

124. Conly J, Stein K. Reduction of vitamin K2 concentrations in human liver associated with the use of broad spectrum antimicrobials. Clin Invest Med 1994;17:531–539.
125. Lipsky JJ. Antibiotic-associated hypoprothrombinemia. J Antimicrob Chemother 1988;21:281–300.
126. Alter HJ, Zvaifler NJ, Rath CE. Interrelationship of rheumatoid arthritis, folic acid, and aspirin. Blood 1971;38: 405–416.
127. Lawrence VA, Loewenstein JE, Eichner ER. Aspirin and folate binding in vivo and in vitro studies of serum binding and urinary excretion of endogenous folate. J Clin Lab Med 1984;103:944–948.
128. Goggin T, Gough H, Bissessar A, et al. A comparative study of the relative effects of anticonvulsant drugs and dietary folate on the red cell folate status of patients with epilepsy. Q J Med 1987;65:911–919.
129. Latham J, Gill DS, Wickramasinghe SN. Effects of phenytoin sodium on doubling time, deoxyuridine suppression, 3H-methotrexate uptake and 57 co-cyanocobalamin uptake in HL60 cells. Clin Lab Haematol 1990;12: 67–75.
130. Botez MI, Joyal C, Maag U, et al. Cerebrospinal fluid and blood thiamine concentrations in phenytoin-treated epileptics. Can J Neurol Sci 1982;9:37–39.
131. Patrini C, Perucca E, Reggiani C, et al. Effects of phenytoin on the in vivo kinetics of thiamine and its phosphoesters in rat nervous tissues. Brain Res 1993;628:179–186.
132. Zangen A, Botzer D, Zangen R, et al. Furosemide and digoxin inhibit thiamine uptake in cardiac cells. Eur J Pharmacol 1998;361:151–155.
133. Levine MA, Pollard HB. Hydrocortisone inhibition of ascorbic acid transport by chromaffin cells. FEBS Lett 1983;158:134–138.
134. Sahud MA, Cohen RJ. Effect of aspirin ingestion on ascorbic acid levels in rheumatoid arthritis. Lancet 1971;1: 937–938.
135. Loh HS, Watters K, Wilson CW, et al. The effects of aspirin on the metabolic availability of ascorbic acid in human beings. J Clin Pharmacol 1973;13:480–486.
136. Coffey G, Wilson CWM. Ascorbic acid deficiency and aspirin induced hematemesis. BMJ 1975;1:208.
137. Dalakas MC, Leon-Monzon ME. Zidovudine-induced mitochondrial myopathy is associated with muscle carnitine deficiency and lipid storage. Ann Neurol 1994;35:482–487.
138. Tein I, DimAuro S, Xie ZW, et al. Valproic acid impairs carnitine uptake in cultured human skin fibroblasts: an in vitro model for pathogenesis of valproic acid-associated carnitine deficiency. Pediatr Res 1993;34:281–287.
139. Wu SP, Shyu MK, Liou HH, Gau CS, Lin CJ. Interaction between anticonvulsants and human placental carnitine transporter. Epilepsia 2004;45:204–210.
140. Aberg F, Appelkvist EL, Broijersen A, et al. Gemfibrozil-induced decrease in serum ubiquinone and alpha- and gamma-tocopherol levels in men with combined hyperlipidaemia. Eur J Clin Invest 1998;28:235–242.
141. Folkers K, Langsjoen P, Willis R, et al. Lovastatin decreases coenzyme Q levels in humans. Proc Natl Acad Sci USA 1990;87:8931–8934.
142. Mortensen SA, Leth A, Agner E, et al. Dose-related decrease of serum coenzyme Q10 during treatment with HMG-CoA reductase inhibitors. Mol Aspects Med 1997;18(Suppl.):S137–144.
143. Ghirlanda G, Oradei A, Manto A, et al. Evidence of plasma CoQ10 lowering effect by HMG-CoA reductase inhibitors: a double-blind, placebo-controlled study. J Clin Pharmacol 1993;33:226–229.
144. DePinieux G, Chariot P, Ammi-Said M, et al. Lipid lowering drugs and mitochondrial function: effects of HMG-CoA reductase inhibitors on serum ubiquinone and blood lactate/pyruvate ratio. Br J Clin Pharmacol 1996;42:333–337.
145. Human JA, Ubbink JB, Jerling JJ, et al. The effect of simvastatin on the plasma antioxidant concentrations in patients with hypercholesterolaemia. Clin Chim Acta 1997;263:67–77.
146. Bargossi AM, Grossi G, Fiorella PL, et al. Exogenous CoQ10 supplementation prevents plasma ubiquinone reduction by HMG-CoA reductase inhibitors. Mol Aspects Med 1994;15(Suppl.):187–193.
147. Shimizu T, Maeda S, Mochizuki H, et al. Theophylline attenuates circulating vitamin B6 levels in children with asthma. Pharmacol 1994;49:392–397.
148. Shimizu T, Maeda S, Arakawa H, et al. Relation between theophylline and circulating vitamin levels in children with asthma. Pharmacol 1996;53:384–389.
149. Delport R, Ubbink JB, Serfontein WJ, et al. Vitamin B6 nutritional status in asthma: the effect of theophylline therapy on plasma pyridoxal-5′-phosphate and pyridoxal levels. Int J Vitam Nutr Res 1988;58:67–72.
150. Laine-Cessac P, Cailleaux A, Allain P. Mechanisms of the inhibition of human erythrocyte pyridoxal kinase by drugs. Biochem Pharmacol 1997;54:863–870.
151. Bartel PR, Ubbink JB, Delport R, et al. Vitamin B6 supplementation and theophylline-related effects in humans. Am J Clin Nutr 1994;60:93–99.

152. Kuwahara H, Noguchi Y, Inaba A, Mizusawa H. Case of an 81 year old woman with theophylline-associated seizures followed by partial seizures due to vitamin B6 deficiency. Clin Neurol 2008;48:125–129.
153. Raskin NH, Fishman RA. Pyridoxine deficiency neuropathy due to hydralazine. N Engl J Med 1964;273:1182–1185.
154. Vidrio H. Interaction with pyridoxal as a possible mechanism of hydralazine hypotension. J Cardiovasc Pharmacol 1990;15:150–156.
155. Krogh-Jensen M, Ekelund S, Svendsen L. Folate and homocysteine status and haemolysis in patients treated with sulphasalazine for arthritis. Scand J Clin Lab Invest 1996;56:421–429.
156. Grieco A, Caputo S, Bertoli A, et al. Megaloblastic anaemia due to sulphasalazine responding to drug withdrawal alone. Postgrad Med J 1986;62:307–308.
157. Logan EC, Williamson LM, Ryrie DR. Sulphasalazine associated pancytopenia may be caused by acute folate deficiency. Gut 1986;27:868–872.
158. Baggott JE, Morgan SL, Ha T, et al. Inhibition of folate-dependent enzymes by non-steroidal anti-inflammatory drugs. Biochem J 1992;282:197–202.
159. Sobczynska-Malefora A, Harrington DJ, Rangarajan S, Kovacs JA, Shearer MJ, Savidge GF. Hyperhomocysteinemia and B-vitamin status after discontinuation of oral anticoagulation therapy in patients with a history of venous thromboembolism. Clin Chem Lab Med 2003;41:1493–1497.
160. Yerby MS. Clinical care of pregnant women with epilepsy; neural tube defects and folic acid supplementation. Epilepsia 2003;44(Suppl. 3):33–40.
161. Kampman MT. Folate status in women of childbearing age with epilepsy. Epilepsy Res 2007;75:52–56.
162. Speidel BD, Meadow SR. Maternal epilepsy and abnormalities of the fetus and newborn. Lancet 1972;300 (7782): 839–843.
163. Kjaer D, Horvath-Puhó E, Christensen J, et al. Antiepileptic drug use, folic acid supplementation, and congenital abnormalities: a population-based case-control study. BJOG 2008;115:98–103.
164. Froscher W, Maier V, Laage M, et al. Folate deficiency, anticonvulsant drugs, and psychiatric morbidity. Clin Neuropharmacol 1995;18:165–182.
165. Kishi T, Fujita N, Eguchi T, et al. Mechanism for reduction of serum folate by antiepileptic drugs during prolonged therapy. J Neurol Sci 1997;145:109–112.
166. Reynolds EH, Trimble MR. Adverse neuropsychiatric effects of anticonvulsant drugs. Drugs 1985;29:570–581.
167. Botez MI, Botez T, Ross-Chouinard A, et al. Thiamine and folate treatment of chronic epileptic patients: a controlled study with the Wechsler IQ scale. Epilepsy Res 1993;16:157–163.
168. Lewis DP, Van Dyke DC, Willhite LA, et al. Phenytoin-folic acid interaction. Ann Pharmacother 1995;29:726–735.
169. Tomson T, Lindbom U, Sundqvist A, et al. Red cell folate levels in pregnant epileptic women. Eur J Clin Pharmacol 1995;48:305–308.
170. Hendel J, Dam M, Gram L, et al. The effects of carbamazepine and valproate on folate metabolism in man. Acta Neurol Scand 1984;69:226–231.
171. Isojarvi JI, Pakarinen AJ, Myllyla VV. Basic hematological parameters, serum gamma-glutamyl-transferase activity, and erythrocyte folate and serum vitamin B12 levels during carbamazepine and oxcarbamazepine therapy. Seizure 1997;6:207–211.
172. Asian K, Bozdemir H, Unsal C, Guvenc B. The effect of antiepileptic drugs on vitamin B12 metabolism. Int J Lab Hematol 2008;30:26–35.
173. Mock DM, Mock NI, Nelson RP, et al. Disturbances in biotin metabolism in children undergoing long-term anticonvulsant therapy. J Pediatr Gastroenterol Nutr 1998;26:245–250.
174. Krause KH, Kochen W, Berlit P, et al. Excretion of organic acids associated with biotin deficiency in chronic anticonvulsant therapy. Int J Vitam Nutr Res 1984;54:217–222.
175. Krause KH, Bonjour JP, Berlit P, et al. Biotin status of epileptics. Ann N Y Acad Sci 1985;447:297–313.
176. Krause KH Bonjour JP, Berlit P, et al. Effect of long-term treatment with antiepileptic drugs on the vitamin status. Drug Nutr Interact 1988;5:317–343.
177. Prasad AS, Lei KY, Moghissi KS, et al. Effect of oral contraceptives on nutrients, III: vitamins B6, B12, and folic acid. Am J Obstet Gynecol 1976;125:1063–1069.
178. Shojania AM. Oral contraceptives: effect of folate and vitamin B12 metabolism. Can Med Assoc J 1982;126: 244–247.
179. Ahmed F, Bamji MS, Iyengar L. Effect of oral contraceptive agents on vitamin nutrition status. Am J Clin Nutr 1975;28:606–615.
180. Bermond P. Therapy of side effects of oral contraceptive agents with vitamin B6. Acta Vitaminol Enzymol 1982;4: 45–54.

181. Villegas-Salas E, Ponce de Leon R, Juarez-Perez MA, et al. Effect of vitamin B6 on the side effects of a low-dose combined oral contraceptive. Contraception 1997;55:245–248.
182. Sanpitak N, Chayutimonkul L. Oral contraceptives and riboflavine nutrition. Lancet 1974;1:836–837.
183. Newman LJ, Lopez R, Cole HS, et al. Riboflavin deficiency in women taking oral contraceptive agents. Am J Clin Nutr 1978;31:247–249.
184. Roe DA, Bogusz S, Sheu J, et al. Factors affecting riboflavin requirements of oral contraceptive users and nonusers. Am J Clin Nutr 1982;35:495–501.
185. Pinto JT, Huang YP, Rivlin RS. Inhibition of riboflavin metabolism in rat tissues by chlorpromazine, imipramine and amitriptyline. J Clin Invest 1981;67:1500–1506.
186. Dutta P, Pinto JT, Rivlin RS. Antimalarial effects of riboflavin deficiency. Lancet 1985;2:1040–1043.
187. Nash AL, Cornish EJ, Hain R. Metabolic effects of oral contraceptives containing 30 micrograms and 50 micrograms of oestrogen. Med J Aust 1979;2:277–281.
188. Rivers JM. Oral contraceptives and ascorbic acid. Am J Clin Nutr 1975;28:550–554.
189. Weininger J, King JC. Effect of oral contraceptive agents on ascorbic acid metabolism in the Rhesus monkey. Am J Clin Nutr 1982;35:1408–1416.
190. Harris AB, Hartley J, Moor A. Reduced ascorbic acid excretion and oral contraceptives. Lancet 1973;2:201–202.
191. McElroy VJ, Schendel HE. Influence of oral contraceptives on ascorbic acid concentrations in healthy, sexually mature women. Am J Clin Nutr 1973;26:191–196.
192. Hudiburgh NK, Milner AN. Influence of oral contraceptives on ascorbic acid and triglyceride status. J Am Diet Assoc 1979;75:19–22.
193. Tyrer LB. Nutrition and the pill. J Reprod Med 1984;29(7 Suppl.):547–550.
194. Opala G, Winter S, Vance C, et al. The effect of valproic acid on plasma carnitine levels. Am J Dis Child 1991;145: 999–1001.
195. Van Wouwe JP. Carnitine deficiency during valproic acid treatment. Int J Vitam Nutr Res 1995;65:211–214.
196. Melegh B, Kerner J, Kispal G, et al. Effect of chronic valproic acid treatment on plasma and urine carnitine levels in children. Acta Paediatr Hung 1987;28:137–142.
197. Farkas V, Bock I, Cseko J, et al. Inhibition of carnitine biosynthesis by valproic acid in rats: the biochemical mechanism of inhibition. Biochem Pharmacol 1996;52:1429–1433.
198. Werner T, Treiss I, Kohlmueller D, et al. Effect of valproate on acylcarnitines in children with epilepsy using ESI-MS/MS. Epilepsia 2007;48:72–76.
199. De Vivo DC, Bohan TP, Coulter DL, et al. L-carnitine supplementation in childhood epilepsy: current perspectives. Epilepsia 1998;39:1216–1225.
200. Hirose S, Mitsudome A, Yasumoto S, et al.Valproate therapy does not deplete carnitine levels in otherwise healthy children. Pediatrics 1998;101:E9.
201. Farhat G, Yamout B, Mikati MA, et al. Effect of antiepileptic drugs on bone density in ambulatory patients. Neurology 2002;58:1348–1353.
202. Hosseinpour F, Ellfolk M, Norlin M, Wikvall K. Phenobarbital suppress vitamin D3 25-hydroxylase expression: a potential new mechanism for drug-induced osteomalacia. Biochem Biophys Res Commun 2007;357:603–607.
203. Bell RD, Pak CY, Zerwekh J, et al. Effect of phenytoin on bone and vitamin D metabolism. Ann Neurol 1979;5: 374–378.
204. Zerwekh JE, Homan R, Tindall R, et al. Decreased serum 24,25-dihydroxyvitamin D concentration during long-term anticonvulsant therapy in adult epileptics. Ann Neurol 1982;12:184–186.
205. Gascon-Barre M, Villeneuve JP, Lebrun LH. Effect of increasing doses of phenytoin on the plasma 25-hydroxyvitamin D and 1,25-dihydroxyvitamin D concentrations. J Am Coll Nutr 1984;3:45–50.
206. Gough H, Goggin T, Bissessar A, et al. A comparative study of the relative influence of different anticonvulsant drugs, UV exposure and diet on vitamin D and calcium metabolism in outpatients with epilepsy. Q J Med 1986;59: 569–577.
207. Hahn TJ, Hendin BA, Scharp CR, et al. Effect of chronic anticonvulsant therapy on serum 25-hydroxycalciferol levels in adults. N Engl J Med 1972;287:900–904.
208. Chung S, Ahn C. Effects of anti-epileptic drug therapy on bone mineral density in ambulatory epileptic children. Brain Dev 1994;16:382–385.
209. Hoikka V, Alhava EM, Karjalainen P, et al. Carbamazepine and bone mineral metabolism. Acta Neurol Scand 1984;70:77–80.
210. Tekgul H, Dizdarer G, Demir N, Ozturk C, Tutuncuoglu S. Antiepileptic drug-induced osteopenia in ambulatory epileptic children receiving a standard vitamin D3 supplement. J Pediatr Endocrinol 2005;18:585–588.

211. Rajantie J, Lamberg-Allardt C, Wilska M. Does carbamazeoine treatment lead to a need of extra vitamin D in some mentally retarded children? Acta Paediatr Scand 1984;73:325–328.
212. Ala-Houhala M, Korpela R, Koivikko M, et al. Long-term anticonvulsant therapy and vitamin D metabolism in ambulatory pubertal children. Neuropediatrics 1986;17:212–216.
213. Keith DA, Gundberg CM, Japour A, et al. Vitamin K-dependent proteins and anticonvulsant medication. Clin Pharmacol Ther 1983;34:529–532.
214. Bengoa JM, Bolt MJ, Rosenberg IH. Hepatic vitamin D 25-hydroxylase inhibition by cimetidine and isoniazid. J Lab Clin Med 1984;104:546–552.
215. Odes HS, Fraser GM, Krugliak P, et al. Effect of cimetidine on hepatic vitamin D metabolism in humans. Digestion 1990;46:61–64.
216. Hatzigeorgiou C, Jackson JL. Hydroxymethylglutaryl-coenzyme A reductase inhibitors and osteoporosis: a meta-analysis. Osteoporos Int 2005;16:990–998.
217. Pérez-Castrillón JL, Vega G, Abad L, et al. Effects of atorvastatin on vitamin D levels in patients with acute ischemic heart disease. Am J Cardiol 2007;99:903–905.
218. Drocourt L, Ourlin JC, Pascussi JM, Maurel P, Vilarem MJ. Expression of CYP3A4, CYP2B6, and CYP2C9 is regulated by the vitamin D receptor pathway in primary human hepatocytes. J Biol Chem 2002;277:25125–25132.
219. Changcheng Z, Assem M, Tay JC, et al. Steroid and xenobiotic receptor and vitamin D receptor crosstalk mediates CYP24 expression and drug-induced osteomalacia. J Clin Invest 2006;116:1703–1712.
220. Rivas P, Górgolas M, García-Delgado R, et al. Evolution of bone mineral density in AIDS patients on treatment with zidovudine/lamivudine plus abacavir or lopinavir/ritonavir. HIV Med 2008;9:89–95.
221. Madeddu G, Spanu A, Solinas P, et al. Bone mass loss and vitamin D metabolism impairment in HIV patients receiving highly active antiretroviral therapy. Q J Nucl Med Mol Imaging 2004;48:39–48.
222. Tebas P, Powderly WG, Claxton S, et al. Accelerated bone mineral loss in HIV-infected patients receiving potent antiretroviral therapy. AIDS 2000;14:F63–F67.
223. Brown TT, Qaqish RB. Antiretroviral therapy and the prevalence of osteopenia and osteoporosis: a meta-analytic review. AIDS 2006;20:2165–2174.
224. Homik J, Suarez-Almazor ME, Shea B, Cranney A, Wells G, Tugwell P. Calcium and vitamin D for corticosteroid-induced osteoporosis. Cochrane Database Syst Rev 1998; CD000952. DOI:10.1002/14651858.CD.00952.
225. Matsuoka LY, Ide L, Wortsman J, et al. Sunscreens suppress cutaneous vitamin D3 synthesis. J Clin Endocrinol Metab 1987;64:1165–1168.
226. Brodie MJ, Boobis AR, Hillyard CJ, et al. Effects of isoniazid on vitamin D metabolism and hepatic monooxygenase activity. Clin Pharmacol Ther 1981;30:363–367.
227. Williams SE, Wardman AG, Taylor GA, et al. Long term study of the effect of rifampicin and isoniazid on vitamin D metabolism. Tubercle 1985;66:49–54.
228. Brodie MJ, Boobis AR, Dollery CT, et al. Rifampicin and vitamin D metabolism. Clin Pharmacol Ther 1980;27: 810–814.
229. Pearson DA. Bone health and osteoporosis: the role of vitamin K and potential antagonism by anticoagulants. Nutr Clin Pract 2007;22:517–544.
230. Sontag TJ, Parker RS. Cytochrome P450 ω-hydroxylase pathway of tocopherol catabolism: novel mechanisms of regulation of vitamin E status. J Biol Chem 2002;277:25290–25296.
231. Manuel-y-Keenoy B, Vinckx M, Vertommen J, van Gaal L, de Leeuw I. Impact of vitamin E supplementation on lipoprotein peroxidation and composition in type 1 diabetic patients treated with atorvastatin. Atherosclerosis 2004;175:369–376.
232. Morrow LE, Grimsely EW. Long-term diuretic therapy in hypertensive patients: effects on serum homocysteine, vitamin B6, vitamin B12, and red blood cell folate concentrations. South Med J 1999;92:866–870.
233. Seligmann H, Halkin H, Rauchfleisch S, et al. Thiamine deficiency in patients with congestive heart failure receiving long-term furosemide therapy: a pilot study. Am J Med 1991;91:151–155.
234. Shimon I, Almog S, Vered Z, et al. Improved left ventricular function after thiamine supplementation in patients with congestive heart failure receiving long-term furosemide therapy. Am J Med 1995;98:485–490.
235. Edelbroek PM, Zitman FG, Schreuder JN, et al. Amitriptyline metabolism in relation to antidepressive effect. Clin Pharmacol Ther 1984;35:467–473.
236. Bell IR, Edman JS, Morrow FD, et al. Vitamin B1, B2, and B6 augmentation of tricyclic antidepressant treatment in geriatric depression with cognitive dysfunction. J Am Coll Nutr 1992;11:159–163.
237. Pinto J, Huang YP, Pelliccione N, et al. Cardiac sensitivity to the inhibitory effects of chlorpromazine, imipramine and amitriptyline upon formation of flavins. Biochem Pharmacol 1982;31:3495–3499.

238. Kishi T. Inhibition of myocardial respiration by psychotherapeutic drugs and prevention by coenzyme Q. In: Yamamura Y, Folkers K, Ito Y (eds). Biomedical and clinical aspects of coenzyme Q. Amsterdam: Elsevier, 1980: 139–154.
239. Okamura N, Ohnishi S, Shimaoka H, Norikura R, Hasegawa H. Involvement of recognition and interaction of carnitine transporter in the decrease of L-carnitine concentration induced by pivalic acid and valproic acid. Pharm Res 2006;23:1729–1735.
240. D'Erasmo E, Ragno A, Raejntroph N, et al. Drug-induced osteomalacia. Recent Prog Med 1998;89:529–533.
241. Spencer H, Lender M. Adverse effects of aluminum-containing antacids on mineral metabolism. Gastroenterology 1979;76:603–606.
242. Shafer RB, Nuttall FQ. Calcium and folic acid absorption in patients taking anticonvulsant drugs. J Clin Endocrinol Metab 1975;41:1125–1129.
243. Reunanen MI, Sotaniemi EA, Hakkarainen HK. Serum calcium balance during early phase of diphenylhydantoin therapy. Int J Clin Pharmacol Biopharm 1976;14:15–19.
244. Xu Y, Hasdhizume T, Shuhart MC, et al. Intestinal and hepatic CYP3A4 catalyze hydroxylation of 1-alpha,25-dihydroxyvitamin D3: implications for drug-induced osteomalacia. Mol Pharmacol 2006;69:56–65.
245. Lukert BP, Raisz LG. Glucocorticoid-induced osteoporosis: pathogenesis and management. Ann Intern Med 1990;112:352–364.
246. Ghishan FK, Walker F, Meneely R, et al. Intestinal calcium transport: effect of cimetidine. J Nutr 1981;111: 2157–2161.
247. Caron P, Gaillard J, Barousse C, et al. Cimetidine treatment of primary hyperparathyroidism. Biomed Pharmacother 1987;41:143–146.
248. Hakanson R, Persson P, Axelson J. Elevated serum gastrin after food intake or acid blockade evokes hypocalcemia. Regul Pept 1990;28:131–136.
249. Stewart CA, Termanini B, Sutliff VE, et al. Iron absorption in patients with Zollinger-Ellison syndrome treated with long-term gastric acid antisecretory therapy. Aliment Pharmacol Ther 1998;12:83–98.
250. Campbell NR, Hasinoff BB, Meddings JB, et al. Ferrous sulfate reduces cimetidine absorption. Dig Dis Sci 1993;38:950–954.
251. Partlow ES, Campbell NR, Chan SC, et al. Ferrous sulfate dos not reduce serum levels of famotidine or cimetidine after concurrent ingestion. Clin Pharmacol Ther 1996;59:389–393.
252. Sturniolo GC, Montino MC, Rossetto L, et al. Inhibition of gastric acid secretion reduces zinc absorption in man. J Am Coll Nutr 1991;10:372–375.
253. Watkins DW, Khalafi R, Cassidy MM, et al. Alterations in calcium, magnesium, iron and zinc metabolism by dietary cholestyramine. Dig Dis Sci 1985;30:477–482.
254. Neuvonen PJ. Interactions with the absorption of tetracyclines. Drugs 1976;11:45–54.
255. Luzecky MH, Burman KD, Schultz R. The syndrome of inappropriate secretion of antidiuretic hormone associated with amitriptyline administration. South Med J 1997;67:495–497.
256. Colgate R. Hyponatraemia and inappropriate secretion of antidiuretic hormone associated with the use of imipramine. Br J Psychiatry 1993;163:819–822.
257. Flint AJ, Crosby J, Genik JL. Recurrent hyponatraemia associated with fluoxetine and paroxetin. Am J Psychiatry 1996;153:134.
258. Goddard C, Paton C. Hyponatreaemia associated with paroxetine. Br Med J 1992;305:1332.
259. Masood GR, Karki SD, Patterson WR. Hyponatraemia with venlafaxine Ann Pharmacother 1998;32:49–51.
260. Bouman WP, Johnson H, Pinner G. Inappropriate antidiuretic hormone secretion and SSRIs. Brit J Psychiatry 1997;170:88–89.
261. Ball C. Fluvoxamine and SIADH. Fluvoxamine and SIADH. Br J Clin Pract 1993;47:62–63.
262. Movig KLL, Leufkens HGM, Lenderink AW, et al. Serotonergic antidepressants associated with an increased risk for hyponatremia in the elderly. Eur J Clin Pharmacol 2002;58:143–148.
263. Dong X, Leppik IE, White J et al. Hyponatremia from oxcarbazepine and carbamazepine. Neurology 2005;65: 1976–1978.
264. Juss JK, Rdhamma AK, Forsyth DR. Tolterodine-induced hyponatraemia. Age Ageing 2005;34:524–525.
265. Raftopoulos H. Diagnosis and management of hyponatremia in cancer patients. Support Care Cancer 2007;15: 1341–1347.
266. Baylis PH, Heath DA. Water disturbances in patients treated with oral lithium carbonate. Ann Intern Med 1978;88: 607–608.
267. Walker RG. Lithium nephrotoxicity. Kidney Int 1993;42:593–598.

268. Shirley DG, Singer DR, Sagnella GA, et al. Effect of a single test dose of lithium carbonate on sodium and potassium excretion in man. Clin Sci 1991;81:59–63.
269. Dorup I, Skajaa K, Clausen T, et al. Reduced concentrations of potassium, magnesium, and sodium-potassium pumps in human skeletal muscle during treatment with diuretics. Br Med J 1988;296:455–458.
270. Malini PL, Strocchi E, Valtancoli G, et al. Angiotensin-converting enzyme inhibitors, thiazide diuretics and magnesium balance: a preliminary study. Magnes Res 1990;3:193–196.
271. Ford ES, Mokdad AH. Dietary magnesium intake in a national sample of U.S. adults. J Nutr 2003;133:2879–2882.
272. Reusz GS, Dobos M, Vasarhelyi B, et al. Sodium transport and bone mineral density in hypercalciuria with thiazide treatment. Pediatr Nephrol 1998;12:30–34.
273. Thurlimann B, Waldburger R, Senn HJ, et al. Plicamycin and pamidronate in symptomatic tumor-related hypercalcemia: a prospective randomized crossover trial. Ann Oncol 1992;3:619–623.
274. Stanton B, Giebisch G, Klein-Robbenhaar G, et al. Effects of adrenalectomy and chronic adrenal corticosteroid replacement on potassium transport in rat kidney. J Clin Invest 1985;75:1317–1326.
275. Seelig MS.Increased need for magnesium with the use of combined oestrogen and calcium for osteoporosis treatment. Magnes Res 1990;3:197–215.
276. Seelig MS. Interrelationship of magnesium and estrogen in cardiovascular and bone disorders, eclampsia, migraine and premenstrual syndrome. J Am Coll Nutr 1993;12:442–458.
277. Vir SC, Love AH. Zinc and copper nutriture of women taking oral contraceptive agents. Am J Clin Nutr 1981;34: 1479–1483.
278. Hinks LJ, Clayton BE, Lloyd RS. Zinc and copper concentrations in leucocytes and erythrocytes in healthy adults and the effect of oral contraceptives. J Clin Pathol 1983;36:1016–1021.
279. Liukko P, Erkkola R, Pakarinen P, et al. Trace elements during 2 years' oral contraception with low-estrogen preparations. Gynecol Obstet Invest 1988;25:113–117.
280. Yunice AA, Czerwinski AW, Lindeman RD. Influence of synthetic corticosteroids on plasma zinc and copper levels in humans. Am J Med Sci 1981;282:68–74.
281. Koskelo EK. Serum selenium in children during anti-cancer chemotherapy. Eur J Clin Nutr 1990;44:799–802.
282. Baum MK, Javier JJ, Mantero-Atienza E, et al. Zidovudine-associated adverse reactions in a longitudinal study of asymptomatic HIV-1-infected homosexual males. J AIDS 1991;4:1218–1226.
283. Solecki TJ, Aviv A, Bogden JD. Effect of a chelating drug on balance and tissue distribution of four essential metals. Toxicol 1984;31:207–216.
284. Schaefer TJ, Wolford RW. Disorders of Potassium. Emerg Med Clin North Am 2005;23:723–747.
285. Oster JR, Materson BJ, Rogers AI. Laxative abuse syndrome. Am J Gastroenterol 1980;74:451–458.
286. Chin RL. Laxative-induced hypokalemia. Ann Emerg Med 1998;32:517–518.
287. Fleming BJ, Genuth SM, Gould AB, et al. Laxative-induced hypokalemia, sodium depletion and hyperreninemia: effects of potassium and sodium replacement on the rennin-angiotensin-aldosterone system. Ann Intern Med 1975;83:60–62.
288. Ritsema GH, Eilers G. Potassium supplements prevent serious hypokalemia in colon cleansing. Clin Radiol 1994;49:874–876.
289. Knobel B, Petchenko P. Hyperphosphatemic hypocalcemic coma caused by hypertonic sodium phosphate (Fleet) enema intoxication. J Clin Gastroenterol 1996;23:217–219.
290. Ehrenpreis ED, Wieland JM, Cabral J, et al. Symptomatic hypocalcemia, hypomagnesemia, and hyperphosphatemia secondary to Fleet's Phospho-Soda colonoscopy preparation in a patient with jejunoileal bypass. Dig Dis Sci 1997;42:858–860.
291. Grosskopf I, Graff E, Charach G, et al. Hyperphosphataemia and hypocalcaemia induced by hypertonic phosphate enema: an experimental study and review of the literature. Hum Exp Toxicol 1991;10:351–355.
292. Reichel H, Deibert B, Geberth S, et al. Frusemide therapy and intact parathyroid hormone plasma concentrations in chronic renal insufficiency. Nephrol Dial Transplant 1992;7:8–15.
293. Ryan MP, Devane J, Ryan MF, et al. Effects of diuretics on the renal handling of magnesium. Drugs 1984;28(Suppl. 1):167–181.
294. Schwinger RH, Erdmann E. Heart failure and electrolyte disturbances. Meth Find Exp Clin Pharmacol 1992;14: 315–325.
295. Quamme GA. Renal magnesium handling: new insights in understanding old problems. Kidney Int 1997;52: 1180–1195.
296. Al-Ghamdi SM, Cameron EC, Sutton RA. Magnesium deficiency: pathophysiologic and clinical overview. Am J Kidney Dis 1994;24:737–752.

297. Kintzel PE. Anticancer Drug-induced Kidney Disorders. Drug Saf 2001;24(1): 19–38.
298. Robertson JI. Diuretics, potassium depletion and the risk of arrhythmias. Eur Heart J 1984;5 (Suppl. A):25–28.
299. Gettes LS. Electrolyte abnormalities underlying lethal and ventricular arrhythmias. Circulation 1992;85(1 Suppl.): I70–176.
300. Clive DM, Stoff JS. Renal syndromes associated with nonsteroidal anti-inflammatory drugs. N Engl J Med 1984;310:563–572.
301. Hottelart C, Achard JM, Moriniere P, et al. Heparin-induced hyperkalemia in chronic hemodialysis patients: comparison of low molecular weight and unfractionated heparin. Artif Organs 1998;22(7): 614–617.
302. Pitt B, Zannad F, Remme WJ, et al. The effect of spironolactone on morbidity and mortality in patients with severe heart failure. Randomized Aldactone Evaluation Study investigators. N Engl J Med 1999;341:709–717.
303. McAdams M. Staffa JA. Dal Pan GJ. The concomitant prescribing of ethinyl estradiol/drospirenone and potentially interacting drugs. Contraception 2007;76(4):278–281.
304. Hussain RM, McIntosh SJ, Lawsone J. Fludrocortisone in the treatment of hypotensive disorders in the elderly. Heart 1996;76:507–509.
305. Bagnis CI, Du Montcel ST, Fonfrede M. Changing electrolyte and acid-basic profile in HIV-infected patients in the HAART era. Nephron Physiol 2006;103:131–138.
306. Izzedine H, Launay-Vacher V, Deray G. Antiviral drug-induced nephrotoxicity. Am J Kidney Dis 2005;45: 804–817.
307. Gearhart MO, Sorg TB. Foscarnet-induced severe hypomagnesemia and other electrolyte disorders. Ann Pharmacother 1993;27:285–289.
308. Anonymous. Morbidity and toxic effects associated with ganciclovir or foscarnet therapy in a randomized cytomegalovirus retinitis trial: studies of ocular complications of AIDS research group, in collaboration with the AIDS Clinical Trials Group. Arch Intern Med 1995;155:65–74.
309. Omar RF, Dusserre N, Desormeaux A, et al. Liposomal encapsulation of foscarnet protects against hypocalcemia induced by free foscarnet. Antimicrob Agents Chemother 1995;39:1973–1978.
310. Jacobson MA, Gambertoglio JG, Aweeka FT, et al. Foscarnet-induced hypocalcemia and effects of foscarnet on calcium metabolism. J Clin Endocrinol Metab 1991;72:1130–1135.
311. Klastersky J, Vanderklen B, Daneau D, et al. Carbenicillin and hypokalemia. Ann Intern Med 1973;78:774–775.
312. Gill MA, DuBe JE, Young WW. Hypokalemiac metabolic alkalosis induced by high-dose ampicillin sodium. Am J Hosp Pharm 1977;34:528–531.
313. Mohr JA, Clark RM, Waack TM, et al. Nafcillin-associated hypokalemia. JAMA 1979;242:544.
314. Nanji AA, Lindsay J. Ticarcillin associated hypokalemia. Clin Biochem 1982;15:118–119.
315. Schlaeffer F. Oxacillin-associated hypokalemia. Drug Intell Clin Pharm 1988;22:695–696.
316. Nanji AA, Denegri JF. Hypomagnesemia associated with gentamicin therapy. Drug Intell Clin Pharm 1984;18: 596–598.
317. Bernardo JF, Murakami S, Branch RA, et al. Potassium depletion potentiates amphotericin-B-induced toxicity in renal tubules. Nephron 1995;70:235–241.
318. Oravcova E, Mistrik M, Sakalova A, et al. Amphotericin B lipid complex to treat invasive fungal infections in cancer patients: report of efficacy and safety in 20 patients. Chemotherapy 1995;41:473–476.
319. Widmer P, Maibach R, Kunzi UP, et al. Diuretic-related hypokalemia. Eur J Clin Pharmacol 1995;49:31–36.
320. Shenfield GM, Knowles GK, Thomas N, et al. Potassium supplements in patients treated with corticosteroids. Br J Dis Chest 1975;69:171–176.
321. Stanton MF, Lowenstein FW. Serum magnesium in women during pregnancy, while taking contraceptives, and after menopause. J Am Coll Nutr 1987;6:313–319.
322. Rolla G, Bucca C, Bugiani M, et al. Hypomagnesemia in chronic obstructive lung disease: effect of therapy. Magnes Trace Elem 1990;9:132–136.
323. Buckley LM, Leib ES, Cartularo KS, et al. Calcium and vitamin D3 supplementation prevents bone loss in the spine secondary to low-dose corticosteroids in patients with rheumatoid arthritis: a randomized, double-blind, placebo-controlled trial. Ann Intern Med 1996;125:961–968.
324. Lems WF, Van Veen GJ, Gerrits MI, et al. Effect of low-dose prednisone (with calcium and calcitriol supplementation) on calcium and bone metabolism in healthy volunteers. Br J Rheumatol 1998;37:27–33.
325. Gennari C. Differential effect on glucocorticoids on calcium absorption and bone mass. Br J Rheumatol 1993;32 (Suppl. 2):11–14.
326. Woo M, Przepiorka D, Ippoliti C, et al. Toxicities of tacrolimus and cyclosporine A after allogenic blood stem cell transplantation. Bone Marrow Transplant 1997;20:1095–1098.
327. Mihatsch MJ, Kyo M, Morozumi K, et al. The side effects of cyclosporine A and tacrolimus. Clin Nephrol 1998;49:356–363.

328. Subbiah V, Tayek JA. Tetany secondary to the use of a proton-pump inhibitor. Ann Intern Med 2002;137:219–220.
329. Boardman PL, Hart FD. Side-effects of indomethacin. Ann Rheum Dis 1967;26:127–132.
330. Leonards JR, Levy G. Gastrointestinal blood loss from aspirin and sodium salicylate tablets in man. Clin Pharmacol Ther 1973;14:62–66.
331. Fleming DJ, Jacques PF, Massaro JM, et al. Aspirin intake and the use of serum ferritin as a measure of iron status. Am J Clin Nutr 2001;74:219–226.
332. Maiorano A, Stallone G, Schena A, et al. Sirolimus interferes with iron homeostasis in renal transplant recipients. Transplantation 2006;82:908–912.
333. Golik A, Zaidenstein R, Dishi V, et al. Effects of captopril and enalapril on zinc metabolism in hypertensive patients. J Am Coll Nutr 1998;17:75–78.
334. Golik A, Modai D, Averbukh Z, et al. Zinc metabolism in patients treated with captopril versus enalapril. Metabolism 1990;39:665–667.
335. O'Connor DT, Strause L, Saltman P, et al. Serum zinc is unaffected by effective captopril treatment of hypertension. J Clin Hypertens 1987;3:405–408.
336. Wester PO. Zinc balance before and during treatment with bendroflumethiazide. Acta Med Scand 1980;208:265–267.
337. Reyes AJ, Leary WP, Lockett CJ, et al. Diuretics and zinc. S Afr Med J 1982;62:373–375.
338. Keen CL, Mark-Savage P, Lönnerdal B, et al. Teratogenic effects of D-penicillamine in rats: relation to copper deficiency. Drug-Nutr Interact 1983;2:17–34.
339. Dastych M, Jezek P, Richtrova M. Effect of penicillamine therapy on the concentration of zinc, copper, iron, calcium, and magnesium in the serum and their excretion in urine. J Gastroenterol 1986;24:157–160.
340. Teherani DK, Altmann H, Tausch G, et al. Zinc levels in blood and urine of rheumatoid arthritis patients after four months treatment with D-penicillamine. Z Rheumatol 1980;39:395–400.
341. Seelig MS. Auto-immune complications of D-penicillamine: a possible result of zinc and magnesium depletion and of pyridoxine inactivation. J Am Coll Nutr 1982;1:207–214.
342. Sandstead HH. Requirements and toxicity of essential trace elements, illustrated by zinc and copper. Am J Clin Nutr 1995;61(Suppl.):621–624.
343. Ravina A, Slezak L, Mirsky N, et al. Reversal of corticosteroid-induced diabetes mellitus with supplemental chromium. Diabet Med 1999;16:164–167.
344. Peretz AM, Neve JD, Famaey JP. Selenium in rheumatic diseases. Semin Arthrit Rheum 1991;20:305–316.
345. Chien LT, Krumdiek CL, Scott CW, et al. Harmful effects of megadoses of vitamins: electroencephalogram abnormalities and seizures induced by intravenous folate in drug-treated epileptics. Am J Clin Nutr 1975;28:51–58.

II CANCER PREVENTION

5 Diet and Childhood Cancer: Preliminary Evidence

Greta R. Bunin and Jaclyn L.F. Bosco

Key Points

- The relationship of mother's, father's, and child's diet with risk of childhood cancer is not well studied and not well understood.
- High intake of cured meats during pregnancy has been consistently associated with increased risk of brain tumors in children. However, other explanations of the observations, such as dietary or other confounders, have not been ruled out.
- Vitamin supplement use may decrease the risk of several childhood cancers; however due to the unidentified critical time period of exposure and specific nutrient(s), these findings must be considered inconclusive.

Abstract

The role of diet in the etiology of cancer in children is not well studied. Based on animal data, epidemiologists hypothesized that gestational exposure to *N*-nitroso compounds increases the risk of brain tumors in children. The hypothesis predicts that cured meats, which contain *N*-nitroso compounds and precursors that form these compounds in vivo, increase risk. The evidence supports an association between frequent eating of cured meat by the mother during pregnancy and increased risk in the child. However, as only dietary factors related to this hypothesis were studied, the possibility exists that correlated dietary characteristics explain the finding. The evidence is not as strong for other aspects of the *N*-nitroso hypothesis, for example, that high vitamin C intake decreases risk because it inhibits synthesis from precursors. Thus, some findings on brain tumors support the *N*-nitroso hypothesis but are also consistent with other mechanisms. Although leukemia is more common than brain tumors in children, fewer studies have addressed the role of diet. Some evidence supports an etiologic role during gestation of foods containing naturally occurring DNA topoisomerase II inhibitors for acute myeloid leukemia with an MLL gene rearrangement in infants, a small subset of childhood leukemia. Recent studies of all childhood leukemia suggest that some aspects of a healthy diet are protective, but replication of these findings is needed. For brain tumors, leukemia, neuroblastoma, and retinoblastoma, protective effects of multivitamin supplements during pregnancy have been observed. Replication and the identification of the critical time period are needed. Although data on the role of diet in childhood cancer are sparse and inconclusive, early evidence suggests that a healthy diet and multivitamins during gestation may reduce risk.

Key Words: Brain tumor; cancer; child; cured meats; DNA topoisomerase II inhibitors; leukemia; *N*-nitroso compounds; multivitamins

5.1. BACKGROUND

Cancer is the most common fatal disease of childhood in the United States. Between ages 1 and 15, only accidents kill more children. Approximately 15 per 100,000 children develop cancer each year or about 7,500 children in the United States *(1)*. This incidence rate indicates that about 1 in 450 children develop cancer before their 15th birthday. The common cancers of childhood are not the

A. Bendich, R.J. Deckelbaum (eds.), *Preventive Nutrition*, Nutrition and Health, DOI 10.1007/978-1-60327-542-2_5,

same as those of later life; leukemia accounts for about one-third of childhood cancers and brain tumors about one-fifth. The other major cancers, in order of frequency, are lymphoma, neuroblastoma, Wilms' tumor, soft tissue sarcoma, osteogenic sarcoma, and retinoblastoma *(2)*. The incidence of childhood cancer increased between the 1970s and the 1990s, but appears to have reached a plateau *(2)*. The same time period had also seen a dramatic improvement in the survival of children with cancer with approximately 80% of these children alive 5 years after diagnosis *(1)*. However, some are left with long-term medical and cognitive problems.

Little is known about the etiology of cancers in children. The medical literature contained few epidemiologic studies of childhood cancer before the 1970s, but since then, the extent of interest and investigation has increased dramatically. Many risk factors have been investigated including genetic abnormalities, parental lifestyle and occupational exposures, and birth characteristics, in addition to aspects of diet that are the focus of this chapter.

The relationship between diet and childhood cancer is not well understood. The possibility that a child's diet or the mother's diet during pregnancy can raise or diminish the risk of these rare cancers at first seems unlikely. The adult cancers most strongly linked with diet, such as stomach and colon, are believed to have latency periods of several decades. In contrast, cancers in children by definition have shorter latencies of no more than 15 years and often less than 5 years. Furthermore, cancers of the digestive tract and of other sites linked to diet in adults rarely occur in children. Therefore, researchers first focused their search for causes on genetic predisposition and exposure to environmental toxins rather than diet.

Diet might act to alter cancer risk in children by mechanisms similar to and different from those proposed for adult cancers. Antioxidants such as vitamin C and beta-carotene may protect against various cancers by their ability to neutralize free radicals and thus prevent oxidative damage to DNA *(3)*. Inadequate intake of folate may encourage malignant transformation of normal cells by altering gene expression and weakening chromosomal structure *(4)*. Exposure to *N*-nitroso compounds may initiate cancer through direct acting or metabolically activated carcinogens in this class of substances. Antioxidants and folate may act in fetuses and children through the same mechanisms as they are hypothesized to act in adults. Carcinogens may also act through the same mechanisms at all ages, but fetuses may be more susceptible to carcinogens, as suggested by animal studies of some *N*-nitroso compounds. Some substances might actually have the opposite effect in fetuses as in adults, as proposed for topoisomerase II inhibitors (*see* Section 5.3). Mechanisms unique to the embryo or fetus may also exist. An excess or deficiency of a dietary component could result in malformation of an organ or subtle cellular changes that increase the organ's susceptibility to cancer. This type of altered development has been proposed as a mechanism leading to cancer in young women after prenatal DES exposure *(5–7)*.

5.2. BRAIN TUMORS

5.2.1. Background

Brain tumors are the second most common type of cancer among children in the United States accounting for about 20% of these cancers. The annual incidence is approximately 4 per 100,000 children under the age of 15 *(2)*. The incidence increased between 1975 and 1989, but the observed trend is thought to reflect not a genuine increase but rather a change in reporting and/or improved diagnosis *(8)*. Since 1990, the incidence has been stable *(2)*. Surgery alone can cure some brain tumors occurring in children, but others require a combination of surgery, radiation, and chemotherapy. With the use of multi-modality therapy, survival from childhood brain tumors has improved to 70% *(2)*, but survivors are often left with neurologic, cognitive, or endocrinologic problems.

Many different histologic types of brain tumor occur in childhood. The major categories are astrocytoma (sometimes referred to as astrocytic glioma) and medulloblastoma (MB) (with histologically similar primitive neuroectodermal tumor (PNET) often included), which account for approximately 50 and 20% of childhood brain tumors, respectively *(9)*. Other types of glioma, ependymoma and oligodendroglioma, comprise another 10% of the total *(9)*. The remaining brain tumors are soft tissue sarcomas, germ cell tumors, and tumors of unspecified type *(9)*.

The patterns of incidence with age and gender differ among the histologic types *(10)*. For example, astrocytoma affects boys and girls with equal frequency, but boys have a higher risk of medulloblastoma. The incidence of astrocytoma peaks between 4 and 8 years of age compared to a peak before age 3 for medulloblastoma. These differences in demographic pattern suggest that the two major categories of childhood brain tumors might differ etiologically. On the other hand, the fact that all tumor types arise in the brain and, thus, share that environment, might imply a common etiology. Perhaps, some etiologic factors are common among different histologic types of brain tumors and others are specific to particular types.

Until recently, epidemiologists generally studied childhood brain tumors as a single entity. If risk factors differed by type of brain tumor, one would expect the studies of all types combined to mostly reflect risk factors for astrocytoma, the most common type. For this reason, studies of all types and astrocytoma are discussed together, and the two studies of medulloblastoma are considered separately.

Most investigations of the relationship between childhood brain tumors and diet have been motivated by the hypothesis that exposure to *N*-nitroso compounds, especially during gestation, increases risk. Preston-Martin et al. first proposed the hypothesis based on animal data and presented the first supporting evidence in 1982 *(11)*.

5.2.2. N-nitroso Hypothesis

The overall category of *N*-nitroso compounds can be broken down into subgroups that include nitrosamines, nitrosamides, and nitrosoureas. *N*-nitroso compounds occur in our environment, as do substances that can combine to form these compounds. Nitrite, nitrogen oxides, and other nitrosating agents can react with nitrogen-containing compounds such as amines, amides, and ureas to form *N*-nitroso compounds. Particularly relevant to the discussion of the relationship between diet and cancer are preformed *N*-nitroso compounds, nitrite, and nitrate, which can be reduced to nitrite in saliva.

Many *N*-nitroso compounds are potent mutagens and animal carcinogens. *N*-nitroso compounds have been found to be carcinogenic in a variety of tissues and organs in 40 animal species. When administered to pregnant animals, some *N*-nitroso compounds induce tumors in the offspring. Of particular relevance to childhood cancer is the fact that some nitrosoureas are potent nervous system carcinogens when given transplacentally *(12)*.

Not only are *N*-nitroso compounds potent carcinogens, but exposure to these compounds is also widespread. *N*-nitroso compounds have been detected in many common products, including cigarette smoke, rubber, cured meats, cosmetics, alcoholic beverages, medications, pesticides, and automobile interiors, and in water, air, and some industrial settings *(12,13)*. Almost all the data on the sources of *N*-nitroso compounds are on the occurrence of nitrosamines. Less is known about the distribution of other *N*-nitroso compounds, including the nitrosoureas that are transplacental nervous system carcinogens in animals.

Although *N*-nitroso compounds occur in the environment, most human exposure is thought to occur via endogenous synthesis from precursors. There is evidence that *N*-nitroso compounds can be synthesized in the stomach and elsewhere in the body *(12)*. Cured meats, baked goods, and cereal contribute most of the nitrite, a NOC precursor, in the diet *(13,14)*. For nitrates, which can be converted to nitrites, vegetables are the main dietary source *(13,14)*.

The endogenous formation of *N*-nitroso compounds induces tumors in animals. When animals are fed *N*-nitroso precursors, for example, nitrite and an amine, a nitrosamine compound is produced in the stomach and tumors result just as from feeding the preformed nitrosamine *(12)*. There are substances that inhibit the formation of *N*-nitroso compounds from precursors in vivo, including vitamin C, vitamin E, selenium, and glutathione *(12)*. In animals, these substances inhibit the formation of *N*-nitroso compounds from precursors and reduce the proportion of animals that develop tumors *(15)*. In some studies, very large doses of vitamin C prevented 100% of tumors *(15)*. In addition to inhibitors, accelerators of the nitrosation reaction are also known and include metal ions, thiocyanate, and certain carbonyl compounds *(12)*.

Based on the animal data, particularly those concerning transplacental carcinogenesis, Preston-Martin et al. hypothesized that exposure to *N*-nitroso compounds during gestation increases the risk of brain tumors in children *(11)*. Children whose mothers frequently ate foods containing *N*-nitroso compounds or nitrite were hypothesized to be at increased risk. The effect of nitrate-rich foods was more difficult to predict; vegetables contribute the majority of nitrates in the diet but also contain vitamin C, an inhibitor of nitrosation. High intakes of vitamins C and E were hypothesized to decrease the risk because of their action as inhibitors of nitrosation reactions.

5.2.3. Maternal Diet

There have now been 10 case–control studies of childhood brain tumors that considered a possible role of pregnancy diet (Table 5.1) *(11,16–25)*. The studies of all types of brain tumors or astrocytoma show a consistent increased risk associated with frequent eating of cured meats; the odd ratios (ORs) for high intake are about 2. Hot dogs, a popular cured meat product queried in most studies, were also associated with increased risk. The only relevant studies with negative findings for cured meats as a group were small *(20,21)* and thus limited in their power to detect associations of low or moderate strength. In contrast to astrocytoma, the weight of the evidence for MB/PNET strongly suggests that they are not associated with cured meat consumption *(17,25)*.

The evidence on the effect of fruit and vegetable consumption during pregnancy is quite limited, although both are relevant to the NOC hypothesis. Fruit as a group, citrus fruit, and individual fruits have not been strongly associated with risk of brain tumors as a group or astrocytoma, although a few non-significant findings have been reported *(18,21)*. For MB/PNET, fruit and fruit juice overall were strongly associated with lower risk in one study, but only modestly so in a follow-up study *(17,24)*. Both of these studies observed significantly decreased risk for only one individual fruit item, namely canned, dried, or frozen peaches and related fruits. The meaning of this finding is not clear. Overall, fruit has not been strongly associated with childhood brain tumors, although the findings for MB/PNET are intriguing.

The evidence linking vegetables with lower risk of brain tumors, like that for fruit, is weak. For all types combined and astrocytoma, some evidence links consumption of vegetables to lower risk; in single studies, a trend of decreasing risk with increasing consumption of vegetables *(19)*, a non-significant decreased risk *(18)*, and decreased risks for some individual vegetables *(21)*. For medulloblastoma, a strong, significant effect for vegetables was observed with individually significant effects of green salad, spinach, and sweet potatoes *(17)*. However, the follow-up study observed no association with

Table 5.1.
Studies of Childhood Brain Tumors and Maternal Consumption of Cured Meats During Pregnancy

Authors (References)	*Sample Size*	*Age (Years)*	*Type(s) of Brain Tumors*	*OR[a] (95% CI)[b] for Maternal Cured Meat Consumption*	
				All Cured Meats	*Hot Dogs*
All brain tumors and astrocytoma					
Preston-Martin et al. *(11)*	209 cases 209 controls	0–14	Tumor of brain or cranial meninges	OR = 2.3[c] High intake	OR = 1.7[c] ≥ 2 times/week
Kuijten et al. *(16)*	163 cases 163 controls	0–14	Astrocytoma	OR = 2.0[c] ≥ 9 times/week	–
Bunin et al. *(18)*	155 cases 155 controls	0–6	Astrocytoma	OR = 1.7 (95% CI 0.8, 3.4) ≥ 5 times/week	OR = 1.9 (95% CI 1.0, 3.7) ≥ 1 time/week
McCredie et al. *(19)*	82 cases 164 controls	0–14	Tumor of brain or cranial nerves	OR = 2.5 (95% CI 1.1, 5.7) ≥ 2.4 times/week	–
Sarasua and Savitz *(20)*	45 cases 206 controls	0–14	Brain tumor	No association	OR = 2.3 (95% CI 1.0, 5.4) > 0 times/week
Cordier et al. *(21)*	75 cases 113 controls	0–15	Brain tumor	ORs 0.6–1.0 for individual cured meats	–
Preston-Martin et al. *(22)*	540 cases 801 controls	0–19	Brain tumor	OR = 2.1 (95% CI 1.3, 3.2) > 7 times/week	OR = 1.4 (95% CI 1.1, 2.0) ≥ 1 times/week
Schymura et al. *(23)*	338 cases 676 controls	0–14	Brain tumor	–	OR = 2.0 (95% CI 1.1, 3.6) 2–3 times/week
Medulloblastoma/primitive neuroectodermal tumor					
Bunin et al. *(17)*	166 cases 166 controls	0–6	MB/PNET[d]	OR = 1.1 (95% CI 0.6, 2.0) ≥ 5 times/week	OR = 1.0 (95% CI 0.6, 1.7) ≥ 1 time/week
Bunin et al. *(24,25)*	315 cases 315 controls	0–6	MB/PNET	OR = 1.0 (95% CI 0.6, 1.5) > 5 times/week	OR = 0.9 (95% CI 0.6, 1.4) ≥ 1 time/week

[a]OR = odds ratio.
[b]95% CI = 95% confidence interval.
[c]95% CI not reported.
[d]MB/PNET = medulloblastoma/primitive neuroectodermal tumor.

vegetables as a group or any individual vegetables *(24)*. In summary, the limited data on vegetables for all tumor types and astrocytoma make interpretation difficult and the two studies of MB/PNET directly conflict.

Vitamin supplements are also sources of vitamin C and other nitrosation inhibitors. Several studies of all brain tumor or astrocytoma *(11,18,20,26)* observed decreased risk with multivitamin use during pregnancy, although the results have not all been statistically significant (Table 5.3). In the studies of

Table 5.2
Studies of Childhood Leukemia and Diet

Authors (References)	*Sample Size*	*Age (Years)*	*Type of Leukemia*	*Dietary Factors Studied*	*Results*
Shu et al. *(46)*	309 cases 618 controls	0–14	Leukemia	Cod liver oil use by child	Cod liver oil OR[a] = 0.3 (95% CI[b] 0.2, 0.7) ALL[c]: OR = 0.4 (95% CI 0.2, 0.9) AML[d]: OR = 0.3 (95% CI 0.1, 1.0)
Peters et al. *(42)*	232 cases 232 controls	0–10	Leukemia	Usual diet of mother, father, child 11 food items	Hot dogs Mother: OR = 2.4 (95% CI 0.7, 8.1) Father: OR = 5.1 (95% CI 1.4, 18) Child: OR = 5.8 (95% CI 2.1, 16.2)
Sarasua and Savitz *(20)*	56 cases 206 controls	0–14	ALL	Mother's and child's diet 5 food items, vitamin supplements	Hot dogs: OR = 0.9 (95%CI 0.4, 1.8) Hamburgers (child): OR = 2.0 (95% CI 0.9, 4.6) Hot dogs, no vitamins (child): OR = 2.9 (95% CI 1.0, 8.6)
Ross et al. *(30)*	84 cases 97 controls	<12.5 months	Leukemia	10 DNAt2[e] inhibitor foods and beverages	DNAt2 inhibitor foods: OR = 1.1 (95% CI 0.5, 2.3) ALL: OR = 0.5 (95% CI 0.2, 1.4) AML: OR = 10.2 (95% CI 1.1, 96)
Jensen et al. *(43)*	138 cases 138 controls	0–14	ALL	76-item FFQ[f] for year prior to pregnancy	Cured meats: OR = 0.7 (95% CI 0.4, 1.1) Vegetables: OR = 0.5 (95% CI 0.3, 0.8) Fruit: OR = 0.7 (95% CI 0.5, 1.0)
Kwan et al. *(45)*	322 cases 403 controls	2–14	Leukemia	Child's intake of 9 foods/food groups, vitamin supplements	Hot dogs/lunch meat: OR = 1.6 (95% CI 0.7, 3.6) Oranges/bananas: OR = 0.5 (95% CI 0.6, 5.3) Orange juice: OR = 0.5 (95% CI 0.3, 0.9)

Table 5.2
(Continued)

Authors (References)	*Sample Size*	*Age (Years)*	*Type of Leukemia*	*Dietary Factors Studied*	*Results*
Spector et al. *(35)*	240 cases 255 controls	<12 months	Acute leukemia	31 foods and beverages	AML MLL+ [g]: OR = 3.2 (95% CI 0.9, 11) AML MLL–[h]: OR = 0.8 (95% CI 0.3, 2.4)
Petridou et al. *(44)*	131 cases 131 controls	1–4	ALL	157-item FFQ for pregnancy	Vegetables OR = 0.7 (95% CI 0.6, 0.9) Fruits: OR = 0.7 (95% CI 0.6, 0.9) Fish and seafood: OR = 0.8 (95% CI 0.6, 0.9)
Dockerty et al. *(41)*	97 cases 303 controls	0–14	ALL	Child's use of vitamin/mineral supplements	Folic acid: OR = 1.0 (95% CI 0.4, 2.8) Iron: OR = 1.1 (95% CI 0.4, 2.8) Multivitamins: OR = 1.0 (95% CI 0.4, 2.8)

[a]OR = odds ratio.
[b]95% CI = 95% confidence interval.
[c]ALL = acute lymphoblastic leukemia.
[d]AML = acute myeloid leukemia.
[e]DNAt2 = DNA topoisomerase II
[f]FFQ = food frequency questionnaire.
[g]MLL + = with a rearrangement involving the mixed-lineage leukemia gene.
[h]MLL – = without a rearrangement involving the mixed-lineage leukemia gene.

medulloblastoma, use of multivitamins close to conception appeared to lower the risk significantly *(17,25)*. One of the studies suggests that longer duration of multivitamin use may protect against brain tumors as a group. The results for vitamin supplementation during gestation suggest protection but whether timing or duration matters and whether the effect varies by tumor type have not been determined.

Although the studies of maternal diet focused on the *N*-nitroso hypothesis, some studies queried a few additional foods. Caffeinated beverages and charcoal broiled foods did not appear to affect risk *(17,18,20)*. Children of mothers who were frequent eaters of non-chocolate candy appeared to be at higher risk in both studies of medulloblastoma *(17,24)*. French fries and chili peppers were associated with medulloblastoma in one study; the other study did not ask about them *(24)*. These findings are based on limited data and require replication.

The studies discussed above were motivated mainly by the *N*-nitroso hypothesis. The *N*-nitroso hypothesis predicts that cured meat is associated with higher risk and fruit, especially those types rich in vitamin C, with lower risk. Vegetables, which contain nitrates and vitamin C, might either raise or lower the risk. Vitamin supplements, which usually contain vitamin C and other inhibitors of NOC formation, would also be expected to lower the risk. The strongest support for the NOC hypothesis comes from the consistent association with maternal cured meat consumption in most studies of all brain tumors or astrocytoma. The evidence linking fruit and vegetable consumption to lower risk of the same groups of brain tumors is weak, as there were no statistically significant findings. The data on vitamin supplements during pregnancy support the NOC hypothesis as most studies showed decreased risk. The results for MB/PNET differ from those for the other entities studied in that cured meats do not seem to affect risk. However, the findings for vitamin supplements in the MB/PNET studies are consistent with the NOC hypothesis and are similar to those for brain tumors overall and astrocytoma. In summary, current evidence on maternal diet and risk of childhood brain tumors as a group or of astrocytoma supports the *N*-nitroso hypothesis to some extent, but important inconsistencies exist.

Inconsistencies with the hypothesis may at least partly result from limitations of the data. The animal data predict that the timing of intake of supplements and foods rich in an inhibitor of nitrosation such as vitamin C may play a role. Eating such foods or taking supplements with cured meat would inhibit the formation of *N*-nitroso compounds and therefore lower the risk. If the cured meat were eaten at one time and the inhibiting supplement or food at another time, the risk would not be as low. The dietary assessments did not ask questions about what foods were eaten together or when during the day supplements were taken and thus cannot address this issue.

Although limitations of the data collected may explain the inconsistencies with the *N*-nitroso hypothesis, other possibilities must also be considered. For example, the association of maternal cured meat consumption with increased risk of childhood brain tumors is compatible with causal exposures other than *N*-nitroso compounds. This is particularly true, as all but one of the studies focused its dietary data collection on components relevant to the NOC hypothesis rather than examining diet in a comprehensive manner. Therefore, a dietary characteristic correlated with frequent cured meat consumption that is the causal exposure could explain the findings. Inconsistencies between the data and the hypothesis might also have occurred by chance as a result of the few studies, most of which had relatively small sample sizes. Similarly, the apparent differences in dietary findings between astrocytoma and medulloblastoma might reflect chance variation rather than distinct etiologies. Future research should examine maternal diet during pregnancy in a comprehensive way, collecting data on macronutrients such as fat and micronutrients such as folate as well as on food components relevant to the *N*-nitroso hypothesis. With these data, epidemiologists will be able to further test the hypothesis as well as more extensively investigate the role of maternal diet.

5.2.4. *Child's Diet*

Few studies have investigated infant or child's diet and the dietary data from each are limited to a small number of food items and vitamin supplements *(11,17,18,20,21,27,28)*. Investigation of the *N*-nitroso hypothesis motivated the studies. The only statistically significant finding that supports the hypothesis comes from the study of medulloblastoma *(17)*in which eating fruit in the first year of life was associated with decreased risk compared to eating no fruit at all. In a study that included all types of brain tumors, McCredie et al. also observed a decreased risk associated with fruit consumption, although it was not statistically significant *(28)*. An effect of fruit was not seen in the study of astrocyoma *(18)*. None of the three studies observed a significantly decreased risk associated with orange juice or vitamin supplements, although a few odds ratios of <1.0 were noted for the latter. The studies collected data on consumption of cured meats, but few infants eat these products and no significant associations were observed. Although no strong findings resulted, more comprehensive examination of infant diet may be a fruitful area of research.

Four studies investigated the child's usual diet before diagnosis and could better address the cured meat question *(11,20,21,27)*. Only one of the four observed a significant association with cured meat consumption *(11)* and, in that study, when child and maternal consumption were analyzed simultaneously, the child's cured meat consumption was not associated with risk. The evidence, then, does not strongly suggest a role for cured meat consumption by the child. However, Sarasua and Savitz found evidence of a possible synergistic effect between cured meats and vitamin supplements *(20)*. They observed large but imprecise odds ratios of about 3–7 for the joint effect of high cured meat consumption and lack of vitamin supplement use. The possible synergistic effect is consistent with the *N*-nitroso hypothesis. Animals fed a *N*-nitroso precursor and a nitrosating agent along with vitamin C produced smaller amounts of *N*-nitroso compounds and developed fewer tumors than those not given vitamin C *(12,15)*. Of the two other studies that investigated vitamin supplement use, one observed an apparent protective effect *(21,27)*. The studies of childhood diet illustrate the need to analyze multiple aspects of diet simultaneously and to analyze the mother's and the child's diet simultaneously.

5.2.5. Summary

For maternal diet during pregnancy, sufficient data exist to conclude that an association exists between frequent consumption of cured meats and risk of astrocytoma. However, the evidence does not yet eliminate the possibility that some correlated aspect of diet is the actual risk factor. The necessity of assessing the child's diet retrospectively makes investigation challenging. Few studies on the child's diet have been conducted and there are no consistent findings. Future research should examine maternal diet during pregnancy in a comprehensive way, collecting data on macronutrients such as fat and micronutrients such as folate as well as on food components relevant to the *N*-nitroso hypothesis. With these data, epidemiologists will be able to further test the hypothesis as well as more extensively investigate the role of maternal diet. In addition, researchers should develop better methods for retrospectively assessing the child's diet, which may also play a role.

5.3. LEUKEMIA

5.3.1. Background

Leukemia accounts for about one-third of all cancer in children under age 15. In the United States, the annual incidence rate of leukemia in children is 5 per 100,000 with about 80% surviving at least 5 years *(2)*. Three-quarters of leukemias in children are classified as acute lymphocytic leukemia (ALL) and 15% as acute myeloid leukemia (AML). Other types of leukemia make up the remaining 10%. Similarly to the different types of brain tumors, different types of leukemia vary in patterns of incidence *(29)*. Acute lymphoblastic leukemia (ALL) in children is more common in males than females and in whites than blacks, and peaks in incidence between ages 3 and 5. The incidence patterns for AML are quite different; male–female and white–black differences are slight and incidence is fairly constant throughout childhood.

5.3.2. Maternal Diet

5.3.2.1. Foods Containing Inhibitors of DNA Topoisomerase II

A hypothesis regarding diet has been put forth specifically for leukemia in infants *(30)*. In about 75% of affected infants, the leukemia cells have a rearrangement involving the *MLL* (mixed-lineage leukemia) gene at chromosome band 11q23. Leukemias that occur after cancer treatment with epipodophyllotoxins, a class of chemotherapeutic agents, also have *MLL* rearrangements. Epipodophyllotoxins inhibit an enzyme called DNA topoisomerase II (DNAt2), which is necessary for DNA replication. If epipodophyllotoxins inhibit DNAt2 and increase the risk of leukemia with *MLL* rearrangements, perhaps other inhibitors of this enzyme also increase the risk of the same leukemia. Inhibitors of DNAt2 include certain flavonoids and quinolones *(31,32)*. Flavonoids, substances found in plants, occur in the diet in fruits, vegetables, herbs, beans, wine, beer, and other plant-derived foods. Quinolones are DNAt2-inhibiting medications, which are used to treat urinary tract infections *(33)*.

Ross and colleagues proposed the specific hypothesis that maternal exposure to DNAt2 inhibitors during pregnancy increases the risk of leukemia with *MLL* rearrangements in infants *(30)*. According to this hypothesis, children of mothers who frequently ate fruits, vegetables, beans, and other plant-derived foods would be at *higher* risk of leukemia. Paradoxically, DNAt2-inhibiting foods and flavonoids have been associated with a decreased risk of some adult cancers *(34)*. Perhaps, as Ross et al. speculate, flavonoids affect fetuses and adults differently. Fetuses are rapidly growing and have high rates of cell division and thus high levels of DNAt2, while adults have much lower rates of cell division and DNAt2 activity. Therefore, Ross et al. suggest that fetuses require high levels of topoisomerase II activity and thus, inhibition may be detrimental.

Two studies have investigated this specific hypothesis for infant leukemia *(30,35)* (Table 5.2). Both studies collected information on the frequency with which mothers ate foods that were believed a priori to contain DNAt2 inhibitors. In the first study, 10 foods and beverages constituted the DNAt2 inhibitor group: beans, fresh vegetables, canned vegetables, fruits, soy, regular coffee, wine, black tea, green tea, and cocoa (as a beverage). In the second study, some additional items were added including onions, apples, berries, red wine, and caffeinated beverages other than coffee. In the first study, although fresh vegetables and coffee were associated with increased risk, Ross et al. observed no association between total consumption of the 10 foods combined *(30)*. However, when the two subgroups of infant leukemia were analyzed separately, the results were quite different. For AML, increasing intake of foods containing topoisomerase II inhibitors was significantly associated with approximately a 10-fold increased risk for high total consumption compared to low total consumption. Additionally, among the individual food items, beans, fresh vegetables, and fruit showed significant trends for AML. Neither the inhibitor-containing foods as a group or any of the composite foods were significantly associated with the risk of ALL.

Nearly a decade after this preliminary study, Ross and colleagues conducted a larger study of the same hypothesis *(35)* using a more contemporary cohort of infants diagnosed with leukemia. In addition, the investigators classified the leukemias as positive or negative for an *MLL* rearrangement. In contrast to the first study, no association was seen for an AML. However, when AML cases were stratified by *MLL* gene status, high consumption of DNAt2 inhibitors tripled the risk of AML in infants who were *MLL* positive. In support of earlier findings, there was little evidence that DNAt2-inhibiting foods increased the risk of ALL. The findings suggest that the effect of DNAt2 may be specific to the *MLL+* subgroup of AML. The investigators also examined the role of fruits and vegetables and found a decreased risk for infant leukemia overall. These results, while intriguing, merit cautious interpretation. Measurement error, confounding by dietary or other exposures, and selection bias may have contributed to the findings. In addition, chance could explain the findings because of the small numbers of subjects in certain categories of intake and leukemia type. Even if confirmed, these findings would not support any advice to pregnant women to avoid DNAt2-inhibiting foods, as infant leukemia is very rare and the *MLL+*AML subgroup even rarer. In addition, the study suggests that fruits and vegetables may decrease risk of infant leukemia overall, even though some fruits and vegetables may increase risk of the MLL+ AML subgroup.

5.3.2.2. Vitamin and/or Mineral Supplements

Several studies have examined multivitamins, folic acid, and other individual vitamin supplements used during pregnancy in relation to leukemia in children (Table 5.3) *(20,36–41)*. Maternal vitamin

Table 5.3
Studies of Childhood Cancer and Maternal Use of Multivitamins and Other Supplements During Pregnancy

Authors (References)	*Sample size*	*Age (years)*	*Type of cancer*	*Findings for use during pregnancy*
Brain tumor				
Preston-Martin et al. *(11)*	209 cases 209 controls	0–14	Tumors of brain and cranial meninges	Vitamin supplement: OR[a]= 0.6[b] (95% CI[c] 0.3, 1.3)
Bunin et al. *(17)*	166 cases 166 controls	0–6	MB/PNET[d]	Multivitamin, ever during pregnancy, no association Multivitamin in first 6 weeks of pregnancy, OR = 0.6 (95% CI 0.3, 1.0)
Bunin et al. *(18)*	155 cases 155 controls	0–6	Astrocytoma	Multivitamin: OR = 0.6 (95% CI 0.2, 1.5)
Sarasua and Savitz *(20)*	45 cases 206 controls	0–14	Brain tumors	Vitamin supplement: OR = 0.7[b] (95% CI 0.3, 1.9)
Preston-Martin et al. *(22)*	540 cases 801 controls	0–19	All brain tumors, subgroups glioma, MB/PNET, other	Multivitamin: OR = 0.7 (95% CI 0.4, 1.1), OR = 0.5 (95% CI 0.4, 0.8), and OR = 0.5 (95% CI 0.4, 0.7) for use during <2, 2 and 3 trimesters, respectively
Preston-Martin et al. *(26)*	1051 cases 1919 controls	0–19	All brain tumors, subgroups glioma, MB/PNET, other	Vitamin supplement, OR = 0.7 (95% CI 0.5, 0.9) and OR = 0.6 (95% CI 0.5, 0.8) for use during 2 and 3 trimesters, respectively
Bunin et al. *(25)*	315 cases 315 controls	0–6	MB/PNET	Multivitamin close to conception: OR = 0.7 (95% CI 0.4, 1.0), no effect of duration
Leukemia				
Robison et al. *(36)*	204 cases 204 controls	0–17	AML[e]	Vitamin or iron supplement: OR = 1.0 (95% CI 0.5, 2.0)
Sarasua and Savitz *(20)*	56 cases 205 controls	0–14	ALL[f]	Vitamin supplement : OR = 0.5[b] (95% CI 0.2, 1.1)
Thompson et al. *(37)*	83 cases 166 controls	0–14	ALL	Folic acid supplement: OR = 0.4 (95% CI 0.2, 0.7), OR = 0.2 (95% CI 0.04, 0.8) and OR = 0.2 (95% CI 0.04, 1.0) for any use, use starting in first 7 weeks of pregnancy, and ≥34 weeks of use, respectively
Shaw et al. *(38)*	789 cases 789 controls	0–14	ALL	Folic acid-containing supplement: OR = 1.0 (95% CI 0.8, 1.2) Taken > 100 times: OR = 0.9 (95% CI 0.7, 1.1)

(*Continued*)

Table 5.3
(Continued)

Authors (References)	*Sample size*	*Age (years)*	*Type of cancer*	*Findings for use during pregnancy*
Jensen et al. *(43)*	138 cases 138 controls	0–14	ALL	Any vitamin supplement in year prior to pregnancy: OR = 0.6 (95% CI 0.2, 1.9) No significant association with any type of supplement in year prior to pregnancy
Ross et al. *(40)*	158 cases 173 controls	0–18	Leukemia in children with Down syndrome	Vitamin use close to conception: OR = 0.6 (95% CI 0.4, 1.0) ALL: OR = 0.5 (95% CI 0.3, 0.9) AML. OR = 0.9 (95% CI 0.5, 1.8)
Dockerty et al. *(41)*	97 cases 303 controls	0–14	ALL	Folic acid-containing supplement: OR = 1.1 (95% CI 0.5, 2.7) Iron-containing supplement: OR = 1.2 (95% CI 0.7, 2.1) Multivitamin: OR = 0.8 (95% CI 0.2, 3.1)
Other cancers				
Bunin et al. *(53)*	182 cases 182 controls	Not stated	Retinoblastoma	Vitamin supplements in first trimester Sporadic non-heritable (unilateral): OR = 0.4 (95% CI 0.2, 0.9) Sporadic heritable (bilateral): OR = 0.3 (95% CI 0.1, 0.9)
Michalek et al. *(55)*	183 cases 372 controls	0–14	Neuroblastoma	Vitamin supplements: OR = 0.5 (95% CI 0.3, 0.7)
Olshan et al. *(54)*	538 cases 504 controls	0–19	Neuroblastoma	Vitamin use OR = 0.7 (95% CI 0.5, 1.1), 0.7 (95% CI 0.5, 1.0), 0.6 (95% CI 0.4, 0.9), 0.6 (95% CI 0.4, 0.9) for use in month before pregnancy, 1st trimester, 2nd trimester, 3rd trimester, respectively.

[a]OR = odds ratio.
[b]Odds ratio and/or 95% confidence interval calculated from data presented in paper.
[c]95% CI = 95% confidence interval.
[d]MB/PNET = medulloblastoma/primitive neuroectodermal tumor.
[e]ALL = acute lymphoblastic leukemia.
[f]AML = acute myeloid leukemia.

supplementation during pregnancy appears to protect against the risk of ALL in children generally and in those with Down syndrome *(20,39,40)*. Studies that focused on the effect of folic acid-containing supplements conflict, with the largest study showing little or no reduction in risk *(37,38,41)*. An effect conferred by a nutrient other than folic acid or an effect at a specific time during pregnancy may explain the different results for vitamin supplements (almost always multivitamins) and for folic acid supplements. Although very limited, the data on AML do not suggest a protective effect of vitamin supplements *(36,40)*. Further research is warranted given the diversity of results from these studies to determine if maternal use of folic acid and multivitamin supplements is protective against leukemia.

5.3.2.3. Foods Containing *N*-Nitroso Compounds

Although animal data have not linked *N*-nitroso compounds with leukemia, the potency of these carcinogens and the possibility of enhanced potency through transplacental exposure suggested investigation. Despite one observation of a fivefold increased risk of leukemia in children of mothers with frequent consumption of hot dogs *(42)*, two other studies refute the association (Table 5.2) *(20,43)*. In addition, investigators have reported negative results for individual and total cured meats, citrus fruit or juice, and interactions of vitamins C or E with cured meats *(20,43)*. Although current evidence does not support the NOC hypothesis, the data are limited as are studies able to examine specific types of leukemia.

5.3.2.4. Other Foods and Nutrients

To our knowledge, only two studies have investigated maternal diet comprehensively in relation to risk of childhood leukemia (Table 5.2). A study in Greece observed that higher maternal consumption of vegetables, fruits, and fish and seafood during pregnancy reduced the risk of ALL, while higher consumption of sugars and meat increased the risk *(44)*. A study in northern California investigated maternal diet in the year prior to pregnancy as an indication of diet at the start of the pregnancy *(43)*. Like the Greek study, this study observed protective effects of high vegetable and fruit consumption. In addition, protein sources appeared to reduce risk, as did high intake of provitamin A carotenoids and the antioxidant glutathione. The two studies suggest that aspects of a healthy diet by women during pregnancy may protect their children against leukemia, but confirmatory studies are required.

5.3.3. Child's Diet

A limited number of studies have investigated the role of child's diet in relation to leukemia (Table 5.2) *(20,41,42,45)*. The evidence on eating cured meats is not consistent. Peters et al. observed elevated risks for eating of cured meats, especially hot dogs, by the child *(42)*. In a small study, cured meat consumption was not linked to risk of leukemia but high cured meat intake in the absence of vitamin supplements was associated with increased risk *(20)*. In a more recent study, Kwan and colleagues observed non-significant weak, but positive associations for regular consumption of hot dogs/lunch meat in 2- to 14-year-olds and in 2- to 5-year-olds compared to rare or no consumption *(45)*. These investigators also observed that children who regularly consumed oranges and bananas during the first 2 years of life reduced the risk of ALL by half, with a similar association observed for orange juice consumption *(45)*. However, one of the previous studies found no association between oranges or orange juice and leukemia risk *(42)*.

Vitamin supplement use by the child was investigated in one study, but no associations were found for folic acid supplements, iron supplements, or multivitamins (Table 5.2) *(41)*. In a study in Shanghai,

China, use of cod liver oil for more than a year appeared to decrease the risk of both ALL and AML *(46)*. Cod liver oil contains vitamins A and D.

The evidence on child's diet is limited and sometimes conflicting. The challenges of assessing diet retrospectively may be greater for child's diet compared to mother's diet. Although adult diet is fairly consistent, a child's diet changes enormously from infancy to young childhood to later childhood. Despite the difficulties, the role of diet during childhood on risk of leukemia as well as other cancers and non-neoplastic diseases should be elucidated.

5.3.4. Summary

Researchers have reported the results of a small number of studies investigating the role of DNAt2-inhibiting foods, *N*-nitroso-related foods, other foods, and vitamin supplementation in relation to childhood leukemia. The studies are few and some of the results conflict. Differences in how dietary intake was assessed, the variation in the time period of consumption, the age group, and different types of controls used to represent the source population that gave rise to the cases could explain the differing results. As in many case–control studies, these had the potential for bias as a result of differential recall between cases and controls and differential participation rates between the cases and controls. Research studies aimed at specific hypotheses as well as those that comprehensively assess diet are needed to further clarify the relation of both maternal and childhood dietary exposures to the risk of childhood leukemia.

5.4. VITAMIN K

The role of vitamin K has been studied in relation to its administration to newborns rather than as a component of diet. In many industrialized countries, newborns are routinely given vitamin K to prevent hemorrhagic disease, unexpected bleeding in previously healthy neonates. In 1990, Golding et al. reported an association between receiving vitamin K as a newborn and development of cancer before age 10 *(47)*. The finding of an approximate doubling of risk arose unexpectedly in a nested case–control study of children born in 1970 in Great Britain. The results of a second study in Great Britain observed an increase in risk of similar magnitude for vitamin K administered intramuscularly but not orally *(48)*. The finding corroborated the first study as in 1970, the year of birth of the children in that study, vitamin K was almost always given intramuscularly. These two studies raised concern because of the high prevalence of exposure among newborns in industrialized countries. Oral vitamin K can be given, but is less effective at preventing hemorrhagic disease. Seeking to confirm or refute the findings, Swedish researchers studied 1.3 million infants born full term after uncomplicated delivery between 1973 and 1989 *(49)*. By record linkage, these children were followed until 1992; approximately 2,350 children in the cohort developed cancer. No increase in risk of cancer overall or of leukemia was observed with exposure to intramuscular vitamin K. Two additional studies, one in the United States and one in Germany, did not observe cancer risk to be associated with vitamin K given to neonates *(50,51)*. When the five studies are considered together, it seems unlikely that vitamin K given neonatally is a major cause of cancer during childhood.

5.5. OTHER CANCERS

A few studies of other childhood cancers have reported isolated findings related to diet. A decreased risk of retinoblastoma was observed in relation to high intakes of vegetables and fruit *(52)*. Similar to the findings for brain tumors and leukemia, use of multivitamins was associated with lower risk of retinoblastoma (Table 5.3) *(53)* and neuroblastoma *(54,55)*. A small study of rhabdomyosarcoma observed an increased risk associated with the child's eating of organ meats *(56)* that was not confirmed in a much larger study (S. Grufferman, personal communication).

5.6. CONCLUSION

Studies of the relationship between diet and risk of childhood cancer are growing in number. The two most common cancers of childhood, leukemia and brain tumor, have been studied to some extent, but our knowledge of the role of diet in the etiologies of these cancers is meager. Nonetheless, the findings suggest that diet does play a role in at least some childhood cancers. For the other cancers of childhood, research on diet has been extremely limited. Future research will elucidate the particulars and extent of the role of diet in relation to the common and less common cancers of childhood.

5.7. RECOMMENDATIONS

Maternal cured meat consumption has been fairly consistently associated with brain tumor risk in children, but whether the association is causal is unclear. Also, the frequency of cured meat consumption that was associated with higher risk varied greatly among the studies. For these reasons, a specific recommendation is not possible or appropriate. However, since cured meats are high in salt and fat, nutritional concerns other than the child's cancer risk, such as keeping one's fat and salt intake within recommendations, require that cured meats be eaten in no more than moderate quantities. Women eating cured meats several times a week or more might wish to reduce their intake during pregnancy.

The early evidence for ALL suggests that a healthy diet or some aspects of a healthy diet may protect an unborn child from developing ALL after birth. The evidence does not justify a recommendation, but supports general recommendations for pregnancy. The data on DNAt2-inhibiting foods raise the possibility that some foods in a healthy diet may increase the risk of a subgroup of infant leukemia. However, fewer than 50 MLL + AMLs occur in the United States each year. The findings for this subset of leukemia may lead to better understanding of its etiology and its treatment, but they should not influence recommendations on diet during pregnancy.

Maternal use of multivitamin supplements during pregnancy has now been associated with decreased risk of leukemia, brain tumors, neuroblastoma, and retinoblastoma in children. Current evidence has not excluded bias or confounding as explanations or determined the duration or timing that confers protection. In addition, the protective nutrient(s) and mechanism have not been identified and may differ among cancers. Some evidence suggests that the periconceptional period may be critical. Currently, medical and public health organizations recommend that women of reproductive age regularly take multivitamins with folate to ensure adequate intake in very early pregnancy to prevent neural tube defects. Early evidence in relation to childhood cancer, although far from conclusive, suggests that the recommendation for prevention of neural tube defects may also protect against childhood cancer.

REFERENCES

1. American Cancer Society. Cancer Facts and Figures 2007. Atlanta, GA: American Cancer Society; 2007.
2. Ries LAG, Melbert D, Krapcho M, Stinchcomb DG, Howlader N, Horner MJ, Mariotto A, Miller BA, Feuer EJ, Altekruse SF, Lewis DR, Clegg L, Eisner MP, Reichman M, Edwards BK (eds). SEER Cancer Statistics Review, 1975–2005. National Cancer Institute. Bethesda, MD, http://seer.cancer.gov/csr/1975_2005/, based on November 2007 SEER data submission, posted to the SEER web site, 2008.
3. Frei B, Stocker R, Ames BN. Antioxidant defenses and lipid peroxidation in human blood plasma. Proc Nat Acad Sci U S A 1988;85:9748–52.
4. Butterworth CE. Folate deficiency and cancer. In: Micronutrients in Health and in Disease Prevention. Bendich A, Butterworth CE (eds). New York: Marcel Dekker, Inc.; 1991.

5. Mittendorf R. Teratogen update: carcinogenesis and teratogenesis associated with exposure to diethylstilbestrol (DES) in utero. Teratology 1995;51:435–45.
6. Nelson KG, Sakai Y, Eitzman B, Steed T, McLachlan J. Exposure to diethylstilbestrol during a critical developmental period of the mouse reproductive tract leads to persistent induction of two estrogen-regulated genes. Cell Growth Differ 1994;5:595–606.
7. McLachlan JA, Newbold RR. Estrogens and development. Environ Health Perspect 1987;75:25–7.
8. Smith MA, Freidlin B, Gloeckler Ries LA, Simon R. Trends in reported incidence of primary malignant brain tumors in children in the United States. J Natl Cancer Inst 1998;90:1269–77.
9. Young JL, Ries LG, Silverberg E, Horm JW, Miller RW. Cancer incidence, survival, and mortality for children younger than 15 years. Cancer 1986;58:598–602.
10. Gurney JG, Severson RK, Davis S, Robison LL. Incidence of cancer in children in the United States: sex-, race-, and 1-year age-specific rates by histologic type. Cancer 1995;75:2186–95.
11. Preston-Martin S, Yu MC, Benton B, Henderson BE. N-Nitroso compounds and childhood brain tumors: a case-control study. Cancer Res 1982;42:5240–5.
12. Lijinsky W. Chemistry and biology of N-nitroso compounds. New York, New York: Cambridge University Press; 1992.
13. National Academy of Sciences. The Health Effects of Nitrate, Nitrite and N-Nitroso Compounds. Washington, DC: National Academy Press; 1981.
14. Howe GR, Harrison L, Jain M. A short diet history for assessing dietary exposure to N-nitrosamines in epidemiologic studies. Am J Epidemiol 1986;124:595–602.
15. Mirvish SS. Experimental evidence for inhibition of N-nitroso compound formation as a factor in the negative correlation between vitamin C consumption and the incidence of certain cancers. Cancer Res 1994;54:1948s–51s.
16. Kuijten RR, Bunin GR, Nass CC, Meadows AT. Gestational and familial risk factors for childhood astrocytoma: results of a case-control study. Cancer Res 1990;50:2608–12.
17. Bunin GR, Kuijten RR, Buckley JD, Rorke LB, Meadows AT. Relation between maternal diet and subsequent primitive neuroectodermal brain tumors in young children. N Engl J Med 1993;329:536–41.
18. Bunin GR, Kuijten RR, Boesel CP, Buckley JD, Meadows AT. Maternal diet and risk of astrocytic glioma in children: a report from the Children's Cancer Group. Cancer Causes Control 1994;5:177–87.
19. McCredie M, Maisonneuve P, Boyle P. Antenatal risk factors for malignant brain tumours in New South Wales children. Int J Cancer 1994;56:6–10.
20. Sarasua S, Savitz DA. Cured and broiled meat consumption in relation to childhood cancer: Denver, Colorado (United States). Cancer Causes Control 1994;5:141–8.
21. Cordier S, Iglesias MJ, Goaster CL, Guyot MM, Mandereau L, Hemon D. Incidence and risk factors for childhood brain tumors in the Ile de France. Int J Cancer 1994;59:776–82.
22. Preston-Martin S, Pogoda JM, Mueller BA, Holly EA, Lijinsky W, Davis RL. Maternal consumption of cured meats and vitamins in relation to pediatric brain tumors. Cancer Epidemiol Biomarkers Prev 1996;5:599–605.
23. Schymura MJ, Zheng D, Baptiste MS, Nasca PC. A case-control study of childhood brain tumors and maternal lifestyle. Am J Epidemiol 1996;143:S8.
24. Bunin GR, Kushi LH, Gallagher PR, Rorke-Adams LB, McBride ML, Cnaan A. Maternal diet during pregnancy and its association with medulloblastoma in children: a Children's Oncology Group study. Cancer Causes Control 2005;16: 877–91.
25. Bunin G, Gallagher PR, Rorke-Adams LB, Robison LL, Cnaan A. Maternal supplement, micronutrient, and cured meat intake during pregnancy and risk of medulloblastoma during childhood: a Children's Oncology Group study. Cancer Epidemiol Biomarkers Prev 2006;15(9):1660–7.
26. Preston-Martin S, Pogoda JM, Mueller BA, et al. Results from an international case-control study of childhood brain tumors: the role of prenatal vitamin supplementation. Environ Health Perspect 1998;106(Suppl. 3): 887–92.
27. Howe GR, Burch JD, Chiarelli AM, Risch HA, Choi BCK. An exploratory case-control study of brain tumors in children. Cancer Res 1989;49:4349–52.
28. McCredie M, Maisonneuve P, Boyle P. Perinatal and early postnatal risk factors for malignant brain tumours in New South Wales children. Int J Cancer 1994;56:11–5.
29. Ross JA, Davies SM, Potter JD, Robison LL. Epidemiology of childhood leukemia, with a focus on infants. Epidemiol Rev 1994;16:243–72.
30. Ross JA, Potter JD, Reaman GH, Pendergrass TW, Robison LL. Maternal exposure to potential inhibitors of DNA topoisomerase II and infant leukemia (United States): a report from the Children's Cancer Group. Cancer Causes Control 1996;7(6):581–90.

31. Austin CA, Patel S, Ono K, Nakane H, Fisher LM. Site-specific DNA cleavage by mammalian DNA topoisomerase II induced by novel flavone and catechin derivatives. Biochem J 1992;282:883–9.
32. Yamashita Y, Kawada SZH. Induction of mammalian topoisomerase II dependent DNA cleavage by nonintercalative flavonoids, genistein and orobol. Biochem Pharmacol 1990;39:737–44.
33. Epstein RJ. Topoisomerases in human disease. Lancet 1988;1:521–4.
34. Steinmetz KA, Potter JD. Vegetables, fruit, and cancer. II. Mechanisms. Cancer Causes Control 1991;2:327–42.
35. Spector LG, Xie Y, Robison LL, et al. Maternal diet and infant leukemia: the DNA topoisomerase II inhibitor hypothesis: a report from the Children's Oncology Group. Cancer Epidemiol Biomarkers Prev 2005;14(3):651–5.
36. Robison LL, Buckley J, Daigle A, et al. Maternal drug use and risk of childhood non-lymphoblastic leukemia among offspring: a report from the Childrens Cancer Study Group. Cancer 1989;64:1169–76.
37. Thompson JR, Gerald PF, Willoughby MLN, Armstrong BK. Maternal folate supplementation in pregnancy and protection against acute lymphoblastic leukaemia in childhood: a case-control study. Lancet 2001;358:1935–40.
38. Shaw AK, Infante-Rivard C, Morrison HI. Use of medication during pregnancy and risk of childhood leukemia (Canada). Cancer Causes Control 2004;15(9):931–7.
39. Wen W, Shu XO, Potter JD, et al. Parental medication use and risk of childhood acute lymphoblastic leukemia. Cancer 2002;95(8):1786–94.
40. Ross JA, Blair CK, Olshan AF, et al. Periconceptional vitamin use and leukemia risk in children with Down syndrome: a Children's Oncology Group study. Cancer 2005;104(2):405–10.
41. Dockerty JD, Herbison P, Skegg DC, Elwood M. Vitamin and mineral supplements in pregnancy and the risk of childhood acute lymphoblastic leukaemia: a case-control study. BMC Public Health 2007;7(147):136.
42. Peters JM, Preston-Martin S, London SJ, Bowman JD, Buckley JD, Thomas DC. Processed meats and risk of childhood leukemia (California, USA). Cancer Causes Control 1994;5:195–202.
43. Jensen CD, Block G, Buffler P, Ma X, Selvin S, Month S. Maternal dietary risk factors in childhood acute lymphoblastic leukemia (United States). Cancer Causes Control 2004;15(6):559–70.
44. Petridou E, Ntouvelis E, Dessypris N, Terzidis A, Trichopoulos D. Maternal diet and acute lymphoblastic leukemia in young children. Cancer Epidemiol Biomarkers Prev 2005;14(8):1935–9.
45. Kwan ML, Block G, Selvin S, Month S, Buffler PA. Food consumption by children and the risk of childhood acute leukemia. Am J Epidemiol 2004;160(11):1098–107.
46. Shu XO, Gao YT, Brinton LA, et al. A population-based case-control study of childhood leukemia in Shanghai. Cancer 1988;62:635–44.
47. Golding J, Paterson M, Kinlen LJ. Factors associated with childhood cancer in a national cohort study. Brit J Cancer 1990;62:304–8.
48. Golding J, Greenwood R, Birmingham K, Mott M. Childhood cancer, intramuscular vitamin K, and pethidine given during labour. Br Med J 1992;305:341–6.
49. Ekelund H, Finnstrom O, Gunnarskog J, Kallen B, Larsson Y. Administration of vitamin K to newborn infants and childhood cancer. Br Med J 1993;307:89–91.
50. Klebanoff MA, Read JS, Mills JL, Shiono PH. The risk of childhood cancer after neonatal exposure to vitamin K. N Engl J Med 1993;329:905–8.
51. von Kries R, Gobel U, Hachmeister A, Kaletsch U, Michaelis J. Vitamin K and childhood cancer: a population based case-control study in Lower Saxony, Germany. Br Med J 1996;313(7051):199–203.
52. Orjuela MA, Titievsky L, Liu X, et al. Fruit and vegetable intake during pregnancy and risk for development of sporadic retinoblastoma. Cancer Epidemiol Biomarkers Prev 2005;14(6):1433–40.
53. Bunin GR, Meadows AT, Emanuel BS, Buckley JD, Woods WG, Hammond GD. Pre- and post-conception factors associated with heritable and non-heritable retinoblastoma. Cancer Res 1989;49:5730–5.
54. Olshan AF, Smith JC, Bondy ML, Neglia JP, Pollock BH. Maternal vitamin use and reduced risk of neuroblastoma. Epidemiology 2002;13:575–80.
55. Michalek AM, Buck GM, Nasca PC, Freedman AN, Baptiste MS, Mahoney MC. Gravid health status, medication use, and risk of neuroblastoma. Am J Epidemiol 1996;143:996–1001.
56. Grufferman S, Wang HH, Delong ER, Kimm SYS, Delzell ES, Falletta JM. Environmental factors in the etiology of rhabdomyosarcoma in childhood. J Natl Cancer Inst 1982;68:107–13.

6 Prevention of Upper Gastrointestinal Tract Cancers

Elizabeth T.H. Fontham and L. Joseph Su

Key Points

- Prevention of both tobacco use and heavy alcohol consumption is a key factor in the primary prevention of upper gastrointestinal tract cancers.
- Infection with *Helicobacter pylori* is a gastric carcinogen. Only about 1–2% of persons infected with *H. pylori* ultimately are diagnosed with gastric cancer; therefore, an understanding of the cofactors in this process is an important area of research. Further, infection with CagA+ *H. pylori* has been associated with an apparent reduced risk of esophageal adenocarcinoma which may complicate treatment decisions.
- Barrett's esophagus is believed to be a premalignant lesion for esophageal adenocarcinoma. Obesity, as determined by elevated body mass index, is also associated with adenocarcinoma of the esophagus, rates of which are increasing in the United States. Maintaining body weight in the normal range throughout adult life is important for the prevention of many cancers, including esophageal adenocarcinoma.
- Consumption of fruits and vegetables is lower than optimal in the United States and increased consumption of these foods will have beneficial effects on the incidence of esophageal and stomach cancers as well as other major chronic diseases.
- Specific micronutrients or combinations of micronutrients may be effective in the prevention of upper gastrointestinal tract cancers, but this has not yet been consistently established in randomized controlled trials.

Key Words: Gastric cancer; esophageal cancer; H. Pylori injection; obesity; tobacco use; alcohol use; fruits and vegetable consumption

6.1. INTRODUCTION

This chapter will focus on lifestyle factors associated with cancers of the esophagus and the stomach. Unlike major cancers such as prostate and breast whose etiologies remain obscure at the present time, hindering primary prevention, cancers of the upper gastrointestinal tract offer well-defined intervention opportunities. Epidemiologic studies have clearly established the important role of alcohol, tobacco, and diet, and recent findings have documented the relation between infection with *Helicobacter pylori* and cancers of the upper gastrointestinal tract. These factors and their interactions will be discussed for cancers of each of these two sites, which together account for approximately 38,000 new cases and 25,000 deaths annually in the United States *(1)*.

6.2. CANCER OF THE ESOPHAGUS

There are two major histologic types of esophageal cancer: squamous cell carcinoma and adenocarcinoma. Squamous cell carcinoma arises in squamous cells that line the esophagus, and it usually occurs in the upper and middle part of the esophagus. Adenocarcinoma begins in the glandular tissue in the lower part of the esophagus at the junction between the esophagus and the stomach. For many

A. Bendich, R.J. Deckelbaum (eds.), *Preventive Nutrition*, Nutrition and Health, DOI 10.1007/978-1-60327-542-2_6,

years, cancer of the esophagus in the United States, and in most areas throughout the world, was virtually synonymous with squamous cell carcinoma *(2)*. Hence, most of the established risk factors for esophageal cancer are specific to this cell type, which comprised the vast majority of cases in studies of this cancer. Recent shifts in the histopathologic cell type have given rise to a rapid increase in the incidence of adenocarcinoma of the esophagus in the United States, particularly among white males *(3,4)*. Because of the increasing importance of esophageal adenocarcinoma, a separate section will consider this entity, which may differ in etiology from squamous cell carcinoma.

6.2.1. Squamous Cell Carcinoma

6.2.1.1. Tobacco and Alcohol Consumption

Both tobacco and heavy alcohol consumption are well-established risk factors for esophageal carcinoma in both men and women. In the United States and other Western countries, over 90% of the risk can be attributed to the individual and joint effects of tobacco and alcohol *(5)*.

An early study by Wynder and Bross *(6)* graphically examined the interaction between alcohol and tobacco and the data suggest a multiplicative effect. Tuyns et al. *(7)* evaluated this relation more formally in data from a case–control study in Brittany. At the highest level of consumption of both alcohol (≥121 g ethanol/day) and tobacco (≥30 g/day), the risk of esophageal cancer was 156 relative to non- or light consumers. The increased risk associated with alcohol consumption appears exponential, whereas increased tobacco smoking appears to yield a more linear increase. Saracci *(8)* estimates that the excess risk because of the interaction of alcohol and tobacco is about 25-fold.

Data from a case–control study in Italy are presented in Table 6.1 *(9)*. Study subjects included 271 male cases and 1,754 male controls with acute illnesses unrelated to tobacco and alcohol consumption. Even with a reference category that included moderate alcohol consumption (<35 drinks/week) by nonsmokers, the estimated odds ratio (OR) of esophageal cancer among heavy smokers (≥25 cigarettes/day, ≥40 year) and very heavy drinkers (≥60 drinks/week) is 22. This report was updated in 1994 to include women *(10)*. Among alcohol drinkers (any vs. none), similar risks were observed for women and men, 3.0 and 4.7, respectively; however, male abstainers had a twofold increased risk, while female nondrinkers had a reduced risk, 0.7, compared with light to moderate drinkers. This study of esophageal cancer fails to support the hypothesis posed by Blume *(11)* that women may be more susceptible to the effects of alcohol, at least for this particular cancer site. An analysis of pooled data from three case–control studies, 1984–1993 in the provinces of Milan and Pordenone *(12)*, 1992–1997 in the provinces of Padua and Pordenone, and the greater Milan area, northern Italy *(13)*, and 1992–1999 in the Swiss Canton of Vaud *(14)*, with 114 female squamous cell esophageal cancer cases and 425 female controls concludes that tobacco smoking and alcohol drinking are the most important risk factors for this neoplasm in women as they are for men in northern Italy and Switzerland *(15)*. An additive interaction (OR = 12.75) was observed for current smokers with heavy drinking (≥3 drinks/day). A 2003 meta-regression analysis pooled eight case–control studies; it also suggested an additive interaction between alcohol drinking of 4+ drinks/day and 30+ cigarettes smoked per day with odds ratio of 12.7 when compared to nondrinkers and nonsmokers *(16)*.

Whether the increased risk of esophageal cancer attributed to alcohol use is a function of the dose of ethanol or whether the type of alcoholic beverage and its other constituents play a role has also been examined in a Japanese study by Hanaoka et al. *(17)* in a study in northern Italy by Zambon et al. *(18)* and in a recent study in Spain by Vioque et al. *(19)*. Their findings confirm those of others that indicate the amount of alcohol consumed, rather than any particular type, is the primary determinant of risk.

Table 6.1
Adjusted Odds Ratios[a] for Cancer of the Esophagus By Alcohol and Tobacco Consumption[b]

	Alcohol (Drinks Per Week)		
Smoking Status	*<35*	*35–59*	*60+*
Nonsmoker	1.0[c]	2.2	2.6
Light	2.1	4.4	5.5
Moderate	4.4	9.7	11.4
Heavy	8.4	18.5	21.8

[a]Adjusted for age, residence, education, and profession.
[b]Adapted from Baron et al. *(8)*.
[c]Reference category.

African Americans and women account for a disproportionate share of mentholated cigarette consumption. Of interest, mentholated brands represented only a fraction (<3%) of total cigarette sales until 1955 *(20)*, but thereafter they increased, peaking at 29% of market share in the late 1970s. Following this changing profile of cigarette consumption was a striking increase in esophageal cancer incidence rates in the 1970s and the 1980s, with a marked apparent impact on rates among African Americans *(21,22)*. Several mechanisms might explain the relationship of mentholated cigarette smoke to esophageal cancer risk and the predominance of squamous cell carcinoma in African Americans. Long-term exposure to menthol attenuates sensations of heat, increasing the potential for thermal damage *(23)*. A study by Hebert et al. also showed an increased risk of squamous cell esophageal cancer associated with smoking mentholated cigarettes in women *(22)*.

Constituents of tobacco smoke have also changed over time *(24)*. The tar yield of cigarettes has decreased over the last few decades *(25)*, which may have influenced the recent trend of declining mortality from upper digestive tract neoplasms in males in several industrialized European countries *(26,27)*. An earlier study in northern Italy reported odds ratio of 2.3 for esophageal cancer in high-tar cigarette smokers compared with low-tar smokers *(28–30)*. The recent analysis of results from two case–control studies, from Italy and Switzerland, involving 395 squamous cell esophageal cancer cases and 1,066 matched controls concludes that there is a direct relationship between tar yield of cigarettes and esophageal cancer. The odds ratios for current smokers compared to never smokers were 4.8 for <20 mg and 5.4 for ≥20 mg tar, respectively, for esophageal cancer based on the brand of cigarette smoked for the longest time *(31)*.

6.2.1.2. Thermal Irritation

Thermal injury as a result of drinking very hot liquids has been suggested to increase risk of esophageal cancer by increasing susceptibility to other carcinogenic exposures *(32–34)*. This hypothesis has some support in both ecologic and analytic studies. Persons living in regions of the world with high rates of esophageal cancer, such as northern Iran and Siberia, are reported to drink excessively hot tea *(35,36)*.

Martinez *(37)* found that more cases than controls reported drinking hot, rather than warm or cold, coffee in Puerto Rico. Both Segi *(38)* and Hirayama *(39)* found an increased risk of esophageal cancer in persons consuming hot tea gruel. In Latin America, several studies have examined the role of maté drinking. DeStefani et al. found a strong association between hot maté consumption and risk of esophageal cancer in Uruguay *(40)*. An earlier case–control study in Brazil failed to find a significant association *(41)*. In 1994, Castelletto et al. examined the role of maté in an Argentinean case–control study *(42)*. They found alcohol, tobacco, and barbecued meat, but not hot maté, to be the primary risks factors. However, hot maté consumption is found to be significantly associated with precancerous lesions of the esophagus in a cross-sectional endoscopic survey in southern Brazil *(43)*.

A study of chronic esophagitis, a precursor lesion for esophageal cancer, in a high-risk region in China lends support to an etiologic role of thermal injury *(44)*. A greater than fourfold excess of mild and moderate esophagitis was found in young persons 15–26 years of age consuming burning-hot beverages [odds ratio (OR) 4.39; 95% confidence interval (CI) 1.72–11.3]. This study design minimizes recall/response bias because case–control status is not known at the time of interview and suggests that this factor may be important at a relatively early stage in the development of this cancer.

6.2.1.3. Nutrition

6.2.1.3.1. Dietary Studies. Fruits and fresh vegetables are consistently associated in studies throughout the world with decreased risk of esophageal cancer, even after controlling for tobacco and alcohol use. Deficiencies of vitamin C, one of several micronutrients contained in fruits and vegetables, have been reported in several areas of the world with exceptionally high rates of esophageal cancer. These include northern Iran *(35)*, Linxian County, China *(45)*,northern and eastern Siberia *(36)*, and Hungary *(46)*, among others. Other dietary deficiencies are also strongly associated with esophageal cancer risk; these include iron, riboflavin, niacin, molybdenum, zinc, folate, and other trace elements *(47–49)*.

The 1961 report by Wynder and Bross noted significantly lower consumption levels of green and yellow vegetables among male cases compared to controls and a nonsignificantly lower consumption level of fruit *(6)*. Potatoes (RR = 0.4, $p < 0.05$)and bananas (RR = 0.3, $p < 0.01$) were determined to be protective in a case–control study in Singapore *(50)*. Frequent consumption of 16 different fruits and vegetables was associated with decreased risk of esophageal cancer in Iran *(51)*. Relative risks for high- vs. low-consumption levels ranged from 0.4 to 0.9 and findings for 10 of the 16 foods were significantly protective. Similar findings were reported in a study of 304 squamous cell esophageal cancer cases and 743 controls in northern Italy *(52)*. High consumption of raw vegetables, citrus fruit, and other fruits are inversely associated with esophageal cancer, ORs = 0.3, 0.4, and 0.5, respectively, after taking the effect of smoking and drinking into account. Increased variety of the fruit and vegetable intake was also found to be significantly associated with lower risk of esophageal cancer in the recent reanalysis of the study *(53)*. A multicenter population-based case–control study of cancers of the esophagus and the gastric cardia in three geographic areas in the United States found that total fruit and vegetable intake is significantly and inversely associated with risk of squamous cell esophageal cancer (OR = 0.90; 95% CI = 0.82, 0.99) *(54)*. Noncitrus fruits (OR = 0.72; 95% CI = 0.55, 0.93) and raw vegetables (OR = 0.75; 95% CI = 0.57, 0.99) were the two only individual categories that were associated with inversed risk of squamous cell esophageal carcinoma. In the same study, fiber, β-carotene, folate, and vitamins C and B6 were significantly inversely associated with esophageal cancer *(55)*. Dietary cholesterol, animal protein, and vitamin B_{12}, in contrast, were significantly associated with esophageal cancer.

A significant inverse trend ($p < 0.001$) was reported between monthly vitamin C consumption and esophageal cancer in white males in New York State *(47)*. A weaker but significant inverse association was observed for vitamin A intake ($p = 0.03$). A fivefold reduction in risk in the highest tertile of fruit and vegetable consumption (>81 times/month) was also found. A report from New York found no association with vitamin C derived from vegetables *(56)*. However, in this study, only 24% of the eligible cases were included and they may not be representative of the total series of cases. A study conducted in Germany with 52 male squamous cell carcinoma patients and 50 randomly selected male controls found that intake of vitamin C greater than 100 mg/day and vitamin E greater than 13 mg/day is associated with reduced risk of esophageal cancer with odds ratios of 0.33 (95% CI = 0.11, 0.92) and 0.13 (95% CI = 0.10, 0.50), respectively, after adjusting for known risk factors such as cigarette smoking and alcohol drinking *(57)*.

Ziegler et al. *(58)* found significant inverse associations between relative risk of esophageal cancer and five indicators of general nutritional status, including total fruit and vegetable consumption (RR *0.5*, p-trend < *0.05)*. This case–control study focused on high-risk black males in Washington, DC. An index of vitamin C intake yielded an estimated relative risk of *0.55* (p-trend < *0.05)* for the highest tertile of consumption. The only other micronutrient significantly inversely associated with risk was riboflavin.

Two case–control studies conducted in the high-risk region of Calvados, France, found a protective effect of vitamin C on esophageal cancer risk *(59,60)*. Approximately threefold significant reductions in risk were observed at the highest level of intake of citrus fruits and of dietary vitamin C. Similarly, DeCarli et al. *(61)* reported a relative risk of 0.3 (0.1–0.6) for high-level fruit consumption and non-significant reductions in risk for high-level vegetable intake. In India, Notani and Jayant *(62)* found a more modest reduction from high-level fruit intake (RR = 0.8, 0.5–1.3), but a significant risk reduction among daily consumers of vegetables (RR = 0.4, 0.2–0.7).

Two 1988 reports support the findings of others indicative of protection from high intake of dietary vitamin C and fresh fruits *(63,64)*. Brown et al. *(63)* found a significant halving in risk in the highest tertile of consumption of citrus fruit, all fruits combined, and dietary vitamin C (p < *0.05)*. A relative risk of 0.4 (0.2–0.8) for high-level consumption of raw vegetables and fresh fruit was found in the California study of Yu et al. *(64)*. Li et al. *(65)* found no reduction in esophageal cancer risk associated with fruit consumption in a high-risk region of China, but a homogeneously low level of intake of fruit in this population makes it a poor one in which to evaluate the association *(66)*. Strong protective effects (p-trend < 0.00 1) associated with consumption of citrus fruits and other fruits were reported by Cheng et al. *(67)*, who conducted a large case–control study in Hong Kong. The proportion of esophageal cancer cases attributable to low-consumption levels of citrus fruits in this population was estimated to be 26%. A retrospective cohort study of esophageal cancer in Linxian, China, reported a significant reduction in risk associated with regular consumption of fresh vegetables, RR 0.66 (0.44–0.99) *(68)*.

A large Italian study of esophageal cancer in lifelong nonsmokers afforded the opportunity to evaluate other risk factors in the absence of residual confounding by tobacco use *(69)*. Although the major risk factor was not unexpectedly alcohol, green vegetables and fresh fruit were associated with significantly reduced relative risks of 0.6 and 0.3, respectively. Similar reductions in risk were associated with β-carotene intake. The estimated relative risk for the combination of high alcohol and low β-carotene was 8.6, with an attributable risk of approximately 45%.

Several dietary factors in addition to fruits and vegetables and their constituent micronutrients have been proposed as candidate protective factors, although the epidemiologic evidence to date is considerably more limited. One such factor is green tea, *Camellia sinensis*. Experimental studies have demonstrated antimutagenic and anticarcinogenic effects, especially in the esophagus *(70–73)*. Findings in a population-based case–control study in China provide some support to this hypothesis *(74)*. After adjustment for confounders including tobacco and alcohol, a significant halving of risk was observed in women drinking green tea (OR = 0.50; 95% CI = 0.30, 0.83) and an inverse dose response was observed. The findings in men were not statistically significant; however, a significant protective effect was observed in both men and women who did not smoke or drink alcohol. Similarly, another study conducted in Yangzhong county, Jiangsu Province in China, also found inverse association between green tea consumption among women *(75)*. However, the number of women who reported drinking tea was too small to reach statistical significance. In contrast, green tea drinking is associated with a nonsignificant increased risk of squamous cell carcinoma among men (OR = 1.37; 95% CI = 0.95, 3.70). Since green tea, as well as other drinks, can be consumed at hot temperatures and since excessively hot fluids have been associated with increased risk of this cancer, the relation between drinking burning-hot fluids was also evaluated. The protective effect of green tea was limited to tea taken at normal temperatures. A more recent case–control study by Ke et al. *(76)* with 1,248 cases and same number of controls in a high-risk region for esophageal cancer in south China examined the relationship between esophageal cancer and drinking Congou tea, a type of black tea grown in China. An inverse association was observed for Congou tea consumption (OR = 0.4, 0.28–0.57) with a strong dose–response relationship ($p < 0.001$) after adjusting for alcohol drinking and smoking. The reduced risk was more pronounced if Congou tea was consumed hot (OR = 0.04, 0.01–0.13) than consumed cold. A population-based matched case–control study in Chuzhou District in Jiangsu Province also found significant reduced risk (OR = 0.13; 95% CI = 0.03, 0.62) of squamous cell esophageal cancer in a multivariate regression model adjusting for history of esophageal lesion, fast eating, utensil cleanup methods, family history of cancer, and *H. pylori* infection *(75)*.

Ginseng, which may be taken as a tea, powder, or as a slice of the root, has also been proposed as a potential anticarcinogen. Unlike the polyphenols in green tea, no specific component or mechanism has been elaborated *(77–78)*. Yun and Choi *(78)* reported a case–control study in Korea where ginseng is commonly used. The relative risk of esophageal cancer associated with ginseng intake was 0.20 (0.09–0.38) after adjustment for tobacco, alcohol, and other confounders. This large reduction in risk was observed in both smokers and nonsmokers. No additional studies have been published to date to confirm this preliminary finding.

6.2.1.3.2. Biochemical Studies. A number of studies have examined biochemical nutritional indicators in blood or tissue, with particular focus on antioxidants. Chen et al. *(79)* collected blood specimens from a sample of the population in 65 different counties in China and correlated the concentration of over 10 different antioxidants with county-specific mortality rates for several cancers, including esophageal cancer. A highly significant inverse relation was found between esophageal cancer rates and both plasma ascorbic acid and selenium in men and selenium in women. A nested case–control study based on the Nutrition Intervention Trials conducted in Linxian, China, examined the relation between baseline serum selenium and the subsequent risk of death from squamous cell esophageal cancer after over 15 years of follow-up (1986–2001) *(80)*. This study found a significant inverse association between baseline serum selenium and death from squamous cell esophageal cancer (RR: 0.83; 95% CI = 0.71, 0.98). Additionally, a case-cohort study design using the same study population examining the pretrial serum vitamin E concentration and esophageal cancer incident suggested that the relative risks for comparisons of the highest to the lowest quartiles of serum α-tocopherol were 0.63 (95% CI = 0.44, 0.91) for squamous cell esophageal cancer *(81)*. Another study in a high-risk region of China found low levels of zinc *(82)*. A population-based case–control study conducted in Washington state *(83)* found no significant difference in nail zinc concentrations in esophageal cancer cases and controls but a large and significant reduction in risk associated with dietary intake of zinc from foods and supplements: OR of *0.5* and 0.1 for the middle and upper tertile of consumption, respectively, trend $p < 0.001$. Other elements in nail tissue associated with esophageal cancer were iron (OR = 2.9 high vs. low levels), calcium (OR 2.6), and cobalt (OR = 1.9). Although this study suggests a number of differences in mineral levels of cases and controls reflecting differences in intake, metabolism, or both, additional investigation is warranted to determine which, if any, of these findings is etiologically meaningful.

6.2.1.3.3. Chemoprevention Studies. Chemoprevention as defined by Sporn and Newton *(84)* is prevention of cancer with pharmacological agents used to inhibit or reverse the process of carcinogenesis. In this relatively new field, which has grown in acceptance since the 1980s, esophageal cancer is one of the first cancer sites for which results from completed trials are available.

Munoz et al. *(85)* reported findings from the first short-term intervention trial in 1985. A total of 610 subjects aged 35–64 in the high-risk region of Huixian, China, were randomized to receive 15 mg (50,000 IU) retinol, 200 mg riboflavin, and 50 mg zinc or placebo once per week for 13.5 months. Five hundred sixty-seven participants completed the trial and underwent endoscopy for histological diagnosis of premalignant lesions of the esophagus (esophagitis, atrophy, dysplasia). The combined treatment had no effect on the prevalence of precancerous lesions of the esophagus. It should be noted, however, that the dose was relatively small and the intervention period short. Micronuclei in exfoliated cells of buccal and esophageal mucosa were evaluated in 170 study subjects from this same trial as an indicator of chromosomal damage *(43)*. No reduction in the prevalence of micronuclei was found in buccal mucosa of subjects after treatment, but a significant reduction in the percentage of micronucleated cells in esophageal mucosa was observed in treated subjects (0.19%) compared to the placebo group (0.31%), $p = 0.04$. In a third report from this same trial, Wahrendorf et al. *(86)* reanalyzed data by blood levels of retinol, riboflavin, and zinc at the beginning and the end of the trial because improvement in blood retinol and zinc levels had been observed in the placebo group as well as the actively treated group. Individuals who had large increases in retinol, riboflavin, and zinc blood levels were more likely to have a histologically normal esophagus at the end of the trial, regardless of the treatment group.

Two large intervention studies conducted in the high-risk population of Linxian, China, were reported in the mid-1990s *(87,88)*. A 6-year randomized trial of daily vitamin/mineral supplementation vs. placebo found no significant reductions in cancer incidence or mortality among adults with preexisting precancerous lesions of the esophagus *(87,89)*. The larger trial in this same area included 29,584 subjects from the general population randomly allocated to combinations of retinol and zinc, riboflavin and niacin, vitamin C and molybdenum, and/or β-carotene, vitamin E, and selenium in doses of one to two times US Recommended Daily Allowances. Significantly reduced total mortality (RR = 0.91, 0.84–0.99) and stomach cancer mortality (RR = 0.79, 0.64–0.99) were observed in those taking β-carotene, vitamin E, and selenium. No significant effects on mortality or cancer incidence, including esophageal cancer, were observed for any of the other vitamin/mineral combinations.

Wang et al. *(90)* evaluated whether any of the vitamin/mineral supplement combinations affected the prevalence of clinically silent precancerous lesions and early invasive cancers of the esophagus and the stomach as determined by endoscopy and biopsy in this same trial. No significant reductions in risk of dysplasia or cancer were observed for any of the supplements, although retinol and zinc were suggestively associated with a lower risk of gastric cancer, OR = 0.38, $p = 0.09$. Similarly, Dawsey et al. *(91)* evaluated the effect of the single vitamin/mineral supplement used in the trial of persons with esophageal dysplasia to see if treatment reduced the prevalence of histological dysplasia or early cancer of the esophagus or the gastric cardia. Modest, nonsignificant risk reductions were observed compared to placebo, (OR = 0.86, 0.54–1.38*)*. The authors concluded that longer interventions with larger number of subjects are required to adequately evaluate the effectiveness of micronutrient supplementation in this high-risk population. In subjects from this same trial, Rao et al. *(92)* evaluated whether epithelial proliferation, an early step in carcinogenesis, was reduced by treatment after 30 months of intervention. The results were similarly inconclusive.

Seven hundred and seventy-eight subjects in the Linxian and Huixian China studies were followed with linear repeated biopsies and histopathology examination for 11 years (1989–2000). Of these, 400 subjects with different severity of esophageal precancerous lesions were randomly divided into two groups for intervention studies with calcium daily and decaffeinated green tea versus placebo for approximately 1 year *(93)*. The study concluded that there was no beneficial effect from either calcium or decaffeinated green tea in alleviating esophageal precancerous lesions and abnormal cell proliferation patterns.

6.2.2. Adenocarcinoma

6.2.2.1. Barrett's Esophagus and Medications

Barrett's esophagus is characterized by the replacement of the lower esophagus, which is normally stratified squamous epithelium, by metaplastic columnar epithelium *(94)*. This condition, attributed to chronic esophageal reflux, is believed to be a premalignant lesion for esophageal adenocarcinoma *(95)*.

Barrett's esophagus displays a similar age, race, and gender distribution as does esophageal adenocarcinoma: it is most common in white males over age 40 *(3,96)*. The reported incidence of esophageal adenocarcinoma in patients with Barrett's is from 30 to over 100 times greater than the rate observed in the general population *(96–99)*.

There also appears to be a familial form of this disease, inherited as an autosomal dominant trait *(100–102)*. Reports of families with the inherited form of Barrett's provide additional support for Barrett's as a precursor lesion.

A related hypothesis has proposed that the use of medications that relax the esophageal sphincter, and thereby promote reflux, may increase the risk of adenocarcinomas of the esophagus and the gastric cardia *(90)*. Histamine H_2 receptor antagonists used routinely for treatment of peptic ulcer and gastroesophageal reflux disease have also been proposed as an etiologic factor *(103)*. In a 1995 report, Chow et al. *(104)* examined the relation between reflux disease and its treatment to risk of adenocarcinomas of the esophagus and the gastric cardia. Significant increased risks of adenocarcinoma were associated with esophageal reflux (OR = 2.1, 1.2–3.6); hiatal hernia (OR = 3.8.1.9–7.6); and esophagitis/esophageal ulcer (5.0, 1.5–16.4). Although a fourfold increased risk was associated with four or more prescriptions for H_2 antagonists, the odds ratio was reduced to 1.5 (0.4–5.4), after adjusting for predisposing conditions. The relation with the use of anticholinergics adjusted for number of conditions was actually inverse: risk decreased with increasing number of prescriptions (p-trend = 0.08). The study findings support the elevated risk of adenocarcinoma conferred by reflux disease but indicate that the mechanism is not strongly related to the treatment of reflux. An interesting finding from a multicenter, population-based case–control study conducted between 1993 and 1995 in three areas of the United States indicates an increased risk of esophageal adenocarcinoma among long-term users of theophylline-containing drugs *(105)*. A hospital-based case–control study in United Kingdom also found that theophylline-containing drugs were significantly associated with esophageal adenocarcinoma *(106)*. The significance of this finding is linked to the rising incidence of asthma and increasing use of asthma medications in the general population and its association with reflux disease. However, a Swedish cohort study followed 92,986 adult patients hospitalized for asthma from 1965 to 1994 for an average of 8.5 years to evaluate their risk of esophagus adenocarcinoma *(107)*. The study findings suggest that the excess risks for the cancer were largely restricted to asthmatic patients who also had a discharge record of gastroesophageal reflux (SIR = 7.5; 95% CI: 1.6–22.0). No association was found for nonreflux asthmatic patients.

6.2.2.2. Tobacco and Alcohol

Two population-based studies of cancers of the esophagus and the gastric cardia conducted in western Washington state (1983–1990) were analyzed to evaluate risk factors for adenocarcinoma compared to squamous cell *(108)*. Use of alcohol and cigarettes was significantly associated with increased risk of both histologic types, but the odds ratios were markedly higher for squamous cell carcinoma. For current smokers of 80+ pack-year compared to nonsmokers, the odds ratios were 16.9 (4.1–6.91) for squamous cell carcinoma and 3.4 (1.4–8.0) for adenocarcinoma. Similarly, for persons who reported drinking 21 or more drinks/week compared to <7/week, the respective odds ratios were 9.5 (4.1–22.3) and 1.8 (1.1–3.1). Population-attributable risk estimates found that cigarette smoking and alcohol together accounted for 87% of the squamous cell carcinomas, while for adenocarcinoma the estimate for cigarettes was 34% and 10% for alcohol consumption of seven or more drinks/week.

Estimates of esophageal adenocarcinoma risk for alcohol and tobacco use by Kabat et al. *(109)* were similar: current smokers, 2.3 (1.4–3.9); 4+ oz of whiskey equivalents per week, 1.9 (1.3–4.3). Brown et al. *(110)* also report that tobacco and alcohol are likely etiologic factors but conferring lower magnitude risk than that associated with squamous cell cancers. The odds ratios at the highest level of smoking (≥40 cigarettes/day) and drinking (≥29 drinks/week) were 2.6 (p-trend < 0.01) and 2.8 (p-trend < *0.05*), respectively. Their study included white men from Atlanta, Detroit, and New Jersey. Significantly increased risks were also found associated with history of ulcer, especially duodenal, and with low social class. The authors note that alcohol and tobacco use, although associated with esophageal adenocarcinoma, does not explain the increased incidence of these tumors in the United States.

In 1997, a multicenter study of esophageal and gastric cancers reported an increased risk of squamous cell carcinoma and adenocarcinoma of the esophagus and adenocarcinomas of all sites in the stomach among smokers *(111)*. Current smokers had a two- to threefold increased risk of adenocarcinomas of the esophagus and the gastric cardia compared to a fivefold increased risk of squamous cell carcinoma of the esophagus. Although risk of these squamous cell tumors declined with duration of smoking cessation, risks of esophageal and gastric cardia adenocarcinomas remained significantly elevated for more than 30 years after cessation. This long lag suggests that the effect of tobacco on these tumors may be on tumor initiation.

More recently, Freedman et al. *(112)* analyzed the data established by the NIH-AARP Diet and Health Study cohort to examine the relationship between esophageal cancer risk and cigarette smoking and alcoholic beverage use. The cohort included 474,606 members of the American Association of Retired Persons resided in six states (California, Florida, Louisiana, New Jersey, North Carolina, and Pennsylvania) and two metropolitan areas (Atlanta and Detroit). Between the initiation of the study in 1995/1996 and the follow-up through 2000, 97 incident cases of ESCC and 205 of esophageal adenocarcinoma were identified. This study found that compared with nonsmokers, current smokers were at increased risk for squamous cell carcinoma [hazard ratio (HR) = 9.27; 95% CI = 4.04, 21.29] and esophageal adenocarcinoma (HR = 3.70; 95% CI = 2.20, 6.22). For drinkers of more than three alcoholic beverages per day, compared with those who drank up to one drink per day, this study found significant associations between alcohol intake and squamous cell esophageal carcinoma risk (HR = 4.93; 95% CI = 2.69, 9.03). However, no significant association with the risk for esophageal adenocarcinoma was found.

6.2.2.3. Obesity and Diet

Two 1995 reports have linked obesity to adenocarcinoma of the esophagus *(108,113)*. A threefold increased risk ($p < 0.01$) was observed at the highest level of body mass index (>26.6 kg/m^2) compared to the lowest in white men *(113)*. No significant associations were found for dietary fat, total calories, meals eaten per day, or consumption of coffee and tea. A protective effect of high intake of raw fruit (OR = 0.4, $p < 0.05$) and vegetable (OR = 0.4, $p < 0.05$) was observed.

Vaughan et al. *(108)* report divergent associations for squamous cell and adenocarcinoma with body mass index. A significantly increased risk of adenocarcinoma was found at the highest decile of body mass index (OR 1.9, 1.1–3.2), whereas body mass was inversely associated with squamous cell carcinoma. The population-attributable risk for body mass index above the 50th percentile was 18% for adenocarcinoma. These observations are consistent with esophageal reflux associated with obesity. These reports were confirmed in a large multicenter population-based case–control study *(114)* in which obesity measured by body mass index was found to be a strong risk factor for esophageal adenocarcinoma and a moderate risk factor for adenocarcinoma of the gastric cardia. The authors suggest that the increasing rates of adenocarcinomas of the esophagus and the cardia may be explained in part by increasing prevalence of obesity in the US population.

Several mechanisms underlying the association between obesity and esophageal adenocarcinoma have been proposed. One is that obesity exacerbates gastroesophageal reflux disease through increased intra-abdominal pressure *(115)*, and the subsequent development of hiatal hernia *(116)* may give rise to encroachment of columnar metaplasia (known as Barrett's esophagus), which is a compensatory response to corrosive effects of digestive enzymes, including gastric and bile acids. Alternatively, central adiposity is also associated with several hormones, such as insulin-like growth factor and adiponectin, which are known to influence carcinogenesis, i.e., cell division, cell death, healing *(116,117)*. A recently nested case–control study within 206,974 members of the Kaiser Permanente multiphasic health checkup cohort found that increasing abdominal diameter ($\geq$25 vs. <20 cm) was strongly associated with an increased risk of esophageal adenocarcinoma (OR = 3.47; 95% CI = 1.29, 9.33) *(116)*. Abdominal diameter was not associated with the risk of cardia adenocarcinomas or esophageal squamous cell carcinomas.

A 2007 report of a meta-analysis of antioxidant intake and risk of esophageal and gastric cardia adenocarcinoma examined data from 10 studies including 1,057 cases of esophageal adenocarcinoma and 644 cases of gastric cardia cancer *(118)*. Summary estimates indicate inverse associations between esophageal adenocarcinoma and higher intake of vitamin C (OR 0.45, 0.39–0.62) and β-carotene (OR = 0.46, 0.36–0.59). Higher intake of vitamin E was not significantly associated with reduced risk (OR = 0.80, 0.63–1.03). Risk of adenocarcinoma of the gastric cardia was inversely associated only with high intake of dietary β-carotene (OR = 0.57, 0.46–0.72).

6.2.2.4. *HELICOBACTER PYLORI*

Peptic ulcer disease and gastric cancer of the antrum and the body have been declining in the past decades *(119,120)*. Gastroesophageal reflux disease, Barrett's esophagus, and esophageal adenocarcinoma rates, in contrast, have increased rapidly, particularly in the Western countries. Recent studies *(121–123)* suggest that this phenomenon may be related to the simultaneous fall in the prevalence of *H. pylori*. A Swedish cohort of 32,906 middle-aged subjects enrolled between 1974 and 1992 were followed through April 1999 *(122)*. *H. pylori* infection was determined through analysis of *H. pylori* antigen in blood samples. A nested case–control study of 49 esophageal cancer cases and 10 matched sets of controls were included in the analysis. Unexpectedly, *H. pylori* seropositivity was present in 22.7% cases and 45.0% of controls. After adjusting for confounding factors, esophageal cancer was associated with *H. pylori* infection (OR = 0.29, 0.12–0.67). The inverse effect between *H. pylori* infection and cancer of the esophagus is stronger for esophageal adenocarcinoma (OR = 0.16) than for squamous cell carcinoma (OR = 0.41). However, the histologic-specific associations did not reach statistical significance because of the small number of cases. A recent meta-analysis of *H. pylori* and esophageal cancer risk concluded that colonization with CagA+ *H. pylori*, a subtype associated with virulence in the stomach, was inversely associated with risk of esophageal adenocarcinoma (OR = 0.41, 0.46–0.68) but there was no association with CagA– *H. pylori* infection. There was no association between *H. pylori* and squamous cell carcinoma of the esophagus *(124)*. These findings and several others have sometimes been reported as *H. pylori* infection "protects" against esophagus cancer *(125,126)*. It is possible that decreased gastroesophageal reflux among *H. pylori*-infected people would reduce the risk for developing an esophageal cancer. Additional information is needed to clarify the nature of the relationship between *H. pylori* infection and esophageal cancer risk. The American College of Gastroenterology guidelines for appropriate testing and treatment of *H. pylori* are given at the end of this chapter and the divergent findings for gastric cancer and esophageal adenocarcinoma are considered.

6.3. CANCER OF THE STOMACH

A steady decline in gastric cancer has been apparent in many countries for the past several decades. The declining rates were first noted in the United States as early as 1930 *(127)* and have persisted into this century *(120)*. Survival rates have not appreciably changed *(120,128)*, therefore, the decline in deaths cannot be attributed to better treatment and prolonged survival but to actual declines in incidence that are now well documented *(129)*. This decline, believed to reflect changes in environmental factors, has been referred to as an "unplanned triumph" since the shifts did not result from active medical or public health intervention and are believed to result from large shifts in food processing and consumption *(130)* as well as a declining prevalence of *H. pylori* infection. It should be noted that the increase in esophageal adenocarcinoma documented in the previous section does include an increase in adenocarcinomas of the gastroesophageal junction and the gastric cardia.

6.3.1. Histologic Types

Adenocarcinomas account for more than 97% of gastric cancers, and studies of etiology are generally limited to this histologic type *(131)*. Building on an earlier observation that gastric carcinomas were often accompanied by features found in intestinal epithelium *(132)*, Lauren *(133)* proposed a classification of adenocarcinomas into two subtypes, "intestinal" and "diffuse." Many, but not all tumors, can thus be classified because some tumors contain characteristics of both types and others neither. Diffuse carcinomas, sometimes referred to as "endemic," tend to occur with similar frequency throughout the world, whereas the distribution of intestinal or "epidemic" type tends to parallel the distribution of overall gastric cancer rates, i.e., this type is relatively more common in areas with high rates and lower where gastric cancers are low *(134)*.

6.3.2. Anatomic Subsites

The stomach is composed of four anatomic regions: the cardia, the fundus, the body, and the antrum, which includes the pylorus. In considering cancer risk factors and prevention, these subsites are often categorized simply as cardia and noncardia cancers. The cardia begins at the gastroesophageal junction and extends distally for several centimeters. Like esophageal adenocarcinomas, rates of adenocarcinoma of the gastric cardia are also increasing and the two cancers share etiologic factors. Because of the close anatomic proximity, there is some misclassification in the assignment of primary site for cancers located in the gastroesophageal junction area.

6.3.3. Risk Factors

6.3.3.1. *HELICOBACTER PYLORI*

Spiral-shaped bacteria in contact with gastric mucosa were first reported about 100 years ago by Pel *(135)* and ignored for the next 90 years. In 1984, Marshall and Warren *(136)* reported isolating these bacteria in cultures of biopsies taken from patients with gastritis and peptic ulcers undergoing endoscopy. By 1994, the International Agency for Research on Cancer, World Health Organization, had determined that infection with *H. pylori* is carcinogenic to humans and declared it a Group I carcinogen based on the large body of epidemiologic research developed during the 11-year period *(137)*.

H. pylori infection is one of the most prevalent infections worldwide, with a range of 20–40% in developed countries and as high as 70–90% in some developing countries *(138,139)*. Prevalence increases with age and no difference in seroprevalence has been found between males and females *(140)*. Socioeconomic status including poor housing conditions, large family size, number of siblings, and low education attainment are predictors of prevalence of infection as well as of gastric cancer *(141,142)*.

Humans are the only known significant reservoir of *H. pylori (143)*. Person-to-person spread is believed to be key in acquisition of infection which generally occurs in childhood. Intrafamilial spread is supported by studies demonstrating a high prevalence of *H. pylori* among family members of an infected individual. A recent study from Greece examined the entire genome of *H. pylori* strains from 32 members of 11 families. The homology of the *H. pylori* genome in members of the same family was striking and strongly supports the hypothesis of transmission from person to person or from a common source *(144)*. Institutionalized populations have also been studied and significantly higher than expected prevalence rates have been observed in both institutionalized adults and children *(145,146)*. Several routes of person-to-person transmission are viable but have not been firmly established to date: both oral–oral and fecal–oral may occur.

H. pylori-contaminated drinking water and sewerage have been demonstrated around the world. Hegarty et al. *(147)* found *H. pylori* in 60% of samples of surface water and 65% of shallow ground water in several states in the United States.

The role of *H. pylori* in gastric carcinogenesis has been explored in correlation and case–control studies, but this approach has yielded equivocal results, largely because of difficulties in determining temporality *(137)*. Cohort studies have provided material for nested case–control analyses that resolved the issue of temporality. *H. pylori* infection was determined by IgG antibodies in serum collected at the time of cohort enrollment which occurred 6–14 years earlier. Forman et al. *(148)* found an approximately threefold increased risk of subsequent gastric cancer in a cohort of Welsh men; Parsonnet et al. *(149)* reported a relative risk of 3.6 (1.8–7.3) in a cohort of men and women in California; and Nomura et al. *(150)* reported a sixfold significantly increased risk in Japanese–American men living in Hawaii.

The mechanisms by which *H. pylori* infection increases gastric cancer risk are the focus of ongoing investigation. *H. pylori* infection, the main cause of chronic gastritis, has been demonstrated to decrease the concentration of ascorbic acid in gastric juice *(151–154)*. *H. pylori* infection is also associated with varying degrees of inflammation *(155)*. In inflammatory states, nitric oxide may be generated and interact with reactive oxygen species forming new cytotoxic compounds *(156,157)*. Thus, *H. pylori* infection has the potential to increase oxidative stress and decrease antioxidant capacity.

The vast majority of individuals infected with *H. pylori* (~80%) manifest no clinical symptoms and fewer than 5% develop gastric cancer *(158)*. Therefore, an understanding of the determinants of cancer development among infected persons is important. Recent research suggests that the interplay of bacterial virulence factors and host susceptibility factors plays a key role in the subsequent development of gastric cancer *(159)*. Cytotoxin-associated antigen (CagA) and vacuolating toxin (VacA) are associated with cancer outcomes. Risks of gastric adenocarcinoma and gastric atrophy, a premalignant condition, have been associated with *cagA*+ and *vacA*+ strains compared to negative strains *(160–162)*. BabA2+-positive strains are also associated with increased gastric cancer risk. BabA is an adherence factor that causes the bacteria to bind to Lewis B blood group antigens on gastric epithelial cells. Host genetic factors that influence the immune response to infection are also believed to play a role. Current research suggests that a pro-inflammatory response to *H. pylori* is associated with increased risk of noncardia gastric cancer. Pro-inflammatory genotypes of tumor necrosis factor-α and *IL-10* have been associated with a doubling of gastric cancer risk *(163)*.

6.3.3.2. Tobacco and Alcohol

Although both tobacco and alcohol use are weakly associated with increased risk of gastric cancer, the strength and magnitude of the association is much less clear than that for esophageal cancer.

Early case–control studies of gastric cancer and alcohol intake were equivocal, with some reporting positive associations *(164,165)* and others none *(166,167)*. Continued study has yielded similar mixed results. Correa et al. *(168)* found twofold elevations in risk of gastric cancer at the highest level of alcohol intake for both whites and blacks in Louisiana. After controlling for other risk factors, wine (OR 2.10, 1.13–3.89) and hard liquor (OR = 1.95, 1.14–3.34) were significantly associated with risk in whites, but not in blacks. A 1990 report of stomach cancer in Los Angeles males also found an increased risk (OR = 3.0, 1.1–8.7) at the highest level of total ethanol intake and significant risks for daily consumption of beer *(169)*. The effect of alcohol was stronger for cancer of the gastric cardia than at other sites.

A twofold increased risk of stomach cancer was found for beer consumption in a German study, but wine and hard liquor were associated with decreased risk *(170)*. This is in contrast to a French study that reported a very large relative risk (6.9, 3.3–14.3) associated with heavy use of red wine *(171)*.

Two cohort studies, however, suggest that alcohol is not an independent risk factor for gastric cancer. Nomura et al. *(172)* found no increased risk of gastric cancer associated with consumption of beer, wine, or hard liquor in Japanese-American men living in Hawaii. Kneller et al. *(173)* likewise found no association for total alcohol or for any specific cancer type.

More consistent findings link smoking to a 1.5- to threefold increased risk of gastric cancer *(168–173)*; however, the overall increased risk has sometimes failed to demonstrate a dose–response *(164,172,174)*. The cohort study by Kneller et al. did find significant increases in risk with both increasing number of cigarettes smoked per day and pack-years of smoking *(173)*. At the highest number of pack-years, the relation of risk was 2.3 (1.23–4.33) and for current use of 30 or more cigarettes/day, the relative risk was 5.8 compared to nonsmokers. Although age at death did not significantly modify risk, the association with smoking was stronger for younger cases. The authors suggest that this finding may reflect a higher proportion of adenocarcinomas of the gastric cardia at younger ages and a stronger relation between smoking and cancers of the cardia than with cancers of other sites in the stomach.

The evidence on tobacco smoke and cancer risk that has accumulated between 1986 and 2002 was recently reviewed by a Work Group assembled by the International Agency for Research on Cancer *(175)*. For stomach cancer, the conclusions were that current smokers were at increased risk for both cardia and noncardia stomach cancers and that the association between tobacco smoking and gastric cancer risk is independent of alcohol consumption and *H. pylori* infection. The absolute risk, however, tends to be higher among smokers who are *H. pylori* positive *(176)*. The proportion of gastric cancer attributable to tobacco smoking, a modifiable risk factor, is about 17 and 11% in men and women, respectively, in developed countries and somewhat lower, 11 and 4%, for men and women in developing countries.

6.3.3.3. Salted, Pickled, and Smoked Foods

Salt has been demonstrated in animal studies to enhance gastric carcinogenesis *(177–180)*. It has been suggested that the action of salt as a gastric mucosal irritant facilitates the action of carcinogens and thus salt acts as a cocarcinogen *(181)*.

Epidemiologic studies also suggest an increased risk of gastric cancer associated with high-salt intake when salted and pickled foods are included in total intake. Death rates throughout regions of Japan *(182)* were found to be correlated with consumption of salted fish and salted vegetables. A geographic correlation has also been demonstrated in China *(183)*. Consumption of salt-cured meats, salted fish, and other salt-preserved foods has been associated with increased risk in case–control studies throughout the world *(184–187)*. Several studies have also reported associations with the addition of salt to foods *(184,188)* or a reported "heavy intake" *(168,189)*.

Many of the strongest findings have been noted in areas of the world where there is a wide range of intake including very high levels, such as in Korea *(186)*. A nested case–control analysis reported by Friedman and Parsonnet *(190)* failed to find evidence that routine salting led to increased risk in a California study population. "Heavy" salt intake in US populations may be quantitatively less than "heavy" intake in other areas of the world and may not be sufficient to demonstrate an increased risk. For example, salted fish and salted vegetables in Japan may contain up to 30% NaCl compared to isotonic saline, which is 0.8% *(181,182,190)*. One of the difficulties in assessing the role of salt is determination of "exposure" where many assessments have been subjective. A step forward in research was the use of a salt taste test *(191)*. Preference for salty taste was objectively assessed using diluted rice cereal (gruel) containing 0.1, 0.3, 0.5, 0.7, and 0.9% NaCl. The five different diluted rice cereals were arranged randomly and the subjects were asked to select a preferred concentration. Cases of gastric cancer were significantly more likely ($p < 0.01$) to prefer the cereals with high-salt concentrations (>0.5%). In this study, persons who were infected with *H. pylori* and who had a high-salt preference had a 10-fold higher risk of early gastric cancer than uninfected persons preferring low salt concentrations ($p < 0.05$).

Beevers et al. *(192)* conducted a preliminary examination of salt intake and *H. pylori* infection as a follow-up to the observation that both high-salt intake and *H. pylori* infection are associated with hypertension. It should be stressed that this study was an ecologic correlation study and therefore should be interpreted cautiously. National *H. pylori* prevalence rates from 10 countries in the EUROGAST study were correlated with national salt excretion levels from the INTERSALT project from the same 10 countries. Statistically significant correlations were found for older men ($r = 0.728$) and women ($r = 0.827$) and in younger men ($r = 0.728$) but not younger women. These findings raise the possibility that salt intake may in some way facilitate *H. pylori* infection. This possible interaction between salt and *H. pylori* infection is also suggested by a 2006 report from a prospective study of gastric cancer in Japan *(193)*. A statistically significant increased risk of stomach cancer was found among consumers at the highest three levels of dietary salt intake compared with the lowest level (<10 g/day). This increased risk was almost threefold for subjects who were infected with *H. pylori* and had atrophic gastritis.

Numerous *N*-nitroso compounds have demonstrated carcinogenicity *(194)*. Based on studies of premalignant lesions of the stomach, it has been hypothesized that intragastric synthesis of *N*-nitroso compounds is a factor in the gastric carcinogenic process *(195)*.

Two studies evaluated factors associated with in vivo nitrosamine formation in humans using the test developed by Ohshima and Bartsch *(196)* that measures urinary excretion of noncarcinogenic *N*-nitrosoproline after ingesting a given dose of proline. Mirvish et al. *(197)* found that men in rural Nebraska who drank water from private wells with a high nitrate content excreted significantly higher *N*-nitrosoproline than did men drinking water with a low nitrate content. Their findings parallel those of a study in Denmark *(198)*. Sierra et al. *(199)* used the nitrosoproline test in children living in high- and low-risk areas for stomach cancer in Costa Rica. They found the concentration excreted by children in the high-risk area significantly greater ($p < 0.04$) compared to children from the low-risk area. They also found that excretion was markedly reduced when ascorbic acid, an inhibitor of nitrosation reactions, was given with the proline.

Associations between gastric cancer and dietary intake of nitrate, nitrite, and preformed nitroso compounds are suggestive *(200–205)*, but the validity of such indices is not well established given the multiple sources, including food, water, and endogenous formation.

6.3.3.4. Fruits and Vegetables

Table 6.2 presents a selected compendium of 31 dietary studies of gastric cancer reported between 1964 and 1995 *(166,168–170,172,173,184,186,189,200–220)*. The strong, consistent inverse association between consumption of fruits and vegetables is abundantly clear. Of the 26 studies described that specifically examined foods and food groups, 24 found a decreased risk of stomach cancer associated with high intake of one or more fruits and vegetables and the vast majority were statistically significant with up to twofold reductions in risk. Only two studies reported an increased risk of gastric cancer associated with fruits *(173)* or vegetables *(185)*, and their findings do little to cast doubt on the apparent protective effect of fruits and vegetables. The findings of Tajima and Tominaga *(185)* stand in contrast to many case–control studies in Japan and elsewhere, and the study by Kneller et al. *(173)* was based on a very limited dietary questionnaire that increases the likelihood of misclassification.

Norat and Riboli conducted a meta-analysis of published case–control and cohort studies of fruit and vegetable consumption *(221)*. The pooled analysis of all studies found that high intake of vegetables is associated with significantly decreased risk of stomach cancer. Fruit intake is associated with a significantly reduced risk of gastric and esophageal cancers. Estimates of the proportion of stomach cancers worldwide that are potentially preventable by increasing fruit and vegetable consumption are up to 50%.

The 2007 World Cancer Research Fund/American Institute for Cancer Research Report – *Food, Nutrition, Physical Activity and the Prevention of Cancer: a Global Perspective (222)* – presents findings from an exhaustive systematic review of available data. This report included a review of a total of 722 publications on stomach cancer and diet. For specific cancers the panel judged the strength of the evidence for individual factors as convincing; probable; limited-suggestive; limited-no conclusion; or substantial effect on risk unlikely. Factors that either increased or decreased risk were considered.

The overall conclusion of the panel was that "food and nutrition play an important role in the prevention and causation of stomach cancer." The panel concluded that consumption of vegetables, allium vegetables in particular, and fruits probably protect against stomach cancer, while salt and salt-preserved foods probably are causally related to stomach cancer. The panel concluded that there was limited but suggestive evidence of a protective effect from pulses (legumes) and foods containing selenium. They also judged that there is limited but suggestive evidence that chili, processed meat, smoked foods, grilled (broiled), and barbequed (charbroiled) animal foods are causes of gastric cancer.

A website address is included in Table 6.3 for accessing much greater detail from the report than is possible in this chapter.

6.3.3.5. Micronutrients

Consumption of fruits and vegetables serve as dietary sources of a plethora of vitamins, minerals, fiber, and less well-studied trace compounds. Many of these are highly correlated with one another, particularly when exposure is based on dietary assessment; therefore, a finding attributed to one may actually reflect the effect of another constituent from the same foods. The strongest findings, therefore, are based on biochemical studies, e. g., blood levels prior to cancer onset, and chemoprevention trials, which actually test the efficacy of specific micronutrients in prevention. The micronutrients believed

Table 6.2
Selected Epidemiologic Studies of Diet and Stomach Cancer Risk

Study (References)	*Population*	*Number of Cases/control or Cohort Size*	*Food or Nutrient*	*Relative Risk, High vs. Low Intake*
Case-control				
Meinsma *(206)*	Holland	340/1060	Vitamin C	Inverse association
			Citrus fruit	$p = 0.1$ males
				$p = 0.001$ females
Higginson *(207)*	United States	93/279	Dairy foods	0.6
			Fresh fruits	Inverse association
			Raw vegetables	Inverse association
Haenszel et al. *(208)*	Japanese in Hawaii	220/440	Tomatoes	0.4 ($p < 0.05$)
			Celery	0.4 ($p < 0.05$)
			Corn	0.5 ($p < 0.05$)
			Onion	0.5 ($p < 0.05$)
			Lettuce	0.8 (NS)
			Western vegetables combined	0.4 ($p < 0.05$)
Graham et al. *(166)*	United States	276/2200	Lettuce	0.64 (trend $p < 0.01$)
Bjelke *(209)*	Norway and United States	162/1394	Vegetable index (Norway)	Inverse association (Norway and United States)
		259/1657	Vitamin C	Inverse association (Norway and United States)
			Fruits and vegetables (United States)	Inverse association (Norway and United States)
Haenszel et al. *(210)*	Japan	783/1566	Fruit	0.7 ($p < 0.05$)
			Plum and pineapple	0.7 ($p < 0.01$)
			Celery	0.6 ($p < 0.01$)
			Lettuce	0.7 ($p < 0.01$)
Correa et al. *(168)*	United States	391/391	Vitamin C	0.50 (trend $p < 0.05$) whites
				0.33 (trend $p < 0.001$) blacks
			Fruit index	0.47 (trend $p < 0.005$) whites
				0.33 (trend $p < 0.001$) blacks
			Vegetable index	0.50 (trend $p < 0.05$) blacks
			Smoked foods	1.98 (trend $p < 0.025$) blacks
Risch et al. *(200)*	Canada	246/246	Vitamin C	0.43 (trend $p = 0.099$)
			Citrus fruit	0.75 (trend $p = 0.006$)
			Nitrite	2.61 (1.61–4.22)
			Carbohydrates	1.53 (1.07–2.18)

Trichopoulos et al. *(211)*	Greece	110/100	Lemons	0.24 (trend $p < 0.01$)
			Oranges	0.33 (trend $p < 0.01$)
			Pasta	3.42 (trend $p < 0.001$)
			Brown bread	0.79 (trend $p < 0.01$)
			Onions	0.68 (trend $p < 0.001$)
Tajima & Tominaja *(185)*	Japan	93/186	Oranges	0.9 (NS)
			Other fruit	1.4 (NS)
			Spinach	2.5 ($p < 0.05$)
			Cabbage	2.2 ($p < 0.01$)
			Green pepper	2.0 ($p < 0.01$)
Jedrychowski et al. *(212)*	Poland	110/110	Fruit	0.3 (0.1–0.6)
			Vegetables	0.6 (0.3–1.4)
LaVecchia et al. *(189)*	Italy	206/474	Vitamin C	0.46 ($p < 0.001$)
			Fruits, index	0.53 (trend $p < 0.01$)
			Citrus fruit	0.58 (trend $p < 0.01$)
			Green vegetable index	0.33 (trend $p < 0.001$)
			Ham	1.6 ($p = 0.04$)
			Polenta	2.32 ($p = 0.007$)
			β-Carotene	0.39 ($p < 0.001$)
You et al. *(213)*	China	564/1131	Vitamin C	0.5 (0.3–0.6)
			Fresh fruit	0.4 (0.3–0.6)
			Fresh vegetables	0.6 (0.4–0.8)
Buiatti et al. *(184)*	Italy	1016/1159	Raw vegetables	0.6 (trend $p < 0.001$)
			Citrus fruits	0.6 (trend $p < 0.001$)
			Other fresh fruits	0.4 (trend $p < 0.001$)
Graham et al. *(56)*	United States	293/293	Raw vegetables	0.43 (0.23–0.78)
			Fruits	No association
			Vitamin C	No association
			Carotene	0.79 (0.63–0.98)
			Sodium	1.51 (1.20–. 91)
			Retinol	1.47 (1.17–1.85)
			Fat	1.37 (1.08–1.74)
Chyou et al. *(201)*	Japanese in Hawaii	111/361	Vegetable index	0.7 (trend $p < 0.001$)
			Fruit index	0.8 (trend $p = 0.20$)
			Nitrite	No association
Buiatti et al. *(202)*	Italy	1016/1159	Vitamin C	0.5 (trend $p < 0.001$)
			α-Tocopherol	0.6 (trend $p < 0.01$)
			β-Carotene	0.6 (trend $p < 0.01$)
			Protein	2.6 (trend $p < 0.001$)
			Nitrites	1.9 (trend $p < 0.001$)
Wu-Williams et al. *(169)*	United States	137/137	Fruit index	0.7 (NS)
			Beef	1.6 (1.0–2.6)

(Continued)

Table 6.2
(Continued)

Study (References)	*Population*	*Number of Cases/control or Cohort Size*	*Food or Nutrient*	*Relative Risk, High vs. Low Intake*
Buiatti et al. *(203)*	Italy	923/1159	Vitamin C	0.5 (0.3–0.6) intestinal type 0.5 (0.3–0.7) diffuse type
			Citrus fruits	0.5 (0.4–0.7) intestinal type 0.6 (0.4–0.9) diffuse type
			Other fresh fruits	0.5 (0.4–0.7) intestinal type 0.4 (0.3–0.6) diffuse type
			Raw vegetables	0.6 (0.4–0.8) intestinal type 0.6 (0.4–0.9) diffuse type
			α-Tocopherol	0.5 (0.3–0.8) intestinal type 0.5 (0.2–0.8) diffuse type
			β-Carotene	0.7 (0.5–0.8) intestinal type 0.6 (0.4–0.7) diffuse type
			Nitrites	1.8 (1.2–2.8) intestinal type 2.8 (1.5–5.0) diffuse type
			Protein	2.4 (1.02–5.6) intestinal type 5.8 (1.8–1.84) diffuse type
Negri et al. *(215)*	Italy	564/6147	Green vegetables	0.4 (0.3–0.6) (trend $p < 0.001$)
			Fruit	0.4 (0.3–0.5) (trend $p < 0.001$)
Boeing et al. *(170)*	Germany	143/579	Vitamin C	0.37 (trend $p < 0.01$)
			Cheese	0.44 (trend $p < 0.01$)
			Processed meat	1.74 (trend $p < 0.01$)
			Whole wheat bread	0.37 (trend $p < 0.001$)
Gonzalez et al. *(216)*	Spain	354/354	Cooked vegetables	0.5 (trend $p = 0.02$)
			Noncitrus fresh fruits	0.6 (trend $p = 0.006$)
			Dried fruits	0.4 (0.2–0.8)
			Meat	0.6 (trend $p = 0.02$)
Hoshiyama & Sasaba *(217)*	Japan	251/483	Fruits	Inverse association
			Raw vegetables	Inverse association
			Pickled vegetables	Increased risk
LaVecchia et al. *(218)*	Italy	723/2024	β-Carotene	0.38 (trend $p < 0.001$)
			Vitamin C	0.53 (trend $p < 0.001$)
			Methionine	2.40 (trend $p < 0.001$)
Lee et al. *(186)*	Korea	213/213		
Hansson et al. *(219)*	Sweden		Total vegetables	0.58 (0.37–0.89) (trend $p = 0.01$)
			Citrus fruits	0.49 (0.29–0.81) (trend $p = 0.004$)

(Continued)

Table 6.2
(Continued)

Study (References)	*Population*	*Number of Cases/control or Cohort Size*	*Food or Nutrient*	*Relative Risk, High vs. Low Intake*
Hansson et al. *(204)*	Sweden		Vitamin C	0.47 (0.30–0.76) (trend $p = 0.003$)
			β-Carotene	0.73 (0.45–1.18) (trend $p = 0.10$)
			Nitrates	0.97 (0.60–1.59) (trend $p = 0.99$)
Lopez-Carrillo et al. *(220)*	Mexico	220/752	Chili peppers (ever, never)	5.49 (2.72–11.06)
			Chili peppers (high vs. none)	17.11 (7.78–37.59)
Gonzales et al. *(205)*	Spain	354/354	Nitrosomines	2.1 (trend $p = 0.007$)
			Fiber	0.35 (trend $p < 0.001$)
			Folate	0.50 (trend $p = 0.008$)
			Vitamin C	0.58 (trend $p = 0.017$)
Cohort				
Nomura et al. *(172)*	Japanese in Hawaii	150/7990	Fruit index	0.8 (0.5–1.3)
			Fried vegetables	0.8 (0.4–1.6)
Kneller et al. *(173)*	United States	75/17,633	Fruit index	1.5 (trend *p*, NS)
			Vegetable index	0.9 (trend *p*, NS)
			Carbohydrates	1.6 (trend $p < 0.05$)

Table 6.3
Website Addresses for Organizational Nutritional Guidelines

American Cancer Society	http://www.cancer.org
National Cancer Institute	http://www.nci.nih.gov
American Heart Association	http://www.americanheart.org
American Diabetes Association	http://www.diabetes.org
World Cancer Research Fund	http://www.wcrf.org

to be most strongly associated with reduced gastric cancer risk based on studies to date are vitamin C, β-carotene, and vitamin E/selenium.

Findings from dietary estimates of intake are also included in Table 6.2. Relatively high consumption of vitamin C and β-carotene is consistently associated with reduced risk of gastric cancer *(64,168,170,189,200,202–204,206,209,214,215)*. Serological assessment also supports a role. Prospective studies that have evaluated vitamin C are scant because vitamin C deteriorates quickly unless specimens are acid stabilized prior to freezing *(223)*. A large well-conducted cohort study, the Basel study *(224)*,did have such material available. Mean plasma vitamin C was significantly lower in persons who died of cancer than in survivors: 47.61 ± 1.78 mol/L vs. 52.76 ± 0.44 mol/L, respectively, $p < 0.01$. The findings were also significant ($p < 0.05$) for persons who subsequently died of stomach cancer and their blood levels were even lower, 42.86 ± 4.88. Low plasma levels of vitamin C were associated with a relative risk of 2.38 for gastric cancer. Low plasma levels of carotene were similarly associated with significantly increased risk of overall mortality from cancer ($p < 0.01$) and cancer of the stomach ($p < 0.01$), with a relative risk of 2.95. No association was observed between plasma levels of vitamin A or E and gastric cancer.

Haenszel et al. *(225)* measured serum micronutrient levels in persons with various premalignant gastric lesions. Carotene levels in both men and women and vitamin E levels in men were significantly lower in subjects with gastric dysplasia than in subjects with normal mucosa or less-advanced lesions.

A recent report from Japan *(226)* evaluated prediagnostic serum selenium and zinc levels and found no excess risk of stomach cancer in those with the lowest levels of selenium (OR = 1.0) or zinc (OR = 1.2).

The most compelling evidence to date for specific micronutrients in chemoprevention of gastric cancer comes from the previously described population trial in China *(88)* that found a significant reduction in stomach cancer mortality among persons taking a combination of β-carotene, vitamin E, and selenium. No reduction in risk was observed among persons taking vitamin C; however, there was no attempt in this trial to eradicate *H. pylon*, which is known to decrease the concentration of ascorbic acid in gastric juice, either by increased oxidation, impaired secretion from blood into the gastric cavity, or both *(150,152–154)*. Another chemoprevention trial conducted in a high-risk population in the Andes mountain area of Colombia was reported in 2000. After 6 years of follow-up study, participants who were treated with either ascorbic acid or β-carotene or anti-*H. pylori* therapy, individually or in combination, were significantly more likely to have regression of gastric premalignant lesions *(191)*.

6.3.3.6. Other Diet

Green tea has been shown to have a protective effect against several types of cancer in animals. Epidemiologic studies also support a protective role but have been complicated by inaccurate measurement by consumption (e. g., self-reported frequency of consumption without cup size, temperature of water, type and amount of tea leaves used, etc.) *(227)*. Three studies of green tea and gastric cancer in China found a protective effect based on the amount of green tea leaves consumed *(228–230)*. Sasazuki et al. *(227)* measured tea polyphenols in plasma of cases and controls in a nested study based on the Japan Public Health Center-based Prospective Study. For men, an increased risk of gastric cancer was associated with high levels of plasma *(–)*-epigallocatechin (EGCG), while for women, a decreased risk of gastric cancer was associated with a high level of another green tea polyphenol, *(–)*-epicatechin-3-gallate (ECG). The authors suggest that cigarette smoking by men may play a role in the observed male–female differences.

6.3.3.7. Obesity

Obesity is a major risk factor for numerous chronic diseases including diabetes, cardiovascular diseases, and cancer. An increased risk of esophageal cancer has been found between body mass index (BMI) as noted earlier in this chapter. A 2008 report of a prospective study from a cohort of approximately 500,000 men and women (NIH-AARP Diet and Health Study) examined BMI and risk of adenocarcinomas of the gastric cardia and noncardia adenocarcinoma *(231)*. Compared to individuals with BMI in the normal range (18.5–25 kg/m^2), a BMI $\geq$35 was associated with a 2.5-fold increased risk of gastric cardia cancer but no increased risk of cancer at other sites in the stomach (0.84, 0.50–1.42). The risk estimates for gastric cardia were very similar to that found for esophageal adenocarcinoma.

6.4. RECOMMENDATIONS

Primary prevention of esophageal, and to a lesser degree stomach, cancer obviously begins with the prevention of tobacco use by teenagers and cessation among addicted adults. Physician prompting and participation in smoking cessation efforts have proven effective. Nicotine replacement in conjunction with behavioral modification may improve the success rate for smokers attempting to quit. More nicotine replacement methods are available now than ever before. These include gum, which was the first type developed; a patch by which nicotine is absorbed through the skin; inhaler; and a nasal spray. These methods can be used to deliver gradually reduced doses of nicotine without exposure to the myriad harmful chemicals contained in tobacco smoke.

Several non-nicotine-containing pharmacologic approaches to cessation have been tried; however, one has been found to have serious adverse side effects associated with such use and others, which are used in the treatment of specific disorders, have not yet been approved by the Food and Drug Administration for use in cessation. These options are expected to develop with additional research.

A reduction in tobacco use by teenagers has proven a persistent challenge because educational programs are offset by well-funded, effective, targeted marketing and promotion by tobacco companies. The terms of the 1998 Tobacco Settlement have the potential to reduce the impact of industry marketing and promotions, particularly if complemented by use of settlement funds for counter-marketing campaigns in those few states that have chosen this approach. Limiting alcohol consumption to moderate levels is particularly important in smokers, and physicians should actively counsel patients accordingly.

Intake of fresh fruits and vegetables in the United States continues to fall short of the recommended "5-A-Day" *(232)*. More recent programs are encouraging even higher levels of consumption—9-A-Day. Increased consumption should continue to be promoted and benefits are expected to accrue in reduced rates of both of these upper digestive tract cancers as well as other epithelial tumors. Effective population-based approaches are important, and since dietary patterns are often established in childhood, promotion of healthy choices in school-based food service programs is an opportunity that should not be missed.

The current dietary recommendations of the American Cancer Society, the American Heart Association, the American Diabetes Association, the National Cancer Institute, and the World Cancer Research Fund are remarkably similar in direction but differ in specificity. The most consistent general recommendations for prevention of the major chronic diseases in this country, including these cancers, are to eat more fruits, vegetables, and grains; eat less high-fat foods and meats; and to be active and stay fit. Website addresses with up-to-date material on dietary recommendations are included for reference in Table 6.3.

The efficacy of vitamin/mineral supplements has not yet been completely established in clinical trials; however, in case–control and cohort studies the individuals in the highest level of intake of specific micronutrients often combine high dietary intake with supplements. With the obvious caution to avoid excessive intake, a multivitamin/mineral supplement, or specific antioxidant supplement, may complement dietary intake, particularly among persons with excessive oxidative stress, such as smokers.

The original European *H. pylori* treatment guidelines were updated in 2000 in a second Maastricht Consensus Report *(233)*. The guidelines recommend a "test-and-treat" approach in adult patients under 45 years of age, presenting in primary care settings with persistent dyspepsia, after excluding those with predominantly gastroesophageal reflux disease symptoms, nonsteroidal anti-inflammatory drug users, and those with alarm symptoms requiring prompt endoscopic investigation (i.e., unexplained weight loss, dysphagia, recurrent vomiting, digestive bleeding or anemia, abnormal physical examination, malabsorption, and concomitant disease with possible digestive involvement).

Additional recommendations from this report include the following:

- The urea breath test or stool antigen test are recommended as diagnostic tests of choice.
- *H. pylori* eradication is recommended in all patients with peptic ulcers, with low-grade, gastric mucosa-associated lymphoid tissue (MALT) lymphoma, with atrophic gastritis and in patients post gastric cancer resection.
- Eradiation is also strongly recommended in infected patients with a first-degree family history of gastric cancer.

The American College of Gastroenterology updated its guidelines for the management of *H. pylori* in 2007 *(234)*. Their recommendations are as follows for testing and treating *H. pylori* infection:

Diagnosis

Established indications

- *H. pylori* testing is indicated in active peptic ulcer disease, a confirmed history of the disease, or low-grade MALT lymphoma.
- Testing should be performed only if the physician plans to treat the disease.
- Of the nonendoscopic tests, urea breath test is the most reliable for diagnosis or for confirming eradication 4 or more weeks after treatment.
- In patients who have not received a PPI (protein pump inhibitor) within 1–2 weeks or an antibiotic or bismuth within 34 weeks of endoscopy, the rapid urease test is inexpensive and accurate.
- Testing for eradication is indicated in patients with *H. pylori*-associated ulcer, persistent dyspeptic symptoms after treatment, or MALT lymphoma, and in those who have undergone resection of early gastric cancer.
- The monoclonal fecal antigen test can be used to establish eradication.

Controversial uses

- Because the relationship between *H. pylori* and GERD is not well defined, testing patients with GERD is not established.
- Testing and treating first-degree relatives of patients with gastric cancer for *H. pylori* may reduce cancer risk.
- Any patient with an ulcer should be treated for *H. pylori*, regardless of the use of NSAIDs/aspirin.

Treatment

Established indications

- The test-and-treat strategy is recommended for patients aged <55 years with uninvestigated dyspepsia and no alarm features (e.g., bleeding, anemia, early satiety, unexplained weight loss).
- Clarithromycin-based triple therapy for 14 days and bismuth quadruple therapy for 10–14 days are accepted first-line treatments.
- Alternative salvage therapies for persistent *H. pylori* infection include bismuth-based quadruple therapy for 7–14 days and levofloxacin-based triple therapy for 10 days.

Controversial uses

- Eradication of *H. pylori* in patients at risk for gastric cancer or with preneoplastic lesions may prevent progression, improve outcomes.
- A small but significant subgroup of patients with functional dyspepsia will have clinical benefit after infection eradication.
- No clinical evidence that *H. pylori* eradication improves/worsens GERD symptoms.

Guidelines have been issued by other countries throughout the world, including Singapore, Japan, and others. Most support the test-and-treat approach in symptomatic patients. Although the spectrum of clinical outcomes associated with *H. pylori* infection is wide ranging, from asymptomatic to gastric cancer, treatment and eradication of infection as recommended above is important because the cofactors that predispose an infected individual to gastric cancer have not yet been established.

REFERENCES

1. Jemal A, Siegel R, Ward E, et al. Cancer statistics, 2008. CA Cancer J Clin 2008;58:71–96.
2. Moses FM. Squamous cell carcinoma of the esophagus. Natural history, incidence, etiology, and complications. Gastroenterol Clin North Am 1991;20:703–16.
3. Blot WJ, Devesa SS, Kneller RW, et al. Rising incidence of adenocarcinoma of the esophagus and gastric cardia. JAMA 1991;265:1287–9.
4. Koch TR. The changing face of esophageal malignancy. Curr Gastroenterol Rep 2003;5:187–91.
5. Day N, Munoz N. Oesophagus. In: Schottenfeld D, Fraumeni JF, Jr., eds. Cancer Epidemiology and Prevention. Philadelphia, PA: WB Saunders, 1982:596–623.
6. Wynder EL, Bross I. A study of etiologic factors in cancer of the esophagus. Cancer 1961;14.
7. Tuyns AJ, Pequignot G, Jensen OM. Esophageal cancer in Ille-et-Vilaine in relation to levels of alcohol and tobacco consumption. Risks are multiplying. Bull Cancer 1977;64:45–60.
8. Saracci R. The interactions of tobacco smoking and other agents in cancer etiology. Epidemiol Rev 1987;9:175–93.
9. Baron AE, Franceschi S, Barra S, et al. A comparison of the joint effects of alcohol and smoking on the risk of cancer across sites in the upper aerodigestive tract. Cancer Epidemiol Biomarkers Prev 1993;2: 519–23.
10. Franceschi S, Bidoli E, Negri E, et al. Alcohol and cancers of the upper aerodigestive tract in men and women. Cancer Epidemiol Biomarkers Prev 1994;3:299–304.
11. Blume SB. Women and alcohol. A review. JAMA 1986;256:1467–70.
12. Fioretti F, Tavani A, La Vecchia C, et al. Histamine-2-receptor antagonists and oesophageal cancer. Eur J Cancer Prev 1997;6:143–6.
13. Franceschi S, Bidoli E, Negri E, et al. Role of macronutrients, vitamins and minerals in the aetiology of squamous-cell carcinoma of the oesophagus. Int J Cancer 2000;86:626–31.
14. Levi F, Pasche C, Lucchini F, et al. Food groups and oesophageal cancer risk in Vaud, Switzerland. Eur J Cancer Prev 2000;9:257–63.
15. Gallus S, Bosetti C, Franceschi S, et al. Oesophageal cancer in women: tobacco, alcohol, nutritional and hormonal factors. Br J Cancer 2001;85:341–5.
16. Zeka A, Gore R, Kriebel D. Effects of alcohol and tobacco on aerodigestive cancer risks: a meta-regression analysis. Cancer Causes Control 2003;14:897–906.
17. Hanaoka T, Tsugane S, Ando N, et al. Alcohol consumption and risk of esophageal cancer in Japan: a case-control study in seven hospitals. Jpn J Clin Oncol 1994;24:241–6.
18. Zambon P, Talamini R, La Vecchia C, et al. Smoking, type of alcoholic beverage and squamous-cell oesophageal cancer in northern Italy. Int J Cancer 2000;86:144–9.
19. Vioque J, Barber X, Bolumar F, et al. Esophageal cancer risk by type of alcohol drinking and smoking: a case-control study in Spain. BMC Cancer 2008;8:221.
20. Historical Trends in the Tobacco Industry, 1925–1967. New York, NY: Maxwell Associates, 1977.
21. Hebert JR, Kabat GC. Menthol cigarettes and esophageal cancer. Am J Public Health 1988;78:986–7.
22. Hebert JR, Kabat GC. Menthol cigarette smoking and oesophageal cancer. Int J Epidemiol 1989;18:37–44.
23. Green BG. Menthol modulates oral sensations of warmth and cold. Physiol Behav 1985;35:427–34.
24. Alberg AJ, Samet JM. Epidemiology of lung cancer. Chest 2003;123:21S–49S.
25. Hoffmann D, Djordjevic MV, Hoffmann I. The changing cigarette. Prev Med 1997;26:427–34.
26. Levi F, Lucchini F, Negri E, et al. Cancer mortality in Europe, 1990–1994, and an overview of trends from 1955 to 1994. Eur J Cancer 1999;35:1477–516.
27. Tang JL, Morris JK, Wald NJ, et al. Mortality in relation to tar yield of cigarettes: a prospective study of four cohorts. BMJ 1995;311:1530–3.
28. Franceschi S, Talamini R, Barra S, et al. Smoking and drinking in relation to cancers of the oral cavity, pharynx, larynx, and esophagus in northern Italy. Cancer Res 1990;50:6502–7.
29. La Vecchia C, Liati P, Decarli A, et al. Tar yields of cigarettes and the risk of oesophageal cancer. Int J Cancer 1986;38:381–5.

30. La Vecchia C, Bidoli E, Barra S, et al. Type of cigarettes and cancers of the upper digestive and respiratory tract. Cancer Causes Control 1990;1:69–74.
31. Gallus S, Altieri A, Bosetti C, et al. Cigarette tar yield and risk of upper digestive tract cancers: case-control studies from Italy and Switzerland. Ann Oncol 2003;14:209–13.
32. Watson W. Carcinoma of oesophagus. Surg Gynecol Obstet 1933;56:884–97.
33. Kwan K. Carcinoma of the esophagus, a statistical study. Chinese Med J 1937;52:237–54.
34. Goldenberg D, Golz A, Joachims HZ. The beverage mate: a risk factor for cancer of the head and neck. Head Neck 2003;25:595–601.
35. Esophageal cancer studies in the Caspian littoral of Iran: results of population studies—a prodrome. Joint Iran-International Agency for Research on Cancer Study Group. J Natl Cancer Inst 1977;59:1127–38.
36. Kolicheva N. Epidemiology of esophagus cancer in the USSR. In: Levin D, ed. Joint USA/USSR Monograph on Cancer Epidemiology in the USA and USSR. Washington, DC: DHHS, 1980.
37. Martinez I. Factors associated with cancer of the esophagus, mouth, and pharynx in Puerto Rico. J Natl Cancer Inst 1969;42:1069–94.
38. Segi M. Tea-gruel as a possible factor for cancer for the esophagus. Gann 1975;66:199–202.
39. Hirayama T. An epidemiological study of cancer of the esophagus in Japan, with special reference to the combined effect of selected environmental factors. Monograph No. 1, Seminar on Epidemiology of Oesophageal Cancer. Bangalore, India, 1971.
40. De Stefani E, Munoz N, Esteve J, et al. Mate drinking, alcohol, tobacco, diet, and esophageal cancer in Uruguay. Cancer Res 1990;50:426–31.
41. Victora CG, Munoz N, Day NE, et al. Hot beverages and oesophageal cancer in southern Brazil: a case-control study. Int J Cancer 1987;39:710–6.
42. Castelletto R, Castellsague X, Munoz N, et al. Alcohol, tobacco, diet, mate drinking, and esophageal cancer in Argentina. Cancer Epidemiol Biomarkers Prev 1994;3:557–64.
43. Munoz N, Victora CG, Crespi M, et al. Hot mate drinking and precancerous lesions of the oesophagus: an endoscopic survey in southern Brazil. Int J Cancer 1987;39:708–9.
44. Wahrendorf J, Chang-Claude J, Liang QS, et al. Precursor lesions of oesophageal cancer in young people in a high-risk population in China. Lancet 1989;2:1239–41.
45. Miller R. Epidemiology. In: Kaplan H, Tschitani P, eds. Cancer in China. New York: Alan R. Liss, 1978:39–57.
46. Rodler I, Zajkas G. Hungarian cancer mortality and food availability data in the last four decades of the 20th century. Ann Nutr Metab 2002;46:49–56.
47. Mettlin C, Graham S, Priore R, et al. Diet and cancer of the esophagus. Nutr Cancer 1981;2:143–7.
48. Chainani-Wu N. Diet and oral, pharyngeal, and esophageal cancer. Nutr Cancer 2002;44:104–26.
49. Chen H, Tucker KL, Graubard BI, et al. Nutrient intakes and adenocarcinoma of the esophagus and distal stomach. Nutr Cancer 2002;42:33–40.
50. De Jong UW, Breslow N, Hong JG, et al. Aetiological factors in oesophageal cancer in Singapore Chinese. Int J Cancer 1974;13:291–303.
51. Cook-Mozaffari PJ, Azordegan F, Day NE, et al. Oesophageal cancer studies in the Caspian Littoral of Iran: results of a case-control study. Br J Cancer 1979;39:293–309.
52. Bosetti C, La Vecchia C, Talamini R, et al. Food groups and risk of squamous cell esophageal cancer in northern Italy. Int J Cancer 2000;87:289–94.
53. Lucenteforte E, Garavello W, Bosetti C, et al. Diet diversity and the risk of squamous cell esophageal cancer. Int J Cancer 2008;123:2397–400.
54. Silvera S, Mayne S, Risch H, et al. Food group intake and risk of subtypes of esophageal and gastric cancer. Int J Cancer 2008;123:852–60.
55. Mayne ST, Risch HA, Dubrow R, et al. Nutrient Intake and Risk of Subtypes of Esophageal and Gastric Cancer. Cancer Epidemiol Biomarkers Prev 2001;10:1055–62.
56. Graham S, Marshall J, Haughey B, et al. Nutritional epidemiology of cancer of the esophagus. Am J Epidemiol 1990;131:454–67.
57. Bollschweiler E, Wolfgarten E, Nowroth T, et al. Vitamin intake and risk of subtypes of esophageal cancer in Germany. J Cancer Res Clin Oncol 2002;128:575–80.
58. Ziegler RG, Morris LE, Blot WJ, et al. Esophageal cancer among black men in Washington, DC. II. Role of nutrition. J Natl Cancer Inst 1981;67:1199–206.
59. Tuyns AJ. Oesophageal cancer in non-smoking drinkers and in non-drinking smokers. Int J Cancer 1983;32:443–4.
60. Tuyns AJ. Cancer risks derived from alcohol. Med Oncol Tumor Pharmacother 1987;4:241–4.

61. Decarli A, Liati P, Negri E, et al. Vitamin A and other dietary factors in the etiology of esophageal cancer. Nutr Cancer 1987;10:29–37.
62. Notani PN, Jayant K. Role of diet in upper aerodigestive tract cancers. Nutr Cancer 1987;10:103–13.
63. Brown LM, Blot WJ, Schuman SH, et al. Environmental factors and high risk of esophageal cancer among men in coastal South Carolina. J Natl Cancer Inst 1988;80:1620–5.
64. Yu MC, Garabrant DH, Peters JM, et al. Tobacco, alcohol, diet, occupation, and carcinoma of the esophagus. Cancer Res 1988;48:3843–8.
65. Li JY, Ershow AG, Chen ZJ, et al. A case-control study of cancer of the esophagus and gastric cardia in Linxian. Int J Cancer 1989;43:755–61.
66. Block G. Vitamin C and cancer prevention: the epidemiologic evidence. Am J Clin Nutr 1991;53:270S–82S.
67. Cheng KK, Day NE, Duffy SW, et al. Pickled vegetables in the aetiology of oesophageal cancer in Hong Kong Chinese. Lancet 1992;339:1314–8.
68. Yu Y, Taylor PR, Li JY, et al. Retrospective cohort study of risk-factors for esophageal cancer in Linxian, People's Republic of China. Cancer Causes Control 1993;4:195–202.
69. Tavani A, Negri E, Franceschi S, et al. Risk factors for esophageal cancer in lifelong nonsmokers. Cancer Epidemiol Biomarkers Prev 1994;3:387–92.
70. Jain AK, Shimoi K, Nakamura Y, et al. Crude tea extracts decrease the mutagenic activity of *N*-methyl-*N*'-nitro-*N*-nitrosoguanidine in vitro and in intragastric tract of rats. Mutat Res 1989;210:1–8.
71. Ito Y, Ohnishi S, Fujie K. Chromosome aberrations induced by aflatoxin B1 in rat bone marrow cells in vivo and their suppression by green tea. Mutat Res 1989;222:253–61.
72. Sasaki Yu F, Imanishi H, Ohta T, et al. Suppressing effect of tannic acid on the frequencies of mutagen-induced sister-chromatid exchanges in mammalian cells. Mutat Res 1989;213:195–203.
73. Stich HF, Rosin MP. Naturally occurring phenolics as antimutagenic and anticarcinogenic agents. Adv Exp Med Biol 1984;177:1–29.
74. Gao Y, McLaughlin J, Blot W, et al. Risk factors for esophageal cancer in Shanghai, China. I. Role of cigarette smoking and alcohol drinking. Int J Cancer 1994;58:192–6.
75. Wang JM, Xu B, Rao JY, et al. Diet habits, alcohol drinking, tobacco smoking, green tea drinking, and the risk of esophageal squamous cell carcinoma in the Chinese population. Eur J Gastroenterol Hepatol 2007;19:171–6.
76. Ke L, Yu P, Zhang ZX, et al. Congou tea drinking and oesophageal cancer in South China. Br J Cancer 2002;86: 346–7.
77. Yun TK, Yun YS, Han IW. Anticarcinogenic effect of long-term oral administration of red ginseng on newborn mice exposed to various chemical carcinogens. Cancer Detect Prev 1983;6:515–25.
78. Yun TK, Choi SY. Preventive effect of ginseng intake against various human cancers: a case-control study on 1987 pairs. Cancer Epidemiol Biomarkers Prev 1995;4:401–8.
79. Chen J, Geissler C, Parpia B, et al. Antioxidant status and cancer mortality in China. Int J Epidemiol 1992;21: 625–35.
80. Wei W-Q, Abnet CC, Qiao Y-L, et al. Prospective study of serum selenium concentrations and esophageal and gastric cardia cancer, heart disease, stroke, and total death. Am J Clin Nutr 2004;79:80–5.
81. Taylor PR, Qiao Y-L, Abnet CC, et al. Prospective Study of Serum Vitamin E Levels and Esophageal and Gastric Cancers. J Natl Cancer Inst 2003;95:1414–6.
82. Thurnham DI, Rathakette P, Hambidge KM, et al. Riboflavin, vitamin A and zinc status in Chinese subjects in a high-risk area for oesophageal cancer in China. Hum Nutr Clin Nutr 1982;36:337–49.
83. Rogers MA, Thomas DB, Davis S, et al. A case-control study of element levels and cancer of the upper aerodigestive tract. Cancer Epidemiol Biomarkers Prev 1993;2:305–12.
84. Sporn MB, Newton DL. Chemoprevention of cancer with retinoids. Fed Proc 1979;38:2528–34.
85. Munoz N, Wahrendorf J, Bang LJ, et al. No effect of riboflavine, retinol, and zinc on prevalence of precancerous lesions of oesophagus. Randomised double-blind intervention study in high-risk population of China. Lancet 1985;2:111–4.
86. Wahrendorf J, Munoz N, Lu JB, et al. Blood, retinol and zinc riboflavin status in relation to precancerous lesions of the esophagus: findings from a vitamin intervention trial in the People's Republic of China. Cancer Res 1988;48:2280–3.
87. Li JY, Taylor PR, Li B, et al. Nutrition intervention trials in Linxian, China: multiple vitamin/mineral supplementation, cancer incidence, and disease-specific mortality among adults with esophageal dysplasia. J Natl Cancer Inst 1993;85:1492–8.
88. Blot WJ, Li JY, Taylor PR, et al. Nutrition intervention trials in Linxian, China: supplementation with specific vitamin/mineral combinations, cancer incidence, and disease-specific mortality in the general population. J Natl Cancer Inst 1993;85:1483–92.

89. Li B, Taylor PR, Li JY, et al. Linxian nutrition intervention trials. Design, methods, participant characteristics, and compliance. Ann Epidemiol 1993;3:577–85.
90. Wang GQ, Dawsey SM, Li JY, et al. Effects of vitamin/mineral supplementation on the prevalence of histological dysplasia and early cancer of the esophagus and stomach: results from the General Population Trial in Linxian, China. Cancer Epidemiol Biomarkers Prev 1994;3:161–6.
91. Dawsey SM, Wang GQ, Taylor PR, et al. Effects of vitamin/mineral supplementation on the prevalence of histological dysplasia and early cancer of the esophagus and stomach: results from the Dysplasia Trial in Linxian, China. Cancer Epidemiol Biomarkers Prev 1994;3:167–72.
92. Rao M, Liu FS, Dawsey SM, et al. Effects of vitamin/mineral supplementation on the proliferation of esophageal squamous epithelium in Linxian, China. Cancer Epidemiol Biomarkers Prev 1994;3:277–9.
93. Wang LD, Zhou Q, Feng CW, et al. Intervention and follow-up on human esophageal precancerous lesions in Henan, northern China, a high-incidence area for esophageal cancer. Gan To Kagaku Ryoho 2002;29(Suppl. 1):159–72.
94. Barrett N. Chronic peptic ulcer of the esophagus and "esophagitis". Br J Surg 1950;38:175–82.
95. Garewal HS, Sampliner R. Barrett's esophagus: a model premalignant lesion for adenocarcinoma. Prev Med 1989;18:749–56.
96. Sjogren RW, Jr., Johnson LF. Barrett's esophagus: a review. Am J Med 1983;74:313–21.
97. Spechler SJ, Robbins AH, Rubins HB, et al. Adenocarcinoma and Barrett's esophagus. An overrated risk? Gastroenterology 1984;87:927–33.
98. Spechler SJ, Goyal RK. Barrett's esophagus. N Engl J Med 1986;315:362–71.
99. Hameeteman W, Tytgat GN, Houthoff HJ, et al. Barrett's esophagus: development of dysplasia and adenocarcinoma. Gastroenterology 1989;96:1249–56.
100. Eng C, Spechler SJ, Ruben R, et al. Familial Barrett esophagus and adenocarcinoma of the gastroesophageal junction. Cancer Epidemiol Biomarkers Prev 1993;2:397–9.
101. Crabb DW, Berk MA, Hall TR, et al. Familial gastroesophageal reflux and development of Barrett's esophagus. Ann Intern Med 1985;103:52–4.
102. Jochem VJ, Fuerst PA, Fromkes JJ. Familial Barrett's esophagus associated with adenocarcinoma. Gastroenterology 1992;102:1400–2.
103. Elder JB, Ganguli PC, Gillespie IE. Cimetidine and gastric cancer. Lancet 1979;1:1005–6.
104. Chow WH, Finkle WD, McLaughlin JK, et al. The relation of gastroesophageal reflux disease and its treatment to adenocarcinomas of the esophagus and gastric cardia. JAMA 1995;274:474–7.
105. Vaughan T, Farrow D, Hansten P, et al. Risk of esophageal and gastric adenocarcinomas in relation to use of calcium channel blockers, asthma drugs, and other medications that promote gastroesophageal reflux. Cancer Epidemiol Biomarkers Prev 1998;7:749–56.
106. Ranka S, Gee J, Johnson I, et al. Non-steroidal anti-inflammatory drugs, lower oesophageal sphincter-relaxing drugs and oesophageal cancer. A case-control study. Digestion 2006;74:109–15.
107. Ye W, Chow WH, Lagergren J, et al. Risk of adenocarcinomas of the oesophagus and gastric cardia in patients hospitalized for asthma. Br J Cancer 2001;85:1317–21.
108. Vaughan T, Davis S, Kristal A, et al. Obesity, alcohol, and tobacco as risk factors for cancers of the esophagus and gastric cardia: adenocarcinoma versus squamous cell carcinoma. Cancer Epidemiol Biomarkers Prev 1995;4: 85–92.
109. Kabat G, Ng S, Wynder E. Tobacco, alcohol intake, and diet in relation to adenocarcinoma of the esophagus and gastric cardia. Cancer Causes Control 1993;4:123–32.
110. Brown L, Silverman D, Pottern L, et al. Adenocarcinoma of the esophagus and esophagogastric junction in white men in the United States: alcohol, tobacco, and socioeconomic factors. Cancer Causes Control 1994;5:333–40.
111. Gammon M, Schoenberg J, Ahsan H, et al. Tobacco, alcohol, and socioeconomic status and adenocarcinomas of the esophagus and gastric cardia. J Natl Cancer Inst 1997;89:1277–84.
112. Freedman ND, Abnet CC, Leitzmann MF, et al. A Prospective Study of Tobacco, Alcohol, and the Risk of Esophageal and Gastric Cancer Subtypes. Am J Epidemiol 2007;165:1424–33.
113. Brown P, Allen AR. Obesity linked to some forms of cancer. W V Med J 2002;98:271–2.
114. Chow WH, Blot WJ, Vaughan TL, et al. Body mass index and risk of adenocarcinomas of the esophagus and gastric cardia. J Natl Cancer Inst 1998;90:150–5.
115. Mayne ST, Navarro SA. Diet, obesity and reflux in the etiology of adenocarcinomas of the esophagus and gastric cardia in humans. J Nutr 2002;132:3467S–70S.
116. Bowers SP, Mattar SG, Smith CD, et al. Clinical and Histologic Follow-Up After Antireflux Surgery for Barrett's Esophagus. J Gastrointest Surg 2002;6:532–9.

117. Hoyo C, Gammon M. Obesity/overweight and risk of esophageal adenocarcinoma. In: McTeirnan A, ed. Cancer Prevention and Management Through Exercise and Weight Control. New York: CRC Press, 2006.
118. Kubo A, Corley DA. Meta-analysis of antioxidant intake and the risk of esophageal and gastric cardia adenocarcinoma. Am J Gastroenterol 2007;102:2323–30; quiz 31.
119. Richter JE, Falk GW, Vaezi MF. Helicobacter pylori and gastroesophageal reflux disease: the bug may not be all bad. Am J Gastroenterol 1998;93:1800–2.
120. Jemal A, Murray T, Samuels A, et al. Cancer Statistics, 2003. CA Cancer J Clin 2003;53:5–26.
121. Chow WH, Blaser MJ, Blot WJ, et al. An inverse relation between cagA+ strains of Helicobacter pylori infection and risk of esophageal and gastric cardia adenocarcinoma. Cancer Res 1998;58:588–90.
122. Henrik Siman J, Forsgren A, Berglund G, et al. Helicobacter pylori infection is associated with a decreased risk of developing oesophageal neoplasms. Helicobacter 2001;6:310–6.
123. Weston AP, Badr AS, Topalovski M, et al. Prospective evaluation of the prevalence of gastric Helicobacter pylori infection in patients with GERD, Barrett's esophagus, Barrett's dysplasia, and Barrett's adenocarcinoma. Am J Gastroenterol 2000;95:387–94.
124. Islami F, Kamangar F. Helicobacter pylori and esophageal cancer risk: a meta-analysis. Cancer Prev Res 2008;1:329–38.
125. Graham DY. Helicobacter pylori is not and never was "protective" against anything, including GERD. Dig Dis Sci 2003;48:629–30.
126. Rauws EA. Should Helicobacter pylori be eradicated before starting long-term proton pump inhibitors? Ital J Gastroenterol Hepatol 1997;29:569–73.
127. Haenszel W. Variation in incidence of and mortality from stomach cancer, with particular reference to the United States. J Natl Cancer Inst 1958;21:213–62.
128. Levin D, Devesa S, Godwin J, et al. Cancer Rates and Risks, 2nd edition. Washington, DC: USDHEW, 1974.
129. Devesa SS, Silverman DT. Cancer incidence and mortality trends in the United States: 1935–1974. J Natl Cancer Inst 1978;60:545–71.
130. Howson CP, Hiyama T, Wynder EL. The decline in gastric cancer: epidemiology of an unplanned triumph. Epidemiol Rev 1986;8:1–27.
131. Nomura A. Stomach. In: Schottenfeld D, Fraumeni I, eds. Cancer Epidemiology and Prevention. Philadelphia, PA: WB Saunders, 1982:624–37.
132. Jarvi O, Lauren P. On the role of heterotopias of the intestinal epithelium in the pathogenesis of gastric cancer. Acta Pathol Microbiol Scand 1951;29:26–44.
133. Lauren P. The Two Histological Main Types of Gastric Carcinoma: Diffuse and So-Called Intestinal-Type Carcinoma. An Attempt at a Histo-Clinical Classification. Acta Pathol Microbiol Scand 1965;64:31–49.
134. Munoz N, Correa P, Cuello C, et al. Histologic types of gastric carcinoma in high- and low-risk areas. Int J Cancer 1968;3:809–18.
135. Pel P. Ziekten van de Maag (Diseases of the Stomach). Amsterdam: De Erven F Bohn, 1899.
136. Marshall BJ, Warren JR. Unidentified curved bacilli in the stomach of patients with gastritis and peptic ulceration. Lancet 1984;1:1311–5.
137. IARC Monographs in the Evaluation of Carcinogenic Risks to Humans. Schistosomes, Liver Flukes and Helicobacter pylon. Lyons, France: World Health Organization, 1994.
138. Megraud F. Epidemiology of Helicobacter pylori infection. Gastroenterol Clin North Am 1993;22:73–88.
139. Parsonnet J. The epidemiology of C. pylon. In: Blaser M, ed. *Campylobacter pyloni*in Gastritis and Peptic Ulcer Disease. New York: Igaku-Shoin, 1989:51–60.
140. An international association between Helicobacter pylori infection and gastric cancer. The EUROGAST Study Group. Lancet 1993;341:1359–62.
141. Ford AC, Forman D, Bailey AG, et al. Effect of sibling number in the household and birth order on prevalence of Helicobacter pylori: a cross-sectional study. Int J Epidemiol 2007;36:1327–33.
142. Goodman KJ, Correa P, Tengana Aux HJ, et al. Helicobacter pylori infection in the Colombian Andes: a population-based study of transmission pathways. Am J Epidemiol 1996;144:290–9.
143. Oderda G. Transmission of Helicobacter pylori infection. Can J Gastroenterol 1999;13:595–7.
144. Roma-Giannikou E, Karameris A, Balatsos B, et al. Intrafamilial spread of Helicobacter pylori: a genetic analysis. Helicobacter 2003;8:15–20.
145. Bohmer CJ, Klinkenberg-Knol EC, Kuipers EJ, et al. The prevalence of Helicobacter pylori infection among inhabitants and healthy employees of institutes for the intellectually disabled. Am J Gastroenterol 1997;92: 1000–4.

146. Malaty HM, Paykov V, Bykova O, et al. Helicobacter pylori and socioeconomic factors in Russia. Helicobacter 1996;1:82–7.
147. Hegarty JP, Dowd MT, Baker KH. Occurrence of Helicobacter pylori in surface water in the United States. J Appl Microbiol 1999;87:697–701.
148. Forman D, Newell DG, Fullerton F, et al. Association between infection with Helicobacter pylori and risk of gastric cancer: evidence from a prospective investigation. BMJ 1991;302:1302–5.
149. Parsonnet J, Friedman GD, Vandersteen DP, et al. Helicobacter pylori infection and the risk of gastric carcinoma. N Engl J Med 1991;325:1127–31.
150. Nomura A, Stemmermann GN, Chyou PH, et al. Helicobacter pylori infection and gastric carcinoma among Japanese Americans in Hawaii. N Engl J Med 1991;325:1132–6.
151. Sobala GM, Schorah CJ, Sanderson M, et al. Ascorbic acid in the human stomach. Gastroenterology 1989;97:357–63.
152. Rathbone BJ, Johnson AW, Wyatt JI, et al. Ascorbic acid: a factor concentrated in human gastric juice. Clin Sci (Lond) 1989;76:237–41.
153. Sobala GM, Pignatelli B, Schorah CJ, et al. Levels of nitrite, nitrate, N-nitroso compounds, ascorbic acid and total bile acids in gastric juice of patients with and without precancerous conditions of the stomach. Carcinogenesis 1991;12:193–8.
154. Ruiz B, Rood JC, Fontham ET, et al. Vitamin C concentration in gastric juice before and after anti-Helicobacter pylori treatment. Am J Gastroenterol 1994;89:533–9.
155. Wyatt JI, Rathbone BJ. Immune response of the gastric mucosa to Campylobacter pylori. Scand J Gastroenterol Suppl 1988;142:44–9.
156. Routledge MN, Wink DA, Keefer LK, et al. Mutations induced by saturated aqueous nitric oxide in the pSP189 supF gene in human Ad293 and E. coli MBM7070 cells. Carcinogenesis 1993;14:1251–4.
157. Radi R, Beckman JS, Bush KM, et al. Peroxynitrite oxidation of sulfhydryls. The cytotoxic potential of superoxide and nitric oxide. J Biol Chem 1991;266:4244–50.
158. Uemura N, Okamoto S, Yamamoto S, et al. Helicobacter pylori infection and the development of gastric cancer. N Engl J Med 2001;345:784–9.
159. Prinz C, Hafsi N, Voland P. Helicobacter pylori virulence factors and the host immune response: implications for therapeutic vaccination. Trends Microbiol 2003;11:134–8.
160. Blaser MJ, Perez-Perez GI, Kleanthous H, et al. Infection with Helicobacter pylori strains possessing cagA is associated with an increased risk of developing adenocarcinoma of the stomach. Cancer Res 1995;55:2111–5.
161. Kuipers EJ, Perez-Perez GI, Meuwissen SG, et al. Helicobacter pylori and atrophic gastritis: importance of the cagA status. J Natl Cancer Inst 1995;87:1777–80.
162. Parsonnet J, Friedman GD, Orentreich N, et al. Risk for gastric cancer in people with CagA positive or CagA negative Helicobacter pylori infection. Gut 1997;40:297–301.
163. El-Omar EM, Rabkin CS, Gammon MD, et al. Increased risk of noncardia gastric cancer associated with proinflammatory cytokine gene polymorphisms. Gastroenterology 2003;124:1193–201.
164. Hirayama T. Epidemiology of stomach cancer. Gann Monogr Cancer Res 1971;11:3–19.
165. Wynder EL, Kmet J, Dungal N, et al. An epidemiological investigation of gastric cancer. Cancer 1963;16:1461–96.
166. Graham S, Schotz W, Martino P. Alimentary factors in the epidemiology of gastric cancer. Cancer 1972;30:927–38.
167. Acheson ED, Doll R. Dietary factors in carcinoma of the stomach: a Study of 100 Cases and 200 Controls. Gut 1964;5:126–31.
168. Correa P, Fontham E, Pickle LW, et al. Dietary determinants of gastric cancer in south Louisiana inhabitants. J Natl Cancer Inst 1985;75:645–54.
169. Wu-Williams AH, Yu MC, Mack TM. Life-style, workplace, and stomach cancer by subsite in young men of Los Angeles County. Cancer Res 1990;50:2569–76.
170. Boeing H, Frentzel-Beyme R, Berger M, et al. Case-control study on stomach cancer in Germany. Int J Cancer 1991;47:858–64.
171. Hoey J, Montvernay C, Lambert R. Wine and tobacco: risk factors for gastric cancer in France. Am J Epidemiol 1981;113:668–74.
172. Nomura A, Grove JS, Stemmermann GN, et al. A prospective study of stomach cancer and its relation to diet, cigarettes, and alcohol consumption. Cancer Res 1990;50:627–31.
173. Kneller RW, McLaughlin JK, Bjelke E, et al. A cohort study of stomach cancer in a high-risk American population. Cancer 1991;68:672–8.
174. Kahn HA. The Dorn study of smoking and mortality among U.S. veterans: report on eight and one-half years of observation. Natl Cancer Inst Monogr 1966;19:1–125.

175. Vineis P, Alavanja M, Buffler P, et al. Tobacco and cancer: recent epidemiological evidence. J Natl Cancer Inst 2004;96:99–106.
176. Tredaniel J, Boffetta P, Buiatti E, et al. Tobacco smoking and gastric cancer: review and meta-analysis. Int J Cancer 1997;72:565–73.
177. Kinosita R. Studies on factors affecting chemical carcinogenesis of mouse stomach. Gann Monogr 1969;8:263–8.
178. Tatematsu M, Takahashi M, Fukushima S, et al. Effects in rats of sodium chloride on experimental gastric cancers induced by N-methyl-N-nitro-N-nitrosoguanidine or 4-nitroquinoline-1-oxide. J Natl Cancer Inst 1975;55:101–6.
179. Kodama M, Kodama T, Suzuki H, et al. Effect of rice and salty rice diets on the structure of mouse stomach. Nutr Cancer 1984;6:135–47.
180. Takahashi M, Kokubo T, Furukawa F, et al. Effects of sodium chloride, saccharin, phenobarbital and aspirin on gastric carcinogenesis in rats after initiation with N-methyl-N'-nitro-N-nitrosoguanidine. Gann 1984;75:494–501.
181. Mirvish SS, Salmasi S, Cohen SM, et al. Liver and forestomach tumors and other forestomach lesions in rats treated with morpholine and sodium nitrite, with and without sodium ascorbate. J Natl Cancer Inst 1983;71:81–5.
182. Sato T, Fukuyama T, Suzuki T. Studies of the causation of gastric cancer. The relation between gastric cancer mortality rate and salted food intake in several places in Japan. Bull Inst Publ Health 1959;8:187–98.
183. Lu JB, Qin YM. Correlation between high salt intake and mortality rates for oesophageal and gastric cancers in Henan Province, China. Int J Epidemiol 1987;16:171–6.
184. Buiatti E, Palli D, Decarli A, et al. A case-control study of gastric cancer and diet in Italy. Int J Cancer 1989;44: 611–6.
185. Tajima K, Tominaga S. Dietary habits and gastro-intestinal cancers: a comparative case-control study of stomach and large intestinal cancers in Nagoya, Japan. Jpn J Cancer Res 1985;76:705–16.
186. Lee JK, Park BJ, Yoo KY, et al. Dietary factors and stomach cancer: a case-control study in Korea. Int J Epidemiol 1995;24:33–41.
187. Ward MH, Lopez-Carrillo L. Dietary factors and the risk of gastric cancer in Mexico City. Am J Epidemiol 1999;149:925–32.
188. Tuyns AJ. Salt and gastrointestinal cancer. Nutr Cancer 1988;11:229–32.
189. La Vecchia C, Negri E, Decarli A, et al. A case-control study of diet and gastric cancer in northern Italy. Int J Cancer 1987;40:484–9.
190. Friedman GD, Parsonnet J. Salt intake and stomach cancer: some contrary evidence. Cancer Epidemiol Biomarkers Prev 1992;1:607–8.
191. Correa P, Fontham ET, Bravo JC, et al. Chemoprevention of gastric dysplasia: randomized trial of antioxidant supplements and anti-helicobacter pylori therapy. J Natl Cancer Inst 2000;92:1881–8.
192. Beevers DG, Lip GY, Blann AD. Salt intake and Helicobacter pylori infection. J Hypertens 2004;22:1475–7.
193. Shikata K, Kiyohara Y, Kubo M, et al. A prospective study of dietary salt intake and gastric cancer incidence in a defined Japanese population: the Hisayama study. Int J Cancer 2006;119:196–201.
194. Bogovski P, Bogovski S. Animal Species in which N-nitroso compounds induce cancer. Int J Cancer 1981;27:471–4.
195. Correa P, Haenszel W, Cuello C, et al. A model for gastric cancer epidemiology. Lancet 1975;2:58–60.
196. Ohshima H, Bartsch H. Quantitative estimation of endogenous nitrosation in humans by monitoring N-nitrosoproline excreted in the urine. Cancer Res 1981;41:3658–62.
197. Mirvish SS, Grandjean AC, Moller H, et al. N-nitrosoproline excretion by rural Nebraskans drinking water of varied nitrate content. Cancer Epidemiol Biomarkers Prev 1992;1:455–61.
198. Moller H, Landt J, Pedersen E, et al. Endogenous nitrosation in relation to nitrate exposure from drinking water and diet in a Danish rural population. Cancer Res 1989;49:3117–21.
199. Sierra R, Chinnock A, Ohshima H, et al. In vivo nitrosoproline formation and other risk factors in Costa Rican children from high- and low-risk areas for gastric cancer. Cancer Epidemiol Biomarkers Prev 1993;2:563–8.
200. Risch HA, Jain M, Choi NW, et al. Dietary factors and the incidence of cancer of the stomach. Am J Epidemiol 1985;122:947–59.
201. Chyou PH, Nomura AM, Hankin JH, et al. A case-cohort study of diet and stomach cancer. Cancer Res 1990;50: 7501–4.
202. Buiatti E, Palli D, Decarli A, et al. A case-control study of gastric cancer and diet in Italy: II. Association with nutrients. Int J Cancer 1990;45:896–901.
203. Buiatti E, Palli D, Bianchi S, et al. A case-control study of gastric cancer and diet in Italy. III. Risk patterns by histologic type. Int J Cancer 1991;48:369–74.
204. Hansson LE, Nyren O, Bergstrom R, et al. Nutrients and gastric cancer risk. A population-based case-control study in Sweden. Int J Cancer 1994;57:638–44.

205. Gonzalez CA, Riboli E, Badosa J, et al. Nutritional factors and gastric cancer in Spain. Am J Epidemiol 1994;139: 466–73.
206. Meinsma I. Nutrition and cancer. Voeding 1964;25:357–65.
207. Higginson J. Etiological factors in gastrointestinal cancer in man. J Natl Cancer Inst 1966;37:527–45.
208. Haenszel W, Kurihara M, Segi M, et al. Stomach cancer among Japanese in Hawaii. J Natl Cancer Inst 1972;49: 969–88.
209. Bjelke E. Epidemiologic studies of cancer of the stomach, colon, and rectum; with special emphasis on the role of diet. Scand J Gastroenterol Suppl 1974;31:1–235.
210. Haenszel W, Kurihara M, Locke FB, et al. Stomach cancer in Japan. J Natl Cancer Inst 1976;56:265–74.
211. Trichopoulos D, Ouranos G, Day NE, et al. Diet and cancer of the stomach: a case-control study in Greece. Int J Cancer 1985;36:291–7.
212. Jedrychowski W, Wahrendorf J, Popiela T, et al. A case-control study of dietary factors and stomach cancer risk in Poland. Int J Cancer 1986;37:837–42.
213. You WC, Blot WJ, Chang YS, et al. Diet and high risk of stomach cancer in Shandong, China. Cancer Res 1988;48:3518–23.
214. Graham S, Haughey B, Marshall J, et al. Diet in the epidemiology of gastric cancer. Nutr Cancer 1990;13:19–34.
215. Negri E, La Vecchia C, Franceschi S, et al. Vegetable and fruit consumption and cancer risk. Int J Cancer 1991;48: 350–4.
216. Gonzalez CA, Sanz JM, Marcos G, et al. Dietary factors and stomach cancer in Spain: a multi-centre case-control study. Int J Cancer 1991;49:513–9.
217. Hoshiyama Y, Sasaba T. A case-control study of single and multiple stomach cancers in Saitama Prefecture, Japan. Jpn J Cancer Res 1992;83:937–43.
218. La Vecchia C, Ferraroni M, D'Avanzo B, et al. Selected micronutrient intake and the risk of gastric cancer. Cancer Epidemiol Biomarkers Prev 1994;3:393–8.
219. Hansson LE, Nyren O, Bergstrom R, et al. Diet and risk of gastric cancer. A population-based case-control study in Sweden. Int J Cancer 1993;55:181–9.
220. Lopez-Carrillo L, Hernandez Avila M, Dubrow R. Chili pepper consumption and gastric cancer in Mexico: a case-control study. Am J Epidemiol 1994;139:263–71.
221. Norat T, Riboli E. Fruit and vegetable consumption and risk of cancer of the digestive tract: meta-analysis of published case-control and cohort studies. IARC Sci Publ 2002;156:123–5.
222. World Cancer Research Fund/American Institute for Cancer Research. Food, Nutrition, Physical Activity, and the Prevention of Cancer: a Global Perspective. Washington DC: AICR, 2007:265–70.
223. Basu T, Schorah C. Vitamin C reserves and requirements in health and disease. Vitamin C in Health and Disease. West Port, CT: AVI, 1982.
224. Stahelin HB, Gey KF, Eichholzer M, et al. Plasma antioxidant vitamins and subsequent cancer mortality in the 12-year follow-up of the prospective Basel Study. Am J Epidemiol 1991;133:766–75.
225. Haenszel W, Correa P, Lopez A, et al. Serum micronutrient levels in relation to gastric pathology. Int J Cancer 1985;36:43–8.
226. Kabuto M, Imai H, Yonezawa C, et al. Prediagnostic serum selenium and zinc levels and subsequent risk of lung and stomach cancer in Japan. Cancer Epidemiol Biomarkers Prev 1994;3:465–9.
227. Sasazuki S, Inoue M, Miura T, et al. Plasma tea polyphenols and gastric cancer risk: a case-control study nested in a large population-based prospective study in Japan. Cancer Epidemiol Biomarkers Prev 2008;17:343–51.
228. Yu GP, Hsieh CC, Wang LY, et al. Green-tea consumption and risk of stomach cancer: a population-based case-control study in Shanghai, China. Cancer Causes Control 1995;6:532–8.
229. Ji BT, Chow WH, Yang G, et al. The influence of cigarette smoking, alcohol, and green tea consumption on the risk of carcinoma of the cardia and distal stomach in Shanghai, China. Cancer 1996;77:2449–57.
230. Mu LN, Lu QY, Yu SZ, et al. Green tea drinking and multigenetic index on the risk of stomach cancer in a Chinese population. Int J Cancer 2005;116:972–83.
231. Abnet CC, Freedman ND, Hollenbeck AR, et al. A prospective study of BMI and risk of oesophageal and gastric adenocarcinoma. Eur J Cancer 2008;44:465–71.
232. Serdula MK, Coates RJ, Byers T, et al. Fruit and vegetable intake among adults in 16 states: results of a brief telephone survey. Am J Public Health 1995;85:236–9.
233. Malfertheiner P, Megraud F, O'Morain C, et al. Current concepts in the management of Helicobacter pylori infection—the Maastricht 2-2000 Consensus Report. Aliment Pharmacol Ther 2002;16:167–80.
234. Chey WD, Wong BC. American College of Gastroenterology guideline on the management of Helicobacter pylori infection. Am J Gastroenterol 2007;102:1808–25.

7 Factors in the Causation of Female Cancers and Prevention

Donato F. Romagnolo and Ivana Vucenik

Key Points

- Cancer statistics in Western and developing countries suggest that geographical differences exist in incidence and mortality. Breast and ovarian cancer are more common in high-income countries. Conversely, endometrial and cervical cancer overall rates are nearly twice as high in middle- to low as in high-income populations. However, changes in risk after migration suggest that environmental factors and diet may play a role in the etiology of these malignancies.
- Dietary patterns appear to modulate the risk of cancers in female carriers of mutations in the susceptibility gene, BRCA1, and selected polymorphisms for CYP1A1, NAT, and GST.
- Protective effects of fruits and vegetables may be related to specific bioactive compounds or specific subgroups of women with family history or differences in ER status.
- Early exposure in life to bioactive food components present in fruits and vegetables may exert protective effects against development of certain cancers later in life.
- Chronic positive energy balance, which can lead to obesity, raises the risk of developing multiple cancers.

Abstract

In this chapter, we discuss risk factors that influence the onset of female cancers. Particular emphasis is given to breast and ovarian cancers and how dietary strategies may help to reduce the risk. Reviewed topics include cancer statistics in Western and developing countries; dietary patterns that modulate the risk of cancers in female carriers of mutations in susceptibility genes and selected polymorphisms; effects of fruits and vegetables on cancer risk. We also discuss the role of diet–hormone receptor status interactions and the impact of energy balance and obesity on cancer risk and prevention. A review of the major intervention trials on diet and breast cancer indicate that dietary fat reduction and increased fruits, vegetable, and fiber intake do not provide the basis for conclusive dietary recommendation. Data are currently lacking concerning the role of healthy dietary patterns and specific nutrients in the etiology of breast cancer. Evidence from cohort and case–control studies suggests that non-starchy vegetables may protect against ovarian and endometrial cancer. There is limited but consistent evidence that carrots, containing high levels of carotenoids, including β- and α-carotene, and certain antioxidants, may be protective against cervical cancer. Results of case–control studies indicate that red meat consumption is positively associated with endometrial cancer due to heme-iron, the generation of *N*-nitroso compounds, and the production of heterocyclic amines and polycyclic aromatic hydrocarbons. The response to certain food components and mutagens is influenced by the interaction with genetic background including mutations in tumor susceptibility genes and polymorphisms for metabolizing genes. Future research is necessary to gain knowledge of the causal relationships between obesity, energy balance, physical activity, and the cancer process. Although caloric restriction is consistently associated with reduced mammary tumor development, it is not clear how it influences the metastatic process.

Key Words: Dietary patterns; female cancers; genetic susceptibility; hormone receptor; obesity; prevention; regional incidence

A. Bendich, R.J. Deckelbaum (eds.), *Preventive Nutrition*, Nutrition and Health, DOI 10.1007/978-1-60327-542-2_7,

7.1. INTRODUCTION

About 10.9 million new cancer cases and 6.7 million cancer deaths occur worldwide every year. *(1–3)*. This toll is projected to grow to 24 million cases and over 16 million deaths annually by 2050 *(4)*. International variations in both cancer incidence and mortality exist among economically developed, high-income countries and developing, low-and-middle income regions of the world *(5)*. However, current evidence points to cancer as a largely preventable disease. The objective of this chapter is to discuss risk factors that influence the onset of female cancers with particular emphasis on breast and ovarian cancers, and how dietary strategies may be used to reduce the risk. Realizing there exists a wealth of scientific information on the subject, to increase focus, we have centered our discussion on selected topics including cancer statistics in Western and developing countries; dietary patterns that modulate the risk of cancers in female carriers of mutations in susceptibility genes and selected polymorphisms; the effects of intake of fruits and vegetables influence on cancer risk; the role of diet–hormone receptor status interactions; and the impact of energy balance and obesity on cancer risk and prevention. Regrettably, many important areas of research could not be discussed or cited due to space limitation. Whenever possible, we formulated recommendations for dietary patterns and intake of specific nutrients with the realization that conflicting evidence exists for certain foods and caution should be exercised when extrapolating scientific data to develop dietary recommendations.

7.2. MULTI-REGIONAL AND MULTI-ETHNIC INCIDENCE OF FEMALE CANCERS

7.2.1. Breast Cancer

Breast cancer is the most common cancer in women worldwide. Approximately 1.15 million cases were recorded in 2002, representing 23% of all female and 11% of overall cancers *(6)*. Incidence rates for 2002 vary internationally by more than 25-fold, ranging from 3.9 cases in Mozambique to 101.1 cases per 100,000 women in the United States. Overall, breast cancer accounts for ~16% of all cancer deaths in adult women *(1,2)* (Table 7.1).

Breast cancer is more common in high-income countries where overall rates are nearly threefold higher than in middle- to low-income regions. However, its incidence is increasing rapidly in middle- and low-income populations residing in Africa, Asia, and Latin America, and in areas where the incidence had previously been low, such as Japan, China, Southern and Eastern Europe *(7)*. Age-adjusted incidence rates range from 75 to 100 per 100,000 women in North America, Northern Europe, and Australia, to less than 20 per 100,000 in parts of Africa and Asia. According to 2006 estimates, about 429,900 cases of breast cancer and 131,900 deaths were diagnosed in Europe, making breast cancer the most frequent cause of tumor death among women in Europe (17.6% of all female cancer deaths) *(8)*. In the United States, rates are higher among white women compared to other ethnic groups, whereas mortality is highest (20–30%) in black women *(9–12)* (Table 7.2).

Breast cancer is more common after menopause. Overall, the risk doubles each decade until menopause, after which it reaches a plateau. However, studies of women who migrated from areas of low to high risk showed that women migrants gained the rate of the host country within one or two generations. These trends indicate that environmental factors, including diet, play an important role in the etiology of this malignancy *(13)*.

Breast cancer can often be detected at early stages through mammography screening. The adequate follow-up of women with a positive result could significantly reduce mortality *(14)*. Survival rates range from more than 90% to less than 50%, depending on the characteristics of the tumor, its size and spread, and the availability of treatment. Average 5-year survival rates range from ~73 to ~57%,

Table 7.1
Leading Sites of New Female Cancer Cases and Deaths Worldwide and by Level of Economic Development

	Estimated New Cases			*Estimated Deaths*		
Worldwide	Breast	1,301,867	(22.8%)	Breast	486,854	(14.0%)
	Cervix uteri	555,094	(9.7%)	Cervix uteri	309,808	(9.3%)
	Ovary	230,555	(4.0%)	Ovary	141,452	(4.3%)
	Corpus uteri	226,787	(4.0%)	Corpus uteri	<100,000	(<3.0%)
	All sites[a]	5,717,275		All sites[a]	3,314,414	
Developed Countries	Breast	679,682	(27.4%)	Breast	203,528	(16.0%)
	Corpus uteri	146,866	(5.9%)	Ovary	66,929	(5.3%)
	Ovary	103,332	(4.2%)	Cervix uteri	42,101	(3.3%)
	Cervix uteri	87,466	(3.5%)	Corpus uteri	<35,000	(<2.8%)
	All sites[a]	2,478,605		All sites[a]	1,272,358	
Developing Countries	Breast	593,233	(18.7%)	Cervix uteri	272,238	(13.5%)
	Cervix uteri	473,430	(15.0%)	Breast	255,576	(12.6%)
	Ovary	123,761	(3.9%)	Ovary	72,433	(3.6%)
	Corpus uteri	<80,000	(<2.5%)	Corpus uteri	<45,000	(2.2%)
	All sites[a]	3,167,802		All sites[a]	2,022,059	

Source: Adapted from Global Cancer Facts and Figures 2007 *(2)*.

Estimates were produced by applying age-specific cancer rates of a defined geographic region (worldwide, developed, and developing countries) from GLOBOCAN 2002 *(4)* to the corresponding age-specific population for the year 2007 from the United Nations population projections (2004 revision). Therefore, estimates for developed and developing countries combined do not sum to worldwide estimates.

[a]Excludes nonmelanoma skin cancer.

respectively, in high and in middle- to low-income countries. Large regional disparities in mortality rates probably reflect significant disparities in detection and management programs *(15)*.

7.2.2. Ovarian Cancer

Ovarian cancer rates are nearly threefold higher in high- than in middle- to low-income regions, and appear to be rising in countries undergoing economic development *(6,7)*. For instance, a fourfold

Table 7.2
Incidence and Mortality of Female Cancers in the United States

Race	Incidence Rates by Race (per 100,000 Women)					Death Rates by Race (per 100,000 Women)				
	Breast	*Ovary*	*Cervix*	*Corpus and Uterus*	*Vulva*	*Breast*	*Ovary*	*Cervix*	*Corpus and Uterus*	*Vulva*
All races	126.1	13.3	8.4	23.4	2.2	25.0	8.8	2.5	4.1	0.5
White	130.6	14.1	8.2	24.3	2.3	24.4	9.2	2.3	3.9	0.5
Black	117.5	10.1	10.8	20.3	1.9	33.5	7.5	4.7	7.1	0.3
Asian/Pacific Islander	89.6	9.8	8.0	16.5	0.7	12.6	4.9	2.2	2.4	0.1
American Indian/Alaska Native	75.0	11.3	6.9	15.3	*	17.1	7.5	3.7	3.1	*
Hispanic	90.1	11.7	13.2	17.8	1.7	15.8	6.0	3.2	3.2	0.3

Source: Adapted from SEER Stat Fact Sheets *(10)*.
* Statistics not shown. Rate based on less than 16 cases for the time interval.

increase in the age-adjusted mortality rate (from 0.9 to 3.6 per 100,000 women) was observed in Japan between 1950 and 1997 *(16)*. Age-adjusted incidence rates range from more than 10 per 100,000 women in Europe and North America, to less than 5 per 100,000 women in parts of Africa and Asia. However, rates are relatively high in some Asian countries including Singapore and the Philippines *(4)*. Ovarian cancer is the eighth most common cancer among women worldwide. In 2008, an estimated 21,650 new cases are expected in the United States where rates are higher among white and Jewish women of Ashkenazi descent *(6,8,9,17)*.

The risk of ovarian cancer increases with age and peaks around 70 years of age, whereas only 10–15% of cases occur before menopause. Ovarian cancer is responsible for more deaths than any other cancer of the female reproductive system. This is due to lack of detectable symptoms at early tumor stages and the fact this malignancy is generally advanced when diagnosed. The 5-year survival rates range from ~30 to 50% *(8)*. Ovarian cancer accounts for about 7% of all cancer incidence and 4% of cancer deaths in women worldwide *(6)*. Changes in risk after migration suggest that environmental factors and diet play a role in the etiology of ovarian cancer *(18)*.

7.2.3. Cervical and Endometrial Cancer

Cervical cancer is the second most commonly diagnosed cancer in women, with an estimated 555,100 new cases worldwide in 2007 occurring mostly in developing countries. In high-income regions, the incidence of cervical cancer and mortality rates have been declining since the 1960s due to widespread implementation of screening and intervention programs *(2,6,19)*.

Overall rates are nearly twice as high in middle- to low as in high-income populations. Age-adjusted incidence rates range from more than 40 per 100,000 women in parts of Africa, South America, and Melanesia to less than 10 per 100,000 women in North America and parts of Asia (Middle East) and Australia *(2,6)*. However, rates are relatively high in India and Bangladesh, and in medically underserved populations mainly due to lack of screening. In the United States, rates are higher among African-American and Hispanic-American than in white women *(8,9)*.

Cervical cancer occurs mostly in women younger than 50 years of age, whereas mortality increases with age *(2,6)*. It is generally accepted that better access to cervical screening programs would be effective in decreasing up to 80% the worldwide incidence and mortality rates for this cancer *(20)*. The vaccination against human *Papilloma* virus is an effective preventive tool that may significantly reduce the occurrence of cervical cancer *(21)*.

Cervical cancers account for nearly 4% of all malignancies (~10% in women) and 4% of all cancer deaths (~9% in women) *(6)*. When detected at early stages, invasive cervical cancer is one of the most successfully treated tumors. The relative 5-year survival rate in the United States is ~92%. For all stages combined, survival rates vary widely from less than 30% in Africa to ~70% in North America and Northern Europe *(2)*.

Endometrial adenocarcinoma is the most common malignant tumor of the female genital tract in developed countries, and the eighth most common cancer in women, and the 17th most common cancer overall *(3–9,22)*. The incidence of endometrial cancer has been increasing in industrialized countries (the United States, Northern Europe, Japan) over the last 20 years, where 80–85% of cases are estrogen-dependent tumors and 15–20% non-estrogen dependent *(3–7)*. In high-income countries, overall rates are nearly fivefold higher than in middle- to low-income regions. Age-adjusted incidence rates range from more than 15 per 100,000 women in North America and parts of Europe to less than 5 per 100,000 women in most of Africa and Asia *(6)*. A 2006 estimate indicated that in Europe endometrial cancer accounted for 10% of cancer incidence (149,300 cases) and 6.2% of cancer deaths (~46,000 cases). Endometrial cancer is the most common gynecologic malignancy in the United States *(22)*. American women over the age of 45 years have a higher risk of endometrial cancer than of cervical cancer. In the United States, rates are higher in white women compared to other ethnic groups, whereas mortality rates are higher in black women *(23)*.

Estrogen is a strong risk factor for endometrial cancer. The risk increases with age, with most diagnoses made after the onset of menopause. The median age of patients at time of diagnosis is 61 years. The incidence is highly dependent on age ranging from 12 to 84 cases per 100,000, respectively, at 40 and 60 years of age *(24,25)*. The 5-year survival rates in the United States are 95, 67, and 23%, if the cancer is diagnosed, respectively, at local, regional, or distant stages *(8)*. The average 5-year survival rate is relatively high, ~73%, although it is lower in middle- (~67%) than in high- (~82%) income countries *(6)*.

7.3. GENE–DIET INTERACTION

7.3.1. Genetic Susceptibility

A genetic contribution to both breast and ovarian cancer is documented by the increased incidence of these tumors in women with a family history. However, mutations in breast and ovarian tumor susceptibility genes account for only 5–10% of breast and ovarian cancer cases. Therefore, the incidence of sporadic breast and ovarian cancer is likely due to other, yet unknown, factors. However, family history conferred by a first-degree relative is a greater relative risk factor for ovarian (~2.1) compared to breast cancer (~3.1) *(26)*.

Autosomal dominant mutations that have been linked to breast and ovarian cancer include the breast cancer (*BRCA1* and *BRCA2*) genes. Breast tumors have also been linked to mutations in the *p53*, *PTEN*, *Chek2*, and *Ataxia telangiectasia* genes. The majority of breast cancers related to BRCA1 mutations occur prior to menopause and tend to be ER-negative (ER–) and progesterone receptor-negative (PR–). In women carrying mutations in susceptibility genes, breast and ovarian cancers tend to occur earlier in life compared to sporadic breast and ovarian tumors. The risk of breast cancer increases with years of reproductive history, postmenopausal hormonal replacement therapy, radiation exposure, alcohol intake, and is reduced by early first pregnancy. The lifetime risk in BRCA1/BRCA2 mutation carriers is 50–85% for breast cancer and 45% for ovarian cancer by age 70 *(27)*. However, BRCA1/BRCA2 mutations do not appear to confer increased risk associated with alcohol consumption, which is a recognized risk factor for breast cancer *(28)*. In individuals that carry mutated BRCA1, cells have one functional and one mutated allele. Therefore, nutritional strategies for BRCA1 mutations carriers should consider opportunities for upregulation of the wild-type BRCA1 allele and prevention of loss of heterozygosity. Conversely, in sporadic breast cancers no somatic BRCA1 mutations have been detected, whereas the frequency of BRCA1 mutations appears to be higher in sporadic ovarian cancers. These observations have led to suggestions that silencing of BRCA1 in breast tissue may occur through epigenetic mechanisms. In this context, diet may play an important preventive role. Low levels of wild-type BRCA1 have been observed in sporadic breast cancers, possibly as a result of hypermethylation of the BRCA1 promoter *(29)*. Therefore, the epigenetic silencing of wild-type BRCA1 and other genes may represent a viable preventative target for dietary intervention.

Studies with a cohort of 83,234 women ranging from 33 to 60 years of age reported that premenopausal women who had a positive family history of breast cancer and consumed five or more servings of fruits and vegetables had a lower risk of breast cancer *(30)*. These effects were attributed to intake of specific carotenoids, vitamins A and C, and total vitamin E. Although this study did not exclude the possibility that multiple components in fruits and vegetables may account for the inverse relationship, it indicated that diet may modify the risk of breast cancer in mutation carriers. Energy intake has been inversely related to breast cancer risk *(31–33)*. With respect to hereditary breast cancer, low-caloric diets through intake restriction *(34)* and diet quality *(35)* have been associated with a reduced incidence of breast cancer in BRCA1 mutation carriers. Conversely, studies reported a positive correlation between risk of breast cancer diagnosed between 30 and 40 years of age and weight gain in BRCA1 mutation carriers *(36)*.

7.3.2. Polymorphisms

Polymorphisms in members of the P450 family of metabolizing enzymes have been shown to influence the effect of diet on cancer risk in genetically predisposed women. With respect to breast cancer, recent studies have shown that coffee consumption was related to a reduced incidence of breast cancer *(37)*. The intake of coffee before 35 years of age was associated with a 64% reduction in the incidence of breast cancer among BRCA1 mutation carriers with at least one variant C allele (AC or CC) of the *CYP1A2* gene, a member of P450 family of metabolizing enzymes *(38)*. The CYP1A2 enzyme participates in the metabolism of estrogen, which is a risk factor in breast carcinogenesis, and the *CYP1A2 CC* genotype has been associated with lower serum estrogen levels *(39)*. One of the plausible mechanisms responsible for the protective effects of coffee is the antioxidant activity of caffeine and its metabolic products, theobromine, and xanthine.

The exposure to heterocyclic amines from well-done meats and fried foods has been shown to increase the risk of breast cancer in genetic polymorphisms of the *N*-acetyltransferase (NAT). Both *NAT1* and *NAT2* variants are involved in the susceptibility to breast cancer, but their impact is not well understood. The NAT1 polymorphism conferred a higher risk of breast cancer in women who consumed well-done meat *(40)*. Increased risk of breast cancer in postmenopausal women was also reported in women who smoked and had the NAT2 slow acetylator genotype *(41)*. The association between the NAT2 intermediate/fast acetylators and total and red meat intake was confirmed in a nested case–control study among postmenopausal women included in the Danish "Diet, Cancer and Health" cohort study *(42)*. This study estimated that a 25 g increment in daily intake of total meat, red meat, or processed meat was associated with a significantly 9, 15, and 23% higher risk of breast cancer. These estimates compared well with those of other studies suggesting a positive association between higher red meat intake ($\sim$100 g/day) in adolescent women and the development of ER+ premenopausal breast tumors *(43)*.

Tobacco smoking may be a risk factor for NAT2 rapid acetylators. Genetic polymorphisms for *NAT1* and *NAT2* could differentially modulate breast cancer risk in presence of glutathione-*S*-transferases (GST)-M1 (*GSTM1*) and GST-T1 (*GSTT1*) null genotypes *(44)*. GSTs are involved in the metabolism of a variety of xenobiotics and lack of these enzymes may lead to accumulation of reactive electrophilic compounds that increase the risk of cancer *(45)*. A significant percentage of the Caucasian population carries a homozygous deletion for GSTs. A 60% increase in the incidence of postmenopausal breast cancer was observed in a cohort of Caucasian women whose *GSTM1* and *GSTT1* genes were deleted *(46)*. The risk was higher for women who consumed meats consistently well- or very well done and carried either one of the null genotypes. The null *GSTT1* genotype conferred a higher risk for postmenopausal breast cancer in women former smokers. Differences in duration and timing of the exposures appeared to influence breast cancer risk. Recent results from the Long Island Breast Cancer Study Project reported a 47% increase among postmenopausal women in relation to a lifetime average intake of grilled/smoked foods, which was more pronounced with low consumption of fruits and vegetables *(47)*. A positive association was observed in ER+/PR+ cases in the highest tertile of benzo[a]pyrene exposure from meat intake when compared with those in the lowest tertile *(48)*. Benzo[a]pyrene is produced during grilling or barbecuing through incomplete combustion of carbon and hydrogen in fat *(49)*.

7.3.3. Isothiocyanates

Isothiocyanates found in cruciferous vegetables have been shown to repress Phase I activating enzymes and induce Phase II conjugating enzymes including GSTs. Through this mechanism, cruciferous vegetables may exert a protective effect against oxidative stress and carcinogenesis induced by reactive metabolites. However, the specific effects of cruciferous vegetables on breast cancer risk remain unclear. The anticarcinogenic properties of indol-3-carbinol (I3C) and diindolylmethane (DIM), secondary metabolites from glucosinolates found in cruciferous vegetables, have been linked to their ability to stimulate the metabolism of estradiol to less potent estrogens *(50)*, induce BRCA1 expression *(51)*, and prevent carcinogen-induced mammary tumors *(52)*. Specific dietary recommendation for BRCA1 mutation carriers included supplementation with 25–50 mg of actual DIM along with 30–40 mg of lycopene (~two glasses of tomato juice/day), 300 μg of 1-selenomethionine, and 340–540 mg of EGCG (epigallocatechin-3-gallate). The latter is a green polyphenol found in green tea (~180 mg/cup) *(53,54)*.

A meta-analysis of 26 published studies revealed an inverse relationship between vegetable consumption and risk of breast cancer *(55)*. In premenopausal women, the risk was marginally inversely associated with consumption of broccoli, but this association was not modified by GSTT1 and GSTM1 phenotypes *(56)*. The same study reported no risk reduction associated with consumption of cruciferous vegetables in postmenopausal women. Recently, the Shanghai Breast Cancer Study examined the effects of cruciferous vegetable intake and genetic polymorphism at the *GSTP1* locus *(57)*. These investigations involved ~3,000 breast cancer cases and indicated that the *GSTP1 Val/Val* genotype was significantly associated with greater breast cancer risk. Interestingly, the association was greater for premenopausal women compared to postmenopausal women with low intake of cruciferous vegetables. White turnip and Chinese cabbage intakes were inversely associated with breast cancer risk. A possible explanation for this association is that the GSTP1 Val/Val phenotype may increase urinary excretion of isothiocianates, thus reducing their levels in the body *(58)*.

7.3.4. Dietary Patterns

Because foods are consumed in associations, it has been difficult to isolate the effects of individual nutrients on cancer risk. Analysis of dietary patterns has been proposed as an alternative approach to studying the association between diet and risk of breast cancer. Results of a study that assessed the influence of diet quality on breast cancer risk in postmenopausal women reported that women who had high scores for the Healthy Eating Index, Recommended Food Score, and alternative Mediterranean Diet Score had a lower risk of ER– breast cancers. Conversely, the same study did not report an association of diet quality with ER+ tumors *(59)*. The 1987–1998 Breast Cancer Detection Demonstration Project investigated dietary patterns in a nationwide sample of ~40,000 US women. This investigation identified three dietary patterns and suggested no significant association between the vegetable–fish/poultry and beef/pork–starch patterns and risk of breast cancer. However, an inverse relationship was reported between invasive breast cancer and the traditional southern diet characterized by high intakes of traditional rural foods including cooked greens, legumes, beans, sweet potatoes, cornbread, cabbage, fried fish and chicken and rice and low intakes of cheese, mayonnaise-salad dressing, alcohol, and salty snacks. This study also reported a significant association between the southern diet and a ~22% lower risk of breast cancer in women without a family history of breast cancer, a body mass index (BMI) <25, and with ER+ and PR+ tumors. The same study suggested that intake of the southern diet in childhood may contribute to the reduced risk of breast cancer observed in manual social class participants compared to non-manual women subjects *(60)*. The mean legume intake in the southern diet was 65% higher compared to the other dietary patterns. Legumes are rich in isoflavones, such as daidzein and genistein, which have been shown to exert antiestrogenic actions. Although some studies reported an inverse relationship between high intake of soybean, which is high in isoflavones, and risk of postmenopausal breast cancer *(61)*, other investigations found no association between breast cancer risk and intake of isoflavones and other flavonoids including flavanones, flavan-3-ols, flavonols, and anthocyanidins *(62)*. These conflicting results may be due to different intake of soya or incorrect assessment of dietary levels. However, the latter study reported a strong, statistically inverse relationship between flavone (i.e., apigenin, luteolin) intake and breast cancer risk. Flavones are found mainly in leafy vegetables, grains, and herbs.

7.3.5. Phytoestrogens

Phytoestrogens are natural compounds found in plants. They are believed to exert protective effects because higher consumption of soybean phytoestrogens is associated with reduced incidence of breast cancer in Asian women compared to women residing in Western countries. However, there are conflicting views concerning recommendations for the use of specific phytoestrogens in breast cancer prevention. A meta-analysis of 18 epidemiological studies published from 1978 through 2004 reported that high soy intake was modestly associated with reduced breast cancer risk *(63,64)*. However, a comprehensive literature review of clinical interventions in postmenopausal women from the European-funded project PhytoHealth generated inconclusive evidence on the effects of isoflavones on breast cancer *(65)*. One potential explanation for these conflicting results is the timing of exposure to isoflavones. For example, compounds found in soy foods, such as genistein, may increase the risk of breast cancer risk if exposure occurs during pregnancy. This effect may be related to the estrogenic effects of genestein, whereas prepubertal exposure may decrease the risk *(66,67)*. The childhood intake of soy was negatively associated with adult mammographic density in a multi-ethnic cohort *(68)* suggesting that adolescent soy exposure may be protective *(69)*. This conclusion finds support in studies reporting that phytoestrogens may function as antiestrogens with premenopausal high circulating levels of estrogens *(70)*. Conversely, phytoestrogens may function as weak estrogens and increase breast cancer risk in postmenopausal women *(71)*. An inverse relationship was reported between soy intake and breast cancer in premenopausal, but not in postmenopausal women *(72)*. Another aspect that deserves consideration is the level of intake of soy products. Levels of circulating isoflavones in women living in Asia may be higher compared to those of women living in Western countries. For example, in women residing in the Netherlands where the risk is high, the median intake and plasma genestein were, respectively, 0.14 mg/day and 4.89 ng/mL *(73)*. In contrast, in Japan, where the incidence of breast cancer is lower, median genistein intake and serum levels were respectively, 22.3 mg/day and 90.2 ng/mL *(74)*. A statistically significant inverse association was reported between plasma genistein (median level 144.5 ng/mL) and the risk of breast cancer in Japan *(75)*. More importantly, the same study reported a 65% reduction in breast cancer in the highest quartile group with a median value of ~350 ng/mL of plasma genestein and a median genestein intake of ~29 mg/day. Therefore, it appears that levels of intake are significantly higher for Asian women compared to women living in Western countries. Furthermore, a recent epidemiological review of soy exposure in Asian women concluded that breast cancer risk was the lowest among women with high intake (>20 mg isoflavones/day) compared to those consuming low (<5 mg isoflavones/day) or intermediate (~10 mg isoflavones/day) levels *(76)*. These studies proposed a protective dose-dependent effect (16% risk reduction/10 mg) of soy isoflavones on risk of breast cancer, and a stronger protective effect if exposure occurred during adolescence than adulthood. Nevertheless, a recent study with a population of British women participating in the European Prospective Investigation into Cancer and Nutrition reported no evidence of an association between dietary isoflavone intake ranging from <1 mg/day to ~32 mg isoflavones/day and risk of breast cancer *(77)*. A possible explanation for the latter results is the large heterogeneity of diets in the British cohort which included vegetarian and non-vegetarian women. Finally, it should be pointed out that the consumption of isoflavone supplements is not recommended for women at high risk of breast cancer, breast cancer survivors, and obese women *(69)*.

7.3.6. Regional Diets

Foods in Western diets have been identified as risk factors for breast cancer. In a population of white women 50–64 years of age comprised in the Cancer Surveillance System of Western Washington, and diagnosed with in situ or invasive breast cancer, red- and high-fat meat intake was significantly associated with an increased risk of postmenopausal breast cancer *(78)*. A study with non-Hispanic white women from the Southwest United States reported a greater risk of breast cancer with Western (high animal products and refined grains) diet compared to Native Mexican (Mexican cheeses, soups, meat dishes, legumes, and tomato-based sauces) and Mediterranean diets. Processing and cooking methods may increase the levels of mammary carcinogens including heterocyclic amines, polyaromatic hydrocarbons, which have been shown to possess carcinogenic properties *(49)*.

The incidence of breast cancer in Chinese populations is about one-third the rate observed in US Caucasian women. Nevertheless, the rate nearly doubles as Chinese women migrate to Western countries, including North America. Dietary practices in Asian populations are known to differ from those of Western societies. A study that examined the impact of intake of fruits and vegetables among Chinese women in Shanghai reported a reduced risk of breast cancer with increasing intake of dark-orange vegetables, Chinese white turnips, and certain dark/green vegetables *(79)*. Compounds contained in dark green/yellow-orange vegetables that may exert a protective effect include α- and β-carotene, folate, and lutein, which have been shown to have anticarcinogenic effects. Food items included in this study were used in 90% of the dishes prepared in Shanghai by deep frying and stir-frying, and 94% of the foods were prepared with soybean oil *(80)*. Supporting evidence for the protective effects of high intake of fruits and vegetables (>3.8 serving/day) in Chinese women against the risk of breast cancer was provided by a more recent case–control studies in Shanghai, China *(81)*.

7.3.7. Fruits and Vegetables

Reports from a cohort study *(82)* indicated no association between intake of fruits and vegetables and the incidence of ovarian cancer. These results have been corroborated by recent reports *(83)* suggesting no clear association between adulthood consumption of fruit and vegetables and risk of ovarian cancer. However, an inverse relationship was suggested between total intake of fruits and vegetables during adolescence and the incidence of ovarian carcinoma *(82)*. These data suggest that early exposure to bioactive food components present in fruits and vegetables may exert protective effects against development of ovarian carcinoma later in life.

The European Prospective Investigation into Cancer and Nutrition (EPIC) study with ~280,000 women between 25 and 70 years of age from 10 European countries reported no association between intake of total vegetable or vegetable subgroups and breast cancer risk *(84)*. This prospective study included more than 3,500 cases of invasive breast cancers. A pooled analysis of eight prospective studies from the United States, Canada, Netherlands, and Sweden including more than 7,000 invasive breast cancer cases among ~350,000 women concluded that fruit and vegetable consumption during adulthood was not associated with reduced breast cancer risk *(85)*. This study examined the effects of total fruits and vegetables, fruits, fruit juices, total fruits, total vegetables, green leafy vegetables, 8 botanically defined fruit and vegetable groups, or 17 specific fruits and vegetables. There was no evidence of interaction between menopausal status and fruits and vegetables groups examined. Similarly, no significant association with breast cancer risk was seen in women residing in Germany, and consuming high intake of fruits and cooked vegetables, whereas an inverse relationship was observed with intake of raw vegetables, total vegetables, and whole-grain products *(86)*. These results contrasted with those of other studies reporting a 25% reduction in breast cancer risk when comparing highest vs lowest consumers *(6)*. Similarly, consumption of 1–2 servings per days of *Brassica* vegetables was associated with a 20–40% lower risk of breast cancer in postmenopausal Swedish women 50–74 years of age *(87)*. A significant protective effect of vegetables, but not fruit, against breast cancer was reported from evaluation of 15 case–control studies *(88)*. Whereas differences may be due to analytical differences and study design, protective effects may be related to specific nutrients in fruits and vegetables or specific subgroups of women with family history *(30)* or differences in ER status *(89)*.

Two randomized dietary intervention trials investigated the role of diet on breast cancer recurrence and survival. The Women's Intervention Nutrition Study (WINS) targeting postmenopausal women reported that a reduction in fat intake was associated with a prolonged survival of women with early-stage ER– breast cancer *(90)*. In contrast, the results of the Woman's Healthy Eating and Living (WHEL) dietary intervention *(91)* that included both pre- and postmenopausal women reported that a diet that was very high in vegetables, fruit, and fiber and low in fat did not reduce additional breast cancer events and mortality among women diagnosed with early-stage breast cancer. The intervention group in the WHEL study consumed 5 vegetable servings and 16 ounces of vegetable juice, 3 servings of fruit, and 30 g of fiber, and reduced fat intake to 15–20%. Conflicting results between the WINS and WHEL study may be due to differences in the intake of calories from fat. Moreover, the WINS trial targeted only postmenopausal women, whereas the WHEL study included both pre-and postmenopausal subjects *(92)*. A recent report from the WHEL project suggested greater survival after breast cancer in physically active women with high vegetable–fruit intake *(93)*. The latter results suggested a positive interaction between dietary pattern and lifestyle factors.

7.4. DIET–HORMONE RECEPTOR STATUS INTERACTION

Diet may differentially impact the etiology of ER+ and ER– breast cancers. The exposure to estrogen has been linked to an increase in the risk of breast cancer *(94)*. High serum estrogen concentrations (~22 vs ~11 pg/mL) have been associated with increased recurrence in women diagnosed with early stage breast cancer *(95)*. A statistical analysis of ~1,400 breast cancer cases and 1,500 controls from the Long Island Breast Cancer Study Project (LIBCSP) indicated an inverse relationship for fruit and vegetable intake, in particular carotenoids (α-carotene, β-carotene, and lutein) and lycopene, among postmenopausal women with ER+ positive tumors compared to premenopausal breast cancers with ER– tumors. Postmenopausal women who consumed more than ~410 μg/day of dietary α-carotene had a 27% reduction in breast cancer incidence. In this study, the percentage of women who had either an ER+/PR+ or ER+/PR– tumor were ~59 and 14%, respectively *(89)*. Data analysis from the Women's Health Initiative Observational Study suggested that dietary α-carotene, β-carotene, and lycopene were inversely associated with the risk of breast cancers that were positive for both ER and PR among postmenopausal women *(96)*. Conversely, an elevated risk of breast cancer was found for premenopausal women with greater red meat intake and ER+ and PR+ status *(97)*. Compounds found in meat that may be responsible for this increased risk include the heterocyclic amine 2-amino-1-methyl-6-phenylimidazo[*4,5*-b]pyridine (PhIP), which is estrogenic in ER+ cells *(98)*.

A study that examined the association between dietary patterns and ER status reported that a prudent diet characterized by high intake of fruits and vegetables, whole grains, low fat dairy products, fish, and poultry may protect against ER– tumors in postmenopausal women *(99)*. These results paralleled those of other investigations reporting an inverse relationship between fruits and vegetable intake in ER– breast cancers *(100)*. Recent studies comprising ~51,000 postmenopausal women from the Swedish Mammography Screening Cohort (SMSC) reported that fiber intake from cereal products was associated with a 50% reduced risk for overall and ER+/PR+ breast tumors among ever-users of postmenopausal hormones *(101)*.

7.5. OBESITY AND ENERGY BALANCE

7.5.1. Obesity and Risk of Female Cancers

Body fatness, overweight, and obesity are pandemic conditions, affecting not only adult subjects, but also young individuals. According to a recent report by the World Cancer Research Fund/American Institute for Cancer Research *(6)*, factors related to food, nutrition, and physical activity that lead to weight gain and obesity may also increase the risk of cancer, whereas caloric restriction and increased physical activity decrease cancer risk, predominantly postmenopausal breast cancer and endometrial cancer.

The chronic excess of energy intake over energy expenditure is a predisposing condition to obesity *(102)*. Although several measurements have been used to define obesity, BMI calculated as weight in kilograms divided by height in meters squared is the most frequently used parameter in epidemiological studies. The World Health Organization (WHO) has proposed a definition of overweight as BMI ranging from 25.0 to 29.9 kg/m^2 and obesity as BMI $\geq 30.0\,kg/m^2$. This classification applies to both sexes and to all adult age groups *(103)*. However, body fat distribution can be assessed using other measurement, including waist and hip circumferences, waist-to-hip ratio, waist-to-thigh ratio, subscapular skinfold, and subscapular to thigh skinfold ratio, which may be better disease risk predictors than BMI alone.

Obesity and overweight are common worldwide and their prevalence is rapidly increasing, thus representing one of the most daunting health challenges of the 21st century. In 2005, it was estimated that 1.6 billion adults worldwide were overweight, of whom 300 million were obese *(104)*. According to estimates by the International Association for the Study of Obesity provided in April 2007, approximately 40–50% of men and 25–35% of women in the European Union were overweight, whereas 15–25% of men and 15–25% of women were obese *(105)*. In the United States, about one-third of the population is overweight, 22% of people are obese, and 24% of the population reports little or no physical activity *(106)*.

Obesity is a risk factor for several chronic diseases and certain types of cancer *(6,107)*. The International Agency for Research on Cancer (IARC) estimates that 25% of cancers worldwide are caused by overweight, obesity, and a sedentary lifestyle *(108)*. Epidemiological studies have shown that obesity is a risk factor for postmenopausal breast cancer and cancers of the endometrium *(107)*. Numerous studies have investigated the complex association between indicators of body size and risk of breast cancer. It is estimated that 30–50% of postmenopausal breast cancer deaths may be attributed to excess body weight (11,000–18,000 per year) *(109)*. However, for premenopausal women, conditions of overweight and obesity lowered the risk of breast cancer, possibly because of a higher frequency of anovulatory menstrual cycles, which may reduce the exposure to estrogens *(110,111)*. In postmenopausal women, the association between BMI and risk of breast cancer was stronger for subjects who did not use hormone-replacement therapy *(107)*. Adult weight gain rather than higher BMI at a younger age has been associated with greater risk of postmenopausal breast cancer. Abdominal adiposity, measured by waist circumference, is positively associated with higher risk of postmenopausal breast cancer, with stronger relationships among non-hormone-replacement therapy users. Conversely, changes in waist circumference have generally not been correlated to risk of breast cancer *(112)*. Interestingly, factors that lead to greater attained height may also contribute to breast and ovarian cancer *(6)*.

Adult weight gain is associated with increased risk of endometrial cancer. Subscapular skinfold parameters have been shown to better predict endometrial cancer risk than waist-to-hip ratio *(107,113)*. Waist circumference, hip circumference, and waist-to-hip ratio were positively correlated with endometrial cancer risk. However, after adjusting for BMI, abdominal body fat was a better predictor for endometrial cancer than gluteofemoral body fat. The association between weight, BMI, and hip circumference and risk of endometrial cancer was stronger for postmenopausal women. In contrast, the association between higher waist circumference and waist-to-hip ratio and endometrial cancer was greater for premenopausal women *(114)*.

7.5.2. Energy Balance and Physical Activity

Dietary energy balance refers to the balance between caloric intake and energy expenditure. Chronic positive energy balance, which can lead to obesity, raises the risk of developing multiple cancers, whereas the state of negative energy balance, as induced by calorie restriction, decreases cancer risk *(115)*. Dietary energy restriction may be a powerful approach for the prevention of cancer and certain chronic diseases. Obesity prevention by caloric restriction is consistently associated with reduced mammary tumor development *(116–118)*.

The term "energy balance" refers to the integrated effects of diet, physical activity, and genetics on growth and body weight over the individual's lifetime *(119)*. Physical activity is defined as skeletal muscle contraction that results in a quantifiable expenditure of energy. The recent 2007 WCRF/AICR Second Report concluded that physical activity protected against several cancers, including postmenopausal breast cancer and endometrial cancer, independently of other factors, such as body fatness, overweight, weight gain, and obesity *(6)*.

7.5.3. Hormonal Regulation

Alterations in circulating hormones modulate the effects of energy balance on cancer risk *(115)*. Women with elevated levels of estrogens and androgens are at increased risk of developing breast cancer *(120)*, whereas elevated estrogen concentrations (unopposed by progesterone) tend to increase the risk of endometrial cancer *(121)*. After menopause, the adipose tissue is the major source of estradiol production *(122)*. Obese postmenopausal women have higher conversion rates of sex hormones and the association between obesity and breast cancer risk in postmenopausal women can largely be explained by increased levels of estrogen *(107,123–125)*. Using the SMSC cohort, Suzuki et al. *(126)* reported that excess endogenous estrogen due to obesity increased the risk of ER+/PR+ postmenopausal breast cancer risk.

For premenopausal women, obesity is associated with a higher frequency of anovulatory cycles and lower levels of circulating sex steroid hormones, which may subsequently reduce breast cancer risk *(6,110,111)*. The "unopposed estrogen" hypothesis proposes that endometrial cancer may develop as a result of the mitogenic effects of estrogens when these are insufficiently counterbalanced by progesterone *(107)*.

High insulin levels have been correlated with insulin resistance syndrome (IRS), obesity at waist circumference, and poor outcome in postmenopausal breast cancer *(127)*. Interventions with early breast cancers have focused on the use of pharmacological compounds such as metmorfin, and lifestyle changes including weight loss and physical activity *(128)*. Interestingly, complications from IRS were important predictors for long-term non-breast cancer mortality in women with early stage breast cancer.

Obesity-induced insulin resistance associated with high systemic levels of insulin-like growth factor 1(IGF-1) increases the risk of breast and endometrial cancer *(129–132)*. Although the involvement of IGF-1 in the cancer process has been documented in vitro *(116,131)*, recent large nested case–control studies failed to see a significant association between IGF factors and breast cancer *(132,133)*. It has been proposed that a condition of high estrogen and low progesterone may promote endometrial tumors by stimulating IGF-1 synthesis while reducing IGFBP-1 *(107,121)*.

The phosphatidylinositol 3-kinase (PI3K)/Akt pathway regulates glucose homeostasis, cell migration, growth, and tumorigenesis. It is a key downstream target of insulin and IGF-I receptors *(134,135)*. The activation of Akt signaling by the PI3K is inhibited by the PTEN factor. A downstream target of Akt is the mammalian target of rapamycin (mTOR), an evolutionary conserved serine/threonine kinase and a key regulator of protein translation and synthesis, centrally involved in cell growth *(136)*. The mTOR pathway integrates changes in nutrients supply, energy, and mitogen signals that regulate cell growth and division. The activation of the PI3K/Akt axis by insulin and IGF-1 leads to activation of mTOR, through inhibition of tuberous sclerosis proteins (TSC1 and TSC2) thus stimulating cell growth and survival. Conversely, energy deprivation (high AMP/ATP ratio) leads to activation of AMPK which directly phosphorylates and activates TSC2 resulting in the inhibition of mTOR. These pathways form a sophisticated system that integrates changes in environmental signals with metabolic cellular responses *(115,131)* (Fig. 7.1). The recurrent activation of Akt in conditions of low AMP/ATP ratio may contribute to repression of TSC2 activity through phosphorylation and activation of mTOR. Therefore, the phosphorylation of TSC2 by Akt or AMPK may lead, respectively, to activation or repression of mTOR, influencing cellular behavior. Recent studies in a mammary tumor model documented that dietary energy restriction-mediated inhibition of carcinogenesis involved the interplay between AMPK and mTOR *(137)*. The activity of mTOR appears to be repressed by dietary energy restriction through activation of AMPK and inactivation of Akt *(138)*. This mechanism provides the molecular basis for understanding the impact of obesity and energy restriction on signal transduction pathways that promote tumor progression, and dietary and pharmacological targets to offset the impact of obesity on the cancer process *(139)*.

A salient characteristic of obesity is elevated levels of circulating leptin, a hormone produced by adipocytes. The human leptin gene promoter harbors multiple transcription regulatory elements that can be activated by insulin *(140)*. Recent findings on the cross talk between mTOR and leptin signaling may have implications for obesity-related conditions such as diabetes, cardiovascular disease,

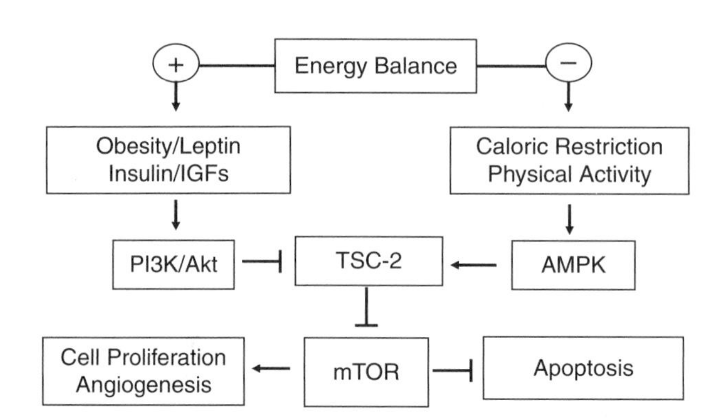

Fig. 1. The interplay between AMPK and mTOR regulates the influence of dietary energy balance on the cancer process. The proposed model suggests that energy restriction and physical activity induce the accumulation of AMPK, which in turn activates TSC (tuberous sclerosis proteins), thus abrogating the negative effects of mTOR (mammalian target of rapamycin) on the apoptotic process. Conversely, conditions of positive energy balance that lead to obesity and increased levels of circulating IGFs and leptin activate the PI3K/AkT pathway leading to abrogation of TSC activity, which in turn leads to activation of mTOR and stimulation of cell proliferation *(131–138)*.

and cancer *(141,142)*. Hyperinsulinemia may induce breast cancer progression through leptin-dependent mechanisms including regulation of the cell cycle, apoptosis, and by modulating the extracellular environment *(143)*. Leptin appears to participate in the regulation of proangiogenic molecules in benign and cancerous endometrial cells and in proliferative processes of the endometrium *(144)*.

Adiponectin is an adipose tissue-secreted hormone with insulin-sensitizing, anti-atherogenic, and anti-diabetic properties. Epidemiological studies have linked low adiponectin levels with increased risk of breast and endometrial cancer *(145–147)*. The circulating levels of adiponectin decrease with increasing body weight and central fat accumulation. Adiponectin levels are regulated by body weight, fat distribution, exercise, and diet. Recently, it has been demonstrated that individuals who adhere to a Mediterranean diet have higher levels of circulating adiponectin *(148)*.

7.6. RECOMMENDATIONS

In November 2007, WCRF/AICR published the Second Expert Report "Diet Nutrition, Physical Activity and Cancer Prevention: a Global Perspective" *(6)*, which represents the most comprehensive review on the evidence of diet, physical activity, and cancer ever completed. The panel judged the epidemiological, experimental, and mechanistic evidence by grading the strength of the causal relationships between food, nutrition and physical activity, and the risk of cancer, and also the relationship of diet and physical activity with weight gain, overweight, and obesity. Results from major intervention trials on diet and breast cancer that focused on dietary fat reduction and increased intake of fruits, vegetable, and fiber did not provide the basis for conclusive dietary recommendation. Data were also lacking concerning the role of healthy dietary patterns and specific nutrients in the etiology of breast cancer. Limited evidence from cohort and case–control studies suggests that non-starchy vegetables protect against ovarian and endometrial cancer. With regards to cervical cancer, there is limited but consistent evidence that carrots, containing high levels of carotenoids, including β- and α-carotene, and certain antioxidants, may protect against cervical cancer. Results of case–control studies suggest that red meat consumption is positively associated with endometrial cancer due to heme-iron, the generation of *N*-nitroso compounds, or the production of heterocyclic amines and polycyclic aromatic hydrocarbons. The cancer risk due to exposure to certain food components and mutagens is influenced by the genetic background as in the case of mutation carriers or in the presence of polymorphisms for metabolizing genes.

Future research is necessary to gain better knowledge of the causal relationships between obesity, energy balance, physical activity, and the cancer process. Although caloric restriction is consistently associated with reduced mammary tumor development, it is not clear how it influences the metastatic process. With respect to breast cancer, breast-feeding appears to be protective for the mother as well as the child. There exists convincing evidence that lactation protects against both premenopausal and postmenopausal breast cancer, whereas limited evidence suggests that lactation protects against cancer of the ovary.

REFERENCES

1. World Health Organization. (2008) World Health Statistics. Geneva, World Health Organization.
2. Garcia, M., Jemal, A., Ward, E. M., et al. (2007) Global Cancer Facts and Figures 2. Atlanta, GA, American Cancer Society.
3. Mackay, J., Jemal, A., Lee, N., and Parkin, D. M. (2006) The Cancer Atlas. Atlanta, GA, American Cancer Society.
4. Ferlay, J., Bray, F., Pisani, P., and Parkin, D. M. (2004) GLOBOCAN 2002: Cancer Incidence, Mortality and Prevalence Worldwide IARC Cancer Base no. 5. [computer program]. Version 2.0. Lyon, IARC.
5. Stewart, B. W. and Kleihues, P. (2003) World Cancer Report. Lyon, IARC/WHO, IARC Press.
6. World Cancer Research Fund/American Institute for Cancer Research. (2007) Food, Nutrition, Physical Activity, and the Prevention of Cancer: a Global Perspective. Washington, DC, AICR.
7. Parkin, D. M., Whelan, S. L., Ferlay, J., and Storm, H. (2005) Cancer Incidence in Five Continents, Vol I to VIII. Lyon, IARC.
8. Ferlay, J., Autier, P., Boniol, M., Heanue, M., Colombet, M., and Boyle, P. (2007) Estimates of the cancer incidence and mortality in Europe in 2006. *Ann Oncol* **18**, 581–2.
9. American Cancer Society. (2008) Cancer Facts and Figures 2008. Atlanta, GA, American Cancer Society.
10. Ries, L. A. G., Melbert, D., Krapcho, M., et al. (eds) (2008) SEER Cancer Statistics Review, 1975–2005. National Cancer Institute. Bethesda, MD, http://seer.cancer.gov/csr/1975_2005/ , based on November 2007 SEER data submission, posted to the SEER web site, 2008.
11. Baquet, C. R., Mishra, S. I., Commiskey, P., Ellison, G. L., and De Shields, M. (2008) Breast cancer epidemiology in blacks and whites: disparities in incidence, mortality, survival rates and histology. *J Natl Med Assoc* **100**, 480–8.
12. Miller, B. A., Chu, K. C., Hankey, B. F., and Ries, L. A. (2008) Cancer incidence and mortality patterns among specific Asian and Pacific Islander populations in the U.S. *Cancer Causes Control* **19**, 227–56.
13. McPherson, K., Steel, C. M., and Dixon, J. M. (2000) ABC of breast diseases. Breast cancer—epidemiology, risk factors, and genetics. *Br Med J* **321**, 624–8.
14. Cokkinides, V., Bandi, P., Siegal, R., Ward, E. M., and Thun, M. J. (2008) Cancer Prevention & Early Detection Facts and Figures 2008. Atlanta, GA, American Cancer Society.
15. Hėry, C., Ferlay, J., Boniol, M., and Autier, P. (2008) Quantification of changes in breast cancer incidence and mortality since 1990 in 35 countries with Caucasian-majority populations. *Ann Oncol* **19**, 1187–94.
16. Tamakoshi, K., Kondo, T., Yatsuya, H., et al. (2001) Trends in the mortality (1950–1997) and incidence (1975–1993) of malignant ovarian neoplasm among Japanese women: analyses by age, time, and birth cohort. *Gynecol Oncol* **83**, 64–71.
17. Chaitchik, S., Ron, I. G., Baram, A., and Inbar, M. (1985) Population differences in ovarian cancer in Israel. *Gynecol Oncol* **21**, 155–60.
18. Henderson, M. M. (1992) International differences in diet and cancer incidence. *J Natl Cancer Inst Monogr* **12**, 59–63.
19. Schorge, J. O., Knowles, L. M., and Lea, J. S. (2004) Adenocarcinoma of the cervix. *Curr Treat Options Oncol* **5**, 119–27.
20. Holcomb, K., and Runowicz, C. D. (2005) Cervical cancer screening. *Surg Oncol Clin N Am* **14**, 777–97
21. Poolman, E. M., Elbasha, E. H., and Galvani, A. P. (2008) Vaccination and the evolutionary ecology of human papilloma virus. *Vaccine* **26**(Suppl. 3), C25–30.
22. Sorosky, I. J. (2008) Endometrial cancer. *Obstet Gynecol* **111**, 436–7.
23. Sherman, M. E., and Devessa, S. S. (2003) Analysis of racial differences in incidence, survival and mortality for malignant tumors of the uterine corpus. *Cancer* **98**, 176–86.

24. Duska, L. R., Garrett, A., Rueda, B. R., Haas, J., Chang, Y., and Fuller, A. F. (2001) Endometrial cancer in woman 40 years old or younger. *Gynecol Oncol* **83**, 388–93.
25. Irvin, W. P., Rice, L. W., and Bereowitz, R. S. (2002) Advances in the management of endometrial adenocarcinoma. A review. *J Reprod Med* **47**, 173–89.
26. Pharoah, P. D., Day, N. E., Duffy, S., Easton, D. F., and Ponder, B. A. (1997) Family history and the risk of breast cancer: a systematic review and meta-analysis. *Int J Cancer* **71**, 800–9.
27. Vogel, V. G. (2008) Epidemiology, genetics, and risk evaluation of postmenopausal women at risk of breast cancer. *Menopause* **15**(4 Suppl.), 782–9.
28. McGuire, V., John, E. M., Felberg, A., et al. (2006) No increased risk of breast cancer associated with alcohol consumption among carriers of BRCA1 and BRCA2 mutations ages <50 years. *Cancer Epidemiol Biomarkers Prev* **15**, 1565–7.
29. Esteller, M., Silva, J. M., Dominguez, G., et al. (2000) Promoter hypermethylation and BRCA1 inactivation in sporadic breast and ovarian tumors. *J Natl Cancer Inst* **92**:564–9.
30. Zhang, S., Hunter, D. J., Forman, M. R., et al. (1999) Dietary carotenoids and vitamins A, C, and E and risk of breast cancer. *J Natl Cancer Inst* **91**, 547–56.
31. Silvera, S. A., Jain, M., Howe, G. R., Miller, A. B., and Rohan, T. E. (2006) Energy balance and breast cancer risk: a prospective cohort study. *Breast Cancer Res Treat* **97**, 97–106.
32. Michels, K. B., and Ekbom, A. (2004) Caloric restriction and incidence of breast cancer. *JAMA* **291**, 1226–30.
33. Malin, A., Matthews, C. E., Shu, X. O., et al. (2005) Energy balance and breast cancer risk. *Cancer Epidemiol Biomarkers Prev* **14**, 1496–501.
34. Nkondjock, A., Robidoux, A., Paredes, Y., Narod, S. A., and Ghadirian, P. (2006) Diet, lifestyle and BRCA-related breast cancer risk among French-Canadians. *Breast Cancer Res Treat* **98**, 285–94.
35. Nkondjock, A., and Ghadirian, P. (2007) Diet quality and BRCA-associated breast cancer risk. *Breast Cancer Res Treat* **103**, 361–9.
36. Kotsopoulos, J., Olopado, O. I., Ghadirian, P., et al. (2005) Changes in body weight and the risk of breast cancer in BRCA1 and BRCA2 mutation carriers. *Breast Cancer Res* **7**, R833–43.
37. Nkondjock, A., Ghadirian, P., Kotsopoulos, J., et al. (2006) Coffee consumption and breast cancer risk among BRCA1 and BRCA2 mutation carriers. *Int J Cancer* **118**, 103–7.
38. Kotsopoulos, J., Ghadirian, P., El-Sohemy, A., et al. (2007) The CYP1A2 genotype modifies the association between coffee consumption and breast cancer risk among BRCA1 mutation carriers. *Cancer Epidemiol Biomarkers Prev* **16**, 912–6.
39. Lurie, G., Maskarinec, G., Kaaks, R., Stanczyk, F. Z., and Le Marchand, L. (2005) Association of genetic polymorphisms with serum estrogens measured multiple times during a 2-year period in premenopausal women. *Cancer Epidemiol Biomarkers Prev* **14**, 1521–7.
40. Krajinovic, M., Ghadirian, P., Richer, C., et al. (2001) Genetic susceptibility to breast cancer in French-Canadians: role of carcinogen-metabolizing enzymes and gene-environment interactions. *Int J Cancer* **92**, 220–5.
41. Ambrosone, C. B., Freudenheim, J. L., Graham, S., et al. (1996) Cigarette smoking, N-acetyltransferase 2 genetic polymorphisms, and breast cancer risk. *JAMA* **276**, 1494–501.
42. Egeberg, R., Olsen, A., Autrup, H., et al. (2008) Meat consumption, N-acetyl transferase 1 and 2 polymorphism and risk of breast cancer in Danish postmenopausal women. *Eur J Cancer Prev* **17**, 39–47.
43. Linos, E., Willett, W. C., Cho, E., Colditz, G., and Frazier, L. A. (2008) Red meat consumption during adolescence among premenopausal women and risk of breast cancer. *Cancer Epidemiol Biomarkers Prev* **17**, 2146–51.
44. Lee, K. M., Park, S. K., Kim, S. U., et al. (2003) *N*-acetyltransferase (NAT1, NAT2) and glutathione *S*-transferase (GSTM1, GSTT1) polymorphisms in breast cancer. *Cancer Lett* **196**, 179–86.
45. García-Closas, M., Kelsey, K. T., Hankinson, S. E., et al. (1999) Glutathione *S*-transferase mu and theta polymorphisms and breast cancer susceptibility. *J Natl Cancer Inst*. **91**, 1960–4.
46. Zheng, W., Wen, W. Q., Gustafson, D. R., Gross, M., Cerhan, J. R., and Folsom, A. R. (2002) GSTM1 and GSTT1 polymorphisms and postmenopausal breast cancer risk. *Breast Cancer Res Treat* **74**, 9–16.
47. Gammon, M. D. and Santella, R. M. (2008) PAH, genetic susceptibility and breast cancer risk: an update from the Long Island Breast Cancer Study Project. *Eur J Cancer* **44**, 636–40.
48. Steck, S. E., Gaudet, M. M., Eng, S. M., et al. (2007) Cooked meat and risk of breast cancer—lifetime versus recent dietary intake. *Epidemiology* **18**, 373–82.
49. Kazerouni, N., Sinha, R., Hsu, C. H., Greenberg, A., and Rothman, N. (2001) Analysis of 200 food items for benzo[a]pyrene and estimation of its intake in an epidemiologic study. *Food Chem Toxicol* **39**, 423–36.

50. Bradfield, C. A. and Bjeldanes, L. F. (1987) Structure-activity relationships of dietary indoles: a proposed mechanism of action as modifiers of xenobiotic metabolism. *J Toxicol Environ Health* **21**, 311–23.
51. Meng, Q., Qi, M., Chen, D. Z., et al. (2000) Suppression of breast cancer invasion and migration by indole-3-carbinol: associated with up-regulation of BRCA1 and E-cadherin/catenin complexes. *J Mol Med* **78**, 155–65.
52. McDougal, A., Gupta, M. S., Morrow, D., Ramamoorthy, K., Lee, J. E., and Safe, S. H. (2001) Methyl-substituted diindolylmethanes as inhibitors of estrogen-induced growth of T47D cells and mammary tumors in rats. *Breast Cancer Res Treat* **66**, 147–57.
53. USDA Report. (2007) Database for the Flavanoids Content of Selected Foods Release 2.1.
54. Kotsopoulos, J. and Narod, S. A. (2000) Towards a dietary prevention of hereditary breast cancer. *Cancer Causes Control* **16**, 125–38.
55. Gandini, S., Merzenich, H., Robertson, C., and Boyle, P. (2000) Meta-analysis of studies on breast cancer risk and diet: the role of fruit and vegetable consumption and the intake of associated micronutrients. *Eur J Cancer* **36**: 636–46.
56. Ambrosone, C. B., McCann, S. E., Freudenheim, J. L., Marshall, J. R., Zhang, Y., and Shields, P. G. (2004) Breast cancer risk in premenopausal women is inversely associated with consumption of broccoli, a source of isothiocyanates, but is not modified by GST genotype. *J Nutr* **134**, 1134–8.
57. Lee, S. A., Fowke, J. H., Lu, W., et al. (2008) Cruciferous vegetables, the GSTP1 Ile105Val genetic polymorphism, and breast cancer risk. *Am J Clin Nutr* **87**, 753–60.
58. Fowke, J. H., Shu, X. O., Dai, Q., et al. (2003) Urinary isothiocyanate excretion, brassica consumption, and gene polymorphisms among women living in Shanghai, China. *Cancer Epidemiol Biomarkers Prev* **12**, 1536–9.
59. Fung, T. T, Hu, F. B., McCullough, M. L., Newby, P. K., Willett, W. C., and Holmes, M. D. (2006) Diet quality is associated with the risk of estrogen receptor-negative breast cancer in postmenopausal women. *J Nutr* **136**, 466–72.
60. Velie, E. M., Schairer, C., Flood, A., He, J. P., Khattree, R., and Schatzkin, A. (2005) Empirically derived dietary patterns and risk of postmenopausal breast cancer in a large prospective cohort study. *Am J Clin Nutr* **82**, 1308–19.
61. Do, M. H., Lee, S. S., Kim, J. Y., Jung, P. J., and Lee, M. H. (2007) Fruits, vegetables, soy foods and breast cancer in pre- and postmenopausal Korean women: a case-control study. *Int J Vitam Nutr Res* **77**, 130–41.
62. Peterson, J., Lagiou, P., Samoli, E., et al. (2003) Flavonoid intake and breast cancer risk: a case–control study in Greece. *Br J Cancer* **89**, 1255–9.
63. Trock, B. J., Hilakivi-Clarke, L., and Clarke, R. (2006) Meta-analysis of soy intake and breast cancer risk. *J Natl Cancer Inst* **98**, 459–71.
64. Hedelin, M., Löf, M., Olsson, M., Adlercreutz, H., Sandin, S., and Weiderpass, E.(2008) Dietary phytoestrogens are not associated with risk of overall breast cancer but diets rich in coumestrol are inversely associated with risk of estrogen receptor and progesterone receptor negative breast tumors in Swedish women. *J Nutr* **138**, 938–45.
65. Cassidy, A., Albertazzi, P., Lise Nielsen, I., et al. (2006) Critical review of health effects of soyabean phyto-oestrogens in post-menopausal women. *Proc Nutr Soc* **65**, 76–92.
66. De Assis, S., and Hilakivi-Clarke, L. (2006) Timing of dietary estrogenic exposures and breast cancer risk. *Ann N Y Acad Sci* **1089**, 14–35.
67. Fritz, W. A., Coward, L., Wang, J., and Lamartiniere, C. A. (1998) Dietary genistein: perinatal mammary cancer prevention, bioavailability and toxicity testing in the rat. *Carcinogenesis* **19**, 2151–8.
68. Maskarinec, G., Takata, Y., Franke, A. A., Williams, A. E., and Murphy, S. P. (2004) A 2-year soy intervention in premenopausal women does not change mammographic densities. *J Nutr* **134**, 3089–94.
69. Duffy, C., Perez, K., and Partridge, A. (2007) Implications of phytoestrogen intake for breast cancer. *CA Cancer J Clin* **57**, 260–77.
70. Bissonauth, V., Shatenstein, B., and Ghadirian, P. (2008) Nutrition and breast cancer among sporadic cases and gene mutation carriers: an overview. *Cancer Detect Prev* **32**, 52–64.
71. Glazier, M. G. and Bowman, M. A. (2001) A review of the evidence for the use of phytoestrogens as a replacement for traditional estrogen replacement therapy. *Arch Intern Med* **161**, 1161–72.
72. Lee, H. P., Gourley, L., Duffy, S. W., Estève, J., Lee, J., and Day, N. E. (1992) Risk factors for breast cancer by age and menopausal status: a case-control study in Singapore. *Cancer Causes Control* **3**, 313–22.
73. Keinan-Boker, L., van Der Schouw, Y. T., Grobbee, D. E., and Peeters, P. H. (2004) Dietary phytoestrogens and breast cancer risk. *Am J Clin Nutr* **79**, 282–8.
74. Yamamoto, S., Sobue, T., Sasaki, S., et al. (2001) Validity and reproducibility of a self-administered food-frequency questionnaire to assess isoflavone intake in a japanese population in comparison with dietary records and blood and urine isoflavones. *J Nutr* **131**, 2741–7.

75. Iwasaki, M., Inoue, M., Otani, T., et al. (2008) Plasma isoflavone level and subsequent risk of breast cancer among Japanese women: a nested case-control study from the Japan Public Health Center-based prospective study group. *J Clin Oncol* **26**, 1677–83.
76. Wu, A. H., Yu, M. C, Tseng, C. C., and Pike, M. C. (2008) Epidemiology of soy exposures and breast cancer risk. *Br J Cancer* **98**, 9–14.
77. Travis, R. C., Allen, N. E., Appleby, P. N., Spencer, E. A., Roddam, A. W., and Key, T. J. (2008) A prospective study of vegetarianism and isoflavone intake in relation to breast cancer risk in British women. *Int J Cancer* **122**, 705–10.
78. Shannon, J., Cook, L. S., and Stanford, J. L. (2003) Dietary intake and risk of postmenopausal breast cancer (United States). *Cancer Causes Control* **14**, 19–27.
79. Malin, A. S., Qi, D., Shu, X-O., et al. (2003) Intake of fruits, vegetables and selected micronutrients in relation to the risk of breast cancer. *Int J Cancer* **105**, 413–8.
80. Dai, Q., Shu, X. O., Jin, F., Gao, Y. T., Ruan, Z. X., and Zheng, W. (2002) Consumption of animal foods, cooking methods, and risk of breast cancer. *Cancer Epidemiol Biomarkers Prev* **11**, 801–8.
81. Shannon, J., Ray, R., Wu, C., et al. (2005) Food and botanical groupings and risk of breast cancer: a case-control study in Shanghai, China. *Cancer Epidemiol Biomarkers Prev* **14**, 81–90.
82. Fairfield, K. M., Hankinson, S. E., Rosner, B. A., Hunter, D. J., Colditz, G. A., and Willett, W. C. (2001) Risk of ovarian carcinoma and consumption of vitamins A, C, and E and specific carotenoids: a prospective analysis. *Cancer* **192**, 2318–26.
83. Koushik, A., Hunter, D. J., Spiegelman, D., et al. (2005) Fruits and vegetables and ovarian cancer risk in a pooled analysis of 12 cohort studies. *Cancer Epidemiol Biomarkers Prev* **14**, 2160–7.
84. van Gils, C. H., Peeters, P. H., Bueno-de-Mesquita, H. B., et al. (2005) Consumption of vegetables and fruits and risk of breast cancer. *JAMA* **293**, 183–93.
85. Smith-Warner, S., Spiegelman, D., Yaun, S. S., et al. (2001) Intake of fruits and vegetables and risk of breast cancer. A pooled analysis of cohort studies. *JAMA* **285**:769–76.
86. Adzersen, K. H., Jess, P., Freivogel, K. W., Gerhard, I., and Bastert, G. (2003) Raw and cooked vegetables, fruits, selected micronutrients, and breast cancer risk: a case-control study in Germany. *Nutr Cancer* **46**, 131–7.
87. Terry, P., Wolk, A., Persson, I., and Magnusson, C. (2001) Brassica Vegetables and breast cancer risk. *JAMA* **285**, 2975–7.
88. Riboli, E. and Norat, T. (2003) Epidemiologic evidence of the protective effect of fruit and vegetables on cancer risk. *Am J Clin Nutr* **78**(Suppl.), 559S–69S.
89. Gaudet, M. M, Britton, J. A., Kabat, G. C., et al. (2004) Fruits, vegetables, and micronutrients in relation to breast cancer modified by menopause and hormone receptor status. *Cancer Epidemiol Biomarkers Prev* **13**, 1485–94.
90. Chlebowski, R. T., Blackburn, G. L., Thomson, C. A., et al. (2006) Dietary Fat reduction and breast cancer outcome: interim efficacy results from the women's intervention nutrition study. *J Natl Cancer Inst* **98**, 1767–76.
91. Pierce, J. P., Natarajan, L., Caan, B. J., et al. (2007) Influence of a diet very high in vegetables, fruit, and fiber and low in fat on prognosis following treatment for breast cancer: the Women's Healthy Eating and Living (WHEL) randomized trial. *JAMA* **298**, 289–98.
92. Nelson, N. (2008) Dietary intervention trial reports no effect on survival after breast cancer. *J Natl Cancer Inst* **100**, 386–7.
93. Pierce, J. P., Stefanick, M. L., Flatt, S. W., et al. (2007) Greater survival after breast cancer in physically active women with high vegetable–fruit intake regardless of obesity. *J Clin Oncol* **25**, 2345–51.
94. Conzen, S. D. (2008) Nuclear receptors and breast cancer. *Mol Endocrinol* **22**, 2215–28.
95. Rock, C. L., Flatt, S. W., Laughlin, G. A., et al. (2008) Women's healthy eating and living study group. Reproductive steroid hormones and recurrence-free survival in women with a history of breast cancer. *Cancer Epidemiol Biomarkers Prev* **17**, 614–20.
96. Cui, Y., Shikany, J. M., Liu, S., Shagufta, Y., and Rohan, T. E. (2008) Selected antioxidants and risk of hormone receptor-defined invasive breast cancers among postmenopausal women in the Women's Health Initiative Observational Study. *Am J Clin Nutr* **87**, 1009–18.
97. Cho, E., Chen, W. Y., Hunter, D. J., et al. (2006) Red meat intake and risk of breast cancer among premenopausal women. *Arch Intern Med* **166**, 2253–9.
98. Lauber, S. N., Ali, S., and Gooderham, N. J. (2004) The cooked food derived carcinogen 2-amino-1-methyl-6-phenylimidazo[4,5-b] pyridine is a potent oestrogen: a mechanistic basis for its tissue-specific carcinogenicity. *Carcinogenesis* **25**, 2509–17.
99. Fung, T. T., Hu, F. B., Holmes, M. D., et al. (2005) Dietary patterns and the risk of postmenopausal breast cancer. *Int J Cancer* **116**, 116–21.

100. Olsen, A., Tjønneland, A., Thomsen, B. L., et al. (2003) Fruits and vegetables intake differentially affects estrogen receptor negative and positive breast cancer incidence rates. *J Nutr* **133**, 2342–7.
101. Suzuki, R., Rylander-Rudqvist, T., Ye, W., Saji, S., Adlercreutz, H., and Wolk, A. (2008) Dietary fiber intake and risk of postmenopausal breast cancer defined by estrogen and progesterone receptor status—a prospective cohort study among Swedish women. *Int J Cancer* **122**, 403–12.
102. Bray, G. A. (2004) Obesity is chronic, relapsing neurochemical disease. *Int J Obes* **28**, 34–38.
103. World Health Organization. (2003) Global Strategy on Diet, Physical Activity and Health—Obesity and Overweight. Geneva, WHO.
104. World Health Organization. (2006) Obesity and Overweight. Fact Sheet no 311. http://www.who.int/mediacentre/factsheets/fs311/en/.
105. International Association for the Study of Obesity. (2007) Adult overweight and obesity in the European Union (EU25). http://www.iotf.org/documents/Europeandatatable_000.pdf.
106. Centers for Disease Control and Prevention (CDC). (2008) Behavioral Risk Factors Surveillance Systems Survey data. Atlanta, Georgia, US Department of Health and Human Services. [updated 2008 Feb 12; cited 2008 Sept 4]. http://www.cdc.gov/NCCdphp/publications/AAG/brfss.htm
107. Pischon, T., Nöthlings, U., and Boeing, H. (2008) Obesity and Cancer. *Proc Nutr Soc* **67**, 128–45.
108. Vainio, H. and Biachini, F.(eds) (2002) Weight Control and Physical Activity. Lyon, IARC Press.
109. Calle, E. E., Rodriguez, C., Walker-Thurmond, K., and Thun, M. J. (2003) Overweight, obesity, and mortality from cancer in a prospective studied cohort of US adults. *N Engl J Med* **348**, 1625–38.
110. Friedenreich, C. M. (2001) Review of anthropometric factors and breast cancer risk. *Eur J Cancer Prev* **10**, 15–32.
111. Carmichael, A. R., and Bates, T. (2004) Obesity and breast cancer: a review of the literature. *Breast* **13**, 85–92.
112. Harvie, M., Hooper, L., and Howell, A. H. (2003) Central obesity and breast cancer: a systematic review. *Obes Rev* **4**, 157–73.
113. Austin, H., Austin, J. M. Jr., Partridge, E. E., Hatch, K. D., and Shingleton, H. M. (1991) Endometrial cancer, obesity, and body fat distribution. *Cancer Res* **51**, 568–72.
114. Friedenreich, C. M., Cust, A., Lahmann, P. H., et al. (2007) Anthropometric factors and risk of endometrial cancer: the European prospective investigation into cancer and nutrition. *Cancer Causes Control* **18**, 399–413.
115. Hursting, S. D., Lashinger, L. M., Colbert, L. H., et al. (2007) Energy balance and carcinogenesis: underlying pathways and targets for intervention. *Curr Cancer Drug Targets* **7**, 325–34.
116. Hursting, S. D., Lavigne, J. A., Berrigan, D., Perkins, S. N., and Barrett, J. C. (2003) Calorie restriction, aging, and cancer prevention: mechanisms of action and applicability to humans. *Annu Rev Med* **54**, 131–52.
117. Thompson, H. J., Zhu, Z., and Jiang, W. (2003) Dietary energy restriction in breast cancer prevention. *J Mammary Gland Biol Neoplasia* **8**, 133–42.
118. Zhou, Z., Jiang, W., McGinley, J. N., Price, J. M., Gao, B., and Thompson, H. J. (2007) Effect of dietary energy restriction on gene regulation in mammary epithelial cells. *Cancer Res* **67**, 12018–25.
119. U.S. Department of Health and Human Services. (2004) National Institutes of Health. Strategic Plan for NIH Obesity Research. A Report of the NIH Obesity Research Task Force. NIH Publication No 04-5493.
120. Key, T., Appleby, P. N., Barnes, I., and Reeves, G. (2002) Endogenous Hormones and Breast Cancer Collaboration Group. Endogenous sex hormones and breast cancer in postmenopausal women: reanalysis of nine prospective studies. *J Natl Cancer Inst* **94**, 606–16.
121. Kaaks, R., Lukanova, A., and Kurzer, M. S. (2002) Obesity, endogenous hormones, and endometrial cancer risk: a synthetic review. *Cancer Epidemiol Biomarkers Prev* **11**, 1531–43.
122. Siiteri, P. K. (1987) Adipose tissue as a source of hormones. *Am J Clin Nutr* **45**, (Suppl.), 277–82.
123. Kaaks, R., Rinaldi, S., Key, T. J., et al. (2005) Postmenopausal serum androgens, oestrogens and breast cancer risk: the European prospective investigation into cancer and nutrition. *Endocr Relat Cancer* **12**, 1071–82.
124. Key, T. J., Appleby, P. N., Reeves, G. K., et al. (2003) Body mass index, serum sex hormones, and breast cancer risk in postmenopausal women. *J Natl Cancer Inst* **95**, 1218–26.
125. Rinaldi, S., Key, T. J., Peeters, P. H., et al. (2006) Anthropometric measures, endogenous sex steroids and breast cancer risk in postmenopausal women: a study within the EPIC cohort. *Int J Cancer* **118**: 2832–39.
126. Suzuki, R., Rylander-Rudqvist, T., Ye, W., Saji, S., and Wolk, A. (2006) Body weight and postmenopausal breast cancer risk defined by estrogen and progesterone receptor status among Swedish women: a prospective cohort study. *Int J Cancer* **119**, 1683–9.
127. Goodwin, P. J., Ennis, M., Bahl, M., et al. (2008) High insulin levels in newly diagnosed breast cancer patients reflect underlying insulin resistance and are associated with components of the insulin resistance syndrome. *Breast Cancer Res Treat* 2008 Apr 25. [Epub ahead of print] PMID: 18437560

128. Ligibel, J. A, Campbell, N., Partridge, A., et al. (2008) Impact of a mixed strength and endurance exercise intervention on insulin levels in breast cancer survivors. *J Clin Oncol* **26**, 907–12.
129. Kaaks, R., and Lukanova, A. (2001) Energy balance and cancer: the role of insulin and insulin-like growth factor-I. *Proc Nutr Soc* **60**, 91–106.
130. Wolf, I., Sadetzki, S., Catalane, R., et al. (2005) Diabetes mellitus and breast cancer. *Lancet Oncol* **6**, 103–11.
131. Hursting, S. D., Nunez, N. P., Varticovski, L., and Vinson, C. (2007) The obesity-cancer link: lessons learned from a fatless mouse. *Cancer Res* **67**, 2391–3.
132. Campbell, K. L., and McTiernan, A. (2007) Exercise and biomarkers for cancer prevention studies. *J Nutr* **137**, 161S–9S.
133. Schernhammer, E. S., Holly, J. M., Hunter, D. J., Pollak, M. N., and Hankinson, S. E. (2006) Insulin-like growth factor-1, its binding proteins (IGFBP-1 and IGFBP-3), and growth hormone and breast cancer risk in The Nurses Health Study II. *Endocr Relat Cancer* **13**, 583–92.
134. Engelman, J. A., Luo, J., and Cantley, L. C. (2006) The evolution of phosphatidylinositol 3-kinases as regulators of growth and metabolism. *Nat Rev Genet* **7**, 606–19.
135. Luo, J., Manning, B. D., and Cantley, L. C. (2003) Targeting the PI3K-Akt pathway in human cancer: rationale and promise. *Cancer Cell* **4**, 257–62.
136. Wullschleger, S., Loewith, R., and Hall, M. N. (2006) TOR signaling in growth and metabolism. *Cell* **124**, 471–84.
137. Jiang, W., Zhu, Z., and Thompson, H. J. (2008) Dietary energy restriction modulates the activity of AMP-activated protein kinase, Akt, and mammalian target of rapamycin in mammary carcinomas, mammary gland, and liver. *Cancer Res* **68**, 5492–9.
138. Moore, T., Beltran, L., Carbajal, S., Traag, J., Hursting, S. D., and DiGiovanni, J. (2008) Dietary energy balance modulates signaling through Akt/mTOR pathways in multiple epithelial tissues. *Cancer Prev. Res* **1**, 65–76.
139. Zhu, Z., Jiang, W., Sells, J. L., Neil, E. S., McGinley, J. N., and Thompson, H. J. (2008) Effect of nonmotorized wheel running on mammary carcinogenesis: circulating biomarkers, cellular processes, and molecular mechanisms in rats. *Cancer Epidemiol Biomarkers Prev* **17**, 1920–9.
140. Bartella, V., Cascio, S., Fiorio, E., Auriemma, A., Russo, A, and Surmacz, E. (2008) Insulin-dependent leptin expression in breast cancer cells. *Cancer Res* **68**, 4919–27.
141. Maya-Montero, C. M. and Bozza, P. T. (2008) Leptin and mTOR: partners in metabolism and inflammation. *Cell Cycle* **7**, 1713–7.
142. Carino, C., Olawaiye, A. B, and Cherfils, S., (2008) Leptin regulation of proangiogenic molecules in benign and cancerous endometrial cells. *Int J Cancer* Sept 17 [Epub ahead of print].
143. Perera, C., Chin, H., Duru, N., and Camarillo, I. (2008) Leptin regulated gene expression in MCF-7 breast cancer cells: mechanistic insights into leptin regulated mammary tumor growth and progression. *J Endocrinol* Aug 20 [Epub ahead of print], 2008
144. Cymbaluk, A., Chudecka-Glaz, A., and Rzepka-Górska, I. (2008) Leptin levels in serum depending on Body Mass Index in patients with endometrial hyperplasia and cancer. *Eur J Obstet Gynecol Reprod Biol* **136**, 74–7.
145. Mantzoros, C. S., Petridou, E., Dessypris, N., et al. (2004) Adiponectin and breast cancer risk. *J Clin Endocrinol Metab* **89**, 1102–7.
146. Petridou, E., Mantzoros, C., Dessypris, N., et al. (2004) Plasma adiponectin concentrations in relation to endometrial cancer: a case control study in Greece. *J Clin Endocrinol Metab* **88**, 993–7.
147. Körner, A., Pazaitou-Panayiotou, K., Kelesidis, T., et al. (2007) Total and high-molecular weight adiponectin in breast cancer: in vitro and in vivo studies. *J Clin Endocrinol Metab* **92**, 1041–48.
148. Mantzoros, C. S., Williams, C. J., Manson, J. E., Meigs, J. B., and Hu, F. B. (2006) Adherence to the Mediterranean dietary pattern is positively associated with plasma adiponectin concentrations in diabetic women. *Am J Clin Nutr* **84**, 328–35.

8 The Role of Nutrition and Diet in Prostate Cancer

Lorelei Mucci and Edward Giovannucci

Key Points

- The more than 30-fold variation in prostate cancer incidence and mortality globally points to a role of lifestyle and dietary factors in the etiology of prostate cancer.
- Prostate cancer is a leading cause of cancer burden among men, although there is heterogeneity in the biological potential of these cancers. As such, in evaluating studies of diet and prostate cancer, one should consider the disease endpoints used, for example, total prostate cancer, high grade or advanced stage prostate cancer, metastatic or fatal disease.
- Screening by prostate-specific antigen (PSA), which began in the United States in the early 1990s, has markedly impacted the results of studies of prostate cancer prevention. Investigations undertaken in populations diagnosed in the PSA era should account for screening practices in the study design or data analysis in order to avoid potential biases.
- Various aspects of diet and energy balance appear to influence prostate cancer at different stages of its development and progression. Some factors (e.g., cruciferous vegetables) may be more important earlier in the disease process, whereas others (e.g., obesity) may impact prostate cancer progression or outcomes.
- Growing evidence suggests that diets rich in tomatoes/lycopene and low in dairy/calcium and processed red meat consumption are associated with a lower risk of prostate cancer, particularly for aggressive disease.
- Although obesity and energy balance appear unrelated to total prostate cancer incidence, they may be linked with an increased risk of biochemical failure and cancer-specific mortality.
- There is more limited evidence of a role for several other dietary factors (fish, soy, zinc) in the etiology and progression of prostate cancer.

Key Words: Fatty acids; migrant studies; obesity; PSA screening; selenium; tomatoes/lycopene

8.1. INTRODUCTION

Prostate cancer represents a considerable public health burden among men. It is a major cause of cancer incidence and mortality, particularly in westernized countries *(1)*. Moreover, prostate cancer is associated with significant impairments in quality of life *(2)*, both from the disease itself and as a consequence of treatment. Prostate cancer prevention would provide the greatest opportunity to reduce suffering from this disease. Tantalizing clues suggest that diet and lifestyle factors could play an important role in the prevention of prostate cancer incidence or progression. Data from experimental models demonstrate direct effects of diet on prostate-specific expression of genes involved in inflammation, angiogenesis, proliferation, and other key pathways involved in cancer progression *(3,4)*. Epidemiological studies have examined an array of dietary and nutritional factors in relation to prostate cancer incidence. In particular, epidemiological studies indicate that higher intake of tomato products

A. Bendich, R.J. Deckelbaum (eds.), *Preventive Nutrition*, Nutrition and Health, DOI 10.1007/978-1-60327-542-2_8,

and lycopene, selenium, and omega-3 fatty acids are associated with lower prostate cancer risk *(5,6)*. Lower levels of these nutrients tend be associated more strongly with advanced rather than localized disease *(6)*, and emerging evidence shows that post-diagnostic diet may influence progression *(7)*, although remarkably few studies have assessed the impact of dietary factors on cancer-specific survival. In this chapter, we summarize the evidence on diet and nutrition in the primary and secondary prevention of prostate cancer. We focus primarily on the factors that have received most scrutiny from human studies.

8.2. GLOBAL BURDEN OF PROSTATE CANCER

8.2.1. Incidence

Prostate cancer is among the most common male cancers, with more than 679,000 cases diagnosed worldwide each year *(1)*. Among men in the United States, who experience the highest rates of this disease globally, the lifetime risk of being diagnosed with prostate cancer is 1 in 6 *(8)*. The burden of prostate cancer shows remarkable worldwide variation (Fig. 8.1), with a more than 60-fold difference in age-adjusted rates between population groups with the highest (African-American men in the United States) and the lowest (Japanese and Chinese men living in their native countries) prostate cancer burden *(1)*. Differences in cancer incidence must be interpreted in the context of diagnostic intensity and screening behaviors across populations. Screening by prostate-specific antigen (PSA), introduced in the early 1990s, has dramatically influenced the incidence and presentation of prostate cancer. Widespread screening in several western countries has led to marked increases in incidence over time (Fig. 8.2), and as a consequence has led to the detection of a significant proportion of latent lesions with questionable clinical importance among tumors diagnosed *(9,10)*. PSA screening has also led to a shift in stage presentation, with concomitant increase in the ratio of localized to advanced disease cases and a decrease in the age at diagnosis *(11)*. Although differences in PSA screening may account for some of the global variation in incidence, geographic differences were apparent already

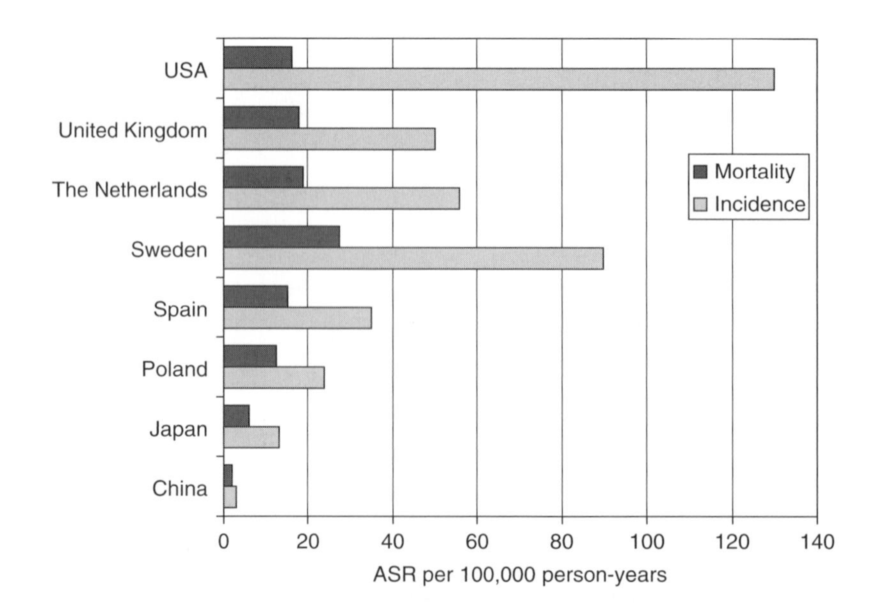

Fig. 8.1. Age-adjusted incidence and mortality rates of prostate cancer among men in selected countries, Ferlay *(1)*.

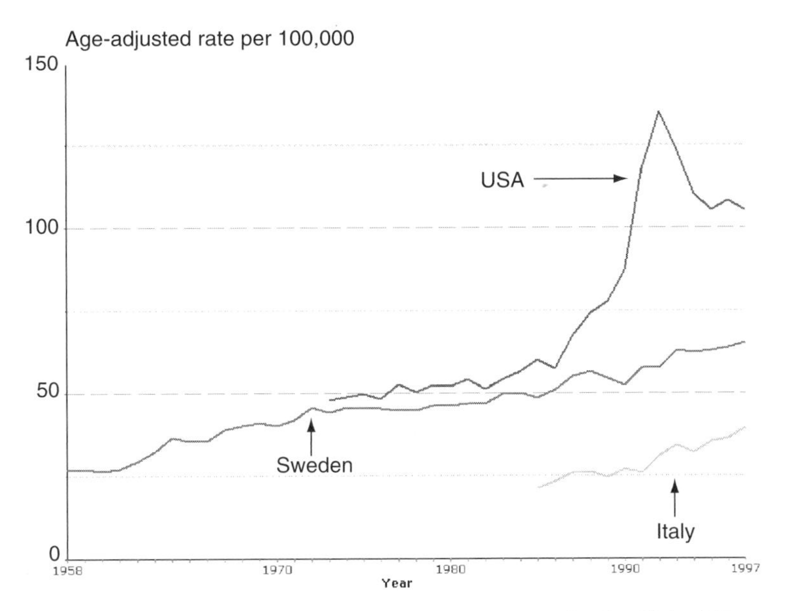

Fig. 8.2. Trends over time in prostate cancer incidence rates among men in the United States, Italy, and Sweden, Ferlay *(1)*.

in the era prior to PSA screening, highlighting a potential role of lifestyle, and particularly dietary patterns to account for the variation in rates.

8.2.2. Mortality

In 2002, 221,000 men died of prostate cancer worldwide, and prostate cancer death is among the most common causes of cancer death among men in developed or westernized countries *(1)*. Notwithstanding the considerable mortality associated with this disease, most men die with and not from their cancers, and many harbor tumors that remain indolent even in the absence of therapy *(12,13)*. The documented benefits of localized treatment for early prostate cancer in reducing mortality *(14)* must be balanced against the small absolute reduction, as well as the attendant effects associated with treatment *(15)*.

Mortality rates are much lower in many Asian countries, consistent with differences in incidence (Fig. 8.1). Even adjusting for age, the mortality rate in the United States for prostate cancer is 18-fold that of China *(1)*. During the past decade, prostate cancer mortality rates have shown some declines. The reasons for this decline remain controversial, but may be attributable at least in part to earlier detection through PSA screening and subsequent earlier treatment *(16)*.

Given the clinical picture of prostate cancer, nutritional and dietary factors hold promise to reduce disease progression and may be an attractive secondary preventive agent for men with low-risk prostate cancer who select active surveillance as primary treatment.

8.2.3. Migrant Studies

Results from migrant studies lend additional support to a role of lifestyle factors in prostate cancer incidence. Prostate cancer rates are greater for immigrants moving from low-risk to high-risk countries compared to those in their native countries *(17,18)*. For example, Japanese immigrants moving to the

United States have prostate cancer rates several fold higher than that of their native counterparts, although still below that of US white men. Interestingly, immigrants typically do not assume the full risk profile of high-risk country natives until the second generation, indicating a possible role of earlier life dietary patterns on prostate cancer risk. Although a shift toward higher incidence rates could also reflect uptake in PSA screening, the observation that prostate cancer mortality rates also increase lends further support for a non-hereditary cause over and beyond any artifactual rise due to screening and enhanced detection.

8.2.4. Influence of PSA Screening on Studies of Prostate Cancer Prevention

Prostate cancer screening using PSA has become extremely common among men, first in the United States during the 1990s and more recently in Europe, Canada, and Australia. Before the widespread use of PSA screening, most prostate cancers were diagnosed clinically, with a relatively high proportion at an advanced stage. In contrast, PSA screening has led to the detection of cancers at an earlier stage *(11)*, with a substantial pool of indolent cancers that would not otherwise have come to light clinically *(10)*.

The uptake of PSA screening has likely resulted in alterations in the observed associations between specific dietary factors and prostate cancer risk over time. First, different dietary factors may impact prostate cancer at various stages of progression, and therefore the associations may differ according to disease subgroups, such as those defined by cancer stage or tumor grade *(6)*. Indeed, it seems unlikely that the factors associated with development of indolent cancers would be similar to those associated with cancers demonstrating biologic potential for aggressive behavior. Since the majority of cancers diagnosed in the PSA era are well-differentiated tumors with low metastatic potential, it seems likely that associations observed in populations in mainly pre-PSA era populations would no longer be observed with respect to total prostate cancer risk.

PSA screening may also be a potential confounder of epidemiological studies. Men who take part in regular screening practices, including PSA screening, tend also to take part in other healthy behaviors, including higher intake of fruits and vegetables *(19)*. Moreover, a prostate cancer diagnosis in the PSA era can often signify that a man happened to have a PSA test at that time. Thus, studies in the PSA era should account for PSA screening practices in their study design or data analysis. For example, advanced prostate cancer or fatal prostate cancer may be an appropriate endpoint, in lieu of total prostate cancer, in studies conducted when PSA screening is widespread.

In this light, we will present summaries of the associations between diet and nutritional factors in relation to total prostate cancer, as well as stratified by disease subtypes or cancer progression.

8.3. DIET AND NUTRITION

A western diet has been proposed as a potentially important prostate cancer risk factor based on international comparisons of prostate cancer mortality rates and the observation that migrants from low- to high-risk geographic areas, as well as their offspring, assume the higher risk profile of their adopted countries. Many aspects of the western diet, including energy imbalance (energy consumption exceeding energy requirements) and high consumption of fat, particularly animal fat and red meat, have been postulated as playing a role in increasing prostate cancer risk. The low rates of prostate cancer in Asia and some Mediterranean countries have pointed toward foods commonly consumed in those populations, including soy and tomato products, in preventing prostate cancer. We summarize the findings for seven major areas of diet and nutrition that have been studied the most extensively in human studies.

8.3.1. *Antioxidants*

8.3.1.1. Lycopene and Tomato-Based Products

Tomatoes are rich in lycopene, a carotenoid with well-documented antioxidant effects. The relationship between tomatoes in prostate cancer prevention has been studied extensively in the epidemiological literature *(20)*, with evidence suggesting a significant benefit associated with higher intake of tomatoes or lycopene. However, the association between tomato products or lycopene and lower prostate cancer risk remains controversial since not all the studies are supportive.

A hypothesized mechanism for cancer prevention by tomatoes/lycopene is via reduction in cellular oxidative stress *(21)*, which can cause chronic inflammation and might, therefore, be related to prostate carcinogenesis *(22)*. In addition, tumor cells produce 10 times the levels of reactive oxygen species compared to normal cells *(23)* leading to potential prostate cancer progression. Antioxidants may lower risk, particularly for advanced prostate cancer, by quenching free radicals and thus ameliorating damage from consequences of chronic inflammation and also by downregulating tumor angiogenesis.

A summary of epidemiological findings on tomatoes and prostate cancer risk was included in a meta-analysis in 2003 *(24)*, including results from 11 case–control and 10 prospective cohort or nested case–control studies that presented data on the intake of tomatoes, tomato products, or dietary lycopene. Compared with men with low intake of tomato products (1st quantile of intake), the relative (RR) of prostate cancer among consumers of higher amounts of raw tomato (5th quantile of intake) was 0.89 (95% CI 0.80–1.00). There is reliable experimental evidence that cooked or processed tomato products, such as tomato sauce, tomato soup, and ketchup, offer more readily bioavailable sources of lycopene than fresh tomatoes *(25)*. Accordingly, some epidemiologic studies have found stronger inverse associations for tomato sauce while reporting weaker results for raw tomato intake *(20)*. The meta-analysis presented a summary RR for intake of cooked tomato products of 0.81 (95% CI 0.71–0.92) comparing extreme categories of intake *(24)*.

The results from cohort studies generally indicate a 25–30% reduction in risk of prostate cancer, whereas dietary-based case–control studies are not supportive of an association *(24)*. For example, the summary RR of prostate cancer from the meta-analysis related to an intake of one serving/day of raw tomato (200 g) was 0.97 (95% CI 0.85–1.10) for the case–control studies and 0.78 (95% CI 0.66–0.92) for cohort studies *(24)*. For cohort studies conducted prior to PSA screening, only a study based in the Netherlands *(26)* found no appreciable association between tomato consumption and prostate cancer risk. However, tomato consumption appeared to be low in this population. In the US-based studies showing associations, the average intake of lycopene may have exceeded that in the Netherlands study by about 10-fold.

The correlation between dietary estimates of lycopene based on food frequency questionnaires and circulating levels measured in blood are relatively low, ranging from 0 to 0.47 *(27)*. The majority of serum or plasma-based studies of lycopene have found protective associations for high lycopene levels with corresponding summary relative risks of 0.55 (95% CI 0.32–0.94) for case–control studies and 0.78 (95% CI 0.61–1.00) for cohort studies *(24)*. A null prospective study of plasma levels and prostate cancer risk was undertaken in a Japanese-American population in Hawaii between 1971 and 1993 *(28)*. Serum lycopene levels were quite low in that population, with median concentration among controls of 134 ng/mL, 60–80% lower than those among men in the study by Hsing et al. *(29)*, the Health Professionals Follow-up Study *(30)* and the Physicians' Health Study *(31)*. However, another recent study conducted in the United States, with higher baseline levels, showed no association between plasma levels of lycopene and risk of total prostate cancer *(32)*. This study included 692 incident prostate cancer cases, although the number of advanced stage cases of prostate cancer was limited as the study was conducted in a population with extensive PSA screening.

Given their hypothesized mechanism, tomatoes and lycopene are also interesting with respect to prostate cancer progression. Epidemiological studies generally point to a stronger reduction in risk of advanced stage or lethal prostate cancer. For example, in the Health Professionals Follow-Up Study, the associations comparing high and low consumption of tomato sauce were 0.75 (95% CI 0.61–0.92) for total prostate cancer and 0.66 (95% CI 0.44–1.00) for advanced stage disease *(6)*. Moreover, the association was more protective for cancers diagnosed prior to the introduction of PSA screening than for later diagnosed cancers. Also in the Health Professionals Follow-Up Study, higher tomato sauce consumption after cancer diagnosis was associated with a lower risk of disease recurrence or progression *(7)*. The idea that lycopene or tomatoes would be associated more strongly with prostate cancer progression was also supported in an analysis from the European Prospective Investigation into Cancer and Nutrition study *(33)* based on 966 total cases and 205 advanced stage cases of prostate cancer. No association was seen for total prostate cancer, but men in the top quintile of plasma lycopene had a significantly reduced risk of advanced stage prostate cancer (RR = 0.40, 95% CI: 0.19–0.88).

Lycopene is found in a number of food items, including tomatoes, salads, soups, pizza, mixed dishes, salsas, ketchup, and juices, and also in some non-tomato items (e.g., watermelon, pink grapefruit). Thus, a dietary questionnaire must include questions about intake of all of these items to capture intake of lycopene. Bioavailability of lycopene from foods is also important, and may vary profoundly across different food items. Lycopene is highly lipophylic. As such, cooking of tomato products disrupts lycopene from binding matrices, and the oil base improves absorption *(20)*. The relatively low correlation between dietary lycopene intake with circulating levels may reflect issues around bioavailability *(20)*.

Whether the association between tomatoes and/or lycopene and prostate cancer in association is causal is a matter of some debate. In many studies, controlling for various dietary and other risk factors do not qualitatively change the observed relative risks. In one case–control study *(34)*, however, a suggestive inverse association observed for cooked tomatoes was substantially attenuated upon adjusting for total fruits or vegetables. In contrast, a larger cohort study *(27)* found little change in the relative risk estimates after controlling for fruit and vegetable intake or for olive oil use.

Although not definitive, the available data suggest that increased consumption of tomato and tomato-based products is associated with lower prostate cancer risk and progression. Whether the effect is driven through lycopene or other aspects of tomatoes remains undetermined. The relationship appears to be stronger for advanced prostate cancer than indolent disease.

8.3.1.2. Selenium

The trace element selenium is not an antioxidant per se, but plays an important role as an essential element for the antioxidant enzyme glutathione peroxidase *(35)* as well as other selenoproteins involved in exerting antitumor effects, including apoptosis and inhibition of cellular proliferation *(36,37)*. Dietary intake of selenium depends on the selenium content of soil in which foods are grown, which varies greatly by geographic area. Ecologic studies have suggested an inverse association between selenium soil content and prostate cancer incidence *(38)*. Because selenium contents in specific foods vary as a function of the selenium content of the soil, epidemiological studies of selenium require biological sampling, primarily measuring levels in blood or toe nails.

Six prospective biomarker studies have reported significant associations between higher levels of selenium and reduced prostate cancer risk *(39–44)*, particularly for advanced disease *(39,40,43)*. Not all epidemiological studies have reported a protective association of selenium, however *(45–47)*. The strongest evidence for selenium comes from a randomized placebo-controlled clinical trial, which found selenium supplementation associated with a 63% prostate cancer reduction *(48,49)*. The association of selenium on risk of prostate cancer, a secondary endpoint for the trial, was apparent already during the first years after randomization. With additional follow-up time, the protective effect of selenium supplementation appeared to be limited to those with low PSA levels at baseline or low selenium levels. Since the activity of some selenoenzymes plateau with higher selenium level *(50)*, the chemopreventive effect of selenium may be greatest in populations with low selenium exposure *(49)*.

Intriguingly, a nationwide program in Finland, a country with historically low selenium content, to supplement soil with selenium has documented no reduction in prostate cancer incidence or mortality (http://www.mtt.fi/english/press/050902.html) despite a marked increase in the selenium status of the general population. Overall, the summary of epidemiological evidence suggests that selenium inhibits prostate carcinogenesis and possibly disease progression, but more study is needed before selenium supplementation can be confidently suggested as a means of primary chemoprevention. The SELECT trial *(51)*, a large randomized intervention trial with selenium (200 mg) and vitamin E (400 IU), could have answered this question definitively. However, the trial was ended 4 years early with the investigators reporting no protective effect of selenium on total prostate cancer incidence at the 200 mg dose (http://www.crab.org/select/). It will be important to follow-up the men in SELECT for prostate cancer progression, though it is unclear if 4 years of exposure will be adequate to document an effect.

8.3.1.3. Vitamin E

The antioxidant properties of vitamin E include its ability to reduce DNA damage and inhibit malignant cellular transformation *(52,53)*. In experimental models, derivatives of vitamin E inhibit growth, induce apoptosis *(54)*, and enhance therapeutic effects in human prostate cancer cells *(55)*.

In humans, studies of vitamin E and prostate cancer are equivocal *(56)*. Results from the Finnish ATBC intervention trial suggested that male smokers randomized to alpha-tocopherol supplements, the most biologically active of the naturally occurring forms of vitamin E, were at significantly lower risk of prostate cancer incidence and cancer-specific mortality *(57)*. The inverse association was observed during the first few years of follow-up suggesting that vitamin E could impact prostate cancer progression. Subsequent follow-up, however, showed no association of vitamin E for prostate cancers diagnosed later in the trial *(58)*.

Epidemiological studies of vitamin E and prostate cancer risk have had mixed results, generally pointing toward no overall association. In the VITamins And Lifestyle (VITAL) study, a cohort study specifically designed to examine supplement use and future cancer risk, a 10-year average intake of supplemental vitamin E was not associated with a reduced prostate cancer risk overall but it was associated with a reduced risk for advanced prostate cancer (regionally invasive or distant metastatic, $n = 123$) (HR 0.43, 95% CI 0.19–1.0 for 10-year average intake $\geq$400 IU/day versus non-use) *(32)*. In a prospective study of plasma vitamin E and prostate cancer mortality, there was a reduced risk associated with higher circulating levels limited to smokers, although the number of cases in the subgroup was small (<30) *(59)*. Other epidemiological studies have similarly found a protective association limited to ever smokers, including prospective studies of vitamin E through supplementation on lethal prostate cancer *(60)* and plasma α-tocopherol levels and aggressive prostate cancer *(31)*. Nested case–control studies *(28,29,46)* using baseline blood specimens have not supported substantial reductions in prostate cancer risk. The major dietary source of vitamin E is γ-tocopherol, and some plasma-based studies have suggested the protective association of vitamin E on prostate cancer is limited to γ-tocopherol *(42,61)*.

Two additional randomized trials among healthy men with lower levels of smoking have recently reported on the findings of vitamin E and prostate cancer incidence. Vitamin E (α-tocopherol, 400 IU) supplementation was the second component of the SELECT trial *(51)*, described above, which reported a slight increased risk of prostate cancer associated with vitamin E. In the Physicians' Health Study II among US male physicians, vitamin E (type, dose) was not associated, positively or negatively, with total cancer incidence *(62,63)*. In both trials, the men will be followed prospectively for cancer progression. These findings are in contrast to those in the Finnish ATBC intervention trial. Of note, in the ATBC trial, all men were smokers and cancers were diagnosed outside the context of PSA screening, and thus, were generally aggressive.

As summarized above, when epidemiologic studies have supported a role for vitamin E, the endpoint has usually been for advanced prostate cancer and often in smokers, who may be prone to oxidative stress. Also of note, in a sample of men in the ATBC trial, those randomized to α-tocopherol had an 11% decrease in vascular endothelial growth factor (VEGF) levels, a cytokine integrally involved in angiogenesis, in contrast to a 10% increase in the placebo group *(64)*. This finding suggests that α-tocopherol may possibly reduce angiogenesis, an important factor for prostate cancer growth.

8.3.2. Energy Balance

Energy balance describes a state of equilibrium between total energy intake versus energy expenditure. In epidemiological studies, the components describing various aspects of energy balance—total energy intake, energy expenditure—are not easily measurable, so measures of body mass and obesity have typically been used as surrogates. Although the etiology and manifestations of obesity are manifold, excess adiposity largely represents a relatively chronic state of overconsumption of energy relative to requirements based on factors such as basal metabolic rate and physical activity. A positive or negative energy balance may influence multiple hormonal pathways that may differentially influence prostate cancer pathogenesis. Moreover, energy balance and obesity may have different effects for disease progression. In the section below, we present a review of the literature on energy intake and obesity, as well as height because to some extent, short stature may represent a relative energy restriction during the growth period.

In a prostate cancer animal model, an energy-restricted diet led to reduced prostate tumor angiogenesis and decreased tumor growth *(4)*, suggesting total energy may be important with respect to disease progression. The epidemiological evidence regarding the effect of total energy intake on prostate cancer risk is inconclusive *(65,66)*, with some studies reporting a positive effect of increased total energy *(67,68)*, while others indicating no association of energy intake with prostate cancer *(69,70)*. In epidemiologic studies, the net result of total energy intake can be difficult to interpret, since its determinants are complex and influenced by body size and physical activity if one is in energy balance, and may be influenced by energy balance if one is gaining or losing weight. Energy intake is also difficult to measure in current epidemiologic studies.

The association between obesity, which may reflect a positive energy imbalance, and prostate cancer risk has been examined in multiple studies and across many populations. The bulk of evidence suggests that excess body weight in adult life is not associated with total prostate cancer incidence *(69,71–82)*. Some studies have suggested a positive association between high body mass index and prostate cancer risk *(83–90)*. A complicating factor is that in some settings, associations with body mass index may reflect lean muscle mass rather than adiposity. For example, a large Swedish study found prostate cancer risk to be more strongly associated with lean body mass than with body mass index *(91)*.

Although obesity plays an unclear role in total prostate cancer incidence, overweight and obese men appear more likely to develop advanced stage or fatal prostate cancer *(92–98)*. In particular, central adiposity has been associated with increased risk of advanced prostate cancer *(99)*. Moreover, among men with prostate cancer, those who were overweight or obese prior to cancer diagnosis were at increased risk of biochemical failure and cancer-specific mortality *(100)*. These findings strongly indicate that excess body weight increases mortality from prostate cancer, although the mechanisms are not completely understood. Obesity could increase risk of prostate cancer mortality through direct causal mechanisms, or indirectly; for example, prostate cancer could be possibly more difficult to diagnose or treat in obese men, leading to delayed diagnosis and suboptimal therapy. Nonetheless, the association between obesity and fatal prostate cancer was observed before the onset of PSA screening, suggesting a true biologic effect of obesity on prostate cancer progression *(83)*.

Issues related to energy balance have been especially difficult to study in earlier life periods, such as childhood and adolescence, during which prostate cancer risk may be influenced. Tallness has been used as a surrogate of high exposure to growth hormones, which are partially influenced by energy balance to the extent that a relative restriction of energy could lead to shorter stature. As has been observed for other cancers, taller height has been a risk factor in some studies of prostate cancer, particularly aggressive disease *(95)*. In a recent study, tallness was not associated with total prostate cancer incidence, but taller men had almost a threefold higher risk of developing fatal prostate cancer *(6)*. Although it is difficult to relate this finding with prevention, the association with height strongly suggests that factors during puberty and adolescence, possibly related to nutritional status, influence prostate carcinogenesis. For example, energy and nutritional restriction resulting in submaximal height may underlie the decreased risk of prostate cancer typically observed in populations that have not experienced economic development.

8.3.3. Fat Intake and Fatty Acids

8.3.3.1. Fat Intake

The results from ecologic studies provided the basis for the initial hypotheses regarding fat intake and prostate cancer, whereby countries with the highest intake of animal products and total fat had the highest rates of prostate cancer *(101)*. Since the 1990s, several analytic epidemiologic studies have been conducted on this topic. Case–control studies have generally shown positive associations with high intake of animal fat and prostate cancer risk *(70,102)*, especially for advanced stage disease *(69)*, although some have shown no association for total fat or saturated fat *(67,103)*.

Findings from prospective cohort studies on the association between dietary fat and prostate cancer are not as consistent. Many have not supported an association *(104–108)*, while other studies have supported a positive association for total dietary fat or fat from animal products *(76)*. In the Malmö Diet and Cancer cohort, where PSA screening was uncommon, intake of total, saturated, or monounsaturated fat was unrelated to total prostate cancer risk *(109)*. A meta-analysis of seven prospective studies found no significant associations between dietary intake of total, saturated, monounsaturated, and polyunsaturated fat and prostate cancer risk *(105)*.

Some prospective studies noted stronger relations for fat intake and risk of advanced or fatal disease than for total prostate cancer *(69,70,110,111)*. Two small studies of men with prostate cancer suggest that high intake of saturated fat at the time of diagnosis is associated with an increased risk of biochemical failure *(112)* and prostate cancer-specific death *(113)*. The stronger findings for advanced disease and progression, if confirmed, suggest that dietary fat may influence late stages of carcinogenesis. However, the EPIC study did not find any association between total, saturated, or monounsaturated fat intake and aggressive prostate cancer defined by advanced stage *(105)*. Moreover, there were suggestions of reduced risk of high-grade prostate cancer associated with increased intake of total fat, monounsaturated, and polyunsaturated fat intake.

8.3.3.2. Fish Intake

Populations with a high consumption of fish, such as Japan and among Eskimos in Alaska, have lower rates of prostate cancer than populations with western food habits, where the fish intake in general is lower *(120–124)*. Fish contain the long-chain marine omega-3 polyunsaturated fatty acids EPA and DHA (described in detail below), which may lower prostate cancer risk and progression *(5)*.

The findings from analytic epidemiologic studies of fish intake and prostate cancer risk and progression have reported inconsistent findings, which may be attributed to the types of fish consumed in different populations as well as the prostate cancer outcome measured. Of four case–control studies, three *(125,126)* reported associations between high intake of fish and lower incidence of prostate cancer. Findings from prospective studies are more mixed *(76,103,108,114,127–129)*, with only two of ten reporting significant inverse associations for total prostate cancer incidence *(128,129)*. In a Swedish twin cohort *(129)*, where men traditionally consume fatty fish such as salmon and herring, moderate consumption of fish was associated with a lower risk of prostate cancer. In the US Health Professionals Follow-up Study (HPFS), consuming fish three or more times per week was associated with a reduced risk of advanced stage or metastatic prostate cancer *(128)*. The Swedish twin study *(129)*, as well the US Physicians' Health Study *(130)* and a Japanese cohort of men *(131)* found higher baseline intake of fish was associated with a reduced risk of cancer-specific mortality among men diagnosed with prostate cancer. Moreover, in the HPFS, regular consumption of fish after prostate cancer diagnosis was associated with a 50% reduction in disease progression *(7)*. The association between intake of fish and reduced prostate cancer progression is intriguing, although not confirmed in all studies *(114)*, including a second Japanese-based study found that high fish consumption was associated with an increased risk of prostate cancer *(132)*.

8.3.3.3. Meat

The weight of evidence from prospective studies suggests no association between intake of meat or fat from animal sources with total prostate cancer risk *(90,105,114)*. Only two prospective studies (of 11) have found positive associations for prostate cancer associated with diets high in meat or animal fat *(76,110)*. The EPIC cohort found no association between fat from animal sources and risk of prostate cancer overall, or for advanced stage disease *(115)*.

There is variability in the nutrient composition of different types of meat, while variations in cooking practices could impact risk of prostate cancer differently. Several studies have focused specifically on red meat, and five cohort studies (of 10) have shown positive associations with prostate cancer associated with high consumption of red meat in the diet *(107,116–118)*. Some cohort studies have focused specifically on processed meats, with perhaps more consistent findings for diets high in processed or cured meats and prostate cancer, particularly advanced disease *(117,119,120)*, although not all studies have confirmed these findings *(107)*. In the grilling of red meat, heterocyclic amine carcinogens form, including 2-amino-1-methyl-6-phenylimidazo[*4,5-b*]pyridine (PhIP), which causes prostate cancer when fed to rats *(22)*. In the Agricultural Health Study, with detailed information on cooking patterns, there was no association with prostate cancer specifically for pan-frying or grilling meats. However, higher intake of well-done meats was associated with 30% greater risk of prostate cancer, and even stronger for advanced disease *(121)*. These findings, while intriguing in light of the experimental data, need to be confirmed in additional populations.

8.3.3.4. Fatty Acids

Several prospective studies have evaluated intake of specific fatty acids, rather than total fat, in relation to prostate cancer. Experimental studies have lent evidence that specific fatty acids from marine and animal sources may have opposite effects on prostate cancer and may play unique roles in development and progression. The hypothesis is that the long-chain *n*–6 polyunsaturated fatty acids (PUFA) metabolites linoleic acid (C18:2 *n*–6) and arachidonic acid (C20:4 *n*–6), which are enriched in animal fat sources, promote carcinogenesis *(133)*. Metabolites of arachidonic acid (**eicosanoids**) such as prostaglandins and leukotrienes inhibit apoptosis *(134–136)*, stimulate cell proliferation *(137)*, promote tumor angiogenesis, and enhance tumor cell adhesion to endothelial cells, and thus may augment prostate cancer's metastatic potential *(138)*. In contrast, the long-chain *n*–3 PUFA, such as eicosapentaenoic acid (**EPA**, C20:5 *n*–3) and docosahexaenoic acid (**DHA**, C22:6 *n*–3) that are found in marine fish, suppress the development of cancer, directly by inhibiting prostate cancer cell growth as well as indirectly through suppressing synthesis of the pro-inflammatory eicosanoids *(139,140)*.

***n*–6 PUFAs**. Epidemiological studies of *n*–6 PUFAs and prostate cancer risk have generally not supported the experimental findings of an increased risk associated with higher *n*–6 PUFAs. Among 20 published studies, only four (three were retrospective case–control studies) reported positive associations for intake or tissue levels of linoleic acid, arachidonic acid, or total *n*–6 PUFAs *(141–143)*. Five of the eight prospective studies actually found inverse associations with higher *n*–6 PUFAs, whereas only one reported a positive association with increased dietary intake. No consistent associations between intake of *(110,120)* or circulating levels *(144–146)* of linoleic acid and prostate cancer risk have been observed. No other *n*–6 fatty acid, including arachidonic acid, was related to prostate cancer risk in any of these studies *(145,146)*. In the Alpha-Tocopherol, Beta-Carotene Cancer Prevention (ATBC) Study, serum LA was not associated with overall prostate cancer, although there was a suggestion of an inverse risk with higher LA levels among men who received α-tocopherol supplementation *(142)*.

***n*–3 PUFAs**. The protective effect of long-chain *n*–3 PUFAs on prostate cancer progression is supported by experimental mouse models whereby animals fed precursors of EPA had reduced prostate tumor recurrence, as well as decreased proliferation and increased apoptosis in recurrent tumor cells *(147)*. Intake of the long-chain *n*–3 fatty acids, measured through dietary assessment and using biomarkers, has been examined to some extent in human studies, although the findings are equivocal. In the HPFS, higher dietary intakes of EPA and DHA were associated with lower total prostate cancer risk as well as advanced disease *(148)*. In the multiethnic cohort, a protective association of higher dietary intake of *n*–3 fatty acid intake was limited to men who were Latino or Caucasian *(107)*. In the Physicians' Health Study, earlier data (1982–1988) showed no significant association between circulating levels of EPA and DHA measured in plasma with prostate cancer risk *(145)*. However, an updated (1982–2000) analysis of these biomarkers measured in whole blood and including a larger number of cases found a significant inverse association for total prostate cancer and a stronger association for aggressive disease *(130)*. An inverse association was confirmed in a population-based study *(149)*, but not in four other studies *(142,143,146,150)*. In a small clinical study of men undergoing radical prostatectomy, EPA and DHA levels measured in the prostate were significantly lower among men diagnosed with advanced versus localized prostate cancer *(151)*.

In western diets, α-linolenic acid is the principal dietary omega *n*–3 fatty acid, a shorter chain *n*–3 whose major contributors are vegetable oils as well as red meat and dairy fat. In a prospective study, higher plasma levels of α-linolenic acid were associated with an increased risk of advanced prostate cancer *(110)*. These findings agree with other plasma-based and dietary-based prospective studies of α-linolenic acid *(129,146,148,150,152)*, particularly for advanced disease, although not all studies have observed an association *(120,153)*. Because α-linolenic acid is a precursor to the longer chain omega-3 fatty acids, the positive association between α-linolenic acid and prostate cancer risk appears discrepant. However, only a small proportion of α-linolenic acid is converted to long the chain omega-3 fatty acids, and thus other metabolites of α-linolenic acid could be deleterious. Interestingly, an increasing ratio of linoleic acid to α-linolenic acid was associated with a lower risk of advanced disease in one prospective study *(145)*.

In western diets, α-linolenic acid is the principal dietary omega *n*–3 fatty acid, a shorter chain *n*–3 whose major contributors are vegetable oils as well as red meat and dairy fat. In a prospective study, higher dietary intake of α-linolenic acid was associated with an increased risk of advanced prostate cancer *(110)*. While these findings agree with other plasma-based and dietary-based prospective studies *(146,148)*, others have reported no association *(120,153)*.

8.3.4. Soy/Phytoestrogens

Given the substantially lower incidence and mortality from prostate cancer in Asia compared to the high rates in western countries, epidemiological attention has turned to dietary practices that are characteristic of these low-risk populations. Traditional Asian diets are notably high in phytoestrogens chiefly from soy-based products *(154,155)*. Phytoestrogens represent a diverse set of compounds in plant foods that share structural similarity with estrogen (17β-estradiol) and exert estrogenic-like effects. For prostate cancer, this mechanism is pertinent since the estrogen-β effects have the potential to inhibit androgen-mediated tumor progression *(156)*.

Phytoestrogens are divided into two main categories: isoflavonoids and lignans. The main source of isoflavonoids is beans, particularly soy beans, and soy products; lignans occur in whole-grain bread, seeds, berries, vegetables, and tea. In populations with high intake of soy products, higher consumption of soy is associated with a lower prostate cancer risk *(77,108,157)*. However, more recent cohort studies among Japanese men living in Hawaii or Japan did not find significant associations between soy consumption and total prostate cancer risk *(132,158)*, although the number of cases was small. In western countries, where soy intake is substantially lower, epidemiological studies evaluating soy and prostate cancer risk are more challenging. However, among Adventist men in California, who regularly consume soy milk, consumption of soy milk more than once per day was associated with a 70% risk reduction in prostate cancer *(159)*. A meta-analysis of two cohort and six case–control studies reported a summary relative risk (95% CI) of 0.7 (0.6–0.8) associated with higher intake of soy food *(160)*. Published since the meta-analysis were findings from the US Multiethnic Cohort Study of men which includes populations with high intake of soy products (Japanese-American) as well as high intake of other legumes (Latinos) *(161)*. Men with the highest intake of soy and/or legumes had a 10% lower risk of total prostate cancer, and an almost 30% lower risk of aggressive disease, defined by high grade or advanced stage. In addition, high intake of soy food was associated with a reduced prostate cancer risk in Japanese men *(162)*.

In western populations, beans and other legumes are the principal source of isoflavonoids, although few epidemiologic studies of prostate cancer have reported on legumes specifically. In the prospective Netherlands cohort, legumes were inversely associated with prostate cancer risk *(26)*, a finding confirmed in some *(125,161,163)* but not all *(164)* studies. Moreover, a Swedish case–control study reported a significant effect modification between dietary intake of phytoestrogens and a gene variant in the promoter of the estrogen receptor-β gene (rs 2, 987, 983–13, 950) *(125)*.

Biomarker studies of phytoestrogens have examined prostate cancer risk in relation to enterolactone and enterodiol, which are converted from the lignans precursors by intestinal flora, as well as the isoflavone-related markers daidzein and genistein, produced by gut microflora from isoflavone precursors. Circulating levels of enterolactone were associated with a reduced risk of prostate cancer in a retrospective study *(165)*, a finding not confirmed in four prospective studies *(166–169)*. For isoflavone biomarkers, high circulating levels of genistein and daidzein were associated with a decreased risk of total prostate cancer in a nested case–control study of Japanese men *(170)*, while a study among

British men found no association with levels of isoflavones in urine and blood. A population-based case–control study in Scottish men found a reduced risk of prostate cancer associated with enterolactone, but not with daidzein or genistein levels *(171)*.

If phytoestrogens help prevent prostate cancer, there are a number of proposed mechanisms. Soy isoflavones may lower risk through estrogenic effects and activation of ERβ, which leads to prostate differentiation, and inhibition of 5-α reductase activity. In addition, soy may inhibit angiogenesis, exert antioxidant activity *(154)*, stimulate apoptosis, and inhibit cell growth *(172,173)* which may influence initiation and progression. Additional research is needed on phytoestrogens and prostate cancer.

8.3.5. Dairy Products, Calcium, and Vitamin D

High intake of dairy products has been associated with an increased risk of prostate cancer in several case–control *(174–177)* and cohort studies *(76,114,175,176–181)*. Positive associations have been observed for total dairy intake, as well as specifically for higher intake of milk *(175,181)*, cheese *(182)*, and yogurt *(181)*. A meta-analysis of 11 cohort studies reported a summary relative risk of 1.11 (95% CI 1.03–1.19) per serving for total dairy, 1.06 (0.91–1.23) for milk, and 1.11 (0.99–1.25) for cheese *(183)*. Most *(114,118,180–182)* but not all *(184,185)* studies published since this meta-analysis have tended to support an association between higher milk or dairy consumption and prostate cancer risk.

Dairy products are common dietary sources of calcium and animal fat, and in some settings dairy products are supplemented with vitamin D. The correlation between dairy foods and these nutrients creates challenges in trying to disentangle the independent effects of dairy, calcium and vitamin D on prostate carcinogenesis. Cohort studies that have tried to parse out effects suggest calcium may be the predominant player in explaining positive associations with prostate cancer. In studies that simultaneously consider dairy intake and calcium, relative risk estimates for dairy are attenuated compared to calcium *(175,180,182)*. In a recent analysis of the European Prospective Investigation into Cancer and Nutrition (EPIC) cohort, dairy protein and dairy calcium were both similarly associated with risk of prostate cancer *(114)*.

Whether specifically calcium increases risk of prostate cancer has been evaluated in several case–control *(174,176)* and cohort *(114,175,177,178,180,186)* studies, with positive associations reported for total or advanced prostate cancer risk. Moreover, very high intake of calcium through diet or supplementation was associated with significant excess risks *(175,177,184)*. Several studies have reported stronger associations between high intake of calcium and risks of aggressive forms of prostate cancer, defined by high grade *(6)*, or advanced or lethal prostate cancer *(175,186)*. Serum levels of calcium and prostate cancer have been explored. In the NHANES, higher circulating levels of calcium were associated with a suggestive increased risk of total cancer incidence, and an increased risk of fatal disease *(187)*, although the number of prostate cancer events was small. Moreover, plasma calcium levels are tightly regulated and it is unclear how this finding related to dietary calcium intake, if at all. Not all studies have confirmed a positive association for calcium *(69,120,184,188)*, while a randomized trial of calcium supplements and colorectal adenoma *(189)* found no positive association with prostate cancer as secondary endpoints, although the follow-up time was short and the cases were primarily PSA-detected, early stage cancers. The link between calcium and prostate cancer risk remains controversial, although the strong findings for aggressive prostate cancer warrant further attention.

One proposed mechanism of calcium and prostate cancer is by suppressing circulating levels of dihydroxyvitamin D ($1,25(OH)_2D$), the bioactive metabolite of vitamin D involved in regulating cellular differentiation and proliferation of many cell types, including prostate epithelia. However, this effect of calcium is modest and the role of circulating $1,25(OH)_2D$ on prostate cancer risk is unclear. Milk has been shown to moderately increase circulating levels of insulin-like growth factor (IGF)-1 *(190)*. None of the studies of dietary or supplemental vitamin D have reported protective effects for prostate cancer *(104,174,176)*. Many studies have evaluated circulating levels of (1,25$(OH)_2$D) and (25(OH)D), a measure of endogenous production and exogenous intake of vitamin D, since sun exposure is a major contributor to vitamin D blood levels. The weight of epidemiological evidence suggests no overall association between circulating vitamin D levels and total prostate cancer risk in populations with adequate vitamin D status *(176,191,192)*. Some have reported inverse associations in subgroups defined by aggressive tumor characteristics *(193–195)* or by genetic variants *(196)*, but these subgroup findings have not been consistent *(197)*. In contrast, in Scandinavian populations with chronically deficient levels, there was an excess risk of prostate cancer associated with the lowest 25(OH)D levels *(198,199)*.

8.3.6. Cruciferous Vegetables

Cruciferous vegetables, such as broccoli, cauliflower, and brussel sprouts, are rich in glucosinolates which may be important for prostate cancer prevention. Isothiocyanates and sulforaphane, the hydrolytic products of glucosinolates, have been shown to upregulate apoptosis, downregulate metastasis and angiogenesis, and suppress inflammatory NFkappa-beta pathways in prostate cancer experimental models *(200)*. Moreover, sulforaphanes induce phase II xenobiotic metabolizing enzymes that protect cells from DNA damage *(201)*. There is considerable experimental data suggesting a benefit of high intake of cruciferous vegetables on prostate cancer *(202–204)*.

Case–control studies have generally reported inverse associations between high consumption of broccoli or total cruciferous vegetables and overall prostate cancer *(157,163,205)*. Results from prospective studies are less consistent. Kirsh et al. *(206)* studied cruciferous vegetables in the screening arm of the PLCO screening trial, to help avoid confounding by PSA screening. During 4 years of follow-up, higher intake of cruciferous vegetables was associated with a reduced risk of total prostate cancer *(206)*, with similar associations for broccoli (RR, 95% CI > 1 serving per week versus less than once per month; 0.5, 95% CI 0.3–0.9) and cauliflower (0.5, 95% CI 0.3–0.9). The inverse association with cruciferous vegetables was also seen in cohorts of men in the United States *(207)* and the Netherlands *(26)* but not in three others studies *(106,208,209)*. In the Health Professionals Follow-Up Study, only a weak non-significant inverse association was found, but this inverse association was strengthened and became significant when long-term use of cruciferous vegetables was considered *(207)*. Three studies specifically evaluated cruciferous intake in relation to extraprostatic prostate cancer *(157,206,207)*. While two found stronger associations for advanced stage disease *(157,206)*, another *(207)* did not.

Among healthy men, a 12-month intervention with a broccoli-rich diet led to changes in inflammation-related pathways that differed as a function of genetic variation in the GSTM1 gene *(210)*. Similarly, in a population-based case–control study, there was a significant interaction between broccoli intake and the GSTM1 genotype on prostate cancer risk, such that men who were GSTM1-positive and high broccoli intake had the greatest reduction in risk *(205)*. These findings for cruciferous vegetables are intriguing in light of the experimental data on sulforaphane and prostate tumorigenesis.

8.3.7. Zinc

Zinc is abundant in a diversity of food sources with both positive (e.g., red meat) and inverse (beans, seafood) associations with prostate cancer and is a component of most multivitamins. Moreover, some specific zinc supplements have doses up to 10 times higher than the recommended daily allowance (11 mg/day). Zinc accumulates in prostate tissue, and prostate epithelial cells contain the highest zinc contents in the human body *(211)*. This essential trace element is important in the normal regulation of prostate cell growth and may play a critical role in DNA and RNA repair *(212)* and preventing prostate tumor invasion *(213)*. At the same time, very high zinc intake is correlated with high circulating levels of IGF-1 *(214)*, which have been associated with advanced stage prostate cancer *(215)*, as well as impaired immune function *(216)*.

Among men in the Health Professionals Follow-up Study, high dose (>100 mg/day) or long-term (>10 years) zinc supplementation was associated with a twofold greater risk of advanced prostate cancer compared to no supplemental zinc *(217)*. A positive finding between zinc supplementation and aggressive prostate cancer was further supported by an Italian case–control study *(218)*. In contrast, daily zinc supplementation was associated with a reduced risk of prostate cancer (defined as total incidence as well as high-grade disease) in a population-based case–control study after controlling for factors including PSA screening *(219)*.

Zinc was further studied as part of a multivitamin–multimineral supplement (vitamin C, beta-carotene, selenium, and zinc) in the randomized Supplementation en Vitamines et Mineraux Antioxydants Trial *(220)*. Among men with normal baseline prostate-specific antigen (PSA) levels (<3 μg/L), supplementation was associated with a reduced risk (0.5, 95% CI 0.3–0.9) of prostate cancer compared with placebo. In contrast, a reduction in risk was not observed among men with elevated PSA levels at baseline, and if anything, supplementation was linked with a suggestive increased risk (1.5, 95% CI 0.9 to 2.7). In the NIH-AARP study of almost 300, 000 men, daily use of a multivitamin was associated with a significant risk of advanced or fatal prostate cancer among men who reported taking a zinc supplement, whereas no association was noted among men who did not take zinc supplement *(221)*.

8.4. RECOMMENDATIONS

One in six men in the United States will be diagnosed with prostate cancer during their lifetime, while 1 in 30 will die of this disease. There is emerging evidence that several dietary and lifestyle factors may alter prostate cancer incidence or progression. However, recommendations for specific changes in these factors for prostate cancer prevention should be considered in the context of promoting overall health among men.

For example, maintaining a healthy weight through prudent dietary choices in combination with exercise may impart substantial benefits for men, by reducing the risk of prostate cancer mortality, as well as risk of other cancers and other chronic diseases. Furthermore, encouraging men to increase intake of specific vegetables (tomato-based products and cruciferous vegetables) as well as fish may also have a positive benefit on reducing other cancers as well as other common diseases, such as cardiovascular disease *(222–224)*. Likewise, reducing intake of red meat, particularly charred or processed meats, may have positive impacts for prostate cancer and overall health. Although the evidence for each of these factors for prostate cancer risk is not definitive, their overall health benefits are clear and it is not inconsequential that most men with prostate cancer will die of other chronic diseases rather than their cancer.

The evidence for soy and phytoestrogens, while intriguing, is not definitive, and their overall role in a healthful diet is not entirely clear. It is true that men in Asia, where soy consumption is generally high, have low rates of prostate cancer, but it is unclear if the high consumption of soy underlies the low cancer rates. On the other hand, an increase of vegetable sources of protein, including soy, in lieu of animal protein may be prudent until more definitive evidence is available.

Avoiding high intake of calcium, particularly through supplementation, may be beneficial for prostate cancer outcomes. In contrast, low calcium intakes could have detrimental effects on bone health, both as a consequence of general aging *(225)* as well as for men with prostate cancer undergoing hormonal therapy *(226)*. However, little evidence suggests that such high intakes as 1, 500 mg/day, which have been associated with greater risk of advanced prostate cancer, are beneficial for general health in middle-age to elderly men. Whether the association is related specifically to calcium or to other components in dairy has not been definitively settled, so consumption of milk and dairy products should be moderate. Similarly, zinc is an essential mineral for immune function and wound healing, but in large quantities may increase risk of aggressive prostate cancer. There is little evidence for any health benefit of zinc at such high intakes. The evidence for high doses of vitamin E is mixed, for both prostate cancer and other health outcomes.

Almost all of the evidence to date has been based on studies that have examined pre-diagnostic diet. Especially in the era of PSA screening, many men are diagnosed at apparently early stages of prostate cancer, but a sizable proportion of these cancers will progress. Since many of the factors discussed have shown associations for advanced stage or fatal prostate cancer, albeit being assessed pre-diagnostically, it is reasonable to make similar recommendations for men diagnosed with prostate cancer, until firmer evidence is in. However, it is important that both the health-care provider and the patient understand this important caveat.

REFERENCES

1. Ferlay J, Bray F, Pisani P, Parkin MX. GLOBOCAN 2002: Cancer Incidence, Mortality and Prevalence Worldwide. Lyon: IARC Press; 2004.
2. Steineck G, Helgesen F, Adolfsson J, et al. Quality of life after radical prostatectomy or watchful waiting. N Engl J Med 2002;347:790–6.
3. El-Bayoumy K, Sinha R. Molecular chemoprevention by selenium: a genomic approach. Mutat Res 2005;591: 224–36.
4. Mukherjee P, Sotnikov AV, Mangian HJ, Zhou JR, Visek WJ, Clinton SK. Energy intake and prostate tumor growth, angiogenesis, and vascular endothelial growth factor expression. J Natl Cancer Inst 1999;91:512–23.
5. Chan JM, Gann PH, Giovannucci EL. Role of diet in prostate cancer development and progression. J Clin Oncol 2005;23:8152–60.
6. Giovannucci E, Liu Y, Platz EA, Stampfer MJ, Willett WC. Risk factors for prostate cancer incidence and progression in the health professionals follow-up study. Int J Cancer 2007;121:1571–8.
7. Chan JM, Holick CN, Leitzmann MF, et al. Diet after diagnosis and the risk of prostate cancer progression, recurrence, and death (United States). Cancer Causes Control 2006;17:199–208.
8. ACS. Cancer Facts & Figures 2008. Atlanta: American Cancer Society; 2008.
9. Ciatto S, Gervasi G, Bonardi R, et al. Determining overdiagnosis by screening with DRE/TRUS or PSA (Florence pilot studies, 1991–1994). Eur J Cancer 2005;41:411–5.
10. Etzioni R, Penson DF, Legler JM, et al. Overdiagnosis due to prostate-specific antigen screening: lessons from U.S. prostate cancer incidence trends. J Natl Cancer Inst 2002;94:981–90.
11. Etzioni R, Gulati R, Falcon S, Penson DF. Impact of PSA screening on the incidence of advanced stage prostate cancer in the United States: a surveillance modeling approach. Med Decis Making 2008;28:323–31.
12. Albertsen PC, Hanley JA, Fine J. 20-year outcomes following conservative management of clinically localized prostate cancer. JAMA 2005;293:2095–101.
13. Johansson JE, Andren O, Andersson SO, et al. Natural history of early, localized prostate cancer. JAMA 2004;291:2713–9.
14. Bill-Axelson A, Holmberg L, Ruutu M, et al. Radical prostatectomy versus watchful waiting in early prostate cancer. N Engl J Med 2005;352:1977–84.
15. Steineck G, Reuter V, Kelly WK, Frank R, Schwartz L, Scher HI. Cytotoxic treatment of aggressive prostate tumors with or without neuroendocrine elements. Acta Oncol 2002;41:668–74.
16. Chu KC, Tarone RE, Freeman HP. Trends in prostate cancer mortality among black men and white men in the United States. Cancer 2003;97:1507–16.
17. Shimizu H, Ross RK, Bernstein L, Yatani R, Henderson BE, Mack TM. Cancers of the prostate and breast among Japanese and white immigrants in Los Angeles County. Br J Cancer 1991;63:963–6.

18. Yu H, Harris RE, Gao YT, Gao R, Wynder EL. Comparative epidemiology of cancers of the colon, rectum, prostate and breast in Shanghai, China versus the United States. Int J Epidemiol 1991;20:76–81.
19. Satia JA, Galanko JA. Demographic, behavioral, psychosocial, and dietary correlates of cancer screening in African Americans. J Health Care Poor Underserved 2007;18:146–64.
20. Giovannucci E. Tomato products, lycopene, and prostate cancer: a review of the epidemiological literature. J Nutr 2005;135:2030S–1S.
21. Heber D, Lu QY. Overview of mechanisms of action of lycopene. Exp Biol Med (Maywood) 2002;227:920–3.
22. Nelson WG, De Marzo AG, Isaacs WB. Prostate cancer. N Engl J Med 2003;349:366–81.
23. Kumar B, Koul S, Khandrika L, Meacham RB, Koul HK. Oxidative stress is inherent in prostate cancer cells and is required for aggressive phenotype. Cancer Res 2008;68:1777–85.
24. Etminan M, Takkouche B, Caamano-Isorna F. The role of tomato products and lycopene in the prevention of prostate cancer: a meta-analysis of observational studies. Cancer Epidemiol Biomarkers Prev 2004;13:340–5.
25. Shi J, Le Maguer M. Lycopene in tomatoes: chemical and physical properties affected by food processing. Crit Rev Food Sci Nutr 2000;40:1–42.
26. Schuurman AG, Goldbohm RA, Dorant E, van den Brandt PA. Vegetable and fruit consumption and prostate cancer risk: a cohort study in The Netherlands. Cancer Epidemiol Biomarkers Prev 1998;7:673–80.
27. Giovannucci E, Rimm EB, Liu Y, Stampfer MJ, Willett WC. A prospective study of tomato products, lycopene, and prostate cancer risk. J Natl Cancer Inst 2002;94:391–8.
28. Nomura AM, Stemmermann GN, Lee J, Craft NE. Serum micronutrients and prostate cancer in Japanese Americans in Hawaii. Cancer Epidemiol Biomarkers Prev 1997;6:487–91.
29. Hsing AW, Comstock GW, Abbey H, Polk BF. Serologic precursors of cancer. Retinol, carotenoids, and tocopherol and risk of prostate cancer. J Natl Cancer Inst 1990;82:941–6.
30. Wu K, Erdman JW, Jr., Schwartz SJ, et al. Plasma and dietary carotenoids, and the risk of prostate cancer: a nested case-control study. Cancer Epidemiol Biomarkers Prev 2004;13:260–9.
31. Gann PH, Ma J, Giovannucci E, et al. Lower prostate cancer risk in men with elevated plasma lycopene levels: results of a prospective analysis. Cancer Res 1999;59:1225–30.
32. Peters U, Leitzmann MF, Chatterjee N, et al. Serum lycopene, other carotenoids, and prostate cancer risk: a nested case-control study in the prostate, lung, colorectal, and ovarian cancer screening trial. Cancer Epidemiol Biomarkers Prev 2007;16:962–8.
33. Key TJ, Appleby PN, Allen NE, et al. Plasma carotenoids, retinol, and tocopherols and the risk of prostate cancer in the European Prospective Investigation into Cancer and Nutrition study. Am J Clin Nutr 2007;86:672–81.
34. Cohen JH, Kristal AR, Stanford JL. Fruit and vegetable intakes and prostate cancer risk. J Natl Cancer Inst 2000;92: 61–8.
35. Combs GF, Jr., Combs SB. The nutritional biochemistry of selenium. Annu Rev Nutr 1984;4:257–80.
36. Menter DG, Sabichi AL, Lippman SM. Selenium effects on prostate cell growth. Cancer Epidemiol Biomarkers Prev 2000;9:1171–82.
37. Redman C, Scott JA, Baines AT, et al. Inhibitory effect of selenomethionine on the growth of three selected human tumor cell lines. Cancer Lett 1998;125:103–10.
38. Rayman MP. The importance of selenium to human health. Lancet 2000;356:233–41.
39. Li H, Stampfer MJ, Giovannucci EL, et al. A prospective study of plasma selenium levels and prostate cancer risk. J Natl Cancer Inst 2004;96:696–703.
40. Yoshizawa K, Willett WC, Morris SJ, et al. Study of prediagnostic selenium level in toenails and the risk of advanced prostate cancer. J Natl Cancer Inst 1998;90:1219–24.
41. Brooks JD, Metter EJ, Chan DW, et al. Plasma selenium level before diagnosis and the risk of prostate cancer development. J Urol 2001;166:2034–8.
42. Helzlsouer KJ, Huang HY, Alberg AJ, et al. Association between alpha-tocopherol, gamma-tocopherol, selenium, and subsequent prostate cancer. J Natl Cancer Inst 2000;92:2018–23.
43. Nomura AM, Lee J, Stemmermann GN, Combs GF, Jr. Serum selenium and subsequent risk of prostate cancer. Cancer Epidemiol Biomarkers Prev 2000;9:883–7.
44. van den Brandt PA, Zeegers MP, Bode P, Goldbohm RA. Toenail selenium levels and the subsequent risk of prostate cancer: a prospective cohort study. Cancer Epidemiol Biomarkers Prev 2003;12:866–71.
45. Goodman GE, Schaffer S, Bankson DD, Hughes MP, Omenn GS. Predictors of serum selenium in cigarette smokers and the lack of association with lung and prostate cancer risk. Cancer Epidemiol Biomarkers Prev 2001;10:1069–76.
46. Hartman TJ, Albanes D, Pietinen P, et al. The association between baseline vitamin E, selenium, and prostate cancer in the alpha-tocopherol, beta-carotene cancer prevention study. Cancer Epidemiol Biomarkers Prev 1998;7:335–40.

47. Peters U, Takata Y. Selenium and the prevention of prostate and colorectal cancer. Mol Nutr Food Res 2008;52: 1261–72.
48. Clark LC, Combs GF, Jr., Turnbull BW, et al. Effects of selenium supplementation for cancer prevention in patients with carcinoma of the skin. A randomized controlled trial. Nutritional Prevention of Cancer Study Group. JAMA 1996;276:1957–63.
49. Duffield-Lillico AJ, Dalkin BL, Reid ME, et al. Selenium supplementation, baseline plasma selenium status and incidence of prostate cancer: an analysis of the complete treatment period of the Nutritional Prevention of Cancer Trial. BJU Int 2003;91:608–12.
50. Neve J. Human selenium supplementation as assessed by changes in blood selenium concentration and glutathione peroxidase activity. J Trace Elem Med Biol 1995;9:65–73.
51. Klein EA, Thompson IM, Lippman SM, et al. SELECT: the next prostate cancer prevention trial. Selenum and Vitamin E Cancer Prevention Trial. J Urol 2001;166:1311–5.
52. Meydani M. Vitamin E. Lancet 1995;345:170–5.
53. Meydani SN, Hayek MG. Vitamin E and aging immune response. Clin Geriatr Med 1995;11:567–76.
54. Gunawardena K, Murray DK, Meikle AW. Vitamin E and other antioxidants inhibit human prostate cancer cells through apoptosis. Prostate 2000;44:287–95.
55. Ripoll EA, Rama BN, Webber MM. Vitamin E enhances the chemotherapeutic effects of adriamycin on human prostatic carcinoma cells in vitro. J Urol 1986;136:529–31.
56. Mucci L, Signorello L, Adami HO. Prostate Cancer. New York: Oxford University Press; 2008.
57. Heinonen OP, Albanes D, Virtamo J, et al. Prostate cancer and supplementation with alpha-tocopherol and beta-carotene: incidence and mortality in a controlled trial. J Natl Cancer Inst 1998;90:440–6.
58. Virtamo J, Pietinen P, Huttunen JK, et al. Incidence of cancer and mortality following alpha-tocopherol and beta-carotene supplementation: a postintervention follow-up. JAMA 2003;290:476–85.
59. Eichholzer M, Stahelin HB, Gey KF, Ludin E, Bernasconi F. Prediction of male cancer mortality by plasma levels of interacting vitamins: 17-year follow-up of the prospective Basel study. Int J Cancer 1996;66:145–50.
60. Chan JM, Stampfer MJ, Ma J, Rimm EB, Willett WC, Giovannucci EL. Supplemental vitamin E intake and prostate cancer risk in a large cohort of men in the United States. Cancer Epidemiol Biomarkers Prev 1999;8:893–9.
61. Huang HY, Alberg AJ, Norkus EP, Hoffman SC, Comstock GW, Helzlsouer KJ. Prospective study of antioxidant micronutrients in the blood and the risk of developing prostate cancer. Am J Epidemiol 2003;157:335–44.
62. Gaziano JM, Glynn RJ, Christen WG et al. Vitamin C and E in the prevention of prostate and total cancer in men: the Physicians' Health Study II RCT. JAMA 2009;301(1):52–62.
63. Lippman SM, Klein EA, Goodman PJ, et al. Effect of selenium and vitamin E on risk of prostate cancer and other cancers: the Selenium and Vitamin E Cancer Prevention Trial (SELECT). JAMA 2009;301:39–59.
64. Woodson K, Triantos S, Hartman T, Taylor PR, Virtamo J, Albanes D. Long-term alpha-tocopherol supplementation is associated with lower serum vascular endothelial growth factor levels. Anticancer Res 2002;22:375–8.
65. Bosland MC, Oakley-Girvan I, Whittemore AS. Dietary fat, calories, and prostate cancer risk. J Natl Cancer Inst 1999;91:489–91.
66. Platz EA. Energy imbalance and prostate cancer. J Nutr 2002;132:3471S–81S.
67. Andersson SO, Wolk A, Bergstrom R, et al. Energy, nutrient intake and prostate cancer risk: a population-based case-control study in Sweden. Int J Cancer 1996;68:716–22.
68. Rohan TE, Howe GR, Burch JD, Jain M. Dietary factors and risk of prostate cancer: a case-control study in Ontario, Canada. Cancer Causes Control 1995;6:145–54.
69. Hayes RB, Ziegler RG, Gridley G, et al. Dietary factors and risks for prostate cancer among blacks and whites in the United States. Cancer Epidemiol Biomarkers Prev 1999;8:25–34.
70. Whittemore AS, Kolonel LN, Wu AH, et al. Prostate cancer in relation to diet, physical activity, and body size in blacks, whites, and Asians in the United States and Canada. J Natl Cancer Inst 1995;87:652–61.
71. Gapstur SM, Gann PH, Colangelo LA, et al. Postload plasma glucose concentration and 27-year prostate cancer mortality (United States). Cancer Causes Control 2001;12:763–72.
72. Giovannucci E, Rimm EB, Stampfer MJ, Colditz GA, Willett WC. Height, body weight, and risk of prostate cancer. Cancer Epidemiol Biomarkers Prev 1997;6:557–63.
73. Greenwald P, Damon A, Kirmss V, Polan AK. Physical and demographic features of men before developing cancer of the prostate. J Natl Cancer Inst 1974;53:341–6.
74. Habel LA, Van Den Eeden SK, Friedman GD. Body size, age at shaving initiation, and prostate cancer in a large, multiracial cohort. Prostate 2000;43:136–43.
75. Kolonel LN. Nutrition and prostate cancer. Cancer Causes Control 1996;7:83–44.

76. Le Marchand L, Kolonel LN, Wilkens LR, Myers BC, Hirohata T. Animal fat consumption and prostate cancer: a prospective study in Hawaii. Epidemiology 1994;5:276–82.
77. Lee MM, Wang RT, Hsing AW, Gu FL, Wang T, Spitz M. Case-control study of diet and prostate cancer in China. Cancer Causes Control 1998;9:545–52.
78. Lund Nilsen TI VL. Anthropometry and prostate cancer risk: a prospective study of 22, 248 Norwegian men. Cancer Causes Control 1999;10.
79. Mills PK, Beeson WL, Phillips RL, Fraser GE. Cohort study of diet, lifestyle, and prostate cancer in Adventist men. Cancer 1989;64:598–604.
80. Putnam SD, Cerhan JR, Parker AS, et al. Lifestyle and anthropometric risk factors for prostate cancer in a cohort of Iowa men. Ann Epidemiol 2000;10:361–9.
81. Schuurman AG, Goldbohm RA, Dorant E, van den Brandt PA. Anthropometry in relation to prostate cancer risk in the Netherlands Cohort Study. Am J Epidemiol 2000;151:541–9.
82. Whittemore AS, Paffenbarger RS, Jr., Anderson K, Lee JE. Early precursors of site-specific cancers in college men and women. J Natl Cancer Inst 1985;74:43–51.
83. Calle EE, Rodriguez C, Walker-Thurmond K, Thun MJ. Overweight, obesity, and mortality from cancer in a prospectively studied cohort of U.S. adults. N Engl J Med 2003;348:1625–38.
84. Cerhan JR, Torner JC, Lynch CF, et al. Association of smoking, body mass, and physical activity with risk of prostate cancer in the Iowa 65+ Rural Health Study (United States). Cancer Causes Control 1997;8:229–38.
85. Chyou PH, Nomura AM, Stemmermann GN. A prospective study of weight, body mass index and other anthropometric measurements in relation to site-specific cancers. Int J Cancer 1994;57:313–7.
86. Lew EA, Garfinkel L. Variations in mortality by weight among 750, 000 men and women. J Chronic Dis 1979;32: 563–76.
87. Snowdon DA, Phillips RL, Choi W. Diet, obesity, and risk of fatal prostate cancer. Am J Epidemiol 1984;120: 244–50.
88. Thompson MM, Garland C, Barrett-Connor E, Khaw KT, Friedlander NJ, Wingard DL. Heart disease risk factors, diabetes, and prostatic cancer in an adult community. Am J Epidemiol 1989;129:511–7.
89. Thune I, Lund E. Physical activity and the risk of prostate and testicular cancer: a cohort study of 53, 000 Norwegian men. Cancer Causes Control 1994;5:549–56.
90. Veierod MB, Laake P, Thelle DS. Dietary fat intake and risk of prostate cancer: a prospective study of 25, 708 Norwegian men. Int J Cancer 1997;73:634–8.
91. Andersson SO, Wolk A, Bergstrom R, et al. Body size and prostate cancer: a 20-year follow-up study among 135006 Swedish construction workers. J Natl Cancer Inst 1997;89:385–9.
92. Efstathiou JA, Bae K, Shipley WU, et al. Obesity and mortality in men with locally advanced prostate cancer: analysis of RTOG 85–31. Cancer 2007;110:2691–9.
93. Gong Z, Agalliu I, Lin DW, Stanford JL, Kristal AR. Obesity is associated with increased risks of prostate cancer metastasis and death after initial cancer diagnosis in middle-aged men. Cancer 2007;109:1192–202.
94. Gong Z, Neuhouser ML, Goodman PJ, et al. Obesity, diabetes, and risk of prostate cancer: results from the prostate cancer prevention trial. Cancer Epidemiol Biomarkers Prev 2006;15:1977–83.
95. MacInnis RJ, English DR. Body size and composition and prostate cancer risk: systematic review and meta-regression analysis. Cancer Causes Control 2006;17:989–1003.
96. Nomura AM. Body size and prostate cancer. Epidemiol Rev 2001;23:126–31.
97. Rodriguez C, Freedland SJ, Deka A, et al. Body mass index, weight change, and risk of prostate cancer in the Cancer Prevention Study II Nutrition Cohort. Cancer Epidemiol Biomarkers Prev 2007;16:63–9.
98. Wright ME, Chang SC, Schatzkin A, et al. Prospective study of adiposity and weight change in relation to prostate cancer incidence and mortality. Cancer 2007;109:675–84.
99. Pischon T, Boeing H, Weikert S, et al. Body size and risk of prostate cancer in the European prospective investigation into cancer and nutrition. Cancer Epidemiol Biomarkers Prev 2008;17:3252–61.
100. Ma J, Li H, Giovannucci E, et al. Prediagnostic body-mass index, plasma C-peptide concentration, and prostate cancer-specific mortality in men with prostate cancer: a long-term survival analysis. Lancet Oncol 2008;9: s1039–47.
101. Rose DP, Boyar AP, Wynder EL. International comparisons of mortality rates for cancer of the breast, ovary, prostate, and colon, and per capita food consumption. Cancer 1986;58:2363–71.
102. Kushi L, Giovannucci E. Dietary fat and cancer. Am J Med 2002;113 (Suppl. 9B):63S–70S.
103. Key TJ, Silcocks PB, Davey GK, Appleby PN, Bishop DT. A case-control study of diet and prostate cancer. Br J Cancer 1997;76:678–87.

104. Chan JM, Pietinen P, Virtanen M, et al. Diet and prostate cancer risk in a cohort of smokers, with a specific focus on calcium and phosphorus (Finland). Cancer Causes Control 2000;11:859–67.
105. Crowe FL, Allen NE, Appleby PN, et al. Fatty acid composition of plasma phospholipids and risk of prostate cancer in a case-control analysis nested within the European Prospective Investigation into Cancer and Nutrition. Am J Clin Nutr 2008;88:1353–63.
106. Hsing AW, McLaughlin JK, Schuman LM, et al. Diet, tobacco use, and fatal prostate cancer: results from the Lutheran Brotherhood Cohort Study. Cancer Res 1990;50:6836–40.
107. Park SY, Murphy SP, Wilkens LR, Henderson BE, Kolonel LN. Fat and meat intake and prostate cancer risk: the multiethnic cohort study. Int J Cancer 2007;121:1339–45.
108. Severson RK, Nomura AM, Grove JS, Stemmermann GN. A prospective study of demographics, diet, and prostate cancer among men of Japanese ancestry in Hawaii. Cancer Res 1989;49:1857–60.
109. Wallstrom P, Bjartell A, Gullberg B, Olsson H, Wirfalt E. A prospective study on dietary fat and incidence of prostate cancer (Malmo, Sweden). Cancer Causes Control 2007;18:1107–21.
110. Giovannucci E, Rimm EB, Colditz GA, et al. A prospective study of dietary fat and risk of prostate cancer. J Natl Cancer Inst 1993;85:1571–9.
111. West DW, Slattery ML, Robison LM, French TK, Mahoney AW. Adult dietary intake and prostate cancer risk in Utah: a case-control study with special emphasis on aggressive tumors. Cancer Causes Control 1991;2:85–94.
112. Strom SS, Yamamura Y, Forman MR, Pettaway CA, Barrera SL, DiGiovanni J. Saturated fat intake predicts biochemical failure after prostatectomy. Int J Cancer 2008;122:2581–5.
113. Meyer F, Bairati I, Shadmani R, Fradet Y, Moore L. Dietary fat and prostate cancer survival. Cancer Causes Control 1999;10:245–51.
114. Allen NE, Key TJ, Appleby PN, et al. Animal foods, protein, calcium and prostate cancer risk: the European Prospective Investigation into Cancer and Nutrition. Br J Cancer 2008;98:1574–81.
115. Crowe FL, Key TJ, Appleby PN, et al. Dietary fat intake and risk of prostate cancer in the European Prospective Investigation into Cancer and Nutrition. Am J Clin Nutr 2008;87:1405–13.
116. Kolonel LN. Fat, meat, and prostate cancer. Epidemiol Rev 2001;23:72–81.
117. Michaud DS, Augustsson K, Rimm EB, Stampfer MJ, Willet WC, Giovannucci E. A prospective study on intake of animal products and risk of prostate cancer. Cancer Causes Control 2001;12:557–67.
118. Rohrmann S, Platz EA, Kavanaugh CJ, Thuita L, Hoffman SC, Helzlsouer KJ. Meat and dairy consumption and subsequent risk of prostate cancer in a US cohort study. Cancer Causes Control 2007;18:41–50.
119. Rodriguez C, McCullough ML, Mondul AM, et al. Meat consumption among Black and White men and risk of prostate cancer in the Cancer Prevention Study II Nutrition Cohort. Cancer Epidemiol Biomarkers Prev 2006;15: 211–6.
120. Schuurman AG, van den Brandt PA, Dorant E, Goldbohm RA. Animal products, calcium and protein and prostate cancer risk in The Netherlands Cohort Study. Br J Cancer 1999;80:1107–13.
121. Koutros S, Cross AJ, Sandler DP, et al. Meat and meat mutagens and risk of prostate cancer in the Agricultural Health Study. Cancer Epidemiol Biomarkers Prev 2008;17:80–7.
122. Dewailly E, Mulvad G, Sloth Pedersen H, Hansen JC, Behrendt N, Hart Hansen JP. Inuit are protected against prostate cancer. Cancer Epidemiol Biomarkers Prev 2003;12:926–7.
123. Nutting PA, Freeman WL, Risser DR, et al. Cancer incidence among American Indians and Alaska Natives, 1980 through 1987. Am J Public Health 1993;83:1589–98.
124. Zhang J, Sasaki S, Amano K, Kesteloot H. Fish consumption and mortality from all causes, ischemic heart disease, and stroke: an ecological study. Prev Med 1999;28:520–9.
125. Hedelin M, Balter KA, Chang ET, et al. Dietary intake of phytoestrogens, estrogen receptor-beta polymorphisms and the risk of prostate cancer. Prostate 2006;66:1512–20.
126. Terry PD, Rohan TE, Wolk A. Intakes of fish and marine fatty acids and the risks of cancers of the breast and prostate and of other hormone-related cancers: a review of the epidemiologic evidence. Am J Clin Nutr 2003;77: 532–43.
127. Andersson SO, Baron J, Wolk A, Lindgren C, Bergstrom R, Adami HO. Early life risk factors for prostate cancer: a population-based case-control study in Sweden. Cancer Epidemiol Biomarkers Prev 1995;4:187–92.
128. Augustsson K, Michaud DS, Rimm EB, et al. A prospective study of intake of fish and marine fatty acids and prostate cancer. Cancer Epidemiol Biomarkers Prev 2003;12:64–7.
129. Terry P, Lichtenstein P, Feychting M, Ahlbom A, Wolk A. Fatty fish consumption and risk of prostate cancer. Lancet 2001;357:1764–6.
130. Chavarro JE, Stampfer MJ, Campos H, Kurth T, Willett WC, Ma J. A prospective study of trans-fatty acid levels in blood and risk of prostate cancer. Cancer Epidemiol Biomarkers Prev 2008;17:95–101.

131. Pham TM, Fujino Y, Kubo T, et al. Fish intake and the risk of fatal prostate cancer: findings from a cohort study in Japan. Public Health Nutr 2008: 1–5.
132. Allen NE, Sauvaget C, Roddam AW, et al. A prospective study of diet and prostate cancer in Japanese men. Cancer Causes Control 2004;15:911–20.
133. Burdge GC, Finnegan YE, Minihane AM, Williams CM, Wootton SA. Effect of altered dietary *n*–3 fatty acid intake upon plasma lipid fatty acid composition, conversion of [13C]alpha-linolenic acid to longer-chain fatty acids and partitioning towards beta-oxidation in older men. Br J Nutr 2003;90:311–21.
134. Ghosh J, Myers CE. Inhibition of arachidonate 5-lipoxygenase triggers massive apoptosis in human prostate cancer cells. Proc Natl Acad Sci U S A 1998;95:13182–7.
135. Ghosh J. Rapid induction of apoptosis in prostate cancer cells by selenium: reversal by metabolites of arachidonate 5-lipoxygenase. Biochem Biophys Res Commun 2004;315:624–35.
136. Pidgeon GP, Kandouz M, Meram A, Honn KV. Mechanisms controlling cell cycle arrest and induction of apoptosis after 12-lipoxygenase inhibition in prostate cancer cells. Cancer Res 2002;62:2721–7.
137. Ghosh J, Myers CE. Arachidonic acid stimulates prostate cancer cell growth: critical role of 5-lipoxygenase. Biochem Biophys Res Commun 1997;235:418–23.
138. Nie D, Krishnamoorthy S, Jin R, et al. Mechanisms regulating tumor angiogenesis by 12-lipoxygenase in prostate cancer cells. J Biol Chem 2006;281:18601–9.
139. Lim K, Han C, Xu L, Isse K, Demetris AJ, Wu T. Cyclooxygenase-2-derived prostaglandin E2 activates beta-catenin in human cholangiocarcinoma cells: evidence for inhibition of these signaling pathways by omega 3 polyunsaturated fatty acids. Cancer Res 2008;68:553–60.
140. Terry PD, Terry JB, Rohan TE. Long-chain (*n*–3) fatty acid intake and risk of cancers of the breast and the prostate: recent epidemiological studies, biological mechanisms, and directions for future research. J Nutr 2004;134: 3412S–20S.
141. Laaksonen DE, Laukkanen JA, Niskanen L, et al. Serum linoleic and total polyunsaturated fatty acids in relation to prostate and other cancers: a population-based cohort study. Int J Cancer 2004;111:444–50.
142. Mannisto S, Pietinen P, Virtanen MJ, et al. Fatty acids and risk of prostate cancer in a nested case-control study in male smokers. Cancer Epidemiol Biomarkers Prev 2003;12:1422–8.
143. Newcomer LM, King IB, Wicklund KG, Stanford JL. The association of fatty acids with prostate cancer risk. Prostate 2001;47:262–8.
144. Chavarro JE, Stampfer MJ, Li H, Campos H, Kurth T, Ma J. A prospective study of polyunsaturated fatty acid levels in blood and prostate cancer risk. Cancer Epidemiol Biomarkers Prev 2007;16:1364–70.
145. Gann PH, Hennekens CH, Sacks FM, Grodstein F, Giovannucci EL, Stampfer MJ. Prospective study of plasma fatty acids and risk of prostate cancer. J Natl Cancer Inst 1994;86:281–6.
146. Harvei S, Bjerve KS, Tretli S, Jellum E, Robsahm TE, Vatten L. Prediagnostic level of fatty acids in serum phospholipids: omega-3 and omega-6 fatty acids and the risk of prostate cancer. Int J Cancer 1997;71:545–51.
147. MacLean CH, Newberry SJ, Mojica WA, et al. Effects of omega-3 fatty acids on cancer risk: a systematic review. JAMA 2006;295:403–15.
148. Leitzmann MF, Stampfer MJ, Michaud DS, et al. Dietary intake of *n*–3 and *n*–6 fatty acids and the risk of prostate cancer. Am J Clin Nutr 2004;80:204–16.
149. Norrish AE, Skeaff CM, Arribas GL, Sharpe SJ, Jackson RT. Prostate cancer risk and consumption of fish oils: a dietary biomarker-based case-control study. Br J Cancer 1999;81:1238–42.
150. Godley PA, Campbell MK, Gallagher P, Martinson FE, Mohler JL, Sandler RS. Biomarkers of essential fatty acid consumption and risk of prostatic carcinoma. Cancer Epidemiol Biomarkers Prev 1996;5:889–95.
151. Freeman VL, Meydani M, Hur K, Flanigan RC. Inverse association between prostatic polyunsaturated fatty acid and risk of locally advanced prostate carcinoma. Cancer 2004;101:2744–54.
152. Norrish AE, Jackson RT, Sharpe SJ, Skeaff CM. Men who consume vegetable oils rich in monounsaturated fat: their dietary patterns and risk of prostate cancer (New Zealand). Cancer Causes Control 2000;11:609–15.
153. Koralek DO, Peters U, Andriole G, et al. A prospective study of dietary alpha-linolenic acid and the risk of prostate cancer (United States). Cancer Causes Control 2006;17:783–91.
154. Messina MJ, Persky V, Setchell KD, Barnes S. Soy intake and cancer risk: a review of the in vitro and in vivo data. Nutr Cancer 1994;21:113–31.
155. Tham DM, Gardner CD, Haskell WL. Clinical review 97: potential health benefits of dietary phytoestrogens: a review of the clinical, epidemiological, and mechanistic evidence. J Clin Endocrinol Metab 1998;83:2223–35.
156. Setlur SR, Mertz K, Hoshida Y, et al. Estrogen-dependent signaling in a molecularly distinct subclass of aggressive prostate cancer. J Natl Cancer Inst 2008.

157. Kolonel LN, Hankin JH, Whittemore AS, et al. Vegetables, fruits, legumes and prostate cancer: a multiethnic case-control study. Cancer Epidemiol Biomarkers Prev 2000;9:795–804.
158. Nomura AM, Hankin JH, Lee J, Stemmermann GN. Cohort study of tofu intake and prostate cancer: no apparent association. Cancer Epidemiol Biomarkers Prev 2004;13:2277–9.
159. Jacobsen BK, Knutsen SF, Fraser GE. Does high soy milk intake reduce prostate cancer incidence? The Adventist Health Study (United States). Cancer Causes Control 1998;9:553–7.
160. Yan L, Spitznagel EL. Meta-analysis of soy food and risk of prostate cancer in men. Int J Cancer 2005;117:667–9.
161. Park SY, Murphy SP, Wilkens LR, Henderson BE, Kolonel LN. Legume and isoflavone intake and prostate cancer risk: the Multiethnic Cohort Study. Int J Cancer 2008;123:927–32.
162. Kurahashi N, Iwasaki M, Sasazuki S, Otani T, Inoue M, Tsugane S. Soy product and isoflavone consumption in relation to prostate cancer in Japanese men. Cancer Epidemiol Biomarkers Prev 2007;16:538–45.
163. Jain MG, Hislop GT, Howe GR, Ghadirian P. Plant foods, antioxidants, and prostate cancer risk: findings from case-control studies in Canada. Nutr Cancer 1999;34:173–84.
164. Hodge AM, English DR, McCredie MR, et al. Foods, nutrients and prostate cancer. Cancer Causes Control 2004;15: 11–20.
165. Hedelin M, Klint A, Chang ET, et al. Dietary phytoestrogen, serum enterolactone and risk of prostate cancer: the cancer prostate Sweden study (Sweden). Cancer Causes Control 2006;17:169–80.
166. Stattin P, Adlercreutz H, Tenkanen L, et al. Circulating enterolactone and prostate cancer risk: a Nordic nested case-control study. Int J Cancer 2002;99:124–9.
167. Stattin P, Bylund A, Biessy C, Kaaks R, Hallmans G, Adlercreutz H. Prospective study of plasma enterolactone and prostate cancer risk (Sweden). Cancer Causes Control 2004;15:1095–102.
168. Kilkkinen A, Virtamo J, Virtanen MJ, Adlercreutz H, Albanes D, Pietinen P. Serum enterolactone concentration is not associated with prostate cancer risk in a nested case-control study. Cancer Epidemiol Biomarkers Prev 2003;12: 1209–12.
169. Ward H, Chapelais G, Kuhnle GG, Luben R, Khaw KT, Binghamk Lack of prospective associations between plasma and urinary phytoestrogens and risk of prostate or colorectal cancer in the European Prospective into Cancer-Norfolk study. Cancer Epidemiol Biomarkers Prev 2008;17:2891–4.
170. Kurahashi N, Iwasaki M, Inoue M, Sasazuki S, Tsugane S. Plasma Isoflavones and Subsequent Risk of Prostate Cancer in a Nested Case-Control Study: the Japan Public Health Center. J Clin Oncol 2008.
171. Heald CL, Ritchie MR, Bolton-Smith C, Morton MS, Alexander FE. Phyto-oestrogens and risk of prostate cancer in Scottish men. Br J Nutr 2007;98:388–96.
172. Bylund A, Saarinen N, Zhang JX, et al. Anticancer effects of a plant lignan 7-hydroxymatairesinol on a prostate cancer model in vivo. Exp Biol Med (Maywood) 2005;230:217–23.
173. Bylund A, Zhang JX, Bergh A, et al. Rye bran and soy protein delay growth and increase apoptosis of human LNCaP prostate adenocarcinoma in nude mice. Prostate 2000;42:304–14.
174. Kristal AR, Cohen JH, Qu P, Stanford JL. Associations of energy, fat, calcium, and vitamin D with prostate cancer risk. Cancer Epidemiol Biomarkers Prev 2002;11:719–25.
175. Tseng M, Breslow RA, Graubard BI, Ziegler RG. Dairy, calcium, and vitamin D intakes and prostate cancer risk in the National Health and Nutrition Examination Epidemiologic Follow-up Study cohort. Am J Clin Nutr 2005;81: 1147–54.
176. Chan JM, Giovannucci E, Andersson SO, Yuen J, Adami HO, Wolk A. Dairy products, calcium, phosphorous, vitamin D, and risk of prostate cancer (Sweden). Cancer Causes Control 1998;9:559–66.
177. Rodriguez C, McCullough ML, Mondul AM, et al. Calcium, dairy products, and risk of prostate cancer in a prospective cohort of United States men. Cancer Epidemiol Biomarkers Prev 2003;12:597–603.
178. Chan JM, Giovannucci EL. Dairy products, calcium, and vitamin D and risk of prostate cancer. Epidemiol Rev 2001;23:87–92.
179. Chan JM, Stampfer MJ, Ma J, Gann PH, Gaziano JM, Giovannucci EL. Dairy products, calcium, and prostate cancer risk in the Physicians' Health Study. Am J Clin Nutr 2001;74:549–54.
180. Mitrou PN, Albanes D, Weinstein SJ, et al. A prospective study of dietary calcium, dairy products and prostate cancer risk (Finland). Int J Cancer 2007;120:2466–73.
181. Kurahashi N, Inoue M, Iwasaki M, Sasazuki S, Tsugane AS. Dairy product, saturated fatty acid, and calcium intake and prostate cancer in a prospective cohort of Japanese men. Cancer Epidemiol Biomarkers Prev 2008;17:930–7.
182. Kesse E, Bertrais S, Astorg P, et al. Dairy products, calcium and phosphorus intake, and the risk of prostate cancer: results of the French prospective SU.VI.MAX (Supplementation en Vitamines et Mineraux Antioxydants) study. Br J Nutr 2006;95:539–45.

183. Huncharek M, Muscat J, Kupelnick B. Dairy products, dietary calcium and vitamin D intake as risk factors for prostate cancer: a meta-analysis of 26, 769 cases from 45 observational studies. Nutr Cancer 2008;60:421–41.
184. Park Y, Mitrou PN, Kipnis V, Hollenbeck A, Schatzkin A, Leitzmann MF. Calcium, dairy foods, and risk of incident and fatal prostate cancer: the NIH-AARP Diet and Health Study. Am J Epidemiol 2007;166:1270–9.
185. Koh KA, Sesso HD, Paffenbarger RS, Jr., Lee IM. Dairy products, calcium and prostate cancer risk. Br J Cancer 2006;95:1582–5.
186. Giovannucci E, Liu Y, Stampfer MJ, Willett WC. A prospective study of calcium intake and incident and fatal prostate cancer. Cancer Epidemiol Biomarkers Prev 2006;15:203–10.
187. Skinner HG, Schwartz GG. Serum calcium and incident and fatal prostate cancer in the National Health and Nutrition Examination Survey. Cancer Epidemiol Biomarkers Prev 2008;17:2302–5.
188. Berndt SI, Carter HB, Landis PK, et al. Calcium intake and prostate cancer risk in a long-term aging study: the Baltimore Longitudinal Study of Aging. Urology 2002;60:1118–23.
189. Baron JA, Beach M, Wallace K, et al. Risk of prostate cancer in a randomized clinical trial of calcium supplementation. Cancer Epidemiol Biomarkers Prev 2005;14:586–9.
190. Heaney RP, McCarron DA, Dawson-Hughes B, et al. Dietary changes favorably affect bone remodeling in older adults. J Am Diet Assoc 1999;99:1228–33.
191. Giovannucci E. Vitamin D and cancer incidence in the Harvard cohorts. Ann Epidemiol 2008;Epub ahead of print.
192. Jacobs ET, Giuliamo AR, Martinez ME, Hollis BW, Reid ME, Marshall JR. Plasma levels of 25-hydroxyvitamin D, 1, 25-dihydroxyvitamin D and the risk of prostate cancer. J Steroid Biochem Mol Biol 2004;89–90:533–7.
193. Corder EH, Guess HA, Hulka BS, et al. Vitamin D and prostate cancer: a prediagnostic study with stored sera. Cancer Epidemiol Biomarkers Prev 1993;2:467–72.
194. Gann PH, Ma J, Hennekens CH, Hollis BW, Haddad JG, Stampfer MJ. Circulating vitamin D metabolites in relation to subsequent development of prostate cancer. Cancer Epidemiol Biomarkers Prev 1996;5:121–6.
195. Platz EA, Leitzmann MF, Hollis BW, Willett WC, Giovannucci E. Plasma 1, 25-dihydroxy- and 25-hydroxyvitamin D and subsequent risk of prostate cancer. Cancer Causes Control 2004;15:255–65.
196. Li H, Stampfer MJ, Hollis BW, et al. A prospective study of plasma vitamin D metabolites, vitamin D receptor polymorphisms, and prostate cancer. PLOS Medicine 2007;4:e103.
197. Ahn J, Peters U, Albanes D, et al. Serum vitamin D concentrations and prostate cancer risk: a nested case control study. J Natl Cancer Inst 2008.
198. Ahonen MH, Tenkanen L, Teppo L, Hakama M, Tuohimaa P. Prostate cancer risk and prediagnostic serum 25-hydroxyvitamin D levels (Finland). Cancer Causes Control 2000;11:847–52.
199. Tuohimaa P, Tenkanen L, Ahonen M, et al. Both high and low levels of blood vitamin D are associated with a higher prostate cancer risk: a longitudinal, nested case-control study in the Nordic countries. Int J Cancer 2004;108: 104–8.
200. Xu C, Shen G, Chen C, Gelinas C, Kong AN. Suppression of NF-kappaB and NF-kappaB-regulated gene expression by sulforaphane and PEITC through IkappaBalpha, IKK pathway in human prostate cancer PC-3 cells. Oncogene 2005;24:4486–95.
201. Verhoeven DT, Verhagen H, Goldbohm RA, van den Brandt PA, van Poppel G. A review of mechanisms underlying anticarcinogenicity by brassica vegetables. Chem Biol Interact 1997;103:79–129.
202. Herman-Antosiewicz A, Singh SV. Checkpoint kinase 1 regulates diallyl trisulfide-induced mitotic arrest in human prostate cancer cells. J Biol Chem 2005;280:28519–28.
203. Hsu JC, Zhang J, Dev A, Wing A, Bjeldanes LF, Firestone GL. Indole-3-carbinol inhibition of androgen receptor expression and downregulation of androgen responsiveness in human prostate cancer cells. Carcinogenesis 2005;26:1896–904.
204. Myzak MC, Hardin K, Wang R, Dashwood RH, Ho E. Sulforaphane inhibits histone deacetylase activity in BPH-1, LnCaP and PC-3 prostate epithelial cells. Carcinogenesis 2006;27:811–9.
205. Joseph MA, Moysich KB, Freudenheim JL, et al. Cruciferous vegetables, genetic polymorphisms in glutathione S-transferases M1 and T1, and prostate cancer risk. Nutr Cancer 2004;50:206–13.
206. Kirsh VA, Peters U, Mayne ST, et al. Prospective study of fruit and vegetable intake and risk of prostate cancer. J Natl Cancer Inst 2007;99:1200–9.
207. Giovannucci E, Rimm EB, Liu Y, Stampfer MJ, Willett WC. A prospective study of cruciferous vegetables and prostate cancer. Cancer Epidemiol Biomarkers Prev 2003;12:1403–9.
208. Key TJ, Allen N, Appleby P, et al. Fruits and vegetables and prostate cancer: no association among 1104 cases in a prospective study of 130544 men in the European Prospective Investigation into Cancer and Nutrition (EPIC). Int J Cancer 2004;109:119–24.

209. Stram DO, Hankin JH, Wilkens LR, et al. Prostate cancer incidence and intake of fruits, vegetables and related micronutrients: the multiethnic cohort study* (United States). Cancer Causes Control 2006;17:1193–207.
210. Traka M, Gasper AV, Melchini A, et al. Broccoli consumption interacts with GSTM1 to perturb oncogenic signalling pathways in the prostate. PLoS ONE 2008;3:e2568.
211. Untergasser G, Rumpold H, Plas E, Witkowski M, Pfister G, Berger P. High levels of zinc ions induce loss of mitochondrial potential and degradation of antiapoptotic Bcl-2 protein in in vitro cultivated human prostate epithelial cells. Biochem Biophys Res Commun 2000;279:607–14.
212. Golovine K, Makhov P, Uzzo RG, Shaw T, Kunkle D, Kolenko VM. Overexpression of the zinc uptake transporter hZIP1 inhibits nuclear factor-kappaB and reduces the malignant potential of prostate cancer cells in vitro and in vivo. Clin Cancer Res 2008;14:5376–84.
213. Ishii K, Usui S, Sugimura Y, et al. Aminopeptidase N regulated by zinc in human prostate participates in tumor cell invasion. Int J Cancer 2001;92:49–54.
214. Holmes MD, Pollak MN, Willett WC, Hankinson SE. Dietary correlates of plasma insulin-like growth factor I and insulin-like growth factor binding protein 3 concentrations. Cancer Epidemiol Biomarkers Prev 2002;11:852–61.
215. Chan JM, Stampfer MJ, Ma J, et al. Insulin-like growth factor-I (IGF-I) and IGF binding protein-3 as predictors of advanced-stage prostate cancer. J Natl Cancer Inst 2002;94:1099–106.
216. Chandra RK. Excessive intake of zinc impairs immune responses. JAMA 1984;252:1443–6.
217. Leitzmann MF, Stampfer MJ, Wu K, Colditz GA, Willett WC, Giovannucci EL. Zinc supplement use and risk of prostate cancer. J Natl Cancer Inst 2003;95:1004–7.
218. Gallus S, Foschi R, Negri E, et al. Dietary zinc and prostate cancer risk: a case-control study from Italy. Eur Urol 2007;52:1052–6.
219. Kristal AR, Stanford JL, Cohen JH, Wicklund K, Patterson RE. Vitamin and mineral supplement use is associated with reduced risk of prostate cancer. Cancer Epidemiol Biomarkers Prev 1999;8:887–92.
220. Meyer F, Galan P, Douville P, et al. Antioxidant vitamin and mineral supplementation and prostate cancer prevention in the SU.VI.MAX trial. Int J Cancer 2005;116:182–6.
221. Lawson KA, Wright ME, Subar A, et al. Multivitamin use and risk of prostate cancer in the National Institutes of Health-AARP Diet and Health Study. J Natl Cancer Inst 2007;99:754–64.
222. He K, Rimm EB, Merchant A, et al. Fish consumption and risk of stroke in men. JAMA 2002;288:3130–6.
223. Joshipura KJ, Ascherio A, Manson JE, et al. Fruit and vegetable intake in relation to risk of ischemic stroke. JAMA 1999;282:1233–9.
224. Michaud DS, Spiegelman D, Clinton SK, Rimm EB, Willett WC, Giovannucci EL. Fruit and vegetable intake and incidence of bladder cancer in a male prospective cohort. J Natl Cancer Inst 1999;91:605–13.
225. Owusu W, Willett WC, Feskanich D, Ascherio A, Spiegelman D, Colditz GA. Calcium intake and the incidence of forearm and hip fractures among men. J Nutr 1997;127:1782–7.
226. Smith MR, Boyce SP, Moyneur E, Duh MS, Raut MK, Brandman J. Risk of clinical fractures after gonadotropin-releasing hormone agonist therapy for prostate cancer. J Urol 2006;175:136–9; Discussion 9.

9 Dietary Supplements and Cancer Risk: Epidemiologic Research and Recommendations

Marian L. Neuhouser and Cheryl L. Rock

Key Points

- Over one-half of all Americans use dietary supplements. Thousands of supplements of multiple combinations of vitamins, minerals, and herbs are available for purchase, but the most commonly used supplements are multivitamins (both with and without minerals) and single supplements of vitamin C, vitamin E, and calcium (with or without vitamin D).
- Dietary supplements can provide a large proportion of total micronutrient intake for many consumers.
- Individuals who smoke should not use β-carotene supplements, as they have been shown to increase lung cancer risk among smokers.
- Vitamin C supplements may reduce the risk of cancers of the GI tract and the bladder.
- Calcium may reduce colon cancer risk.

Key Words: Vitamin; minerals; dietary supplements; cancer

9.1. INTRODUCTION

Millions of Americans use dietary supplements *(1,2)*. Data from national nutrition surveys suggest that approximately 52% of all American adults use dietary supplements *(1,3)* with substantially higher use among certain population subgroups such as cancer survivors *(4–7)*. One reason for the high prevalence of use is the 1994 passage of Public Law 103–417, the Dietary Supplement and Health Education Act (DSHEA) *(8)*. This legislation discontinued the premarket safety evaluations for ingredients used in supplements, placed the burden of proof of product safety on the US Food and Drug Administration (FDA) instead of the manufacturers and permitted limited nutrition support statements without prior approval from the FDA *(8)*. These regulations resulted in an exponential increase in the number and variety of dietary supplements available for over-the-counter purchase, including those that consumers believe may prevent chronic diseases such as cancer *(9–11)*.

The medical community is conflicted about the efficacy of dietary supplements in relation to cancer prevention. Randomized controlled trials of supplements have yielded some entirely unexpected findings. β-carotene, for which observational studies consistently linked higher dietary intake with reduced risk for lung cancer, was found to increase the incidence of lung cancer in two large clinical trials *(12,13)*. Selenium, which was hypothesized to reduce the risk of non-melanomatous skin cancers among persons with previous basal cell or squamous cell carcinomas, had no effect on basal cell cancers, increased the risk of squamous and total non-melanoma skin cancers and reducedthe risk

A. Bendich, R.J. Deckelbaum (eds.), *Preventive Nutrition*, Nutrition and Health, DOI 10.1007/978-1-60327-542-2_9,

of several other cancers *(14)*. Selenium supplementation was subsequently tested in a large clinical trial for the primary prevention of prostate cancer, in which beneficial effects on incidence were not observed *(15)* (www.crab.org/select). Some observational studies suggest that calcium supplements are associated with an increased risk of prostate cancer (particularly for aggressive cancers) *(16–20)*, but others have noted a reduced risk for cancer with use of vitamin D; often calcium and vitamin D are part of the same supplement *(21)*. Data are inconsistent regarding calcium supplementation and risk of colon cancer; observational data note protective associations *(22–26)* while randomized controlled trials show no effect *(27)*. These results are juxtaposed with a 2002 report claiming that most Americans do not obtain sufficient vitamins from the diet to prevent chronic disease, including certain cancers, and advised that all adults should use a daily multivitamin supplement *(28,29)*. Conversely, an NIH State-of-the-Science Panel concluded that current evidence, especially results from randomized controlled trials, does not support the usefulness of dietary supplements for cancer prevention *(30)*. These opposing views from the medical community and unexpected results from intervention studies underscore the need for a synthesis of the research literature on supplements and cancer prevention.

This chapter describes epidemiologic evidence to support or refute associations of dietary supplement use with cancer risk. As an introduction, we provide a definition of dietary supplements and briefly review potential biological mechanisms whereby dietary supplements could prevent cancer. We also present methodologic considerations important for understanding epidemiologic studies on supplement use and cancer risk. The majority of the chapter is devoted to synthesizing results from studies that have provided data on supplement use and cancer risk. We then discuss issues relevant to this research, with emphasis on problems in assessment of supplement use and potential confounding factors. Lastly, we provide information on new studies in this area and give our recommendations about the use of dietary supplements to prevent cancer.

9.1.1. Definition of Dietary Supplements in the United States

Dietary supplements as defined by the Dietary Supplement and Health Education Act (DSHEA) are vitamins, minerals, herbs or other botanicals, and amino acids which are sold as capsules, gelcaps, powders, or liquids and are intended to supplement the diet through increased dietary intake *(8)*. These categories are not mutually exclusive as many supplements contain mixtures of 10 or more micronutrients, herbs, and other potentially bioactive compounds. Highly fortified food products and items intended as the only component of a meal (e.g., meal replacement beverages) are not classified as dietary supplements.

9.1.2. Hypothesized Mechanisms of Effect

There is extensive evidence that high consumption of plant foods is associated with lower risk of human cancers *(31–35)*. A comprehensive review of diet and cancer concluded that the current evidence demonstrates a protective effect of vegetable consumption, and less definitively, fruit consumption, against almost all major cancers *(31)*. The mechanisms underlying these associations are complex and likely involve numerous compounds and multiple biochemical pathways *(36,37)*. Included among potential agents from plant foods are a variety of vitamins (e.g., vitamin C, vitamin E, folate) and minerals (e.g., calcium, selenium), as well as a myriad of bioactive compounds such as carotenoids (α-carotene, β-carotene, β-cryptoxanthin, lycopene, lutein, and zeaxanthin) and flavonoids (e.g., quercetin, naringenin). Whether dietary supplements containing micronutrients and bioactive compounds found in plant foods would be effective chemopreventive agents is of considerable public health interest. A particularly important point is that the bioavailability of nutrients from supplements has not been completely characterized; for some nutrients the bioavailability is increased from supplements, while for others it is decreased *(38,39)*. For example, the bioavailability of folic acid is greater in supplement form, compared to natural food forms (i.e., fruits and vegetables), but the bioavailability of supplemental calcium is dependent on the form (i.e., calcium citrate versus calcium carbonate *(38–40)*.

Laboratory studies provide evidence for mechanisms whereby micronutrients commonly found in dietary supplements could prevent, or promote, cancer. Much attention has focused on nutrients and bioactive food components with antioxidant properties: carotenoids, vitamin C, vitamin E, and selenium *(15,41–46)*. There are many potentially relevant functions for antioxidants, including protection of cell membranes and DNA from oxidative damage, scavenging and reduction of *N*-nitroso compounds, and serving as cofactors or structural components for enzymes whose substrates are reactive nitrogen or reactive oxygen species *(41,46–50)*. Vitamin E is a non-specific chain-breaking antioxidant, while vitamin C is a strong intracellular and extracellular antioxidant via its electron donor capabilities *(51)*. Selenium is a structural component of the glutathione peroxidases, which comprise both the intracellular and extracellular antioxidant defense systems *(51)*.

Micronutrients have preventive properties apart from their potential antioxidant capabilities. Vitamin A (i.e., retinol) plays a role in the differentiation of normal epithelial cells and the maintenance of intercellular communication through gap junctions, thus repressing the processes leading to abnormal cell replication *(52,53)*. The retinoic acid receptors, RAR and RXR, regulate gene expression of numerous enzymes and proteins and retinoids upregulate the synthesis of natural killer cells and cytokines involved in the inflammatory response *(54)*. Carotenoids exhibit antioxidant activity in vitro; however, laboratory evidence strongly suggests that the retinoid-like activities of the carotenoid metabolites are the more important mechanisms by which they may inhibit the progression of carcinogenesis in the human biological system *(55,56)*. Similar to the effects of retinoids, carotenoids influence cell growth regulation, including the inhibition of growth and malignant transformation and the promotion of apoptosis in transformed cells *(53,57–60)*. Several carotenoids (β-carotene, canthaxanthin, and lycopene) and excentric cleavage products of β-carotene have been shown to inhibit both estrogen receptor (ER)-positive and ER-negative breast tumor cell growth in vitro *(58,59)*. Carotenoids also inhibit estrogen signaling of 17β-estradiol and thus inhibit estrogen-induced cell proliferation *(60)*. Vitamin C enhances the immune response and connective tissue integrity primarily via its role as a cofactor or co-substrate for enzymes involved in biosynthesis of collagen, catecholamines, and mixed function oxidases *(51)*. Vitamin E has strong antiproliferative effects on cultured human tumor cells, possibly mediated by its influence on important cell-signaling pathways including TGF-β, c-Jun, and the mitogen-activated protein kinase signaling pathway *(61)*. Folate may be related to cancer risk because inadequate amounts of the vitamin may increase hypomethylation of DNA, with subsequent loss of the normal controls on gene expression *(62,63)*. Low folate status may also impair DNA repair capacity, a noted risk factor for human cancers *(64)*. Calcium could reduce colon cancer risk by binding bile acids *(65)* or by regulating colorectal epithelial cell proliferation *(66)*. Evidence is suggestive, but not conclusive, that vitamin D reduces risk of colon, breast, prostate, and other cancers by its crucial role in maintaining calcium homeostasis, but perhaps more importantly by regulation of gene transcription, *(67)* inhibiting epithelial cell proliferation and enhancing apoptosis and cellular differentiation *(68–70)*. Selenium may block the clonal expansion of early malignant cells by modulation of cell cycle proteins and apoptotic proteins, in addition to its antioxidant functions *(71,72)*. Conversely, iron may increase risk of cancer because it enhances the growth of transformed cells and acts as a pro-oxidant, thereby increasing carcinogenic DNA changes and general oxidative stress *(73)*. Improved understanding of cancer biology will be needed for identifying the cellular and molecular processes that can be affected by vitamin and mineral supplementation.

9.1.3. Prevalence of Supplement Use in the United States

In the United States, consumption of dietary supplements is widespread and has been increasing over the past 15 years. In NHANES 1999–2000, 52% of adults reported taking a dietary supplement in the past month *(1)*. In multivariate analysis, women (versus men), non-Hispanic whites (versus non-Hispanic blacks or Mexican-Americans), and those with a higher level of education, lower body mass index, and higher level of physical activity were associated with a greater likelihood of reporting use of dietary supplements *(1,2)*. Importantly, many consumers take multiple supplements either alone or in combination with prescription or over-the-counter medications *(3)*. For example, in a large cohort study of dietary supplement use and cancer risk in Western Washington State, 32% of the 76,072 cohort participants reported daily use of five or more supplements over the previous 10 years *(74)*. It is important to note, though, that usage patterns among a group of habitual supplement users may differ from the general population. Dietary supplements comprise a substantial portion of Americans' out-of-pocket medical expenditures estimated at over $20 billion/year *(30)*.

9.1.4. Objectives of this Chapter

This chapter presents human observational and experimental data on the use of dietary supplements and cancer risk. To examine the role of supplement use, we included only studies that presented findings on multivitamins or nutrients from supplements separate from food. Studies on total intake of nutrients (diet plus supplements) are not presented because it is not possible in such studies to separate effects of other bioactive compounds present in foods from those of the specific micronutrients of interest. Hereafter, we use the term "dietary supplement" to include both vitamin and mineral supplements, and we use the term "multivitamin" to refer to a one-a-day type multivitamins, which typically also contain minerals.

9.2. BACKGROUND IMPORTANT IN INTERPRETING PUBLISHED STUDIES

9.2.1. Micronutrient Intakes from Foods and Supplements Differ Markedly

Dietary supplements can provide a large proportion of total intakes of some micronutrients, and therefore, the variability in total micronutrient intake attributable to supplements can overwhelm that from foods. For example, in the Women's Health Initiative (WHI) *(75)*, supplement users obtained from 50 to 70% of their total retinol, vitamin C, and vitamin E from supplements, and the median dose from supplements was generally greater than that from foods (Table 9.1) *(76)*. Additional data from WHI show that antioxidant supplement use increased over time such that compared to 1993–1994, the odds of using single supplements of vitamin C and vitamin E in 1998 were 1.37 and 2.10, respectively *(77)*. These results were remarkably similar to results among men enrolled in a large chemoprevention trial for the primary prevention of prostate cancer. Forty-four percent of the men enrolled in the trial reported use of a multivitamin on a regular basis and approximately one-third used high-dose single supplements of vitamin C or vitamin E. Among supplement users, nutrient intake from supplements contributed to about half of total β-carotene and folate intakes and approximately 60% of vitamins A, C, D and 90% of vitamin E intakes (Fig. 9.1) *(78)*. These data illustrate the point that for many nutrients (such as vitamin E), the dose available from supplements (typically

Table 9.1
Vitamin and Mineral Supplement Use Among 16,747 Participants in the Women's Health Initiative (WHI) *(76)*

Nutrient	*Taking Supplement-Containing Nutrient (%)*	*Intake from Supplements Among Supplement Users (Median)*	*Intake from Foods (Median)*	*Supplement Intake as % of Total Intake Among Supplement Users (Mean)*	*Supplement Intake as % of Total Intake Among all Participants (Mean)*
Retinol, mcg	43.5	2250	439	53.6	35.2
β-carotene, mcg	42.6	4500	3264	37.1	24.3
Vitamin C, mg	53.1	200	95	51.4	33.7
Vitamin E, mg	53.2	30	7	71.3	46.7
Folate, mcg	44.0	400	245	41.1	26.9
Calcium, mg	51.5	500	672	30.9	20.3
Iron, mg	37.0	18	13	33.4	21.9
Selenium, mcg	32.5	20	89	11.3	7.4

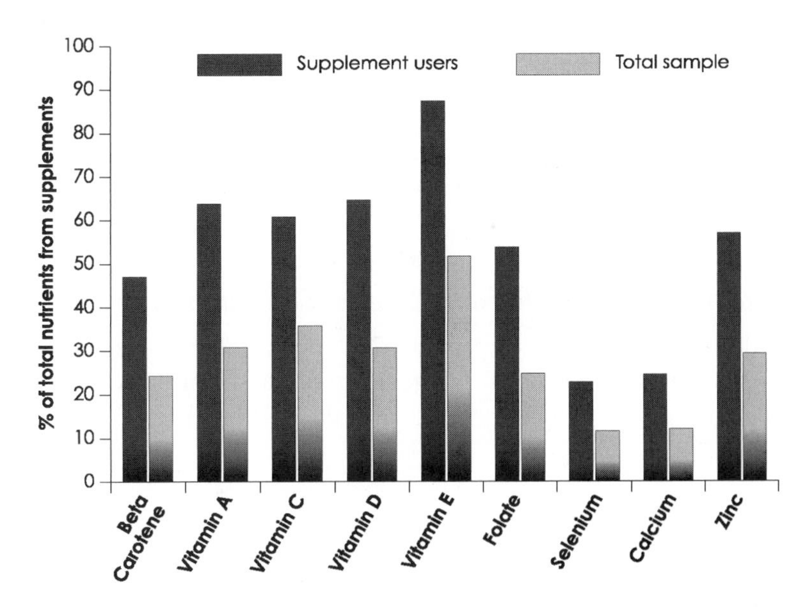

Fig. 9.1. Nutrient intake among supplement users ($n = 9263$) and complete sample ($n = 15{,}387$) of men enrolled in the prostate cancer prevention trial *(78)*.

200–1000 mg) is many times larger than can possibly be obtained from foods (about 8–10 mg). Therefore, in many observational studies of cancer risk, the highest levels of intake of many micronutrients could only be obtained from supplements. It follows that some of the significant findings on vitamin or mineral intake were actually detecting an association between supplement use and cancer risk. However, because many of the earlier published studies do not present findings separately for nutrients from foods versus nutrients from supplements, the studies presented here are likely a subset of those that could address associations of supplements with cancer risk. Studies published within the last 10 years have more consistently presented results separately for nutrients from food versus supplements.

An additional important point with regard to micronutrients from food versus supplements is that the chemical isomers in supplements may differ from those in food. This is particularly important for vitamin E. The natural isomer most abundant in nature (RRR-α-tocopherol) is preferentially retained, compared to synthetic α-tocopherol, which consists of eight diastereoisomers (DL-α-tocopherol) *(79)*. Although the principal vitamin E isomer in foods is γ-tocopherol, which may be more effective at trapping reactive nitrogen species in cell culture studies *(80)*, it is found in only very small concentrations in the human biological system.

9.2.2. The Effects of Micronutrients in Multivitamins Cannot Be Isolated

Findings from investigations that measure multivitamin use are a particular analytical challenge because one cannot isolate the potential effect of the micronutrient of interest (e.g., vitamin A, vitamin D) from all the other vitamins and minerals in the supplement. To investigate micronutrients from supplements, study samples must have sufficient numbers of participants using single supplements (e.g., a capsule containing only vitamin A) to separate out the effects of these micronutrients from multivitamin use. We suggest that findings on micronutrients that are seldom taken as single supplements (e.g., vitamin A, thiamin, zinc) almost certainly reflect the use of multivitamins and are thus confounded by other constituents. Therefore, we only present observational studies on multivitamins and the following micronutrients: vitamin C, vitamin E, and calcium. These micronutrients represent the most commonly used single supplements *(1,74)*, and therefore, it is at least plausible that studies of these micronutrients had enough users to isolate their effect from that of multivitamins. There are insufficient epidemiologic data addressing herbal supplements and cancer risk to be included in this chapter.

9.2.3. Issues of Study Design as Related to Research on Vitamin Supplements

Below we briefly describe three study designs used in epidemiologic research, with additional comments on their strengths and weaknesses when used to study associations of dietary supplements use with cancer risk.

9.2.3.1. Randomized Controlled Trials (RCT)

In RCT, the investigators allocate the exposure (i.e., the supplement) at random. Participants are then followed over a period of time to assess the occurrence of a specified disease outcome. Assuming the sample is sufficiently large, the experimental design of an RCT provides a very high degree of assurance about the validity of the results because both known and unknown characteristics of the intervention and control groups will be identical, thereby eliminating the biases of observational studies, which are described below. However, RCTs are too costly to conduct multiple trials for different types of cancers, nutrients, doses, combinations, or for long periods of time. The latter point is particularly important because carcinogenesis is a process that takes place over a period of many years; so short-term trials often do not yield definitive information *(27)*. In addition, in many trials, the participants are selected to be at high risk for the cancer of interest (e.g., smokers in studies of lung cancer) *(12,81)* limiting the generalizability of the findings. Finally, prevention trials in humans cannot test an agent with known risk.

9.2.3.2. Cohort Studies

In cohort studies, the supplement use of a group of disease-free participants is measured. This group is then followed to assess the occurrence of multiple disease outcomes. Cohorts are attractive for studies of supplement use because they can assess the effects of many types, doses, and combinations of supplements, with multiple cancers. Their primary limitation is that the exposure is self-selected, so that investigators must measure and control for factors (such as diet or exercise) likely to confound supplement–cancer associations. In addition, some cohort studies have inadequate statistical power to test associations of supplement use with cancer, because supplement use was relatively rare when these cohorts were established *(82)*, although many cohorts have updated their dietary assessment strategies to include assessment of supplement use *(74,83,84)*.

9.2.3.3. Case–Control Studies

In case–control studies, participants are selected based on whether they do (case) or do not (control) have cancer. The groups are then compared to see whether supplement use varies by disease status. Selection bias can occur if supplement users are more likely than non-users to agree to be controls in a study. This bias is probable, because supplement users are more interested in health issues than non-users *(85)*, and therefore more likely to participate in a research study. Selection bias may, therefore, lead to finding a protective effect of a supplement, which may be erroneous. Recall bias can occur if individuals with cancer remember and report their use of supplements differently than controls, which can result in either an overestimation or underestimation of the association of supplement use with cancer risk. In addition, since the assessment typically measures behaviors that occurred right before the cancer diagnosis, it is difficult to make inferences about cancer risk per se since carcinogenesis is such a lengthy process.

9.3. REVIEW OF STUDIES ON VITAMIN SUPPLEMENTS AND CANCER RISK

9.3.1. Methods

Below we summarize the published literature on vitamin supplements and cancer risk, organized by supplement type and by study design. We present results grouped by supplement type because an evaluation of which vitamin supplements are associated with cancer at all sites is important for the development of public health recommendations for the prevention of cancer as a whole. We used the MEDLINE® database of the National Library of Medicine to identify epidemiologic studies on dietary supplement use and cancer risk. We limited our review in several ways. Because few studies presented findings on vitamin supplements before 1985, we searched the database from 1980 onward (up to July, 2008). We included a paper only if there were at least 50 cancer endpoints (cases or deaths) in adults. We did not review studies of precancerous conditions (e.g., colorectal adenomas or leukoplakia). We present all data on multivitamins and other commonly used single supplements: vitamin C, vitamin E, calcium. Additionally, we give data from RCTs of β-carotene and selenium because the unexpected results from these trials have strongly influenced public health recommendations for these nutrients. Further, due to the recent interest in folic acid and the potential that heavy supplement use may increase, rather than decrease cancer risk, we include a short section on folic acid. As noted above, because many other vitamins and minerals (e.g., vitamin A, zinc) are generally only obtained from multivitamin pills, we do not present results for those nutrients.

9.3.2. Randomized, Controlled Trials of β-Carotene

Table 9.2 gives results from eight randomized controlled trials that have examined either supplemental β-carotene alone or β-carotene combined with other supplemental nutrients *(12,13,86–94)*. These trials were motivated both by observational epidemiology, animal experiments, and mechanistic studies of carcinogenesis. There are strong and consistent findings of protective associations for fruit and vegetable consumption (the dominant dietary sources of carotenoids), total carotenoid intake, and serum carotenoid concentration on cancer incidence, in particular, for cancer of the lung. However, the intervention trials of β-carotene supplements, either alone or combined with other agents, did not support protective effects for incident cancer, cancer mortality, or total mortality. In two studies of persons at high risk of lung cancer, due to smoking or asbestos exposure [the Beta-Carotene and Retinol Efficacy Trial (CARET) and the Alpha-Tocopherol Beta-Carotene Cancer Prevention Study

Table 9.2
Randomized Controlled Trials of β-Carotene (or Combinations Containing β-Carotene) and Risk of Cancer

Study Name (Ref)	*Agent*	*Endpoint(s)*	*Cases*	*Relative Risk for Supplementation*
Skin Cancer Prevention Study *(86,87)*	50 mg β-carotene	Cancer mortality		
		All sites	82	0.8 (0.5–1.3)[a]
		Incidence		
		Non-melanoma skin cancer	1,952	1.0 (0.9–1.2)[a]
		Basal cell	651	1.0 (0.9–1.2)[a]
		Squamous cell	132	1.2 (0.9–1.7)[a]
Nutrition Intervention Trials in Linxian, China *(88)*	15 mg β-carotene, 50 mcg selenium, and 30 mg α-tocopherol	Cancer mortality		
		All sites	792	0.9 (0.8–1.0)
		Esophageal	360	1.0 (0.8–1.2)
		Stomach	331	0.8 (0.6–1.0)
		Cardia	253	0.8 (0.6–1.0)
		Non-cardia	78	0.7 (0.5–1.1)
Nutrition Intervention Trials in Linxian, China *(89)*	High-dose (2–3 times the RDA) multivitamin with minerals and 15 mg β-carotene	Cancer mortality		
		All sites	176	1.0 (0.7–1.3)[a]
		Esophageal	82	0.8 (0.5–1.3)[a]
		Stomach	77	1.2 (0.9–1.9)[a]
		Incidence		
		All sites	448	1.0 (0.9–1.2)[a]
		Esophageal	251	0.9 (0.7–1.2)[a]
		Stomach	177	1.2 (0.9–1.6)[a]
The Alpha-Tocopherol, Beta-Carotene Study Intervention period results *(12)*	20 mg β-carotene	Cancer mortality		
		Lung	564	1.2 (ns)[b]
		Other	552	1.0 (ns)[b]
		Incidence		
		Lung	876	1.2 (1.0–1.4)
		Four cancers	149–250	No associations

Table 9.2
(Continued)

Study Name (Ref)	*Agent*	*Endpoint(s)*	*Cases*	*Relative Risk for Supplementation*
The Alpha-Tocopherol, Beta-Carotene Study Post-Intervention Follow-Up *(91)*	N/A	Cancer mortality		
		All sites	1847	
		1993–1996		1.11 (1.03–1.21)[a]
		1996–1999		1.07 (0.99–1.15)[a]
		Incidence		
		Lung		
		1993–1996	498	1.17 (0.98–1.39)[a]
		1996–1999	539	0.97 (0.82–1.15)[a]
		Colon		
		1993–1996	92	1.06 (0.70–1.60)[a]
		1996–1999	113	1.88 (0.82–1.15)[a]
		All other sites		
		1993–1999	1878	No associations
Physicians' Health Study *(90)*	50 mg β-carotene on alternate days	Cancer mortality		1.0 (0.9–1.2)[a]
		All sites	766	1.0 (0.9–1.1)[a]
		Incidence		
		Nine cancers	50–1,047	No associations
MRC/BHF Heart Protection Study *(92)*	600 mg vitamin C, 250 mg vitamin C, 20 mg β-carotene	Cancer mortality		
		All sites	704	1.04 (0.90–1.21)
		Incidence		
		All sites	1617	0.98 (0.89–1.08)
		Lung	301	1.1 (n.s.)
		Stomach	116	1.0 (.n.s)
		Prostate	290	0.9 (n.s)
Beta-Carotene and Retinol Efficacy Trial or CARET *(13)*	30 mg β-carotene and 25,000 IU vitamin A	Cancer mortality		
		Lung cancer	254	1.5 (1.1–2.0)
		Incidence		
		Lung cancer	388	1.3 (1.0–1.6)
		Prostate cancer	300	No associations
		Other cancers	730	No associations
Women's Health Study *(93)*	50 mg β-carotene on alternate days	Cancer mortality		
		All sites	59	1.11 (0.67–1.85)
		Incidence		
		All sites	747	1.03 (0.89–1.18)
SU.VI.MAX Study *(94)*	120 mg vitamin C, 30 mg vitamin E, 6 mg β-carotene, 100 μ selenium, 20 mg zinc	Cancer mortality		
		All sites	174	0.77 (0.57–1.00)
		Men	103	0.63 (0.63–0.93)
		Women	71	1.03 (0.64–1.63)
		Incidence		
		All sites	562	0.90 (0.76–1.06)
		Men	212	0.69 (0.53–0.91)
		Women	350	1.04 (0.85–1.29)

[a]Adjusted relative risk, see original studies for details.
[b]Non-significant.

(ATBC)], β-carotene significantly increased lung cancer incidence by 20 and 30% *(12,13)*, while one trial conducted in a low-risk population of male physicians and another among female health professionals found no effect of β-carotene on lung cancer incidence *(90)* or total cancer incidence and mortality *(93)*. In CARET, β-carotene also significantly increased lung cancer mortality and cardiovascular disease mortality *(13)*. CARET and ATBC continued follow-up with study participants for several years for cancer and mortality endpoints *(91,95)*. ATBC reported that even after the cessation of the β-carotene intervention, the relative risk for mortality in the β-carotene arm remained elevated for up to 6 years during the post-trial follow-up. The majority of the excess deaths were due to cardiovascular events *(91)*. In terms of cancer incidence in the ATBC follow-up period, there were no statistically significant differences in cancer incidence between β-carotene recipients and non-recipients at any site with the exception of a late effect of β-carotene being associated with an 88% increased risk of colorectal cancer *(91)*. In CARET, one of the most striking results was that fruit and vegetable consumption provided protection against lung cancer risk among these heavy smokers, but only among those taking the placebo. There was no protective association for persons taking the CARET study vitamins *(96)*. In essence, the high-dose β-carotene plus retinol supplements had the apparent effect of negating any benefit from a high fruit and vegetable diet *(96)*.

A British trial in 20,536 adults investigated whether a combined supplement of β-carotene, vitamin C, and vitamin E could reduce the incidence of cardiovascular disease. Cancer was a secondary endpoint of this trial; there were no statistically significant differences in cancer incidence or mortality at any cancer site between the treatment groups *(92)*. In a trial examining skin cancer, there was no effect of β-carotene on any type of skin cancer or cancer mortality overall *(86)*. Two trials in China examined β-carotene combined with other micronutrients. In one trial focused on cancers of the upper digestive tract, supplementation of 15 mg β-carotene with selenium and vitamin E resulted in a near statistically significant 20% reduction in mortality from stomach cancer *(88)*. In a trial of persons with esophageal dysplasia, supplementation with 15 mg β-carotene plus high-dose multivitamins with minerals had no effect on total, esophageal or stomach cancer incidence or mortality *(89)*. In the SU.VI.MAX Study, subgroup analysis revealed that supplementation with a combination of fairly low doses of vitamin C, vitamin E, β-carotene, selenium, and zinc (versus placebo) was associated with lower mortality in men but not women *(94)*. The investigators suggested that this differential finding may be attributed to lower baseline status, especially for β-carotene, in the men enrolled in that trial. In men with a normal level of prostate-specific antigen (PSA) participating in that trial, vitamin supplementation was associated with a significant reduction in the incidence of prostate cancer (HR 0.52, 95% CI 0.29, 0.92) *(97)*. An additional subset analysis of the SU.VI.MAX results revealed that the antioxidant supplement increased the risk of skin cancer in women, but not men *(98)*.

The fairly consistent negative or null findings from these clinical trials of supplemental β-carotene are perplexing, given the strong and consistent protective effects of dietary and serum carotenoids found in observational studies *(31,32,99–102)*. However, convincing evidence now exists that β-carotene supplements should be contraindicated in smokers or by others at risk for lung cancer via occupational exposures to lung carcinogens *(31)*.

9.3.3. Randomized, Controlled Trials of Selenium

The Nutritional Prevention of Cancer Trial was designed to test whether selenium supplementation would prevent recurrence of non-melanoma skin cancers, among persons living in parts of the United States where soil selenium levels are low *(103)*. Compared to those in the placebo group there was a 25% reduction in total cancer incidence and a 40% reduction in cancer mortality. The selenium supplementation reduced prostate and colon cancers by about half, but it statistically significantly increased squamous cell and total non-melanoma skin cancer by 25% and 17%, respectively, compared to placebo *(14,103,104)*.

These findings generated considerable clinical and scientific interest in selenium's cancer preventive properties *(105,106)*, which resulted in the development of the Selenium and Vitamin E Cancer Prevention Trial (SELECT). This trial was designed to test whether selenium (200 μg/day l-selenomethionine) and/or vitamin E (400 mg/day dl-α-tocopherol) on prostate cancer incidence in >35,000 men enrolled at >400 sites in the United States *(107)*. In 2008, the active treatment (supplementation) in this trial was halted because the supplements, taken alone or together for an average of 5 years, did not prevent prostate cancer (www.crab.org/select). The study was designed to detect a 25% reduction in prostate cancer, and it was evident that effect was unlikely to be achieved even with continuation of treatment, based on the data examined at the 5-year time point. Further, there were two findings that caused some concern, although these findings were not statistically significant. Slightly more cases of prostate cancer in men taking only vitamin E and slightly more cases of diabetes in men taking only selenium were observed. Most of the men enrolled in SELECT will be followed for about 3 more years, so that their health status can be monitored (www.crab.org/select).

9.3.4. Randomized, Controlled Trials of Vitamin E

The effects of vitamin E supplementation on cancer incidence and mortality have been examined in four randomized, controlled trials. In the ATBC Study (discussed above), male smokers were assigned to receive 50 mg/day dl-α-tocopherol with or without the β-carotene, and no effects on cancer incidence were observed in association with vitamin E supplementation *(12)*. Supplementation with a combination of β-carotene, vitamin E (30 mg/day), and selenium was associated with a reduction in cancer rates (RR 0.87, 95% CI 0.75, 1.00), especially stomach cancer (RR 0.79, 95% CI 0.64, 0.99), in the general population of Linxian County, China *(88)*. In the Heart Outcomes Prevention Evaluation (HOPE) and HOPE-The Ongoing Outcomes (HOPE-TOO) studies, cancer incidence and cancer mortality were examined as primary outcomes in a target group of 9,541 and 7,030 individuals, respectively *(108)*. Study participants were $\geq$55 years of age with vascular disease or diabetes mellitus, randomized to vitamin E supplementation (400 mg/day RRR-α-tocopherol) or placebo. There were no differences in cancer incidence and cancer deaths across the two study arms, during a median duration of follow-up of 7 years. In the Women's Health Study, 39,876 women $\geq$45 years were prescribed vitamin E supplements (600 mg every other day RRR-α-tocopherol with or without aspirin and followed for an average of 10.1 years) *(109)*. No effects on cancer incidence or overall mortality were observed in association with vitamin E supplementation.

9.3.5. Multivitamins

Multivitamins are the most commonly used dietary supplement in the United States and elsewhere *(1,74,110)*, as evidenced by the many studies that have reported associations of multivitamin use with cancer incidence or mortality *(25,45,83,84,88,111–139)*. While only a few studies have examined all cancers sites, results are not particularly consistent or promising. One randomized controlled trials of multivitamins is completed *(139)* and at least one other is in progress *(140)*. The completed trail was conducted in a resource-poor area of China where much of the population had nutritionally inadequate diets *(88,139)*. The overall results demonstrated a reduced risk of mortality of any cancer and specifically from cancers of the stomach, gastric cardia, and esphogus *(88)*. However, there was no association with lung cancer mortality *(139)* and no association with incidence of other cancers of the upper digestive tract *(88)*. Results from the PHSII Trial testing a standard multivitamin versus placebo on various health outcomes, including cancer, among male physicians aged 50 years and older will be available in 2012 *(140)*.

Among observational studies examining multivitamin use and cancers of the oral cavity, esophagus, pharynx, and stomach *(112–114,123,127,129,132,133)*, one found a statistically significant 70% increased risk for oral cancer *(123)*, but among the other studies there were no significant associations or consistent trends. The studies of breast cancer were inconsistent: several cohort studies found a relative risk near the null value of 1.0 *(109,115,126,138)*, but one of these cohort studies suggested a modest effect modification by alcohol intake *(109)*. Only one cohort study, conducted among US female nurses, reported a weak, non-significant inverse association of multivitamin use with breast cancer risk when the supplements were used for 5–9 years *(126)*. Case–control studies of multivitamin use and breast cancer risk have been inconsistent; one study suggested an inverse association of multivitamin use with breast cancer risk, while another case–control study found a statistically significant 30% increased risk *(116)*; still another case–control study reported a non-significant 27% reduced risk *(130)*. A case–control study in China found no association between multivitamin use and breast cancer risk *(141)*. Of the observational studies that examined multivitamin and colorectal cancer, five out of seven studies of colon cancer found reduced risks *(24–26,117,124,125,135)* though only two findings of 50 and 75% reduced risks were statistically significant *(25,124)*. Additionally, three large cohort studies reported reduced risk of colon cancer, but only for long-term use of multivitamins (>10 years) *(126,137,142,143)*. One of these found the colon cancer reduced risk only among men *(137)*, while rectal cancer risk was reduced among women taking multivitamins *(136)*. Two studies of multivitamins and cervical cancer found statistically significant 40 and 60% reduced risks *(118,119)*. One study of bladder cancer found a statistically significant 39% reduced risk *(120)* while another reported no association *(128)*. One study of prostate cancer reported no associations of multivitamins with disease risk *(131)*, while two studies reported increased risk of prostate cancer among users of multivitamins *(134,144)*, particularly increased risk for advanced or fatal prostate cancer *(134)*. Two studies of non-Hodgkins lymphoma found a near doubling in risk for women who used multivitamins over a 10-year period, but no association for men *(83)*. Multivitamins were shown to have no association with lung cancer risk in a large cohort study *(45)* and an increased risk for women in a pooled analysis from eight cohort studies *(43)*.

Studies of multivitamins do not support a protective effect of multivitamins for cancers of the upper digestive tract (oral, pharyngeal, esophageal, and stomach), lung, or breast. Evidence is not at all consistent for associations with cancers of the colon, bladder, and cervix while there may be increased risks of non-Hodgkin's lymphoma in women and prostate cancer in men. An Executive Summary from the 2006 NIH Conference "Multivitamin/Mineral Supplements and Chronic Disease Prevention" noted that further research is necessary before it can be concluded that the multivitamins are unequivocally harmful or beneficial in relation to cancer prevention *(30)*. This conclusion is consistent with those in a recent report from the Women's Health Initiative *(145)*. In this cohort of 161,808 postmenopausal women who were followed for 8 years, on average, use of multivitamins neither increased nor decreased risk of eight cancers (invasive breast, colon, endometrium, lung, kidney, ovary, stomach, and bladder) *(145)*. The WHI study of multivitamins and cancer risk is one of the largest to date and endpoints were ascertained via medical records with physician review, which strengthens the view that the results and conclusions are valid. Health-care providers should remain attuned to additional research in the area so as to provide the best clinical information to patients seeking advice about multivitamin use.

9.3.6. *Vitamin C*

Many studies have reported associations of supplemental vitamin C with cancer incidence or mortality *(22,25,42,43,45,83,88,111–115,118–121,123,126–132,141,146–165)*. Four studies examined all cancer sites combined, and none found either large or statistically significant associations *(23,111,146,165)*. Several studies examined cancers of the upper digestive tract. The single randomized controlled trial found no associations with cancers of the upper digestive tract *(88)*. Of 15 observational studies, three found statistically significant protective associations: a 50% reduced risk of pharyngeal *(113)*, a 30% reduced risk of oral plus pharyngeal *(114)*, and one found a halving in risk, but the results were statistically significant only among the non-Hispanic white study participants *(132)*.Two studies found a non-significant 25% reduced risk for esophageal cancers *(123,166)*. The studies of stomach cancer found consistent inverse associations with vitamin C, but the confidence intervals around the point estimates all included the null value of 1.0 *(127,129,147,164)*. Eight studies of breast cancer found no statistically significant associations or consistent trends *(115,126,130,146,148–150,164)*. Of the six studies that examined lung cancer, one found a statistically significant 60% reduced risk for men and a non-significant 40% increased risk for women *(151)*, while results of other studies were inconsistent and none were statistically significant *(43,45,146,152,161)*. Of the studies that examined colon cancer, one found a statistically significant 40% reduced risk *(25)*, two found non-statistically significant reduced risks *(22,146)*, and one a non-statistically significant reduced risk for women only *(153)*. Both studies of cervical cancer found statistically significantly reduced risks, 50% for in situ *(119)* and 30% for invasive *(118)*. Results from the six studies that examined bladder cancer were inconsistent. One study found a statistically significant 40% reduced risk for men only *(146)*. Of the remaining studies, one found a statistically significant 60% reduction *(120)*, two found a non-significant 30–50% reduced risk *(128,154)* while another found a 50% reduced risk in women and a 20% increased risk in men, both of which were non-statistically significant *(155)*. Only one cohort study reported a non-significant 25% increased risk for bladder cancer mortality associated with use of vitamin C supplements *(158)*. The single study of ovarian cancer reported a statistically significant halving in risk associated with use of supplemental vitamin C *(159)*. Studies of basal cell carcinoma, prostate cancer, and non-Hodgkin's lymphoma did not find any statistically significant associations with vitamin C use *(42,83,121,131,160)*.

Studies of vitamin C do not support a protective effect for total, breast, prostate, or lung cancers. There is modest evidence for a protective effect for cancers of the upper digestive tract, cervix, ovary, bladder, and colon.

9.3.7. *Vitamin E*

Numerous observational studies have examined associations of supplemental vitamin E with cancer incidence or mortality *(22,25,42,43,45,83,88,111,114,115,118–121,123,126,127,129–131,141,146–150,153,156–159,162–165,167–171)*. Only four studies report associations with all cancer sites combined; two found non-statistically significant 20% *(88,146)* reductions in risk, and one found a non-statistically significant 60% reduction in risk *(111)*. One of these studies was a randomized controlled trial that combined vitamin E with β-carotene and selenium, so its result cannot be attributed to vitamin E alone *(88)*. None of the nine studies of breast cancer found statistically significant associations or consistent trends *(115,126,130,141,146,148–150,164)*, although one found a 20% reduction in breast cancer risk with vitamin E supplement use among women with low dietary vitamin E intake (OR 0.8, 95% CI 0.6, 1.0). Of the three studies on lung cancer, only the study among non-smokers found a significant 50% reduced risk for both men and women *(168)*. Importantly, the Alpha-Tocopherol Beta-Carotene Study, which was a randomized trial that included one study arm designed specifically to test whether vitamin E could prevent lung cancer in smokers found no effect *(12)*. However, an unexpected finding in this trial was a statistically significant 30% reduced risk for prostate cancer in the vitamin E arm of the trial, as noted above *(167)*. One cohort study reported a modest increased risk of late stage prostate cancer with vitamin E, *(171)* but another cohort study and a case–control study found non-statistically significant 24 and 30% reduced risks *(131,160)*. A large cohort in Western Washington State found no overall association of vitamin E with prostate cancer risk when assessed an average 10-year use of ≥400 IU versus no use. However, risk of advanced disease (defined as regionally invasive or distant metastasis) was significantly decreased by 64% with long-term vitamin E use *(172)*.

Twelve studies examined associations of vitamin E with upper digestive tract cancers. Of the two randomized controlled trials, the trial examining the vitamin E, selenium, and β-carotene combined found a near statistically significant 20% reduction in stomach cancer *(88)*, while the other trial of vitamin E alone found a non-statistically significant 30% increased risk *(173)*. Of the three studies that examined oral cancer alone, or oral plus pharyngeal cancer, two found statistically significant protective effects ranging from 50 to 70% reduced risk *(114,169)*. The remaining studies of upper digestive tract cancers found either no associations *(127)* or non-statistically significant protective associations *(123,129,147)*. Of the eight studies that examined colon cancer, two observational studies found statistically significant protective associations of 50 and 60% reduced risks *(22,25)* while the others found only modest, non-statistically significant inverse associations. For example, a randomized controlled trial found a non-significant 20% reduced risk *(173)* and three out of four observational studies found small reduced risks in women but not men *(146,153,170)*, while one reported non-significant reduced risks of similar magnitude for both men and women *(84)*. Vitamin E supplements significantly reduced ovarian cancer risk by 57% and cervical cancer risk by 40–60% in the few studies of gynecological cancers *(119,159)*. Two studies of bladder cancer found 50–60% reduced risks associated with vitamin E supplementation *(120,158)*.

Studies of vitamin E do not support an association with breast or smoking-related lung cancers. The SELECT study results are not consistent with some of the evidence from the observational studies suggesting that vitamin E supplementation may be protective against prostate cancer, particularly with long-term use. There is modest evidence for protective associations of vitamin E with cancers of the upper digestive tract, colon, bladder, and the female reproductive tract.

9.3.8. Calcium

Calcium and cancer incidence has been examined in numerous studies, both clinical trials and observational studies *(16,19,21,23–26,129,156,174–185)*. Some of these studies examine calcium alone, others examine calcium plus vitamin D while others investigate calcium or vitamin D from other types of supplements; thus, investigating the exposure from calcium or vitamin D separately or together is extremely difficult. In a randomized controlled trial of calcium, calcium plus vitamin D or placebo conducted in postmenopausal women, total cancer incidence at all common cancer sites was significantly lower among the women taking calcium plus vitamin D, but the number of cases was so small that inferences about the effectiveness are unclear *(21)*. Of the studies that examined colon cancer, two found a statistically significant reduced risk *(26,174)*, three found non-statistically significant 20 and 30% reduced risks *(23–25)* and two reported no association *(178,183)*. A single report for rectal cancer found a near statistically significant 24% reduced risk *(175)*. Among secondary outcomes in the Women's Health Initiative randomized controlled trial of calcium plus vitamin D or placebo, colon cancer incidence did not vary between the intervention and placebo groups *(27)*. The data on prostate cancer are inconsistent; several cohort studies have reported large (as high as 300%) and statistically significant and increased risks *(16,19)*. Results from other studies are not consistent; for example, the exceptionally wide confidence intervals in one study *(20)* and the small number of cases in the highest quintile of calcium intake in the other *(19)* suggest that we view these findings with caution. Notably, a case–control reported no association of supplemental calcium with prostate cancer risk *(131)*, but a later publication from that study suggested a modest association of use of calcium supplement with increased risk for metastatic prostate cancer *(177)*. Results from a randomized trial and large cohort did not reveal any significant associations (either protective or harmful) for use and calcium supplements and prostate cancer risk *(184,185)* There were non-statistically significant associations of calcium supplementation with cancers of the endometrium *(156)* and ovary *(176)* and inconsistent associations with cancers of the upper digestive tract *(129)*. Calcium and calcium plus vitamin D are of interest to many clinicians and researchers in relation to breast cancer risk. However, results from a large clinical trial and a large cohort of postmenopausal women reveal no associations of calcium or calcium plus vitamin D with breast cancer risk *(180,182)*.

Studies of calcium supplementation (with or without vitamin D) and cancer are limited. There is modest evidence for a protective association of calcium supplementation with colon cancer risk in observational studies, but these findings are not confirmed by randomized controlled trials. A few studies have noted increased risk of prostate cancer when calcium supplements are used, but the findings are inconsistent. While there is tremendous scientific interest in vitamin D and cancer risk *(186)*, vitamin D is seldom used as a single supplement. More often, it is used in combination with calcium. Thus, it is not possible at this time to draw any conclusions about vitamin D as a single supplement and overall cancer risk.

9.3.9. Folic Acid

Folic acid is an important water-soluble B vitamin that primarily functions as a methyl donor. The folate pathway is involved in key biochemical reactions related to nucleotide synthesis, DNA repair, DNA methylation (which is related to gene silencing), and prevention of DNA strand breaks *(187–190)*. Several case–control and cohort studies have demonstrated an inverse association of folic acid supplementation via single supplements or with that obtained from multivitamins with reduced risk of colon cancer, or a precursor of colon cancer—colorectal adenomas *(124,135)*. While the work presented in this chapter is primarily focused on cancer endpoints, it is worthwhile noting that numerous studies have been published on the associations of supplemental folic acid with surrogate endpoint biomarkers including aberrant crypt foci, DNA methylation, and dysplasias *(64,190)*. While many of these studies initially appeared promising with regard to folic acid supplementation and cancer risk reduction, more recently it has become apparent that caution must be applied to these findings *(191,192)*. In 2007, results were published from a double-blinded, placebo-controlled trial of 1 mg folic acid/day (with or without 325 mg/day aspirin) or placebo (with or without 325 mg/day aspirin) where the primary trial endpoint was recurrence of colorectal adenomas, a known risk factor for colon cancer *(193)*. The folic acid supplements did not reduce the risk of recurrence of colorectal adenoma; rather in participant follow-up, it was noted that the folic acid supplementation was associated with a more than twofold increased risk of three or more adenomas *(193)*. These results are somewhat consistent with findings from the HOPE-2 Trial (Heart Outcomes Prevention Evaluation) which was a randomized, double-blind, placebo-controlled trial of a single pill of 2.5 mg folic acid plus 50 mg vitamin B_6 plus 1 mg vitamin B_{12} or placebo taken daily for an average of 5 years. Secondary outcomes in the trial were cancers of the colon, lung, breast, prostate, and skin (melanoma) and there were no differences in cancer incidence or cancer mortality between the intervention and placebo groups cancers of the upper digestive tract. More recent data suggest that a dual role for folic acid in supplemental form may exist in relation to carcinogenesis. For persons with an existing colorectal adenoma, folic acid supplements may accelerate the growth of cancers. On the other hand, for persons with especially low folate status, supplements may prevent DNA damage, which is a clear precursor to cancer *(191)*. Additional research is clearly needed before recommendations can be made about folic acid and cancer prevention *(192,194)*.

9.4. DISCUSSION

The results from this chapter indicate that the associations of dietary supplement use with cancer risk are complex and generally inconsistent. The strongest findings are from the RCTs that indicate β-carotene can increase incidence of lung cancer in smokers. No supplements appear to be definitively related to breast cancer. Cancers of the gastrointestinal tract appear to be inversely associated with the supplements examined herein, with perhaps the most consistent associations seen between multivitamins, vitamin E, calcium and colon cancer, and vitamin C with both upper and lower gastrointestinal cancers. Notably, when subgroup analysis is conducted with consideration of the dietary intake of micronutrients or baseline nutritional status, supplementation appears more likely to confer benefits rather than null effects or increased risk. However, in the recently published study of multivitamin use in the Women's Health Initiative, stratified analyses by fruit and vegetable intake did not support the notion that those with poor diets may receive benefit from multivitamins *(145)*.

There are a number of methodologic problems in much of the epidemiologic research on dietary supplement use. These limitations are important to consider before drawing conclusions from this chapter. Here we discuss three important issues: (1) measurement error in assessment of supplement use, (2) importance of a time-integrated measure, and (3) supplement use as a marker for behaviors that may alter cancer risk.

9.4.1. Measurement Error in Assessment of Supplement Use

Much recent research supports the notion that systematic bias in dietary self-report is a common problem in observational studies *(195,196)*. Because individuals have a strong tendency to underreport their dietary intake, any observed associations with disease outcomes often become attenuated or null *(195,197)*. Even less is known about the measurement properties of instruments used to assess dietary supplement use and the extent to which supplement use is misreported. Epidemiologic studies typically use personal interviews or self-administered questionnaires to obtain information on three to five general classes of multiple vitamins and on single supplements, the dose of single supplements, and sometimes frequency and/or duration of use. Considerable effort has been expended toward validating dietary supplement collection instruments *(39,198–202)*.

In 1998, results were published from a validation study comparing supplement data collected in a telephone interview and from a brief self-administered questionnaire with data derived from a detailed in-person interview and transcription of the labels of supplement bottles (i.e., a gold standard) among adult supplement users in Washington State ($n = 104$). Correlation coefficients comparing average daily supplemental vitamin and mineral intake from the interview or questionnaire to the gold standard ranged from 0.8 for vitamin C to 0.1 for iron *(200)*. These results suggest that commonly used epidemiologic methods of assessing supplement use may incorporate significant amounts of error in estimates of some nutrients. The effect of this type of non-differential measurement error is to attenuate measures of association, which could obscure many significant associations of supplement use with cancer.

In 2003, a validity study of a very extensive and detailed 24-page instrument of dietary supplement use was compared to an array of nutritional biomarkers. In a sample of 220 adults aged

50–74 years, there were modest correlations of self-reported use of supplemental intake with serum concentrations of vitamin C ($r = 0.29$) and β-carotene ($r = 0.31$) and a very good correlation with serum vitamin E ($r = 0.69$) *(199)*. The high correlation with serum vitamin E might suggest that this self-administered supplement questionnaire was quite accurate with the lower correlation for the other biomarkers due to other influences on those serum measures. Still, there are limitations with these self-reported measures. Until recently, few reliable databases existed that included reliable ingredient information for the thousands of dietary supplements available for purchase in the United States *(39,203)*. A current and ongoing project between the USDA's Nutrient Data Laboratory, the Office of Dietary Supplements, and the National Center for Health Statistics will provide for the first time a Dietary Supplement Ingredient Database (DSID) *(204)*. This database will include analytic data for common vitamin and mineral preparations, as opposed to simply the label ingredient information, and will greatly improve the ability to accurately assess nutrient exposure from dietary supplements *(203,204)*.

9.4.2. Importance of a Time-Integrated Measure of Supplement Use

Investigators studying diet and chronic diseases usually wish to measure an individual's long-term nutrient intake because the induction and latent periods for these diseases are long. However, many studies only asked participants about their current use of vitamin and mineral supplements, or only obtained information about supplement use at one point in time. Potential sources of variability in supplement use over time include changes in (1) the type of multivitamin used, (2) number of years the supplement was taken, (3) formulations of multivitamins or dose of single supplements, and (4) frequency of taking supplements.

Investigators in Washington State conducted a mailed survey to examine the relationship between current and long-term (10-year) supplement use ($n = 325$ adults) *(202)*. Estimates of current daily intakes for supplemental micronutrients were roughly twice that of average daily intake over the past 10 years. Correlations between current intake and long-term intake from supplements alone were 0.77, 0.75, and 0.65 for vitamin C, E, and calcium, respectively *(202)*. This type of measurement error may also have contributed to many of the null associations in this chapter.

9.4.3. Supplement Use as a Marker for Cancer-Related Behavior

Observational studies on supplement use can be seriously compromised by confounding because supplement use is strongly related to other factors that affect cancer risk. Supplement users are more likely than non-users to be female, non-Hispanic white, better educated, affluent, non-smokers, light drinkers, and to consume diets lower in fat and higher in fiber and some micronutrients *(1,2,9, 205–207)*. However, potential confounding variables include those specific to cancer risk such as preventive screening, use of potentially chemopreventive agents, and diet-related attitudes and behavior *(85,208)*.

Demographic and health-related characteristics of high-dose supplement users were assessed as part of a cohort study of dietary supplement users and cancer risk in Western Washington. Among women, those who had a mammogram in the previous 2 years were 60% more likely to be users of calcium, 50% more likely to use multivitamins, and 20 and 40% more likely to use vitamins C and E, respectively, than women who did not have a mammogram. Among men, those who had a PSA test within the previous 2 years were about 1.5 times more likely to be users of vitamin E, 40% more likely to be users of multivitamins and vitamin C than men who did not get a PSA test *(74,85)*.

For both men and women, there was a strong positive association between having a sigmoidoscopy, using NSAIDs and using multivitamins and single supplements. High-dose supplement users were statistically significantly more likely to be of normal weight, be non-smokers, exercise regularly and eat five or more servings of fruits and vegetables per day *(74)*. These results are similar to previous results from a random-digit-dial survey to monitor cancer risk behavior in adults in Washington State ($n = 1{,}449$) *(85)*.

These relationships could confound studies of supplement use and cancer risk in complex ways. For example, male supplement users were more likely to have had a PSA test, which is associated with increased diagnosis of prostate cancer. Thus, supplement users could appear to have a higher incidence of prostate cancer. However, if early diagnosis of prostate cancer by PSA reduces mortality, supplement users could appear to have lower prostate cancer mortality. Health beliefs influence cancer risk through behavior such as diet and exercise. For example, in a previous prospective study, it was reported that belief in a connection between diet and cancer was a statistically significant predictor of changes to more healthful diets over time *(209)*. In cohort studies, the increasing healthfulness of supplements users' diets and other health practices over time could result in a spurious positive association between supplement use and chronic disease.

In theory, control in analyses for demographics and health-related behavior adjusts for these confounding factors. However, absence of residual confounding cannot be assured, especially if important confounding factors are unknown, assessed with error, not assessed at all, or not included in the analyses. Therefore, many of the observational studies of supplement use and cancer risk may actually be assessing healthy behaviors in general; it is very hard to disentangle these exposures *(85,206,210,211)*.

9.4.4. Future Research

Despite the large number of studies reviewed in this report, the published studies to date on dietary supplements and cancer risk are far from definitive. Research on dietary supplements must continue before public health recommendations can be formulated. The National Institutes of Health appears to be committed to research on dietary supplements. The 1994 DSHEA legislation mandated the creation of the Office of Dietary Supplements (ODS) *(8)*. The ODS supports research, sponsors workshops and consensus conferences and disseminates information about dietary supplements to researchers, clinicians, and consumers. One goal of some of the ODS-funded research is to establish a network of Dietary Supplement Research Centers *(212)*. The Office of Dietary Supplements can be accessed at http://dietary-supplements.info.nih.gov/. The website contains links to several of the dietary supplement consensus conferences, which have occurred over the last several years.

Several large projects funded by the National Institutes of Health have been reported recently. Specifically with regard to supplements, the Women's Health Initiative tested among other outcomes such as fracture risk whether a combined dose of calcium and vitamin D would reduce the incidence of colorectal or breast cancer in postmenopausal women, but no associations were reported for either cancer outcome *(27,180)*. The Selenium and Vitamin E for the Prevention of Prostate Cancer (SELECT) ended early due to the slight increased risk of diabetes and prostate cancer. The Physicians' Health Study II is a randomized trial of β-carotene, vitamin E, vitamin C, and multivitamins among healthy, male physicians to test whether these supplements will reduce the incidence of total and prostate cancers, as well as cardiovascular disease and eye diseases *(140)*. These and other investigations have provided important data on specific dietary supplements in relation to cancer risk that will be useful to both scientists and clinicians.

9.5. RECOMMENDATIONS

Increasing numbers of Americans are using dietary supplements. Health professionals are confronted regularly with questions regarding the efficacy of these compounds. Physicians must remain informed about research showing efficacy, harm, or no effect of dietary supplements in relation to cancer prevention.

Most supplement users believe that these compounds improve their health. In a small study on motivations and beliefs of supplement users ($n = 104$), we found that supplement users believed that multivitamins helped them feel better (41%), vitamin C prevents colds and/or flu (76%), and vitamin E and calcium prevent chronic disease (60–80%). Many participants felt that foods could not supply adequate amounts of certain nutrients, indicating that advice to "eat a balanced diet" would not reduce their motivation to take dietary supplements *(10)*.

Given the large numbers of Americans taking dietary supplements, we believe that it is important to formulate recommendations regarding their use. We believe the following recommendations regarding dietary supplements and cancer risk are consistent with the literature and are appropriate based on current knowledge:

- Results of randomized controlled trials clearly indicate that cigarette smokers, or other individuals at high risk for lung cancer, should not take β-carotene supplements. Healthy adults will receive no benefit from β-carotene supplementation.
- A daily multivitamin and mineral pill is likely neither harmful nor beneficial. If a multivitamin is used, doses should generally not exceed the daily value without specific clinical evidence suggesting the need to restore adequate status and medical monitoring for adverse effects.
- Vitamin C supplementation may reduce the risk of some cancers, particularly those of the GI tract and the bladder. Anecdotal evidence indicates that doses as high as 1 g/day are safe *(213)*. However, pharmacokinetic evidence suggests that there is a ceiling effect of oral vitamin C on plasma levels *(214)*. Therefore, higher doses of C may not offer an increase in benefit.
- There is evidence from observational studies that calcium supplements may reduce risk of colon cancer. However, results from the Women's Health Initiative randomized trial did not support the observational data *(27)*. Given the evidence that calcium might prevent age-related fractures, use of calcium supplements in the range of 500–1,000 mg/ day may be prudent for many Americans, particularly those with inadequate calcium intake from dietary sources *(215)*.
- There is emerging and compelling data to suggest a preventive role for vitamin D in relation to breast, prostate, and colon cancers *(21,215–218)*. In addition to cutaneous exposure to sunlight, many Americans obtain vitamin D either from milk products or multivitamins. Despite the increasing number of publications of animal and in vitro studies of the potential role for vitamin D in carcinogenesis, few consumers use single supplements of vitamin D that are not combined with calcium *(1)*. Thus, at this point in time it is difficult to disentangle the potential associations for vitamin D from that of other micronutrients concurrently ingested.
- There is conflicting evidence demonstrating that folic acid can both increase and decrease colon cancer risk *(124,191,193,194)*. Limitations of the currently published studies are that many of the outcomes are intermediate endpoints (e.g., adenomas). Further research is needed in this area.
- There is little epidemiologic research on other vitamin (e.g., vitamin A, vitamin B_1, B_2, B_6, and B_{12}), mineral (e.g., chromium, copper, magnesium, iron, zinc), and herbal supplements and cancer. Additionally, due to the risk of toxicity, we do not recommend high-dose supplementation of vitamins A, particularly for women of childbearing potential *(219)*. Given the possibility that iron increases cancer risk *(73,220)*, we would advise against large doses of this mineral for the purposes of preventing cancer. No recommendations are possible regarding other minerals.

Americans need a strong message that there are many bioactive compounds in foods, especially in fruits and vegetables, which likely play an important role in the prevention of cancer and other diseases *(31)*. Dietary supplements cannot replace the benefits obtained from eating a diet high in fruit and vegetables, nor can they reverse the damage caused by a low-fiber, high-fat diet. Finally, health-care providers should always ask patients about the use of dietary supplements. As noted in this report, many Americans use multiple supplements on a regular basis *(4,74)*. Clinicians must be aware of several issues including the potential for supplement–drug interactions, high monetary expenditures for supplements, and the tempting possibility for patients to replace (or substitute) important healthy behaviors, such as maintaining or losing weight, engaging in physical activity, eating a low-fat/high fruit and vegetable diet, and smoking cessation, with a dietary supplement pill. However, published evidence suggests that users of dietary supplements adhere to healthy lifestyle behaviors *(85)*. Additional research results will provide clinicians with information that may be useful in formulating public health recommendations about dietary supplements and cancer prevention.

REFERENCES

1. Radimer K, Bindewald B, Hughes J, et al. Dietary supplement use by US adults: data from the National Health and Nutrition Examination Survey, 1999–2000. Am J Epidemiol 2004;160:339–49.
2. Rock CL. Multivitamin-multimineral supplements: who uses them? Am J Clin Nutr 2007;85:277S–9S.
3. Kaufman DW, Kelly JP, Rosenberg L, et al. Recent patterns of medication use in the ambulatory adult population of the United States. JAMA 2002;287:337–44.
4. Patterson RE, Neuhouser ML, Hedderson MM, et al. Changes in diet, physical activity, and supplement use among adults diagnosed with cancer. J Am Diet Assoc 2003;103:323–8.
5. Burstein HJ, Gelber S, Guadagnoli E, et al. Use of alternative medicine by women with early-stage breast cancer. N Engl J Med 1999;340:1733–9.
6. Newman V, Rock CL, Faerber S, et al. Dietary supplement use by women at risk for breast cancer recurrence. The Women's Healthy Eating and Living Study Group. J Am Diet Assoc 1998;98:285–92.
7. Wiygul JB, Evans BR, Peterson BL, et al. Supplement use among men with prostate cancer. Urology 2005;66:161–6.
8. Dietary Supplement and Health Education Act of 1994 Public Law 103–417, Washington, DC 103rd Congress, 1994.
9. Neuhouser ML. Dietary supplement use by American women: challenges in assessing patterns of use, motives and costs. J Nutr 2003;133:1992S–6S.
10. Neuhouser ML, Patterson RE, Levy L. Motivations for using vitamin supplements. J Am Diet Assoc 1999;99:851–4.
11. Satia-Abouta J, Kristal AR, Patterson RE, et al. Dietary supplement use and medical conditions—the VITAL study. Am J Prev Med 2003;24:43–51.
12. The Alpha-Tocopherol Beta Carotene Cancer Prevention Study Group. The effect of vitamin E and beta carotene on the incidence of lung cancer and other cancers in male smokers. N Engl J Med 1994;330: 1029–35.
13. Omenn GS, Goodman GE, Thornquist MD, et al. Effects of a combination of beta carotene and vitamin A on lung cancer and cardiovascular disease. N Engl J Med 1996;334:1150–5.
14. Duffield-Lillico AJ, Slate EH, Reid ME, et al. Selenium supplementation and secondary prevention of nonmelanoma skin cancer in a randomized trial. J Natl Cancer Inst 2003;95:1411–81.
15. Klein EA. Selenium and Vitamin E Cancer Prevention Trial. Ann New York Acad Sci 2004;1031:234–41.
16. Giovannucci E, Rimm EB, Wolk A, et al. Calcium and fructose intake in relation to risk of prostate cancer. Cancer Res 1998;58:442–7.
17. Giovannucci E, Liu Y, Platz EA, et al. Risk factors for prostate cancer incidence and progression in the Health Professionals Follow-Up Study. Int J Cancer 2007;121.
18. Chan JM, Stampfer MJ, Ma J, et al. Dairy products, calcium, and prostate cancer risk in the Physician's Health Study. Am J Clin Nutr 2001;74:549–54.
19. Rodriguez C, McCullough M, Mondul A, et al. Calcium, dairy products, and risk of prostate cancer in a prospective cohort of United States men. Cancer Epidemiol Biomarkers Prev 2003;12:597–603.
20. Giovannucci E, Liu Y, Stampfer MJ, et al. A prospective study of calcium intake and incident and fatal prostate cancer. Cancer Epidemiol Biomarkers Prev 2006;15:203–10.

21. Lappe JM, Travers-Gustafson D, Davies KM, et al. Vitamin D and calcium supplementation reduces cancer risk: results of a randomized trial. Am J Clin Nutr 2007;85:1586–91.
22. Bostick RM, Potter JC, McKenzie DR, et al. Reduced risk of colon cancer with high intake of vitamin E: The Iowa Women's Health Study. Cancer Res 1993;53:4230–7.
23. Bostick RM, Potter JD, Sellers TA, et al. Relation of calcium, vitamin D, and dairy food intake to incidence of colon cancer among older women. Am J Epidemiol 1993;137:1302–17.
24. Wu K, Willett WC, Fuchs CA, et al. Calcium intake and risk of colon cancer in women and men. J Natl Cancer Inst 2002;94:437–46.
25. White E, Shannon JS, Patterson RE. Relationship between vitamin and calcium supplement use and colon cancer. Cancer Epidemiol Biomarkers Prev 1997;6:769–74.
26. McCullough ML, Robertson AS, Rodriguez C, et al. Calcium, vitamin D, dairy products, and risk of colorectal cancer in the Cancer Prevention Study II Nutrition Cohort (United States). Cancer Causes Control 2003;14: 1–12.
27. Wactawski-Wende J, Kotchen JM, Anderson GL, et al. Calcium plus vitamin D supplementation and the risk of colorectal cancer. The New England Journal of Medicine 2006;354:684–96.
28. Fletcher RH, Fairfield KM. Vitamins for chronic disease prevention in adults—Clinical Applications. JAMA 2002;287:3127–9.
29. Fairfield KM, Stampfer M. Vitamin and mineral supplements for cancer prevention: issues and evidence. Am J Clin Nutr 2007;85:289S–92S.
30. NIH State-of-the-Science Panel. National Institutes of Health State-of-the-Science Conference Statement: Multivitamin/mineral supplements and chronic disease prevention. Ann Intern Med 2006;145:364–71.
31. World Cancer Research Fund/American Institute of Cancer Research. Food, nutrition, physical activity and the prevention of cancer: a global perspective: American Institute for Cancer Research, 2007.
32. Kolonel LN, Hankin JH, Whittemore AS, et al. Vegetables, fruits, legumes and prostate cancer: a multiethnic case-control study. Cancer Epidemiol Biomarkers Prev 2000;9:795–804.
33. Kristal AR, Lampe JW. *Brassica* vegetables and prostate cancer risk: a review of the epidemiological evidence. Nutr Cancer 2002;42:1–9.
34. Murillo G, Mehta RG. Cruciferous vegetables and cancer prevention. Nutr Cancer 2001;41:17–28.
35. Lin J, Zhang SM, Cook NR, et al. Dietary intakes of fruit, vegetables, and fiber, and risk of colorectal cancer in a prospective cohort of women (United States). Cancer Causes Control 2005;16:225–33.
36. Lampe JW. Health effects of vegetables and fruit: assessing mechanisms of action in human experimental studies. Am J Clin Nutr 1999;70:475s–90s.
37. Bonnesen C, Eggleston IM, Hayes JD. Dietary indoles and isothiocyanates that are generated from cruciferous vegetables can both stimulate apoptosis and confer protection against DNA damage in human colon cell lines. Cancer Res 2001;61:6120–30.
38. Yetley EA. Multivitamin and multimineral dietary supplements: definitions, characterization, bioavailability, and drug interactions. Am J Clin Nutr 2007;85:269S–76S.
39. Dwyer JT, Holden J, Andrews K, et al. Measuring vitamins and minerals in dietary supplements for nutrition studies in the USA. Analytical and Bioanalytical Chemistry 2007;389:37–46.
40. Hendrich S, Fisher K. What do we need to know about active ingredients in dietary supplements? Summary of workshop discussion. J Nutr 2001;131:1387S–8S.
41. Konopacka M, Rzeszowska-Wolny J. Antioxidant vitamins C, E and β-carotene reduce DNA damage before as well as after γ-ray irradiation of human lymphocytes in vitro. Mutat Res 2001;491:1–7.
42. Fung TT, Hunter DJ, Spiegelman D, et al. Vitamins and carotenoids intake and the risk of basal cell carcinoma of the skin in women (United States). Cancer Causes Control 2002;13:221–30.
43. Cho E, Hunter DJ, Spiegelman D, et al. Intakes of vitamins A, C and E and folate and multivitamins and lung cancer: A pooled analysis of 8 prospective studies. Int J Cancer 2006;118:970–8.
44. Michaud DS, Pietinen P, Taylor PR, et al. Intakes of fruits and vegetables, carotenoids and vitamins A, E, C in relation to the risk of bladder cancer in the ATBC cohort study. Br J Cancer 2002;87:960–5.
45. Slatore CG, Littman AJ, Au DH, et al. Long-term use of supplemental multivitamins, vitamin C, vitamin E, and folate does not reduce the risk of lung cancer. Am J Respir Crit Care Med 2008;177:524–30.
46. Møller P, Loft S. Oxidative DNA damage in human white blood cells in dietary antioxidant intervention studies. Am J Clin Nutr 2002;36:303–10.
47. Cooke MS, Evans MD, Mistry N, et al. Role of dietary antioxidants in the prevention of in vivo oxidative DNA damage. Nutr Res Rev 2002;15:19–41.

48. Thompson HJ. DNA oxidation products, antioxidant status, and cancer prevention. J Nutr 2004;134:3186S–7S.
49. Keum YS, Yu S, Change PP-J, et al. Mechanism of action of sulforaphane: inhibition of p38 mitogen-activated protein kinase isoforms contributing to the induction of antioxidant response element-mediated heme oxygenase-1 in human hepatoma HepG2 cells. Cancer Res 2006;66:8804–13.
50. Hayes JD, McMahon M. Molecular basis for the contribution of the antioxidant responsive element to cancer chemoprevention. Cancer Lett 2001;174:103–13.
51. Institute of Medicine, Food and Nutrition Board. Dietary Reference Intakes for Vitamin C, Vitamin E, Selenium and Carotenoids. Washington, DC: National Academy Press, 2000.
52. Hossian MZ, Wiliens LR, Mehta PP, et al. Enhancement of gap junctional communication by retinoids correlates with their ability to inhibit neoplastic transformation. Carcinogenesis 1989;10:1743–8.
53. Torres AG, Borojevic R, Trugo NMF. β-Carotene is accumulated, metabolized, and possibly converted to retinol in human breast carcinoma cells (MCF-7). Int J Vitam Nutr Res 2004;74:171–7.
54. Dietary Reference Intakes for Vitamin A, Vitamin K, Arsenic, Boron, Chromium, Copper, Iodine, Iron, Manganese, Molybdenum, Nickel, Silicon, Vanadium, and Zinc. Washington, DC: National Academy Press, 2001.
55. Krinsky NI. The antioxidant and biological properties of the carotenoids. Ann N Y Acad Sci 1998;854:443–7.
56. Bertram JS. Carotenoids and gene regulation. Nutr Rev 1999;57:182–91.
57. Cui Y, Lu Z, Bai L, et al. β-Carotene induces apoptosis and up-regulates peroxisome proliferator-activated receptor γ expression and reactive oxygen species production in MCF-7 cancer cells. Eur J Cancer 2007;43:2590–601.
58. Prakash P, Russell RM, Krinsky NI. In vitro inhibition of proliferation of estrogen-dependent and estrogen-independent human breast cancer cells treated with carotenoids or retinoids. J Nutr 2001;131:1574–80.
59. Tibaduiza EC, Fleet JC, Russell RM, et al. Excentric cleavage products of beta-carotene inhibit estrogen receptor positive and negative breast tumor cell growth in vitro and inhibit activator protein-1-mediated transcriptional activation. J Nutr 2002;132:1368–75.
60. Hirsch K, Atzmon A, Danilenko M, et al. Lycopene and other carotenoids inhibit estrogenic activity of 17 β-estradiol and genistein in cancer cells. Breast Cancer Res Treat 2007;104:221–30.
61. Kline K, Yu W, Sanders BG. Vitamin E: mechanisms of action as tumor cell growth inhibitors. J Nutr 2001;131: 161S–3S.
62. Harnack L, Jacobs DR, Nicodemus K, et al. Relationship of folate, vitamin B-6, vitamin B-12, and methionine intake to incidence of colorectal cancers. Nutr Cancer 2002;43:152–8.
63. Purohit V, Abdelmalek MF, Barve S, et al. Role of S-adenosylmethionine, folate, and betaine in the treatment of alcoholic liver disease: summary of a symposium. Am J Clin Nutr 2007;86:14–24.
64. Goode EL, Ulrich CM, Potter JD. Polymorphisms in DNA repair genes and associations with cancer risk. Cancer Epidemiol Biomarkers Prev 2002;11:1513–30.
65. Alberts DS, Rittenbaugh C, Story JA, et al. Randomized, double-blinded, placebo-controlled study of effect of wheat bran fiber and calcium on fecal bile acids in patients with resected adenomatous colon polyps. J Natl Cancer Inst 1996;88:81–92.
66. Bostick RM, Potter JD, Fosdick L, et al. Calcium and colorectal epithelial cell proliferation: a preliminary randomized, double-blinded, placebo-controlled clinical trial. J Natl Cancer Inst 1993;85:132–41.
67. Welsh J, Wietzke J. Impact of the vitamin D3 receptor on growth-regulatory pathways in mammary gland and breast cancer. J Steroid Biochem Mol Biol 2003;83:85–92.
68. Holick MF. Vitamin D. In: Stipanuk MH, ed. Biochemical and Physiological Aspects of Human Nutrition. Philadelphia, PA: W.B. Saunders, 2000:624–36.
69. Guzey M, Kitada S, Reed JC. Apoptosis induction by 1α,25-Dihydroxyvitamin D_3 in prostate cancer. Mol Cancer Ther 2002;1:667–77.
70. Pike JW, Meyers M, Watanuki M, et al. Perspectives on mechanisms of gene regulation by 1,25-dihydroxyvitamin D3 and its receptor. J Steroid Biochem Mol Biol 2007;103:389–95.
71. Venkateswaran V, Klotz LH, Fleshner NE. Selenium modulation of cell proliferation and cell cycle biomarkers in human prostate carcinoma cell lines. Cancer Res 2002;92:2540–5.
72. Brigelius-Flohe R, Banning A. Part of the series: from dietary antioxidants to regulators in cellular signaling and gene regulation. Sulforaphane and selenium, partners in adaptive response and prevention of cancer. Free Radic Res 2006;40:775–87.
73. Choi J-Y, Neuhauser ML, Barnett MJ, et al. Iron intake, oxidative stress-related genes (*MnSOFandMPO)* and prostate cancer risk in CARET cohort. Carcinogenesis 2008;29:964–70.
74. White E, Patterson RE, Kristal AR, et al. VITamins And Lifestyle Cohort Study: study design and characteristics of supplement users. Am J Epidemiol 2004;159:83–93.

75. Women's Health Initiative Study Group. Design of the Women's Health Initiative Clinical Trial and Observational Study. Control Clin Trials 1998;19:61–109.
76. Patterson RE, Kristal AR, Tinker LF, et al. Measurement characteristics of the Women's Health Initiative food frequency questionnaire. Ann Epidemiol 1999;9:178–87.
77. Shikany JM, Patterson RE, Agurs-Collins T, et al. Antioxidant supplement use in Women's Health Initiative participants. Prev Med 2003;36:379–87.
78. Neuhouser ML, Kristal AR, Patterson RE, et al. Dietary supplement use in the Prostate Cancer Prevention Trial: implications for prevention trials. Nutr Cancer 2001;39:12–8.
79. Traber MG, Burton GW, Ingold KU, et al. RRR-and SRR-alpha-tocopherols are secreted without discrimination in human chylomicrons, but RRR-alpha-tocopherol is preferentially secreted in very low density lipoproteins. J Lipid Res 1990;31:675–85.
80. Giovannucci EL. γ–Tocopherol: a new player in prostate cancer prevention? J Natl Cancer Inst 2000;92:1966–7.
81. Omenn GS, Goodman GE, Thornquist MD, et al. The β-Carotene and Retinol Efficacy Trial (CARET) for chemoprevention of lung cancer in high risk populations: smokers and asbestos-exposed workers. Cancer Res 1994;54: 2038s–43s.
82. Patterson RE, White E, Kristal AR, et al. Vitamin supplements and cancer risk: a review of the epidemiologic evidence. Cancer Causes Control 1997;8:786–802.
83. Zhang SM, Giovannucci EL, Hunter DJ, et al. Vitamin supplement use and the risk of Non-Hodgkin's Lymphoma among women and men. Am J Epidemiol 2001;153:1056–63.
84. Wu K, Willett WC, Chan JM, et al. A prospective study on supplemental vitamin E intake and risk of colon cancer in women and men. Cancer Epidemiol Biomarkers Prev 2002;11:1298–304.
85. Patterson RE, Neuhouser ML, White E, et al. Cancer-related behavior of vitamin supplement users. Cancer Epidemiol Biomarkers Prev 1998;7:79–81.
86. Greenberg RE, Baron JA, Stukel TA, et al. A clinical trial of beta carotene to prevent basal-cell and squamous-cell cancers of the skin. N Engl J Med 1990;323:789–95.
87. Greenberg ER, Baron JA, Karagas MR, et al. Mortality associated with low plasma concentration of beta carotene and the effect of oral supplementation. JAMA 1996;275:699–703.
88. Blot WJ, Li JY, Taylor PR, et al. Nutrition intervention trials in Linxian, China: supplementation with specific vitamin/mineral combinations, cancer incidence, and disease-specific mortality in the general population. J Natl Cancer Inst 1993;85:1483–92.
89. Li J, Taylor PR, Li B, et al. Nutrition intervention trials in Linxian, China: multiple vitamin/mineral supplementation, cancer incidence, and disease-specific mortality among adults with esophageal dysplasia. J Natl Cancer Inst 1993;85:1492–8.
90. Hennekens CH, Buring JE, Manson JE, et al. Lack of effect of long-term supplementation with beta carotene on the incidence of malignant neoplasms and cardiovascular disease. N Engl J Med 1996;334:1145–9.
91. The Alpha-Tocopherol Beta Carotene Cancer Prevention Study Group. Incidence of cancer and mortality following α-tocopherol and β-carotene supplementation. A postintervention follow-up. JAMA 2003;290:476–85.
92. Heart Protection Study Collaborative Group. MRC/BHF Heart Protection Study of antioxidant vitamin supplementation in 20,536 high-risk individuals: a randomised placebo-controlled trial. Lancet 2002;360:23–33.
93. Lee IM, Cook NR, Manson JE, et al. β-carotene supplementation and incidence of cancer and cardiovascular disease: the Women's Health Study. J Natl Cancer Inst 1999;91:1202–6.
94. Hercberg S, Galan P, Preziosi P, et al. The SU.VI.MAX study: a randomized, placebo-controlled trial of the health effects of antioxidant vitamins and minerals. Arch Int Med 2004;164:2335–42.
95. Goodman GE, Thornquist MD, Balmes J, et al. The Beta-Carotene and Retinol Efficacy Trial: incidence of lung cancer and cardiovascular disease mortality during 6-year follow-up after stopping β-carotene and retinol supplements. J Natl Cancer Inst 2004;96:1743–50.
96. Neuhouser ML, Patterson RE, Thornquist MD, et al. Fruits and vegetables are associated with lower lung cancer risk only in the placebo arm of the β-Carotene and Retinol Efficacy Trial (CARET). Cancer Epidemiol Biomarkers Prev 2003;12:350–8.
97. Meyer F, Galan P, Douville P, et al. Antioxidant vitamin and mineral supplementation and prostate cancer prevention in the SU. VI. MAX trial. Int J Cancer 2005;116:182–186.
98. Hercberg S, Ezzedine K, Guinot C, et al. Antioxidant supplementation increases the risk of skin cancers in women but not in men. J Nutr 2007;137:2098–105.
99. Michaud DS, Feskanich D, Rimm EB, et al. Intake of specific carotenoids and risk of lung cancer in 2 prospective US cohorts. Am J Clin Nutr 2000;72:990–7.

100. Feskanich D, Ziegler RG, Michaud DS, et al. Prospective study of fruit and vegetable consumption and risk of lung cancer among men and women. J Natl Cancer Inst 2000;92:1812–23.
101. Norrish AE, Jackson RT, Sharpe SJ, et al. Prostate cancer and dietary carotenoids. Am J Epidemiol 2000;151: 119–23.
102. Wu K, Erdman JW, Schwartz SJ, et al. Plasma and the dietary carotenoids, and the risk of prostate cancer: a nested case-control study. Cancer Epidemiol Biomarkers Prev 2004;13:260–9.
103. Clark LC, Combs GF, Turnbull BW, et al. Effects of selenium supplementation for cancer prevention in patients with carcinoma of the skin. A randomized controlled trial. Nutritional Prevention of Cancer Study Group. JAMA 1996;276:1957–63.
104. Clark LC, Dalkin B, Krongrad A, et al. Decreased incidence of prostate cancer with selenium supplementation: results of a double-blind cancer prevention trial. Br J Urol 1998;81:730–4.
105. Platz EA, Helzlsour KJ. Selenium, zinc and prostate cancer. Epidemiol Rev 2001;23:93–101.
106. Helzlsouer KJ, Huang H-Y, Alberg AJ, et al. Association between α-tocopherol, γ-tocopherol, selenium and subsequent prostate cancer. J Natl Cancer Inst 2000;92:2018–23.
107. Lippman SM, Goodman PJ, Klein EA, et al. Designing the selenium and vitamin E cancer prevention trial (SELECT). J Natl Cancer Inst 2005;97:94–102.
108. The HOPE and HOPE-Too Trial Investigators, Lonn E, Bosch J, et al. Effects of long-term vitamin E supplementation on cardiovascular events and cancer: a randomized controlled trial. JAMA 2005;293:1338–47.
109. Lee I-M, Cook NR, Gaziano JM, et al. Vitamin E in the primary prevention of cardiovascular disease and cancer. The Women's Health Study: a randomized controlled trial. JAMA 2005;294:56–65.
110. Watkins ML, Erickson JD, Thun MJ, et al. Multivitamin use and mortality in a large prospective study. Am J Epidemiol 2000;152:149–62.
111. Losonczy KG, Harris TB, Havlik RJ. Vitamin E and vitamin C supplement use and risk of all-cause and coronary heart disease mortality in older persons: the Established Populations for Epidemiologic Studies of the Elderly. Am J Clin Nutr 1996;64:190–6.
112. Brown LM, Swanson CA, Gridley G, et al. Adenocarcinoma of the esophagus: role of obesity and diet. J Natl Cancer Inst 1995;87:104–9.
113. Rossing MA, Vaughan TL, McKnight B. Diet and pharyngeal cancer. Int J Cancer 1989;44:593–7.
114. Gridley G, McLaughlin JK, Block G, et al. Vitamin supplement use and reduced risk of oral and pharyngeal cancer. Am J Epidemiol 1992;135:1083–92.
115. Hunter DJ, Manson JE, Colditz GA, et al. A prospective study of the intake of vitamins C, E, and A and the risk of breast cancer. N Engl J Med 1993;329:234–40.
116. Ewertz M, Gill C. Dietary factors and breast-cancer risk in Denmark. Int J Cancer 1990;46:779–84.
117. Martinez ME, Giovannucci EL, Colditz GA, et al. Calcium, vitamin D, and the occurrence of colorectal cancer among women. J Natl Cancer Inst 1996;88:1375–82.
118. Ziegler RG, Brinton LA, Hamman RF, et al. Diet and the risk of invasive cervical cancer among white women in the United States. Am J Epidemiol 1990;132:432–45.
119. Ziegler RG, Jones CH, Brinton LA, et al. Diet and the risk of in situ cervical cancer among white women in the United States. Cancer Causes Control 1991;2:17–29.
120. Bruemmer B, White E, Vaughan TL, et al. Nutrient intake in relation to bladder cancer among middle-aged men and women. Am J Epidemiol 1996;144:485–95.
121. Hunter DJ, Colditz GA, Stampfer MJ, et al. Diet and risk of basal cell carcinoma of the skin in a prospective cohort of women. Ann Epidemiol 1992;2:231–9.
122. Kirkpatrick CS, White E, Lee JAH. Case-control study of malignant melanoma in Washington State. II. Diet, alcohol, obesity. Am J Epidemiol 1994;139:869–80.
123. Barone J, Tailoi E, Hebert JR, et al. Vitamin supplement use and risk for oral and esophageal cancer. Nutr Cancer 1992;18:31–41.
124. Giovannucci E, Stampfer M, Colditz GA. Multivitamin use, folate and colon cancer in women in the Nurses' Health Study. Ann Intern Med 1998;129:517–24.
125. Fuchs CS, Willett WC, Colditz GA, et al. The influence of folate and multivitamin use on the familial risk of colon cancer in women. Cancer Epidemiol Biomarkers Prev 2002;11:227–34.
126. Zhang S, Hunter DJ, Forman M. Dietary carotenoids and vitamins A, C and E and risk of breast cancer. J Natl Cancer Inst 1999;91:547–56.
127. Jacobs EJ, Connell CJ, McCullough ML, et al. Vitamin C, vitamin E, and multivitamin supplement use and stomach cancer mortality in the Cancer Prevention Study II cohort. Cancer Epidemiol Biomarkers Prev 2002;11:35–41.

128. Michaud DS, Spiegelman D, Clinton SK, et al. Prospective study of dietary supplements, macronutrients, and risk of bladder cancer in US men. Am J Epidemiol 2000;152:1145–53.
129. Mayne ST, Risch HA, Dubrow R, et al. Nutrient intake and risk of subtypes of esophageal and gastric cancer. Cancer Epidemiol Biomarkers Prev 2001;10:1055–62.
130. Moorman PG, Ricciuti MF, Millikan RC, et al. Vitamin supplement use and breast cancer in a North Carolina population. Public Health Nutr 2001;4:821–7.
131. Kristal AR, Stanford LL, H CJ, et al. Vitamin and mineral supplement use is associated with reduced risk of prostate cancer. Cancer Epidemiol Biomarkers Prev 1999;8:887–92.
132. Brown LM, Swanson CA, Gridley G, et al. Dietary factors and the risk of squamous cell esophageal cancer among black and white men in the United States. Cancer Causes Control 1998;9:467–74.
133. Farrow DC, Vaughan TL, Berwick M, et al. Diet and nasopharyngeal cancer in a low-risk population. Int J Cancer 1998;78:675–9.
134. Lawson KA, Wright ME, Subar AF, et al. Multivitamin use and risk of prostate cancer in the National Institutes of Health-AARP Diet and Health Study. J Natl Cancer Inst 2007;99.
135. Zhang SM, Moore SC, Lin J, et al. Folate, vitamin B6, multivitamin supplements, and colorectal cancer risk in women. Am J Epidemiol 2006;163:108–15.
136. Hu J, Mery L, Desmeules M, et al. Diet and vitamin or mineral supplementation and risk of rectal cancer in Canada. Acta Oncol 2007;46:342–54.
137. Hu J, Morrison H, DesMeules M, et al. Diet and vitamin or mineral supplementation and risk of colon cancer by subsite in Canada. Eur J Cancer Prev 2007;16:275–91.
138. Ishitani K, Lin J, Manson JE, et al. A prospective study of multivitamin supplement use and risk of breast cancer. Am J Epidemiol 2008;167:1197–1206.
139. Kamangar F, Qiao YL, Yu B, et al. Lung cancer chemoprevention: a randomized, double-blind trial in Linxian, China. Cancer Epidemiol Biomarkers Prev 2006;15:1562–4.
140. Christen WG, Gaziano JM, Hennekens CH. Design of Physicians' Health Study II—a randomized trial of beta-carotene, vitamins E and C, and multivitamins, in prevention of cancer, cardiovascular disease, and eye disease, and review of results of completed trials. Ann Epidemiol 2000;10:125–34.
141. Dorjgochoo T, Shrubsole MJ, Shu XO, et al. Vitamin supplement use and risk for breast cancer: the Shanghai Breast Cancer Study. Breast Cancer Res Treat 2008;111:269–78.
142. Jacobs EJ, Connell CJ, Patel AV, et al. Multivitamin use and colon cancer mortality in the Cancer Prevention Study II cohort (United States). Cancer Causes Control 2001;12:927–34.
143. Jacobs EJ, Conell CJ, Chao A, et al. Multivitamin use and colorectal cancer incidence in a US cohort: does timing matter? Am J Epidemiol 2003;158:621–8.
144. Stevens VL, McCullough ML, Diver WR, et al. Use of multivitamins and prostate cancer mortality in a large cohort of US men. Cancer Causes Control 2005;16:643–50.
145. Neuhouser ML, Wassertheil-Smoller S, Thomson C, et al. Multivitamin use and risk of cancer and cardiovascular disease in the Women's Health Initiative Cohorts. Arch Int Med 2009;169:294–304.
146. Shibata A, Paganini-Hill A, Ross PK, et al. Intake of vegetables, fruits, beta-carotene, vitamin C and vitamin supplements and cancer incidence among the elderly: a prospective study. Br J Cancer 1992;66:673–9.
147. Hansson L, Nyren O, Bergstrom R, et al. Nutrients and gastric cancer risk. A population-based case control study in Sweden. Int J Cancer 1994;57:638–44.
148. Rohan TE, Howe GR, Friedenreich CM, et al. Dietary fiber, vitamins A, C, and E, and risk of breast cancer: a cohort study. Cancer Causes Control 1993;4:29–37.
149. Kushi LJ, Fee RM, Sellers TA, et al. Intake of Vitamins A, C, E and postmenopausal breast cancer. Am J Epidemiol 1996;144:165–74.
150. Freudenheim JL, Marshall JR, Vena JE, et al. Premenopausal breast cancer risk and intake of vegetables, fruits, and related nutrients. J Natl Cancer Inst 1996;88:340–8.
151. Le Marchand L, Yoshizawa CN, Kolonel LN, et al. Vegetable consumption and lung cancer risk: a population-based case-control study in Hawaii. J Natl Cancer Inst 1989;81:1158–64.
152. Jain M, Burch JD, Howe GR, et al. Dietary factors and risk of lung cancer: results from a case-control study, Toronto, 1981–1985. Int J Cancer 1990;86:33–8.
153. Wu HA, Paganini-Hill A, Ross RK, et al. Alcohol, physical activity and other risk factors for colorectal cancer: a prospective study. Br J Cancer 1987;55:687–94.
154. Steineck G, Hagman U, Gerhardsson M, et al. Vitamin A supplements, fried foods, fat and urothelial cancer, A case-referent study in Stockholm in 1985–87. Int J Cancer 1990;45:1006–11.

155. Nomura AMY, Kolonel LN, Hankin JH, et al. Dietary factors in cancer of the lower urinary tract. Int J Cancer 1991;48:199–205.
156. Barbone F, Austin H, Partridge EE. Diet and endometrial cancer: a case-control study. Am J Epidemiol 1993;137: 393–409.
157. Jacobs EJ, Connell CJ, Patel AV, et al. Vitamin C and vitamin E supplement use and colorectal cancer mortality in a large American Cancer Society cohort. Cancer Epidemiol Biomarkers Prev 2001;10:17–23.
158. Jacobs EJ, K HA, Briggs PJ, et al. Vitamin C and vitamin E supplement use and bladder cancer mortality in a large cohort of US men and women. Am J Epidemiol 2002;156:1002–10.
159. Fleischauer AT, Olson SH, Mignone L, et al. Dietary antioxidants, supplements, and risk of epithelial ovarian cancer. Nutr Cancer 2001;40:92–8.
160. Schuurman A, Goldbohm RA, Brants HAM, et al. A prospective cohort study on intake of retinol, vitamins C and E, and carotenoids and prostate cancer risk (Netherlands). Cancer Causes Control 2002;13: 573–82.
161. Lee DH, Jacobs Jr DR. Interaction among heme iron, zinc, and supplemental vitamin C intake on the risk of lung cancer: Iowa Women's Health Study. Nutr Cancer 2005;52:130–7.
162. Kirsh VA, Hayes RB, Mayne ST, et al. Supplemental and dietary vitamin E, β-carotene, and vitamin C intakes and prostate cancer risk. J Natl Cancer Inst 2006;98:245–54.
163. Brzozowska A, Kaluza J, Knoops KTB, et al. Supplement use and mortality: the SENECA study. Eur J Nutr 2008;47:131–7.
164. Cui Y, Shikany JM, Liu S, et al. Selected antioxidants and risk of hormone receptor-defined invasive breast cancers among postmenopausal women in the Women's Health Initiative Observational Study. Am J Clin Nutr 2008;87: 1009–18.
165. Messerer M, Hakansson N, Wolk A, et al. Dietary supplement use and mortality in a cohort of Swedish men. Br J Nutr 2008;99:626–31.
166. Mayne ST, Cartmel B, Baum M, et al. Randomized trial of supplemental β-carotene to prevent second head and neck cancer. Cancer Res 2001;61:1457–63.
167. Heinonen OP, Albanes D, Virtamo J, et al. Prostate cancer and supplementation with α-tocopherol and β-carotene: incidence and mortality in a controlled trial. J Natl Cancer Inst 1998;90:440–6.
168. Mayne ST, Janerich DT, Greenwald P, et al. Dietary beta-carotene and lung cancer risk in U.S. nonsmokers. J Natl Cancer Inst 1994;86:33–8.
169. Day GL, Blot WJ, Austin DF, et al. Racial differences in risk of oral and pharyngeal cancer: alcohol, tobacco, and other determinants. J Natl Cancer Inst 1993;88:340–8.
170. Slattery ML, Edwares SL, Anderson K, et al. Vitamin E and colon cancer: Is there an association? Nutr Cancer 1998;30:201–6.
171. Chan J, Stampher MJ, Ma J, et al. Supplemental vitamin E intake and prostate cancer risk in a large cohort of men in the United States. Cancer Epidemiol Biomarkers Prev 1999;8:893–9.
172. Peters U, Littman AJ, Kristal AR, et al. Vitamin E and selenium supplementation and risk of prostate cancer in the VITamins And Lifestyle (VITAL) study cohort. Cancer Causes Control 2008;19:75–87.
173. The ATBC Cancer Prevention Study Group. The alpha-tocopherol, beta-carotene lung cancer prevention study: design, methods, participant characteristics, and compliance. Ann Epidemiol 1994;4:1–10.
174. Kampman E, Goldbohm RA, van den Brandt PA, et al. Fermented dairy products, calcium, and colorectal cancer in the Netherlands cohort study. Cancer Res 1994;54:3186–90.
175. Zheng W, Anderson KE, Kushi LH, et al. A prospective cohort study of intake of calcium, vitamin D, and other micronutrients in relation to incidence of rectal cancer among postmenopausal women. Cancer Epidemiol Biomarkers Prev 1998;7:221–5.
176. Goodman MT, Wu AH, Tung K-H, et al. Association of dairy products, lactose, and calcium with the risk of ovarian cancer. Am J Epidemiol 2002;156:148–57.
177. Kristal AR, Cohen JH, Qu P, et al. Associations of energy, fat, calcium and vitamin D with prostate cancer risk. Cancer Epidemiol Biomarkers Prev 2002;11:719–25.
178. Marcus PM, Newcomb PA. The association of calcium and vitamin D, and colon and rectal cancer in Wisconsin women. Int J Epidemiol 1998;27:788–93.
179. Shin M-Y, Holmes MD, Hankinson SE, et al. Intake of dairy products, calcium, and vitamin D and risk of breast cancer. J Natl Cancer Inst 2002;94:1301–11.
180. Chlebowski RT, Johnson KC, Kooperberg C, et al. Calcium plus vitamin d supplementation and the risk of breast cancer. J Natl Cancer Inst 2008;100:1581–91.

181. Knight JA, Lesosky M, Barnett H, et al. Vitamin D and reduced risk of breast cancer: a population-based case-control study. Cancer Epidemiol Biomarkers Prev 2007;16:422–9.
182. McCullough ML, Rodriguez C, Diver WR, et al. Dairy, calcium, and vitamin D intake and postmenopausal breast cancer risk in the Cancer Prevention Study II Nutrition Cohort. Cancer Epidemiol Biomarkers Prev 2005;14: 2898–904.
183. Flood A, Peters U, Chatterjee N, et al. Calcium from diet and supplements is associated with reduced risk of colorectal cancer in a prospective cohort of women. Cancer Epidemiol Biomarkers Prev 2005;14: 126–32.
184. Koh KA, Sesso HD, Paffenbarger Jr RS, et al. Dairy products, calcium and prostate cancer risk. Br J Dermatol 2006;95:1582–5.
185. Baron JA, Beach M, Wallace K, et al. Risk of prostate cancer in a randomized clinical trial of calcium supplementation. Cancer Epidemiol Biomarkers Prev 2005;14:586–9.
186. Davis CD, Hartmuller V, Freedman DM, et al. Vitamin D and cancer: current dilemmas and future needs. Nutr Rev 2007;65:S71–4.
187. Reed MC, Nijhout HF, Neuhouser ML, et al. A mathematical model gives insights into nutritional and genetic aspects of folate-mediated one-carbon metabolism. J Nutr 2006;136:2653–61.
188. Pool-Zobel BL, Bub A, Liegibel UM, et al. Mechanisms by which vegetable consumption reduces genetic damage in humans. Cancer Epidemiol Biomarkers Prev 1998;7:891–9.
189. Pool-Zobel BL, Abrahamse SL, Collins AR, et al. Analysis of DNA strand breaks, oxidized bases, and glutathione *S*-transferase P1 in human colon cells from biopsies. Cancer Epidemiol Biomarkers Prev 1999;8:609–14.
190. Lin X, Tascilar M, Lee WH, et al. GSTP1 CpG island hypermethylation is responsible for the absence of GSTP1 expression in human prostate cancer cells. Am J Path 2001;159:1815–26.
191. Kim Y-I. Folic acid supplementation and cancer risk: Point. Cancer Epidemiol Biomarkers Prev 2008;17.
192. Ulrich CM, Potter JD. Folate supplementation: too much of a good thing? Cancer Epidemiol Biomarkers Prev 2006;15:189–93.
193. Cole BF, Baron JA, Sandler RS, et al. Folic acid for the prevention of colorectal adenomas: a randomized clinical trial. JAMA 2007;297:2351–9.
194. Ulrich CM. Folate and cancer prevention—Where to next? Counterpoint. Cancer Epidemiol Biomarkers Prev 2008;17:2226–30.
195. Neuhouser ML, Tinker L, Shaw PA, et al. Use of recovery biomarkers to calibrate nutrient consumption self-reports in the Women's Health Initiative. Am J Epidemiol 2008;167:1247–59.
196. Subar A, Kipnis V, Troiano RP, et al. Using intake biomarkers to evaluate the extent of dietary misreporting in a large sample of adults: the OPEN Study. Am J Epidemiol 2003;158:1–13.
197. Prentice RL. Measurement error and results from analytic epidemiology: dietary fat and breast cancer. J Natl Cancer Inst 1996;88:1738–47.
198. Satia JA, King IB, Morris JS, et al. Toenail and plasma levels as biomarkers of selenium exposure. Ann Epidemiol 2006;16:53–58.
199. Satia-Abouta J, Patterson RE, King IB, et al. Reliability and validity of self-report of vitamin and mineral supplement use in the VITamins and Lifestyle Study. Am J Epidemiol 2003;157:944–54.
200. Patterson RE, Kristal AR, Levy L, et al. Validity of methods used to assess vitamin and mineral supplement use. Am J Epidemiol 1998;148:643–9.
201. Patterson RE, Levy L, Tinker LF, et al. Evaluation of a simplified vitamin supplement inventory developed for the Women's Health Initiative. Public Health Nutr 1999;2:273–6.
202. Patterson RE, Neuhouser ML, White E, et al. Measurement error from assessing use of vitamin supplements at one point in time. Epidemiology 1998;9:567–9.
203. Dwyer JT, Picciano MF, Raiten DJ. Food and dietary supplement databases for What We Eat in America-NHANES. J Nutr 2003;133:624S–34S.
204. Roseland J, Holden JM, Andrews KW, et al. Dietary supplement ingredient database (DSID): Preliminary USDA studies on the composition of adult multivitamin/mineral supplements. J Food Composition Anal 2008;21:S69–77.
205. Block G, Cox G, Madans J, et al. Vitamin supplement use, by demographic characteristics. Am J Epidemiol 1988;127:297–309.
206. Lyle BJ, Mares-Perlman JA, Klein BEK, et al. Supplement users differ from nonusers in demographic, lifestyle, dietary and health characteristics. J Nutr 1998;128:2355–62.
207. Hoggatt KJ, Bernstein L, Reynolds P, et al. Correlates of vitamin supplement use in the United States: data from the California Teachers Study cohort. Cancer Causes Control 2002;13:735–40.

208. Patterson RE, Kristal AR, Lynch JC, et al. Diet-cancer related beliefs, knowledge, norms and their relationship to healthful diets. J Nutr Educ 1995;27:86–92.
209. Patterson RE, Kristal AR, White E. Do beliefs, knowledge, and perceived norms about diet and cancer predict dietary change? Am J Public Health 1996;86:1394–400.
210. Harnack L, Block G, Subar A, et al. Associations of cancer prevention-related nutrition knowledge, beliefs and attitudes to cancer prevention dietary behavior. J Am Diet Assoc 1997;97:957–65.
211. Jasti S, Siega-Riz AM, Bentley ME. Dietary supplement use in the context of health disparities: cultural, ethnic and demographic determinants of use. J Nutr 2003;133:2010S–3S.
212. Costello RB, Coates P. In the midst of confusion lies opportunity: fostering quality science in dietary supplement research. J Am Coll Nutr 2001;20:21–5.
213. Diplock AT. Safety of antioxidant vitamins and beta-carotene. Am J Clin Nutr 1995;62:1510S–6S.
214. Blanchard J, Tozer TN, Rowland M. Pharmacokinetic perspectives in megadoses of ascorbic acid. Am J Clin Nutr 1997;66:1165–70.
215. Holick MF. Vitamin D: importance in the prevention of cancers, type 1 diabetes, heart disease, and osteoporosis. Am J Clin Nutr 2004;79:362–71.
216. Holick MF. Vitamin D: the underappreciated D-lightful hormone that is important for skeletal and cellular health. Curr Opin Endocrinol Diabetes 2002;9:87–98.
217. Giovannucci E, Liu Y, Rimm EB, et al. Prospective study of predictors of vitamin D status and cancer incidence and mortality in men. J Natl Cancer Inst 2006;98:451–9.
218. Garland CF, Gorham ED, Mohr SB, et al. Vitamin D and prevention of breast cancer: Pooled analysis. J Steroid Biochem Mol Biol 2007;103:708–11.
219. Kaegi E. Unconventional therapies for cancer: 5. Vitamins A, C and E. Can Med Assoc J 1998;158:1483–8.
220. Bird CL, Witte JS, Swendseid ME, et al. Plasma ferritin, iron intake, and the risk of colorectal polyps. Am J Epidemiol 1996;144:34–41.

III Cardiovascular Disease Prevention

10 *N*–3 Fatty Acids from Fish and Plants: Primary and Secondary Prevention of Cardiovascular Disease

William E. Connor and Sonja L. Connor

Key Points

- Animal studies, epidemiological studies, and clinical trials have shown that fish and fish oil may reduce sudden death by preventing cardiac arrhythmias.
- *N*–3 fatty acids have an antithrombotic effect through the diminution of thromboxane A_2 that produces platelet aggregation and vasoconstriction.
- EPA and DHA have been shown to inhibit atherosclerosis, probably because of their suppression of cellular growth factors that inhibit the proliferation of smooth muscle cells.
- *N*–3 fatty acids lower very-low-density lipoprotein (VLDL) and triglyceride through depression of synthesis of triglyceride in the liver. *N*–3 fatty acids also suppress postprandial lipemia, which reduces chylomicron remnants that are atherogenic.
- Fish oil does not adversely affect glucose control in patients with diabetes.
- *N*–3 fatty acids have been uniformly associated with a mild decrease in systolic blood pressure and at times a decrease in diastolic blood pressure.
- The intake of *n*–3 fatty acids should be increased to prevent deaths from coronary heart disease.

Key Words: Algal oil; arrhythmia; coronary heart disease; fish oil; glucose; platelet; thrombosis; triglycerides

10.1. INTRODUCTION

Fish and fish oils contain the very-long-chained and highly polyunsaturated *n*–3[1] fatty acids which are derived from phytoplankton, the base of the food chain in the oceans, lakes, and rivers *(1)*. Phytoplankton synthesize the *n*–3 fatty acids, eicosapentaenoic (20:5) (EPA) and docosahexaenoic acids (22:6) (DHA), which are subsequently incorporated into fish, shellfish, and sea mammals. The plants synthesize an *n*–3 fatty acid, linolenic acid (18:3), that can be converted by the body to EPA and more slowly to DHA *(2,3)*. The *n*–3 fatty acids have profound biological and biochemical effects in the body. Despite a wealth of scientific information [a review listed over 120 references about cardiovascular effects alone *(4)*], clinical interest in *n*–3 fatty acids has not been high in the United States despite considerable attention to their use in Europe and Japan. This chapter will focus on the considerable and underappreciated potential benefits of the *n*–3 fatty acids in cardiovascular disease.

In the 1950s, it was discovered that polyunsaturated vegetable oils containing *n*–6 linoleic acid had a pronounced plasma cholesterol-lowering effect, yet the mechanism of this action has remained obscure *(1)*. In those early days, it was noted that fish oil, which was also polyunsaturated, had a

A. Bendich, R.J. Deckelbaum (eds.), *Preventive Nutrition*, Nutrition and Health, DOI 10.1007/978-1-60327-542-2_10,

similar hypocholesterolemic effect. No mention was made of the fact that fish oil contained very-long-chain *n*–3 fatty acids (C20:5 and C22:6) and that these might act differently than the *n*–6 fatty acid of vegetable oils, such as linoleic acid (C18:2). These early data about fish oil lay fallow until the pioneering observations of Dyerberg and Bang focused special attention on the *n*–3 fatty acids, eicosapentaenoic acid (EPA, 20:5) and docosahexaenoic acid (DHA, 22:6), found in marine oils *(5)*. They observed a lower coronary mortality among the Greenland Eskimos whose diet was especially rich in marine oils compared to Danish people eating a high-saturated-fat diet *(6)*. Later it was found that not only were these *n*–3 fatty acids cholesterol lowering, but, in addition, they had a profound plasma triglyceride-lowering effect, especially in hypertriglyceridemic patients *(7–9)*. Over two decades of research in humans, animals, perfused organs, and tissue cultures have firmly documented the mechanisms of the hypolipidemic actions of these *n*–3 fatty acids from fish and, furthermore, have demonstrated that these fatty acids have many other beneficial effects in cardiovascular disease.

This chapter will focus on seven different areas of research, which will help to answer the question about the potential benefits of *n*–3 fatty acids from fish oil on primary and secondary prevention of cardiovascular disease. These are listed below and will be discussed in detail subsequently. We first will discuss how *n*–3 fatty acids will prevent further events in those patients who already have coronary heart disease (secondary prevention). This is especially relevant to staving off fatal arrhythmias of the heart and thrombosis. Then we will discuss how *n*–3 fatty acids might prevent coronary disease in healthy individuals, especially in those with certain risk factors.

Secondary Prevention of Coronary Heart Disease:

- Antiarrhythmic actions
- Thrombosis

Primary Prevention of Coronary Heart Disease:

- Experimental animal studies to inhibit the growth of atherosclerotic plaques
- Lipid and lipoprotein disorders
- Diabetes mellitus
- Hypertension

10.2. ANTIARRHYTHMIC ACTIONS

10.2.1. Animal Studies

Sudden death from ventricular arrhythmias is a much-dreaded complication in patients with coronary heart disease. Several experimental studies have addressed this problem with the use of *n*–3 fatty acids from fish oil. McLennan et al. used coronary artery ligation in the rat to produce an in vivo model of ventricular fibrillation and myocardial infarction *(10)*. They found that the number of ventricular ectopic beats and duration of tachycardia or fibrillation was increased when the rats were fed sheep kidney fat (a saturated fat) when compared to rats fed tuna fish oil, a rich source of *n*–3 fatty acids. The rats fed tuna fish oil had a significantly reduced incidence and severity of arrhythmias. In another animal study, ventricular fibrillation was prevented by fish oil during both the occlusion of the coronary artery and the reperfusion *(11)*.

In other experiments, Hallaq et al. *(12)* have used isolated neonatal cardiac myocytes (from hearts of 1-day-old rats) as a model for the study of cardiac arrhythmogenic factors that are modified by *n*–3 fatty acids. They incubated isolated myocytes (for 3–5 days) in a culture medium enriched with arachidonic acid (AA) or EPA (20:5 *n*–3). The AA-enriched myocytes developed a toxic cytosolic calcium concentration on exposure to ouabain, whereas EPA-enriched myocytes preserved physiologic calcium levels. An increase of EPA in the membrane phospholipids was demonstrated with a small reduction in arachidonic acid in myocytes fed EPA. A second study by the same researchers further indicated the mechanism of action of the fish oil fatty acids in preventing the arrhythmias of these isolated myocytes *(13)*. It was found that *n*–3 fatty acids prevented a calcium-depleted state in the myocytes caused by the l-type calcium channel blocker nifedipine. The protective effects of the *n*–3 fatty acids appeared to result from their modulatory effects on nifedipine-sensitive l-type calcium channels. In a recent study, dogs were given intravenously pure EPA, DHA, or α-linolenic acid *(14)*. Tests were performed in a dog model of sudden death. With infusion of EPA, five of seven dogs did not have a fatal ventricular arrhythmia ($p < 0.02$); with DHA and α-linolenic acid, six of eight dogs in each group were protected ($p < 0.004$ for each). These studies indicated a definite beneficial effect of dietary *n*–3 fatty acids on the heart; both marine and plant sources of these fatty acids prevented cardiac arrhythmias in animals.

10.2.2. Population Studies

The epidemiological data about dietary *n*–3 fatty acids and coronary disease are extensive and go back to the initial observations of Dyerberg and Bang, who found a much lower rate of coronary heart disease in the Greenland Eskimos compared with Danes *(6)*. They deduced by means of extensive studies that it was the *n*–3 fatty acid content of the Eskimo diet that inhibited the atherosclerotic disease despite the fact that the Eskimo diet was a high-cholesterol, high-fat diet *(5)*. Their dietary fat, instead of being pathogenic, was protective since it was derived from the seas (fish, seal, etc.) and contained *n*–3 fatty acids. Furthermore, autopsy studies revealed that atherosclerosis in Alaskan Eskimos was much less than atherosclerosis found in Caucasians living in Alaska *(15)*.

A number of studies correlating fish consumption (providing *n*–3 fatty acids) and the mortality from coronary heart disease have been carried out *(4)*. In Dutch men the mortality from coronary heart disease was more than 50% lower among those who consumed at least 30 g of fish per day than among those who did not eat fish *(16)*. In the MRFIT trial, *n*–3 fatty acid consumption correlated inversely with all-cause mortality and coronary mortality *(17)*. Even in the Harvard Health Professionals Follow-up Study of 51,529 men, the consumption of one to two servings of fish per week was associated with a lower incidence of coronary heart disease *(18)*. Some of these studies will be described in greater detail later.

More recently, fatty fish consumption was associated with prevention of cardiac arrest from ventricular fibrillation in coronary patients. Ventricular fibrillation is the cause of death in most patients with coronary heart disease and it accounts for 20–30% of people whose first indication of coronary disease is cardiac arrest. A study from the University of Washington compared the effects of eating fish with the incidence of cardiac arrest *(19)*. There was a 50% reduction in the risk of cardiac arrest in people who consumed at least one fatty fish meal per week. A typical fatty fish would be salmon. Other fatty fish include sardines, mackerel, and Chilean sea bass. Even those who consumed a less fatty fish, such as tuna, had benefit because all fish and shellfish contain the beneficial *n*–3 fatty acids.

This protection against cardiac arrest occurred from the *n*–3 fatty acids (EPA and DHA) in the fat of fish. The effects of eating fish were reflected biochemically in the fatty acids of the red blood cells. If the red blood cells had a relatively low level of *n*–3 fatty acids, 3.3% of total fatty acids, there was a much greater risk of cardiac arrest than in those individuals whose red blood cell *n*–3 fatty acids were 5% or more of the total fatty acids. In other words, there was a 70% reduction in the risk of cardiac arrest in those people with the higher red blood cell *n*–3 fatty acid content *(19)*. Likewise, the 20,551 men aged 40–84 years in the US Physicians Health Study had a 52% reduction in the risk of sudden cardiac death in those who consumed fish at least once a week *(20)*.

Data from 76,283 women in the Nurses' Health Study showed that a daily intake of 1.1 g or more of α-linolenic acid (18:3 *n*–3) protected against fatal ischemic heart disease and that this protection probably resulted from an antiarrhythmic effect of α-linolenic acid *(21)*. However, the protective effect of α-linolenic acid did not extend to nonfatal myocardial infarction, for which there was a nonsignificant trend for an effect.

10.2.3. Clinical Trials

These epidemiological data are buttressed by a randomized controlled clinical trial in 2,333 men who had recovered from myocardial infarction and who were then asked to increase their intake of fatty fish or take fish oil *(22)*. There was a 29% reduction in the 2-year all-cause mortality in subjects advised to eat fatty fish and there was also a reduction in deaths from ischemic heart disease, but no reduction in nonfatal myocardial infarction. This was the first intervention trial in which all-cause mortality was reduced in a coronary intervention program. One likely reason for the reduction in coronary mortality was the decrease in cardiac arrest as documented by Siscovick et al. *(19)*. Men who ate fatty fish at least once a week had a 50% reduction in cardiac arrest, which probably resulted from the antiarrhythmic action of the *n*–3 fatty acids discussed earlier.

In a second clinical trial, 223 patients with angiographically proven coronary artery disease received about 1.5 g/day of *n*–3 fatty acids from fish oil concentrate for 2 years *(23)*. The progression of coronary disease was significantly but modestly decreased compared to the control group. Thus, the evidence becomes stronger and stronger that even some fish (or *n*–3 fatty acid consumption from fish oil) on a consistent basis (at least one serving a week) may prevent many deaths from coronary heart disease.

However, the most recent and largest test of fish oil in coronary patients occurred in Italy as the GISSI-Prevenzione Trial in 11,324 patients who had survived a recent myocardial infarction *(24)*. These patients were divided into four groups and were randomly assigned daily supplements of 1 g *n*–3 polyunsaturated fatty acids and 300 mg vitamin E or no supplements (control). The amount of fish oil utilized provided approximately 0.85 g of EPA and DHA in ethyl esters in a ratio of 1:2. There was no placebo control group. The duration of the supplementation was 3½ years. In this massive trial, treatment with the fish oil *n*–3 polyunsaturated fatty acids, but not vitamin E, significantly lowered the risk of death (14–20% less) and of cardiovascular death (17–30% less, dependent on the type of statistical analyses). The deaths and cardiovascular events included nonfatal myocardial infarction and stroke. Plasma triglycerides were lowered in the *n*–3 polyunsaturated fatty acid-treated patients.

This 1999 study was reanalyzed in 2002 *(25)*. It was shown that the risk of sudden death was significantly prevented by *only* 3 months of treatment with fish oil and there was a 67% reduction in the overall deaths. This benefit continued for 3½ years to the end of the study. It is important to recognize that the benefit from fish oil occurred early. Instead of dying from sudden death after the onset of a myocardial infarction, fish oil-treated men survived in greater numbers.

Several conclusions may be drawn from this study. Fish oil was given over the relatively long time period of 3½ years and it was safe. The authors expected a greater reduction in deaths as occurred in the DART study in Wales (29%). They attributed their less-positive, but still positive, results to a number of coexisting factors. One problem was the relatively low death rate in the control group. To be noted is the fact that all of the patients were consuming the Mediterranean diet that in itself is associated with a lowered death rate from coronary heart disease, perhaps in part from more antioxidant consumption and more fish. A second possible problem was the ratio of EPA to DHA, which in most other studies was 1.5–2.0 but which in this study was 0.5. EPA may be the more active of the two *n*–3 polyunsaturated fatty acids in its clinical effects; i.e., a direct antagonist of the cyclooxygenase enzyme that produces thromboxane A_2 in platelets. It was felt that the reduction in death rate occurred largely because of the prevention of sudden death from cardiac arrhythmias. This is the first large-scale trial proving that a low dose of fish oil (*n*–3 polyunsaturated fatty acids) over a 3½ year period of time saved lives. The evidence about the cardioprotective effects of fish oil fatty acids is steadily mounting.

The Lyon Diet Heart Study, a randomized secondary prevention trial (following a first myocardial infarction), showed a remarkable 76% reduction in risk of cardiac death and nonfatal heart attack after 4 years *(26,27)*. Subjects followed a diet high in canola oil and canola oil margarine (high in the plant *n*–3 fatty acid α-linolenic acid). The diet was also low in fat and high in complex carbohydrate and fiber. Very recent data indicate that α-linolenic acid may have a direct effect on cardiac arrhythmias, while other cardiovascular effects are likely mediated through the synthesis of EPA and DHA *(2)*. Conversion of α-linolenic acid to eicosapentaenoic acid is fairly rapid and measurable within a few days after the dietary ingestion of linolenic acid and consists of desaturation, elongation, and further desaturation, with the rate-limiting enzyme step being the first desaturation step brought about by Δ^6-desaturase.

Even in Japan, which has a low coronary death rate and a high intake of fish, a clinical trial of fish oil had a positive result *(28)*. There were two groups of hypercholesterolemic patients. One group received 1,800 mg of EPA per day as an ethyl ester plus a statin drug. The control group received only a statin. The EPA + statin group had a 19% reduction in major coronary events compared to the statin-only group, $p = 0.011$. Note that this purified fish oil preparation did not contain any DHA, unlike other clinical trials of fish oil. The EPA group did have more hemorrhagic manifestations. The all-cause deaths were similar in the two groups.

The clinical prevention of sudden cardiac death by *n*–3 polyunsaturated fatty acids and the mechanism of this prevention are summarized as follows *(29)*. Both the clinical and animal studies showing the antiarrhythmic effects of *n*–3 PUFA convey exactly the same message: namely that fish oil fatty acids are a powerful but simple modality to prevent the 300,000 episodes of sudden death occurring in the United States annually. The mechanism of the antiarrhythmic actions is to modulate ion channels so as to stabilize the cardiac myocytes electrically. Fatty acids act to inhibit the fast, voltage-dependant sodium current and the l-type calcium currents. It was also suggested that the electrical activity in brain neurons is similarly modulated by the *n*–3 fatty acids, quieting action on all excitable tissues, including the neurons of the central nervous system.

Are there contraindications to the use of fish oil to prevent cardiac arrhythmia? The answer to that question comes from the clinical trials in patients with implanted defibrillators, which seemed to be ideal to test further the fish oil hypothesis. The results from all three studies did not suggest benefit *(30–32)*. In fact, in one study the incidence of ventricular tachycardia was greater in the fish oil group compared to the control group given olive oil *(31)*. We conclude that patients with implanted defibrillators should not be given fish oil.

10.3. THROMBOSIS

N–3 fatty acids from fish oil have invariably had an antithrombotic effect, particularly a diminution in thromboxane A_2 that produces platelet aggregation and vasoconstriction *(1,33,34)*. Platelet reactivity and adhesion were, therefore, considerably reduced after fish oil ingestion *(35)*. There have also been reductions of PAI-1, fibrinogen, TPA and increases in platelet survival and bleeding time *(4)*. Enhanced fibrinolysis has also been observed *(4)*. Perhaps even more significant was a study in baboons showing that *n*–3 fatty acids eliminated both vascular thrombus formation and vascular lesions after vascular injury *(36)*. The baboons treated with fish oil showed decreases in thrombus formation at sites of surgical carotid endarterectomy. The intake of α-linolenic acid either has no effect or leads to decreased platelet aggregation when compared with linoleic acid *(37,38)*. The function of the endothelium, important in both thrombosis and atherosclerosis, is affected by *n*–3 fatty acids *(33)*. The production of prostacyclin is enhanced and endothelial-derived relaxation factor (EDRF), or nitric oxide, which is depressed in atherosclerotic disease, is greatly increased by the *n*–3 fatty acids of fish oil *(39–41)*. Eight patients with coronary artery disease received 1.8 g/day EPA for 6 weeks *(42)*. EPA had beneficial effects on both nitric oxide-dependent and nondependent forearm vasodilatation. The production of prostacyclin was increased in rats fed 2.5 and 5% linseed oil *(43)*. Fish oil supplementation has improved arterial compliance in diabetic subjects *(44)*. Fish oil supplementation also inhibits norepinephrine-medicated vasoconstriction that also increases arterial compliance *(45)*. Fifteen obese people with insulin resistance were fed four diets of 4 weeks each. The 20 g/day α-linolenic acid (flaxseed oil) diet *(46)* caused a marked rise in arterial compliance; however, insulin sensitivity and HDL cholesterol decreased and LDL oxidizability increased.

10.4. EXPERIMENTAL ATHEROSCLEROSIS AND FISH OIL

When menhaden oil (a fish oil product) was incorporated in atherogenic diets fed to rhesus monkeys, aortic plaques and their cholesterol content were much less than in the nonfish-oil-fed groups *(47)*. Since the plasma lipid levels were roughly similar in control groups, the inhibition of atherosclerosis may have involved other mechanisms operative in the vessel wall itself. Carotid atherosclerosis was likewise inhibited. Pigs fed an atherogenic diet had much less coronary atherosclerosis when given cod liver oil containing the *n*–3 fatty acids *(48)*. There is good evidence that EPA and DHA from fish oil are even incorporated into advanced human atherosclerotic plaques *(49)*. They are present in complicated plaques as components of cholesterol esters and phospholipids. The incorporation of EPA and DHA from the plasma lipoproteins into the plaques is detectable within a week of fish oil feeding. Perhaps the inhibition of atherosclerosis occurs because EPA and DHA inhibit cellular growth in the arterial wall *(50)*. Atherosclerosis cannot develop even after injury and the influx of low-density lipoprotein (LDL) cholesterol and cholesterol ester unless there is also a cellular reaction. Two important cells in atherosclerosis are smooth muscle cells and macrophages. Because of the suppression of cellular growth factors by *n*–3 fatty acids, proliferation of smooth muscle cells was inhibited *(51)*. Likewise, macrophage infiltration into the vessel wall was lessened by *n*–3 fatty acids *(47)*. Even the initial lesion of atherosclerosis—the fatty streak—develops less under the influence of dietary *n*–3 fatty acids *(47)*.

10.5. EFFECTS ON THE PLASMA LIPIDS AND LIPOPROTEINS

A major effect of dietary *n*–3 fatty acids from fish oil is on the plasma levels of lipids and lipoproteins *(8,52)*. As will be shown, the science in this area is very clear: *n*–3 fatty acids in practical doses (<7 g/day) lower the plasma very-low-density lipoprotein (VLDL) and triglyceride levels through depression of synthesis of triglyceride in the liver. *N*–3 fatty acids from fish oil also suppress postprandial lipemia, the chylomicron remnants of which are considered atherogenic. As VLDL concentrations decrease, LDL cholesterol rises, possible transiently. HDL cholesterol does not change. Like the drug gemfibrozil, *n*–3 fatty acids may cause an increase in LDL as they lower the plasma triglyceride concentration in some hyperlipidemic states such as familial combined hyperlipidemia as will be discussed in detail later. It is unclear whether *n*–3 fatty acids from plants (α-linolenic acid) have these actions. α-Linolenic acid from flaxseed oil has not been shown to lower the plasma triglyceride levels except in very large amounts (38 g/day) *(53)*.

Theoretically, the ideal nutritional program to reduce the plasma lipid and lipoprotein concentrations maximally would be a very-low-cholesterol and saturated-fat diet, which would upregulate the LDL receptor and reduce LDL plasma concentrations, combined with a diet containing fish oil, which would suppress VLDL production and lower plasma triglyceride concentrations. This point of view is buttressed by a well-controlled dietary study of fish oil and saturated fat fed as isolated variables. Saturated fat raised the plasma LDL cholesterol levels and fish oil lowered the plasma triglyceride levels *(54)*.

10.5.1. Effects of Fish Oil on Normal Subjects

Several recent reviews have documented that *n*–3 fatty acids from fish have a great effect on plasma lipids and lipoproteins, even in normal subjects *(1,8,52)*. The principal action is on the plasma triglyceride and VLDL concentrations. This hypolipidemic action is well illustrated in a study of 12 healthy adults (six men and six women) who were given three different diets fed in random order for 4 weeks each: a saturated control diet, a salmon diet containing considerable amounts of *n*–3 fatty acids, and a vegetable oil diet high in *n*–6 fatty acids *(51)*. Both the salmon diet and the vegetable oil diet decreased the plasma cholesterol similarly, from 188 to 162 mg/dL. Both diets reduced LDL, from 128 to 108 mg/dL. HDL cholesterol levels were not changed by the salmon oil diet. The salmon diet decreased VLDL cholesterol levels and the changes in plasma triglyceride were most striking, from 76 to 50 mg/dL. The polyunsaturated vegetable oils did not lower VLDL and triglyceride levels.

10.5.2. Studies in Hyperlipidemic Patients

Because of the hypolipidemic effect of *n*–3 fatty acids from fish oil in normal subjects, it seemed most reasonable to test their effects in hyperlipidemic patients *(9)*. The two groups of hyperlipidemic patients selected for study were characterized by hypertriglyceridemia, since the depression of the plasma triglyceride and VLDL appeared to be a unique effect of *n*–3 fatty acids from fish oil.

Twenty hypertriglyceridemic patients volunteered for the study (8 men and 12 women). Ten of the patients presented with increased levels of both VLDL and LDL, consistent with the type II-B phenotype. Their mean plasma lipid levels at time of entry were 337 mg/dL for cholesterol and 355 mg/dL for triglyceride. Clinically, many of these patients had familial combined hyperlipidemia, a disorder characterized by a strong disposition to the development of coronary heart disease and by overproduction of lipoproteins, particularly VLDL.

The other 10 patients had apparent type V hyperlipidemia, as characterized by increased chylomicrons and greatly increased VLDL levels in the fasting state. Their mean plasma lipid levels at entry were 514 mg/dL for cholesterol and 2,874 mg/dL for triglyceride. Four of the type V patients had concomitant, noninsulin-dependent diabetes mellitus, and two had adult-onset, insulin-dependent diabetes mellitus. Their insulin doses and diabetic control remained constant throughout the study despite the salmon oil.

Both overproduction of VLDL and impaired clearance of the remnants of chylomicron and VLDL metabolism characterize the type V phenotype. Clinically, type V patients have the "chylomicronemia" syndrome, which is characterized by episodes of abdominal pain from enlargement of abdominal viscera (hepatomegaly and splenomegaly) and by episodes of acute pancreatitis. These patients also suffer from eruptive xanthomata, neuropathy, and lipemia retinalis. Although LDL levels are low in patients with fasting chylomicronemia (type V), the presence of the atherogenic remnant particles predisposes them to the development of atherosclerotic complications, including coronary heart disease.

Special care was taken to make certain that the patients were in steady-state conditions before entry. Steady state was defined as a constancy of body weight and diet, and an absence of any residual hypolipidemic drug effect. Most of the patients had not been receiving any hypolipidemic drugs just prior to the study. In the patients previously given drugs, these were discontinued, and plasma lipid levels were monitored until predrug levels were attained.

Two different control diets were used for the two groups of hypertriglyceridemic patients, depending on the phenotype of hyperlipidemia. Patients with combined hyperlipidemia (type II-B) received their usual low-cholesterol (100 mg/day), low-fat (20–30% of total calories) diet. Subsequent dietary periods for these patients consisted of a fish oil diet for 4 weeks followed, in some patients, by a 4-week period of a diet high in vegetable oil containing a predominance of *n*–6 fatty acids. Both of these diets were balanced for cholesterol content (approximately 250 mg/day) and contained 30% of calories as fat. The diets in all periods were eucaloric, such that the subjects neither gained nor lost weight.

For patients with fasting chylomicronemia (type V), the control diet consisted of a very-low-fat diet (5%) in order to lower plasma triglyceride levels maximally. The next dietary interval contained fish oil at 20 or 30% of total calories. Finally, a high *n*–6 vegetable oil diet was also provided, which contained 20–30% of total calories as fat and 200–300 mg of cholesterol per day. Both the fish oil and the vegetable oil diets were initially used cautiously in the patients with fasting chylomicronemia (type V) in order to minimize the risk of hepatosplenomegaly, abdominal pain, and acute pancreatitis.

The salmon oil diet provided about 20 g/day of *n*–3 fatty acids for a 2,600-kcal intake, with 30% of total calories as fat. On the other hand, the vegetable oil diet provided about 47 g of the *n*–6 polyunsaturated fatty acid, linoleic acid. Thus, the fish oil diets actually provided 43–64% less total polyunsaturated fatty acids than the vegetable oil diet, gram for gram.

The fish oil diet decreased the plasma LDL cholesterol levels in the patients with combined hyperlipidemia (type II-B) by 26 mg/dL. Of individual lipoprotein cholesterol changes, the decline of VLDL cholesterol was the most striking; but LDL and HDL cholesterol also decreased. The plasma triglyceride changes were even greater than the cholesterol changes with the fish oil diet. The plasma triglyceride level decreased from 334 to 118 mg/dL. This occurred largely because of the change in VLDL triglyceride, which was lowered from 216 to 55 mg/dL.

The highly polyunsaturated *n*–6 vegetable oil diet had a much weaker effect on VLDL cholesterol and triglyceride. LDL values were similar, but in contrast, HDL cholesterol was higher after the vegetable oil diet. Plasma apolipoprotein changes reflected the lipoprotein lipid changes. In the type II-B patients, there were significant reductions in apo B and C-III levels in the fish oil period, which paralleled the declines in LDL and VLDL levels.

In the type V patients, effects of the fish oil diet were even more striking (Figs. 10.1 and 10.2). With consumption of the very-low-fat control diet, their initial plasma lipid levels declined considerably but still remained greatly elevated. Many of these patients still had milky-appearing plasma, with chylomicrons present in the fasting state. The first change to occur in these patients after the fish oil diet was the virtual disappearance of fasting chylomicronemia, which had been present in five of the patients. During the fish oil diet, total plasma triglyceride decreased from a control value of 1,353 to 281 mg/dL, a drop of 79% (Fig. 10.1). VLDL triglyceride decreased similarly, from 1,087 to 167 mg/dL. Plasma cholesterol levels declined into the normal range after the fish diet, from 373 to

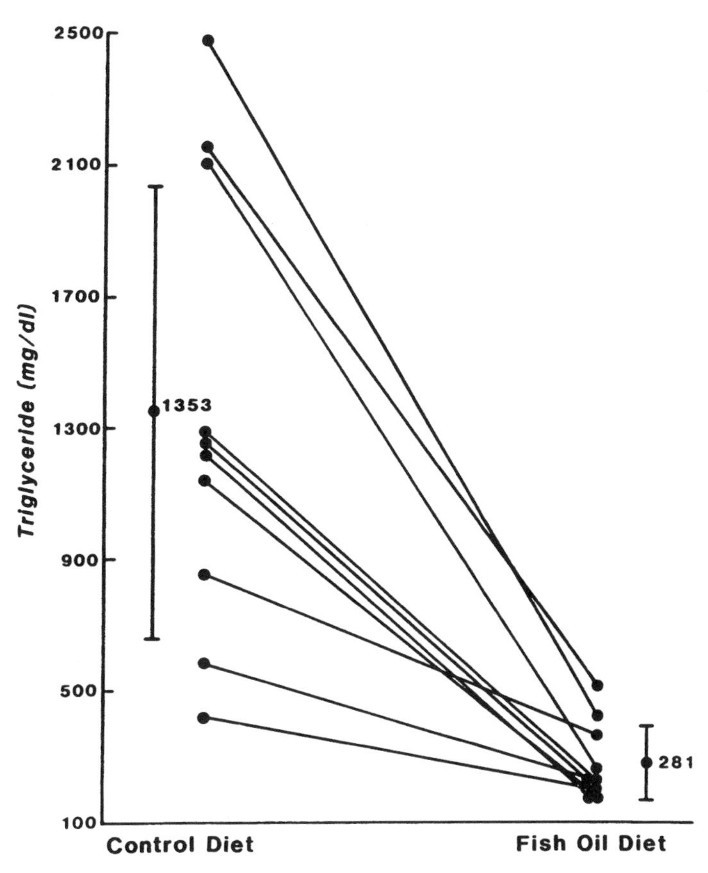

Fig. 10.1. The changes in plasma triglyceride levels in the 10 type V patients: control diet versus fish oil diet. To convert triglyceride from milligrams per deciliter to millimoles per liter, multiply by 0.0113.

207 mg/dL (Fig. 10.2). Most of this total plasma cholesterol decrease occurred as the result of marked changes in the amount of VLDL cholesterol, which decreased from 270 to 70 mg/dL. Of interest was the 48% concomitant rise of LDL cholesterol, from the low value of 84 to 125 mg/dL. Apolipoprotein levels changed to reflect the altered lipoprotein lipid levels. Apo A-1 levels did not change, whereas apo B, C-III, and E all decreased significantly.

When the *n*–6-rich vegetable oil replaced the fish oil in the diets of eight patients, all patients with fasting chylomicronemia (type V) had increases in plasma triglyceride levels within 3–4 days. After 10–14 days of the *n*–6 vegetable oil feeding, the mean plasma triglyceride values rose 198%, and VLDL triglyceride increased from 171 to 550 mg/dL. Plasma cholesterol also increased, from 195 to 264 mg/dL. LDL cholesterol levels, on the contrary, were decreased 28% by the vegetable oil diet: another indication that the metabolic abnormality of the type V phenotype was worsening. Because of enhanced hypertriglyceridemia and the risk of development of abdominal pain typical of this type V disorder, the vegetable oil feeding period was discontinued prematurely in all type V patients *(9)*.

10.5.3. Implications of the Fish Oil Studies in Hypertriglyceridemic Patients

In the 20 hypertriglyceridemic patients, fish oil incorporated in the diet led to an even more profound hypolipidemic effect than had been observed in normal subjects. The plasma triglyceride levels decreased in each of the 20 patients, a 79% decrease in the type V patients and a 64% decrease in

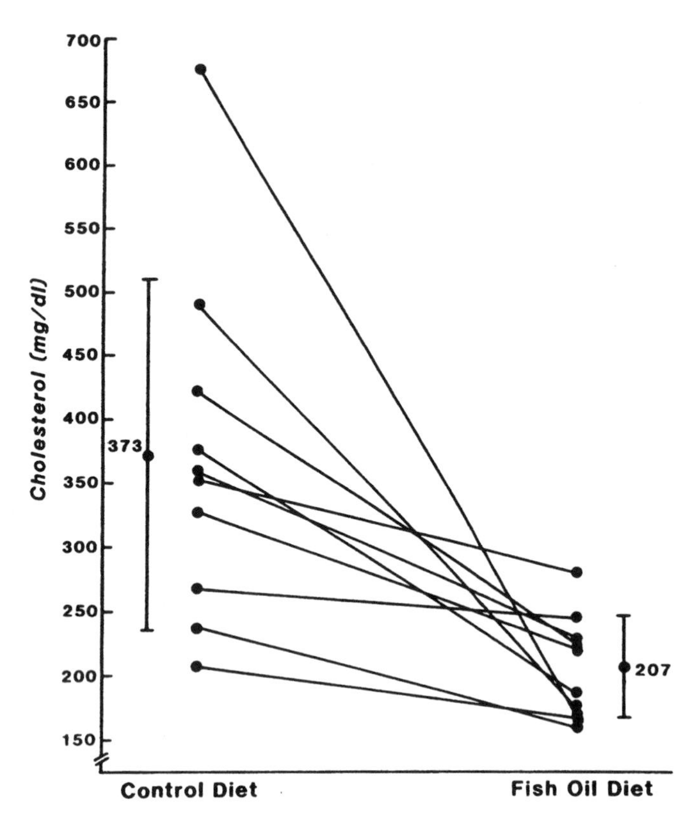

Fig. 10.2. The changes in plasma cholesterol levels in the 10 type V patients: control diet versus fish oil diet. To convert cholesterol from milligrams per deciliter to millimoles per liter, multiply by 0.026.

the patients with combined hyperlipidemia (type II-B); plasma cholesterol levels decreased 45 and 27%, respectively. In the 12 normal subjects previously investigated *(55)*, decreases were less for plasma triglyceride (38%) and much less for plasma cholesterol (14%). Apparently, the greater the hypertriglyceridemia, the greater the reductions brought about by dietary fish oil—in plasma lipids and especially in VLDL.

These results may have considerable therapeutic importance for patients with severe and moderate hypertriglyceridemia. The only dietary treatment to date for severely hypertriglyceridemic patients with fasting chylomicronemia (type V) has been the very severe and therapeutically difficult restriction of dietary fat to between 5 and 10% of total calories in an effort to approach normal plasma triglyceride levels. Americans find this possible to do on a short-term basis but very difficult on a long-term basis because they are accustomed to eating higher quantities of fat, i.e., approximately 40% of total calories. Hitherto, all fatty foods have been contraindicated in patients with fasting chylomicronemia (type V). The findings of this study suggest that some fatty, and even high-cholesterol, foods (i.e., fish or even shellfish) containing marine *n*–3 fatty acids are quite appropriate for ingestion and may produce further triglyceride lowering over and above that which results from the very-low-fat diet.

Other studies in familial combined hyperlipidemia and in type IV hyperlipidemia have shown increases in LDL and apo B, while plasma VLDL and triglyceride values decline *(8,56,57)*. Such LDL increases have also occurred in type IV patients given the drug gemfibrozil. Perhaps this is an expected physiological action when hypertriglyceridemia is being corrected. Should the LDL levels become abnormally high after either drugs or fish oil, then further therapy of the LDL specifically

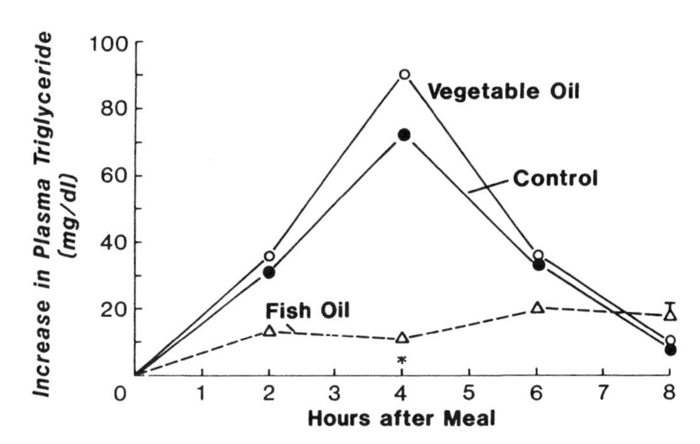

Fig. 10.3. The increase in plasma triglyceride levels following the ingestion of 50 g of fat. Saturated fat test meal given during the saturated fat diet (closed circles); vegetable oil test meal given during the vegetable oil diet (open circles); salmon oil test meal given during the salmon oil diet (triangles). Mean fasting triglyceride levels were 72 ± 19, 76 ± 37, and 46 ± 11 mg/dL before the saturated fat, vegetable and, salmon oil, respectively, test meals were administered.

is warranted (i.e., bile acid-binding resins or one of the statins such as lovastatin). Fish oil has also produced plasma cholesterol and triglyceride lowering in type III patients and in familial hypercholesterolemia *(58)*.

10.5.4. Reduction of Postprandial Lipemia After Fatty Meals

It has been observed that fish oils markedly decreased the usual chylomicronemia that follows fatty meals *(59,60)*. In other words, fat tolerance was greatly improved (*see* Fig. 10.3). This improvement could result from diminished absorption, slower synthesis, and slower entry of chylomicrons into the circulation or, alternatively, from a more rapid removal of the chylomicrons that do appear in the circulation. There is no evidence for diminished absorption, and fat balance studies have not shown increased fat excretion in stools after dietary periods enriched with fish oil. Whether reduced chylomicron production or enhanced removal of chylomicrons is responsible has not yet been completely clarified. Fish oil feeding produces smaller VLDL particle size in animals compared to vegetable oil feeding. Smaller VLDL would have, then, an enhanced catabolism. Perhaps, after a background diet of fish oil, chylomicrons are smaller in size and hence more rapidly catabolized, the result being a much flatter fat tolerance curve *(59)*.

10.5.5. The Mechanism of the Hypolipidemic Effects of Fish Oil

How *n*–3 fatty acids exert their effects to decrease the levels of plasma triglyceride and cholesterol has been tested in humans in two different sets of experiments: (a) the inhibition by fish oil of the usual hypertriglyceridemia that inevitably results when a high-carbohydrate (CHO) diet is suddenly fed to humans and (b) the effects of fish oil on apo B, VLDL, and LDL production rates and turnovers.

10.5.5.1. Fish oil and the Inhibition of Carbohydrate-Induced Hypertriglyceridemia

The well-known phenomenon of carbohydrate-induced hypertriglyceridemia is a physiologic response. In this model, VLDL triglyceride synthesis is stimulated as the dietary CHO intake abruptly increases. The increased VLDL synthesis leads to hypertriglyceridemia, which may persist for many weeks. If *n*–3 fatty acids do inhibit VLDL synthesis, then the usual CHO-induced hypertriglyceridemia should not occur when fish oil is incorporated into the high-CHO diet.

Seven mildly hypertriglyceridemic, but otherwise healthy, subjects (ages 22–54 years) were fed three different experimental diets *(61)*. Each was composed of a liquid formula plus three bran muffins per day to supply fiber. The baseline diet contained 45% of calories from CHO. The high-carbohydrate (high-CHO) diets were then divided into control and fish groups, both containing 15, 10, and 75% of calories as fat, protein, and CHO, respectively. In the baseline and high-CHO control diets, a blend of peanut oil and cocoa butter provided the fat, which was replaced by fish oil, in the form of a commercially available marine lipid concentrate, in the high-CHO fish oil diet. The total amount of fish oil consumed per day was 50 g (in a 3,000-kcal diet), equivalent to approximately 3.3 tablespoons of oil. This amount provided 8.5 g of EPA and 5.5 g of DHA.

The three experimental diets were fed in three different sequences in the Clinical Research Center (Fig. 10.4). In the first sequence, the high-CHO control diet preceded the high-CHO fish oil diet (Fig. 10.4a). In the second sequence, the high-CHO diet was given for 20 days instead of 10 in order to demonstrate that the hypertriglyceridemia did not spontaneously resolve after the first 10 days. It was then followed by the fish oil diet (Fig. 10.4b). In the third sequence, the fish oil was fed first with the high-CHO diet for 25 days and then removed to permit the effects of the high CHO to be manifest for the next 15 days (Fig. 10.4c). Three subjects were studied with the first sequence, and two subjects each were studied with the second and third sequences.

In all seven subjects, the high-CHO control diet increased the plasma triglyceride levels over the baseline diet from 105 to 194 mg/dL *(61)*. The magnitude of the CHO-induced hypertriglyceridemia correlated significantly with each individual's baseline triglyceride levels. The rise in plasma triglyceride levels was complete by day 5 and resulted almost entirely from an increase in the VLDL triglyceride fraction, which more than doubled during the control diet from 69 to 156 mg/dL (Fig. 10.4). Although the total plasma cholesterol levels did not change, VLDL cholesterol levels approximately doubled from 18 to 34 mg/dL; and HDL cholesterol was reduced from 49 to 41 mg/dL.

When the fat of the high-CHO control diet was replaced isocalorically with fish oil, the elevated plasma triglyceride concentration was reduced from 194 to 75 mg/dL, a decrease of 61%. This decrease usually occurred within 3 days (Fig. 10.4a). Once again, changes in VLDL triglyceride levels were largely responsible for this effect (156–34 mg/dL) (Fig. 10.5). Total cholesterol levels decreased insignificantly during the high-CHO fish oil diet—from 172 to 153 mg/dL—primarily because of the drop in VLDL cholesterol levels (34–12 mg/dL) *(61)*.

The hypertriglyceridemia persisted even when the period of CHO induction was prolonged from 10 to 20 days and did not significantly decrease until fish oil was incorporated into the diet (Fig. 10.4b). When the high-CHO fish oil diet followed the baseline diet, the plasma triglyceride level did not rise, but when the high-CHO control diet was fed subsequently, the level increased (Fig. 10.4c). The high-CHO control diet decreased the levels of apo B and increased apo C-III concentrations; apo A-1 and E levels did not change. The high-CHO fish oil diet decreased apo A-1 and apo C-III levels; apo B and E concentrations did not change.

The incorporation of corn oil in place of fish oil into the high-CHO regimen failed to prevent the induced hypertriglyceridemia. For the three subjects who participated in this study, the average triglyceride levels were as follows: baseline, 93 ± 23 mg/dL; high-CHO control, 196 ± 58 mg/dL; high-CHO corn oil, 215 ± 90 mg/dL; and high-CHO fish oil, 86 ± 10 mg/dL.

Dietary fish oil not only prevented but also rapidly reversed the dietary, CHO-induced elevations in plasma triglyceride and VLDL levels, whereas the *n*–6 fatty acid-rich corn oil had no effect at all. Since the primary difference between corn oil and commercial fish oil preparation is the *type* of polyunsaturated fatty acids present (corn oil, 57% 18:2 *n*–6 linoleic acid; the commercially available

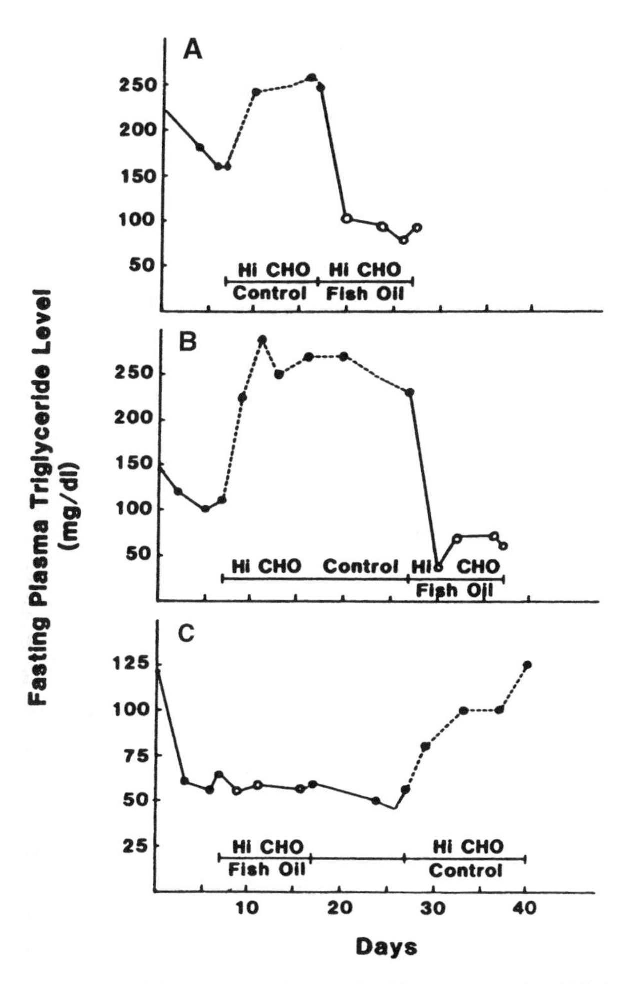

Fig. 10.4. The effects of the baseline diet, high-carbohydrate (Hi-CHO) control, and high-carbohydrate (Hi-CHO) fish oil diets on plasma triglyceride levels in three subjects. We see the reversal of carbohydrate-induced hypertriglyceridemia by dietary fish oil (**a**), the persistence of the hypertriglyceridemia (throughout 20 days) and the subsequent reversal by fish oil (**b**), and the prevention of carbohydrate-induced hypertriglyceridemia by fish oil (**c**).

fish oil preparation, 32% *n*–3 fatty acids), the difference in effect is due to the *n*–3 fatty acids in the fish oil. This finding implied a probable inhibitory effect of *n*–3 fatty acids on hepatic VLDL production.

10.5.5.2. Fish Oil and the Synthesis and Turnover of Apo B, VLDL, and LDL

The hypothesis that *n*–3 fatty acids probably reduced VLDL levels by inhibiting VLDL synthesis was supported by studies designed to elucidate further mechanisms of the hypotriglyceridemic effect of *n*–3 fatty acids. Dietary fish oil affected either the synthesis or the removal of VLDL. The rates of flux and turnover of VLDL triglyceride were measured after injection of ^{3}H-glycerol into people studied under two dietary protocols, one containing fish oil and the other containing fats typical of the American diet *(62)*. This technique permitted the calculation of both synthetic and removal rates of VLDL.

Ten male subjects were selected on the basis of having a wide range of fasting plasma triglyceride concentrations, from 34 to 4,180 mg/dL, so that the hypothesis about the mechanism of action of dietary fish oils could be tested in subjects with greatly different pool sizes of plasma triglyceride.

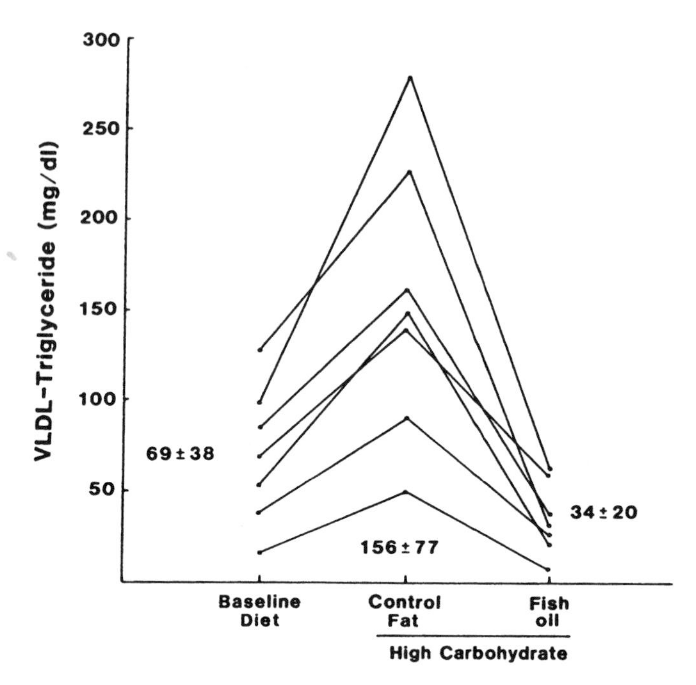

Fig. 10.5. The effects of the high-carbohydrate control and fish oil diets on the plasma VLDL triglyceride levels in the seven subjects.

Liquid formula diets containing 15–20% fat, 65–75% CHO, and 10–15% protein were fed during both the control and the fish oil dietary periods. The two diets differed only in the type of fat they contained. In the control diet, a blend of cocoa butter and peanut oil (1:2) was incorporated into the formulas. The fish oil diet containing the commercial preparation was taken in three divided doses daily and was not mixed into the formulas. The principal difference between the two diets was the higher content of linoleic acid (18:2 *n*–6) in the control diet and the presence of *n*–3 fatty acids in the fish oil diet. The former diet contained virtually no *n*–3 fatty acids, whereas the latter provided about 17 g/day of these highly polyunsaturated fatty acids.

The experimental diets were consumed for a period of 3–5 weeks before the actual VLDL turnover procedure was conducted. This time was needed for the plasma triglyceride levels to stabilize, particularly in the subjects whose triglyceride levels were above normal. Seven subjects consumed the control diet first, followed by the fish oil diet; in the remaining three, the order was reversed. The order in which the diets were administered did not affect the results.

The isocaloric substitution of fish oil for the control vegetable fat produced the expected significant reductions in the total and lipoprotein lipid levels in all 10 subjects. Total cholesterol levels for all 10 subjects fell from 195 to 144 mg/dL, a reduction of 22%. Decreases in VLDL levels accounted for most of the drop in plasma cholesterol (83–21 mg/dL). LDL cholesterol levels did not change significantly, whereas HDL cholesterol concentrations fell from 31 to 24 mg/dL. All of these changes were evident in both the normal and the hypertriglyceridemic groups.

After the administration of ^{3}H-glycerol and its incorporation into the triglyceride of VLDL, the decay curves were analyzed by computer models so that VLDL synthesis and turnover could be calculated. The incorporation of *n*–3 fatty acids into the diet caused a 72% decrease in the VLDL triglyceride pool size (11.4–3.2 g; $p < 0.025$). The decreased pool size was associated with a 45% reduction in the VLDL triglyceride synthetic rate (23–12.6 mg/h/IW; $p < 0.005$) and a 45% decrease in the residence time of VLDL triglyceride in the plasma (5.8–3.2 h; $p < 0.005$). The reciprocal of the residence time is the fractional catabolic rate (FCR), which was increased by 65% (0.23–0.38 h^{-1}; $p < 0.005$). There was a significant rise in the cholesterol/triglyceride ratio in VLDL during the fish oil interval (0.18–0.25; $p < 0.05$). Finally, the ratio of the fast to the slow synthetic pathways did not change with fish oil feeding. The same trends were seen in both normal and hypertriglyceridemic patients. Similar results have also been found by a slightly different dietary plan and with the labeling of VLDL apo B with ^{125}I *(63)*. There was a striking reduction of VLDL synthesis and enhanced turnover.

Direct evidence that the hepatic synthesis of triglyceride and VLDL is suppressed by *n*–3 fatty acids from fish oil has been seen in three in vitro studies of the perfused rat liver and in studies of liver cells from rats and rabbits in primary culture *(64–66)*. In all of these studies, triglyceride synthesis was reduced. In one, enhanced ketone body production resulted; in the others there was a diversion of *n*–3 fatty acids from triglyceride synthesis into phospholipid synthesis *(66)*. When the net results of the human and animal studies are taken together, the evidence is very strong that suppression of VLDL and triglyceride synthesis is a primary mechanism for the hypolipidemic effects of *n*–3 fatty acids, coupled with an increased fractional catabolic rate of VLDL.

10.5.6. Fish Oil and LDL Turnover

Labeled LDL turnover studies have been carried out in normal subjects given fish oil. It was demonstrated that there was a decreased synthesis of LDL and a tendency for an increased fractional catabolic rate *(67)*. Spady and colleagues have shown enhancement of LDL receptor activity after the administration of fish oil in the rat *(68)*. This latter result fits in very well with the increased fractional catabolic rate observed in the normal human subjects. It seems very clear that the *n*–3 fatty acids from fish oil have major effects on all of the major lipoproteins with the exception of HDL.

10.5.7. Summary and Conclusions: Fish Oil Effects on Plasma Lipids and Lipoproteins

The *n*–3 fatty acids from fish oil and fish have been shown to have a remarkable effect on the synthesis and clearance of triglyceride-rich lipoproteins, especially VLDL and chylomicrons. Even LDL synthesis and clearance have been affected. Because of these significant effects on lipoprotein synthesis and clearance, beneficial effects of fish oil have been demonstrated in a variety of hyperlipidemic states, especially those with conditions such as hypertriglyceridemia and chylomicronemia. Therapeutic implications for fish oil are especially positive in type V, type IV, and type III hyperlipidemia. A similar effectiveness has been shown in hypertriglyceridemic diabetic patients without affecting glucose homeostasis *(69)*.

Difficulty in interpreting the effects of fish oil in various hyperlipidemic patients has occurred because of vastly different experimental conditions *(8)*. In some studies, fish oil was simply added as a supplement to the usual diet in doses of 8–16 g/day. In the control period, a placebo oil such as olive oil or safflower oil was not always utilized. In other studies, there was the customary diet plus the use of an appropriate placebo oil. Furthermore, various kinds of fish oils have been utilized, some containing a considerable amount of cholesterol and saturated fat. Newer fish oils have less-saturated fatty acids, have higher concentrations of *n*–3 fatty acids, and a lower cholesterol content.

Some conclusions have emerged from the wide variety of studies, most of which have not been controlled for caloric and body weight stability. Fish oil is most effective when the diet is well controlled. In these studies, LDL lowering as well as profound VLDL and triglyceride lowering in normal subjects and in a wide variety of hyperlipidemic states has usually occurred. In our experience, this lowering of plasma cholesterol levels has occurred in patients with elevated plasma triglyceride concentrations (types V, II-a, II-B, III, and IV hyperlipidemia), with the most dramatic results occurring in patients with fasting chylomicronemia (type V) who do not tolerate any other kind of dietary fat *(8,9,52,70)*. In the literature and in our experience, HDL levels have not been greatly affected by fish oil. Clearly the use of fish oil in hyperlipidemia must be individualized as to both use and dosage. Lower doses of fish oil (8–15 g/day) or 2.4–4.5 g of *n*–3 fatty acids (EPA and DHA) particularly lower the plasma triglyceride levels *(8)*.

Why the plasma LDL and apo B have at times increased after fish oil administration when at the same time the plasma VLDL and triglyceride have decreased is one of the most challenging questions and may relate to fundamental aspects of VLDL–LDL metabolism. Normally, LDL is derived from two sources: conversion from VLDL and direct synthesis from the liver. The catabolism of VLDL is likewise in two directions through intermediate-density lipoprotein (IDL). This lipoprotein may be removed by the apo E receptor in the liver or converted to LDL. The animal experiments of Huff and Telford suggest why, in some instances, fish oil might increase LDL *(71)*. Turnover studies in the miniature pig revealed that fish oil feeding increased the proportion of VLDL being converted to LDL. Apparently, the *n*–3 fatty acids of fish oil produce a smaller VLDL particle, which is more likely to be converted to LDL. In this pig study, LDL concentrations, however, did not increase because the direct synthesis of LDL was reduced more by fish oil than the increase in LDL from VLDL. These pig studies await confirmation in humans. They do explain why LDL may increase in some humans fed fish oil: more VLDL is converted to LDL and direct LDL synthesis does not decrease, thus adding up to more LDL. LDL turnover studies have shown decreased production of LDL in normal humans given large amounts of salmon oil versus vegetable oil *(67)*. In this study, the plasma LDL decreased after *n*–3 fatty acids.

10.6. FISH OIL IN DIABETIC PATIENTS

In diabetic patients there is enhanced risk for vascular disease so that the use of fish oil might be particularly desirable if glucose control is not disturbed. The literature is controversial with regard to the effects of *n*–3 fatty acids in diabetic patients *(72,73)*. In type I insulin-dependent diabetic patients, there is universal agreement that glucose control is not hampered with fish oil supplementation and that the beneficial effects on the plasma lipids and lipoproteins have been demonstrated. In type II adult-onset, noninsulin-dependent diabetic patients, the results have been somewhat conflicting, possibly because such patients are very susceptible to the caloric load imposed. In most studies, the plasma triglyceride and VLDL concentrations have declined, but some studies have shown the deterioration in glucose homeostasis. This literature has recently been reviewed by Heine and colleagues *(72)*. Most of the studies have been short term and, in some, caloric control has been somewhat distorted by the administration of calorie-dense fish oil without there being a suitable placebo. The addition of fish oil to the usual diet would be hypercaloric, thereby disturbing glucose control. When there is attention to the caloric content of the supplement, then there are beneficial effects on the plasma triglyceride and VLDL without disturbing glucose homeostasis as illustrated in the following experiments.

These problems and objections were considered in an experimental design of a study in 16 adult-onset diabetic patients who were randomized to a double-blind, placebo-controlled crossover study *(69)*. The subjects of the study were overweight and most were receiving hypoglycemic agents. There was a 3-month stabilization baseline period in which they were given a eucaloric, lower fat (30% of the calories from fat), high-complex carbohydrate diet with 55% from carbohydrate. This was followed by two 6-month intervention periods in which the subjects continued on the same diet and received a supplement of 15 g/day of either olive oil or fish oil. The fish oil contained 6 g/day of *n*–3 fatty acids. The end points of the study were plasma lipid and lipoprotein concentrations and glucose homeostasis. The plasma triglyceride concentrations were much lower with the fish oil preparation versus olive oil

(260 versus 449 mg/dL). VLDL cholesterol as well as VLDL triglyceride was lower. The total plasma cholesterol was unchanged. There was a significant increase in LDL cholesterol, as has been mentioned previously, when hypertriglyceridemic individuals are given fish oil, from 117 to 145 mg/dL. HDL cholesterol did not change.

However, the effects of fish oil on glucose homeostasis revealed no difference from the 6-month period of olive oil administration. Body weights were unchanged. Fasting glucose levels were 172 and 178 mg/dL, olive oil versus fish oil. Another measure of diabetic control, hemoglobin A-1C, revealed no difference as did also the 24-h urinary glucose excretion, the plasma C-peptide, and the 24-h urinary C-peptide.

In view of the extremely high mortality from coronary heart disease in adult-onset diabetic patients, this hypolipidemic action of fish oil was of interest because diabetic control did not deteriorate and there were significantly beneficial plasma lipid–lipoprotein effects. The other actions of the *n*–3 fatty acids from fish oil in inhibiting the development of atherosclerosis, in preventing thromboxane A_2 formation, in increasing endothelial-derived relaxing factor, and in inhibiting platelet-derived growth factor would all be additional reasons for postulating a therapeutic benefit from the use of fish oil in diabetic patients.

10.7. HYPERTENSION

N–3 fatty acids in 11 studies have been uniformly associated with a mild decrease in systolic blood pressure uniformly and at times a decrease in diastolic blood pressure, particularly in the upright position *(4)*. This has especially occurred in mild hypertensives *(4,74,75)*. A randomized controlled trial was conducted to study the effects of fish intake and weight loss on blood pressure in medication-treated overweight hypertensives *(76)*. Combining a daily fish meal with weight loss resulted in additive decreases in ambulatory blood pressure and decreases in heart rate. The suggested mechanism has been an attenuation in the responses of forearm vascular resistance and blood flow to angiotensin, i.e., less vascular reactivity. Since the decreases in both systolic and diastolic pressures are not great even in the best studies (4.6 and 3.0 mmHg) *(75)*, fish oil cannot be regarded as a single treatment modality for hypertension. However, when used for other purposes, the mild blood pressure-lowering effect of *n*–3 fatty acids would certainly provide an added benefit.

10.8. CONCLUSIONS

The *n*–3 fatty acids from fish, fish oil, and plants greatly inhibit the atherosclerotic process, coronary thrombosis, and cardiac arrhythmias by a variety of actions and should be considered as an important therapeutic modality in patients with already established coronary heart disease and to prevent coronary disease in highly susceptible people.

10.9. RECOMMENDATIONS

The intake of *n*–3 fatty acids from fish and plants should definitely be increased to prevent coronary heart disease *(2,77)*. This could be best in the form of two to three fish meals per week in the context of a low-fat diet. Fish, of course, could be substituted for meat in the diet. An intake of α-linolenic acid of 2 g/day or 1% of energy has been suggested *(77)*. At the same time, the diet should be reduced in fat content to 20% of the total calories with a high-carbohydrate, high-fiber intake. The intake of cholesterol should be limited to 100 mg/day.

Table 10.1 provides the fat content and *n*–3 fatty acid content for a wide variety of fish and shellfish *(78)*. All fish and shellfish contain the *n*–3 fatty acids even when the fat content is rather low as it is in

Table 10.1
Fat and *n*–3 Fatty Acid (Eicosapentaenoic Acid and Docosahexaenoic Acid) Content of Seafood and Fish Oils

Seafood (100 g, Edible Portion, Raw)	*Fat (g)*	*n–3 Fatty Acids (EPA and DHA)(g)*
Fish		
Anchovy, European	4.8	1.4
Bass, striped	2.3	0.8
Bluefish	4.2	0.8
Carp	5.6	0.3
Catfish, channel, wild	2.8	0.4
Catfish, channel, farmed	7.6	0.3
Cod, Atlantic	0.7	0.2
Cod, Pacific	0.6	0.2
Flatfish (flounder and sole)	1.2	0.2
Haddock	0.7	0.2
Halibut, Atlantic and Pacific	2.3	0.4
Herring, Atlantic	9.0	1.6
Herring, Pacific	13.9	1.7
Mackerel, Atlantic	13.9	2.3
Mullet, striped	3.8	0.3
Ocean perch, Atlantic	1.6	0.3
Pike, Walleye	1.2	0.3
Pompano, Florida	9.5	0.6
Sablefish	15.3	1.4
Salmon, Atlantic, farmed	13.4	2.0
Salmon, Atlantic, wild	6.3	1.4
Salmon, Chinook	10.4	2.0
Salmon, pink	3.5	1.0
Salmon, Sockeye	8.6	1.2
Sardines, in sardine oil[a]	15.5	3.3
Sardines, in oil, drained	11.5	1.0
Sardines, in tomato sauce, drained	10.5	1.4
Snapper, mixed species	1.3	0.3
Sturgeon, mixed species	4.0	0.3
Swordfish	4.0	0.6
Trout, mixed species	6.6	0.7
Trout, rainbow, farmed	5.4	1.7
Trout, rainbow, wild	3.5	0.6
Tuna, light, canned in water, drained	0.8	0.3
Tuna, fresh, yellowfin	1.0	0.2
Crustaceans		
Crab, Dungeness	1.0	0.3
Crayfish, mixed species	1.0	0.1
Lobster, Northern	0.9	0.0
Shrimp, mixed species	1.7	0.5

Seafood (100 g, Edible Portion, Raw)	*Fat (g)*	*n–3 Fatty Acids (EPA and DHA) (g)*
Mollusks		
Abalone, mixed species	1.0	Trace
Clam, mixed species	1.0	0.1
Mussel, blue	2.2	0.4
Octopus, common	1.0	0.2
Oyster, Pacific	2.3	0.7
Scallop, mixed species	0.8	0.2
Squid, mixed species	1.4	0.5

[a]Analysis by the Atherosclerosis Research Laboratory, Portland, OR.

shellfish. The lower the fat content, the higher the percentage of *n*–3 fatty acids present in a given fish or shellfish. The goal of this recommendation is to produce an increased content of the *n*–3 fatty acids, EPA and DHA, in the blood and tissues of the body. This will occur if there is regular consumption of fish and shellfish (at least to the extent of 200–300 g/week) such that these *n*–3 fatty acids will be present in the diet.

With wild fish stocks being depleted, consumers will more and more find only farmed fish on the markets, especially salmon and catfish. An important question is the EPA and DHA content of farmed fish. Preliminary data from our laboratory indicate that EPA and DHA is 50–75% lower in farmed catfish. Farmed and wild salmon have similar amounts of EPA and DHA. Additional analyses are needed to confirm these data.

Table 10.2 provides the *n*–3 fatty acid (α-linolenic acid) content from various oils and foods *(76)*. Daily consumption of 15 g (1 tablespoon) canola oil or canola oil margarine provides as much as 1 g α-linolenic acid/day. Other rich dietary sources of α-linolenic acid include English walnuts and flaxseed oil. α-Linolenic acid is also a prominent fatty acid of green leafy vegetables; however, because these vegetables have such a low fat content, the net amount of α-linolenic acid ingested from these sources is small. Human milk but not cow's milk is a good source of the *n*–3 fatty acids including α-linolenic acid, EPA, and DHA. The amounts in human milk (2.2% of total fatty acids) provide a good basis for considering similar amounts in the diets of children and adults. In recognition of the importance of *n*–3 fatty acids in human nutrition, the producers of infant formulas are including soy oil, an excellent source of α-linolenic acid, as one of the fat ingredients instead of corn and coconut oils that are poor sources. DHA and arachidonic acid (*n*–6) have now been included in many infant formulas. Likewise, intravenous fat preparations use soy oil instead of safflower oil that is poor in *n*–3 fatty acids.

Should it be desired to ascertain what the long-term intake of fish and shellfish has been in the past, there are excellent markers to document this point. The markers would be the measurement of the *n*–3 fatty acids in the plasma which would reflect a more immediate intake, their measurement in red blood cells which, because of the greater half-life of these cells, would reflect the intake over a longer period of time, and finally biopsies of the adipose tissue whose fatty acids would reflect the intake over many months and years *(79)*.

For the intensive treatment of various forms of hyperlipidemia as well as the production of an antithrombotic state, fish oils would need to be utilized in addition to the consumption of fish. The dose of fish oil might well be from 4 to 15 g/day, titrated according to the end point desired, and the content of EPA + DHA in the fish oil preparation utilized.

Table 10.2
α-Linolenic Acid Content of Various Oils and Foods

Food	*Linolenic Acid (wt%)*
Oils	
Flaxseed oil	53.3
Canola oil	9.1
Soy oil	6.8
Corn oil	1.2
Olive oil	0.8
Safflower oil	0.0
Chocolate	0.1
Coconut oil	0.0
Nuts	
English walnuts	9.1
Hazelnuts	0.9
Cashews	0.6
Almonds	Trace
Peanuts, dry roasted	Trace
Green, leafy vegetables, raw (edible portion)	
Brussels sprouts	0.20
Kale	0.13
Spinach	0.12

For people who are unable to consume fish or shellfish, the use of fish oil would again be advisable. For primary prevention, 2–3 g/day would be desirable. Higher doses, as noted above, should be used for secondary prevention and the attainment of discrete end points of plasma lipid and lipoprotein levels and platelet function.

A variety of fish oils are available commercially. Most include vitamin E as an antioxidant. They range from 1 g capsules containing EPA + DHA in the amount of 300 mg per capsule to highly purified ethyl esters of EPA and DHA, 900 mg in a 1 g capsule (Lovaza®). The more concentrated preparation could be used to treat hypertriglyceridemia. Fish oils are generally well tolerated when consumed after a meal.

The American Heart Association has recommended two fatty fish meals per week for the US population and 1 g of EPA + DHA for patients with coronary disease. If healthy individuals are unable to consume two meals of fatty fish per week, one to two fish oil capsules per day might achieve the same benefit.

REFERENCES

1. Goodnight SH Jr., Harris WS, Connor WE, Illingworth DR. Polyunsaturated fatty acids, hyperlipidemia and thrombosis. *Arteriosclerosis*. 1982;2:87–113.
2. Connor WE. α-Linolenic acid in health and disease. *Am J Clin Nutr*. 1999;69:827–828.
3. Voss A, Reinhart M, Sankarappa S, Sprecher H. The metabolism of 7,10,13,16,19-docosapentaenoic acid to 4,7,10,16,19-docosahexaenoic acid in rat liver is independent of 4-desaturase. *J Biol Chem*. 1991;266:19995–20000.
4. Connor WE. *N*–3 fatty acid and heart disease. In: Kritchevsky D, Carroll KK, eds. *Nutrition and Disease Update: Heart Disease*. Champaign, IL: American Oil Chemists' Society Press, 1994;7–42.

5. Bang HO, Dyerberg J, Hyorne N. The composition of food consumed by Greenlandic Eskimos. *Acta Med Scand.* 1973;200:69–73.
6. Bang HO, Dyerberg J. Lipid metabolism and ischemic heart disease in Greenland Eskimos. In: Draper HH, ed. *Advanced Nutrition Research*, vol 3. New York: Plenum Press, 1980;1–32.
7. Connor WE. Hypolipidemic effects of dietary *n*–3 fatty acids in normal and hyperlipidemic humans: effectiveness and mechanisms. In: Simopoulos AP, ed. *Health Effects of Polyunsaturated Fatty Acids in Seafoods*. Chapter 10. Orlando, FL: Academic Press, 1986;173–210.
8. Harris WS. Fish oils and plasma lipid and lipoprotein metabolism in humans: a critical review. *J Lipid Res.* 1989;30: 785–807.
9. Phillipson BE, Rothrock DW, Connor WE, Harris WS, Illingworth DR. The reduction of plasma lipids, lipoproteins and apoproteins in hypertriglyceridemic patients by dietary fish oil. *N Engl J Med.* 1985;312:1210.
10. McLennan PL, Abeywardena MY, Charnock JS. Influence of dietary lipids on arrhythmias and infarction after coronary artery ligation in rats. *Can J Physiol Pharmacol.* 1985;63:1411–1417.
11. McLennan PL, Abeywardena MY, Charnock JS. Dietary fish oil prevents ventricular fibrillation following coronary artery occlusion and reperfusion. *Am Heart J.* 1988;116:706–717.
12. Hallaq H, Sellmayer A, Smith TW, Leaf A. Protective effect of eicosapentaenoic acid on ouabain toxicity in neonatal rat cardiac myocytes. *Proc Natl Acad Sci.* 1990;87:7834–7838.
13. Hallaq H, Smith TW, Leaf A. Modulation of dihydropyridine-sensitive calcium channels in heart cells by fish oil fatty acids. *Proc Natl Acad Sci.* 1992;89:1760–1764.
14. Billman GE, Kang JX, Leaf A. Prevention of sudden cardiac death by dietary pure omega-3 polyunsaturated fatty acids in dogs. *Circulation.* 1999;99:2452–2457.
15. Newman WP, Propst MT, Rogers DR, Middaugh JP, Strong JP. Atherosclerosis in Alaska natives and non-natives. *Lancet.* 1993;341:1056–1057.
16. Kromhout D, Bosschieter EB, Coulander C. The inverse relation between fish consumption and 20-year mortality from coronary heart disease. *N Engl J Med.* 1985;312:1205–1209.
17. Dolecek TA, Grandits G. Dietary polyunsaturated fatty acids and mortality in the Multiple Risk Factor Intervention Trial (MRFIT). *World Rev Nutr Diet.* 1991;66:205–216.
18. Ascherio A, Rimm EB, Stampfer MJ, Giovannucci EL, Willett WC. Dietary intake of marine *n*–3 fatty acids, fish intake, and the risk of coronary disease among men. *N Engl J Med.* 1995;332:977–982.
19. Siscovick DS, Raghunathan TE, King I, Weinmann S, Wicklund KG, Albright J, et al. Dietary intake and cell membrane levels of long-chain *n*–3 polyunsaturated fatty acids and the risk of primary cardiac arrest. *JAMA.* 1995;274:1363–1367.
20. Albert CM, Hennekens CH, O'Donnell CJ, Ajani UA, Carey VJ, Willett WC, Ruskin JN, Manson JE. Fish consumption and risk of sudden cardiac death. *JAMA.* 1998;279:23–28.
21. Hu FB, Stampfer MJ, Manson JE, Rimm EB, Wolk A, Colditz GA, Hennekens CH, Willett WC. Dietary intake of alpha-linolenic acid and risk of fatal ischemic heart disease among women. *Am J Clin Nutr.* 1999;69: 890–897.
22. Burr ML, Fehily AM, Gilbert JF, Rogers S, Holliday RM, Sweetnam PM, Elwood PC, Deadman NW Effects of changes in fat, fish, and fibre intakes on death and myocardial reinfarction: diet and reinfarction trial (DART). *Lancet.* 1989;2:757–762.
23. von Schacky C, Angerer P, Kothny W, Theisen K, Mudra H. The effect of dietary omega–3 fatty acids on coronary atherosclerosis. A randomized, double-blind, placebo-controlled trial. *Ann Intern Med.* 1999;130:554–562.
24. GISSI-Prevenzione Investigators. Dietary supplementation with *n*–3 polyunsaturated fatty acids and vitamin E after myocardial infarction: results of the GISSI-Prevenzione trial. *Lancet.* 1999;354:447–455.
25. GISSI II—Marchioli R, Barzi F, and Bomba E, et al. Early protection against sudden death by *n*–3 polyunsaturated fatty acids after myocardial infarction: time-course analysis of the results of the Gruppo Italiano per Studio della Sopravvivenza nell' Infarto Miocardico (GSSI)-Prevenzione. *Circulation.* 2002;105:1897–1903
26. de Lorgeril M, Salen P, Martin J-L, Mamelle N, Monjaud I, Touboul P, Delaye J. Effect of a Mediterranean type of diet on the rate of cardiovascular complications in patients with coronary artery disease. *J Am Coll Cardiol.* 1996;28: 1103–1108.
27. de Logeril M, Salen P, Martin J-L, Monjaud I, Delaye J, Mamelle N. Mediterranean diet, traditional risk factors, and the rate of cardiovascular complications after myocardial infarction. *Circulation.* 1999;99:779–785.
28. Yokoyama M, Origasa H, Matsuzaki M, Matsuzawa Y, Saito Y, Ishikawa Y, Oilawa S, Sasaki J, Hishida H, Itakura H, Kita T, Kitabatake A, Nakaya N, Sakata T, Shimada K, Shirato K. Effects of eicosapentaenoic acid on major coronary events in hypercholesterolaemic patients (JELIS): a randomized open-label, blinded endpoint analysis. *Lancet.* 2007;369:1090–1098.

29. Leaf A, Kang JS, Billman GE. Clinical prevention of sudden cardiac death by *n*–3 polyunsaturated fatty acids and mechanism of prevention of arrhythmias by *n*–3 fish oils. *Circulation*. 2003;107:2646–2652.
30. Leaf A, Albert CMJosephson M, Steinhaus D, Kluger J, Kang JX, Cox B, Zhang H, Schoenfeld D. Prevention of fatal arrhythmias in high-risk subjects by fish oil *n*–3 fatty acid intake. *Circulation*. 2005;112:2762–2768.
31. Raitt MH, Connor WE, Morris C, Kron J, Halperin B, Chugh SS, McClelland J, Cook J, MacMurdy K, Swenson R, Connor SL, Gerhard G, Kraemer DF, Oseran D, Marchant C, Calhoun D, Shnider R, McAnulty J. Fish oil supplementation and risk of ventricular tachycardia and ventricular fibrillation in patients with implantable defibrillators: a randomized controlled trial. *JAMA*. 2005;293:2884–2991.
32. Brouwer IA, Zock PL, Camm AJ, Bocker D, Hauer RN, Wever EF, Dullemeijer C, Ronden JE, Katan MB, Lubinski A, Buschler H, Schouten EG. Effect of fish oil on ventricular tachyarrhythmia and death in patients with implantable cardioverter defibrillators: the study on Omega-3 Fatty Acids and Ventricular Arrhythmia (SOFA) randomized trial. *JAMA*. 2006;295:2613–2619.
33. Leaf A, Weber PC. Cardiovascular effects of *n*–3 fatty acids. *N Engl J Med*. 1988;318:549–557.
34. Andriamampandry MD, Leray C, Freund M, Cazenave JP, Gachet C. Antithrombotic effects of (*n*–3) polyunsaturated fatty acids in rat models of arterial and venous thrombosis. *Thromb Res*. 1999;93:9–16.
35. Goodnight SHE Jr., Harris WS, Connor WE. The effects of dietary *n*–3 fatty acids upon platelet composition and function in man: a prospective, controlled study. *Blood*. 1981;58:880–885.
36. Harker LA, Kelly AB, Hanson SR, Krupski W, Bass A, Osterud B, FitzGerald GA, Goodnight SH, Connor WE. Interruption of vascular thrombus formation and vascular lesion formation by dietary *n*–3 fatty acids in fish oil in non human primates. *Circulation*. 1993;87:1017–1029.
37. Mutanen M, Freese R. Polyunsaturated fatty acids and platelet aggregation. *Curr Opin Lipidol*. 1996;7:14–19.
38. Allman MA, Pena MM, Pang D. Supplementation with flaxseed oil versus sunflowerseed oil in healthy young men consuming a low fat diet: effects on platelet composition and function. *Eur J Clin Nutr*. 1995;49:169–178.
39. DeCaterina R, Giannessi D, Mazzone A, Bernini W, Lazzerini G, Maffei S, Cerri M, Salvatore L, Weksler B. Vascular prostacyclin is increased in patients ingesting ω–3 polyunsaturated fatty acids before coronary artery bypass graft surgery. *Circulation*. 1990;82:428–438.
40. Shimokawa H, Vanhoutte PM. Dietary *n*–3 fatty acids and endothelium-dependent relaxation in porcine coronary arteries. *Am J Physiol*. 1989;256:H968–H973.
41. Nordoy A, Hatcher L, Goodnight S, FitzGerald GA, Connor WE. Effects of dietary fat content, saturated fatty acids and fish oil on eicosanoid production and hemostatic parameters in normal men. *J Lab Clin Med*. 1994;123: 914–920.
42. Tagawa H, Shimokawa H, Tagawa T, Kuroiwa-Matsumoto M, Hirooka Y, Takeshita A. Long-term treatment with eicosapentaenoic acid augments both nitric oxide-mediated and non-nitric oxide-mediated endothelium-dependent forearm vasodilatation in patients with coronary artery disease. *J Cardiovasc Pharmacol*. 1999;33:633–640.
43. Rupp H, Turcani M, Ohkubo T, Maisch B, Brilla CG. Dietary linolenic acid-mediated increase in vascular prostacyclin formation. *Mol Cell Biochem*. 1996;162:59–64.
44. McVeigh GE, Brennan GM, Cohn JN, Finkelstein SM, Hayes RJ, Johnston GD. Fish oil improves arterial compliance in non-insulin-dependent diabetes mellitus. *Arterioscler Thromb*. 1994;14:1425–1429.
45. Chin JFP, Gust AP, Nestel PJ, Dart AM. Marine oils dose-dependently inhibit vasoconstriction of forearm resistance vessels in humans. *Hypertension*. 1993;21:22–28.
46. Nestel PJ, Pomeroy SE, Sasahara T, Yamashita T, Liang YL, Dart AM, Jennings GL, Abbey M, Cameron JD. Arterial compliance in obese subjects is improved with dietary plant *n*–3 fatty acid from flaxseed oil despite increased LDL oxidizability. *Arterioscler Thromb Vasc Biol*. 1997;17:1163–1170.
47. Davis HR, Bridenstine RT, Vesselinovitch D, Wissler RW. Fish oil inhibits development of atherosclerosis in Rhesus monkeys. *Arteriosclerosis*. 1987;7:441–449.
48. Weiner BH, Ockene IS, Levine PH, Cuénoud HF, Fisher M, Johnson BF, Daoud AS, Jarmolych J, Hosmer D, Johnson MH, Natale A, Vaudreuil C, Hoogasian JJ. Inhibition of atherosclerosis by cod-liver oil in a hyperlipidemic swine model. *N Engl J Med*. 1986;315:841–846.
49. Rapp JH, Connor WE, Lin DS, Porter JM. Dietary eicosapentaenoic acid (EPA) and docosahexaenoic acid (DHA) from fish oil: their incorporation into advanced human atherosclerotic plaques. *Arterioscler Thromb*. 1991;11:903–911.
50 Fox PL, DeCorleto PE. Fish oils inhibit endothelial cell production of platelet-derived growth factor-like protein. *Science*. 1988;241:453–456.
51. Kaminski WE, Jendraschak E, Kiefl R, Von Schacky C. Dietary ω–3 fatty acids lower levels of platelet-derived growth factor mRNA in human mononuclear cells. *Blood*. 1993;81:1871–1879.
52. Harris WS. *N*–3 fatty acids and serum lipoproteins: human studies. *Am J Clin Nutr*. 1997;65:1645S–1654S.

53. Singer P, Berger I, Wirth M, Godicke W, Jaeger W, Voigt S. Slow desaturation and elongation of linoleic and α-linolenic acids as a rationale of eicosapentaenoic acid-rich diet to lower blood pressure and serum lipids in normal, hypertensive and hyperlipemic subjects. *Prostaglandins Leukot Med.* 1986;24:173–193.
54. Nordoy A, Hatcher LF, Ullmann DL, Connor WE. The individual effects of dietary saturated fat and fish oil upon the plasma lipids and lipoproteins in normal men. *Am J Clin Nutr.* 1993;57:634–639.
55. Harris WS, Connor WE, McMurry MP. The comparative reduction of the plasma lipids and lipoproteins by dietary polyunsaturated fats: salmon oil versus vegetable oils. *Metabolism.* 1983;32:179.
56. Friday KE, Failor RA, Childs MT, Bierman EL. Effects of *n*–3 and *n*–6 fatty acid-enriched diets on plasma lipoproteins in heterozygous familial hypercholesterolemia. *Arterioscler Thromb.* 1991;11:47–54.
57. Failor RA, Childs MT, Bierman EL. The effects of *n*–3 and *n*–6 fatty acid-enriched diets on plasma lipoproteins and apoproteins in familial combined hyperlipidemia. *Metabolism.* 1988;37:1021–1028.
58. Illingworth DR, Schmidt E. The influence of dietary *n*–3 fatty acids on plasma, lipids and lipoproteins. *Ann N Y Acad Sci.* 1993;676:60–69.
59. Harris WS, Connor WE, Alam N, Illingworth DR. The reduction of postprandial triglyceride in humans by dietary *n*–3 fatty acids. *J Lipid Res.* 1988;29:1451–1460.
60. Weintraub MS, Zechner R, Brown A, Eisenberg S, Breslow JL. Dietary polyunsaturated fats of the ω–6 and ω-3 series reduce postprandial lipoprotein levels: chronic and acute effects of fat saturation on postprandial lipoprotein metabolism. *J Clin Invest.* 1988;82:1884–1893.
61. Harris WS, Connor WE, Inkeles SB, Illingworth DR. Dietary *n*–3 fatty acids prevent carbohydrate-induced hypertriglyceridemia. *Metabolism.* 1984;33:1016–1019.
62. Harris WS, Connor WE, Illingworth DR, Rothrock DW, Foster DM. Effect of fish oil on VLDL triglyceride kinetics in man. *J Lipid Res.* 1990;31:1549–1558.
63. Nestel PJ, Connor WE, Reardon MR, Connor S, Wong S, Boston R. Suppression by diets rich in fish oil of very low density lipoprotein production in men. *J Clin Invest.* 1984;74:82.
64. Wong SH, Nestel PH, Trimble RP, Storer BG, Illman RJ, Topping DL. The adaptive effects of dietary fish and safflower oil on lipid and lipoprotein metabolism in perfused rat liver. *Biochem Biophys Acta.* 1983;792:103.
65. Wong S, Reardon M, Nestel P. Reduced triglyceride formation from long chain polyenoic fatty acids in rat hepatocytes. *Metabolism.* 1985;34:900–905.
66. Benner KG, Sasaki A, Gowen DR, Weaver A, Connor WE. The differential effect of eicosapentaenoic acid and oleic acid on lipid synthesis and VLDL secretion in rabbit hepatocytes. *Lipids.* 1990;25:534–540.
67. Illingworth DR, Harris WS, Connor WE. Inhibition of low density lipoprotein synthesis by dietary *n*–3 fatty acids in humans. *Arteriosclerosis.* 1984;4:270–275.
68. Ventura MA, Woollett LA, Spady DK. Dietary fish oil stimulates hepatic low density lipoprotein transport in the rat. ‘ *J Clin Invest.* 1989;84:528–537.
69. Connor WE, Prince MJ, Ullman D, Riddle M, Hatcher L, Smith FE, Wilson D. The hypotriglyceridemic effect of fish oil in adult-onset diabetes without adverse glucose control. *Ann NY Acad Sci.* 1993;683:337–440.
70. Schmidt EB, Kristensen SD, DeCaterina R, Illingworth DR. The effects of *n*–3 fatty acids on plasma lipids and lipoproteins and other cardiovascular risk factors in patients with hyperlipidemia. *Atherosclerosis.* 1993;103:107–121.
71. Huff MW, Telford DE. Dietary fish oil increases the conversion of very low density lipoprotein B to low density lipoprotein. *Arteriosclerosis.* 1989;9:58–66.
72. Heine RJ. Dietary fish oil and insulin action in humans. *Ann N Y Acad Sci.* 1993;683:110–121.
73. Connor WE. Diabetes, fish oil and vascular disease. *Ann Intern Med.* 1995;123:950–952, 1995.
74. Toft I, Bonaa KH, Ingebretsen OC, Nordoy A, and Jenssen T. Effects of *n*–3 polyunsaturated fatty acids on glucose homeostasis and blood pressure in essential hypertension: a randomized, controlled trial. *Ann Intern Med.* 1995;123:911–918.
75. Bonaa KH, Bjerve KS,Straumme B, Gram IT, Thelle D. Effect of eicosapentaenoic and docosahexaenoic acids on blood pressure in hypertension: a population-based intervention trial from the Tromso study. *N Engl J Med.* 1990;322: 795–801.
76. Bao DQ, Mori TA, Burke V, Puddey IB, Beilen LJ. Effects of dietary fish and weight reduction on ambulatory blood pressure in overweight hypertensives. *Hypertension.* 1998;32:710–717.
77. de Deckere EA, Korver O, Verschuren PM, Katan MB. Health aspects of fish and *n*–3 polyunsaturated fatty acids from plant and marine origin. *Eur J Clin Nutr.* 1998;52:749–753.
78. http://www.nal.usda.gov/fnic/foodcomp/search / Accessed 10/21/08.
79. Leaf DA, Connor WE, Barstad L, Sexton G. Incorporation of dietary *n*–3 fatty acids into the fatty acids of human adipose tissue and plasma lipid classes. *Am J Clin Nutr.* 1995;62:68–73.

11 Cardiovascular Effects of *Trans* Fatty Acids

Ronald P. Mensink

Key Points

- *Trans* fatty acids are found in hydrogenated fats and in fats from ruminants.
- The intake of *trans* fatty acids in the United States is in decline and is around 2% of energy.
- Tissue levels of *trans* fatty acids reflect dietary intakes.
- *Trans* fatty acids have an adverse effect on the serum lipoprotein profile.
- Epidemiological studies show a positive relationship between the intake of *trans* fatty acids from partially hydrogenated fats and cardiovascular risk.
- Keep the intake of *trans* fatty acids below 1% of energy.

Key Words: Cardiovascular disease; endothelial function; hemostasis; inflammation; LDL oxidation; metabolism; serum lipoproteins; *trans* fatty acids

11.1. INTRODUCTION

The position of the double bond in the carbon chain of a fatty acid is indicated in several ways. When counted from the carboxyl end (–COOH) of the molecule, the position (x) is denoted by the "Δ–x"-nomenclature, while the "n–x" classification is used when counting starts from the methyl end (–CH_3). Thus, "n–3" means that the double bond is located at the third carbon atom from the methyl end. These double bonds can have either the *cis* or the *trans* configuration. *Cis* means that the two carbon atoms adjacent to the double bound point in the same direction, whereas *trans* means that the two carbon atoms are located at opposite sides of the double bond. As an example, α-linolenic acid (C18:3 n–3), which belongs to the n–3 family, and one of its *trans* isomers (C18:3 n–3 Δ–9*c*,12*c*,15*tr*) are shown in Fig 11.1. These two molecules are the so-called "geometrical isomers."

Trans fatty acids in the diet are derived from several sources. First, they are formed, to small extents, from polyunsaturated fatty acids by bacteria in the first stomach (rumen) of ruminant animals.

Second, *trans* fatty acids are produced by industrial hydrogenation and deodorization of vegetable oils. Hydrogenation is used to change the chemical, physical, and sensory characteristics of a vegetable oil to make the oil suitable for the production of foods such as margarines, shortenings, and biscuits. When all double bonds are hydrogenated, a saturated fatty acid is formed. However, the *cis* double bond may also isomerize into a *trans* double bond without net uptake of hydrogen. Furthermore, double bonds can migrate along the molecule, which results in the formation of positional isomers. Therefore, through hydrogenation, many different molecules can be formed (Fig. 11.2), although *trans* monounsaturated fatty acids are the most prevailing molecules.

Trans polyunsaturated fatty acids—in particular those from α-linolenic acid—can be produced during deodorization, the last step of the refining process of vegetable oils. Finally, heating and frying of oils at high temperatures may also result in the formation of *trans* polyunsaturated bonds.

A. Bendich, R.J. Deckelbaum (eds.), *Preventive Nutrition*, Nutrition and Health, DOI 10.1007/978-1-60327-542-2_11,

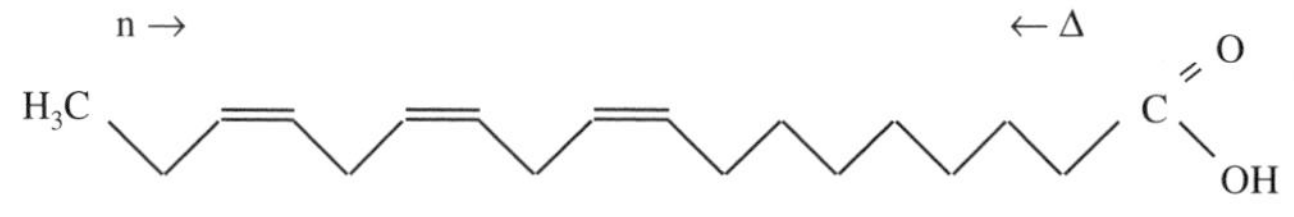

n → ← Δ

H_3C ... C(=O)OH

Cis α-linolenic acid
C18:3n-3 Δ9c,12c,15c

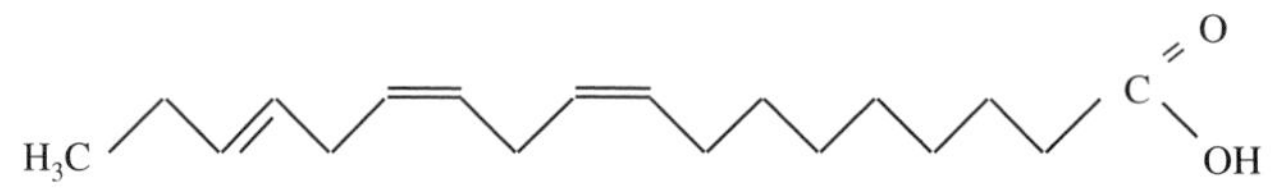

H_3C ... C(=O)OH

Trans α-linolenic acid
C18:3n-3 Δ9c,12c,15tr

Fig. 11.1. Chemical structure of all-*cis* C18:3 *n*–3 and 9-*cis*, 12-*cis*, 15-*trans* C18:3 *n*–3 (C18:3 *n*–3 Δ9*c*,12*c*,15*tr*).

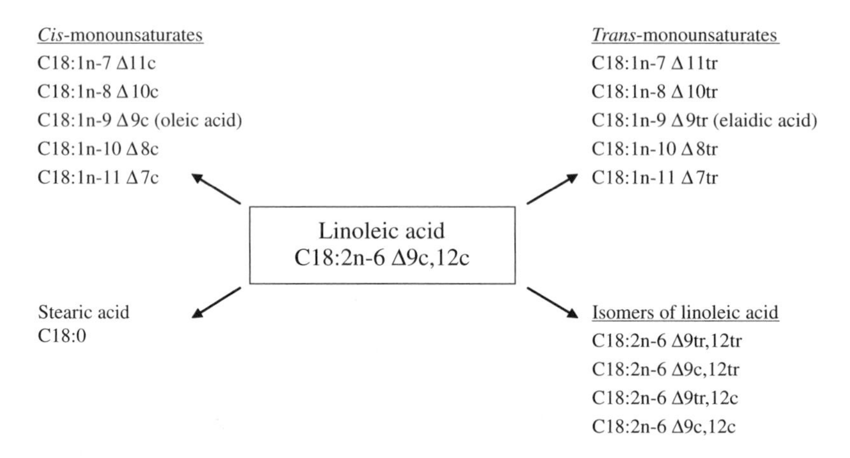

Fig. 11.2. Potential conversions of linoleic acid into its positional and geometrical isomers.

The beneficial effects of *cis* fatty acids on health differ from those of *trans* fatty acids. This chapter now discusses the effects of *trans* fatty acids on various risk parameters for coronary heart disease (CHD), such as lipids and lipoproteins, low-density lipoprotein (LDL) oxidation, and hemostasis. The metabolism of *trans* fatty acids and their effects on the desaturation and elongation of other fatty acids are also addressed.

11.2. *TRANS* FATTY ACIDS IN FOODS

Trans fatty acids in margarines and dairy products are mainly found in triacylglycerols; in meat, *trans* fatty acids are found in phospholipids. Intakes of *trans* fatty acids in the United States are declining. From 1980 to 1982, it was estimated that daily intakes were 3.0 and 2.8% of energy in men and women, respectively. During 2000–2002, these values were 2.3% for men and 2.2% for women *(1)*. For comparison, the intake of saturated fatty acids in 2000–2002 was about 11% of energy. Due to stringent regulations concerning the labeling of *trans* fatty acids, the rapid reformulation of food

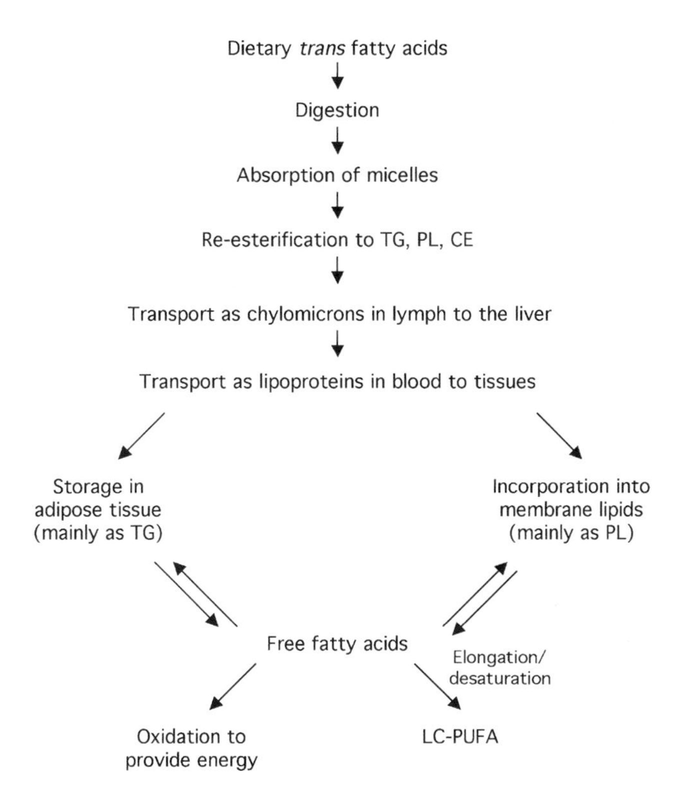

Fig. 11.3. Simplified scheme of *trans* fatty acid metabolism in humans. TG, triacylglycerols; PL, phospholipids; CE, cholesteryl esters; LC-PUFA, long-chain polyunsaturated fatty acids.

products by the industry, and increased consumer awareness, it is expected that current intakes of *trans* fatty acids will be lower.

Major contributors to total *trans* fatty acid intake are fried foods and snacks, stick margarines, bakery products, and, to a minor extent, dairy products. About 85% of all *trans* fatty acids in the diet are *trans* monounsaturated fatty acids, mostly C18:1 *trans*. When derived from ruminant fats, the double bond is mainly at the Δ–11 position (C18:1 Δ–11 *trans*, which is called vaccenic acid). This is because these *trans* isomers are formed through the specific action of bacteria in the rumen. In industrially hydrogenated oils, *trans* isomers are formed chemically and therefore the double bond varies more or less randomly between positions Δ–8 and Δ–13. The remaining 15% of *trans* fatty acid intake is mainly provided by mono-*trans* isomers of linoleic and α-linolenic acid; a few di-*trans* isomers are also present.

11.3. METABOLISM

11.3.1. Digestion, Absorption, and Incorporation into Blood Lipids

The metabolic pathways of *trans* fatty acids after intake are shown in Fig. 11.3. After ingestion, triacylglycerols are hydrolyzed by pancreatic lipase into free fatty acids, sn2-monoglycerides, and, to a small extent, diglycerides. Phospholipids are split by pancreatic phospholipases, which remove the fatty acid from the sn1 or the sn2 position. As a result, lysophospholipids and free fatty acids are formed.

The breakdown products of triacylglycerol and phospholipids are incorporated into micelles and absorbed by the enterocyte. Hydrolysis and absorption of oleic acid and its positional and geometrical isomers are comparable. Within the enterocyte, the free fatty acids are used for the formation of triacylglycerols, cholesteryl esters, and phospholipids. After esterification, lipids are used for the formation of chylomicrons, which transport the fatty acids through the lymph to the liver. In the liver, fatty acids can be incorporated into very-low-density lipoproteins (VLDLs). The incorporation of *trans* isomers of C18:1 Δ–9 into VLDL, but also in other lipoproteins like LDL and high-density lipoproteins (HDLs), is lower than that of C18:1 Δ–9*c*. As any other fatty acid, *trans* fatty acids are also incorporated into tissue phospholipids, stored in adipose tissue, desaturated and elongated into longer chain polyunsaturated fatty acids (LC-PUFA), and oxidized *(3–7)*.

11.3.2. Tissue Levels

As humans do not make *trans* fatty acids, tissue levels reflect dietary intakes. The *trans* fatty acid content of adipose tissue for US subjects is about 4% of total fatty acids *(7)*. Approximately 70% of the *trans* fatty acids are C18:1 isomers with double bonds at the Δ8, 9, 10, 11, 12, or 13 positions. *Trans* C16:1 isomers, which originate mainly from dairy fat, are also found in adipose tissue. About 20% of the *trans* isomers are *trans* C18:2 isomers. Both mono-*trans* (Δ9*c*,12*t* and Δ9*t*,12*c*) and di-*trans* (Δ9*t*,12*t*) C18:2 isomers and several *trans* C18:3 isomers can be detected in adipose tissue. *Trans* C18:1 and C18:2 fatty acids have been found in human kidney, brain, heart, liver, aorta, jejunum, and human milk *(8)*, and *trans* C18:3 can be detected in platelets and human milk *(9,10)*.

11.3.3. Conversion

Human liver microsomal complexes contain three different desaturation enzymes: Δ–9, Δ–6, and Δ–5 desaturases, which insert double bonds at the Δ–9, Δ–6, and Δ–5 position of the fatty acid molecule, respectively. Furthermore, fatty acids can be elongated by addition of a two-carbon unit. In this way, stearic acid, linoleic acid, and α-linolenic acid are converted into their longer chain metabolites (Fig. 11.4), which play an important role in many physiological processes. As chain elongation and desaturation always occur at the carboxyl end of the fatty acid molecule, the position of the first double bond at the methyl end will not change during conversion reactions.

Trans fatty acids are desaturated by the same enzymes as *cis* fatty acids. Except for Δ–8, Δ–9, and Δ–10 *trans* isomers, C18:1 positional isomers are good substrates for Δ–9 desaturase *(11)*. Consequently, *cis,trans* and *trans,cis* fatty acids can be formed. Some *trans* isomers of C18:1 are also substrates for Δ–6 and Δ–5 desaturases *(12)*. In addition, C18:1 *trans* isomers can be elongated into C20 and C22 fatty acids.

Trans linoleic acid can be converted into *trans* isomers of arachidonic acid (C20:4 *n*–6), although at a lower rate than *all-cis* C18:2 *(13)*. *Trans* isomers of α-linolenic acid can be converted into *trans* isomers of docosahexaenoic acid (C22:6 *n*–3) *(14)*.

Because *trans* and *cis* fatty acids are converted by the same desaturase and elongase enzymes, competition between fatty acids exists. In vitro studies do suggest that both the *cis* and the *trans* monoenoic positional isomers inhibit Δ–6 and –5 desaturation of linoleic and α-linolenic acids, and their longer chain metabolites. In contrast, C18:1 Δ–9*tr* increased Δ–9 desaturation of stearic acid, while C18:1 Δ–11*tr* and C18:2 Δ–9*tr*,12*tr* had no effects on Δ–9 desaturase activity *(15,16)*. It is difficult to extrapolate these in vitro findings to the in vivo situation. In fact, in humans no effect of *trans* α-linolenic acid was found on the conversion of linoleic acid *(17)*. Therefore, it seems that in adult volunteers consuming a diet adequate in essential fatty acids, the present intake of *trans* isomers does not affect desaturase activity.

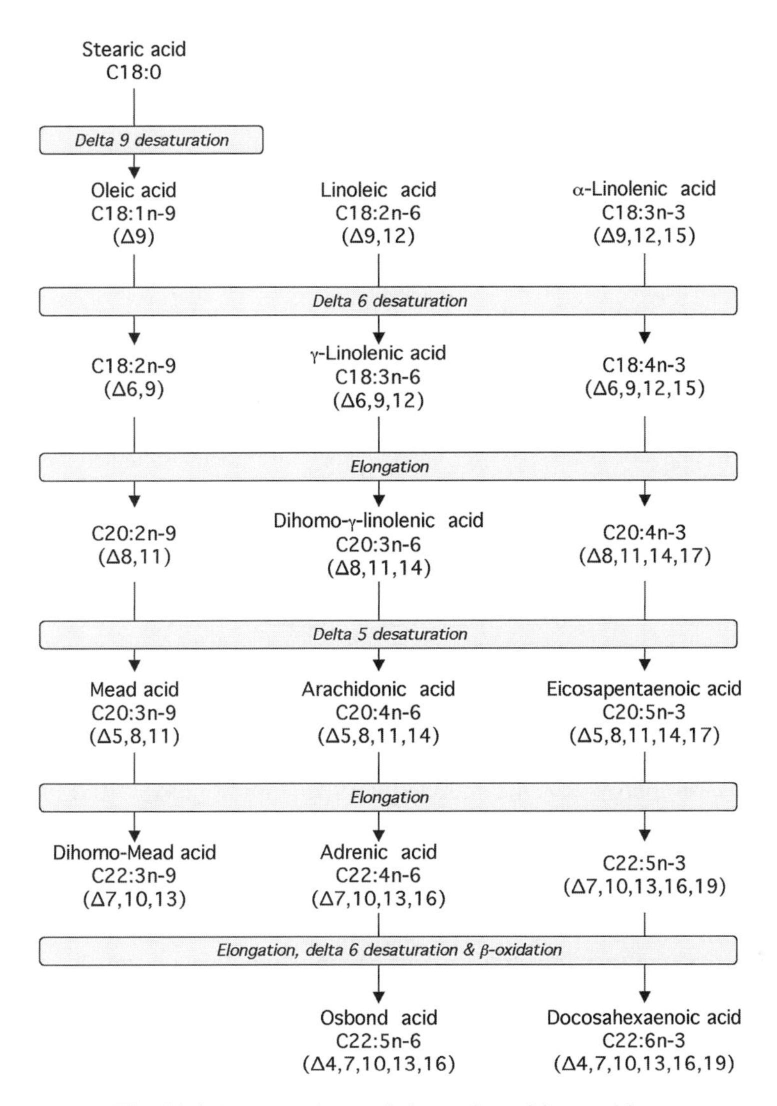

Fig. 11.4. Desaturation and elongation of fatty acids.

11.3.4. Oxidation

Only a few human studies have been carried out to examine the postprandial oxidation of *trans* fatty acids. As compared to their *cis* counterparts, *trans* isomers of oleic and α-linolenic acids were oxidized to the same extent, while isomerization increased the postprandial oxidation of linoleic acid *(18,19)*.

11.4. EFFECTS OF *TRANS* MONOUNSATURATED FATTY ACIDS ON SERUM LIPIDS AND LIPOPROTEINS

11.4.1. Serum Total, LDL, and HDL Cholesterol

Effects of dietary *trans* monounsaturated fatty acids on serum cholesterol concentrations have been investigated since the early 1960s. Most studies found increased serum total cholesterol levels in humans consuming partially hydrogenated vegetable oils. Results, however, were not

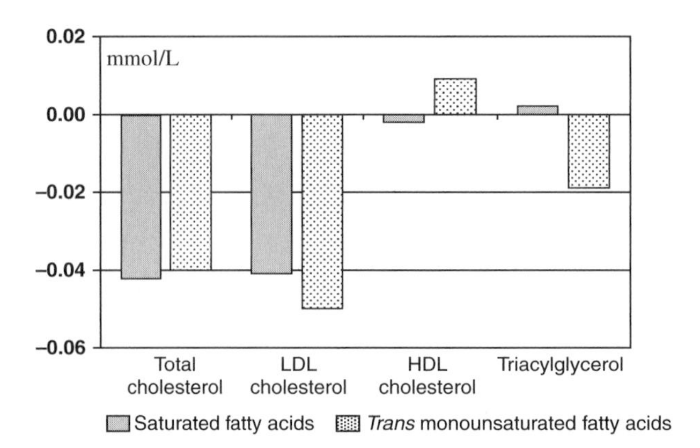

Fig. 11.5. Effects of exchanging 1% of energy from saturated or *trans* monounsaturated fatty acids for oleic acid on serum total, LDL, and HDL cholesterol and on triacylglycerol concentrations.

uniform. In addition, these earlier studies did not examine the relationship between dietary *trans* monounsaturated fatty acids and the distribution of cholesterol over the different lipoproteins. In one of the first controlled intervention studies that specifically examined the effects of *trans* monounsaturated fatty acids on serum LDL and HDL cholesterol, 25 men and 34 women were fed a diet high in oleic acid, high in *trans* isomers of oleic acid, or high in a mixture of saturated fatty acids lauric and palmitic acid *(20)*. Each diet was fed for 3 weeks. The level of *trans* monounsaturated fatty acids in the *trans* diet was 11% of energy. Results showed that *trans* monounsaturated fatty acids significantly raised serum total and LDL cholesterol levels by 0.26 and 0.37 mmol/L, respectively, and lowered HDL cholesterol levels by 0.17 mmol/L compared to oleic acid. These effects were confirmed by later studies that tested lower intakes of *trans* monounsaturated fatty acids. Recently, a meta-analysis was performed that combined the results of all recent studies on the effects of *trans* fatty acids on serum lipoproteins. It was estimated that each additional percent of dietary energy as *trans* monounsaturated fatty acids at the expense of oleic acid results in an increase in LDL cholesterol levels of 0.050 mmol/L and a decrease in HDL cholesterol levels of 0.009 mmol/L (Fig. 11.5) *(21)*.

Consumption of *trans* monounsaturated fatty acids also decreases LDL particle size *(22)*. This is another reason to keep the intake of *trans* fatty acids as low as possible, as small, dense LDL is a risk marker for CHD. Similarly, an elevated fasting triacylglycerol concentration is a risk marker for CHD. It has been estimated that each additional percent of energy as *trans* monounsaturated fatty acids at the expense of oleic acid increases triacylglycerol levels by 0.019 mmol/L *(21)*.

Finally, effects of *trans* monounsaturated fatty acids on lipoprotein(a) [Lp(a)] have been studied. Lp(a) is an LDL particle with an extra glycoprotein [apoprotein(a)], attached through a disulfide link. High levels of Lp(a) increase the risk for CHD. Therefore, effects of *trans* monounsaturated fatty acids on Lp(a) have been investigated in several intervention studies. From these studies it can be concluded that—when compared with saturated fatty acids—*trans* monounsaturated fatty acids raise Lp(a), in particular in subjects who already have increased Lp(a) levels *(21)*.

11.4.2. Conclusion

Trans monounsaturated fatty acids do have adverse effects on the serum lipoprotein profile, as they raise serum LDL cholesterol, Lp(a), and triacylglycerol levels, while HDL cholesterol levels and LDL particle size decrease. *Trans* polyunsaturated fatty acids—though less extensively studied—also seem to have unfavorable effects on the serum lipoprotein profile *(23)*.

11.4.3. Mechanism

The mechanisms underlying the effects of *trans* fatty acids on serum lipoprotein concentrations are poorly understood. It has been hypothesized that cholesteryl ester transfer protein (CETP) is involved. CETP transfers cholesteryl esters from HDL to the apolipoprotein B-containing lipoproteins LDL and VLDL in exchange for triacylglycerol. Indeed, an increased transfer of cholesteryl ester transfer activity has been found in volunteers who consumed a *trans* fatty acid-enriched diet *(24)*.

Lecithin:cholesterol acyltransferase (LCAT) has also been suggested to be involved. LCAT, which is bound to HDL, esterifies free cholesterol from tissues by the transfer of an acyl group from phosphatidylcholine (PC). Since a significant part of the dietary *trans* fatty acids is incorporated into PC *(6)*, a few studies have investigated the effect of *trans* fatty acids on LCAT. In one study it was shown that *trans* fatty acids decreased LCAT activity *(25)*, while another study did not demonstrate an effect of *trans* fatty acids on LCAT activity *(26)*.

In the liver, fatty acids from chylomicrons are esterified to free cholesterol by acyl-CoA:cholesterol acyltransferase (ACAT). The affinity of ACAT differs for different fatty acids. A low affinity of the enzyme for a particular fatty acid increases the proportion of free cholesterol in the liver, which decreases LDL receptor activity. Since the LDL receptor removes LDL from plasma, a decreased LDL receptor activity results in higher plasma LDL concentrations. Thus, through this pathway, dietary fatty acids can influence the plasma LDL cholesterol concentration. In hamsters, *trans* monounsaturated fatty acids indeed decreased hepatic cholesteryl ester concentrations relative to oleic acid *(27)*. This suggests that the affinity of ACAT is lower for *trans* monounsaturated fatty acids as compared to oleic acid, which may result in higher free cholesterol levels in the liver. In addition, the plasma LDL cholesterol concentration was increased and the LDL receptor activity was decreased on the diet high in *trans* monounsaturated fatty acids. *Trans* monounsaturated fatty acids may therefore affect plasma LDL cholesterol via the pathway as described above. This mechanism, however, has not been investigated in man. It is expected that a decreased flux of LDL cholesterol from the blood to cells leads to an increased de novo cholesterol synthesis, simply because cells do need cholesterol. However, in healthy volunteers, de novo cholesterol synthesis was not different after consumption of a diet high in hydrogenated or unhydrogenated corn oil *(28)*. In another study, however, exchanging partially hydrogenated fat for palmitic acid in the diet did increase endogenous cholesterol synthesis *(29)*. Thus, the precise mechanism by which *trans* fatty acids influence serum lipoprotein concentrations is unknown.

11.5. EFFECTS OF *TRANS* FATTY ACIDS ON OTHER RISK FACTORS FOR CORONARY HEART DISEASE

11.5.1. LDL Oxidation

Free radicals initiate the oxidation of unsaturated fatty acids in the LDL particle, which may ultimately lead to the formation of an atherosclerotic plaque. As the fatty acid composition of the LDL particle reflects the fatty acid composition of the diet, effects of increased intakes of *trans* fatty acids on the in vitro susceptibility of LDL to oxidation have been investigated in several intervention trials. However, no effects were found after consumption of *trans* monounsaturated fatty acids as compared to other fatty acids *(28)*. Thus, at least in vitro, *trans* fatty acids have no impact on LDL oxidizability.

11.5.2. Hemostasis

Only a few human studies have investigated the effects of *trans* fatty acids on markers for platelet aggregation, coagulation, and fibrinolysis—three determinants of hemostatic function. These studies have yielded conflicting results, which may be partly due to the many different methods to measure hemostatic function. Therefore, it seems that *trans* fatty acids do not have a major impact on hemostatic function *(30–32)*.

11.5.3. Low-Grade Systemic Inflammation

Recent epidemiological studies have suggested that *trans* fatty acids from partially hydrogenated oils are related to parameters that reflect a low-grade proinflammatory state. In women, *trans* fatty acid intake was positively associated with an increased activity of the tumor necrosis factor (TNF) system. Only in women with an increased body mass index, *trans* fatty acid intake correlated positively with plasma concentrations of interleukin-6 (IL-6) and C-reactive protein (CRP) *(33)*. In another study in women, *trans* fatty acid intake was positively associated with an increased activity of the TNF system and increased plasma concentrations of interleukin-6 and C-reactive protein *(34)*. Comparable conclusions were drawn from a study in patients with established heart disease *(35)*.

Results of dietary intervention studies are however less consistent than those of the epidemiological studies. In one study it was found that *trans* fatty acids increased plasma concentrations of IL-6 and C-reactive protein, as compared with oleic acid but not as compared with stearic acid *(36)*. In subjects with moderately increased LDL cholesterol levels, the production of IL-6 and TNF-α by cultured mononuclear cells was increased after consumption of a diet rich in *trans* fatty acids, as compared with linoleic acid. Effects of saturated fatty acids were in between *(37)*. Other intervention studies, however, did not suggest that *trans* fatty acids increase the proinflammatory response *(38)*.

11.5.4. Endothelial Cell Function

Epidemiological studies have also suggested that *trans* fatty acids also impair endothelial function *(34)*. An increase in plasma E-selectin concentrations was also observed in a dietary intervention study when *trans* fatty acids replaced oleic acid, saturated fatty acids, or carbohydrates *(36)*. Compared with saturated fatty acids, consumption of *trans* fatty acids also impaired endothelial function, as reflected by a reduction in brachial artery flow-mediated vasodilatation *(39)*.

11.6. EPIDEMIOLOGICAL STUDIES

Various approaches have been used in epidemiological studies to examine the relationship between the intake of *trans* fatty acids and the risk for CHD. Because high serum LDL cholesterol levels are positively associated with CHD and high levels of HDL cholesterol are antiatherogenic, some epidemiological studies focused on the relation between the proportion of *trans* fatty acids in tissues, which reflects dietary intakes, or the dietary *trans* fatty acid intake and the lipid profile *(40–42)*. Altogether, these epidemiological studies confirmed the dietary intervention studies, which found that *trans* fatty acids have a negative impact on the lipid profile. However, these studies did not address the question of whether high intakes of *trans* fatty acids were truly related to the risk for CHD. This has been examined in several case–control studies. In these studies, the proportion of *trans* fatty acids in tissues or the dietary *trans* fatty acid intake of case subjects (e.g., subjects with CHD) has been compared with that of control subjects (e.g., subjects without CHD). Some studies have suggested higher total *trans* fatty acid intakes in cases as compared to controls, while other studies have not found any effects of *trans* fatty acids on CHD risk *(21,43)*. However, case–control studies have the disadvantage that they are sensitive to information bias, selection bias, and confounding. Four prospective cohort studies, which suffer less from these disadvantages, have examined the relationship between the *trans* fatty acid intake and the risk for CHD. In the Nurses' Health Study *(44)*, the relative risk (RR) of CHD

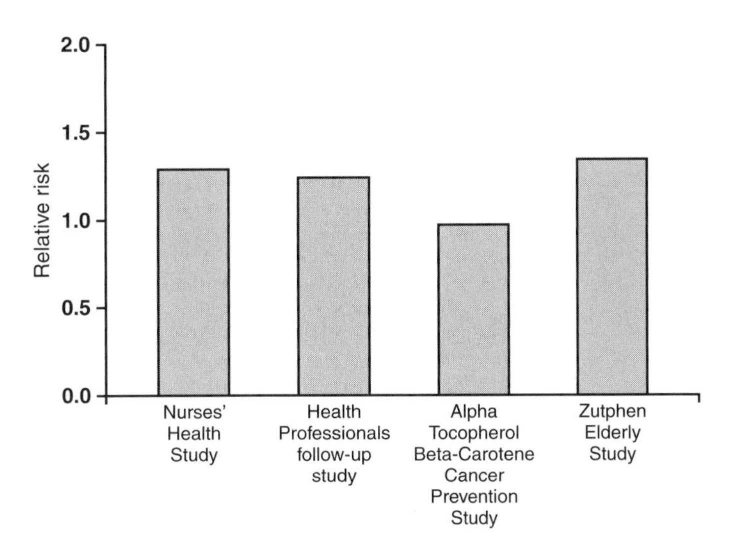

Fig. 11.6. Relative risk of coronary heart disease of the middle vs. the lowest quintile *(44–46)* or tertile *(47)* from four different prospective cohort studies.

was 1.33 for the highest quintile of *trans* fatty acid intake (median intake: 2.8% of energy) relative to the lowest quintile (median intake: 1.3% of energy). This means that women who consume 2.8% of energy as*trans* fatty acids per day have a 33% higher risk for CHD than women with a *trans* fatty acid intake of 1.3% of energy. These results were confirmed by three other prospective cohort studies *(45–47)*, shown in Fig. 11.6.

In these epidemiological studies, the focus was on *trans* fatty acids from partially hydrogenated oils. Some prospective cohort studies have also looked at relationships between intakes of *trans* fatty acids from ruminants and risk for CHD. In these studies, no significant positive associations with CHD risk were identified. It is impossible to conclude if the absence of an association was due to the lower levels of intake of ruminant *trans* fatty acids, to differences in metabolic effects of these two types of *trans* fatty acids, or to effects of other constituents in ruminant-derived products *(21)*. Whatever be the explanation, the impact of the current level of intake of *trans* fatty acids from ruminant sources on cardiovascular health seems limited.

In summary, results from case–control studies are inconsistent. For prospective cohort studies, results are more uniform: a high intake of *trans* fatty acids from partially hydrogenated oils is associated with an increased risk of CHD.

11.7. CONCLUSIONS AND RECOMMENDATIONS

The digestion and absorption of *trans* monounsaturated fatty acids are comparable with those of their *cis*isomers. However, incorporation of various *trans* monoenoic isomers into lipid classes and lipoproteins is lower than that of the *cis* isomers. Furthermore, both *trans* monounsaturated and *trans* polyunsaturated fatty acids are desaturated and elongated into longer chain metabolites. At present intakes, *trans* isomers do not affect desaturase activity in adults consuming a diet adequate in essential fatty acids. As compared to their *cis* counterparts, *trans* isomers of oleic and α-linolenic acids are oxidized to the same extent, whereas isomerization increased the postprandial oxidation of linoleic acid. However, these processes have mainly been investigated in animal and in vitro studies. More human in vivo studies are necessary.

Most studies on the effects of *trans* fatty acids and health have focused on *trans* monounsaturated fatty acids from partially hydrogenated oils, and effects of *trans* polyunsaturated fatty acids are hardly known. *Trans* monounsaturated fatty acids increase serum LDL cholesterol, triacylglycerols, and Lp(a) concentrations and decrease those of serum HDL cholesterol as compared with oleic acid. The exact underlying mechanisms for these effects are not known. Furthermore, *trans* monounsaturated fatty acids seem to have no major impact on LDL oxidizability, platelet aggregation, coagulation, and fibrinolysis. Prospective epidemiological studies showed a positive association between *trans* fatty acid intakes and the risk for CHD. Based on these results, a reduction in the *trans* fatty acid intake of less than 1% of energy is more than justified *(48)*.

REFERENCES

1. Lee S, Harnack L, Jacobs DR Jr, Steffen LM, Luepker RV, Arnett DK. Trends in diet quality for coronary heart disease prevention between 1980–1982 and 2000–2002: The Minnesota Heart Survey. J Am Diet Assoc 2007;107: 213–22.
2. Emken EA. Do *trans* acids have adverse health consequences? In: Nelson G, ed. Health Effects of Dietary Fatty Acids. Champaign, IL: American Oil Chemists Society, 1991:245–60.
3. Emken EA, Dutton HJ, Rohwedder WK, Rakoff H, Adlof RO. Distribution of deuterium-labeled *cis* and *trans*-12-octadecemoic acids in human plasma and lipoprotein lipids. Lipids 1980;15:864–71.
4. Emken EA, Adlof RO, Rohwedder WK, Gulley RM. Incorporation of deuterium-labeled *trans*- and *cis*-13-octadecenic acids in human plasma lipids. J Lipid Res 1983;24:34–46.
5. Emken EA, Rohwedder WK, Adlof RO, DeJarlais WJ, Gulley RM. Absorption and distribution of deuterium-labeled *trans*- and *cis*-11-octadecenoic acid in human plasma and lipoprotein lipids. Lipids 1986;21:589–95.
6. Emken EA, Adlof RO, Rohwedder WK, Gulley RM. Incorporation of *trans*-8- and *cis*-8-octadecenoic acid isomers in human plasma and lipoprotein lipids. Lipids 1989;24:61–9.
7. Kris-Etherton PM, ed. *Trans* fatty acids and coronary heart disease risk. Am J Clin Nutr 1995;62:655S–708S.
8. Heckers H, Korner M, Tuschen TW, Melcher FW. Occurrence of individual *trans*-isomeric fatty acids in human myocardium, jejunum and aorta in relation to different degrees of atherosclerosis. Atherosclerosis 1977;28: 389–98.
9. Sébédio JL, Vermunt SH, Chardigny JM, et al. The effect of dietary *trans* alpha-linolenic acid on plasma lipids and platelet fatty acid composition: the Transline study. Eur J Clin Nutr 2000;54:104–13.
10. Chardigny JM, Wolff RL, Mager E, Sébédio JL, Martine L, Juanéda P. *Trans* mono- and polyunsaturated fatty acids in human milk. Eur J Clin Nutr 1995;49:523–31.
11. Mahfouz MM, Valicenti AJ, Holman RT. Desaturation of isomeric *trans*-octadecenoic acids by rat liver microsomes. Biochim Biophys Acta 1980;618:1–12.
12. Pollard MR, Gunstone FD, James AT, Morris LJ. Desaturation of positional and geometric isomers of monoenoic fatty acids by microsomal preparations from rat liver. Lipids 1980;15:306–14.
13. Beyers EC, Emken EA. Metabolites of *cis, trans*, and *trans, cis* isomers of linoleic acid in mice and incorporation into tissue lipids. Biochim Biophys Acta 1991;1082:275–84.
14. Grandgirard A, Piconneaux A, Sébédio JL, O'Keefe SF, Semon E, Le Quere JL. Occurrence of geometrical isomers of eicosapentaenoic and docosahexaenoic acids in liver lipids of rats fed heated linseed oil. Lipids 1989;24: 799–804.
15. Cook H, Emken EA. Geometric and positional fatty acid isomers interact differently with desaturation and elongation of linoleic and linolenic acids in cultured glioma cells. Biochem Cell Biol 1990;68:653–60.
16. Rosenthal MD, Whitehurst MC. Selective effects of isomeric *cis* and *trans* fatty acids on fatty acyl $\Delta 9$ and $\Delta 6$ desaturation by human skin fibroblasts. Biochim Biophys Acta 1983;753:450–9.
17. Scrimgeour CM, Macvean A, Fernie CE, et al. Dietary *trans* α-linolenic acid does not inhibit D5- and D6-desaturation of linoleic acid in man. Eur J Lipid Sci Technol 2001;103:341–9.
18. DeLany JP, Windhauser MM, Champagne CM, Bray GA. Differential oxidation of individual dietary fatty acids in humans. Am J Clin Nutr 2000;72:905–11.
19. Bretillon L, Chardigny JM, Sebedio JL, et al. Isomerization increases the postprandial oxidation of linoleic acid but not alpha-linolenic acid in men. J Lipid Res 2001;42:995–7.
20. Mensink RP, Katan MB. Effect of dietary *trans* fatty acids on high-density and low-density lipoprotein cholesterol levels in healthy subjects. N Engl J Med 1990;323:439–45.

21. Mozaffarian D, Katan MB, Ascherio A, Stampfer MJ, Willett WC. *Trans* fatty acids and cardiovascular disease. N Engl J Med 2006;354:1601–13.
22. Mauger JF, Lichtenstein AH, Ausman LM, et al. Effect of different forms of dietary hydrogenated fats on LDL particle size. Am J Clin Nutr 2003;78:370–5.
23. Vermunt SH, Beaufrere B, Riemersma RA, et al. Dietary *trans* alpha-linolenic acid from deodorised rapeseed oil and plasma lipids and lipoproteins in healthy men: the Transline study. Brit J Nutr 2001;85:387–92.
24. Gatto LM, Sullivan DR, Samman S. Postprandial effects of dietary *trans* fatty acids on apolipoprotein(a) and cholesteryl ester transfer. Am J Clin Nutr 2003;77:1119–24.
25. Subbaiah PV, Subramanian VS, Liu M. *Trans* unsaturated fatty acids inhibit lecithin: cholesterol acyltransferase and alter its positional specificity. J Lipid Res 1998;39:1438–47.
26. Van Tol A, Zock PL, Van Gent T, Scheek LM, Katan MB. Dietary *trans* fatty acids increase serum cholesteryl ester transfer protein activity in man. Atherosclerosis 1995;115:129–34.
27. Woollett LA, Daumerie CM, Dietschy JM. *Trans*-9-octadecenoic acid is biologically neutral and does not regulate the low density lipoprotein receptor as the *cis* isomer does in the hamster. J Lipid Res 1994;35:1661–73.
28. Cuchel M, Schwab US, Jones PJ, et al. Impact of hydrogenated fat consumption on endogenous cholesterol synthesis and susceptibility of low-density lipoprotein to oxidation in moderately hypercholesterolemic individuals. Metabolism 1996;45:241–7.
29. Sundram K, French MA, Clandinin MT. Exchanging partially hydrogenated fat for palmitic acid in the diet increases LDL-cholesterol and endogenous cholesterol synthesis in normocholesterolemic women. Eur J Nutr 2003;42:188–94.
30. Mutanen M, Aro A. Coagulation and fibrinolysis factors in healthy subjects consuming high stearic or *trans* fatty acid diets. Thromb Haemost 1997;77:99–104.
31. Turpeinen A, Wubert J, Aro A, Lorenz R, Mutanen M. Similar effects of diets rich in stearic acid or *trans*-fatty acids on platelet function and endothelial prostacyclin production in humans. Arterioscler Thromb Vasc Biol 1998;18:316–22.
32. Sanders TA, Oakley FR, Crook D, Cooper JA, Miller GJ. High intakes of *trans* monounsaturated fatty acids taken for 2 weeks do not influence procoagulant and fibrinolytic risk markers for CHD in young healthy men. Br J Nutr 2003;89:767–76.
33. Mozaffarian D, Pischon T, Hankinson SE, et al. Dietary intake of *trans* fatty acids and systemic inflammation in women. Am J Clin Nutr 2004;79:606–12.
34. Lopez-Garcia E, Schulze MB, Meigs JB, et al. Consumption of *trans* fatty acids is related to plasma biomarkers of inflammation and endothelial dysfunction. J Nutr 2005;135:562–6.
35. Mozaffarian D, Rimm EB, King IB, Lawler RL, McDonald GB, Levy WC. *Trans* fatty acids and systemic inflammation in heart failure. Am J Clin Nutr 2004;80:1521–5.
36. Baer DJ, Judd JT, Clevidence BA, Tracy RP. Dietary fatty acids affect plasma markers of inflammation in healthy men fed controlled diets: a randomized crossover study. Am J Clin Nutr 2004;79:969–73.
37. Han SN, Leka LS, Lichtenstein AH, Ausman LM, Schaefer EJ, Meydani SN. Effect of hydrogenated and saturated, relative to polyunsaturated, fat on immune and inflammatory responses of adults with moderate hypercholesterolemia. J Lipid Res 2002;43:445–52.
38. Mensink RP. Effects of products made from a high-palmitic acid, *trans*-free semiliquid fat or a high-oleic acid, low-*trans* semiliquid fat on the serum lipoprotein profile and on C-reactive protein concentrations in humans. Eur J Clin Nutr 2008;62:617–24.
39. De Roos NM, Bots ML, Katan MB. Replacement of dietary saturated fatty acids by *trans* fatty acids lowers serum HDL cholesterol and impairs endothelial function in healthy men and women. Arterioscler Thromb Vasc Biol 2001;21:1233–7.
40. Hudgins LC, Hirsch J, Emken EA. Correlation of isomeric fatty acids in human adipose tissue with clinical risk factors for cardiovascular disease. Am J Clin Nutr 1991;53:474–82.
41. Troisi RJ, Willett WC, Weiss ST. *Trans*-fatty acid intake in relation to serum lipid concentrations in adult men. Am J Clin Nutr 1992;56:1019–24.
42. Siguel E, Lerman RH. *Trans*-fatty acid patterns in patients with angiographically documented coronary artery disease. Am J Cardiol 1993;71:916–20.
43. Sun Q, Ma J, Campos H, Hankinson SE, Manson JE, Stampfer MJ, Rexrode KM, Willett WC, Hu FB. A prospective study of *trans* fatty acids in erythrocytes and risk of coronary heart disease. Circulation 2007;115:1858–65.
44. Oh K, Hu FB, Manson JE, Stampfer MJ, Willett WC. Dietary fat intake and risk of coronary heart disease in women: 20 years of follow-up of the nurses' health study. Am J Epidemiol. 2005;161:672–9.
45. Pietinen P, Ascherio A, Korhonen P, et al. Intake of fatty acids and risk of coronary heart disease in a cohort of Finnish men. Am J Epidemiol 1997;145:876–7.

46. Ascherio A, Rimm EB, Giovannucci EL, Spiegelman D, Stampfer M, Willett WC. Dietary fat and risk of coronary heart disease in men: cohort follow up study in the United States. Br Med J 1996;313:84–90.
47. Oomen CM, Ocke MC, Feskens EJ, van Erp-Baart MA, Kok FJ, Kromhout D. Association between *trans* fatty acid intake and 10-year risk of coronary heart disease in the Zutphen Elderly Study: a prospective population-based study. Lancet 2001;357:746–51.
48. Lichtenstein AH, Appel LJ, Brands M, Diet and lifestyle recommendations revision 2006: a scientific statement from the American Heart Association Nutrition Committee. Circulation 2006;114:82–96

12 Antioxidant and B-vitamins and Atherosclerosis

Juliana Hwang-Levine, Wendy J. Mack, and Howard N. Hodis

Key Points

- Atherosclerosis is an interlinked inflammatory free-radical process.
- Observational studies suggest that antioxidant and B-vitamins may reduce cardiovascular disease.
- Randomized controlled trials have not substantiated a reduction in cardiovascular disease with antioxidant or B-vitamin supplementation.
- Randomized controlled trials raise a level of concern for the safety of certain antioxidant vitamin supplementation under certain conditions.
- Current data do not support the use of antioxidant or B-vitamin supplementation for the prevention or treatment of cardiovascular disease.
- Consumption of fruits and vegetables high in antioxidant and B-vitamins appears to be an important component of a healthy dietary intake and lifestyle.

Key Words: Antioxidant vitamins; atherosclerosis; β-carotene; B-vitamins; cardiovascular disease; folic acid; observational studies; randomized controlled trials; vitamin B_6; vitamin B_{12}; vitamin C; vitamin E

12.1. INTRODUCTION

Cardiovascular disease (CVD) is the main cause of death in Western countries. It is well accepted that high intake of fruits and vegetables has a significant role in the prevention of many chronic diseases such as CVD, cancer, and degenerative brain disorders *(1)*. Due to the complexity of the nutrient content of fruits and vegetables, identification of specific nutrients responsible for cardioprotection is challenging. Collectively, the likely substances that provide the beneficial effects of fruits and vegetables include antioxidant vitamins and B-vitamins.

Atherosclerosis, the main cause of CVD, is a chronic inflammatory response in the walls of arteries. Atherosclerosis is primarily a process that affects medium and large-size arteries. The initial evidence of atherosclerosis is the fatty streak which is due to the accumulation of lipid-laden foam cells in the intimal layer of the artery. The fatty streak develops into the fibrous plaque which is the characteristic feature of atherosclerosis. Eventually the plaque invades the vessel lumen and obstructs blood flow *(2)*.

It is widely recognized that there is a strong association between oxidative stress, inflammation, and endothelial dysfunction at the onset of atherosclerosis *(3)*. This indicates a potential role for antioxidant and anti-inflammatory intervention in CVD. Experimental evidence indicates thatoxidative

A. Bendich, R.J. Deckelbaum (eds.), *Preventive Nutrition*, Nutrition and Health, DOI 10.1007/978-1-60327-542-2_12,

damage as well as inflammatory processes may be involved with atherogenic-promoting processes, such as endothelial damage *(4)*, and that antioxidants and anti-inflammatory agents can control oxidative damage and inflammation in atherosclerosis lesions *(5,6)*.

Studies in humans indicate an association between blood measures of oxidative processes, inflammatory markers, and atherosclerosis *(7)*. These data are paralleled by observational studies, including arterial imaging studies *(8,9)* that have demonstrated an inverse association between use of antioxidant vitamin and B-vitamin supplementation and atherosclerosis and CVD. However, data from randomized controlled trials (RCTs) have been less consistent, and the safety of certain supplements has even been questioned in some studies. In general, data from RCTs have not translated into clinical benefit by failing to reduce atherosclerosis and cardiovascular morbidity and mortality *(10)*.

This chapter will review the current evidence from observational studies and RCTs for an antiatherogenic effect of antioxidant vitamin and B-vitamin supplementation in humans.

12.2. REACTIVE OXYGEN SPECIES, ANTIOXIDANTS, AND ATHEROSCLEROSIS

Reactive oxygen species (ROS) such as superoxide (O_2^-) are increased in atherosclerosis *(11)*. In the arterial wall, activated NADH/NAD(P)H oxidase, xanthine oxidase, and 12/15 lipoxygenase reduce oxygen to superoxide ($O_2^{-}\cdot$) as well as disrupt the mitochondrial respiratory chain *(3)*. Evidence suggests that membrane-bound NAD(P)H oxidases are the major source of O_2^- generation and that both NAD(P)H oxidase-derived $O_2^{-}\cdot$ and mitochondrial-derived $O_2^{-}\cdot$ represent the bulk of the superoxide radical in the vasculature *(12)*.

ROS at physiological levels function as homeostatic signaling molecules that regulate cell growth and adaptive responses. At higher concentrations ROS cause cellular injury and death *(13)*. ROS peroxidize lipid components, thereby injuring cell membranes and nuclei (hydroxyl radicals). ROS also interact with endogenous vasoreactive mediators such as nitric oxide (NO) formed by endothelial cells. Increased ROS formation promotes endothelial dysfunction by reducing NO bioavailability *(14)*. ·rapidly scavenges NO-generating peroxynitrite, which is a highly reactive radical that facilitates lipid peroxidation *(15)*, endothelial injury, and atherosclerosis *(16)*.

Lipid peroxidation, a free-radical-related process, underlies many of the proinflammatory and vascular cell changes associated with atherogenesis *(17)*. Certain steps of arterial wall cellular metabolism that lead to atheroma formation are modulated by low-density lipoprotein (LDL)-derived oxidation products. The type and proportion of oxidation products are dependent on the radical-generating system *(18)*.

Excess accumulation of LDL in the extracellular subendothelial space of arteries leads to LDL oxidation. Oxidized LDL (Ox-LDL) is important in the initiation and progression of atherosclerosis as Ox-LDL acquires properties that make it more atherogenic than normal unmodified LDL. Ox-LDL stimulates the secretion of many cytokines by endothelial and smooth muscle cells such as monocyte chemoattractant protein-1 (MCP-1) and macrophage colony-stimulating factor (MCS), thereby attracting monocytes and T lymphocytes to the endothelium. Ox-LDL also stimulates expression of endothelial adhesion molecules that bind monocytes and lymphocytes which then migrate into the vascular wall, where monocytes differentiate into macrophages. In addition, Ox-LDL inhibits motility of tissue macrophages decreasing their egress from the arterial wall, essentially trapping them in the subendothelial space. This exacerbates the atherogenic process as the retained macrophages further oxidize LDL. Oxidation of LDL results in the rapid uptake of LDL through several different macrophage scavenger receptors leading to foam cell formation through macrophagecholesteryl ester enrichment. The macrophages then coalesce into a fatty streak which develops into more advanced lesions called plaques. Ox-LDL may contribute to the progression of fatty streaks to plaques by stimulating endothelial and smooth muscle cell secretion of cytokines such as macrophage colony-stimulating factor (M-CSF) and granulocyte-macrophage colony-stimulating factor (GM-CSF), which all contribute to cellular proliferation of a variety of cell types in the arterial wall. Ox-LDL is also immunogenic, eliciting autoantibody formation and reactive T lymphocytes in the arterial wall. The aggregation of macrophages and white cells within the arterial wall results in the expression of a variety of inflammatory and immune mediators such as tumor necrosis factor, leukotrienes, interleukins, and gamma interferon, contributing to the progression of atherosclerosis through inflammatory pathways.

Antioxidant enzymes catalyze the conversion of ROS into less reactive species or scavenge ROS in specific compartments of the cell *(19)*. In the pathogenesis of atherosclerosis, there is an increase in ROS production or inadequate antioxidant activity, resulting in an imbalance between oxidant stress and antioxidant capability *(11)*. As such, emphasis has been placed on antioxidant interventions for atherosclerosis. Different categories of antioxidant defenses have been described and are subdivided into nonenzymatic and enzymatic antioxidants *(10)*. The major nonenzymatic antioxidants are ascorbic acid (vitamin C), α-tocopherol (vitamin E), carotenoids (β-carotene, lycopene, etc.), uric acid, bilirubin, and ubiquinol-10. Enzymatic antioxidants include superoxide dismutase (SOD), catalase, glutathione, and peroxidase *(20,21)*. Vitamin E, vitamin C, and carotenoids (precursors to vitamin A) are the most studied natural antioxidants.

Vitamin E plays an essential role inhibiting oxidative stress. Vitamin E is comprised of numerous isomeric forms, of which α-tocopherol is the most biologically active *(22)*. As the most lipid-soluble vitamin, vitamin E is the major vitamin component of cell membranes and circulating lipoproteins. Vitamin E protects against lipid peroxidation, scavenges free radicals, and inhibits LDL oxidation, all of which are believed to play a crucial role in the development and progression of atherosclerosis *(23,24)*.

Vitamin C is an important water-soluble vitamin found primarily in fruits and vegetables *(25,26)*. The biological role of vitamin C is mainly related to its capability to function as a reducing agent. Ascorbate and dehydroascorbate are both active forms of vitamin C; ascorbate is readily oxidized to dehydroascorbate. Vitamin C acts as a hydrogen donor, reversing oxidation and inactivating free radicals thereby protecting proteins and lipids *(27)*.

Vitamin A is a lipid-soluble vitamin that has three active forms: retinol, retinal, and retinoic acid. Carotenoids (α-carotene, β-carotene, and β-cryptoxanthin) are precursors to vitamin A. Of the carotenoids, β-carotene is the most important due to its antioxidant role in quenching singlet oxygen, thereby preventing lipid peroxidation and oxidative damage to DNA *(28,29)*.

Experimental studies using the antioxidants, probucol and butylated hydroxytoluene (BHT), offer compelling but indirect evidence that LDL oxidation is an important contributor to atherogenesis *(5,6)*. Treatment of endothelial cells with oxidants results in augmented expression of cytokine-mediated formation of adhesion molecules (vascular cell adhesion molecule, VCAM and intercellular adhesion molecule, ICAM) *(30)* that are inhibited by antioxidants involving redox-sensitive transcriptional or post-transcriptional factors *(31)*. Similarly, the induction of factors, such as monocyte chemotactic peptide, is evoked by Ox-LDL and lipid peroxidation products through oxidant-sensitive transcriptional factors *(31)*. Vitamin E has been shown to have a modulatory role on signal transduction events associated with the regulation of smooth muscle cell proliferation in addition to its direct antioxidant activity (i.e., inhibition of lipid peroxidation and free-radical generation), although an indirect action related to modulating the cellular redox state is plausible *(32)*. Mechanisms to prevent lipid peroxidation and protein oxidation have been the focus of many antioxidant studies.

12.3. B-VITAMINS, INFLAMMATION, AND ATHEROSCLEROSIS

The origin of the hypothesis that B-vitamins may have a role in CVD was the observation of premature atherosclerosis in adolescents with homocysteinuria due to a genetic disorder in the cystathionine-β-synthase gene *(33)*. Since this original observation, many studies have found an association between increased CVD risk and elevated blood levels of total homocysteine (tHcy) *(34,35)*. It has been shown that tHcy causes inflammation as well as endothelial dysfunction, platelet activation, and oxidation of LDL *(36)*.

Elevated levels of tHcy are a result of an imbalance between tHcy production and metabolism. Deficiencies of folic acid, vitamin B_{12}, and vitamin B_6 result in an increase in tHcy concentrations since these B-vitamins are co-factors for the conversion of homocysteine to methionine *(36)*. Multivitamin users have lower levels of tHcy as compared with nonusers *(37)*.

Interest in B-vitamins has expanded to their potential capability of improving endothelial function by exerting anti-inflammatory effects associated with atherosclerosis *(38)*. The anti-inflammatory effects of folic acid are in part exerted by lowering plasma tHcy concentration and by a direct effect on endothelial nitric oxide (eNOS) expression *(39)*. Moreover, folic acid has been associated with improvement in a number of cardiovascular markers such as endothelial function, arterial stiffness, blood pressure, and pro-thrombotic activity *(40,41)*.

12.4. OBSERVATIONAL STUDIES

12.4.1. Ecological Studies

Early ecological studies reported an inverse association between consumption of antioxidant vitamin-containing fruits and vegetables and CVD. Data from the US Department of Agriculture and vital statistics from 1964 to 1978 indicated an inverse correlation between fruit and vegetable consumption and CVD mortality, which was especially true for foods rich in vitamin C *(42)*. A similar correlation was observed between increased ascorbate production and reduction in coronary heart disease (CHD) mortality in the United States between 1958 and 1978 *(43)*. In Scotland and England, fresh fruit and vegetable consumption was inversely associated with CVD mortality *(44,45)*. Vitamin C from fresh fruit and vegetable consumption was inversely related with CHD mortality *(44,45)*. Although these observations are important, results from ecological studies are limited since they are susceptible to large interpretative biases from the inability to link CVD risk with food consumption in specific individuals. In addition, these studies were not adjusted for concomitant CVD risk factors.

12.4.2. Epidemiological Studies

12.4.2.1. Vitamin E Dietary and Supplementary Intake

Six large-scale prospective cohort studies have investigated the relationship between vitamin E intake and CVD as shown in Table 12.1: Nurses' Health Study *(46)*, Health Professional Follow-up Study *(47)*, Finnish Mobile Clinic Study *(48)*, Established Populations for Epidemiologic Studies of the Elderly *(49)*, Iowa Women's Health Study *(50,51)*, and Rotterdam Study *(52)*. The Iowa Women's Health Study *(50,51)* and the Established Populations for Epidemiologic Studies of the Elderly *(49)* found decreased CHD risk associated with dietary intake of vitamin E. The Nurses' Health Study *(46)*, Health Professional Follow-up Study *(47)*, Finnish Mobile Clinic Study *(48)*, and Rotterdam Study *(52)* found that supplementary vitamin E intake was significantly associated with lower CHD risk.

The Nurses' Health Study *(46)* was the largest of these studies, with 87,245 female nurses 34–59 years of age who were free of CVD at the beginning of the study and completed dietary questionnaires

Table 12.1
Epidemiological Studies of Dietary and Supplementary Vitamin Intake

Study	*Group*	*Gender/Age (years)*	*Vit E*	*Vit C*	*β-car*	*B-vit*
Nurses' Health Study	87,245 American nurses	Women 34–59	x	x	x	x
Health Professional Follow-up Study	39,910 American health professionals	Men 40–75	x	x	x	x
Finnish Mobile Clinic Study	5,133 Finns	Men and women 30–69	x	x	x	–
Established Populations for Epidemiologic Studies of the Elderly	11,178 Americans	Men and women 67–105	x	x	–	–
Iowa Women's Health Study	34,486 post-menopausal American women	55–69	x	x	x	–
Rotterdam Study	4,802 Dutch	Men and women 55–95	x	x	x	–
First National Health and Nutrition Examination Survey (NHANES I)	11,348 Americans	Men and women 25–74	–	x	–	x
Alameda County Study	3,119 Americans	Men and women 16 and older	–	x	–	–
Western Electric Study	1,556 Americans	Men 40–55	–	x	x	–
Swedish Women's Study	1,462 Swedish	Women 38–60	–	x	–	–
British Department of Health and Nutrition Examination Survey Epidemiologic Follow-up Study	730 British	Men and women 65 and older	–	x	–	–
Massachusetts Health Care Panel Study	1,299 Americans	MEN and women 66 and older	–	–	x	–

that assessed their consumption of nutrients and supplementary antioxidant vitamins. During follow-up of 8 years, 552 cases of CHD were documented, including 437 nonfatal myocardial infarctions (MIs) and 115 coronary deaths. Compared with women in the lowest quintile of vitamin E intake (<3.5 IU/day), those in the top quintile (>21.5 IU/day) had a relative risk (RR) of CHD of 0.66 (95% confidence interval [CI], 0.50–0.87, $p < 0.001$) after adjustment for age, smoking, and a number of additional CVD risk factors. The CHD benefit was attributable primarily to supplementary vitamin E intake, because dietary sources alone were not associated with significant reductions in CHD. Those who used supplementary vitamin E for less than 2 years had no significant reduction in CHD risk (RR = 0.86, 95% CI, 0.52–1.43). However, use of supplementary vitamin E for 2 years or more was associated with a significant reduction in CHD risk after adjustment for CVD risk factors and other antioxidant vitamins (RR = 0.59, 95% CI, 0.38–0.91). Women taking at least 100 IU/day of vitamin E supplements (i.e., not multivitamins) had a RR of CHD of 0.57 (95% CI, 0.41–0.78) compared with nonusers. There was no trend toward a greater decrease in CHD risk with increasing daily intake of supplementary vitamin E; RR was 0.56 (95% CI, 0.21–1.51) for intake of 100–250 IU/day; 0.56 (95% CI, 0.33–0.96) for intake of 300–500 IU/day; and 0.58 (95% CI, 0.24–1.42) for intake greater than 600 IU/day.

In the Health Professionals Follow-up Study, 39,910 US male health professionals, 40–75 years of age, originally free of CVD, completed dietary questionnaires that assessed daily dietary and supplementary antioxidant intake of vitamin C, carotene, and vitamin E *(47)*. During 4 years of follow-up, 667 cases of CHD were documented, including 360 coronary artery bypass grafts (CABG) or percutaneous transluminal coronary angioplasties (PTCA), 201 nonfatal MIs, and 106 fatal MIs. Compared with men in the lowest quintile of vitamin E intake (median intake = 6.4 IU/day), those in the top quintile (median intake = 419 IU/day) had a RR of CHD of 0.64 (95% CI, 0.49–0.83, $p < 0.003$) after adjustment for age, smoking, body mass index, hypertension and additional CVD risk factors, and a RR of CHD of 0.60 (95% CI, 0.44–0.81, $p < 0.01$) with further adjustments for vitamin C and carotene intake. Benefit was attributable primarily to supplementary vitamin E intake since dietary sources alone were not associated with significant reductions in CHD. For men consuming 60 IU/day vitamin E, the RR of CHD was 0.64 (95% CI, 0.49–0.83) compared with those consuming <7.5 IU/day. Maximal reduction in risk was seen with an intake of 100–249 IU/day vitamin E, with no further decrease in risk at higher intakes. Men ingesting at least 100 IU/day vitamin E supplements for 2 years or longer had a RR of CHD of 0.63 (95% CI, 0.47–0.84) compared with nonusers after controlling for multivitamin use. Men ingesting at least 100 IU/day of vitamin E supplements not as multivitamins had a RR of CHD of 0.75 (95% CI, 0.61–0.93) compared with nonusers.

The Finnish Mobile Clinic Study was conducted among 5,133 Finnish men and women *(48)* 30–69 years of age who were initially free of CVD. Daily dietary and supplementary antioxidant vitamin intake was assessed with questionnaires at baseline. During the mean 14-year follow-up period, there were 186 CHD deaths in men and 58 CHD deaths in women. In the 2,748 men, those in the highest versus lowest tertile of vitamin E intake had a RR for CHD death of 0.68 (95% CI, 0.42–1.11, p-value for trend = 0.01) after adjustment for CVD risk factors. In the 2,385 women, those in the highest versus lowest tertile of vitamin E intake had a RR for CHD death of 0.35 (95% CI, 0.14–0.88, p-value for trend < 0.01) after adjustment for CVD risk factors. Individuals who died of CHD consumed more dairy products and less vegetable, fruits, and margarine than did the survivors. The adjusted RR for CHD mortality among the 3% of men and women who used supplements containing vitamin E or vitamin C versus those individuals who did not use supplements was 0.55 (95% CI, 0.18–1.73).

The Established Populations for Epidemiologic Studies of the Elderly was a prospective study designed to determine the effect of vitamin E and vitamin C supplement intake (not part of a multivitamin) on CHD mortality in 11,178 men and women aged 61–105 years *(49)*. During the

6–9 years of follow-up, there were 3,490 deaths, including 1,101 CHD deaths. Vitamin E supplementary intake was associated with a RR for CHD mortality of 0.59 (95% CI, 0.37–0.93) as well as a RR for all-cause mortality of 0.73 (95% CI, 0.58–0.91) after adjustment for CVD risk factors. However, the dosages of the vitamin supplements and consistency of their use were not assessed.

The Iowa Women's Health Study followed 34,486 post-menopausal women for up to 7 years *(50,51)*. The women, 55–69 years of age and initially free of CVD, completed questionnaires that assessed daily dietary and supplementary antioxidant vitamin A, E, and C intake. During follow-up, 242 women died of CHD. In the overall cohort, vitamin E intake from food and supplements was not associated with a lower risk of CHD death. However, vitamin E intake from food was inversely associated with CHD death among the subgroup of 21,809 women who did not consume vitamin supplements. Compared with women in the lowest quintile of vitamin E intake (4.91 IU/day) those in the top quintile (9.64 IU/day) had a RR for CHD death of 0.38 (95% CI, 0.18–0.80, $p < 0.004$) after adjustment for age, total energy intake, and a variety of additional CVD risk factors. Unlike the Nurses' Health Study *(46)* and the Health Professionals Follow-up Study *(47)*, supplementary intake of vitamin E was not associated with lower CHD risk. Intake of foods rich in vitamin E, such as margarine, nuts and seeds, and mayonnaise or creamy salad dressings, was inversely associated with a lower risk for CHD death.

The Rotterdam Study *(52)* consisted of 4,802 Dutch men and women 55–95 years old who were initially free of MI. Dietary levels of vitamin C, vitamin E, and β-carotene were assessed by a semi-quantitative food frequency questionnaire. During a 4-year follow-up period, 124 subjects had an MI. Dietary intake of vitamin E was not associated with risk of MI.

12.4.2.2. Vitamin C Dietary and Supplementary Intake

Eleven large-scale prospective cohort studies, Nurses' Health Study *(46)*, Health Professional Follow-up Study *(47)*, Finnish Mobile Clinic Study *(48)*, Established Populations for Epidemiologic Studies of the Elderly *(49)*, Iowa Women's Health Study *(50)*, Rotterdam Study *(52)*, First National Health and Nutrition Examination Survey (NHANES I) *(53)*, Alameda County Study *(54)*, Western Electric Study *(55)*, Swedish Women's Study *(56)*, and British Department of Health and Nutrition Examination Survey Epidemiologic Follow-up Study *(57)*, have examined the relationship between vitamin C intake and CVD (Table 12.1). The results from these studies have been less consistent than those for vitamin E, and the majority of the studies do not support a relationship between vitamin C intake and CVD protection. Nine of these studies *(46–50,52,53,55,56)* found no CVD risk reduction from vitamin C intake. After adjustment for other vitamin and multivitamin intake, both the Nurses' Health Study *(46)* and the Health Professionals Follow-up Study *(47)* showed no CVD benefit from vitamin C intake. In fact, in the Health Professionals Follow-up Study, men in the top quintile of vitamin C intake (median intake = 1,162 mg/day) compared to the lowest quintile (median intake = 92 mg/day) had a nonsignificantly increased risk for CHD (RR = 1.25, 95% CI, 0.68–1.16) *(47)*. In addition, in the Iowa Women's Health Study, women in the top quintile of vitamin C intake (196.3 mg/day) compared to the lowest quintile (87.3 mg/day) also had a nonsignificant increased risk for CHD mortality (RR = 1.43, 95% CI, 0.75–2.70) *(50)*.

The First National Health and Nutrition Examination Survey Epidemiologic Follow-up Study was the only large-scale prospective cohort study to find a significant CVD benefit from vitamin C intake *(53)*. This study involved a median follow-up of 10 years in 11,348 men and women 15–74 years old. In the cohort of 4,479 men, there were 1,069 verified deaths and in the 6,869 women there were 740 verified deaths. Men who had a dietary intake of vitamin C of 50 mg/day or more and who consumed vitamin C supplements on a regular basis had a CVD standardized mortality ratio of 0.58 (95% CI,

0.41–0.78). Women with the same vitamin C intake had a CVD standardized mortality ratio of 0.66 (95% CI, 0.53–0.82). Comparisons were made relative to all US whites, for whom the standardized mortality rate was defined to be 1.00. The average vitamin C supplementary intake among users was 800 mg/day. The inverse association between vitamin C intake and CVD mortality appeared to be explained by supplementary intake since vitamin C dietary intake alone was not significantly associated with reduced CVD mortality. Caution in the interpretation of these results is warranted, however, since the use of other vitamin supplements or multivitamins and possible confounding by other cardiovascular risk factors were not considered in the data analyses.

In the Finnish Mobile Clinic Study, women in the highest versus lowest tertile of vitamin C intake had a RR of CHD death of 0.49 (95% CI, 0.24–0.98, $p = 0.06$) after adjustment for CVD risk factors; men had a RR of 1.00 (95% CI, 0.68–1.45, $p = 0.94$) *(48)*. In the Established Populations for Epidemiologic Studies of the Elderly, vitamin C supplementary versus no supplementary intake was associated with a RR of CHD mortality of 0.99 (95% CI, 0.74–1.33) after adjustment for CVD risk factors *(49)*. In the Rotterdam Study, there was no association between tertiles of dietary vitamin C intake and risk of MI in age- and sex-adjusted models *(52)*. In the same study, in multivariate-adjusted logistic regression analyses, vitamin C intake was significantly inversely associated with peripheral arterial disease in women (highest versus lowest quartile: RR = 0.64, 95% CI, 0.48–0.89, $p = 0.006$) and a 100 mg increase in intake was associated with a 0.013 ankle-arm systolic blood pressure index increase (95% CI, 0.001–0.025).

In the Alameda County Study *(54)*, 3,119 men and women 16 years of age and older were followed for up to 10 years. The subjects completed dietary questionnaires that assessed daily dietary and supplementary antioxidant vitamin intake. During follow-up, 276 deaths occurred. Vitamin C intake above versus below 50 mg/day was not significantly associated with mortality from cancer, CVD, or total mortality.

In the Western Electric Study, 1,556 men 40–55 years old were followed for 24 years *(55)*. Although there was a RR of 0.75 (95% CI, 0.52–1.07) for CHD death in the highest tertile of vitamin C intake (>112 mg/day) versus the lowest tertile (<83 mg/day), this was not statistically significant. In a 12-year follow-up study of 1,462 Swedish women 38–60 years old, vitamin C intake estimated from 24-h dietary recalls was not associated with CVD mortality, MI, or stroke after controlling for CV risk factors *(56)*.

In the British Department of Health and Social Security Nutritional Survey, 730 British men and women 65 years and older were followed for up to 20 years *(57)*. The subjects were initially free of CVD and completed 7-day dietary records. During follow-up, there were 124 deaths from stroke and 182 deaths from CHD. Compared with subjects in the lowest tertile of vitamin C intake (27.9 mg/day), those in the top tertile (>44.9 mg/day) had a RR of death from stroke of 0.50 (95% CI, 0.30–0.80, $p < 0.003$) after adjustment for age, sex, and CVD risk factors. In contrast to the stroke results, there was no significant association with CHD mortality (RR = 0.8, 95% CI, 0.6–1.2).

12.4.2.3. β-Carotene Dietary and Supplementary Intake

Seven large-scale prospective cohort studies, Nurses' Health Study *(46)*, Health Professional Follow-up Study *(47)*, Finnish Mobile Clinic Study *(48)*, Iowa Women's Health Study *(50)*, Rotterdam Study *(52)*, Western Electric Study *(58)*, and Massachusetts Health Care Panel Study *(59)*, have investigated the relationship between β-carotene intake and CVD (Table 12.1). The reported relationships between β-carotene intake and CVD have been inconsistent. Of the seven reported prospective cohort studies, only three have shown a statistically significant inverse relationship between β-carotene intake and CVD *(47,59,60)*. Although women in the highest versus lowest quintile of β-carotene intake had a RR of 0.78 (95% CI, 0.59–1.03) for CHD risk in the Nurses' Health Study *(46)*, this did not reach statistical significance after adjustment for age, smoking, and other CHD risk factors.

In the Health Professionals Follow-up Study, men in the top quintile of carotene intake (dietary and supplementary median intake = 19,034 IU/day) versus the lowest quintile (median intake = 3,969 IU/day) had a lower risk for CHD (RR = 0.71, 95% CI, 0.53–0.86, $p = 0.03$) *(47)*. Although this inverse relationship between carotene intake and CHD risk was not found among those men who had never smoked (RR = 1.09, 95% CI, 0.66–1.79, $p = 0.64$), an inverse association was apparent among current smokers (RR = 0.30, 95% CI, 0.11–0.82, $p = 0.02$) and former smokers (RR = 0.60, 95% CI, 0.38–0.94, $p = 0.04$).

In the Finnish Mobile Clinic Study, women in the highest versus lowest tertile of β-carotene intake had a RR of CHD death of 0.62 (95% CI, 0.30–1.29, $p = 0.60$) after adjustment for CHD risk factors; men had a RR of 1.02 (95% CI, 0.70–1.48, $p = 0.36$) *(48)*.

In the Iowa Women's Health Study, no association was found between CHD mortality and vitamin A, retinol or carotenoids, from dietary or supplementary intake alone or from dietary and supplementary intake combined *(50)*.

In the Rotterdam Study, during a 4-year follow-up period, 124 subjects had an MI. Risk of MI for the highest compared with the lowest tertile of β-carotene intake was 0.55 (95% CI, 0.34–0.83, $p = 0.013$) adjusted for age, sex, body mass index, pack-years, income, education, alcohol intake, energy-adjusted intakes of vitamin C and E, and use of antioxidant vitamin supplements, suggesting an inverse relation with risk of MI *(52)*.

Although the RR for CHD death was 0.79 (95% CI, 0.60–1.04) in middle-aged men in the highest (>15.9 mg/day) versus the lowest tertile (<2.9 mg/day) of β-carotene intake in the Western Electric Study, this did not reach statistical significance *(58)*.

The Massachusetts Health Care Panel Study examined the association between consumption of carotene-containing fruits and vegetables and CVD mortality in a prospective cohort of 1,299 Massachusetts residents 66 years old and older initially free of CVD *(59)*. During 4.75 years of follow-up, 161 cases of CVD mortality were documented, including 48 confirmed fatal MI. Compared with men and women in the lowest quartile of carotene-containing fruit and vegetable consumption (<0.8 servings/day), those in the top quartile (2.05 servings/day) had a RR of CVD death of 0.59 (95% CI, 0.37–0.94, $p = 0.014$) after adjustment for age, smoking, cholesterol intake, alcohol consumption, and additional CVD risk factors, and a RR of 0.27 (95% CI, 0.10–0.74, $p = 0.005$) for fatal MI.

12.4.2.4. B-Vitamin Dietary and Supplementary Intake

Three large-scale prospective cohort studies, Nurses' Health Study *(61)*, Health Professional Follow-up Study *(62)*, and NHANES I *(63)*, have examined the relationship between B-vitamin intake and CVD (Table 12.1). The results from these studies have been inconsistent and the majority of the studies do not support a relationship between folate and B-vitamin intake and CVD protection. The Nurses' Health Study *(61)* and the Health Professional Follow-up Study *(62)* primarily examined the relationship between B-vitamin intake and stroke.

In the Nurses' Health Study *(61)*, 1,140 incident cases of stroke were identified. No association between the intake of folate and incidence of stroke was observed; the RR for the highest quintile of dietary folate intake compared to the lowest quintile was 1.01 (95% CI, 0.79–1.29, $p = 0.80$).

In the Health Professional Follow-up Study *(62)*, folate, vitamin B_6, and vitamin B_{12} intake (including multivitamin consumption) were assessed through dietary questionnaires. Intake of folate was associated with a significantly lower risk of ischemic but not hemorrhagic stroke. The RR of ischemic stroke was 0.71 (95% CI, 0.52–0.96, $p = 0.05$) for men in the highest quintile of intake compared with those in the lowest quintile. Intake of vitamin B_{12}, but not vitamin B_6, was also inversely associated with risk of ischemic stroke.

In NHANES I *(63)*, 9,764 US men and women 25–74 years old were free of CVD at baseline. Dietary questionnaires that assessed daily dietary and supplementary folate intake were assessed and 926 incident strokes and 3,758 incident CVD events were documented over 19 years of follow-up. The RR for stroke was 0.79 (95% CI, 0.63–0.99, $p = 0.03$) and the RR for CVD was 0.86 (95% C, 0.78–0.95, $p < 0.001$ for trend) in the highest versus lowest quartile of dietary folate intake after adjustment for established CVD risk factors and other dietary factors.

12.4.3. Studies Measuring Serum Antioxidant Concentrations

Overall, studies examining the relationship between serum antioxidant concentrations and CVD have been inconclusive *(64–84)*. Specifically, studies examining the relationship between serum β-carotene, vitamin A, and vitamin C concentrations and CVD have been inconsistent and those between serum vitamin E concentrations and CVD have not consistently confirmed the observational studies examining dietary and supplementary intake of vitamin E and CVD.

12.4.3.1. Cross-Sectional Studies

In the cross-sectional World Health Organization/Multinational Monitoring Project of Trends and Determinants of Cardiovascular Disease Study (WHO/MONICA), multiple groups of approximately 100 healthy men aged 40–49 years from 16 different worldwide regions differing sixfold in age-standardized CHD mortality were compared on plasma levels of vitamins A, C, E, and carotene *(65,66)*. Independent of lipid levels, the plasma vitamin E concentration was the strongest inverse factor in relation to cross-cultural CHD mortality rates. This study also indicated that the low vitamin E plasma levels had a greater impact on CHD mortality than classical risk factors, such as hypercholesterolemia or hypertension. Across all 16 regions, where LDL-cholesterol (LDL-C) levels ranged from 3.36 mmol/L (130 mg/dL) to 4.91 mmol/L (190 mg/dL), moderate associations between CHD mortality rates and plasma cholesterol ($r^2 = 0.29$, $p = 0.03$) and diastolic blood pressure ($r^2 = 0.25$, $p = 0.05$) were found; a considerably stronger inverse association was found for lipid-standardized vitamin E plasma levels ($r^2 = 0.62$, $p = 0.0003$). Lipid-standardized vitamin A ($r^2 = 0.24$, $p = 0.05$), vitamin C ($r^2 = 0.11$, $p = 0.22$), and carotene ($r^2 = 0.04$, $p = 0.48$) plasma levels showed weaker or no associations with CHD mortality rates. In 12 out of the 16 regions with "common" blood pressure and plasma LDL-C levels ranging from 3.88 mmol/L (150 mg/dL) to 4.40 mmol/L (170 mg/dL), the absolute levels of vitamin E ($r^2 = 0.63$, $p = 0.002$) and lipid-standardized vitamin E ($r^2 = 0.73$, $p = 0.0004$) showed strong inverse correlations with CHD mortality rates. Additionally, vitamin C plasma levels showed a moderately strong statistically significant inverse correlation with CHD mortality rates among the 12 regions ($r^2 = 0.41$, $p = 0.03$). Lipid-standardized vitamin A ($r^2 = 0.16$, $p = 0.19$) and carotene ($r^2 = 0.21$, $p = 0.14$) plasma levels showed no association with CHD mortality rates among the 12 regions. In regions with low and medium CHD risk (Italy, Switzerland and Northern Ireland), the average vitamin E plasma levels among healthy middle-aged males were 26–28 μmol/L; in regions with the highest CHD risk (Finland and Scotland), vitamin E plasma levels were 20–21.5 μmol/L *(66)*. On average, vitamin E plasma levels were about 25% lower ($p < 0.01$) in regions with high CHD risk versus regions of low-to-medium risk and there was only a small overlap in the distributions of vitamin E plasma levels. This indicates that there may be a threshold of CHD risk from vitamin E plasma levels below 25–30 μmol/L *(66)*. A similar threshold for CHD risk may be operational for vitamin C plasma levels below 23–51 μmol/L *(66)*. The WHO/MONICA study is limited by lack of individual control of confounders. Data from a smaller cross-sectional survey *(67)* of four European regions did not support the hypothesis that plasma antioxidant concentrations explain regional differences in CHD mortality.

In a cross-sectional survey *(68)* of 595 individuals 50–84 years old conducted in an urban population of India, a significant inverse association between plasma vitamin A, vitamin C, vitamin E, and β-carotene levels and prevalence of CAD (defined as presence of angina or diagnosis of MI) was found. The vitamin A and E levels remained inversely related to CAD after adjustment for other CVD risk factors. The adjusted OR for CAD between the lowest (<11.8 μmol/L) and highest (>19.2 μmol/L) quintiles of plasma vitamin levels in 523 subjects without CAD and 72 subjects with CAD was significant only for plasma vitamin E levels (OR = 2.53, 95% CI, 1.11–5.31).

The relation between serum ascorbic acid level and the prevalence of CVD was analyzed among 6,624 US men and women 40–74 years of age enrolled in the National Health and Nutrition Examination Survey *(69)*. Compared with individuals in the lowest ($\leq$0.4 mg/dL) versus the highest ($\geq$1.1 mg/dL) tertile of serum vitamin C concentrations, there was a 27% decreased prevalence of CHD (OR = 0.73, 95% CI, 0.59–0.90) and a 26% decreased prevalence of stroke (OR = 0.74, 95% CI, 0.56–0.97) after adjustment for CVD risk factors obtained from demographic and historical information. Other CVD risk factors such as plasma cholesterol levels and other serum antioxidant levels were not controlled for in the analyses. In addition, misclassification of CVD may have occurred since this diagnosis was self-reported.

The interpretation of these cross-sectional studies is limited for at least two reasons: (1) intake, and thus plasma vitamin levels may change after a CVD event and (2) determination of prevalent disease cannot separate whether the associations are with incident events or survival from the event.

12.4.3.2. Case–Control Studies

Case–control and prospective cohort studies have yielded mixed results and in general have not confirmed the inverse relationship between serum vitamin antioxidant levels and CVD found in the cross-sectional studies measuring serum antioxidant concentrations. Three nested case–control studies conducted within large prospective cohort studies have been reported *(70–72)*. In two of these studies, no association between serum vitamin E or vitamin A concentrations and subsequent CVD mortality was found *(70,71)*. However, in the nested case–control studies from the Netherlands *(70)* and eastern Finland *(71)*, blood samples were collected at baseline and vitamin assays were determined 7–10 years after sampling. Blood samples were frozen at –20°C, a temperature at which antioxidant vitamins are unstable and undergo degradation. Thus, it is probable that the measured vitamin E levels were inaccurate. In the nested case–control study from Washington County, Maryland, blood samples were properly frozen at –70°C and analyzed for vitamin levels 16 years after collection *(72)*. Although no association was found between serum α-tocopherol levels and risk for MI in the total sample, a protective association for MI was suggested with higher serum levels of α-tocopherol in the subgroup of individuals with serum cholesterol levels $\geq$6.21 mmol/L ($\geq$240 mg/dL). In addition, an increasing risk for MI was found with decreasing serum β-carotene concentrations ($p = 0.02$) with a similar trend found with decreasing levels of lutein ($p = 0.09$). However, the excess risk for MI associated with low serum carotenoid concentrations was limited to current smokers.

In a case–control study of 110 cases of angina pectoris and 394 controls selected from a sample of 6,000 men aged 35–54 years, a significant inverse association was found between plasma vitamin E levels and angina pectoris *(73)*. Plasma vitamin levels were determined at the time the cohort was identified. Vitamin E levels remained independently and inversely associated with the risk of angina pectoris after adjustment for age, smoking, blood pressure, lipids, and weight. The adjusted OR for angina pectoris between the lowest compared to highest quintiles of vitamin E plasma levels was 2.68

(95% CI, 1.07–6.70, $p = 0.02$). Adjusted plasma levels of vitamin C, vitamin A, and carotene were inversely but not significantly associated with angina pectoris. Several double-blind RCTs examining the effect of vitamin E administration on angina pectoris have yielded mixed results in trials with subjects free of CHD at baseline and followed up for incident angina pectoris as well as in trials with subjects with stable angina pectoris *(85–87)*.

12.4.3.3. Prospective Cohort Studies

At least four prospective cohort studies, the Basel Prospective Study *(64)*, the Lipid Research Clinics Coronary Primary Prevention Trial and Follow-up Study *(74)*, the Prospective Population Study of Men from Eastern Finland *(75)*, and the Zutphen Elderly Study *(88)*, have reported the relationship between serum vitamin E, vitamin C, and β-carotene concentrations and CVD risk. The Basel Prospective Study examined the relationship between baseline serum vitamin C, vitamin E, and β-carotene concentrations and subsequent CVD mortality in 2,974 male Swiss pharmaceutical company employees aged 41–59 years, initially without CVD *(64)*. During 12 years of follow-up, 553 men died, including 132 from ischemic heart disease (IHD) and 31 from stroke. Compared with men in the highest quartile of serum carotene (α-carotene plus β-carotene) concentration, those in the lowest quartile (<23 μmol/L) had a RR of CVD mortality of 1.53 (95% CI, 1.07–2.20, $p < 0.024$) after adjustment for age, smoking, blood pressure, and cholesterol. Compared with men in the highest quartile of serum vitamin C concentration, those in the lowest quartile (<22.7 μmol/L) had a RR of CVD mortality of 1.25 (95% CI, 0.77–2.01, $p = 0.38$). Independently, low serum concentrations of carotene and vitamin C were not associated with death from stroke, RR of 2.07 (95% CI, 0.78–5.46, $p = 0.14$) and 1.28 (95% CI, 0.40–4.09, $p = 0.34$), respectively. However, the risk of death from stroke was significantly increased in subjects who were in the lowest quartiles of both carotene (<0.23 μmol/L) and vitamin C (<22.7 μmol/L) plasma levels relative to those in the highest quartiles of both carotene and vitamin C (RR = 4.17 (95% CI, 1.68–10.33), $p = 0.002$). CVD mortality was not associated with serum vitamin E concentrations. This latter finding may result from the fact that the median serum vitamin E concentration in the cohort was 35 μmol/L, greater than the presumed threshold for CHD risk of 25–30 μmol/L, as suggested by the WHO/MONICA cross-sectional study *(66)*. This also may have been true for the vitamin A levels.

The relationship between serum carotenoid concentration and subsequent CHD events (nonfatal MI and CHD death) was prospectively assessed in 1899 hyperlipidemic men assigned to the placebo arm of the Lipid Research Clinics Coronary Primary Prevention Trial *(74)*. Serum carotenoid levels and smoking status were available for 1,883 men aged 40–59 years. After 13 years of follow-up, men in the highest quartile of serum β-carotene concentration (>3.16 μmol/L) had a RR for CHD of 0.64 (95% CI, 0.44–0.92, $p = 0.01$) compared with those in the lowest quartile (<2.33 μmol/L) after adjustment for age, smoking, high-density lipoprotein cholesterol (HDL-C), LDL-C, and other known CHD risk factors. This finding was stronger among the 441 men who had never smoked in whom the RR was 0.28 (95% CI, 0.11–0.73, p-trend = 0.06). For the 679 current smokers, the RR for CHD was 0.78 (95% CI, 0.44–1.34, p-trend = 0.04). This contrasts to other reports where carotene intake was more strongly associated with risk in smokers.

The relationship between serum vitamin C concentration and subsequent MI was prospectively assessed in 1605 randomly selected men in eastern Finland with an average age of 54 years who were initially free of CVD *(75)*. After up to 8.75 years of follow-up, there were 70 fatal or nonfatal MIs. Compared with men in the highest quintile of plasma vitamin C concentration (>64.8 μmol/L) men in the lowest quintile (<11.4 μmol/L) had a RR of fatal or nonfatal MI of 4.03 (95% CI, 1.74–9.36, $p = 0.001$) after adjustment for age, season, and examination year. However, after adjustment for a number of additional CVD risk factors, the RR was substantially diminished to 2.08 (95% CI, 0.82–5.30) and was no longer statistically significant.

In the Zutphen Elderly Study *(88)*, 559 men with a mean age of 72 years initially free of chronic disease were followed for 15 years. Carotenoids, α- and γ-tocopherols, and vitamin C were assessed in relation to CVD mortality; 197 men died from CVD. After adjustment for age, smoking, and other potential lifestyle and dietary confounders, the RR for CVD for a 1 standard deviation increase in intake was 0.81 (95% CI, 0.66–0.99) for α-carotene and 0.80 (95% CI, 0.66–0.97) for β-carotene. Carrots were the primary source of α- and β-carotenes and their consumption was related to a lower risk of death from CVD. Intake of carotenoids other than α- and β-carotenes, vitamin C, and α- and γ-tocopherols were not associated with CVD mortality.

Taken together, the studies relating serum antioxidant vitamin concentrations with CVD have been inconsistent and inconclusive. These studies further complicate the inconsistencies seen with the vitamin C dietary intake studies and in general have failed to confirm the apparent protective effect of high dietary and supplementary intake of vitamin E. This seemingly contradistinction could result from the nutrients associating with CVD through mechanisms not reflected by blood levels.

12.4.4. Studies Measuring Blood B-vitamin and tHCY Concentrations

Over the past decade, many studies have used plasma tHcy levels as a surrogate for B-vitamin intake since B-vitamins are important direct or indirect co-factors in the metabolism of tHcy and supplementation with these vitamins reduces plasma tHcy levels. Therefore, many studies not only evaluate plasma folate and B-vitamin concentrations but tHcy levels as well *(89–99)*. In general, studies examining the relationship between blood levels of B-vitamins and CVD have been mixed whereas those between blood levels of tHcy and CVD have been more consistent.

In the Canadian Nutrition Survey *(89)*, 5,056 men and women aged 35–79 years with no history of CHD were followed for 15 years for CHD mortality, with a total of 165 CHD deaths. Serum folate levels were significantly inversely associated with risk of fatal CHD with a rate ratio for individuals in the lowest serum folate level category (<6.8 nmol/L [*3*ng/mL]) compared with the highest category (>13.6 nmol/L [*6*ng/mL]) of 1.69 (95% CI, 1.10–2.61).

In NHANES I, 2006 individuals aged 25–74 years were studied *(90)*. After adjusting for age, race, sex, education, diabetes, history of heart disease, systolic blood pressure, body mass index, hemoglobin level, cigarette smoking, and alcohol intake, participants with a plasma folate concentration ≤9.2 nmol/L had a RR for ischemic stroke of 1.37 (95% CI, 0.82–2.29) compared to participants with a plasma folate concentration >9.2 nmol/L. Moreover, whites with a plasma folate concentration ≤9.2 nmol/L had a RR for ischemic stroke of 1.18 (95% CI, 0.67–2.08), whereas blacks had a RR of 3.60 (95% CI, 1.02–12.71) compared to their counterparts with plasma folate levels >9.2 nmol/L. Therefore, this study suggested that a folate concentration ≤9.2 nmol/L may be a risk factor for ischemic stroke, especially in blacks.

In the Kuopio Ischemic Heart Disease Risk Factor Study (KIHD) *(91)*, 734 men aged 46–64 years were followed for 5 years and 3 months. Serum folate levels were measured at baseline. During follow-up, six (2.5%) men with higher serum folate concentrations (highest tertile, >11.3 nmol/L) and 28 (5.7%) men with lower serum folate (lowest two tertiles, <11.3 nmol/L) developed an acute coronary event ($p = 0.008$). Adjusting for age, examination years, and plasma lycopene concentration, the RR for an acute coronary event was 0.31 (95% CI, 0.11–0.90, $p = 0.031$) in men with higher serum folate concentrations compared with men with lower serum folate levels.

In the Framingham Heart Study, 1,041 elderly subjects (418 men and 623 women aged 67–96 years) were assessed by ultrasonography for the degree of stenosis of the extracranial carotid arteries, and plasma folate, vitamin B_{12}, and vitamin B_6 levels were determined. Subjects were classified into two categories according to the degree of carotid artery stenosis, 0–24% and 25–100%. Plasma concentrations of folate and pyridoxal-5′-phosphate (PLP) (the active form of vitamin B_6) and the level of folate intake were inversely associated with carotid artery stenosis after adjustment for age, sex, and other risk factors *(92)*.

In the Physicians' Health Study *(93)*, 14,916 male physicians aged 40–84 years with no prior history of MI or stroke provided plasma at baseline and were followed for 7.5 years. Samples from 333 men who subsequently developed MI and their paired controls matched for age and smoking were analyzed for folate and vitamin B_6. Men with the lowest 20% of folate levels (<2.0 ng/mL) had a RR of 1.4 (95% CI, 0.9–2.3) for MI compared with those in the top 80% of folate levels. For men with the lowest 20% of vitamin B_6 values, the RR was 1.5 (95% CI, 1.0–2.2) for MI compared with those in the top 80% of vitamin B_6 values. When both folate and vitamin B_6 were included in a model with CVD risk factors, the RR of MI for low as compared with high levels of folate was 1.3 (95% CI, 0.8–2.1) and for PLP 1.3 (95% CI, 0.9–2.1). tHcy did not add significant predictive value to these results, except in the first half of the follow-up interval where men with the top 5% of plasma tHcy values had an almost threefold increased risk for MI.

In the Atherosclerosis Risk in Communities (ARIC) study *(94)*, middle-aged men and women were followed for 3.3 years. CHD incidence was associated positively ($p < 0.05$) with plasma tHcy concentrations in women but not men and CHD was inversely associated ($p < 0.05$) with plasma folate levels (women only), plasma PLP levels (both sexes), and B-vitamin supplementation (women only).

Serum was analyzed for tHcy from 229 men who died from IHD who were age-matched with 1,128 control subjects in the British United Provident Association (BUPA) *(95)*, a study of 21,520 men aged 35–64 years without an initial history of IHD. Serum tHcy levels were significantly higher in men who died of IHD than in men who did not (mean, 13.1 versus 11.8 μmol/L; $p < 0.001$). The risk of IHD among men in the highest quartile of serum tHcy levels was 3.7 times (or 2.9 times after adjusting for other risk factors) the risk among men in the lowest quartile (95% CI, 1.8–4.7). There was a continuous dose–response relationship with risk increasing by 41% (95% CI, 20–65%) for each 5 μmol/L increase in the serum tHcy level. After adjustment for apolipoprotein B levels and blood pressure, this estimate was 33% (95% CI, 22–59%) for each 5 μmol/L increase in the serum tHcy level.

In the Rotterdam Study *(96)*, tHcy levels were evaluated in relation to atherosclerosis and symptomatic CVD in a random sample of 630 men and women from 7,983 subjects aged 55 years and older. Carotid plaques and common carotid artery intima-media thickness (CCA IMT) were assessed by ultrasonography, lower extremity (peripheral) artery atherosclerosis was determined by the ankle-brachial index (ABI, the ratio of the ankle to arm systolic blood pressure), and the prevalence of CVD was assessed by history of MI or stroke. In subjects aged 55–74 years, elevated plasma tHcy levels (≥18.6 μmol/L) was associated with a thicker CCA IMT, a lower ABI and an increased prevalence of CVD (OR = 3.0, 95% CI, 1.5–6.1) after adjusting for sex and age. In subjects aged 75 years and older, plasma tHcy levels were not associated with atherosclerosis and CVD.

In the British Regional Heart Study *(97)*, serum was collected and stored from 5,661 men aged 40–59 years. During subsequent follow-up there were 141 incident cases of stroke among men with no history of stroke at screening. Serum tHcy concentration was measured in 107 case and 118 control men (matched for age group and town) who did not develop a stroke or MI during follow-up. Serum tHcy concentrations were significantly higher in cases than controls (geometric mean 13.7 (95% CI, 12.7–14.8) versus 11.9 (95% CI, 11.3–12.6) μmol/L; $p = 0.004$). There was a graded increase in the RR of stroke in the second, third, and fourth quartiles of the tHcy distribution (OR = 1.3, 1.9, 2.8, respectively, p-for-trend = 0.005) relative to the first quartile. Adjustment for age, town, social class, body mass index, hypertensive status, cigarette smoking, forced expiratory volume, packed-cell volume, alcohol intake, diabetes, HDL-C, and serum creatinine did not attenuate the association. This study therefore suggested that tHcy is a strong and independent risk factor for stroke.

In a Finnish population-based study *(98)*, 7,424 men and women aged 40–64 years initially free of atherosclerotic vascular disease at baseline were followed for 9 years, and 134 male and 131 female cases with either MI or stroke were identified. For each case, a control subject was selected belonging to the same sex and 5-year age group. The mean serum tHcy concentration of male cases and controls was 9.99 μmol/L and 9.82 μmol/L at baseline, respectively, and that of female cases and controls 9.58 μmol/L and 9.24 μmol/L, respectively. The differences between cases and controls were not statistically significant. This study found no significant association between serum tHcy levels and atherosclerotic vascular disease, MI, or stroke in logistic regression analyses.

In a multicenter case–control study in Europe *(99)*, 750 cases with documented vascular disease (MI, angina pectoris, stroke, peripheral vascular disease, or coronary artery or carotid artery stenosis) and 800 control subjects matched for age and sex were compared. Plasma was obtained within 1 year of the CVD event. Plasma tHcy levels (before and after methionine loading), red cell folate, vitamin B_{12}, and vitamin B_6 were analyzed. Plasma tHcy concentrations greater than the 80th percentile for control subjects either fasting (12.1 μmol/L) or after a methionine load (38.0 μmol/L) were associated with an elevated risk of vascular disease independent of traditional vascular risk factors (smoking, hypertension, hypercholesterolemia). Moreover, concentrations of red cell folate below the 10th percentile (<513 nmol/L) and concentrations of vitamin B_6 below the 20th percentile (<23.3 nmol/L) were also associated with increased risk of vascular disease. It was concluded from this study that lower levels of folate and vitamin B_6 confer an increased risk for atherosclerosis.

Meta-analysis of 27 studies relating plasma tHcy levels with CVD showed that levels of tHcy are positively associated with CVD risk *(100)*. Increasing plasma tHcy level is an independent graded risk factor for atherosclerotic vascular disease. The odds ratio (OR) for coronary artery disease (CAD) for each 5 μmol/L tHcy increment is 1.6 (95% CI, 1.4–1.7) for men and 1.8 (95% CI, 1.3–1.9) for women. The OR for cerebrovascular disease for each 5 μmol/L tHcy increment is 1.5 (95% CI, 1.3–1.9). Peripheral arterial disease also shows a strong association with plasma tHcy levels. Increased folic acid intake reduces tHcy levels, therefore it is assumed that lowering tHcy levels will decrease CVD.

In 1996, the Food and Drug Administration issued a regulation requiring all cereal grain and flour products to be fortified with folic acid to reduce the risk of neural tube defects in newborns. The fortification of cereal grain and flour products with folic acid was associated with a substantial improvement in folate status and reduction in plasma tHcy levels in a population of middle-aged and older adults *(37,101)*.

12.5. COMPLETED RANDOMIZED CONTROLLED TRIALS

Although the conclusions from epidemiological studies concerning the CVD protective effects of antioxidant vitamins and B-vitamins have been mixed, there was enough consistency in the data and potential benefit in reducing disease risk to warrant conduct of clinical trials. Epidemiological studies are unable to completely control for confounding variables that could affect disease outcome and bias estimates of associations between antioxidant vitamins and B-vitamins and CVD risk. It is not possible to conclusively determine whether the association of CVD with dietary assessment or serum measurements of vitamins represent a true association or are confounded by other dietary or lifestyle practices, which themselves are protective. Additionally, the protective effects seen in association studies may be a result of other components of the foods or a combination of antioxidants of these other food components. The epidemiological data examining the relationship between antioxidant vitamin and B-vitamin intake and CVD are limited in providing definitive answers about whether these vitamins

Table 12.2
Randomized Controlled Trials of Supplementary Vitamin Intake with Cardiovascular Disease as a Secondary Outcome

Study	*Group*	*Gender/Age (years)*	*Vit E*	*Vit C*	*β-car*	*B-vit*
Alpha-Tocopherol Beta-Carotene Cancer Prevention Study (ATBC)	29,133 Finnish	Men and women 50-69	x	–	x	–
Lixian Study	29,584 Chinese	Men and women 40–69	x	x	x	–
Beta-Carotene and Retinol Efficacy Trial (CARET)	18,314 American	Men and women 45–74	–	–	x	–
Skin Cancer Prevention Study (SCPS)	1,720 American	Men and women 27–84	–	–	x	–
Physicians' Health Study (PHS)	22,071 American physicians	Men 40–84		–	x	–
Womens' Health Study (WHS)	39,876 American	Women 45 and older	–	–	x	–
Nambour Skin Cancer Prevention Trial (NSCPT)	1,621 Australian	Men and women 20–69	–	–	x	–

are protective for CVD. Unbiased estimates of the efficacy of vitamins as therapeutic or preventive agents can only be obtained from RCTs.

To date, at least 24 RCTs of antioxidant vitamin, folate, and B-vitamin supplementation examining CVD events and atherosclerosis progression have been published (Tables 12.2–12.4). Of these 24 RCTs, 7 were designed to determine whether antioxidant vitamin supplementation was effective in preventing cancer with CVD as a secondary outcome *(102–111)* (Table 12.2). These trials were conducted in individuals predominantly asymptomatic for CVD at baseline. The remaining 17 RCTs were specifically designed to evaluate CVD outcomes with antioxidant vitamin, folate, and B-vitamin supplementation (Tables 12.3 and 12.4). Of these 17 trials, 10 have been conducted in individuals with established CVD at baseline *(112–121)*, 2 have been conducted in individuals without established CVD *(122,123)*, and 5 have been arterial imaging trials to measure the progression of atherosclerosis *(124–128)* (Table 12.4).

12.5.1. Randomized Controlled Trials with CVD as a Secondary Outcome

12.5.1.1. Vitamin E

The Alpha-Tocopherol Beta-Carotene Cancer (ATBC) prevention study was a randomized, double-blind, placebo-controlled trial designed to determine whether daily α-tocopherol, β-carotene, or both would reduce the incidence of lung and other cancers *(102)*. A total of 29,133 male Finnish smokers 50–69 years of age were randomized in a 2 × 2 factorial design to α-tocopherol (50 mg/day)

Table 12.3
Randomized Controlled Trials of Supplementary Vitamin Intake with Cardiovascular Disease as a Primary Outcome

Study	*Group*	*Age (years)*	*Vit E*	*Vit C*	*β-car*	*B-vit*
Secondary Prevention Trials						
Cambridge Heart Antioxidant Study (CHAOS)	2,002 patients with coronary artery disease	Ave 61.8	x	–	–	–
Gruppo Italiano per lo Studio della Sopravvivenza nell'Infarto Miocardico Acuto Prevenzione Trial (GISSI)	11,324 patients with myocardial infarction	Ave 59.4	x	–	–	–
Heart Outcomes Prevention Evaluation (HOPE)	9,541 patients with cardiovascular disease or risk factors	≥55	x	–	–	–
Secondary Prevention with Antioxidants of Cardiovascular Disease in End-Stage Renal Disease (SPACE)	196 hemodialysis patients with pre-existing cardiovascular disease	40–75	x	–	–	–
Heart Protection Study (HPS)	20,536 patients with cardiovascular disease or risk factors	40–80	x	x	x	–
Vitamin Intervention for Stroke Prevention (VISP)	3,680 patients with nondisabling stroke	≥35	–	–	–	x
Heart Outcomes Prevention Evaluation 2 (HOPE-2)	9,541 patients with cardiovascular disease or risk factors	≥55	x	–	–	x
Netherlands Study	593 patients with stable CAD	55–75	–	–	–	x
Norwegian Vitamin Trial (NORVIT)	3,749 patients with acute myocardial infarction	30–85	–	–	–	x
Women's Antioxidant and Folic Acid Cardiovascular Study (WAFACS)	5,442 women with history of CVD or 3 or more risk factors	≥42	–	–	–	x
Primary Prevention Trials						
Primary Prevention Project (PPP)	4,495 patients with at least one cardiovascular risk	≥50	x	–	–	–
Supplementation en Vitamines et Minéraux Antioxydants (SU.VI.MAX)	13,017 men and women	35-60	x	x	x	–

Table 12.4
Arterial Imaging Randomized Controlled Trials

Study	*Group*	*Age (y)*	*Vit E*	*Vit C*	*β-car*	*B-vit*
Coronary Angiographic Trials						
HDL-Atherosclerosis Treatment Study (HATS)	160 patients with clinical coronary disease and stenosis	51–57	x	x	x	–
Women's Angiographic Vitamin and Estrogen (WAVE)	423 post-menopausal women with coronary stenosis and luminal stenosis	ave 65	x	x	–	–
B-Mode Ultrasound Trials						
Antioxidant Supplementation in Atherosclerosis Prevention Study (ASAP)	520 men and post-menopausal women with serum cholesterol ≥5 mM/L	45–69	x	x	–	–
Study to Evaluate Carotid Ultrasound Changes in Patients Treated with Ramipril and Vitamin E (SECURE)	732 patients with cardiovascular disease or risk factors	≥55	x	–	–	–
Vitamin E Atherosclerosis Prevention Study (VEAPS)	353 men and women with ≥130 mg/dL LDL-cholesterol	≥40	x	–	–	–

alone, β-carotene (20 mg/day) alone, both α-tocopherol and β-carotene, or placebo. After a median follow-up of 6.1 years of randomized treatment, there was no overall benefit of either supplement on CVD. Relative to the individuals who did not receive α-tocopherol, those assigned to α-tocopherol experienced a 16% nonsignificant reduction in death from ischemic stroke (RR = 0.84, 95% CI, 0.59–1.19) but a statistically significant 50% increase in death from hemorrhagic stroke (RR = 1.5, 95% CI, 1.02–2.20). Individuals assigned to α-tocopherol had a 5% nonsignificant reduction in death from IHD (RR = 0.95, 95% CI, 0.85–1.06) but a 2% nonsignificantly higher overall mortality (RR = 1.02, 95% CI, 0.95–1.09) compared with those individuals who did not receive α-tocopherol. Regarding the primary trial endpoint, the β-carotene group experienced a statistically significant 18% increase in the incidence of lung cancer relative to the placebo group (RR = 1.18, 95% CI, 1.03–1.36). The incidence of lung cancer was not affected by α-tocopherol supplementation.

In the subset of 1,862 men randomized to ATBC who had a previous MI, the first major coronary event after randomization was determined *(103)*. In this subset of subjects, 424 nonfatal MI and fatal CHD cases occurred during follow-up. Relative to subjects who received placebo (438 subjects), those assigned to α-tocopherol (466 subjects) experienced a 38% significant reduction in nonfatal MI (RR = 0.62, 95% CI, 0.41–0.96). However, there was a 33% nonsignificant increase in fatal CHD cases in the α-tocopherol group (RR = 1.33, 95% CI, 0.84–2.05). All events combined, nonfatal MI and fatal CHD, were nonsignificantly reduced 10% in subjects assigned to the α-tocopherol group (RR = 0.90, 95% CI, 0.67–1.22). Subjects assigned to the α-tocopherol plus β-carotene group experienced a 58% significant increase in death from CHD relative to the placebo group (RR = 1.58, 95%

CI, 1.05–2.40). All events combined, nonfatal MI and fatal CHD, were nonsignificantly increased 14% in the α-tocopherol plus β-carotene group (RR = 1.14, 95% CI, 0.87–1.51). Nonfatal MI was nonsignificantly reduced 14% in the α-tocopherol + β-carotene group (RR = 0.86, 95% CI, 0.58–1.26).

In the subset of 27,271 men randomized to ATBC who did not have a previous MI, the first major coronary event after randomization was determined *(104)*. In this subset of subjects, 2,111 nonfatal MIs and fatal CHD occurred during follow-up. Relative to subjects who received placebo (6,849 subjects), those assigned to the α-tocopherol group (6,820 subjects) experienced a 4% nonsignificant increase in nonfatal MI (RR = 1.04, 95% CI, 0.89–1.22). Fatal CHD was nonsignificantly decreased 10% in the α-tocopherol group (RR = 0.90, 95% CI, 0.75–1.08). All events combined, nonfatal MI and fatal CHD, were nonsignificantly reduced 2% in subjects assigned to the α-tocopherol group (RR = 0.98, 95% CI, 0.87–1.10). Subjects assigned to the α-tocopherol plus β-carotene group experienced a 6% nonsignificant decrease in death from CHD, relative to the placebo group (RR = 0.94, 95% CI, 0.79–1.13). All events combined, nonfatal MI and fatal CHD, were nonsignificantly decreased 3% in the α-tocopherol plus β-carotene group (RR = 0.97, 95% CI, 0.86–1.09). Nonfatal MI was nonsignificantly reduced 1% in the α-tocopherol plus β-carotene group (RR = 0.99, 95% CI, 0.84–1.16).

The Linxian study was a randomized, double-blind, placebo-controlled trial designed to determine whether daily intake of four combinations of nine individual vitamin–mineral supplements reduces overall or cancer mortality or incidence of cancer *(105)*. Subjects 40–69 years of age were recruited from four Linxian (Chinese) communities, and 29,584 individuals were randomized to placebo or retinol (5,000 IU) and zinc (22.5 mg); riboflavin (3.2 mg) and niacin (40 mg); vitamin C (120 mg) and molybdenum (30 μg); and vitamin E (30 mg), β-carotene (15 mg), and selenium (Se) (50 μg). Linxian Province was chosen because inhabitants of this area have one of the highest esophageal/gastric cancer rates in the world, as well as a low intake of several micronutrients. After a follow-up of 5.25 years of randomized treatment, total mortality was significantly reduced 9% among those receiving vitamin E, β-carotene, and Se supplementation relative to placebo (RR = 0.91, 95% CI, 0.84–0.99, $p = 0.03$). This reduction mostly resulted from lower cancer mortality, especially stomach cancer. There was also a 10% nonsignificant reduction in cerebrovascular mortality (RR = 0.90, 95% CI, 0.86–1.07). Because vitamin E, β-carotene, and Se were used in combination, it was not possible to determine which supplement or supplements contributed to the lower mortality rates. The other three combination regimens of retinol and zinc, riboflavin and niacin, and vitamin C and molybdenum had no significant effects on mortality rates. The relevance of these results to Western populations is unclear.

12.5.1.2. Vitamin C

In the Linxian study, there was no significant effect on mortality rates in the vitamin C plus molybdenum supplement group *(105)*. In a smaller, separate study of 538 subjects 52–97 years old, there was no reduction in total mortality at 6 months in those individuals randomized to 200 mg/day vitamin C versus placebo *(106)*.

12.5.1.3. β-Carotene

In the ATBC prevention study (male smokers), subjects randomized to β-carotene experienced an 8% significant increase in total mortality (RR = 1.08, 95% CI, 1.01–1.16) and a 12% significant increase in death from IHD (RR = 1.12, 95% CI, 1.01–1.25) compared with those who did not receive β-carotene *(102)*. In addition, relative to those individuals who did not receive β-carotene, those assigned to β-carotene experienced a 23% nonsignificant increase in death from ischemic stroke

(RR = 1.23, 95% CI, 0.86–1.76) and a 17% nonsignificant increase in death from hemorrhagic stroke (RR = 1.17, 95% CI, 0.80–1.70). In addition to apparently increasing all of the CVD mortality endpoints, β-carotene also significantly increased the incidence of lung cancer and nonsignificantly increased prostate, colon, rectal, and stomach cancers.

In the ATBC prevention study subset *(103)* of men who had a previous MI, subjects assigned to the β-carotene group (461 subjects) experienced a 75% significant increase in death from CHD compared with men in the placebo group (438 subjects) (RR = 1.75, 95% CI, 1.16–2.64). All events combined, nonfatal MI and fatal CHD, were nonsignificantly increased 11% in subjects assigned to the β-carotene group relative to the placebo group (RR = 1.11, 95% CI, 0.84–1.48). In the β-carotene group, nonfatal MI was nonsignificantly reduced 33% relative to the placebo group (RR = 0.67, 95% CI, 0.44–1.02).

In the ATBC prevention study subset *(104)* of men who did not have a previous MI, subjects assigned to the β-carotene only group (6,821 subjects) experienced a 1% nonsignificant reduction in death from CHD relative to those men randomized to the placebo group (6,849 subjects) (RR = 0.99, 95% CI, 0.83–1.19). All events combined, nonfatal MI and fatal CHD, were nonsignificantly increased 3% in subjects assigned to the β-carotene group relative to the placebo group (RR = 1.03, 95% CI, 0.91–1.16). In the β-carotene group, nonfatal MI was nonsignificantly increased 6% relative to the placebo group (RR = 1.06, 95% CI, 0.90–1.24).

In the Linxian study, β-carotene supplementation was used in combination with vitamin E and Se *(105)*. The results of this combination of supplements on total mortality, CVD mortality, and cancer are summarized in Section 5.1.1.

The Beta-Carotene and Retinol Efficacy Trial (CARET) was a randomized, double-blind, placebo-controlled trial designed to determine whether daily treatment with a combined supplement containing β-carotene and vitamin A would reduce the incidence of lung cancer, the primary trial endpoint *(107)*. A total of 18,314 men and women 45–74 years of age with a high risk for lung cancer from asbestos and/or smoking exposure were randomized to placebo or a combined supplement of β-carotene (30 mg/day) and vitamin A (25,000 IU/day) in the form of retinyl palmitate. After a mean follow-up of 4 years of randomized treatment, CARET was terminated early because of a 28% significantly increased risk for lung cancer in the combined supplement group compared with the placebo group (RR = 1.28, 95% CI, 1.04–1.57). In the combined supplement group, overall mortality was significantly increased 17% (RR = 1.17, 95% CI, 1.03–1.33, $p = 0.02$) and death from CVD was nonsignificantly increased 26% (RR = 1.26, 95% CI, 0.99–1.61) compared with the placebo group.

The Skin Cancer Prevention Study (SCPS) was a randomized, double-blind, placebo-controlled trial designed to determine the effect of daily β-carotene supplementation on all-cause mortality and mortality from CVD and cancer in subjects with at least one biopsy-proven basal cell or squamous cell skin cancer *(108)*. A total of 1,188 men and 532 women with a mean age of 63 years were randomized to placebo or β-carotene (50 mg/day) supplementation. After a median follow-up of 4.3 years of randomized treatment, subjects randomized to β-carotene supplementation showed a 3% nonsignificant increase in all-cause mortality (RR = 1.03, 95% CI, 0.82–1.30) and a 16% nonsignificant increase in CVD mortality (RR = 1.16, 95% CI, 0.82–1.64) compared with the placebo group. In subgroup analyses, there was no evidence of a protective effect of β-carotene supplementation on mortality in subjects with initial plasma β-carotene levels below the median concentration or among subjects classified by smoking history.

The Physicians' Health Study (PHS) was a randomized, double-blind, placebo-controlled trial designed to test whether aspirin and β-carotene supplementation could prevent cancer and CVD *(109)*. Using a 2 × 2 factorial design, 22,071 US male physicians 40–84 years old without a history of cancer (except nonmelanomatous skin cancer) and CVD were randomized to aspirin (325 mg/day) plus β-carotene placebo, β-carotene (50 mg every other day) plus aspirin placebo, both active agents, or both placebos. At the beginning of the study, 11% of the subjects were current smokers and 39% were former smokers. The aspirin portion of the study was terminated early because of a 44% statistically significant ($p < 0.001$) reduction in risk for a first MI *(129)*. After a mean follow-up of 12 years of randomized treatment, subjects randomized to β-carotene supplementation showed no statistically significant overall benefit or harm with respect to cancer, CVD events, CVD mortality, or total mortality *(109)*. However, there were certain trends consistent with the ATBC prevention study *(102–104)*, CARET *(107)*, and SCPS *(108)* in the current smokers who were randomized to β-carotene supplementation. Current smokers randomized to β-carotene supplementation had an 8% greater risk for MI (RR = 1.08, 95% CI, 0.80–1.48), 18% greater risk for stroke (RR = 1.18, 95% CI, 0.83–1.67), 15% greater risk for CVD events (nonfatal MI, nonfatal stroke, and CVD death) (RR = 1.15, 95% CI, 0.93–1.43), 13% greater risk for CVD mortality (RR = 1.13, 95% CI, 0.80–1.61), and 5% greater risk for death from all causes (RR = 1.05, 95% CI, 0.86–1.29) compared with placebo subjects; all RRs were nonsignificant *(109)*.

The Women's Health Study was a randomized, double-blind, placebo-controlled trial using a 2 × 2 × 2 factorial design to test aspirin (100 mg every other day), vitamin E (600 IU every other day), and β-carotene (50 mg every other day) in the prevention of cancer and CVD in 39,876 women aged 45 years or older *(110)*. The β-carotene component was terminated early after a median intervention of 2.1 years primarily because of the null findings of β-carotene and cancer incidence after 12 years of randomized treatment in the companion PHS. Compared with placebo, there were no statistically significant differences in the incidence of cancer, CVD, or total mortality after the 2.1-year treatment period or after an additional 2-year follow-up. MI was nonsignificantly reduced 16% in the β-carotene group relative to the placebo group (RR = 0.84, 95% CI, 0.56–1.27), whereas stroke was nonsignificantly increased 42% (RR = 1.42, 95% CI, 0.96–2.10) and death from CVD nonsignificantly increased 17% (RR = 1.17, 95% CI, 0.54–2.53) in the β-carotene group relative to the placebo group.

The Nambour Skin Cancer Prevention Trial (NSCPT) was a randomized, double-blind, placebo-controlled trial using a 2 × 2 factorial design to test β-carotene (30 mg/day) and 15-plus sunscreen in the prevention of basal and squamous cell carcinomas in 1,621 residents 20–69 years old of Nambour in southeast Queensland, Australia *(111)*. After 4.5 years of intervention, there were no significant differences in the incidence of new skin cancers between the treatment groups. During the course of the trial, there were 11 deaths among the 801 participants randomized to β-carotene supplementation and 21 deaths among the 820 participants randomized to placebo, resulting in a nonsignificant 50% reduction in overall mortality (RR = 0.50, 95% CI, 0.24–1.03). Although there were few CVD deaths (6 in the β-carotene group and 12 in the placebo group), they were nonsignificantly reduced 50%.

12.5.2. Randomized Controlled Trials with CVD as the Primary Outcome

12.5.2.1. Secondary Prevention Trials

12.5.2.1.1. Antioxidant Vitamin Supplementation. The Cambridge Heart Antioxidant Study (CHAOS) was a randomized, double-blind, placebo-controlled, secondary prevention trial designed to determine whether α-tocopherol would reduce CVD risk *(112)*. The primary trial outcomes were a combined endpoint of CVD death and nonfatal MI and nonfatal MI alone. A total of 2002 subjects with angiographically proven CAD were randomized to α-tocopherol (1,035 subjects) and placebo (967 subjects). The first 546 subjects assigned to the α-tocopherol group received 800 IU/day; the remainder received 400 IU/day. After a median follow-up of 510 days, subjects assigned to the α-tocopherol group experienced a 47% significant reduction in the CVD death and nonfatal MI combined endpoint (RR = 0.53, 95% CI, 0.34–0.83, $p = 0.005$). The beneficial effect on the composite endpoint resulted from a 77% significant reduction in the risk for nonfatal MI relative to placebo (RR = 0.23, 95%

CI, 0.11–0.47). However, there was an 18% nonsignificant increase in CVD death in the α-tocopherol group (RR = 1.18, 95% CI, 0.62–1.27). In addition, total mortality was nonsignificantly greater in the α-tocopherol group than in the placebo group (3.5 versus 2.7%, $p = 0.31$). The pattern of a significant reduction in nonfatal MI and a nonsignificant increase in CVD death in CHAOS is similar to the results from the ATBC substudy and remains unexplained *(103)*.

The Gruppo Italiano per lo Studio della Sopravvivenza nell'Infarto Miocardico Prevenzione Trial (GISSI) was an open-label trial conducted in 1,665 women and 9,659 men with an average age of 59.4 years who had a recent (≤3 months) MI *(113)*. Subjects were randomly assigned according to a 2 × 2 factorial design to receive either synthetic vitamin E 300 mg daily or no supplement and either an *n*–3 polyunsaturated fatty acid (*n*–3 PUFA) supplement or no supplement for a mean of 3.5 years. The primary trial outcome was a composite of death, nonfatal MI, and stroke. Intention-to-treat analyses were done according to the factorial design (two way) and by treatment group (four way). Vitamin E had no beneficial effect in either analysis on the primary trial endpoint relative to the other treatment groups. By the two-way analysis, a total of 730 of the 5,660 subjects randomized to vitamin E (12.9%) and 770 of the 5,664 subjects randomized to the control group (13.6%) had a primary outcome event, yielding a nonsignificant risk reduction of 5% (RR = 0.95, 95% CI, 0.86–1.05). CVD death was also nonsignificantly decreased 6% relative to the control group (RR = 0.94, 95% CI, 0.81–1.10). By the four-way analysis, a total of 371 of the 2,830 subjects randomized to vitamin E alone (13.1%) and 414 of the 2,828 subjects randomized to the control group (14.6%) had a primary outcome event, yielding a nonsignificantly reduced risk of 11% (RR = 0.89, 95% CI, 0.77–1.03). CVD death was significantly decreased 20% relative to the control group in the four-way analysis (RR = 0.80, 95% CI, 0.65–0.99). On the other hand, both the two-way and the four-way analyses indicated a significant reduction of 10–15% in the primary trial outcome (RR = 0.85, 95% CI, 0.74–0.99) and a significant reduction of 17–30% in CVD death (RR = 0.70, 95% CI, 0.71–0.97) by *n*–3 PUFA relative to the other treatment groups.

The Heart Outcomes Prevention Evaluation (HOPE) Study was a randomized, double-blind, placebo-controlled trial conducted in 2,545 women and 6,996 men 55 years of age or older who had either established CVD or diabetes in addition to one other CVD risk factor *(114)*. Subjects were randomly assigned according to a 2 × 2 factorial design to receive either natural vitamin E 400 IU daily or matching placebo and either Ramipril (an angiotensin-converting enzyme inhibitor) or matching placebo for a mean of 4.5 years. The primary trial outcome was a composite of MI, stroke, and death from cardiovascular causes. A total of 772 (16.2%) of the 4,761 subjects randomized to vitamin E and 739 (15.5%) of the 4,780 subjects randomized to placebo had a primary outcome event (RR = 1.05, 95% CI, 0.95–1.16, $p = 0.33$). There were also no significant differences between vitamin E and placebo in the components of the composite primary trial outcome with a nonsignificant 17% increase in stroke (RR = 1.17, 95% CI, 0.95–1.42, $p = 0.13$) and 5% increase in CVD death (RR = 1.05, 95% CI, 0.90–1.22, $p = 0.54$). On the other hand, relative to subjects randomized to placebo, subjects receiving Ramipril had a 22% significant reduction in the primary trial outcome (RR = 0.78, 95% CI, 0.70–0.86, $p < 0.001$) *(130)*.

The Secondary Prevention with Antioxidants of Cardiovascular Disease in End-Stage Renal Disease (SPACE) was a randomized, double-blind, placebo-controlled trial conducted in 61 women and 135 men 40–75 years of age with pre-existing CVD *(115)*. All participants were chronic hemodialysis patients. Subjects were randomly assigned to either natural vitamin E 800 IU daily or matching placebo for a median of 519 days. The primary trial outcome was a composite of fatal and nonfatal MI, stroke, peripheral vascular disease, and unstable angina. Fifteen (16%) of the 97 subjects randomized to vitamin E and 33 (33%) of the 99 subjects randomized to placebo had a primary outcome event, a significant reduction in risk of 54% (RR = 0.46, 95% CI, 0.27–0.78, $p = 0.014$). Five (5.1%) of the 97 subjects randomized to vitamin E and 17 (17.2%) of the 99 subjects randomized to placebo had a MI, a significant reduction in risk of 70% (RR = 0.30, 95% CI, 0.11–0.78 $p = 0.016$). CVD mortality was nonsignificantly reduced 39% relative to placebo (RR = 0.61, 95% CI, 0.28–1.30).

The Heart Protection Study (HPS) was a randomized, double-blind, placebo-controlled trial conducted in 5,082 women and 15,454 men 40–80 years old who had either established CAD, other occlusive vascular disease, or diabetes *(116)*. Subjects were randomly assigned according to a 2 × 2 factorial design to receive either antioxidant vitamin supplementation (synthetic vitamin E 600 mg, vitamin C 250 mg, and β-carotene 20 mg daily) or matching placebo and either simvastatin or matching placebo for a mean of 5 years. The primary outcomes were fatal and nonfatal vascular events. There was no significant difference in all-cause mortality between subjects randomized to antioxidant vitamin supplementation (1,446 of 10,269 subjects [14.1%]) and to placebo (1,389 of 10,267 subjects [13.5%]), for an increased risk of 4% (RR = 1.04, 95% CI, 0.97–1.12, $p = 0.30$). Similarly, there were no significant differences in deaths resulting from vascular or nonvascular causes. A total of 878 (8.6%) of the 10,269 subjects randomized to antioxidant vitamin supplementation and 840 (8.2%) of the 10,267 subjects randomized to placebo died from a vascular cause, for an increased risk of 5% (RR = 1.05, 95% CI, 0.95–1.15, $p = 0.30$). There was no significant difference in the number of subjects having a nonfatal MI or coronary death: 1,063 (10.4%) of 10,269 subjects randomized to antioxidant vitamin supplementation versus 1,047 (10.2%) of 10,267 subjects randomized to placebo, for an increased risk of 2% (RR = 1.02, 95% CI, 0.94–1.11, $p = 0.7$). There was no significant difference in the number of subjects who had a fatal or nonfatal stroke, 511 (5%) of 10,269 subjects randomized to antioxidant vitamin supplementation versus 518 (5%) of 10,267 subjects randomized to placebo, for a decreased risk of 1% (RR = 0.99, 95% CI, 0.87–1.12, $p = 0.80$). On the other hand, relative to subjects randomized to placebo, subjects receiving simvastatin had a significant reduction of 13% in all-cause mortality (RR = 0.87, 95% CI, 0.91–0.94, $p = 0.0003$) and a 17–27% significant reduction in CVD mortality and vascular events *(131)*.

12.5.2.1.2. B-Vitamin Supplementation. In the Vitamin Intervention for Stroke Prevention (VISP) *(132)*, 3,680 adults with nondisabling cerebral infarction were randomized in a double-blind, placebo-controlled trial. All participants received best medical and surgical care plus a daily multivitamin containing the US Food and Drug Administration's reference daily intakes of other vitamins; patients were randomly assigned to receive once-daily doses of a high-dose formulation ($n = 1{,}827$) containing 25 mg of pyridoxine, 0.4 mg of cobalamin, and 2.5 mg of folic acid or a low-dose formulation ($n = 1{,}853$) containing 200 μg of pyridoxine, 6 μg of cobalamin, and 20 μg of folic acid. The mean reduction of the plasma tHcy level was 2 μmol/L greater in the high-dose group than in the low-dose group but there was no treatment effect on any endpoint. The unadjusted RR for any stroke, CHD event, or death was 1.0 (95% CI, 0.8–1.1) with 18.0% of subjects in the high-dose group and 18.6% of subjects in the low-dose group having these events during the 2 years of randomized treatment. The risk of ischemic stroke during the 2 years of randomized treatment was 9.2% for the high-dose group and 8.8% for the low-dose group (RR = 1.0; 95% CI, 0.8–1.3, $p = 0.80$). There was a persistent and graded association between baseline plasma tHcy level and outcomes. This study suggested that a moderate reduction of tHcy with B-vitamin supplementation after nondisabling cerebral infarction had no effect on vascular outcomes during 2 years of follow-up.

The Heart Outcomes Prevention Evaluation 2 (HOPE-2) was a randomized, double-blind, placebo-controlled trial of 5,522 men and women 55 years of age or older with known CVD or diabetes mellitus and at least one additional vascular risk factor *(133)*. Subjects received a daily supplement of 2.5 mg of folic acid, 50 mg of vitamin B_6, and 1 mg of vitamin B_{12} or matching placebo for an average of 5 years. Although mean plasma tHcy levels decreased by 2.4 μmol/L in the vitamin supplementation group and increased by 0.80 μmol/L in the placebo group, there was no statistically significant difference between treatment groups in the primary trial endpoint of CVD death, MI, and stroke. Primary trial events occurred in 519 subjects (18.8%) randomized to the vitamin therapy group and 547 subjects (19.8%) randomized to the placebo group (RR = 0.95, 95% CI, 0.84–1.07, $p = 0.41$). As compared with placebo, vitamin supplementation did not significantly decrease CVD death (RR = 0.96, 95% CI, 0.81–1.13) or MI (RR = 0.98, 95% CI, 0.85–1.14). Fewer subjects randomized to vitamin supplementation (4%) than placebo (5.3%) had a stroke (RR = 0.75, 95% CI, 0.59–0.97). However, more subjects in the vitamin supplementation group (9.7%) than the placebo group (7.9%) were hospitalized for unstable angina (RR = 1.24, 95% CI, 1.04–1.49).

In an open-label study *(134)* conducted in the Netherlands of 593 subjects with stable CAD who were followed for 24 months, 300 subjects were randomized to folic acid (0.5 mg/day) and 293 served as untreated controls. At baseline, all subjects had been on statin therapy for a mean of 3.2 years. In subjects treated with folic acid, plasma tHcy levels decreased by 18% from 12.0 ± 4.8 to 9.4 ± 3.5 μmol/L, whereas the tHcy levels were unaffected in the control group ($p < 0.001$ between treatment groups). The primary trial endpoint (all-cause mortality and a composite of vascular events) occurred in 31 (10.3%) of subjects in the folic acid group and in 28 (9.6%) subjects in the control group (RR = 1.05, 95% CI, 0.63–1.75). This study indicated that statin and folic acid treatment in subjects with stable CAD did not reduce the trial endpoint relative to statins alone.

The Norwegian Vitamin Trial (NORVIT) *(120)* was a randomized, double-blind, placebo-controlled trial of 3,749 men and women who had an acute MI within 7 days before randomization. Subjects were randomized in a 2 × 2 factorial design to receive one of the following four daily treatments: 0.8 mg of folic acid, 0.4 mg of vitamin B_{12}, and 40 mg of vitamin B_6; 0.8 mg of folic acid and 0.4 mg of vitamin B_{12}; 40 mg of vitamin B_6; or placebo. The primary trial endpoint during a median follow-up of 40 months was a composite of recurrent MI, stroke, and sudden death attributed to CAD. The mean plasma tHcy level was lowered by 27% among subjects randomized to folic acid plus vitamin B_{12}, but this treatment had no significant effect on the primary trial endpoint (RR = 1.08, 95% CI, 0.93–1.25, $p = 0.31$). Treatment with vitamin B_6 was not associated with any significant benefit with regard to the primary trial endpoint (RR = 1.14, 95% CI, 0.98–1.32, $p = 0.09$). In the group randomized to folic acid, vitamin B_{12}, and vitamin B_6, there was a trend toward an increased risk of the primary trial endpoint (RR = 1.22, 95%CI, 1.00–1.50, $p = 0.05$). This study indicated that B-vitamin supplementation not only failed to lower the risk of recurrent CVD after an initial MI, but in fact may be harmful.

The ongoing randomized, double-blind, placebo-controlled Women's Antioxidant Cardiovascular Study (WACS) *(135)*, a 2 × 2 × 2 factorial design of three antioxidant vitamins (C, E, and β-carotene), was expanded to a four-group factorial trial, the Women's Antioxidant and Folic Acid Cardiovascular Study (WAFACS) *(121)* to study folic acid, vitamin B_6, and vitamin B_{12}. WAFACS includes 5,442 US health professional women aged 42 years or older with either a history of CVD or three or more coronary risk factors. Women were randomized to a combination pill containing 2.5 mg of folic acid, 50 mg of vitamin B_6, and 1 mg of vitamin B_{12} or a matching placebo for 7.3 years; 406 women in the active group and 390 women in the placebo group experienced a CVD event. Subjects receiving B-vitamin supplementation had similar risk for the composite CVD primary endpoint (MI, stroke, coronary revascularization, or CVD mortality) as did women receiving placebo (RR = 1.03, 95% CI, 0.90–1.19, $p = 0.65$). Additionally, the primary components of the combined endpoint were not significantly different between treatment groups: MI (RR = 0.87, 95% CI, 0.63–1.22, $p = 0.42$), stroke (RR = 1.14, 95% CI, 0.82–1.57, $p = 0.44$), and CVD mortality (RR = 1.01, 95% CI, 0.76–1.35, $p = 0.93$). In a subgroup analysis, mean plasma tHcy level was decreased by 18.5% (95% CI, 12.5–24.1%, $p < 0.001$) in the B-vitamin supplementation group ($n = 150$) over that in the placebo group ($n = 150$), for a difference of 2.27 μmol/L (95% CI, 1.54–2.96 μmol/L). After 7.3 years of randomized treatment, supplementation with a combination of B-vitamins did not reduce CVD events among high-risk women despite a significant decrease in tHcy levels.

The Swiss Heart Study (SHS) *(136)* was a randomized, double-blind placebo-controlled trial of 553 subjects who had undergone angioplasty of at least one significant coronary stenosis. Participants were randomly assigned to receive a combination of folic acid (1 mg/day), vitamin B_{12} (400 μg/day), and vitamin B_6 (10 mg/day) ($n = 272$) or placebo ($n = 281$) for 6 months. The composite primary endpoint was defined as death, nonfatal MI and need for repeat revascularization evaluated at 6 months and 1 year. After a mean follow-up of 11 months, the composite endpoint was significantly lower in subjects treated with tHcy-lowering therapy relative to placebo (15.4% versus 22.8%) (RR = 0.68, 95% CI, 0.48–0.96, $p = 0.03$). The reduction in the composite endpoint was primarily due to a reduced rate of target lesion revascularization (9.9% versus 16.0%) (RR = 0.62, 95% CI, 0.40–0.97, $p = 0.03$). A nonsignificant trend was seen in fewer deaths (1.5% versus 2.8%) (RR = 0.54, 95% CI, 0.16–1.70, $p = 0.27$) and nonfatal MI (2.6% versus 4.3%), RR of 0.60 (95% CI, 0.24–1.51; $p = 0.27$) with tHcy-lowering therapy relative to placebo treatment. These findings remained unchanged after adjustment for potential confounders. In this study, tHcy-lowering therapy with B-vitamins resulted in a significant decrease in the incidence of major adverse events after percutaneous coronary intervention.

The effect of B-vitamin supplementation was studied in Germany and Netherlands, among 636 subjects who had undergone coronary artery stenting in a double-blind, multicenter trial *(137)*. The subjects were randomly assigned to 1 mg of folic acid, 5 mg of vitamin B_6, and 1 mg of vitamin B_{12} intravenously followed by daily oral doses of 1.2 mg of folic acid, 48 mg of vitamin B_6, and 60 μg of vitamin B_{12} for 6 months, or to placebo. The angiographic endpoints (minimal luminal diameter, late loss, and restenosis rate) were assessed at 6 months by means of quantitative coronary angiography. At follow-up, the mean minimal luminal diameter was significantly smaller in the B-vitamin supplementation group than in the placebo group (1.59 ± 0.62 mm versus 1.74 ± 0.64 mm, $p = 0.008$) and the extent of late luminal loss was greater (0.90 ± 0.55 mm versus 0.76 ± 0.58 mm, $p = 0.004$). The restenosis rate was greater in the B-vitamin group than in the placebo group (34.5% versus 26.5%, $p = 0.05$) and a higher percentage of subjects in the B-vitamin group required repeated target vessel revascularization (15.8% versus 10.6%, $p = 0.05$). This study suggested that B-vitamin supplementation had adverse effects on the risk of restenosis after percutaneous coronary intervention.

To date, four RCTs *(138–141)* have reported the effect of B-vitamin supplementation on CVD and all-cause mortality among subjects with end-stage renal failure. Hyperhomocysteinemia is frequently detected in subjects with end-stage renal disease and in dialysis patients. In addition, patients with end-stage renal disease have greater CVD risk than individuals with normal renal function. However, in all four RCTs reporting the effect of B-vitamin supplementation on CVD in patients with end-stage renal disease, there was no statistically significant benefit or harm of B-vitamin supplementation on CVD or all-cause mortality.

While experimental and epidemiological evidence supports a plausible role for B-vitamin supplementation with lowering of tHcy levels in the prevention of CVD, collectively, RCTs do not provide clear evidence of such an effect in tens of thousands of participants with pre-existing CVD or renal disease. RCTs in subjects without pre-existing CVD have not been reported to date. It has been proposed that RCTs of tHcy lowering should be conducted in subjects who have a persistent excess prevalence of hyperhomocysteinemia or in populations located in regions where mandatory fortification with folic acid has not been implemented. However, in the few RCTs where this has been accomplished *(120,134,140,141)* studies have not yielded differential results from those of the collective RCT data.

12.5.2.2. Primary Prevention Trials

The Primary Prevention Project (PPP) was an open-label trial conducted in 2,583 women and 1,912 men with a mean age of 64.4 years without pre-existing CVD *(122)*. Subjects were randomly assigned according to a 2 × 2 factorial design to receive either synthetic vitamin E 300 mg daily or no supplement and either low-dose aspirin or no aspirin for a mean of 3.6 years. The primary trial outcome was a composite of CVD, nonfatal MI, stroke, and nonfatal stroke. A total of 56 (2.5%) of the 2,231 subjects randomized to vitamin E and 53 (2.3%) of the 2,264 subjects randomized to placebo had a primary trial outcome event (RR = 1.07, 95% CI, 0.74–1.56). There were also no significant differences between vitamin E and placebo in the components of the composite primary trial outcome with a 14% reduction in CVD death (RR = 0.86, 95% CI, 0.49–1.52), 1% increase in nonfatal MI (RR = 1.01, 95% CI, 0.56–2.03), and 56% increase in nonfatal stroke (RR = 1.56, 95% CI, 0.56–2.03). On the other hand, subjects receiving low-dose aspirin had a significant 24% reduction in CVD death relative to subjects not receiving aspirin (RR = 0.76, 95% CI, 0.31–0.99).

Supplementation en Vitamines et Mineraux Antioxydants (SU.VI.MAX) *(123)* was a randomized, double-blind, placebo-controlled primary prevention trial. In total, 13,017 French adults were randomized, including 7,876 women aged 35–60 years and 5,141 men aged 45–60 years. All participants took a single daily capsule of a combination of 120 mg of ascorbic acid, 30 mg of vitamin E, 6 mg of β-carotene, 100 μg of selenium, and 20 mg of zinc or a placebo. Median follow-up was 7.5 years. There were no major differences detected between treatment groups in ischemic CVD incidence (134 subjects [2.1%] in the intervention group versus 137 subjects [2.1%] in the placebo group) or all-cause mortality (76 subjects [1.2%] in the intervention group versus 98 subjects [1.5%] in the placebo group). A significant interaction between sex and treatment group effects on cancer incidence was found ($p = 0.004$). Sex-stratified analysis showed a protective effect of antioxidants in men (RR = 0.69, 95% CI, 0.53–0.91) but not in women (RR = 1.04, 95% CI, 0.85–1.29). A similar trend was observed for all-cause mortality (RR = 0.63, 95% CI, 0.42–0.93 in men versus RR = 1.03, 95% CI, 0.64–1.63 in women, $p = 0.11$ for interaction).

12.5.3. Arterial Imaging Trials

12.5.3.1. Coronary Angiographic Trials

The HDL-Atherosclerosis Treatment Study (HATS) was a randomized, double-blind, placebo-controlled trial of 160 subjects (13% female; average age 53 years) with CAD and a low HDL-C level (≤35 mg/dL for men, ≤40 mg/dL for women) *(124)*. Subjects with angiographically documented CAD were randomly assigned to one of four regimens: antioxidants (synthetic vitamin E 800 IU, vitamin C 1,000 mg, natural β-carotene 25 mg, and Se 100 μg daily), simvastatin–niacin plus antioxidants, simvastatin–niacin, or placebos. After 3 years of intervention, standardized coronary angiograms were performed and compared with the baseline coronary angiograms also obtained under standardized conditions. The primary trial endpoints were angiographic evidence of a change in coronary artery stenosis and the occurrence of a cardiovascular event, death, MI, stroke, or revascularization. The simvastatin and niacin dosages were titrated to obtain an LDL-C goal of <90 mg/dL and to raise HDL-C by at least 10 mg/dL over baseline. The mean per subject change in lesion percent diameter stenosis (%S) was +3.9%S with placebos, –0.4%S with simvastatin–niacin alone ($p < 0.001$ for the comparison with the placebo group), +1.8%S with antioxidants alone ($p = 0.16$ for the comparison with the placebo group) and +0.7%S with simvastatin–niacin plus antioxidants ($p = 0.004$ for the comparison with the placebo group). The frequency of clinical event endpoints was 24% in the placebo group, 3% in the simvastatin–niacin group, 21% in the antioxidant group, and 14% in the simvastatin–niacin plus antioxidant group. Only the frequency of clinical events in the simvastatin–niacin group significantly differed from the placebo group ($p = 0.003$).

The Women's Angiographic Vitamin and Estrogen (WAVE) trial was a double-blind, placebo-controlled trial conducted in 423 post-menopausal women (average age of 65 years) with angiographically established CAD *(125)*. Subjects were randomly assigned according to a 2 × 2 factorial design to receive either vitamin E 400 IU plus vitamin C 500 mg daily or matching placebo and either conjugated equine estrogen 0.625 mg daily (plus medroxyprogesterone acetate 2.5 mg daily for women with a uterus) or matching placebo. The mean interval between the baseline and follow-up coronary angiograms was 2.8 years. The primary clinical trial endpoint was based on the change in the mean minimum lumen diameter of all qualifying coronary arterial segments and on the incidence of MI and death. The mean minimum lumen diameter (in millimeters/year [mm/y]) worsened in all four groups, –0.0010 mm/y with placebos, –0.048 mm/y with hormone therapy alone, –0.042 mm/y with antioxidant therapy alone, and –0.046 mm/y with antioxidant therapy plus hormone therapy; none of the differences compared with placebo were significant. The frequency of the composite clinical endpoint was 2.8% in the placebo group, 4.9% in the hormone therapy group, 6.7% in the antioxidant group, and 6.5% in the antioxidant therapy plus hormone therapy group. The frequency of clinical events was not significantly different amongst groups ($p = 0.54$). However, all-cause mortality was significantly higher in women assigned to antioxidant vitamin supplementation compared with women assigned to antioxidant vitamin placebo, 16 versus 6 (hazard ratio = 2.6, 95% CI, 1.1–7.2, $p = 0.047$).

12.5.3.2. B-Mode Ultrasound Trials

The Antioxidant Supplementation in Atherosclerosis Prevention (ASAP) study was a double-blind, placebo-controlled trial conducted in Eastern Finland in 264 post-menopausal women and 256 men 45–69 years old with serum cholesterol ≥194 mg/dL *(126)*. Subjects were randomly assigned according to a 2 × 2 factorial design to receive twice daily either vitamin E 136 IU, slow-release vitamin C 250 mg, a combination of both vitamins or placebo for 3 years. The primary trial endpoint was progression of CCA IMT determined every 6 months from serial high-resolution B-mode ultrasonograms. In women, there were no significant differences in progression rates among the treatment groups: 0.015 mm/y in women who received only vitamin E, 0.017 mm/y in women who received only vitamin C, 0.016 mm/y in women who received the vitamin combination, and 0.016 mm/y in women randomized to placebo. In men, the rates of progression were 0.018 mm/y in men who received only vitamin E, 0.017 mm/y in men who received only vitamin C, 0.011 mm/y in men who received both vitamins, and 0.020 mm/y in men randomized to placebo. The progression rate was significantly less in the men who were randomized to both vitamins compared to all men ($p = 0.009$) or compared with men who received placebo ($p = 0.008$). The treatment effect of the vitamin combination was limited to hypercholesterolemic men who smoked. Baseline plasma α-tocopherol and total ascorbate levels were lowest among men who smoked relative to men who did not smoke or to women who did or did not smoke.

The Study to Evaluate Carotid Ultrasound Changes in Patients Treated with Ramipril and Vitamin E (SECURE) *(127)* was a substudy of HOPE *(114)*. SECURE was a double-blind, placebo-controlled trial conducted in 172 women and 560 men 55 years of age or older who had either established CVD or diabetes in addition to one other cardiovascular risk factor. Subjects were randomly assigned according to a 3 × 2 factorial design to receive either natural vitamin E 400 IU daily or matching placebo and either ramipril 2.5 mg/day or 10 mg/day or matching placebo for a mean of 4.5 years. The primary trial endpoint was progression of the mean maximum carotid artery IMT determined every 6 months from serial high-resolution B-mode ultrasonograms. Carotid IMT progression rates did not differ between subjects randomized to vitamin E and those randomized to placebo, 0.0174 mm/y versus 0.0180 mm/y, respectively. On the other hand, the mean progression rate among subjects randomized to ramipril was significantly less than the progression rate of subjects randomized to placebo, 0.021 mm/y for the placebo group, 0.0180 mm/y for the ramipril group receiving 2.5 mg/day, and 0.0137 mm/y for the ramipril group receiving 10 mg/day ($p = 0.033$).

The Vitamin E Atherosclerosis Prevention Study (VEAPS) was a randomized, double-blind, placebo-controlled trial conducted in 185 women and 168 men $\geq$40 years old with LDL-C $\geq$130 mg/dL and no clinical signs or symptoms of CVD *(128)*. Subjects were randomly assigned to either synthetic vitamin E 400 IU daily or matching placebo for 3 years. The primary trial endpoint was progression of CCA IMT *(142–145)* determined every 6 months from serial high-resolution B-mode ultrasonograms. Although on-trial plasma vitamin E levels were significantly increased and several markers indicated a significant reduction in LDL oxidation in the vitamin E group, there was no difference in the average rates of CCA IMT progression between the two treatment groups. In fact, the rate of progression in the vitamin E group was twice that of the placebo group, 0.0040 mm/y versus 0.0023 mm/y ($p = 0.08$).

12.5.4. Nutri-genomic Trials

An evolving area of clinical medicine is that of pharmacogenetics, in which administration of specific therapies to maximize benefit and to avoid risk are based upon an individual's genetic profile. The relevance of this field intersects with that of vitamin E supplementation and diabetes mellitus. It has been described that diabetic individuals with a specific haptoglobin genotype (Hp 2-2) have a greater risk for vascular complications than those with the other haptoglobin genotypes (Hp 1-1 and Hp 2-1) *(146,147)*. In several studies, it has been reported that diabetics with the Hp 2-2 genotype who receive vitamin E supplementation have an approximate 50% reduction in cardiovascular events relative to those diabetics with the Hp 2-2 genotype who do not receive vitamin E supplementation *(147,148)*. With vitamin E supplementation, the cardiovascular event rate in diabetics with the Hp 2-2 genotype is reduced to the cardiovascular event rates of diabetics with the Hp 1-1 and Hp 2-1 genotypes. In contrast, vitamin E supplementation appears to have no effect on the cardiovascular event rates in diabetics with the Hp 1-1 and Hp 2-1 genotypes.

12.5.5. Summary of Randomized Controlled Trials of Antioxidant Vitamin and B-Vitamin Supplementation

Despite enormous experimental evidence indicating the beneficial effects of antioxidant and B-vitamin intake on cardiovascular events, the majority of RCTs conducted to date have failed to demonstrate the efficacy of vitamin supplementation in reducing CVD.

Observational studies have reported an inverse correlation between vitamin intake and CVD, frequently at doses above the recommended daily allowance *(46,47,149)*. In contrast, many large intervention trials have found little or no effect on atherosclerosis to a modest increase in mortality with high-dose vitamin use *(114,117,150)*. The most mentioned reason for the difference between observational and interventional studies is that subjects who use vitamin supplementation generally have a healthier lifestyle (better diet, more exercise activity, and less smoking) versus those who choose not to use vitamins. An alternate and equally plausible explanation for the disparity between observational studies and RCTs is the timing of intervention. Given that CVD is a multifactorial indolent progressive process in which clinical events occur 50–70 years after initiation of the process, the relatively short length of intervention in RCTs, (average 2–7 years) as well as the time frame in which the intervention has been initiated in relation to the atherosclerotic process, may be insufficient or too late to intervene on disease progression and for the vitamins to play a meaningful role. An additional explanation for the disparity between observational and intervention studies may be the large difference in patient populations studied. Participants in RCTs are usually medically well managed and vitamin supplementation may not offer or have further perceptible benefit.

Publication of no less than 24 RCTs examining the effects of vitamin supplementation on CVD events and atherosclerosis progression have yielded consistent, conflicting and occasionally, somewhat troublesome data. In general, when one examines the vitamin E data (Tables 12.2–12.4), there is a general null effect of this antioxidant supplement on CVD events and on the progression of atherosclerosis. In fact, when one examines the data carefully, there are some troublesome results. In ATBC, there was no overall benefit of vitamin E supplementation on CVD. However, subjects who received vitamin E had a significantly greater risk of death from hemorrhagic stroke *(102)*. Similarly, in HOPE-2 *(151)*, vitamin E supplementation caused an increased risk for heart failure.

In CHAOS, although there was a reduction in risk from nonfatal MI, there was a nonsignificantly increased risk for both CVD death and total mortality in the vitamin E group *(112)*. Similar to CHAOS, the ATBC substudy *(103)* in men with previous MI showed the same pattern of a significant reduction in nonfatal MI but a nonsignificant increase in CVD death. Interestingly, the increase in risk for CVD death was not apparent in the ATBC substudy of men without a previous MI *(104)*. The cohort of ATBC was comprised of individuals who smoked *(102)*. In the arterial imaging trials, VEAPS *(128)* showed a nonsignificantly greater progression rate of subclinical atherosclerosis in healthy men and women receiving vitamin E relative to placebo, whereas ASAP *(126)* and SECURE *(127)* showed no effect of vitamin E supplementation relative to placebo. However, there have been trial results indicating a significant reduction of CVD events with vitamin E supplementation as in CHAOS *(112)* and the smaller SPACE trial *(115)*. CHAOS and SPACE are similar in that both of these trials used a higher dosage of vitamin E (in CHAOS a lower dosage was subsequently used) relative to most of the other trials and they were conducted for the shortest time periods relative to other trials. SPACE, like CHAOS did not show a reduction in CVD mortality with vitamin E supplementation. The one trial that did show a significant reduction in CVD mortality was GISSI *(113)*. However, this finding was restricted to the four-way analysis, comparing 371 (13.1%) of the 2,830 subjects randomized to vitamin E alone to 414 (14.6%) of the 2,828 subjects randomized to the control group. Neither the two-way nor the four-way analysis showed a significant reduction in the primary trial outcome composite of death, nonfatal MI, and stroke.

In the two randomized controlled trials that studied vitamin C alone *(105,106)*, there was no evidence that vitamin C was effective in reducing CVD events or mortality. No adverse effects were observed.

The results from the ATBC *(102–104)*, Linxian *(105)*, CARET *(107)*, SCPS *(108)*, PHS *(109)*, WHS *(110)*, and NSCPT *(111)* trials provided no support for a beneficial effect of β-carotene in the prevention of CVD in contradistinction to the beneficial effects inferred from observational studies. In fact, taken together, the results of these trials are troubling since subjects, particularly smokers randomized to β-carotene supplementation, had increased rates of cancer, CVD events, CVD mortality, and overall mortality. In particular, in the ATBC substudy of subjects who had a previous MI, those randomized to β-carotene supplementation had a 75% significantly greater risk of CVD mortality compared with placebo *(103)*. Those subjects who received both vitamin E and β-carotene supplementation had a 58% significantly greater risk of CVD mortality than those receiving placebo *(103)*. These trials indicate that β-carotene is not the primary component responsible for the lower risks of cancer and CVD associated with the high intakes of fruits and vegetables. Although one could argue that β-carotene should still be tested in other populations, there is little support for efficacy, and good reason for concern for the safety of β-carotene supplementation for the prevention of cancer and CVD in certain populations, such as smokers.

The five RCTs of B-vitamin supplementation *(117,120,121,132,134)* consistently demonstrated that there is no effect of B-vitamin supplementation in secondary prevention for CHD. The effects on stroke prevention are mixed. Whereas HOPE-2 showed a significant reduction of stroke in individuals without pre-existing stroke, VISP, a trial specifically designed to determine the effect of B-vitamin supplementation on the secondary prevention of stroke was null. Three of the five RCTs of B-vitamin supplementation in secondary prevention of CHD were mainly randomized in the United States and Canada *(117,121,132)* where folate fortification of cereal grain and flour products is mandatory, while the other two trials *(120,134)* were conducted in areas of non-fortification with folate suggesting that dietary fortification with folate does not reduce the power of these trials. Despite a significant decrease in plasma tHcy levels in these trials, B-vitamin supplementation failed to lower CVD risk in patients with pre-existing vascular disease.

The serial arterial imaging trials provided important information regarding the effects of antioxidant supplementation on the progression of atherosclerosis. The coronary angiographic trials, HATS *(124)* and WAVE *(125)*, showed that cocktails of antioxidants including vitamin E, vitamin C, β-carotene, and Se have no beneficial effect on the progression of established coronary artery atherosclerosis. In fact, in HATS, when antioxidant therapy was combined with simvastatin–niacin, the coronary angiographic and clinical event benefits of simvastatin–niacin alone were diminished. In terms of the angiographic endpoint, this adverse interaction between the antioxidant cocktail and the simvastatin–niacin therapy was significant ($p = 0.02$). This interaction appeared to be the result of a significant blunting of the increase in the HDL_2 atherosclerosis protective subfraction by the antioxidant therapy *(124)*.

In contrast to coronary angiography, which is a measurement endpoint for established late-stage atherosclerosis, B-mode carotid artery ultrasound is used to measure arterial wall thickness, the earliest anatomical manifestation of atherosclerosis *(152)*. Three completed trials, VEAPS *(128)*, SECURE *(127)*, and ASAP *(126)*, showed that vitamin E had no effect on the progression of subclinical atherosclerosis. In fact, in VEAPS *(128)*, subjects who were randomized to vitamin E had a nonsignificant twofold greater progression rate of subclinical atherosclerosis than those subjects randomized to placebo. This effect was seen in men and women. ASAP also showed that vitamin C alone had no effect on the progression of subclinical atherosclerosis *(126)*. In contrast to the null effects of vitamin E and vitamin C alone, ASAP showed that a combination of vitamin E and vitamin C had a significant effect in reducing the progression of subclinical atherosclerosis *(126)*. However, this effect was limited to males who smoked and had high serum cholesterol levels. Relative to the other groups of subjects in ASAP (nonsmoking men and women and smoking women), smoking men had lower baseline serum vitamin C and E levels. Along with other trial results, such as those from SPACE *(115)*, ASAP indicated that individuals who are vitamin deficient may derive anti-atherosclerosis benefits from antioxidant vitamin supplementation. Although serum vitamin levels were not reported from SPACE, it is generally believed that chronic dialysis patients have low intracellular vitamin concentrations and experience a much greater level of oxidative stress than most other populations studied with the possible exception of smokers.

It remains to be seen whether administration of antioxidant vitamin and B-vitamin supplementation in relation to earlier stages of atherosclerosis (perhaps during childhood or young adulthood) will be effective in reducing atherosclerosis progression. By the time atherosclerosis is well established, it is possible that antioxidant vitamin and B-vitamin supplementation is ineffective in reducing progression of atherosclerosis and CVD events. Animal studies have predominantly demonstrated antioxidants to be effective in preventing atherosclerosis when atherosclerosis is in its earliest stages. There are very little data indicating that antioxidant vitamin and B-vitamin supplementation is therapeutically effective in treating atherosclerosis once it is established.

Epidemiological studies probably reflect long-term lifestyle behaviors, such as high dietary and supplementary intake of vitamins over many years, perhaps even during the early formative years of atherosclerosis development. In contrast, RCTs have investigated vitamin intervention for a limited period of time and in most instances in populations that already have some degree of atherosclerosis. Although an inverse association with vitamin intake found in epidemiological studies may have resulted from uncontrolled confounding, the divergent results of observational studies and RCTs alternatively suggest that antioxidant vitamins and B-vitamins may have different effects in preventing atherosclerosis, reducing atherosclerosis progression or affecting the rupture of atherosclerosis plaques. Before final conclusions are reached regarding the effects of antioxidant vitamin and B-vitamin supplementation on CVD, questions raised by the completed RCTs need to be addressed.

12.6. ONGOING RANDOMIZED CONTROLLED TRIALS

Several RCTs are ongoing. These trials are testing vitamin supplementation in both primary and secondary prevention of CVD. The Supplementation with Folate, vitamin B_6, and B_{12} and/or OMega-3 fatty acids (SU.FOL.OM3) study is a randomized, double-blind, placebo-controlled secondary prevention trial testing supplementation with folate, vitamin B_6 and B_{12}, and/or omega-3 fatty acids for 5 years on recurrent IHD events in subjects with atherosclerosis in the coronary or cerebral arteries; 3,000 subjects between 45 and 80 years old have been recruited for this trial *(153)*.

The Folic Acid for Vascular Outcome Reduction In Transplantation (FAVORIT) Trial is designed to evaluate whether lowering plasma tHcy levels using B-vitamin supplementation will reduce CVD events in renal transplant recipients. This population of individuals is at high risk for CVD; 4,000 renal transplant recipients have been recruited and randomized to multivitamins that include either high-dose or low-dose folic acid (5 or 0 mg), vitamin B_6 (50 or 1.4 mg), and vitamin B_{12} (1,000 or 2 μg) *(154)*.

The Vitamin to Prevent Stroke (VITATOPS) Trial is designed to determine whether B-vitamin supplements (folic acid 2 mg; vitamin B_6 25 mg; vitamin B_{12} 500 μg) reduce the risk of stroke and other serious vascular events in patients with recent stroke or transient ischemic attacks of the brain or eye. A total of 8,000 patients were recruited between 2000 and 2004 from multiple centers for this randomized, double-blind, placebo-controlled trial *(155)*.

The Western Norway B-vitamin Intervention Trial (WENBIT) is a secondary prevention trial *(156)* being conducted in a cohort of subjects recruited from a population without mandatory folic acid food fortification. From 1999 to 2004, a total of 3,090 subjects with a median age of 62 years (20.5% women) undergoing coronary angiography for stable angina pectoris were recruited. Subjects were assigned to four groups receiving daily oral treatment with (a) folic acid 0.8 mg, vitamin B_{12} 0.4 mg, and vitamin B_6 40 mg, (b) folic acid and vitamin B_{12}, (c) vitamin B_6 alone, and (d) placebo.

The B-Vitamin Atherosclerosis Intervention Trial (BVAIT) is a randomized, double-blind, placebo-controlled serial arterial imaging trial of 506 men and post-menopausal women (39%) without clinical evidence of CVD or diabetes mellitus with an average age of 61.4 years. Subjects were randomized to a daily combination pill of folic acid 5 mg, vitamin B_6 50 mg, and vitamin B_{12} 0.4 mg or matching placebo for an average 3 years. Every 6 months CCA IMT was measured and the rate of progression of subclinical atherosclerosis will be determined between the two treatment groups. BVAIT is unique in that it is the only trial testing B-vitamin therapy under primary prevention conditions in a healthy general population of subjects.

12.7. RECOMMENDATIONS

Although observational data suggest that antioxidant vitamins and B-vitamins may reduce CVD risk and basic research provides plausible mechanisms for such an effect, these data have not been substantiated by completed RCTs. Additionally, results from these completed RCTs raise a level of concern for the safety of certain antioxidant vitamin supplementation under certain conditions. In the final analysis, current data do not support the use of antioxidant vitamin or B-vitamin supplementation for the prevention or treatment of CVD. Guidelines for the use of antioxidant vitamin and B-vitamin supplementation for the prevention or treatment of CVD will have to await results from ongoing RCTs and assessment of the effects of antioxidant vitamin and B-vitamin supplementation in certain populations, such as those with low serum vitamin levels and those with chronic disease states that may take a toll on the body's antioxidant capacity, such as diabetics and other special populations. Regardless of the results of these RCTs, however, consumption of fruits and vegetables high in antioxidant vitamins and B-vitamins appears to be an important component of a healthy dietary intake and lifestyle.

REFERENCES

1. Ignarro, L. J., Balestrieri, M. L., and Napoli, C. (2007) Nutrition, physical activity, and cardiovascular disease: an update. *Cardiovasc Res* **73**, 326–40.
2. Faxon, D. P., Fuster, V., Libby, P., Beckman, J. A., Hiatt, W. R., Thompson, R. W., Topper, J. N., Annex, B. H., Rundback, J. H., Fabunmi, R. P., Robertson, R. M., Loscalzo, J., and American Heart Association. (2004) Atherosclerotic Vascular Disease Conference: Writing Group III: pathophysiology. *Circulation* **109**, 2617–25.
3. Madamanchi, N. R., Vendrov, A., and Runge, M. S. (2005) Oxidative stress and vascular disease. *Arterioscler Thromb Vasc Biol* **25**, 29–38.
4. Rong, J. X., Rangaswamy, S., Shen, L., Dave, R., Chang, Y. H., Peterson, H., Hodis, H. N., Chisolm, G. M., and Sevanian, A. (1998) Arterial injury by cholesterol oxidation products causes endothelial dysfunction and arterial wall cholesterol accumulation. *Arterioscler Thromb Vasc Biol* **18**, 1885–94.
5. Björkhem, I., Henriksson-Freyschuss, A., Breuer, O., Diczfalusy, U., Berglund, L., and Henriksson, P. (1991) The antioxidant butylated hydroxytoluene protects against atherosclerosis. *Arterioscler Thromb* **11**, 15–22.
6. Hodis, H. N., Chauhan, A., Hashimoto, S., Crawford, D. W., and Sevanian, A. (1992) Probucol reduces plasma and aortic wall oxysterol levels in cholesterol fed rabbits independently of its plasma cholesterol lowering effect. *Atherosclerosis* **96**, 125–34.
7. Stringer, M. D., Görög, P. G., Freeman, A., and Kakkar, V. (1989) Lipid peroxides and atherosclerosis. *BMJ* **298**, 281–4.
8. Azen, S. P., Qian, D., Mack, W. J., Sevanian, A., Selzer, R. H., Liu, C. R., Liu, C. H., and Hodis, H. N. (1996) Effect of supplementary antioxidant vitamin intake on carotid arterial wall intima-media thickness in a controlled clinical trial of cholesterol lowering. *Circulation* **94**, 2369–72.
9. Hodis, H. N., Mack, W. J., LaBree, L., Cashin-Hemphill, L., Sevanian, A., Johnson, R., and Azen, S. P. (1995) Serial coronary angiographic evidence that antioxidant vitamin intake reduces progression of coronary artery atherosclerosis. *JAMA* **273**, 1849–54.
10. Thomson, M. J., Puntmann, V., and Kaski, J. C. (2007) Atherosclerosis and oxidant stress: the end of the road for antioxidant vitamin treatment? *Cardiovasc Drugs Ther* **21**, 195–210.
11. Griendling, K. K., and FitzGerald, G. A. (2003) Oxidative stress and cardiovascular injury: part II: animal and human studies. *Circulation* **108**, 2034–40.
12. Harrison, D., Griendling, K. K., Landmesser, U., Hornig, B., and Drexler, H. (2003) Role of oxidative stress in atherosclerosis. *Am J Cardiol* **91**, 7A–11A.
13. Lum, H., and Roebuck, K. A. (2001) Oxidant stress and endothelial cell dysfunction. *Am J Physiol Cell Physiol* **208**, C719–41.
14. Harrison, D. G. (1997) Endothelial function and oxidant stress. *Clin Cardiol* **20**, II-11–7.
15. Bonetti, P. O., Lerman, L. O., and Lerman, A. (2003) Endothelial dysfunction: a marker of atherosclerotic risk. *Arterioscler Thromb Vasc Biol* **23**, 168–75.
16. Nedeljkovic, Z. S., Gokce, N., and Loscalzo, J. (2003) Mechanisms of oxidative stress and vascular dysfunction. *Postgrad Med J* **79**, 195–9.
17. Uchida, K. (2007) Lipid peroxidation and redox-sensitive signaling pathways. *Curr Atheroscler Rep* **9**, 216–21.

18. Torzewski, M., and Lackner, K. J. (2006) Initiation and progression of atherosclerosis—enzymatic or oxidative modification of low-density lipoprotein? *Clin Chem Lab Med* **44**, 1389–94.
19. Griendling, K. K., Sorescu, D., and Ushio-Fukai, M. (2000) NAD(P)H oxidase, role in cardiovascular biology and disease. *Circ Res* **86**, 494–501.
20. Urso, M. L., and Clarkson, P. M. (2003) Oxidative stress, exercise, and antioxidant supplementation. *Toxicology* **189**, 41–54.
21. Bonomini, F., Tengattini, S., Fabiano, A., Bianchi, R., and Rezzani, R. (2008) Atherosclerosis and oxidative stress. *Histol Histopathol* **23**, 381–90.
22. Traber, M. G. (2007) Heart disease and single-vitamin supplementation. *Am J Clin Nutr* **85**, 293S–9S.
23. Singh, U., Devaraj, S., and Jialal, I. (2005) Vitamin E, oxidative stress, and inflammation. *Annu Rev Nutr* **25**, 151–74.
24. Meydani, M. (2004) Vitamin E modulation of cardiovascular disease. *Ann N Y Acad Sci* **1031**, 271–9.
25. Nishikimi, M., and Yagi, K. (1996) Biochemistry and molecular biology of ascorbic acid biosynthesis. *Subcell Biochem* **25**, 17–39.
26. Haytowitz, D. B. (1995) Information from USDA's nutrient data bank. *J Nutr* **125**, 1952–5.
27. Tolbert, B. M. (1985) Metabolism and function of ascorbic acid and its metabolites. *Int J Vitam Nutr Res Suppl* **27**, 121–68.
28. Di Mascio, P., Kaiser, S., and Sies, H. (1989) Lycopene as the most efficient biological carotenoid singlet oxygen quencher. *Arch Biochem Biophys* **274**, 532–8.
29. Tapiero, H., Townsend, D. M., and Tew, K. D. (2004) The role of carotenoids in the prevention of human pathologies. *Biomed Pharmacother* **58**, 100–10.
30. Berliner, J. A., and Heineicke, J. W. (1996) The role of oxidized lipoproteins in atherogenesis. *Free Rad Biol Med* **20**, 707–27.
31. Suzuki, Y. J., Forman, H. J., and Sevanian, A. (1997) Oxidants as stimulators of signal transduction. *Free Radic Biol Med* **22**, 269–85.
32. Ricciarelli, R., Tasinato, A., Clément, S., Ozer, N. K., Boscoboinik, D., and Azzi, A. (1998) Alpha-Tocopherol specifically inactivates cellular protein kinase C alpha by changing its phosphorylation state. *Biochem J* **334**, 243–9.
33. McCully, K. S. (1969) Vascular pathology of homocysteinemia: implications for the pathogenesis of arteriosclerosis. *Am J Pathol* **56**, 111–28.
34. Heart Study Collaboration. (2002) Homocysteine Studies Collaboration. *JAMA* **288**, 15–22.
35. Wald, D. S., Law, M., and Morris, J. K. (2002) Homocysteine and cardiovascular disease: evidence on causality from a meta-analysis. *BMJ* **325**, 1202–6.
36. Mangoni, A. A., and Jackson, S. H. (2002) Homocysteine and cardiovascular disease: current evidence and future prospects. *Am J Med* **112**, 556–65.
37. Malinow, M. R., Nieto, F. J., Kruger, W. D., Duell, P. B., Hess, D. L., Gluckman, R. A., Block, P. C., Holzgang, C. R., Anderson, P. H., Seltzer, D., Upson, B., and Lin, Q. R. (1997) The effects of folic acid supplementation on plasma total homocysteine are modulated by multivitamin use and methylenetetrahydrofolate reductase genotypes. *Arterioscler Thromb Vasc Biol* **17**, 1157–62.
38. Papatheodorou, L., and Weiss, N. (2007) Vascular oxidant stress and inflammation in hyperhomocysteinemia. *Antioxid Redox Signal* **9**, 1941–58.
39. Moens, A. L., Vrints, C. J., Claeys, M. J., Timmermans, J. P., Champion, H. C., and Kass, D. A. (2008) Mechanisms and potential therapeutic targets for folic acid in cardiovascular disease. *Am J Physiol Heart Circ Physiol* **294**, H1971–7.
40. Stehouwer, C. D., and van Guldener, C. (2003) Does homocysteine cause hypertension? *Clin Chem Lab Med* **41**, 1408–11.
41. Sauls, D. L., Arnold, E. K., Bell, C. W., Allen, J. C., and Hoffman, M. (2007) Pro-thrombotic and pro-oxidant effects of diet-induced hyperhomocysteinemia. *Thromb Res* **120**, 117–26.
42. Verlangieri, A. J., Kapeghian, J. C.,el-Dean, S., and Bush, M. (1985) Fruit and vegetable consumption and cardiovascular mortality. *Med Hypotheses* **16**, 7–15.
43. Ginter, E. (1979) Decline in coronary mortality in United States and vitamin C. *Am J Clin Nutr* **32**, 511–2.
44. Acheson, R. M., and Williams, D. R. R. (1993) Does consumption of fruit and vegetables protect against stroke? *Lancet* **1**, 1191–3.
45. Armstrong, B. K., Mann, J. I., Adelstein, A. M., and Eskin, F. (1975) Commodity consumption and ischemic heart disease mortality, with special reference to dietary practices. *J Chron Dis* **28**, 455–69.
46. Stampfer, M. J., Hennekens, C. H., Manson, J. E., Colditz, G. A., Rosner, B., and Willett, W. C. (1993) Vitamin E consumption and the risk of coronary disease in women. *N Engl J Med* **328**, 1444–9.

47. Rimm, E. B., Stampfer, M. J., Ascherio, A., Giovannucci, E., Colditz, G. A., and Willett, W. C. (1993) Vitamin E consumption and the risk of coronary heart disease in men. *N Engl J Med* **328**, 1450–6.
48. Knekt, P., Reunanen, A., Järvinen, R., Seppänen, R., Heliövaara, M., and Aromaa, A. (1994) Antioxidant vitamin intake and coronary mortality in a longitudinal population study. *Am J Epidemiol* **139**, 1180–9.
49. Losonczy, K. G., Harris, T. B., and Havlik, R. J. (1996) Vitamin E and vitamin C supplement use and risk of all-cause and coronary heart disease mortality in older persons. *Am J Clin Nutr* **64**, 190–6.
50. Kushi, L. H., Folsom, A. R., Prineas, R. J., Mink, P. J., Wu, Y., and Bostick, R. M. (1996) Dietary antioxidant vitamins and death from coronary heart disease in postmenopausal women. *N Engl J Med* **334**, 1156–62.
51. Yochum, L. A., Folsom, A. R., and Kushi, L. H. (2000) Intake of antioxidant vitamins and risk of death from stroke in postmenopausal women. *Am J Clin Nutr* **72**, 476–83.
52. Klipstein-Grobusch, K., Geleijnse, J. M., den Breeijen, J. H., Boeing, H., Hofman, A., Grobbee, D. E., and Witteman, J. C. (1999) Dietary antioxidants and risk of myocardial infarction in the elderly. *Am J Clin Nutr* **69**, 261–6.
53. Enstrom, J. E., Kanim, L. E., and Klein, M. A. (1992) Vitamin C intake and mortality among a sample of the United States population. *Epidemiology* **3**, 194–202.
54. Enstrom, J. E., Kanim, L. E., and Breslow, L. (1986) The relationship between vitamin C intake, general health practices, and mortality in Alameda County, California. *Am J Public Health* **76**, 1124–30.
55. Pandey, D. K., Shekelle, R., Selwyn, B. J., Tangney, C., and Stamler, J. (1995) Dietary vitamin C and beta-carotene and risk of death in middle-aged men. *Am J Epidemiol* **142**, 1269–78.
56. Lapidus, L., Andersson, H., Bengtsson, C., and Bosaeus, I. (1986) Dietary habits in relation to incidence of cardiovascular disease and death in women: a 12-year follow-up of participants in the population study of women in Gothenburg, Sweden. *Am J Clin Nutr* **44**, 444–8.
57. Gale, C. R., Martyn, C. N., Winter, P. D., and Cooper, C. (1995) Vitamin C and risk of death from stroke and coronary heart disease in cohort of elderly people. *BMJ* **310**, 1563–6.
58. Shekelle, R. B., Lepper, M., Liu, S., Maliza, C., Raynor, W. J. J., Rossof, A. H., Paul, O., Shryock, A. M., and Stamler, J. (1981) Dietary vitamin A and risk of cancer in the Western Electric study. *Lancet* **2**, 1185–90.
59. Gaziano, J. M., Manson, J. E., Branch, L. G., Colditz, G. A., Willett, W. C., and Buring, J. E. (1995) A prospective study of consumption of carotenoids in fruits and vegetables and decreased cardiovascular mortality in the elderly. *Ann Epidemiol* **5**, 225–60.
60. Klipstein-Grobusch, K., Launer, L. J., Geleijnse, J. M., Boeing, H., Hofman, A., and Witteman, J. C. (2000) Serum carotenoids and atherosclerosis. *Atherosclerosis* **148**, 49–56.
61. Al-Delaimy, W. K., Rexrode, K. M., Hu, F. B., Albert, C. M., Stampfer, M. J., Willett, W. C., and Manson, J. E. (2004) Folate intake and risk of stroke among women. *Stroke* **35**, 1259–63.
62. He, K., Merchant, A., Rimm, E. B., Rosner, B. A., Stampfer, M. J., Willett, W. C., and Ascherio, A. (2004) Folate, vitamin B6, and B12 intakes in relation to risk of stroke among men. *Stroke* **35**, 169–74.
63. Bazzano, L. A., He, J., Ogden, L. G., Loria, C., Vupputuri, S., Myers, L., and Whelton, P. K. (2002) Dietary intake of folate and risk of stroke in US men and women: NHANES I epidemiologic follow-up study. *Stroke* **33**, 1183–8.
64. Gey, K. F., Stähelin, H. B., and Eichholzer, M. (1993) Poor plasma status of carotene and vitamin C is associated with higher mortality from ischemic heart disease and stroke. *Clin Investig* **71**, 3–6.
65. Gey, K. F., Puska, P., Jordan, P., and Moser, U. K. (1991) Inverse correlation between plasma vitamin E and mortality from ischemic heart disease in cross-cultural epidemiology. *Am J Clin Nutr* **53**, 326S–34S.
66. Gey, K. F., Brubacher, G. B., and Stähelin, H. B. (1987) Plasma levels of antioxidant vitamins in relation to ischemic heart disease and cancer. *Am J Clin Nutr* **45**, 1368–77.
67. Riemersma, R. A., Oliver, M., Elton, R. A., Alfthan, G., Vartiainen, E., Salo, M., Rubba, P., Mancini, M., Georgi, H., and Vuilleumier, J. P. (1990) Plasma antioxidants and coronary heart disease: vitamins C and E, and selenium. *Eur J Clin Nutr* **44**, 143–50.
68. Singh, R. B., Ghosh, S., Niaz, M. A., Singh, R., Beegum, R., Chibo, H., Shoumin, Z., and Postiglione, A. (1995) Dietary intake, plasma levels of antioxidant vitamins, and oxidative stress in relation to coronary artery disease in elderly subjects. *Am J Cardiol* **76**, 1233–8.
69. Simon, J. A., Hudes, E. S., and Browner, W. S. (1998) Serum ascorbic acid and cardiovascular disease prevalence in U.S. adults. *Epidemiology* **9**, 316–21.
70. Kok, F. J., de Bruijn, A. M., Vermeeren, R., Hofman, A., van Laar, A., de Bruin, M., Hermus, R. J., and Valkenburg, H. A. (1987) Serum selenium, vitamin antioxidants, and cardiovascular mortality: a 9-year follow-up study in the Netherlands. *Am J Clin Nutr* **45**, 432–68.
71. Salonen, J. T., Salonen, R., Penttilä, I., Herranen, J., Jauhiainen, M., Kantola, M., Lappeteläinen, R., Mäenpää, P. H., Alfthan, G., and Puska, P. (1985) Serum fatty acids, apolipoproteins, selenium and vitamin antioxidants and the risk of death from coronary artery disease. *Am J Cardiol* **56**, 226–31.

72. Street, D. A., Comstock, G. W., Salkeld, R. M., Schüep, W., and Klag, M. J. (1994) Serum antioxidants and myocardial infarction: are low levels of carotenoids and alpha-tocopherol risk factors for myocardial infarction? *Circulation* **90**, 1154–61.
73. Riemersma, R. A., Wood, D. A., Macintyre, C. C., Elton, R. A., Gey, K. F., and Oliver, M. F. (1991) Risk of angina pectoris and plasma concentrations of vitamins A, C, and E and carotene. *Lancet* **337**, 1–5.
74. Morris, D. L., Kritchevsky, S. B., and Davis, C. E. (1994) Serum carotenoids and coronary heart disease. *JAMA* **272**, 1439–41.
75. Nyyssönen, K., Parviainen, M. T., Salonen, R., Tuomilehto, J., and Salonen, J. T. (1997) Vitamin C deficiency and risk of myocardial infarction: prospective population study of men from eastern Finland. *BMJ* **314**, 634–8.
76. Ramirez, J., and Flowers, N. C. (1980) Leukocyte ascorbic acid and its relationship to coronary artery disease in man. *Am J Clin Nutr* **33**, 2079–87.
77. Kok, F. J., van Poppel, G., Melse, J., Verheul, E., Schouten, E. G., Kruyssen, D. H., and Hofman, A. (1991) Do antioxidants and polyunsaturated fatty acids have a combined association with coronary atherosclerosis? *Atherosclerosis* **86**, 85–90.
78. Halevy, D., Thiery, J., Nagel, D., Arnold, S., Erdmann, E., Höfling, B., Cremer, P., and Seidel, D. (1997) Increased oxidation of LDL in patients with coronary artery disease is independent from dietary vitamins E and C. *Arterioscler Thromb Vasc Biol* **17**, 1432–7.
79. Iribarren, C., Folsom, A. R., Jacobs, D. R. J., Gross, M. D., Belcher, J. D., and Eckfeldt, J. H. (1997) Association of serum vitamin levels, LDL susceptibility to oxidation, and autoantibodies against MDA-LDL with carotid atherosclerosis: a case-control study. *Arterioscler Thromb Vasc Biol* **17**, 1171–7.
80. Bonithon-Kopp, C., Coudray, C., Berr, C., Touboul, P. J., Fève, J. M., Favier, A., and Ducimetière, P. (1997) Combined effects of lipid peroxidation and antioxidant status on carotid atherosclerosis in a population aged 59–71 years. *Am J Clin Nutr* **65**, 121–7.
81. Devaraj, S., Tang, R., Adams-Huet, B., Harris, A., Seenivasan, T., de Lemos, J. A., and Jialal, I. (2007) Effect of high-dose alpha-tocopherol supplementation on biomarkers of oxidative stress and inflammation and carotid atherosclerosis in patients with coronary artery disease. *Am J Clin Nutr* **86**, 1392–8.
82. Aldred, S., Sozzi, T., Mudway, I., Grant, M. M., Neubert, H., Kelly, F. J., and Griffiths, H. R. (2006) Alpha tocopherol supplementation elevates plasma apolipoprotein A1 isoforms in normal healthy subjects. *Proteomics* **6**, 1695–703.
83. Buijsse, B., Feskens, E. J., Schlettwein-Gsell, D., Ferry, M., Kok, F. J., Kromhout, D., and de Groot, L. C. (2005) Plasma carotene and alpha-tocopherol in relation to 10-y all-cause and cause-specific mortality in European elderly. *Am J Clin Nutr* **82**, 879–86.
84. Carpenter, K. L., Kirkpatrick, P. J., Weissberg, P. L., Challis, I. R., Dennis, I. F., Freeman, M. A., and Mitchinson, M. J. (2003) Oral alpha-tocopherol supplementation inhibits lipid oxidation in established human atherosclerotic lesions. *Free Radic Res* **37**, 1235–44.
85. Rapola, J. M., Virtamo, J., Haukka, J. K., Heinonen, O. P., Albanes, D., Taylor, P. R., and Huttunen, J. K. (1996) Effect of vitamin E and beta carotene on the incidence of angina pectoris: a randomized, double-blind, controlled trial. *JAMA* **275**, 693–8.
86. Gillilan, R. E., Mondell, B., and Warbasse, J. R. (1997) Quantitative evaluation of vitamin E in the treatment of angina pectoris. *Am Heart J* **93**, 444–9.
87. Anderson, T. W., and Reid, D. B. (1974) A double-blind trial of vitamin E in angina pectoris. *Am J Clin Nutr* **27**, 1174–8.
88. Buijsse, B., Feskens, E. J., Kwape, L., Kok, F. J., and Kromhout, D. (2008) Both alpha- and beta-carotene, but not tocopherols and vitamin C, are inversely related to 15-year cardiovascular mortality in Dutch elderly men. *J Nutr* **138**, 344–50.
89. Morrison, H. I., Schaubel, D., Desmeules, M., and Wigle, D. T. (1996) Serum folate and risk of fatal coronary heart disease. *JAMA* **275**, 1893–6.
90. Giles, W. H., Kittner, S. J., Anda, R. F., Croft, J. B., and Casper, M. L. (1995) Serum folate and risk for ischemic stroke. *Stroke* **26**, 1166–70.
91. Voutilainen, S., Lakka, T. A., Porkkala-Sarataho, E., Rissanen, T., Kaplan, G. A., and Salonen, J. T. (2000) Low serum folate concentrations are associated with an excess incidence of acute coronary events. *Eur J Clin Nutr* **54**, 424–8.
92. Selhub, J., Jacques, P. F., Bostom, A. G., D'Agostino, R. B., Wilson, P. W., Belanger, A. J., O'Leary, D. H., Wolf, P. A., Schaefer, E. J., and Rosenberg, I. H. (1995) Association between plasma homocysteine concentrations and extracranial carotid-artery stenosis. *N Engl J Med* **332**, 286–91.
93. Chasan-Taber, L., Selhub, J., Rosenberg, I. H., Malinow, M. R., Terry, P., Tishler, P. V., Willett, W., Hennekens, C. H., and Stampfer, M. J. (1996) A prospective study of folate and vitamin B6 and risk of myocardial infarction in US physicians. *J Am Coll Nutr* **15**, 136–43.

94. Folsom, A. R., Nieto, F. J., McGovern, P. G., Tsai, M. Y., Malinow, M. R., Eckfeldt, J. H., Hess, D. L., and Davis, C. E. (1998) Prospective study of coronary heart disease incidence in relation to fasting total homocysteine, related genetic polymorphisms, and B vitamins. *Circulation* **98**, 204–10.
95. Wald, N. J., Watt, H. C., Law, M. R., Weir, D. G., McPartlin, J., and Scott, J. M. (1998) Homocysteine and ischemic heart disease: results of a prospective study with implications regarding prevention. *Arch Intern Med* **158**, 862–7.
96. Bots, M. L., Launer, L. J., Lindemans, J., Hofman, A., and Grobbee, D. E. (1997) Homocysteine, atherosclerosis and prevalent cardiovascular disease in the elderly. *J Intern Med* **242**, 339–47.
97. Perry, I. J., Refsum, H., Morris, R. W., Ebrahim, S. B., Ueland, P. M., and Shaper, A. G. (1995) Prospective study of serum total homocysteine concentration and risk of stroke in middle-aged British men. *Lancet* **346**, 1395–8.
98. Alfthan, G., Pekkanen, J., Jauhiainen, M., Pitkäniemi, J., Karvonen, M., Tuomilehto, J., Salonen, J. T., and Ehnholm, C. (1994) Relation of serum homocysteine and lipoprotein(a) concentrations to atherosclerotic disease in a prospective Finnish population based study. *Atherosclerosis* **106**, 9–19.
99. Robinson, K., Arheart, K., Refsum, H., Brattström, L., Boers, G., Ueland, P., Rubba, P., Palma-Reis, R., Meleady, R., Daly, L., Witteman, J., and Graham, I. (1998) Low circulating folate and vitamin B6 concentrations: risk factors for stroke, peripheral vascular disease, and coronary artery disease. *Circulation* **97**, 437–43.
100. Boushey, C. J., Beresford, S. A., Omenn, G. S., and Motulsky, A. G. (1995) A quantitative assessment of plasma homocysteine as a risk factor for vascular disease: probable benefits of increasing folic acid intakes. *JAMA* **274**, 1049–57.
101. Jacques, P. F., Selhub, J., Bostom, A. G., Wilson, P. W., and Rosenberg, I. H. (1999) The effect of folic acid fortification on plasma folate and total homocysteine concentrations. *N Engl J Med* **340**, 1449–54.
102. The Alpha-Tocopherol, Beta Carotene Cancer Prevention Study Group. (1994) The effect of vitamin E and beta carotene on the incidence of lung cancer and other cancers in male smokers. *N Engl J Med* **330**, 1029–35.
103. Rapola, J. M., Virtamo, J., Ripatti, S., Huttunen, J. K., Albanes, D., Taylor, P. R., and Heinonen, O. P. (1997) Randomised trial of alpha-tocopherol and beta-carotene supplements on incidence of major coronary events in men with previous myocardial infarction. *Lancet* **349**, 1715–20.
104. Virtamo, J., Rapola, J. M., Ripatti, S., Heinonen, O. P., Taylor, P. R., Albanes, D., and Huttunen, J. K. (1998) Effect of vitamin E and beta carotene on the incidence of primary nonfatal myocardial infarction and fatal coronary heart disease. *Arch Intern Med* **158**, 668–75.
105. Blot, W. J., Li, J. Y., Taylor, P. R., Juo, W., Dawsey, S., Wang, G. Q., Yang, C. S., Zheng, S. F., Gail, M., Li, G. Y., Yu, Y., Liu, B., Tangrea, J., Sun, Y., Liu, F., Fraumeni, J. F., Zhang, Y., and Li, B. (1993) Nutrition intervention trials in Linxian, China: supplementation with specific vitamin/mineral combinations, cancer incidence, and disease-specific mortality in the general population. *J Natl Cancer Inst* **85**, 1483–92.
106. Wilson, T. S., Datta, S. B., Murrell, J. S., and Andrews, C. T. (1973) Relation of vitamin C levels to mortality in a geriatric hospital: a study of the effect of vitamin C administration. *Age Ageing* **2**, 163–71.
107. Omenn, G. S., Goodman, G. E., Thornquist, M. D., Balmes, J., Cullen, M. R., Glass, A., Keogh, J. P., Meyskens, F. L., Valanis, B., Williams, J. H., Barnhart, S., and Hammar, S. (1996) Effects of a combination of beta carotene and vitamin A on lung cancer and cardiovascular disease. *N Engl J Med* **334**, 1150–5.
108. Greenberg, E. R., Baron, J. A., Karagas, M. R., Stukel, T. A., Nierenberg, D. W., Stevens, M. M., Mandel, J. S., and Haile, R. W. (1996) Mortality associated with low plasma concentration of beta carotene and the effect of oral supplementation. *JAMA* **275**, 699–703.
109. Hennekens, C. H., Buring, J. E., Manson, J. E., Stampfer, M., Rosner, B., Cook, N. R., Belanger, C., LaMotte, F., Gaziano, J. M., Ridker, P. M., Willett, W., and Peto, R. (1996) Lack of effect of long-term supplementation with beta carotene on the incidence of malignant neoplasms and cardiovascular disease. *N Engl J Med* **223**, 1145–9.
110. Lee, I. M., Cook, N. R., Manson, J. E., Buring, J. E., and Hennekens, C. H. (1999) Beta-carotene supplementation and incidence of cancer and cardiovascular disease. *J Natl Cancer Inst* **91**, 2102–6.
111. Green, A., Williams, G., Neale, R., Hart, V., Leslie, D., Parsons, P., Marks, G. C., Gaffney, P., Battistutta, D., Frost, C., Lang, C., and Russell, A. (1999) Daily sunscreen application and beta-carotene supplementation in prevention of basal-cell and squamous-cell carcinomas of the skin: a randomised controlled trial. *Lancet* **354**, 723–9.
112. Stephens, N. G., Parsons, A., Schofield, P. M., Kelly, F., Cheeseman, K., and Mitchinson, M. J. (1996) Randomised controlled trial of vitamin E in patients with coronary disease. *Lancet* **347**, 781–6.
113. GISSI-Prevenzione Investigators (Gruppo Italiano per lo Studio della Sopravvivenza nell'Infarto miocardico). (1999) Dietary supplementation with *n*–3 polyunsaturated fatty acids and vitamin E after myocardial infarction. *Lancet* **354**, 447–55.
114. Yusuf, S., Dagenais, G., Pogue, J., Bosch, J., and Sleight, P. (2000) Vitamin E supplementation and cardiovascular events in high-risk patients. *N Engl J Med* **342**, 154–60.

115. Boaz, M., Smetana, S., Weinstein, T., Matas, Z., Gafter, U., Iaina, A., Knecht, A., Weissgarten, Y., Brunner, D., Fainaru, M., and Green, M. S. (2000) Secondary prevention with antioxidants of cardiovascular disease in endstage renal disease (SPACE): randomised placebo-controlled trial. *Lancet* **356**, 1213–8.
116. Heart Protection Study Collaborative Group. (2002) MRC/BHF Heart Protection Study of antioxidant vitamin supplementation in 20,536 high-risk individuals: a randomised placebo-controlled trial. *Lancet* **360**, 23–33.
117. Lonn, E., Yusuf, S., Arnold, M. J., Sheridan, P., Pogue, J., Micks, M., McQueen, M. J., Probstfield, J., Fodor, G., Held, C., Genest, J. J., and Investigators. (2006) Homocysteine lowering with folic acid and B vitamins in vascular disease. *Engl J Med* **354**, 1567–77.
118. Bostom, A. G., Selhub, J., Jacques, P. F., and Rosenberg, I. (2001) Power Shortage: clinical trials testing the "homocysteine hypothesis" against a background of folic acid-fortified cereal grain flour. *Ann Intern Med* **135**, 133–7.
119. Hankey, G. J., Eikelboom, J. W., Loh, K., Tang, M., Pizzi, J., Thom, J., and Yi, Q. (2005) Sustained homocysteine-lowering effect over time of folic acid-based multivitamin therapy in stroke patients despite increasing folate status in the population. *Cerebrovasc Dis* **19**, 110–6.
120. Bønaa, K. H., Njølstad, I., Ueland, P. M., Schirmer, H., Tverdal, A., Steigen, T., Wang, H., Nordrehaug, J. E., Arnesen, E., Rasmussen, K., and NORVIT Trial Investigators. (2006) Homocysteine lowering and cardiovascular events after acute myocardial infarction. *N Engl J Med* **354**, 1578–88.
121. Albert, C. M., Cook, N. R., Gaziano, J. M., Zaharris, E., MacFadyen, J., Danielson, E., Buring, J. E., and Manson, J. E. (2008) Effect of folic acid and B vitamins on risk of cardiovascular events and total mortality among women at high risk for cardiovascular disease: a randomized trial. *JAMA* **299**, 2027–36.
122. de Gaetano, G., and Collaborative Group of the Primary Prevention Project (PPP). (2001) Low-dose aspirin and vitamin E in people at cardiovascular risk: a randomised trial in general practice. *Lancet* **357**, 89–95.
123. Hercberg, S., Galan, P., Preziosi, P., Bertrais, S., Mennen, L., Malvy, D., Roussel, A. M., Favier, A., and Briançon, S. (2004) The SU.VI.MAX study: a randomized, placebo-controlled trial of the health effects of antioxidant vitamins and minerals. *Arch Intern Med* **164**, 2335–42.
124. Brown, B. G., Zhao, X. Q., Chait, A., Fisher, L. D., Cheung, M. C., Morse, J. S., Dowdy, A. A., Marino, E. K., Bolson, E. L., Alaupovic, P., Frohlich, J., and Albers, J. J. (2001) Simvastatin and niacin, antioxidant vitamins, or the combination for the prevention of coronary disease. *N Engl J Med* **345**, 1583–92.
125. Waters, D. D., Alderman, E. L., Hsia, J., Howard, B. V., Cobb, F. R., Rogers, W. J., Ouyang, P., Thompson, P., Tardif, J. C., Higginson, L., Bittner, V., Steffes, M., Gordon, D. J., Proschan, D., Younes, N., and Verter, J. I. (2002) Effects of hormone replacement therapy and antioxidant vitamin supplements on coronary atherosclerosis in post-menopausal women: a randomized controlled trial. *JAMA* **288**, 2432–40.
126. Salonen, J. T., Nyyssönen, K., Salonen, R., Lakka, H. M., Kaikkonen, J., Porkkala-Sarataho, E., Voutilainen, S., Lakka, T. A., Rissanen, T., Leskinen, L., Tuomainen, T. P., Valkonen, V. P., Ristonmaa, U., and Poulsen, H. E. (2000) Antioxidant Supplementation in Atherosclerosis Prevention (ASAP) study: a randomized trial of the effect of vitamins E and C on 3-year progression of carotid atherosclerosis. *J Intern Med* **248**, 377–86.
127. Lonn, E., Yusuf, S., Dzavik, V., Doris, C., Yi, Q., Smith, S., Moore-Cox, A., Bosch, J., Riley, W., Teo, K., and Investigators. (2001) Effects of ramipril and vitamin E on atherosclerosis. *Circulation* **103**, 919–25.
128. Hodis, H. N., Mack, W. J., LaBree, L., Mahrer, P. R., Sevanian, A., Liu, C. R., Liu, C. H., Hwang, J., Selzer, R. H., Azen, S. P., and VEAPS Research Group. (2002) Alpha-tocopherol supplementation in healthy individuals reduces low-density lipoprotein oxidation but not atherosclerosis. *Circulation* **106**, 1453–9.
129. Steering Committee of the Physicians' Health Study Research Group. (1989) Final report on the aspirin component of the ongoing Physicians' Health Study. *N Engl J Med* **321**, 129–35.
130. Yusuf, S., Sleight, P., Pogue, J., Bosch, J., Davies, R., and Dagenais, G. (2000) Effects of an angiotensin-converting-enzyme inhibitor, ramipril, on cardiovascular events in high-risk patients. *N Engl J Med* **342**,145–53.
131. Heart Protection Study Collaborative Group. (2002) MRC/BHF Heart Protection Study of cholesterol lowering with simvastatin in 20,536 high-risk individuals: a randomised placebo-controlled trial. *Lancet* **360**, 7–22.
132. Toole, J. F., Malinow, M. R., Chambless, L. E., Spence, J. D., Pettigrew, L. C., Howard, V. J., Sides, E. G., Wang, C. H., and Stampfer, M. (2004) Lowering homocysteine in patients with ischemic stroke to prevent recurrent stroke, myocardial infarction, and death. *JAMA* **291**, 565–75.
133. Ray, J. G., Kearon, C., Yi, Q., Sheridan, P., Lonn, E., and Heart Outcomes Prevention Evaluation (HOPE) 2 Investigators. (2007) Homocysteine-lowering therapy and risk for venous thromboembolism: a randomized trial. *Ann Intern Med* **146**, 761–7.
134. Liem, A., Reynierse-Buitenwerf, G. H., Zwinderman, A. H., Jukema, J. W., and van Veldhuisen, D. J. (2003) Secondary prevention with folic acid: effects on clinical outcomes. *J Am Coll Cardiol* **41**, 2105–13.

135. Manson, J. E., Gaziano, J. M., Spelsberg, A., Ridker, P. M., Cook, N. R., Buring, J. E., Willett, W. C., and Hennekens, C. H. (1995) A secondary prevention trial of antioxidant vitamins and cardiovascular disease in women: rationale, design, and methods. *Ann Epidemiol* **5**, 261–9.
136. Schnyder, G., Roffi, M., Flammer, Y., Pin, R., and Hess, O. M. (2002) Effect of homocysteine-lowering therapy with folic acid, vitamin B12, and vitamin B6 on clinical outcome after percutaneous coronary intervention: a randomized controlled trial. *JAMA* **288**, 973–9.
137. Lange, H., Suryapranata, H., De Luca, G., Börner, C., Dille, J., Kallmayer, K., Pasalary, M. N., Scherer, E., and Dambrink, J. H. (2004) Folate therapy and in-stent restenosis after coronary stenting. *N Engl J Med* **350**, 2673–81.
138. Righetti, M., Serbelloni, P., Milani, S., and Ferrario, G. (2006) Homocysteine-lowering vitamin B treatment decreases cardiovascular events in hemodialysis patients. *Blood Purif* **24**, 379–86.
139. Righetti, M., Ferrario, G. M., Milani, S., Serbelloni, P., La Rosa, L., Uccellini, M., and Sessa, A. (2003) Effects of folic acid treatment on homocysteine levels and vascular disease in hemodialysis patients. *Med Sci Monit* **9**, 19–24.
140. Zoungas, S., McGrath, B. P., Branley, P., Kerr, P. G., Muske, C., Wolfe, R., Atkins, R. C., Nicholls, K., Fraenkel, M., Hutchison, B. G., Walker, R., and McNeil, J. J. (2006) Cardiovascular morbidity and mortality in the Atherosclerosis and Folic Acid Supplementation Trial (ASFAST) in chronic renal failure: a multicenter, randomized, controlled trial. *J Am Coll Cardiol* **47**, 1108–16.
141. Wrone, E. M., Hornberger, J. M., Zehnder, J. L., McCann, L. M., Coplon, N. S., and Fortmann, S. P. (2004) Randomized trial of folic acid for prevention of cardiovascular events in end-stage renal disease. *J Am Soc Nephrol* **15**, 420–6.
142. Pignoli, P., Tremoli, E., Poli, A., Oreste, P., and Paoletti, R. (1986) Intimal plus medial thickness of the arterial wall: a direct measurement with ultrasound imaging. *Circulation* **74**, 1399–406.
143. Mack, W. J., LaBree, L., Liu, C., Selzer, R. H., and Hodis, H. N. (2000) Correlations between measures of atherosclerosis change using carotid ultrasonography and coronary angiography. *Atherosclerosis* **150**, 371–9.
144. Hodis, H. N., Mack, W. J., LaBree, L., Selzer, R. H., Liu, C. R., Liu, C. H., and Azen, S. P. (1998) The role of carotid arterial intima-media thickness in predicting clinical coronary events. *Ann Intern Med* **128**, 262–9.
145. Selzer, R. H., Hodis, H. N., Kwong-Fu, H., Mack, W. J., Lee, P. L., Liu, C. R., and Liu, C. H. (1994) Evaluation of computerized edge tracking for quantifying intima-media thickness of the common carotid artery from B-mode ultrasound images. *Atherosclerosis* **111**, 1–11.
146. Levy, A. P., Hochberg, I., Jablonski, K., Resnick, H. E., Lee, E. T., Best, L., Howard, B. V., and Strong Heart Study. (2002) Haptoglobin phenotype is an independent risk factor for cardiovascular disease in individuals with diabetes. *J Am Coll Cardiol* **40**, 1984–90.
147. Milman, U., Blum, S., Shapira, C., Aronson, D., Miller-Lotan, R., Anbinder, Y., Alshiek, J., Bennett, L., Kostenko, M., Landau, M., Keidar, S., Levy, Y., Khemlin, A., Radan, A., and Levy, A. P. (2008) Vitamin E supplementation reduces cardiovascular events in a subgroup of middle-aged individuals with both type 2 diabetes mellitus and the haptoglobin 2-2 genotype: a prospective double-blinded clinical trial. *Arterioscler Thromb Vasc Biol* **28**, 341–7.
148. Levy, A. P., Gerstein, H. C., Miller-Lotan, R., Ratner, R., McQueen, M., Lonn, E., and Pogue, J. (2004) The effect of vitamin E supplementation on cardiovascular risk in diabetic individuals with different haptoglobin phenotypes. *Diabetes Care* **27**, 2767.
149. Rimm, E. B., Willett, W. C., Hu, F. B., Sampson, L., Colditz, G. A., Manson, J. E., Hennekens, C., and Stampfer, M. J. (1998) Folate and vitamin B6 from diet and supplements in relation to risk of coronary heart disease among women. *JAMA* **279**, 359–64.
150. Salonen, R. M., Nyyssönen, K., Kaikkonen, J., Porkkala-Sarataho, E., Voutilainen, S., Rissanen, T. H., Tuomainen, T. P., Valkonen, V. P., Ristonmaa, U., Lakka, H. M., Vanharanta, M., Salonen, J. T., and Poulsen, H. E. (2003) Six-year effect of combined vitamin C and E supplementation on atherosclerotic progression. *Circulation* **107**, 947–53.
151. Lonn, E., Bosch, J., Yusuf, S., Sheridan, P., Pogue, J., Arnold, J. M., Ross, C., Arnold, A., Sleight, P., Probstfield, J., and Dagenais, G. R. (2005) Effects of long-term vitamin E supplementation on cardiovascular events and cancer: a randomized controlled trial. *JAMA* **293**, 1338–47.
152. Blankenhorn, D. H., and Hodis, H. N. (1994) Arterial imaging and atherosclerosis reversal. *Arterioscler Thromb* **14**, 177–92.
153. Galan, P., de Bree, A., Mennen, L., Potier de Courcy, G., Preziozi, P., Bertrais, S., Castetbon, K., and Hercberg, S. (2003) Background and rationale of the SU.FOL.OM3 study: double-blind randomized placebo-controlled secondary prevention trial to test the impact of supplementation with folate, vitamin B6 and B12 and/or omega-3 fatty acids on the prevention of recurrent ischemic events in subjects with atherosclerosis in the coronary or cerebral arteries. *J Nutr Health Aging* **7**, 428–35.
154. Bostom, A. G., Carpenter, M. A., Kusek, J. W., Hunsicker, L. G., Pfeffer, M. A., Levey, A. S., Jacques, P. F., McKenney, J., and FAVORIT Investigators. (2006) Rationale and design of the Folic Acid for Vascular Outcome Reduction In Transplantation (FAVORIT) trial. *Am Heart J* **152**, 448.e1–7.

155. The VITATOPS Trial Study Group. (2002) The VITATOPS (Vitamins to Prevent Stroke) trial: rationale and design of an international, large, simple, randomised trial of homocysteine-lowering multivitamin therapy in patients with recent transient ischaemic attack or stroke. *Cerebrovasc Dis* **13**, 120–6.
156. Bleie, Ø., Refsum, H., Ueland, P. M., Vollset, S. E., Guttormsen, A. B., Nexo, E., Schneede, J., Nordrehaug, J. E., and Nygård, O. (2004) Changes in basal and postmethionine load concentrations of total homocysteine and cystathionine after B vitamin intervention. *Am J Clin Nutr* **80**, 641–8.

13 B Vitamins in the Prevention of Cognitive Decline and Vascular Dementia

Kristen E. D'Anci and Irwin H. Rosenberg

Key Points

- Cognitive decline in aging is associated with nutritional status, most notably with low folate and vitamin B12 and with elevated homocysteine.
- Homocysteine levels can be lowered through nutritional supplementation with B vitamins including folic acid, vitamin B12, and vitamin B6.
- Blood levels of folate are more strongly associated with cognitive function.
- Intervention trials with B vitamins have shown mixed findings with respect to cognitive performance.
- Supplementation with adequate doses of B vitamins appears to be most effective in preventing cognitive decline in individuals with low nutrient intake and status.

Abstract

As the number of elderly continues to increase worldwide, age-related neurological disorders, such as Alzheimer's disease and vascular dementia, are a growing concern. In some cases, vascular dementia and cognitive decline in aging are associated with nutritional status and elevated homocysteine, suggesting that improving nutritional status can play a meaningful role in the prevention of cognitive impairment. The research described in this chapter represents current understanding on the relationships of folate and vitamin B12 nutritional status with cognitive function and dementia in adults and elderly. Low B vitamin status is associated with increased homocysteine levels and there is evidence that insufficient B vitamin intake is associated with lower cognitive scores in comparison to adequate intake. However, higher rates of cognitive decline have been reported with high levels of folate and folic acid intake in adults, and memory performance may be impaired with high folate intake in individuals with low vitamin B12 status. Overall, studies reported lower folate blood levels and a higher prevalence of deficiency among subjects with dementia. In general, vitamin B12 serum levels were lower in patients with dementia relative to nondemented individuals; however, this relationship was not as consistent as that for folate. Subsequent to mandatory folic acid fortification, stroke mortality has decreased at a greater rate in the United States and Canada, suggesting a positive effect of fortification on cerebrovascular health. Interventions with folic acid and with combinations of B vitamins were able to improve cognitive function or prevent decline, especially in subjects with low nutrient status. As with the data for blood nutrient levels, evidence that vitamin B12 treatment improves cognitive function is conflicting and less positive.

Key Words: Aging; Alzheimer's disease; B vitamins; cerebrovascular disease; cognition; dementia; folate; folic acid; vitamin B6; vitamin B12

13.1. INTRODUCTION

As the number of elderly in the United States and globally continues to increase, age-related neurological disorders, such as Alzheimer's disease and vascular dementia, are a growing concern. The loss of memory, emotional changes, and impairments in general cognitive functioning frequently result in social isolation, loss of independence, and institutionalization. However, cognitive decline is not an inevitable consequence of growing old. Indeed, although some forms of cognitive disorders may

A. Bendich, R.J. Deckelbaum (eds.), *Preventive Nutrition*, Nutrition and Health, DOI 10.1007/978-1-60327-542-2_13,

have a genetic component, cognitive decline is influenced by nutritional factors and may be secondary to nutritionally mediated conditions such as diabetes or vascular disease. As such, there is a strong need to identify modifiable nutritional factors that regulate the proper maintenance of brain function to facilitate healthy aging.

The relationship between diet and vascular disease is well established. The nutritional foundation of cardiovascular disease prevention is a diet high in fruits and vegetables and fiber and low in saturated fat. One of the prevailing risk factors for cardio- and cerebrovascular disease is elevated blood levels of homocysteine. Plasma homocysteine may be considered a functional indicator of B vitamin status, including that of folate and vitamin B12 and, to a lesser extent, vitamin B6. High plasma homocysteine concentrations can be largely attributed to inadequate status of these vitamins *(1)*. Data from several laboratories indicate that plasma homocysteine increases with age independent of vitamin status and that hyperhomocysteinemia is highly prevalent in the elderly. Several studies have shown consistent and strong relationships between homocysteine concentration, heart disease, and other vascular outcomes including cerebrovascular disease *(2,3)*.

Cerebrovascular disease is a potentially preventable cause of cognitive impairment in old age. All of its manifestations can lead to vascular cognitive impairment. Vascular cognitive impairment ranges in severity from subtle neuropsychological deficits to frank dementia and frequently coexists with and possibly contributes to other neurodegenerative conditions such as Alzheimer's disease. Mildly elevated plasma homocysteine (homocysteinemia) has been implicated as a powerful, and potentially modifiable, risk factor for cognitive dysfunction, cerebrovascular disease, and Alzheimer's dementia. Being in the upper quartile of the population for plasma homocysteine doubles an individual's risk of developing dementia. Other research shows that moderate hyperhomocysteinemia is an independent risk factor for stroke *(4)* and may predict the severity of cerebral atherosclerosis for patients with cerebral infarction *(5)*. Plasma homocysteine concentrations are increased in stroke patients *(6)*, and homocysteine concentration is independently associated with an increased likelihood of nonfatal stroke *(7)*. However, whether homocysteine is actually a cause of or simply a marker for these conditions has not been established.

13.2. B VITAMINS, HOMOCYSTEINE, AND BRAIN FUNCTION

Possible biochemical interpretations of the putative effects of low B vitamin status and high homocysteine on cognitive decline can be made on the basis of one-carbon metabolism (Fig. 13.1). Folate serves as a carrier of one-carbon groups for the methylation cycle. In this cycle, methionine with its available methyl group is activated by adenosine triphosphate to form *S*-adenosylmethionine, which is the universal methyl donor in a multitude of methyl transfer reactions including many that are of vital importance to central nervous function. Through the transfer of its methyl group, *S*-adenosylmethionine is converted to *S*-adenosylhomocysteine, which is hydrolyzed to homocysteine. Homocysteine can regenerate methionine for an additional methylation cycle by acquiring a new methyl group from methyl tetrahydrofolate in a reaction that is catalyzed in all tissues by vitamin B12-requiring methionine synthase *(9)*. Excess intracellular homocysteine can also be removed from the methylation pathway by conversion to cystathionine in the *trans*-sulfuration pathway or through export into circulation *(9)*. It has been proposed that cognitive impairment in the elderly is due in part to vasotoxic effects of homocysteine and/or to impaired methylation reactions in brain tissue *(10)*. While it has yet to be determined if hyperhomocysteinemia is a cause of vascular disease or indicative of some other physiological change leading to vascular damage, current data demonstrate that homocysteine is strongly associated with cognitive dysfunction in aging *(11–13)*.

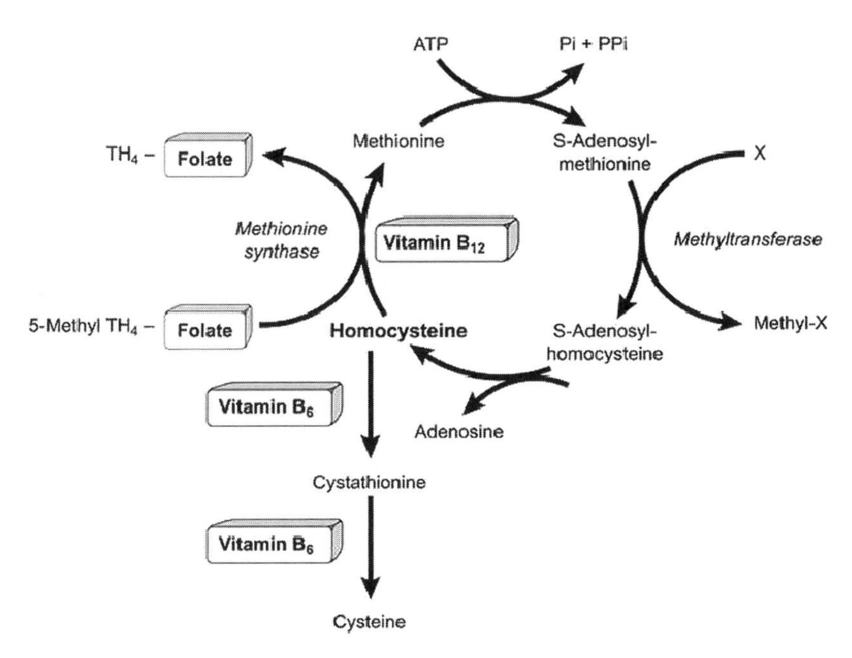

Fig. 13.1. Use of *S*-adenosylmethionine as the methyl donor for biological methylation reactions results in the formation of *S*-adenosylhomocysteine. Homocysteine is formed from the hydrolysis of *S*-adenosylhomocysteine. Homocysteine may be remethylated to form methionine by a folate-dependent reaction that is catalyzed by methionine synthase, a vitamin B12-dependent enzyme. Alternately, homocysteine may be metabolized to cysteine in reactions catalyzed by two vitamin B6-dependent enzymes. (Adapted from *(8)*)

13.3. B VITAMIN DEFICIENCY AND COGNITIVE IMPAIRMENT

It has long been known that severe deficiency of vitamins such as niacin, vitamin B12, and thiamine causes cognitive impairment *(2)* and that replacement of deficient nutrients can prevent or ameliorate those forms of cognitive impairment that are caused by deficiency (i.e., *14*). While more severe vitamin deficiencies or congenital defects are not common in the United States, these milder sub-clinical B vitamin deficiencies are prevalent in the elderly *(12,15,16)*. Decreased intestinal absorption and poor appetite contribute to these sub-clinical deficiencies. Studies suggest that even moderately low or sub-clinical levels of B vitamins are associated with cognitive impairment, dementia, and other psychiatric disorders *(17–19)*.

Extensive research supports the hypothesis that B vitamin deficiencies moderate cognitive brain functioning through effects on cerebrovascular health, DNA synthesis, and neurotransmitter metabolism *(2,20)*. Vitamin B12 and folate are closely linked in the methylation process, such that a B12 deficiency can lead to a secondary folate deficiency through a decrease in the retention of folate. It is proposed that low levels of B12 are linked with peripheral neuropathy and subacute combined degeneration of the spinal cord, whereas folate may have more of a role in cognition and mood, perhaps through effects on serotonin, dopamine, and noradrenergic systems *(20,21)*.

13.4. DIETARY B VITAMINS AND COGNITION

In a seminal study, healthy, independently living elderly individuals with sub-clinical malnutrition (i.e., low dietary intake of protein and selected vitamins including vitamin B12) scored lower on tests of verbal memory and nonverbal abstract reasoning than did their peers with normal intake *(22)*. However, subsequent studies have yielded conflicting results with respect to nutrient intake and cognitive function. In a prospective cohort study *(23)* examining the relation between nutritional intake and daily functioning, dietary intakes were not associated with a change in functional decline over a 6-month period in nursing home residents. In a retrospective case–control study *(24)* comparing patients with Alzheimer's disease with healthy controls, cases and controls were asked to recall their past food consumption using a food-frequency questionnaire during three age periods: 20–39, 40–59, and 60 or more years of age. It was found that those with Alzheimer's disease had lower mean dietary intakes of vitamin B6 and folate than controls in the over-60 age group, but not in younger age groups. However, there was no relationship between either folate intake and homocysteine levels or homocysteine levels and cognitive status. Finally, adding to the complexity of these findings, some research *(25)* shows a slower decline in cognitive test performance over a 6-year period in subjects with high vitamin B12 intake but faster decline among subjects with a high folate intake (>400 μg/day) from either food sources or supplements.

13.5. B VITAMIN BLOOD LEVEL ASSOCIATIONS WITH COGNITION AND DEMENTIA

Poor B vitamin status and/or high homocysteine is associated with poorer cognitive performance *(26–30)*. Previous studies have found that patients with dementia, especially those with Alzheimer's disease, have lower serum concentrations of B vitamins *(12,31,32)*; moreover, serum levels of these micronutrients are associated with the severity of the disease *(33)*. Several cross-sectional studies have found that patients with Alzheimer's dementia had significantly higher levels of serum total homocysteine than did age-matched hospitalized controls *(34)* and healthy community-dwelling elderly individuals *(34,35)*. In a longitudinal study, higher levels of homocysteine in Alzheimer's patients were associated with greater progression of hippocampal atrophy as measured by medial temporal lobe thickness, as well as a similar trend in Mini Mental State Evaluation score decline *(36)*. More recent studies have found inconsistent results with respect to the predictive value of high homocysteine or low B vitamin status and cognitive performance in the elderly *(37–38)*.

13.6. VITAMIN B12

A number of studies have investigated a potential correlation between serum vitamin B12 levels and cognitive function or diagnosis of several types of dementia and cognitive impairment *(39)*. Most of these studies have focused on Alzheimer's disease. Based on longitudinal studies, serum vitamin B12 levels did not affect the risk of developing Alzheimer's disease or dementia. The existing evidence from studies that implemented a cognitive function assessment instrument did not support any correlation between serum vitamin B12 levels and cognitive function. Among cross-sectional studies, there was a tendency for vitamin B12 serum levels to be lower in patients with Alzheimer's disease or other types of dementia, which in certain studies reached statistical significance. However, this trend was not consistent. Finally, an inverse relationship between vitamin B12 levels and duration of Alzheimer's disease has been reported. In general, evidence from longitudinal cohort and case–control studies suggests that there is no significant association between blood concentrations or the dietary intake of vitamin B12 and cognitive test performance or the progression of Alzheimer's disease. Although some studies reported higher vitamin B12 blood concentrations to be associated with better cognitive test performance, no consistent pattern of association with a particular cognitive domain has been reported.

13.7. FOLATE

With respect to folate, Miller and colleagues *(40)* showed that red blood cell folate levels are predictive of homocysteine levels. Elevated homocysteine levels, in turn, are associated with poorer performance on several cognitive tasks. In this study, however, there was no clear relationship between folate status and cognitive performance. Teunissen and colleagues *(41)* found that serum levels of homocysteine were negatively correlated with verbal learning and memory at baseline testing only, whereas higher serum folate levels were associated with better delayed recall performance. Elevated homocysteine was associated with poorer functioning on several cognitive tasks looking at immediate recall, attention, and performance during a 6-year follow-up period and there were no further associations with folate status. Morris and coworkers *(42)* reported similar findings with respect to folate and recall. Folate status was positively correlated with recall performance. Moreover, elevated levels of homocysteine were associated with poor recall. When looking at folate and homocysteine levels, low folate levels in combination with higher levels of homocysteine (above the 80th percentile) were associated with significantly poorer performance than low folate or lower homocysteine levels alone *(42)*.

Recent research shows an association with folate levels and mild cognitive impairment and some forms of dementia. In these studies, people with the lowest serum levels of folate were at greater risk for Alzheimer's disease, mild cognitive impairment, and dementia. Snowdon and coworkers *(33)* examined atrophy in the brains of deceased nuns with Alzheimer's disease and compared these data with serum folate levels that had been determined earlier in the nuns' lives. They found an inverse relationship between folate status and severity of atrophy. This effect was seen even in participants without significant atherosclerosis or brain infarcts, suggesting further that the role of folate is not limited to its relationship with homocysteine and homocysteine's putative vascular effects. While there is abundant evidence to indicate an association between folate status and the development or progression of cognitive decline or dementia, a causative role is not clear-cut. On the one hand, some data suggest that folate inadequacy precedes onset of cognitive impairment or dementia. On the other hand, there is evidence to suggest that dietary intakes of folate are lower in people with Alzheimer's disease. In contrast to the data on vitamin B12, the majority of studies evaluating blood folate concentrations reported a positive association between low folate levels and poor cognitive test performance.

13.8. FORTIFICATION WITH FOLIC ACID AND CEREBROVASCULAR EFFECTS

A number of countries have instituted mandatory folic acid fortification of wheat flour and cereal grains, with the aim of reducing neural tube defects in developing fetuses. Fortification has not been implemented in many countries, particularly in Europe, due in part to concerns about the potential for high levels of folic acid intake to exacerbate the neurological consequences of a vitamin B12 deficiency. Indeed, some of the research discussed in this chapter suggests that high levels of folate or folic acid in people with low vitamin B12 status have negative effects on memory and cognition. However, in the United States, since the initiation of fortification, positive effects have been seen with respect to increased blood folate levels and reduced homocysteine levels *(43–45)*. As described above, homocysteine levels are associated with severity of cerebrovascular disease, including stroke and cerebral atherosclerosis. Furthermore, low folate and high homocysteine are known risk factors for vascular disease, and population-wide reduction of these risk factors may be proposed to reduce incidence of vascular disease, stroke, and subsequent vascular dementia. Supporting this proposal, there has been a reduction in cardiovascular and stroke mortality in the United States and Canada since the introduction of folic acid fortification *(46)*. It remains to be seen whether folic acid fortification is associated with a change in the incidence of vascular dementia and cognitive decline.

13.9. INTERVENTION TRIALS WITH B VITAMINS

High-dose supplementation with B vitamins, particularly in combination, can decrease homocysteine levels and produce changes in cognitive performance, but these changes are not clear-cut *(47)*. For example, Bryan and colleagues *(48)* provided women with 750 μg folate for 35 days and conducted cognitive and mood tests pre- and post-supplementation. They found that folate significantly improved performance in a speed-of-processing task for women with initially lower folate levels. In younger women, they found that folate significantly enhanced recall for those with initially lower folate levels. However, the effects of folate supplementation were not consistent across the study or across different age groups. Nilsson and colleagues *(49)* studied vitamin B12 and folic acid supplementation in older patients with mild-to-moderate dementia. Performance on cognitive tasks was measured before and after 2 months supplementation with 5 mg/day folic acid and 1 mg/day cyanocobalamin. They found that in patients with elevated homocysteine, vitamin supplementation decreased homocysteine levels and patients improved on measures of attention, memory, and orientation. In contrast, Sommer and colleagues *(50)* examined 20 mg/day folic acid supplementation for 10 weeks in patients with dementia and reported that those in the folic acid condition performed worse relative to controls on some tasks and that there were no group differences for other tasks. High-dose supplementation with a combination of vitamins B12, B6, and folic acid was not effective in preventing cognitive decline in Alzheimer's disease *(51)* or in individuals at risk for cardiovascular disease *(52)*. For most of the trials, no effect on cognitive decline was found for vitamin B12 given alone *(47)*. Additionally, some studies showed worsening on cognitive functioning following B12 intervention *(53,54)*. Although data from a trial which supplemented older individuals with folic acid for 3 years show reductions in homocysteine and slower decline relative to controls *(55)*, intervention trials have, in general, shown little benefit of vitamin supplementation on brain function. Nevertheless, cognitive improvement may be seen with supplementation in individuals with initially low nutrient levels *(52,56)*. Some evidence suggests that lowering homocysteine levels with vitamin supplementation reduces the risk of cardiovascular disease *(13,57)*, but other research has shown no benefit of lowering homocysteine *(58–60)*.

There are several considerations in interpreting the findings on vitamin supplementation and cognition. First, there is no standard dose of vitamin or duration of treatments that is recommended for this type of trial. Second, many trials may have had too few participants to sufficiently determine an effect on cognitive performance. Third, in people with dementia or other cognitive impairment, the severity of impairment alone is of importance. For example, Nilsson and colleagues excluded severely demented patients from their study because the patients were unable to complete testing either before or after supplementation. As a correlative, the ability of vitamin supplementation to offset existing cognitive decline is somewhat dependent upon the duration and consequent irreversibility of cognitive decline. Finally, the relationship between nutrient status and homocysteine status should be considered. Again, referring to the Nilsson study *(49)*, patients with elevated homocysteine showed cognitive improvements following vitamin supplementation, whereas those with normal levels of homocysteine, but who did have mild-to-moderate dementia, did not. Similarly, in women at high risk for cardiovascular disease, there was no evidence for preventing cognitive decline in those with normal nutrient intakes; however, some benefits were seen with supplementation in those with initially low levels of nutrient intakes.

13.10. CONCLUSIONS

The evidence for an association between elevated plasma homocysteine and cognitive dysfunction is compelling. Elevated plasma total homocysteine has been reproducibly linked to diseases of the aging brain including subtle age-related cognitive decline, cerebrovascular disease and stroke, vascular dementia, and Alzheimer's disease. Nevertheless, it is unclear whether elevated homocysteine mediates or is otherwise associated with vascular brain aging. Epidemiological data provide the basis for a hypothesis implicating homocysteine as a mediator of vascular and neuronal pathology. If this is true, then homocysteine-lowering therapies could reduce the incidence of cognitive decline, stroke, and dementia. The evidence for this is not consistent. Highlighting the importance of prevention, some research shows that supplementation with B vitamins, while successful at lowering homocysteine, is not beneficial in mediating cognitive function in individuals with irreversible compromised brain function. However, the role of homocysteine lowering in the long-term prevention of cognitive decline, whether via promotion of vascular health or via direct neural effects, has yet to be adequately established.

The research described in this chapter represents current understanding on the relationships of folate and vitamin B12 nutritional status with cognitive function and dementia in adults and elderly. There is evidence that insufficient B vitamin intake is associated with lower cognitive scores in comparison to adequate intake. Low B vitamin status is associated with increased homocysteine levels. However, higher rates of cognitive decline have been reported with high levels of folate and folic acid intake in adults, and memory performance may be impaired with high folate intake in individuals with low vitamin B12 status. Overall, studies reported lower folate blood levels and a higher prevalence of deficiency among subjects with dementia. In general, vitamin B12 serum levels were lower in patients with dementia relative to nondemented individuals; however, this relationship was not as consistent as that for folate. Stroke mortality has decreased at a greater rate in the United States and Canada subsequent to mandatory folic acid fortification, suggesting a positive effect of fortification on cerebrovascular health. Interventions with folic acid and with combinations of B vitamins were able to improve cognitive function or prevent decline, albeit inconsistently, especially in subjects with low nutrient status. As with the data for blood nutrient levels, evidence that vitamin B12 treatment improves cognitive function is conflicting and less positive.

Overall, there is support for the concept that diets rich in B vitamins, especially folate and vitamin B12, are beneficial in maintaining brain function in aging. Some smaller experimental trials show that supplementation with vitamin B12 and folic acid may prevent cognitive decline or improve cognitive function to some degree, particularly in individuals with low baseline levels of these nutrients. To date, intervention trials indicate that supplementation with B vitamins for a period of 1.5–3 years, even while adequate to reduce circulating levels of homocysteine, may not be sufficient to prevent cognitive decline. Furthermore, some research indicates that high levels of supplementation with folic acid may have negative consequences on cognitive performance in vitamin B12-deficient individuals.

13.11. RECOMMENDATIONS

Adhering to a diet rich in whole grains, leafy greens, as well as lean meats and low-fat dairy may confer the greatest benefit in prevention of decline and maintaining cognitive function. Moreover, following such a diet should be considered a life-long goal rather than an immediate action to treat current decline. Folic acid supplementation is beneficial in those with folate-deficient diets, but may be harmful at high doses in the presence of low vitamin B12 status. Low B12 status is more prevalent in elders than previously known and steps for prevention need to be considered. Given the high prevalence of populations in the low and marginal folate and vitamin B12 status, the results presented here support the concern that cognitive function and risk of age-related decline represent a nutritionally modifiable public health problem.

REFERENCES

1. Selhub, J., Jacques, P.F., Wilson, P.W., Rush, D., and Rosenberg, I.H. (1993) Vitamin status and intake as primary determinants of homocysteinemia in an elderly population. *JAMA* **270**, 2693–2698.
2. Rosenberg, I.H. and Miller, J.W. (1992) Nutritional factors in physical and cognitive functions of elderly people. *Am J Clin Nutr* **55**, 1237S–1243S.
3. Smith, A.D. (2008) The worldwide challenge of the dementias: a role for B vitamins and homocysteine? *Food Nutr Bull* **29**, S143–S172.
4. Bostom, A.G., Rosenberg, I.H., Silbershatz, H., Jacques, P.F., Selhub, J., D Agostino, R.B., Wilson, P.W., and Wolf, P.A. (1999) Nonfasting plasma total homocysteine levels and stroke incidence in elderly persons: the Framingham Study. *Ann Intern Med* **131**, 352–355.
5. Yoo, J.H., Chung, C.S., and Kang, S.S. (1998) Relation of plasma homocyst(e)ine to cerebral infarction and cerebral atherosclerosis. *Stroke* **29**, 2478–2483.
6. Brattstrom, L,E,, Hardevo, J.E., and Hultverg, B.L. (1984) Moderate homocysteinemia—a possible risk factor for arteriosclerotic cerebrovascular disease. *Stroke* **15**, 1012–1016.
7. Giles, W.H., Croft, J.B., Greenlund, K.J., Ford, E.S., and Kittner, S.J. (1998) Total homocyst(e)ine concentration and the likelihood of nonfatal stroke: results from the Third National Health and Nutrition Examination Survey, 1988–1994. *Stroke* **29**, 2473–2477.
8. Selhub, J. and Miller, J.W. (1992) The pathogenesis of homocysteinemia: interruption of the coordinate regulation by S-adenosylmethionine of the remethylation and transsulfuration of homocysteine. *Am J Clin Nutr* **55**, 131–138.
9. Selhub, J. (1999) Homocysteine metabolism. *Ann Rev Nutr* **19**, 217–246.
10. Troen, A. and Rosenberg, I. (2005) Homocysteine and cognitive function. *Semin Vasc Med* **5**, 209–214.
11. Morris, M.S., Jacques, P.F., Rosenberg, I.H., Selhub, J., Bowman, B.A., Gunter, E.W., Wright, J.D., and Johnson, C.L. (2000) Serum total homocysteine concentration is related to self-reported heart attack or stroke history among men and women in the NHANES III. *J Nutr* **130**, 3073–3076.
12. Selhub, J., Bagley, L.C., Miller, J., and Rosenberg, I.H. (2000) B vitamins, homocysteine, and neurocognitive function in the elderly. *Am J Clin Nutr* **71**, 614S-620S.
13. Wald, D.S., Law, M., and Morris, J.K. (2002) Homocysteine and cardiovascular disease: evidence on causality from a meta-analysis. *Brit Med* **325**, 1202–1208.
14. van Asselt, D.Z., Pasman, J.W., van Lier, H.J., Vingerhoets, D.M., Poels, P.J., Kuin, Y., Blom, H.J., and Hoefnagels, W.H. (2001) Cobalamin supplementation improves cognitive and cerebral function in older, cobalamin-deficient persons. *J Gerontol A Biol Sci Med Sci* **56**, M775–779.
15. Joosten, E., van den Berg, A., Riezler, R., Naurath, H.J., Lindenbaum, J., Stabler, S.P., and Allen, R.H. (1993) Metabolic evidence that deficiencies of vitamin B-12 (cobalamin), folate, and vitamin B-6 occur commonly in elderly people. *Am J Clin Nutr* **58**, 468–476.
16. Lindenbaum, J., Rosenberg, I.H., Wilson, P.W., Stabler, S.P., and Allen, R.H. (1994) Prevalence of cobalamin deficiency in the Framingham elderly population. *Am J Clin Nutr* **60**, 2–11.
17. Bell, I.R., Edman, J.S., Marby, D.W., Satlin, A., Dreier, T., Liptzin, B., and Cole, J.O. (1990) Vitamin B12 and folate status in acute geropsychiatric inpatients: affective and cognitive characteristics of a vitamin nondeficient population. *Biol Psychiatry* **27**, 125–137.
18. Riggs, K.M., Spiro, A., III, Tucker, K., and Rush, D. (1996) Relations of vitamin B-12, vitamin B-6, folate, and homocysteine to cognitive performance in the Normative Aging Study. *Am J Clin Nutr* **63**, 306–314.
19. Tucker, K.L., Riggs, K.M., and Spiro, A.I. (1999) Nutrient intake is associated with cognitive function: The Normative Aging Study (Abstract). *Gerontologist* **39**, 149.
20. Bottiglieri, T., Laundy, M., Crellin, R., Toone, B.K., Carney, M.W., and Reynolds, E.H. (2000) Homocysteine, folate, methylation, and monoamine metabolism in depression. *J Neurol Neurosurg Psychiatr* **69**, 228–232.
21. Alpert, M., Silva, R.R., and Pouget, E.R. (2003) Prediction of treatment response in geriatric depression from baseline folate level: interaction with an SSRI or a tricyclic antidepressant. *J Clin Psychopharmacol* **23**, 309–313.
22. Goodwin, J.S., Goodwin, J.M., and Garry, P.J. (1983) Association between nutritional status and cognitive functioning in a healthy elderly population. *JAMA* **249**, 2917–2921.
23. Deijen, J.B., Slump, E., Wouters-Wesseling, W., De Groot, C.P., Gallè, E., and Pas, H. (2003) Nutritional intake and daily functioning of psychogeriatric nursing home residents. *J Nutr Health Aging* **7**, 242–246.

24. Mizrahi, E.H., Jacobsen, D.W., Debanne, S.M., Traore, F., Lerner, A.J., Friedland, R.P., and Petot, G.J. (2003) Plasma total homocysteine levels, dietary vitamin B6 and folate intake in AD and healthy aging. *J Nutr Health Aging* **7**, 160–165.
25. Morris, M.S., Jacques, P.F., Rosenberg, I.H., and Selhub, J. (2007) Folate and vitamin B-12 status in relation to anemia, macrocytosis, and cognitive impairment in older Americans in the age of folic acid fortification. *Am J Clin Nutr* **85**, 193–200.
26. Clarke, R., Birks, J., Nexo, E., Ueland, P.M., Schneede, J., Scott, J., Molloy, A., and Evans, J.G. (2007) Low vitamin B-12 status and risk of cognitive decline in older adults. *Am J Clin Nutr* **86**, 1384–1391.
27. Haan, M.N., Miller, J.W., Aiello, A.E., Whitmer, R.A., Jagust, W.J., Mungas, D.M., Allen, L.H., and Green, R. (2007) Homocysteine, B vitamins, and the incidence of dementia and cognitive impairment: results from the Sacramento Area Latino Study on Aging. *Am J Clin Nutr.* **85**, 511–517.
28. Kado, D.M., Karlamangla, A.S., Huang, M.H., Troen, A., Rowe, J.W., Selhub, J., and Seeman, T.E. (2005) Homocysteine versus the vitamins folate, B6, and B12 as predictors of cognitive function and decline in older high-functioning adults: MacArthur Studies of Successful Aging. *Am J Med* **118**, 161–167.
29. Quadri, P., Fragiacomo, C., Pezzati, R., Zanda, E., Forloni, G., Tettamanti, M., and Lucca, U. (2004) Homocysteine, folate, and vitamin B-12 in mild cognitive impairment, Alzheimer disease, and vascular dementia. *Am J Clin Nutr* **80**, 114–122.
30. Tucker, K.L., Qiao, N., Scott, T., Rosenberg, I., and Spiro, A. 3rd. (2005) High homocysteine and low B vitamins predict cognitive decline in aging men: the Veterans Affairs Normative Aging Study. *Am J Clin Nutr* **82**, 627–635.
31. Ikeda, T., Furukawa, Y., Mashimoto, S., Takahashi, K., and Yamada, M. (1990) Vitamin B12 levels in serum and cerebrospinal fluid of people with Alzheimer's disease. *Acta Psychiatr Scand* **82**, 327–329.
32. Karnaze, D.S. and Carmel, R. (1987) Low serum cobalamin levels in primary degenerative dementia. Do some patients harbor atypical cobalamin deficiency states? *Arch Intern Med* **147**, 429–431.
33. Snowdon, D.A., Tully, C.L., Smith, C.D., Riley, K.P., and Markesbery, W.R. (2000) Serum folate and the severity of atrophy of the neocortex in Alzheimer disease: findings from the Nun Study. *Am J Clin Nutr* **71**, 993–998.
34. Joosten, E., Lesaffre, E., Riezler, R., Ghekiere, V., Dereymaeker, L., Pelemans, W., and Dejaeger, E. (1997) Is metabolic evidence for vitamin B-12 and folate deficiency more frequent in elderly patients with Alzheimer's disease? *J Gerontol A Biol Sci Med Sci* **52**, M76–79.
35. Clarke, R., Woodhouse, P., Ulvik, A., Frost, C., Sherliker, P., Refsum, H., Ueland, P.M., and Khaw, K.T. (1998a) Variability and determinants of total homocysteine concentrations in plasma in an elderly population. *Clin Chem* **44**, 102–107.
36. Clarke, R., Smith, A.D., Jobst, K.A., Refsum, H., Sutton, L., and Ueland, P.M. (1998b) Folate, vitamin B12, and serum total homocysteine levels in confirmed Alzheimer disease. *Arch Neurol* **55**, 1449–1455.
37. Ellinson, M., Thomas, J., and Patterson, A. (2004) A critical evaluation of the relationship between serum vitamin B, folate and total homocysteine with cognitive impairment in the elderly. *J Hum Nutr Diet* **17**, 371–383.
38. Ravaglia, G., Forti, P., Maioli, F., Muscari, A., Sacchetti, L., Arnone, G., Nativio, V., Talerico, T., and Mariani, E. (2003) Homocysteine and cognitive function in healthy elderly community dwellers in Italy. *Am J Clin Nutr* **77**, 668–673.
39. Raman,G., Tatsioni, A., Chung, M., Rosenberg, I.H., Lau, J., Lichtenstein, A.H., and Balk, E.M. (2007) Heterogeneity and lack of good quality studies limit association between folate, vitamins B-6 and B-12, and cognitive function. *J Nutr* **137**, 1789–1794.
40. Miller, J.W., Green, R., Ramos, M.I., Allen, L.H., Mungas, D.M., Jagust, W.J., and Haan, M.N. (2003) Homocysteine and cognitive function in the Sacramento Area Latino Study on Aging. *Am J Clin Nutr* **78**, 441– 447.
41. Teunissen, C.E., Blom, A.H., Van Boxtel, M.P., Bosma, H., de Bruijn, C., Jolles, J., Wauters, B.A., Steinbusch, H.W., and de Vente, J. (2003) Homocysteine: a marker for cognitive performance? A longitudinal follow-up study. *J Nutr Health Aging* **7**, 153–159.
42. Morris, M.S., Jacques, P.F., Rosenberg, I.H., Selhub, J.; National Health and Nutrition Examination Survey. (2001) Hyperhomocysteinemia associated with poor recall in the third National Health and Nutrition Examination Survey. *Am J Clin Nutr* **73**, 927–933.
43. Choumenkovitch, S.F., Jacques, P.F., Nadeau, M.R., Wilson, P.W., Rosenberg, I.H., and Selhub, J. (2001) Folic acid fortification increases red blood cell folate concentrations in the Framingham study. *J Nutr* **131**, 3277–3280.
44. Jacques, P.F., Selhub, J., Bostom, A.G., Wilson, P.W., and Rosenberg, I.H. (1999) The effect of folic acid fortification on plasma folate and total homocysteine concentrations. *N Engl J Med* **340**, 1449–1454.
45. Ganji, V. and Kafai, M.R. (2006) Trends in serum folate, RBC folate, and circulating total homocysteine concentrations in the United States: analysis of data from National Health and Nutrition Examination Surveys, 1988–1994, 1999–2000, and 2001–2002. *J Nutr* **136**, 153–158.

46. Yang, Q., Botto, L.D., Erickson, J.D., Berry, R.J., Sambell, C., Johansen, H., and Friedman, J.M. (2006) Improvement in stroke mortality in Canada and the United States, 1990 to 2002. *Circulation* **113**, 1335–1343.
47. Balk, E.M., Raman, G., Tatsioni, A., Chung, M., Lau, J., and Rosenberg, I.H. (2007) Vitamin B6, B12, and folic acid supplementation and cognitive function: a systematic review of randomized trials. *Arch Intern Med* **167**, 21–30.
48. Bryan, J., Calvaresi, E., and Hughes, D. (2002) Short-term folate, vitamin B12 or vitamin B6 supplementation slightly affects memory performance but not mood in women of various ages. *J Nutr* **132**, 1345–1356.
49. Nilsson, K., Gustafson, L., and Hultberg, B. (2001) Improvement of cognitive functions after cobalamin/folate supplementation in elderly patients with dementia and elevated plasma homocysteine. *Int J Geriatr Psychiatry* **16**, 609–614.
50. Sommer, B.R., Hoff, A.L., and Costa, M. (2003) Folic acid supplementation in dementia: a preliminary report. *J Geriatr Psychiatry Neurol* **16**, 156–159.
51. Aisen, P.S., Schneider, L.S., Sano, M., Diaz-Arrastia, R., van Dyck, C.H,, Weiner, M.F., Bottiglieri, T., Jin, S., Stokes, K.T., Thomas, R.G., Thal, L.J.; Alzheimer Disease Cooperative Study. (2008) High-dose B vitamin supplementation and cognitive decline in Alzheimer disease: a randomized controlled trial. *JAMA* **300**, 1774–83.
52. Kang, J.H., Cook, N., Manson, J., Buring, J.E., Albert, C.M., and Grodstein, F. (2008) A trial of B vitamins and cognitive function among women at high risk of cardiovascular disease. *Am J Clin Nutr* **88**, 1602–1610.
53. Hvas, A.M., Juul, S., Lauritzen, L., Nexo, E., and Ellegaard, J. (2004) No effect of vitamin B12 treatment on cognitive function and depression: a randomized placebo controlled study. *J Affect Disord* **81**, 269–273.
54. Eussen, S.J., de Groot, L.C., Joosten, L.W., Bloo, R.J., Clarke, R., Ueland, P.M., Schneede, J., Blom, H.J., Hoefnagels, W.H., and van Staveren, W.A. (2006) Effect of oral vitamin B-12 with or without folic acid on cognitive function in older people with mild vitamin B-12 deficiency: a randomized, placebo-controlled trial. *Am J Clin Nutr* **84**, 361–370.
55. Durga, J., van Boxtel, M.P., Schouten, E.G., Kok, F.J., Jolles, J., Katan, M.B., and Verhoef, P. (2007) Effect of 3-year folic acid supplementation on cognitive function in older adults in the FACIT trial: a randomised, double blind, controlled trial. *Lancet* **369**, 208–216.
56. Fioravanti, M., Ferrario, E., Massaia, M., Cappa, G., Rivolta, G., Grossi, E., and Buckley, A.E. (1997) Low folate levels in the cognitive decline of elderly patients and the efficacy of folate as a treatment for improving memory deficits. *Arch Gerontol Geriatr* **26**, 1–13.
57. Schnyder, G., Roffi, M., Pin, R., Flammer, Y., Lange, H., Eberli, F.R., Meier, B., Turi, Z.G., and Hess, O.M. (2001) Decreased rate of coronary restenosis after lowering of plasma homocysteine levels. *N Engl J Med* **345**, 1593–1600.
58. Bonaa, K.H., Njolstad, I., Ueland, P.M., Schirmer, H., Tverdal, A., Steigen, T., Wang, H., Nordrehaug, J.E., Arnesen, E., Rasmussen, K.; NORVIT Trial Investigators. (2006) Homocysteine lowering and cardiovascular events after acute myocardial infarction. *N Engl J Med* **354**, 1578–1588.
59. Lonn, E., Yusuf, S., Arnold, M.J., Sheridan, P., Pogue, J., Micks, M., McQueen, M.J., Probstfield, J., Fodor, G., Held, C., Genest, J. Jr; Heart Outcomes Prevention Evaluation (HOPE) 2 Investigators. (2006) Homocysteine lowering with folic acid and B vitamins in vascular disease. *N Engl J Med* **354**, 1567–1577.
60. Toole, J.F., Malinow, M.R., Chambless, L.E., Spence, J.D., Pettigrew, L.C., Howard, V.J., Sides, E.G., Wang, C.H., and Stampfer, M. (2004) Lowering homocysteine in patients with ischemic stroke to prevent recurrent stroke, myocardial infarction, and death: the Vitamin Intervention for Stroke Prevention (VISP) randomized controlled trial. *JAMA* **291**, 565–575.

IV Diabetes and Obesity

14 The Women's Health Initiative: Lessons for Preventive Nutrition

Cynthia A. Thomson, Shirley A.A. Beresford, and Cheryl Ritenbaugh

Key Points

- The Women's Health Initiative (WHI) is the largest primary prevention study of diet ever undertaken among postmenopausal women in the United States.
- Postmenopausal women represent a subpopulation with unique health issues and variable diet and nutrition influences on health status.
- Excess body weight is a significant health risk for cardiovascular disease, breast cancer, colorectal cancer, and type 2 diabetes in postmenopausal women.
- The diet intervention aimed at reducing total fat to 20% of kilocalories was not associated with reduced risk for breast or colorectal cancer. Likewise, neither cardiovascular disease nor diabetes was significantly reduced in the WHI Diet Modification study population. Those who reported the highest dietary fat intake at baseline or reported the greatest reduction in dietary fat showed trends toward reduced risk for several of these chronic diseases.
- Supplementation with calcium and vitamin D did not reduce risk for colorectal cancer nor was it associated with a significant reduction in fracture risk, although women with lower baseline serum vitamin D levels did have some protection against fractures with combined calcium and vitamin D supplementation.

Key Words: Cardiovascular disease; cancer; clinical trial; diet; dietary adherence

14.1. INTRODUCTION

Toward the end of the 1980s, considerable evidence was accumulating from observational studies and some short-term trials with nondisease endpoints that identified factors that might benefit women in their postmenopausal years. The balance of randomized trials prior to that point had been focused on men and men's health. The confluence of scientific discovery and political will led to the design and ultimate funding of what was to be known as the Women's Health Initiative (WHI). Two main factors in this category were hormone replacement therapy (hormone therapy), suggested to prevent cardiovascular disease, the number one cause of death among women, and total dietary fat reduction for purposes of reducing risk for breast and colorectal cancers, also major causes of death and disability among women. These became the two, overlapping, main trials that were the core of WHI.

The rationale, sample size estimates, and intervention methodology for each of the main trials were developed separately, and different eligibility criteria were applied to each. Figure 14.1 illustrates how the two trials actually overlapped once the recruitment for all of the studies within WHI was complete.

The Women's Health Initiative (WHI) was the largest combined clinical and observational trial ever undertaken among postmenopausal women in the United States *(1)*. The study was originally founded in 1992 out of the Office of the Director of NIH, then Bernadine Healy; subsequently, the funding was moved to the National Heart, Lung and Blood Institute (NHLBI) of the National Institutes of Health,

A. Bendich, R.J. Deckelbaum (eds.), *Preventive Nutrition*, Nutrition and Health, DOI 10.1007/978-1-60327-542-2_14,

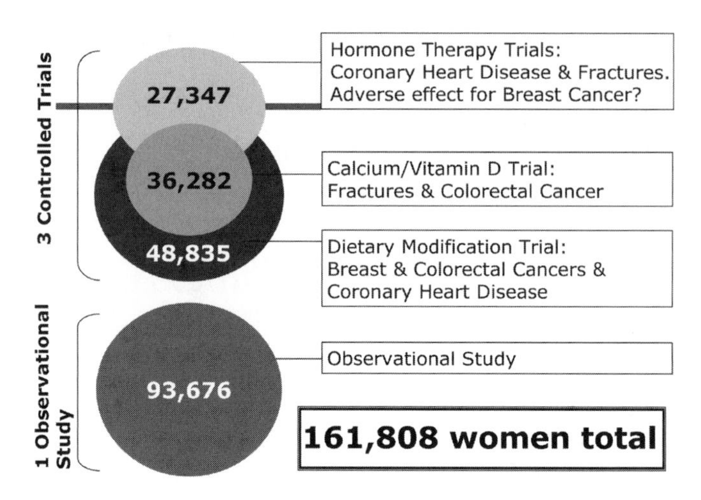

Fig 14.1. Women's Health Initiative (WHI) study design.

US Department of Health and Human Services *(2)*. From its inception, the WHI sought to answer significant research questions to inform clinical practice for postmenopausal women over the age of 50 years. Specifically the hypotheses to be tested focused on critical causes of morbidity and mortality in this segment of the population: (a) the role of hormone therapy to prevent cardiovascular disease and osteoporotic fractures with the potential risk of increasing breast cancer; (b) dietary fat reduction to prevent breast and colorectal cancer and possibly cardiovascular disease; and (c) supplemental vitamin D and calcium to prevent both colorectal cancer and hip and other fractures. The study design for this complex clinical trial is illustrated in Fig. 14.1. Briefly, at study onset, women were screened for eligibility to either the low-fat dietary modification trial or the hormone therapy trial, although subjects could participate in both. At the end of the first year, the third clinical trial, studying supplemental calcium and vitamin D, was added and again subjects were screened for eligibility and randomized to supplementation or placebo; participation in the alternate trials was permitted if the subject met eligibility requirements for the specific trial(s) of interest.

The primary focus of this chapter is the Diet Modification (DM) trial because of its relevance to preventive nutrition. The hormone and calcium/vitamin D trials will be briefly described along with results from the calcium/vitamin D trial; for detailed results from the hormone trial, the reader is referred to the primary outcome papers (for citations, *see* Table 14.2). In addition to the clinical trials, women not interested or ineligible for these trials could elect to enroll in the WHI Observational Study (OS), where demographic, clinical, and lifestyle data were collected periodically and prospectively for use in epidemiological analyses of multiple research questions of relevance in this population of postmenopausal women.

14.1.1. Scientific Rationale for the DM Trial

Ever since the international rates of dietary fat consumption and mortality from breast cancer were found to be strongly associated *(3)*, scientists have worked to test this hypothesis in epidemiological studies, with conflicting results *(4–6)*. Some of the discrepancies in findings from different studies may be partially explained by the limited range of fat intake in individual studies compared to the international comparative studies and also by imperfect adjustment for measurement error and cross-cultural differences in nutrient databases associated with dietary assessment. By 1993, the time was therefore right to evaluate the hypothesis with a carefully conducted randomized controlled trial. The choice of age groups for this trial was based on statistical power considerations, taking into account the logistics of recruiting very large numbers of women and the incidence of breast cancer by age group *(1)*.

The intervention was developed based on behavioral research that had demonstrated that group sessions resulted in greater behavioral change as compared to lower intensity methods of changing dietary behavior. The feasibility of implementing such a program was evaluated in a series of behavior change trials, including the Women's Health Trial *(7)* and the Women's Health Initiative Feasibility Study in Minority Populations *(8)*. These studies showed that not only could women change their dietary fat intake by nearly 50%, but that the changes they made had some impact on the dietary behavior of their husbands also *(9)*.

The primary endpoints, breast and colorectal cancers, of the DM Trial of the WHI were selected based on the literature available prior to 1992, and scientific evidence relating these outcomes to dietary fat intake, evidence which is summarized below.

14.1.1.1. Breast Cancer

The literature has been summarized in Prentice et al. *(10,11)*. Briefly, animal studies in rats and mice showed that higher tumor rates were associated with the high-fat feeds *(12)*. Migrant studies supported the international dietary comparisons by showing that women born in Japan but moving to the United States during their adult life had rates of breast cancer closer to those of the United States than to those of Japan *(13)*. Reviews of observational studies, using meta-analytic methodology, came to different conclusions. Those restricted to case–control studies estimated a significant excess risk associated with higher fat consumption *(14)*. A meta-analysis that combined case–control and cohort studies also estimated a significantly positive risk *(15)*. On the other hand, a meta-analysis restricted to prospective observational studies did not find a significant association between dietary fat and risk of breast cancer *(16)*. One explanation that had been suggested for the inconsistency among the epidemiological studies was the different methods of dietary assessment used *(17)*.

14.1.1.2. Colorectal Cancer

The literature has been briefly summarized in Beresford et al. *(18)*. Like the international studies of breast cancer mortality, Carroll et al. *(3)* and Prentice and Sheppard *(5)* demonstrated that colorectal cancer mortality was about one-third lower in countries with 50% lower fat intake than that of the US population. Again, supporting these observations, studies found that women migrating from countries with low-fat consumption to countries with high-fat consumption experienced the higher colorectal cancer rates of their new country *(19,20)*. The preponderance of evidence from within-country observational studies *(5,21–24)* suggested that dietary fat was a risk factor for, and fruit, vegetables, and grains were protective against, risk of colorectal cancer.

14.1.1.3. Cardiovascular Disease

Howard et al. *(25)* have recently summarized the evidence. Both observational studies and randomized trials have identified strong associations between serum cholesterol [especially low-density lipoprotein cholesterol (LDL-C) level] and other cardiovascular disease (CVD) risk factors. Increased serum cholesterol and LDL-C have also consistently been associated with increased intake of saturated and *trans*fatty acids and dietary cholesterol. Unsaturated fatty acids, plant proteins, grains, and fibers have been associated with reduced blood cholesterol levels. A direct relationship between dietary intake of saturated fatty acids and rates of CVD has been found in several observational studies

(26–31). Early dietary intervention trials investigated not only total fat reduction but also more specifically the substitution of one kind of fat for another, or emphasizing one diet type over another. These trials demonstrated cardiovascular risk reduction associated with poly- and monounsaturated rather than saturated fat *(32–34)* and with Mediterranean-type or very-low-fat pattern rather than the typical US diet *(35–37)*. Because this outcome (cardiovascular disease) was a secondary outcome in the DM trial, the intervention developed for the WHI focused on cancer risk reduction through reduced total fat and increased fruits, vegetables, and grains, assuming this would likewise benefit cardiovascular risk reduction as well.

14.1.2. Eligibility Criteria: Clinical Trials (CT) and Observational Study (OS)

As detailed in the design paper, all women enrolled in the WHI were between the ages of 50 and 79 years at the time of randomization and were required to meet specific eligibility criteria, which varied somewhat across the clinical trials and the OS, prior to enrollment *(1)*. Overall eligibility required study subjects to be postmenopausal, provide written informed consent, and to plan to remain in the general geographical location for a minimum of 3 years. In addition, to be considered eligible for the OS, women had to have a predicted survival of 3 years or more, report no dependency on alcohol or drugs, and no dementia or mental illness, all factors that might preclude obtaining reliable study data. Eligibility for all parts of the clinical trial (CT) included no history of breast cancer (including recent clear mammography), no other cancer in previous 10 years, no myocardial infarct, stroke, or TIA in previous 6 months, and no severe or chronic hepatitis. Women also had to demonstrate a body mass index greater than 18 kg/m^2, hematocrit over 32%, platelet count above 75,000, blood pressure of less than 200/105 mmHg; in addition, women were ineligible if currently taking corticosteroids.

Additional exclusion criteria for the hormone trial included endometrial hyperplasia or cancer, malignant melanoma, history of pulmonary emboli or deep vein thrombosis, bleeding disorders, elevated triglycerides, abnormal gynecological exam, previous osteoporotic fracture treated with hormones, or use of anticoagulants or tamoxifen. Women also had to be willing to discontinue hormone therapy in order to be randomized into the treatment or placebo arms of the HT trial; women who had not had a hysterectomy had to be willing to have endometrial evaluations regularly.

The calcium/vitamin D trial further excluded women with a history of renal calculi or hypercalcemia, currently using corticosteroids, or taking vitamin D supplementation above 600 IU/day. Supplementation below this level, as is common in multivitamin supplements, was not an exclusion criterion nor was the use of supplemental calcium. It is important to note that all hormone or CaVitD clinical trial participants had to complete a baseline eligibility visit and successfully complete run-in evaluation on placebo (HT and CaVitD) prior to randomization into the clinical trial arms.

The dietary eligibility criteria for the DM trial were based on an initial food frequency screening for dietary fat intake; women who reported an intake of <32% total energy as dietary fat or total caloric intake <600 or >5,000 kcal/day were ineligible. Other exclusion criteria included therapeutic diet restriction in conflict with dietary assignments including diabetic or low-salt diets, reported intolerance of high-fiber intake, history of Type 1 diabetes mellitus, colorectal or breast cancer, reported intake of >10 meals outside home weekly, and inability to complete a 4-day food record.

14.1.3. Recruitment

Over 160,000 women were recruited overall between fall 1993 and summer 1999. Forty clinical centers from 24 states across the United States participated in recruitment. Ten centers were targeted to enhance minority recruitment including Atlanta, GA, Arizona (Tucson, Phoenix), Birmingham, AL, Chicago-Rush, IL, Detroit, MI, Honolulu, HI, La Jolla, CA, Medlantic, DC, Miami, FL, and

San Antonio, TX. In addition, three sites were responsible for collecting more detailed data and clinical assessment of bone health using dual X-ray absorptiometry (DEXA) (Arizona, Pittsburgh, and Birmingham).

The vast majority of study participants were ultimately recruited through direct mail campaigns. However, women were more likely to respond when they had already learned of the study through other channels, such as newspaper and television advertisements, local media reports, public service announcements, health fairs, brochures placed in local health clinics, pharmacies, beauty salons, libraries, churches and clinics, mailings to local health-care providers, and "name a friend" where additional women were recruited by participants currently active in the study.

14.1.4. Baseline Characteristics of Study Sample

Final enrollment numbers included 93,676 in OS, 27,347 in HT trial, 48,836 in DM, and 36,282 in CaVitD, indicating that the DM and HT trials achieved recruitment goals, while the CaVitD attained 81% of goal and the OS 93%. The majority of women recruited into WHI, both CT and OS, were between the ages of 60–69 years, white, achieved an education beyond high school, were married, and overweight. On average, between 15 and 20% of women screened for study participation entered the CT, while approximately 24% entered the OS. Recruitment rates were lower for women over age 70 years as well as Native Americans and Asians *(38)*. An estimated 10% of women reported a current smoking habit, the majority consumed 1–7 drinks/week and reported low physical activity levels, over 60% of women took at least one dietary supplement including almost 25% who reported regular use of calcium supplementation. Clinical tests of lipids, glucose/insulin, and select nutrients were completed for the random 8% subsample of WHI subjects across all treatment groups. Results of the clinical tests suggested that the women had borderline hypercholesterolemia (mean of 220 mg/dL across all study groups), mean triglycerides ranging from 131 to 144 mg/dL; and fasting glucose of 94–102 mg/dL *(38)*. These indicators along with the elevated BMI and waist circumference of the average study participant suggest that mild-to-moderate metabolic abnormalities were common in the study population.

14.1.5. Data Collection and Time Points

The amount of data collected for the WHI study is massive and the infrastructure to successfully complete the trial is extremely complex *(39)*. In addition to numerous questionnaires focused on demographic, lifestyle, and clinical characteristics, biosamples were also collected on the entire study sample with analysis of lipids, nutrients, and select cardiovascular risk factors completed on a random 8% subsample at baseline (38-Appendix tables) and Table 14.1.

Procedures for collecting clinical measurement data are included in Table 14.1. In addition, interested researchers can enter the National Heart, Lung, Blood Institute (NHLBI) WHI web site for more detailed information including PDF copies of all study forms, procedures, and protocols [http://www.nhlbi.nih.gov/whi *(40)*].

14.1.6. Progress to Date

To date, over 450 manuscripts have been published in the peer-reviewed literature using data collected in the context of the WHI CT and OS. The published manuscripts describing the primary hypotheses tested within the HT, DM, and CaVitD trials are summarized in Table 14.2 and were published in the *Journal of the American Medical Association* and the *New England Journal of Medicine* between 2002 and 2006. A comprehensive listing of all study publications and manuscripts in pro-

Table 14.1
WHI Clinical Measurement Procedures

Clinical Measurement	*Procedures*	*Data Entry Form*
Resting pulse	Measured after 5-min rest at radial artery. Recorded as number of beats in 30 s and then multiplied by 2	*Form 80* Physical Measurements
Blood pressure	Measured over brachial artery using stethoscope bell and mercury manometer. Cuff size determined by standardized arm circumference measurement. After a 5-min rest, maximal inflation level determined and two blood pressure measurements taken with 30-s rests in between. Systolic value (Phase I) recorded at the 1st of 2nd or more Korotkoff sounds. Diastolic (Phase V) recorded when the last rhythmic sound heard. Recorded in mmHg to nearest even digit, rounded up	*Form 80* Physical Measurements
Height	Wall-mounted stadiometer used. Measured at end-inspiration with shoes removed. Recorded to nearest one-tenth centimeter, rounded up	*Form 80* Physical Measurements
Weight	Calibrated balance beam or digital scale used. Measured with shoes, heavy clothing, and pocket contents removed. Recorded to nearest one-tenth kilogram, rounded up	*Form 80* Physical Measurements
Waist and hip circumferences	Measured with extra layers of clothes removed (nonbinding undergarments only) at horizontal plane: waist at level of natural waist (narrowest part of torso) at end-expiration; hips at site of maximum extension of buttocks. Recorded to nearest half-centimeter, rounded up	*Form 80* Physical Measurements
Grip strength	Calibrated Jamar hand-grip dynamometer used. Staff demonstration and participant submaximal trial followed by two measurements in dominant arm with staff coaching for maximal performance. Recorded to nearest kilogram, rounded up	*Form 90* Functional Status
Chair stand	Stopwatch and straight-backed, nonpadded, flat, armless chair used. If participant able to complete single chair stand without using arms, then two 15-s trials of repeated chair stands performed with arms folded across chest with a 1- to 2-min rest in between trials. Chair stands counted as participant arose; counts for both trials recorded	*Form 90* Functional Status
Timed walk	Gait course marked with end lines 6 m apart and "Xs" marked just beyond each end line. Stopwatch used. Participant used ambulatory aids as needed and walked course at usual speed until "X" was reached. Second trial performed to "X" at the other end. Recorded as number of seconds (to nearest one-tenth second) from start to when foot completely crossed the end line	*Form 90* Functional Status

Clinical breast examination	Performed by licensed clinician. Inspection of both breasts with participant in sitting position. Systematic palpation of all four quadrants of both breasts and axillae with participant in supine position. Findings recorded, including follow-up of abnormalities	*Form 84* Clinical Breast Exam
Pelvic examination	Performed by licensed clinician. Inspected external genitalia, pelvic structures (with valsalva), vaginal mucosa, and cervix. Bimanual exam performed with and without valsalva, followed by rectovaginal exam. Findings recorded, including follow-up of abnormalities	*Form 81* Pelvic Exam
Pap smear	Performed by licensed clinician. Exocervical scraping obtained using wooden spatula and endocervical scraping obtained using cytobrush. Two smears fixed on slide and sent to local lab. Findings and lab report recorded	*Form 92* Pap Smear
Endometrial aspiration	Performed by licensed clinician. Flexible aspirator used to obtain endometrial tissue from all uterine surfaces. Tip of aspirator cut off and allowed to fall into formalin bottle with specimen. Sent to local lab. Findings and lab report recorded	*Form 82* Endometrial Aspiration
Transvaginal uterine ultrasound	Performed at local ultrasound center if WHI clinician unable to do endometrial aspiration (e.g., due to cervical stenosis). Recorded thickness of endometrial stripe and any significant pathology	*Form 83* Transvaginal Uterine Ultrasound
Cognitive assessment	Modified Mini Mental Status Examination (3MSE) used. Items asked in order, using standardized strategies and probes. Recorded participant responses to items and computer scored after data entry	*Form 39* Cognitive Assessment
Electrocardiogram	MAC-PC electrocardiograph used. 12-Lead ECG recorded for 5 s with participant in supine position after standard skin preparation, limb lead placement, and chest lead placement using a Heart Square. Digitized ECG recording transmitted to ECG Reading Center	*Form 86* ECG
Mammogram	Performed at local mammography center by standard low-dose radiation technique. Radiologist report recorded as negative, benign, probably benign, suspicious, or highly suggestive of malignancy, using ACR classification system. Follow-up of abnormal findings also recorded	*Form 85* Mammography
Bone density scan	Dual-energy X-ray absorptiometry (DXA) bone mineral density using Hologic QDR-2000 and standard scanning and analysis procedures for measuring hip, AP lumbar spine, and whole-body bone density	*Form 87* Bone Density Scan

cess as well as guidelines for paper proposal submission is available on the NHLBI-WHI Scientific Resources web site at http://www.nhlbi.nih.gov/whi *(40)*.

In addition to publications, over 150 ancillary studies have been funded, which utilize WHI data, biosamples, and/or related resources. The ancillary study topics of relevance to preventive nutrition include such areas as diet and age-related eye disease, body composition and breast and bone density, vitamin supplements and cognitive change, gene–environmental interactions in colorectal

Table 14.2
Publication of Main WHI Trial Findings

1. The Writing Group for the WHI Investigators. (2002) Risks and benefits of estrogen plus progestin in healthy post-menopausal women: principal results of the Women's Health Initiative randomized controlled trial. *JAMA* **288**(3), 321–33.
2. The Women's Health Initiative Steering Committee. (2004) Effects of Conjugated equine estrogen in postmenopausal women with hysterectomy: The Women's Health Initiative Randomized Controlled Trial. *JAMA* **291**, 1701–12.
3. Beresford, S., Johnson, K., Ritenbaugh, C., Lasser, N., Snetselaar, L., Black, H., Anderson, G., Assaf, A., Bassford, T., Bowen, D., Brunner, R., Brzyski, R., Caan, B., Chlebowski, R., et al. (2006) Low-fat dietary pattern and risk of colorectal cancer: The Women's Health Initiative randomized controlled dietary modification Trial. *JAMA* **295**, 643–54.
4. Howard, B.V., Van Horn, L., Hsia, J., Manson, J.E., Stefanick, M.L., Wassertheil-Smoller, S., Kuller, L.H., LaCroix, A.Z., Langer, R.D., Lasser, N.L., Lewis, C.E., Limacher, M.C., Margolis, K.L., Mysiw, W.J., et al. (2006) Low-fat dietary pattern and risk of cardiovascular disease: The Women's Health Initiative randomized controlled dietary modification trial. *JAMA* 295, 655–66.
5. Prentice, R.L., Caan, B., Chlebowski, R.T., Patterson, R., Kuller, L.H., Ockene, J.K., Margolis, K.L., Limacher, M.C., Manson, J.E., Parker, L.M., Paskett, E., Phillips, L., Robbins, J., Rossouw, J.E., et al. (2006) Low-fat dietary pattern and risk of invasive breast cancer: The Women's Health Initiative randomized controlled dietary modification trial. *JAMA* 295:629–42.
6. Wactawski-Wende, J., Kotchen, J.M., Anderson, G.L., Assaf, A.R., Brunner, R.L., O'Sullivan, M.J., Margolis, K.L., Ockene, J.K., Phillips, L., Pottern, L., Prentice, R.L., Robbins, J., Rohan, T.E., Sarto, G.E., et al. (2006) Calcium plus vitamin D supplementation and the risk of colorectal cancer. *N Engl J Med* **354**(7), 684–96.
7. Jackson, R.D., LaCroix, A.Z., Gass, M., Wallace, R.B., Robbins, J., Lewis, C.E., Bassford, T., Beresford, S.A., Black, H.R., Blanchette, P., Bonds, D.E., Brunner, R.L., Brzyski, R.G., Caan, B., et al. (2006) Calcium plus vitamin D supplementation and the risk of fractures. *N Engl J Med* **354**(7), 669–83.

cancer prevention, serum fatty acids and ischemic stroke, biochemical and anthropometric heterogeneity and obesity, selenium and colorectal cancer risk, and choline/betaine intake and chronic disease. In addition, two dietary measurement validation studies have been completed to develop calibration equations for energy and protein misreporting in the WHI population: the Nutritional Biomarkers Study and the Nutrition and Physical Activity Assessment Study. These ancillary studies frequently include assessment of additional research biomarkers beyond the core analytes funded under the parent WHI study. Investigators interested in proposing an ancillary study are encouraged to partner with WHI investigative team members. Details are available on the WHI operations web site at http://www.nhlbi.nih.gov/whi *(40)*.

14.2. WHI DIET MODIFICATION STUDY

14.2.1. Study Hypotheses

The Women's Health Initiative (WHI) Diet Modification (DM) trial was initiated to test two primary hypotheses:

1. The intervention diet (described below), as compared to usual diet, adhered to over a period of 9 years, would significantly reduce the risk for breast cancer among postmenopausal women.

2. The intervention diet (described below), as compared to usual diet, adhered to over a period of 9 years, would significantly reduce the risk for colorectal cancer among postmenopausal women.

The secondary hypothesis was the following:

1. The intervention diet, as compared to usual diet, would result in a significantly lower incidence of cardiovascular disease.

14.2.2. Study Population

The WHI dietary modification (DM) trial recruited 48,835 healthy postmenopausal women across 40 clinical sites nationally into a randomized controlled trial between 1993 and 1999. The details of the DM trial design have been previously published *(41)*. The inclusion criteria for the diet trial required enrolled women to be between the ages of 50 and 79 years of age; exclusion criteria included cancer diagnosis in previous 10 years, any previous history of breast or colorectal cancer, and predicted life span of less than 3 years. In addition, women completed a food frequency questionnaire at baseline to determine if their dietary fat intake accounted for less than 32% of total energy intake; if so, the women were excluded from the DM trial. The demographic and clinical characteristics of the DM study population have been previously described. Generally the study participants were white (~20% minority), well-educated, married, retired or unemployed, and were overweight with a mean body mass index (BMI) of just over 27 kg/m^2 and with significant abdominal obesity with a mean waist circumference of 89 cm. Only 7% were current tobacco smokers. The average physical activity levels were estimated on self-report at 10 METS/day *(38)*.

14.2.3. Dietary Intervention

Study women were randomly assigned in a 40:60 distribution to either a low-fat diet or their usual/control diet (n = 19,541 and 29,294, respectively) (Fig. 14.2) *(42,43)*. The low-fat diet consisted of 20% of energy as fat. As time went on, the intervention further focused on reducing saturated

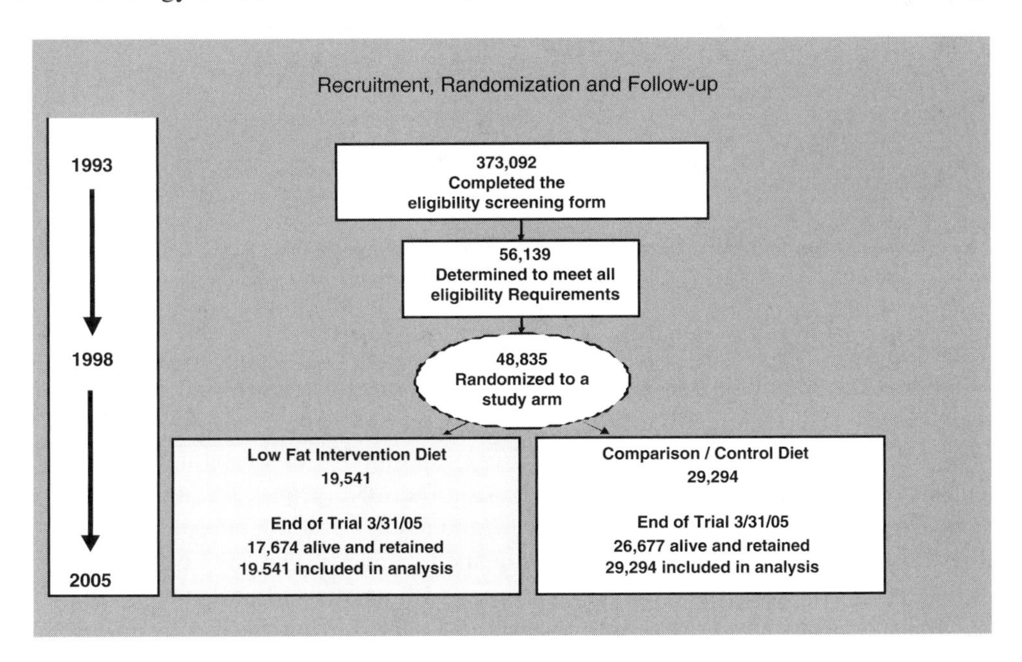

Fig. 14.2. WHI DM trial: recruitment, randomization, and follow-up.

fat intake to <7% kcal, increasing intake of vegetables and fruit to five or more savings daily and increasing servings of grains to >6 servings with an emphasis (through education) on whole-grain foods. No weight loss component was included.

14.2.3.1. Dietary Counseling

In order to achieve and sustain the dietary change goals of the WHI DM low-fat diet intervention, study subjects participated in a standardized counseling program that provided women with one individual counseling session with a registered dietitian/nutritionist, in which they received a fat gram target based on consuming 20% of reported calories as fat. This was followed by 18 small groups (8–15 women) counselling sessions during the first 12 months on study and 4 sessions/year during the following years, ending in summer 2004 *(41)*. The education and counseling regarding fat intake focused on limiting grams of fat. A number of behavioral change theoretical models and tools were applied during the counseling process to promote dietary adherence including self-monitoring, motivational interviewing, goal setting (daily fat gram goal), targeted messaging, and tailored feedback (Fig. 14.3) *(42)*. Further, given the emphasis of the study on minority ethnic/racial group recruitment/representation, all written materials were evaluated for acceptance and translation among special populations.

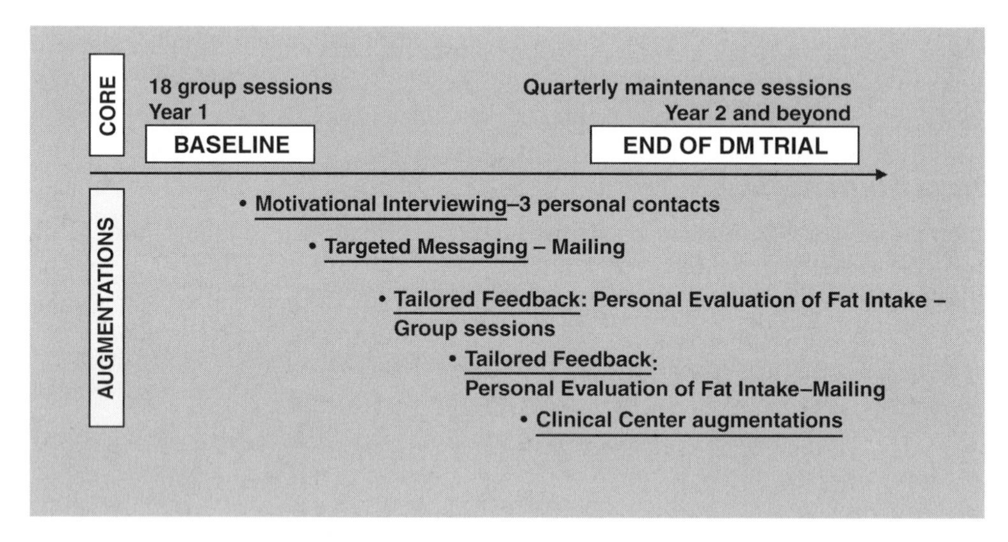

Fig. 14.3. WHI dietary intervention strategies.

14.2.4. Trial Duration, Data Collection, and Time Points

The WHI DM trial continued for 12 years with a mean follow-up period of 8.1 years. Dietary intake was measured at baseline and year 1 for all study participants using the WHI food frequency questionnaire; follow-up measurements using the FFQ continued with a rotating one-third of the population completing the FFQ annually (100% study sample over 3 years), through year 9. In addition, all subjects provided a 4-day food record at baseline and a 4.6% random subsample of the population completed a single 24-h recall of dietary intake at years 3, 6, and 9. The FFQ was the primary dietary assessment method for the DM trial *(44)*, but the repeat 24-h recalls and/or diet records are being used to describe consistencies or differences in diet–outcome associations derived from the different diet data collection methods.

In addition to repeated measures of dietary intake, blood samples for biospecimen acquisition were collected on all DM study subjects at baseline and years 1, 3, and 6. A random subsample of DM trial subjects have had their biosamples evaluated for key nutrients including plasma carotenoids, serum (OH)vitamin D, as well as health indicators such as glucose, insulin, and lipids as part of the core analyte subsample.

Outcomes including breast and colorectal cancer, cardiovascular disease, and stroke were collected on study participants through biannual telephone calls from trained study outcome assessors at the 40 clinical sites across the United States *(45)*. Once a self-report of one of the key health outcomes was collected, corresponding medical records were collected locally, adjudicated locally, and were sent to the study coordinating center in Seattle to verify the self-report and local adjudication. The medical records were then reviewed by centrally trained medical doctor adjudicators as has been previously described *(47)*. Self-reports of hospitalizations and major health events were also followed by collection of medical records; for these, adjudication was limited to the local site.

14.2.5. Was Dietary Change Achieved?

The two randomized groups began the study with the same dietary intake patterns. Figure 14.4 illustrates the mean differences in dietary intake from baseline to year 3 in the DM intervention group only. Data are presented for select nutrients and food groups which were targeted for behavioral modification. A the 3-year time point, total energy as fat was reduced by 25.4% as compared to baseline percentage energy consumed as fat. Dietary change was apparent not only for total fat intake but also for intake of saturated fat, vitamin E (high in vegetable oils, nuts/seeds), and red meat. While total energy restriction was not a dietary goal (nor was physical activity addressed in the trial), caloric intake was a slight 4.1% lower, on average, among women randomized to the low-fat diet at year 3 as compared to baseline *(46)*.

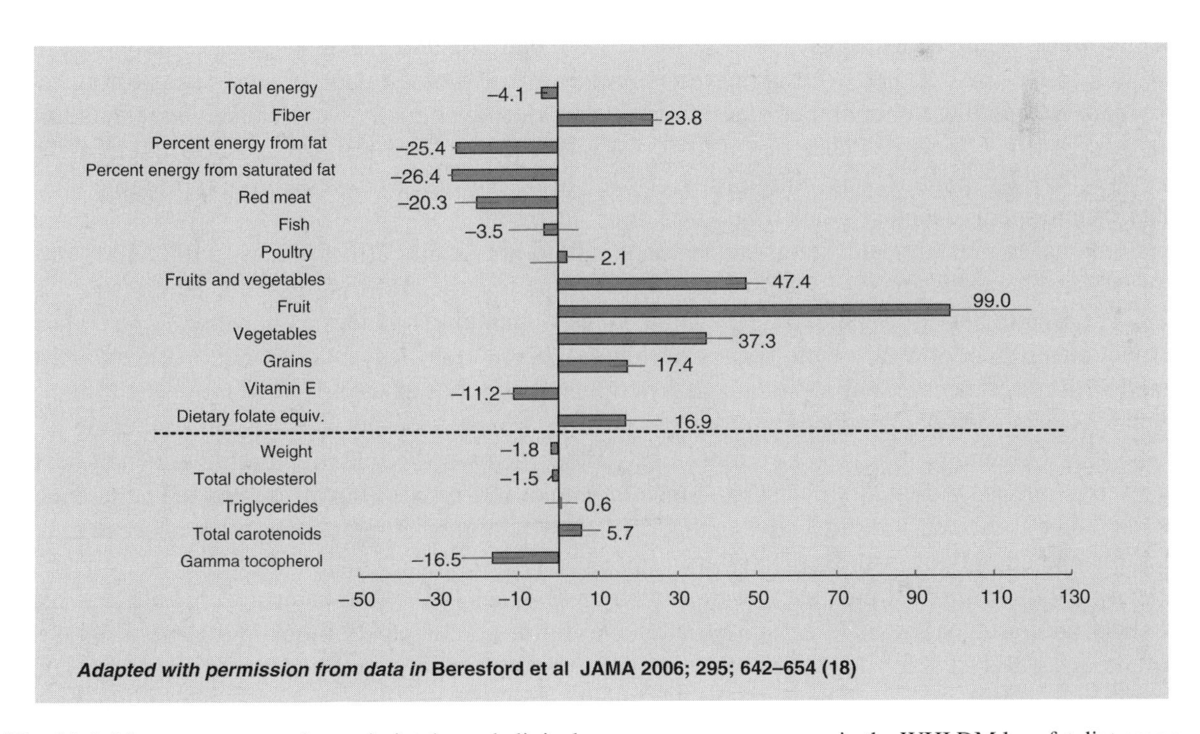

Fig. 14.4. Mean percentage change in intake and clinical measures among women in the WHI DM low-fat diet group.

In addition to changes in self-reported dietary intake, select biomarkers were also evaluated among a randomly selected 6% subsample of the DM intervention and control women. These analyses showed that the net difference in serum cholesterol favored the intervention by –3.3%, LDL by –3.5%, and as is common with diets low in saturated fat, HDL decreased slightly, while total triglycerides showed an increase in the DM intervention diet group. γ-Tocopherol (vitamin E) decreased by 0.21 μg/dL. In relation to fruit and vegetable intake, total plasma carotenoids increased only slightly by an average 0.04 μg/dL in the intervention diet group. In addition to diet-related biomarkers, serum estradiol levels were also significantly reduced, while sex hormone-binding globulin increased in the study population randomized to the low-fat diet group *(10)*.

Several publications have evaluated factors associated with adherence to the low-fat diet. Factors reported to be associated with poorer adherence to the DM diet include advanced age, minority ethnicity, lower SES, and obesity *(47)*. Among older WHI participants at Eastern state sites of WHI, adherence was best among assertive women, those who had valued a low-fat diet for many years of adult life, and those who felt they had acquired the requisite knowledge and skills for dietary change; in contrast, nonadherent women were unable to effectively resist emotional eating and were more concerned with negative response from others regarding their participation in a low-fat diet plan *(48)*. Of note, similar factors explained nonadherence to the NCEP dietary guidelines for elevated cholesterol in WHI OS study women *(49)*, with the addition of being married, current smoking, and lower physical activity. When a subsample ($n = 100$) of DM trial participants was queried regarding the definition of healthy eating, these women defined healthy eating in terms of greater intake of fruits and vegetables, followed by higher whole-grain intake, lower fat and lower intake meat/protein. Of interest, 39% reported "healthy" as equivalent to "balanced" *(50)*. In this sample of women, those who demonstrated a capacity to maintain the WHI dietary goals also valued the behavior "don't overeat," while nonmaintainers described healthy eating as "consistent/patterned" eating. These results suggest that restricted total intake that is not perceived as inflexible day-to-day may be a significant factor in achieving and maintaining the WHI eating pattern. The use of self-monitoring tools, including food diaries, Fat Scan, keeping track goals, Quick Scan, Picture Tracker, and eating pattern changes, varied among women assigned to the low-fat diet, with approximately half of the women using the self-monitoring tools provided. Overall, the total number of days of self-monitoring annually was inversely associated with total percentage of energy intake as fat *(51)*. Sample self-monitoring tools are shown here as their adoption for other patient counseling programs should be considered as a method to promote dietary adherence to specific dietary goals (Figs. 14.5 and 14.6).

Functional or mental health status was also thought to potentially affect success with dietary modification. In an analysis among 13,277 of the 19,542 women in the DM trial, baseline administration of the SF-36 Health Survey, a standardized, validated instrument that includes eight subscale assessments of functional status and well-being, suggested that adherence to the low-fat intervention plan was associated with select measures of physical and functional health. For example, greater physical functioning was associated with attendance at more group counseling sessions as well as self-monitoring of fat intake *(52)*. Self-monitoring was also more likely in women who scored higher on the mental health, social functioning, and vitality measures. Since fat intake self-monitoring also predicted adherence to the WHI diet, these data are suggestive in that patients with higher SF-36 may be more successful with adopting and complying with medically prescribed changes in dietary behaviors.

Early results from WHI indicate that most women assigned to the low-fat arm were able to achieve the daily fat gram goal by making appropriate lower fat or nonfat substitutions in their food intake or eliminating added fats such as gravies, sauces, salad dressings, etc. generally *(47)*. Fat was also reduced by reducing dessert, meat, and dairy intake. Of interest, high-fat bread and salty snack food intake was reduced only slightly, with an average reduction of only 2 g daily. And while added fats and meat con-

Women's Health Initiative -- Keeping Track of Goals

Name:________________	Fat Gram Goal: _______	Session #______	Group #_______
Fat Gram Score: ____________	**F/V Score:** ___________	**Grain Score:** __________	

One Fruit/Vegetable Serving = 1/2 cup cut up fruit or vegetable • 1/2 cup potatoes • 1 medium piece of fruit 1 cup raw leafy vegetable • 3/4 cup juice • 1/4 cup dried fruit

One Grain Serving = 1 slice bread • 1/2 bagel • 1/2 cup cooked cereal or grits • 1 cup cold cereal 8 crackers • 3 cups popcorn • 1/2 cup pasta, rice, or noodles • 1/2 cup cooked dried peas or beans

Date:________ F/V Serv: □□□□□□□□□ = ___ Grain Serv: □□□□□□□□□ = ___ Fat Grams: Breakfast = ______ Lunch = ______ Dinner = ______ Snacks = ______ **Total Fat Grams** = ______	Date:________ F/V Serv: □□□□□□□□□ = ___ Grain Serv: □□□□□□□□□ = ___ Fat Grams: Breakfast = ______ Lunch = ______ Dinner = ______ Snacks = ______ **Total Fat Grams** = ______	**Goals / Notes.....**
Date:________ F/V Serv: □□□□□□□□□ = ___ Grain Serv: □□□□□□□□□ = ___ Fat Grams: Breakfast = ______ Lunch = ______ Dinner = ______ Snacks = ______ **Total Fat Grams** = ______	Date:________ F/V Serv: □□□□□□□□□ = ___ Grain Serv: □□□□□□□□□ = ___ Fat Grams: Breakfast = ______ Lunch = ______ Dinner = ______ Snacks = ______ **Total Fat Grams** = ______	**Goals / Notes.....**
Date:________ F/V Serv: □□□□□□□□□ = ___ Grain Serv: □□□□□□□□□ = ___ Fat Grams: Breakfast = ______ Lunch = ______ Dinner = ______ Snacks = ______ **Total Fat Grams** = ______	Date:________ F/V Serv: □□□□□□□□□ = ___ Grain Serv: □□□□□□□□□ = ___ Fat Grams: Breakfast = ______ Lunch = ______ Dinner = ______ Snacks = ______ **Total Fat Grams** = ______	**Goals / Notes.....**

Fig. 14.5. DM monitoring tools: keeping track of goals.

tributed the most fat calories in the diet initially and were reduced the most in the DM low-fat diet arm during year 1, these foods were also returned to the diet during year 2 more than other fat sources; this suggests that reducing intake of these foods is challenging for the study participants for a long term.

Additional predictors of dietary change included higher level of education, younger age, attending more of the dietary counseling sessions as well as reporting a more optimistic approach to life *(53)*.

14.2.6. The WHI DM Trial and Body Weight

The WHI study had no dietary goal related to total energy intake nor did the counseling include efforts to promote weight loss among the women on study. An early reduction in energy intake resulted in the small (~2 kg average) loss of body weight shown in the low-fat diet group during the first year of study. Of interest, the initial weight loss during year 1 was followed by a pattern of slow, steady

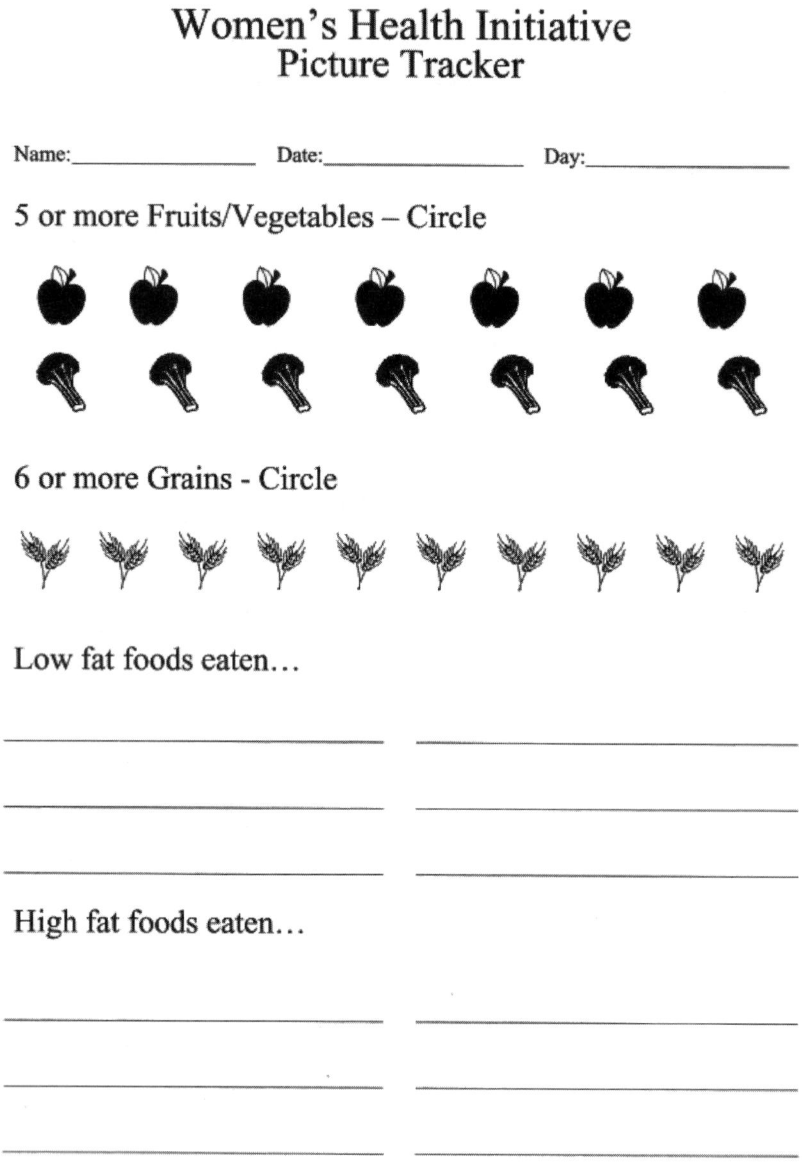

Women's Health Initiative
Picture Tracker

Name:__________ Date:__________ Day:__________

5 or more Fruits/Vegetables – Circle

6 or more Grains - Circle

Low fat foods eaten…

High fat foods eaten…

Fig. 14.6. WHI DM monitoring tools: picture tracker.

weight regain that paralleled the pattern seen in control group women throughout the study so that, on average, the intervention group women had returned to a mean body weight just above the initial values for body weight at the 8.1 year time point *(54)*. This pattern of steady body weight gain over several years in both diet groups after year 1 was predominantly in women under age 60 years. In older women, control group assignment was associated with an initial stable body weight followed by a steady decline over time. The intervention provided an initial weight loss during year one, a slight regain during the next several years, and then a steady decline such that women over the age of 70 years had a net loss resulted in a net loss of body weight at year 8 of the WHI trial. These differential patterns of weight change over time in relation to diet group assignment as well as age are likely to have clinical relevance in terms of health outcomes in this population as outcomes continue to be

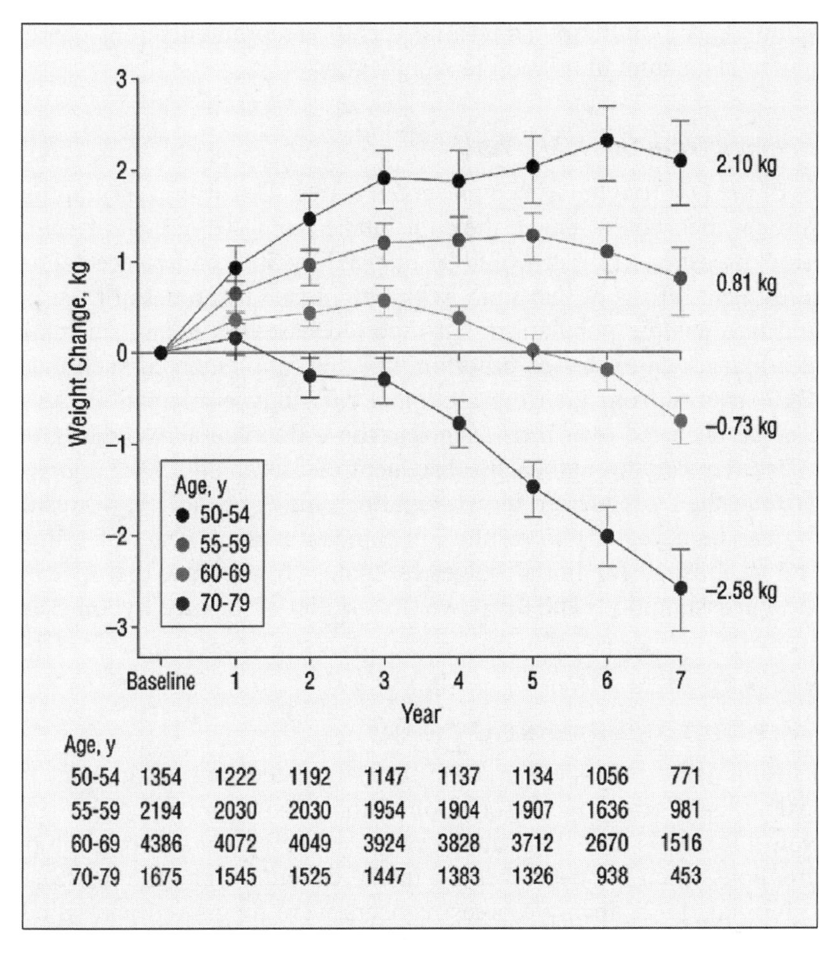

ARCHIVES OF
INTERNAL MEDICINE

Fig. 14.7. Change in body weight among WHI DM participants over time and by age group.

collected longitudinally. Figure 14.7 illustrates the age-related change in body weight and includes data for all WHI clinical trial participants randomized to the placebo arm of any of the three WHI CTs *(55)*.

In addition, body weight changes were also evaluated in relation to assignment to the intervention versus placebo arm of the calcium/vitamin D clinical trial within WHI. While this was not an a priori hypothesis of WHI, mounting scientific evidence in the later years of the trial suggesting an inverse association between calcium intake and body weight supported evaluation of this relationship. Caan et al. reported that randomization to calcium (1,000 mg)/cholecalciferol (440 IU) resulted in a small (–0.13 kg) but statistically significant difference in weight gain at year 3 *(57)*. Weight control was significant only among women who reported a total calcium intake from diet and supplement of <1,200 mg at study initiation, suggesting that the beneficial effects are dependent on baseline nutrient status and that supplementation of well-nourished women does not provide additional protection against weight gain in later life. The analysis evaluated the combined exposure to both dietary and supplemental calcium; based on the outcomes, no specific recommendations can be made regarding diet versus supplemental calcium in relation to weight control.

14.2.7. Key Outcomes from the DM Trial

The hypotheses of the DM trial have been presented above. The design assumption was that women assigned to the low-fat diet would reduce their fat intake from 40 to 20% of energy as fat and that after a 9-year period, breast cancer risk would be reduced by 50% and colorectal cancer risk by 30% *(1)*. However, recruitment efforts at that time, with heightened awareness of the problems of eating excess fat, resulted in a sample population that reported baseline dietary fat intake that was much lower than desired. Indeed, an exclusion criterion was then implemented such that women consuming diets with <32% energy from fat were ineligible for study participation. As anticipated, some regression to the mean occurred over time such that the difference across diet groups was maximal at year 1 (10.7%) and reduced through each subsequent measurement to 9.5 and 8.1% at years 3 and 6, respectively *(56)*. Further, a reduction in participating clinics from 44 to 40 resulted in heightened recruitment during the final years; this resulted in a mean average years of follow-up of 8.1 years rather than 9 years established for initial power estimates. In the end, the difference achieved was only 70% of the design assumptions and the power to test diet-associated hypotheses was significantly reduced.

14.2.7.1. Low-Fat Diet and Breast Cancer

The DM trial results were published in 2006 and suggested that assignment to the low-fat diet resulted in a nonsignificant protective association with invasive breast cancer risk (RR 0.91, 95% confidence interval (CI) 0.83–1.01). This number was, however, what would be predicted from the initial study assumptions when the decreased fat intake difference and the decreased follow-up time were considered. In total, 1,727 invasive breast cancers were diagnosed and adjudicated during the study period, representing 3.5% of the DM population *(10)*. When 4-day food records were used to estimate change in fat intake as a percentage of total energy, a subgroup analysis showed that among women who reported fat intake at the uppermost quartile (>36.8% fat kcal) and reduced dietary fat by an average of 12.2% over the course of the study, invasive breast cancer risk was reduced by 22% (RR 0.78, 95% CI 0.64–.095) ($p = 0.04$). A separate subgroup analysis indicated that the low-fat diet may have proven beneficial in reducing estrogen receptor-positive tumors that coexpressed progesterone receptor negativity. In this group, the RR was 0.64 (95% CI 0.49–0.84); no association was found for any other hormone receptor subgroup. Of interest, when the association between dietary fat and breast cancer outcomes were assessed in a case-control subsample of 1836 WHI DM control group women using the two dietary measurement instruments, the data showed a significant association for Food Record Data (RR 1.82) but a nonsignificant association using the FFQ diet data (RR 0.67) *(57)*. In addition, a subsequent analysis of the low-fat dietary associations with all cancer sites suggested that the low-fat diet was potentially protective against ovarian cancer *(58)*.

14.2.7.2. Low-Fat Diet and Colorectal Cancer Risk

Over the course of the DM trial (8.1 years), 480 women were diagnosed with invasive colorectal cancer. Despite a significant reduction in dietary fat intake, which averaged 10.7% at year 1 and 8.1% at year 6, there was no significant difference in colorectal cancer rates for women randomized to the low-fat or comparison diets (HR 1.08, 95% CI 0.90–1.29) *(18)*. While not statistically significant, the data were suggestive of elevated risk for proximal colon cancers (HR 1.25, 95% CI 0.96–1.61); associations between the low-fat intervention and protection from rectal cancer were also nonsignificant, but the total number of cases was small ($n = 117$). Analysis of dietary factors previously reported to be associated with colorectal cancer in the published literature showed that none was associated with increased (energy, fat, red meat, alcohol) or decreased (fruit/vegetables, whole grains, carotenoids, folate, or dietary calcium) risk of colorectal cancer in the WHI DM trial *(18)*.

14.2.7.3. Low-Fat Diet and Cardiovascular Disease Risk

A number of cardiovascular-related diagnoses were evaluated in the context of the WHI DM trial. In relation to hypertension, the low-fat diet was associated with a mean decrease in systolic and diastolic blood pressure of –0.2 and –0.3 mmHg, respectively. Stroke, a common consequence of poorly controlled blood pressure, occurred in a total of 1,076 women or 2% of the DM population. Stroke risk was not associated with low-fat dietary assignment *(25)*. Blood pressure was not modified in relation to treatment assignment for the CaVitD trial subjects *(59)*.

Overall there were 2,549 coronary heart disease diagnoses/events (myocardial infarction, bypass, stent, angioplasty) *(25)*. In relation to heart disease risk, factor VIIC was reduced, on average, by 4.9% or more from baseline to year 3 in the intervention diet versus control group subjects but total cardiovascular disease rates were not significantly reduced with the low-fat diet (RR 0.98; 85% CI 0.92–1.05) nor were CHD or CHD death *(25)*. With the mildly reduced RR for a number of cardiovascular-related measures, assessment of cardiovascular risk using the Framingham risk score suggested that the WHI low-fat diet resulted in a net 3–4% risk reduction *(25)*.

14.2.7.4. Total Mortality and Global Index

When comparing global index for the intervention subjects ($n = 2{,}051$ annualized cases) and comparison group subjects ($n = 3{,}207$ annualized cases), no significant difference was demonstrated. Similarly, mortality rates were also comparable across diet groups, suggesting that the low-fat diet also did not modify these more global indicators of health status over the 8 year stud period *(56)*.

14.2.8. Role of Low-Fat Diet in Modifying Other Health Risks

The DM cohort (along with the Observational Study (OS) cohort) with the repeated measures of diet and lifestyle factors as well as adjudicated and self-reported health outcomes has provided an unprecedented opportunity to evaluate the role of diet in modifying the risk of a number of health outcomes. Additionally the calcium and vitamin D trial provides similar opportunities. The results of numerous analyses are summarized in Table 14.3, where association being evaluated, analytical cohort, sample size, and risk ratios are described.

14.3. CLINICAL APPLICATIONS

The outcomes of the DM study within WHI did not provide statistically significant support for a low-fat diet in reducing risk of breast cancer or colorectal cancer. Nor was the diet intervention associated with reduced risk of cardiovascular disease, obesity, or diabetes. However, there are a

Table 14.3

Summary of WHI Diet, Nutrition Research Findings from the DM and Calcium/vitamin D Clinical Trials and WHI Observational Study (as of fall 2008)

First Author, Journal, Year	*Sample Population*	*Dietary/ Anthropometric Measure*	*Outcome Measure*	*Key Findings*	*Comment*
Overall health					
Brunner, R.L., et al., *J Am Diet Assoc*, 2008 *(67)*	33,067 Females (50–79 years), supplement = 1,000 mg Ca and 400 IU VitD	Calcium intake	Physical functioning	No significant difference between supplement versus placebo group	Study findings did not support Vit D/Ca protecting against physical function decline in pop
Ma, Y., et al., *Nutrition*, 2008 *(68)*	1,958 postmenopausal women, observation group	Dietary fiber intake	Hs-CRP. IL-6, TNF-αR2	Inverse association with dietary fiber intake and IL-6 and TNF-αR2 (*p*-value = 0.01, 0.002 total fiber, respectively)	No association between hs-CRP and dietary fiber in postmenopausal women
Caan, B., et al., *Arch Intern Med*, 2007 *(55)*	36,282 Females (50–79 years), supplement = 1,000 mg Ca and 400 IU VitD	Calcium intake	Change in body weight	Supplement group = mean wt. loss of 0.13 kg, 95% CI 0.21–0.05, *p*-value = 0.001	VitD/Ca supplement showed small effect on prevention of wt. gain, primarily observed in women with inadequate Ca intake
Howard, B.V., et al., *JAMA*, 2006 *(54)*	48,835 postmenopausal DM group (40% intervention/60% obs.)	Dietary fat, fruit, vegetable and grain intake	Change in body weight	Intervention group = mean wt. loss of 2.2 kg, *p*-value <0.001 after 1 year; 0.4 kg, *p*-value 0.01 at 7.5 years	Wt. loss was greatest for women who decreased overall energy from fat. Similar but lesser trend seen for increasing fruits, vegetables, and fiber

Cancer					
Breast					
Cui, Y. et al., *Am J Clin Nutr*, 2008 *(69)*	2,509 breast cancer cases within WHI observational group (84,805)	Dietary α- and β-carotene, lycopene, vitamin E and vitamin C	Invasive breast cancer	Dietary α- and β-carotene, lycopene inversely associated with ER + PR + breast cancer	Vitamin E (regardless of source) and dietary vitamin C not associated with decreased risk of any type of breast cancer. Supplemental vitamin C showed weak association with/ overall breast cancer risk
Prentice, R.L. et al., *JAMA*, 2006 *(10)*	48,835 postmenopausal DM group (40% intervention/60% obs.)	Dietary fat, fruit, vegetable and grain intake	Invasive breast cancer	655 breast cancer cases observed in the intervention group versus 1,072 in the comparison group (HR 0.91, CI 0.83–1.01)	Low-fat diet did not show a statistically significant reduction in breast cancer; however, nonsignificant trend suggests decreased risk with longer follow-up
Colorectal					
Kabat, G.C. et al., *Cancer Causes Control*, 2008 *(70)*	1,476 colorectal cancer cases within WHI (158,800)	Glycemic index/glycemic load (GI/GL)	Colorectal cancer	Total carbohydrate intake, GI, GL, intake of sugars and fiber showed no association with colorectal cancer	Analyses by cancer subsite yielded null results with exception of rectal cancer GL(HR between highest and lowest quartiles 1.84, CI 0.95–3.56)
Wactawski-Wende, J., et al., *N Eng J Med*, 2006 *(71)*	36, 282 females (50–79 years), supplement = 1,000 mg Ca and 400 IU vitamin D	Calcium and vitamin D intake	Colorectal cancer	168 cancer cases in the supplement group versus 154 in the comparison group (HR 1.08, CI 0.86–1.34, $p = 0.51$)	Calcium/vitamin D supplementation for 7 years had no effect on incidence of colorectal cancer in postmenopausal women

(*Continued*)

Table 14.3
(Continued)

First Author, Journal, Year	*Sample Population*	*Dietary/ Anthropometric Measure*	*Outcome Measure*	*Key Findings*	*Comment*
Beresford, S.A., et al., *JAMA*, 2006 *(18)*	48,835 postmenopausal DM group (40% intervention/60% obs.)	Dietary fat, fruit, vegetable, and grain intake	Colorectal cancer	201 cancer cases in the intervention group versus 279 in the comparison group (HR 1.08, CI 0.90–1.29)	Low-fat dietary pattern did not reduce risk of colorectal cancer. However, potential interaction observed between baseline aspirin use and estrogen–progestin use ($p = 0.01$ for each)
Ovarian					
Prentice, R.L., et al., *J Natl Cancer Inst*, 2007 *(58)*	48,835 postmenopausal DM group (40% intervention/60% obs.)	Dietary fat, fruit, vegetable, and grain intake	Ovarian cancer	Ovarian cancer rates statistically lower in the intervention group versus the comparison group ($p = 0.03$)	For the first 4 years, intervention and comparison group ovarian cancer risk was similar; however, over the next 4.1 years, intervention group showed decreased risk of ovarian cancer
Cardiovascular disease					
Hsia, J., et al., *Circulation*, 2007 *(72)*	36,282 females (50–79 years), supplement = 1,000 mg Ca and 400 IU VitD	Calcium and vitamin D intake	Myocardial infarction (MI) or coronary heart disease death, stroke	MI or coronary heart disease death confirmed in 499 in supplement group versus 475 in the comparison group (HR 1.04, CI 0.92–1.18). Stroke confirmed in 362 supplement group versus 377 comparison group (HR 1.04, CI 0.82–1.10)	Women with higher calcium intake at baseline were at no higher risk for coronary event or stroke ($p = 0.91$ and 0.14, respectively) if assigned to supplement group

Howard, B.V. et al., *JAMA*, 2006 *(25)*	48,835 postmenopausal DM group (40% intervention/60% obs.)	Dietary fat, fruit, vegetable, and grain intake	Fatal and nonfatal CHD, stroke, and CVD	The numbers who developed CHD, stroke, and CVD (annualized incidence rates) were 1,000, 434, and 1,357 in the intervention and 1,549, 642, and 2,088 in the comparison group, respectively. The diet had no significant effects on incidence of CHD (HR 0.97, CI 0.90–1.06), stroke (HR 1.02, CI 0.90–1.15), or CVD (HR 0.98, CI 0.92–1.05)	Trends toward greater reductions in CHD risk were observed in those with lower intakes of saturated fat or trans fat or higher intakes of vegetables/fruits
Bone health					
Jackson, R.D. et al., *N Eng J Med*, 2006 *(73)*	36,282 females (50–79 years), supplement = 1,000 mg Ca and 400 IU VitD	Calcium and vitamin D intake	Hip fractures, bone density, and kidney stones	Hip bone density was 1.06% higher in the calcium plus vitamin D group than in the placebo group ($p < 0.01$)	Among healthy postmenopausal women, calcium/vitamin D supplementation resulted in a small but significant improvement in hip bone density, did not significantly reduce hip fracture, and increased the risk of kidney stones

(*Continued*)

Table 14.3
(Continued)

First Author, Journal, Year	*Sample Population*	*Dietary/ Anthropometric Measure*	*Outcome Measure*	*Key Findings*	*Comment*
Diabetes mellitus					
Tinker, L.F., et al., *Arch Intern Med*, 2008 *(74)*	48,835 postmenopausal DM group (40% intervention/60% obs.)	Dietary fat, fruit, vegetable, and grain intake	Diagnosis of oral or injection-controlled diabetes mellitus	Incident-treated diabetes was reported by 1,303 intervention participants and 2,039 comparison participants (HR 0.96, CI 0.90–1.03; $p = 0.25$). Weight loss occurred in the intervention group, with a difference between intervention and comparison groups of 1.9 kg after 7.5 years ($p < 0.001$)	A low-fat dietary pattern among postmenopausal women showed no evidence of reducing diabetes risk. Trends toward reduced incidence were greater with greater decreases in total fat intake and weight loss
De Boer, I.H., et al., *Diabetes Care*, 2008 *(75)*	33,951 females (50–79 years) self-reported, nondiabetic supplement = 1,000 mg Ca and 400 IU VitD versus placebo (total group = 36,282)	Calcium and vitamin D intake	Diagnosis of oral or injection-controlled diabetes mellitus	2,291 women were newly diagnosed with diabetes. The hazard ratio for incident diabetes associated with calcium/vitamin D treatment was 1.01 (CI 0.94–1.10) based on intention to treat	Calcium plus vitamin D supplementation did not reduce the risk of developing diabetes over 7 years of follow-up in this randomized, placebo-controlled trial

Obesity					
Lou, J., et al., *Br J Cancer*, 2008 *(76)*	138,503 postmenopausal women	Prospective study looking at weight and waist-to-hip ratio	Diagnosis of pancreatic cancer	251 women were diagnosed with pancreatic cancer. Women in the highest quintile of waist-to-hip ratio had a 70% (95% CI 10–160%) excess risk of pancreatic cancer compared to women in the lowest quintile	Obesity, especially central adiposity, increases risk of pancreatic cancer in postmenopausal women
Kabat, G.C., et al., *Am J Epidemiol*, 2008 *(77)*	161,809 postmenopausal women; smokers versus nonsmokers	Prospective study looking at weight and waist-to-hip ratio	Diagnosis of lung cancer	1,365 women were diagnosed with lung cancer. Baseline BMI was inversely associated with lung cancer risk in current smokers and nonsmokers (HR 0.61, CI 0.40–0.94); waist circumference was positively associated with lung cancer risk (HR 1.56, CI 0.91–2.69) in current smokers and (HR 1.50, CI 0.98–2.31) in nonsmokers	These findings suggest that in smokers, BMI is inversely associated with lung cancer risk and that waist circumference is positively associated with risk

(*Continued*)

Table 14.3
(Continued)

First Author, Journal, Year	*Sample Population*	*Dietary/ Anthropometric Measure*	*Outcome Measure*	*Key Findings*	*Comment*
Luo, J., et al., *Am J Epidemiol*, 2007 *(78)*	140,057 postmenopausal women	Prospective study looking at weight change and waist-to-hip ratio	Diagnosis of renal cell carcinoma	269 women were diagnosed with renal cell carcinoma. Waist-to-hip ratio was associated with renal cell carcinoma risk in highest versus lowest quartile (RR 1.8, CI 1.2–2.5). Women who experienced weight cycling more than 10 times were at 2.6 times greater risk for developing lung cancer versus women whose weight remained stable (CI 1.6–4.2)	Obesity, specifically central adiposity, is associated with increased risk of renal cell carcinoma among postmenopausal women. This research may also indicate that weight cycling is independently associated with an increased risk of renal cell carcinoma in this population
Morimoto, L.M., et al., *Cancer Causes Control*, 2002 *(79)*	85,917 postmenopausal women in the observational group	Body weight, lifetime weight history	Diagnosis of breast cancer	1,030 women were diagnosed with invasive breast cancer. Among HRT nonusers, heavier women (baseline BMI >31.1) had an elevated risk of postmenopausal breast cancer, RR 2.52, CI 1.62–3.93, compared to slimmer women (BMI < 22.6)	Increasing BMI appears to be most pronounced among younger postmenopausal women. Change in BMI since age 18, maximum BMI, and weight were also associated with breast cancer in HRT nonusers. Lifetime weight gain is also a strong predictor of breast cancer. Waist-to-hip ratio was not related to postmenopausal breast cancer risk

McTiernan, A., et al., *Obesity*, 2006 *(80)*	267 postmenopausal women randomly selected from the Diet Modification Group	None	Association between physical activity, body weight, and sex hormones	BMI was positively associated with estrone ($\beta = 0.031$, $p < 0.001$), estradiol ($\beta = 0.048$, $p < 0.001$), free estradiol ($\beta = 0.062$, $p < 0.001$), free testosterone ($\beta = 0.017$, $p = 0.02$), and prolactin ($\beta = 0.012$, $p = 0.02$) and negatively associated with SHBG ($\beta = -0.02$, $p = 0.001$). Total physical activity was negatively associated with concentrations of estrone, estradiol, and androstenedione ($\beta = -0.006$, -0.007, and -0.005, respectively, all $p \leq 0.05$). Women with high BMI/low physical activity had the highest mean estrone concentration of 28.8 pg/mL (p trend $<$ 0.001)	The associations between overweight/obesity, sex hormones, and sedentary lifestyle on breast cancer risk may in part be explained by this research

(*Continued*)

Table 14.3
(Continued)

First Author, Journal, Year	*Sample Population*	*Dietary/ Anthropometric Measure*	*Outcome Measure*	*Key Findings*	*Comment*
McTigue, K., et al., *JAMA*, 2006 *(81)*	90,185 postmenopausal women from observation group		Mortality, coronary heart disease, HTN, DM	Extreme obesity prevalence differed with race/ethnicity, from 1% among Asian and Pacific Islanders to 10% among black women. All-cause mortality rates per 10,000 person-years were 68.39 (CI 65.26–71.68) for normal body mass index, 71.16 (CI 67.68–74.82) for overweight, 84.47 (CI 78.90–90.42) for obesity 1, 102.85 (CI 92.90–113.86) for obesity 2, and 116.85 (CI 103.36–132.11) for extreme obesity	Obesity set as a BMI of 30 or greater may lead to misinterpretation of individuals and populations at risk

Howard, B., et al., *Int J Obes Relat Metab Disord*, 2004 *(82)*	Blood samples randomly obtained from 3,389 (60% white, 20% black, 12% Hispanic, and 8% Asian/Pacific Islander) postmenopausal women from the entire observational study (161,809)		Insulin concentrations and resistance	Insulin resistance and insulin concentrations were independent predictors of increases in weight in white women ($p = 0.002$ and 0.004, respectively) and in the combined group ($p = 0.027$ and 0.039, respectively). Those in the highest quartile of insulin resistance gained 0.4 kg in 3 years and those in the lowest quartile lost 0.06 kg. A significant interaction between obesity and insulin resistance was observed ($p = 0.002$ for white women and 0.032 for the whole group)	Insulin resistance appears to be a predictor of weight gain in postmenopausal women, except among the most obese women

number of lessons learned along the way that will help to advance diet–disease association research for years to come *(56)*. The lessons learned include the following:

- Self-report of dietary intake using a food frequency questionnaire may not be a good estimate of actual intake. Having a more detailed dietary record or multiple recalls could provide stronger results. Also having additional biomarker analyses to measure adherence and guide individual change in diet as well as to estimate measurement error within the total population would strengthen future study designs. While the cost associated with such an undertaking was prohibitive during the development of WHI, new and emerging automated dietary assessment methods may reduce costs considerably. The recently completed nutritional biomarker studies should also help to advance our understanding of measurement error in diet research *(60)*.
- Adopting a low-fat diet late in life may be an insufficient "exposure" to reduce or reverse detrimental effects of lifelong eating. Studies targeting women of younger age should be considered. However, there is no indication from subgroup analysis that women in the 50–60-year age group had any different results in terms of low-fat diet and health outcomes than the women recruited between the ages of 70 and 79 years.
- The absolute change in dietary fat (and possibly fruits and vegetables) and duration of dietary fat restriction may have been insufficient to modify disease risk. For example, despite a concerted effort to screen out women with lower fat diets, many women, at the time of study entry, had dietary fat intakes below the average for women of the same age in the general population, thus the actual reduction in exposure over time was not as great as was initially predicted. Further, the study was terminated short of the original design estimates and adherence was lessened over time (as has been demonstrated in numerous dietary intervention trials).
- Weight control was not included in the original study design. Weight control is difficult to achieve on an individual basis, let alone in the large group counseling setting; the absolute size of the study sample made more individualized diet counseling impossible; had a weight loss component been proposed, it is unlikely that significant weight loss would have occurred or been maintained. The fact that the low-fat diet resulted in a lessened weight gain over the study period among women randomized to the low-fat diet is promising, but it was insufficient to significantly modify disease risk for obesity-related diagnoses.
- The target study outcomes—breast and colorectal cancer as well as cardiovascular disease risk—are modifiable through physical activity, and yet no physical activity component was included in the trial design. Again, costs and ability to attain and sustain significant increases in physical activity in older women influenced the decision not to focus on activity as a behavioral goal. Further, physical activity levels did not change significantly during the study. Certainly, if physical activity had increased and been maintained at a higher level as adjuvant to the low-fat diet, there would have been a greater likelihood of risk reduction.
- In the calcium and vitamin D supplementation trial, study participants were permitted to continue their own independent use of calcium and up to 600 IU of vitamin D daily. This protocol design resulted in some overlap in exposure to these nutrients in the supplemented versus placebo group and may have contributed to the null findings found.
- Ideally, adjudication of additional health outcomes beyond those involved in testing of the primary and secondary hypotheses would allow for more accurate assessment of diet–disease associations (i.e., additional cancer sites, arthritis, autoimmune diseases, etc), particularly for rare diseases where a large sample size, such as the WHI study provides, would be needed in order to achieve adequate statistical power.
- Translation of research findings can be challenging, especially given the multiple interventions, study arms, and the inclusion of an observational cohort and clinical trials within the overall study design. On the other hand, having both study types within the larger study population is unprecedented and allows for a thorough comparison of diet–disease hypothesis testing in the context of the two parallel approaches employed in diet research.

14.4. RECOMMENDATIONS

The Women's Health Initiative is the largest trial ever undertaken to determine the role of a low-fat diet in reducing risk for common chronic diseases of postmenopausal women. In addition, it was the first study to prospectively test in a randomized, placebo-controlled study design the association between supplemental calcium and vitamin D and osteoporotic fracture and/or colorectal cancer risk in a large sample of postmenopausal women. While the paramount finding of the WHI was the result of the HT trial, indicating an increased risk for cardiovascular disease *(61)* and breast cancer among postmenopausal women taking hormones *(62)*, the dietary modification trial has and will continue to inform clinical care for women in this age group for years to come.

Despite evidence that the diet intervention designed to achieve a low-fat diet did not modulate breast cancer risk in the overall WHI DM trial population, there were some promising results that the diet was efficacious among women with the highest baseline fat intake who adhered to the diet. Further, the low-fat diet was associated with a significant reduction in ovarian cancer risk *(58)*. These two findings suggest that postmenopausal women should be provided with appropriate dietary assessment and proper counseling to adopt an eating plan that is low in fat and high in fruits, vegetables, and grains in an effort to reduce these risks. Further, while the reduction in total fat was not protective against cardiovascular disease, it has subsequently been reported that such an eating plan, specifically reduced in saturated fat and in addition to weight loss in overweight women, should be advised as per the ATP III guidelines *(63)*.

While the the results of the calcium/vit D supplementation trial where null, these findings should not deter clinicians from assessing dietary intake of these nutrients to promote adequacy for optimal bone health. In fact, numerous studies support a protective role for both calcium and vitamin D in reducing osteoporosis and related fracture risk *(64–66)*. The subgroup of women with lower serum vitamin D levels did demonstrate reduced fracture risk with supplementation, suggesting that the "healthy" volunteer effect was a significant effect modifier for this CT within WHI.

Importantly, the WHI is only just beginning to inform clinical care for aging women. As the population of older women is increasing steadily in the United States, so will our need to better understand the role of lifestyle factors, including diet, in reducing disease risk and promoting improved health. The wealth of clinical, lifestyle, demographic, and genetic data generated from this large, longitudinal study will continue to impact clinical knowledge well beyond the initial proposed hypotheses.

14.5. ACKNOWLEDGMENTS

The following dietitians and nutritionists provided support to the implementation of the dietary modification study of WHI: Brooke Adachi; Tanya Agurs-Collins; Gayle Alleman; Joli Allen; Linda Antinoro; Christine Armes; Grace Arriola; Frani Averbach; Heather Backer; Kelly Bailey; Denise Barratt; Mary Paula Baumann; Michelle Bean; Marlene Beddome; Margaret Beesley; Lorna Belden; Randi Belhumeur; Connie Belk; Pat Bergen; Eleanor Betz; Kathy Bierl; Cindy Bjerk; Kerri Bland; Mary Pat Bolton; Charlotte Bragg; Anne Brettschneider; Amy A. Brewer; Mireille Bright-Gbebry; Marilyn Brill; Julie Burgess-Bottom; Deborah Burman; Jill Burns; Elizabeth Burrows; Kathy Burton; Darcy Bushnell; Bette Caan; Arlene Caggiula; Louise Calderera; Virginia Cantrell; Sara Carrion; Michelle Berry Casavale; Aimee Cassulo; Dana Chamberlain; Teresa Clark; Carolyn Clifford; Rosemary Clifford; Dorsey Cokkinias; Amy Coleman; Patricia Collette; Lisa Cooper; Betsy Costello; Janice Cox; Christi Coy; Martha Sue Dale; Dorothy Delessio; Stephanie Dellner; Jan Depper; Janet DiIulio; Nora DiLaura; Pia DiMarzio; Panayiota Doering; Linda Doroshenko; Eva Dugger; Carolyn Ehret; Louise Engar; Jackie Enlund; Margaret Farrell; Jennifer Fine; Darlene Fontana; Amy Ford; Gail Frank; Trish Freed; Ann Gallagher; Karen Gantt; Angela Gardner; Linda Gay; Judy Gerber;

Niki Gernhofer; Vicki Gobel; Sue Goldman; Nancy Goyings; Barbara Graf; Theresa Greco; Lynne Gregor; Peggy Gregson; Adriann Grosz; Andrea Guastamacchio; Valerie Haack; Pamela Haines; Beth Hall; Susan Hall; Mary Hankey; Judith Hartman; Kathy Hattaway; Holly Henry; Judith Hinderlitter; Mary Lynne Hixson; Elizabeth Hoelscher; Donna Hollinger; Leah Hopkins; Gerrilyn Hopper; Heather Horowitz; Kathy Hrovat; Anne Hubbell; Sharon Innes; Sharon Jackson; Carolyn Jennings; Lynn Johnson; Marian Johnson; Bobbette Jones; Donna Toms Jones; Kirsten Jordahl-Nielsen; Sarah Josef; Cheryl Kaplan; Stephanie Keck; Karen Kedrowski; Mark Kestin; Laura Kinzel; Lee-Ann Klein; Heidi Koempel; Sandy Kowal; Jen Kring; Alan Kristal; Brenda Krpata; Danielle Lacina; Phil Lacher; Mary Lamon-Smith; Patti Laqua; Mary Larez; Lynne Larson; Vera I. Lasser; Claire LeBrun; Elisa Lee; Sandy Lemanski; Judy Levin; Barbara Lewin; Ilona Lichty; Marlene Llano; Denise Londergan; Leticia Lopez; MaryAnn Lueck; Pat Lyman; Sally Mackey; Katherine Malville-Shipan; Elizabeth Marchese; Nicole Marrari; Shannon Martin; Barbara Mascitti; Alice McCarley; Gayle McCartney; Carol McDermott; Ann McDonald; Allison McKenny; Kathy McManus; Joan McPherson; Becky Meehan; Renee Melton; Mary Messier; Carole Milas; Audrey Miller; Connie Mobley; Debra Mohns; Carol Molfetta; Deborah Monsegue; Margaret Moonan; Ann Gowan Morrissey; Yasmin Mossavar-Rahmani; Dru Mueller; Mary Mueller; Ann Mylod; Sue Nelson; Marian Neuhouser; Alice Nicolai; Doris Nicolas-Mir; Patricia Ochoa; Deborah Ocken; Annette Okuma; Monica Olsen; Pat Ormond; Christina Padilla; Ruth Patterson; Julia Paulk; Margaret F. Pedersen; Elaine Percival; Lois Peterson; Marie Phipps; Jacqui Pirofalo; Joan Pluess; Rosa Polanco; Megan Porter; Carolyn Prosak; Suzanne Proulx; Diane Quigley; Janet Raines; Susan Rauth; Suzanne Rebro; Cyndi Reeser; Cheryl Ritenbaugh; Connie Roberts; Ramona Robinson-O'Brien; Angela Rodriguez; Nancy Saal; Angelique Salonga; Ethel Sanders; Monica Schoenberger; Lisa Sewell; Julie Shapero; Angie Sharp-Jacobson; Audrey Shweky; Karen Silva; Nancy Simpson; Elise Sinagra; Bali Singh; Donna Sivertsen; Beth Smith; Karen Smith; Nancy Smith; Tracy Smith; Linda Snetselaar; Jennifer Soboslai; Oehme Soule; Karen Southern; Jackie St. Cyr; Sachiko St. Jeor; Lorraine G. Staats; Jennifer Summers; Susie Swenson; Joyce Sydell; Houra Taheri; Marisa Tanko; Kimberly Thedford; Nancy Tiger; Robin Thomas; Alice Thomson; Cyndi Thomson; Lesley Tinker; Maribel Tobar; Cynthia Tucker; Nicki Turner; Alice Valoski; Linda Van Horn; Lucy Villarreal; Eileen Vincent; Shirley Vosburg; Carol Walsh; Christie Wayment; Donna J. Weiss; Maria Welch; Andrea Wenger; Julie West; Catherine Willeford; Lois Wodarski; Dolores Wolongevicz; Judy Wylie-Rosett; Alexandra Wynne; Heidi Zahrt.

REFERENCES

1. Women's Health Initiative Study Group. (1998) Design of the Women's Health Initiative clinical trial and observational study. *Control Clin Trials* **19**, 61–109.
2. Rossouw, J.E., Finnegan, L.P., Harlan, W.R., Pinn, V.W., Clifford, C., McGowan, J.A. (1995) The Evolution of the Women's Health Initiative: Perspectives from the NIH. *J Am Med Womens Assoc* **50**, 50–5.
3. Carroll, K.K. (1975) Experimental evidence of dietary factors and hormone-dependent cancers. *Cancer Res* **35**, 3374–83.
4. Willett, W.C. (1997) Fat, energy and breast cancer. *J Nutr* **127**(5 Suppl.), 921S–3S.
5. Wolk, A., Bergström, R., Hunter, D., Willett, W., Ljung, H., Holmberg, L., Bergkvist, L., Bruce, A., Adami, H.O. (1998) A prospective study of association of monounsaturated fat and other types of fat with risk of breast cancer. *Arch Intern Med* **158**(1), 41–45.
6. Prentice, R.L., Sheppard, L. (1990) Dietary fat and cancer: consistency of the epidemiologic data and disease prevention that may follow from a practical reduction in fat consumption. *Cancer Causes Control* **1**, 81–97.
7. Insull, W., Henderson, M.M., Prentice, R.L., et al. (1990) Results of a randomized feasibility study of a low-fat diet. *Arch Intern Med* **150**, 421–27.

8. Fouad, M.N., Corbie-Smith, G., Curb, D., Howard, B.V., Mouton, C., Simon, M., Talavera, G., Thompson, J., Wang, C.Y., White, C., Young, R. (2004) Special populations recruitment for the Women's Health Initiative: successes and limitations. *Control Clin Trials* **25**(4), 335–52.
9. White, E., Hurlich, M., Thompson, R.S., Woods, M.N., Henderson, M.M., Urban, N., Kristal, A. (1991) Dietary changes among husbands of participants in a low-fat dietary intervention. *Am J Prev Med* **7**, 319–25.
10. Prentice, R.L., Caan, B., Chlebowski, R.T., Patterson, R., Kuller, L.H., Ockene, J.K., Margolis, K.L., Limacher, M.C., Manson. J.E., Parker, L.M., Paskett, E., Phillips, L., Robbins, J., Rossouw, J.E., Sarto, G.E., Shikany, J.M., Stefanick, M.L., Thomson, C.A., Van Horn, L., Vitolins, M.Z., Wactawski-Wende, J., Wallace, R.B., Wassertheil-Smoller, S., Whitlock, E., Yano, K., Adams-Campbell, L., Anderson, G.L., Assaf, A.R., Beresford, S.A., Black, H.R., Brunner, R.L., Brzyski, R.G., Ford, L., Gass, M., Hays, J., Heber, D., Heiss, G., Hendrix, S.L., Hsia, J., Hubbell, F.A., Jackson, R.D., Johnson, K.C., Kotchen, J.M., LaCroix, A.Z., Lane, D.S., Langer, R.D., Lasser, N.L., Henderson, M.M. (2006) Low-fat dietary pattern and risk of invasive breast cancer: The Women's Health Initiative randomized controlled dietary modification trial. *JAMA* **295**, 629–42.
11. Prentice, R.L., Kakar, F., Hursting, S., Sheppard, L., Klein, R., Kushi, L.H. (1988) Aspects of the rationale for the Women's Health Trial *J Natl Cancer Inst* **80**, 802–14.
12. Ip, C. (1990) Quantitative assessment of fat and calorie as risk factors in mammary carcinogenesis in an experimental model. *Prog Clin Biol Res* **346**, 107–17.
13. Tominaga, S. and Kuroishi, T. (1997) An ecological study on diet/nutrition and cancer in Japan. *Int J Cancer* **S10**, 2–6.
14. Howe, G.R., Hirohata, T., Hislop, T.G., et al. (1990) Dietary factors and risk of breast cancer: combined analysis of 12 case-control studies. *J Natl Cancer Inst* **82**, 561–9.
15. Boyd, N.F., Stone, J., Vogt, K.N., Connelly, B.S., Martin, L.J., Minkin, S. (2003) Dietary fat and breast cancer risk revisited: a meta-analysis of the published literature. *Br J Cancer* **89**, 1672–85.
16. Hunter, DJ., Spiegelman, D., Adami, H.O., et al. (1996) Cohort studies of fat intake and the risk of breast cancer: a pooled analysis. *N Engl J Med* **334**, 356–61.
17. Bingham, S.A., Luben, R., Welch, A., Wareham, N., Khaw, K.T., Day, N. (2003) Are imprecise methods obscuring a relation between fat and breast cancer? *Lancet* **362**, 212–14.
18. Beresford, S.A., Johnson, K.C., Ritenbaugh, C., Lasser, N.L., Snetselaar, L.G., Black, H.R., Anderson, G.L., Assaf, A.R., Bassford, T., Bowen, D., Brunner, R.L., Brzyski, R.G., Caan, B., Chlebowski, R.T., Gass, M., Harrigan, R.C., Hays, J., Heber, D., Heiss, G., Hendrix, S.L., Howard, B.V., Hsia, J., Hubbell, F.A., Jackson, R.D., Kotchen, J.M., Kuller, L.H., LaCroix, A.Z., Lane. D.S., Langer, R.D., Lewis, C.E., Manson, J.E., Margolis, K.L., Mossavar-Rahmani, Y., Ockene, J.K., Parker, L.M., Perri, M.G., Phillips, L., Prentice, R.L., Robbins, J., Rossouw, J.E., Sarto, G.E., Stefanick, M.L., Van Horn, L., Vitolins, M.Z., Wactawski-Wende, J., Wallace, R.B., Whitlock, E. (2006) Low-fat dietary pattern and risk of colorectal cancer: The Women's Health Initiative randomized controlled dietary modification trial. *JAMA* **295**, 643–54.
19. McMichael, A.J., Giles, G.G. (1988) Cancer in migrants to Australia: extending the descriptive epidemiological data. *Cancer Res* **48**, 751–6.
20. Thomas, D.B., Karagas, M.R. (1987) Cancer in first and second generation Americans. *Cancer Res* **47**, 5771–6.
21. Steinmetz, K.A., Potter, J.D. (1993) Food-group consumption and colon cancer in the Adelaide case-control study, I: vegetables and fruit. *Int J Cancer* **53**, 711–19.
22. Steinmetz, K.A., Kushi, L.H., Bostick, R.M., Folsom, A.R., Potter, J.D. (1994) Vegetables, fruit, and colon cancer in the Iowa Women's Health Study. *Am J Epidemiol* **139**, 1–15.
23. Howe, G.R., Benito, E., Castelleto, R., et al. (1992) Dietary intake of fiber and decreased risk of cancers of the colon and rectum: evidence from the combined analysis of 13 case-control studies. *J Natl Cancer Inst* **84**, 1887–96
24. Trock, B., Lanza, E., Greenwald, P. (1990) Dietary fiber, vegetables, and colon cancer: critical review and metaanalyses of the epidemiologic evidence. *J Natl Cancer Inst* **82**, 650–61.
25. Howard, B.V., Van Horn, L., Hsia, J., Manson, J.E., Stefanick, M.L., Wassertheil-Smoller, S., Kuller, L.H., LaCroix, A.Z., Langer, R.D., Lasser, N.L., Lewis, C.E., Limacher, M.C., Margolis, K.L., Mysiw, W.J., Ockene, J.K., Parker, L.M., Perri, M.G., Phillips, L., Prentice, R.L., Robbins, J., Rossouw, J.E., Sarto, G.E., Schatz, I.J., Snetselaar, L.G., Stevens, V.J., Tinker, L.F., Trevisan, M., Vitolins, M.Z., Anderson, G.L., Assaf, A.R., Bassford, T., Beresford, S.A., Black, H.R., Brunner, R.L., Brzyski, R.G., Caan, B., Chlebowski, R.T., Gass, M., Granek, I., Greenland, P., Hays, J., Heber, D., Heiss, G., Hendrix, S.L., Hubbell, F.A., Johnson, K.C., Kotchen, J.M. (2006) Low-fat dietary pattern and risk of cardiovascular disease: The Women's Health Initiative randomized controlled dietary modification trial. *JAMA* **295**, 655–66.
26. Keys, A. (1983) From Naples to seven countries-a sentimental journey. *Prog Biochem Pharmacol* **19**, 1–30.
27. Oh, K., Hu, F.B., Manson, J.E., Stampfer, M.J., Willett, W.C. (2005) Dietary fat intake and risk of coronary heart disease in women: 20 years of follow-up of the Nurses' Health Study. *Am J Epidemiol* **161**, 672–9.

28. Liu, S., Stampfer, M.J., Hu, F.B., et al. (1999) Whole grain consumption and risk of coronary heart disease: results from the Nurses' Health Study. *Am J Clin Nutr* **70**, 412–9.
29. Liu, S., Manson, J.E., Lee, I.-M., et al. (2000) Fruit and vegetable intake and risk of cardiovascular disease: the Women's Health Study. *Am J Clin Nutr* **72**, 922–8.
30. Fung, T.T., Stampfer, M.J., Manson, J.E., Rexrode, K.M., Willett, W.C., Hu, F.B. (2004) Prospective study of major dietary patterns and stroke risk in women. *Stroke* **35**, 2014–9.
31. Kris-Etherton, P.M., Harris, W.S., Appel, L.J. (2002) American Heart Association Nutrition Committee. Fish consumption, fish oil, omega-3 fatty acids, and cardiovascular disease [published correction appears in Circulation. 2003;**107**:512]. *Circulation* **21**, 2747–57.
32. Dayton, S., Pearce, M.L., Hashimoto, S., Cixon, W.J., Tomlyasu, U. (1969) A controlled trial of a diet high in unsaturated fat for preventing complications of atherosclerosis. *Circulation* **60**, S111–63.
33. Leren, P. (1970) The Oslo Diet–Heart Study: eleven year report. *Circulation* **42**, 935–42.
34. Turpeinen, O., Karvonen, M.J., Pekkarinen, M., Miettinen, M., Elosuo, R., Paavilainen, E. (1979) Dietary prevention of coronary heart disease: the Finnish Mental Hospital Study. *Int J Epidemiol* **8**, 99–118.
35. de Lorgeril, M., Salen, P., Martin, J.L., Monjaud, I., Delaye, J., Mamelle, N. (1999) Mediterranean diet, traditional risk factors and the rate of cardiovascular complications after myocardial infarction: final report of the Lyon Diet Heart Study. *Circulation* **99**, 779–85.
36. de Lorgeril, M., Renaud, S., Mamelle, N., et al. (1994) Mediterranean alpha-linolenic acid-rich diet in secondary prevention of coronary heart disease [published correction appears in Lancet. 1995;**345**:738]. *Lancet* **343**, 1454–9.
37. Ornish, D., Scherwitz, L.W., Billings, J.H., et al. (1998) Intensive lifestyle changes for reversal of coronary heart disease. *JAMA* **80**, 2001–7.
38. Hays, J., Hunt, J.R., Hubbell, A., Anderson, G.L., Limacher, M., Allen, C., Rossouw, J.E. (2003) The Women's Health Initiative recruitment methods and results. *Ann Epidemiol* **13**, S18–77.
39. Anderson, G.L., Manson, J., Wallace, R., Lund, B., Hall, D., Davis, S., Shumaker, S., Wang, C.Y., Stein, E., Prentice, R.L. (2003) Implementation of the Women's Health Initiative Study Design. *Ann Epidemiol* **13**, S5–17.
40. National Heart, Lung and Blood Institute, WHI Scientific Results at: https:// www.nhlbi.nih.gov/.whi / Accessed January 26, 2009 24, 2008.
41. Ritenbaugh, C., Patterson, R.E., Chlebowski, R.T., Caan, B., Fels-Tinker, L., Howard, B., Ockene, J. (2003) The women's health initiative dietary modification trial: overview and baseline characteristics of participants. *Ann Epidemiol* **13**, S87–97.
42. Beresford, S.A., Caan, B.J., Tinker, L. Women's health initiative: dietary results and clinical outcomes, Food and Nutrition Conference and Exposition, Honolulu, HI, September, 2006.
43. Tinker, L.F., Burrows, E.R., Henry, H., Patterson, R., Rupp, J., Van Horn, L. (1996) The Women's Health Initiative: overview of the nutrition components. In: Krummel, D.A., Kris-Etherton, P.M., eds. *Nutrition and women's health.* Gaithersburg, MD: Aspen Publishers, pp. 510–42.
44. Patterson, R.E., Kristal, A.R., Tinker, L.F., Carter, R.A., Bolton, M.P., Agurs-Collins, T. (1999) Measurement characteristics of the Women's Health Initiative food frequency questionnaire. *Ann Epidemiol* **9**, 178–87.
45. Curb, J.D., McTiernan, A., Heckbert, S.R., Kooperber, C., Stanford, J., Nevitt, M., Johnson, K.C., Proulx-Burns, L., Pastore, L., Criqui, M., Daugherty, S. (2003) Outcomes ascertainment and adjudication methods in the Women's Health Initiative. *Ann Epidemiol* **13**, S122–8.
46. Women's Health Initiative Study Group. (2004) Dietary adherence in the Women's Health Initiative Dietary Modification Trial. *J Am Dent Assoc* **104**, 654–8.
47. Patterson, R.E., Kristal, A., Rodabough, R., Caan, B., Lillington, L., Mossavar-Rahmani, Y., Simon, M.S., Snetselaar, L., Van Horn, L. (2003) Changes in food sources of dietary fat in response to an intensive low-fat dietary intervention: early results from the Women's Health Initiative. *J Am Dent Assoc* **103**, 454–60.
48. Kearney, M.H., Rosal, M.C., Ockene, J.K., Churchill, L.C. (2002) Influences on older women's adherence to a low-fat diet in the Women's Health Initiative. *Psychosom Med* **64**, 450–7.
49. Hsia, J., Rodabough, R., Rosal, M.C., Cochrane, B., Howard, B.V., Snetselaar, L., Frishman, W.H., Stefanick, M.L. (2002) Compliance with National Cholesterol Education Program dietary and lifestyle guidelines among older women with self-reported hypercholesterolemia. The Women's Health Initiative. *Am J Med* **113**, 384–92.
50. Hopkins, S., Burrows, E., Bowen, D.J., Tinker, L.F. (2001) Differences in eating pattern labels between maintainers and nonmaintainers in the Women's Health Initiative. *J Nutr Educ* **33**, 278–83.
51. Mossavar-Rahmani, Y., Henry, H., Rodabough, R., Bragg, C., Brewer, A., Freed, T., Kinzel, L., Pedersen, M., Soule, C.O., Vosburg, S. (2004) Additional self-monitoring tools in the dietary modification component of the Women's Health Initiative. *J Am Dent Assoc* **104**, 76–85.

52. Tinker, L.F., Perri, M.G., Patterson, R.E., Bowen, D.J., McIntosh, M., Parker, L.M., Sevick, M.A., Wodarski, L.A. (2002) The effects of physical and emotional status on adherence to a low-fat dietary pattern in the Women's Health Initiative. *J Am Dent Assoc* **102**, 789–800.
53. Tinker, L.F., Rosal, M.C., Young, A.F., Perri, M.G., Patterson, R.E., Van Horn, L., Assaf, A.R., Bowen, D.J., Ockene, J., Hays, J., Wu, L.(2007) Predictors of dietary change and maintenance in the Women's Health Initiative Dietary Modification Trial. *J Am Dent Assoc* **107**, 1155–65.
54. Howard, B.V., Manson, J.E., Stefanick, M.L., Beresford, S.A., Frank, G., Jones, B., Rodabough, R.J., Snetselaar, L., Thomson, C.A., Tinker, L., Vitolins, M., Prentice, R. (2006) Low-fat dietary pattern and weight change over 7 years: The Women's Health Initiative Dietary Modification Trial. *JAMA* **295**, 39–49.
55. Caan, B., Neuhouser, M., Aragaki, A., Lewis, C.B., Jackson, R., LeBoff, M.S., Margolis, K.L., Powell, L., Uwaifo, G., Whitlock, E., Wylie-Rosett, J., LaCroix, A. (2007) Calcium plus vitamin D supplementation and the risk of postmenopausal weight gain. *Arch Intern Med* **167**, 893–902.
56. Prentice, R.L., Anderson, G.L. (2007) The Women's Health Initiative: Lesson's Learned. *Annual Rev Public Health* **29**, 131–50.
57. Freedman, L.S., Potischman, N., Kipnis, V., Midthune, D., Schatzkin, A., Thompson, F.E., Troiano, R.P., Prentice, R., Patterson, R., Carroll, R., Subar, A.F. (2006) A comparison of two dietary instruments for evaluating the fat–breast cancer relationship. *Int J Epidemiol* **35**, 1011–21.
58. Prentice, R.L., Thomson, C.A., Caan. B., Hubbell, F.A., Anderson, G.L., Beresford, S.A., Pettinger, M., Lane, D.S., Lessin, L., Yasmeen, S., Singh, B., Khandekar, J., Shikany, J.M., Satterfield, S., Chlebowski, R.T. (2007) Low-fat dietary pattern and cancer incidence in the Women's Health Initiative Dietary Modification Randomized Controlled Trial. *J Natl Cancer Inst* **99**, 1534–43.
59. Margolis, K.L., Ray, R.M., Van Horn, L., Manson, J.E., Allison, M.A., Black, H.R., Beresford, S.A., Connelly, S.A., Curb, J.D., Grimm, R.H., Kotchen, T.A., Kuller, L.H., Wassertheil-Smoller, S., Thomson, C.A., Torner. J.C., for the Women's Health Initiative Investigators. (2008) Effect of Calcium and Vitamin D Supplementation on Blood Pressure. The Women's Health Initiative Randomized Trial. *Hypertension* **52**, 847–55.
60. Neuhouser, M.L., Tinker, L., Shaw, P.A., Schoeller, D., Bingham, S.A., Van Horn, L., Beresford, S.A., Caan, B., Thomson, C.A., Satterfield, S., Kuller, L., Heiss, G., Smit, E., Sarto, G., Ockene, J., Stefanick, M.L., Assaf, A., Runswick, S., Prentice, R.L. (2008) Use of recovery biomarkers to calibrate nutrient consumption self-reports in the Women's Health Initiative. *Am J Epidemiol* **167**, 1247–59.
61. Manson, J.E., Hsia, J., Johnson, K.C., Rossouw, J.E., Assaf, A.R., Lasser, N.L., Trevisan, M., Black, H.R., Heckbert, S.R., Detrano, R., Strickland, O.L., Wong, N.D., Crouse, J.R., Stein, E., Cushman, M.; Women's Health Initiative Investigators. (2003) Estrogen plus progestin and the risk of coronary heart disease. *N Engl J Med* **349**(6), 523–34
62. Chlebowski, R.T., Hendrix, S.L., Langer, R.D., Stefanick, M.L., Gass, M., Lane, D., Rodabough, R.J., Gilligan, M.A., Cyr, M.G., Thomson, C.A., Khandekar, J., Petrovitch, H., McTiernan, A.; WHI Investigators. (2003) Influence of estrogen plus progestin on breast cancer and mammography in healthy postmenopausal women: The Women's Health Initiative Randomized Trial. *JAMA* **89**(24), 3243–53.
63. Grundy, S.M., Cleeman, J.I., Merz, N.B., Brewer, H.T., Clark, L.B., Hunninghake, D.B., Pasternak, R.C., Smith, S.C., Stone, N.J.; for the Coordinating Committee of the National Cholesterol Education Program. (2004). Implications of recent clinical trials for the National Cholesterol Education Program Adult treatment Panel III Guidelines. *Circulation* **110**, 227–39.
64. Lewiecki, E.M. (2008) Prevention and treatment of postmenopausal osteoporosis. *Obstet Gynecol Clin North Am* **35**(2), 301–15.
65. Jackson, R.D. and Shidham, S. (2007) The role of hormone therapy and calcium plus vitamin D for reduction of bone loss and risk for fractures: lessons learned from the Women's Health Initiative. *Curr Osteoporos Rep* **5**, 153–9.
66. Dawson-Hughes, B., Bischoff-Ferrari, H.A. (2007) Therapy of osteoporosis with calcium and vitamin D. *J Bone Miner Res* **22**(Suppl. 2), V59–63.
67. Brunner, R.L., Cochrane, B., Jackson, R.D., Larson, J., Lewis, C., Limacher, M., Rosal. M., Shumaker, S., Wallace, R.; Women's Health Initiative Investigators. (2008) Calcium, vitamin D supplementation, and physical function in the Women's Health Initiative. *J Am Dent Assoc* **108**, 1472–9.
68. Ma, Y., Hébert, J.R., Li, W., Bertone-Johnson, E.R., Olendzki, B., Pagoto, S.L., Tinker, L., Rosal, M.C., Ockene, I.S., Ockene, J.K., Griffith, J.A., Liu, S. (2008) Association between dietary fiber and markers of systemic inflammation in the Women's Health Initiative Observational Study. *Nutrition* **24**, 941–9.
69. Cui, Y., Shikany, J.M., Liu, S., Yasmeen, S., Rohan, T.E. (2008) Selected antioxidants and risk of hormone receptor-defined invasive breast cancers among postmenopausal women in the Women's Health Initiative Observational Study. *Am J Clin Nutr* **87**, 1009–18.

70. Kabat, G.C., Shikany, J.M., Beresford, S.A., Caan, B., Neuhouser, M.L., Tinker, L.F., Rohan, T.E. (2008) Dietary carbohydrate, glycemic index, and glycemic load in relation to colorectal cancer risk in the Women's Health Initiative. *Cancer Causes Control* **19**, 1291–98.
71. Wactawski-Wende, J., Kotchen, J.M., Anderson, G.L., Assaf, A.R., Brunner, R.L., O'Sullivan, M.J., Margolis, K.L., Ockene, J.K., Phillips, L., Pottern, L., Prentice, R.L., Robbins, J., Rohan, T.E., Sarto, G.E., Sharma. S., Stefanick, M.L., Van Horn, L., Wallace, R.B., Whitlock, E., Bassford, T., Beresford, S.A., Black, H.R., Bonds, D.E., Brzyski, R.G., Caan, B., Chlebowski, R.T., Cochrane, B., Garland, C., Gass, M., Hays, J., Heiss, G., Hendrix, S.L., Howard, B.V., Hsia, J., Hubbell, F.A., Jackson, R.D., Johnson, K.C., Judd, H., Kooperberg, C.L., Kuller, L.H., LaCroix, A.Z., Lane, D.S., Langer, R.D., Lasser, N.L., Lewis, C.E., Limacher, M.C., Manson, J.E.(2006) Women's Health Initiative Investigators. Calcium plus vitamin D supplementation and the risk of colorectal cancer. *N Engl J Med* **354**, 684–96.
72. Hsia, J., Heiss, G., Ren, H., Allison, M., Dolan, N.C., Greenland, P., Heckbert, S.R., Johnson, K.C., Manson, J.E., Sidney, S., Trevisan, M., Women's Health Initiative Investigators. (2007) Calcium/vitamin D supplementation and cardiovascular events. *Circulation* **115**, 846–54.
73. Jackson, R.D., LaCroix, A.Z., Gass, M., Wallace, R.B., Robbins, J., Lewis, C.E., Bassford, T., Beresford, S.A., Black, H.R., Blanchette, P., Bonds, D.E., Brunner, R.L., Brzyski, R.G., Caan, B., Cauley, J.A., Chlebowski, R.T., Cummings, S.R., Granek, I., Hays, J., Heiss, G., Hendrix, S.L., Howard, B.V., Hsia, J., Hubbell, F.A., Johnson, K.C., Judd, H., Kotchen, J.M., Kuller, L.H., Langer, R.D., Lasser, N.L., Limacher, M.C., Ludlam, S., Manson, J.E., Margolis, K.L., McGowan, J., Ockene, J.K., O'Sullivan, M.J., Phillips, L., Prentice, R.L., Sarto, G.E., Stefanick, M.L., Van Horn, L., Wactawski-Wende, J., Whitlock, E., Anderson, G.L., Assaf, A.R., Barad, D.; Women's Health Initiative Investigators. (2006) Calcium plus vitamin D supplementation and the risk of fractures. *N Engl J Med* **354**, 669–83.
74. Tinker, L.F., Bonds, D.E., Margolis, K.L., Manson, J.E., Howard, B.V., Larson, J., Perri, M.G., Beresford, S.A., Robinson, J.G., Rodriguez, B., Safford, M.M., Wenger, N.K., Stevens, V.J., Parker, L.M. (2008) Low-fat dietary pattern and risk of treated diabetes mellitus in postmenopausal women: the Women's Health Initiative randomized controlled dietary modification trial. *Arch Intern Med* **168**, 1500–11.
75. de Boer, I.H., Tinker, L.F., Connelly, S., Curb, J.D., Howard, B.V., Kestenbaum. B., Larson, J.C., Manson, J.E., Margolis, K.L., Siscovick, D.S., Weiss, N.S.; Women's Health Initiative Investigators. (2008) Calcium plus vitamin D supplementation and the risk of incident diabetes in the Women's Health Initiative. *Diabetes Care* **31**, 701–7.
76. Luo, J., Margolis, K.L., Adami, H.O., LaCroix, A., Ye, W.; Women's Health Initiative Investigators. (2008) Obesity and risk of pancreatic cancer among postmenopausal women: the Women's Health Initiative (United States). *Br J Cancer* **99**, 527–31.
77. Kabat, G.C., Kim, M., Hunt, J.R., Chlebowski, R.T., Rohan, T.E. (2008) Body mass index and waist circumference in relation to lung cancer risk in the Women's Health Initiative. *Am J Epidemiol* **168**, 158–69.
78. Luo, J., Margolis, K.L., Adami, H.O., Lopez, A.M., Lessin, L., Ye, W.; for the Women's Health Initiative Investigators. (2007) Body size, weight cycling, and risk of renal cell carcinoma among postmenopausal women: The Women's Health Initiative (United States). *Am J Epidemiol* **166**,752–9.
79. Morimoto, L.M., White, E., Chen, Z., Chlebowski, R.T., Hayes, J., Kuller, L., Lopez, A.M., Manson, J., Margolis, K.L., Muti, P.C., Stefanick, M.L., McTiernan, A. (2002) obesity, body size and risk of post-menopausal breast cancer: the Women's Health Initiative (United States). *Cancer Causes Control* **13**, 741–51.
80. McTiernan, A., Wu, L., Chen, C., Chlebowski, R., Mossavar-Rahmani, Y., Modugno, F., Perri, M.G., Stanczyk, F.Z., Van Horn, L., Wang, C.Y., Women's Health Initiative Investigators. (2006) Relation of BMI and physical activity to sex hormones in postmenopausal women. *Obesity (Silver Spring)* **14**, 1662–77.
81. McTigue, K., Larson, J.C., Valoski, A., Burke, G., Kotchen, J., Lewis, C.E., Stefanick, M.L., Van Horn, L., Kuller, L. (2006) Mortality and cardiac and vascular outcomes in extremely obese women. *JAMA* **296**, 79–86.
82. Howard, B.V., Adams-Campbell, L., Allen, C., Black, H., Passaro, M., Rodabough, R.J., Rodriguez, B.L., Safford, M., Stevens, V.J., Wagenknecht, L.E. (2004) Insulin resistance and weight gain in postmenopausal women of diverse ethnic groups. *Int J Obes Relat Metab Disord* **28**, 1039–47.

15 Role of Nutrition in the Pathophysiology, Prevention, and Treatment of Type 2 Diabetes and the Spectrum of Cardiometabolic Disease

Cristina Lara-Castro and W. Timothy Garvey

Key Points

- The overlapping syndromes of prediabetes and metabolic syndrome place individuals at increased risk of Type 2 diabetes and cardiovascular disease, and in aggregate, these disorders comprise the spectrum of cardiometabolic disease. Treatment of these prediabetic states is critical for reducing patient suffering and social costs attributable to the increasing prevalence of diabetes worldwide.
- Insulin resistance is central to the pathophysiology of Type 2 diabetes and metabolic syndrome. While obesity can exacerbate insulin resistance, it is abnormal lipid accumulation in visceral adipose tissue, and in skeletal muscle cells and hepatocytes, that appear to be more potent and independent mediators of cardiometabolic disease.
- Cardiometabolic disease can be effectively treated or prevented using nutrition as a component of lifestyle therapy, which, given the underlying pathophysiology, will need to augment insulin sensitivity, enhance insulin secretion, and/or ameliorate cardiovascular risk factors. This can be accomplished via hypocaloric feeding or altered macronutrient composition of the diet.
- Hypocaloric diets resulting in 5–10% weight loss are effective over a wide spectrum of caloric composition ranging from low carbohydrate to low fat.
- Isocaloric diets can also improve insulin sensitivity and cardiovascular risk factors particularly if enriched in monounsaturated fat or fiber with reduced intake of saturated fat.
- Genome-wide association studies have confirmed multiple susceptibility loci (i.e., gene-based single nucleotide polymorphisms—SNPs) for Type 2 diabetes and obesity and present a powerful paradigm for the identification and study of nutrient–gene interactions, provided that investigators measure diet as an environmental variable.

Key Words: Cardiometabolic disease; fiber; metabolic syndrome; monounsaturated fatty acids; prediabetes; susceptibility genes; Type 2 diabetes; weight loss

15.1. THE BURDEN OF DIABETES, PREDIABETES, AND METABOLIC SYNDROME

Diabetes encompasses a group of complex, chronic, and progressive diseases that are primarily defined on the basis of hyperglycemia. Two major types of diabetes are recognized and include Type 1 diabetes and Type 2 diabetes. Type 1 diabetes, which accounts for ~5% of diabetes cases, is caused mainly by immune-mediated pancreatic B-cell destruction leading to absolute insulin deficiency. Type 2 diabetes, which accounts for the majority of patients (90–95%), features a multifactorial pathogenesis involving defects in both insulin action and insulin secretion as a result of a complex interaction of genetic and environmental influences. The prevalence of diabetes has increased

A. Bendich, R.J. Deckelbaum (eds.), *Preventive Nutrition*, Nutrition and Health, DOI 10.1007/978-1-60327-542-2_15,

Table 15.1
Criteria for Diagnosis of Prediabetes and Diabetes Mellitus

- Normoglycemia
  - o Fasting glucose <100 mg/dL (<5.6 mM)
  - o 2-h postchallenge <140 mg/dL (<7.8 mM)
- Abnormal glucose tolerance—prediabetes
  - 1] Impaired fasting glucose (IFG)
    - o Fasting glucose 100–125 mg/dL (5.6–.0 mM)
  - 2] Impaired glucose tolerance (IGT)
    - o 2-h postchallenge 140–199 mg/dL (7.8–11.1 mM)
- Diabetes mellitus
  - o Fasting glucose ≥ 126 mg/dL (7.0 mM)
  - o 2-h postchallenge ≥ 200 mg/dL (11.1 mM)

Postchallenge indicates 75 g oral glucose. Note that metabolic syndrome (*see* Table 15.2) is also a prediabetic state.

in industrialized and developing countries alike; recent estimates indicate that there were 171 million individuals with diabetes in the world in the year 2000 and that this number is projected to increase to 366 million in 2030. In the United States alone, about 23.6 million people have diabetes, and, of those, 17.9 million are diagnosed and 5.7 million are undiagnosed *(1)*. This overall increase in diabetes prevalence primarily reflects an increase in Type 2 diabetes due in part to the epidemic increase in overweight and obesity occurring in world populations over that same time period. This is a concerning trend, given that diabetes imposes a significant public health burden and large resource demands on health-care systems. Diabetes is the main cause of kidney failure, nontraumatic limb amputation, and new-onset blindness in US adults, and patients with diabetes have cardiovascular disease rates and stroke incidence about two to four times higher than adults without diabetes.

The progression to overt diabetes is heralded by a transitional prediabetic state intermediate between normal glycemia and the physiology of healthy individuals and the hyperglycemia and aberrant metabolism that characterizes diabetes. The prediabetic state is defined by one or both of two diagnostic entities: (i) prediabetes is defined on the basis of elevated fasting and 2-h postprandial glucose concentrations (*see* Table 15.1) and (ii) metabolic syndrome is a trait complex that underscores the fact that the pathophysiology of diabetes involves much more than abnormal glucose homeostasis (*see* Table 15.2). Both prediabetes and metabolic syndrome markedly increase risk of future diabetes and also greatly augment the risk for cardiovascular disease even in those individuals who do not go on to develop frank diabetes *(2)*. The progression of abnormal glucose tolerance and metabolic syndrome to Type 2 diabetes and cardiovascular disease is schematically depicted in Fig. 15.1. In order to reduce the burden of diabetes, action must be taken to prevent the development of prediabetes and metabolic syndrome, as well as the progression of these conditions to diabetes.

15.2. DIABETES PROGRESSION

15.2.1. Prediabetes and Abnormalities of Carbohydrate Metabolism

It is important to identify subjects at risk for diabetes because lifestyle therapy and pharmacological interventions are able to effectively prevent (or delay) Type 2 diabetes. The prediabetic state results

Table 15.2
Clinical Criteria for Diagnosis of Metabolic Syndrome

(A) NCEP Adult Treatment Panel III (ATP III)
Three or more of the following risk factors:

Risk factor	*Defining Level*
• Waist circumference	
Men	>102 cm (>40 in.)
Women	>88 cm (>35 in.)
• Triglycerides	≥150 mg/dL (≥1.7 mmol/L)
• HDL cholesterol	
Men	<40 mg/dL (1.03 mmol/L)
Women	<50 mg/dL (1.29 mmol/L)
• Blood pressure	≥130/≥85 mmHg
• Fasting glucose	≥100 mg/dL (5.6 mM)

(B) World Health Organization
Insulin resistance, identified by one of the following:
- Type 2 diabetes
- Impaired fasting glucose (IFG)
- Impaired glucose tolerance (IGT)

Plus any two of the following:
- Antihypertensive medication and/or high blood pressure (≥140 mmHg systolic or ≥90 mmHg diastolic)
- Plasma triglycerides ≥150 mg/dL (≥1.7 mmol/L)
- HDL cholesterol <35 mg/dL (<0.9 mmol/L) in men or <39 mg/dL (<1.0 mmol/L) in women
- BMI >30 kg/m^{2}and/or waist:hip ratio >0.9 in men and >0.85 in women
- Urinary albumin excretion rate ≥20 μg/min or albumin:creatinine ratio ≥30 mg/g

(C) International Diabetes Federation 2005
Central obesity, race/ethnicity specific
- Waist circumference ≥94 cm for men of European descent and ≥80 cm for women of European descent; ≥90 cm for men of southeast Asia and ≥80 cm for women of southeast Asia; ≥85 cm for men of Japan and ≥90 cm for women of Japan.

Plus any two of the following four factors:
- Raised triglyceride level: ≥150 mg/dL (1.7 mmol/L), or specific treatment for this lipid abnormality
- Reduced HDL cholesterol: <40 mg/dL (1.03 mmol/L) in males and <50 mg/dL (1.29 mmol/L) in females, or specific treatment for this lipid abnormality
- Raised blood pressure: systolic BP ≥ 130 mmHg or diastolic BP ≥ 85 mmHg, or treatment of previously diagnosed hypertension
- Raised fasting plasma glucose ≥100 mg/dL (5.6 mmol/L), or previously diagnosed Type 2 diabetes.

from defects in both insulin action and insulin secretion, which are the same abnormalities that sustain hyperglycemia in Type 2 diabetes. There exists a high degree of individual variability in insulin sensitivity in nondiabetic populations, and relative insulin resistance is a trait that is detectable early in life, even in childhood. Glucose tolerance remains normal as long as the pancreatic β cells can mount exaggerated insulin secretory responses that compensate for insulin resistance through hyperinsulinemia. However, in certain individuals, the β cells begin to fail after years of metabolic stress and can no longer fully compensate for the degree of prevailing insulin resistance. The exhaustion in β cells involves both a defect in glucose sensing necessary for the insulin secretory response and a diminution in β-cell mass. These individuals develop abnormal glucose tolerance with glucose levels that do not meet criteria for diabetes but are nevertheless too high to be considered normal. There are two types of abnormal glucose tolerance, and either or both qualify individuals as having prediabetes. Patients with impaired fasting glucose (IFG) have fasting glucose levels but <126 mg/dL (7.0 mmol/L), and patients with impaired glucose tolerance (IGT) have 2-h values during a 75-g oral glucose tolerance test (OGTT) of 140 mg/dL (7.8 mmol/L) but <200 mg/dL (11.1 mmol/L). (Note that confirmed fasting glucose values of 126 mg/dL or greater and/or 2-h values of 200 mg/dL or more would qualify individuals for the diagnosis of diabetes). Table 15.1outlines the diagnostic criteria for prediabetes advocated by the American Diabetes Association *(3)*. Prediabetes is becoming more common worldwide, and the US Department of Health and Human Services estimates that about one in four US adults aged 20 years or older—or 57 million people—had prediabetes in 2007. Up to half of patients with prediabetes are likely to develop Type 2 diabetes within 10 years, unless they take steps to prevent or delay diabetes. Thus it is critically important to develop optimal strategies for the management of prediabetes and the prevention of Type 2 diabetes.

15.2.2. Insulin Resistance, Obesity, and Metabolic Syndrome

As discussed above, insulin resistance antedates the development of diabetes and helps maintain the diabetes state. In addition to placing increased demands on pancreatic β cells, relative insulin resistance is associated with multiple clinical and anthropometric traits, a trait cluster referred to as metabolic syndrome (also known as insulin resistance syndrome, syndrome X). Metabolic syndrome is the result of a pathophysiological process responsible for the co-occurrence of multiple cardiometabolic risk factors that may include elevated blood pressures, abnormal glucose intolerance, upper body fat distribution, generalized obesity, dysfibrinolysis, inflammation, and dyslipidemia characterized by high triglycerides, low HDL cholesterol, and small dense LDL particles. Metabolic syndrome places individuals at increased risk, not only for future Type 2 diabetes but also for atherosclerotic diseases. In order to identify patients for more aggressive risk factor management and cardiovascular disease surveillance, diagnostic criteria for metabolic syndrome have been devised as delineated in Table 15.2 *(4–6)*. These diagnostic schemes have been proposed by (i) the Adult Treatment Panel III (ATPIII) of the National Cholesterol Education Project, which reflects a relative emphasis on cardiovascular disease risk factors and is most widely used in the United States; (ii) the World Health Organization, which requires the presence of abnormal glucose tolerance; and (iii) the International Diabetes Federation, which takes into account racial/ethnic differences in the waist circumference associated with cardiometabolic risk. It is important to consider, however, that these criteria exhibit poor sensitivity for identifying patients with insulin resistance and dyslipidemia, and do not take into account the impact of age, race, and ethnicity *(7,8)*. Furthermore, the pathophysiological process responsible for metabolic syndrome can be operative, placing individuals at increased risk of diabetes and cardiovascular disease, even when diagnostic criteria are not fully satisfied. Nevertheless, patients who meet defined criteria for the diagnosis of metabolic syndrome or prediabetes are clearly at increased risk for future diabetes and cardiovascular disease, and these diagnostic entities are valuable tools for identifying patients most deserving of aggressive interventions to prevent cardiometabolic disease. Indeed, cardiometabolic risk is further augmented in patients who have a combination of any two or more of IFG, IGT, or metabolic syndrome, compared with those who have only one of these diagnostic entities. For this reason, the American College of Endocrinologists (ACE) has singled out patients with two or more of IFG, IGT, and metabolic syndrome for more aggressive lifestyle intervention and the

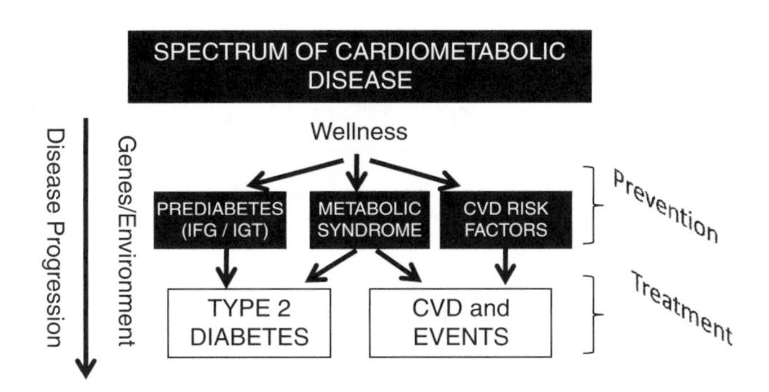

Fig. 15.1 The spectrum of progression of cardiometabolic disease.

consideration of antidiabetic drug therapy *(2)*. This spectrum of cardiometabolic disease is illustrated Fig. 15.1.

Obesity is associated with insulin resistance. A positive relationship has been observed between measures of general adiposity such as body mass index (BMI) and measures of insulin resistance such as maximally stimulated glucose uptake rates using the hyperinsulinemic euglycemic clamp *(9)*. However, there is a high degree of individual variability such that lean or obese individuals can be either relatively insulin sensitive or insulin resistant. In fact, the correlation coefficient (i.e., R^2) associating BMI with clamp measures of insulin sensitivity indicates that only ~10% of individual variability in insulin sensitivity can be explained by differences in BMI *(10)*. Thus, while obesity is an important factor exacerbating insulin resistance, individual variation in insulin sensitivity largely exists independent of the degree of generalized obesity. A more important factor is where fat is located. Accumulation of fat in the omental or the visceral compartment (i.e., upper body fat distribution) and accumulation of intracellular triglyceride in muscle cells and in hepatocytes have been associated with insulin resistance, independent of the degree of general adiposity. Multiple authors have demonstrated that relative redistribution of fat to the omental compartment has profound adverse effects on insulin sensitivity, although increased abdominal subcutaneous fat also contributes to insulin resistance. These observations are relevant to our current understanding of metabolic syndrome pathophysiology and the development of the cardiometabolic trait cluster. Accumulation of omental lipid alters secretion of multiple adipokines and proinflammatory cytokines from adipose tissue, including adiponectin, resistin, leptin, free fatty acids, PAI-1, TNF-α, IL-6, MCP-1, and others. These factors enter the circulation and act as hormones to alter metabolism in multiple tissues. For example, adiponectin can enhance insulin action in skeletal muscle and suppress foam cell formation; thus, the decline in adiponectin with omental fat accumulation likely contributes to insulin resistance in muscle and accelerated atherogenesis in the vascular wall *(11,12)*. Increased circulating concentrations of free fatty acids induce insulin resistance in muscle and augment vascular reactivity to endogenous pressors. Increased release of MCP-1 and other factors contributes to macrophage infiltration in adipose tissue, and the resulting inflammation is associated with increased systemic release of proinflammatory cytokines from adipose tissue that could contribute to insulin resistance and vascular wall inflammation *(13)*. In this way, omental adipose tissue could orchestrate the development of the cardiometabolic trait complex via effects of the secreted adipose tissue factors on key organ systems relevant to metabolism and atherogenesis. The evidence suggests that these secreted factors induce skeletal muscle insulin resistance and may constitute a mechanistic link between increased ventral fat, insulin resistance, and the expression of metabolic syndrome.

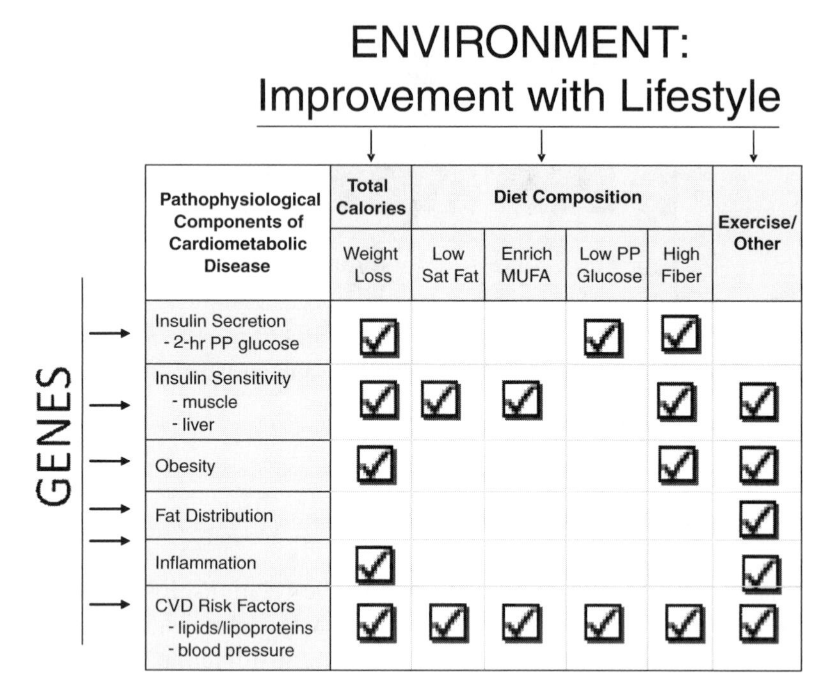

Fig. 15.2 Differential effects of lifestyle changes (environment) on pathophysiological components of cardiometabolic disease.

15.3. NUTRITIONAL STRATEGIES FOR THE TREATMENT OF PREDIABETIC STATES AND THE PREVENTION OF TYPE 2 DIABETES

Given the pathophysiology of prediabetes and metabolic syndrome, preventive nutrition therapy will be effective in halting the progression to Type 2 diabetes to the extent that the diet enhances insulin secretory capacity, ameliorates insulin resistance, and/or improves the cardiometabolic risk cluster profile. These traits are genetically determined, and nutrient–gene interactions will determine the efficacy of diet in ameliorating these processes. This chapter will consider these aspects of nutrition therapy, including effects of hypocaloric diets that produce weight loss, alterations in macronutrient composition (fat versus carbohydrate), changes in the types of dietary fat or carbohydrate consumed, as well as a short discussion on specific medications that act through nutrition-related mechanisms and have been proven useful in the prevention of Type 2 diabetes. The ability of diet to reverse metabolic defects and cardiovascular risk factors is depicted in Fig. 15.2.

15.3.1. Weight Loss

15.3.1.1. Hypocaloric Diets

Both acute energy restriction and chronic weight loss have been shown to improve glucose metabolism in insulin-resistant and Type 2 diabetic individuals. Sustained weight loss of 5–10% of initial body weight can significantly decrease fasting blood glucose, insulin levels, hemoglobin A1c concentrations, and medication requirements in obese patients with Type 2 diabetes. In nondiabetic individuals who are overweight and insulin resistant, both hyperinsulinemia and insulin resistance substantially improve following weight loss, whereas in insulin-sensitive overweight individuals, lesser improvement may be observed. Enhanced insulin sensitivity following weight loss is partially related to the loss of total fat and highly correlated with the loss of visceral and intramyocellular fat *(10,14)*. In these studies, it is important to emphasize that negative energy balance produces weight loss regardless of the macronutrient composition of the diet (i.e., low-carbohydrate versus low-fat diets) *(15–17)*. While various diet plans can emphasize factors that affect hunger, satiety, and adherence, caloric reduction is the essential component and a sine qua non of the weight loss.

When considering the metabolic benefits of diet and weight loss, lifestyle intervention resulting in modest weight loss can dramatically prevent progression toward Type 2 diabetes in high-risk individuals. This has been consistently demonstrated in several lifestyle intervention trials conducted in different populations. The Finnish Diabetes Prevention Study *(18)* was one of the first controlled, randomized studies to show that Type 2 diabetes is preventable with lifestyle intervention. In this study, 522 middle-aged, overweight subjects with IGT were randomized to a standard-care control group or an intensive lifestyle intervention group for 3 years. The intervention goals were to reduce body weight by 5%, reduce dietary to <30% of total calories and saturated fat to <10% of calories, increase dietary fiber to 15 g/1,000 kcal, and promote moderate-intensity physical activity for 30 min/day. The risk of diabetes was reduced by 58% in the intensive lifestyle intervention group compared with the control group. These results have been reproduced by the Diabetes Prevention Program (DPP), in which 3,234 subjects with abnormal glucose tolerance were randomized to a control group, metformin treatment, or a lifestyle intervention comparable to that on the Finnish study *(19)*. Over a 4-year period, lifestyle intervention produced a modest degree of weight loss (5–7%) and a 58% reduction in the progression to Type 2 diabetes compared with the control group. Lifestyle intervention was also superior to metformin therapy for diabetes prevention. Importantly, the efficacy of lifestyle changes in decreasing incident diabetes was similar in Caucasians, African-Americans, Hispanic Americans, Asian-Americans, and Native Americans. A third lifestyle intervention was conducted in Asians. The Da Qing study *(20)* recruited 577 people with IGT who were relatively lean (mean BMI 25.8 kg/m^2) in contrast to the more obese patients enrolled in the Finnish study or the DPP. Participants in this study were randomized to one of four intervention protocols (diet alone, exercise alone, diet–exercise combined, or none). The 6-year incidence of Type 2 diabetes was lower in the three intervention groups (41–46%) compared with the control group (68%). The lifestyle intervention in all three of these important studies featured a reduced-calorie diet that was low in total and saturated fat, and emphasized high-fiber foods and fresh fruits and vegetables. Although the independent contributions of diet and exercise in the prevention of Type 2 diabetes were not rigorously assessed, the relative degree of diabetes prevention in all three studies was directly correlated with the degree of sustained weight loss.

15.3.1.2. The Complex Relationship between Obesity and Insulin Resistance: A Mechanistic Hypothesis

We have established above that general obesity can explain only a delimited portion of individual variability in insulin sensitivity in cross-sectional studies. This would seem to be at odds with the observation that weight loss consistently leads to clinically significant improvements in insulin resistance, as well as the strong epidemiological link between obesity and the development of Type 2 diabetes *(21)*. One hypothesis that could resolve this paradox is that it is not changes in general obesity per se that largely mediate benefits of weight loss or the epidemiological association with diabetes; rather, some other metabolic concomitant of the obese state is directly responsible. The fact that only moderate weight reduction, as opposed to achieving ideal body weight, yields maximal metabolic benefits of weight loss already hints at this possibility since this represents a disconnect between degree of obesity and insulin sensitivity.

There are two possible mechanistic explanations for the apparent paradox. First, changes in omental fat mass and functional adipokine secretion may directly influence insulin sensitivity in a manner that is only indirectly related to overall adiposity. Second, fat accumulation in other compartments such as intramyocellular or intrahepatocellular fat has been shown to be highly correlated with insulin resistance *(22)*, although the mechanisms underlying this interrelationship have not been elucidated. It remains possible that fluctuations in these ectopic depots, which may or may not track with overall adiposity in different clinical settings, are a direct mediator of insulin resistance. To support this contention, we have shown that a short-term hypocaloric state (i.e., 3–5 days on a very-low-calorie diet) is sufficient to cause significant improvements in insulin sensitivity associated with dramatic ~50% decrements in intramyocellular lipid, yet accompanied by only small changes in body weight *(10)*. Similarly, rapid improvements in glucose tolerance are observed following bariatric surgery, before any significant reductions in fat mass are realized. Thus, the relationship between insulin sensitivity and obesity is complex, and determinants other than changes in general adiposity may be in large part responsible for diet-induced improvements in insulin resistance.

15.3.2. Alterations in Macronutrient Composition: Low-Fat Versus High-Fat Diets

Changes in macronutrient composition have been used to promote weight loss and enhance insulin sensitivity, independent of an overall reduction in calorie ingestion. During the last decade, popular low-carbohydrate diets in particular have been widely used and received a great deal of attention in the popular press. These diets limit the amount and composition of carbohydrates and are characterized by an increase in dietary fat, to achieve a degree of unintentional calorie reduction through a blunting of the appetite. This can range from a proper mix or "zone" of complex carbohydrates that reduces postprandial serum insulin (the Zone diet, the South Beach diet) to extreme high-fat diets that induce satiety by causing ketogenesis and reducing gastrointestinal motility (the Atkins diet).

Low-carbohydrate (low-carb) diets can also be termed high-fat diets since it is impractical to make up the carbohydrate caloric deficit with dietary protein. Low-carb, high-fat diets were first described by Banting in 1863 *(23)*. Studies evaluating these diets can be difficult to compare since investigators have employed variable duration, diet composition parameters and end points, methodologies (in particular those used to assess insulin sensitivity), and nonrandomized design. Importantly, the composition of dietary fat (saturated versus polyunsaturated and monounsaturated) is not carefully described in most instances. Nevertheless, it is clear that high-fat diets can be used to effectively achieve short-term weight loss. Furthermore, these studies have generally shown no consistent detrimental effects of high-fat intake on insulin sensitivity over a broad range of dietary fat content, including several randomized studies using the hyperinsulinemic glucose clamp technique or frequently sampled intravenous glucose tolerance test (FSIGT) to quantify insulin sensitivity *(24,25)*. One dissenting study by Lovejoy et al. *(26)* used the clamp technique to show that a 3-week high-fat diet (50% fat, 35% carbohydrate, and 15% protein) did induce relative insulin resistance compared with an isocaloric low-fat diet (20% fat, 55% carbohydrate, and 15% protein); however, this could be explained by a higher proportion of saturated fatty acids in the high-fat diet in this study.

Two randomized controlled trials that compared the longer term effects (6 months to 1 year) of traditional low-fat diets versus ad libitum high-fat diets have received a great deal of attention *(15,16)*. Foster et al. *(15)* compared the efficacy of the low-carbohydrate (initially restricted to 20 g carbohydrate/day), high-fat "Atkins" diet against a conventional low-fat (25% of calories), low-calorie diet (1,200–1,500 kcal/day in females and 1,500–1,800 kcal/day in males) in otherwise healthy obese subjects (mean BMI = 34 kg/m^2). The high-fat diet produced a greater weight loss than the low-fat diet after 6 months (6.7 versus 2.7 kg) but at 1 year the amount of weight loss was not significantly different between the two groups (4.3 versus 2.5 kg). About 40% of the 63 randomized subjects did not finish the study. Insulin sensitivity was assessed using the quantitative insulin sensitivity check index, which is an index based on fasting glucose and insulin concentrations. This is a suboptimal measure of insulin sensitivity since the index is affected by insulin secretion, which can vary independently from insulin sensitivity among individuals. Despite this limitation, the index was interpreted to show an increase in insulin sensitivity at 6 months but no change from baseline at 1 year in both dietary subgroups, with no significant differences between the subgroups. LDL and total cholesterol was lower at 3 months in the low-fat diet subgroup, while HDL cholesterol was higher and triglycerides lower at 1 year in the high-fat diet subgroup. In the second study, Samaha et al. *(16)* compared the effects of a low-carbohydrate (30 g/day), high-fat diet versus a low-fat (30 g/day), NHLBI diet designed to create a caloric deficit of 500 kcal/day. Their subjects were severely obese (mean BMI 43 = kg/m^2) and most were African-Americans, hypertensive, and characterized by either Type 2 diabetes or metabolic syndrome. In this 6-month study, subjects on the high-fat diet lost more weight than those on the low-fat diet; however, the amount of weight loss was low (5.8 versus 1.9 kg) and the dropout rate was again very high, particularly in the high-fat diet group (47 versus 33% in the low-fat diet group), indicative of pervasive noncompliance. LDL and HDL cholesterol was not affected by the diets although triglycerides were lowered in the high-fat diet group. The authors also emphasized that the high-fat diet led to greater improvements in insulin sensitivity than the low-fat diet group, but these effects were minimal and the authors again used a suboptimal index based on fasting glucose and insulin levels as a measure of insulin sensitivity. On balance, the studies published to date indicate that high-fat diets can be used safely to effect short-term weight loss, without adversely affecting cardiovascular risk factors such as blood pressure and circulating lipids.

An alternative approach to an ad libitum high-fat diet is an ad libitum high-carbohydrate diet consisting of high-fiber foods with low caloric density. This approach can also be used effectively to promote weight loss. An example is the EatRight® program employed at the University of Alabama at Birmingham *(27)*. This program emphasizes the ingestion of large quantities of high-bulk, low-energy-density foods (primarily vegetables, fruits, high-fiber grains, and cereals) and moderation in high-energy-density foods (meats, cheeses, sugars, and fats). This approach produces equal satiety at reduced energy intake compared with a high-fat diet comprised of energy-dense foods. EatRight participants lose an average of 6.3–8.2 kg by the end of the 12-week program, and, overall, 53% of participants maintain their reduced weight or continue to lose weight 2 years later, while only 23% regain all their lost weight.

The few studies that have compared high-fat versus low-fat diets cannot be construed as indicating that high-fat diets are more effective in achieving weight loss than low-fat, high-fiber diets. For example, in addition to high rates of dropout and noncompliance, the studies by Foster et al. and Samaha et al. did not control for the types or composition of fat or carbohydrates in the diets, which could have affected study end points. As discussed below, variations in the types of fats in the high-fat diets (e.g., amount of monounsaturated fatty acids) or types of carbohydrate in the low-fat diets (e.g., complex versus simple carbohydrate, ratio of carbohydrate to fiber), or other unknown factors, may have influenced study parameters and contributed to inconsistent results with respect to weight loss, insulin sensitivity, and lipid levels. Regardless, the extent of weight loss is directly related to reduced caloric ingestion. Previous metabolic studies have shown that when the energy content of a hypocaloric diet is stable, macronutrient composition does not influence weight loss. Additional research is needed to explore differential effects of low- and high-fat diets on weight loss, appetite behavior, satiety, insulin sensitivity, and cardiovascular disease risk factors in both the short and the long term. In particular, very little is known regarding long-term safety of high-fat diets and their long-term effects on metabolism and cardiovascular disease risk.

15.3.3. Pharmacological Agents That Affect Nutrient Availability

In addition to lifestyle interventions, pharmacological agents designed to interfere with macronutrient metabolism have also been used as adjunctive therapies in the prevention of Type 2 diabetes. Recent reports have shown that dietary fat absorption can be targeted pharmacologically to decrease the progression toward Type 2 diabetes in prediabetic individuals. The XENDOS study (Xenical in the Prevention of Diabetes in Obese Subjects Study) *(28)* used orlistat, a gastrointestinal lipase inhibitor currently approved for the management of obesity, as an addition to lifestyle changes in the prevention of Type 2 diabetes. The patients ($N = 3{,}305$) were randomly assigned to lifestyle plus placebo intervention or lifestyle changes plus orlistat (120 mg three times daily). Compared with lifestyle changes alone, the addition of orlistat resulted in a greater reduction in the incidence of Type 2 diabetes over 4 years (9.0% in the placebo group versus 6.2% in the orlistat group, $p = 0.0032$) and produced greater weight loss in a clinically representative obese population. Orlistat was safe and well tolerated during the length of the study. Interestingly, despite similar weight loss in subjects with IGT and normal glucose tolerance, the difference in diabetes incidence was detectable only in the IGT subgroup, emphasizing the importance of metabolic risk stratification identifying high-risk individuals.

Drugs affecting carbohydrate digestion, targeted to a decrease in postprandial hyperglycemia, have also proven successful in preventing Type 2 diabetes. The α-glucosidase inhibitor acarbose helps control glycemia in diabetes by impeding intestinal digestion of polysaccharides and reducing postprandial glycemic excursions. In a multicenter diabetes prevention trial, patients with IGT were randomly allocated to 100 mg acarbose or placebo three times daily over a period of 3 years (STOP-NIDDM trial) *(29)*. The primary end point was development of diabetes on the basis of a yearly oral glucose tolerance test (OGTT). Despite high attrition rates, fewer subjects in the acarbose arm developed Type 2 diabetes compared to those in placebo (32 versus 42%; $p = 0.0015$), corresponding to a 25% decrease in incident diabetes compared with the placebo-control group. Additional observations were that acarbose significantly increased reversion of impaired glucose tolerance to normal glucose tolerance and was associated with a decrease in cardiovascular disease events. The data from the XENDOS and STOP-NIDDM trials indicate that drug-induced decrements in the absorption of dietary fat and carbohydrate can effectively prevent progression to Type 2 diabetes in high-risk individuals with abnormal glucose tolerance.

15.3.4. Alterations in Composition of Dietary Fat

15.3.4.1. Composition of Dietary Fat: Saturated Versus Polyunsaturated

Investigators have examined whether the composition of dietary fat, independent of total fat consumption, can modulate insulin sensitivity and cardiovascular risk factors. With respect to saturated fat, epidemiological studies show that high intake of total and saturated fat intake is associated with insulin resistance, and this relationship has components that are both dependent and independent of increased body adiposity *(30)*. Multiple cross-sectional studies have similarly found that high intake of both saturated and *trans*fatty acids is associated with hyperinsulinemia and with increased risk of Type 2 diabetes, independent of body adiposity *(30,31)*. High consumption of polyunsaturated fatty acids (PUFA) does not appear to have the same adverse effects and may even result in an increase in insulin sensitivity. For example, Summers et al. studied the effect of substituting dietary saturated fat with polyunsaturated fat on insulin sensitivity in healthy, obese, and Type 2 diabetic subjects *(32)*. Their findings demonstrated that an isocaloric diet enriched in polyunsaturated fat resulted in both an increase in insulin sensitivity assessed by glucose clamp and a lowering of LDL cholesterol, when compared with a diet rich in saturated fatty acids. However, it was not possible in this study to conclude whether it was the increase in dietary PUFA or the decrease in saturated fat that produced the relative benefits in the PUFA diet subgroup. In any event, diets enriched in polyunsaturated fat have not consistently been shown to improve insulin sensitivity, and long-term intervention trials have not been conducted. Discrepancies in the short-term studies are often attributable to the failure to control for dietary fatty acid and carbohydrate composition (e.g., amount of monounsaturated fatty acid), total calories, physical activity, and population characteristics such age, gender, and adiposity.

15.3.4.2. Monounsaturated Fatty Acids (MUFA)

Beneficial effects of a high-MUFA diet on glycemic control in Type 2 diabetes have been demonstrated in a meta-analysis of randomized trials using isoenergetic high-MUFA diets *(33)*. More recently, short-term intervention studies in healthy volunteers have shown that the isocaloric substitution of MUFA for saturated fat, or even substituting MUFA for carbohydrates, can have positive effects on insulin sensitivity. Similar results were obtained in a 3-month trial that evaluated insulin sensitivity in healthy volunteers receiving diets varying in fatty acid composition (omega-3 PUFA versus MUFA versus saturated fat) *(34)*. It was the MUFA-enriched diet that led to significant increases in insulin sensitivity, and this effect was greater when the total amount of fat was modest (<37% of calories). In addition, a Mediterranean-style diet high in MUFA has been shown effective in improving insulin sensitivity and associated cardiometabolic risk in predominantly European populations, including improvements in lipid profile, whole body inflammation, and endothelial dysfunction, independent of changes in body weight *(35)*. A diet enriched in MUFA has also been used as a strategy to maintain metabolic balance after weight loss. Due et al. *(36)* have recently compared the effect of three ad libitum diets, different in type and amount of fat and carbohydrate, on insulin resistance, glucose tolerance, and weight loss maintenance. In this 6-month study, participants were assigned to one of three diets after volunteers had lost 8% or more of their body weight: (1) MUFA diet—moderate in fat (35–45% of energy) and high in monounsaturated fatty acids (>20% of energy); (2) LF diet—low-fat diet (20–30% of energy); and (3) control diet—35% of energy as fat (>15% of energy as saturated fatty acids). After 6 months, the MUFA diet reduced fasting glucose (–3.0%), insulin levels (–9.4%), and the homeostasis model assessment of insulin resistance score (–12.1%). In contrast, these metabolic variables were adversely affected (i.e., increased) in subjects consuming the control diet or the LF diet. These and other studies indicate that a diet enriched in monounsaturated fat has a more favorable effect on glucose homeostasis than does the typical Western diet in the short term and may also be superior in maintaining reduced body weight after a period of weight loss. For these reasons, current dietary recommendations of the American Diabetes Association in essence allow diabetic patients to substitute MUFAs for carbohydrates up to a combined intake of MUFA plus carbohydrates that accounts for 60–70% of total calories *(37)*.

15.3.4.3. Omega-3 and Omega-6 Fatty Acids and Conjugated Linoleic Acids (CLAs)

While evidence suggests that omega-3-FA from fish or fish oil may help prevent heart disease, the effects of omega-3 and omega-6 FA on glucose homeostasis are inconsistently reported. Human studies have shown that enrichment of omega-3 FA in skeletal muscle tissue is associated with enhanced insulin sensitivity, whereas a high ratio of phospholipid omega-6/omega-3 in muscle membrane lipids is associated with increased fasting insulin levels and relative body weight. However, these associations do not necessarily imply causality with respect to dietary ingestion, which can be confirmed only by intervention studies. Short-term diet intervention studies in subjects with and without Type 2 diabetes have shown that there is no effect of omega-3 fatty acids on insulin sensitivity. Interventions in these studies involved dietary fish oil supplementation (3–6 g/day), with a duration that ranged from 2 to 12 weeks *(38)*. Longer term intervention studies are scarce and have also failed to show significant improvements in insulin sensitivity in healthy and diabetic patients. For example, a 6-month randomized controlled trial in diabetic patients evaluated the effects of a moderate supplementation of fish oil (2.7 g/day for the first two months and 1.7 g/day for the remaining 4 months) on glucose control and lipid metabolism. Fish oil had a significant hypotriglyceridemic effect, without any change in glucose control and no change in peripheral glucose utilization measured by hyperinsulinemic clamp *(34)*.

More recently, investigators have hypothesized that conjugated linoleic acids (CLA) can modulate human adiposity and insulin sensitivity. This is based on studies in rodents showing that mixed CLA isomers may reverse insulin resistance and may exert beneficial effects on glucose metabolism in diabetes *(39)*. However, human metabolic studies have provided conflicting results. Obese subjects receiving 3.4 g/day of the *trans*-10, *cis*-12 CLA isomer lost weight and waist circumference after 12 weeks but surprisingly became more insulin resistant. In contrast, subjects receiving 3.4 g/day of a mixture of CLA isomers, *cis*-9, *trans*-11 and *trans*-10, *cis*-12, during the same period experienced no effects on body weight or insulin sensitivity from baseline *(40)*. Additional trials are needed to assess whether specific CLAs affect human adiposity and insulin action.

In conclusion, isocaloric diets enriched in saturated and *trans*fatty acids may increase insulin resistance, while monounsaturated fat appears to improve insulin sensitivity. On the other hand, metabolic studies using diets high in PUFA have provided conflicting conclusions on the effect of these compounds in glucose metabolism. Further studies are necessary to determine the long-term effects of PUFA-rich diets in insulin sensitivity, lipid metabolism, and body fat distribution. Determining the long-term benefits of diets varying in dietary fat composition, as well as the adherence and compliance to changes in dietary fatty acids, is of prime importance in trying to translate these changes to a "real-world" setting.

15.3.5. Alterations in Composition of Dietary Carbohydrate: The Importance of Fiber

As observed for fat, the composition of dietary carbohydrates, as opposed to the total amount of carbohydrate calories, may potentially influence body weight and insulin sensitivity. The glycemic index (GI) has been established to physiologically classify carbohydrates based on postmeal glycemic responses, since this is not always predictable based on simple versus complex carbohydrate chemical structure. In terms of weight management, the data are equivocal as to whether high-GI diets promote weight gain. A recent evidence-based report from the World Health Organization found that the only carbohydrate-relevant dietary factor, which convincingly protected against weight gain and obesity, was a high dietary fiber intake *(41)*. In addition, popular "low-carb" diets attempt to lower dietary GI to prevent high insulin secretory responses, since insulin is considered the direct culprit as an appetite stimulant. This is despite any rigorous supportive data and in the face of convincing contrary evidence that central administration of insulin acts to suppress appetite and reduce energy intake in primates and rodents *(42)*. For the most part, studies on the effects of low- versus high-glycemic-index foods have assessed short-term effects of foods or liquid meals on hunger, satiation, satiety, and short-term energy intake. Overall, results from these investigations indicate that consumption of high-GI carbohydrates has less of an effect to suppress appetite and a diminished ability to induce satiety than foods with lower GI. The implication is that long-term consumption of high-GI diets may lead to energy overconsumption and, therefore, promote weight gain and/or the maintenance of excess body weight. However, well-controlled randomized trials of high quality have failed to find any effect of a high- versus low-GI diets on body weight, body fat, or insulin levels. Furthermore, because multiple dietary, physiological, and environmental factors affect GI, its validity as a meaningful way to characterize food on a consistent basis has been questioned and its implementation in nutritional recommendations is problematic.

With respect to effects on insulin sensitivity, observational studies suggest that diets enriched in simple carbohydrates and/or fructose can be associated with low plasma HDL cholesterol and insulin resistance, whereas diets supplying higher amounts of complex carbohydrates and fiber are associated with increased insulin sensitivity. Consistent with this idea, studies in rodents demonstrate that high-sucrose and high-fructose diets induce insulin resistance. However, controlled intervention studies in humans indicate that high-sucrose or high-fructose diets do not have these same consequences and do not worsen insulin sensitivity when assessed by hyperinsulinemic clamps *(43)*.

On balance, available data suggest that dietary fiber, rather than available carbohydrates or dietary GI per se, is directly responsible for any pronounced effects of carbohydrates on insulin sensitivity in humans. In a randomized crossover study comparing isocaloric high- versus low-GI diets, there was no observed benefit of the low-GI diet on insulin sensitivity *(44)*. However, a low-GI diet with a greater amount of fiber and whole-grain products seemed to improve glycemic and insulin responses and lowered the risk of Type 2 diabetes *(45)*, indicating that the fiber content in low-GI foods may play a role in their metabolic effects. In support of this contention, studies on the effects of dietary intake of fiber, particularly whole-grain foods, have been fairly consistent in demonstrating an effect to enhance insulin sensitivity. In the Insulin Resistance Atherosclerosis study, which included 978 adults with normal or impaired glucose tolerance, whole-grain intake was significantly associated with insulin sensitivity, as assessed by the minimal model analysis of the frequently sampled intravenous glucose tolerance test. The analyses indicated that the fiber and magnesium content of whole-grain foods accounted for some of their effects on insulin sensitivity *(46)*. In the Framingham Offspring Cohort, whole-grain intake was inversely associated with BMI, LDL cholesterol, and fasting insulin, and this was largely attributable to dietary fiber content and magnesium *(47)*. Finally, a randomized controlled experiment employing hyperinsulinemic euglycemic clamps showed that a whole-grain diet improved insulin sensitivity over baseline, while a refined-grain diet had no effects over a 6-week period *(48)*.

15.4. GENETIC BASES OF TYPE 2 DIABETES

While uncommon monogenic forms of diabetes have been identified, Type 2 diabetes in the vast majority of patients is a polygenic disorder with complex gene–gene and gene–environment interactions. Further, accumulating evidence indicates that susceptibility genes confer only small-to-moderate relative risk, without any single gene exerting major effects. Since Type 2 diabetes is a heterogeneous disorder resulting from combined defects in both insulin action and insulin secretion, genetic factors will predictably confer insulin resistance or β-cell secretory failure as subphenotypes. Extensive studies assessing multiple candidate genes have met with limited success, and genome-wide linkage scans have localized regions on several chromosomes harboring T2DM susceptibility genes *(49,50)*; however, with few exceptions, most identified genes in these studies confer small-to-moderate risk and have yielded inconsistent results in replication efforts *(51)*. On the other hand, genome-wide association studies (GWAS), which assess association between phenotypes and interspersed single nucleotide polymorphisms (SNP) among unrelated individuals, may represent a breakthrough in the search for genes contributing to diabetes and other polygenic diseases. In 2007, the GWAS approach identified multiple gene-based SNPs associated with diabetes, which have been independently confirmed in GWAS studies of several Caucasian populations. In addition, several of these genes had been previously implicated as diabetes susceptibility genes in candidate gene studies or in family studies employing linkage analyses. These combined observations add great credence that these SNPs actually mark susceptibility genes or at least loci for diabetogenic genes. Diabetes-associated SNPs have included genes that influence insulin secretion (*KCNJ11, TCF2, TCF7L2, HHEX-IDE, SLC30A8*, and *CDKAL1*): *PPARγ*, which had been shown in candidate gene studies to affect insulin sensitivity and body weight; *WFS1*, which is mutated in Wolfram syndrome; *IGF2BP2*, which encodes the binding protein for IGF2; and *FTO*, which had previously been identified as an obesity gene. At present, there is an intense search to prove that the genes containing these SNPs are the actual genes conferring diabetes risk, that these genes have a role in racial/ethnic groups other than northern European Caucasians, and to elucidate the biochemical mechanisms by which these genes could impair glucose tolerance.

One of the diabetes genes identified in the GWAS studies is the transcription factor 7-like 2 gene (*TCF7L2*) located on chromosome 10q. Two intronic SNPs in linkage disequilibrium (rs12255372 and rs7903146) were shown to be associated with an ~twofold increase in risk for T2DM in an Icelandic population and were strongly correlated with a microsatellite marker that was previously found to be linked to T2DM *(51)*. The diabetes association with the two SNPs at the *TCF7L2* locus was replicated in Danish and US Caucasian cohorts *(52–54)*. Subsequent studies addressed the mechanisms by which *TCF7L2* could impair glucose tolerance. Homozygosity for the at-risk allele for rs12255372 was shown to be associated with impaired insulin secretion *(55,56)* without any effects on insulin resistance or body weight. *TCF7L2* is a transcription factor that regulates the WNT signaling pathway and the secretion of incretin hormones. While the molecular mechanisms by which *TCF7L2* impairs insulin secretion remain to be elucidated, clearly this gene could confer diabetes risk via a reduction in insulin secretory responses.

Given the ability of diet to influence metabolism and prevent progression to diabetes in high-risk patients with IGT, it is almost certain that diet will interact with susceptibility genes in the development of diabetes. However, little information is available regarding nutrient–gene interactions in the pathogenesis of diabetes. One reason for the lack of current data is that diet and other environmental factors are rarely measured in whole-genome scan studies that can require large numbers of volunteers to achieve sufficient power. The success and increasing application of SNP-based GWAS studies offer exciting possibilities to include diet as a covariable in the association between SNPs and diabetes-related phenotypes. These results hold promise for the notion of "personalized medicine," where individual diets would be tailored for optimal interaction with susceptibility alleles to enhance outcomes regarding cardiometabolic disease.

As an example of nutrient–gene interaction, we have reported an exon 6 splice donor polymorphism for the uncoupling protein 3 gene (*UCP3*), with an allelic frequency of ~10% in individuals with African ancestry. This polymorphism results in the loss of a splice junction and truncation of the 6th and final transmembrane-spanning domain of the UCP3 protein, which resides in the mitochondrial inner membrane. Heterozygosity for the splice donor polymorphism is associated with increased respiratory quotient and decreased rates of basal lipid oxidation and confers increased risk for severe obesity. Thus, the exon 6 splice donor polymorphism appears to influence macronutrient fuel partitioning away from fat and in favor of carbohydrate to maintain the basal metabolic rate. Conceivably, this polymorphism could have arisen in Africa as a thrifty gene to alter *UCP3* function in a manner that promotes fat storage during food abundance and increases survivability during famine. This "thrifty" mechanism, however, may adversely lead to progressive obesity and diabetes in an alternative environment with chronic exposure to a high-fat diet. The polymorphism was detected in Gullah-speaking African-Americans living on barrier islands and coastal communities in South Carolina, who are fairly uniform in their consumption of a high-fat diet *(57)*. In this way, the *UCP3* polymorphism could interact with diet as an obesity susceptibility gene in Africans and African diasporic populations. Indeed, increased RQ and low-fat oxidation have previously been shown to be risk factors for future weight gain and the development of obesity *(58)*.

15.5. CONCLUSION AND RECOMMENDATIONS

Nutrition is central to prevention and treatment across the spectrum of cardiometabolic disease, which includes diabetes, cardiovascular disease, prediabetes, and metabolic syndrome. Lifestyle therapy including diet can effectively prevent progression of prediabetes and metabolic syndrome to overt diabetes and cardiovascular disease events. Since many patients are overweight, nutritional therapy should be targeted to achieve 5–10% loss in body weight, which can be effectively and safely accomplished using either low-carbohydrate or low-fat diets. It is important to note, however, that the safety of low-carbohydrate (i.e., high-fat), weight-loss diets has not been established much over 1 year in duration, and evidence suggests that this macronutrient composition may adversely affect cardiometabolic disease if patients are isocaloric. While moderate weight loss is largely responsible for cardiometabolic benefits, increments in insulin sensitivity may be more directly related to the loss of particular fat depots (i.e., intramyocellular, intrahepatocellular, or omental adipose tissue) rather than the loss of total body weight per se. Alterations in dietary composition per se can also facilitate weight loss and/or increase insulin sensitivity. Diets enriched in saturated fats can induce insulin resistance, while fat substitution with monounsaturated fats can enhance insulin sensitivity. Therefore, in assessing the safety of low-carbohydrate, high-fat diets, it is important to consider the composition of ingested fat (i.e., percent that is monounsaturated and polyunsaturated). High-fiber, high-carbohydrate diets comprised of foods with low caloric density can similarly be used for effective weight reduction and to ameliorate insulin resistance. While some data suggest that low-glycemic-index diets are most advantageous in this regard, these effects may have more to do with increments in dietary fiber rather than differences in available carbohydrates. For prevention and treatment of cardiometabolic disease, lifestyle therapy will need to either increase insulin secretion, enhance insulin sensitivity, or ameliorate cardiovascular disease risk factors. These traits are polygenic and represent metabolic phenotypes influenced by disease susceptibility genes. Thus, the impact of diet undoubtedly entails nutrient–gene interactions; however, knowledge in this area is lacking partly due to the fact that environment is rarely quantitatively assessed in whole-genome scan studies. Recent success using the genome-wide association approach (GWAS) presents a powerful paradigm for studies defining nutrient–gene interactions in populations and generating information relevant to personalized medicine in individuals. Effective nutritional treatment of prediabetes and metabolic syndrome is critical if we are to reduce the worldwide burden of patient suffering and health-care costs caused by diabetes.

REFERENCES

1. Definition and Diagnosis of diabetes mellitus and intermediate hyperglycemia: report of a WHO/IDF consultation. World Health Organization Report, 2006.
2. Garber AJ, Handelsman Y, Einhorn D, et al. Diagnosis and management of prediabetes in the continuum of hyperglycemia: when do the risks of diabetes begin? A consensus statement from the American College of Endocrinology and the American Association of Clinical Endocrinologists. Endocr Pract 2008;14:933–46.
3. American Diabetes Association. Diagnosis and classification of diabetes mellitus.. Diabetes Care 2008;31:S55–60.
4. Grundy SM, Cleeman JI, Daniels SR, et al. Diagnosis and management of the metabolic syndrome: an American Heart Association/National Heart, Lung, and Blood Institute Scientific Statement. Circulation 2005;112:2735–52.
5. Alberti KG, Zimmet PZ. Definition, diagnosis and classification of diabetes mellitus and its complications. Part 1: diagnosis and classification of diabetes mellitus provisional report of a WHO consultation. Diabet Med 1998;15: 539–53.
6. InternationalDiabetes Federation: The IDF consensus worldwide definition of the metabolic syndrome. April 14, 2005 Available at www.idf.org/webdata/docs/IDFMetasyndrome definition.pdf.
7. Garvey WT, Kwon S, Zheng D, Shaughnessy S, Wallace P, Pugh K, Jenkins AJ, Klein RL, Liao Y. The Effects of Insulin Resistance and Type 2 Diabetes Mellitus on Lipoprotein Subclass Particle Size and Concentration Determined by Nuclear Magnetic Resonance. Diabetes 2003;52:453–62.

8. Liao Y, Kwon S, Shaughnessy S, Wallace P, Hutto A, Jenkins AJ, Klein RL, Garvey WT. Critical evaluation of ATP III criteria in identifying insulin resistance with dyslipidemia. Diabetes Care 2004;27: 978–83.
9. Lara-Castro C, Garvey WT. Diet, insulin resistance, and obesity: zoning in on data for Atkins dieters living in South Beach.J Clin Endocrinol Metab 2004;89:4197–205.
10. Lara-Castro C, Newcomer BR, Rowell J, et al. Effects of short-term very low-calorie diet on intramyocellular lipid and insulin sensitivity in nondiabetic and type 2 diabetic subjects. Metabolism 2008;57:1–8.
11. Lara-Castro C, Fu Y, Chung BH, Garvey WT. Adiponectin and the metabolic syndrome: mechanisms mediating the risk of metabolic and cardiovascular disease. Curr Opin Lipidol 2007;18:263–70.
12. Tian L, Luo N, Klein RL, Chung BH, Garvey WT, Fu Y. Adiponectin reduces lipid accumulation in macrophage foam cells. Atherosclerosis, 2009;202:152–61.
13. Hotamisligil GS. Inflammation and metabolic disorders. Nature 2006;444:860–7.
14. Goodpaster BH, Kelley DE, Wing RR, Meier A, Thaete FL. Effects of weight loss on regional fat distribution and insulin sensitivity in obesity. Diabetes 1999;48:839–47.
15. Foster GD, Wyatt HR, Hill JO, et al. A randomized trial of a low-carbohydrate diet for obesity. N Engl J Med 2003;348:2082–90.
16. Samaha FF, Iqbal N, Seshadri P, et al. A low-carbohydrate as compared with a low-fat diet in severe obesity. N Engl J Med 2003;348:2074–81.
17. Nordmann AJ, Nordmann A, Briel M, Keller U, Yancy WS Jr, Brehm BJ, Bucher HC. Effects of low-carbohydrate vs low-fat diets on weight loss and cardiovascular risk factors: a meta-analysis of randomized controlled trials. Arch Intern Med 2006;166:285–293.
18. Tuomilehto J, Lindstrom J, Eriksson JG, et al. Prevention of type 2 diabetes mellitus by changes in lifestyle among subjects with impaired glucose tolerance. N Engl J Med 2001;344:1343–50.
19. Diabetes Prevention Program Research Group. Reduction in the incidence of Type 2 Diabetes with lifestyle intervention or metformin. *N Engl J Med* 2002;346:393–403.
20. Pan XR, Li GW, Hu YH, et al. Effects of diet and exercise in preventing NIDDM in people with impaired glucose tolerance. The Da Qing IGT and Diabetes Study. Diabetes Care 1997;20:537–44.
21. Must A, Spadano J, Coakley EH, Field AE, Colditz G, Dietz WH. The disease burden associated with overweight and obesity. *JAMA* 1999;282:1523–9.
22. Perseghin G, Scifo P, De Cobelli F, et al. Intramyocellular triglyceride content is a determinant of in vivo insulin resistance in humans: a 1H-13C nuclear magnetic resonance spectroscopy assessment in offspring of type 2 diabetic parents. *Diabetes* 1999;48:1600–6.
23. Banting W. Letter on Corpulence, Addressed to the Public, 3rd ed. London: Harrison, 1864.
24. Borkman M, Campbell LV, Chisholm DJ, Storlien LH. Comparison of the effects on insulin sensitivity of high carbohydrate and high fat diets in normal subjects. J Clin Endocrinol Metab 1991;72:432–7.
25. Swinburn BA, Boyce VL, Bergman RN, Howard BV, Bogardus C. Deterioration in carbohydrate metabolism and lipoprotein changes induced by modern, high fat diet in Pima Indians and Caucasians. J Clin Endocrinol Metab 1991;73:156–65.
26. Lovejoy JC, Windhauser MM, Rood JC, de la Bretonne JA. Effect of a controlled high-fat versus low-fat diet on insulin sensitivity and leptin levels in African-American and Caucasian women. Metabolism 1998;47:1520–4.
27. Weinsier RL. EatRight Lose Weight: Seven Simple Steps. Birmingham, AL: Oxmoor House, 1997.
28. Torgerson JS, Hauptman J, Boldrin MN, Sjostrom L. XENical in the prevention of diabetes in obese subjects (XENDOS) study: a randomized study of orlistat as an adjunct to lifestyle changes for the prevention of type 2 diabetes in obese patients. Diabetes Care 2004;27:155–61.
29. Chiasson JL, Josse RG, Gomis R, Hanefeld M, Karasik A, Laakso M. Acarbose for prevention of type 2 diabetes mellitus: the STOP-NIDDM randomised trial. Lancet 2002;359:2072–7.
30. Mayer-Davis EJ, Monaco JH, Hoen HM, et al. Dietary fat and insulin sensitivity in a triethnic population: the role of obesity. The Insulin Resistance Atherosclerosis Study (IRAS). Am J Clin Nutr 1997;65:79–87.
31. Parker DR, Weiss ST, Troisi R, Cassano PA, Vokonas PS, Landsberg L. Relationship of dietary saturated fatty acids and body habitus to serum insulin concentrations: the Normative Aging Study. Am J Clin Nutr 1993;58: 129–36.
32. Summers LK, Fielding BA, Bradshaw HA, et al. Substituting dietary saturated fat with polyunsaturated fat changes abdominal fat distribution and improves insulin sensitivity. Diabetologia 2002;45:369–77.
33. Garg A. High-monounsaturated-fat diets for patients with diabetes mellitus: a meta-analysis. Am J Clin Nutr 1998;67:577S–82S.

34. Vessby B, Unsitupa M, Hermansen K, et al. Substituting dietary saturated for monounsaturated fat impairs insulin sensitivity in healthy men and women: The KANWU Study. Diabetologia 2001;44:312–9.
35. Esposito K, Marfella R, Ciotola M, et al. Effect of a Mediterranean-style diet on endothelial dysfunction and markers of vascular inflammation in the metabolic syndrome: a randomized trial. J Am Medical Assoc 2004;292:1440–6.
36. Due A, Larsen TM, Mu H, Hermansen K, Stender S, Astrup A. Comparison of 3 ad libitum diets for weight-loss maintenance, risk of cardiovascular disease, and diabetes: a 6-mo randomized, controlled trial. Am J Clin Nutr 2008;88: 1232–41.
37. Bantle JP, Wylie-Rosett J, Albright AL, et al. Nutrition recommendations and interventions for diabetes: a position statement of the American Diabetes Association. Diabetes Care 2008;31(Suppl. 1):S61–78.
38. Rivellese AA, De Natale C, Lilli S. Type of dietary fat and insulin resistance. Ann N Y Acad Sci 2002;967:329–35.
39. Brown JM, McIntosh MK. Conjugated linoleic acid in humans: regulation of adiposity and insulin sensitivity. J Nutr 2003;133:3041–6.
40. Riserus U, Arner P, Brismar K, Vessby B. Treatment with dietary *trans*10*cis*12 conjugated linoleic acid causes isomer-specific insulin resistance in obese men with the metabolic syndrome. Diabetes Care 2002;25:1516–21.
41. WHO. Diet, Nutrition, and the Prevention of Chronic Diseases. Tech Rep Ser 916. Geneva: World Health Organization, 2003.
42. Schwartz MW, Figlewicz DP, Baskin DG, Woods SC, Porte D Jr. Insulin in the brain: a hormonal regulator of energy balance. Endocr Rev 1992;13:387–414.
43. Daly M. Sugars, insulin sensitivity, and the postprandial state. Am J Clin Nutr 2003;78:865S–72S.
44. Kiens B, Richter EA. Types of carbohydrate in an ordinary diet affect insulin action and muscle substrates in humans. Am J Clin Nutr 1996;63:47–53.
45. Hu FB, van Dam RM, Liu S. Diet and risk of Type II diabetes: the role of types of fat and carbohydrate. Diabetologia 2001;44:805–17.
46. Liese AD, Roach AK, Sparks KC, Marquart L, D'Agostino RB Jr, Mayer-Davis EJ. Whole-grain intake and insulin sensitivity: the Insulin Resistance Atherosclerosis Study. Am J Clin Nutr 2003;78:965–71.
47. McKeown NM, Meigs JB, Liu S, Saltzman E, Wilson PW, Jacques PF. Carbohydrate nutrition, insulin resistance, and the prevalence of the metabolic syndrome in the Framingham Offspring Cohort. Diabetes Care 2004;27:538–46.
48. Pereira MA, Jacobs DR Jr, Pins JJ, et al. Effect of whole grains on insulin sensitivity in overweight hyperinsulinemic adults. Am J Clin Nutr 2002;75:848–55.
49. Hoffmann K, Mattheisen M, Dahm S, et al. A German genome-wide linkage scan for type 2 diabetes supports the existence of a metabolic syndrome locus on chromosome 1p36.13 and a type 2 diabetes locus on chromosome 16p12.2. Diabetologia 2007;50:1418–22.
50. Hsueh WC, Silver KD, Pollin TI, et al. A genome-wide linkage scan of insulin level derived traits: the Amish Family Diabetes Study. Diabetes 2007;56:2643–8.
51. Grant SF, Thorleifsson G, Reynisdottir I, et al. Variant of transcription factor 7-like 2 (TCF7L2) gene confers risk of type 2 diabetes. Nat Genet 2006;38:320–3.
52. Scott LJ, Mohlke KL, Bonnycastle LL, et al. A genome-wide association study of type 2 diabetes in Finns detects multiple susceptibility variants. Science 2007;316:1341–5.
53. Zeggini E, Weedon MN, Lindgren CM, et al. Replication of genome-wide association signals in UK samples reveals risk loci for type 2 diabetes. Science 2007;316:1336–41.
54. Florez JC, Manning AK, Dupuis J, et al. A 100 K genome-wide association scan for diabetes and related traits in the Framingham Heart Study: replication and integration with other genome-wide datasets. Diabetes 2007;56:3063–74.
55. Florez JC, Jablonski KA, Bayley N, et al. TCF7L2 polymorphisms and progression to diabetes in the Diabetes Prevention Program. N Engl J Med 2006;355:241–50.
56. Munoz J, Lok KH, Gower BA, et al. Polymorphism in the transcription factor 7-like 2 (TCF7L2) gene is associated with reduced insulin secretion in nondiabetic women. Diabetes 2006;55:3630–4.
57. Argyropoulos G, Brown AM, Willi SM, et al. Effects of mutations in the human uncoupling protein 3 gene on the respiratory quotient and fat oxidation in severe obesity and type 2 diabetes. J Clin Invest 1998;102:1345–51.
58. Ravussin E. Metabolic differences and the development of obesity. Metabolism 1995;44:12–4.

16 Nutrition, Metabolic Syndrome, and Diabetes in the Senior Years

Barbara Stetson and Sri Prakash L. Mokshagundam

Key Points

- Prevention or delay of the development of long-term complications of high blood glucose and related metabolic abnormalities and improvement of quality of life are key issues for nutritional considerations in diabetes care.
- Preventive care can substantially add to quality of life in those living with diabetes, including those in their senior years. In the United States, 44% of persons with self-reported diagnosed diabetes are of age 65 or older and 18% are over age 75.
- Prevention of the development of diabetes in those with risk factors such as "metabolic syndrome" is highly relevant to the care of older adults. Over 40% of adults over the age of 70 years have the metabolic syndrome.
- Older adults with diabetes may be functionally limited by the presence of hypoglycemia. Factors that may play a role in the increased risk of hypoglycemia in the senior years include poor nutritional status, cognitive dysfunction, polypharmacy, and comorbid illnesses.
- Diabetes prevalence-related comorbidities such as diabetic retinopathy, cardiovascular disease, peripheral vascular disease, and congestive heart failure may result in decreased usual activity and limit activities of daily living, including transportation, shopping for food, and ability to read food labels and restaurant menus.
- Given the high rates of depression in the diabetes population and in the senior years, careful assessment of depressive symptomology and its impact on dietary intake, diabetes self-care, and health outcomes is critical.

Key Words: Blood glucose; diabetes; hypoglycemia; metabolic syndrome; older adults; quality of life

16.1. INTRODUCTION

The prevalence of obesity is rising so rapidly in so many countries that the World Health Organization (WHO) has declared that there is now a global epidemic of obesity. Internationally, emergence of new cases of diabetes mellitus (diabetes) parallels the increases seen in Western countries and are increasing even more quickly in Asia. The risks of type 2 diabetes in these countries tend to increase at levels of body mass index generally classified as nonobese in Caucasian westerners *(1)*. These worldwide changes are due to an accelerated prevalence of obesity, today's predominance of sedentary lifestyle, and the rapidly growing population of older adults *(2)*.

In the United States (US) the estimated number of individuals with diabetes is approximately 23.6 million, of whom 5.7 million are undiagnosed (National Diabetes Information Clearinghouse—http://diabetes.niddk.nih.gov/). Type 2 diabetes disproportionately affects minority populations, including African Americans, Hispanics, Native Americans, Asian Americans, and Pacific Islanders. Risk factors for diabetes that are specific to these populations include genetic, behavioral, and lifestyle factors *(3)*.

A. Bendich, R.J. Deckelbaum (eds.), *Preventive Nutrition*, Nutrition and Health, DOI 10.1007/978-1-60327-542-2_16,

16.2. PREVALENCE OF DIABETES IN OLDER ADULTS

The aging of America is also contributing to the increasing numbers of cases of diabetes, as diabetes prevalence increases with age. In developing countries, the majority of people with diabetes are between 45 and 64 years of age. In developed countries, the majority of people with diabetes are age 65 or older. In the United States, the oldest of the large baby boomer cohort are now approaching age 60, and increasing numbers will soon join these ranks. The prevalence of diabetes in individuals over 60 years is estimated at 23.8% (12.2 million). It is also estimated that there are 536,000 new cases of diabetes diagnosed in subjects over 60 years in the United States. The Third National Health and Nutrition Examination Survey (NHANES III) included information on type 2 diabetes and included persons age 75 and older. Extrapolation from this nationally representative sample indicates that in the United States, 44% of persons with self-reported diagnosed diabetes are age 65 or older and 18% are over age 75 *(4)*.

16.3. BURDEN OF DIABETES

Type 2 diabetes exerts a tremendous economic burden, accounting for over 100 billion dollars in annual health-care expenditures in the United State and 28% of the Medicare budget for older Americans *(5)*. In NHANES III, among persons with type 2 diabetes over age 65, 21% reported being in poor health and 35% reported having at least one hospitalization in the preceding year *(6)*.

Diabetes is a chronic disease that leads to a variety of micro and macrovascular complications that affect almost all systems in the body. While the primary abnormality in diabetes, elevated blood glucose level, remains largely asymptomatic, the consequences of sustained elevation in blood glucose are potentially devastating. Seniors with diabetes have higher rates of functional disability and comorbid conditions relative to those without diabetes.

Cardiovascular disease (CVD) is the most frequent and costly complication of type 2 diabetes. A recent review indicates that when cardiovascular events are stratified by diabetes status, relative risk for men is twice and for women is threefold of gender-matched nondiabetics. Among all CVD events, diabetes accounted for 56% of events in men and 78% of events in women. A number of diabetes-related risk factors have been associated with CVD. Epidemiological studies have also suggested that postchallenge hyperglycemia is a risk factor for cardiovascular disease. Albuminuria in diabetics has been shown to have a CVD risk that is four to five times compared to diabetics without albuminuria, suggesting that these should be targets of preventive strategies in persons with diabetes *(7)*.

16.4. GOALS OF DIABETES TREATMENT

The management of diabetes requires a combination of lifestyle interventions and medications. Diabetes is often a progressive disease requiring changing therapeutic strategies. Thus, primary, secondary, and tertiary prevention efforts are all key to public health goals. Clinically, the interaction between lifestyle changes and medications must be carefully considered across the preventive care continuum. This chapter will address secondary and tertiary nutrition prevention, beginning with issues relevant to persons already diagnosed with diabetes and then addressing those with risk factors and metabolic syndrome.

Dietary intervention to maintain optimal glycemic control is a key component of management of those diagnosed with diabetes. The aims of diabetes treatment are to (a) decrease/prevent the development of long-term complications of high blood glucose and related metabolic abnormalities, (b) improve the quality of life of individuals with diabetes, and (c) treat or prevent the development of symptoms of high or low blood glucose.

16.5. DIAGNOSIS AND CLASSIFICATION OF DIABETES

The American Diabetes Association (ADA) and the WHO criteria for the diagnosis of diabetes *(8–10)* emphasize fasting blood glucose levels *(8)* (*see* Table 16.1 for the specific ADA diagnostic criteria). In addition to the occurrence of chronically elevated blood glucose levels, isolated postchallenge hyperglycemia is a particularly common problem among older adults who have abnormal glucose tolerance. The oral glucose tolerance test is less reproducible than the fasting plasma glucose levels and hence is used less often in routine clinical practice. The ADA diagnostic criteria were developed for general use and apply broadly to all age groups. No specific ADA guidelines exist for the senior years.

16.5.1. Typologies of Diabetes in Older Adults

The proper classification of diabetes is important in setting goals for nutritional management. Diabetes is broadly classified into type 1 and type 2 diabetes.

The majority of older adults with diabetes have type 2 diabetes, which is characterized by two defects—insulin resistance and defective insulin secretion *(11)*. The majority of individuals with type 2 diabetes are obese. However, in the older population the proportion of subjects with type 2 diabetes who are underweight increases and could be as high as 20%. This is particularly true in the nursing home population (*see* Section 16.10). Type 2 diabetes results from a combination of insulin resistance, increased hepatic glucose production, and defective insulin secretion *(12)*. Insulin resistance is generally considered the early defect in type 2 diabetes. Insulin resistance is often present in nondiabetic relatives of individuals with type 2 diabetes and in persons with impaired glucose tolerance. Several studies have also demonstrated defective insulin secretion in these at-risk individuals.

The exact mechanism of insulin resistance in type 2 diabetes is unclear. A variety of genetic and environmental factors lead to decreased insulin sensitivity. Of importance to this chapter, obesity and decreased physical activity have been known to decrease insulin sensitivity. Aging is associated with a change in body composition with increase in fat mass and decrease in muscle mass *(13,14)*. This could be partly responsible for the increase in insulin resistance with aging. Aging is also associated with a decline in insulin secretion, particularly a blunting of the first-phase insulin secretion *(15)*. First-phase insulin secretion is an important determinant of postchallenge blood glucose levels. Age-related changes in health behaviors such as increased sedentary lifestyles may also further

Table 16.1
American Diabetes Association Criteria for Diagnosis of Diabetes

	Normal	*Prediabetes*	*Diabetes Mellitus*
Fasting blood glucose (mg/dL)	<100	101–125	>125
2-h postglucose (mg/dL)	<140	141–199	≥200

compound these changes. Type 2 diabetes is a progressive disorder. The progression of the clinical picture with increasing blood glucose levels, requiring increasing doses of medications, is mainly due to a progressive decline in β-cell function. When β-cell function is markedly reduced, exogenous insulin will be necessary to regulate blood glucose levels.

Type 1 diabetes is an autoimmune disorder resulting from cell-mediated and antibody-mediated destruction of β-cells of the islets *(16)*. Insulin is required for the management of type 1 diabetes. Failure to treat with insulin results in the development of an acute metabolic complication—diabetic ketoacidosis. Although type 1 diabetes most commonly occurs in the first three decades of life, it can develop at any age, even in the senior years. The basic underlying mechanism of disease is autoimmune destruction of the pancreatic islets. Circulating islet cell antibodies can be demonstrated in the majority of individuals, especially in the first few years after diagnosis. In addition to new onset type 1 diabetes, older adults may have preexisting type 1 diabetes. Type 1 diabetes, particularly of long duration, is often very "brittle" with wide fluctuations in blood glucose levels and episodes of recurrent and severe hypoglycemia.

16.6. ESTABLISHING MEDICATION AND NUTRITIONAL MANAGEMENT GOALS IN OLDER ADULTS WITH DIABETES

Once the type of diabetes is established, medication and nutritional management goals should be developed. In addition to tailoring the nutritional recommendation to assist glycemic control, consideration of other important risk factors is critical. Obesity, dyslipidemia, hypertension, and insulin resistance are important and often overlapping factors warranting consideration when planning dietary interventions for older adults with type 2 diabetes. Avoidance of hypoglycemia, particularly recurrent and/or severe hypoglycemia, is a major consideration in type 1 diabetes. Lifestyle interventions that have been recommended for the management of diabetes have positive effects on both insulin secretion and insulin resistance. Aggressive lifestyle intervention can prevent the progression of impaired glucose tolerance to diabetes and could decrease the dose and number of medications for the management of type 2 diabetes.

16.6.1. Medication Use and Glycemic Control

The major aim of treating diabetes is to decrease the rate of micro and macrovascular disease associated with elevated blood glucose. Two landmark trials have served as the basis for current recommendations for the management of blood glucose levels in diabetes. The Diabetes Control and Complications Trial (DCCT) was conducted in adults with type 1 diabetes and compared intensive insulin treatment using multiple insulin injections or an insulin pump to conventional treatment using twice daily injections of intermediate- and short-acting insulin over a follow-up period of 7 years *(17,18)*. The results showed significant reduction in risk of all microvascular disease endpoints in the intensively treated group. However, the DCCT did not include older adults and did not have the statistical power to analyze benefits on macrovascular risk reduction. The clear demonstration of a relationship between glycemic control, measured by reduction in hemoglobin A1c, and improved outcomes indicates that similar outcomes would be expected in adults in their senior years. A downside to tight control was indicated by the findings of higher risk of hypoglycemia in the intensively treated group. Given the burden of potential hypoglycemia and potential impact on quality of life in an older adult with a limited prognosis, the cost–benefit ratio of intensive glycemic control versus hypoglycemia risk must be carefully considered.

The United Kingdom Prospective Diabetes Study (UKPDS) was a long-term study of a variety of treatment options in adults with type 2 diabetes *(18)*. The important findings of the UKPDS

Table 16.2
Factors to Consider in Management of Diabetes in Older Adults

1. Individualize glycemic goals. The goal, in most cases, should usually include the standard A1c target of <7%. Consider a higher goal, if appropriate, based on the following factors:
   - patient preference
   - diabetes severity
   - life expectancy
   - functional status and social support
2. Keep therapy as simple and inexpensive as possible
3. Encourage diabetes education of the patient and primary caregivers, with the reminder that such education is covered by Medicare
4. Treat hypertension and dyslipidemia to decrease cardiovascular risk
5. Screen for depression and offer therapy promptly if the diagnosis is made
6. Maintain an updated medication list and monitor regularly for adverse drug effects
7. Screen annually for cognitive impairment and other geriatric syndromes (e.g., urinary incontinence, pain, injurious falls)

Source: Modified from Olson DE and Norris SL *(107)*.

can be summarized as follows: (a) A reduction of 1% in hemoglobin A1c results in ~22% reduction in microvascular complications; (b) reduction in microvascular complications with reduction in hemoglobin A1c is observed, irrespective of the type of intervention; (c) glycemic control in type 2 diabetes worsens over time and necessitates changes in medication, irrespective of initial management approach; (d) in a subgroup of subjects treated with metformin, there was a significant reduction in macrovascular disease. The DCCT did not include older adults and there were few older participants in the UKPDS. Although the UKPDS included an older population, subgroup analysis of the older age group is not available. Hence, the applicability of these studies to older adults is limited.

The ADA goal for glycemic control is to have HbA1c levels ≤7.0%. In addition the ADA recommends individualization of glycemic control goals with an HbA1c goal of <6% when feasible. The ADA goals are general and written to broadly apply to all persons with diabetes. However, older adult persons with diabetes require special consideration and require reassessment of goals. While the general principles of diabetes care remain the same, it is now well recognized that in managing the older adult, additional factors need to be considered and the goals modified accordingly. The 2003 California Healthcare Foundation/American Geriatric Society (AGS) guidelines have helped clarify some of these concerns and several concepts have been subsequently incorporated in the ADA guidelines. These guidelines have emphasized the need for individualizing diabetes care, aggressively addressing cardiovascular risk factors, and stressed glycemic control in preventing microvascular complications. The AGS guidelines also recognize the importance of comorbidities that are common in the older adults with diabetes and have significant impact on the ability to maintain strict glycemic control. These include depression, cognitive impairment, urinary incontinence, falls, pain, and polypharmacy *(19)*.

In addition to glycemic goals, several metabolic and cardiovascular risk factors must also be considered due to the higher rates of cardiovascular disease and its substantial impact on morbidity and mortality in persons with diabetes (*see* Table 16.2). The high risk of cardiovascular morbidity and mortality in diabetes is due to a variety of factors. These include overall blood glucose control, glycemic fluctuation, postprandial blood glucose levels, high LDL cholesterol, low HDL cholesterol, elevated serum triglycerides, blood pressure, and altered coagulation profile. In addition, systemic inflammation, a prooxidant state, and endothelial dysfunction play a significant role. Recommendations to focus on maintaining optimal control of blood pressure in persons with diabetes are based on the positive results of such control as demonstrated in the UKPDS, Hypertension Optimal Treatment (HOT) trial, and Arterial Blood Pressure Control in Diabetes (ABCD) studies *(20–22)*. Additionally, the benefits of lowering total and LDL cholesterol have also been demonstrated in intervention studies. Reducing LDL-C has been shown to decrease cardiovascular events to 80 years of age, but studies are needed to explore the efficacy of lipid-lowering therapy in individuals older than 80 years. The magnitude of delay in progression of atherosclerotic disease in response to screening-guided therapy has not been well delineated. The cost-effectiveness of lipid screening in the older diabetes population, as well as the subgroups that would benefit most from such screening, needs to be further studied. Based on ongoing clinical trial data, the guidelines may be modified to recommend that an LDL-C not greater than 70 mg/dL is the target level for special high-risk populations. In the PROVE-IT study, a subgroup analysis showed less benefit in patients over the age of 65 compared with younger patients. The TNT trial excluded patients over the age of 75 years. Hence, the applicability of these data to the population with diabetes in their senior years remains unclear. Optimal implementation of the current guidelines and the use of available agents will ultimately depend on expanding the knowledge base of health-care providers and may require far-reaching educational programs that change the way that risk factor management is viewed by caregivers and patients alike *(23–35)*.

Any nutritional approach to the management of diabetes must specifically address the issues related to cardiovascular risk. Cardiovascular risk reduction in diabetes is achieved through a combination of lifestyle changes and pharmacological interventions that address the multiple risk factors. A general outline of lifestyle and pharmacological approaches is shown in Table 16.3.

Taken together, these lifestyle and pharmacological interventions can play a major role in the management of cardiovascular risk reduction in persons with diabetes. The recommended goals for management of weight, blood pressure, and lipids are outlined in Table 16.4.

16.7. GENERAL DIABETES DIETARY RECOMMENDATIONS

The general goals of nutritional recommendations for the management of diabetes include the following:

1. Achieve and maintain blood glucose levels as outlined above
2. Achieve and maintain optimum lipid levels
3. Achieve and maintain reasonable body weight. This would include weight loss, if overweight, and weight gain, if undernourished
4. Prevent acute complications
5. Maintain overall health

The general recommendations for macronutrient intake in diabetic diet are shown in Table 16.5. The carbohydrate composition (amount and type) of the diet has been a focus of many recommendations and subject of recent controversy. There are no well-designed studies that have compared different dietary approaches. In a study of a high-carbohydrate (60%), low-fat (25%) diet, compared to a low-carbohydrate (35%), high-monounsaturated-fat (50%) diet *(36)*, plasma glucose, triglyceride, and VLDL cholesterol were lower in the subjects in the low-carbohydrate/high-monosaturated-fat diet group. However, the use of high-fat/low-carbohydrate diets could lead to more hypoglycemic episodes and ketosis.

Higher protein diets have been recommended and have been popular for weight loss. An empirical review of studies of high- versus low-protein diets found that short-term (<6 months), high-protein diets may help people lose more weight and body fat, due in part to increased satiety *(37)*. However, the long-term effects of high-protein diets and their efficacies in persons with diabetes have not been

Table 16.3
Lifestyle and Pharmacological Approaches to Risk Factor Management

Risk Factor	*Lifestyle Intervention*	*Pharmacological Intervention*
HbA1c	Diet and exercise	Insulin-sensitizing agents, insulin secretagogues, insulin
Postprandial glucose/glycemic excursion	Carbohydrate content of meals (amount, type, timing, personal response to CHO based on postprandial SMBG feedback)	Repaglinide (Prandin), nateglinide (Starlix), sitagliptin, exenatide, pramlintide, short-acting insulins (insulin lis-pro, insulin aspart, regular insulin)
LDL cholesterol	Low-cholesterol diet Exercise	Statins Bile acid-binding agents Niacin
Triglyceride	Low-fat diet Exercise/weight loss	Gemfibrozil/fenofibrate Niacin (long acting) Omega-3 fatty acids
Low HDL cholesterol	Exercise Smoking cessation	Niacin Gemfibrozil/fenofibrate
High blood pressure	Low-sodium diet Exercise Weight loss	Variety of antihypertensive agents (ace inhibitors preferred)
Procoagulant state	Exercise/weight loss	Aspirin
Proinflammatory state	Diet (e.g., increased proportion of less-refined CHO, increased vegetable and fruit intake, reduced saturated fat,), exercise, weight loss, medication	Aspirin

well tested. Higher protein content has been shown to increase risk of development and progression of diabetic nephropathy. It has been suggested that a high-protein diet (2 g/kg of body weight) may be contraindicated for persons with poorly controlled diabetes or complications *(38)*. Low-carbohydrate diets are not recommended in the management of diabetes. Dietary carbohydrate is the major contributor to postprandial glucose concentration and an important source of energy, water-soluble vitamins and minerals, and fiber. Thus, in agreement with the National Academy of Sciences-Food and Nutrition Board, a recommended range of carbohydrate intake is 45–65% of total calories. In addition, because the brain and the central nervous system have an absolute requirement for glucose as an energy source, restricting total carbohydrate to <130 g/day is not recommended *(39)*. Monitoring carbohydrate intake, whether by carbohydrate counting, using the exchange system, or experience-based estimation, remains a key component of managing diabetes. The use of complex carbohydrates is preferred. The use of lower fat content in the diet is based on the need to restrict caloric intake, improve lipid levels, and assist weight loss. While the use of vitamin supplements is not generally recommended for subjects with type 2 diabetes, the ADA recommends multivitamins for older adults.

Table 16.4
Recommended Assessment and Management Goals for CVD Risk Factors in Diabetes

Parameter	*Frequency*	*Goal*
HbA1c	3–4 months	<7% (<6.5%)[a]
Fasting blood glucose	2–7 times a week[b]	80–120 mg/dL[a] (90–130 mg/dL)
Postprandial blood glucose	2–7 times a week[b]	<40 mg/dL[a] (<180 mg/dL)
LDL cholesterol	Annual. 3–4 months, if abnormal	<100 mg/dL[c]
HDL cholesterol	Annual. 3–4 months, if abnormal	>45 mg/dL
Triglycerides	Annual. 3–4 months, if abnormal	<200 mg/dL
Systolic blood pressure	Each visit	<130 mmHg
Diastolic blood pressure	Each visit	<80 mmHg
Microalbuminuria	Annual	Normal or no progression
Eye examination	Annual	Normal or no progression
Neurological examination	Annual	Normal or no progression

[a]Recommended by AACE/ACE.
[b]No clear recommendation (need to individualize).
[c]Optional goal of <70 mg/dL in highest risk individuals.

Table 16.5
Macronutrient Content of General Diabetes Diet

Carbohydrates	45–65% of total caloric intake
Protein	12–20% of caloric intake <30% of total caloric intake
Fat	<200 mg/day of cholesterol Saturated and polyunsaturated fat <10% of total caloric intake (2–40 g/day)

There has been no convincing evidence to recommend the routine use of antioxidants, vitamin E, or C in subjects with diabetes.

16.7.1. Balancing Diet and Medication

The interaction of diet and medication is of particular importance in the management of diabetes. Insulin and drugs that increase insulin secretion are likely to induce hypoglycemia if meals are not taken at appropriate times. Erratic eating habits might require readjustment of medications, either dose, timing, or both. This may be of particular concern in the hospitalized older patients with diabetes. Poor eating habits might also necessitate change to medications that are less likely to cause hypoglycemia when used alone. Metformin (Glucophage) and the thiazolidenediones (pioglitazone and rosiglitazone) are least likely to cause hypoglycemia when used alone. The problem of unwanted weight gain is another issue for consideration for many overweight individuals. Insulin and the thiazolidenediones, particularly when used in combination, are most likely to result in weight gain. The deleterious effect of weight gain in overweight older adults must be evaluated in tandem with the potential benefits of improved glycemic control. Loss of appetite may occur with Metformin, exenatide (a GLP-1 receptor agonist), and pramlintide (an amylin analog) and would be of concern in the undernourished persons with diabetes. Due to concerns of lactic acidosis, metformin may not be appropriate for persons with predisposing conditions such as heart failure, liver problems, and renal insufficiency.

16.8. BODY WEIGHT AND FUNCTIONAL STATUS IN OLDER ADULTS WITH DIABETES

16.8.1. Overweight and Obesity

Overweight not only is an important risk factor for the development of diabetes but also has a significant impact on diabetes progression and the development of complications *(40)*. Only recently has the problem of obesity been systematically examined in older adults. Obesity appears to be common in older adults until the eighth decade of life and then declines in the oldest old. Data from NHANES III indicate that type 2 diabetes strongly increases in prevalence with increasing overweight in older as well as younger adults. Personal risks for diabetes were observed to be stronger for obese younger adults but still substantially elevated in older adults with odds ratio of 3.4 (95% CI, 1.1–8.3) for the most obese men over age 55 and 5.8 (95% CI, 4.2–7.4) for the most obese women over age 55. In a study of 3-year mortality in community-dwelling older adults, unintentional weight loss and underweight BMI were associated with elevated mortality. Overweight or obesity and intentional weight loss were not associated with mortality. These results are consistent with other findings and suggest that undernutrition may pose greater mortality risk in seniors than do obesity or intentional weight loss *(41)*. This may be an issue particularly for the oldest old and those who have impaired functional status. One study found that at least 21% of nursing home patients with type 2 diabetes were underweight *(42)*.

16.9. SPECIAL NUTRITION INTERVENTION SITUATIONS IN DIABETES PREVENTIVE NUTRITION

16.9.1. Hospitalization

A variety of systemic problems that affect glycemic control in the hospital setting have been identified. Barriers that may impact an individual's nutrition status and subsequently affect glycemic control include poor appetite, inability to eat, increased nutrient, and calorie needs due to catabolic stress, variation in diabetes medications, and the possible need for enteral or parenteral nutrition support. Proper timing of meals and the relation to medications are important. Insulin should be administered immediately before or after a meal. Due to the wide heterogeneity in the hospital population, individualization of nutrition recommendations is key to improving outcomes. The common practice of ordering "ADA Diet" is strongly discouraged as the ADA does not endorse any specific diet. The consistent carbohydrate meal planning system is encouraged. For this system to be effective, it is important that nursing and nutrition services coordinate their services.

Management of hyperglycemia in the hospital setting has gained increasing attention over the last few years. Hyperglycemia is common among hospitalized subjects and has been shown to be associated with higher mortality and morbidity in a variety of studies. This is particularly relevant to the older population, since they are more likely to be admitted to the hospital and have higher rates of diabetes. The American Association of Clinical Endocrinologists (AACE) and ADA have recommended glycemic targets for patients with hyperglycemia, depicted in Table 16.6 *(43,44)*. While randomized controlled studies in the intensive care unit support the glycemic goal of <110 mg/dL in the ICU, there are no trials that clearly justify the recommended goal blood glucose levels in the non-ICU setting *(45)*. The key areas of focus to improve inpatient glycemic control are the following:

Table 16.6
Summary of ADA and AACE Glycemic Control Target Recommendations

Location	*ADA Recommendation*	*AACE Recommendation*
ICU	As close to 110 mg/dL as possible; generally <180 mg/dL	<110 mg/dL
General ward	As close to 90–130 mg/dL as possible; postprandial glucose <180 mg/dL	Premeal <110 mg/dL, postmeal <180 mg/dL

1. Establishing screening criteria for appropriate referral to a registered dietitian
2. Identifying nutrition-related issues in clinical pathways and patient care plans
3. Implementing and maintaining standardized diet orders such as consistent carbohydrate menus
4. Integrating blood glucose monitoring results with nutrition care plans
5. Using standing orders for diabetes education and diabetes MNT as appropriate
6. Standardizing discharge follow-up orders for MNT and diabetes education postdischarge when necessary *(46,47)*

Patients requiring clear or full liquid diets should receive 200 g carbohydrate/day in equally divided amounts at meal and snack times. Liquids should not be sugar free. Patients require carbohydrate and calories, and sugar-free liquids do not meet these nutritional needs. For tube feedings, either a standard enteral formula (50% carbohydrate) or a lower carbohydrate content formula (33–40% carbohydrate) may be used. Calorie needs for most patients are in the range of 25–35 kcal/kg every 24 h. Care must be taken not to overfeed patients because this can exacerbate hyperglycemia. After surgery, food intake should be initiated as quickly as possible. Progression from clear liquids to full liquids to solid foods should be completed as rapidly as tolerated *(48)*.

16.9.2. Long-Term Care

Residents of long-term care facilities may face additional or unique problems. They tend to be often underweight and it is not necessary to make any caloric restrictions in these subjects. Low body weight has been associated with higher mortality and morbidity in these subjects. Restricting food choices may lead to poor overall nutritional status in these subjects and has not been shown to improve glycemic control. Hence, the use of “no concentrated sugar,” “no sugar added,” or “liberal diabetic diet” is discouraged *(49,50)*.

16.9.3. Enteral and Parenteral Nutrition

Enteral and parenteral nutrition might pose additional challenges in the management of patients with diabetes. While the glycemic goals for individuals receiving enteral and parenteral nutrition are the same as glycemic goals for the general population of diabetics, achievement of normoglycemia may be more difficult in patients who are acutely ill. There is evidence that poor glycemic control in subjects on parenteral or enteral nutrition is related to poor outcomes. It is estimated that up to 30% of patients who receive parenteral nutrition have diabetes. Many of these patients have no previous history of diagnosed diabetes and develop diabetes due to stress-induced increases in counterregulatory hormones and cytokines.

The relative value of high-carbohydrate versus high-fat enteral feeds for persons with diabetes has been debated *(51)*. The most widely used commercial enteral preparations for individuals with diabetes provide, but 1 calorie/mL, 40% (CHOICEdm TF; Novartis Medical Nutrition) to 34% (Glucerna; Abbott Laboratories, Inc.) carbohydrate and 43% (CHOICEdm TF) to 49% (Glucerna) fat. They also

Table 16.7
ATP III Criteria for Diagnosis of Metabolic Syndrome

Risk Factor	*Defining Level*
Abdominal obesity (waist circumference)	>102 cm (>40 in.)
Men	>88 cm (>35 in.)
Women	
Triglyceride	>150 mg/dL
HDL cholesterol	
Men	<40 mg/dL
Women	<50 mg/dL
Blood pressure	<130/85 mmHg
Fasting blood glucose	>110 mg/dL

Note: Diagnosis is established if three or more risk factors are present.

have high monounsaturated fatty acids (MUFA; 35% of kilocalories in Glucerna). MUFA has been shown to be beneficial in improving lipid profile, glycemic control, and lowering insulin level *(52)*. CHOICEdmTF has a higher content of medium-chain triglycerides and has no fructose. The use of insulin or oral agents in persons receiving enteral nutrition should be tailored to match the timing of feeds. Parenteral nutrition fluids are high in carbohydrate and derive only fewer calories from fat. In persons with diabetes, particularly in less severely stressed individuals, the proportion of carbohydrate may be decreased but is still very high. The usual rate of glucose infusion is 4–5 g/kg body weight and lipid infusion of 1–1.5 g/kg body weight. This requires adequate use of insulin to maintain normoglycemia *(53)*. Insulin infusion not only maintains glycemic control but also prevents protein breakdown and promotes protein synthesis.

16.10. PREVENTIVE NUTRITION ISSUES AND DIABETES RISK FACTORS AND METABOLIC SYNDROME

The term "metabolic syndrome" refers to a cluster of abnormalities that were initially described by Reaven to include hypertension, dyslipidemia, abnormal blood glucose, and abdominal obesity. Insulin resistance was considered to be a central and pathogenetic abnormality in this syndrome. Over the last 20 years, a number of other clinical and laboratory features have been proposed to be components of this syndrome. Some of these features included high c-reactive protein, nonalcoholic fatty liver disease, low plasminogen activator inhibitor 1, high fibrinogen, polycystic ovarian disease, and low adiponectin levels. In 2001, the National Cholesterol Education Program–Adult Treatment Panel III (NCEP–ATP III) recommended the use of the metabolic syndrome in cardiovascular risk assessment. The NCEP–ATP III defined metabolic syndrome as shown in Table 16.7 *(54)*. A number of other organizations including the WHO, the International Diabetes Federation, the AACE, and the American Heart Association/National Heart Lung and Blood Institute have offered different diagnostic criteria for metabolic syndrome. However, the most widely used definition in the Unites States is the one offered by NCEP–ATP III. The prevalence of the metabolic syndrome as defined by the NCEP–ATP III in the NHANES III increases with age. Over 40% of adults over the age of 70 have the metabolic syndrome *(55)*.

The exact significance of the metabolic syndrome has been a subject of controversy. The main significance appears to be the ability of the syndrome to identify individuals who are at high risk of developing cardiovascular disease and/or diabetes *(56–58)*.

Clinical trials have not been designed specifically to test the effect of lipid and blood pressure interventions in participants with the metabolic syndrome. Diabetes may be considered as a good model of the metabolic syndrome, because 85% of diabetic subjects have the metabolic syndrome, and there are clear data that lipid and blood pressure interventions work very well in those with diabetes. However, there are no specific studies addressing this in older adults. Many critics, including members of a joint panel of the ADA and the European Association for the Study of Diabetes, have questioned the utility of the metabolic syndrome in clinical practice, citing the lack of agreement in diagnostic criteria, the absence of a clear understanding of an underlying pathophysiologic mechanism linking the multiple components of the syndrome, and the variable risk associated with individual syndrome components *(59)*.

While proponents of the continued use of the metabolic syndrome in clinical practice agree with many of the criticisms, they argue that its recognition in clinical practice is useful. The use of the metabolic syndrome may encourage health-care providers to look for other risk factors once one risk factor is found and more importantly encourage weight loss (through a program of diet and exercise) and increased physical activity rather than always prescribing a different drug for each medical condition. The management of the metabolic syndrome is primarily aimed at reducing cardiovascular risk. In individuals with prediabetes, the aim would be to prevent progression to type 2 diabetes.

16.10.1. Prevention of Type 2 Diabetes—Targeting Prediabetes

Several large, randomized clinical trials have demonstrated the efficacy of both lifestyle and pharmacological interventions in preventing the progression of impaired fasting glucose or impaired glucose tolerance. In the landmark Diabetes Prevention Program (DPP), progression to diabetes was reduced by 65% in the lifestyle intervention group and by 31% in the metformin group *(60)*. Over 50% of participants in the study had three or more components of the metabolic syndrome. In older participants, lifestyle intervention had an even greater impact than metformin *(61)*. Among participants age 60 and older, lifestyle intervention reduced the risk of development of diabetes by 71%. Participants in the 60–85-year-old age group were the most likely to achieve weight loss (5–7% loss in body weight was achieved with dietary change and activity) and physical activity goals. No age differences were noted in the reduction of caloric intake. Older participants receiving metformin, the medication intervention arm of the study, did not experience benefits to the degree as younger, heavier participants. Diabetes incidence rates over time fell with increasing age, while in the metformin group, the youngest participants showed the lowest diabetes incidence. A trend was noted that metformin had lower effectiveness relative to lifestyle change with increased participant age, despite comparable to superior medication adherence and greater weight loss in the older metformin group participants. Age differences in response to pharmacologic treatment and physiology, as well as behavior, all likely played a role in the findings. Of note, DPP participants were community dwelling, relatively healthy, and free of significant physical limitations and frailty; however, the intensive lifestyle intervention was modified to accommodate participants who developed limitations over the course of the study *(61)*.

In the DREAM trial, both rosiglitazone and metformin were shown to reduce progression to type 2 diabetes in nondiabetic adults with impaired fasting glucose and/or impaired glucose tolerance. However, rosiglitazone use was associated with weight gain *(62)*, adding to growing concern about the potential cardiovascular adverse effects of thiazolidenediones. Further, metformin is not yet approved by the FDA for use in the nondiabetic population, dampening widespread use of these medications.

Another large study prospectively tracked at-risk older adults and highlighted the relationship between lifestyle variables and metabolic syndrome and the feasibility and benefits of lifestyle change. Participants were 3,051 men aged 60–75 without diabetes or history of CHD, who had been followed as participants in the British Regional Heart Study and completed a 20-year follow-up. Metabolic syndrome variables and lifestyle factors, including dietary intake, were also assessed. One fourth of the sample had metabolic syndrome. After adjusting for demographics and other lifestyle factors, variable associations with metabolic syndrome were as follows: BMI had the most significant, positive association; physical activity was negatively associated, total dietary fat and alcohol intake showed no significant association; no significant association for saturated or polyunsaturated fat and PS ratio. Carbohydrate intake was associated with greater odds of metabolic syndrome; participants eating a low-fat, high-carbohydrate diet had the greatest odds of having metabolic syndrome, while those eating a low-carbohydrate, high-fat diet showed no increase in having metabolic syndrome. The authors cautioned against emphasizing total fat reduction in older adults given the risk of increased carbohydrate intake, which could adversely impact lipids and the development of metabolic syndrome.

These large prospective studies highlight that change in lifestyle, even at older ages, influenced metabolic syndrome variables, with changes observed within 3 years of changes in behavior. Weight loss, irrespective of baseline BMI, was associated with lower risk. Increased leisure time activity was also associated with lower likelihood of metabolic syndrome *(63)*. Findings suggest that modest weight loss and exercise is appropriate for seniors at risk for type 2 diabetes *(64)*.

The effect of lifestyle intervention on blood pressure, dyslipidemia, and central obesity in the setting of metabolic syndrome in individuals with normal blood glucose has not been studied. Currently, dyslipidemia and hypertension in subjects with the metabolic syndrome are managed similarly to those without metabolic syndrome.

16.11. CLINICAL ISSUES IMPACTING OLDER ADULTS WITH AND AT RISK FOR DIABETES

16.11.1. Dietary Habits of Older Adults with Diabetes

Of the few systematic studies of dietary intake in older adults with diabetes, both the type of food consumed and the pattern of eating behaviors emerge as important influences on nutritional intake. An Australian study of adults over age 65 found that both diabetic and nondiabetic age-matched subjects' typical dietary habits exceeded recommended levels of dietary fat and provided inadequate fiber. Only 6% of the diabetic subjects consumed a diet with at least 50% carbohydrate and less than 30% fat *(65)*. This suggests that recommendations to increase fiber and decrease fat may be appropriate for many older adults with diabetes.

Persons with diabetes must follow a diet that incorporates healthy food choices and spacing of meals to be consistent with exogenous insulin use and physical activity, with the goal of maintaining euglycemia. Unhealthful snacking is one area that threatens optimal nutrition in the senior years. A random telephone survey of 335 community-dwelling US adults age 55 and older found that 98% of older adults reported snacking at least once each day, with evening being the most common time and nearly all snacking occurring at home *(66)*. Taste outranked nutrition as a snack selection criteria. Fruits were popular but were chosen less often than other less healthful snacks. This suggests that the context of snacking is an important consideration influencing choices made when snacking and should be addressed when providing nutritional guidance to older patients with diabetes. Concrete suggestions for replacing highly processed, high-fat snack foods with fruits and vegetables and other nutritious snacks may assist older adults in the selection of healthier snacks (*see* Ref. *(67)* for examples). Patients may be encouraged to incorporate healthy alternatives into shopping lists and keep them in the home as replacements for preferred but unhealthy items. Evening activities that are alternatives to snacking may also be encouraged.

16.11.2. Hypoglycemia in Older Adults

Hypoglycemia is a major limiting factor in the management of diabetes. The incidence of hypoglycemia is relatively high in older compared to younger adults. A variety of factors may play a role in the increased risk of hypoglycemia in the senior years. These include poor nutritional status, cognitive dysfunction, polypharmacy, and comorbid illnesses. Except in the severely malnourished, poor dietary intake by itself does not lead to hypoglycemia. The most common cause of hypoglycemia remains the use of blood glucose-lowering agents. Drugs that increase insulin secretion and insulin itself can cause hypoglycemia. A major finding of the DCCT, which was conducted with young, healthy adults, was that the major deleterious health consequence of tight blood glucose control in persons with type 1 diabetes is hypoglycemia. Drugs that enhance insulin sensitivity (thiazolidenediones), decrease hepatic glucose production (metformin), or decrease carbohydrate absorption (α-glucosidase inhibitors) have very low risk of hypoglycemia, except when used in combination with insulin or an insulin secretagogue. When medication use creates problems of consistent hypoglycemia, patients must learn how to avoid and manage hypoglycemic episodes. Older adults taking insulin who have high variability in blood glucose levels, exhibit very low average blood glucose concentrations, have had diabetes for a long duration, have a low body mass index, or who have high levels of vigorous physical activity may be at particular risk of severe hypoglycemia *(68)*.

16.11.3. Self-Monitoring and Dietary Treatment of Hypoglycemia

Frequent self-monitoring of blood glucose levels provides specific information that may serve as feedback for guiding decisions about moment-to-moment treatment needs, thus helping individuals to anticipate or prevent severely low glucose levels. However, frequent blood glucose testing may be perceived as expensive, inconvenient, or painful. Unfortunately, rather than performing frequent blood glucose testing, many individuals simply rely on their symptoms or estimates about their blood glucose levels when deciding what to eat or how vigorously to exercise or whether to operate a motor vehicle *(69)*. By increasing the frequency of blood glucose testing (at least four times per day for persons taking insulin) and making informed decisions about when to eat additional carbohydrate (e.g., eat 15 g of carbohydrate to raise blood glucose levels about 45 mg/dL) or to identify personal sources of vigorous physical activity contributing to low blood glucose levels, patients may learn to prevent severe hypoglycemia. Educating patients about the importance of always carrying glucose tabs or gel or fast acting carbohydrate snacks or placing them in various locations such as the car or relative's homes may also aid in the treatment of mild-to-moderate hypoglycemic episodes. Recommendations for management of hypoglycemia in older adults are presented in Table 16.8.

16.11.4. Hypoglycemia Unawareness and Treatment of Hypoglycemia

Many individuals develop the syndrome of hypoglycemia unawareness, in which the warning symptoms that indicate that hypoglycemia is developing (e.g., tremulousness, tachycardia) are decreased or not detected. Without these warning symptoms, individuals are not able to take actions such as eating to prevent continued reductions in blood glucose levels and severe hypoglycemic episodes may result. Following episodes of hypoglycemia, counterregulatory hormone stores may not be available, and thresholds for symptoms of hypoglycemia may shift to lower glucose concentrations. Thus, patients with recurrent hypoglycemia may be particularly at risk for unawareness and for severely low hypoglycemic episodes. Failure to test blood glucose levels regularly can contribute to the problem of hypoglycemia unawareness. This cycle is particularly problematic for older adults who are highly physically active or who skip meals, do not eat sufficient quantities of food to match their insulin

Table 16.8
Dietary Management of Hypoglycemia

1. Check blood glucose level by glucose monitor
2. If blood glucose is less than 60 mg/dL or symptomatic, treat with 15 g of carbohydrate (½ cup juice, ½ cup regular soft drink, glucose gel)
3. Repeat blood sugar reading in 15 min after treatment and again after 60 min
4. Repeat step 2 until blood glucose is >60 mg/dL
5. If meals are due within 60 min, eat meal now
6. If meals are not due within 60 min, follow the glucose treatment with a snack containing carbohydrate and one protein (cheese and crackers, peanut butter and crackers, skim milk and crackers, or a small sandwich)
7. If blood glucose <40 mg/dL and/or subject is stuporous, confused, or unresponsive, give 1 amp of D50W as IV push and start D10W at 60 cc/h. Check blood glucose every 5 min and repeat till blood glucose >60 mg/dL or till awake. Give oral carbohydrate once awake

doses, or consume a high-fat diet, which delays carbohydrate absorption and is not accounted for in the timing of insulin administration.

Alcohol consumption, while not typically problematic when consumed in moderation, can pose risks for hypoglycemia in older adults taking insulin. In particular, the major risk of alcohol-related hypoglycemia is in persons in a fasting state and those who are alcohol dependent. The disinhibiting effect of alcohol poses the risk of hypoglycemia unawareness, making blood glucose monitoring essential. The potential for a delayed risk of hypoglycemia the morning after evening alcohol intake should also be emphasized *(70)*. The problem of patient hypoglycemia unawareness should be considered if the patient's HbA1c is low (e.g., <6.0), and if s/he describes inability to detect counterregulatory autonomic symptoms (e.g., tremulousness, pounding heart, anxiety, queasy stomach, sweating, flushed face) when blood glucose levels are low *(71)*. Potential barriers to blood glucose testing or adequate food consumption such as financial constraints, fear of pain, depression, or feelings of being overwhelmed by diabetes should be assessed.

Structured psychoeducational intervention and print materials to promote reduced hypoglycemia unawareness have been developed and systematically evaluated for nearly 25 years in the Blood Glucose Awareness Training (BGAT) program developed by Cox and colleagues *(69)*. The BGAT program focuses on improving the accuracy of patients' detection and interpretation of relevant blood glucose symptoms and other internal and external cues. Prospective, controlled studies including long-term follow-up indicate that training in BGAT results in improved accuracy of recognition of current blood glucose levels, improved detection of hypoglycemia in individuals with hypoglycemia unawareness, improved judgment regarding when to treat low blood glucose levels, reduced occurrence of severe hypoglycemia, improved judgment about not driving while hypoglycemic, reduction in the rate of motor vehicle violations, and better-preserved counterregulatory hormonal response during intensive insulin *(69,72)*. Overall, intensive training to promote blood glucose awareness can have significant and sustained benefits and aid in more consistent dietary management of seniors who take insulin.

16.11.5. Self-Management Behaviors

Few studies of diabetes self-management education (DSME) have focused on older adults. Intervention guidelines have been based on expert consensus rather than on empirical evidence *(73)*. Some older adults may have limited knowledge and/or understanding of diabetes care. Self-management steps such as home-based blood glucose monitoring, planning for meals and adjusting food intake based on blood glucose levels, and when (and how to adjust) insulin or take oral diabetes agents may be influenced by cognitive functioning, physical status, and personal preferences and resources. Physical barriers to optimal diabetes dietary intake may include swallowing difficulties, poor dentition, decreased thirst or appetite, and influence of medications on taste. Psychosocial influences may include limited financial resources, difficulties with transportation, and limited social support. Given the variability in the older adult population, it has been argued that individual self-management intervention may be preferable to group-based intervention. However, several studies have demonstrated positive outcomes with group intervention with seniors with diabetes. A simplified self-care regimen is optimal *(73)*.

16.12. DIABETES LIFESTYLE CHANGE AND PHYSICAL LIMITATIONS

Diabetes-related comorbidities such as diabetic retinopathy, cardiovascular disease, peripheral vascular disease, and congestive heart failure may result in decreased usual activity and limit activities of daily living (ADLs) and instrumental activities in daily living (IADLs), including limited transportation options, limited ability to shop for food or ability to read restaurant menus. Self-monitoring may also be influenced by reduced visual acuity. Diminished fine motor skills may also impact the ability to functionally conduct a finger stick, conduct the steps necessary for using the glucometer, and read the results. Due to such limitations, many older patients may require alternative choices of meters or assistance in blood glucose testing *(74)*. Comorbid health conditions increase polypharmacy, so pill boxes and mediplanners may also be helpful in simplifying medication management.

16.13. PSYCHOSOCIAL AND BEHAVIORAL ISSUES RELATED TO SELF-CARE AND DIETARY INTAKE IN OLDER ADULTS WITH DIABETES

16.13.1. Depression

The past decade has seen the emergence of a large and empirically based literature documenting the prevalence and impact of depression in persons living with diabetes *(75)*. A systematic review found a mean prevalence rate of depression in 23.4% of persons with diabetes. It appears that adults with diabetes who are depressed have average HbA1c levels 0.5–1.0% higher relative to those who are nondepressed. The depression–diabetes link may be particularly salient for older adults. Data from the Epidemiological Catchment Area study of more than 18,000 adults conducted in five sites found depressive symptoms in 15% of adults over age 65 and lifetime rate of depression in 2% of women and 3% of men *(76)*. The prevalence of depression in primary care geriatric clinic populations is estimated to be about 5%. In nursing homes, estimates of depression have ranged from 15 to 25% at any given point, with an incidence of 13% per year *(77,78)*. In older adults with diabetes, rates of depression are likely even higher. In a study of Medicare claimants, major depression was significantly more prevalent in those with diabetes and those with both diabetes and depression had more treatment visits, more time in inpatient stays, and had higher medical costs, even after excluding mental health treatment-related services *(79)*.

Prevalence and associations with depressive symptoms was examined in adults over age 65 with diabetes living in rural communities in the ELDER (Evaluating Long-term Diabetes Self-Management Among Elder Rural Adults) and Hispanic Established Population for the Epidemiologic Study of the Elderly (EPESE) studies. In ELDER, nearly 16% of the sample had depressive symptomology, regardless of ethnic group. Depressive symptoms were more common in those with lower functional status, lower income, more chronic health conditions, lower education, unmarried, and in women *(80)*.

In EPESE, death rates were substantially higher when a high level of depressive symptoms was comorbid with diabetes, cardiovascular disease, hypertension, stroke, and cancer. The odds of having died among persons with diabetes with high levels of depressive symptoms were three times that of diabetics without high levels of depressive symptoms. Hence, an interaction between depression and diabetes and the prevalence of other risk factors greatly increased absolute risk of mortality in these large studies of older adults.

In depressed seniors, indirect self-destructive behavior, such as not eating and medication nonadherence, may be more common than overt self-harming gestures such as suicide attempts and is associated with decreased survival. Many older adults might consider depression to be a normal part of aging and may not report their symptoms to a health-care provider. Health-care providers may also attribute some depressive symptoms to old age or other physical ailments or mood disturbance may be less prominent than multiple somatic complaints. Some older patients with depression may present with "failure to thrive" rather than specific complaints *(81)*. In a large HMO study of 4,463 patients with diabetes with an average age of 63 years, major depression was associated with poor adherence to self-care reflected by poor diet, inactivity, and lower medication adherence. No differences between depressed and nondepressed participants were observed for preventive health behaviors such as screenings for retinopathy and microalbumin and foot care and blood glucose self-monitoring *(82)*. Given the high rates of depression in this population, careful assessment of depressive symptomology and its impact on dietary intake, related aspects of diabetes self-care, and health outcomes are critical.

Recidivism of depression in persons with diabetes must also be considered when planning interventions for adults in their senior years, since their rates appear to be higher than in the general population *(83)*. Following treatment for depression, nonremission in adults with type 2 diabetes appears to be associated with lower adherence to blood glucose monitoring, higher HbA1c levels, and higher body weight *(84)*. Age differences in depression treatment response also appear to be relevant to the consideration of psychosocial issues in preventive care with older adults with diabetes. A multicenter, double-blind, placebo-controlled trial examined depression treatment in a maintenance treatment trial of adults with diabetes who had participated in a Sertraline treatment trial and had achieved depression recovery. Participants aged 55 and younger and those over age 55 were compared with regard to differences in time to depression recurrence. Younger participant group had significant increase in time to recurrence with Sertraline; however, the older group showed no intervention effect, due to a high placebo response rate. This suggests that older persons with diabetes may have unique issues with regard to optimal approaches for long-term management of depression *(85)*. Further, it is unclear if depression treatment actually improves metabolic control or self-care behaviors in depressed adults with diabetes. In the Pathways Study, participants with diabetes were randomized to depression intervention consisting of pharmacotherapy, problem-solving treatment, or both, or routine primary care. Symptoms of depression were reduced but HbA1c did not show improvement *(86)*. The enhanced depression care did not result in improvements in diabetes self-care behaviors including optimal dietary intake, physical activity, smoking cessation, and adherence to medications *(87)*. Results were similar for both younger and older (over age 60) study participants *(88)*.

16.13.2. Social Support

Social isolation is an established risk factor for morbidity and mortality in numerous disease states, with the largest body of literature linking it to CVD, a common outcome of diabetes *(89)*. One avenue in which social support may impact outcomes in chronic diseases such as diabetes is through its impact on self-care behaviors. Recently widowed persons may have limited cooking skills or access to shopping. Such persons may also be depressed and withdraw from usual daily activities, including social, food-related activities such as dining out or even preparing regular meals.

Interestingly, studies of the impact of social support and health behavior in persons with diabetes indicate substantial gender differences. Higher levels of social support have been associated with improved glycemic control in women with type 2 diabetes. However, several studies have found that a high level of perceived support is associated with less diabetes control in men, and it may be that forms of support that are satisfactory to men may reinforce patterns of eating, drinking, and exercise that are inconsistent with optimal diabetes self-care *(90)*. In older adults who continue to live with a spouse, husband's food preference, regardless of nutritional content, is often the best predictor of family meals that are eaten *(91)*. This indicates that not only is social isolation an important influence on dietary intake, but day-to-day social support and interactions among family and friends may both positively and negatively impact dietary intake in older persons with diabetes. Thus, it is imperative to evaluate the social context of patients' food purchases and dietary intake patterns.

Diabetes education programs that include the older adult's spouse may help to promote optimal social support for healthful dietary and lifestyle changes. Substantial improvements were found for diabetes knowledge, psychosocial functioning, and metabolic control in a 6-week diabetes education program for male diabetes patients aged 65–82 years of age, that included their spouses *(92)*.

These studies highlight that to provide optimal care, one must consider the social support system, provide the senior with encouragement and reinforcement for self-care behaviors, and provide instrumental assistance and technical advice as needed.

16.13.3. Cognitive Dysfunction

Both cross-sectional and prospective studies associate diabetes with cognitive decline in older adults *(93)*. Cognitive function is an important consideration since impaired function can impact comprehension of health-related information, including instructions regarding self-care such as procedures for conducting blood glucose testing, making adjustments in dietary intake, and taking multiple medications *(94)*. A number of studies have indicated that older adults with both type 1 and type 2 diabetes have impaired cognitive function compared with age-matched groups, and the level of impairment worsens with increases in hyperglycemia. Tightened blood glucose control, even in the absence of normoglycemia, can result in cognitive improvements in older persons with diabetes *(95,96)*.

Studies of cognitive functioning in diabetes in the senior years are confounded by the influence of other comorbid medical conditions, making it challenging to distinguish the independent contributions of diabetes. Despite inconsistent findings in studies comparing cognition in older adults with and without diabetes, overall it appears that diabetes is associated with impairment in more complex facets of cognitive functioning such as psychomotor efficiency and verbal memory. Influences likely impacting functioning include duration of illness, blood glucose control, and age. However, a review of studies of older adults with diabetes examined cognitive functioning and self-management behavior found few associations. Evaluation of minor cognitive impairment and diabetes self-management task performance in a small sample of older adults also found no significant influence on performance *(97,98)*. At present, it remains unclear how and at what threshold cognitive impairment influences acquisition of new diabetes self-care demands.

If frequent self-monitoring and adherence to specific dietary guidelines are within the cognitive abilities of the older patient with diabetes, then attainment of tight blood glucose control may be a reasonable goal. However, intensive self-management may not always be realistic for many cognitively impaired seniors. Unfortunately, despite the potential physical benefits, intensive management and tight control may require so many day-to-day demands that this may be difficult to practically achieve. Cognitive dysfunction can make adherence to dietary recommendations particularly difficult. For example, older adults who are cognitively impaired may not remember structured mealtimes that are coordinated with their insulin regimen. Difficulty in following a complicated meal plan, such as carbohydrate counting or using a sliding scale to match insulin units to intake, may make such treatment regimens too overwhelming to be practical. For some older individuals, a concrete, structured meal plan can help minimize ambiguity regarding their diabetes diet. Such plans may be made in conjunction with a diabetes educator or a dietitian. Helpful strategies include using cues in the environment such as regular meals, setting alarms, providing written information with large print and pictures, training videotapes, and assessing comprehension and skill by asking for demonstrations. In addition, provision of home-based caretakers or meal services may also assist the older person with cognitive impairment or significant physical disability or other barriers to obtain access to optimal nutrition that is consistent with diabetes goals.

16.13.4. Personal Beliefs and Nutrition in Older Adults with Diabetes

Older patients' personal views of diabetes appear to substantially impact their levels of diabetes self-care. A study of adults over age 60 with type 2 diabetes found that perceptions regarding the cause of diabetes, treatment effectiveness, and seriousness of one's diabetes were all significantly associated with quality of life and negative affect. Beliefs regarding treatment effectiveness were particularly predictive of dietary intake and physical activity *(99)*. Studies of women reflecting heterogeneous ethnic groups and socioeconomic status indicate that they conceptualize diabetes in terms of their own cognitive explanatory models, which are not necessarily congruent with the way in which health providers conceptualize diabetes *(100)*.

An interview study was conducted with adults with type 2 diabetes over age 65 to assess their experiences with diabetes and their health goals and practices. The majority were African American (79%) and female (57%) and had diabetes-related comorbidities. Interview synthesis indicated that this sample focused their health-care goal outcomes from a functional and social perspective, rather than a biomedical focus. Medical professions were cited as influential (by 43% of participants); however, friends, family, and media also appeared to be critical sources of influence in shaping expectations and care goals. Participants' personal diabetes care goals predominantly centered on activities of daily living and maintaining their independence and avoiding becoming a burden to family (71%). Friends and family experiences with medical conditions and events shaped goals (50%). Findings also suggested that these older participants did not differentiate between the relative importance of different treatment goals and the prevention of different health complications. Diet and exercise were underemphasized relative to medication management *(101)*. Effective interactions with older patients should consider their experiences and influences and focus on quality of life and patient definitions of their health-care goals. Clearly communicating priorities in self-care tasks may facilitate adherence to medical goals and optimal care.

Many older adults may not be aware of the link between diabetes and cardiovascular disease and may benefit from enhanced education to promote awareness. A study of 1,109 adults over age 45 with diabetes randomly sampled from a large US mixed model managed care organization. Adults age 65 and older were compared to those 45–64 years of age. Older adults had longer duration of diabetes and higher rates of heart disease and cardiovascular comorbidities but were less likely to relate these health comorbidities to their diabetes *(102)*.

Collaborative provider–patient discussions regarding self-care, and personal preferences and goals, and awareness of personal risk factors, may enhance patient motivation and treatment engagement. Integrating patient values and beliefs about health and quality of life, considering available emotional and social support and financial resources, and addressing problem solving regarding goals and treatment adjustments may best promote long-term regimen adherence *(94)*.

16.14. CULTURAL ISSUES AND NUTRITION IN DIABETES

Research also suggests that ethnic minority and older adults may have culturally unique health-related perspectives that are not effectively targeted by traditionally delivered health promotion interventions *(103)*. While being healthy appears to be important and a general awareness of what to do to stay healthy is evident, operational definitions of health in these populations are often somewhat different than that typically used in prevention and health promotion efforts. For example, focus group studies with underserved ethnic minorities found that a prevailing belief was that better health behaviors could build resistance to acute illnesses and keep them healthy but that chronic diseases such as diabetes were due to fate and heredity and beyond their individual control. In general, participants did not appear to make the cognitive "link" between chronic disease prevention and the importance of diet, physical activity, and weight control. Most participants expressed an interest in "doing better" but were not able to specify how such healthful changes might be made.

Qualitative evaluations of cultural influences on diabetes reveal the complexity of psychosocial influences on diabetes lifestyle change and why traditional health provider perspective-based dietary interventions with minority persons often fail. An interview-based study of 20 middle-aged Mexican American women with type 2 diabetes revealed that their personal understanding and interpretation of their diabetes was most heavily based on their family's experiences and on community influences *(100)*. From the participants' perspectives, the severity of their diabetes was indicated by being treated with insulin injections and the provider being vigilant, while treatment with oral medications and the perception that providers had a lax attitude were taken to mean that the diabetes was not severe. Having diabetes was also viewed as a confusing, silent illness and provider provision of information was often viewed as insufficient. Participant comments revealed that many found that provider comments were predominantly focused on negative aspects of their behavior, were confrontational, and at times, petty or demeaning. Provider focus on positive gains to be made with behavior change, reinforcement for accomplishments, and avoiding pejorative terms (e.g., obese) may go a long way in engaging many patients and enhancing a more collaborative relationship.

The strong influence of family and culture on adherence to a diabetes diet and lifestyle change is also evident in focus group-based qualitative research with African American women with type 2 diabetes. Factors influencing optimal diabetes diet and physical activity behaviors were evaluated in a study of 70 southern, predominantly rural African American women, of whom 65% were age 55 and older *(104)*. These women described the psychological impact of diabetes as being stronger than the physical impact, and the psychological issues reported included feelings of nervousness, fatigue, worrying, and having feelings of dietary deprivation, including craving for sweets. Participants reported considerable life stress other than diabetes, particularly having a multi-caregiver role. Family members complaining about and resistant toward healthy food preparation methods was common. Positive family support for diabetes was evident in the form of instrumental support from adult daughters or other female family members or friends to older, single, or widowed women. In addition, spirituality and religiosity emerged as a main theme in all groups. Spirituality was largely viewed as a primary source of emotional support, a positive influence on diabetes, and a contributor to quality of life. This study exemplifies the importance of incorporating family and the church in self-care behaviors of many southern African American women.

These findings demonstrate why consideration of the social and cultural context of older adult's lives is critical for the development of interventions to promote diet and lifestyle change. Collaborative patient–provider care and family-centered and church-based approaches may offer more appropriate avenues for efforts to promote optimal diabetes care and maximize the effective delivery of dietary interventions for many seniors with diabetes who are frequently overrepresented in ethnic minority populations.

16.15. QUALITY OF LIFE AND DIABETES DIET IN OLDER ADULTS

In considering prescribing a diabetes diet or when addressing issues related to dietary adherence, it is important to consider the impact of dietary change on the individual's quality of life. Diabetes itself poses numerous challenges and the many lifestyle demands are among the most difficult for patients. Both type 1 and type 2 diabetes appear to have an impact on health-related quality of life *(105)*. Diabetes-specific quality-of-life issues are associated with overall well-being, and dietary restrictions and daily hassles related to diabetes care are significantly associated with treatment satisfaction as well as with general well-being. A liberalized diet and flexible insulin therapy are among the diabetes-related factors most associated with favorable quality of life in persons with type 1 diabetes *(106)*.

For some seniors with diabetes, functional status is an important quality-of-life care consideration. For individuals with limited cognitive function or multiple comorbidities, the primary goal of care may be achieving a satisfactory quality of life rather than aggressive treatment regimens. In other cases, for those who are robust, goals and care may be similar to those for younger persons with diabetes. The AGS guidelines for improving care of older persons address issues that are more common in the older diabetes population, including depression, pain, falls with injury, and declining functional status. Consistent with research, the guidelines suggest that goals be developed based on the functional status and personal desires *(107)*. Of note, studies of functional status and health outcomes in older adults with diabetes have found that persons with low functional level (defined by three or more limitations in IADLs or ADLs) did not benefit from aggressive intervention aimed at blood glucose control *(108)*.

As highlighted in the previous sections, psychosocial and functional influences play a substantial role in the dietary and lifestyle behaviors of older adults with diabetes.

16.16. BEHAVIOR CHANGE INTERVENTION STUDIES OF SELF-MANAGEMENT AND LIFESTYLE CHANGE IN OLDER ADULTS WITH DIABETES

Few studies of diabetes self-management education (DSME) have focused on older adults. Intervention guidelines have been based on expert consensus rather than on empirical evidence *(73)*. It is unclear if strategies for DSME that are effective with younger patients are optimal for older adults. Consideration of issues already discussed in this chapter highlights the unique self-management issues facing many seniors with diabetes. Generational cohort influences may influence readiness for receipt of detailed self-care information. Many of the oldest of older adults may have limited knowledge or understanding of diabetes care. Self-management steps may be influenced by cognitive functioning, physical status, and personal preferences and resources. Physical barriers to optimal diabetes dietary intake may include swallowing difficulties, poor dentition, decreased thirst or appetite, and influence of medications on taste. Psychosocial influences may include limited financial resources, difficulties in transportation, and limited social support. Providing DSME content over multiple contacts and use of memory cues such as large print handouts and cues to use at home can be helpful. A simplified self-care regimen is optimal with goals not only to maintain diabetes control but also to have good quality of life *(73)*. Many clinicians advocate individual DSME for older adults; however, several studies have indicated that group education is effective. As can be seen from the few examples of controlled group diabetes education programs specifically developed for older adults, participants in intensive programs have tended to be younger, from the baby boomer generation, limiting generalizability to the larger geriatric diabetes population.

An example of one of the few theoretically based lifestyle change intervention programs developed specifically for older adults with diabetes is the "Sixty-something . . ." study *(109)*. Principles of SCT were used to develop a 10-session self-management training program, with 102 adults over 60 years of age with type 2 diabetes. Subjects were randomly assigned to immediate or delayed intervention conditions. The intervention taught problem-solving skills and strategies for enhancing self-efficacy for overcoming personal barriers to adhering to their diabetes diet and other aspects of the diabetes regimen. The immediate intervention produced greater reductions in caloric intake and percent of calories from fat, greater weight loss and increases in the frequency of blood glucose testing compared to delayed controls. Improvements were generally maintained at 6-month follow-up. Results from subjects receiving the delayed intervention closely approximated those for the immediate intervention subjects.

Successful lifestyle change was also achieved in the Mediterranean Lifestyle Program, a randomized intervention trial with 279 postmenopausal women with type 2 diabetes aimed at reducing CHD risk factors. The intervention component consisted of a 3-day retreat and 6 months of weekly meetings with an emphasis on a Mediterranean low-saturated-fat diet, moderate physical activity, stress management, smoking cessation, and group support. Behavior change was greater in participants receiving the intervention than in the usual care control group with greater changes in eating patterns, stress management, activity level, and smoking cessation. Intervention resulted in improvements in BMI, HBA1c, plasma fatty acids, and quality of life. Feasibility of this intensive approach to lifestyle change is an important issue for consideration; participants in the Mediterranean diet arm had a higher rate of attrition relative to those in the control condition *(110)*.

Chodosh and colleagues conducted a meta-analysis of self-management intervention studies that compared outcomes with usual care or control condition for diabetes, osteoarthritis, or hypertension. Twenty diabetes studies were examined for impact on HbA1c. Pooled effect size was –0.36 in favor of self-management intervention, indicating lower HbA1c in treatment groups. For fasting blood glucose outcomes, the pooled effect size for 14 studies was –0.28, equating to a decrease of 0.95 mmol/L in blood glucose level. No statistically significant differences were observed in weight change between intervention and control conditions in the 13 studies reporting this outcome. Only three studies primarily focused on diet and education; these had a pooled effect size of –0.062 for HbA1c. Post hoc analyses aimed at understanding key, essential elements of the self-management intervention programs did not support any key program aspects *(111)*.

16.17. DIET-SPECIFIC INTERVENTION IN OLDER ADULTS WITH DIABETES

A randomized nutrition education intervention was developed for older adults with type 2 diabetes. Participants were age 65 and older and without functional or cognitive impairment. The 10-week intervention was based on Social Cognitive Theory and meaningful learning approaches aimed at minimizing the presentation of too much information and maximizing learning by breaking information down into small pieces and successively adding upon each concept across the intervention. Strategies included limiting the content introduced at each session, meaningfully organizing the content, integrating preexisting knowledge with new information, and modeling and in-session practice regarding decision making related to reading food labels, food purchasing, meal planning, and using the information for diabetes self-care. Goal setting, self-monitoring, and feedback were also utilized. The intervention resulted in improved glycemic control *(112)*, greater total knowledge, positive outcome expectancies, decision-making skills, and reduced self-management barriers *(113)*.

16.18. WEIGHT LOSS INTERVENTION IN OLDER ADULTS WITH DIABETES

Weight loss issues that must be considered for older adults with diabetes include the impact of restrictions on quality of life and potential loss of lean muscle mass from decreased protein intake. In research undertaken with younger adults with diabetes, weight loss programs that combine diet, physical activity, and theoretically guided behavior change techniques have been shown to be the most effective over the short term *(3)*. Hypocaloric diets, which can improve glucose tolerance and lipid levels, may also be appropriate for older, obese persons with diabetes. Behavioral weight control interventions with persons with type 2 diabetes have found that even reductions of approximately 10% of weight loss can decrease hypertension and lipid abnormalities and improve glycemic control, with improvements related to the magnitude of weight loss *(3)*.

The Look AHEAD trial is the first large, multisite clinical trial with type 2 diabetes that compared intensive weight loss intervention to support and education. This randomized controlled trial was conducted at multiple centers, with 5,145 adults aged 45–74 with type 2 diabetes. Group and individual meetings focused on intensive lifestyle change with weight loss and maintenance as the primary goal. Restriction of caloric intake was the primary approach with a goal of limiting total fat calories to 30%, with a maximum of 10% saturated fat and minimum of 15% from protein. Portion controlled diets were prescribed as were structured meal plans. Home-based exercise with gradually increasing goals was utilized. Walking was encouraged although other moderate-intensity activities could be chosen. A toolbox approach using algorithms and assessments of progress was used. After the initial 6-month intervention period, toolbox options included Orlistat and/or advanced behavioral approaches for participants with difficulties in achieving goals. At 1-year follow-up, intervention participants had lost an average of 8.6% body weight, with greater weight loss, increased cardiovascular fitness, and improved cardiovascular risk factors relative to participants in the control condition. Use of glucose-lowering medications decreased in intervention participants and increased in control group participants. Use of antihypertensive medications increased in control group participants but remained unchanged in the intervention group. Use of lipid-lowering medications increased in both intervention and control groups, with smaller increases in intervention participants. Findings highlight that substantial weight loss is feasible in persons with type 2 diabetes. Notably, even participants on insulin lost an average of 7.6% of bodyweight from baseline. Follow-up of trial participants is planned for up to 11.5 years and will provide evidence regarding ongoing risk factor improvements and maintenance of health gains *(114)*.

16.19. IMPARTING DIABETES DIETARY INFORMATION TO OLDER ADULTS

In order to impart diabetes diet information in a fashion that will lead to actual changes in behavior and maintenance of these changes, it is critical to consider the psychosocial and cultural influences that are present for each individual patient. Simple and concrete statements such as "eat less fat" or "eat less food" or "get more walking in each day" may promote learning and minimize failure. Nutrition information is best presented in sequenced manageable steps that can then be individualized to the patient's setting. Simple tip sheets and problem-solving approaches discussed in earlier sections may also be helpful. The National Diabetes Education Program (NDEP) has prepared materials adapted from those used in the DPP for use by primary care providers for middle-age and older adults. These materials address motivational approaches that consider readiness for change, behavioral relapse. Materials can help to assess a patient's personal readiness for change and help in setting up a walking program. This toolkit, the NDEP GAMEPLAN (Goals, Accountability, Monitoring, Effectiveness, Prevention through a Lifestyle of Activity and Nutrition), is copyright free and contains health-care provider information with background information and patient handouts that may be copied (http://ndep.nih.gov/diabetes/pubs/GPToolkit.pdf). For provider information, prediabetes risk factors are presented along with diagnostic decision trees, screening approaches are reviewed, DPP findings are summarized, and commonly asked primary care provider questions are addressed *(64)*.

It is also important to be mindful of the range of functioning in older adults. Older adults of the World War II generation have tended to be characterized as somewhat reverential toward physicians and the health-care system. However, baby boomers, who are now entering the realm of older adulthood, tend to differ from previous generations and tend to have high expectations of their health providers and desire additional information. This generation of "new" older adults tends to want a collaborative relationship with their health-care provider and desire additional information including resources such as self-help publications, Internet, video, and audiotapes. They demand convenience, expect hard evidence of quality and expertise, can be skeptical of advice at face value, and are often willing to explore alternative therapies *(111)*. In order to meet the needs of the range of older adults with diabetes, it is clear that a "one size fits all" approach will not be effective. Rather issues related to culture and ethnicity and generational cohort must be considered.

16.20. RECOMMENDATIONS

1. Establish the type of diabetes and medication regimen in order to appropriately integrate dietary goals.
2. Consider the importance of cardiovascular risk—including obesity and lipids—in developing diabetes dietary goals and routine assessment.
3. Work with the patient to set a goal of achieving and maintaining reasonable body weight. For obese seniors, moderate weight loss may achieve dramatic results and exercise may greatly enhance dietary intervention. Maintenance of behavior change and weight loss is critical. For underweight adults, focus on promotion of optimal nutritional intake and functional status.
4. Educate older adults with diabetes about the rationale for diet and lifestyle change and link to health outcomes; promote self-efficacy for change.
5. Consider the risk of hypoglycemia for older adults taking insulin—particularly those with poor nutritional status, cognitive dysfunction, polypharmacy and comorbid illness. Encourage frequent self-monitoring and dietary self-treatment and preventive strategies.
6. Assess seniors' specific dietary patterns such as food choices, quantity eaten, and unplanned snacking and the lifestyle contexts in which they occur.
7. Address psychosocial issues that may influence dietary intake, including depression, social support, cognitive status, attitudes and perceptions, and the impact of the diabetes regimen on quality of life.
8. Address and intervene within individuals' cultural context.
9. Provide a collaborative relationship with each patient, offer resources and provide concrete, behavioral strategies to promote behavior change.
10. When appropriate, provide self-help materials including Internet and National Diabetes Education Program resources (e.g., from the www.ada.org, CDC http://www.cdc.gov/diabetes/, AADE www.aadenet.org, or NIH http://www2.niddk.nih.gov/.

REFERENCES

1. Seidell J. Obesity, insulin resistance and diabetes—a worldwide epidemic. Br J Nutr 2000;83(Suppl. 1):S5–S8.
2. Visscher TL, Seidell JC. The public health impact of obesity. Annu Rev Public Health 2001;22:355–375.
3. Wing RR, Goldstein MG, Acton KJ, Birch LL, Jakicic JM, Sallis JF Jr, Smith-West D, Jeffery RW, Surwit RS. Behavioral science research in diabetes: lifestyle changes related to obesity, eating behavior, and physical activity. Diabetes Care 2001;24(1):1–2.
4. King H, Aubert RE, Herman WH. Global burden of diabetes 1995–2025: prevalence, numerical estimates, and projections. Diabetes Care 1998;21 (9):1414–1431.
5. Ratner R. Type 2 diabetes mellitus: the grand overview. Diabetic Med 1998;15 (Suppl. 4):S4–S7.

6. Shorr RI, Franse LV, Resnick HE, DiBari M, Johnson KC, Pahor M. Glycemic control of older adults with type 2 diabetes: findings from the Third National Health and Nutrition Examination Survey, 1988–1994. J Am Geriatr Soc 2000;48 (3):264–267.
7. Howard BV, Magee MF. Diabetes and cardiovascular disease. Curr Atheroscler Rep 2000;2 (6):476–481.
8. Report of the Expert Committee on the diagnosis and classification of diabetes mellitus. Diabetes Care 1997;20: 1183–1197.
9. Organization WH. Definition, diagnosis and classification of diabetes mellitus and its complications: report of a WHO consultation. Part 1, Diagnosis and Classification of Diabetes Mellitus (WHO/NCD/NCS/99.2). Geneva: World Health Org; 1999
10. The Expert Committee on the Diagnosis and Classification of Diabetes. Follow-up report on the diagnosis of diabetes mellitus. Diabetes Care 2003;26:3160–3167.
11. Morley J. An overview of diabetes mellitus in older persons. Clin Geriatr Med 1999;15(2):211–224.
12. Dagogo-Jack S, Santiago JV. Pathophysiology of type 2 diabetes and modes of action of therapeutic interventions. Arch Intern Med 1997;157(16):1802–1817.
13. Elahi D, Muller DC. Carbohydrate metabolism in the elderly. Eur J Clin Nutr 2000;54(Suppl. 3):S112–S120.
14. Beaufrere B, Morio B. Fat and protein redistribution with aging: metabolic considerations. Eur J Clin Nutr 2000;54 (Suppl. 3):S48–S53.
15. Chiu KC, Lee NP, Cohan P, Chuang LM. Beta cell function declines with age in glucose tolerant Caucasians. Clin Endocrinol (Oxf) 2000;53(5):569–575.
16. Falorni A, Kockum I, Sanjeevi CB, Lernmark A. Pathogenesis of insulin-dependent diabetes mellitus. Baillieres Clin Endocrinol Metab 1995;9(1):25–46.
17. Anonyms. The diabetes control and complications trial research group. N Engl J Med 1993; 329 (14): 977–986.
18. Turner RC, Holman RR. Lessons from UK prospective diabetes study. Diabetes Res Clin Pract 1995;28:S151–S157.
19. Brown AF, Mangione CM, Saliba D, Sarkisian CA. Guidelines for improving the care of the older person with diabetes mellitus. J Am Geriatr Soc 2003;51 (5 Suppl.):S265–S280.
20. Adler AI, Stratton IM, Neil HA, Yudkin JS, Matthews DR, Cull CA, Wright AD, Turner RC, Holman RR. Association of systolic blood pressure with macrovascular and microvascular complications of type2 diabetes (UKPDS 36): prospective observational study. BMJ 2000;321(7258):412–419.
21. Hansson L, Zanchetti A, Carruthers SG, Dahlof B, Elmfeldt D, Julius S, Menard J, Rahn KH, Wedel H, Westerling S. Effects of intensive blood-pressure lowering and low-dose aspirin in patients with hypertension: principal results of the Hypertension Optimal Treatment (HOT) randomized trial. HOT Study Group. Lancet 1998;351(9118):1755–1762.
22. Villarosa IP, Bakris GL. The appropriate blood pressure control in diabetes (ABCD) trial. J Hum Hypertens 1998;12(9):653–655.
23. Goldberg RB, Mellies MJ, Sacks FM, Moye LA, Howard BV, Howard WJ, Davis BR, Cole TG, Pfeffer MA, Braunwald E. The CARE Investigators. Cardiovascular events and their reduction with pravastatin in diabetic and glucose-intolerant myocardial infarction survivors with average cholesterol levels: subgroup analyses in the cholesterol and recurrent events (CARE) trial. Circulation 1998;8(98):2513–2519.
24. Haffner SM. The Scandinavian Simvastatin Survival Study (4S) subgroup analysis of diabetic subjects: Implications for the prevention of coronary heart disease. Diabetes Care 1997;20(4):469–471.
25. Helmy T, Patel AD, Alameddine F, Wenger NK. Management strategies of dyslipidemia in the elderly: review. Arch Gen Med 2005;7 (4):8.
26. Sever PS, Dahlöf B, Poulter NR, et al. Prevention of coronary and stroke events with atorvastatin in hypertensive patients who have average or lower-than-average cholesterol concentrations, in the Anglo-Scandinavian Cardiac Outcomes Trial—Lipid Lowering Arm (ASCOT-LLA): a multicentre randomised controlled trial. Lancet 2003;361: 1149–1158.
27. Shepherd J, Blauw GJ, Murphy MB, et al. Pravastatin in elderly individuals at risk of vascular disease (PROSPER): a randomised controlled trial. Lancet 2002;360:1623–1630.
28. Cannon CP, Braunwald E, McCabe CH, et al. Intensive versus moderate lipid lowering with statins after acute coronary syndromes. N Engl J Med 2004;350:1495–1504.
29. LaRosa JC, Grundy SM, Waters DD, et al. Treating to New Targets (TNT) Investigators. Intensive lipid lowering with atorvastatin in patients with stable coronary disease. N Engl J Med 2005;352:1425–1435.
30. Group. HPSC. MRC/BHF Heart Protection Study of cholesterol lowering with simvastatin in 20,536 high-risk individuals: a randomised placebo-controlled trial. Lancet 2002;360:7–22.
31. Downs JR, Clearfield M, Weis S, et al. Primary prevention of acute coronary events with lovastatin in men and women with average cholesterol levels: results of AFCAPS/TexCAPS. Air Force/Texas Coronary Atherosclerosis Prevention Study. JAMA 1998;279:1615–1622.

32. Shepherd J, Cobbe SM, Ford I, et al. Prevention of coronary heart disease with pravastatin in men with hypercholesterolemia. West of Scotland Coronary Prevention Study Group. N Engl J Med 1995;333:1301–1307.
33. Sacks FM, Pfeffer MA, Moye LA, et al. The effect of pravastatin on coronary events after myocardial infarction in patients with average cholesterol levels. Cholesterol and Recurrent Events Trial investigators. N Engl J Med 1996;335:1001–1009.
34. Hunt D, Young P, Simes J, et al. Benefits of pravastatin on cardiovascular events and mortality in older patients with coronary heart disease are equal to or exceed those seen in younger patients: results from the LIPID trial. Ann Intern Med 2001;134:931–940.
35. Anonyms. Randomised trial of cholesterol lowering in 4444 patients with coronary heart disease: the Scandinavian Simvastatin Survival Study (4S). Lancet 1994;344:1383–1389.
36. Garg A, Bonanome A, Grundy SM, Zhang ZJ, Unger RH. Comparison of a high-carbohydrate diet with a high-monosaturated fat diet in patients with non-insulin-dependent diabetes mellitus. N Engl J Med 1988;319(13):829–834.
37. Halton TL, Hu FB. The effects of high protein diets on thermogenesis, satiety and weight loss: a critical review. J Am Coll Nutr 2004;23:373–385.
38. Salahudeen AK. Halting progression of renal failure: consideration beyond angiotensin II inhibition. Nephrol Dial Transplant 2002;17(11):1871–1875.
39. Sheard NF, Clark NG, Brand-Miller JC, Franz MJ, Pi-Sunyer FX, Mayer-Davis E, Kulkarni K, Geil P. Dietary carbohydrate (Amount and Type) in the prevention and management of diabetes: a statement by the American diabetes association. Diabetes Care 2004;27(9):2266–2271.
40. Jovanovic L, Gondos B. Type 2 diabetes: the epidemic of the new millennium. Ann Clin Lab Sci 1999;29(1):33–42.
41. Locher JL, Roth DL, Ritchie CS, Cox K, Sawyer P, Bodner EV, Allman RM. Body mass index, weight loss, and mortality in community-dwelling older adults. J Gerontol A Biol Sci Med Sci 2007;62:1389–1392.
42. Mooradian AD, Osterweil D, Petrawek D, Morley JE. Diabetes Mellitus in elderly nursing home patients. J Am Geriatr Soc 1988;36:391–396.
43. Garber AJ, Moghissi ES, Bransome ED Jr, et al. American College of Endocrinology position statement on inpatient diabetes and metabolic control. Endocr Pract 2004;10:77–82.
44. Clement S, Braithwaite SS, Magee MF, Ahmann A, Smith EP, Schafer RG, Hirsch IB; The American Diabetes Association Diabetes in Hospitals Writing Committee. Management of diabetes and hyperglycemia in hospitals. Diabetes Care 2004;27:553–591.
45. Inzucchi SE. Management of hyperglycemia in the hospital setting. N Engl J Med 2006;355(18):1903–1911.
46. Swift CS, Boucher JL. Nutrition therapy for the hospitalized patient with diabetes. Endocr Pract 2006;12 (Suppl. 3):61–67.
47. Boucher JL, Swift CS, Franz MJ, Kulkarni K, Schafer RG, Pritchett E, Clark NG. Inpatient management of diabetes and hyperglycemia: implications for nutrition practice and the food and nutrition professional. J Am Diet Assoc 2007;107(1):105–111.
48. Association AD. Diabetes nutrition recommendations for health care institutions (Position Statement). Diabetes Care 2004;27(Suppl. 1):S55–S57.
49. Coulston AM, Mandelbaum D, Reaven GM. Dietary management of nursing home residents with non-insulin-dependent diabetes mellitus. Am J Clin Nutr 1990;51:67–71.
50. Tariq SH, Karcic E, Thomas DR, Thomson K, Philpot C, Chapel DL, Morley JE. The use of a no-concentrated-sweets diet in the management of type 2 diabetes in nursing homes. J Am Diet Assoc 2001;101:1463–1466.
51. Wright J. Total parenteral nutrition and enteral nutrition in diabetes. Curr Opin Clin Nutr Metab Care 2000;3:5–10.
52. Garg A. High-MUFA diets for patients with DM: a meta-analysis. Am J Clin Nutr 1998;67 (Suppl 3):577S–582S.
53. Hongsermeier T, Bistrian BR. Evaluation of a practical technique of determining insulin requirements in diabetic patients receiving total parenteral nutrition. J Parenter Enter Nutr 1993;17:16–19.
54. Expert Panel on Detection E, and Treatment of High Blood Cholesterol in Adults. Executive summary of the third report of the National Cholesterol Education Program (NCEP) Expert Panel on Detection, Evaluation, and Treatment of High Blood Cholesterol in Adults (Adult Treatment Panel III). JAMA 2001;285:2486–2497.
55. Ford ES, Giles WH, Dietz WH. Prevalence of the metabolic syndrome among US adults: findings from the third National Health and Nutrition Examination Survey. JAMA 2002;287:356–359.
56. Stern MP, Williams K, Gonzalez-Villalpando C, Hunt KJ, Haffner SM. Does the metabolic syndrome improve identification of individuals at risk of type 2 diabetes and/or cardiovascular disease? Diabetes Care 2004;27:2676–2681.
57. Ninomiya JK, L'Italien G, Criqui MH, Whyte JL, Gamst A, Chen RS. Association of the metabolic syndrome with history of myocardial infarction and stroke in the Third National Health and Nutrition Examination Survey. Circulation 2004;109:42–46.

58. Lorenzo C, Okoloise M, Williams K, Stern MP, Haffner SM. The metabolic syndrome as predictor of type 2 diabetes: the San Antonio Heart Study. Diabetes Care 2003;26:3153–3159.
59. Kahn R, Buse J, Ferrannini E, Stern M. The metabolic syndrome: time for a critical appraisal: joint statement from the American Diabetes Association and the European Association for the Study of Diabetes. Diabetes Care 2005;28:2289–2304.
60. Group. DPPR. Reduction in the incidence of type 2 diabetes with lifestyle intervention or metformin. N Engl J Med 2002;346:393–403.
61. Crandall J, Schade D, Ma Y, Fujimoto WY, Barrett-Conner E, Fowlder S, Dagogo-Jack S, Andres R. for the Diabetes Prevention Program Research Group. The influence of age on the effects of lifestyle modification and metformin in prevention of diabetes. J Gerontol 2006;61A (10):1075–1081.
62. DREAM (Diabetes REduction Assessment with ramipril and rosiglitazone Medication) Trial Investigators GH, Yusuf S, Bosch J, Pogue J, Sheridan P, Dinccag N, Hanefeld M, Hoogwerf B, Laakso M, Mohan V, Shaw J, Zinman B, Holman RR. Effect of rosiglitazone on the frequency of diabetes in patients with impaired glucose tolerance or impaired fasting glucose: a randomised controlled trial. Lancet 2006; 368 (9541): 1096–1105.
63. Wannamethee SG, Shaper AG, Whincup PH. Modifiable lifestyle factors and the metabolic syndrome in older men: effects of lifestyle changes. J Am Geriatr Soc 2006;S4:1909–1914.
64. Kelly JM, Marrero DG, Gallivan J, Leontos C, Perry S. Diabetes prevention: a GAMEPLAN for success. Geriatrics 2004;59 (7):26.
65. Horwath CC, Worsley A. Dietary habits of elderly persons with diabetes. J Am Diet Assoc 1991;91(5):553–557.
66. Cross AT, Babicz D, Cushman LF. Snacking habits of senior Americans. J Nutr Elder 1995;14 (2–3):27–38.
67. James-Enger K. Express lane. Diabetic cooking snack savvy. Diabetes Forecast 2002;55(6):73–75.
68. Janssen MM, Snoek FJ, de Jongh RT, Casteleijn S, Deville W, Heine RJ. Biological and behavioural determinants of the frequency of mild, biochemical hypoglycaemia in patients with type 1 diabetes on multiple injection therapy. Diabetes Metab Res Rev 2000;16(3):157–163.
69. Cox DJ, Gonder-Frederick L, Polonsky W, Schlundt D, Kovatchev B, Clarke W. Blood Glucose Awareness Training (BGAT-2): Long-term benefits. Diabetes Care 2001;24(4):637–642.
70. Meeking DR, Cavan DA. Alcohol ingestion and glycemic control in patients with insulin-dependent diabetes mellitus. Diabet Med 1997;14(4):279–283.
71. Bolli GB. How to ameliorate the problem of hypoglycemia in intensive as well as nonintensive treatment of type 1 diabetes. Diabetes Care 1999;22(Suppl. 2):B43–B52.
72. Wild D, von Maltzahn R, Brohan E, Christensen T, Clauson P, Gonder-Frederick L. A critical review of the literature on fear of hypoglycemia in diabetes: implications for diabetes management and patient education. Patient Educ Couns 2007;68:10–15.
73. Suhl E, Bonsignore P. Diabetes self-management education for older adults: general principles and practical application. Diabetes Spectr 2006;19(4):234–240.
74. American Association of Diabetes Educators. Special considerations for the education and management of older adults with diabetes. Diabetes Educ 2000; 26 (1): 37–39.
75. Lustman PJ, Penckofer SM, Clouse RE. Recent advances in understanding depression in adults with diabetes. Curr Diab Rep 2007;4:114–122.
76. Fombonne E. Increased rates of depression: update of epidemiological findings and analytic problems. Acta Psychiatr Scand I994;90:145–156.
77. Consensus Panel. Diagnosis and treatment of depression in late life. JAMA 1992; 268 (8): 1018–1024.
78. Slater SL, Katz IR. Prevalence of depression in the aged: Formal calculations versus clinical facts. J Am Geriatr Soc 1995;43:778–779.
79. Finkelstein EA, Bray JW, Chen H, Larson MJ, Miller K, Tompkins C, Keme A, Manderscheid R. Prevalence and costs of major depression among elderly claimants with diabetes. Diabetes Care 2003;26:415–420.
80. Bell RA, Smith SL, Arcury TA, Snively BM, Stafford JM, Quandt SA. Prevalence and correlates of depressive symptoms among rural older African Americans, Native Americans, and Whites with diabetes. Diabetes Care 2005;28(4):823–829.
81. Sarkisian CA, Lachs MS. "Failure to thrive" in older adults. Ann Intern Med 1996;124:1072–1078.
82. Lin EHB, Katon W, Von Korff M, Rutter C, Simon GE, Oliver M, Ciechanowski P, Ludman EJ, Bush T, Young B. Relationship of depression and diabetes self-care, medication adherence, and preventive care. Diabetes Care 2004;27(9):2154–2160.
83. Lustman P, Griffith L, Freedland K, Clouse R. The course of major depression in diabetes. Gen Hosp Psychiatry 1997;19:138–143.

84. Lustman P, Freedland K, Griffith LS. Clouse RE. Predicting response to cognitive behavior therapy of depression in type 2 diabetes. Gen Hosp Psychiatry 1998;20:302–306.
85. Williams MM, Clouse RE, Nix BD, Rubin EH, Sayuk GS, McGill JB, Gelenberg AJ, Ciechanowski PS, Hirsch IB, Lustman PJ. Efficacy of sertraline in prevention of depression recurrence in older versus younger adults with diabetes. Diabetes Care 2007;30(4):801–806.
86. Katon W, Von Korff M, Ciechanowski P, Russo J, Lin E, Simon G, Ludman E, Walker E, Bush T, Young G. Behavioral and clinical factors associated with depression among individuals with diabetes. Diabetes Care 2004;27(4):914–920.
87. Lin EHB, Katon W, Rutter C, Simon GE, Ludman EJ, Von Korff M, Young B, Oliver M, Ciechanowski PC, Kinder L, Walker E. Effects of enhanced depression treatment on diabetes self-care. Ann Fam Med 2006;4(1):46–53.
88. Williams JW, Katon W, Lin EHB, Noel PH, Worchel J, Cornell J, Harple L, Fultz BA, Junkeler E, Mka VA, Unutzer J. for the IMPACT Investigators. The effectiveness of depression care management on diabetes-related outcomes in older patients. Ann InternMed 2004;140(12):1015–1025.
89. Orth-Gomer K. International epidemiological evidence for a relationship between social support and cardiovascular disease. In: Shumaker SA, Czajkowski SM, eds. Social Support and Cardiovascular Disease. New York: Plenum Press; 1994.
90. Kaplan RM, Hartwell SL. Differential effects of social support and social network on physiological and social outcomes in men and women with type II diabetes mellitus. Health Psychol 1987;6:387–398.
91. Weidner G, Healy AB, Matarazzo JD. Family consumption of low fat foods: stated preference versus actual consumption. J ApplSoc Psychol 1985;15:773–779.
92. Gilden JL, Hendryz M, Casia C, Singh SP. The effectiveness of diabetes education programs for older patients and their spouses. J Am Geriatr Soc 1989;37(11):1023–1054.
93. Bennett DA. Diabetes and change in cognitive function. Arch Intern Med 2000;160(2):141–143.
94. Jack L Jr, Airhihenbuwa CO, Namageyo-Funa A, Owens MD, Vinicor F. The psychosocial aspects of diabetes care. Geriatrics 2004;59(5):26–32.
95. Bent N, Rabitt P, Metcalfe D. Diabetes mellitus and the rate of cognitive ageing. Br J Clin Psychol 2000;39(4):349–362.
96. Meneilly GS, Cheung E, Tessier D, Yakura C, Tuokko H. The effect of improved glycemic control on cognitive functions in the elderly patient with diabetes. Gerontologist 1993;48 (4):M117–M121.
97. Asimakopoulou K, Hampson SE. Cognitive Function in Type 2 Diabetes: Relationship to Diabetes Self-Management. Guilford, Surrey, UK: University of Surrey; 2001.
98. Asimakopoulou K, Hampson SE. Cognitive functioning and self-management in older people with diabetes. Diabetes Spectr 2002;15(2):116–121.
99. Hampson SE, Glasgow RE, Foster LS. Personal models of diabetes among older adults: relationship to self-management and other variables. Diabetes Educ 1995;21(4):300–307.
100. Alcozer F. Secondary analysis of perceptions and meanings of type 2 diabetes among Mexican American women. Diabetes Educ 2000;26(5):785–795.
101. Huang ES, Gorawara-Bhat R, Chin MH. Self-reported goals of older patients with type 2 diabetes. J Am Geriatr Soc 2005;53:306–311.
102. O'Connor PJ, Desai JR, Solberg LI, Rush WA, Bishop DB. Variation in diabetes care by age: opportunities for customization. BMC Fam Pract 2003;4(16):http://www.biomedcentral.com/1471-2296/1474/1416.
103. White SL, Maloney SK. Promoting healthy diets and active lives to hard-to-reach groups: market research study. Public Health Rep 1990;105(3):224–231.
104. Samuel-Hodge CD, Headen SW, Skelly AH, Ingram AF, Keyserling TC, Jackson EJ, Ammerman AS, Elasy TA. Influences on day to day self-management of type 2 diabetes among African American women: spirituality, the multi-caregiver role, and other social context factors. Diabetes Care 2000;23(7):928–933.
105. Jacobson AM. Quality of life in patients with diabetes mellitus. Semin Clin Neuropsychiatry 1997;2(1):82–93.
106. Bott U, Muhlhauser I, Overmann J, Berger M. Validation of a diabetes-specific quality of life scale for patients with type 1 diabetes. Diabetes Care 1998;21(5):757–769.
107. Olson DE, Norris SL. Overview of AGS guidelines for the treatment of diabetes mellitus in geriatric populations. Geriatrics 2004;59(4):18–24.
108. Blaum CS, Ofstedal MB, Langa KM, Wray LA. Functional status and health outcomes in older Americans with diabetes mellitus. J Am Geriatr Soc 2003;51(6):745–753.
109. Glasgow RE, Toobert DJ, Hampson SE, Brown JE, Lewinsohn PM, Donnelly J. Improving self-care among older patients with type II diabetes: The "Sixty Something..." Study. Patient Educ Couns 1992;19(1):61–74.
110. Toobert DJ, Glasgow RE, Stryker LA, Barrera M Jr, Radcliffe JL, Wander RC, Bagdade JD. Biologic and quality of life outcomes from the Mediterranean lifestyle program. Diabetes Care 2003;26(8):2288–2293.

111. Clark B. Older, Sicker, smarter, and redefining quality: the older consumer's quest for service. In: Dychtwald K, ed. Healthy Aging. Challenges and Solutions. Gaithersburg MD: Aspen Publishers, Inc.; 1999.
112. Miller CA, Edwards L, Kissling G, Sanville L. Nutrition Education improves metabolic outcomes among older adults with diabetes mellitus: Results from a randomized controlled trial. Prev Med 2002;34:252–259.
113. Miller CA, Edwards L, Kissling G, Sanville L. Evaluation of a theory-based nutrition intervention for older adults with diabetes mellitus. J Am Diet Assoc 2002;102(8):1069–1081.
114. Group TLAR. Reduction in weight and cardiovascular disease risk factors in individuals with type 2 diabetes. Diabetes Care 2007; 30 (6): 1374–1383.

17 Adipokines, Nutrition, and Obesity

Melissa E. Gove and Giamila Fantuzzi

Key Points

- White adipose tissue (WAT) is a complex and highly active secretory organ, sending out and responding to signals that modulate appetite, energy expenditure, insulin sensitivity, endocrine and reproductive systems, bone metabolism, inflammation, and immunity.
- Obesity is characterized by an increase in fat mass, increased macrophage infiltration of WAT, and an abnormal adipokine and cytokine production, contributing to generation of a state of low-grade chronic inflammation.
- Leptin, adiponectin, visfatin, resistin, and adipsin are adipokines since they are predominantly secreted from adipocytes. The cytokines IL-6 and TNF-α are also secreted from WAT but are not considered adipokines since adipocytes are not the primary source of these molecules.
- Leptin, adiponectin, resistin, and visfatin, as well as IL-6 and TNF-α, provide an important link between obesity, insulin resistance, and inflammatory disorders.
- The effects of nutrients on adipokines and cytokines secreted from WAT are still highly controversial and warrant further research.

Abstract

The traditional view of white adipose tissue (WAT) existing solely as an energy reservoir is no longer true. Although predominantly consisting of adipocytes, WAT is also composed of preadipocytes, endothelial cells, fibroblasts, and macrophages located in the stromovascular fraction. As a result, WAT is a complex and highly active secretory organ capable of modulating appetite, energy expenditure, insulin sensitivity, endocrine and reproductive systems, bone metabolism, inflammation, and immunity. The release of adipokines and cytokines by WAT is one of the most important ways this tissue influences physiological and pathological processes throughout the body. Adipokines are biologically active mediators released from adipocytes and include proteins such as leptin, adiponectin (APN), resistin, adipsin, and visfatin, as well as factors traditionally considered as cytokines such as interleukin (IL)-6, tumor necrosis factor (TNF)-α, and others. Obesity is characterized by an increase in fat mass, which is associated with a state of chronic inflammation with macrophage infiltration of WAT and abnormal production of adipokines and cytokines. Leptin, APN, resistin, visfatin, IL-6, and TNF-α help provide a link between obesity and development of insulin resistance, impaired glucose tolerance, and Type 2 diabetes (T2D) commonly associated with obesity. How specific nutrients affect production and secretion of adipokines and WAT-derived cytokines is currently controversial

Key Words: Adipokines; cytokines; insulin resistance; white adipose tissue; obesity

17.1. INTRODUCTION

17.1.1. Adipose Tissue as an Endocrine Organ

In mammals, adipose tissue consists of brown adipose tissue (BAT) and white adipose tissue (WAT). BAT is involved in heat generation and is a minor component of human adipose tissue; most of the adipose tissue in mammals is WAT, the subject of this chapter *(1)*. Traditionally, it has been widely accepted that the basic functionality of white adipocytes is to accumulate and release free fatty acids

A. Bendich, R.J. Deckelbaum (eds.), *Preventive Nutrition*, Nutrition and Health, DOI 10.1007/978-1-60327-542-2_17,

(FFA). Adipocytes store triglycerides (TG) during times of calorie abundance and facilitate retrieval of TG during periods of calorie deficit, predominantly serving as an energy source *(2)*. During fasting there is a dynamic equilibrium between the release of FFA into the circulation and their uptake and oxidation by peripheral tissues, most importantly skeletal muscle. Therefore, the size of adipose tissue stores increases in times of positive energy balance and decreases when energy expenditure exceeds intake *(3)*. In a lean environment, adipocytes remain small and efficiently store FFA as TG, which are then mobilized during periods of caloric need to generate ATP *(4)*. In an environment of overnutrition and/or decreased energy expenditure, surplus energy is deposited in adipose tissue as TG, available to be utilized for energy in times of caloric restriction *(5)*.

These traditional functions, however, are no longer considered the sole functions of WAT. Although adipocytes are the main cellular component of WAT, several other cell types, including preadipocytes, endothelial cells, fibroblasts, macrophages, and other leukocytes, reside in the stromovascular fraction of this tissue *(6)*, which contains a well-organized vasculature and is highly innervated. These characteristics link WAT to the whole body and make it an essential tissue in the regulation of several physiological functions *(2)*. As a result, WAT is currently considered as a complex and highly active secretory organ, sending out and responding to signals that modulate appetite, energy expenditure, insulin sensitivity, endocrine and reproductive systems, bone metabolism, inflammation, and immunity *(7,8)*.

Release of adipokines by WAT is one of the most important ways for this tissue to influence physiological and pathological processes throughout the body. Adipokines are biologically active mediators released from adipocytes and include proteins such as leptin, adiponectin (APN), resistin, adipsin, and visfatin, as well as factors traditionally considered as cytokines such as interleukin (IL)-6, tumor necrosis factor (TNF)-α, chemokines, and others. Adipokines will be described in detail below.

17.2. OBESITY: WHAT HAPPENS TO ADIPOSE TISSUE?

The incidence of obesity and associated disorders is dramatically increasing worldwide.Although there are complex genetic and environmental components to the development of obesity, this condition is largely due to a positive imbalance between food intake and energy expenditure *(5)*. Obesity is characterized by increased fat mass caused by the combination of increase in the size of preexisting adipocytes and de novo adipocyte differentiation *(2)*. This increase in fat mass is associated with a chronic state of inflammation, characterized by abnormal production of adipokines and proinflammatory cytokines, leading to the induction of inflammation, that can be detected through increased circulating levels of the acute-phase protein C-reactive protein (CRP) and the activation of proinflammatory signaling pathways *(6,7,9)*.

Chronic activation of the immune system in obesity likely contributes to the insulin resistance (IR), impaired glucose tolerance, and Type 2 diabetes (T2D) commonly associated with obesity *(9)*. Insulin resistance is defined as a diminished ability of the cell to respond to insulin, usually compensated for by hyperinsulinemia, which contributes to the subsequent β-cell failure in T2D. In adipose tissue, IR leads to an increase in lipolysis and subsequent release of FFA into the circulation. The increased availability and utilization of FFA by skeletal muscle contributes to the development of skeletal muscle IR and increase in hepatic glucose production *(9)*. The role of individual adipokines and cytokines secreted from WAT in the development of IR will be discussed below.

Adipose tissue of obese individuals contains a large number of macrophages compared with WAT of lean persons. Infiltration of WAT by macrophages is directly correlated with adiposity and adipocyte size in both human subjects and mice *(10)*. Adipose tissue of lean individuals contains approximately

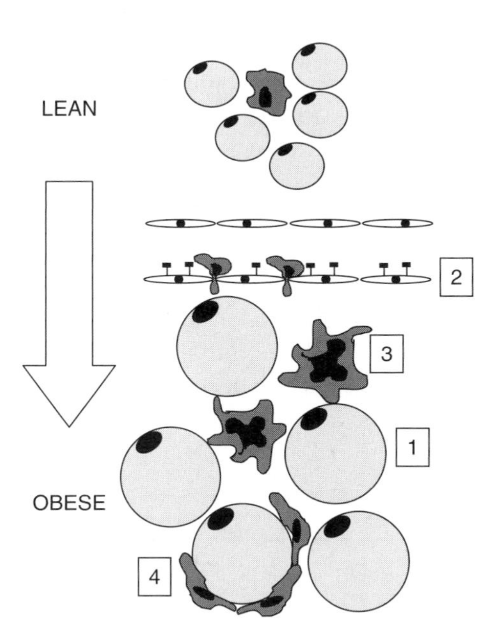

Fig. 17.1. Adipose tissue in lean and obese individuals. In adipose tissue of obese individuals, (1) enlarged adipocytes produce high levels of leptin and cytokines. Production of these factors (2) leads to upregulation of adhesion molecules on vascular endothelial cells, favoring transmigration of monocytes from the circulation to adipose tissue. Transmigration is facilitated by increased production of chemokines by adipose tissue. Increased infiltration of monocytes leads to (3) accumulation of macrophages in adipose tissue in obesity. These cells become activated and release a variety of proinflammatory factors, leading to amplification of the vicious inflammatory cycle. Enlarged adipocytes (4) become susceptible to cell death and are phagocytosed by macrophages, which surround dead adipocytes forming crown-like structures.

10% macrophages, whereas the macrophage content of WAT of obese individuals can be as high as 50% of the total number of cells *(4)*. This increase in macrophage infiltration could represent the cause and/or the consequence of the low-grade inflammation seen in obesity *(11)*. Although preadipocytes have been reported to differentiate into macrophages *(12)*, bone marrow chimera studies in mice have demonstrated that WAT macrophages are bone marrow-derived, indicating that macrophages present in obese individuals originate from circulating monocytes infiltrating WAT *(10,13)*. The chemokine monocyte chemoattractant protein (MCP)-1, which recruits monocytes to the site of inflammation, is expressed and secreted by WAT *(8)*. Both circulating and WAT levels of MCP-1 are increased in obesity, suggesting that MCP-1 may be involved in recruiting circulating monocytes in obese individuals *(14)*. It has also been proposed that factors secreted by hypertrophied, mature adipocytes can activate endothelial cells present in WAT, which subsequently favors monocyte adhesion and migration, leading to macrophage infiltration *(11)*. These recruited macrophages are largely responsible for the locally produced TNF-α and an important part of the production of IL-6 and inducible nitric oxide synthase (iNOS) *(11)* in obese individuals (*see* Fig. 17.1).

Adipose tissue is distributed throughout the body in specific depots, each depot being important in maintaining homeostasis and efficient energy utilization *(15)*. However, in obesity, where there is a surplus of nutrients, WAT expands to accommodate this nutrient overload to areas that are not designed to store large amounts of lipids and are more susceptible to the toxic effects of excess fat *(5)*. Accumulation of fat occurs subcutaneously and/or around organs such as liver or pancreas, which is referred to as visceral WAT *(5,16)*. Visceral adipose tissue is an active participant in obesity-related diseases and is a primary contributor to the development of T2D and IR *(9,16)*. Subcutaneous adipose tissue does not appear to be as important a mediator. A possible explanation for these functional differences is the anatomical location of visceral *versus* subcutaneous WAT *(16)*. In addition, compared to subcutaneous WAT, visceral WAT and its infiltrating macrophages produce higher levels of proinflammatory cytokines, such as TNF-α and IL-6, both of which have been implicated in the development of T2D and IR *(8,16)*.

17.3. ADIPOKINES AND CYTOKINES

As mentioned above, adipokines are bioactive peptides predominantly secreted by adipocytes that are capable of acting at both the local and the systemic level *(8)*. Leptin, APN, resistin, and visfatin provide an important link between obesity, IR, and inflammatory disorders *(6)* (*see* Table 17.1). Other cytokines are also secreted from WAT. However, although some are produced by adipocytes, they are not considered adipokines because adipocytes are not the main source of these mediators *(6)*.

Table 17.1
Major Adipokines

Adipokine	*Source*	*Biological Activities*
Leptin	**Adipocytes** Fundus of the stomach Skeletal muscle Liver Lymphocytes Placenta	Anorexigenic Proinflammatory Chemotactic Antiapoptotic Stimulates endothelial cell growth
Adiponectin	**Adipocytes** Hepatocytes Osteoclasts Synovial fibroblasts Skeletal muscle cells	Anti-inflammatory Proinflammatory Increases FA oxidation Increases glucose utilization Protects vasculature
Resistin	Adipocytes[a] Monocytes/macrophages[a]	Proinflammatory
Visfatin	**Visceral adipocytes** Neutrophils Macrophages	Antiapoptotic
Adipsin	**Adipocytes**	Homologue of complement factor D

The table lists individual members of the adipokine family, their cellular/tissue source (in bold the cells that produce the highest levels), and their major biological activities.
[a]Monocytes/macrophages are the major source of resistin in humans and adipocytes in rodents.

17.3.1. Leptin

Years before the discovery of leptin, a genetic defect that caused a severely obese phenotype due to overeating and decreased energy expenditure was observed in mice that had a genetic defect in the *ob* gene; as a result these obese mice carrying the mutation were called *ob/ob* mice. A series of parabiotic experiments suggested that *ob/ob* mice did not produce a satiety factor but were able to respond to that factor from a parabiotic mouse. Similar parabiotic experiments performed with *db/db* mice, which display a phenotype similar to the *ob/ob* mice due to a mutation in the *db* gene, led to the hypothesis that the *db* gene encoded for the *ob* receptor *(17)*. Nearly three decades later, leptin was discovered in 1994 by Friedman et al. *(18)*. This16-kDa polypeptide encoded by the *ob* gene was termed leptin from the Greek word *leptos*, meaning thin *(17,19)*.

Leptin is the most well-characterized adipokine to date. Although it is predominantly produced by adipocytes, leptin is also produced by the fundus of the stomach, skeletal muscle, liver, lymphocytes, and placenta, but at much lower levels than WAT *(19,20)*. Circulating serum levels and mRNA expression of leptin in WAT are strongly associated with body mass index (BMI) and fat mass. In addition, leptin mRNA expression is higher in subcutaneous adipose tissue than in visceral adipose tissue in humans *(8,11)*, leading to serum levels of leptin to be 2–3 times higher in women than in men, even when adjusted for age and BMI *(6)*.

Leptin exerts its effects by binding to its receptor (Ob–R). There are five alternatively spliced isoforms of Ob–R: Ob–Ra, Ob–Rb, Ob–Rc, Ob–Rd, and Ob–Re, each capable of forming homodimers in the presence and absence of its ligand *(17,20)*. Activation of OB–R results from a ligand-induced conformational change rather than from dimerization *(17)*. However, only the long form, Ob–Rb, contains an intracellular region capable of downstream signaling upon leptin binding. Like other class-1 cytokine receptors, Ob–Rb requires the activation of Janus tyrosine kinase 2 (JAK2) for propagation of leptin signaling. Leptin binding to Ob–Rb stimulates autophosphorylation of JAK2, which leads to further phosphorylation of the three intracellular tyrosine residues 985, 1,077, and 1,138 *(21,22)*. When activated, each tyrosine residue recruits specific downstream signaling proteins such as SH2-domain containing phosphatase-2 (SHP-2), signal transducer and activator of transcription (STAT)-5, and STAT-3, all of which mediate leptin's signaling *(23)*.

The main function of leptin is an inhibitory effect on appetite and therefore leptin is an anorexigenic molecule *(17)*. Leptin helps maintaining long-term control of adiposity and regulates adaptive metabolic changes in response to nutritional modifications. Leptin also regulates short-term energy intake by modulating meal size according to changes in energy balance *(24)*. In agreement with these functions of leptin, the highest expression of Ob–Rb is found in the feeding centers of the hypothalamus, including the arcuate, dorsomedial, ventromedial, and premammillary nuclei *(20,25)*. Within the arcuate nucleus, Ob–Rb is found in two distinct populations of "first-order" neurons *(24)*. One population of neurons synthesizes orexigenic, appetite-stimulating, neuropeptide Y (NPY) and agouti-related peptide (AgRP) and the other synthesizes anorexigenic, appetite-inhibiting, proopiomelanocortin (POMC) *(24,25)*. Leptin decreases appetite and increases energy expenditure by activating POMC neurons and inhibits NPY/AgRP neurons, suppressing expression of orexigenic neuropeptides *(22)*. When energy stores are in excess and leptin activity is high, leptin stimulates the production of POMC and inhibits the production of NPY/AgRP. However, when energy stores are low, decreased leptin activity stimulates appetite by suppressing synthesis of anorectic POMC and increases the expression of orexigenic peptides *(22)*. In addition to its critical role in the homeostatic control of food intake, mostly controlled by the hypothalamus, leptin also acts in the cortex and limbic areas of the brain, where it regulates cognitive and hedonic responses to feeding *(26)*.

In addition to being obese, mice and humans with leptin or leptin-receptor deficiency have other endocrine abnormalities including infertility, alterations in bone metabolism, and dysfunction of the immune system. In fact, with the onset of starvation or acute fasting, leptin levels fall dramatically out of proportion with fat mass, signaling to the brain to initiate the adaptive response to starvation *(17,27)*. These neuroendocrine changes associated with a dramatic decrease in leptin include a decrease in reproductive hormones to limit procreation, a fall in thyroid hormones to conserve metabolism, and an increase in stimulation of the hypothalamus–pituitary–adrenal axis to mobilize needed energy stores *(17,27)*. Interestingly, mice and humans with a leptin or leptin-receptor deficiency live in a state of perceived starvation *(17)*. Other endocrine alterations associated with the lack of leptin include suppression of immune function and increased bone development *(8)*. Leptin's role in regulating the immune system has been fueled largely by early observations of thymus atrophy in *db/db* mice. Leptin protects T lymphocytes from apoptosis and regulates T-cell activation and proliferation *(7)*. Leptin also influences cytokine production by immune cells, stimulates endothelial cell growth and angiogenesis, and accelerates wound healing *(8,17)*. Most neuroendocrine and immune alterations associated with fasting and/or starvation are normalized by the administration of leptin. Thus, leptin's activities are not restricted to control of appetite and energy expenditure; in fact, this adipokine regulates a variety of biological functions and is a critical factor in maintaining body homeostasis.

17.3.2. Adiponectin

Adiponectin is secreted predominantly from adipocytes and circulates around 5–30 μg/mL in healthy humans, making it the highest circulating adipokine *(7)*. Adiponectin is also expressed and secreted from other cell types, such as cytokine-stimulated hepatocytes, osteoclasts, synovial fibroblasts, and skeletal muscle cells *(28–30)*. Adiponectin was characterized in 1995 and 1996 by four independent groups using four different methods *(8)*. The structure of APN is a 247-amino-acid protein that consists of an amino-terminal signal sequence, a variable region, a collagenous domain, and a carboxy-terminal globular domain *(31)*. Adiponectin is a close homologue of C1q, a complement protein, and TNF-α, sharing structural similarities but not sequence similarities with these molecules.

In its basic form, APN exists as a low-molecular-weight (LMW) trimer, which further associates with other trimers to form middle-molecular-weight (MMW) hexamers and high-molecular-weight (HMW) oligomers. These HMW oligomers form bouquet-like structures through disulphide bonds located within the collagenous domains of each monomer *(31,32)*. All three forms of APN are found circulating in serum, but whether the different forms of APN exhibit differential biological actions remains unclear. A globular form of APN may also exist. Leukocyte elastase, which is secreted from activated monocytes and neutrophils, cleaves APN and generates the globular domain that is capable of forming a trimer *(7,33,34)*. Globular APN, however, remains to be shown to exist physiologically in vivo. To add to the complexity, hydroxylation and glycosylation of the lysines in the collagenous domain are necessary for complete biologic activity *(35)*.

APN exerts the majority of its effects by signaling through its receptors. Two APN receptors have been identified to date, AdipoR1 and AdipoR2. AdipoR1 is ubiquitously expressed but is predominately found in skeletal muscle, whereas AdipoR2 is abundantly expressed in the liver *(36)*. In a fasted state, mRNA expression of AdipoR1 and AdipoR2 increases in the liver and skeletal muscle; this is quickly restored to normal levels upon feeding. Mouse studies have shown hypoinsulinemia to increase mRNA levels of APN receptors in skeletal muscle, restoring mRNA levels to normal following insulin treatment *(37)*. The opposite is seen in hyperglycemic and hyperinsulinemic conditions, which decrease mRNA expression of both APN receptors *(33)*.

AdipoR1 and AdipoR2 contain seven transmembrane domains that are structurally and functionally different to other G protein receptors, having the N-terminus located in the cytoplasm and the C-terminus located externally *(33)*. In vitro studies have shown AdipoR1 to be a high-affinity receptor for globular APN and a low-affinity receptor for full-length APN, while AdipoR2 is an intermediate-affinity receptor for both globular and full-length APN. APN acts through the activation of signaling molecules such as peroxisome proliferator-activated receptor (PPAR)-α, AMP-activated protein kinase (AMPK), and mitogen-activated protein kinase (p38 MAPK) *(38)*. The activation of PPAR-α has been previously shown to be involved in APN-stimulated fatty acid (FA) oxidation but not for glucose uptake, while AMPK and MAPK activation is likely involved in both FA oxidation and glucose uptake *(38)*. However, as stated earlier, whether the different forms of APN initiate the same signals remains controversial *(31)*.

In addition to receptor-dependent effects, the high circulating concentration of APN suggests that low-affinity receptor-independent activities might be critical in the biological functions of this protein. APN binds to and inhibits the activity of the growth factors heparin-binding epidermal growth factor (HB-EGF), basic fibroblast growth factor (bFGF), and platelet-derived growth factor (PDGF), all of which are involved in tissue repair *(1,39)*. APN also binds to a diverse number of chemokines (such as CCL2, CCL5, and CXCL12), displacing them from their interaction with proteoglycans in the extracellular matrix *(40)*. APN can also bind calreticulin, which opsonizes apoptotic cells and facilitates their clearance by macrophages *(41)*. Finally, APN interacts with collagen I, III, and IV in the injured vascular wall and myocardium *(42)*. Therefore, APN acts in a complex way, both receptor dependent and -independent, to regulate a variety of biological functions involved in metabolism, tissue repair, cell proliferation, and inflammation, among others.

Unlike leptin, APN levels decrease with the increase in fat mass that is observed in obesity. The inverse correlation between APN and IR has been well established both in vitro and in vivo in animal and human studies *(35,43)*. Although the mechanisms are not completely clear, hypoadiponectinemia occurs in IR subjects that are either lipodystrophic or obese, indicating that APN is directly involved in regulating glucose homeostasis and hepatic insulin sensitivity *(32)*. APN improves insulin sensitivity by various mechanisms *(43)*. In the liver, APN activates AMPK, which enhances insulin sensitivity by increasing FA oxidation, downregulating gluconeogenic enzymes, and increasing glucose-6-phosphate biosynthesis *(43,44)*. Activation of PPAR-α by APN further increases insulin sensitivity by increasing FA oxidation, and thereby decreasing FA synthesis in the liver *(33)*. In muscle, APN stimulates phosphorylation of acetyl-CoA carboxylase, FA oxidation, glucose utilization, and lactate production *(8,33,43)*. As in the liver, these effects are seen in parallel with the activation of AMPK and PPAR-α by APN and are eliminated when inhibiting AMPK and PPAR-α signaling *(33,44–46)*. Treatment with thiazolindinediones (TZD), drugs that activate PPAR-γ and that are commonly used to treat for T2D and IR, increases APN levels and subsequently improves glucose tolerance and insulin sensitivity. PPAR-γ is a ligand-activated transcription factor involved in the regulation of adipocyte differentiation and numerous adipocyte genes. Studies using mice and adipocyte cultures have shown that the activation of PPAR-γ increases APN mRNA levels in adipose cells *(35,47)*, further strengthening the link between adipocyte biology and IR.

Reduced levels of APN in obesity also contribute to endothelial dysfunction observed in cardiovascular diseases. In the early stages of cardiovascular disease, endothelial cell activation occurs via inflammatory stimuli such as TNF-α, which induce the expression of vascular disease adhesion molecule-1 (VCAM-1), endothelial-leukocyte adhesion molecule-1 (E-selectin), and intracellular adhesion molecule-1 (ICAM-1), a crucial event in the development of vascular disease *(48,49)*. Adiponectin reverses these deleterious effects of TNF-α on endothelial dysfunction by hindering the ability of TNF-α to activate the proinflammatory transcription factor nuclear factor κB (NFκB) and the subsequent induction of adhesion molecules *(48,50)*. In addition, APN inhibits foam cell formation and smooth muscle cell migration, both of which play an important role in the development of atherosclerosis *(51)*. These effects of APN are also seen in vivo in apolipoprotein E (apoE)-deficient mice, a well-established model for atherosclerosis. When treated with adenovirus-derived APN, apoE-deficient mice did not develop severe atherosclerosis. This is due to APN's ability to attenuate the endothelial inflammatory response and macrophage foam cell formation by inhibiting lipoprotein lipase and class-A-scavenger receptor *(52)*. In addition, in vivo models of myocardial ischemia have shown that APN administration protects the heart from ischemia-reperfusion injury by inhibiting NO and superoxide production and thus protects tissues from nitrative and oxidative stress *(53)*. In conclusion, similar to leptin, APN is also a pleiotropic, adipocyte-derived molecule critical in maintaining homeostasis of a variety of tissues and processes.

17.3.3. Resistin

Resistin was discovered independently by three different groups using three different methods *(54–56)*. This 12-kDA polypeptide is the product of the *retn* gene and belongs to a unique family of cysteine-rich C-terminal domain proteins called resistin-like molecules (RELM) *(8)*. In rodents, resistin is produced almost exclusively from adipocytes *(57)*. However, there is controversy surrounding this adipokine, since some studies have shown human adipose tissue to express resistin and others did not find its presence or detected it at very low levels in human adipocytes. In fact, it is believed that monocytes and macrophages, rather than adipocytes, are the predominant producers of resistin in humans *(11)*.

Homcomb et al. *(56)* first described the gene family and its specificity to adipose tissue in 2000 *(56)*. Shortly after, Steppan et al. *(54)* discovered elevated serum levels of resistin in rodent models of genetic and diet-induced obesity, suggesting a role in obesity-related dysfunction. Further investigation indicated that administration of recombinant resistin to healthy wild-type mice impaired glucose homeostasis and insulin sensitivity compared to vehicle-injected mice. In addition, neutralization of resistin by injection of antibodies into diet-induced obese mice decreased blood glucose levels and improved insulin sensitivity *(57)*. Further animal studies using resistin-deficient mice showed a decrease in fasting glucose, improved glucose tolerance, and enhanced insulin sensitivity compared to wild-type mice. These resistin-deficient mice have reduced fasting glucose after a high-fat diet compared to their weight-matched controls *(8,58)*. In vitro studies have shown the ability of resistin to suppress insulin-stimulated glucose uptake, which is subsequently reversed after anti-resistin antibody treatment in cultured 3T3-L1 adipocytes *(8,20)*. Although there are some conflicting data, in rodents, resistin is generally considered a link to the development of hyperglycemia and IR observed in obesity *(11)*.

Most of the work linking resistin to obesity-related disorders has been conducted in mouse models. However, the human resistin protein is only 55% homologous to its mouse counterpart, suggesting that resistin may not be evolutionarily well conserved across species *(6)*. As a result, the role of resistin in humans remains largely unclear. In addition, there is no correlation between resistin mRNA expression, body weight, obesity, and IR in humans *(20)*. Although there is no clear link between resistin, IR, and T2D in humans, the proinflammatory properties of resistin suggest a role in inflammatory processes *(59)*. In primary human macrophages, resistin expression is induced by the proinflammatory cytokines IL-1, IL-6, and TNF-α in combination with stimulation with endotoxin *(6)*. In these studies it appears that resistin is able to be induced by and in turn to induce the proinflammatory cytokines IL-6 and TNF-α via activation of the NFκB pathway *(60)*. Resistin also upregulates the adhesion molecules VCAM-1 and ICAM-1 in human aortic endothelial cells. This induction of adhesion molecules by resistin is inhibited by APN, which also inhibits TNF-α's ability to induce adhesion molecules on the vascular endothelium *(48,49)*. In addition, resistin stimulates smooth muscle cell proliferation in the human aorta, another physiological process necessary in the development of atherosclerosis *(58)*. Thus, in humans, resistin may be more critical to the regulation of vascular function and inflammatory responses than as a factor controlling metabolic processes.

17.3.4. Visfatin

Although visfatin is considered a recently identified adipokine, the peptide was actually first identified in 1994 by Samal et al. *(61)* to be highly expressed in bone marrow, liver, and muscle tissue. This 52-kDa protein was isolated and characterized from a human peripheral blood lymphocyte cDNA library and termed pre-B-cell colony-enhancing factor (PBEF) *(61)*. Since the initial discovery, PBEF has been implicated in several inflammatory disorders, such as acute lung failure, sepsis, inflammatory bowel disease, and rheumatoid arthritis *(62)*.

This molecule was rediscovered in 2005 as visfatin *(63)* and was named such because it was found to be expressed predominantly in visceral adipose tissue. In both genetic and diet-induced animal models of obesity, visfatin expression is increased in visceral adipose tissue *(58)*. With visfatin rediscovery, Fukuhara et al. demonstrated insulin-like effects of visfatin in vitro as well as in vivo when administered to wild-type mice by binding to and activating the insulin receptor *(63)*. Since then the authors have been questioned and some of their findings retracted *(59)*. More recent data suggest that serum visfatin increases with progressive β-cell deterioration observed in T2D *(64)*. Interestingly, circulating levels of visfatin increase in overweight and obese individuals, with an even larger increase seen in those with metabolic syndrome compared to subjects that do not fulfill the criteria for complete metabolic syndrome diagnosis *(65)*.

In addition to visceral WAT, visfatin expression is also upregulated in activated neutrophils and macrophages *(6)*. Visfatin inhibits apoptosis of neutrophils and upregulates the expression of the proinflammatory cytokines IL-1β, IL-6, and TNF-α in human monocytes, as well as in mice treated with recombinant visfatin *(6,59)*.

17.3.5. Adipsin

Adipsin was first discovered in 1983 by Spiegelman et al. *(66)* as being specifically induced during adipocyte differentiation in the adipocyte cell line 3T3-F442A. This 28-kDa protein shares sequence homology with the serine proteases family. Murine adipsin shares the greatest sequence homology with human complement factor D, the rate-limiting enzyme in the alternative pathway of the complement system *(67)*. Adipsin is synthesized and secreted by epididymal fat, cultured fat cells, and by the sciatic nerve *(68)*.

In genetic and diet-induced obese mice, mRNA levels of adipsin in WAT are 50–100-fold lower compared with wild-type mice. Circulating protein levels are also decreased in animal models of obesity *(67)*. However, adipsin levels are either increased or unchanged in human obesity *(7)*.

17.3.6. IL-6

IL-6 is a proinflammatory cytokine that circulates in multiple glycosylated forms ranging from 22 to 27 kDa in size. It exerts its effects by binding to the IL-6 receptor, which belongs to the class I family of cytokine receptors, the same group to which the leptin receptor also belongs *(8,22)*. IL-6 is produced by many cell types, including fibroblasts, endothelial cells, monocytes, and adipocytes *(11)*. In fact, IL-6 levels are increased in both the serum and WAT of obese individuals *(11)*. Both adipocytes and stromovascular cells within WAT are capable of producing IL-6; in the absence of acute inflammation, WAT contributes approximately 15–30% of total circulating IL-6 *(7,58)*. However, most of the IL-6 secreted from WAT originate from endothelial cells, monocytes, and macrophages located in the stromovascular fraction *(58)*. Visceral WAT secretes 2–3 times more IL-6 compared to subcutaneous WAT, which may explain why central obesity has such a deleterious role in obesity-related disorders *(8,58)*. Both WAT expression and circulating levels of IL-6 decrease in parallel with weight loss and in patients undergoing bariatric surgery *(59)*.

IL-6 is a multifunctional cytokine and targets several different cell and tissue types *(11)*. IL-6 was one of the first cytokines to be linked to and considered to be a predictor for development of IR and cardiovascular disease *(59)*. In fact, high levels of IL-6 in adipose tissue have been positively correlated with circulating CRP, an important risk factor for cardiovascular disease *(58)*. In addition, peripheral administration of IL-6 decreases insulin signaling by reducing the expression of insulin receptor substrate (IRS) components through activation of the JAK/STAT pathway *(8)*. How this occurs remains unclear at this time, but there are data suggesting the participation of protein kinases and tyrosine phosphatases, as well as the possible interaction of suppressor of cytokine signaling (SOCS) with the insulin receptor *(58)*. There is, however, some controversy surrounding the role of IL-6 in IR. In addition to its negative role in regulating IR, IL-6 also increases FA oxidation both in vitro and in vivo by activating AMPK in skeletal muscle and WAT, thus improving insulin sensitivity in these tissues *(58,59)*. Mice lacking functional IL-6 are insulin resistant, which is reversed upon IL-6 replacement therapy, suggesting that this molecule aids in preventing IR rather than precipitating it *(8,11,59)*. Clearly, IL-6 is a potent proinflammatory cytokine and is found at high levels in obesity; however, its role in obesity-related diseases is still controversial.

17.3.7. TNF-α

TNF-α is a proinflammatory cytokine that is mainly produced by macrophages and lymphocytes *(11)*. This 26-kDa transmembrane protein is cleaved into a 17-kDa active protein and initiates its effects by signaling through type I and type II TNF-α receptors *(8)*. Although predominantly produced by immune cells, TNF-α can also be produced by WAT and acts as a link between adiposity and the development of IR. These observations are supported by the high levels of TNF-α found in both serum and WAT of obese humans and mice *(8,9)*. However, TNF-α acts locally at the site of WAT through autocrine or paracrine mechanisms rather than systemically, inducing IR as well as the production of leptin and IL-6 *(6,9)*. TNF-α expression has a strong positive correlation with the degree of obesity, represented by BMI, and a negative correlation with lipoprotein lipase activity *(69)*. Animal studies have shown that complete abrogation of functional TNF-α or lack of its two receptors significantly improves insulin sensitivity in multiple models of rodent obesity *(70)*.

In addition, TNF-α inhibits insulin action and insulin receptor signaling in vitro in cultured human adipocytes *(69)*. TNF-α directly inhibits insulin signaling by activating serine kinases that increase serine phosphorylation of IRS-1, making it a poor substrate for the insulin receptor and incapable of initiating proper signaling *(71)*. Insulin signaling is also impaired due to TNF-α's ability to increase serum FFA, which induce IR in several different tissues *(8)*. This strong association between TNF-α and IR makes it a likely target for IR therapy. However, although anti-TNF-α therapy is effective in chronic diseases such as rheumatoid arthritis and inflammatory bowel disease, neutralizing TNF-α activity in diabetic patients has not shown to increase insulin sensitivity *(7)*.

17.4. NUTRITION, ADIPOKINES, AND CYTOKINES

Most mammalian genes display a circadian rhythm in their expression. These daily oscillations in gene expression occur as the result of the daily light/dark cycle, feeding behavior, and sleeping pattern, in addition to other environmental and physiological cues *(72)*. Although originally thought to reside only in the hypothalamic suprachiasmatic nucleus, circadian clocks exist and are active in peripheral tissues as well. Zvonic et al. found the presence of active circadian clocks in adipose tissue, further linking circadian dysfunction with the development of obesity *(73)*. In addition, many adipokines and cytokines secreted from adipose tissue have a circadian rhythm. Leptin has a circadian rhythm in both gene expression and protein secretion, with peaks in expression occurring during the sleep phase of the sleep–wake cycle in humans *(74)*. Studies conducted in wild-type C57BL/6 mice have demonstrated a circadian rhythm for APN, resistin, visfatin, and adipsin expression *(75)*. In addition, secretion of IL-6 and TNF-α also has a strong circadian pattern *(73)*. However, in the altered metabolic state of obesity, the rhythmic expression of some adipokines and cytokines is altered. For example, the diurnal expression of APN and resistin is decreased in obese KK mice and even more so in obese diabetic KK-A^y mice *(75)*, but the rhythmic expression of leptin remains unchanged *(72)*.

Depending on the size and nutrient composition of a meal, a variety of changes in hormonal levels can occur within minutes and hours after food ingestion. This has prompted the investigation into whether adipokines and cytokines are modulated postprandially *(76)*. Data on this subject, however, are controversial. In the case of leptin, studies in humans and mice have shown that acute fasting and exercise lead to a rapid reduction in plasma concentrations of leptin, while overfeeding results in acute elevation of leptin levels *(77)*. Poppitt et al. have demonstrated a decrease in leptin levels after a high-fat meal in lean individuals and shown that an increase in saturation of the fatty meal did not further alter the response of the adipokine *(78)*. However, interpretation of the postprandial response of leptin is somewhat difficult due to the diurnal changes in leptin mentioned above *(78)*.

Data surrounding the impact of a high-lipid diet on circulating levels of APN in humans are even more controversial. Esposito et al. showed that both nondiabetic and diabetic individuals demonstrated a decrease in APN concentrations from baseline after consumption of a high-fat meal *(79)*. However, other studies show no alterations in APN levels following high-fat feeding in either nondiabetic or diabetic populations *(76,80)*. In addition, IL-6 levels increase in response to a high-fat feeding, while there are no alterations in TNF-α production *(80)*. Overall, the available data suggest that, although feeding and specific nutrients may acutely alter the expression of adipokines and cytokines, these changes are generally of small magnitude compared to the long-term effect of excessive or deficient nutrition, exercise, infection, or inflammation.

17.5. CONCLUSIONS AND RECOMMENDATIONS

The purpose of WAT existing solely as an energy source is no longer true. Research has indicated that WAT is an important mediator of physiological and pathological processes, largely because of the release of adipokines and cytokines. We now know that WAT has a dominant role in maintaining normal lipid and glucose metabolism and that dysfunction of WAT observed in obesity has a direct impact on lipid and glucose homeostasis *(4)*. How this dysfunction occurs in WAT is still largely unknown. Further understanding of the mechanisms of how adipokines and cytokines are involved in the development of metabolic disorders will help find treatment options for obesity, T2D, and IR.

REFERENCES

1. Wang Y, Lam KS, Xu JY, Lu G, Xu LY, Cooper GJ, and Xu A. (2005) Adiponectin inhibits cell proliferation by interacting with several growth factors in an oligomerization-dependent manner. *J Biol Chem* **280**, 18341–7.
2. Wang P, Mariman E, Renes J, and Keijer J. (2008) The secretory function of adipocytes in the physiology of white adipose tissue. *J Cell Physiol* **216**, 3–13.
3. Al-Hasani H, and Joost HG. (2005) Nutrition-/diet-induced changes in gene expression in white adipose tissue. *Best Pract Res Clin Endocrinol Metab* **19**, 589–603.
4. Guilherme A, Virbasius JV, Puri V, and Czech MP. (2008) Adipocyte dysfunctions linking obesity to insulin resistance and type 2 diabetes. *Nat Rev Mol Cell Biol* **9**, 367–77.
5. Sethi JK, and Vidal-Puig AJ. (2007) Thematic review series: adipocyte biology. Adipose tissue function and plasticity orchestrate nutritional adaptation. *J Lipid Res* **48**, 1253–62.
6. Tilg H, and Moschen AR. (2006) Adipocytokines: mediators linking adipose tissue, inflammation and immunity. *Nat Rev Immunol* **6**, 772–83.
7. Fantuzzi G. (2005) Adipose tissue, adipokines, and inflammation. *J Allergy Clin Immunol* **115**, 911–9; quiz 20.
8. Kershaw EE, and Flier JS. (2004) Adipose tissue as an endocrine organ. *J Clin Endocrinol Metab* **89**, 2548–56.
9. Lorenzo M, Fernandez-Veledo S, Vila-Bedmar R, Garcia-Guerra L, De Alvaro C, and Nieto-Vazquez I. (2008) Insulin resistance induced by tumor necrosis factor-alpha in myocytes and brown adipocytes. *J Anim Sci* **86**, E94–104.

10. Weisberg SP, McCann D, Desai M, Rosenbaum M, Leibel RL, and Ferrante AW, Jr. (2003) Obesity is associated with macrophage accumulation in adipose tissue. *J Clin Invest* **112**, 1796–808.
11. Bastard JP, Maachi M, Lagathu C, Kim MJ, Caron M, Vidal H, Capeau J, and Feve B. (2006) Recent advances in the relationship between obesity, inflammation, and insulin resistance. *Eur Cytokine Netw* **17**, 4–12.
12. Charriere G, Cousin B, Arnaud E, Andre M, Bacou F, Penicaud L, and Casteilla L. (2003) Preadipocyte conversion to macrophage. Evidence of plasticity. *J Biol Chem* **278**, 9850–5.
13. Curat CA, Miranville A, Sengenes C, Diehl M, Tonus C, Busse R, and Bouloumie A. (2004) From blood monocytes to adipose tissue-resident macrophages: induction of diapedesis by human mature adipocytes. *Diabetes* **53**, 1285–92.
14. Sartipy P, and Loskutoff DJ. (2003) Monocyte chemoattractant protein 1 in obesity and insulin resistance. *Proc Natl Acad Sci U S A* **100**, 7265–70.
15. Frayn KN, Karpe F, Fielding BA, Macdonald IA, and Coppack SW. (2003) Integrative physiology of human adipose tissue. *Int J Obes Relat Metab Disord* **27**, 875–88.
16. Hamdy O, Porramatikul S, and Al-Ozairi E. (2006) Metabolic obesity: the paradox between visceral and subcutaneous fat. *Current diabetes Rev* **2**, 367–73.
17. Fantuzzi G, and Faggioni R. (2000) Leptin in the regulation of immunity, inflammation, and hematopoiesis. *J Leukoc Biol* **68**, 437–46.
18. Zhang Y, Proenca R, Maffei M, Barone M, Leopold L, and Friedman JM. (1994) Positional cloning of the mouse obese gene and its human homologue. *Nature* **372**, 425–32.
19. Juge-Aubry CE, Henrichot E, and Meier CA. (2005) Adipose tissue: a regulator of inflammation. *Best Pract Res Clin Endocrinol Metab* **19**, 547–66.
20. Meier U, and Gressner AM. (2004) Endocrine regulation of energy metabolism: review of pathobiochemical and clinical chemical aspects of leptin, ghrelin, adiponectin, and resistin. *Clin Chem* **50**, 1511–25.
21. Buettner C, Pocai A, Muse ED, Etgen AM, Myers MG Jr, and Rossetti L. (2006) Critical role of STAT3 in leptin's metabolic actions. *Cell Metab* **4**, 49–60.
22. Robertson SA, Leinninger GM, and Myers MG Jr. (2008) Molecular and neural mediators of leptin action. *Physiol Behav* **94**, 637–42.
23. Gong Y, Ishida-Takahashi R, Villanueva EC, Fingar DC, Munzberg H, and Myers MG Jr. (2007) The long form of the leptin receptor regulates STAT5 and ribosomal protein S6 via alternate mechanisms. *J Biol Chem* **282**, 31019–27.
24. Valassi E, Scacchi M, and Cavagnini F. (2008) Neuroendocrine control of food intake. *Nutr Metab Cardiovasc Dis* **18**, 158–68.
25. Munzberg H, and Myers MG, Jr. (2005) Molecular and anatomical determinants of central leptin resistance. *Nat Neurosci* **8**, 566–70.
26. Rosenbaum M, Sy M, Pavlovich K, Leibel RL, and Hirsch J. (2008) Leptin reverses weight loss-induced changes in regional neural activity responses to visual food stimuli. *J Clin Invest* **118**, 2583–91.
27. Chan JL, and Mantzoros CS. (2005) Role of leptin in energy-deprivation states: normal human physiology and clinical implications for hypothalamic amenorrhoea and anorexia nervosa. *Lancet* **366**, 74–85.
28. Ehling A, Schaffler A, Herfarth H, Tarner IH, Anders S, Distler O, Paul G, Distler J, Gay S, Scholmerich J, Neumann E, and Muller-Ladner U. (2006) The potential of adiponectin in driving arthritis. *J Immunol* **176**, 4468–78.
29. Berner HS, Lyngstadaas SP, Spahr A, Monjo M, Thommesen L, Drevon CA, Syversen U, and Reseland JE. (2004) Adiponectin and its receptors are expressed in bone-forming cells. *Bone* **35**, 842–9.
30. Delaigle AM, Jonas JC, Bauche IB, Cornu O, and Brichard SM. (2004) Induction of adiponectin in skeletal muscle by inflammatory cytokines: in vivo and in vitro studies. *Endocrinology* **145**, 5589–97.
31. Oh DK, Ciaraldi T, and Henry RR. (2007) Adiponectin in health and disease. *Diabetes Obes Metab* **9**, 282–9.
32. Trujillo ME, and Scherer PE. (2005) Adiponectin—journey from an adipocyte secretory protein to biomarker of the metabolic syndrome. *J Intern Med* **257**, 167–75.
33. Kadowaki T, and Yamauchi T. (2005) Adiponectin and adiponectin receptors. *Endocr Rev* **26**, 439–51.
34. Waki H, Yamauchi T, Kamon J, Kita S, Ito Y, Hada Y, Uchida S, Tsuchida A, Takekawa S, and Kadowaki T. (2005) Generation of globular fragment of adiponectin by leukocyte elastase secreted by monocytic cell line THP-1. *Endocrinology* **146**, 790–6.
35. Chandran M, Phillips SA, Ciaraldi T, and Henry RR. (2003) Adiponectin: more than just another fat cell hormone? *Diabetes Care* **26**, 2442–50.
36. Guzik TJ, Mangalat D, and Korbut R. (2006) Adipocytokines - novel link between inflammation and vascular function? *J Physiol Pharmacol* **57**, 505–28.
37. Tsuchida A, Yamauchi T, Ito Y, Hada Y, Maki T, Takekawa S, Kamon J, Kobayashi M, Suzuki R, Hara K, Kubota N, Terauchi Y, Froguel P, Nakae J, Kasuga M, Accili D, Tobe K, Ueki K, Nagai R, and Kadowaki T. (2004)

Insulin/Foxo1 pathway regulates expression levels of adiponectin receptors and adiponectin sensitivity. *J Biol Chem* **279**, 30817–22.

38. Yamauchi T, Kamon J, Ito Y, Tsuchida A, Yokomizo T, Kita S, Sugiyama T, Miyagishi M, Hara K, Tsunoda M, Murakami K, Ohteki T, Uchida S, Takekawa S, Waki H, Tsuno NH, Shibata Y, Terauchi Y, Froguel P, Tobe K, Koyasu S, Taira K, Kitamura T, Shimizu T, Nagai R, and Kadowaki T. (2003) Cloning of adiponectin receptors that mediate antidiabetic metabolic effects. *Nature* **423**, 762–9.
39. Fayad R, Pini M, Sennello JA, Cabay RJ, Chan L, Xu A, and Fantuzzi G. (2007) Adiponectin deficiency protects mice from chemically induced colonic inflammation. *Gastroenterology* **132**, 601–14.
40. Masaie H, Oritani K, Yokota T, Takahashi I, Shirogane T, Ujiie H, Ichii M, Saitoh N, Maeda T, Tanigawa R, Oka K, Hoshida Y, Tomiyama Y, and Kanakura Y. (2007) Adiponectin binds to chemokines via the globular head and modulates interactions between chemokines and heparan sulfates. *Exp Hematol* **35**, 947–56.
41. Takemura Y, Ouchi N, Shibata R, Aprahamian T, Kirber MT, Summer RS, Kihara S, and Walsh K. (2007) Adiponectin modulates inflammatory reactions via calreticulin receptor-dependent clearance of early apoptotic bodies. *J Clin Invest* **117**, 375–86.
42. Okamoto Y, Arita Y, Nishida M, Muraguchi M, Ouchi N, Takahashi M, Igura T, Inui Y, Kihara S, Nakamura T, Yamashita S, Miyagawa J, Funahashi T, and Matsuzawa Y. (2000) An adipocyte-derived plasma protein, adiponectin, adheres to injured vascular walls. *Horm Metab Res* **32**, 47–50.
43. Mlinar B, Marc J, Janez A, and Pfeifer M. (2007) Molecular mechanisms of insulin resistance and associated diseases. *Clin Chim Acta* **375**, 20–35.
44. Combs TP, Berg AH, Obici S, Scherer PE, and Rossetti L. (2001) Endogenous glucose production is inhibited by the adipose-derived protein Acrp30. *J Clin Invest* **108**, 1875–81.
45. Yamauchi T, Kamon J, Minokoshi Y, Ito Y, Waki H, Uchida S, Yamashita S, Noda M, Kita S, Ueki K, Eto K, Akanuma Y, Froguel P, Foufelle F, Ferre P, Carling D, Kimura S, Nagai R, Kahn BB, and Kadowaki T. (2002) Adiponectin stimulates glucose utilization and fatty-acid oxidation by activating AMP-activated protein kinase. *Nat Med* **8**, 1288–95.
46. Diez JJ, and Iglesias P. (2003) The role of the novel adipocyte-derived hormone adiponectin in human disease. *Eur J Endocrinol* **148**, 293–300.
47. Boden G, Cheung P, Mozzoli M, and Fried SK. (2003) Effect of thiazolidinediones on glucose and fatty acid metabolism in patients with type 2 diabetes. *Metabolism* **52**, 753–9.
48. Ouchi N, Kihara S, Arita Y, Maeda K, Kuriyama H, Okamoto Y, Hotta K, Nishida M, Takahashi M, Nakamura T, Yamashita S, Funahashi T, and Matsuzawa Y. (1999) Novel modulator for endothelial adhesion molecules: adipocyte-derived plasma protein adiponectin. *Circulation* **100**, 2473–6.
49. Ouchi N, Kihara S, Arita Y, Okamoto Y, Maeda K, Kuriyama H, Hotta K, Nishida M, Takahashi M, Muraguchi M, Ohmoto Y, Nakamura T, Yamashita S, Funahashi T, and Matsuzawa Y. (2000) Adiponectin, an adipocyte-derived plasma protein, inhibits endothelial NF-kappaB signaling through a cAMP-dependent pathway. *Circulation* **102**, 1296–301.
50. Goldstein BJ, and Scalia R. (2004) Adiponectin: a novel adipokine linking adipocytes and vascular function. *J Clin Endocrinol Metab* **89**, 2563–8.
51. Arita Y, Kihara S, Ouchi N, Maeda K, Kuriyama H, Okamoto Y, Kumada M, Hotta K, Nishida M, Takahashi M, Nakamura T, Shimomura I, Muraguchi M, Ohmoto Y, Funahashi T, and Matsuzawa Y. (2002) Adipocyte-derived plasma protein adiponectin acts as a platelet-derived growth factor-BB-binding protein and regulates growth factor-induced common postreceptor signal in vascular smooth muscle cell. *Circulation* **105**, 2893–8.
52. Okamoto Y, Kihara S, Ouchi N, Nishida M, Arita Y, Kumada M, Ohashi K, Sakai N, Shimomura I, Kobayashi H, Terasaka N, Inaba T, Funahashi T, and Matsuzawa Y. (2002) Adiponectin reduces atherosclerosis in apolipoprotein E-deficient mice. *Circulation* **106**, 2767–70.
53. Tao L, Gao E, Jiao X, Yuan Y, Li S, Christopher TA, Lopez BL, Koch W, Chan L, Goldstein BJ, and Ma XL. (2007) Adiponectin cardioprotection after myocardial ischemia/reperfusion involves the reduction of oxidative/nitrative stress. *Circulation* **115**, 1408–16.
54. Steppan CM, Bailey ST, Bhat S, Brown EJ, Banerjee RR, Wright CM, Patel HR, Ahima RS, and Lazar MA. (2001) The hormone resistin links obesity to diabetes. *Nature* **409**, 307–12.
55. Kim KH, Lee K, Moon YS, and Sul HS. (2001) A cysteine-rich adipose tissue-specific secretory factor inhibits adipocyte differentiation. *J Biol Chem* **276**, 11252–6.
56. Holcomb IN, Kabakoff RC, Chan B, Baker TW, Gurney A, Henzel W, Nelson C, Lowman HB, Wright BD, Skelton NJ, Frantz GD, Tumas DB, Peale FV Jr, Shelton DL, and Hebert CC. (2000) FIZZ1, a novel cysteine-rich secreted protein associated with pulmonary inflammation, defines a new gene family. *EMBO J* **19**, 4046–55.
57. Steppan CM, and Lazar MA. (2004) The current biology of resistin. *J Intern Med* **255**, 439–47.
58. Antuna-Puente B, Feve B, Fellahi S, and Bastard JP. (2008) Adipokines: the missing link between insulin resistance and obesity. *Diabetes Metab* **34**, 2–11.

59. Tilg H, and Moschen AR. (2008) Inflammatory mechanisms in the regulation of insulin resistance. *Mol Med* **14**, 222–31.
60. Lago F, Dieguez C, Gomez-Reino J, and Gualillo O. (2007) Adipokines as emerging mediators of immune response and inflammation. *Nat Clin Pract* **3**, 716–24.
61. Samal B, Sun Y, Stearns G, Xie C, Suggs S, and McNiece I. (1994) Cloning and characterization of the cDNA encoding a novel human pre-B-cell colony-enhancing factor. *Mol Cell Biol* **14**, 1431–7.
62. Luk T, Malam Z, and Marshall JC. (2008) Pre-B cell colony-enhancing factor (PBEF)/visfatin: a novel mediator of innate immunity. *J Leukoc Biol* **83**, 804–16.
63. Fukuhara A, Matsuda M, Nishizawa M, Segawa K, Tanaka M, Kishimoto K, Matsuki Y, Murakami M, Ichisaka T, Murakami H, Watanabe E, Takagi T, Akiyoshi M, Ohtsubo T, Kihara S, Yamashita S, Makishima M, Funahashi T, Yamanaka S, Hiramatsu R, Matsuzawa Y, and Shimomura I. (2005) Visfatin: a protein secreted by visceral fat that mimics the effects of insulin. *Science* **307**, 426–30.
64. Lopez-Bermejo A, Chico-Julia B, Fernandez-Balsells M, Recasens M, Esteve E, Casamitjana R, Ricart W, and Fernandez-Real JM. (2006) Serum visfatin increases with progressive beta-cell deterioration. *Diabetes* **55**, 2871–5.
65. Filippatos TD, Derdemezis CS, Kiortsis DN, Tselepis AD, and Elisaf MS. (2007) Increased plasma levels of visfatin/pre-B cell colony-enhancing factor in obese and overweight patients with metabolic syndrome. *J Endocrinol Invest* **30**, 323–6.
66. Spiegelman BM, Frank M, and Green H. (1983) Molecular cloning of mRNA from 3T3 adipocytes. Regulation of mRNA content for glycerophosphate dehydrogenase and other differentiation-dependent proteins during adipocyte development. *J Biol Chem* **258**, 10083–9.
67. White RT, Damm D, Hancock N, Rosen BS, Lowell BB, Usher P, Flier JS, and Spiegelman BM. (1992) Human adipsin is identical to complement factor D and is expressed at high levels in adipose tissue. *J Biol Chem* **267**, 9210–3.
68. Cook KS, Min HY, Johnson D, Chaplinsky RJ, Flier JS, Hunt CR, and Spiegelman BM. (1987) Adipsin: a circulating serine protease homolog secreted by adipose tissue and sciatic nerve. *Science* **237**, 402–5.
69. Hotamisligil GS. (1999) The role of TNFalpha and TNF receptors in obesity and insulin resistance. *J Intern Med* **245**, 621–5.
70. Uysal KT, Wiesbrock SM, and Hotamisligil GS. (1998) Functional analysis of tumor necrosis factor (TNF) receptors in TNF-alpha-mediated insulin resistance in genetic obesity. *Endocrinology* **139**, 4832–8.
71. Hotamisligil GS. (2003) Inflammatory pathways and insulin action. *Int J Obes Relat Metab Disord* **27 Suppl 3**, S53–5.
72. Ptitsyn AA, and Gimble JM. (2007) Analysis of circadian pattern reveals tissue-specific alternative transcription in leptin signaling pathway. *BMC Bioinformatics* **8 Suppl 7**, S15.
73. Zvonic S, Ptitsyn AA, Conrad SA, Scott LK, Floyd ZE, Kilroy G, Wu X, Goh BC, Mynatt RL, and Gimble JM. (2006) Characterization of peripheral circadian clocks in adipose tissues. *Diabetes* **55**, 962–70.
74. Bray MS, and Young ME. (2007) Circadian rhythms in the development of obesity: potential role for the circadian clock within the adipocyte. *Obes Rev* **8**, 169–81.
75. Ando H, Yanagihara H, Hayashi Y, Obi Y, Tsuruoka S, Takamura T, Kaneko S, and Fujimura A. (2005) Rhythmic messenger ribonucleic acid expression of clock genes and adipocytokines in mouse visceral adipose tissue. *Endocrinology* **146**, 5631–6.
76. Imbeault P. (2007) Environmental influences on adiponectin levels in humans. *Appl Physiol Nutr Metab* **32**, 505–11.
77. Matarese G. (2000) Leptin and the immune system: how nutritional status influences the immune response. *Eur Cytokine Netw* **11**, 7–14.
78. Poppitt SD, Leahy FE, Keogh GF, Wang Y, Mulvey TB, Stojkovic M, Chan YK, Choong YS, McArdle BH, and Cooper GJ. (2006) Effect of high-fat meals and fatty acid saturation on postprandial levels of the hormones ghrelin and leptin in healthy men. *Eur J Clin Nutr* **60**, 77–84.
79. Esposito K, Nappo F, Giugliano F, Di Palo C, Ciotola M, Barbieri M, Paolisso G, and Giugliano D. (2003) Meal modulation of circulating interleukin 18 and adiponectin concentrations in healthy subjects and in patients with type 2 diabetes mellitus. *Am J Clin Nutr* **78**, 1135–40.
80. Poppitt SD, Keogh GF, Lithander FE, Wang Y, Mulvey TB, Chan YK, McArdle BH, and Cooper GJ. (2008) Postprandial response of adiponectin, interleukin-6, tumor necrosis factor-alpha, and C-reactive protein to a high-fat dietary load. *Nutrition* **24**, 322–9.

18 Diet, Obesity, and Lipids: Cultural and Political Barriers to Their Control in Developing Economies

Henry Greenberg, Anne Marie Thow, Susan R. Raymond, and Stephen R. Leeder

Key Points

- There are cultural barriers to risk factor control in developing economies.
- There are political barriers to risk factor control in developing countries.
- Urbanization dominates the influence on the transition to the current state of cardiovascular risk.
- Global health organizations need to change goals and structures.
- The North Karelia project of the 1970s is an effective roadmap for developing future programs.

Abstract

Serum lipids and their societal counterparts, diet and nutritional behavior, are major risk factors for atherosclerotic vascular disease. We posit that there are two complex barriers to carrying out effective policies to combat hyperlipidemia and obesity in developing economies. The first is that chronic diseases and their risk factors are not on the agenda of the global health community. The second, and the focus of this chapter, is that the barriers are political and cultural, arenas rarely entered by global health assistance organizations. In order to do this, the global health community needs to change its orientation and its structure.

The most obvious barriers are the lifestyle transitions induced by urbanization, such as diet, use of leisure time, exercise or the lack thereof, and employment demands. New global trade relationships and their priorities are also important barriers. Educational systems and health-care delivery systems also are not oriented to focus on the epidemic of chronic illness.

The Finnish North Karelia project, begun in the 1970s, is a helpful roadmap, and its design and priorities can play an important role in the future.

Key Words: Culture; global health assistance; hyperlipidemia; nutrition; obesity; politics; urbanization

18.1. INTRODUCTION

Cardiovascular disease is a major contributor to morbidity and mortality in developing countries *(1)*. Serum lipids and obesity *(2–4)* are major risk factors for the development and progression of atherosclerotic cardiovascular disease. While genetic determinants play a role, diet, nutrition, and pharmacological intervention are the cornerstones of current interventional practice. Drugs, especially statins, are effective in lipid control, but they are expensive, possess side effects, and, most importantly, are not begun until clinical disease is suspected either because elevated lipid levels or other prominent risk factors are present or because overt disease is discovered. Nutritional control of lipids and of body mass index is the key preventive and public health goal.

A. Bendich, R.J. Deckelbaum (eds.), *Preventive Nutrition*, Nutrition and Health, DOI 10.1007/978-1-60327-542-2_18,

Population-level nutritional behavior depends on both personal and environment factors. There is growing evidence of the importance of creating environments that support healthy nutritional behavior changes, where public policy plays a critical role. This chapter will examine cultural and political barriers to cardiovascular disease prevention through improved nutrition, in the context of complex new policy challenges and the expanded role of non-government forces. To achieve effective political change, the role of public health professionals needs to be reassessed and must adapt to these emerging challenges.

18.1.1. Barriers to Action

In developing economies, there are two complex barriers to establishing and carrying out effective policies to combat hyperlipidemia and obesity. First, chronic disease is not on the priority agenda of any global health assistance organization, and not on the agenda at all in most, the WHO 2005 recognition of the dominance of chronic illness notwithstanding *(5)*. The focus on communicable disease both from within country and from the assistance agencies dims the impetus for public policy directed toward chronic disease prevention. Until systems for monitoring and managing hypertension, diabetes, as well as hyperlipidemia are established, any intervention requiring ongoing monitoring of behavior as well as drug compliance cannot be successful. Establishing such systems requires policy commitment which is often lacking in developing nations. This is an unfortunate reality that we *(6,7)* and others *(8)* have discussed previously.

The second barrier is the focus of this chapter. The impediments to establishing and carrying out effective policies to combat hyperlipidemia and obesity, as well as other risk factors, are cultural and political, and hence are even more complex than technical policy formulation. Commercial pressures, trade and other economic forces, and cultural expectations, particularly those emerging from recent urban transformations, divert efforts to combat the risk factors for chronic disease. To prevent the young from developing hyperlipidemia or obesity, or to slow or reverse the patterns in older people requires changes in social and political organization as well as in individual behavior.

18.2. THE PROBLEMS OF CULTURE AND POLITICS

In its simplest presentation, the problem universally is to reduce caloric intake and enhance caloric expenditure through increased physical activity while ensuring that calories from saturated and trans fats are kept within acceptable guideline ranges. The advice is simple to give; the necessary steps are straightforward and easily understood; and the outcomes are readily measured.

Converting advice to action, of course, is the true challenge. As author Gordon Dickson has observed, “Some people like my advice so much that they frame it upon the wall instead of using it.” The barriers to implementation of effective public health advice are, in no small measure, in the political and cultural dimensions of change. In many situations they are often not easy to define, and, where they can be defined, surmounting them requires sustained input from a wide array of participants.

Urbanization provides a powerful lens through which to illustrate the intersection between the engines of economic progress, the pressures of culture, and the problems of hyperlipidemia and obesity. The urban transformation is occurring almost everywhere, and with great rapidity *(9)*. Demographic projections show that the world is well along the path of a massive shift from countryside to city or to a megacity environment. Moreover, even in emerging economies urban income is higher than rural income, despite widening economic inequalities. This translates to disposable income coupled to broad product access.

With urbanization comes contact with new foods, new expectations, new family structures, and new exposures. High-energy fast food is cheap, available, chic, and tasty. Urban work schedules disrupt traditional family life and one of the first casualties is the family meal. Television, Internet, and video games are ubiquitous and command attention, eroding time for physical activity, which itself is not a planned part of daily life and, in any event, may not be accommodated by readily accessible facilities. Newly urbanized children see a future with more excitement, uncertainty, and variety than they did in the countryside and the anchors to tradition break quickly.

Urbanization is not the sole energy force fueling cultural shape shifting that breeds obesity, of course. Democratization spurs wider choices for people, including diets; and school curricula and public health initiatives often do not keep up with the expanding realities of public health needs. As education becomes more important for individual employment and national economic growth, the inclusion of physical education in school curricula declines. The falloff in breast feeding and increased use of infant formula may participate in the promotion of CVD risk factors, although the field suffers from a lack of randomized data *(10)*.

The list of positive economic development trends with negative effects on public health is long. The length of the list, however, does not leave public health without positive strategic options.

There are many governmental agencies whose policies impinge upon diet, obesity, and lipids (Table 18.1). An important contribution of the WHO study on macroeconomics and health was the recommendation that each country that is able to do so establish a macroeconomic commission on health *(11)*. It would require that all government agencies include health, particularly public health, considerations in the formulation of public policy. Examples include redesigning the subsidy policies of the department of agriculture, tilted so as to favor vegetable oils rather than animal fats. Such policies are credited with positive CVD health impacts in Poland *(12)* and Finland *(13)*. Other strategies include integrating diet and nutrition programs in school curricula which most likely has a long-term

Table 18.1
Disconnects Between Cultural Change and Policy Rigidity Create Barriers to Effective Program Action

What Cultures and Conditions Change. . .
- "Modernization" of preferred food types
- Erosion of image of traditional foods
- Increased pace and pressures of lifestyles
- Erosion of family mealtime and traditions
- Increase in employment that is sedentary
- Increase in technology as a measure of modernity
- Increase in use of TV and video options as recreation

What Current Policies Do Not Change. . .
- Agricultural subsidies
- Requirements for nutritional information on packaged foods
- Food advertising requirements and standards
- "Silos" of policy formulation with nutrition as the policy orphan
- Investment policies for urban planning
- K-12 education curriculum
- Youth physical education investments
- Priority of nutrition within public health education
- Trade policies on food commodities and good processing investments

impact, although short-term impact is unclear. Where they exist, ministries of youth and sport have a clear role, but need to be viewed through the health prism to gain adequate budgets *(14)*. Urban planning needs to incorporate facilities for physical recreational activity and outdoor safety needs to be insured so that facilities will be used. The use of taxing power can be used effectively for social engineering and can play many roles in shaping a country's consumption and utilization of food stuffs.

Such broad, yet integrated solutions to the overarching cultural effects of economic change, however, must operate through political structures. Engaging political actors to move public health initiatives forward is not easy. Imposing political interventions is not easy. Not only is change usually opposed by whatever structures are entrenched, food production and consumption are everywhere embedded in cultural, economic, and political infrastructure. Such change is made all the more difficult because the factual basis for choosing strategies for action that are worth political risk are often untested or unconfirmed. Any change will engender pushbacks, and major changes will engender major pushbacks. High-energy fast food has many allies, including consumers, farmers, national and multinational corporations. Moreover, food outlets of all types employ a large number of low-skilled young urbanites, whose employment is in the larger economic and societal interests. Attempting to curtail this will meet resistance that claims an urban public good as its rationale for objection.

Trade liberalization is another political and cultural reality that is increasingly affecting diets in the developing world. World Trade Organization (WTO) membership is an imperative in the global economic community, along with participation in a rising number of bilateral and regional trade agreements. These are driving an agenda of liberalization in developing countries which incorporate facilitation of imports, promotion of commodity exports, and decreased support and protection for domestic production and industry. These processes have changed nutritional incentives by influencing agricultural production toward cash cropping, facilitating investment in food processing, and enhancing the influence of global food processors through marketing *(15–17)*.

Adopting an outward-looking WTO-led liberalization agenda is also likely to decrease inward-looking political willingness and capacity for implementing changes to policy designed to improve population nutrition. The WTO enforces member's commitments to its agenda, limiting the ability of countries to effect "interventionalist" national trade, agricultural, and fiscal policies to promote consumption of healthy food, for example, agricultural subsidies for healthy oil for either domestic or export markets. While there is some precedent for health being considered as an interpretive principle in WTO dispute settlement *(18)*, the transferability to nutrition is limited by the lack of "cause and effect" pathways such as the relatively direct health implications of importing building material containing asbestos. In addition, food is one of the most highly traded commodities and the influence of the food industry on policy is now quite substantial.

There are other arenas that impinge on nutrition in which change will come only with engagement of the political leadership and only with great effort and transformational leadership. Educational establishments are large, effective political units nearly everywhere and unless time, money, and budget allocations support desired change, it is unlikely to occur, anywhere. Urban development in much of the world is done with little planning and adding considerations for bike paths, playgrounds, and public safety requires a sophistication, and budget, that is often lacking. The recent American Heart Association (AHA) *(19)* scientific statement on Population-Based Prevention of Obesity scours the literature and recognizes very little of promise.

18.2.1. Strategies for Solutions

Having laid out this bleak argument, on top of a paucity of funding for chronic disease, can there be a silver lining? Yes, but the global health assistance community will need new spectacles to see it.

The AHA position piece *(19)* mentions the Finnish North Karelia *(13)* project of the 1970s but does not really see it for what it is. It is an early roadmap. The North Karelia project was fundamentally a political initiative, sparked by a petition from provincial representatives who were incensed about the extremely high rates of cardiovascular disease in their province *(20)*. Not only this but also its central strategies were based on multi-sectoral policy initiatives. Attributes and results of North Karelia include (a) the authority of accurate, believable, powerful, and specific data accepted by the population and the government; (b) demonstration that upstream changes can have a quickly responsive downstream effect; (c) realization that broad perception of a problem can bring together effectively many government and non-government agencies; and (d) understanding that public health intervention can play an important role is national economic development. Also, Finland developed a cold climate rapeseed variety that permitted substitution of vegetable oil for animal fat without a national economic burden, an underappreciated but important component of the project *(13)*. This scenario is not unlike the resolution the Clinton Foundation developed with the soft drink manufacturers to remove sweetened drinks from schools in that they had other products to fill the vending machines *(21)*. In both settings the underpinnings of success were established by either government or corporate activity that permitted change to occur without jeopardizing economic positions.

Today's developing world has recently evolved characteristics that make a North Karelia approach feasible in these newly fertile environments. First, the infectious and other diseases of childhood are diminishing in import, leaving public health and medical capacity to concentrate on other health issues. Second, as life expectancy increases and fertility falls, populations age, and age matters—in the economy as consumers and at the ballot box as voters. Third, as economies shift from labor-intensive agrarian to service-oriented urban, the individual worker/employee and his/her skills accrue value and their health matters to employers. Fourth, HIV/AIDS treatment requires a system for managing chronic disease not unlike what is needed for hypertension, diabetes, or obesity/lipid disorders. These arguments need to be carried to those with public authority. This lobbying is a new, key role for the new health assistance organization.

Hence there are possible scenarios for developing countries to adopt that will begin an effective assault on changing how their people eat. Since chronic disease risk factor management ultimately requires individual participation, whether eating differently or stopping smoking, an approach to obesity and diet should be part of a broader comprehensive program that begins far upstream of public health. The strategy will need to begin with cultural and political communication. Given the complexity of asking people to change behavior and take medicine for asymptomatic diseases, a likely starting point for any program will be a public information campaign to set the stage. Not only will assistance organizations need to understand how to do this, but have to understand how to get the public health establishment and other governmental departments to develop and fund such campaigns.

Other collaborators in the fight against obesity will be the educational establishment and the medical community. Both will need to change what they do and their priorities in order to develop collaboration with non-medical sectors (such as economics, behavioral science, finance) whose interest, skills, and influence are central to changing the way people think and act. They will not be able to change without the budget support to augment the school curriculum or to have time and staff to set up monitoring programs for chronic disease. Even after changes in the professional attitudes and goals, the culture in the educational, medical, and public health professions will need to develop a capacity to collaborate across disciplines and economic sectors.

No doubt the challenge of change to the agricultural community and the food industry will be considerable, probably even greater than that faced by the educational, medical, and public health communities. If agriculture subsidies need to be changed there will be intense political jockeying required. The failure of the Doha round of WTO negotiations points to the intensity of the problems.

The food industry is powerful and a major economic contributor to development; it will cooperate only if the coalition-building strategy accommodates its long-term interests along with those of all other sectors that must find the middle ground of compromise.

To create a political environment in which these issues can be raised in the corridors of power with any reasonable expectation of implementation, political operatives and lobbyists will need to be on staff. If the government cannot be persuaded that these interventions, from information programs to bike paths, are necessary steps in the transition to contemporary public health goals, and do so with budgetary support and interdepartmental cooperation, little will get done. The consensus- and coalition-building roles of the assistance organizations will be their critical component.

There is one further programmatic role for the new assistance organization. The message has to go to the specific community at risk—rural farmers in Tanzania; new urban immigrants in Mumbai; nomadic herdsman in Mongolia; emerging Chinese middle-class factory workers with money in their pockets. There will be a pressing need for an anthropologist on staff. Policy will need to be culture specific *(22)*.

18.3. CONCLUSION AND RECOMMENDATION

The public health community needs a long time frame in which to intervene and then judge progress. Fortunately, partial or gradual responses generate partial or graded, but clearly visible, results. Stroke rates fall (blood pressure intervention), pneumonia hospitalizations decrease (smoking cessation), and diabetes incidence drops (weight loss/change in diet) rapidly enough with even partial risk factor control. If measurements are done well, and this is an essential part of any program, the results can begin a feedback loop for additional support. Tying changes in health measurements to economic metrics such as absenteeism or hospital admissions can be used to bolster ongoing support.

The key in all of this will be finding the central compromises that will allow multi-sectoral coalitions to exist within the cultures of nations and or organizations. There is irony here, of course. Politics—one of the barriers to change—is precisely what the health sector will need to master. For politics is, in fact, the art of compromise.

REFERENCES

1. Leeder S, Raymond S, Greenberg H. A Race against Time; The challenge of cardiovascular disease in developing economies. New York: Earth Institute, Columbia University, 2004.
2. Jensen MK, Chiuve SE, Rimm EB, et al. Obesity, behavioral lifestyle factors, and risk of acute coronary events. Circulation 2008;117: 3062–3069.
3. Bogers RP, Bemelmans WJE, Hoogenveen RT, et al. Association of overweight with increased risk of coronary heart disease partly independent of blood pressure and cholesterol levels. Arch Intern Med 2007;167:1720–1728.
4. Gelber RP, Gaziano JM, Orav EJ, et al. Measures of obesity and cardiovascular risk among men and women. J Am Coll Cardiol 2008;52:605–615.
5. WHO. Preventing chronic disease: a vital investment. Geneva: World Health Organization, 2005.
6. Greenberg H, Raymond SU, Leeder SR. Cardiovascular disease and global health: threat and opportunity. Health Aff 2005;24:W5–W31 (Web exclusive).
7. Greenberg H, Raymond SU, Leeder SR. Global health assistance for chronic illness: a look at the practical. Prog Cardiovasc Dis 2008;51:90–97.
8. Chopra M, Galbraith S, Darnton-Hill I. A global response to a global problem: the epidemic of overnutrtion. Bull WHO 2002;80:952–958.
9. Montgomery MR. The urban transformation of the developing world. Science 2008;319:761–764.
10. Diet, Nutrition, and the Prevention of Chronic Diseases. WHO technical report series-916. Geneva: World Health Organization, 2003.
11. WHO. Report of the Commission on Macroeconomics and Health: investing in health for economic development. Geneva: World Health Organization (Macroeconomic commission), 2001.

12. Zatonski WA, Willett W. Changes in dietary fat and declining coronary heart disease in Poland: population based study. BMJ 2005;331:187–188.
13. Vartiainen E, Jousilahti P, Alfhtan G. et al. Cardiovascular risk factor changes in Finland, 1972–1997. Intl J Epidemiol 2000;29:49–56.
14. Amusa LO, Toriola AL, Onyewadume IU, Dhaliwai HS. Perceived barriers to sport and recreation participation in Botswana. Afr J Phys Health Educ Recreation Dance 2008;14:115–129.
15. Rayner G, Hawkes C, Lang T, et al. Trade liberalization and the diet transition: a public health response. Health Promot Int 2006;21(Suppl. 1):67–74.
16. Hawkes C, Thow AM. Implications of CAFTA-DR for the nutrition transition in Central America. Rev Panam Salud Publica 2008;24:345–360.
17. Hawkes C. Uneven dietary development: linking the policies and processes of globalization with the nutritional transition, obesity, and diet-related chronic diseases. Global Health 2006;2(4):1–18.
18. Bloche MG, Jungman ER. Health policy and the World Trade organization. In Kawachi and Wamala, eds, Globalization and Health. New York: Oxford University Press, 2007.
19. Kumanyika SK, Obarzanek E, Stettler N, et al. Population-based prevention of obesity. Circulation 2008;118:428–464.
20. Finland National Public Health Institute 2006. North Karelia Project. Available at http://www.ktl.fi/portal/265. Accessed 27 September 2008.
21. Burros M, Warner M. Bottlers agree to a school ban on sweet drinks. New York Times, p. A-1, May 4, 2006.
22. Janes CR. Going global in century XXI: medical anthropology and the new primary health care. Human Organization 2004;63:457–471.

V

Prevention of Major Disabilities; Improvement in Health Outcomes

19 Diet, Osteoporosis, and Fracture Prevention: The Totality of the Evidence

Robert P. Heaney

Key Points

- Bone health requires total nutrition. This is because the integrity of bone tissue depends on the integrity of its cells, which like most other tissues, needs a broad array of macro- and micronutrients. Additionally, calcium and protein play key structural roles, since the bulk of the bony material is made up of these substances.
- Bone turns over relatively slowly. Thus the effects of inadequate nutrition on bone are often delayed and the structural properties of bone tend to reflect past nutrition more than current intakes.
- Calcium is a threshold nutrient. The minimum daily requirement is the intake at which bony response plateaus. To ensure reaching this threshold, calcium intake should be 1,500 mg/day both during growth and once again after age 50. Risk of osteoporotic hip and other non-spine fractures can be reduced by 30–50% with life-long calcium intakes in this range.
- Vitamin D is produced predominantly in the skin. Recommended daily oral intakes are sufficient only to prevent the most extreme bony manifestations of vitamin D deficiency. Optimal vitamin D status is ensured by serum 25(OH)D values ≥80 nmol/L (32 ng/mL). Lower values are associated with impaired regulation of calcium absorption and increased osteoporotic fracture risk. Daily utilization of vitamin D may be as high as 4,000 IU (100 μg). For most elderly individuals a daily oral dose of 1,000–2,000 IU is necessary to sustain adequate serum 25(OH)D concentrations.
- Protein, once thought to be potentially harmful to bone when ingested in large quantities, is now best understood as complementary to calcium. Together the two nutrients provide the bulk constituents of bony material. To achieve the full benefit of either, the intake of the other must be adequate as well. Protein intakes that optimize bony response are uncertain, but appear from available data to be above 1.0 g/kg/day.
- Recovery from hip fracture can be substantially improved with aggressive attention to the nutritional status of hip fracture patients, with special emphasis on repairing the protein malnutrition common in such patients.
- Even though typical magnesium intakes are below the RDA (310 mg/day and 400 mg/day for women and men, respectively), there appear to be few skeletal consequences of the shortfall. Supplemental magnesium does not improve calcium absorption in individuals consuming typical diets and has no recognized effect on calcium balance.
- Vitamin K, zinc, manganese, and copper are involved in various aspects of bone matrix formation, but it is not known whether deficiency of any of them contributes to the development or severity of typical osteoporosis.

Key Words: Calcium; parathyroid hormone (PTH) and hip fracture; phosphorus; protein; sodium, vitamin D; vitamin K

19.1. INTRODUCTION

19.1.1. Nutrition in the Osteoporotic Fracture Context

Osteoporosis is currently defined as a condition of skeletal fragility due to decreased bone mass and to microarchitectural deterioration of bone tissue, with consequent increased risk of fracture. The condition is multifactorial in pathogenesis. Nutrition affects bone health in two distinct ways. First, bone tissue deposition, maintenance, and repair are the result of cellular processes, which are

A. Bendich, R.J. Deckelbaum (eds.), *Preventive Nutrition*, Nutrition and Health, DOI 10.1007/978-1-60327-542-2_19,

as dependent on nutrition as are the corresponding processes of any other tissue. The production of bone matrix, for example, requires the synthesis and post-translational modification of collagen and an array of other proteins. Nutrients involved in these cellular activities include not only the amino acid building blocks of the protein itself, but vitamins C, D, and K, and the minerals phosphorus, copper, manganese, and zinc. Additionally, the regulation of calcium homeostasis through modulation of bone resorption requires normal magnesium nutrition. Second, the skeleton serves as a very large nutrient reserve for two elements, calcium and phosphorus. The size of that reserve (ultimately equivalent to the strength of the skeletal structures) will be dependent in part upon the daily balance between absorbed intake and excretory loss of these two minerals; although after infancy, calcium is more often rate-limiting than phosphorus.

Strength in bone, as in most engineering structures, is dependent not only on its massiveness, but also on the arrangement of its material in space, and on the intrinsic strength of its component material. In bone, that material strength is influenced over long periods of use by the accumulation of unrepaired fatigue damage, as well as by the prevalence and location of remodeling loci, which are locally weak until fully repaired. All three factors play a role in most low trauma fractures, and it is not often possible to say which may be the most important in any given case.

Bone mass and density are themselves influenced by many factors. The three most commonly found to be limiting in industrialized nations are physical activity, gonadal hormones, and nutrition. In adults of these nations the nutrients most apt to be in short supply are calcium and vitamin D. Calcium intake, specifically, may be inadequate for the straightforward reason that it is low; however, even when statistically "normal," it may still be inadequate because of subnormal absorption *(1)* or greater than normal excretory loss *(2–3)*.

Because calcium and vitamin D are the nutrients most likely to be limiting in the industrialized nations, much of the following discussion will focus on these two nutrients. It is necessary to stress at the outset that both function mainly as nutrients, not as drugs. Hence their beneficial effects will be confined to individuals in whom intake of either is insufficient. Also, calcium is not an isolated nutrient; it occurs in foods along with other nutrients; and it has been shown that diets low in calcium tend also to be nutritionally poor in other respects as well *(4)*. Thus, while it is necessary to deal with nutrients one by one in an analysis such as this, it is useful to bear in mind that the disorders in our patients are likely to be more complex.

19.2. CALCIUM

The primitive function of the skeleton is to serve as a source and as a sink for calcium and phosphorus, that is, as a reserve to offset shortages and as a place for safely storing dietary surpluses, at least after periods of depletion. This reserve feature is expressed, for example, in laboratory animals such as cats, rats, and dogs, which, when placed on low calcium intakes, will reduce bone mass as needed to maintain near constancy of calcium levels in the extracellular fluid *(5)*. This activity is mediated by parathyroid hormone (PTH) *(6)* and involves the destruction of microscopic volumes of bone, not the leaching of calcium from bone.

Reserves, by their nature, are designed to tide organisms over external shortages. When intake is inadequate, the reserve is the first component of the body content to be depleted. With most nutrients this depletion of the reserve has no detectable impact upon the health or functioning of the organism. Only after the reserve is exhausted and the metabolic pool begins to be depleted does clinical disease express itself. For some nutrients (e.g., vitamin A, energy), the reserve can be quite large, and the latent period may last many months. But for others (e.g., the water-soluble vitamins), the reserve may be very small and detectable dysfunction develops quickly when intake drops.

Calcium is a unique nutrient in that, over the course of evolution of the higher vertebrates, the calcium reserve acquired a second role, namely internal stiffening and rigidity—what is today the most apparent feature of the skeleton. Calcium is the only nutrient with a reserve that possesses such a major, if secondary, function, and the size of the reserve is unusually large, relative to the cellular and extracellular metabolic pools of calcium. As a result, dietary insufficiency virtually never directly impairs tissue functions that are dependent on calcium, at least in ways we now recognize. However, since bone strength is a function of bone mass, it follows inexorably that any decrease whatsoever in the size of the calcium reserve—any decrease in bone mass—will produce a corresponding decrease in bone strength. We literally walk about on our calcium reserve. It is this unique feature of calcium nutrition which is a major part of the linkage of calcium and bone status.

19.2.1. Ascertaining the Requirement for Calcium

Calcium functions as a threshold nutrient, much as does iron. This means that, below some critical value, the nutrient effect (bone mass for calcium and hemoglobin mass for iron) will be limited by available supplies, while above that value, i.e., the "threshold," no further benefit for that particular function will accrue from additional increases in intake. This biphasic relationship is illustrated in Fig. 19.1, in which the intake–effect relationship is depicted first schematically (1A), and then (1B) as exemplified by bone calcium data derived from a growing animal model. In panel 1B the effect of the nutrient is expressed directly as the amount of bone calcium an animal is able to accumulate from any given intake. The minimum requirement can be defined as the intake at which the curve first becomes flat.

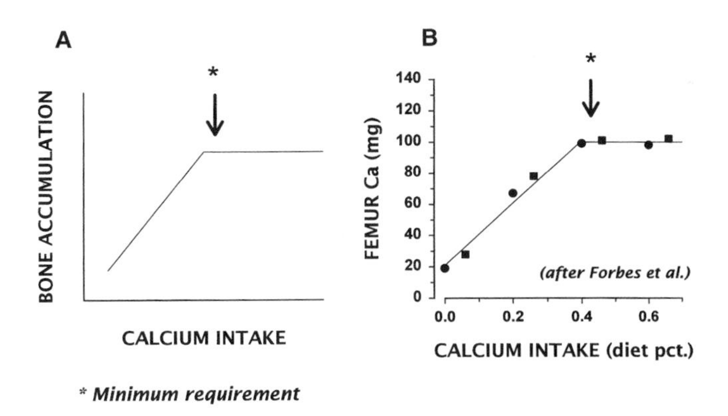

Fig. 19.1. Threshold behavior of calcium intake. (**a**) Theoretical relationship of bone accumulation to intake. Below a certain value—the threshold—bone accumulation is a linear function of intake (the ascending line); in other words, the amount of bone that can be accumulated is limited by the amount of calcium ingested. Above the threshold (the horizontal line), bone accumulation is limited by other factors and is no longer related to changes in calcium intake. (**b**) Actual data from two experiments in growing rats, showing how bone accumulation does, in fact, exhibit a threshold pattern. (Redrawn from data in Forbes RM et al. *(7)*.) (Copyright Robert P. Heaney, 1992. Reproduced with permission)

But the same basic relationship holds throughout life, even when bone may be undergoing some degree of involution. The threshold concept is generalized to all life situations in Fig. 19.2a, which shows schematically what the intake/retention curves look like during growth, maturity, and involution. In brief, the plateau occurs at a positive value for retention during growth, at zero retention in the

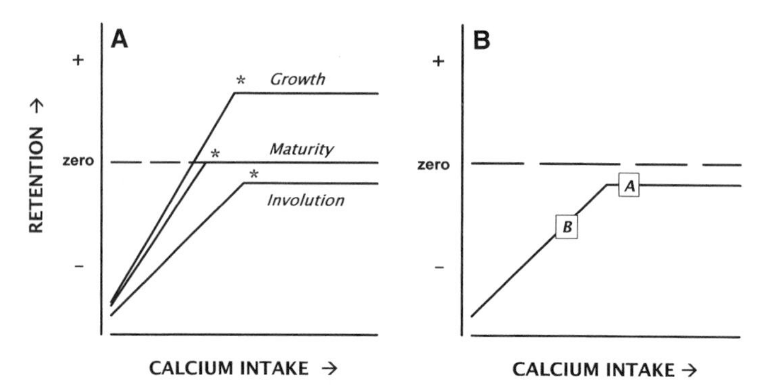

Fig. 19.2. (**a**) Schematic calcium intake and retention curves for three human life stages. Retention is greater than zero during growth, zero at maturity, and may be negative during involution. (**b**) The involution curve only. Point **B** designates an intake below the maximal calcium retention threshold, and Point **A**, an intake above the threshold. (Copyright Robert P. Heaney, 1998. Reproduced with permission)

mature individual, and sometimes at a negative value in the elderly. (Available evidence suggests that the plateau during involution is negative in the first 3–5 years after menopause, rises to zero for the next 10–15 years, and then becomes increasingly negative in the old elderly.) At all life stages the best representation of the minimum requirement is the intake value just at or above the effect threshold of Figs. 19.1 and 19.2.

In Fig. 19.2b, which shows only the involutional curve, there are two points located along the curve, one below (B) and one above (A) the threshold. At *A*, calcium retention is negative for reasons intrinsic to the skeleton (e.g., disuse), while at *B*, involutional effects are compounded by inadequate intake, which makes the balance more negative than it needs to be. Point B (or below) is probably where most older adults in the industrialized nations would be situated today (*see* below). The goal of calcium nutrition in this life stage is to move them to point *A* and thereby to make certain that insufficient calcium intake is not aggravating any underlying bone loss.

There has been much uncertainty and confusion in recent years about what that intake may be for various ages and physiological states. With the 1994 Consensus Development Conference on Optimal Calcium Intake *(8)* and the DRIs released by the Food and Nutrition Board of the Institute of Medicine *(9)*, the bulk of that confusion has been resolved. The evidence for the intakes recommended by these expert panels is summarized both in their respective reports and in comprehensive reviews of the relationship of nutrition and osteoporosis *(10,11)* and only the highlights will be mentioned in ensuing sections of this chapter.

It is worth noting, however, that the recommendations of both panels, while expressed in quantitative terms, are basically qualitative, and can be summarized as follows: Contemporary calcium intakes in the United States, by both men and women, are too low for optimal bone health. The most persuasive of the evidence leading to this conclusion came in the form of several randomized controlled trials showing both reduction in age-related bone loss and reduction in fractures following augmentation of prevailing calcium intakes *(12–19)*, For technical reasons relating to bone remodeling biology *(20)*, randomized controlled trials are not well suited to dose ranging. Hence, while the evidence was persuasive that prevailing intakes were too low, recommended levels in several cases involve ranges, and are clearly prudential judgments, centered of necessity on intakes employed in the trials concerned. The consensus panel's recommendations, the DRIs of the Institute of Medicine, and the corresponding 1989 RDAs *(21)* are set forth in Table 19.1.

Table 19.1
Various Estimates of the Calcium Requirement in Women

Age	*1989 RDA*	*NIH*[a]	*1997 DRI (AI)*[b]
1–5	800	800	500/800
6–10	800	800–1,200	800/1,300
11–24	1,200	1,200–1,500	1,300/1,000
Pregnancy/lactation	1,200	1,200–1,500	1,000
24–50/65	800	1,000	1,000/1,200
65–	800	1,500	1,200

[a]Recommendations for women as proposed by the Consensus Development Conference on Optimal Calcium Intake *(8)*.
[b]"AI" refers to "Adequate Intake" *(9)*, a value which, in this context, is equivalent to an average requirement. The corresponding RDA could be 20–30% higher, i.e., 1,000 in children, 1,600 in adolescents, 1,200 in adults out to age 50, 1,200 during pregnancy and lactation, and 1,450 in those over age 50. The presence of two values reflects the fact that the age categories for the DRIs overlapped those of the NIH.

19.2.2. Primary Prevention: The Acquisition of Genetically Programmed Bone Mass

The human skeleton contains at birth approximately 25–30 g calcium and, at maturity in women, 1,000–1,200 g. All of this difference must come in by way of the diet. Further, unlike other structural nutrients such as protein, the amount of calcium retained is always substantially less than the amount ingested. This is both because absorption efficiency is relatively low even during growth, and because calcium is lost daily through shed skin, nails, hair, and sweat, as well as in urine and unreclaimed digestive secretions. Except during infancy and the pubertal growth spurt, when retention may be as much as 25–30% of intake, only about 4–8% of ingested calcium is retained each day during most of the growth period. This inefficient retention is not so much because ability to build bone is limited but because the primitive calcium intake to which human physiology is adapted was high. An intestinal absorptive barrier is a protection against calcium surfeit, and inefficient cutaneous and renal retention reflect primitive environmental abundance. At primitive intakes, 4–8% retention was entirely adequate.

When ingested calcium is less than optimal, the balance between bone formation and resorption, normally positive during growth, falls toward zero. This occurs because PTH augments bone resorption at the endosteal–trabecular surface of growing bones in order to sustain the level of ionized calcium in the extracellular fluid, which has to meet the demands of ongoing mineralization at the periosteum and growth plates. In other words, what is not provided by the diet is taken from the calcium (skeletal) reserve. When these demands exceed the amount of calcium absorbed from the diet plus that released from growth-related bone modeling, more PTH is secreted, and resorption increases still further, until balance becomes zero or even negative. Growth in bone size continues, however, and a limited quantity of mineral now has to be redistributed over an expanding structural volume.

Net bone accumulation during growth will be greater as calcium intake increases, but only to the point where endosteal–trabecular resorption is due solely to the genetic program governing the shaping of bone and is not being driven by the body's need for calcium. Above that level, as depicted in the data of Fig. 19.1b, further increases in calcium intake will produce no further bony accumulation. The intake required to achieve the full genetic program, and thus to assure peak bone mass, is the intake that corresponds to the beginning of the plateau region in Fig. 19.1. This value will be different for different stages of growth, in part because growth rates are not constant and also because, as body size increases, obligatory calcium losses through skin and excreta increase as well *(22)*.

When the many published reports of calcium balance studies during growth are combined, it is possible to make out in humans the pattern of plateau behavior found in laboratory animals and then, from the aggregated data, to estimate the intake values that correspond to the threshold *(22,23)*. Fig. 19.3 represents one example of the relationship between intake and retention, combining the results of many published studies of calcium balance derived from a subset of the adolescents whose balances were assembled by Matkovic *(24)*. It clearly shows the plateau type of behavior that both animal studies and theoretical considerations predict. It also shows that, at intakes less than the plateau threshold, daily storage is less than optimal, i.e., accumulation of bone is limited by intake. Any such limiting intake must therefore be considered inadequate.

As can be seen from Fig. 19.3, the daily threshold value for adolescents is about 1,500–1,600 mg. Best available estimates for the value of this daily threshold from balance studies performed at other stages of growth are 1,400 mg in children and 1,000 mg in young adults out to age 30 (*see* below). The dual intake balance studies of Jackman et al. *(23)* found that 1,300 mg/day in adolescents was the lowest intake consistent with their model for the retention plateau.

These values, based on analysis of balance data, are buttressed by several randomized controlled trials of calcium supplementation in children and adolescents *(15,16,25,26)* and by a longitudinal observational study in young adults *(27)*. The controlled trials demonstrated that bone gain during growth was greater when intake was elevated above the 1989 RDA (800 mg/day). Some of the gain seen within the first 6–12 months of augmented calcium intake represents a phenomenon known as the remodeling transient *(20)*, which, while it confers improved bone strength in its own right, confounds estimates of the requirement since it reflects mainly the transition between two bone remodeling steady states. Nevertheless, computer modeling of the transient in these trials indicates that not all of the additional gain can be explained solely as a transient. Thus these data suggest that the 1989 RDAs lie on the ascending portion of the threshold curves of Fig. 19.2 rather than on the plateau, as they should. This explains the upward revisions reflected in the recommendations of the NIH Optimal Intake panel and of the Food and Nutrition Board.

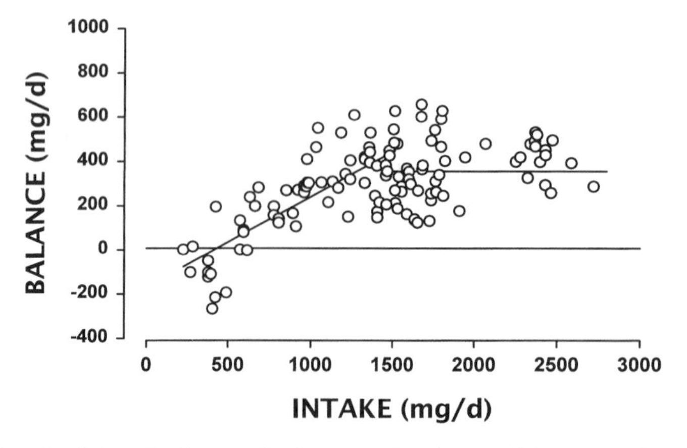

Fig. 19.3. The relationship of calcium intake, on the horizontal axis, to calcium retention (balance), on the vertical axis, for a subset of the adolescents described by Matkovic and Heaney *(22)*. Note that, despite the "noisiness" that is inevitable in measurements of balance in humans, there is clear evidence of an intake plateau, as observed in the animal experiments of Fig. 19.1. Note also that, for this age, the threshold of the plateau occurs at about 1,500 mg Ca/day. (Copyright Robert P. Heaney, 1992. Reproduced with permission)

Despite the evidence just cited, pointing to relatively high intake thresholds for maximal calcium retention during growth, epidemiologic data indicate that differences in peak bone mass related to habitual calcium intake are less than might have been anticipated. This discrepancy suggests that individuals with habitually lower intakes may nevertheless catch up to some extent. This seems to be the case in the controlled trial of Matkovic et al. *(28)*, in which not only did catch-up seem to occur in the unsupplemented adolescent girls, but the calcium benefit itself was most apparent in the tallest subjects, i.e., those with the greatest growth need for calcium.

Recker et al. *(27)*, in a longitudinal study of young adults, showed prospectively that bone augmentation continues into the third decade. Bone mass gains in this study ranged from 0.5% per year for the forearm to 1.25% per year for total body bone mineral. The single most important correlate of the rate of bone accumulation was calcium intake. This study, while it lacked the inferential power of a randomized controlled trial, nevertheless had an advantage over such trials in that it studied individuals on their self-selected intakes, i.e., at a steady state for bone remodeling, and thus avoided the confounding effect of the remodeling transient. The rate of bone accumulation in this study was inversely proportional to age, with the best estimate of the age at which the rate reached zero being approximately 29–30 years. Thus the window of opportunity to achieve the full genetic program (i.e., the chance for catch-up) appears to remain at least partly open until about age 30.

19.2.3. Secondary Prevention: The Conservation of Acquired Bone Mass

Studies of calcium requirement in mature, but still premenopausal women, have, in general, yielded results compatible with the newer recommendations of Table 19.1. Welten et al. *(29)*, in a meta-analysis of studies in this age group, concluded that calcium intake was positively associated with bone mass. Heaney et al. *(30)*, in a study of estrogen-replete women ingesting their habitual calcium intakes, found zero calcium balance at a mean intake slightly under 1,000 mg/day. By contrast, Nordin et al. *(2)* found a figure closer to 600 mg/day and Recker et al. *(31)* in a prospective study of bone mass in premenopausal women found no detectable bone loss over a 2-year period on an estimated mean calcium intake of 651 mg. (Corresponding RDAs, meeting the requirement of 95% of all individuals, would be between ~800 mg/day and ~1,200 mg/day.)

In brief, while there may be other health reasons for maintaining an even higher calcium intake during the mature years, bone health seems to be supported adequately by an intake in the range of 800–1,200 mg/day; lower intakes may lead to premenopausal bone loss or failure to achieve peak mass, or both.

19.2.4. Menopause

Estrogen has a bewildering variety of actions. In bone it seems to adjust the bending setpoint of the mechanical feedback loop that regulates bone mass. (The bending set point is the amount of bending a bone experiences during loading that is sufficient to trigger a bone remodeling response.) Accordingly, whenever women lose ovarian hormones, either naturally at menopause or earlier as a result of anorexia nervosa or athletic amenorrhea, the skeleton appears to sense that it has more bone than it needs, and hence, during ongoing continuous remodeling, allows resorption to carry away more bone than formation replaces. (Precisely the same change occurs when men lose gonadal hormones for any reason.) This amounts to raising the set point of the mechanical feedback loop that functions to maintain bone bending under load within safe limits. This downward adjustment in bone mass due to gonadal hormone lack varies somewhat from site to site across the skeleton, but at the spine amounts to approximately 12–15% of the bone a woman had prior to menopause *(32)*.

The importance of this phenomenon in a discussion of nutrient effects is to distinguish menopausal bone loss from nutrient deficiency loss and to stress that menopausal loss, which is due mainly to absence of gonadal hormones, not to nutrient deficiency, cannot be substantially influenced by diet. Almost all of the published studies of calcium supplementation within 5 years following menopause failed to prevent bone loss. Even Elders et al., who employed a calcium intake in excess of 3,100 mg/day succeeded only in slowing menopausal loss, not in preventing it *(33)*. Only a few reports, such as the study of Aloia *(19)*, contain clear evidence for a benefit of a high calcium intake at this life stage, and even here, estrogen produced a greater effect. Nevertheless, one can find in many of the published reports evidence of small calcium effects at even this life-stage, and it may be that, in any group of early menopausal women, there are some whose calcium intake is so inadequate that they are losing bone for two reasons (estrogen lack *plus* calcium insufficiency).

Important as menopausal bone loss is, it is only a one-time, downward adjustment, and, if nutrition is adequate, the loss continues for only a few years, after which the skeleton comes into a new steady state (although at a somewhat lower bone mass). It is in this context that the importance of achieving a high peak skeletal mass during growth becomes apparent. One standard deviation for lumbar spine bone mineral content in normal women is about 12–15% of the young adult mean, and for total body bone mineral, about 10–12%. Hence a woman who has a bone mass at least one standard deviation above the mean can sustain the 12–15% menopausal loss and still end up with about as much bone as the average woman has before menopause. By contrast, a woman at or under one standard deviation below the young adult mean premenopausally drops to two standard deviations below the mean as she crosses menopause and is therefore, by the WHO criteria *(34)*, already osteopenic and verging on frankly osteoporotic.

As noted, the menopausal bone mass adjustment theoretically stops with a loss of about 15% at the spine, but this is true only so long as calcium intake is adequate. In this regard, it is important to note that estrogen has non-skeletal effects as well, i.e., it improves intestinal calcium absorption and renal calcium conservation *(32,35,36)*. As a result, an estrogen-deficient woman has a higher calcium requirement, and unless she raises her calcium intake after menopause, she will continue to lose bone after the estrogen-dependent quantum has been lost, even if the same diet would have been adequate to maintain her skeleton before menopause. In other words, early in the menopausal period, her bone loss is mainly (or entirely) because of estrogen withdrawal, while later it will be because of inadequate calcium intake. Figure. 19.4 assembles, schematically, the set of factors contributing to bone loss in the postmenopausal period. The figure shows both the self-limiting character of the loss due to estrogen deficiency and the usually slower, but progressive loss due to nutritional deficiency (if present). Unlike the estrogen-related loss which mostly plays itself out in 3–6 years, an ongoing calcium deficiency loss will continue to deplete the skeleton indefinitely for the remainder of a woman's life, that is, unless calcium intake is raised to a level sufficient to stop it. Furthermore, since both absorption efficiency *(35)* and calcium intake *(37)* decline with age, the degree of calcium shortfall actually tends to worsen with age.

Thus it is important for a woman to increase her calcium intake after menopause. Both the 1984 NIH consensus conference on osteoporosis and the 1994 Consensus Conference on Optimal Calcium Intake *(8)* recommended intakes of 1,500 mg/day for estrogen-deprived postmenopausal women. The "Adequate Intakes" of the Food and Nutrition Board *(9)* for everyone over age 50, when translated into RDA format (Table 19.1), are nearly identical (1,450 mg/day). It may be that the optimal intake is somewhat higher still (*see* below), but median intakes in the United States for women of this age are

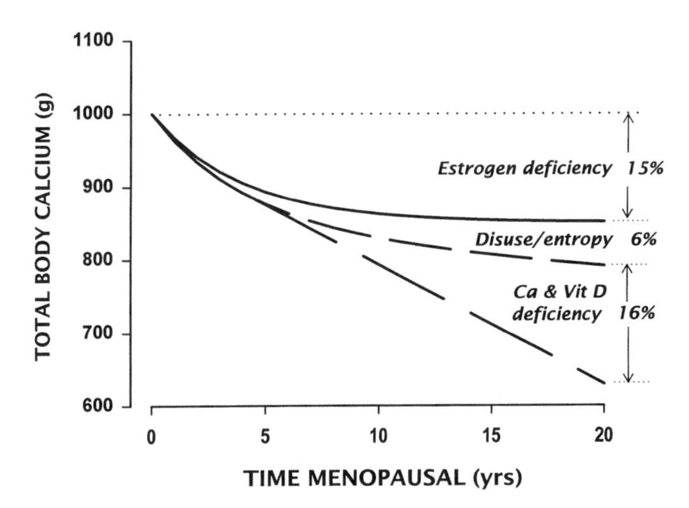

Fig. 19.4. Partition of age-related bone loss in a typical postmenopausal woman with an inadequate calcium intake. Based on a model described in detail elsewhere *(32)*. (Copyright Robert P. Heaney, 1990. Reproduced with permission)

in the range of 500–600 mg/day *(37,38)*, and if the bulk of them could be raised even to 1,500 mg/day, the impact on skeletal health would be considerable.

19.2.5. Senescence

Age-related bone loss occurs in both sexes, regardless of gonadal hormone levels, generally starting about age 50. However, it is obscured in the years immediately following menopause in women by the substantially larger effect of estrogen withdrawal (*see* Fig. 19.4). It probably occurs, however, even in estrogen-treated women, at about the same rate as in men. This rate is generally reported to be on the order of 0.3–1.0% per year during the sixth and seventh decades, and then accelerates with advancing age. For example, loss from the hip in the control subjects of the study by Chapuy et al. *(13)*, at an average age of 84, was 3% per year. Age-related loss involves both cortical and trabecular bone and can come about by several mechanisms: disuse, remodeling errors, and nutritional deficiency, summarized in Fig. 19.4.

While nutrient deficiency is clearly only a part of the total problem, nevertheless it is common. That the 3% loss in the control subjects of Chapuy et al. was related to their nutritional status is indicated by the fact that this loss was completely obliterated in the calcium and vitamin D supplemented women. Intestinal calcium absorption efficiency declines with age *(35)*, at the same time as nutrient intake itself generally declines *(37)*, the result is that the diet of aging individuals becomes more and more inadequate. McKane et al. *(39)* showed that the high PTH levels and abnormal PTH secretory dynamics typically found in elderly women are due to calcium deficiency, and that PTH function can be entirely normalized by calcium intakes of 2,400 mg/day. In a somewhat younger group of elderly subjects, Peacock et al. *(40)* showed that elevating calcium intake from ~550 mg/day to ~1,300 mg/day effectively obliterated bone loss at hip and total body over a 4-year treatment period.

It is in this age group also that the most dramatic and persuasive evidence for fracture prevention by high calcium intakes has been produced *(13,17,41,42)*. This is partly because most fragility fractures rise in frequency with age, and hence the opportunity to see a fracture benefit (if one exists) is greater then. Chapuy et al. *(13)* showed a reduction in hip fracture risk of 43% by 18 months after starting supplementation with calcium plus vitamin D, and a 32% reduction in other extremity fractures.

Dawson-Hughes et al. *(41)* produced a greater than 50% reduction in all non-vertebral fractures by 36 months of supplementation with both calcium and vitamin D. Chevalley et al. *(17)* in another study in elderly women, showed that, even when vitamin D was given to both groups, extra calcium reduced femoral bone loss *and* vertebral fracture incidence. And Recker et al. *(42)* in a 4-year, randomized controlled trial in elderly women (mean age 73) demonstrated that a calcium supplement reduced both age-related bone loss and incident vertebral fractures. Their subjects had all received a multivitamin supplement as had the subjects of Chevalley et al. *(17)*; hence, the effect in the calcium-supplemented groups of both studies can be attributed to the extra calcium alone.

These findings do not mean that vitamin D is unimportant in this age group (*see* below). It is likely that intakes of both calcium and vitamin D are inadequate in most elderly individuals, and the unrecognized prevalence of combined deficiency has made it difficult to study the actual requirements of either nutrient in this age group.

The calcium intake achieved in the Chapuy study was about 1,700 mg/day, 1,400 mg/day in the Chevalley study, 1,300 mg/day in the Dawson-Hughes study, and about 1,600 mg/day in the Recker study. These values are in the range of the intake earlier found by Heaney et al. *(30)* to be the mean requirement for healthy estrogen-deprived older women (1,500–1,700 mg/day). All these studies are, therefore, congruent with the recommendations of Table 19.1.

It has been generally considered that the anti-fracture benefit of calcium observed in these controlled trials was due to protection against calcium deficiency-induced bone loss. However, the fracture benefit is substantially larger than the bone mass difference would predict and the fracture reduction is apparent within a very few months of starting calcium (and vitamin D) supplementation *(13,41,43)*—well before much effect on bone mass could have occurred. A better explanation seems to be reduction in the excessively high remodeling activity that is common in postmenopausal women and in the elderly generally. Remodeling activity is emerging as an important risk factor for fragility, and any agency that reduces remodeling seems to reduce fracture risk, independently of any effect on bone mass *(43)*.

An important feature of these controlled trials in already elderly individuals was that bone mass was low in both treated and control groups at the start of the study, and while a significant difference in fracture rate was produced by calcium supplementation, even the supplemented groups had what would have to be considered an unacceptably high fracture rate. What these studies do not establish is how much lower the fracture rate might have been if a high calcium intake had been provided for the preceding 20–30 years of these women's lives. The studies of Matkovic et al. *(44)* and Holbrook et al. *(45)*, although not randomized trials, strongly suggest that the effect may be larger than has been found with treatment started in the seventh to ninth decades of life. Both of these observational studies reported a hip fracture rate that was roughly 60% lower in elderly persons whose habitual calcium intakes had been high. While findings from observational studies such as these had not been considered persuasive in the absence of proof from controlled trials, the now many controlled trials showing a skeletal benefit of added calcium have removed that uncertainty.

Despite the massiveness and consistency of the body of evidence linking calcium intake to bone mass and fracture risk, the publication in 2006 of the results of the calcium-D arm of the Women's Health Initiative (WHI) trial *(46)* seemed to reopen the entire question, casting sufficient doubt so as to result in a substantial decline in calcium supplement sales in the 6 months following publication. In this randomized, double-blind, placebo-controlled trial, 36,282 women over age 50 were randomized either to a treatment containing 1,000 mg calcium (as $CaCO_3$) plus 400 IU vitamin D_3 or to placebo. In intention-to-treat analysis, there was a statistically significant reduction in age-related bone loss, a non-significant (12%) reduction in hip fracture, and a 30%, significant reduction in women not using personal calcium supplements. When analysis was confined to those who were treatment-adherent, there was a statistically significant, 29% reduction in hip fractures. However, since treatment adherence may be associated with other factors that might have been the actual cause of the lower fracture risk, such "per protocol" analysis is generally considered less persuasive than intention-to-treat analysis.

It is important to dissect apart these findings, if for no other reason than because the WHI report has been considered in some sectors to refute the larger body of prior evidence, simply because of WHI's vastly larger sample size. Three crucial differences between WHI and all preceding studies stand out. First, in WHI there was no low calcium intake control group. Mean calcium intake in the cohort prior to entry was between 1,100 and 1,200 mg/day, i.e., effectively at the AI value of the Institute of Medicine. In terms of Figs. 19.1 and 19.2, both the placebo and the treated groups were already on the response plateau (or close to it) prior to taking any supplemental calcium. Hence little or no response should have been expected. Second, baseline vitamin D status was very low (mean serum 25(OH)D was 42 nmol/L (17 ng/mL), and the effective vitamin D dose (~200 IU/day after factoring in compliance) would not have been sufficient to raise serum 25(OH)D by more than 5 nmol/L (~2 ng/mL). Hence calcium absorption efficiency would have been impaired *(47)*. Finally, observed hip fracture risk was about half of what had been anticipated from Medicare data, at the design phase of the project.

Hence, rather than refuting the conclusions of prior work, the findings of WHI (low absolute fracture rate and little response to extra calcium beyond an already relatively high intake) are precisely consistent with current understanding of how calcium intake relates to bone status. The observed protection against bone loss and the non-significant fracture risk reduction can best be explained as a partial response due to the fact that, while mean intake on randomization was close to adequate, roughly half would have had lower than average intakes and would have been in a position to show at least partial response.

In brief, the totality of the evidence indicates clearly that an adequate calcium intake is essential for bone health and underscores the importance of achieving at least the 1,200–1,500 mg target figure for the elderly. At the same time it must be stressed, once again, that osteoporosis is a multifactorial condition, and that removing one of the pathogenic factors (i.e., insuring an adequate calcium intake) cannot be expected to eradicate all osteoporotic fractures.

19.2.6. Nutrient–Nutrient Interactions: Factors that Influence the Requirement

There are several nutritional factors which influence or have been proposed to influence the calcium requirement. These are probably less important than once thought, since at calcium intakes in the ranges currently recommended, interactions tend to have negligible impact on the calcium economy. Nevertheless, the issue of interactions continues to arise and so will be addressed briefly here.

The principal interacting nutrients are sodium, protein, caffeine, and fiber. Fiber and caffeine influence calcium absorption *(48–51)* and typically exert relatively minor effects, while sodium and protein influence urinary excretion of calcium *(51,52)* and can be of much greater significance for the calcium economy when calcium intakes are low. The net effects of phosphorus and fat in humans are minor to non-existent.

The basis for the differing importance of these nutritional factors on the calcium economy is illustrated in Fig. 19.5, which partitions the variance in calcium balance observed in 560 balances in healthy middle-aged women consuming typical intakes, and studied in the author's laboratory. As Fig. 19.5 shows, only 11% of the variance in balance among these women is explained by differences in their actual calcium intakes. By contrast, absorption efficiency explains about 15%, while urinary losses explain more than half.

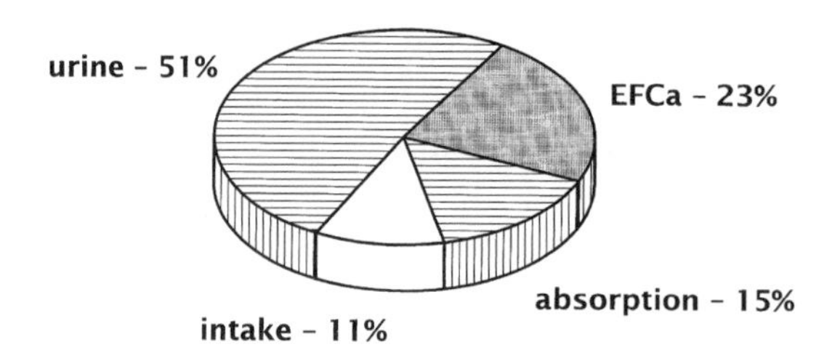

Fig. 19.5. Partition of variance in calcium balance in normal women among the input–output processes involved in calculation of balance. (Copyright Robert P. Heaney, 1994. Reproduced with permission)

19.2.6.1. Influences on Intestinal Absorption of Calcium

19.2.6.1.1. Fiber. The effect of fiber is variable, and generally small. Many kinds of fiber have no influence at all on absorption, such as the fiber in green, leafy vegetables *(10)*. The fiber in wheat bran, by contrast, reduces absorption of co-ingested calcium, although except for extremes of fiber intake *(53)*, the overall effect is generally relatively small. Often lumped together with fiber are associated plant food constituents such as phytate and oxalate. Both can reduce the availability of any calcium contained in the same food, but, unlike bran, generally do not affect co-ingested calcium from other foods. For example, for equal ingested loads, the calcium of beans is only about half as available as the calcium of milk *(54)*, while the calcium of spinach and rhubarb is nearly totally unavailable *(55)*. For spinach and rhubarb, the inhibition is mostly due to oxalate. For common beans, phytate is responsible for about half the interference, and oxalate, the other half. Even so, the effects of phytate and oxalate are highly variable from food to food. There is a sufficient quantity of both anti-absorbers in beans to complex all the calcium also present, and yet absorptive interference is only half what might be expected.

19.2.6.1.2. Caffeine. Often considered to have a deleterious effect on the calcium economy, caffeine actually has the smallest effect of the known interacting nutrients *(49)*. A single cup of brewed coffee causes deterioration in calcium balance of ~3 mg *(50,51,56)*, mainly by reducing absorption of calcium *(50)*. The effect is probably on active transport, although this is not known for certain. This small effect is more than adequately offset by a tablespoon or two of milk *(50,56)*.

19.2.6.2. Influences on Renal Conservation of Calcium

19.2.6.2.1. Protein and Sodium. As noted, the effects of protein and of sodium can be substantial *(2,3,51)*. Both nutrients can increase urinary calcium loss across the full range of their own intakes, from very low to very high—so it is not a question of harmful effects of an *excess* of these nutrients. Sodium and calcium share the same transport system in the proximal tubule, and every 2,300 mg sodium excreted by the kidney pulls 20–60 mg of calcium out with it. And every gram of typical protein (whether from animal or vegetable sources) metabolized in adults causes an increment in urine calcium loss of about 1 mg. This latter effect is probably due to excretion of the sulfate load produced in the metabolism of sulfur-containing amino acids (and is thus a kind of endogenous analog of the acid-rain problem).

Although much of the literature in this field stresses the role of sodium, as such, it is important to recognize that most dietary sodium is in the form of table salt, sodium chloride, and that the often ignored anion plays an important role in these interactions. This issue is too complex for exhaustive treatment here, and it will be sufficient only to note that sodium bicarbonate does not have the same hypercalciuric effect as sodium chloride *(57)*, and that potassium bicarbonate completely obliterates

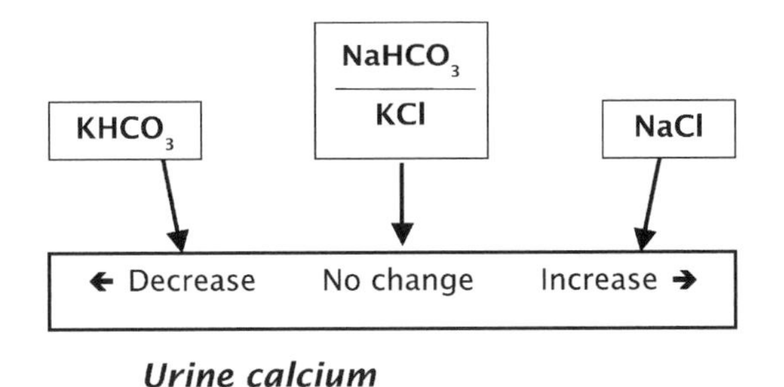

Fig. 19.6. Effects of various sodium and potassium salts on urine calcium. (Copyright Robert P. Heaney, 2003. Used with permission)

the calciuria of a high sodium chloride intake *(58)*, Fig. 19.6 summarizes the interplay of sodium, potassium, and their respective anions, on urine calcium.

At low salt and protein intakes, the minimum calcium requirement for an adult premenopausal female may be as little as 450 mg/day *(52)*, whereas if her intake of both nutrients is high, she may require as much as 2,000 mg/day to maintain calcium balance. A forceful illustration of the importance of sodium intake is provided by the report of Matkovic et al. *(59)*, showing that urine calcium remains high in adolescent girls on calcium intakes too low to permit bone gain. The principal determinant of urinary calcium in such young women is sodium intake *(60)*, not calcium intake.

Differences in protein and sodium intake from one national group to another are perhaps part of the explanation why studies in different countries have shown sometimes strikingly different calcium requirements. At the same time, one usually finds a positive correlation between calcium intake and bone mass within each national range of intakes *(61)*. Hence although sodium (and protein) intake differences between cultures obscure the calcium effect, they do not obliterate it.

For diets high in calcium, as would have been the case for our hunter–gatherer ancestors, high protein and possibly high sodium intakes could have been handled by the body perfectly well. At the low intakes that prevail today, an individual's absorptive performance is close to maximal. Augmented loss from increased sodium or protein intake cannot be offset by increasing extraction from the diet, both because there is less there to extract, and because extraction efficiency is already at the upper end of its possible range. By contrast, at intake levels typical of those that prevailed during hominid evolution, intestinal absorption is predominantly passive, and the full range of absorptive adaptation is available to offset increased excretory or cutaneous losses. In brief, these nutrients create problems for the calcium economy of contemporary adult humans mainly because we typically have calcium intakes that are low relative to those of pre-agricultural humans, and sodium intakes that are high.

19.2.6.2.2. Acid Ash Residue. The acid/alkaline ash characteristic of the diet may also be important, although the quantitative relationship of this diet feature to the calcium requirement has been less fully explored to date. Nevertheless, it has clearly been shown that substitution of metabolizable anions (e.g., bicarbonate or acetate) for fixed anions (e.g., chloride) in various test diets will lower obligatory urinary calcium loss substantially *(62,63)*. This suggests that primarily vegetarian diets create a lower calcium requirement and provides a further explanation for the seemingly lower requirement in many non-industrialized populations. However, it is not yet clear whether, within a population, vegetarians have higher bone mass values than omnivores, and some data suggest they may actually have less dense skeletons, possibly because of the often very low calcium levels of such diets *(64,65)*.

19.2.6.2.3. Phosphorus. Phosphorus is commonly believed to reduce calcium absorption, but the evidence for that effect is scant to non-existent, and there is much contrary evidence. In analysis of

567 metabolic balances performed in healthy middle-aged women studied on their usual diets, variation in phosphorus intake over a nearly sixfold range had no detectable effect on calcium absorption efficiency *(66)*. And in a controlled metabolic study, Spencer found no effect of even large increments in phosphate intake on overall calcium balance at low, normal, and high intakes of calcium *(67)*. In adults, Ca:P ratios ranging from 0.2 to above 2.0 are without effect on calcium balance, at least so long as adjustments are made for calcium intake *(51,66)*.

What phosphorus does is depress urinary calcium loss and elevate digestive juice secretion of calcium, by approximately equal amounts, with little or no net effect on balance *(68)*. While it is true that stoichiometric excesses of phosphate will tend to form complexes with calcium in the chyme, various calcium phosphate salts have been shown to exhibit absorbability similar to other calcium salts, and phosphate is, of course, a principal anion of the major food source of calcium (dairy products). In any case, phosphate itself is more readily absorbed than calcium (by a factor of 2–5 ×), and at intakes of both nutrients in the range of their respective RDAs, absorption will leave a stoichiometric excess of calcium in the ileum, not the other way about. This explains the seeming paradox that high calcium intakes can block phosphate absorption (as in management of end-stage renal disease), while achievably high phosphate intakes have little or no effect on calcium absorption.

19.2.6.2.4. Aluminum. Although not in any proper sense a nutrient, aluminum, in the form of Al-containing antacids, also exerts significant effects on obligatory calcium loss in the urine *(69)*. By binding phosphate in the gut, these substances reduce phosphate absorption, lower integrated 24-h serum phosphate levels, and thereby elevate urinary calcium loss. (This is the opposite of the more familiar hypocalciuric effect of oral phosphate supplements.) Therapeutic doses of Al-containing antacids can elevate urine calcium by 50 mg/day or more.

19.2.7. Calcium Sources

The best calcium sources are, of course, foods. In a modern, Western diet, food items that provide more than 100 mg of calcium per serving are limited to dairy products (with the exception of cottage cheese), greens of the mustard family (collards, kale, mustard), calcium-set tofu, sardines, and a few nuts (especially hazelnuts and almonds). Smaller amounts of calcium are ubiquitous in many leafy vegetables, but with the exception of shellfish, calcium levels are low in most meats, poultry, or fish. As noted earlier, the calcium of beans is only about half as available as the calcium of milk, and the calcium of high oxalate vegetables (such as spinach and rhubarb) is almost completely unavailable. Figure 19.7 displays the *available* calcium in a variety of foods. "Available" represents the product of the fractional absorbability of the calcium in a food and its total calcium content. It is thus the actual, gross[1] amount of calcium a particular food delivers into the blood of the absorbing subject.

In general, most diets without dairy products have total calcium nutrient densities under 20 mg Ca/100 kCal, and *available* calcium densities lower still. Since total energy intake for adult American women is in the range of 1,400–1,800 kCal/day, it follows that most diets low in dairy products will be low in calcium—probably 300–400 mg Ca/day or less—far short of levels currently considered optimal.

In part as a response to this dilemma, the Surgeon General, in his 1988 report on Nutrition and Health *(70)*, recommended judicious, low-level calcium fortification of many items in the food chain. An increasing number of fortified foods is becoming available each year—ranging from fruit juice, to

[1] In the process of digestion substantial quantities of calcium enter the gut in the form of digestive secretions and sloughed mucosa. For this reason, net absorption is always less than gross.

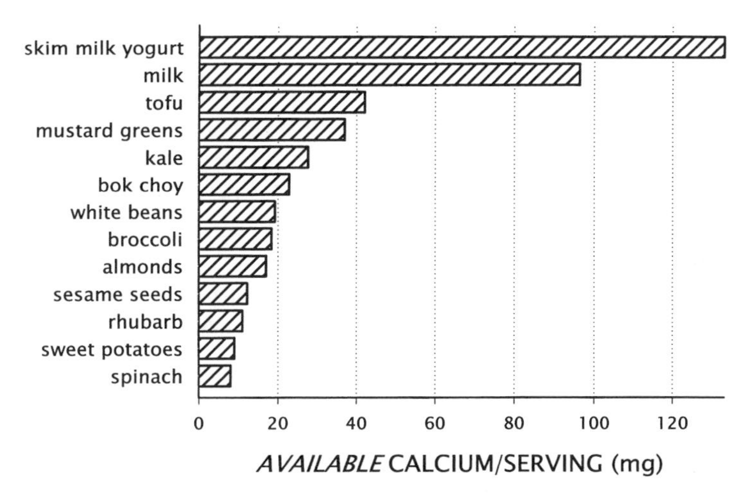

Fig. 19.7. Available calcium per serving for a variety of foods. "Available calcium" is the product the calcium content of a food and its fractional absorbability. (Copyright Robert P. Heaney, 1998. Reproduced with permission)

bread, to breakfast cereals, to potato chips, to rice. Where bioavailability of the calcium in these foods has been ensured, these foods should be useful adjuncts in the attempt to improve calcium intake at the population level. However, satisfactory bioavailability does not automatically follow from simply adding calcium to a product. Recent entries into the soy and rice beverage markets, as well as several of the calcium-fortified orange juices, exhibit very poor physical characteristics *(71)*. Soy beverage fortified to the calcium load of cow milk has been shown to deliver only 75% of the amount of calcium as cow milk *(72)* when the fortificant is optimally suspended in the beverage, and as little as 30% when (as is usually the case) the fortificant has settled to the bottom of the carton *(71)*.

The principal calcium supplement in the US market is calcium carbonate, available as such, or as oyster shell or dolomite. When the tablet is competently formulated, so that it disintegrates in the stomach, or when the supplement is chewed, the carbonate is quite well absorbed and generally very well tolerated. There is no requirement for gastric acid, per se, so long as the carbonate salt is taken with meals. Calcium citrate and calcium citrate malate (CCM) are also good sources, but they tend to be more expensive *(73)* and tend to require more tablets to deliver the same quantity of calcium. CCM has been reported in some studies to exhibit higher absorption efficiency than calcium carbonate *(41)* and the study of Peacock et al. *(40)*, using CCM as the calcium source, reported one of the largest calcium effects on age-related bone loss observed to date. In the rare case in which the carbonate seems to be not well tolerated, these other sources provide useful alternatives.

Divided doses enhance absorption from both supplements and foods, since absorption fraction is an inverse function of load size. All calcium sources (including food) interfere with iron absorption when the two nutrients are ingested at the same meal. However, single-meal studies miss the body's up-regulation of iron absorption in the face of need, and chronic feeding studies have revealed no deterioration of iron status in subjects consuming high calcium diets. Perhaps of greater relevance, Matkovic and his colleagues have convincingly shown that adolescent girls are able to increase total body iron stores normally in the presence of 1,600 mg calcium intake *(74)*. This is a particularly reassuring finding since females in this age group are among the most vulnerable to iron deficiency in the United States today. However, if an adult is iron-deficient (e.g., as a result of severe blood loss) and is taking an iron supplement, it may be best if the meal at which the iron is taken not contain a large amount of calcium (food or supplement).

19.3. VITAMIN D

Vitamin D facilitates active transport of calcium across the intestinal mucosa, at least partly by inducing the formation of a calcium transport complex in intestinal mucosal cells. This function is particularly important for adaptation to low calcium intakes. Absorption also occurs passively, probably mainly by way of paracellular diffusion. This route is not dependent on vitamin D and is not well studied. The proportion of absorption by the two mechanisms varies with intake and is not well characterized in humans; at high calcium intakes (above 2,400 mg/day) gross absorption fraction approaches 10–15% of intake. Under these circumstances it is likely that active transport contributes relatively little to the total absorbed load. Nevertheless, it is generally considered that vitamin D status enables regulation of absorptive performance and that it thereby influences the calcium requirement. The quantitative importance of active transport of calcium at prevailing calcium intakes is illustrated in Fig. 19.8, which plots the actual net quantity of calcium absorbed as a function both of calcium intake and of active absorption. As can be seen, some degree of active absorption is needed to offset excretory losses even at calcium intakes as high as 50 mmol (2,000 mg)/day.

A major storage form of the vitamin at prevailing inputs is 25-hydroxyvitamin D [*25* (OH)D], and its plasma level is generally regarded as the best clinical indicator of vitamin D status. Although usually considered to be about three orders of magnitude less potent than calcitriol in promoting active transport in animal receptor assays, there is growing evidence that it may possess physiological functions in its own right *(75,76)*, and in the only human dose–response studies performed to date, 25(OH)D was found to have a molar potency in the range of 1/100 to 1/125 that of $1,25(OH)_2D_3$ *(77,78)*, not the 1/2,000 figure usually considered to reflect relative 25(OH)D activity.

Vitamin D status commonly deteriorates in the elderly, whose plasma 25(OH)D levels are generally lower than in young adults *(79)*. This difference is due partly to decreased solar exposure, partly to

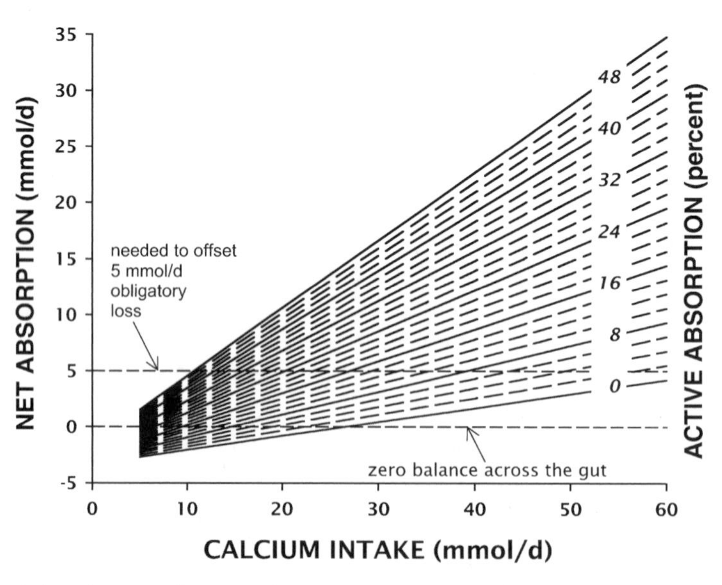

Fig. 19.8. Relationship of vitamin D-mediated, active calcium absorption, calcium intake, and net calcium gain across the gut. Each of the contours represents a different level of active absorption above a baseline passive absorption of 12.5%. (The values along each contour represent the sum total of passive and variable active absorption.) The horizontal dashed lines indicate 0 and 5 mmol/day net absorption, respectively. The former is the value at which the gut switches from a net excretory to a net absorptive mode, and the latter is the value needed to offset typical urinary and cutaneous losses in mature adults. (Copyright Robert P. Heaney, 1999. Reproduced with permission)

decreased efficiency of skin vitamin D synthesis, and partly to decreased intake of milk, the principal dietary source of the vitamin. Moreover, the elderly exhibit other abnormalities of the vitamin D endocrine system which may further impair their ability to adapt to reduced calcium intake. These include decreased responsiveness of the renal 1-α-hydroxylase to parathyroid hormone *(80)* and, possibly, decreased mucosal responsiveness to calcitriol *(81)*.

For all these reasons there is a growing consensus that the requirement for ingested vitamin D rises with age *(9)*, and a body of data which strongly suggest that relative vitamin D deficiency plays a role in several components of the osteoporosis syndrome. While a trial by Lips et al. *(82)* of vitamin D supplementation noted no fracture reduction, others have demonstrated clear benefits from supplementing vitamin D in the elderly. One example is the finding by Heikinheimo et al. *(83)*, in a randomized controlled trial, of significant reduction in all fractures in an elderly Finnish population given a single annual injection of 150,000–300,000 IU vitamin D each fall (equivalent to ~400–800 IU/day). The impressive fracture reductions noted in the trials of Chapuy et al. *(13)* and Dawson-Hughes et al. *(41)* may also have been due in part to the vitamin D supplementation that was a component of both trials. Most recently, in a large prospective study of British elderly subjects, vitamin D at an average dose of ~820 IU/day, reduced typical osteoporotic fracture risk by 33% *(84)*.

The foregoing studies (as well as others *(85,86)*) strongly suggest that vitamin D insufficiency is prevalent in the middle-aged and elderly of Northern Europe and North America. In one study of North Italian centenarians, severe vitamin D deficiency was virtually universal *(87)*. However, except in this latter study, frank osteomalacia was not reported to be a significant feature of the clinical status of study participants. This old criterion for true vitamin D deficiency is much too insensitive to be clinically useful today, since it misses most of the cases of clinically important vitamin D deficiency.

How the vitamin D requirement ought to be defined is another matter. Holick has shown that it takes an intake of at least 600 IU/day, from all sources, to sustain serum 25(OH)D levels in healthy young adults *(88)*, and the doses of vitamin D used in the studies summarized above also suggest that an intake in the range of 500–800 IU/day is required for full expression of the known skeletal effects of vitamin D in adults. This is substantially above the 1989 RDA of only 200 IU for adults *(21)*, The current DRIs for vitamin D *(9)* reflect this realization, if only to a limited extent. They included an increase from 200 to 400 IU for adults aged 50–70, and to 600 IU for those over age 70. Vieth has presented evidence that the requirement may be higher still *(89)*. In a meta-analysis of published anti-fracture trials, Bischoff-Ferrari et al. *(90)* showed that a minimum of 700–800 IU/day was required to reduce hip fracture risk.

My colleagues and I have shown in controlled dosing studies that the body normally utilizes ~4,000 IU (100 μg)/day, most of which must be coming from cutaneous sources *(91)*. This same study showed that serum 25(OH)D rose by about 0.7 nmol/L for every 1 μg daily input of cholecalciferol. Other studies *(82,89)* provide estimates of rates of increase ranging up to 1.2 nmol/L per microgram daily input. Taking a value between these extremes (1 nmol/L per microgram), it follows that an individual with a serum 25(OH)D of 50 nmol/L will typically need 30 μg (1,200 IU) cholecalciferol daily to achieve a stable level of 80 nmol/L.

Ascertaining the vitamin D requirement is complicated by the mixed input of cutaneous and dietary vitamin D. Since skin production under ambient conditions is largely unknown, it is difficult to estimate how much must be provided orally in those who are housebound or excluded from skin exposure to solar radiation, but it could be as much as 4,000 IU (100 μg)/day.

The foregoing discussion has focused primarily on effects of vitamin D on bone and on the calcium economy, which is appropriate, as the skeleton is the focus of this chapter. However, no treatment of vitamin D today would be complete without at least mentioning the broad array of tissue effects of the vitamin. Briefly, 25(OH)D is converted intracellularly to calcitriol in a wide variety of tissues, by locally expressed 1-α-hydroxylases. Within the cell this calcitriol serves as one component of the complex that unlocks key response information in the genome, thereby facilitating actions as diverse as macrophage function, infection resistance, postural stability, blood pressure, cancer expression, and a variety of autoimmune diseases. Optimal blood levels, enabling these functions, are in excess of 80 nmol/L, and probably closer to 120–150 nmol/L (the latter figure is probably close to the primitive level which obtained during the millennia of hominid evolution). Several recent reviews summarize this information *(92–95)*.

19.4. PROTEIN

Two seemingly contradictory facts seem well established with respect to calcium and protein: (1) protein intake can increase urinary calcium loss *(51,96,97)* and (2) protein aids recovery from hip fracture *(98,99)* and slows age-related bone loss *(99,100)*. For the most part, the studies establishing these diverse and to some extent contradictory effects have been performed by varying only the nutrient concerned. For example, the calciuric effects of protein have been demonstrated most clearly in studies in which purified protein or protein hydrolysates were used, with each gram of protein resulting in an rise in urinary calcium excretion of approximately 1 mg. Spencer, however, observed that, when the protein was fed as ground beef, urinary calcium did not rise, with the difference in response being due, presumably, to the fact that the meat contained substantial quantities of phosphorus *(101)*.

Kerstetter et al. *(102)* have reported that high protein intakes enhance calcium absorption, an effect which would counter a calciuric effect (were there to be one). Not everyone has been able to reproduce this finding in chronic feeding studies *(66)*, and it may be, to the extent that the phenomenon is operative, that it applies only acutely. More recently Roughead et al. *(103)* in a controlled feeding study have reported no effect of high and low protein intakes from meat (117 g/day vs. 68 g/day) on either calcium balance or urinary calcium excretion, a finding consistent with that of Spencer et al. *(101)* and Heaney *(66)*, but at odds with that of Kerstetter *(102)*.

The very reproducible calciuric effect of pure protein or amino acids had led, several years ago, to the tentative conclusion that high protein intakes might be deleterious for the skeleton. However, not only do the studies involving food sources of protein, such as those just cited, fail to support that conclusion, but epidemiological studies, such as those from the Framingham osteoporosis cohort *(100)* indicate instead that age-related bone loss in postmenopausal women is *inversely* related to protein intake, not *directly*, as might have been predicted from the calciuric effect.

In this connection it is of interest to examine the interaction of protein and calcium intakes in the study of Dawson-Hughes and colleagues *(104)*. In their randomized controlled trial of calcium supplementation, the bone gain associated with calcium supplementation was confined to individuals in the highest tertile of protein intake, while in the placebo group there was a non-significant trend toward worsening bone status as protein intake rose. This latter effect is what would be predicted if there were some degree of protein-induced calciuria without an offsetting increase in absorbed calcium. Protein intake in this study spanned only a relatively narrow range and was not randomly assigned to the subjects, and so these results cannot be considered final. Nevertheless they do exhibit two interesting features: (1) high protein intake clearly did not block the positive effect of calcium and (2) most of the protein was from animal sources (as was true for the Framingham osteoporosis cohort, as well). This latter point provides no support for the hypothesis that animal foods (as contrasted with vegetable protein sources), by increasing urinary calcium loss, artificially elevate the calcium requirement. In the author's cohort of mid-life women, protein intake was positively associated with calcium balance *(105)*. Specifically, the positive association of calcium intake to calcium balance, reported earlier *(106)*, was found only in women with protein intakes above 62 g/day (i.e., about 1.0 g/kg/day, an intake above the current RDA of 0.8 g/kg/day).

The importance of achieving an adequate protein intake is at least twofold. First, protein is a bulk constituent of bone. Because of extensive post-translational modification of the collagen molecule (e.g., cross-linking, hydroxylation) many of the amino acids released in bone resorption cannot be recycled. Hence, bone turnover requires a continuing supply of fresh protein. Second, protein elevates serum IGF-1 *(107,108)*, which is trophic for bone *(108)*. For both reasons a diet inadequate in protein would be expected to reduce the bony response to calcium. Thus, it may be tentatively concluded that protein intakes in the individuals concerned were suboptimal. To the extent that this is true, and that both of the bulk constituents of bone (calcium and protein) are ingested at suboptimal intakes, then it follows that the true effect of neither nutrient can be discerned in studies that do not ensure full repletion of the other.

The IGF-1 response is of particular interest. IGF-1 rises with protein intake, but above certain protein intakes, no further increase in IGF-1 can be produced. This effect is analogous to calcium retention, which rises at suboptimal calcium intakes, but which plateaus at or above the individual's calcium intake requirement. For both nutrients a rise in retention (for calcium) or a rise in IGF-1 (for protein) can be taken as evidence that the pre-supplement intake of the corresponding nutrient was suboptimal. The fact that IGF-1 has not reached its plateau at intakes in the range of current RDAs *(106)* has led some to conclude that the current RDA for protein is set too low, particularly for the elderly. In any case, it is important to understand this relationship of nutrient intakes in various studies to the respective plateaus. Different studies would be expected to show more or less of a beneficial effect of increasing one or both nutrient intakes depending on whether the pre-supplement intake was below or above the respective threshold level.

19.5. VITAMIN K

The chemistry and physiology of vitamin K have been extensively reviewed elsewhere *(109,110,111)*. In brief, vitamin K is necessary for the gamma-carboxylation of glutamic acid residues in a large number of proteins. Most familiar are those related to coagulation, in which seven vitamin K-dependent proteins are involved in one way or another. The gamma-carboxyglutamic acid residues in the peptide chain bind calcium, either free or on the surface layers of crystals, and have been thought to function in varying ways including catalysis of the coagulation cascade, inhibition of mineralization (as in urine), and generation of osteoclast chemotactic signals.

Three vitamin K-dependent proteins are found in bone matrix: osteocalcin (bone gla-protein—BGP), matrix gla-protein, and protein S. Only BGP is unique to bone. There is also a kidney gla-protein (nephrocalcin), which may be involved in renal reabsorption of calcium. BGP binds avidly to hydroxyapatite and is chemotactic for bone-resorbing cells. Roughly 30% of the synthesized BGP is not incorporated into matrix, but is released instead into the circulation, where, like alkaline phosphatase, it can be measured and used as an indicator of bone turnover. In vitamin K deficiency, such as would occur with coumarin anticoagulants, serum BGP levels decline, and the degree of carboxylation of the circulating BGP falls dramatically. While it would seem therefore that vitamin K deficiency would have detectable skeletal effects, they have been very hard to find. Rats reared and sustained to adult life under near total suppression of BGP gamma-carboxylation show only minor skeletal defects, mostly related to abnormalities in the growth apparatus *(109)*. In aging humans, the problem of detecting skeletal abnormalities is compounded by the fact that the bulk of the skeleton was formed prior to the onset of any deficiency, and thus bone tends to be an insensitive indicator of current nutritional stresses.

Various vitamin K-related abnormalities have been described in association with osteoporosis, but their significance to skeletal status remains unclear. Women with low dietary vitamin K intakes have significantly lower values for BMD at hip and spine than do those with higher intakes *(112)*, and greater risk for hip fracture *(113)*. Circulating vitamin K and menaquinone levels are low in hip fracture patients *(114)*. BGP is under-carboxylated in osteoporotics, and this defect responds to relatively small doses of vitamin K. However, maximal suppression of undercarboxylation has been reported to require in excess of 1,000 μg of vitamin K per day *(115)*. Finally, urine calcium has been reported to be high in osteoporotics and to fall on administering vitamin K *(116)*.

Whether or not vitamin K is important for bone health, serum vitamin K levels are indicators of general nutritional status, and it may simply be that the observation of low vitamin K levels in osteoporotics, especially in those with hip fracture, is mainly a reflection of the often poor nutrition of these individuals.

19.6. MAGNESIUM

The adult female RDA for magnesium was 280 mg/day in the 1989 RDAs *(21)* and was revised upward to 320 mg/day in the 1997 DRIs *(9)*. Only about 25% of adult US females achieve this level of intake on any given day. Average intakes tend to be in the range of 70–80% of the RDA. While severe magnesium deficiency is a well-described syndrome *(117)*, interfering both with PTH secretion and PTH action on bone, it is uncertain whether mild departures from the RDA have any adverse effect or even whether the RDA needs to be as high as it is now set. There is, as well, a widespread popular belief that magnesium is necessary for optimal calcium absorption. However, the many studies establishing the benefit of supplemental calcium described earlier achieved their effect without adding magnesium to the diets of their subjects. Furthermore, Spencer, in a series of careful metabolic studies, showed that a tripling of magnesium intake had no effect on absorption efficiency for calcium *(118)*. Thus there is no clear evidence that supplemental magnesium materially aids the prevention or treatment of osteoporosis. Moreover, magnesium salts, when used as a component of a combined supplement tablet (e.g., as in dolomite), displace calcium and make it more difficult (i.e., more pills are required) to get sufficient calcium by this route.

However, two sets of observations point to a niche role for magnesium. First, an unknown, but probably small, proportion of patients with osteoporosis have silent celiac disease as a contributory factor in their disease. These individuals commonly have some degree of magnesium deficiency. Since the underlying problem in such cases is asymptomatic, it is usually unrecognized, and hence untreated. For that reason, there may well be a small group of osteoporotic patients who would benefit from supplemental magnesium (as well as from calcium and vitamin D). Second, Sahota et al. *(119,120)* have shown that patients with low vitamin D status commonly exhibit a blunted PTH response to the consequent low absorbed calcium intake, and that such patients exhibit positive magnesium tolerance tests and respond to supplemental magnesium. Whether such supplementation improves their bone status is unclear. Nevertheless, the demonstration of subclinical magnesium deficiency in such patients suggests that additional magnesium intake might be beneficial overall.

19.7. TRACE MINERALS

Several trace minerals, notably zinc, manganese, and copper, are essential metallic co-factors for enzymes involved in synthesis of various bone matrix constituents. Ascorbic acid (along with zinc) is needed for collagen cross-linkage. In growing animals, diets deficient in these nutrients produce clear-cut skeletal abnormalities *(121)*. Additionally, zinc deficiency is well known to produce growth retardation and other abnormalities in humans. But it is not known whether significant deficiencies of these elements develop in previously healthy adults, or at least, if they do, whether such deficiencies contribute detectably to the osteoporosis problem. Copper deficiency is reported to be associated with osteoporotic lesions in sheep, cattle, and rats *(122)*. Copper has not been much studied in connection with human osteoporosis, but in one study in which serum copper was measured, levels were negatively correlated with lumbar spine BMD, even after adjusting for body weight and dietary calcium intake *(123)*.

Copper deficient animals develop reduced collagen cross-links, a factor that is known to weaken bone strength. Oxlund has reported reduced extractable cross-links in the bone of osteoporotic patients *(124)*, but it is not known whether copper deficiency was the cause.

In one four-way, randomized trial, copper, as a part of a trace mineral cocktail including also zinc and manganese, slowed bone mineral loss in postmenopausal women, when given either with or without supplemental calcium *(125)*. There appeared to be a small additional benefit from the extra trace minerals; however, the only statistically significant effect in this study was associated with the calcium supplement. This could mean that trace mineral deficiency plays no role in osteoporosis, but it could also mean that not all of the women treated suffered from such deficiency. In fact, since both osteoporotic and age-related bone loss are multifactorial, and since there is no known way to select subjects for inclusion on the basis of presumed trace mineral need, one would presume that only some of the subjects in such a study might be deficient. Thus the suggestive findings of this study have to be considered grounds for further exploration of this issue.

19.8. NUTRITION AND HIP FRACTURE

Nutrition enters into the hip fracture problem in two ways: in predisposing to fracture and in recovery from the assault of the injury and its repair. Fractures in the old elderly, and particularly hip fractures, are concentrated in institutionalized persons with multiple disabilities. The osteoporotic elderly, generally, are known to have depleted lean body mass and fat mass, and, when studied, have been found to have low circulating values for several key nutritional indicator variables, from serum albumin to ferritin and vitamins A, D, and K *(98)*. Survival 2 years after injury is four times higher in patients with serum albumin values above 3.5 g/dL than in patients with values below 3.0 g *(98)*. Additionally, patients with hip fracture often have low calcium intakes, and in the majority of studies evaluating the matter, dietary calcium earlier in life is inversely associated with hip fracture risk. In brief, hip fracture is a problem concentrated in multiply compromised individuals, and the prospect of successfully intervening to reduce risk has proved daunting even to contemplate.

However, one aspect of the problem is partly amenable to change. The relative malnutrition of patients suffering hip fracture and coming to hospital for repair contributes significantly to the often unsatisfactory outcomes for this common fracture (i.e., 15–20% excess mortality; 50% institutionalization of the survivors). Delmi et al. *(98)*, in a randomized trial of a protein-based nutrient supplement given to patients newly hospitalized for hip fracture, found that only 26% of unsupplemented individuals had outcomes classified as “good” at 6 months after injury, while nearly 60% of supplemented individuals had “good” outcomes. The investigators noted that the hospital diets offered the unsupplemented individuals were nutritionally adequate, but were frequently unconsumed, while the investigators ensured the ingestion of the supplement. This is not an isolated observation; others *(126)* had earlier found qualitatively similar benefit from nutritional supplementation in such patients. Additionally, protein supplements after fracture have been shown to retard bone loss in the contralateral hip *(99)*. The consistency of these findings constitutes a challenge to the health professions to apply these basic nutritional principles in the management of their patients.

19.9. NUTRITION AND GLUCOCORTICOID-INDUCED OSTEOPOROSIS

Glucocorticoid-induced osteoporosis (GIO) can be one of the most disabling and rapidly progressing forms of the disorder. Severe compression fractures can occur in as little as 6 months after starting corticosteroid therapy. The reasons are not entirely clear. Corticosteroids depress virtually all anabolic activities in the body, including osteoblastic new bone formation. Excessive bone resorption often occurs as well, which, when taken together with depressed bone formation, leads to negative bone balance. However, the rate of progression of GIO is too rapid to be explained on bone mass grounds alone. It is likely that accelerated remodeling, particularly in vertebral cancellous bone, directly compromises trabecular strength to an extent out of proportion to the loss in mass. Whatever the ultimate mechanism, current standards of practice call for co-therapy with a remodeling suppressor whenever glucocorticoids are prescribed for a chronic condition.

However, pharmacologic prophylaxis of GIO does not negate the need for nutritional measures as well. Vitamin D status and calcium absorption are commonly low in patients taking corticosteroids, and both defects aggravate the tendency to increased remodeling and bone resorption. Sufficient vitamin D_3 should be given to raise the serum 25(OH)D level to 80 nmol/L (32 ng/ml), and calcium intake should be at least 1,500 mg/day, and perhaps as much as 2,500 mg/day. These measures alone are not sufficient, but pharmacologic prophylaxis alone is insufficient as well. Both are necessary.

19.10. RECOMMENDATION

Calcium intake should be high throughout life: 1,500 mg/day during growth and at least that much in the elderly. Foods are the best sources, but given caloric restriction, that means low fat milk and yogurt, for the most part, as well as the widespread utilization of calcium-fortified foods. Supplements are convenient and often necessary, but should not be a substitute for a national nutritional policy or for a good diet. The elderly are commonly vitamin D deficient as well as calcium-deprived. Conscious efforts must be made to ensure a daily intake of at least 600–1,000 IU. The old elderly often suffer some degree of global undernutrition in addition to their specific deficiencies of calcium and vitamin D. Given the common isolation of elderly living alone, this is not an easy problem to solve. At very least we must make an effort to feed them after they develop fractures.

REFERENCES

1. Heaney RP, Recker RR. Distribution of calcium absorption in middle-aged women. Am J Clin Nutr 1986;43:299–305.
2. Nordin BEC, Polley KJ, Need AG, Morris HA, Marshall D. The problem of calcium requirement. Am J Clin Nutr 1987;45:1295–1304.
3. Nordin BEC, Need AG, Morris HA, Horowitz M. Sodium, calcium and osteoporosis In: Burckhardt P, Heaney RP, eds. Nutritional Aspects of Osteoporosis, Vol. 85. New York: Raven Press, 1991;279–295.
4. Barger-Lux MJ, Heaney RP, Packard P, Lappe JM, Recker RR. Nutritional correlates of low calcium intake. Clin Appl Nutr 1992;2:39–44.
5. Gershon-Cohen J, Jowsey J. The relationship of dietary calcium to osteoporosis. Metabolism 1964;13:221–226.
6. Jowsey J, Raisz LG. Experimental osteoporosis and parathyroid activity. Endocrinol 1968;82:384–396.
7. Forbes RM, Weingartner KE, Parker HM, Bell RR, Erdman JW Jr. Bioavailability to rats of zinc, magnesium and calcium in casein-, egg- and soy protein-containing diets. J Nutr 1979;109:1652–1660.
8. NIH Consensus Conference: Optimal Calcium Intake. JAMA 1994;272:1942–1948.
9. Dietary Reference Intakes for Calcium, Magnesium, Phosphorus, Vitamin D, and Fluoride. Food and Nutrition Board, Institute of Medicine. Washington, DC: National Academy Press, 1997.

10. Heaney RP. Nutritional factors in osteoporosis. Ann Rev Nutr 1993;13:287–316.
11. Heaney RP. Calcium, dairy products, and osteoporosis. J Am Coll Nutr 2000;19:83S–99S.
12. Dawson-Hughes B, Dallal GE, Krall EA, Sadowski L, Sahyoun N, Tannenbaum S. A controlled trial of the effect of calcium supplementation on bone density in postmenopausal women. N Engl J Med 1990; 323: 878–883.
13. Chapuy MC, Arlot ME, Duboeuf F, et al. Vitamin D_3 and calcium to prevent hip fractures in elderly women. N Engl J Med 1992;327:1637–1642.
14. Reid IR, Ames RW, Evans MC, Gamble GD, Sharpe SJ. Effect of calcium supplementation on bone loss in postmenopausal women. N Engl J Med 1993;328:460–464.
15. Johnston CC Jr, Miller JZ, Slemenda CW, et al. Calcium supplementation and in-creases in bone mineral density in children. N Engl J Med 1992;327:82–87.
16. Lloyd T, Andon MB, Rollings N, et al. Calcium supplementation and bone mineral density in adolescent girls. JAMA 1993;270:841–844.
17. Chevalley T, Rizzoli R, Nydegger V, et al. Effects of calcium supplements on femoral bone mineral density and vertebral fracture rate in vitamin D-replete elderly patients. Osteoporos Int 1994;4:245–252.
18. Reid IR, Ames RW, Evans MC, Gamble GD Sharpe SJ. Long-term effects of calcium supplementation on bone loss and fractures in postmenopausal women: a randomized controlled trial. Am J Med 1995;98:331–335.
19. Aloia JF, Vaswani A, Yeh JK, Ross PL, Flaster E, Dilmanian FA. Calcium supplementation with and without hormone replacement therapy to prevent postmenopausal bone loss. Ann Intern Med 1994;120:97–103.
20. Heaney RP. The bone remodeling transient: implications for the interpretation of clinical studies of bone mass change. J Bone Miner Res 1994;9:1515–1523.
21. Recommended Dietary Allowances, 10th edition. Washington, DC: National Acad. Press, 1989.
22. Matkovic V, Heaney RP. Calcium balance during human growth. Evidence for threshold behavior. Am J Clin Nutr 1992;55:992–996.
23. Jackman LA, Millane SS, Martin BR, Wood OB, McCabe GP, Peacock M, Weaver CM. Calcium retention in relation to calcium intake and postmenarcheal age in adolescent females. Am J Clin Nutr 1997;66:327–333.
24. Matkovic V. Calcium metabolism and calcium requirements during skeletal modeling and consolidation of bone mass. Am J Clin Nutr 1991;54:245S-260S.
25. Chan GM, Hoffman K, McMurray M. The effect of dietary calcium supplementation on pubertal girls' growth and bone mineral status. J Bone Miner Res 1991;6:S240.
26. Cadogan J, Eastell R, Jones N, Barker ME. Milk intake and bone mineral acquisition in adolescent girls: randomised, controlled intervention trial. BMJ 1997;315:1255–1260.
27. Recker RR, Davies KM, Hinders SM, Heaney RP, Stegman MR, Kimmel DB. Bone gain in young adult women. JAMA 1992;268:2403–2408.
28. Matkovic V, Badenhop-Stevens N, Landoll J, Goel P, Li B. Long term effect of calcium supplementation and dairy products on bone mass of young females. J Bone Miner Res 2002;19(Suppl. 1):S172 (Abstract #1200).
29. Welten DC, Kemper HCG, Post GB, van Staveren WA. A meta-analysis of the effect of calcium intake on bone mass in females and males. J Nutr 1995;2802–2813.
30. Heaney RP, Recker RR, Saville PD. Menopausal changes in calcium balance performance. J Lab Clin Med 1978;92:953–963.
31. Recker RR, Lappe JM, Davies KM, Kimmel DB. Change in bone mass immediately before menopause. J Bone Miner Res 1992;7:857–862.
32. Heaney RP. Estrogen-calcium interactions in the postmenopause: a quantitative description. Bone Miner 1990;11: 67–84.
33. Elders PJM, Netelenbos JC, Lips P, et al. Calcium supplementation reduces vertebral bone loss in perimenopausal women: a controlled trial in 248 women between 46 and 55 years of age. J Clin Endocrinol Metab 1991; 73:533–540.
34. Kanis JA, Melton LJ III, Christiansen C, Johnston CC, Khaltaev N. The diagnosis of osteoporosis. J Bone Miner Res 1994;9:1137–1141.
35. Heaney RP, Recker RR, Stegman MR, Moy AJ. Calcium absorption in women: relationships to calcium intake, estrogen status, and age. J Bone Miner Res 1989;4:469–475.
36. Nordin BEC, Need AG, Morris HA, Horowitz M, Robertson WO. Evidence for a renal calcium leak in postmenopausal women. J Clin Endocrinol Metab 1991;72:401–407.
37. Carroll MD, Abraham S, Dresser CM. Dietary intake source data: US, 1976–1980. Vital & Health Statistics, Serv. 11-NO. 231, DRHS. Publ. No. (PHS) 83-PHS, March 1983. Washington, DC: Gov. Printing Office.

38. Alaimo K, McDowell MA, Briefel RR, et al. Dietary intake of vitamins, minerals, and fiber of persons ages 2 months and over in the United States: Third National Health and Nutrition Examination Survey, Phase 1, 1988–1991. Advance data from vital and health statistics; no.258. Hyattsville, MD: National Center for Health Statistics, 1994.
39. McKane WR, Khosla S, Egan KS, Robins SP, Burritt MF, Riggs BL. Role of calcium intake in modulating age-related increases in parathyroid function and bone resorption. J Clin Endocrinol Metab 1996;81:1699–1703.
40. Peacock M, Liu G, Carey M, McClintock R, Ambrosius W, Hui S, Johnston CC Jr. Effect of calcium or 25OH vitamin D_3 dietary supplementation on bone loss at the hip in men and women over the age of 60. J Clin Endocrinol Metab 2000;85:3011–3019.
41. Dawson-Hughes B, Harris SS, Krall EA, Dallal GE. Effect of calcium and vitamin D supplementation on bone density in men and women 65 years of age or older. N Engl J Med 1997;37:670–676.
42. Recker RR, Hinders S, Davies KM. Correcting calcium nutritional deficiency prevents spine fractures in elderly women. J Bone Miner Res 1996;11:1961–1966.
43. Heaney RP. Is the paradigm shifting? Bone 2003;33:457–465.
44. Matkovic V, Kostial K, Simonovic I, Buzina R, Brodarec A, Nordin BEC. Bone status and fracture rates in two regions of Yugoslavia. Am J Clin Nutr 1979;32:540–549.
45. Holbrook TL, Barrett-Connor E, Wingard DL. Dietary calcium and risk of hip fracture: 14-year prospective population study. Lancet 1988;2:1046–1049.
46. Jackson RD, LaCroix AZ, Gass M, Wallace RB, Robbins J, Lewis CE, Bassford T, Beresford SAA, Black HR, Blanchette P, Bonds DE, Brunner RL, Brzyski RG, Caan B, Cauley JA, Chlebowski RT, Cummings SR, Granek I, Hays J, Heiss G, Hendrix SL, Howard BV, Hsia J, Hubbell FA, Johnson KC, Judd H, Kotchen JM, Kuller LH, Langer RD, Lasser NL, Limacher MC, Ludlam S, Manson JE, Margolis KL, McGowan J, Ockene JK, O'Sullivan MJ, Phillips L, Prentice RL, Sarto GE, Stefanick ML, Van Horn L, Wactawski-Wende J, Whitlock E, Anderson GL, Assaf AR, Barad D. Calcium plus vitamin D supplementation and the risk of fractures. N Engl J Med 2006; 354:669–683.
47. Heaney RP, Dowell MS, Hale CA, Bendich A. Calcium absorption varies within the reference range for serum 25-hydroxyvitamin D. J Am Coll Nutr 2003;22:142–146.
48. Pilch SM, ed. Physiological effects and health consequences of dietary fiber. Prepared for the Center for Food Safety and Applied Nutrition, Food and Drug Administration under Contract No. FDA 223-84-2059 by the Life Sciences Research Office, Federation of American Societies for Experimental Biology. 1987. Available from FASEB Special Publications Office, Bethesda, MD.
49. Heaney RP. Effects of caffeine on bone and the calcium economy. Food Chem Toxicol 2002;40:1263–1270.
50. Barger-Lux MJ, Heaney RP. Caffeine and the calcium economy revisited. Osteoporos Int 1995;5:97–102.
51. Heaney RP, Recker RR. Effects of nitrogen, phosphorus, and caffeine on calcium balance in women. J Lab Clin Med 1982;99:46–55.
52. Nordin BEC, Need AG, Morris HA, Horowitz M. The nature and significance of the relationship between urinary sodium and urinary calcium in women. J Nutr 1993;123:1615–1622.
53. Weaver CM, Heaney RP, Martin BR, Fitzsimmons ML. Human calcium absorption from whole wheat products. J Nutr 1991;121:1769–1775.
54. Weaver CM, Heaney RP, Proulx WR, Hinders SM, Packard PT. Absorbability of calcium from common beans. J Food Sci 1993;58:1401–1403.
55. Weaver CM, Heaney RP, Nickel KP, Packard PT. Calcium bioavailability from high oxalate vegetables: Chinese vegetables, sweet potatoes, and rhubarb. J Food Sci 1997;62:524–525.
56. Barrett-Connor E, Chang JC, Edelstein SL. Coffee-associated osteoporosis offset by daily milk consumption. JAMA 1994;271:280–283.
57. Morris RC Jr, Frassetto LA, Schmidlin O, Forman A, Sebastian A. Expression of osteoporosis as determined by diet-disordered electrolyte and acid-base metabolism. In: Burckhardt P, Dawson-Hughes B, Heaney RP, eds. Nutritional Aspects of Osteoporosis. New York: Academic Press, 2001;357–378.
58. Sellmeyer DE, Schloetter M, Sebastian A. Potassium citrate prevents increased urine calcium excretion and bone resorption induced by a high sodium chloride diet. J Clin Endocrinol Metab 2002;87:2008–2012.
59. Matkovic V, Fontana D, Tominac C, Goel P, Chesnut CH III. Factors that influence peak bone mass formation: a study of calcium balance and the inheritance of bone mass in adolescent females. Am J Clin Nutr 1990;52:878–888.
60. Matkovic V, Ilich JZ, Andon MB, et al. Urinary calcium, sodium, and bone mass of young females. Am J Clin Nutr 1995;62:417–425.
61. Lau EMC, Cooper C, Woo J. Calcium deficiency—a major cause of osteoporosis in Hong Kong Chinese. In: Burckhardt P, Heaney RP, eds. Nutritional Aspects of Osteoporosis, Vol. 85. New York: Raven Press, 1991;175–180.

62. Berkelhammer CH, Wood RJ, Sitrin MD. Acetate and hypercalciuria during total parenteral nutrition. Am J Clin Nutr 1988;48:1482–1489.
63. Sebastian A, Harris ST, Ottaway JH, et al. Improved mineral balance and skeletal metabolism in postmenopausal women treated with potassium bicarbonate. N Engl J Med 1994;330:1776–1781.
64. Barr SI, Prior JC, Janelle KC, Lentle BC. Spinal bone mineral density in premenopausal vegetarian and nonvegetarian women: cross-sectional and prospective comparisons. J Am Diet Assoc 1998;98:760–765.
65. Chiu J-F, Lan S-J, Yang C-Y, Wang P-W, Yao W-J, Su I-H, Hsieh C-C. Long-term vegetarian diet and bone mineral density in postmenopausal Taiwanese women. Calcif Tissue Int 1997;60:245–249.
66. Heaney RP. Dietary protein and phosphorus do not affect calcium absorption. Am J Clin Nutr 2000;72:758–761.
67. Spencer H, Kramer L, Osis D, Norris C. Effect of phosphorus on the absorption of calcium and on the calcium balance in man. J Nutr 1978;108:447–457.
68. Heaney RP, Recker RR. Determinants of endogenous fecal calcium in healthy women. J Bone Miner Res 1994;9:1621–1627.
69. Spencer H, Kramer L, Norris C, Osis D. Effect of small doses of aluminum-containing antacids on calcium and phosphorus metabolism. Am J Clin Nutr 1982;36:32–40.
70. The Surgeon General's Report on Nutrition and Health. DHHS (PHS) Publication No.88-50211, 1988.
71. Heaney RP, Rafferty K, Bierman J. Not all calcium-fortified beverages are equal. Nutr Today 2005 40(1):39–44.
72. Heaney RP, Dowell MS, Rafferty K, Bierman J. Bioavailability of the calcium in fortified soy imitation milk, with some observations on method. Am J Clin Nutr 2000;71:1166–1169.
73. Heaney RP, Dowell MS, Bierman J, Hale CA, Bendich A. Absorbability and cost effectiveness in calcium supplementation. J Am Coll Nutr 2001;20:239–246.
74. Ilich-Ernst JZ, McKenna AA, Badenhop NE, Clairmont AC, Andon MB, Nahhas, RW, Goel P, Matkovic V. Iron status, menarche, and calcium supplementation in adolescent girls. Am J Clin Nutr 1998;68:880–887.
75. Barger-Lux MJ, Heaney RP, Lanspa SJ, Healy JC, DeLuca HF. An investigation of sources of variation in calcium absorption physiology. J Clin Endocrinol Metab 1995;80:406–411.
76. Devine A, Wilson SG, Dick IM, Prince RL. Effects of vitamin D metabolites on intestinal calcium absorption and bone turnover in elderly women. Am J Clin Nutr 2002;75:283–288.
77. Colodro IH, Brickman AS, Coburn JW, Osborn TW, Norman AW. Effect of 25-hydroxy-vitamin D_3 on intestinal absorption of calcium in normal man and patients with renal failure. Metabolism 1978;27:745–753.
78. Heaney RP, Barger-Lux MJ, Dowell MS, Chen TC, Holick MF. Calcium absorptive effects of vitamin D and its major metabolites. J Clin Endocrinol Metab 1997;82:4111–4116.
79. Francis RM, Peacock M, Storer JH, Davies AEJ, Brown WB, Nordin BEC. Calcium malabsorption in the elderly: the effect of treatment with oral 25-hydroxyvitamin D_3. Eur J Clin Invest 1983;13:391–396.
80. Slovik DM, Adams JS, Neer RM, Holick MF, Potts JT Jr. Deficient production of 1,25-dihydroxyvitarnin D in elderly osteoporotic patients. N Engl J Med 1981;305:372–374.
81. Francis RM, Peacock M, Taylor GA, Storer JR, Nordin BEC. Calcium malabsorption in elderly women with vertebral fractures: evidence for resistance to the action of vitamin D metabolites on the bowel. Clin Sci 1984;66: 103–107.
82. Lips P, Graafmans WC, Ooms ME, Bezemer D, Bouter LM. Vitamin D supplementation and fracture incidence in elderly persons. Ann Intern Med 1996;124:400–406.
83. Heikinheimo RJ, Inkovaara JA, Harju EJ, et al. Annual injection of vitamin D and fractures of aged bones. Calcif Tissue Int 1992;51:105–110.
84. Trivedi DP, Doll R, Khaw KT. Effect of four monthly oral vitamin D_3 (cholecalciferol) supplementation on fractures and mortality in men and women living in the community: randomised double blind controlled trial. Br Med J 2003;326:469–474.
85. Thomas MK, Lloyd-Jones DM, Thadhani RI, Shaw AC, Deraska DJ, Kitch BT, Vamvakas EC, Dick IM, Prince RL, Finkelstein JS. Hypovitaminosis D in medical inpatients. N Engl J Med 1998;338:777–783.
86. LeBoff MS, Kohlmeier L, Hurwitz S, Franklin J, Wright J, Glowacki J. Occult vitamin D deficiency in postmenopausal US women with acute hip fracture. JAMA 1999;281:1505–1511.
87. Passeri G, Pini G, Troiano L, Vescovini R, Sansoni P, Passeri M, Gueresi P, Delsignore R, Pedrazzoni M, Franceschi C. Low vitamin D status, high bone turnover and bone fractures in centenarians. J Clin Endocrinol Metab 2003;88:5109–5115.
88. Holick MF. Sources of vitamin D: diet and sunlight. In: Burckhardt P, Heaney RP, eds. Challenges of Modern Medicine, Nutritional Aspects of Osteoporosis Vol. 7, Proceedings of 2nd International Symposium on Osteoporosis, Lausanne, May 1994. Rome, Italy: Ares-Serono Symposia Publications, 1995;289–309.

89. Vieth R. Vitamin D supplementation, 25-hydroxyvitamin D levels, and safety. Am J Clin Nutr 1999;69:842–856.
90. Bischoff-Ferrari HA, Dawson-Hughes B, Baron JA, Burckhardt P, Li R, Spiegelman D, Specker B, Orav JE, Wong JB, Staehelin HB, O'Reilly E, Kiel DP, Willett WC. Calcium intake and hip fracture risk in men and women: a meta-analysis of prospective cohort studies and randomized controlled trials. Am J Clin Nutr 2007:86:1780–1790.
91. Heaney RP, Davies KM, Chen TC, Holick MF, Barger-Lux MJ. Human serum 25-hydroxy-cholecalciferol response to extended oral dosing with cholecalciferol. Am J Clin Nutr 2003;77:204–210.
92. Holick MF, Chen TC. Vitamin D deficiency: a worldwide problem with health consequences. Am J Clin Nutr 2008;87:1080S–1086S.
93. Holick MF. Vitamin D deficiency. N Engl J Med 2007;357:266–281.
94. Spina CS, Tangpricha V, Uskokovic M, Adorinic L, Maehr H, Holick MF. Vitamin D and cancer. Anticancer Res 2006;26(4A):2515–2524.
95. Holick MF. High prevalence of vitamin D inadequacy and implications for health. Mayo Clin Proc 2006;81:353–373.
96. Johnson NE, Alcantara EN,Linkswiler HM. Effect of protein intake on urinary and fecal calcium and calcium retention of young adult males. J Nutr 1970;100:1425–1430.
97. Chu JY, Margen S, Costa FM. Studies in calcium metabolism. II. Effects of low calcium and variable protein intake on human calcium metabolism. Am J Clin Nutr 1975;28:1028–1035.
98. Delmi M, Rapin C-H, Bengoa J-M, Delmas PD, Vasey H, Bonjour J-P. Dietary supplementation in elderly patients with fractured neck of the femur. Lancet 1990;335:1013–1016.
99. Schürch M-A, Rizzoli R, Slosman D, Vadas L, Vergnaud P, Bonjour J-P. Protein supplements increase serum insulin-like growth factor-I levels and attenuate proximal femur bone loss in patients with recent hip fracture. Ann Intern Med 1998;128:801–809.
100. Hannan MT, Tucker KL, Dawson-Hughes B, Cupples LA, Felson DT, Kiel DP. Effect of dietary protein on bone loss in elderly men and women: The Framingham Osteoporosis Study. J Bone Miner Res 2000;15: 2504–2512.
101. Spencer H, Kramer L, Osis D. Effect of a high protein (meat) intake on calcium metabolism in man. Am J Clin Nutr 1978;31:2167–2180.
102. Kerstetter JE, O'Brien KO, Insogna KL. Dietary protein affects intestinal calcium absorption. Am J Clin Nutr 1998;68:859–865.
103. Roughead ZK, Johnson LK, Lykken GI, Hunt JR. Controlled high meat diets do not affect calcium retention of bone status in healthy postmenopausal women. J Nutr 2003;133:1020–1026.
104. Dawson-Hughes B, Harris SS. Calcium intake influences the association of protein intake with rates of bone loss in elderly men and women. Am J Clin Nutr 2002;75:773–779.
105. Heaney RP. Effects of protein on the calcium economy. In: Burckhardt P, Heaney RP, Dawson-Hughes B, eds. Nutritional Aspects of Osteoporosis 2006. Amsterdam: Elsevier Inc., 2007;191–197.
106. Heaney RP, Recker RR, Saville PD. Calcium balance and calcium requirements in middle-aged women. Am J Clin Nutr 30:1603–1611, 1977.
107. Heaney RP, McCarron DA, Dawson-Hughes B, Oparil S, Berga SL, Stern JS, Barr SI, Rosen CJ. Dietary changes favorably affect bone remodeling in older adults. J Am Diet Assoc1999;99:1228–1233.
108. Bonjour JP, Schüch MA, Chevalley T, Ammann P, Rizzoli R. Protein intake, IGF-1 and osteoporosis. Osteoporos Int 1997;7(Suppl. 3):S36–S42.
109. Price PA. Role of vitamin-K-dependent proteins in bone metabolism. Ann Rev Nutr 1988;8:565–583.
110. Szulc P, Delmas PD. Is there a role for vitamin K deficiency in osteoporosis? In: Burckhardt P, Heaney RP, eds. Challenges of Modern Medicine, Nutritional Aspects of Osteoporosis, Vol. 7, Proceedings of 2nd International Symposium on Osteoporosis, Lausanne, May 1994). Ares-Serono Publications, Rome, Italy, 1995;7:357–366.
111. Cashman KD, O'Connor E. Does high vitamin K_1 intake protect against bone loss in later life? Nutr Rev 2008;66: 532–538.
112. Booth SL, Broe KE, Gagnon DR, Tucker KL, Hannan MT, McLean RR, Dawson-Hughes B, Wilson PWF, Cupples LA, Kiel DP. Vitamin K intake and bone mineral density in women and men. Am J Clin Nutr 2003; 77:512–516.
113. Feskanich D, Weber P, Willett WC, Rockett H, Booth SL, Colditz GA. Vitamin K intake and hip fractures in women: a prospective study. Am J Clin Nutr 1999;69:74–79.
114. Hodges SJ, Pilkington MJ, Stamp TCB, et al. Depressed levels of circulating menaquinones in patients with osteoporotic fractures of the spine and femoral neck. Bone 1991;12:387–389.
115. Binkley NC, Krueger DC, Kawahara TN, Engelke JA, Chappell RJ, Suttie JW. A high phylloquinone intake is required to achieve maximal osteocalcin γ-carboxylation. Am J Clin Nutr 2002;76:1055–1060.

116. Knapen MHJ, Hamulyak K, Vermeer C. The effect of vitamin K supplementation on circulating osteocalcin (bone gla protein) and urinary calcium excretion. Ann Intern Med 1989;111:1001–1005.
117. Shils ME. Magnesium. In: Shils ME, Olson JA, Shike M, eds. Modern Nutrition in Health and Disease, 8th Edition, Vol.1. Philadelphia, PA: Lea & Febiger, 1994;164–184.
118. Spencer H, Fuller H, Norris C, Williams D Effect of magnesium on the intestinal absorption of calcium in man. J Am Coll Nutr 1994;13:485–492.
119. Sahota O, Mundey MK, San P, Godber IM, Lawson N, Hosking DJ. The relationship between vitamin D and parathyroid hormone: calcium homeostasis, bone turnover, and bone mineral density in postmenopausal women with established osteoporosis. Bone 2004;35:312–319.
120. Sahota O, Mundey MK, San P, Godber IM, Hosking DJ. Vitamin D insufficiency and the blunted response in established osteoporosis: the role of magnesium deficiency. Osteoporos Int 2006;17:1013–1021.
121. Mertz W, ed. Trace Elements in Human and Animal Nutrition, 5th edition. San Diego, CA: Academic Press, Inc., 1987.
122. Strain JJ. A reassessment of diet and osteoporosis—possible role for copper. Med Hypotheses 1988;27:333–338.
123. Howard G, Andon M, Bracker M, Saltman P, Strause L. Low serum copper, a risk factor additional to low dietary calcium in postmenopausal bone loss. J Trace Elem Exp Med 1992;5:23–31.
124. Oxlund H, Barckman M, Ørtoft G, Andreasen TT. Reduced concentrations of collagen cross-links are associated with reduced mechanical strength in bone. J Bone Miner Res 1995;10:S179.
125. Strause L, Saltman P, Smith K, Andon M. The role of trace elements in bone metabolism. In: Burckhardt P, Heaney RP, eds. Nutritional Aspects of Osteoporosis, Serono Symposia Publication Vol. 85. New York: Raven Press, 1991;223–233.
126. Bastow MD, Rawlings J, Allison SP. Benefits of supplementary tube feeding after fractured neck of femur. Br Med J 1983;287:1589–1592.

20 Gastric Acid Secretions, Treatments, and Nutritional Consequences

Ronit Zilberboim and Adrianne Bendich

Key Points

- Gastric acid provides an important selective advantage today as when vertebrates evolved. With the eradication of one acid-related gastrointestinal (GI) disease there may be an increase in the incidence of another acid-related GI disease, demonstrating the importance of acid balance.
- Certain gastroesophageal diseases require chronic acid-suppressive therapy.
- Nutritional intervention at the initiation of acid-suppressive therapy could prevent adverse nutritional consequences.
- The effect of acid-suppressive therapy on both vitamin and mineral status is observed only after many years of therapy and recent clinical evidence is available to demonstrate such effects.
- Although the use of proton pump inhibitors (PPI) for the diagnosis and treatment of gastrointestinal disorders is effective, step-down strategies for acid-suppressive therapies should be developed and explored in support of updated indication.

Key Words: Achlorhydria; acid-suppression therapies; calcium and iron metabolism; community-acquired pneumonia; gastric secretions; gastroesophageal reflux disease; *Helicobacter pylori*; histamine receptor antagonists; vitamin B_{12}; proton pump inhibitors

20.1. INTRODUCTION

The objectives of this chapter are to provide an overview of the roles of gastric acid in health and disease, the use of acid-suppressive therapies to mitigate prevalent excess acid symptoms and/or diseases, and to review potential detrimental consequences related to their use. Particular emphasis is placed on the nutritional consequences of gastric acid imbalance as it is related to vitamin and mineral status. Finally, reduced gastric acid is also associated with disturbance in the natural bacterial balance throughout the gastrointestinal tract and the resulting gastrointestinal bacterial overgrowth is reviewed.

20.2. GASTRIC ACID

Gastric acid secretion is believed to have developed in the evolution of vertebrates and to provide selective advantages. Acid balance in the stomach is precisely regulated as acid is needed to facilitate digestion and absorption on one hand while too much acid can damage the mucosa and cause ulcers. Stomach acid aids in protein digestion and is thought to be involved in the absorption of minerals such as calcium and iron, as well as vitamins such as B_{12}. In addition, acid in the stomach has a central role in prevention of bacterial overgrowth and therefore protection against enteric infection. Gastric acid is also involved in upper GI motility and finally may modulate feeding behavior *(1,2)*.

A. Bendich, R.J. Deckelbaum (eds.), *Preventive Nutrition*, Nutrition and Health, DOI 10.1007/978-1-60327-542-2_20,

20.2.1. Regulation and Secretion of Gastric Acid

The stomach is organized as vertical tubules and contains two functional elements, the oxyntic and pyloric glands (Fig. 20.1). The oxyntic and pyloric glands are organized in the stomach in three areas: the fundus, corpus, and the antrum. The oxyntic gland is located in the fundus and the corpus which consists of the majority of the stomach's area (80%). The remaining section of the stomach (20%), the antrum, is covered with the pyloric gland. In terms of acid secretion, the parietal cells which are located only in the fundus and the corpus area secrete acid via the hydrogen potassium adenosine triphosphatase pump (H^+K^+ATPase/ATPase), also known as the proton pump. The pyloric area contains gastrin cells (also called G cells) which produce gastrin, a gastric acid secretion stimulator. D cells, that secrete somatostatin, the main inhibitor of acid secretion, are located throughout the stomach both on oxyntic and pyloric glands. Enterochromaffin-like (ECL) cells secrete histamine and are only located in the oxyntic area. It is estimated that human stomach contains 1×10^9 parietal cells and 9×10^6 gastrin cells *(1)*.

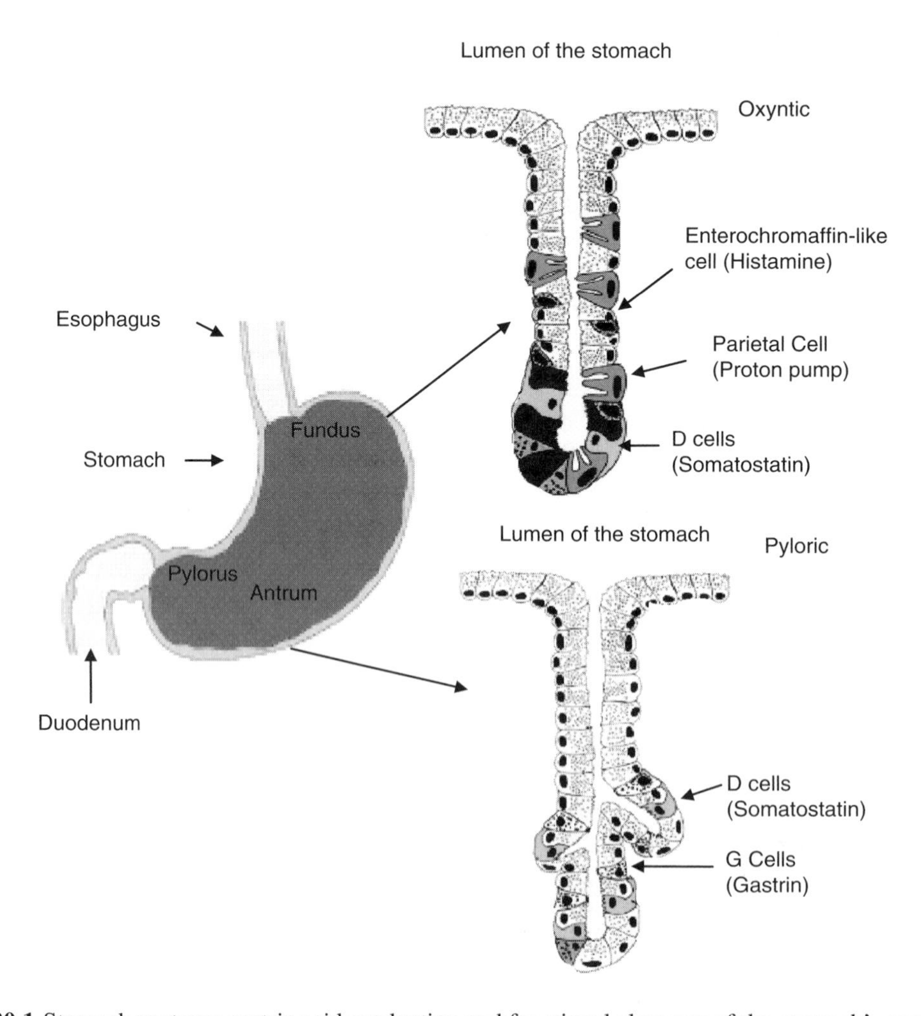

Fig. 20.1 Stomach anatomy gastric acid production and functional elements of the stomach's mucosa.

Gastric acid secretions are stimulated both centrally and peripherally. Central stimulation occurs via peptides that are produced in the gut and can signal the brain directly and indirectly *(3)*. Peripherally, gastric acid secretion is regulated via hormonal, neural, and biological agents; the main stimulants include gastrin (hormonal), histamine (paracrine), and acetylcholine (neurocrine). In response to food, gastrin functions primarily by releasing histamine from ECL cells via cholecystokinin-2 (CCK2). The released histamine diffuses to the parietal cells where it stimulates acid secretion via activation of histamine H_2-receptors. In addition, gastrin acts directly on the parietal cell by binding to CCK2 and induces release of cytosolic calcium which ultimately leads to the activation of the proton pump. Acetylcholine functions both by stimulating parietal cells as well as through the inhibition of somatostatin release. Lastly, gastric acid secretions can be modulated by infection with *Helicobacter pylori* (HP). The main inhibitor of gastric acid secretion is somatostatin which acts to attenuate acid secretion *(1,3)*. The active parietal cell secretes hydrochloric acid (HCl) at a concentration of 160 mM/L or pH of 0.8. The acid extrudes through the mucus layer based on the pressure generated during the secretion (~17 mmHg) *(1)*. Importantly, for acid to be secreted from the parietal cell, physiological processes that include exocytosis of the tubulovesicle followed by endocytosis after the termination must occur. The ATPase pump transforms from a sequestered state in the inactive cytoplasmic tubulovesicles to a docking stage followed by priming and fusion with the apical plasma membrane where it is in the active state. The cycle is completed once the stimulus has decayed and the proton pump is internalized via endocytosis and acid secretion is terminated. A detailed review of the early signaling events that ultimately result in acid secretion by parietal cells, recycling of the proton pump and the role of the cytoskeleton in support of the active acid secretion can be found in Yao and Forte *(4)*.

20.2.2. Diseases of Gastric Acid Secretion

Several diseases disturb gastric secretions and both over- and under-secretion are observed in clinical practice. While hyposecretion is observed in atrophic and autoimmune gastritis, hypersecretion is manifested in the Zollinger–Ellison syndrome (ZES), duodenal ulcer, and certain types of gastric ulcer. Individuals with HP infection also suffer from disturbances in acid secretion and interestingly colonization elicits both secretion extremes; hypersecretion and hyposecretion have been reported *(1)*. Table 20.1 summarizes diseases associated with both hyper- and hyposecretion, their prevalence, and consequences.

20.2.2.1. Gastroesophageal Reflux Disease (GERD)

Gastroesophageal reflux disease (GERD) was defined by the Montreal consensus as "a condition which develops when the reflux of the stomach content causes troublesome symptoms and/or complications" *(5)*. This definition is accepted as the operational definition for GERD by the American Gastroenterological Association (AGA) *(6)*. GERD tends to be chronic, lifelong condition that may require continuous management including symptom relief and complication management *(7)*. GERD is most prevalent in North America and in Europe and between 10 and 29% of adults have this disease *(8)*. Reflux disease is one of the most frequent disorder seen in primary care, as well as in secondary referral centers and therefore has a high economic impact *(9)*. GERD affects the esophagus primarily, and the major symptoms are heartburn and regurgitation. With regard to duration, it was estimated, based on a survey of 1,000 people that suffer from heartburn in the US, that while about 15% of those surveyed typically experience heartburn for less than 1 year, about 30% experience it for over 10 years *(10)*.

The GERD population has been sub-classified into several distinct groups including a non-erosive reflux disease (NERD) group. Importantly, a large portion of the NERD population has normal

Table 20.1
Diseases Related to Perturbation in Acid Secretion

Disease[a]	*Acid secretion status*	*Prevalence/populations*	*Acid perturbation mechanism and consequences*	*References*
Zollinger–Ellison syndrome (ZES)	Hypersecretion	Rare condition: 1–3 cases/year/million	• Gastrinoma, neuroendocrine tumor that causes ectopic secretion of gastrin leading to gastric acid hypersecretion • Chronic use of H_2RA and/or PPIs • Severe refractory peptic diseases and frequent complications • 60–90% of tumors are malignant	*(23,99)*
Chronic HP infection	Hypersecretion (*see* hyposecretion below)	10–15% of HP cases; antral predominant (located in the antrum – lower part of the stomach)	• Increase in the production of gastric acid perhaps due to reduced somatostatin • Duodenal ulcer mainly due to the perturbations in acid secretion	*(1)*
Duodenal ulcer	Hypersecretion	In children the prevalence is 1 out of 2,500 hospital admissions and only about 30% was HP related *(100)*. In US adults it is estimated that up to 20% are idiopathic (HP negative; NSAIDs negative) *(101)*	• Idiopathic (HP negative; NSAIDs negative); increased basal and stimulated acid production and/or reduced somatostatin	*(1,100, 102–104)*
Gastric ulcer	Hyposecretion to normal acid secretion	Very rare; may be related to other diseases	• Altered gastric mucosal defense • Oxyntic mucosa inflammation with reduced parietal mass cells	*(1)*

Autoimmune gastritis	Hyposecretion	General population about 2%; prevalence increases with age (up to 12% for 80 years old); In diabetes type 1 5–10%	• High serum gastrin and low pepsinogen I concentrations • ECL cells undergo hyperplasia which may progress to dysplasia and gastric carcinoid tumor (4–9% of patients) • Affect the partial cells in the corpus and fundus location of the stomach • Vitamin B_{12} deficiency (megaloblastic pernicious anemia) • Iron deficiency anemia (in 20–40% of patients with autoimmune gastritis)	*(105)*
Chronic atrophic gastritis/gastric cancer	Hyposecretion		• Chronic atrophic gastritis is a precursor of intestinal gastric cancer • Initiation of the carcinogenic process in most cases related to HP infection	*(99,106,107)*
Acute HP infection	Hyposecretion		• Decrease in acid secretion is thought to facilitate survival of the organism and colonization of the stomach • Protection against GERD, Barrett's esophagus, and esophageal adenocarcinoma	*(1,108)*

(*Continued*)

Table 20.1 (Continued)

Disease[a]	*Acid secretion status*	*Prevalence/populations*	*Acid perturbation mechanism and consequences*	*References*
			• Nutritional consequences related to vitamin B_{12} malabsorption	
Chronic HP infection pan-gastritis	Hyposecretion		• Direct inhibition of parietal cells probably due to a constituent from the bacteria • Inhibition of expression of portion of the ATPase pump • Indirectly changing hormonal, pancreatic or neural regulatory functions • Stimulation of somatostatin thus inhibition of gastrin, histamine, and acid secretion • Due to the inflammation products there is a functional inhibition of the parietal cells; alternatively products of the HP itself inhibit the partial cell from secreting acid	

[a]Other disease states that have been associated with hypersecretion such as renal failure are not reviewed here.

physiological acid exposure, but interestingly, they are more sensitive to acid than those with erosive reflux disease *(11)*. Unlike duodenal and gastric ulcers where excessive acid is associated with the etiology of the disease, the pathogenesis in GERD is mainly related to acid in the wrong place. Notably, similar symptoms, in terms of frequency and severity, are experienced in both erosive and non-erosive patients. There are, however, several complications associated with GERD that are encountered in clinical practice including reflux esophagitis and Barrett's esophagus *(9)* and further, both nighttime and supine reflux are associated with severe esophagitis or complicated GERD.

20.2.2.2. Infections with *H. Pylori* (HP)

H. pylori (HP) is a human pathogen that has been linked with sanitation and water quality in most but not all studies *(12)*. Successful eradication of the bacteria is associated with improved health outcomes including reduced peptic ulcer risk and reduced peptic ulcer recurrence *(13)*. HP affects half of the world's population and is considered the major cause of acute gastritis, chronic gastritis, gastroduodenal ulceration and is recognized as a carcinogen for its role in gastric carcinogenesis *(13,14)*. Further, HP infection is implicated with a number of extra gastric diseases *(15)*. Infection with HP often occurs in childhood, and once established, can persist lifelong if untreated. Overall, the prevalence of HP infection is higher in developing countries when compared to developed countries, but within a region it can vary by ethnicity and socioeconomic factors. It is commonly understood that prevalence of infection is decreasing in many countries due to improvements in sanitation and living standards and the relatively recent movement of populations from rural to urban settings *(12)*. Bacterial and host factors contribute to the degree and severity of inflammation. With regard to the bacteria, HP strains may differ in their motility, adherence to gastric epithelial cells as well as the chemicals that they produce to help their survival in the acidic environment or synthesis of the cytotoxin that they produce. Varied host immune responses to the pathogen have been documented *(15)*. The interaction between host genetic factors and susceptibility to HP has not been elucidated yet *(16)*. Most HP-infected individuals suffer from reduced gastric acid secretion due to atrophy, inactivation of the ATPase and direct inhibition of the acid secretion. In contrast, hypergastrinemia causes acid hypersecretion in individuals with antral gastritis. Table 20.1 describes the consequences of both hyper and hyposecretion as related to HP infection. Eradication of HP restores somatostatin and gastrin and thus acid secretion.

20.2.2.3. Ulcers

There are four types of gastric ulcers based on location, acid secretion, and concomitant association with duodenal ulcer. Most gastric ulcers are thought to be due to altered gastric mucosal defense despite normal to low basal and stimulated gastric acid secretions. Currently gastric ulcer is managed by inhibiting gastric acid secretion and the removal of the assaulting agent (nonsteroidal anti-inflammatory drugs or management of HP infection). Increased basal and stimulated acid production is often seen with the development of duodenal ulcers, and acid control is central to the management of these patients. Before the invention of acid-suppressive therapies histamine H_2-receptor antagonists (H_2RAs) and proton pump inhibitors (PPIs), antacids were commonly used. However, since the ulcer was not cured by the antacids recurrence and complication often resulted. Consequently, surgery procedures such as vagotomy that aimed to denervate the acid-producing areas of the stomach or other more extensive ulcer surgeries such as subtotal gastric restriction were used. After the development of acid-suppression therapies especially PPIs, it was possible to use these medications that allowed the duodenal ulcer to heal and further to prevent recurrence. Importantly, healing is correlated with duration of acid-suppressive therapy, and longer periods of time are usually recommended. HP infection is considered the root cause for most duodenal ulcers. Increased parietal cell mass due to suppression of somatostatin has been documented in HP-infected duodenal ulcer patients *(1)*. Patients with idiopathic ulcers are usually maintained on PPI therapy, and higher doses may be required to control acid secretion and prevent ulcer relapse *(17)*. Prevalence of these diseases is summarized in Table 20.1.

20.3. PHARMACOLOGICAL INTERVENTIONS RELATED TO EXCESS GASTRIC ACID

Too much acid and especially acid in the wrong place is associated with several diseases that may require pharmacologic intervention, mainly in the form of acid suppression. There are numerous interventions and these are considered safe and effective. However, it is important to note that their use pattern has shifted from acute to chronic. Currently, leading the way in acid-suppressive therapies are proton pump inhibitors (PPIs) and histamine-2 receptor antagonists (H_2RAs). Antacids, in contrast, act on acid that is already in the stomach and when dosed appropriately can provide rapid and effective acid neutralization; however, their use is limited as these do not control acid secretion and only provide temporarily relief.

In the past, antacids, compounds that neutralize acid that is already in the stomach, were used for the management of acid hypersecretion. Currently two classes of systemic drugs are used to prevent gastric acid secretion: H_2RAs and PPIs. These medications are widely prescribed due to their beneficial therapeutic ability to reduce gastric acid secretion regardless of stimulus *(1)*. Over 82 million prescriptions for PPIs and over 19 million prescriptions for H_2RAs are written yearly worldwide *(18)*. PPI therapy is popular among doctors and patients and importantly, patients are using PPIs on a long-term and continuous basis *(19)*. It is increasingly common to take these drugs on a chronic basis to prevent recurrent gastroesophageal reflux (GERD), avoid potential complications such as the precancerous condition, Barrett's esophagus, strictures, and to prevent complications due to chronic use of nonsteroidal anti-inflammatory drugs (NSAIDs). It has been reported that about two-thirds of PPI prescribing in the United Kingdom is for long-term therapy *(20)*. Additional indications also exist; for example, in Japan, several PPIs are officially approved for the management of HP infection *(21)*.

20.3.1. Antacids

Antacids such as calcium carbonate ($CaCO_3$), aluminum hydroxide ($Al(OH)_3$), sodium bicarbonate ($NaHCO_3$), and magnesium hydroxide ($Mg(OH)_2$) or combinations of these salts were employed throughout most of the twentieth century to manage acid hypersecretion. These temporarily buffer gastric acid and provide a transient increase in stomach pH that has been associated with relief from pain. For many decades antacids were the only treatment available for acid imbalance diseases. Consequently, a large number of antacid compounds were recognized as safe and effective as reflected in the first over the counter (OTC) monograph that was produced by the FDA in 1974. Due to the large number of potential compounds used in antacid formulations, these combinations needed to comply with two main requirements. The first was commercial in nature and required a minimum amount of each of the compounds in a formulation mixture of 25%. The second condition was related to efficacy and required 5 milliequivalents of acid neutralization capacity as a minimum acid neutralization capacity per dose. The antacids have several important limitations. Due to their local mode of action, it is difficult to maintain the buffering for a prolonged period of time and repeated dosing with the antacid is needed *(22)*.

20.3.2. *Histamine H_2-Receptor Antagonists (H_2RAs)*

The first selective H_2-receptor antagonist, cimetidine, was developed and commercialized by Smith Kline and French in the late 1970s. It became commercially available under the brand name Tagamet. Several other analogues including ranitidine (Zantac), famotidine (Pepcid), and nizatidine (Axid) soon followed. It was found later that this class of antagonist only blocked one of the parietal cell secretagogues, the histamine receptor and abolished gastrin but did not affect acetylcholine, indicating that gastrin action was mediated via neuroendocrine cells *(23)*. H_2-receptor antagonists work as competitive inhibitors and concerns related to rebound as a result of its mode of action surfaced soon after their introduction to the marketplace. As acid is secreted by the proton pump, the use of H_2RAs significantly reduced acid secretion but without actually totally stopping it. Despite advantages that may be associated with partial acid suppression, the H_2RAs class use dramatically deceased as newer interventions that were able to totally block acid release became commercially available. Besides the incomplete acid suppression, there were several studies that documented a rebound effect (acid hypersecretion) and tachyphylaxis/tolerance with the H_2RAs after prolonged use. Hypersecretion has been demonstrated with and without a meal. Increased sensitivity or up regulation of the H_2 receptor to histamine stimulation as a result of the chronic competitive inhibition is considered the main mechanism by which rebound develops, and perhaps to the development of tolerance as well. The development of tolerance to H_2RAs is perceived as loss of efficacy *(24)*.

20.3.3. *Proton Pump Inhibitors (PPIs)*

In the late 1980s, the first proton pump inhibitor (PPI), omeprazole, was introduced. Several other related moieties later became available; all work on the same cellular mechanism. PPIs inhibit gastric acid secretion independent of gastric acid stimulators including histamine, gastrin, or acetylcholine. The PPI class is the standard drug currently used for acid-related gastrointestinal diseases. PPIs are powerful and the most effective acid secretion inhibitors due to their direct effect on the proton pump. PPIs provide a more sustained increase in gastric pH relative to H_2RAs which only affect the histamine receptors thus allowing for some acid secretion *(21)*. Most PPIs are taken in the form of prodrugs. To prevent protonation in the stomach prior to duodenum absorption, PPIs are formulated with an acid resistant coating. PPIs inhibit acid secretion via the attachment of their active form to the proton pump ATPase. The bond between the PPI and the ATPase prevents the final stage of acid secretion, the proton transport across the membranes from the parietal cell.

Chemically PPIs are either substituted pyridylmethylsulfinyl benzimidazole or imidazopyridine derivatives. The pyridine and the imidazole are heterocyclic moieties forming the PPI prodrug that can be protonated as a function of the conditions in the acid space of the parietal cells. The protonation of both the pyridine and the imidazole is a prerequisite for their conversion to the active drug. Consequently, the activation of PPIs is dependent on the pH and when the pH is below 2.5 there is greater activation. At the same time this also delays the initiation of acid suppression *(25)*. The rate of activation, which differs across the PPI category, depends on the pH in the acid space and the chemical properties of the heterocyclic rings. The activated drug forms one or more irreversible disulfide bonds with the cysteine(s) in the proton pump. Therefore, the duration of their effects is longer than expected based on their blood levels. PPIs accumulate in acidic spaces of stimulated parietal cells where their concentration can increase up to 1,000 times that of the blood. Following the first protonation, the PPI binds to the ATPase allowing for the second protonation to occur. Importantly, the second interaction location on the ATPase differs for the different PPI drugs (dependent on the pH). These properties also highlight the need to consume the PPIs shortly after the initiation of a meal when the proton pump is activated *(23)*. Uptake of PPIs by other cells such as HP, colonic epithelial cells, and several others has also been reported *(21)*. These cells may also contain ATPases that are slightly different than the parietal cell ATPase.

Clinically available PPIs are omeprazole, S-omeprazole, lansoprazole, pantoprazole, and rabeprazole, and currently in development is tenatoprazole *(25)*. PPIs are capable of raising intragastric pH by several units and reduce hydrogen ion concentration (acidity) by hundreds to thousands fold *(26)*. The dosage forms and the indications associated with the available PPIs are tabulated in Table 20.2.

20.3.3.1. Medical Therapy for Gastroesophageal Reflux Disease

Medical therapy is based on the classical view of GERD and is focused on inhibition of acid production with PPIs dominating the treatment algorithm *(27)*. Further, increasing pH to over 4 especially during the night has been recommended. The recent AGA position paper concluded that empirical therapy with PPIs is appropriate initial management for patients with uncomplicated heartburn. Further, there are data to support treatment with antisecretory agents, and that PPIs are more effective than H_2RAs. However, it also highlights that despite lack of data to support PPI intervention twice a day or nocturnal dose with H_2RAs, these approaches are widely used in clinical practice *(6)*. For example, Fass recently reported that during a period of 7 years (1997–2004) there was a 50% increase in double dose PPI use in patients with GERD *(28)*. Another GERD algorithm, one that emphasizes the importance of lifestyle changes to help control the triggers, is also available *(9)*. This European algorithm focuses both on the latest clinical practice experience and is based on findings that shows that the majority of typical GERD patients have no evidence of erosive esophagitis. Further, it highlights the prevalence of the documented treatment failure in this population. Many GERD patients that fail once a day PPI, and despite lack of data, are prescribed high dose of PPI but still fail to respond, thus making this population treatment failure. There is only limited information to support other interventions as first line of therapy including over the counter (OTC) antacids in combination with H_2RA or a PPI or lifestyle changes. As for lifestyle changes, there are ample studies that show an association between weight and GERD, however, definitive studies that demonstrate long-term success are lacking *(29–32)*. For example, a meta-analysis of OTC medicines, it was concluded that for infrequent post-prandial symptoms antacids and antacids in combination with alginates are quite effective *(33)*. Overall, it is expected that PPIs will still prominently be recommended, however, the use of antacids as alginate–antacid combinations are likely to increase *(9)*.

20.3.3.2. Medical Therapy for HP Infection

The first-line treatment of HP infection is a triple therapy which combines PPI, clarithromycin, and amoxicillin or metronidazole. However, due to increased bacterial resistance, clarithromycin-containing triple therapies do not reliably produce a high (at least 80%) cure rate *(34)*. The second line of therapy, the quadruple therapy, consists of bismuth +metronidazole +tetracycline +PPI, is effective when resistance to metronidazole is low. A third line of therapy for patients that failed both the triple and the quadruple therapy uses PPI and amoxicillin together with levofoxacin. Each line of therapy produces efficacy of about 70% *(35)*. An emerging alternative strategy is sequential therapy where the antibiotics are administered in sequence. The initial treatment phase includes PPI and amoxicillin followed by PPI, clarithromycin, and tinidazole *(36)*. These emerging therapies are considered superior alternatives to the triple therapy for treatment of patients infected with resistant strains of HP. Eradication has been difficult due to both recurrence, resistance to clarithromycin and resistant strains of HP *(37)*. Recurrence rates of HP infection remain high in developing countries, and in certain populations in developed countries *(12)*. Based on a recent meta-analysis, it was concluded that in

Table 20.2
Indications, Dosage, and Elimination Routes for PPIs

Brand Name/Chemical Name/Manufacturer	*Indication/Dose (mg/day) and duration/recommended duration for specific indication*	*Dosage form/Manufacturer directions*	*Half-life (hours)/Elimination routes/Protein bound (%)*
Aciphex/rabeprazole sodium/Eisai *(42)*	Healing of erosive or ulcerative GERD: 20 mg/day; for 4–8 weeks Maintenance of healing of erosive or ulcerative GERD: 20 mg/day Treatment of symptomatic GERD: 20 mg/day for 4 weeks Healing of duodenal ulcers: 20 mg/day for up to 4 weeks	Form: Delayed-release, enteric-coated tablets containing 20 mg of rabeprazole sodium Directions: Take with or without food	Half-life: 1–2 h. Elimination urine 90%; Remainder in feces Protein bound 96.3%

(*Continued*)

Table 20.2
(Continued)

Brand Name/Chemical Name/Manufacturer	*Indication/Dose (mg/day) and duration/recommended duration for specific indication*	*Dosage form/Manufacturer directions*	*Half-life (hours)/Elimination routes/Protein bound (%)*
Prevacid/Lansoprazole /TAP *(43)*	Duodenal ulcer (Short-term treatment): 15 mg/day for 4 weeks Duodenal ulcer (maintenance of healed): 15 mg/day HP eradication to reduce the risk of duodenal ulcer recurrence triple therapy: 30 mg plus amoxicillin 1 g and clarithromycin 500 mg twice daily for 10 or 14 days HP eradication to reduce the risk of duodenal ulcer recurrence dual therapy: 30 mg plus amoxicillin 1 g (three times daily for 14 days) Benign gastric ulcer: 30 mg/day for up to 8 weeks NSAIDs-associated gastric ulcer, healing: 30 mg/day for 8 weeks NSAIDs-associated gastric ulcer, risk reduction: 15 mg/day for up to 8 weeks Short-term treatment of symptomatic GERD: 15 mg/day for up to 8 weeks Short-term treatment of erosive esophagitis: 30 mg/day for up to 8 weeks Pediatric (1–11 years) short-term treatment of symptomatic GERD and short-term treatment of erosive esophagitis: 15 mg/day for up to 12 weeks for ≤30 kg; 30 mg/day for up to 12 k for 30 kg Adolescent (12–17 years) Short-term treatment of symptomatic GERD Non-erosive GERD: 15 mg/day for up to 8 weeks Erosive Esophagitis: 30 mg/day for up to 8 weeks Maintenance of healing of erosive esophagitis: 15 mg/day Pathological hypersecretory conditions including ZES: 60 mg/day	Form: Prevacid is supplied in delayed-release capsules, in delayed-release orally disintegrating tablets for oral administration and in a packet for delayed-release oral suspension. Each form is available in either 15 or 30 mg strengths. Directions: Take before eating	Half-life: 1.5 h Elimination urine: 1/3; Feces: 2/3 Protein bound 97%

Prilosec/Omeprazole (delayed-release capsules) Omeprazole magnesium (for delayed-release oral suspension)/ AstraZeneca *(109)*	Short-term treatment of active duodenal ulcer: 20 mg once daily for 4 weeks. Some patients may require an additional 4 weeks HP eradication to reduce the risk of duodenal ulcer recurrence: *Triple therapy*: Prilosec 20 mg, Amoxicillin 1,000 mg, and Clarithromycin 500 mg. Each drug twice daily for 10 days *Dual therapy*: PRILOSEC 40 mg (once daily for 14 days), Clarithromycin 500 mg (three times daily for 14 days) Gastric ulcer: 40 mg once daily for 4 to 8 weeks GERD: 20 mg once daily for 4–8 weeks Maintenance of healing of erosive esophagitis: 20 mg once daily Pathological hypersecretory conditions: 60 mg (varies with individual patient) once daily Pediatric patients (1–16 years of age) GERD and maintenance of healing of erosive esophagitis — Weight / Dose: 5–10 kg / 5 mg 10–20 kg / 10 mg ≥20 kg / 20 mg	Form: Prilosec delayed-release capsules: 10 mg, 20 mg, and 40 mg Prilosec for delayed-release oral suspension: 2.5 or 10 mg Directions: Delayed-release capsules should be taken before eating and swallowed whole. Packets for delayed-release oral suspension should be emptied into a container containing water (5 mL water for a 2.5 mg packet and 15 mL of water for a 10 mg packet), stirred, allowed to thicken for 2–3 min, and stirred and drank within 30 min. If any material remains after drinking, add more water, stir and drink immediately. Once daily	Half-life: 0.5–1 h Elimination urine: 77%; Remainder in feces Protein bound: 95%

(*Continued*)

Table 20.2
(Continued)

Brand Name/Chemical Name/Manufacturer	*Indication/Dose (mg/day) and duration/recommended duration for specific indication*	*Dosage form/Manufacturer directions*	*Half-life (hours)/Elimination routes/Protein bound (%)*
Protonix/Pantoprazole sodium/Wyeth *(110)*	Treatment of erosive esophagitis: 40 mg/day for up to 8 weeks Maintenance of healing of erosive esophagitis : 40 mg/day Pathological hypersecretory conditions including ZES: 40 mg twice daily for as long as clinically indicated. Doses as high as 240 mg/day have been administered	Form: Delayed-release tablet available in two strengths. Each delayed-release tablet contains 45.1 or 22.6 mg of pantoprazole sodium sesquihydrate (equivalent to 40 or 20 mg pantoprazole, respectively) Directions: Take with or without food	Half-life: 1 h Elimination urine: 71%; Feces: 18% Protein bound: 98%
Nexium/esomeprazole magnesium/ AstraZeneca LP *(44)*	GERD, healing of erosive esophagitis: 20 or 40 mg/day for up to 4–8 weeks GERD, maintenance of healing of erosive esophagitis: 20 mg/day Symptomatic GERD: 20 mg/day for 4 weeks Risk reduction of NSAIDs-associated gastric ulcer: 20 or 40 mg/day for up to 6 months HP eradication to reduce the risk of duodenal ulcer recurrence, triple therapy: 40 mg/day for 10 d plus amoxicillin 1,000 mg twice daily for 10 days and clarithromycin 500 mg twice daily for 10 days Adolescent Use 12- to 17-year olds, short-term treatment of GERD: 20 or 40 mg/day for up to 8 wk Pathological hypersecretory conditions including ZES: 40 mg twice daily	Form: Nexium is supplied in delayed-release capsules and in packets for a delayed-release oral suspension. Each delayed-release capsule contains 20 or 40 mg of esomeprazole. Each packet of Nexium. For delayed-release oral suspension contains 20 or 40 mg of esomeprazole, in the form of the same enteric-coated granules used in Nexium delayed-release capsules. Directions: At least 1 h before meals	Half-life: 40 mg dose: 1.5 h 20 mg dose: 1.2 h Elimination urine: 80%; Remainder in feces Protein bound: 97%

Zegerid/Omeprazole sodium bicarbonate/ Santarus *(45)*	Short-term treatment of active duodenal ulcer: 20 mg/day for 4 weeks Benign gastric ulcer: 40 mg/day for 4–8 weeks Symptomatic GERD (with no esophageal erosions): 20 mg/day for up to 4 weeks GERD with erosive esophagitis: 20 mg/day for up to 4–8 weeks Maintenance healing of erosive esophagitis: 20 mg/day Reduction of risk of upper gastrointestinal bleeding in critically ill patients : 40 mg initially followed by 40 mg 6–8 h later and 40 mg/day thereafter for 14 days (oral suspension only)	Form: Immediate-release capsules and unit-dose packets as powder for oral suspension. Each capsule contains either 40 or 20 mg of omeprazole and 1,100 mg of sodium bicarbonate. Packets of powder for oral suspension contain either 40 or 20 mg of omeprazole and 1,680 mg of sodium bicarbonate. Directions: On an empty stomach at least 1 h before a meal	Half-life: 1 h Elimination urine: 77%; Remainder in feces Protein bound: 95%

GERD—gastroesophageal reflux disease; ZES—Zollinger–Ellison Syndrome; NSAID–nonsteroidal anti-inflammatory drugs; HP–*H. pylori*.

developing countries recurrence is related to infection with a new strain while in developed countries it is due to re-infection with the same strain *(38)*.

20.3.3.3. Medical Therapy for Gastroesophageal Complication Induced by Nonsteroidal Anti-inflammatory Drugs (NSAIDs)

Both prescription and OTC nonsteroidal anti-inflammatory drugs (NSAIDs) are commonly used for relief of pain and inflammation associated with musculoskeletal injury and arthritis. It is well known that the use of NSAIDs is associated with upper GI complications such as GI bleeding, ulcer perforation, and symptomatic peptic ulcer disease. These upper GI complications develop in about 1–2% of NSAIDs users; non-users in contrast are at a lower risk for developing such complications *(39)*. To overcome the complications of NSAIDs, several gastroprotective strategies including certain drugs and drug combinations have been used. Gastroprotective therapies include cyclooxygenase-2 (COX-2) inhibitors, or cotherapy with PPIs or misoprostol, and new strategies including vitamin C in combination of aspirin *(40)*. Notably, commonly used gastroprotective strategies reduce the risk of upper GI complications in NSAIDs users. The combination of PPI with COX-2 has been shown to enhance the gastroprotective efficacy relative to COX-2 alone. In contrast, it was reported that H_2RAs did not reduce the gastrointestinal complication risk while on NSAIDs *(39)*. PPIs are considered as primary gastroprotective agents in gastrointestinal complications in cases of NSAIDs, and their use significantly increased with the withdrawal of COX-2 *(41)*.

20.3.4. Drug–Drug Interactions

There are certain drugs that are affected by co-administration with PPIs and, at times, readjustment of the PPI and/or the co-administered drug may be required. Notably, PPIs and drugs that are metabolized by cytochromes in the liver including certain cytochromes (CYP2C19) may be affected due to co-administration. Patients receiving warfarin and PPIs (including Aciphex, Prevacid, Nexium, and Zegerid), concomitantly may experience increased risk of bleeding due to increases in the international normalized ratio (INR) of platelets and increase in prothrombin time *(42–45)*. PPIs may also interfere with the bioavailability of drugs that are dependent on gastric pH. For example, Prevacid, Nexium, and Zegerid decrease plasma levels of the human immunodeficiency virus (HIV) protease inhibitor atazanavir, which depends on an acidic gastric pH for absorption *(43–45)*. PPIs should not be co-administered with atazanavir. In addition, PPIs should be taken 30 min prior to sucralfate, an older medication for the treatment of GERD and active duodenal ulcers, that has a buffering capacity because it has been shown that absorption of Prevacid and omeprazole is delayed and bioavailability reduced when administered with sucralfate *(43)*.

Increases in plasma concentrations of both PPI and antibiotic drugs have been observed when these are taken concomitantly. This is especially important as the standard treatment for HP infection may include the combination of these drugs. For example, increased plasma concentrations of rabeprazole and 14-hydroxyclarithromycin may be experienced by individuals administered combinations of rabeprazole (Aciphex), amoxicillin, and clarithromycin *(42)*. Similarly, increases in plasma levels of omeprazole, clarithromycin, and 14-hydroxyclarithromycin may occur when omeprazole (in the drug Zegerid) and clarithromycin are co-administered *(45)*.

Doubling of esomeprazole activity level may result from concomitant administration of esomeprazole (Nexium) and a combined inhibitor of certain cytochromes including CYP2C19 and CY3A4, such as voriconazole. Although not normally required, dose adjustment upward may be required in patients with Zollinger–Ellison's Syndrome (ZES) *(44)*.

Zegerid may prolong the elimination of drugs that are metabolized by oxidation in the liver, including diazepam, warfarin, and phenytoin. Also, when administered concomitantly with the immunosuppressive drug, tacrolimus (a macrolide), Zegerid may increase serum levels of tacrolimus *(45)*. Recently there is some evidence that suggests that omeprazole may negatively affect cardiovascular events in patients who also take clopidogrel. It was hypothesized that omeprazole competes with liver enzymes that metabolize clopidogrel (Plavix) resulting in enhanced platelet activity (reduced efficacy of the clopidogrel) for patients who are taking the anti-platelet drug, clopidogrel *(46)*. However, it appears that this is not a class effect. In contrast to the reported negative omeprazole–clopidogrel drug interaction, the intake of other PPIs like pantoprazole or esomeprazole was not associated with impaired response to clopidogrel *(47)*.

20.4. NUTRITIONAL CONSEQUENCES OF REDUCED GASTRIC ACID

Long-term decrease in gastric acid has been associated with several effects on nutritional status.

20.4.1. Vitamin B_{12} and Gastric Acid

Acid and pepsin are needed for the digestion of protein and the release of the vitamin B_{12} from dietary sources where vitamin B_{12} is usually protein bound. Gastric acid is central to the release of vitamin B_{12}, and therefore reduced gastric acid may result in reduced B_{12} bioavailability. Despite strong biological plausibility, mixed data have been reported in relation to the usage of acid-suppressive therapies and vitamin B_{12} status. There are only a few studies where reduced vitamin B_{12} status was clearly demonstrated.

Vitamin B_{12} (cobalamin) deficiency is prevalent in the elderly with potential irreversible hematological and neurological consequences affecting sensory and motor function if unrecognized and untreated. More subtle effects have also been described, including osteopenia, neurocognitive impairment, and increased vascular disease risk associated with elevated homocysteine *(48)*. Prevalence ranging from 3 to 40% was reported *(49)*. Others suggest that between 10 and 20% of the population is affected, and importantly that about half of those affected are clinically symptomatic *(50,51)*. The consequences of reduced vitamin B_{12} absorption may take years to develop because the body is able to reutilize vitamin B_{12} by enterohepatic recirculation, and most patients will not manifest signs and symptoms of vitamin B_{12} deficiency until at least 3 years of no vitamin B_{12} absorption *(51)*. Other markers of vitamin B_{12} deficiency such as homocysteine and methylmalonic acid levels could be used to demonstrate earlier deficiency manifestation, therefore allowing for earlier diagnosis and thus enable corrective actions before complications develop *(49)*.

Most cases of vitamin B_{12} deficiency may be due to either reduced intake or malabsorption. Populations at risk for vitamin B_{12} deficiency include strict vegetarians as vitamin B_{12} is found primarily in animal-based foods. The majority of dietary vitamin B_{12} is found in meat, poultry, fish, eggs, and dairy products where it is bound to protein. Reduced intake may be due to lifestyle choices and may be related to economic status *(52)*.

Several studies ascertained the relationship between acid-suppressive therapies and vitamin B_{12} status. Some but not all studies showed a reduced vitamin B_{12} status with increased duration of acid-suppressive therapy. A dose–response reduction in vitamin B_{12} absorption has been demonstrated in healthy individuals that were treated with omeprazole for 2 weeks. Absorption of vitamin B_{12} was reduced about fourfold after ingestion of 20 mg omeprazole (from 3.2 to 0.9%), and a further reduction in vitamin B_{12} absorption was observed (from 3.6 to 0.4%) after doubling the omeprazole dose *(53)*. Additional data to support the association between acid-suppressive therapies and vitamin B_{12} status come from a retrospective case–control state-wide Medicaid drug database. This study identified

125 patients who started vitamin B_{12} supplementation (via injection) during the study period as cases and 500 matched controls. Chronic acid-suppressive therapy was defined as either H_2RAs or PPI use for 10 of the 12 previous months. The odds ratio for the initiation of vitamin B_{12} therapy was significantly higher for users of acid suppressives; while 18% of the users of acid-suppressive drugs initiated vitamin B_{12} therapy, compared to 11% of the control group who initiated vitamin B_{12} therapy. It was concluded that patients initiating vitamin B_{12} therapy were more likely to be receiving long-term acid-suppressive therapy *(51)*. Diminished serum vitamin B_{12} was demonstrated based on a study of 542 elderly patients who were PPI users but not those who used H_2RAs. Twenty eight percent were using H_2RAs only and 26% were using PPI only, and the average duration was 18 months. The vitamin B_{12} status of the 141 PPI users was evaluated as a function of time and further the effect of duration of use of PPI was evaluated with and without vitamin B_{12} supplementation. Importantly, concomitant oral B_{12} supplementation slowed but did not prevent the decline in vitamin B_{12} status seen when long-term PPI therapy was used. Finally, a recent small, long-term study of PPI use and vitamin B_{12} status showed a significant trend toward reduced vitamin B_{12} levels with increased duration of PPI therapy. Patients with gastrinoma (ZES) and without gastrinoma that were acid hypersecretors and were treated with lansoprazole (PPI) with an average treatment of 140 months were studied. Overall vitamin B_{12} went down from a median between 890 and 950 pg/mL in the first few years to 495 pg/mL, a drop of 46% over 12 years. Ten percent had low serum vitamin B_{12} and 31% had normal vitamin B_{12} but had other symptoms related to vitamin B_{12} deficiency. Vitamin B_{12} therapy reduced vitamin B_{12} deficiency markers and confirmed the effect of PPI *(54)*. Importantly, as expected, evidence of vitamin B_{12} appears only after many years on PPI therapy.

Other populations at risk for vitamin B_{12} deficiency include patients that suffer from acid imbalance. Achlorhydric patients (gastric restriction or vagotomy), those who have atrophic gastritis, those who are infected with HP, and those who chronically use acid-suppressive therapy may be affected.

20.4.2. Vitamin C and Gastric Acid

Vitamin C (ascorbic acid) is a water soluble antioxidant and a reducer of carcinogenic nitrite. It is plentiful in fruits and vegetables. In addition to the many well-established roles of vitamin C, there are epidemiological, experimental, and animal data that indicate that diets that are high in fruits and vegetables (a source of vitamins and minerals) may reduce the risk of gastric cancer. The results from randomized trials are not consistent, however *(55)*. In addition, antioxidants may also protect against the development of esophageal adenocarcinoma, despite different etiologies in the development of gastric relative to esophageal cancers. Reduced vitamin C levels both in plasma and in the affected tissues of individuals who suffer from esophageal adenocarcinoma relative to matched healthy individuals suggest a role for oxidative stress in the pathogenesis and progression of esophageal adenocarcinoma (*see* Chapter 6) *(56)*.

Ascorbic acid is present in small concentration in human saliva and it is actively secreted into the gastric lumen. In the upper part of the stomach, ascorbic acid is rapidly oxidized even under fasting conditions, therefore limiting its antioxidant capability. Ascorbic acid is thought to have a critical role in the management of toxic nitrogen species *(57)*. Its primary evolutionary benefits may stem from its functions related to nitrite and iron chemistry *(26)*.

McColl's team demonstrated that in healthy volunteers, 40 mg dose of omeprazole (PPI) significantly reduced fasting gastric levels of vitamin C, and that the effect on the active form, ascorbic acid was even more profound. Therefore, individuals that are on omeprazole therapy have a limited capability to reduce nitrite levels, absorb iron, and a reduced antioxidant potential. This compromised capability may manifest itself in diseases such as iron deficiency and iron deficiency anemia as well as negative gastric outcomes as discussed above. While on omeprazole and regardless of the HP status, there was a reduction in the active form of vitamin C, ascorbic acid. Total vitamin C, however, was reduced only in the HP positive subgroup. Reduced serum concentration of vitamin C was also noted as a function of HP status *(58,59)*. These results indicate that the effect of omeprazole therapy is further complicated by HP status *(59,60)*.

20.4.3. Calcium and Gastric Acid

Calcium is the most common mineral in the human body and 99% of the body's total calcium is present in the skeleton and teeth. Calcium has many critical biological roles and is necessary to sustain life. Calcium's functions stem directly from its chemical properties including intermediate binding affinity, ionic radius, and coordination number which together allow flexibility in its interactions *(61)*. Interestingly, calcium deficiency does not manifest itself as shortage of calcium for cellular or physiological processes rather as a decrease in the calcium reservoir itself, the bones. Calcium is the most studied mineral in relationship to human health (*see* Chapter 19) *(62)*.

Calcium (Ca^{+2}) is absorbed in the intestine via two mechanisms: an active, transcellular and a passive, paracellular absorption. The active, classical epithelial transcellular transport occurs with the help of the calcium-binding protein calbindin and calcium ATPase pump. Active transport occurs primarily in the duodenum and upper jejunum (closer to the stomach). In this process, calcium is transferred across membranes into the extracellular space. The amount of calcium absorbed is dependent on the bioavailability of the calcium in the diet and the capacity for absorption in the intestine (which is regulated at least in part by 1, 25 dihydroxy vitamin D). The second absorption mechanism is a passive, paracellular process that allows calcium to permeate across gap junctions. This process occurs through the length of the intestine. The amount of calcium transferred is dependent on the lumen's permeability which is influenced by 1, 25 dihydroxy vitamin D and calcium concentration *(63)* as well as the concentration of "free" calcium. The majority of ingested calcium is not absorbed and it is excreted in the feces.

Stomach pH may affect the intestinal calcium absorption in several ways. First, the solubility of the calcium salts is dependent on their physical and chemical properties and in particular solubility is a function of pH. In addition, pH may affect both active and passive calcium transport. In terms of the active transport the pore structure that is associated with the process may be affected due to the change in the confirmation of the proteins and in particular it is suspected that the ionic amino acids like glutamate act as a "pH sensor." Similarly, it has been suggested that pH affects the tight junctions that control the passive absorption from the lumen into the interstitial spaces *(64)*.

The long-term effects of acid-suppressive therapy, and in particular that of PPIs due to their more effective acid suppression, on bone homeostasis have not been extensively studied *(65)*. One way in which acid-suppression therapy may affect calcium absorption is based on the reduced stomach acidity; however, due to complex and confounding effects such as source of calcium and co-ingestion with food, it has been difficult to demonstrate such an effect. In vitro calcium solubility is dependent on pH, however, there are data that demonstrated that in vitro solubility does not truly represent human physiology as it relates to calcium absorption. Bo-Linn and colleagues demonstrated that calcium solubility was not a prerequisite for absorbability *(66)*, and Heaney et al. *(67)*, showed that when solubility increased by a factor of over 100, absorbability only doubled. Further, in the presence of a meal, bioequivalent calcium absorption from calcium salts that are opposite in their neutral pH solubility, calcium citrate, and calcium carbonate, also demonstrated that calcium solubility was not a prerequisite for efficient absorption *(68)*. Calcium absorption has been shown to be greatly enhanced by consumption with a meal, and interestingly, ingestion of food is usually followed by a 2–3 units increase in the stomach pH (from about 1 or 2 to well over 4). Despite stimulation of acid secretion due to consumption of a meal, the food's large buffering capacity results in a net increase in pH, an effect that holds even for small meals such as breakfast *(69)*. Therefore, the role of food and that of gastric acid (or pH) in calcium absorption appears to be conflicted and inconclusive.

20.4.3.1. Gastric Acid and Calcium Absorption

The effect of achlorhydria or acid-suppressive therapy on calcium absorption was evaluated in several studies. Two small studies, one with achlorhydric individuals (Recker *(70)*) and one with healthy volunteers on omeprazole (40 mg/day) (Serfaty-Lacrosniere *(71)*), showed that when calcium was taken with a meal, there was no difference in calcium absorption. The Serfaty-Lacrosniere *(71)* study demonstrated that omeprazole did not negatively affect calcium absorption when the calcium was part of a meal (milk and cheese). The Recker *(70)* study, however, showed the importance of the meal when calcium was provided as calcium carbonate or calcium citrate. Notably, without a meal, calcium from calcium carbonate was absorbed to a much smaller degree in achlorhydric (0.047) compared to normal individuals (0.225). For calcium citrate, however, calcium absorption was not dependent on co-ingestion with a meal and the fraction absorbed was found to be 0.243 for the normal individuals and doubled to 0.453 for the achlorhydric. In contrast to these two studies where reduced stomach acidity did not seem to affect calcium absorption, there were two studies where calcium absorption was affected by the use of acid-suppressive therapy. O'Connell *(72)* evaluated the effect of 20 mg of omeprazole on calcium absorption in a placebo-controlled study. Of importance in this study, calcium was provided as isotope-labeled calcium carbonate, at 500 mg of elemental calcium without a meal. Omeprazole significantly impaired calcium absorption; in fact, calcium absorption was reduced from 9.1 (on placebo) to 3.5% (on omeprazole). Finally, in a small placebo-controlled study with eight healthy male volunteers, Graziani and colleagues evaluated the effect of a high dose of omeprazole (60 mg/day; 20 mg every 8 h) on calcium absorption and urinary calcium excretion. One thousand mg of elemental calcium were provided in the form of food (milk and cheese) in a single meal. The results indicate that after ingestion of the test meal, calcium plasma level increased in the placebo, but did not change when volunteers were taking omeprazole. Calcium balance was also affected as was seen by urinary calcium excretion. In the placebo, urine calcium level was higher than after omeprazole where calcium excretion over a 24 h period was 33% lower *(73)*.

20.4.3.2. Gastric Acid Suppression and Fractures: Epidemiological Studies

While calcium absorption studies provide conflicting data, directional consistency was revealed based on one earlier small and three recent large epidemiological studies *(74–77)*. These studies demonstrate the effect of acid-suppressive therapy on increased risk of fracture. Increased risk for hip factures was first reported for the H_2RA cimetidine in a small case–control study with 356 men with a radiologically confirmed hip fracture and 402 controls. Significant increased risk with an odds ratio of 2.5 was reported *(77)*. This early finding was followed by several large epidemiological studies showing similar effects. Vestergaard et al. *(75)* in a case–control study were the first to demonstrate that PPI use was associated with increased risk of fractures. Subjects were sampled from the Danish population and eligible cases consisted of 124,655 individuals who suffered a fracture in the calendar year 2000. The control group of 373,962 were matched by gender and age from the same database. Exposure and variable confounders included antacids and acid-suppressive therapies and several other drugs that may interfere with calcium absorption or resorption. Interestingly, while H_2RAs reduced the risk of any fracture (OR 0.88) and hip fractures (OR 0.69), PPIs significantly increased the risk of fractures (OR 1.18, 1.45, and 1.60 for overall, hip and spine fractures). Notably, due to the method by which the data were collected the population was not restricted by age *(75)*. Similar observations were also reported based on a very large nested case–control study with 13,556 cases of hip fracture and 135,386 controls in the United Kingdom. While studying the effects of long-term PPI use, Yang et al. showed that both men and women PPI users over 50 years old were at increased risk of hip fracture and that the risk increased with increased duration and dose of PPI *(74)*. Adjusted odds ratio (AOR) was 1.44 for more than 1 year of PPI therapy, up to 2.65 for longer term and higher dose. Similarly when comparing duration of PPI therapy, AOR increased from 1.22 to 1.58 for 1–4 years, respectively. The relationship between duration of exposure to PPIs and osteoporosis-related fractures was also explored by Targownik *(76)*. In a retrospective cohort based on medical claims data in Manitoba, Canada, patients with hip, vertebral, and wrist fractures who were over 50 years old were matched with three controls based on age, sex, and other comorbidities and AOR was calculated. Overall, there were 15,792 fracture cases and 47,289 controls and follow-up for over 5 years for about a 1/3 of the cohort. Exposure to PPIs for over 7 years was associated with increased overall fracture risk; OR of 1.6 for hip fracture after 5 years increased to 4.55 after 7 years of PPI use.

20.4.3.3. Gastric Acid Suppression and Fractures: Prospective Studies

There are now two prospective studies that also support the other epidemiological findings. The first prospective study examined the association between acid-suppressive therapy and skeletal outcomes using two large cohorts, one of 5,755 men and the other of 5,339 women over the age of 65 *(18)*. There were a total of 822 men and 753 women that were either PPI or H_2RA users. Baseline characteristics revealed that dietary elemental calcium intake was slightly over 700 mg, and that about 50% of the women and 35% of the men were taking calcium supplements. After multivariate analysis, when the combination of therapies (either PPI or H_2RAs) was examined, a lower cross-sectional bone mineral density (BMD) was observed for men, and interestingly, there were no significant BMD differences among women. Total bone loss at the hip was also only significant in men who used PPI or H_2RAs. In terms of fracture outcomes, there was a 34% significant increased risk of nonspine fracture in the women PPI users and a 20% increase when looking at PPI or H_2RAs users. There was also an association of a slight increased risk of nonspine fracture among men who were not taking calcium supplements. It is important to note that in the women's cohort, mean duration of PPI therapy was 1.8 years and that of H_2RAs was 3.6 years; there was a threefold increase in the number who used PPI therapy during the duration of the study (average 4.9 years). In the men's cohort there was a 50% increase in PPI use during the study (average 4.6 years). Yu et al. estimated that one extra nonspine fracture would be expected for every 10 women treated with PPIs for 5 years and concluded that the use of PPIs in older women, and perhaps older men with low calcium intake, may be associated with a modestly increased risk of nonspine fractures. In the second prospective study, Roux et al. *(79)* also demonstrated increased risk of vertebral fractures in postmenopausal women recruited from five European centers. About 2,400 women were recruited from 1999 to 2001 and were followed for a mean 6.1 years for the Osteoporosis and Ultrasound Study; data from 1,211 women were used for vertebral fracture and 1,346 for nonvertebral fracture analysis. Radiographs were preformed in the beginning and in final visit and omeprazole exposure and potential cofounders were also ascertained. In the multivariate analysis, omeprazole was a significant and independent predictor of fractures with a relative risk of 3.1, and importantly, it was independent of age *(78)*.

It has been suggested that one way to explain the mechanism by which the PPIs increase the risk of fractures is related to their activity on the osteoclast ATPase. As described above, the PPIs exert their activity via their attachment to the ATPase and prevention of acid secretion. While PPIs primarily act on the gastric parietal cell ATPase, PPIs can also exert some activity on the bone's osteoclast ATPase activity. Localization of the PPIs activity may be related to the general structure as well as the different sub units of the ATPase in the parietal cell and osteoclast in bone *(78)*.

Overall, most studies consistently show an association of low magnitude between long-term PPI use and increased risk of all fractures and/or hip fractures. Further, increased PPI dose and/or duration also consistently contributed to the increased risk. Although these studies may not be as conclusive as placebo-controlled intervention studies, they illustrate that more research is needed into the effects of long-term use of PPIs. As summarized by Richards and Goltzman *(79)*, clinicians should help patients understand risk-benefit and especially after long-term use. There may be situations where reduced dose and/or finite duration should be considered *(79)*. In summary, the totality of the data begins to indicate that the use of acid-suppressive therapy, and in particular PPIs consistently albeit to a small degree, increases the risk of fractures. Currently, perhaps due to the relatively short duration, or the small number of people who were actually on acid-suppressive therapies, the prospective study did not provide enough data to support reduced BMD as a cause for the fracture findings.

20.4.4. Iron Absorption and Gastric Acid

Iron absorption is highly regulated and is primarily dependent on the level of iron stores and erythropoietic activity. Iron in the diet can be divided into two forms, heme and nonheme. The heme iron is in the ferrous form (Fe^{+2}), whereas most of the nonheme iron is in the ferric (Fe^{+3}) form. There is a separate transporter protein for nonheme iron, and the ferric form requires reduction (in the presence of acid) to the ferrous form before it can be transported across the intestinal lumen and absorbed *(80)*. Heme iron is relatively more bioavailable and is only slightly influenced by dietary factors. In contrast nonheme iron absorption is dependent on its solubility and interactions with other meal components. Gastric acid is needed to maintain nonheme iron in solution in the ferrous form and therefore decreased stomach acid may lead to impaired iron absorption *(81,82)*. Ascorbic acid in the stomach enhances absorption of nonheme iron, primarily through reduction of ferric iron to the ferrous form *(83)*. Other organic acids, such as citric, lactic, and malic acid also enhance nonheme iron absorption. The level of gastric acid in the stomach also affects the potential to absorb iron; achlorhydria has been associated with increased risk of iron deficiency and is seen most frequently in the elderly *(84–86)*.

20.4.4.1. Gastric Acid Suppression and Iron Absorption

A small study using radioactively labeled iron (nonheme) and cimetidine (H_2RA) showed a reduction in the percent of iron absorbed that was dose dependent. While iron absorption was reduced by 28% when 300 mg of cimetidine was administered, iron absorption was reduced by 42 and 65% with the increase in cimetidine to 600 and 900 mg, respectively. Similarly, antacid (1.2 g aluminum hydroxide and 1.2 g of magnesium hydroxide and 90 mg simethicone) also reduced iron absorption by 52% *(85)*.

In contrast, in patients with abnormally high absorption of iron (hereditary hemochromatosis (HH)), reduced iron absorption is advantageous. Recent data indicate that long-term use of PPIs in patients with HH reduced the requirement for maintenance phlebotomy. PPI-induced suppression of gastric acid reduced dietary iron absorption in the patients and there was a significant reduction in the volume of blood needed to be removed annually while on the PPI. It was concluded that administration of a PPI to patients with HH can inhibit the absorption of nonheme iron from a test meal and the habitual diet *(87)*.

20.4.5. *Nutritional Consequences Summary*

Increasing evidence suggests that acid-suppressive therapies, and in particular long-term PPI use, are not without risk. Consequences include reduced mineral and vitamin absorption. In some cases, such as in the increased risk of fractures, the association between acid-suppressive therapies and adverse events is based primarily on observational studies *(74–76)*. While these observational studies use large databases that carefully document both the cases based on disease codes and medication usages, it is important to note that such associations are prone to confounders such as other conditions, lifestyle influences, and other OTC medication intake. Further, no conclusive data are available based on short-term intervention studies that evaluate the effect of acid-suppressive therapies on calcium absorption. Finally, in the cases of vitamin B_{12} and iron much smaller cohorts have been used to demonstrate the effect. Despite these limitations it is important to note that these adverse events have occurred and that risk-benefit analysis should always consider the potential for adverse nutritional effects.

20.4.6. *Infectious Diseases*

Several lines of defense mechanisms exist to protect the internal environment from potential toxins and microbial pathogens. Structural integrity combined with protective agents help to maintain the gastrointestinal tract. In the case of pathogenic assault, there are local GI factors that are mobilized and, in addition if needed, systemic immune responses are available to help prevent the pathogenic invasion. Gastric acid is one of the protective agents available to help protect against microbial pathogens reaching the intestine *(88)*. At a low pH (below pH 3), gastric content is capable of killing bacteria within 15 min *(89)*. The potential for increased susceptibility to infections as a complication of acid-suppressive therapy and in particular PPIs has been suggested in the past based on reduced stomach acid or hypochlorhydria. Data are currently accumulating with regard to increased risk for both upper and lower GI tract infections.

20.4.6.1. Community-Acquired Pneumonia Case–Control and Prospective Studies

Gulmez et al. *(90)* reported increased risk of community-acquired pneumonia in PPI users in a population-based study in Denmark. There were 7,642 cases based on well-defined hospital discharge diagnosis with 34,176 control subjects matched by age and sex. Logistic regression was used to control confounders. It was reported that the adjusted odds ratio of the association between current use of PPI with community-acquired pneumonia was moderate, at 1.5, and statistically significant. However, the association was not PPI dose dependent. Further there was no association with H_2 receptor antagonists or with past use of PPIs *(90)*. A prospective study with children also demonstrated increased risk for community-acquired pneumonia in children who were on gastric acid-suppression therapy. The incidence of community-acquired pneumonia prior to the study was similar across the groups. Children who were referred to centers with common GERD-related symptoms and received either ranitidine (H_2 receptor antagonist) or omeprazole (PPI) and others that visited the center for routine examinations were included in the study. Data were from 186 subjects of which 95 were in the control group and 91 received acid-suppression therapy (47 on ranitidine and 44 on omeprazole). Importantly, except for GERD, these children were healthy and had no co-morbidities. The study found that 2% of the control children suffered from pneumonia in the follow-up period, 12% of those on the acid-suppression therapy suffered from pneumonia *(91)*.

20.4.6.2. *Clostridium Difficile*-Associated Disease and Gastric Suppression

Clostridium difficile is a gram-positive anaerobic spore-forming bacterium that is strongly associated with hospital-acquired diarrhea in developing countries. The clinical results of *C. difficile*-associated disease (CDAD) are related to the kind and amount of toxins that the *C. difficile* produces. The most common risk factor for *C. difficile* infection is prior use of antibiotics; after the loss of commensal bowel flora opportunistic colonization with *C. difficile* often follows *(92)*. Importantly, despite being asymptomatic, the disease may not occur unless the normal flora is disturbed. Although the spores are resistant to acid, the vegetative stage is acid sensitive so overall the impact of decrease in gastric acid is biologically plausible *(93)*. Loss of normal gastric acidity has been associated with colonization of the gastric and duodenal portion of the upper gastrointestinal tract *(94)*. Despite the plausible connection between the role of gastric acid in preventing *C. difficile* colonization, the specific mechanism has not been clearly elucidated. In the hospital setting, *C. difficile* in considered the main cause of colitis and is associated with increased mortality and morbidity and increased length of stay, therefore of economic significance *(93)*.

Most, but not all, studies indicate increased risk for *C. difficile* with the use of acid suppression. In a US study carried out in a nursing home setting, the use of PPIs was associated with increased risk of CDAD. This was a 347 bed long-term care facility and there were 25 cases of CDAD during the study period and 28 other patients were selected as control. All subjects that had CDAD infection used antibiotics several months prior to the onset of the CDAD (from 2 to 5 months) and 60% were on PPIs. Only 32% of the patients in the control group were PPI users *(92)*. In a retrospective case–control study in the United Kingdom, it was demonstrated that recent use of PPIs was associated with a 2.5 odds ratio increased risk of *C. difficile* diarrhea. The cases included 176 inpatients that were positive for *C. difficile*. Controls were selected and matched from the hospital population *(93)*. Dial and colleagues reported several studies all indicating increased risk of CDAD with use of acid-suppression medicines *(95–97)*. In a 2004 article *(96)*, 1,187 inpatients in a Montreal teaching hospital who received antibiotics in the previous 9 months and who also received acid-suppression therapy were compared to those that did not receive acid-suppression therapy. The odds ratio for the association between *C. difficile* and PPIs use was significant at 2.1. In contrast to PPIs, there was no increased risk with the use of H_2RAs. In another case–control study *(97)*, conducted in a separate Montreal teaching hospital, there were 94 cases of positive *C. difficile* and 94 controls. Again the users of PPIs had a higher odds ratio of 2.7 for the development of *C. difficile*. Again in this population there was no association of *C. difficile* illness while on H_2RAs *(96)*. Interestingly, about 50% of the patients receiving antibiotics were also prescribed PPIs and another 10% receiving H_2RAs; however, it was not possible to ascertain the reason for the prescription of the acid suppressers. It has been suggested recently that acid-suppression therapy is overused in hospitals and that many patients continue to take the medication even 3 months after discharge *(98)*. In 2005 Dial et al. *(97)* confirmed previous results in a population-based case–control study with a United Kingdom population based on the General Practice Database. There were 1,672 cases of *C. difficile* and 10 matching controls were selected for each case. The adjusted odds ratio was 2.9 for the association of PPI use and *C. difficile* infection *(97)*.

20.5. CONCLUSIONS

Acid-suppressive therapies and in particular PPIs have transformed our ability to regulate gastric acid secretion and overcome diseases associated with hypersecretion. Treatment failures, however, illustrate that not all gastroesophageal diseases are caused by hypersecretion, and further, there are data indicating that these interventions especially PPIs, when used chronically, are associated with negative nutritional and other negative consequences. Further, our increasing knowledge related to the role of HP infection in acid secretion creates opportunities to manage the root cause of some symptoms such that prevalence of more severe outcomes may be reduced.

The important negative consequences of long-term therapy with PPIs is a valid concern as there are no studies of use beyond 10 years and only a limited number of studies with relatively few patients have examined other possible effects of long-term gastric acid-suppression treatments.

20.6. RECOMMENDATIONS

Long-term use of acid-suppressive therapy should be evaluated at least annually, and step-down algorithms should be considered. Clinicians should monitor patients on chronic acid-suppressive therapy for vitamin and mineral status and initiate supplementation before deficiencies develop. HP infection should be ascertained and treated as there are numerous negative consequences associated with this infection.

REFERENCES

1. Schubert ML, Peura DA. Control of gastric acid secretion in health and disease. Gastroenterology 2008;134(7): 1842–1860.
2. Pohl D, Fox M, Fried M, Goke B, Prinz C, Monnikes H, et al. Do we need gastric acid? Digestion 2008;77(3–4): 184–197.
3. Schubert ML. Gastric secretion. Curr Opin Gastroenterol 2007;23(6):595–601.
4. Yao X, Forte JG. Cell biology of acid secretion by the parietal cell. Annu Rev Physiol 2003;65:103–131.
5. Vakil N, van Zanten SV, Kahrilas P, Dent J, Jones R. The Montreal definition and classification of gastroesophageal reflux disease: a global evidence-based consensus. Am J Gastroenterol 2006;101(8):1900–1920.
6. Kahrilas PJ, Shaheen NJ, Vaezi MF, Hiltz SW, Black E, Modlin IM, et al. American Gastroenterological Association Medical Position Statement on the management of gastroesophageal reflux disease. Gastroenterology 2008;135(4):1383–1391.
7. Fiedorek S, Tolia V, Gold BD, Huang B, Stolle J, Lee C, et al. Efficacy and safety of lansoprazole in adolescents with symptomatic erosive and non-erosive gastroesophageal reflux disease. J Pediatr Gastroenterol Nutr 2005;40(3): 319–327.
8. Dent J, El Serag HB, Wallander MA, Johansson S. Epidemiology of gastro-oesophageal reflux disease: a systematic review. Gut 2005;54(5):710–717.
9. Tytgat GN, McColl K, Tack J, Holtmann G, Hunt RH, Malfertheiner P, et al. New algorithm for the treatment of gastro-oesophageal reflux disease. Aliment Pharmacol Ther 2008;27(3):249–256.
10. Shaker R, Castell DO, Schoenfeld PS, Spechler SJ. Nighttime heartburn is an under-appreciated clinical problem that impacts sleep and daytime function: the results of a Gallup survey conducted on behalf of the American Gastroenterological Association. Am J Gastroenterol 2003;98(7):1487–1493.
11. Thoua NM, Khoo D, Kalantzis C, Emmanuel AV. Acid-related oesophageal sensitivity, not dysmotility, differentiates subgroups of patients with non-erosive reflux disease. Aliment Pharmacol Ther 2008;27(5):396–403.
12. Bruce mg, Maaroos HI. Epidemiology of Helicobacter pylori infection. Helicobacter 2008;13(Suppl. 1):1–6.
13. Kwok A, Lam T, Katelaris P, Leong RW. Helicobacter pylori eradication therapy: indications, efficacy and safety. Expert Opin Drug Saf 2008;7(3):271–281.
14. Kandulski A, Selgrad M, Malfertheiner P. Helicobacter pylori infection: a clinical overview. Dig Liver Dis 2008;40(8):619–626.
15. Permin H, Andersen LP. Inflammation, immunity, and vaccines for Helicobacter infection. Helicobacter 2005;10(Suppl. 1):21–25.
16. Chmiela M, Michetti P. Inflammation, immunity, vaccines for Helicobacter infection. Helicobacter 2006;11(Suppl. 1):21–26.
17. McColl KE. How I manage H. pylori-negative, NSAID/aspirin-negative peptic ulcers. Am J Gastroenterol 2009;104(1):190–193.
18. Yu EW, Blackwell T, Ensrud KE, Hillier TA, Lane NE, Orwoll E, et al. Acid-suppressive medications and risk of bone loss and fracture in older adults. Calcif Tissue Int 2008;83(4):251–259.
19. Raghunath AS, O'Morain C, McLoughlin RC. Review article: the long-term use of proton-pump inhibitors. Aliment Pharmacol Ther 2005;22(Suppl. 1):55–63.

20. Raghunath AS, Hungin AP, Mason J, Jackson W. Helicobacter pylori eradication in long-term proton pump inhibitor users in primary care: a randomized controlled trial. Aliment Pharmacol Ther 2007;25(5):585–592.
21. Masaoka T, Suzuki H, Hibi T. Gastric epithelial cell modality and proton pump inhibitor. J Clin Biochem Nutr 2008;42(3):191–196.
22. Scheindlin S. A century of ulcer medicines. Mol Interv 2005;5(4):201–206.
23. Shin JM, Vagin O, Munson K, Kidd M, Modlin IM, Sachs G. Molecular mechanisms in therapy of acid-related diseases. Cell Mol Life Sci 2008;65(2):264–281.
24. Gillen D, McColl KE. Problems related to acid rebound and tachyphylaxis. Best Pract Res Clin Gastroenterol 2001;15(3):487–495.
25. Shin JM, Cho YM, Sachs G. Chemistry of covalent inhibition of the gastric (H+, K+)-ATPase by proton pump inhibitors. J Am Chem Soc 2004;126(25):7800–7811.
26. McColl KE. Effect of proton pump inhibitors on vitamins and iron. Am J Gastroenterol. 2009;104:S5–S9.
27. Tytgat GN, Heading RC, Muller-Lissner S, Kamm MA, Scholmerich J, Berstad A, et al. Contemporary understanding and management of reflux and constipation in the general population and pregnancy: a consensus meeting. Aliment Pharmacol Ther 2003;18(3):291–301.
28. Fass R. Persistent heartburn in a patient on proton-pump inhibitor. Clin Gastroenterol Hepatol 2008;6(4):393–400.
29. Hampel H, Abraham NS, El Serag HB. Meta-analysis: obesity and the risk for gastroesophageal reflux disease and its complications. Ann Intern Med 2005;143(3):199–211.
30. Fraser-Moodie CA, Norton B, Gornall C, Magnago S, Weale AR, Holmes GK. Weight loss has an independent beneficial effect on symptoms of gastro-oesophageal reflux in patients who are overweight. Scand J Gastroenterol 1999;34(4):337–340.
31. Ruhl CE, Everhart JE. Overweight, but not high dietary fat intake, increases risk of gastroesophageal reflux disease hospitalization: the NHANES I Epidemiologic Followup Study. First National Health and Nutrition Examination Survey. Ann Epidemiol 1999;9(7):424–435.
32. Fisher BL, Pennathur A, Mutnick JL, Little AG. Obesity correlates with gastroesophageal reflux. Dig Dis Sci 1999;44(11):2290–2294.
33. Tran T, Lowry AM, El Serag HB. Meta-analysis: the efficacy of over-the-counter gastro-oesophageal reflux disease therapies. Aliment Pharmacol Ther 2007;25(2):143–153.
34. Graham DY, Shiotani A. New concepts of resistance in the treatment of Helicobacter pylori infections. Nat Clin Pract Gastroenterol Hepatol 2008;5(6):321–331.
35. Rokkas T, Sechopoulos P, Robotis I, Margantinis G, Pistiolas D. Cumulative H. pylori eradication rates in clinical practice by adopting first and second-line regimens proposed by the Maastricht III consensus and a third-line empirical regimen. Am J Gastroenterol 2009;104(1):21–25.
36. Vakil N. H. pylori treatment: new wine in old bottles? Am J Gastroenterol 2009;104(1):26–30.
37. Vakil N. Helicobacter pylori treatment: is sequential or quadruple therapy the answer? Rev Gastroenterol Disord 2008;8(2):77–82.
38. Niv Y. H pylori recurrence after successful eradication. World J Gastroenterol 2008;14(10):1477–1478.
39. Targownik LE, Metge CJ, Leung S, Chateau DG. The relative efficacies of gastroprotective strategies in chronic users of nonsteroidal anti-inflammatory drugs. Gastroenterology 2008;134(4):937–944.
40. Konturek PC, Kania J, Hahn EG, Konturek JW. Ascorbic acid attenuates aspirin-induced gastric damage: role of inducible nitric oxide synthase. J Physiol Pharmacol 2006;57(Suppl. 5):125–136.
41. Sun SX, Lee KY, Bertram CT, Goldstein JL. Withdrawal of COX-2 selective inhibitors rofecoxib and valdecoxib: impact on NSAID and gastroprotective drug prescribing and utilization. Curr Med Res Opin 2007;23(8):1859–1866.
42. Physician's Desk Reference: PDR—Aciphex Tablets (Eisai). PDR Electronic Library. 2008. 6-18-2008.
43. Physician's Desk Reference: PDR—Prevacid Delayed-Release Capsules (TAP). PDR Electronic Library. 2008. 6-18-2008.
44. Physician's Desk Reference: PDR—Nexium Delayed-Release Capsules (AstraZeneca LP). PDR Electronic Library. 2008. 6-18-2008.
45. Physician's Desk Reference: PDR—Zegerid Powder for Oral Solution (Santarus). PDR Electronic Library. 2008. 6-18-2008.
46. Gilard M, Arnaud B, Cornily JC, Le Gal G, Lacut K, Le Calvez G, et al. Influence of omeprazole on the antiplatelet action of clopidogrel associated with aspirin: the randomized, double-blind OCLA (Omeprazole CLopidogrel Aspirin) study. J Am Coll Cardiol 2008;51(3):256–260.
47. Siller-Matula JM, Spiel AO, Lang IM, Kreiner G, Christ G, Jilma B. Effects of pantoprazole and esomeprazole on platelet inhibition by clopidogrel. Am Heart J 2009;157(1):148–5.

48. Green R. Is it time for vitamin B-12 fortification? What are the questions? Am J Clin Nutr 2009;89(2):712S–716S.
49. Dharmarajan TS, Kanagala MR, Murakonda P, Lebelt AS, Norkus EP. Do acid-lowering agents affect vitamin B12 status in older adults? J Am Med Dir Assoc 2008;9(3):162–167.
50. Lechner K, Fodinger M, Grisold W, Puspok A, Sillaber C. Vitamin B12 deficiency. New data on an old theme. Wien Klin Wochenschr 2005;117(17):579–591.
51. Force RW, Meeker AD, Cady PS, Culbertson VL, Force WS, Kelley CM. Ambulatory care increased vitamin B12 requirement associated with chronic acid suppression therapy. Ann Pharmacother 2003;37(4):490–493.
52. Allen LH. Causes of vitamin B12 and folate deficiency. Food Nutr Bull 2008;29(2 Suppl.):S20–S34.
53. Marcuard SP, Albernaz L, Khazanie PG. Omeprazole therapy causes malabsorption of cyanocobalamin (vitamin B12). Ann Intern Med 1994;120(3):211–215.
54. Hirschowitz BI, Worthington J, Mohnen J. Vitamin B12 deficiency in hypersecretors during long-term acid suppression with proton pump inhibitors. Aliment Pharmacol Ther 2008;27(11):1110–1121.
55. Liu C, Russell RM. Nutrition and gastric cancer risk: an update. Nutr Rev 2008;66(5):237–249.
56. Fountoulakis A, Martin IG, White KL, Dixon MF, Cade JE, Sue-Ling HM, et al. Plasma and esophageal mucosal levels of vitamin C: role in the pathogenesis and neoplastic progression of Barrett's esophagus. Dig Dis Sci 2004;49(6):914–919.
57. Pietraforte D, Castelli M, Metere A, Scorza G, Samoggia P, Menditto A, et al. Salivary uric acid at the acidic pH of the stomach is the principal defense against nitrite-derived reactive species: sparing effects of chlorogenic acid and serum albumin. Free Radic Biol Med 2006;41(12):1753–1763.
58. Henry EB, Carswell A, Wirz A, Fyffe V, McColl KE. Proton pump inhibitors reduce the bioavailability of dietary vitamin C. Aliment Pharmacol Ther 2005;22(6):539–545.
59. Mowat C, Carswell A, Wirz A, McColl KE. Omeprazole and dietary nitrate independently affect levels of vitamin C and nitrite in gastric juice. Gastroenterology 1999;116(4):813–822.
60. Mowat C, Williams C, Gillen D, Hossack M, Gilmour D, Carswell A, et al. Omeprazole, Helicobacter pylori status, and alterations in the intragastric milieu facilitating bacterial N-nitrosation. Gastroenterology 2000;119(2):339–347.
61. Heaney RP. Bone as the calcium nutrient reserve. In: Weaver CM, Heaney RP, eds. Calcium in Human Health. Totowa, NJ: Humana Press Inc., 2006:7–12.
62. Weaver CM, Heaney RP. Introduction. In: Weaver CM, Heaney RP, eds. Calcium in Human Health. Totowa, NJ: Humana Press Inc., 2006:1–3.
63. Awumey EM, Bukoski RD. Cellular functions and fluxes of calcium. In: Weaver CM, Heaney RP, eds. Calcium in Human Health. Totowa, NJ: Humana Press Inc., 2006:13–35.
64. Hoenderop JG, Nilius B, Bindels RJ. Calcium absorption across epithelia. Physiol Rev 2005;85(1):373–422.
65. Wright MJ, Proctor DD, Insogna KL, Kerstetter JE. Proton pump-inhibiting drugs, calcium homeostasis, and bone health. Nutr Rev 2008;66(2):103–108.
66. Bo-Linn GW, Davis GR, Buddrus DJ, Morawski SG, Santa AC, Fordtran JS. An evaluation of the importance of gastric acid secretion in the absorption of dietary calcium. J Clin Invest 1984;73(3):640–647.
67. Heaney RP, Recker RR, Weaver CM. Absorbability of calcium sources: the limited role of solubility. Calcif Tissue Int 1990;46(5):300–304.
68. Heaney RP, Dowell SD, Bierman J, Hale CA, Bendich A. Absorbability and cost effectiveness in calcium supplementation. J Am Coll Nutr 2001;20(3):239–246.
69. Simonian HP, Vo L, Doma S, Fisher RS, Parkman HP. Regional postprandial differences in pH within the stomach and gastroesophageal junction. Dig Dis Sci 2005;50(12):2276–2285.
70. Recker RR. Calcium absorption and achlorhydria. N Engl J Med 1985;313(2):70–73.
71. Serfaty-Lacrosniere C, Wood RJ, Voytko D, Saltzman JR, Pedrosa M, Sepe TE, et al. Hypochlorhydria from short-term omeprazole treatment does not inhibit intestinal absorption of calcium, phosphorus, magnesium or zinc from food in humans. J Am Coll Nutr 1995;14(4):364–368.
72. O'Connell MB, Madden DM, Murray AM, Heaney RP, Kerzner LJ. Effects of proton pump inhibitors on calcium carbonate absorption in women: a randomized crossover trial. Am J Med 2005;118(7):778–781.
73. Graziani G, Como G, Badalamenti S, Finazzi S, Malesci A, Gallieni M, et al. Effect of gastric acid secretion on intestinal phosphate and calcium absorption in normal subjects. Nephrol Dial Transplant 1995;10(8):1376–1380.
74. Yang YX, Lewis JD, Epstein S, Metz DC. Long-term proton pump inhibitor therapy and risk of hip fracture. JAMA 2006;296(24):2947–2953.
75. Vestergaard P, Rejnmark L, Mosekilde L. Proton pump inhibitors, histamine H2 receptor antagonists, and other antacid medications and the risk of fracture. Calcif Tissue Int 2006;79(2):76–83.

76. Targownik LE, Lix LM, Metge CJ, Prior HJ, Leung S, Leslie WD. Use of proton pump inhibitors and risk of osteoporosis-related fractures. CMAJ 2008;179(4):319–326.
77. Grisso JA, Kelsey JL, O'Brien LA, Miles CG, Sidney S, Maislin G, et al. Risk factors for hip fracture in men. Hip Fracture Study Group. Am J Epidemiol 1997;145(9):786–793.
78. Roux KJ, Crisp ML, Liu Q, Kim D, Kozlov S, Stewart CL, et al. Nesprin 4 is an outer nuclear membrane protein that can induce kinesin-mediated cell polarization. Proc Natl Acad Sci U S A 2009.
79. Richards JB, Goltzman D. Proton pump inhibitors: balancing the benefits and potential fracture risks. CMAJ 2008;179(4):306–307.
80. Zimmermann MB, Hurrell RF. Nutritional iron deficiency. Lancet 2007;370(9586):511–520.
81. Brody T. Inorganic nutrients. Nutritional Biochemistry. London: Academic Press, 1999:693–878.
82. Hallberg L. Perspectives on nutritional iron deficiency. Annu Rev Nutr 2001;21:1–21.
83. Hurrell RF, Reddy MB, Juillerat M, Cook JD. Meat protein fractions enhance nonheme iron absorption in humans. J Nutr 2006;136(11):2808–2812.
84. Cook JD, Brown GM, Valberg LS. The effect of achylia gastrica on iron absorption. J Clin Invest 1964;43:1185–1191.
85. Skikne BS, Lynch SR, Cook JD. Role of gastric acid in food iron absorption. Gastroenterology 1981;81(6):1068–1071.
86. Bezwoda W, Charlton R, Bothwell T, Torrance J, Mayet F. The importance of gastric hydrochloric acid in the absorption of nonheme food iron. J Lab Clin Med 1978;92(1):108–116.
87. Hutchinson C, Geissler CA, Powell JJ, Bomford A. Proton pump inhibitors suppress absorption of dietary non-haem iron in hereditary haemochromatosis. Gut 2007;56(9):1291–1295.
88. Howden CW, Hunt RH. Relationship between gastric secretion and infection. Gut 1987;28(1):96–107.
89. Tennant SM, Hartland EL, Phumoonna T, Lyras D, Rood JI, Robins-Browne RM, et al. Influence of gastric acid on susceptibility to infection with ingested bacterial pathogens. Infect Immun 2008;76(2):639–645.
90. Gulmez SE, Holm A, Frederiksen H, Jensen TG, Pedersen C, Hallas J. Use of proton pump inhibitors and the risk of community-acquired pneumonia: a population-based case-control study. Arch Intern Med 2007;167(9):950–955.
91. Canani RB, Cirillo P, Roggero P, Romano C, Malamisura B, Terrin G, et al. Therapy with gastric acidity inhibitors increases the risk of acute gastroenteritis and community-acquired pneumonia in children. Pediatrics 2006;117(5):e817-e820.
92. Al Tureihi FI, Hassoun A, Wolf-Klein G, Isenberg H. Albumin, length of stay, and proton pump inhibitors: key factors in Clostridium difficile-associated disease in nursing home patients. J Am Med Dir Assoc 2005;6(2):105–108.
93. Cunningham R, Dale B, Undy B, Gaunt N. Proton pump inhibitors as a risk factor for Clostridium difficile diarrhoea. J Hosp Infect 2003;54(3):243–245.
94. Thorens J, Froehlich F, Schwizer W, Saraga E, Bille J, Gyr K, et al. Bacterial overgrowth during treatment with omeprazole compared with cimetidine: a prospective randomised double blind study. Gut 1996;39(1):54–59.
95. Dial MS. Proton pump inhibitor use and enteric infections. Am J Gastroenterol. 2009;104:S10–S16.
96. Dial S, Alrasadi K, Manoukian C, Huang A, Menzies D. Risk of Clostridium difficile diarrhea among hospital inpatients prescribed proton pump inhibitors: cohort and case-control studies. CMAJ 2004;171(1):33–38.
97. Dial S, Delaney JA, Barkun AN, Suissa S. Use of gastric acid-suppressive agents and the risk of community-acquired Clostridium difficile-associated disease. JAMA 2005;294(23):2989–2995.
98. Parente F, Cucino C, Gallus S, Bargiggia S, Greco S, Pastore L, et al. Hospital use of acid-suppressive medications and its fall-out on prescribing in general practice: a 1-month survey. Aliment Pharmacol Ther 2003;17(12):1503–1506.
99. Berna MJ, Hoffmann KM, Serrano J, Gibril F, Jensen RT. Serum gastrin in Zollinger-Ellison syndrome: I. Prospective study of fasting serum gastrin in 309 patients from the National Institutes of Health and comparison with 2229 cases from the literature. Medicine (Baltimore) 2006;85(6):295–330.
100. Elitsur Y, Lawrence Z. Non-Helicobacter pylori related duodenal ulcer disease in children. Helicobacter 2001;6(3):239–243.
101. Laine L, Hopkins RJ, Girardi LS. Has the impact of Helicobacter pylori therapy on ulcer recurrence in the United States been overstated? A meta-analysis of rigorously designed trials. Am J Gastroenterol 1998;93(9):1409–1415.
102. Aro P, Storskrubb T, Ronkainen J, Bolling-Sternevald E, Engstrand L, Vieth M, et al. Peptic ulcer disease in a general adult population: the Kalixanda study: a random population-based study. Am J Epidemiol 2006;163(11):1025–1034.
103. Jang HJ, Choi MH, Shin WG, Kim KH, Chung YW, Kim KO, et al. Has peptic ulcer disease changed during the past ten years in Korea? A prospective multi-center study. Dig Dis Sci 2008;53(6):1527–1531.
104. Gisbert JP, Blanco M, Mateos JM, Fernandez-Salazar L, Fernandez-Bermejo M, Cantero J, et al. H. pylori-negative duodenal ulcer prevalence and causes in 774 patients. Dig Dis Sci 1999;44(11):2295–2302.
105. De Block CE, De Leeuw IH, Van Gaal LF. Autoimmune gastritis in type 1 diabetes: a clinically oriented review. J Clin Endocrinol Metab 2008;93(2):363–371.

106. Weck MN, Brenner H. Prevalence of chronic atrophic gastritis in different parts of the world. Cancer Epidemiol Biomarkers Prev 2006;15(6):1083–1094.
107. Kuipers EJ. Proton pump inhibitors and Helicobacter pylori gastritis: friends or foes? Basic Clin Pharmacol Toxicol 2006;99(3):187–194.
108. Ruscin JM, Page RL, Valuck RJ. Vitamin B(12) deficiency associated with histamine(2)-receptor antagonists and a proton-pump inhibitor. Ann Pharmacother 2002;36(5):812–816.
109. AstraZeneca. Prilosec Package Insert (omeprazole and omeprazole magnesium). Package Insert. 2009.
110. Physician's Desk Reference: PDR—Protonix Tablets (Wyeth). PDR Electronic Library. 2008. 6-18-2008.

21 Nutritional Antioxidants, Dietary Carbohydrate, and Age-Related Maculopathy and Cataract

Chung-Jung Chiu and Allen Taylor

Key Points

- Age-related cataract (ARC) and maculopathy (ARM) are two major causes of blindness worldwide.
- There are several important reasons to study relationships between risk for ARC/ARM and nutrition.
- Oxidative stress is associated with compromises to the lens and retina, and abundant literature indicates that antioxidants may ameliorate the risk for ARM and probably ARC.
- Accumulation of glycation and glycoxidation endproducts in drusen and cataracts indicates that dietary carbohydrate is related to risk for ARC and ARM.
- Dietary pattern analysis will give further insight into dietary effects as a whole on risk for ARC and ARM.
- Further efforts should include identifying specific nutrient(s) in foods, defining optimal levels of the nutrient(s), and determining if, and the age when, supplementation should be begun.

Abstract

Loss of vision is the second greatest fear, next to death, among the elderly. Age-related cataract (**ARC**) and maculopathy (**ARM**) are two major causes of blindness worldwide. There are several important reasons to study relationships between risk for ARC/ARM and nutrition: (1) because it is likely that the same nutritional practices that are associated with prolonged eye function will also be associated with delayed age-related compromises to other organs and perhaps, aging in general; (2) surgical resources are insufficient to provide economic and safe surgeries for cataract and do not provide a widely practicable cure for ARM; (3) there will be considerable financial savings and improvements in quality of life if health is extended, particularly given the rapidly growing elderly segment of our population.

It is clear that oxidative stress is associated with compromises to the lens and retina. Recent literature indicates that antioxidants may ameliorate the risk for ARM and probably ARC. Given the association between oxidative damage and age-related eye debilities, it is not surprising that over 70 studies have attempted to relate antioxidant intake to risk for ARC and ARM. This chapter will review epidemiological literature about ARC and ARM with emphasis on roles for vitamins C and E and carotenoids. Since glycation and glycoxidation are major molecular insults which involve an oxidative stress component, we also review for the first time new literature that relates dietary carbohydrate intake to risk for ARC and ARM. To evaluate dietary effects as a whole, several studies have tried to relate dietary patterns to risk for ARC. We will also give some attention to this emerging research.

Proper nutrition, possibly including use of antioxidant supplements for the nutritionally impoverished, along with healthy life styles, eliminating smoking and obesity, appears to provide the least costly and most practical means to delay ARC and ARM.

Given the promise of salutary effects of optimal nutrition on visual function, further studies should be devoted to identifying the most effective nutritional and environmental strategy to prevent or delay the development and progress of ARM/ARC. The efforts should include identifying the right nutrient(s) or foods, defining useful levels of the nutrient(s), and determining the age when the supplementation should be begun.

A. Bendich, R.J. Deckelbaum (eds.), *Preventive Nutrition*, Nutrition and Health, DOI 10.1007/978-1-60327-542-2_21,

Key Words: Aging; carbohydrate; dietary pattern; evidence-based medicine; fat; glycemic index; healthy eating index; lens; lutein; nutrition; retina; vitamin C; vitamin E; zeaxanthin

21.1. INTRODUCTION

Age-related cataract (ARC) and maculopathy (ARM) are the major causes of blindness among the elderly population throughout the world *(1–3)*. In the United States, the prevalence of visually significant ARC increases from approximately 5% at age 65 years to about 50% for persons older than 75 years *(4,5)* and costs associated with cataract account for 10% of the medical budget. The prevalence of ARM also increases dramatically with age and is over 10% among people >80 years old *(6)*. Within two decades, it is anticipated that the prevalence of both ARC and ARM will increase by 50% as the number of elderly Americans increases *(6,7)*. A delay in these two major age-related disabilities would not only enhance the quality of life for much of the world's older population but also substantially reduce the economic burden which is imposed by this debility.

Various risk factors including oxidative stress resulting from sunlight exposure and smoking *(8–11)*, etc., along with or independent of glycation (nonenzymatic modification of cellular constituents by reaction with sugars and oxidation of these moieties) or their sequelae, play important roles in the pathogenesis of ARC and ARM (reviewed in *(12)* and *(6,7,13)*). Manipulating the antioxidant balance to prevent ARC and ARM through diet or supplementation has attracted much interest. In this review of the epidemiological literature, we describe the evidence for a link between ARC, ARM, nutritional antioxidants, carbohydrate, and dietary patterns.

21.2. EPIDEMIOLOGICAL STUDIES REGARDING ANTIOXIDANTS AND CATARACT

Within the last two decades, considerable effort has been dedicated to determining whether antioxidants can diminish the risk for cataract (Tables 21.1 and 21.2). We first review data from the intervention trials (Table 21.2). Next we review the observational studies, including describing novel results which relate carbohydrate nutrition to risk for ARC. This topic has not been reviewed previously. We then move to a summary of the observational data regarding foods and nutrients in relation to ARC. The overall impression created by the observational data is that nutrient intake or blood nutrient levels are related to risk for cataract and that nutrition might be exploited to diminish the risk for this debility. Upon first inspection, the data from the intervention trials are less encouraging (Table 21.2), but newer studies add confidence to observations that cataract is delayed in persons with adequate supplies of dietary antioxidants.

21.2.1. Intervention Trials

Ten randomized, double-blinded, placebo-controlled intervention trials have assessed the effect of vitamin supplements on cataract risk, with mixed results (Table 21.2). Each of these listed specific formulations of vitamins and minerals, including over the counter multivitamins used in the Italian-American Clinical Trial of Nutritional Supplements and Age-Related Cataract (CTNS) *(14)*. The AREDS, VECAT, ATBC, Women's Health Study trials (including β-carotene trial and vitamin E trial), and APC all found no statistically significant association *(15–20)*. The Linxian, Physician's Health Study, and REACT trials each found primarily protective effects of use of specific nutrient supplements

Table 21.1
Design and Sample Size of Major Epidemiological Studies Relating Risk for Cataract to Nutritional Antioxidants

Study	*Design*	*Sample size*	*References*
Age-Related Eye Disease Study (AREDS) (2001)	Intervention	$n = 4,596$	*(15)*
Health Professionals Follow-up Study (HPFS) (1999)	Prospective	$n = 36,644$	*(64)*
Nurses' Health Study (NHS) (1999)	Prospective	$n = 73,956$	*(54)*
NHS (1999)	Prospective	$n = 77,466$	*(63)*
Women's Health Study (WHS) (2004)	Intervention	$n = 39,876$	*(23)*
REACT (2002)	Intervention	$n = 445$	*(24)*
Physicians' Health Study (PHS) (2003)	Intervention	$n = 20,968$	*(26)*
WHS (2008)	Prospective	$n = 35,551$	*(51)*
WHS (2008)	Prospective	$n = 39,876$	*(22)*
Italian-American Clinical Trial of Nutritional Supplements and Age-Related Cataract (CTNS) (2008)	Intervention	$n = 1,020$	*(14)*
Blue Mountains Eye Study (2000)	Retrospective	$n = 2,900$	*(42)*
POLA study (2000)	Retrospective	$n = 2,584$	*(67)*
Dherani et al. (2008)	Retrospective	$n = 1,443$	*(41)*
CTNS (2005)	Retrospective	$n = 1,020$	*(38)*
Gale et al. (2001)	Retrospective	$n = 372$	*(43)*
Gritz et al. (2006)	Intervention	$n = 798$	*(21)*
NHS (1992)	Prospective	$n = 50,828$	*(36)*
Italian-American Cataract Study (1991)	Retrospective	$n = 1,477$	*(44)*
Jacques and Chylack (1991)	Retrospective	$n = 112$	*(28)*
Nutrition and Vision Project (NVP) (1997)	Retrospective	$n = 294$	*(29)*
NVP (2001)	Retrospective	$n = 478$	*(16)*
NVP (2004)	Prospective	$n = 407$	*(18)*
NVP (2005)	Retrospective	$n = 408$	*(18)*
Beaver Dam Eye Study (BDES) (2008)	Prospective	$n = 4,926$	*(47)*
Knekt et al. (1992)	Prospective	$n = 141$	*(62)*
Blue Mountains Eye Study (BMES) (2001)	Retrospective	$n = 385$	*(68)*
Lens Opacities Case–Control Study (1991)	Retrospective	$n = 1,380$	*(30)*
Lens Opacities Case–Control Study (1995)	Retrospective	$n = 1,380$	*(56)*
Barbados Eye Study (1997)	Retrospective	$n = 4,314$	*(69)*
Longitudinal Study of Cataract (1998)	Prospective	$n = 764$	*(46)*

(*Continued*)

Table 21.1
(Continued)

Study	*Design*	*Sample size*	*References*
BDES (1999)	Prospective	$n = 1{,}354$	*(37)*
BDES (1999)	Prospective	$n = 252$	*(59)*
Nutritional Factors in Eye Disease Study (1994)	Retrospective	$n = 2{,}152$	*(31)*
BDES (1995)	Retrospective	$n = 1{,}919$	*(32)*
Nutritional Factors in Eye Disease Study (1995)	Retrospective	$n = 400$	*(57)*
BDES (1996)	Prospective	$n = 3{,}220$	*(71)*
BDES (2000)	Prospective	$n = 2{,}434$	*(39)*
Visual Impairment Project (VIP) (1999)	Retrospective	$n = 3{,}271$	*(58)*
Vitamin E, Cataract and Age-Related Maculopathy Trial (2004)	Intervention	$n = 1{,}193$	*(20)*
AREDS (2006)	Prospective	$N = 4{,}590$	*(70)*
India-US Case–Control Study (1989)	Retrospective	$n = 1{,}990$	*(33)*
Vitamin E and Cataract Prevention Study (VECAT) (1999)	Prospective	$n = 1{,}111$	*(60)*
Nourmohammadi et al. (2008)	Retrospective	$N = 88$	*(49)*
Robertson et al. (1989)	Retrospective	$n = 304$	*(34)*
Rodríguez-Rodríguez et al.	Retrospective	$N = 177$	*(50)*
Rouhiainen et al. (1996)	Prospective	$n = 410$	*(61)*
Physicians' Health Study (1994)	Prospective	$n = 17{,}744$	*(55)*
Linxian cataract studies (1993)	Intervention	$n = 2{,}141$ (multivitamin/mineral supplement in trial 1); $n = 3{,}249$ (retinol/zinc, riboflavin/niacin, ascorbic acid/ molybdenum, and selenium/α-tocopherol/ β-carotene in trial 2)	*(19)*
BMES (2008)	Prospective	$n = 2{,}464$	*(40)*
Tarwadi et al. (2004)	Retrospective	$n = 243$	*(48)*
Tavani et al. (1996)	Retrospective	$n = 913$	*(53)*
NVP (2002)	Retrospective	$n = 492$	*(17)*
Alpha-tocopherol Beta-carotene (ATBC) Study (1998)	Intervention	$n = 28{,}934$	*(25)*
Valero et al. (2002)	Retrospective	$n = 677$	*(35)*
Baltimore Longitudinal Study on Aging (1993)	Retrospective	$n = 671$	*(45)*
VIP (2006)	Retrospective	$n = 2{,}322$	*(65)*
JPHC (2007)	Prospective	$n = 35{,}186$	*(127)*

Table 21.2
Comparison of Intervention Trials on Cataract

Study 1	*Linxian*
	Sperduto et al. (1993) (19)
Design	Two randomized, double-masked trials with end-of-trial eye examinations in nutritionally deprived Chinese subjects
Subjects	Trial 1: 2,141 subjects aged 45–74 years; Trial 2: 3,249 subjects aged 45–74 years
Evaluation of opacity	The lens Opacities Classification System II (128). A person was judged to have a nuclear cataract if either eye had a grade ≥NII, a cortical cataract if either eye had a grade ≥ CII, and a posterior subcapsular cataract if either eye had a grade ≥PI
Baseline opacity	Does not apply
Supplement	Multivitamin/mineral supplement or matching placebo in trial 1; factorial design to test the effect of four different vitamin/mineral combinations in trial 2 (retinol/zinc, riboflavin/niacin, ascorbic acid/molybdenum, and selenium/α-tocopherol/β-carotene)
Duration	5–6 years
Result	In the first trial, there was a statistically significant 36% reduction in the prevalence of nuclear cataract for persons aged 65–74 years who received the supplements. In the second trial, the prevalence of nuclear cataract was significantly lower in persons receiving riboflavin/niacin compared with persons not receiving these vitamins. Again, persons in the oldest group, 65–74 years, benefited the most (44% reduction in prevalence). No treatment effect was noted for cortical cataract in either trial. Although the number of posterior subcapsular cataracts was very small, there was a statistically significant deleterious effect of treatment with riboflavin/niacin
Conclusion	Increased β-carotene intake and supplement use is not related to diminished risk for nuclear, cortical, or posterior subcapsular cataract. Increased riboflavin intake or supplement use is related to diminished risk for nuclear cataract. Increased multivitamin supplement use or antioxidant index intake is not related to diminished risk for cortical cataract
Study 2	*ATBC*
	Teikari et al. (1998) (25)
Design	Randomized, double-blind, placebo-controlled, 2 × 2 factorial trial in male smokers
Subjects	28,934 Finnish male smokers 50–69 years of age
Evaluation of opacity	Cataract extraction, ascertained from the National Hospital Discharge Registry
Baseline opacity	Does not apply
Supplement	α-Tocopherol (Vit E) 50 mg/day, β-carotene 20 mg/day
Duration	5–8 years (median 5.7 years)
Result	Neither α-tocopherol (RR, 0.91; 95% CI: 0.74, 1.11) nor β-carotene (RR, 0.97; 95% CI: 0.79, 1.19) supplementation affected the incidence of cataract surgery
Conclusion	Supplementation with α-tocopherol or β-carotene does not affect the incidence of cataract extractions among male smokers

(*Continued*)

Table 21.2 (Continued)

Study 3	*AREDS*
	AREDS (2001) (15)
Design	A randomized, placebo-controlled, clinical trial
Subjects	$n = 4{,}596$ US subjects; median: 68 years; range 55–80 years old
Evaluation of opacity	An extension of the Wisconsin System for Subjective Classifying Cataracts from photographs. The area of lens involvement is used to assess the severity of cortical and PSC opacities. Optical density of nuclear opacity is graded against a series of seven standard photographs. Progression of lens opacity (nuclear, cortical, and/or PSC) is defined as follows: nuclear: 1.5 unit increase from baseline on a scale ranging from 0.9 to 6.1; cortical: 10% absolute increase in area from baseline within the central 5-mm diameter circle of the lens; PSC: 5% absolute increase in area from baseline within the central 5-mm diameter circle of the lens. Only relatively advanced cataracts were rated as "lens events." The "lens event rates" were relatively high, ranging from 0.7% for cortical to 21.3% for mixed
Baseline opacity	Nuclear: <2.0: 37%, 2.0–3.9: 48%, ≥4.0: 15%; cortical: <0.1: 48%, 0.1–4.9: 41%, ≥5.0: 11 %; PSC: <0.1: 90%, 0.1–4.9: 8%, ≥5.0: 2%
Supplement	Vit C: 500 mg, Vit E: 400 IU, β-carotene: 15 mg, zinc (zinc oxide): 80 mg, copper (cupric oxide): 2 mg: 66% of participants took Centrum® (containing an RDA amount of Vit C, E, β-carotene, and zinc). This may have reduced the difference in vitamin concentrations between the treated and placebo groups
Duration	6.3 years of AREDS supplement use
Result	No significant protective or adverse effect was found
Conclusion	Use of a high-dose formulation of vitamin C, vitamin E, and β-carotene in a relatively well-nourished older adult cohort had no apparent effect on the 7-year risk of development or progression of age-related nuclear, cortical, or PSC lens opacities or visual acuity loss
Study 4	*REACT*
	Chylack et al. (2002) (24)
Design	A randomized, placebo-controlled, clinical trial
Subjects	297 (US + UK) randomized; before the start of the study, the UK cohort was relatively poorer in nutritional status; US age: mean 65 years, UK age: mean 68 years, $P \leq 0.005$
Evaluation of opacity	Standardized digital lens images (CASE 2000 CCD) graded using computer-assisted (objective) image analysis: cortical: % area opaque (0–100), PSC: % area opaque (0–100), nuclear dip: ln:sd (10–36), scatter density anterior 1 mm and posterior 1 mm: sd (200–1000); the primary endpoint was the percentage increase in pixels opaque: used as a continuous variable; more advanced cataract progression was found in the UK cohort than in the US cohort
Baseline opacity	Nuclear dip (pixel density): mean: 16.7; LOCS III nuclear score: mean: 2.4; LOCS III cortical score: mean: 1.8; LOCS III PSC score: mean: 0.58
Supplement	Vit C: 750 mg, Vit E: 600 IU, β-carotene: 18 mg
Duration	3 years of supplement use
Result	There was about half as much change among users of the supplement for 3 years. The protection was significant only in the US cohort (which had earlier opacities) and in the overall cohort (US + UK) at 3 years; at no time it was significant in the UK cohort alone
Conclusion	Dietary use of the vitamins C, E, and β-carotene for 3 years produced a small deceleration in progression of age-related mixed cataract

Study 5	*Physicians' Health Study* Christen et al. (2003) (26)
Design	Randomized, double-masked, placebo-controlled trial
Subjects	Male US physicians aged 40–84 years ($n = 22{,}071$)
Evaluation of opacity	Does not apply
Baseline opacity	Does not apply
Supplement	50 mg β-carotene
Duration	Alternate days for 12 years
Result	Christen et al. found a 27% reduced rate of extraction (*RR*, 0.73; *CI*: 0.53–1) between men who are current smokers and who did vs did not take the β-carotene supplement 50 mg on alternate days, in a prospective study of 20,968 physicians
Conclusion	Randomized trial data from a large population of healthy men indicated no overall benefit or harm of 12 years of β-carotene supplementation on cataract or cataract extraction. However, among current smokers at baseline, β-carotene appeared to attenuate their excess risk of cataract by about one-fourth
Study 6	*Women's Health Study* Christen et al. (2004) (23)
Design	Randomized, double-masked, placebo-controlled trial
Subjects	Female US health professionals aged 45 years or older (n = 39,876)
Evaluation of opacity	Does not apply
Baseline opacity	Does not apply
Supplement	50 mg β-carotene on alternate day
Duration	Alternate days for 2.1 years
Result	There were 129 cataracts in the β-carotene group and 133 in the placebo group (*RR*, 0.95; 95% *CI*: 0.75–1.21). For cataract extraction, there were 94 cases in the β-carotene group and 89 cases in the placebo group (*RR*, 1.04; 95% *CI*: 0.78–1.39). Subgroup analyses suggested a possible beneficial effect of β-carotene in smokers
Conclusionm	These randomized trial data from a large population of apparently healthy female health professionals indicate that 2 years of β-carotene treatment has no large beneficial or harmful effect on the development of cataract during the treatment period
Study 7	*VECAT* McNeil et al. (2004) (20)
Design	A randomized, placebo-controlled, clinical trial
Subjects	1,193 subjects with early or no cataract, aged 55–80 years; mean: 65 years
Evaluation of opacity	The eye with the more advanced opacity; Nidek EAS 1000; Wilmer lens grading system: cortical: (integral) 0(0), 0.5(1/16), 1(1/8), 2(1/4), 3(1/2), 4($\geq \frac{1}{2}$); nuclear: 0–4.9; PSC: multiply the vertical and horizontal dimensions measured by the calibration of the slit beam height The cutoff points for defining cataract: nuclear: 2.0, cortical: 2.0, PSC: greater or equal to 1 mm^2 The criteria for cataract progression: cortical: increase more than 1; nuclear: increase more than 0.5; PSC: increase 1 mm^2 or more

(*Continued*)

Table 21.2
(Continued)

Baseline opacity	Wilmer grade: cortical: median = 0, mean = 0.5, range: 0–4.0; nuclear: median, mean = 1.5, range: 0.5–4.5; digital grade: nuclear: mean = 8.015 (95% *CI*: 7.99–8.03)
Supplement	Vit E: 500 IU in soybean oil encapsulated in gelatin
Duration	4 years of Vit E supplement use
Result	For cortical cataract the 4-year cumulative incidence rate was 4.5% among those randomized to vitamin E and 4.8% among those randomized to placebo ($P = 0.87$). For nuclear cataract the corresponding rates were 12.9 and 12.1% ($P = 0.77$), for PSC cataract they were 1.7 and 3.5% ($P = 0.08$), while for any of these forms of cataract they were 17.1 and 16.7%, respectively. Progression of cortical cataract was seen in 16.7% of the vitamin E group and 18.4% of the placebo group ($P = 0.76$). Corresponding rates for nuclear cataract were 11.4 and 11.9% ($P = 0.84$), while those of any cataract were 16.5 and 16.7%, respectively. There was no difference in the rate of cataract extraction between the two groups ($P = 0.87$)
Conclusion	Vitamin E given for 4 years at a dose of 500 I/U daily did not reduce the incidence of or progression of nuclear, cortical, or posterior subcapsular cataracts. These findings do not support the use of vitamin E to prevent the development or slow the progression of age-related cataracts
Study 8	*APC Study* Gritz et al. (2006) (21)
Design	A randomized, placebo-controlled, clinical trial
Subjects	798 subjects, aged 35–50 years, and best-corrected visual acuity of 20/40 or better in both eyes and no history of diabetes mellitus, intraocular surgery, radiation therapy, corticosteroid therapy, or active use of vitamin supplements
Evaluation of opacity	Slit lamp biomicroscopy was performed following dilatation. Each cataract was graded on nuclear color and opalescence, cortical opacity, and posterior subcapsular opacity, in 0.1 units, with the grader referring to the LOCS III standard transparency as needed. Three ophthalmologists each graded cataracts in both eyes of all patients, masked to the other observer's scores. Two teams of three ophthalmologists graded the same subjects over the course of the study
Baseline opacity	The baseline LOCS III scores for all cataract subtypes were comparable between the two groups. At baseline, slightly more than 73% of eyes had significant nuclear opalescence and almost 59% of eyes had significant nuclear color changes (defined as a grade of $\geq$2.0 on the LOCS III scale). Significant cataract of at least one category of cataract, as per LOCS III definitions, was present in at least one eye of 79% of the subjects. Despite the degree of cataract present, 84.2% of right eyes had acuity of 20/20 or better and a similar condition existed in left eyes. Over 84% of all eyes were within 1 diopter of emmetropia
Supplement	Vitamin C, 500 mg; vitamin A (β-carotene from commercially grown algae in the all trans form with small amounts of other carotenoids like lycopene), 25,000 IU (15 mg); vitamin E (RRR-α-tocopherol from soya oil with small amounts of other tocopherols), 400 IU

Duration	5 years
Result	Treatment groups were comparable at baseline. There was high compliance with follow-up and study medications. There was progression in cataracts. There was no significant difference between placebo and active treatment groups for either the primary or secondary outcome variables
Conclusion	Antioxidant supplementation with β-carotene, vitamins C and E did not affect cataract progression in a population with a high prevalence of cataract whose diet is generally deficient in antioxidants
Study 9	*WHS* Christen et al. (2008) (22)
Design	A randomized, placebo-controlled, clinical trial
Subjects	37,675 female health professionals aged 45 years or older without a diagnosis of cataract at baseline
Evaluation of opacity	On each annual follow-up questionnaire, women were asked if they had a new diagnosis of cataract or had undergone cataract extraction since completion of the last questionnaire. After the report of a cataract event, written consent identifying the treating ophthalmologist(s) or optometrist(s) was obtained. Ophthalmologist(s) and optometrist(s) were contacted by mail and were asked to complete a cataract questionnaire supplying information about the presence of lens opacities, date of diagnosis, visual acuity loss, cataract extraction, other ocular abnormalities that could explain visual acuity loss, cataract type (nuclear sclerosis, cortical, posterior subcapsular), and cause (including age related, traumatic, congenital, inflammatory, or surgery or steroid induced). Alternatively, ophthalmologist(s) and optometrist(s) were allowed to provide the requested information by supplying photocopies of the relevant medical records
Baseline opacity	NA
Supplement	Vitamin E (600 IU of α-tocopherol [National Source Vitamin E Association, Washington, DC] every other day)
Duration	Average of 9.7 years of treatment and follow-up
Result	There was no significant difference between vitamin E and placebo groups in the incidence of cataract (1,159 vs 1,217 cases; *RR*, 0.96; 95% *CI*: 0.88–1.04). In subgroup analyses of subtypes, there were no significant effects of vitamin E on the incidence of nuclear (1,056 vs 1,127 cases; *RR*, 0.94; 95% *CI*: 0.87–1.02), cortical (426 vs 461 cases; *RR*, 0.93; 95% *CI*: 0.81–1.06), or posterior subcapsular cataract (357 vs 359 cases; *RR*, 1.00; 95% *CI*: 0.86–1.16). Results were similar for extraction of cataract and subtypes. There was no modification of the lack of effect of vitamin E on cataract by baseline categories of age, cigarette smoking, multivitamin use, or several other possible risk factors for cataract
Conclusion	These data from a large trial of apparently healthy female health professionals with 9.7 years of treatment and follow-up indicate that 600 IU natural-source vitamin E taken every other day provides no benefit for age-related cataract or subtypes

(*Continued*)

Table 21.2 (Continued)

Study 10	*CTNS Study* CTNS Study Group (2008) (14)
Design	A randomized, placebo-controlled, clinical trial
Subjects	1,020 participants (mean age, 68 ± 5 years [range, *55–75*]) with early or no cataract
Evaluation of opacity	Slit lamp and retroillumination lens photographs of each eligible eye were taken at the qualification visit, the randomization visit, and then annually, using specially modified slit-lamp (model SL-6E, Topcon Corp., Tokyo, Japan) and retroillumination (Neitz Instrument Co., Ltd., Tokyo, Japan) cameras. Photographs were graded in a masked fashion using standardized grading procedures. Best-corrected VA (BCVA) was measured at each visit according to the Early Treatment Diabetic Retinopathy Study protocol
Baseline opacity	355 subjects (69.6%) in the placebo group and 355 subjects (69.6%) in the treatment group had early cataract
Supplement	Centrum, a multivitamin/mineral supplement containing US RDI levels of nutrients
Duration	Mean follow-up time was 9.0 ± 2.4 years
Result	There was a decrease in total lens events in participants assigned to the multivitamin/mineral formulation compared with those assigned to the placebo (*HR*, 0.82; 95% *CI*: 0.68–0.98; $P = 0.03$). Nuclear events were significantly less common (*HR*, 0.66; 95% *CI*: 0.50–0.88; $P = 0.004$) and PSC events significantly more common (*HR*, 2.00; 95% *CI*: 1.35–2.98; $P < 0.001$) in participants taking the multivitamin/mineral formulation than in those assigned to the placebo. No statistically significant treatment effects were seen for cortical opacities, moderate VA loss, or cataract surgery
Conclusion	Lens events were less common in participants who took the multivitamin/mineral formulation, but treatment had opposite effects on the development or progression of nuclear vs PSC opacities, the two most visually important opacity subtypes

among subjects from certain subgroups of age, gender, or nationality *(21–23)*. The Italian-American Clinical Trial of Nutritional Supplements and Age-Related Cataract (CTNS) found that lens events were less common in participants who took the multivitamin/mineral formulation, Centrum. However, while the 9-year treatment delayed nuclear opacification, it was associated with more rapid progress of PSC opacities *(14)*. Interestingly, there were no effects on visual acuity in this study.

Although the AREDS *(15)* did not allow for the conclusions that use of nutrient supplements decreases the risk for the progress of opacities, several elements of the design of the study suggest further analyses would be worthwhile: (a) 66% of participants took Centrum® (containing an RDA amount of vitamins C, E, β-carotene, and zinc). This may have reduced the difference in vitamin concentrations between the treated and placebo groups. (b) More sensitive grading methods were employed. It appears that the chances of finding relationships might be greater if one studies younger age groups, in which the opacities are at early stages, particularly cohorts in which supplement use can be evaluated over a long period of time (i.e., 10 years). For example, in the Nutrition and Vision Project (NVP) sub-study of the Nurses' Health Study (NHS) it was observed that use of supplements for 10 or more years, begun relatively early in life (starting before 50 years of age), is required to obtain the benefit of diminishing chances for incidence or progress of lens opacities *(16–18)*. A biochemical corollary of this data is that, like aggregation of protein (which in fact biochemically describes a cataract), once the process has begun it is difficult to stop.

21.2.1.1. Antioxidant Supplements and Nuclear Cataract

Two *(14,19)* of the six intervention trials that examined separately for nuclear cataract *(14,15,19–22)* found support for the hypothesis that use of antioxidant supplements is related to diminished risk for nuclear cataract.

Sperduto et al. *(19)* found multivitamin and riboflavin and niacin supplements are associated with 43 and 55% reduced risk of nuclear cataract, respectively, in 66- to 71-year-old subjects in the Linxian study (Table 21.2). Use of multivitamins had no effect in the entire cohort but the riboflavin and niacin supplements were associated with salutary effects for nuclear cataract in the complete cohort. The authors caution that additional research is needed in less nutritionally deprived populations before these findings can be translated into nutritional recommendations for the general public *(19)*.

The CTNS followed 1,020 participants, 55–75 years old and with early or no cataract, for an average of 9.0 ± 2.4 years *(14)*. There was a decrease in total lens events in participants assigned to Centrum, a multivitamin/mineral supplement containing US RDI levels of nutrients, compared with those assigned to the placebo (hazard ratio [*HR*], 0.82; 95% confidence interval [*CI*]: 0.68–0.98; $P = 0.03$). While nuclear events were significantly less common (*HR*, 0.66; 95% *CI*: 0.50–0.88; $P = 0.004$), PSC events were significantly more common (*HR*, 2.00; 95% *CI*: 1.35–2.98; $P < 0.001$) in participants taking the multivitamin/mineral formulation than in those assigned to the placebo and no statistically significant treatment effects were seen for cortical opacities, moderate VA loss, or cataract surgery.

21.2.1.2. Antioxidant Supplements and Cortical Cataract

None of the six intervention studies that have examined risk for cortical cataract showed an effect of use of antioxidants on this type of opacity *(14,15,20,19,21,22)* (Table 21.2). The trial supplementations included multivitamin/mineral, retinol/zinc, riboflavin/niacin, ascorbic acid/molybdenum, and selenium/α-tocopherol/β-carotene in the Linxian study *(19)*, Centrum® in the CTNS *(14)*, vitamin E in the Australian study and WHS *(20,22)*, and vitamin C, E, and β-carotene in the AREDS and APC *(15,21)*.

21.2.1.3. Antioxidant Supplements and Posterior Subcapsular Cataract

Five intervention studies *(14,15,19,21,22)* examined the risk for posterior subcapsular cataract (PSC) in relation to use of nutrient supplements. Although the number of PSC cataracts was very small, Sperduto et al. found that use of riboflavin and niacin supplements was associated with an over 2-fold increased rate of PSC (*RR*, 2.64; *CI*: 1.31–5.35) in an intervention study of 3,249 Chinese subjects *(19)*. Similarly, the CTNS found PSC events significantly more common (*HR*, 2.00; 95% *CI*: 1.35–2.98; $P < 0.001$) in participants taking Centrum than in those assigned to the placebo *(14)*. However, there was no effect on visual acuity.

21.2.1.4. Antioxidant Supplements and Mixed Cataract

Four intervention studies evaluated the effect on mixed cataract *(14,22,23,24)*. Chylack et al. *(24)* found that use of a supplement containing vitamin C, vitamin E, and β-carotene diminished risk for mixed cataract in US subjects, but not in less well-nourished British subjects (Table 21.2). The US subjects had earlier opacities and were younger (US age: mean 65 years, UK age: mean 68 years). In the Women's Health Study no large beneficial or harmful effect was observed after 2.1 years of use of β-carotene supplements *(23)* or after an average of 9.7 years of use of vitamin E on alternate days *(22)*. The CTNS found that total lens events were less common in participants who took

Centrum than in those assigned to the placebo but treatment had opposite effects on the development or progression of nuclear vs PSC opacities *(14)*.

21.2.1.5. Antioxidant Supplements and Cataract Extraction

Of the five intervention studies *(14,20,23,25,26)*, only the Women Health Study *(26)* found that use of antioxidant supplements is related to diminished risk for cataract extraction. In that study β-carotene supplement use was associated with a 27% reduced rate of cataract extraction (*RR*, 0.73; *CI*: 0.53–1) among male current smokers.

21.2.2. Studies Regarding Vitamin C

Biochemical data indicate potential anticataractogenic and cataractogenic roles for vitamin C, but epidemiological literature is far more supportive of the former *(16,17,24,27–38)*. Vitamin C supplement use and elevated blood levels of vitamin C were inversely associated with at least one type of cataract in many epidemiological studies *(16,17,23,28–38)*. The optimal level of vitamin C intake seems to be more than 130 mg, with intakes of >300 mg providing no added benefit *(16)*. It appears that supplement intake must be maintained for 10 years to obtain this benefit *(16,17,39)*.

21.2.2.1. Vitamin C and Nuclear Cataract

Ten *(16,29–32,35,37,38,40,41)* of the 19 studies *(15,16,18,21,29–32,35,37,38,40–47)* found that increased vitamin C intake, supplement use, or blood levels are related to diminished risk for nuclear cataract (Fig. 21.1). It appears that long-term duration *(16,17,39)* is needed to obtain this benefit and that diabetes might affect this association *(31)*.

21.2.2.2. Vitamin C and Cortical Cataract

Fourteen studies examined relationships between vitamin C and risk for cortical cataract *(15,17,21,28,30,31,35,40–45,47)*. Four studies found that increased vitamin C intake, supplement use, or blood levels were related to diminished risk for cortical cataract *(17,28,35,41)*. Similarly, Jacques and Chylack found low vitamin C and limited fruits or vegetable intake were associated with an increased risk of cortical cataracts *(28)*. Taylor et al. found a role for vitamin C in diminishing the risk of cortical cataracts in women aged <60 years *(17)*. Only one study found that increased vitamin C supplement use was related to increased risk for cortical cataract in non-diabetics and decreased the risk in diabetics *(31)*. The remaining nine studies showed no effect, including the AREDS and APC intervention *(15,21,30,40,42–45,47)*.

21.2.2.3. Vitamin C and PSC Cataract

Twelve studies examined relationships between vitamin C and risk for PSC cataract *(15,17,21,28,30,35,38,40–43,47)*. Three studies found that increased vitamin C intake or elevated blood levels were related to diminished risk for PSC cataract *(28,35,38)*. Consistently, Jacques and Chylack found that low plasma vitamin C, low vitamin C intake, and fewer fruits or vegetable intake were associated with higher odds of PSC cataract *(28)*. Ferrigno et al. found high plasma vitamin C levels were associated with a protective effect on PSC cataract in the cohort in the Italian-American Trial of Nutritional Supplements and Age-Related Cataract *(38)*. The remaining nine studies showed no effect, including the AREDS and APC intervention studies *(15,17,21,30,40–43,47)*.

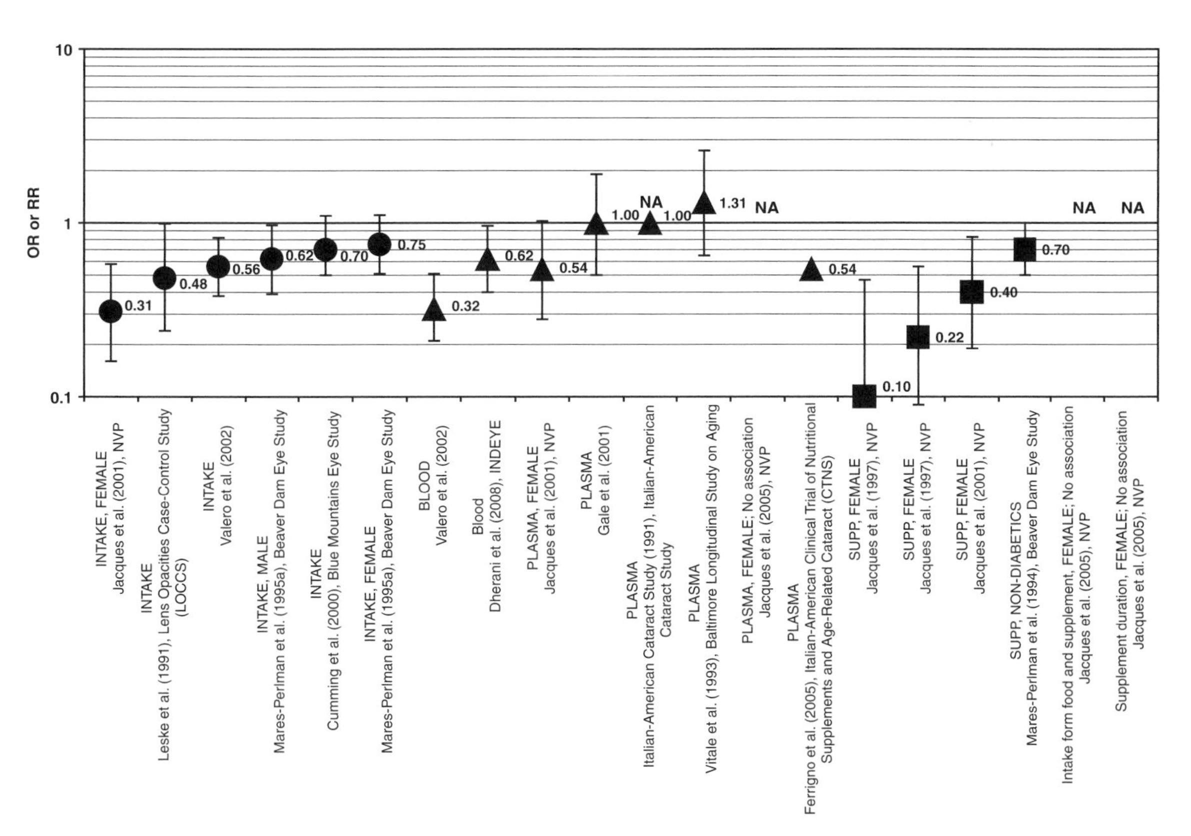

Fig. 21.1. Retrospective observational studies on vitamin C and nuclear cataract. Risk ratio or odds ratio, for high vs low intake (with or without supplements), blood levels, or supplement use, in log scale (•: diet, ▲: blood, ▪: supplement, **NA**: *OR* or *RR* not available)

21.2.2.4. Vitamin C and Mixed Cataract

Nine studies examined relationships between vitamin C and risk for mixed cataract *(24,30,33,41,48–52)*. Five studies found that increased vitamin C supplement use or blood levels were related to diminished risk for mixed cataract, including the REACT intervention *(24,33,41,50,52)*. In contrast, Leske et al. found no effect *(30)*. Tarwadi et al. found lower plasma vitamin C levels in all patients and lowest in low-income patients *(48)*. Nourmohammadi et al. found that patients had a moderately lower ascorbic acid concentration in serum than the control group, which was not statistically significant. Christen et al. followed 35,551 women for a mean of 10 years and found no association between dietary vitamin C (with or without supplements) and cataract *(49)*.

21.2.2.5. Vitamin C and Cataract Extraction

Eight studies examined relationships between vitamin C and risk for cataract extraction *(15,34,36,40,52–55)*. Two studies, Robertson et al. in a retrospective study *(34)* and Hankinson et al. in a prospective study *(36)*, found vitamin C supplementation was associated with a reduced risk of cataract extraction. Yoshida et al. found that a higher vitamin C intake from foods only was significantly associated with a reduced risk of cataract extraction in women (*OR* = 0.64; 95% *CI*: 0.41–0.94), but not in men (*OR* = 0.70; 95% *CI*: 0.44–1.20), in a 5-year prospective population-based analysis using data from a cohort of over 30,000 Japanese residents recruited to the Japan Public Health

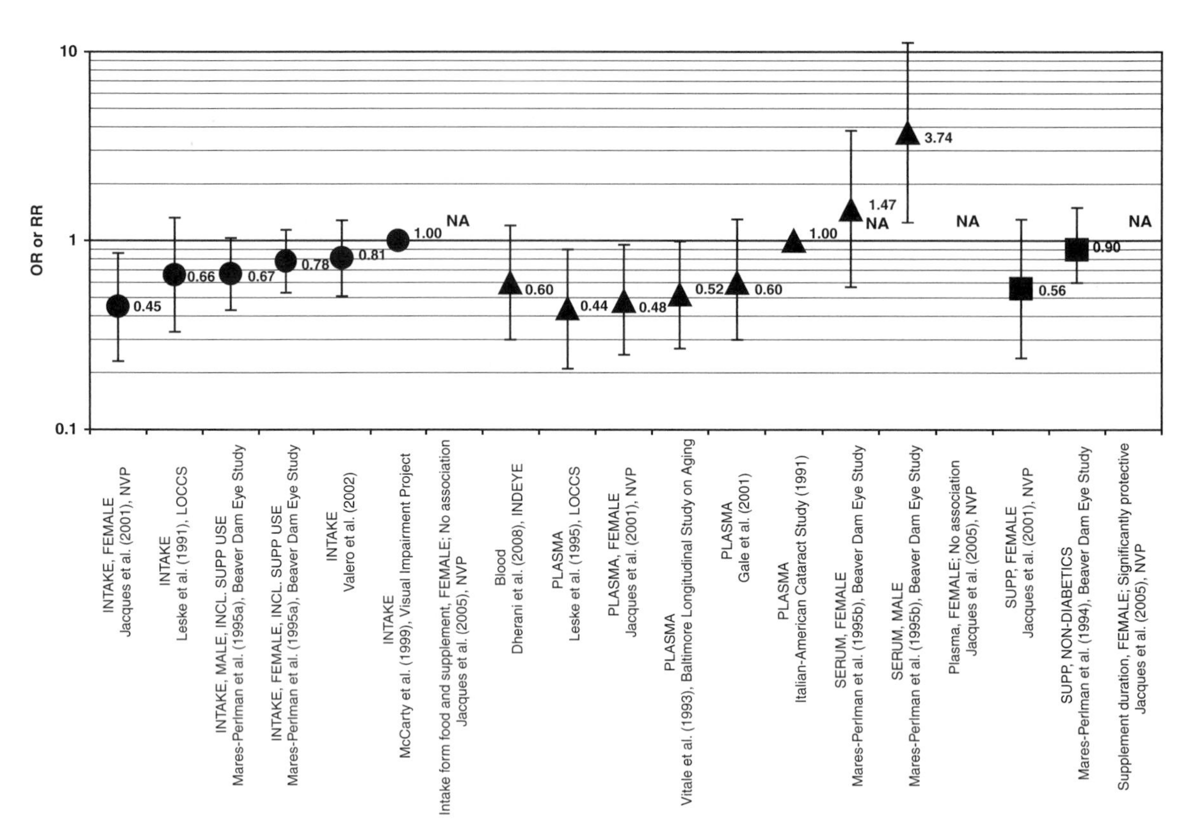

Fig. 21.2. Retrospective observational studies on vitamin E and nuclear cataract. Risk ratio or odds ratio, for high vs low intake (with or without supplements), blood levels, or supplement use, in log scale (•: diet, ▲: blood, ▪: supplement, **NA**: *OR* or *RR* not available)

Center-based Prospective Study (JPHC Study) cohort I *(52)*. The five remaining studies showed no effect, including the AREDS intervention *(15,40,53–55)*.

21.2.3. Studies Regarding Vitamin E

Vitamin E supplements often provide significantly higher levels of this nutrient than can be obtained in the diet. A variety of studies regarding relationships between vitamin E supplement use, dietary intake and plasma levels, and risk for various forms of cataract showed mixed results.

21.2.3.1. Vitamin E and Nuclear Cataract

Twenty-three studies examined relationships between vitamin E and risk for nuclear cataract *(15,16,18,20–22,30–32,35,37,40,41,43–47,56–60)*. Five studies, including our Nutrition and Vision Project, found that increased vitamin E intake, supplement use, or blood levels are related to diminished risk for nuclear cataract or cataract progress *(16,18,45,46,56)*. Only one study found that increased vitamin E blood level in women is related to increased risk for nuclear cataract *(57)*. Of the remaining studies, two studies were suggestive of decreased risk *(31,32)* and the remaining 15 studies showed no effect, including the AREDS, VECAT, APC, and WHS interventions *(15,20–22,30,35,37,40,41,43,44,47,58–60)* (Fig. 21.2).

Jacques et al. found a 55% reduced odds of nuclear cataract (*RR*, 0.45; *CI*: 0.23–0.86) in women with the highest (>90 mg/day) vs lowest (<12 mg/day) vitamin E intake *(16)*. Data from Mares-Perlman et al. were suggestive of a 33% reduced rate of cataract (*RR*, 0.67; *CI*: 0.43–1.03; *P* for trend = 0.03) in men with the highest vs lowest vitamin E intake *(32)*.

Plasma levels of vitamin E were also related to risk of nuclear cataract. Leske et al. found a 56% reduced rate of cataract (*RR*, 0.44; *CI*: 0.21–0.9) in subjects with the highest vs lowest plasma level of vitamin E *(16)*. Jacques et al. found a 52% reduced rate of nuclear opacity (*RR*, 0.48; *CI*: 0.25–0.95) in women with the highest vs lowest vitamin E plasma level *(16)*. Vitale et al. found a 48% reduced rate of nuclear cataract (*RR*, 0.52; *CI*: 0.27–0.99) in subjects with the highest vs lowest plasma vitamin E level in a retrospective study of 671 subjects *(45)*. Similar to the data noted above, but in a prospective study of 764 subjects, Leske et al. found a 42% reduced rate of cataract (*RR*, 0.58; *CI*: 0.36–0.94) in subjects with the highest vs lowest plasma vitamin E level *(46)*. In contrast, Mares-Perlman et al. found a 3.7 times increased rate of nuclear cataract (*RR*, 3.74; *CI*: 1.25–11.2) between men with the highest vs lowest serum vitamin E level *(57)*.

Data from Jacques et al. also found a statistically significant trend (*P* = 0.03) for a relation between the period of vitamin E supplement use and decreased odds for opacities *(16)*. In subsequent work, they found that use of vitamin E supplements for more than 10 years is associated with decreased progress of nuclear opacities in a study of 405 subjects *(18)*. Mares-Perlman et al. were suggestive of a 10% reduced rate of nuclear cataract (*RR*, 0.9; *CI*: 0.6–1.5; *P* for trend = 0.02) in users vs nonusers of vitamin E supplements *(31)*. Leske et al. found a 57% reduced rate of nuclear cataract (*RR*, 0.43; *CI*: 0.19–0.99) in users vs nonusers of vitamin E supplements in a prospective study of 764 subjects *(46)*.

21.2.3.2. Vitamin E and Cortical Cataract

Twenty studies examined relationships between vitamin E and risk for cortical cataract *(15,17,20–22,28,30,31,35,38,40,41,43–45,47,56,57,60,61)*. Three studies found that increased vitamin E intake, supplement use, or blood levels were related to diminished risk for cortical cataract *(30,60,61)*. One study found that high plasma vitamin E levels were associated with increased prevalence of cortical cataract *(38)*. The remaining 16 studies showed no effect, including the AREDS, VECAT, APC, and WHS interventions *(15,17,20–22,28,31,35,40,41,43–45,47,56,57)*.

Leske et al. found a 41% reduced rate of cortical cataract (*RR*, 0.59; *CI*: 0.35–0.99) in subjects with the highest vs lowest vitamin E intake *(30)*. Rouhiainen et al. found a 73% reduced rate of cortical cataract (*RR*, 0.27; *CI*: 0.08–0.83) in subjects with the highest vs lowest plasma vitamin E level in a prospective study of 410 subjects *(61)*. Nadalin et al. found a 56% reduced rate of cortical cataract (*RR*, 0.44; 95% *CI*: 0.25–0.77) in subjects of users vs nonusers of vitamin E supplements in a prospective study of 1,111 subjects *(60)*. But, Ferrigno et al. found that high plasma vitamin E levels were associated with increased prevalence of cortical cataract (*OR* = 1.99; 95% *CI*: 1.02-3.90) *(38)*. They explained that this unexpected association could be due to unadjusted confounding *(38)*.

21.2.3.3. Vitamin E and PSC Cataract

Fourteen studies examined relationships between vitamin E and risk for PSC cataract *(15,17,21,22,28,30,35,38,40,41,43,47,56,58)*. One study of 3,271 subjects in Australia found that increased vitamin E intake was related to 47% increased risk for PSC cataract (*RR*, 1.47; *CI*: 1.04–2.09) *(58)*. Ferrigno et al. in the Italian-American Clinical Trial of Nutritional Supplements and Age-Related Cataract also found that high plasma vitamin E levels were associated with increased prevalence of PSC (*OR* = 3.27; 95% *CI*: 1.34–7.96). The remaining 12 studies showed no effect, including the AREDS, APC, and WHS interventions *(15,17,21,22,28,30,35,40,41,43,47,56)*.

21.2.3.4. Vitamin E and Mixed Cataract

Eight studies examined relationships between vitamin E and risk for mixed cataract *(22,24,30,38,41,49,51,56)*. Five studies found that increased vitamin E intake or supplement use was related to diminished risk for mixed cataract, including the REACT intervention *(24,30,41,49,51)*. One study found that high plasma vitamin E levels were associated with increased prevalence of mixed cataract *(38)*. The remaining two studies showed no effect *(22,56)* (*see* Tables 21.1 and 21.2 for LOCCS and WHS intervention data). Leske et al. found a 42% reduced rate of mixed cataract (*RR*, 0.58; *CI*: 0.37–0.93) in subjects with the highest vs lowest vitamin E intake *(30)*. Chylack et al. found a reduced rate of mixed cataract in US subjects but not in the less than well-nourished UK subjects in the REACT intervention of 445 subjects *(24)* (*see* Tables 21.1 and 21.2). They concluded that daily use of vitamins C, E, and β-carotene for 3 years produced a small deceleration in progression of age-related mixed cataract.

Nourmohammadi et al. in a small-scale study (88 patients and healthy controls) found that the mean serum concentration of α-tocopherol in patients (9.16 ± 2.53 μg/mL) with cataract was lower than in the control group ($P < 0.001$) *(49)*. Dherani et al., in a cross-sectional study of people aged ≥50 years identified from a household enumeration of 11 randomly sampled villages in north India, found that higher blood level of α-tocopherol is associated with decreased odds of cataract (*OR* for the highest (≥20.84 μmol/L) compared with the lowest (<17.02 μmol/L) tertile was 0.58, 95% *CI*: 0.36–0.94; *P* for trend = 0.04) *(41)*. However, in analysis of continuous data, no significant inverse association was found *(41)*. Christen et al., in a prospective study of 35,551 women followed for a mean of 10 years in the WHS, found that comparing women in the extreme quintiles, the multivariate *RR* of cataract was 0.86 (95% *CI*: 0.74–1.00; test for trend, $P = 0.03$) for vitamin E from food plus supplements *(51)*. However, the results from foods only were not significant (*RR*, 0.92; 95% *CI*: 0.80–1.06; test for trend, $P = 0.39$) *(51)*. Ferrigno et al. in the Italian-American Clinical Trial of Nutritional Supplements and Age-Related Cataract also found that high plasma vitamin E levels were associated with increased prevalence of mixed cataract ($OR = 1.86$; 95% *CI*: 1.08–3.18) *(38)*.

21.2.3.5. Vitamin E and Cataract Extraction

Ten studies examined relationships between vitamin E and risk for cataract extraction *(15,20,25,34,36,40,53–55,62)*. Tavani et al. found a 50% reduced rate of cataract extraction (*RR*, 0.5; *CI*: 0.3–1.0) in subjects with the highest vs lowest vitamin E intake in a retrospective study of 913 subjects *(53)*. Robertson et al. found a 56% reduced rate of cataract extraction (*RR*, 0.44; *CI*: 0.24–0.77) in subjects who used vitamin E supplements vs those who did not in a retrospective study of 304 subjects *(34)*. The remaining eight studies showed no effect, including the AREDS, VECAT, and ATBC interventions *(15,20,25,36,40,54,55,62)*. Tan et al., following 2,464 persons aged ≥49 years for either 5 or 10 years in the Blue Mountain Eye Study, found that the third ($OR = 1.60$; 95% *CI*: 1.04–2.45; *P* for trend = 0.127) and highest ($OR = 1.55$; 95% *CI*: 1.02–2.38; *P* for trend = 0.127) quintile of vitamin E intake (foods plus supplements) at baseline compared with the lowest quintile was associated with an increased risk for cataract surgery *(40)*. However, the authors interpreted the results as a non-significant association.

21.2.4. Studies Regarding Carotenoids

The major carotenoids which are found in the lens and have been related to lens health are lutein, zeaxanthin, and riboflavin. Levels of β-carotene, the best known carotenoid because of its importance as a vitamin A precursor, are vanishingly low in the lens. Lutein and zeaxanthin are important carotenoid components of the human diet, and several investigations suggest that elevated intake of foods rich in lutein, such as spinach, is related to decreased risk for cataract. Previous study results are mixed for lutein and zeaxanthin as well as for other carotenoids.

21.2.4.1. Total carotenoids

21.2.4.1.1. Total Carotenoids and Nuclear Cataract. Four studies examined relationships between total carotenoids and risk for nuclear cataract *(16,18,57,63)*. One study found that increased serum levels of total carotenoids in women are related to a 3.95-fold increased rate of cataract (*RR*, 3.95; *CI*: 1.65–9.47) *(57)* but, curiously, to a 45% reduced (*RR*, 0.55; *CI*: 0.2–1.49; *P* for trend = 0.002) risk in men. The three remaining studies showed no effect *(16,18,63)*.

21.2.4.1.2. Total Carotenoids and Cortical Cataract. Three studies examined relationships between total carotenoids and risk for cortical cataract *(17,28,57)*. Of these, one study found that increased blood total carotenoid levels are related to 90% diminished risk for cortical cataract *(28)*. Of the remaining two studies, one study was suggestive of an increased risk with high carotenoid blood levels in women *(57)* and the other showed no effect *(17)*.

21.2.4.1.3. Total Carotenoids and PSC Cataract. All three studies that examined relationships between total carotenoids and risk for PSC cataract found statistically significant data to support the hypothesis that increased blood levels of total carotenoids or intake are related to diminished risk for PSC cataract *(17,28,63)*.

Taylor et al. found an 81% reduced rate of PSC cataract (*RR*, 0.19; *CI*: 0.05–0.68; *P* for trend = 0.01) in women who never smoked and who had the highest vs lowest carotenoid intake *(17)*. In the same study, they also found a 59% reduced rate of PSC cataract (*RR*, 0.41; *CI*: 0.17–0.99) in women with the highest vs lowest plasma level of carotenoids *(17)*. Chasan-Taber et al. *(63)* found a 31% reduced rate of PSC cataract (*RR*, 0.69; *CI*: 0.49–0.98) between women with the highest vs lowest carotenoid intake in a prospective study of 77,466 subjects.

21.2.4.1.4. Total Carotenoids and Cataract Extraction. Four studies examined relationships between total carotenoids and risk for cataract extraction *(36,54,63,64)*. Hankinson et al. found a 27% reduced rate of cataract extraction (*RR*, 0.73; *CI*: 0.55–0.97) between women with the highest vs lowest carotenoid intake in a prospective study of 50,828 subjects *(36)*, and Chasan-Taber et al. found a 15% reduced rate of extraction (*RR*, 0.85; *CI*: 0.72–1) between women with the highest and lowest carotenoid intake in a prospective study of 77,466 subjects *(63)*. Brown et al. *(64)* were suggestive of a 15% reduced rate of extraction (*RR*, 0.85; *CI*: 0.68–1.07; *P* for trend = 0.05) between men with the highest and lowest carotenoid intake in a prospective study of 36,644 subjects *(54)*.

21.2.4.2. Lutein and zeaxanthin

21.2.4.2.1. Lutein/Zeaxanthin and Nuclear Cataract. Ten studies examined relationships between lutein/zeaxanthin (LZ) and risk for nuclear cataract *(16,18,32,37,41,43,57,59,63,65)*. Jacques et al. in the NVP found a 48% reduced rate of nuclear cataract (*RR*, 0.52; *CI*: 0.29–0.91) between women with the highest and lowest LZ intake *(16)*. The Beaver Dam Eye Study found a 60% reduced rate of nuclear cataract (*RR*, 0.4; *CI*: 0.2–0.8) between subjects (age < 65) with the highest and lowest lutein intake and that persons in the highest quintile of lutein intake in the distant past were half as likely (*RR*, 0.5; *CI*: 0.3–0.8, *P* for trend = 0.002) to have an incident of nuclear cataract as persons in the lowest quintile of intake in a prospective study of 1,354 subjects *(37)*. Similarly, there was a suggestion of a 27% reduced rate of nuclear cataract (*RR*, 0.73; *CI*: 0.5–1.06; *P* for trend = 0.02) in women with the highest vs lowest lutein intake in a retrospective study of 1,919 subjects *(32)*. In contrast, Mares-Perlman et al. found a 4.1-fold increased rate of nuclear cataract (*RR*, 4.09; *CI*:

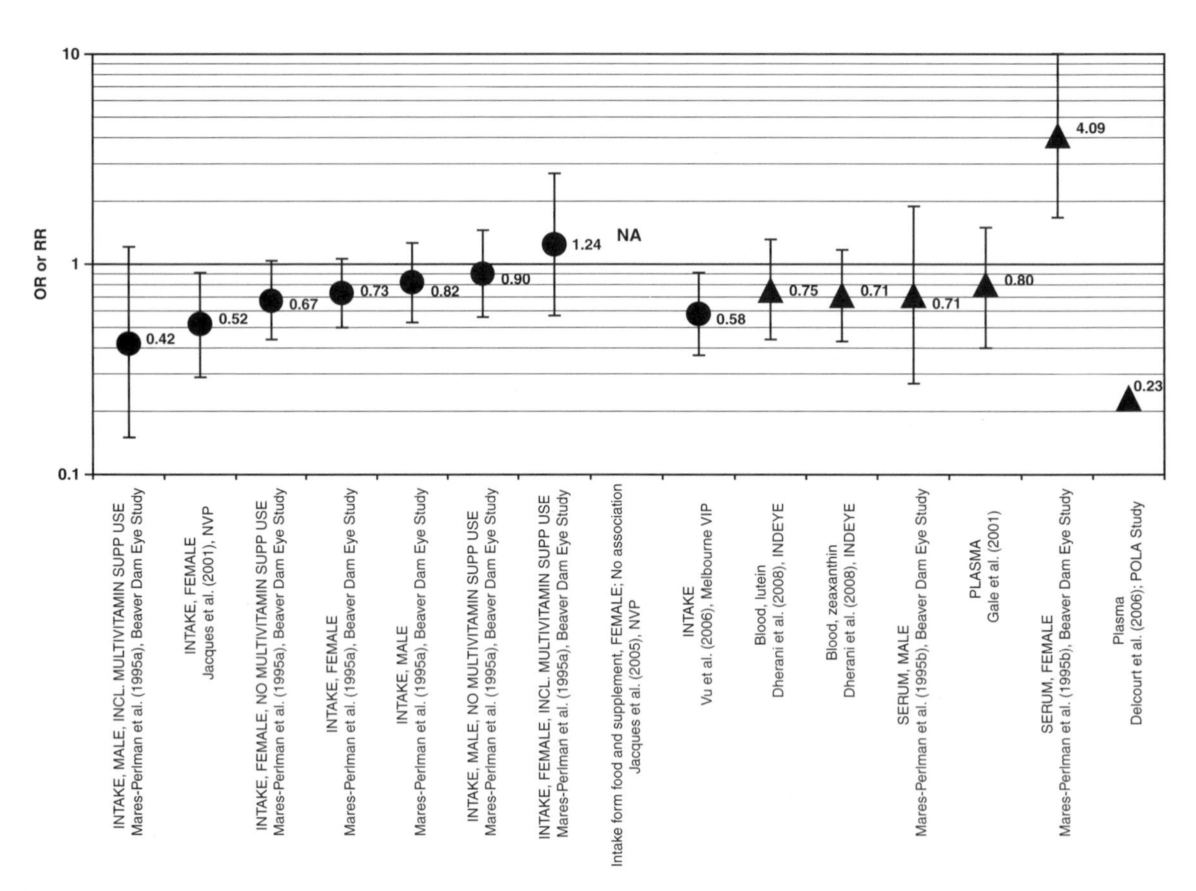

Fig. 21.3. Retrospective observational studies on lutein/zeaxanthin and nuclear cataract. Risk ratio or odds ratio, for high vs low intake (with or without supplements), blood levels, or supplement use, in log scale (•: diet, ▲: blood, ▪: supplement, **NA**: *OR* or *RR* not available)

1.67–10.03; *P* for trend = 0.006) in women with the highest vs lowest serum level of lutein in a retrospective study of 400 subjects *(57)*. Vu et al. in a cross-sectional analysis of 2,322 subjects from the Melboure Visual Impairment Project found that the *OR*s for nuclear cataract were 0.67 (0.46–0.96) and 0.60 (0.40–0.90) for every 1-mg increase in crude and energy-adjusted daily LZ intake, respectively *(65)*. The *OR*s (95% *CI*) for those in the top quintile of crude LZ intake was 0.58 (0.37–0.92; *P* = 0.023 for trend), and it was 0.64 (0.40–1.03) for energy-adjusted LZ intake (*P* = 0.018 for trend) *(65)*. The remaining five studies showed no association *(18,41,43,59,63)* (Fig. 21.3).

21.2.4.2.2. Lutein and Zeaxanthin and Cortical Cataract. Five studies examined relationships between LZ blood levels or intake and cortical cataract *(17,41,43,57,65)*. Three studies showed no effect *(17,43,65)*. Mares-Perlman et al. were suggestive of a 4.8-fold increased rate of cortical cataract (*RR*, 4.84; *CI*: 0.83–28.1; *P* for trend = 0.03) in men with the highest vs lowest lutein serum level *(57)*. Dherani et al., in a cross-sectional study of people aged ≥50 years identified from a household enumeration of 11 randomly sampled villages in north India, found that, in analysis of continuous data, a significant inverse association with cortical cataract was found for blood lutein level, but not for zeaxanthin *(41)*. However, no significant results were found in tertile analyses for either lutein or zeaxanthin *(41)*.

21.2.4.2.3. Lutein and Zeaxanthin and PSC Cataract. Five studies examined relationships between LZ and risk for PSC cataract *(17,41,43,63,65)*. Gale et al. found a 50% reduced rate of PSC cataract (*RR*, 0.5; *CI*: 0.2–1; *P* for trend = 0.012) in subjects with the highest vs lowest plasma level of lutein in a retrospective study of 372 subjects *(43)* and Chasan-Taber et al. found a 32% reduced rate of PSC cataract (*RR*, 0.68; *CI*: 0.48–0.97) in women with the highest vs lowest LZ intake *(63)*. The remaining three studies showed no effect *(17,41,65)*.

21.2.4.2.4. Lutein and Zeaxanthin and Mixed Cataract. Three studies examined relationships between LZ and risk for mixed cataract *(41,50,51)*. In a cross-sectional study, Rodríguez-Rodríguez et al. investigated the relationship between the intake of antioxidant nutrients monitored for 7 consecutive days using a "precise individual weighing" method and cataracts in 177 institutionalized elderly people (61 men and 116 women) aged ≥65 years *(50)*. The results showed that subjects who consumed >3.29 mg/day (95 percentile) of lutein were less likely to have cataracts (*OR* = 0.086; *CI*: 0.007–1.084; *P* < 0.05) than those whose consumption was <0.256 mg/day (5 percentile). In men, high intakes of zeaxanthin seemed to provide a protective effect against the problem (*OR* = 0.96; *CI*: 0.91–0.99; *P* < 0.05) *(50)*. Dherani et al., in analysis of continuous data, found significant inverse associations with mixed cataract for both lutein and zeaxanthin levels in blood *(41)*. However, in tertile analyses, a significant association was found for zeaxanthin but not for lutein *(41)*. Christen et al., in a prospective study of 35,551 women followed for a mean of 10 years in the WHS, found that comparing women in the extreme quintiles, the multivariate *RR*of cataract was 0.82 (95% *CI*: 0.71–0.95; test for trend, *P* = 0.045) for LZ intake *(51)*.

21.2.4.2.5. Lutein and Zeaxanthin and Cataract Extraction. Two studies examined relationships between lutein/zeaxanthin intake and risk for cataract extraction *(63,64)*. Data from Brown et al. were suggestive of a 19% reduced rate of extraction (*RR*, 0.81; *CI*: 0.65–1.01; *P* for trend = 0.03) in men with the highest vs lowest LZ intake in a prospective study of 36,644 subjects *(64)*. But, the other study showed no effect *(63)*.

21.2.4.3. β-Carotene

21.2.4.3.1. β-Carotene and Nuclear Cataract. It is likely that β-carotene levels in the lens are below detection limits, puzzling the data which relate β-carotene to cataract. Of the 14 studies that examined relationships between β-carotene and risk for nuclear cataract *(15,16,19,21,32,37,40,41,43,45,57,59,63,66)*, only three found statistically significant data to support the hypothesis that increased β-carotene intake or supplement use is related to diminished risk for nuclear cataract *(16,32,41)*. Jacques et al. found a 48% reduced rate of nuclear cataract (*RR*, 0.52; *CI*: 0.28–0.97) in women with the highest vs lowest β-carotene intake *(16)*. Mares-Perlman et al. found a 71% reduced rate of nuclear cataract (*RR*, 0.29; *CI*: 0.1–0.84) in men with the highest vs lowest β-carotene intake *(32)*but a 2.8-fold higher rate of nuclear cataract (*RR*, 2.8; *CI*:1.21–6.48) in women with the highest vs lowest serum β-carotene level *(57)*. Dherani et al., in analysis of continuous data, found a significant inverse association with nuclear cataract for blood level of β-carotene; however, in tertile analyses, no significant association was found *(41)*. Of the remaining 11 studies, one study was suggestive of diminished risk (*P* for trend = 0.033) *(43)* and the others showed no effect, including the AREDS and Linxian Intervention Studies *(15,19,21,37,40,45,57,59,63,66)*.

21.2.4.3.2. β-Carotene and Cortical Cataract. Nine studies examined relationships between β-carotene and risk for cortical cataract *(15,17,19,21,40,41,43,45,57)*. Data from Mares-Perlman et al. were suggestive of a 72% reduced rate of cortical cataract (*RR*, 0.28; *CI*: 0.06–1.24; *P* for trend = 0.05) in men but opposing effects in women with the highest vs lowest serum β-carotene level *(57)*. The remaining eight studies showed no effect, including the AREDS and Linxian Intervention Study *(15,17,19,21,40,41,43,45)*.

21.2.4.3.3. β-Carotene and PSC Cataract. Eight studies examined relationships between β-carotene and risk for PSC cataract *(15,17,19,21,40,41,43,63)*. Taylor et al. found a 72% reduced rate of cataract (*RR*, 0.28; *CI*: 0.08–0.96; *P* for trend = 0.02) in women who never smoked and who had the highest vs lowest β-carotene intake *(17)*. This is corroborated by observations from Chasan-Taber et al. that suggest a 32% reduced rate of PSC cataract (*RR*, 0.68; *CI*: 0.48–0.97) in women with the highest vs lowest β-carotene intake in a prospective study of 77,466 subjects *(63)*. The remaining six studies showed no effect, including the AREDS and Linxian Intervention Study *(15,19,21,40,41,43)*.

21.2.4.3.4. β-Carotene and Mixed Cataract. Five studies indicate relationships between β-carotene and risk for mixed cataract *(23,24,41,48,51)*. Chylack et al. found that use of supplements containing β-carotene (along with vitamins C and E) diminished risk for mixed cataract in US subjects but not the less well-nourished British subjects in an intervention study of 445 subjects *(24)*. Data from Tarwadi and Agte suggested lower plasma levels of β-carotene in all patients and lowest levels of β-carotene in low-income patients *(48)*. Christen et al., in the β-carotene component of the trial in the Women's Health Study, found that 2 years of β-carotene treatment has no large beneficial or harmful effect on the development of cataract during the treatment period *(23)*. Similarly, in a prospective study of 35,551 women followed for a mean of 10 years in the WHS, they found no significant associations between β-carotene intake from foods alone or from foods plus supplements and risk for mixed cataract *(51)*. Dherani et al., in analysis of continuous data, found a significant inverse association with mixed cataract for blood levels of β-carotene. However, in tertile analyses, no significant association was found *(41)*.

21.2.4.3.5. β-Carotene and Cataract Extraction. Eight studies examined relationships between β-carotene and risk for cataract extraction *(15,25,26,40,53,62–64)*. Christen et al. found a 27% reduced rate of extraction (*RR*, 0.73; *CI*: 0.53–1) in men who are current smokers and who did vs did not take the β-carotene supplement 50 mg on alternate days in a prospective study of 20,968 physicians *(26)*. The seven remaining studies showed no effect, including the AREDS and ATBC intervention studies *(15,25,40,53,62–64)*.

21.2.4.4. α-Carotene

21.2.4.4.1. α-Carotene and Nuclear Cataract. Nine studies examined relationships between α-carotene and risk for nuclear cataract *(16,18,32,37,41,43,57,59,63)*. Gale et al. found a 50% reduced rate of nuclear cataract (*RR*, 0.5; *CI*: 0.3–0.9; *P* for trend = 0.006) in subjects with the highest vs lowest plasma α-carotene levels *(43)*. Mares-Perlman et al. found a 39% reduced rate of nuclear cataract (*RR*, 0.61; *CI*: 0.39–0.95; *P* for trend = 0.04) in men but not in women with the highest vs lowest α-carotene intake *(32)*but found a 2.62 times increased rate of cataract (*RR*, 2.62; *CI*: 1.12–6.13) in women but not in men with the highest vs lowest serum α-carotene level in a retrospective study of 400 subjects *(57)*. Dherani et al. found a significant inverse association with nuclear cataract for the second tertile compared with the first tertile of blood level of α-carotene (*OR* = 0.63; 95% *CI*: 0.43, 0.93, *P* for trend = 0.2) in the Indian cohort. However, results for the third tertile or analyses of continuous data showed no significant associations *(41)*. The remaining five studies showed no effect *(16,18,37,59,63)*.

21.2.4.4.2. α-Carotene and Cortical Cataract. Four studies have examined relationships between α-carotene and risk for cortical cataract *(17,41,43,57)*. Dherani et al. found a significant inverse association with cortical cataract for the second tertile compared with the first tertile of blood level of α-carotene (*OR* = 0.45; 95% *CI*: 0.27, 0.75, *P* for trend = 0.7). However, results for the third tertile or analyses of continuous data showed no significant associations *(41)*. The remaining three studies found no statistically significant data *(17,43,57)*.

21.2.4.4.3. α-Carotene and PSC Cataract. Four studies have examined relationships between α-carotene and risk for PSC cataract *(17,41,43,63)*. Data from the NVP were suggestive of a 71% reduced rate of cataract (*RR*, 0.29; *CI*: 0.08–1.05; *P* for trend = 0.02) in women with the highest vs lowest α-carotene intake who never smoked *(17)*. The remaining three studies found no effect *(41,43,63)*.

21.2.4.4.4. α-Carotene and Mixed Cataract. One of the two studies *(41,51)* that examined relationships between α-carotene and risk for mixed cataract found statistically significant associations *(41)*. Dherani et al. found a significant inverse association with mixed cataract for the third tertile compared with the first tertile of blood level of α-carotene (*OR* = 0.69; 95% CI: 0.50, 0.95, *P* for trend = 0.05) *(41)*.

21.2.4.4.5. α-Carotene and Cataract Extraction. Of the two studies that examined relationships between α-carotene intake and risk for cataract extraction neither found statistically significant associations *(63,64)*.

21.2.4.5. Retinol/vitamin A

21.2.4.5.1. Retinol/Vitamin A and Nuclear Cataract. Four studies have examined relationships between retinol/vitamin A and risk for nuclear cataract *(40,41,47,67)*. Delcourt et al. found that higher plasma retinol levels were associated with decreased risk for nuclear cataract (*RR*, 0.75; *CI*: 0.66–0.86) in the cohort of the Pathologies Oculaires Liees a l'Age (POLA) study in southern France *(67)*. Tan et al., following 2,464 persons aged ≥49 years for either 5 or 10 years in the Blue Mountain Eye Study, found that the highest quintile of vitamin A intake (foods plus supplements) at baseline compared with the lowest quintile was associated with a suggestively reduced risk for nuclear cataract (*OR* = 0.66; 95% CI: 0.42–1.03; *P* for trend = 0.056) *(40)*. However, the authors interpreted the results as a non-significant association. Dherani et al., in both tertile and continuous analyses, found significant inverse associations with nuclear cataract for blood levels of retinol *(41)*. Klein et al. found no evidence of an association between vitamin A supplement use and incident nuclear cataract risk in a prospective study of the Beaver Dam Eye Study *(47)*.

21.2.4.5.2. Retinol/Vitamin A and Cortical Cataract. Three studies have examined relationships between retinol/vitamin A and risk for cortical cataract *(40,41,47)*. Klein et al. found a small protective effect for cortical cataract by vitamin A supplement use (*OR* = 0.42; 95% CI: 0.24, 0.73) in a prospective study of the Beaver Dam Eye Study *(47)*. The remaining two studies found no statistically significant data *(40,41)*.

21.2.4.5.3. Retinol/Vitamin A and PSC Cataract. None of the three studies that examined relationships between retinol/vitamin A and risk for PSC cataract found a significant association *(40,41,47)*.

21.2.4.5.4. Retinol/Vitamin A and Mixed Cataract. Three studies examined relationships between retinol/vitamin A and risk for cortical cataract *(41,49,67)*. Delcourt et al. found that higher plasma retinol levels were associated with decreased risk for mixed (*RR*, 0.75; *CI*: 0.66–0.86) cataract in the POLA cohort study in southern France *(67)*. Dherani et al., in both tertile and continuous analyses, found significant inverse associations with mixed cataract for blood level of retinol *(41)*. Nourmohammadi et al. did not find a significant association between serum retinol levels and risk of mixed cataract *(49)*.

21.2.4.5.5. Retinol/Vitamin A and Cataract Extraction. Two studies examined relationships between retinol/vitamin A and risk for cataract surgery *(40,67)*. Delcourt et al. found that higher plasma retinol levels were associated with decreased risk for cataract extraction (*RR*, 0.75; *CI*: 0.66–0.86) in the POLA cohort study in southern France *(67)*. However, Tan et al., following 2,464 persons aged ≥49 years for either 5 or 10 years in the Blue Mountain Eye Study, found no association between vitamin A intake (foods plus supplements) at baseline and risk for cataract surgery *(40)*.

21.2.4.6. Riboflavin

21.2.4.6.1. Riboflavin and Nuclear Cataract. Eight studies examined relationships between riboflavin and risk for nuclear cataract *(16,19,30–32,42,68)*. Five studies found statistically significant data to support the hypothesis that increased riboflavin intake or supplement use is related to diminished risk for nuclear cataract, including the Linxian Intervention Study *(16,19,32,42)*. One study was suggestive of decreased risk in non-diabetic subjects (*RR*, 0.8; *CI*: 0.5–1.3; *P* for trend = 0.02) *(31)*. The two remaining studies found no effect *(30,68)*.

Jacques et al. found a 63% reduced rate of cataract (*RR*, 0.37; *CI*: 0.19–0.73; *P* for trend = 0.03) in women with the highest vs lowest riboflavin intake *(16)*. Cumming et al. found a 50% reduced rate of nuclear cataract (*RR*, 0.5; *CI*: 0.3–0.9; *P* for trend = 0.01) in subjects with the highest vs lowest riboflavin intake *(42)*, and Mares-Perlman et al. found a 44 and 33% reduced rate of nuclear cataract (*RR*, 0.56; *CI*: 0.36–0.87; *P* for trend = 0.009) in women (*RR*, 0.67; *CI*: 0.46–0.98; *P* for trend = 0.026), respectively, with the highest vs lowest riboflavin intake *(32)*. Mares-Perlman et al. also found suggestions of a 20% reduced rate of cataract (*RR*, 0.8; *CI*: 0.5–1.3; *P* for trend = 0.02) in non-diabetics with the highest vs lowest riboflavin supplements *(31)*. Sperduto et al. found a 55% reduced rate of nuclear cataract (*RR*, 0.45; *CI*: 0.31–0.64) in 65- to 74-year-old poorly nourished Chinese who took supplements containing riboflavin and niacin *(19)*. In all the subjects in this portion of the Linxian study, there was a 41% reduction in risk for nuclear cataract (*RR*, 0.59; *CI*: 0.45–0.79) in users of this supplement *(19)*. Jacques et al. found that higher riboflavin intake may reduce the progression of age-related lens opacification *(18)*.

21.2.4.6.2. Riboflavin and Cortical Cataract. Five studies examined relationships between riboflavin and risk for cortical cataract *(17,19,31,30,42)*. Leske et al. found a 41% reduced rate of cortical cataract (*RR*, 0.59; *CI*: 0.36–0.97) in subjects with the highest vs lowest riboflavin intake *(30)*. Cumming et al. found a 30% reduced rate of cortical cataract (*RR*, 0.7; *CI*: 0.5–1) in subjects with the highest vs lowest riboflavin intake *(42)*. In contrast, Mares-Perlman et al. were suggestive of a 70% increased rate of cortical cataract (*RR*, 1.7; *CI*: 0.9–3.1; *P* for trend = 0.05) in non-diabetics with the highest vs lowest riboflavin supplements *(31)*. The remaining two studies found no effect, including the Linxian Intervention Study *(17,19)*.

21.2.4.6.3. Riboflavin and PSC Cataract. Four studies examined relationships between riboflavin and risk for PSC cataract *(17,19,30,42)*. In the three observational studies no effect was observed *(17,30,42)*. This is contrast with the Linxian Intervention Study which found a 2.6 times increased rate of PSC cataract (*RR*, 2.64; *CI*: 1.31–5.35) *(19)* (Tables 21.1 and 21.2).

21.2.4.6.4. Riboflavin and Mixed Cataract. Two studies examined relationships between riboflavin intake and risk for mixed cataract *(30,48)*. One did not find a statistically significant association *(30)*. The other study found that plasma levels of riboflavin were subnormal in all patients and lowest in low-income patients *(48)*.

21.2.5. Studies Regarding Multivitamins and Antioxidant Indices

In addition to commercial over the counter multivitamin supplement use, this section includes results from studies which used “antioxidant indices” that approximate the combined effects of the multiple antioxidants that are contained in the diet on risk for cataract. However, since single nutrients appear to have strong influences on the indices, its usefulness has been questioned.

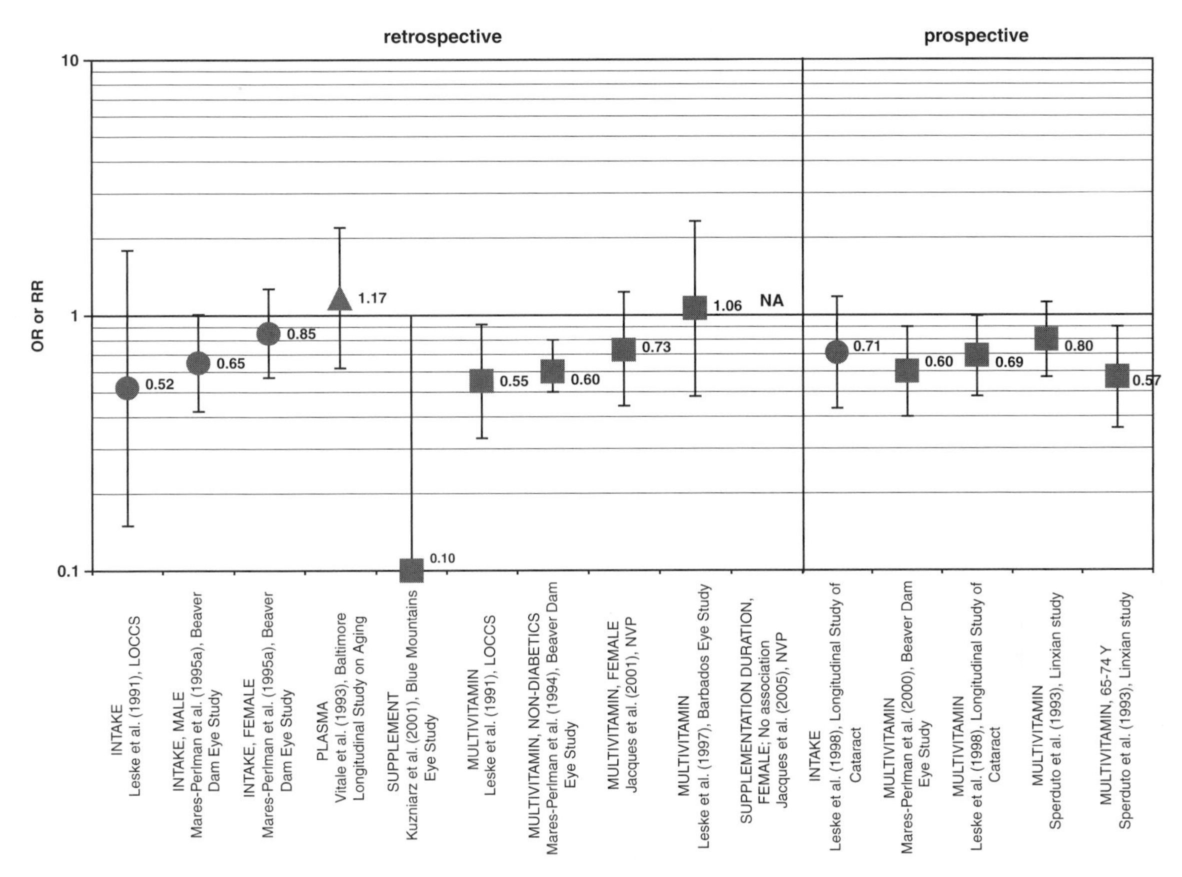

Fig. 21.4. Retrospective observational studies on multivitamins/antioxidant index and nuclear cataract. Risk ratio or odds ratio, for high vs low intake (with or without supplements), blood levels, or supplement use, in log scale (•: diet, ▲: blood, ▪: supplement, **NA**: *OR* or *RR* not available)

21.2.5.1. Multivitamins, Antioxidant Indices, and Nuclear Cataract

Sixteen studies examined relationships between multivitamins or antioxidant index and risk for nuclear cataract *(14,16,19,21,30–32,39,40,45–47,68–70)*. Nine studies found statistically significant data to support the hypothesis that increased multivitamin supplement use or antioxidant index intake is related to diminished risk for nuclear cataract, including the Linxian Intervention Study *(14,16,19,30–32,39,40,45,46,68–70)*. Kuzniarz et al. found a 90% reduced rate of nuclear cataract (*RR*, 0.1; *CI*: 0–1; *P* for trend = 0.02) in subjects who did vs did not use multivitamin supplements *(68)*. Leske et al. *(30)* found a 45% reduced rate of nuclear cataract (*RR*, 0.55; *CI*: 0.33–0.92) in users of multivitamin supplements in the Lens Opacities Case–Control Study and also found a 31% reduced rate of nuclear cataract (*RR*, 0.69; *CI*: 0.48–0.99) in users of multivitamin supplements in a prospective study *(46)*. Mares-Perlman et al. found a 40% reduced rate of nuclear cataract (*RR*, 0.6; *CI*: 0.5–0.8) in non-diabetics with the highest vs lowest multivitamin supplements *(31)* and also found a 40% reduced rate of cataract (*RR*, 0.6; *CI*: 0.4–0.9) in subjects who did vs did not use multivitamins for more than 10 years *(39)*. Sperduto et al. found a 43% reduced rate of nuclear cataract (*RR*, 0.57; *CI*: 0.36–0.9) in 65- to 74-year-old Chinese people who used multivitamins for 6 years *(21)* (Tables 21.1, 21.2 and Fig.21.4). In addition, data from Mares-Perlman et al. were suggestive of a 35% (*RR*, 0.65; *CI*: 0.42–1.01) reduced risk of nuclear cataract in the highest vs lowest quintile of multivitamin supplement use (Mares-Perlman et al.). The AREDS in a prospective study (median follow-up, 6.3 years) showed that Centrum use, adjusted for propensity score and other covariates, was protective for nuclear opacity events ($OR = 0.75$; 95% *CI*: 0.61–0.91, $P = 0.004$) *(70)*. The findings were corroborated by a randomized, double-masked, placebo-controlled clinical trial, the CTNS. In the trial 1,020 participants, aged 55–75 years, were followed for an average of 9.0 years. Nuclear events were significantly less common (*HR*, 0.66; 95% *CI*: 0.50–0.88; $P = 0.004$). However, in that trial PSC events were significantly more common (*HR*, 2.00; 95% *CI*: 1.35–2.98; $P < 0.001$) in participants taking Centrum than in those assigned to the placebo *(14)*. Tan et al., following 2,464 persons aged ≥49 years for either 5 or 10 years in the Blue Mountain Eye Study, found that an above-median intake of combined antioxidants (vitamins C and E, β-carotene, and zinc from foods plus supplements) at baseline was associated with a reduced risk of incident nuclear cataract ($OR = 0.51$; 95% *CI*: 0.34, 0.76) *(40)*. The remaining six studies showed no effect *(16,21,32,45,47,69)*.

21.2.5.2. Multivitamins, Antioxidant Indices, and Cortical Cataract

Fourteen studies examined relationships between use of multivitamin supplements or antioxidant indices and risk for cortical cataract *(14,17,19,21,28,30,31,39,40,45,47,69–71)*. Five studies found statistically significant data to support the hypothesis that increased multivitamin supplement use or antioxidant index intake or blood levels is related to diminished risk for cortical cataract *(28,30,39,47,71)*. One study found statistically significant data indicating that increased multivitamins supplement use is related to increased risk for cortical cataract *(31)*. The remaining eight studies showed no effect, including the Linxian Intervention Study *(14,17,19,21,40,45,69,70)*.

Jacques and Chylack found an 84% reduced rate of cortical cataract ($OR = 0.16$) in subjects with the highest vs lowest antioxidant index in a retrospective study of 112 subjects *(28)*. Leske et al. found a 58% reduced rate of cortical cataract (*RR*, 0.42; *CI*: 0.18–0.97) in subjects with the highest vs lowest antioxidant intake index and found a 48% reduced rate of cataract (*RR*, 0.52; *CI*: 0.36–0.75) in subjects who did vs did not use multivitamin supplements *(30)*. In contrast, Mares-Perlman et al. *(31)* found a 60% increased rate of cortical cataract (*RR*, 1.6; *CI*: 1.1–2.1) in non-diabetics with the highest vs lowest multivitamin supplements in a retrospective study of 1,862 subjects. However, Mares-Perlman et al. *(71)* found a 30% reduced rate of cortical cataract incidence (*RR*, 0.7; *CI*: 0.5–1) and a 20% reduced rate of cataract progress (*RR*, 0.8; *CI*: 0.6–1) in users vs nonusers of multivitamin supplements in a prospective study of 3,220 subjects. In a more recent work, Mares-Perlman et al. *(39)* found a 60% reduced rate of cortical cataract (*RR*, 0.4; *CI*: 0.2–0.8; *P* for trend = 0.002) in subjects who took multivitamin supplements for more than 10 years in a prospective study of 2,434 subjects. Klein et al. found a small protective effect for cortical cataracts by multivitamins on incident nuclear cataract risk ($OR = 0.77$; 95% CI: 0.62, 0.95) in a prospective study of the Beaver Dam Eye Study *(47)*.

21.2.5.3. Multivitamins and Antioxidant Indices, and PSC Cataract

Twelve studies examined relationships between multivitamins or antioxidant index and risk for PSC cataract *(14,17,19,21,28,30,31,33,39,40,47,70)*. Three studies found statistically significant data to support the hypothesis that increased multivitamin supplement use or antioxidant index intake or blood antioxidant index is related to diminished risk for PSC cataract *(28,30,33)*. However, the CTNS 9-year trial of 1,020 participants, aged 55–75 years, showed that PSC events were significantly more common (*HR*, 2.00; 95% *CI*: 1.35–2.98; $P < 0.001$) in persons taking Centrum than in those assigned to the placebo *(14)*. The remaining eight studies showed no effect, including the Linxian Intervention Study *(17,19,21,31,39,40,47,70)*.

Jacques and Chylack found that adjusted prevalence for PSC cataract was decreased by 80% (*RR*, 0.16; *CI*: 0.04–0.82) for persons with high antioxidant index scores, based upon combined intake of vitamins C, E, and carotenoids *(28)*. Leske et al. found a 60% reduced rate of PSC cataract (*RR*, 0.4; *CI*: 0.21–0.77) among users of multivitamin supplements *(30)*, and Mohan et al. found a 77% reduced rate of PSC cataract (*RR*, 0.23; *CI*: 0.06–0.88) in subjects with the highest vs lowest plasma antioxidant index in a retrospective study of 1,990 subjects *(33)*.

21.2.5.4. Multivitamins and Antioxidant Indices, and Mixed Cataract

Four studies examined relationships between multivitamins or antioxidant index and risk for mixed cataract *(14,30,33,70)*. Leske et al. found a 61% reduced rate of cataract (*RR*, 0.39; *CI*: 0.19–0.8) in subjects with the highest vs lowest antioxidant intake index and also found a 30% reduced rate of mixed cataract (*RR*, 0.7; *CI*: 0.51–0.97) in multivitamin supplement users *(30)*. The AREDS, in a prospective study (median follow-up, 6.3 years), showed that Centrum® use, adjusted for propensity score and other covariates, was associated with a reduction in "any" lens opacity progression ($OR = 0.84$; 95% *CI*: 0.72–0.98, $P = 0.025$). Results from the CTNS also showed that there was a decrease in total lens events in participants assigned to Centrum compared with those assigned to the placebo (*HR*, 0.82; 95% *CI*: 0.68–0.98; $P = 0.03$). In contrast, Mohan et al. found an 87% increased rate of mixed cataract (*RR*, 1.87; *CI*: 1.29–2.69) in subjects with the highest vs lowest plasma antioxidant index in a retrospective study of 1,990 subjects *(33)*.

21.2.5.5. Multivitamins and Antioxidant Indices, and Cataract Extraction

Six studies examined relationships between use of multivitamins or antioxidant indices and risk for cataract extraction *(14,36,40,54,55,62)*. Knekt et al. found a 62% reduced rate of cataract extraction (*RR*, 0.38; *CI*: 0.15–1) in subjects with the highest vs lowest levels of serum vitamin E and β-carotene *(62)*. Seddon et al. found a 27% reduced rate of cataract extraction (*RR*, 0.73; *CI*: 0.54–0.99) in male physicians who did vs did not use multivitamins, but not vitamin C or E supplements alone, in a prospective study of 17,744 subjects *(55)*. The remaining four studies showed no effect *(14,36,40,54)*.

21.3. EPIDEMIOLOGICAL STUDIES RELATING DIETARY PATTERNS, INDICES, CARBOHYDRATE INTAKE TO RISK FOR AGE-RELATED CATARACT AND MACULOPATHY

Most of the information about nutrition and age-related eye diseases described in this chapter was derived from "single-nutrient" studies. Such studies tend to ignore interactions and correlations (multicollinearity) between nutrients, to be unable to detect small effects from single nutrients, and to be of less value for dietary advice. Studies derived from dietary indices, such as Healthy Eating Index (HEI), or empirical dietary patterns using either cluster analysis or factor analysis will offer an opportunity to address these issues *(72)*. However, little has been done in this field. Using the USDA Food Guide Pyramid as a dietary pattern our NVP found that women in the highest quartile category of HEI scores were significantly less likely to have nuclear opacities than those in the lowest category ($OR = 0.47$; 95% *CI*: 0.26–0.84), especially among nonusers of supplemental vitamin C ($OR = 0.23$; 95% *CI*: 0.10–0.52) *(73)*.

Biochemical and recent epidemiological studies indicate that elevated exposure to carbohydrate may also be deleterious to eye tissues and that it would be useful to ask if dietary carbohydrate intake

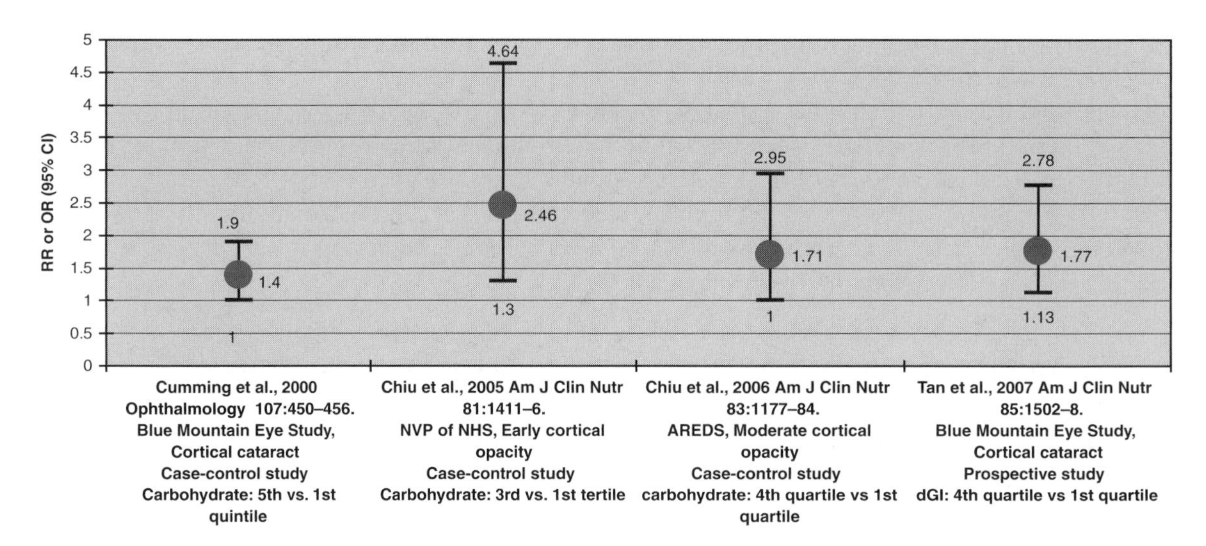

Fig. 21.5. Observational studies on dietary carbohydrate and cortical cataract. **dGI**, dietary glycemic index

is related to odds for age-related eye diseases *(74,75)*. The dietary glycemic index (GI) is the weighted average of GI from individual foods *(76,77)* and can thus be considered as an index of dietary carbohydrate consumption which can be used to define a pattern of intake of carbohydrate. Several studies indicate that the dietary GI is of value in studying the relations between dietary carbohydrate and diseases *(77)*. As carbohydrate-containing foods account for the most frequently consumed foods and provide the largest proportion of energy, quantitative indices based on physiological responses to carbohydrate intake will simplify the classification of carbohydrate foods, and be helpful for people, especially diabetics, to select foods for dietary management.

Our data from the NVP of the NHS show that higher total carbohydrate intake is associated with the increased odds for cortical opacities and higher dietary GI is related to higher odds for early age-related macular degeneration (AMD) *(75,78)*. The AMD findings were further confirmed and expanded in both cross-sectional and prospective studies in the AREDS cohort *(79–81)* and in a recent replication study in the Blue Mountain Eye Study (BMES) *(82)* (Fig. 21.5). Inconsistent with the findings from the NVP *(75)*, the AREDS *(79)*, and an earlier cross-sectional study in the BMES *(42)*, a prospective study in the BMES found that higher dietary GI was associated with risk of cortical cataract but not with nuclear or PSC cataract and no association between GL or carbohydrate quantity and any cataract subtype *(83)* (Fig. 21.6).

21.4. EPIDEMIOLOGICAL STUDIES REGARDING ANTIOXIDANTS AND ARM

Accumulating evidence indicates that oxidative stress, including glycoxidation, plays an important role in the pathogenesis of ARM *(84)*. Much effort has been engaged in the research of modulating antioxidant balance by appropriate diet or supplementation with specific micronutrients to prevent or delay the development or progression of ARM. For example, the AREDS *(85)*, a major clinical trial, suggests that use of antioxidant supplements (vitamins C (500 mg) and E (400 IU), β-carotene (15 mg), and zinc (80 mg of zinc as zinc oxide)) offers considerable benefit with respect to progress to advanced AMD (a 28% reduced risk (*OR* = 0.72; 99% *CI*: 0.52–0.98)) in a mean follow-up of 6.3 years in persons older than 55 years. Some other observational studies and intervention trials (Table 21.3) corroborate these data and indicate that further work should be done in order to extend the AREDS results. A recent prospective study from the Rotterdam group indicated that an above-median intake

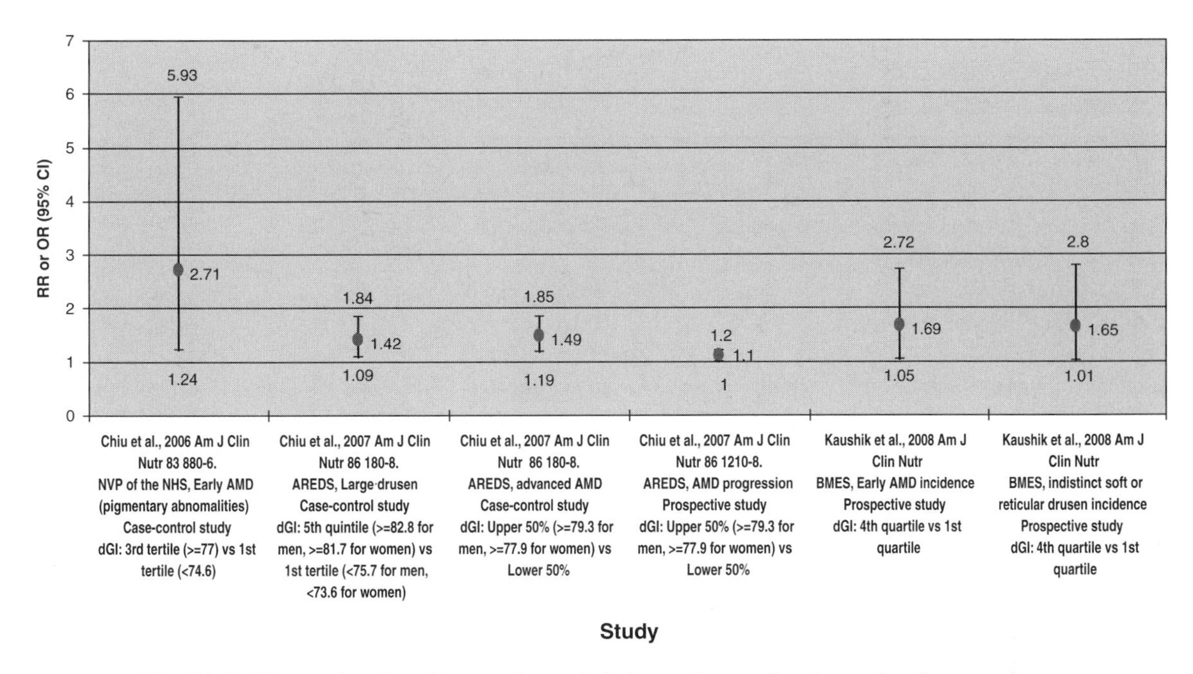

Fig. 21.6. Observational studies on glycemic index and age-related macular degeneration

of all four nutrients, β-carotene (3.6 mg), vitamin C (114 mg), vitamin E (13 IU), and zinc (9.6 mg), was associated with a 35% reduced risk (*HR*, 0.65; 95% *CI*: 0.46–0.92) of AMD in a mean follow-up of 8.0 years in persons older than 55 years *(86)*. This result was independent of supplement use. The results imply that dietary intake of much lower levels of the micronutrients which were used in the AREDS formulation can substantially reduce the risk of developing AMD.

21.4.1. Intervention Trials

Nine randomized, double-blinded, placebo-controlled intervention trials have assessed the effect of vitamin or micronutrient supplements on ARM risk *(85,87–96)*. The AREDS found that supplementation of vitamin C (500 mg), vitamin E (400 IU), β-carotene (15 mg), and zinc (80 mg of zinc as zinc oxide) reduced the risk of AMD progression in persons older than 55 years *(85)*. A recent meta-analysis raised concerns about the safety of taking a vitamin E supplement higher than 400 IU/day *(87)*. However, review of that analysis and available data indicates that there is little reason for concern about intakes of 400 IU/day *(88)*. Nevertheless, the safety of pharmacologic doses of vitamin E, alone or combined with other supplements, seems to warrant further study. The National Eye Institute is sponsoring a new study (AREDS II) to further advance our understanding of the role of the vitamins and minerals used in the AREDS formulation as well as potential benefit of lutein/zeaxanthin and ω-3 long-chain polyunsaturated fatty acids in delaying the progression of vision loss *(89)*. Results derived from AREDS also provide valuable data regarding the natural history of AMD, which will increase our ability to identify high-risk populations and facilitate further research applications. For example, a simplified five-step severity scale has been developed to provide convenient risk categories for development of advanced AMD that can be determined by clinical examination or by less demanding photographic procedures than used in the AREDS *(97)*, while a nine-step severity scale derived from the system was also developed to provide convenient risk categories and acceptable reproducibility for research purpose *(98)*.

Table 21.3
Design and Sample Size of Major Epidemiological Studies Relating Risk for Age-Related Maculopathy to Nutritional Antioxidants

Study	*Design*	*Sample size*	*References*
AREDS (Age-Related Eye Disease Study Research Group) (2001)	Intervention	3,640	*(85)*
AREDS (2007)	Retrospective	4,519	*(113)*
Belda et al. (1999)	Retrospective	40	*(118)*
Bartlett et al. (2007)	Intervention	25	*(95)*
Blumenkranz et al. (1986)	Retrospective	49	*(99)*
Cardinault et al. (2005)	Retrospective	55	*(117)*
Nurses' Health Study (NHS) and Health Professionals Follow-up Study (HPFS) (2004)	Prospective	77,562 women 40,866 men	*(111)*
NHS and HPFS	Prospective	71,494 women and 41,564 men	*(126)*
Physicians' Health Study (PHS)(1999)	Prospective	21,120	*(108)*
POLA study (1999)	Retrospective	2,584	*(109)*
EDCCS (Eye Disease Case Control Study) (1993)	Retrospective	1,036	*(101)*
Blue Mountains Eye Study (BMES) (2002)	Prospective	2,335	*(110)*
Gale et al. (2003)	Retrospective	380	*(124)*
The First National Health and Nutrition Examination Survey (NHANES I) (1988)	Retrospective	3,082	*(100)*
Ishihara et al. (1997)	Retrospective	101	*(103)*
Kaiser et al. (1995)	Intervention	20	*(91)*
Beaver Dam Eye Study (BDES) (2008)	Prospective	4,926	*(47)*
BDES (1995)	Retrospective	334	*(115)*
BDES (1996)	Prospective	1,968	*(106)*
The Third National Health and Nutrition Examination Survey (NHANES III) (2001)	Retrospective	8,222	*(123)*
Newsome et al. (1988)	Intervention	151	*(90)*
Parisi et al. (2008)	Intervention	27	*(96)*
Sanders et al. (1993)	Retrospective	130	*(114)*
Eye Disease Case–Control Study (1994)	Retrospective	876	*(102)*
Simonelli et al. (2002)	Retrospective	94	*(104)*
BMES (1997)	Retrospective	312	*(116)*
Snellen et al. (2002)	Retrospective	138	*(122)*
Stur et al. (1996)	Intervention	112	*(92)*
BMES (2008)	Prospective	2,454	*(112)*
VECAT (2002)	Intervention	1,193	*(94)*
ATBC study (1998)	Intervention	29,133	*(93)*
Trieschmann et al. (2007)	Prospective	136	*(125)*
BDES (1998)	Prospective	1,586	*(107)*
Rotterdam Study (2005)	Prospective	4,765	*(86)*
Baltimore Longitudinal Study of Aging (BLSA) (1994)	Retrospective	976	*(105)*

Bartlett et al. in a double-masked randomized controlled trial, following 25 subjects (placebo (n = 10) or active (n = 15) groups) for 9 months, found that 6 mg of lutein supplementation in combination with other antioxidants (retinol 750 μg, vitamin C 250 mg, vitamin E 34 mg, zinc 10 mg, copper 0.5 mg) is not beneficial *(95)*. Parisi et al. followed 27 patients with nonadvanced AMD and visual acuity ≥0.2 logarithm of the minimum angle of resolution for 12 months in a randomized controlled trial. Fifteen patients (mean age, 69.4 ± 4.31 years) took daily supplements of vitamin C (180 mg), vitamin E (30 mg), zinc (22.5 mg), copper (1 mg), lutein (10 mg), zeaxanthin (1 mg), and astaxanthin (4 mg); 12 similarly aged patients had no dietary supplementation during the same period. The authors found that the supplementation was beneficial only for a selective dysfunction in the central retina (0–5°) but not for the more peripheral (5–20°) retinal areas *(96)*.

Newsome et al. found that a high-dose zinc supplement (oral zinc sulfate 100 mg twice daily; 5.3 RDA) significantly reduced the risk of vision loss after a follow-up of 12–24 months *(90)*. However, four trials, including VECAT and ATBC, found no statistically significant effect *(91–94)* (Table 21.4).

21.4.2. Vitamin C and ARM

Seventeen studies have assessed the association between vitamin C intake or supplement use and risk for ARM *(47,86,99–113)*. Fourteen studies did not find a significant association between vitamin C and risk for ARM *(86,99–103,105–109,111–113)*.

Three studies found significant results, but in opposite direction *(47,104,110)*. Simonelli et al. found that serum vitamin C levels were significantly lower ($P < 0.05$) in subjects with late ($N = 29$) vs early ($N = 19$) in a group of Italian patients, suggesting a benefit of maintaining elevated levels of vitamin C *(104)*. In contrast, Flood et al. found higher vitamin C intake from diet and supplements was associated with increased risk for early ARM (drusen or pigmentary abnormalities and absence of AMD) (intake in the fifth quintile, OR = 2.3; 95% CI: 1.3–4.0; P for trend = 0.002) in a prospective study (n = 2,335) of the Blue Mountain Eye Study *(110)*. Klein et al., in a prospective study of the Beaver Dam Eye Study, found that vitamin C supplement use was associated with increased odds of late AMD *(47)*. However, this association may reflect advice by eye care providers to take such supplements when earlier lesions of AMD were found. Van Leeuwen et al. did not find that subjects with a higher dietary intake of vitamin C alone had reduced risk of AMD (per 1 – SD increase, RR, 1.02; 95% CI: 0.94–1.10) but they found that an above-median intake of all four nutrients, β-carotene, vitamin C, vitamin E, and zinc, was associated with a 35% reduced risk (HR, 0.65; 95% CI: 0.46–0.92) of AMD in a prospective study (N = 4,170, mean follow-up of 8 years) of the Rotterdam Study *(86)*.

21.4.3. Vitamin E and ARM

Vitamin E supplements often provide significantly higher levels of this nutrient than can be obtained in the diet. A variety of studies examined relationships between vitamin E supplement use, dietary intake and plasma levels, and risk for ARM. The results from the observational studies were mixed.

Twelve of the 20 studies that have assessed this association found no significant results *(86,99,101,102,106,108,111–117)*. Klein et al., in a prospective study of the Beaver Dam Eye Study, found that vitamin E supplement use was associated with increased odds of late AMD *(47)*. However, this association may reflect advice by eye care providers to take such supplements when earlier lesions of AMD were found. The remaining seven studies found significantly protective effects under certain circumstances *(86,103–105,107,109,118)*. West et al. found that subjects with plasma α-tocopherol levels in high quartile had lower risk of AMD (OR = 0.43; 95% CI: 0.25–0.73) in a case–control study of the Baltimore Longitudinal Study of Aging (BLSA), but they did not find significant association

Table 21.4
Design and Sample Size of Major Randomized Controlled Clinical Trials on Age-Related Maculopathy

Study 1	Newsome et al. (1998) (90)
Subject age	Mean: 68 years; range 42–89 years
Number	151
Place	United States
Evaluation of outcome	Visual acuity was tested using current glasses plus the pinhole as an estimation of best-corrected visual acuity and eye charts with an illumination of about 150 foot-candles on eye chart
Baseline ARM	Fundus characteristics, including drusen, visual acuity, geographic atrophy, pigment mottling, retinal pigment epithelial detachments, were similar in the two groups
Supplement	100 mg zinc sulfate with food twice daily (40% bioavailability; 5.3 times the 15-mg RDA)
Duration	6-Month supply at entry and at subsequent 6-month intervals for a follow-up of 12–24 months
Conclusion	This is the first controlled oral intervention study to show a positive, if limited, treatment effect in macular degeneration
Study 2	Kaiser et al. (1995) (91)
Subject age	>65 years
Number	20
Place	Switzerland
Evaluation of outcome	Visual acuity
Baseline ARM	Early stage of AMD
Supplement	Visaline
Duration	6 months
Conclusion	The effect of the treatment was not statistically different between the two admittedly small groups in terms of visual and retinal acuity, color vision, and contrast sensitivity. Despite the lack of such measurable differences in this short study, the patients' own subjective assessments, however, were much better in the Visaline-treated group. Due to the short duration of the observation time, we cannot comment on a possible long-term effect
Study 3	Stur et al. (1996) (92)
Subject age	Mean: 71.5 years
Number	112
Place	Austria
Evaluation of outcome	1. Visual acuity 2. Contrast sensitivity 3. Color discrimination 4. Retinal grating acuity 5. Wisconsin grading system for macular drusen
Baseline ARM	AMD and exudative lesions in one eye and a visual acuity of better than 20/40 and macular degeneration without any exudative lesion in the second eye
Supplement	Oral zinc sulfate
Duration	2 years
Conclusion	Oral zinc substitution has no short-term effect on the course of age-related macular degeneration in patients who have an exudative form of the disease in one eye

Study 4	Teikari et al. (1998) (93)
Subject age	Smoking males aged 50–69 years
Number	29,133
Place	Finland
Evaluation of outcome	Photographic evidence of ARM at final follow-up
Baseline ARM	No evaluation
Supplement	α-Tocopherol (50 mg/day) β-Carotene (20 mg/day)
Duration	5–8 years
Conclusion	No beneficial effect of long-term supplementation with α-tocopherol or β-carotene on the occurrence of ARM was detected among smoking males
Study 5	AREDS (2001) (85)
Subject age	55–80 years
Number	3,640
Place	United States
Evaluation of outcome	1. Photographic assessment of progression to or treatment for advanced AMD 2. At least moderate visual acuity loss from baseline (≥15 letters)
Baseline ARM	Participants had extensive small drusen, intermediate drusen, large drusen, noncentral geographic atrophy, or pigment abnormalities in one or both eyes, or advanced AMD or vision loss due to AMD in one eye. At least one eye had best-corrected visual acuity of 20/32 or better
Supplement	(1) Antioxidants (vitamin C, 500 mg; vitamin E, 400 IU; and β-carotene, 15 mg); (2) zinc, 80 mg, as zinc oxide and copper, 2 mg, as cupric oxide; (3) antioxidants plus zinc; or (4) placebo
Duration	Average follow-up: 6.3 years
Conclusion	Persons older than 55 years should have dilated eye examinations to determine their risk of developing advanced AMD. Those with extensive intermediate size drusen, at least one large druse, noncentral geographic atrophy in one or both eyes, or advanced AMD or vision loss due to AMD in one eye, and without contraindications such as smoking should consider taking a supplement of antioxidants plus zinc such as that used in this study
Study 6	VECAT Taylor et al. (2002) (94)
Subject age	55–80 years
Number	1,193
Place	Australia
Evaluation of outcome	The clinical assessment of AMD was performed with 90° and 78° diopter lenses and was graded according to the international classification. One-frame simultaneous stereophotographs of the macula were taken with a Nidek 3-DX fundus camera (Nidek, Japan) with Kodachrome 64° ASA color film. These photographs were graded independently for AMD by trained graders according to the international classification. Sets of circles were used to estimate drusen size and the area affected with drusen or abnormalities in retinal pigmentary. At the end of the study the initial and final photographs were reassessed for any change with a "side by side" comparison in a masked and randomized fashion. The grading of retinal photographs was performed over 5 years

(*Continued*)

Table 21.4
(Continued)

	Primary outcome: development of early age-related macular degeneration in retinal photographs. Other measures included alternative definitions of age-related macular degeneration, progression, changes in component features, visual acuity, and visual function
Baseline ARM	18% Early AMD and 0.5% late AMD were found in both groups
Supplement	Vitamin E 500 IU daily
Duration	4 years (1995–1996–2000)
Conclusion	Daily supplement with vitamin E supplement does not prevent the development or progression of early or later stages of age-related macular degeneration
Study 7	LAST Richer et al. (2004) (129)
Number	90
Place	United States
Evaluation of outcome	Mean eye macular pigment optical density
Baseline ARM	Atrophic ARMD patients were referred by ophthalmologists at two Chicago-area veterans medical facilities
Supplement	Group 1 received lutein 10 mg (L); in Group 2, a lutein 10 mg/antioxidants/vitamins and minerals broad-spectrum supplementation formula (L/A); and in Group 3, a maltodextrin placebo (P)
Duration	12 months
Conclusion	In this study, visual function is improved with lutein alone or lutein together with other nutrients. Further studies are needed with more patients, of both genders, and for longer periods of time to assess long-term effects of lutein or lutein together with a broad spectrum of antioxidants, vitamins, and minerals in the treatment of atrophic age-related macular degeneration
Study 8	Bartlett et al. (2007) (95)
Number	25
Place	United Kingdom
Evaluation of outcome	Contrast sensitivity (CS) was measured using a Pelli-Robson chart (Clement Clarke International, Edinburgh Way, Harlow, Essex CM20 2TT, UK) and scored per letter
Baseline ARM	There was no significant difference in baseline visual acuity (VA) between active (0.20 ± 0.28) and placebo (0.08 ± 0.15) groups ($t = 1.229$, $P = 0.229$). The baseline CS scores were 1.43 ± 0.20 and 1.36 ± 0.20 log units for the placebo and active groups, respectively. Both groups fell below the normal CS score reported for this age group, which is 1.65 log units and is repeatable to within ± 0.15 log units
Supplement	6 mg lutein, 750 μg retinol equivalents, 250 mg vitamin C, 34 mg vitamin E, 10 mg zinc, and 0.5 mg copper
Duration	9 months
Conclusion	The results suggest that 6 mg of lutein supplementation in combination with other antioxidants is not beneficial for this group

Study 9	Parisi et al. (2008) (96)
Number	27
Place	Italy
Evaluation of outcome	Multifocal electroretinogram (mfERG) recordings are an electrophysiologic method of evaluating the function of localized retinal or macular areas
Baseline ARM	At baseline, we observed highly significant reductions of N1–P1 response amplitude densities (RADs = N1–P1 response amplitude densities, nV/deg^2) of R1 and R2 in treated AMD [T-AMD] group and nontreated AMD [NT-AMD] group when compared with healthy controls (1-way analysis of variance $P < 0.01$). N1–P1 RADs of R3–R5 observed in T-AMD and NT-AMD were not significantly different ($P > 0.05$) from controls. No significant differences ($P > 0.05$) were observed in N1–P1 RADs of R1–R5 between T-AMD and NT-AMD at baseline (0–2.5° (R1), 2.5–5° (R2), 5–10° (R3), 10–15° (R4), and 15–20° (R5))
Supplement	Vitamin C (180 mg), vitamin E (30 mg), zinc (22.5 mg), copper (1 mg), lutein (10 mg), zeaxanthin (1 mg), and astaxanthin (4 mg; AZYR SIFI, Catania, Italy)
Duration	12 months
Conclusion	In nonadvanced AMD eyes, a selective dysfunction in the central retina (0–5°) can be improved by the supplementation with carotenoids and antioxidants. No functional changes are present in the more peripheral (5–20°) retinal areas

with severe AMD (geographic atrophy or neovascular features) *(105)*. Ishihara et al. found serum vitamin E-alpha levels tended to be lower in the patient group than in the control group *(103)*. Vanden Langenberg et al. found that vitamin E intake from past diet (without supplement) in the fifth quintile had lower risk of large drusen (the highest vs lowest quintile of intake, *OR* = 0.4; 95% *CI*: 0.2–0.9; *P* for trend = 0.04) in a prospective study (n = 1,586) of the Beaver Dam Eye Study cohort, but they did not find significant associations with either baseline diet (without supplement), past or baseline intake (including supplements) *(107)*. Neither did they find any significant association with pigmentary abnormalities, a precursor of ARM *(107)*. Belda et al. found serum vitamin E levels negatively correlated with severity of AMD ($N = 25$) ($r = -0.815$, $P < 0.001$) *(118)*. Delcourt et al. found subjects with a higher fasting blood α-tocopherol–lipid ratio had reduced risk of late AMD (neovascular AMD or geographic atrophy) (the highest vs. lowest quintile of the ratio, *OR* = 0.18; 95% *CI*: 0.05–0.67, *P* for trend = 0.004) in the POLA study *(109)*, but they did not find any significant association with fasting blood α-tocopherol (not the ratio) levels. Simonelli et al. found serum vitamin E levels and vitamin E/cholesterol ratios were significantly lower ($P < 0.05$) in subjects with late ARM ($n = 29$) than in early ARM ($n = 19$) and in older control subjects ($N = 24$) in an Italian study *(104)*. Van Leeuwen et al. found that subjects with a higher dietary intake of vitamin E alone had reduced risk of AMD (per 1 – SD increase, *RR*, 0.92; 95% *CI*: 0.84–1.00) and an above-median intake of all four nutrients, β-carotene, vitamin C, vitamin E, and zinc, was associated with a 35% reduced risk (*HR*, 0.65; 95% *CI*: 0.46–0.92) of AMD in a prospective study (n = 4,170, mean follow-up of 8 years) of the Rotterdam Study.

21.4.4. Carotenoids and ARM

Despite the presence of some 30 or more carotenoids in human serum, only lutein and zeaxanthin are found in the retina. The only source of lutein and zeaxanthin is diet, especially from leafy green vegetables *(119)*. Several investigations suggest that elevated intake of foods rich in lutein and zeaxanthin, such as spinach, is related to the level of these carotenoids in the macula *(120,121)*. Despite there having been many studies, no clear patterns have appeared as yet regarding associations between AMD and intake of lutein/zeaxanthin, as well as other carotenoids.

21.4.4.1. Total carotenoids

Three of the six studies which related total carotenoids to ARM found no significant results *(103,111,117)*, including a prospective study of 77,562 women from the Nurses' Health Study and 40,866 men from the Health Professionals Follow-up Study *(111)*. The other three studies found a protective role of total carotenoids on ARM *(101,102,104)*. The EDCCS Group found that fasting blood total carotenoids >5th quintile was associated with lower risk for neovascular AMD compared with all controls (the highest vs lowest quintile of blood level, $OR = 0.34$; 95% *CI*: 0.21–0.55, P for trend < 0.0001) or controls without large drusen (the highest vs lowest quintile of blood level, $OR = 0.35$; 95% *CI*: 0.20–0.61, P for trend $= 0.0001$) *(101)*. They also reported that intake of carotenoids >5th quintile was associated with lower risk for exudative AMD (the highest vs lowest quintile of intake, $OR = 0.57$; 95% *CI*: 0.35–0.92, P for trend $= 0.02$) *(102)*. Simonelli et al. found that serum total carotenoid levels were significantly lower ($P < 0.01$) in subjects with late ARM ($N =$ 29) than in early ARM ($N = 19$) in a small group of Italian patients *(104)*.

21.4.4.2. Lutein and zeaxanthin

Eighteen studies have related lutein and zeaxanthin to risk for AMD *(86,101,102,104,106,107,110–115,117,122–126)*. Seven studies indicated beneficial effects due to consuming higher level of lutein/zeaxanthin but 11 studies, including five prospective studies from the Beaver Dam Eye Study *(107)*, the Blue Mountain Eye Study *(110)*, the Nurses' Health Study, the Health Professionals Follow-up Study *(111,126)*, and the Rotterdam Study, did not find significant results *(86,104,106,107,110,111,114,115,117)*.

The EDCCS reported that subjects with fasting blood lutein and zeaxanthin >80th percentile had reduced risk of neovascular AMD (>80th percentile vs $\leq$20th percentile of blood level, $OR = 0.3$; 95% *CI*: 0.2–0.6, P for trend $= 0.0001$) *(101)*, and Seddon et al. reported that intake of lutein and zeaxanthin >5th quintile had reduced risk of exudative AMD (the highest vs lowest quintile of intake, $OR = 0.43$; 95% *CI*: 0.2–0.7, P for trend < 0.001) *(102)*. Snellen et al. found a clear inverse dose–response relationship between lutein/zeaxanthin intake and occurrence of neovascular AMD *(122)*.

Gale et al. reported that zeaxanthin, but not lutein or lutein/zeaxanthin, may protect against AMD (*OR*of AMD for the lowest vs highest third of plasma zeaxanthin = 2.0, 95% *CI*: 1.0–4.1) *(124)*. Using the Third National Health and Nutrition Examination Survey (NHANES III) data, Mares-Perlman et al. found that in the youngest age groups who were at risk for developing early onset (ages 40–59 years) or late onset of (ages 60–79 years) AMD, higher levels of lutein and zeaxanthin in the diet were related to lower odds for pigmentary abnormalities (the highest vs lowest quintile of intake, $OR =$ 0.1; 95% *CI*: 0.1–0.3) and of late ARM (the highest vs lowest quintile of intake, $OR = 0.1$; 95% *CI*: 0.0–0.9) *(123)*. However, inverse relations of lutein and zeaxanthin in the diet or serum to any form of AMD were not observed overall. Interestingly, Mares-Perlman et al. suggested that relationships of lutein and zeaxanthin to AMD may be influenced by age and race *(123)*. In a case–control study using the AREDS baseline data, higher dietary intake of lutein/zeaxanthin was independently associated with decreased likelihood of having neovascular AMD, geographic atrophy, and large or extensive intermediate drusen *(113)*. In a population-based prospective study in the BMES cohort, Tan et al. found that higher dietary lutein and zeaxanthin intake reduced the risk of long-term incident AMD *(112)*.

21.4.4.3. β-Carotene

Fifteen studies tried to relate β-carotene to AMD *(40,86,101,102,104–107,110,111, 113–117)*. Results from the EDCCS support a protective role for β-carotene on advanced AMD *(101,102)*.

The EDCCS reported that subjects with fasting blood β-carotene >80th percentile had lower odds of neovascular AMD (>80th percentile vs $\leq$ 20th percentile of blood level, $OR = 0.3$; 95% *CI*: 0.2–0.5, *P* for trend < 0.0001) *(101)* and intake of β-carotene >5th quintile was related to reduced odds of exudative AMD (the highest vs lowest quintile of intake, $OR = 0.59$; 95% *CI*: 0.4–0.96, *P* for trend = 0.03) *(102)*.

Van Leeuwen et al. did not find that subjects with a higher dietary intake of β-carotene alone had reduced risk of AMD (per 1 – SD increase, *RR*, 1.00; 95% *CI*: 0.94–1.06) but they found that an above-median intake of all four nutrients, β-carotene, vitamin C, vitamin E, and zinc, was associated with a 35% reduced risk (*HR*, 0.65; 95% *CI*: 0.46–0.92) of AMD in a prospective study ($N = 4{,}170$, mean follow-up of 8 years) of the Rotterdam Study *(86)*. However, in a cross-sectional analysis of the AREDS baseline data, Chiu et al. demonstrated that consuming diets which provide low dGI and higher intakes of vitamins C and E, zinc, lutein/zeaxanthin, docosahexaenoic acid (DHA), and eicosapentaenoic acid (EPA) were associated with the greatest reduction in risk for prevalent drusen and advanced AMD, while dietary β-carotene did not affect these relationships (in press). In a prospective study of the AREDS cohort, Chiu et al. found that AREDS supplement users consuming higher (upper 50% or $\geq$2.2 mg/day) β-carotene from foods had over 50% increased risk for advanced AMD (*HR*, 1.52; 95% CI: 1.10, 2.11) (in revision). Similarly, in an independent analysis in the BMES, Tan et al. found that higher β-carotene intake (from foods only or from foods plus supplements) was associated with an increased risk of late AMD *(112)*.

The remaining 11 studies did not find significant results *(104–107,110–117)*.

21.4.4.4. α-Carotene

Ten studies have related α-carotene to ARM *(86,101,102,106,107,110,111,114,115,117)*. Eight studies did not find significant results *(86,102,106,110,111,114,115,117)*.

The EDCCS reported that subjects with fasting blood α-carotene >80th percentile had lower odds of neovascular AMD (>80th percentile vs $\leq$20th percentile of blood level, $OR = 0.5$; 95% *CI*: 0.3–0.8, *P* for trend = 0.003) *(101)*, but they did not find any significant association between intake of α-carotene and odds of exudative AMD *(102)*. Vanden Langenberg et al. found that higher α-carotene from past diet was associated with lower occurrence of large drusen (the highest vs lowest quintile of intake, $OR = 0.52$; 95% *CI*: 0.3–1.0, *P* for trend = 0.02), while they did not find significant associations between intake (both past and baseline diet) of α-carotene and pigmentary abnormalities or between baseline diet and large drusen *(107)*.

21.4.4.5. Vitamin A and retinol

Fifteen studies related vitamin A or retinol to ARM *(47,62,99,100,102–105,107,109–114)*. Of these, only four studies found significant results *(47,86,100,102,107)*.

Using the First National Health and Nutrition Examination Survey (NHANES I) data, Goldberg et al. found that more frequent intake of fruits and vegetables rich in vitamin A was associated with lower prevalence of AMD; however, they did not find any significant association with intake of vitamin A by 24-h recall data *(100)*. The EDCCS found that intake of vitamin A (including supplement) was associated with decreased risk of exudative AMD, but they did not find associations either with vitamin A intake (without supplement), status of use of vitamin A supplement, total retinol intake, or retinol intake without supplements *(102)*. Vanden Langenberg et al. found that the higher pro-A carotenoids from either past or baseline diet, the lower the occurrence of large drusen (the highest vs lowest quintile of intake, past diet: $OR = 0.53$; 95% CI: 0.3–1.0 ($P > 0.05$), P for trend = 0.03; baseline diet: $OR = 0.45$; 95% CI: 0.2–1.0, P for trend = 0.03), while they did not find significant associations between intake (both past and baseline diet) of pro-A carotenoids and pigmentary abnormalities *(107)*.

Klein et al., in a prospective study of the Beaver Dam Eye Study, found that vitamin A supplement use was associated with increased odds of late AMD *(47)*. However, this association may reflect advice by eye care providers to take such supplements when earlier lesions of AMD were found.

21.4.4.6. Lycopene

Ten studies related lycopene to ARM *(86,101,102,104,106,110,111,114,115,117)*. Three studies found significant results *(104, 115, 117)*.

Mares-Perlman et al. found that subjects in the lowest quintile of serum lycopene were twice as likely as all others combined to have AMD ($OR = 2.2$; 95% CI: 1.1–4.5) in a nested case–control study within the Beaver Dam Eye Study *(115)*. Simonelli et al. found that serum lycopene levels were significantly lower ($P < 0.05$) in late AMD ($n = 29$) and early AMD ($n = 19$) than controls ($n = 46$) in an Italian study *(104)*. Cardinault et al. found that concentrations in serum and lipoparticle (LDL and HDL) fractions of lycopene were significantly lower ($P < 0.05$) in AMD patients ($n = 34$) than controls ($n = 21$).

21.4.4.7. Cryptoxanthin and ARM

Ten studies examined the association between cryptoxanthin and ARM *(86,101,102,104,106,110,111,114,115,117)*. Two studies found significant results to support the hypothesis that high blood cryptoxanthin levels are protective against ARM *(101,104)* and one study was suggestive of a protective role of β-cryptoxanthin intake on the risk of neovascular ARM *(111)*. The remaining seven studies found no significant results *(86,102,106,110,114,115,117)*.

The EDCCS reported that subjects with fasting blood cryptoxanthin levels >80th percentile had reduced odds of neovascular AMD (>80th percentile vs ≤20th percentile of blood level, $OR = 0.4$; 95% CI: 0.2–0.6, P for trend = 0.0001) *(101)*. Similarly, Simonelli et al. found that subjects with late ARM ($N = 29$) had serum β-cryptoxanthin levels which were significantly lower ($P < 0.05$) than those with early ARM ($N = 19$) and controls ($N = 46$) *(104)*. However, Cho et al., in a prospective study from the Nurses' Health Study ($N = 77{,}562$) and the Health Professionals Follow-up Study ($N = 40{,}866$), obtained data that were suggestive of a protective role of β-cryptoxanthin intake on the risk of neovascular ARM (the highest quintile vs lowest of intake, $OR = 0.68$; 95% CI: 0.45–1.02, P for trend = 0.03), while they did not find any significant association with early ARM *(111)*.

21.4.5. Studies Regarding Multivitamins, Antioxidant Indices, and ARM

Seven studies examined the association between multivitamin or antioxidant index and risk for ARM *(47,101,102,106,108,111,122)*. One study found that a fasting blood antioxidant index was associated with reduced risk of AMD *(101)*. The remaining six studies did not find any significant association *(47,102,106,108,111,122)*.

In a cross-sectional analysis of the AREDS baseline data, Chiu et al. demonstrated that consuming diets which provide low dGI and higher intakes of vitamins C and E, zinc, lutein/zeaxanthin, DHA, and EPA were associated with the greatest reduction in risk for prevalent drusen and advanced AMD, while dietary β-carotene did not affect these relationships (in press). In a prospective study of the AREDS cohort, Chiu et al. found that AREDS supplement users consuming higher (upper 50%

or $\geq$2.2 mg/day) β-carotene from foods had over 50% increased risk for advanced AMD (*HR*, 1.52; 95% *CI*: 1.10, 2.11) (in revision).

21.5. CONCLUSIONS AND CONSIDERATIONS FOR FUTURE RESEARCH

It is surprising that the lens, which is metabolically quiescent, and the retina, which is among the most metabolically active tissues in the body share many molecular and environmental disease-related risk factors. Oxidative stress and its sequelae are clearly involved in the etiology of ARC and ARM. Given these associations, it is not surprising that over 70 studies have attempted to relate antioxidant intake to risk for ARC and ARM. These studies make it clear that consistently good nutrition, based upon diets which are rich in fruits and vegetables, with good sources of ω-3 fats and lower readily digested carbohydrates, coupled with healthy lifestyles, which preclude known risk factors for eye diseases such as smoking, overweight, poverty, and high simple carbohydrate must be begun early in life and/or when ARC or ARM are at an early stage in order to obtain the benefit of prolonged sight later in life. Since the same dietary practice also optimizes many aspects of health, including CVD and immune function, there should be good incentive to do so. In the absence of good nutrition, i.e., among the nutritionally impoverished, supplementation appears to offer benefit with respect to reducing risk for AMD and cataract. Fortunately, some of these changes, such as moving to lower glycemic index diets, are easily achieved with minor diet-behavior modification.

In terms of identifying high-risk populations, it will be helpful to clarify the interactions between nutrition and other risk factors. Further studies are warranted to determine how genetic factors interact with nutrients to affect the risk of AMD. Present results indicate that begun early in life, dietary intake of about 250 mg/day vitamin C and 90 mg/day vitamin E, 3 mg/day lutein should provide sufficient reserves to provide eye health benefit. These levels are considerably higher than the current recommended daily intakes. The lutein recommendation is based upon the observational data from the NVP *(16)*. Because the bioavailability of ascorbate may decrease with age, slightly higher intakes might be required in the elderly. The ongoing AREDS 2 intervention trials regarding lutein and zeaxanthin should yield qualitative and quantitative information with regard to the benefit of these carotenoids, especially for the prevention of AMD. Because intake or plasma measurements are highly variable and the effects of diet are likely to be cumulative, studies should be performed for at least 10 years on populations for which long-term dietary records or multiple measures of plasma nutrient levels are available. Identifying the best means to quantify nutrient intake would also be of advantage. The new definitions and refinements of techniques to quantify progress of ARC and ARM will also be helpful. Because delaying progress of early lesions also results in retardation of compromises of visual function, early endpoints are preferable. Greater ability to compare among studies would also be invaluable with respect to analyzing data from multiple studies. In terms of fully utilizing available information and maximizing statistical power, using the eye, instead of person, as the unit of analysis provides more power to studies.

In conclusion, the data from the diverse epidemiological studies suggest that there are many common insults and mechanistic compromises that are associated with aging and that proper nutrition started early in life (a time yet to be defined) and maintained through maturity may address some of these compromises and provide for extended youthful function during older age. Indeed, proper nutrition, possibly including use of antioxidant supplements for those at risk, along with healthy lifestyles may provide the least costly and most practical means to delay ARC and ARM. In addition to proper nutrition, epidemiologic data would suggest that avoiding stress (i.e., smoking, obesity, excess sunlight exposure, and diets which are rich in simple sugars) may also be protective.

21.6. ACKNOWLEDGMENTS

Any opinions, conclusions, or recommendations expressed in this publication are those of the authors and do not necessarily reflect the views or policies of the U.S. Department of Agriculture, nor does mention of trade names, commercial products, or organizations imply endorsement by the US Government.

Financial support for this project has been provided by R01-13250 and R03-EY014183-01A2 from the National Institutes of Health; and by grants from the Johnson and Johnson Focused Giving Program.

REFERENCES

1. Kupfer C. The conquest of cataract: a global challenge. Trans Ophthalmol Soc U K 1985;104(Pt 1):1–10.
2. Schwab L. Cataract blindness in developing nations. Int Ophthalmol Clin 1990;30(1):16–8.
3. World Health Organization. Use of intraocular lenses in cataract surgery in developing countries: memorandum from a WHO meeting. Bull World Health Org 1991;69(6):657–66.
4. Klein BE, Klein R, Linton KL. Prevalence of age-related lens opacities in a population: The Beaver Dam Eye Study. Ophthalmology 1992;99:546–52.
5. Leibowitz HM, Krueger DE, Maunder LR, et al. The Framingham Eye Study monograph: An ophthalmological and epidemiological study of cataract, glaucoma, diabetic retinopathy, macular degeneration, and visual acuity in a general population of 2631 adults, 1973–1975. Surv Ophthalmol 1980;24(Suppl):335–610.
6. Friedman DS, O'Colmain BJ, Munoz B, et al. Prevalence of age-related macular degeneration in the United States. Arch Ophthalmol 2004;122:564–72.
7. Siegal M, Chiu CJ, Taylor A. Antioxidant status and risk for cataract. In: Bendich A, Deckelbaum RJ, eds. Preventive Nutrition: The comprehensive guide for health professionals. Third ed. Totowa, NJ: Humana Press Inc.; 2005: 463–503.
8. McCarty CA, Taylor HR. A review of the epidemiologic evidence linking ultraviolet radiation and cataracts. Dev Ophthalmol 2002;35:21–31.
9. West ES, Schein OD. Sunlight and age-related macular degeneration. Int Ophthalmol Clin 2005;45:41–7.
10. Robman L, Taylor H. External factors in the development of cataract. Eye 2005;19:1074–82.
11. Klein R, Peto T, Bird A, Vannewkirk MR. The epidemiology of age-related macular degeneration. Am J Ophthalmol 2004;137:486–95.
12. Taylor A, ed. Nutritional and environmental influences on the eye. Boca Raton, FL: CRC Press; 1999.
13. Chiu CJ, Taylor A. Nutritional antioxidants and age-related cataract and maculopathy. Exp Eye Res 2007;84: 229–45.
14. Clinical Trial of Nutritional Supplements and Age-Related Cataract Study Group, Maraini G, Sperduto RD, et al. A randomized, double-masked, placebo-controlled clinical trial of multivitamin supplementation for age-related lens opacities. Clinical trial of nutritional supplements and age-related cataract report no. 3. Ophthalmology 2008;115: 599–607.e1.
15. Age-Related Eye Disease Study Research Group. A randomized, placebo-controlled, clinical trial of high-dose supplementation with vitamins C and E and beta carotene for age-related cataract and vision loss: AREDS report no. 9. Arch Ophthalmol 2001;119(10):1439–52.
16. McNeil JJ, Robman L, Tikellis G, Sinclair MI, McCarty CA, Taylor HR. Vitamin E supplementation and cataract: randomized controlled trial. Ophthalmology 2004;111:75–84.
17. Teikari JM, Rautalahti M, Haukka J, et al. Incidence of cataract operations in Finnish male smokers unaffected by a tocopherol or b carotene supplements. J Epidemiol Community Health 1998;52:468–72.
18. Christen W, Glynn R, Sperduto R, Chew E, Buring J. Age-related cataract in a randomized trial of beta-carotene in women. Ophthalmic Epidemiol 2004;11:401–12.
19. Gritz DC, Srinivasan M, Smith SD, et al. The Antioxidants in Prevention of Cataracts Study: effects of antioxidant supplements on cataract progression in South India. Br J Ophthalmol 2006;90:847–51.
20. Christen WG, Glynn RJ, Chew EY, Buring JE. Vitamin E and age-related cataract in a randomized trial of women. Ophthalmology 2008;115:822–9.
21. Sperduto RD, Hu TS, Milton RC, et al. The Linxian cataract studies. Two nutrition intervention trials. Arch Ophthalmol 1993;111(9):1246–53.

22. Christen WG, Manson JE, Glynn RJ, et al. A randomized trial of beta carotene and age-related cataract in US physicians. Arch Ophthalmol 2003;121:372–8.
23. Chylack LT, Jr., Brown NP, Bron A, et al. The Roche European American Cataract Trial (REACT): a randomized clinical trial to investigate the efficacy of an oral antioxidant micronutrient mixture to slow progression of age-related cataract. Ophthalmic Epidemiol 2002;9(1):49–80.
24. Jacques PF, Chylack LTJ, Hankinson SE, et al. Long-term nutrient intake and early age-related nuclear lens opacities. Arch Ophthalmol 2001;119:1009–19.
25. Taylor A, Jacques PF, Chylack LT, Jr., et al. Long-term intake of vitamins and carotenoids and odds of early age-related cortical and posterior subcapsular lens opacities. Am J Clin Nutr 2002;75(3):540–9.
26. Jacques PF, Taylor A, Moeller S, et al. Long-term nutrient intake and 5-year change in nuclear lens opacities. Arch Ophthalmol 2005;123:517–26.
27. Taylor A, ed. Nutritional and environmental influences on the risk for cataract. New York: CRC Press; 1999.
28. Jacques PF, Chylack LT, Jr. Epidemiologic evidence of a role for the antioxidant vitamins and carotenoids in cataract prevention. Am J Clin Nutr 1991;53(1 Suppl):352S–5S.
29. Jacques PF, Taylor A, Hankinson SE, et al. Long-term vitamin C supplement use and prevalence of early age-related lens opacities. Am J Clin Nutr 1997;66(4):911–6.
30. Leske MC, Chylack LT, Jr., Wu SY. The Lens Opacities Case-Control Study. Risk factors for cataract. Arch Ophthalmol 1991;109(2):244–51.
31. Mares-Perlman JA, Klein BE, Klein R, Ritter LL. Relation between lens opacities and vitamin and mineral supplement use. Ophthalmology 1994;101(2):315–25.
32. Mares-Perlman JA, Brady WE, Klein BE, et al. Diet and nuclear lens opacities. Am J Epidemiol 1995;141(4):322–34.
33. Mohan M, Sperduto RD, Angra SK, et al. India-US case-control study of age-related cataracts. India-US Case- Control Study Group. Arch Ophthalmol 1989;107(5):670–6.
34. Robertson JM, Donner AP, Trevithick JR. Vitamin E intake and risk of cataracts in humans. Ann N Y Acad Sci 1989;570:372–82.
35. Valero MP, Fletcher AE, De Stavola BL, Vioque J, Alepuz VC. Vitamin C is associated with reduced risk of cataract in a Mediterranean population. J Nutr 2002;132(6):1299–306.
36. Hankinson SE, Stampfer MJ, Seddon JM, et al. Nutrient intake and cataract extraction in women: a prospective study. BMJ 1992;305(6849):335–9.
37. Lyle BJ, Mares-Perlman JA, Klein BE, Klein R, Greger JL. Antioxidant intake and risk of incident age-related nuclear cataracts in the Beaver Dam Eye Study. Am J Epidemiol 1999;149(9):801–9.
38. Ferrigno L, Aldigeri R, Rosmini F, Sperduto RD, Maraini G. The Italian-American Cataract Study Group. Associations between plasma levels of vitamins and cataract in the Italian-American Clinical Trial of Nutritional Supplements and Age-Related Cataract (CTNS): CTNS Report #2. Ophthalmic Epidemiol 2005;12:71–80.
39. Mares-Perlman JA, Lyle BJ, Klein R, et al. Vitamin supplement use and incident cataracts in a population-based study. Arch Ophthalmol 2000;118(11):1556–63.
40. Tan AG, Mitchell P, Flood VM, et al. Antioxidant nutrient intake and the long-term incidence of age-related cataract: the Blue Mountains Eye Study. Am J Clin Nutr 2008;87:1899–905.
41. Dherani M, Murthy GV, Gupta SK, et al. Blood levels of vitamin C, carotenoids and retinol are inversely associated with cataract in a North Indian population. Invest Ophthalmol Vis Sci 2008;49:3328–35.
42. Cumming RG, Mitchell P, Smith W. Diet and cataract: the Blue Mountains Eye Study. Ophthalmology 2000;107(3):450–6.
43. Gale CR, Hall NF, Phillips DI, Martyn CN. Plasma antioxidant vitamins and carotenoids and age-related cataract. Ophthalmology 2001;108:1992–8.
44. The Italian-American Cataract Study Group. Risk factors for age-related cortical, nuclear, and posterior subcapsular cataracts. Am J Epidemiol 1991;133:541–53.
45. Vitale S, West S, Hallfrisch J, et al. Plasma antioxidants and risk of cortical and nuclear cataract. Epidemiology 1993;4(3):195–203.
46. Leske MC, Chylack LT, Jr., He Q, et al. Antioxidant vitamins and nuclear opacities: the longitudinal study of cataract. Ophthalmology 1998;105(5):831–6.
47. Klein BE, Knudtson MD, Lee KE, et al. Supplements and age-related eye conditions the beaver dam eye study. Ophthalmology 2008;115:1203–8.
48. Tarwadi K, Agte V. Linkages of antioxidant, micronutrient, and socioeconomic status with the degree of oxidative stress and lens opacity in Indian cataract patients. Nutrition 2004;20:261–7.

49. Nourmohammadi I, Modarress M, Khanaki K, Shaabani M. Association of serum alpha-tocopherol, retinol and ascorbic acid with the risk of cataract development. Ann Nutr Metab 2008;52:296–8.
50. Rodríguez-Rodríguez E, Ortega RM, López-Sobaler AM, Aparicio A, Bermejo LM, Marín-Arias LI. The relationship between antioxidant nutrient intake and cataracts in older people. Int J Vitam Nutr Res 2006;76: 359–66.
51. Christen WG, Liu S, Glynn RJ, Gaziano JM, Buring JE. Dietary carotenoids, vitamins C and E, and risk of cataract in women: a prospective study. Arch Ophthalmol 2008;126:102–9.
52. Yoshida M, Takashima Y, Inoue M, et al. Prospective study showing that dietary vitamin C reduced the risk of age-related cataracts in a middle-aged Japanese population. Eur J Nutr 2007;46:118–24.
53. Tavani A, Negri E, La Vecchia C. Food and nutrient intake and risk of cataract. Ann Epidemiol 1996;6:41–6.
54. Chasan-Taber L, Willett WC, Seddon JM, et al. A prospective study of vitamin supplement intake and cataract extraction among U.S. women. Epidemiology 1999;10(6):679–84.
55. Seddon JM, Christen WG, Manson JE, et al. The use of vitamin supplements and the risk of cataract among US male physicians. Am J Public Health 1994;84(5):788–92.
56. Leske MC, Wu SY, Hyman L, et al. Biochemical factors in the lens opacities. Case-control study. The Lens Opacities Case-Control Study Group. Arch Ophthalmol 1995;113(9):1113–9.
57. Mares-Perlman JA, Brady WE, Klein BE, et al. Serum carotenoids and tocopherols and severity of nuclear and cortical opacities. Invest Ophthalmol Vis Sci 1995;36(2):276–88.
58. McCarty CA, Mukesh BN, Fu CL, Taylor HR. The epidemiology of cataract in Australia. Am J Ophthalmol 1999;128(4):446–65.
59. Lyle BJ, Mares-Perlman JA, Klein BE, et al. Serum carotenoids and tocopherols and incidence of age-related nuclear cataract. Am J Clin Nutr 1999;69(2):272–7.
60. Nadalin G, Robman LD, McCarty CA, Garrett SK, McNeil JJ, Taylor HR. The role of past intake of vitamin E in early cataract changes. Ophthalmic Epidemiol 1999;6(2):105–12.
61. Rouhiainen P, Rouhiainen H, Salonen JT. Association between low plasma vitamin E concentration and progression of early cortical lens opacities. Am J Epidemiol 1996;144(5):496–500.
62. Knekt P, Heliovaara M, Rissanen A, Aromaa A, Aaran RK. Serum antioxidant vitamins and risk of cataract. BMJ 1992;305(6866):1392–4.
63. Chasan-Taber L, Willett WC, Seddon JM, et al. A prospective study of carotenoid and vitamin A intakes and risk of cataract extraction in US women. Am J Clin Nutr 1999;70(4):509–16.
64. Brown L, Rimm EB, Seddon JM, et al. A prospective study of carotenoid intake and risk of cataract extraction in US men. Am J Clin Nutr 1999;70(4):517–24.
65. Vu HT, Robman L, Hodge A, McCarty CA, Taylor HR. Lutein and zeaxanthin and the risk of cataract: the Melbourne visual impairment project. Invest Ophthalmol Vis Sci 2006;47:3783–6.
66. Delcourt C, Carrière I, Delage M, Barberger-Gateau P, Schalch W, POLA Study Group. Plasma lutein and zeaxanthin and other carotenoids as modifiable risk factors for age-related maculopathy and cataract: the POLA Study. Invest Ophthalmol Vis Sci 2006;47:2329–35.
67. Delcourt C, Cristol JP, Tessier F, Leger CL, Michel F, Papoz L. Risk factors for cortical, nuclear, and posterior subcapsular cataracts: the POLA study. Am J Epidemiol 2000;151:497–504.
68. Kuzniarz M, Mitchell P, Cumming RG, Flood VM. Use of vitamin supplements and cataract: the Blue Mountains Eye Study. Am J Ophthalmol 2001;132(1):19–26.
69. Leske MC, Wu SY, Connell AM, Hyman L, Schachat AP. Lens opacities, demographic factors and nutritional supplements in the Barbados Eye Study. Int J Epidemiol 1997;26(6):1314–22.
70. Milton RC, Sperduto RD, Clemons TE, Ferris FL, Age-Related Eye Disease Study Research Group. Centrum use and progression of age-related cataract in the Age-Related Eye Disease Study: a propensity score approach. AREDS report No. 21. Ophthalmology 2006;113:1264–70.
71. Mares-Perlman JA, Brady WE, Klein BEK, Klein R, Palta M. Supplement use and 5-year progression of cortical opacities. Invest Ophthalmol Vis Sci 1996;37:S237.
72. Newby PK, Muller D, Tucker KL. Associations of empirically derived eating patterns with plasma lipid biomarkers: a comparison of factor and cluster analysis methods. Am J Clin Nutr 2004;80:759–67.
73. Moeller SM, Taylor A, Tucker KL, et al. Overall adherence to the dietary guidelines for Americans is associated with reduced prevalence of early age-related nuclear lens opacities in women. J Nutr 2004;134:1812–9.
74. Stitt AW. The Maillard Reaction in Eye Diseases. Ann NY Acad Sci 2005;1043:582–97.
75. Chiu CJ, Morris MS, Rogers G, et al. Carbohydrate intake and glycemic index in relation to the odds of early cortical and nuclear lens opacities. Am J Clin Nutr 2005;81:1411–6.

76. Jenkins DJ, Wolever TM, Taylor RH. Glycemic index of foods: a physiological basis for carbohydrate exchange. Am J Clin Nutr 1981;34:362–6.
77. Jenkins DJA, Kendall CWC, Augustin LSA, et al. Glycemic index: overview of implications in health and disease. Am J Clin Nutr 2002;76 (suppl):266S–73S.
78. Chiu CJ, Hubbard LD, Armstrong J, et al. Dietary glycemic index and carbohydrate in relation to early age-related macular degeneration. Am J Clin Nutr 2006;83:880–6.
79. Chiu CJ, Milton RC, Gensler G, Taylor A. Dietary carbohydrate and glycemic index in relation to cortical and nuclear lens opacities in the Age-Related Eye Disease Study. Am J Clin Nutr 2006;83:1177–84.
80. Chiu CJ, Milton RC, Gensler G, Taylor A. Association between dietary glycemic index and age-related macular degeneration in the Age-Related Eye Disease Study. Am J Clin Nutr 2007;86:180–8.
81. Chiu CJ, Milton RC, Klein R, Gensler G, Taylor A. Dietary carbohydrate and progression of age-related macular degeneration, a prospective study from the Age-Related Eye Disease Study. Am J Clin Nutr 2007;86:1210–8.
82. Kaushik S, Wang JJ, Flood V, et al. Dietary glycemic index and the risk of age-related macular degeneration. Am J Clin Nutr 2008;88:1104–10.
83. Tan J, Wang JJ, Flood V, et al. Carbohydrate nutrition, glycemic index, and the 10-y incidence of cataract. Am J Clin Nutr 2007;86:1502–8.
84. Mares-Perlman J, Klein R. Diet and age-related macular degeneration. In: Taylor A, ed. Nutritional and environmental influences on the eye. Boca Raton, FL: CRC Press;1999:181–214.
85. Age-Related Eye Disease Study Research Group. A randomized, placebo-controlled, clinical trial of high-dose supplementation with vitamins C and E, beta carotene, and zinc for age- related macular degeneration and vision loss: AREDS report no. 8. Arch Ophthalmol 2001;119:1417–36.
86. van Leeuwen R, Boekhoorn S, Vingerling JR, et al. Dietary Intake of Antioxidants and Risk of Age-Related Macular Degeneration. JAMA 2005;294:3101–7.
87. Miller ERr, Pastor-Barriuso R, Dalal D, Riemersma RA, Appel LJ, Guallar E. Meta-analysis: high-dosage vitamin E supplementation may increase all-cause mortality. Ann Intern Med 2005;142:37–46.
88. Hathcock JN, Azzi A, Blumberg J, et al. Vitamins E and C are safe across a broad range of intakes. Am J Clin Nutr 2005;81:736–45.
89. Chew EY, Clemons T. Vitamin E and the Age-Related Eye Disease Study supplementation for age-related macular degeneration. Arch Ophthalmol 2005;123:395–6.
90. Newsome DA, Swartz M, Leone NC, Elston RC, Miller E. Oral zinc in macular degeneration. Arch Ophthalmol 1988;106:192–8.
91. Kaiser HJ, Flammer J, Stumpfig D, Hendrickson P. Visaline in the treatment of age-related macular degeneration: a pilot study. Ophthalmologica 1995;209:302–5.
92. Stur M, Tittl M, Reitner A, Meisinger V. Oral zinc and the second eye in age-related macular degeneration. Invest Ophthalmol Vis Sci 1996;37:1225–35.
93. Teikari JM, Laatikainen L, Virtamo J, et al. Six-year supplementation with alpha-tocopherol and beta-carotene and age-related maculopathy. Acta Ophthalmol Scand 1998;76:224–9.
94. Taylor HR, Tikellis G, Robman LD, McCarty CA, McNeil JJ. Vitamin E supplementation and macular degeneration: randomised controlled trial. BMJ 2002;325(7354):11–4.
95. Bartlett HE, Eperjesi F. Effect of lutein and antioxidant dietary supplementation on contrast sensitivity in age-related macular disease: a randomized controlled trial. Eur J Clin Nutr 2007;61:1121–7.
96. Parisi V, Tedeschi M, Gallinaro G, et al. Carotenoids and antioxidants in age-related maculopathy italian study: multifocal electroretinogram modifications after 1 year. Ophthalmology 2008;115:324–33.
97. Age-Related Eye Disease Study (AREDS) Research Group. A simplified severity scale for age-related macular degeneration: AREDS Report No. 18. Arch Ophthalmol 2005;123:1570–4.
98. Age-Related Eye Disease Study Group. The Age-Related Eye Disease Study severity scale for age-related macular degeneration: AREDS Report No. 17. Arch Ophthalmol 2005;123:1484–98.
99. Blumenkranz MS, Russell SR, Robey MG, Kott-Blumenkranz R, Penneys N. Risk factors in age-related maculopathy complicated by choroidal neovascularization. Ophthalmology 1986;93:552–8.
100. Goldberg J, Flowerdew G, Smith E, Brody JA, Tso MOM. Factors associated with age-related macular degeneration. Am J Epidemiol 1988;128(4633):700–10.
101. EDCCS (Eye Disease Case Control Study). Antioxidant status and neovascular age-related macular degeneration. Eye Disease Case-Control Study Group. Arch Ophthalmol 1993;111:104–9.
102. Seddon JM, Ajani UA, Sperduto RD, et al. Dietary carotenoids, vitamins A, C, and E, and advanced age-related macular degeneration. Eye Disease Case-Control Study Group. JAMA 1994;272(7279):1413–20.

103. Ishihara N, Yuzawa M, Tamakoshi A. Antioxidants and angiogenetic factor associated with age-related macular degeneration (exudative type). Nippon Ganka Gakkai Zasshi 1997;101:248–51.
104. Simonelli F, Zarrilli F, Mazzeo S, et al. Serum oxidative and antioxidant parameters in a group of Italian patients with age-related maculopathy. Clin Chim Acta 2002;320:111–5.
105. West S, Vitale S, Hallfrisch J, et al. Are antioxidants or supplements protective for age-related macular degeneration? Arch Ophthalmol 1994;112:222–7.
106. Mares-Perlman J, Klein R, Klein BEK, et al. Association of zinc and antioxidant nutrients with age-related maculopathy. Arch Ophthalmol 1996;114:991–7.
107. VandenLangenberg GM, Mares-Perlman JA, Klein R, Klein BE, Brady WE, Palta M. Associations between antioxidant and zinc intake and the 5-year incidence of early age-related maculopathy in the Beaver Dam Eye Study. Am J Epidemiol 1998;148:204–14.
108. Christen WG, Ajani UA, Glynn RJ, et al. Prospective cohort study of antioxidant vitamin supplement use and the risk of age-related maculopathy. Am J Epidemiol 1999;149:476–84.
109. Delcourt C, Cristol JP, Tessier F, Leger CL, Descomps B, Papoz L. Age-related macular degeneration and antioxidant status in the POLA study. POLA Study Group. Pathologies Oculaires Liees a l'Age. Arch Ophthalmol 1999;117(10):1384–90.
110. Flood V, Smith W, Wang JJ, Manzi F, Webb K, Mitchell P. Dietary antioxidant intake and incidence of early age-related maculopathy: the Blue Mountains Eye Study. Ophthalmology 2002;109:2272–8.
111. Cho E, Seddon JM, Rosner B, Willett WC, Hankinson SE. Prospective study of intake of fruits, vegetables, vitamins, and carotenoids and risk of age-related maculopathy. Arch Ophthalmol 2004;122:883–92.
112. Tan JS, Wang JJ, Flood V, Rochtchina E, Smith W, Mitchell P. Dietary antioxidants and the long-term incidence of age-related macular degeneration: the Blue Mountains Eye Study. Ophthalmology 2008;115:334–41.
113. Age-Related Eye Disease Study Research Group, SanGiovanni JP, Chew EY, et al. The relationship of dietary carotenoid and vitamin A, E, and C intake with age-related macular degeneration in a case-control study: AREDS Report No. 22. Arch Ophthalmol 2007;125:1225–32.
114. Sanders TAB, Haines AP, Wormald R, Wright LA, Obeid O. Essential fatty acids, plasma cholesterol, and fat-soluble vitamins in subjects with age-related maculopathy and matched control subjects. Am J Clin Nutr 1993;57(7114): 428–33.
115. Mares-Perlman JA, Brady WE, Klein R, et al. Serum antioxidants and age-related macular degeneration in a population-based case-control study. Arch Ophthalmol 1995;113(12):1518–23.
116. Smith W, Mitchell P, Rochester C. Serum beta carotene, alpha tocopherol, and age-related maculopathy: the Blue Mountains Eye Study. Am J Ophthalmol 1997;124:838–40.
117. Cardinault N, Abalain JH, Sairafi B, et al. Lycopene but not lutein nor zeaxanthin decreases in serum and lipoproteins in age-related macular degeneration patients. Clin Chim Acta 2005;357:34–42.
118. Belda JI, Roma J, Vilela C, et al. Serum vitamin E levels negatively correlate with severity of age-related macular degeneration. Mechanisms of Ageing & Development 1999;107(2):159–64.
119. Mangels AR, Holden JM, Beecher GR, Forman MR, Lanza E. Carotenoid content of fruits and vegetables: an evaluation of analytic data. J Am Diet Assoc 1993;93:284–96.
120. Landrum JT, Bone RA, Joa H, Kilburn MD, Moore LL, Sprague KE. A one year study of the macular pigment: The effect of 140 days of a lutein supplement. Exp Eye Res 1997;65:57–62.
121. Hammond BR, Jr., Johnson EJ, Russell RM, et al. Dietary modification of human macular pigment density. Invest Ophthalmol Vis Sci 1997;38:1795–801.
122. Snellen EL, Verbeek AL, Van Den Hoogen GW, Cruysberg JR, Hoyng CB. Neovascular age-related macular degeneration and its relationship to antioxidant intake. Acta Ophthalmol Scand 2002;80:368–71.
123. Mares-Perlman JA, Fisher AI, Klein R, et al. Lutein and zeaxanthin in the diet and serum and their relation to age-related maculopathy in the third national health and nutrition examination survey. Am J Epidemiol 2001;153: 424–32.
124. Gale CR, Hall NF, Phillips DI, Martyn CN. Lutein and zeaxanthin status and risk of age-related macular degeneration. Invest Ophthalmol Vis Sci 2003;44:2461–5.
125. Trieschmann M, Beatty S, Nolan JM, et al. Changes in macular pigment optical density and serum concentrations of its constituent carotenoids following supplemental lutein and zeaxanthin: the LUNA study. Exp Eye Res 2007;84: 718–28.
126. Cho E, Hankinson SE, Rosner B, Willett WC, Colditz GA. Prospective study of lutein/zeaxanthin intake and risk of age-related macular degeneration. Am J Clin Nutr 2008;87:1837–43.

127. Yoshida M, Takashima Y, Inoue M, et al. Prospective study showing that dietary vitamin C reduced the risk of age-related cataracts in a middle-aged Japanese population. Eur J Nutr 2007 Jan 30; [Epub ahead of print] 2007.
128. Chylack LT, Leske MC, McCarthy D, Khu P, Kashiwagi T, Sperduto R. Lens opacities classification system II (LOCS II). Arch Ophthalmol 1989;107(4125):991–7.
129. Richer S, Stiles W, Statkute L, et al. Double-masked, placebo-controlled, randomized trial of lutein and antioxidant supplementation in the intervention of atrophic age-related macular degeneration: the Veterans LAST study (Lutein Antioxidant Supplementation Trial). Optometry 2004;75:216–30.

22 Micronutrients and Immunity in Older People

John D. Bogden and Donald B. Louria

Key Points

- Placebo-controlled clinical trials, despite their limitations, are the best approach for studying effects of micronutrients on immunity.
- High doses of some single nutrients may improve immunity in relatively short time periods—weeks to months – but persistence of these effects is not known at this time. High doses of other micronutrients may adversely affect immunity.
- Some micronutrients may interfere with the beneficial effects of other micronutrients on immunity; this effect will depend on relative doses.
- Low-to-moderate-dose multivitamin/mineral supplements may require considerable time (6 months to 1 year or more) before they enhance immune functions and reduce susceptibility to infectious diseases, and the timing of their effects may differ in men and women.
- High- and even low-dose micronutrient supplements may enhance immunity even in the absence of evidence of underlying deficiencies.
- Long-term ingestion of single nutrient supplements, especially at high doses, may have beneficial and/or adverse effects on immunity and other outcomes.
- Micronutrient supplements are not a substitute for a good diet and regular exercise but rather are a complementary measure.

Key Words: Elderly; immunity; infection; micronutrients; minerals; vitamins

22.1. INTRODUCTION

Aging has been described as a group of processes that promote vulnerability to challenges, thereby increasing the likelihood of death. Since there is evidence that depressed immunity can increase the risk of death, it is likely that changes in immunity with age are key factors in the aging process.

Theories of aging include the free radical, programmed senescence, and immunologic theories (*1*). Evidence for the immunologic theory of aging is based largely on the well-described changes with age that occur in various species that have been studied, including man, and on observations from cross-sectional studies that demonstrate an association between maintenance of good immune function and longevity (*1*,*2*). A limitation of this theory is that it lacks the universality of other theories, such as the free radical theory of aging, since it is not applicable to lower organisms that do not have well-developed immune systems. Of course, the complexity of aging may require the use of more than one theory to understand it, and the various theories are not necessarily independent of one another. For

A. Bendich, R.J. Deckelbaum (eds.), *Preventive Nutrition*, Nutrition and Health, DOI 10.1007/978-1-60327-542-2_22,

example, there is evidence that demonstrates that antioxidant nutrients that reduce free radical damage can improve immunity in older people (*3*), suggesting that the free radical and immunologic theories may be complementary.

22.2. AGING AND IMMUNITY

22.2.1. General Changes in Immunity with Aging

It is useful to distinguish between primary changes that develop due to the age-dependent intrinsic decline of immunity and secondary changes that are the result of "environmental" factors such as prescription and nonprescription drug use, physical activity, and diet. In fact, Lesourd and Mazari (*4*) have suggested that secondary rather than primary changes in immunity with age are more likely to explain the increased incidence and severity of infectious diseases in older people.

Changes in immunity with aging include inhibited T-lymphocyte functions, decreased antibody production and responses, increased autoimmune activity with compromised self–non-self discrimination, and greater heterogeneity in immunologic responses (*5–9*). With regard to the latter, depressed T-cell function is the most common and may begin as early as the sixth decade. However, T-cell dysfunction is neither inevitable nor predictable. For example, we (*8*) measured delayed hypersensitivity skin test responses in 100 people aged 60–89. We found that although 41% were anergic to a panel of seven skin test antigens and an additional 29% were "relatively anergic," responding to only one of the seven antigens, the remaining 30% were reactive, responding to two or more of the skin test antigens, often with sizable reactions.

22.2.2. Specific Changes in Immunity with Aging

22.2.2.1. Involution of the Thymus

The most striking changes in immunity with increasing age are inhibited T-cell functions (Table 22.1). These are likely related to the well-known involution of the thymus (*10*). The differentiation process by which stem cells become T lymphocytes occurs in this organ. It is a two-lobed structure in mammals, located in the thorax above the heart. There are several stages in the process by which immature stem cells (pre-T cells) become mature T cells. These include migration to the thymus, where some cells are stimulated to grow and others die; differentiation, in which the mature

Table 22.1
Some Specific Changes in Immunity with Aging

Involution of the thymus
Decreased thymic hormone concentrations
Decreased delayed hypersensitivity skin test responses
Decreased interleukin-2 secretion
Decreased lymphocyte proliferative responses to mitogens
Lower antibody titers after vaccination
Increased serum autoantibodies
Increased soluble interleukin-2 receptors
Reduced phagocytosis by polymorphonuclear leukocytes
Reduced intracellular killing by polymorphonuclear leukocytes

phenotype of T cells develops in the thymus, including surface expression of accessory molecules; positive selection, in which self-major histocompatibility complex (MHC)-restricted T cells are selected and other cells rejected; and negative selection, which ensures that surviving mature T cells are self-tolerant. The selective survival or death of cells results in a self-MHC-restricted, self antigen-tolerant, mature T-cell population (*10*).

The thymus is the principal site of T-cell maturation. Involution with age occurs soon after puberty. Since some maturation of T cells continues throughout adult life, it is likely that a remnant of the thymus or some other tissue continues to effect T-cell maturation (*10*). However, since memory T cells have a long life span of 20 years or more (*10*), the involution of the thymus does not cause compromised immunity in young adults but is likely to contribute to depressed immunity as the time since thymic involution increases.

The involution of the thymus prior to the peak reproductive years suggests that this process may provide an evolutionary advantage. One hypothesis is that involution provides a net benefit, since it reduces the risk of autoimmune reactions (*11*). According to this theory, the increased risk of cancer or infectious diseases due to depressed cellular immunity is a detriment that is offset by a reduced risk of autoimmune disease that accompanies thymic involution. Although attractive, this theory of immunologic "trade-offs" as an adaptation to aging requires additional supporting evidence.

An alternative hypothesis has been proposed by Siskind (*12*), who suggests that adaptation to environmental pathogens occurs early in life and, thereafter, relative constancy of immune function rather than adaptability may be most beneficial. He further speculates that efforts to modify cellular immunity in later life, e.g., by pharmacological or nutritional means, may do more harm than good. Though interesting, this hypothesis is not widely supported and not consistent with the known association between good cellular immunity and reduced morbidity and mortality in older people.

22.2.2.2. T-LYMPHOCYTE FUNCTIONS

Changes in T lymphocytes with aging include a shift in relative percentages of subpopulations and qualitative changes in cell surface receptors (*13*). In comparison to T cells from younger people, cells of the elderly are deficient in in vitro production of certain T-cell growth factors such as interleukin-2 (IL-2) and have a decreased ability to bind and respond to it (*14–17*). McMurray (*15*) has outlined evidence that implicates nutrient-mediated effects at virtually every step in the development and expression of T-cell immunity, from direct effects on the thymus and thymic hormone production through T-cell maturation and distribution, antigen reactivity, lymphokine production, and even composition of the T-cell membrane.

Delayed hypersensitivity skin test (DTH) responses involve T-lymphocyte proliferation, production of IL-2 and other lymphokines, and infiltration of the test site with mononuclear cells, resulting 24–72 h later in induration and erythema; it is the T-cell parameter that is most consistently and profoundly affected by nutritional status (*15*). Reduced DTH is also the immune parameter most consistently associated in older people with increased infectious disease morbidity and mortality from all causes, as found by Meakins et al. (*18*) and Christou et al. (*19*) for surgery patients and Wayne et al. (*20*) and Roberts-Thomson et al. (*21*) for initially healthy people aged 60 or older.

In their investigation, Christou et al. (*19*) studied the relationship between presurgery DTH responses and postsurgical sepsis-related death in 245 subjects with a median age of 67 years and a range of 24–98 years. Initially anergic subjects experienced significantly more postsurgical mortality than those who were reactive. Since all the subjects had gastrointestinal cancers that prompted the decision to operate, it could be argued that the initial severity of the disease increased both the incidence of anergy and the risk of dying postoperatively. Thus, initial disease severity could explain the

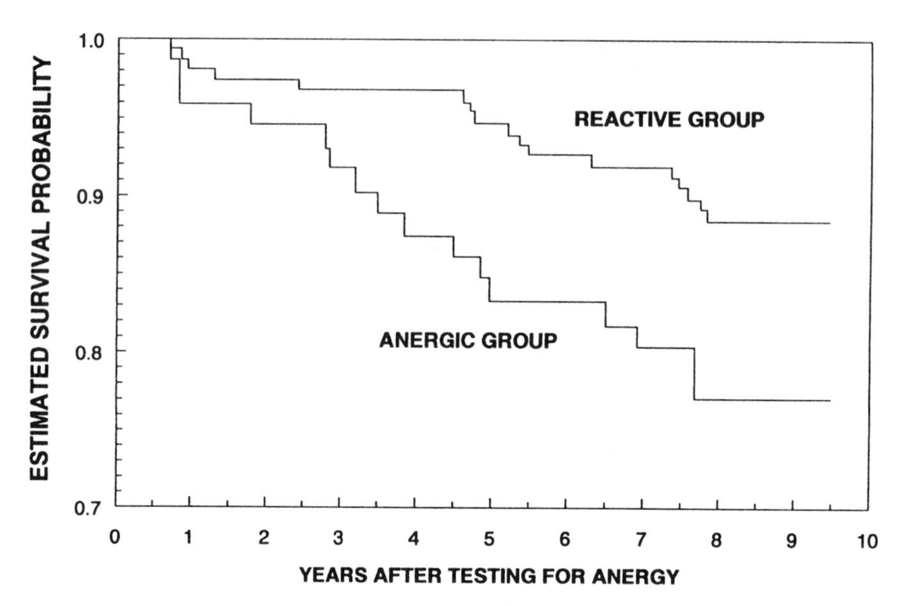

Fig. 22.1. Kaplan–Meier curves of all-cause mortality for initially anergic and reactive older people during 10-year follow-up period. Subjects ($n = 273$) were aged 60 or older and apparently healthy at enrollment. Participants were considered anergic if their responses to each of four skin test antigens were less than 5 mm of induration. Most of the excess mortality in the anergic group occurred within 5 years of enrollment. Adapted from Wayne et al. (*20*).

apparent strong relationship between preoperative DTH responses and postsurgical mortality. However, the study of Wayne et al. (*20*) did not have this confounder since they looked prospectively at healthy adults over a 10-year time period. In this investigation, the authors followed 273 initially healthy subjects aged 60 or older with no history of serious medical problems. DTH responses to four recall antigens were measured at enrollment. Anergy (failure to respond to any skin test antigen) at enrollment in the study was associated with a significantly increased risk of dying in the 10-year follow-up period (Fig. 22.1). At the end of 10 years, 11% of the initially reactive subjects had died compared to 22% of the anergic participants. The study demonstrates that anergy to skin test antigens, even when present in healthy older people, is associated with subsequently increased all-cause mortality. The authors also found a 2½-fold increase in cancer mortality in the initially anergic group in comparison to the reactive group. However, this was not statistically significant, probably because of the relatively small number of cancer deaths observed.

Evidence for the decline in T-cell function with age includes a considerable number of studies that demonstrate reduced lymphocyte proliferative responses (LPR) to mitogens or antigens, as well as depressed DTH responses to recall antigens (*7,16–23*). Indeed, these two measures of T-cell function have been the most widely studied functional tests done in conjunction with the assessment of the effects of nutritional intervention on immunity. A problem with lymphocyte proliferative responses to mitogens is the considerable variability of these assays, even in laboratories with rigid quality control procedures.

There is some evidence for changes in T-lymphocyte subsets with aging, in particular decreases in CD4+, increases in CD8+ cells, and decreases in the CD4+/CD8+ ratio (*9*). There is also evidence that lymphocyte subsets are altered in older people who are ill. For example, Markewitz et al. (*24*) have found that immunosuppression in cardiopulmonary bypass surgery patients aged 55 or older is associated with decreases in CD4+ T cells and increases in CD8+ T cells. Higa et al. (*25*) have found that increases in CD8+ T cells predict a longer period of recovery after onset of acute herpetic pain during herpes zoster infection. The increased incidence of this disease in older people is thought to be due to the depressed cellular immunity that occurs with age (*14*).

Measurement of lymphocyte subsets is a key component in the evaluation of immune function (*26,27*). Knowledge of lymphocyte subset numbers (cells/mL and percent of total) allows determination of relationships between immune functions and the number and percentage of cells responsible for these functions. This can permit distinguishing between effects due to increased numbers of a particular subgroup of cells versus enhanced activity by the same number of cells. The latter could be related to antigen-binding capacity per cell. Indeed, changes in antigen-binding capacity per cell could be a mechanism by which micronutrients influence immune functions. However, the effects of changes in antigen-binding capacity per cell on declining immunity with age are largely unexplored.

22.2.2.3. Other Immune System Changes with Aging

There is some evidence for a decline in B-cell functions with age, although it is likely related at least in part to the T-cell dependence of B-cell functions. Older people vaccinated with tetanus toxoid, varicella zoster, or hepatitis B antigens demonstrate reduced antibody production as well as a greater percentage of nonresponders compared to young adults. This may also be true after pneumococcal and influenza virus immunization, though the evidence is not as convincing (*28*).

Perskin and Cronstein (*29*) have reported that aging produces alterations in neutrophil plasma membrane viscosity that may result in compromised neutrophil function and increased susceptibility to infection with specific pyogenic bacteria. This is consistent with studies of Nagel et al. (*30*), Shoham-Kesary and Gershon (*31*), and Corberand et al. (*32*) that suggest compromised in vitro activity of neutrophils from older people. Depending on the microorganism studied, the neutrophil activity that was depressed was phagocytosis or intracellular killing.

A review by Makinodan et al. (*33*) suggests that although antigen-responsive cells such as B cells, monocytes, and killer cells are vulnerable to aging, T cells are clearly the most vulnerable. This is the reason that many of the studies of nutrition, immunity, and aging have focused on T-cell functions.

Sen et al. (*28*) have published an insightful review that distinguishes between an increased incidence versus greater severity of infectious diseases in older people. For example, they report an increased case–fatality ratio for bacterial meningitis and pneumococcal pneumonia in older people, and an increased incidence of diseases such as urinary tract infections and varicella zoster. Other diseases such as influenza virus infection and gram-negative sepsis are both more frequent and more severe in older people. They suggest that in addition to changes in immunity with age, urinary tract, respiratory tract, and neurological changes may contribute to the increase in infectious disease morbidity and mortality in older people.

Relationships among the interleukins, their receptors, and immunity have been widely discussed in the recent literature. Of particular interest in the elderly is interleukin-2 (IL-2), since its production is decreased in older people (*9*). Interestingly, soluble IL-2 receptor (IL-2R) levels are higher in older than in younger adults (*34*) and it has been suggested that this may be a factor in the decline of cellular immunity with age, since high serum concentrations of soluble IL-2R may compete with and decrease IL-2 binding to T-cell IL-2 receptors and thereby compromise immunity (*35,36*). We have previously found that serum IL-2R concentrations are relatively lower in physically active older people compared with sedentary seniors and that exercise/physical activity habits and multivitamin supplementation may interact to influence soluble serum IL-2R concentrations (*37*). We have also verified the higher levels of soluble IL-2R in older people (unpublished data).

The immunity of the very old may in part explain their survival to an advanced age. The oldest people, including centenarians studied by Sansoni et al. (*38*), tend to have well-preserved immune functions, such as natural killer cell activity, that are often better than those aged 50–80 years. In addition, those above age 90 tend to have lower serum autoantibody concentrations than those in the 60–80 year range (*14,39*). Thus, preserved immune functions and reduced autoimmunity appear to be associated with the ability to live to age 90 and beyond.

22.3. MICRONUTRIENTS AND IMMUNITY

22.3.1. Nutrition, Immunity, and Aging

Scrimshaw and SanGiovanni (*40*) have noted that infections, no matter how mild, can adversely affect nutritional status, which in turn can compromise immunity and exacerbate the effects of infection. They discuss evidence for the effects of various micronutrient deficiencies on immunity, including β-carotene; pyridoxine; folic acid; pantothenate; vitamins A, B_{12}, C, D, and E; the trace elements iron, zinc, copper, and magnesium. In general, cell-mediated and nonspecific immune functions are more sensitive to single micronutrient deficiencies than humoral immunity.

Fraker (*41*) has noted that the immune system is a large "organ" composed of the blood, spleen, lymphatic system, thymus, and other components. In addition, millions of new immune system cells are produced daily. Its large size and high cellular turnover combine to make the immune system a major user of nutrients. Thus, it is not surprising that some aspects of immunity are very sensitive to nutritional deficiencies.

One key question is whether the decline in immunity with aging is due, at least in part, to nutritional deficiencies and/or increased requirements. Another possibility is that micronutrient supplementation might improve immunity even in the absence of an underlying "deficiency," defined by factors such as low circulating nutrient concentrations or consistently low intakes.

Human studies of protein-calorie malnutrition (PCM) in underdeveloped countries or in hospitalized adults demonstrate a causal association between undernutrition and secondary immunodepression that results in diminished resistance to infectious diseases (*14,15,42,43*). This association is consistent enough to permit the use of DTH in medical and surgical patients as a predictor of clinical prognosis (*19*). Thus, there appears to be little doubt that severe malnutrition has a major impact on resistance to disease that is mediated in part through the immune system. There is also evidence that moderate to marginal undernutrition may compromise immunity (*44,45*).

McMurray (*15*) has noted that dietary deficiencies, both moderate and severe, of specific nutrients profoundly alter cell-mediated immune responses in humans and experimental animals. Diets with inadequate contents of calories, proteins, vitamin A, pyridoxine, biotin, or zinc can result in depressed production of thymic hormones critical for T-lymphocyte differentiation. Reduced numbers and depressed in vitro function of T cells have also been reported in experimental deficiencies of zinc, copper, iron, and vitamins A and E. Depressed DTH responses are a consistent result of dietary inadequacies of protein, pyridoxine, iron, zinc, and vitamins A and C.

A comprehensive review by Beisel (*46*) extensively examined the literature on single nutrients and immunity up to 1982. The water-soluble vitamins that appear to be most critical for maintaining immunity are vitamin B_6, folate, vitamin B_{12}, and vitamin C. Among the lipid-soluble micronutrients, vitamins A and E appear to exert the most significant impacts. Recent studies have shown that vitamin D is also an important immune modulator. Trace metals that exert substantial influences on immune functions are iron, zinc, selenium, and copper (*15,46*).

Since the variability in immune responses increases with aging, subgroups that have impaired immunity because of nutrient deficiencies are more likely to be observed in the elderly than in other age groups. In addition, when episodes of nutritional vulnerability overlap with suboptimal immune function, an adverse synergistic interaction is possible (*15*). These factors make it more rewarding to study nutrition/immunity relationships in older rather than in younger adults. In 1985, Beisel (*47*) noted that individual studies of immunity in humans have not been systematic or comprehensive; despite the passage of 24 years, this continues to be an accurate observation. The lack of comprehensive studies is no doubt related to the very considerable expense that would be incurred in studying multiple immune responses in a sizeable number of older people.

It can be useful to compare relationships between nutrition and immunity in healthy older people with these relationships in diseases in which immune functions are compromised. In the case of HIV infection, we (*48*) have found that compromised nutritional and antioxidant status begins early in the course of infection and may contribute to disease progression. This observation can be compared to the decline in cellular immunity that begins in many older people in the fifth or sixth decade and is associated with a reduced life expectancy (*2,5,20*).

22.3.2. Cross-Sectional Studies on Micronutrient Nutrition and Immunity

Goodwin and Garry (*49*) compared immunological functions of healthy elderly New Mexico residents consuming higher than RDA levels (5× RDA or greater) of micronutrients with similar individuals not taking supplements. Micronutrients evaluated were vitamins A, C, D, and E, the B vitamins, iron, calcium, and zinc. There was no significant difference between the two groups in DTH responses or in vitro lymphocyte proliferative responses to mitogens. The authors suggested that the immune-enhancing properties of high doses of vitamins might be the result of a nonspecific adjuvant effect that does not persist with time.

More recently the same authors (*50*) studied 230 healthy older men and women to determine if subclinical micronutrient deficiencies could contribute to the depressed immunity found in many of the elderly. Immune functions studied included DTH responses, in vitro lymphocyte proliferative responses to PHA, lymphocyte counts, and levels of serum autoantibodies. Spearman correlation coefficients were calculated to assess associations between blood micronutrient concentrations and selected immune functions. The authors also compared subjects with the lowest responses to those with the highest. There were no significant associations between low serum micronutrient concentrations and immune functions, and the authors suggested that subtle nutrient differences did not appear to contribute to the immunodeficiency of aging. However, the population sample studied was relatively affluent and people taking prescription drugs or daily over-the-counter medications, as well as those with a serious medical problem, were excluded. Thus, the study may have excluded those subjects who might benefit most from micronutrient supplements.

Kawakami et al. (*51*) studied 155 healthy subjects aged 20–99 years and suggested that cell-mediated immunity was reduced as a result of malnutrition.

In a more recent study (*52*) we examined relationships between immunity and dietary and serum antioxidants, B vitamins, essential trace metals, and serum homocysteine in 65 older men and women aged 53–86 years. Subjects who had used vitamin or mineral supplements in the preceding 3 months were excluded. Soluble serum interleukin-2 receptor concentrations (sIL-2R) were positively associated with body mass index and serum concentrations of homocysteine, and negatively associated with serum β-carotene and dietary lycopene. This is a logical result that suggests that systemic inflammation, for which elevated sIL-2R levels are a marker, is increased by obesity and factors that increase serum homocysteine but reduced by dietary antioxidants. In a multiple regression model, the above four factors and serum vitamin B_6 concentrations explained 52% of the variability in sIL-2R. The percentage of subjects with anergy to a panel of seven recall skin antigens was 25%, and these responses were negatively associated with T-helper cell number, suggesting the reduced numbers of the latter as a factor that may have contributed to the anergy of the subjects. T-helper cell numbers were positively associated with serum copper, and natural killer cell numbers were positively associated with dietary folate and vitamin B_6. These results document significant relationships between micronutrient nutrition and immunity, and suggest that IL-2 may be influenced by dietary antioxidants and B vitamins, including those that modify homocysteine metabolism.

The above studies were not attempts to intervene by provision of micronutrient supplements but were assessments of associations between the subjects' usual intakes or blood concentrations and selected immune functions. Variables that cannot be controlled in cross-sectional studies may mask associations between nutritional factors and immunity, especially since immunity is likely to be dependent on a number of factors, only one of which is nutritional status. Such studies are valuable as a way to identify nutrients for more intensive study but can only provide statistical associations that may not be cause/effect relationships. The latter can be assessed by standard placebo-controlled clinical trials.

22.3.3. Clinical Trials of Single Micronutrients

Several clinical trials have been conducted during the last two decades. These have included depletion/repletion studies in young volunteers and provision of micronutrient supplements to older people who did not appear to have preexisting deficiencies. Jacob et al. (*53*) studied the effects of short-term ascorbate depletion on immunity and other factors in young adult males confined to a metabolic ward. Ascorbate depletion was achieved using daily doses of 5–20 mg/day, while repletion was achieved with doses of 60 (the RDA at that time) to 250 mg/day. Although lymphocyte proliferative responses to mitogens were not affected by ascorbate depletion/repletion, delayed dermal hypersensitivity skin test (DTH) responses to a panel of seven recall antigens were markedly depressed by ascorbate depletion. Repletion for 28 days at either 60 or 250 mg/day did not restore the mean antigen score to the predepletion level, though there was some improvement in induration in three of the eight men studied. These results suggest that DTH is more sensitive to ascorbate depletion than mitogen responses. They further suggest that the repletion period was of insufficient duration to produce a return of DTH to baseline levels and/or the repletion doses were not large enough.

Fuller et al. (*54*) studied the effect of β-carotene supplementation on the UV radiation-induced photosuppression of DTH in 24 young adult males, aged 19-39 years. They found that exposure to a UV-A/B light source over a 16-day period significantly reduced DTH responses in a control (placebo) group to 39% of the initial values but did not induce significant reductions in a group given 30 mg β-carotene per day. This group repeated the study in elderly men and found that UV suppression of DTH was again prevented with β-carotene supplementation, though, as could be anticipated, there was more variability in DTH responses in the older people compared to young adults (*55,56*).

Watson et al. (*57*) investigated the effects of β-carotene on lymphocyte subpopulations in male and female subjects with a mean age of 56 years. β-Carotene was given at doses of 15, 30, 45, or 60 mg/day for 2 months. Using monoclonal antibodies to identify lymphocyte subsets, they found that the percentages of T-helper and natural killer cells, as well as cells with IL-2 and transferrin receptors, were increased in a dose-related fashion. There were no significant effects of β-carotene on T-suppressor cells. However, the number of subjects in each treatment group was only 3-5; thus, further investigation is needed to confirm these findings.

Santos et al. (*58*) found that men participating in the Physicians' Health Study who consumed 50 mg of β-carotene on alternate days for an average of 12 years had significantly greater natural killer cell activity than controls given placebos. Surveillance by natural killer cells is considered to be protective against the development of cancer. However, two large intervention trials have found an association between high doses of β-carotene and the development of lung cancer in cigarette smokers (*59,60*).

The role of the immune system in leading to the development of lung cancer in these studies is not known.

Talbott et al. (*61*) in a pilot study investigated the impact of pyridoxine on lymphocyte responses in 15 older (aged 65-81 years) mostly female subjects and found that administration of a pharmacologic dose (50 mg/day) of pyridoxine hydrochloride significantly increased in vitro lymphocyte proliferative responses to phytohemagglutin (PHA), pokewood mitogen, and *Staphylococcus aureus*.

Meydani et al. (*62*) have reported that vitamin B_6 deficiency impairs IL-2 production and lymphocyte proliferation in older adults. Each of these measurements was reduced by about 50% by depletion, while repletion with near RDA levels of B6 eventually increased values to about the baseline levels. Although only eight subjects were studied, this investigation supports a number of other studies that suggest that vitamin B_6 may play a key role in immune responses (*63*).

In another study, Meydani et al. (*64*) gave older people 50, 200, or 800 mg of vitamin E daily for 4–5 months. This resulted in improved antibody titers to hepatitis B vaccine and enhanced DTH responses, especially in the group consuming 200 mg of vitamin E per day. This suggests 200 mg as a recommended dose, although lower doses may be equally effective when administered for longer periods of time. In a more recent study, Pallast et al. (*65*) investigated the effects of 6 months of supplementation of healthy older men (aged 65–80 years) with vitamin E at doses of 50 and 100 mg daily for 6 months. There was a dose-related trend of increased DTH responses, especially in those subjects with initially low responses, suggesting that there are subgroups of older people that might benefit most from vitamin E supplements.

There has been considerable interest in the potential for zinc to improve immune functions in older people. It is clear that severe zinc deficiency in animals and people, e.g., as found in the disease acrodermatitis enteropathica, can greatly compromise cellular immunity and lead to the development of life-threatening opportunistic infections (*66*). There are also reports of significant associations between plasma or cellular zinc concentrations and immune functions such as DTH responses in older people (*8,67*). However, more recent studies of the impact of zinc supplementation on immunity in older people have not been encouraging. They have either demonstrated no beneficial effect of zinc supplements on immunity or an adverse effect even when the supplements contained modest doses of zinc in the range of 15–25 mg/day (*68,69*). In the absence of an underlying deficiency, use of zinc supplements by older people, especially at doses that exceed 15 mg/day, is more likely to adversely affect immunity than improve it. The effects of zinc on immunity in the elderly have been reviewed (*70*).

Doherty et al (*71*) studied the effect of moderate (1.5 mg/kg) versus high (6.0 mg/kg) zinc supplementation on mortality in 141 young children in Bangladesh with protein-energy malnutrition and weight-for-age *z* scores of about –4.6. Mortality was significantly greater in the high-dose group, with sepsis a frequent contributing factor. The results suggest that high-dose zinc supplementation may contribute to increased risk of sepsis and mortality in severely malnourished children. Although this study involved only very young children, aged 6 months to 3 years, it suggests caution in the use of high-dose zinc supplements by any age group.

In another study conducted in 56 *Shigella flexneri*-infected children in Bangladesh, Rahman et al (*72*) administered 20 mg of zinc daily in combination with a multivitamin. Controls received only the multivitamin. They found that adjunct therapy with zinc during acute shigellosis significantly improved seroconversion to shigellacidal antibody responses and also increased the percentages of B lymphocytes and plasma cells in the blood. Given the widespread presence of zinc and other nutrient deficiencies in Bangladesh, the wider applicability of these results to other populations is uncertain.

Martineau et al (*73*) studied the effect of a single high dose of 2500 μg of vitamin D on antimycobacterial immunity in 192 adults living in London who had been exposed to tuberculosis patients. The vitamin D supplement significantly improved in vitro antimycobacterial immunity in the study subjects.

22.3.4. Clinical Trials of Combinations of Micronutrients

The studies discussed above focused on the effects of relatively large doses of individual micronutrients on immune functions. There have been a limited number of published placebo-controlled trials of the effects of multivitamin/mineral supplements on immune functions in older people.

In the first of these studies, we investigated the effects of zinc given in combination with a multivitamin on immune functions in 63 older people (*74*). All subjects received a standard multivitamin/mineral supplement that contained all the essential micronutrients except zinc. In addition, subjects received 15 mg or 100 mg of zinc or a placebo. Daily consumption of the multivitamin/mineral supplement for 1 year was associated with enhanced DTH and mitogen responses, but these effects were reduced and delayed by ingestion of 15 and especially 100 mg of zinc each day. These data suggest that interactions among micronutrients may influence their effects on immunity and that some individual micronutrients, even at modest doses, may have unexpected adverse effects. The adverse impact of zinc is consistent with other previously cited studies that indicate that zinc supplements in healthy older people either do not improve immunity or adversely affect it (*68,69*). Penn et al. (*75*) studied the effects of a supplement containing vitamin C (100 mg), vitamin A (8,000 IU), and vitamin E (50 mg) on immune functions; the supplement or a placebo was administered for 28 days to the 30 elderly subjects studied. All were patients who had been hospitalized for at least 3 months. The number and the percent of CD4+ and CD8+ T cells were significantly increased in the supplemented group but not in a placebo group. Proliferative responses of lymphocytes to the mitogen PHA were also significantly increased in the supplemented group by 64-283% but were not affected by the placebo. There was biochemical evidence of deficiencies of vitamins A, C, and/or E in 5–47% of the supplemented subjects at enrollment into the study. Thus, it is possible that the improvement in cellular immunity in these subjects with short-term administration of vitamins A, C, and E was due to correction of underlying deficiencies that are more likely to be present in hospitalized than in independently living older people. These results suggest that this group of micronutrients may be particularly important for enhancement of immune responses in older people.

In another study, Chavance et al. (*76*) enrolled 218 subjects aged 60 or older who were living independently and had not used any vitamin supplements for at least the prior 3 months. They were given a low-dose multivitamin or placebo for 4 months. No clinical or laboratory assessments of immune function were conducted. The authors found no significant effects of supplementation on the incidence of infections; however, effects on the duration of each infection or the total number of days of infection were not assessed. As suggested by the authors, the failure to find any significant effects on the incidence of infections may be due to the short duration of supplementation. This is consistent with our results, which suggest that periods of supplementation of more than 6 months are required before significant improvements in immune functions occur in older people.

We also conducted a randomized, placebo-controlled, double-blind trial of the effects of RDA-level micronutrient supplementation on plasma vitamin and trace metal concentrations and immune functions in independently living healthy older subjects (*77*). The over-the-counter micronutrient supplement used in the study contained RDA levels of each of the essential vitamins and low-to-moderate doses of minerals.

Of the 65 subjects enrolled, 56 (86%) completed the 1-year study. About two-third were females. As expected, there were no statistically significant effects of the placebo on plasma micronutrient concentrations. In contrast, the data for the micronutrient supplement group show statistically significant increases at 6 and/or 12 months for plasma concentrations of ascorbate, β-carotene, folate, vitamin B_6, and α-tocopherol. These data verify compliance of the study subjects and demonstrate that supplementation with RDA levels of the latter micronutrients can increase their plasma concentrations in older people.

Table 22.2
Delayed Hypersensitivity Skin Test Responses of Placebo and Micronutrient Groups

Subgroup and Response Type	*Placebo Group*			*Micronutrient Group*		
	0 month	*6 month*	*12 month*	*0 month*	*6 month*	*12 month*
All subjects						
Positive responses	1.65 0.30	1.42 + 0.25	1.73 + 0.29	1.45 + 0.25[a]	1.76 + 0.27[ab]	2.38 + 0.33[b]
Total induration (mm)	5.37 + 1.02	4.76 + 0.93	5.80 + 0.95	5.21 + 0.98[a]	5.73 + 0.94[ab]	8.40 + 1.25[b]
Males						
Positive responses	2.93 + 0.60	1.93 + 0.30	2.50 + 0.78	1.64 + 0.33[a]	2.59 + 0.43[ab]	2.86 + 0.53[b]
Total induration (mm)	8.86 + 1.91	6.36 + 1.29	8.88 + 2.51	6.23 + 1.15	8.85 + 1.58	10.91+ 2.08
Females						
Positive responses	1.18 + 0.29	1.24 + 0.31	1.45 + 0.27	1.33 + 0.36[a]	1.25 + 0.29[ab]	2.08 + 0.42[b]
Total induration (mm)	4.08 + 1.09	4.17 + 1.16	4.67 + 0.83	4.58 + 1.41[a]	3.83 + 0.95[a]	6.86 + 1.49[b]

Mean + SE; $n = 26$ for placebo group (7 males, 19 females), $n = 29$ for micronutrient group (11 males, 18 females). Positive responses are the mean number of antigens eliciting a response from a total of seven antigens. Total induration is the sum of the indurations of all positive responses. Within groups, values in the same row with different letter superscripts are significantly different, $P < 0.05$ (Wilcoxon signed-rank test).

Table 22.2 contains the data on DTH for all study subjects combined and for males and females separately. For induration in the placebo group, there were no statistically significant differences between the 0- and 6-month results, 0- and 12-month results, or 6- and 12-month data. Similar results were obtained for the analyses of the data for the placebo group on the number of positive responses.

For the micronutrient supplement group, there was also no significant difference for the data on induration at 0 and 6 months. However, there was a statistically significant difference between 0- and 12-month induration results ($p = 0.005$). There was an increase in induration between 6 and 12 months, but this did not achieve statistical significance ($p = 0.056$). Similar trends were observed for the individual skin test antigens.

Similar results were also obtained for the number of positive responses in the micronutrient treatment group. The mean number of positive responses in the placebo group increased by only 4.8% between 0 and 12 months, and induration by 8.0%. In contrast, in the micronutrient supplement group, the mean number of positive responses increased by 64% and induration by 61% between 0 and 12 months. These data provide strong evidence for the enhancement of DTH after 1 year of micronutrient supplementation.

The results also suggest that some enhancement of DTH responses occurred sooner (at 6 months) in the male subjects than in the females (Table 22.2). The male subjects had significantly greater DTH responses than the females at enrollment; this is consistent with previous data that suggest that DTH responses in males may differ from those in females (*78*). The diets of the male subjects differed from the females, being higher in energy intake as well as intake of individual micronutrients, and it is possible that this factor may have interacted with micronutrient supplementation to influence DTH responses.

There was an increase between 0 and 12 months in the number of subjects in the placebo group with low blood concentrations of some of the micronutrients measured, specifically β-carotene, retinol, folate, and vitamin B_6. This trend differed significantly from the micronutrient group, for which the number of low values changed very little between 0 and 12 months. Thus, the improvement in skin test responses in the micronutrient group is not due to the correction of underlying micronutrient deficiencies for the nine micronutrient concentrations that we determined in blood, at least as defined by current guidelines for low circulating concentrations. The increased number of low values in the placebo group suggests that older people who do not take vitamin supplements for a year may have an increased risk of developing one or more low concentrations, particularly for vitamin B_6, folate, and β-carotene.

Our data suggest that enhancement of immune functions in older subjects by low-dose micronutrient supplementation takes approximately 1 year. These results also suggest that the diets of older people are inadequate in one or more micronutrients and/or that the current RDAs for one or more micronutrients may be too low to support optimal immunity in older adults. For optimal responses, they required the RDA level of the vitamins in the supplement in addition to the amounts in their food.

It could be argued that a 60% increase in DTH responses over a 1-year period is only a mean increase of about 5% per month. However, this increase far exceeds the decline in DTH responses per year that occurs with aging and thus may completely prevent it. These results suggest that older subjects who take a "one-a-day"-type multivitamin supplement faithfully for at least 6-12 months may experience a substantial improvement in measures of cellular immunity such as DTH responses. It is possible that more rapid and/or larger increases in DTH responses would occur if higher doses of micronutrients were used.

Girodon et al. (*79*) studied the effects of trace element and vitamin supplementation on immunity and infections in institutionalized subjects aged 65 and older in France. Subjects ($n = 725$) received daily for 2 years a placebo, a trace element supplement containing 20 mg zinc and 100 μg selenium, a vitamin supplement with 120 mg vitamin C, 6 mg β-carotene, and 15 mg of vitamin E, or both the vitamin and trace element supplements. DTH responses were not significantly influenced by any treatment, but antibody responses to influenza vaccine were improved in the groups given zinc and selenium, and the incidence of respiratory tract infection was marginally lower ($p = 0.06$) in these groups. The vitamin and trace element supplements also reduced the prevalence of underlying deficiencies of these nutrients. Since these were institutionalized subjects with a high frequency of low blood micronutrient concentrations, the applicability of these results to healthy independently living people is uncertain. Nevertheless, this large study provides the first evidence that selenium may be a key nutrient in the maintenance of immunity in older people.

Winkler et al. (*80*) found that daily ingestion of a micronutrient supplement containing minerals and vitamins reduced the incidence and severity of the symptoms of infection with the common cold. However, the reductions in symptom "scores" were only about 19% and the duration of infection was not reduced. Because the study subjects also ingested a supplement containing "probiotic bacteria," the influence of the micronutrient supplement alone cannot be determined. The study subjects were young adults; thus, the applicability of the results to the elderly is uncertain.

22.3.5. Limitations of Current Knowledge

The above studies that focused on the effects of multivitamin/mineral supplements on immune functions, in combination with the short-term higher dose single nutrient studies such as those of Meydani et al. (*62,64*), Watson (*57*), and Talbott (*61*), provide solid evidence that micronutrient supplements can enhance immune functions in older people, but data on effects on the incidence and prevalence of infectious diseases are quite limited. Despite the evidence provided by these studies, we do not know if long-term daily use of multivitamin/mineral supplements will enhance immune functions and reduce the incidence and severity of infectious diseases in older people beyond the 1–2 year duration of the longest studies done to date. This is an unfortunate gap in our knowledge, because millions of older Americans currently consume a multivitamin/mineral supplement daily, either alone or in combination with one or more single nutrients at higher doses (*81,82*). This situation is in part the result of the limited objectives of all previously completed studies. All of the single nutrient studies have been of short duration, usually using high doses of one micronutrient, given to a relatively small number of subjects. Most of these studies have not assessed the impact of single nutrient supplementation on the incidence of infectious diseases, a limitation related to the small number of subjects enrolled in these studies and their short duration, with a consequent lack of statistical power to assess disease incidence. The studies on multivitamin and/or trace element supplements also have limitations: (1) The study of Chavance et al. (*76*) was of only 4 months duration. Although, this study assessed the impact of multivitamin supplementation on the incidence of infectious diseases, it did not include any measures of immune function. (2) The study of Penn et al. (*75*) was only 1 month in duration and included only older people who had been hospitalized for at least 3 months. (3) Our studies (*74,77*) assessed DTH responses, lymphocyte proliferative responses to mitogens, and natural killer cell activity, but we could not examine other measures of immunity or clinical outcomes and the period of supplementation was limited to 1 year. (4) The 2-year study of Girodon et al. (*79*) included only institutionalized subjects.

Thus, additional studies of micronutrient/immunity/disease relationships are required, in particular studies that focus on clinical outcomes.

22.4. FACTORS THAT CAN INFLUENCE NUTRITION–IMMUNITY RELATIONSHIPS

Factors that may influence micronutrient/immunity relationships in older people include gender, stress, disease, physical activity and exercise, obesity, and food choices.

In our 1994 study (*77*) of the effects of low-dose micronutrient supplements on immunity in older people, improvements in DTH responses occurred sooner in the males than the females. Although the reason for this is not known, one possibility is that the higher intake of micronutrients from food in the men results in a larger total micronutrient intake.

There are a considerable number of reports that psychological and physiological stress in experimental animals and people can depress cellular immune functions (*83,84*), though it is beyond the scope of this chapter to assess these studies in any detail. As an example, death of a spouse has been associated with depressed immune functions (*81*). However, virtually all studies of relationships between stress and immunity have not adequately assessed nutritional factors that may be altered by stress and in most cases have completely ignored nutrition. Physical and psychological stress can modify food intake in animals and people, and thus studies of stress/immunity relationships are usually confounded by nutritional factors that have not been adequately evaluated.

There is considerable evidence that physical activity/exercise patterns can influence immunity (*85–89*). In general, the data suggest that very strenuous exercise can acutely depress immunity. For example, various studies have found that participants in marathons have a significantly increased risk of respiratory infections in the 1–2-week period following the race (*88,89*). Chronic overtraining has also been associated with depressed immunity (*87*). In contrast, regular moderate exercise appears to enhance immune functions (*87*). One hypothesis is that regular exercise contributes to the maintenance of muscle mass, and muscle is the source of a key nutrient, glutamine, required by lymphocytes (*90*). In addition, alterations in cytokine levels as a result of regular exercise may also be a factor (*91,92*).

In a review, Nieman (*93*) concluded that infection risk following intensive exercise is likely related to acute nonpersistent changes in immunity. However, unless the athlete exceeds his or her usual training limits, immunocompromise is unlikely, though further research is needed to confirm this conclusion. In general, studies of macronutrient or micronutrient supplements in combination with exercise have shown no effects on immunity. For example, in a randomized trial of 112 elderly (mean = 79.2 ± 5.9 years) men and women, Paw et al. (*94*) reported that exercising twice per week improved DTH skin test responses to recall antigens, but consumption of micronutrient-enriched foods (25–100% of the RDA for various micronutrients) did not enhance the effect of exercise.

Stallone (*95*) has outlined studies that indicate that excess body weight in humans or experimental animals is associated with impairments in host defense mechanisms. Definitive studies have not been done, but there are data suggesting both beneficial and detrimental effects of weight loss on immunity. In experimental animals, it is well known that chronically reduced energy intake without malnutrition can profoundly ameliorate the detrimental effects of aging on immunity and can increase mean and maximum life span (*96*).

The well-established importance of some micronutrients such as zinc in the maintenance of immune function suggests that choices of foods high in these micronutrients may be beneficial, but this has not been validated in well-controlled studies.

Because of the evolutionary development of humans as hunter-gatherers who consumed foods but not supplements, it has been argued that appropriate food choices are sufficient to achieve optimal health, including optimal immunity. There is a substantial body of evidence that supports wise food choices, including diets high in fruit and vegetable intake and low in saturated fat, as key factors in preventing some chronic diseases. However, arguments based on evolution are compromised by two factors: first, that evolution has programmed humans and other species to live through our peak reproductive years, but not necessarily beyond them, and second, that our preagricultural ancestors had intakes of some nutrients (e.g., calcium, iron, and zinc) much higher than those of modern humans (*97*). Olshansky et al. (*98*) have used the term "manufactured time" to describe the use of prescription drugs and other methods to increase the odds of living beyond our reproductive years. Thus, it should not be surprising that micronutrient supplements may be particularly beneficial to the immune and other organ systems of older people.

Goodwin (*99*) has suggested that the relationship between depressed cellular immune function and subsequently increased mortality may be due to compromised immunity being a marker for clinically latent diseases or poor overall physiologic function. However, impaired immunity may also contribute to a reduced ability to defend against infections, cancers, and perhaps cardiovascular heart disease.

22.5. SOME RECENT REVIEWS

In a recent review, Bogden and Oleske (*100*) concluded that adverse effects from high-dose micronutrient supplements may occur in HIV-1-infected patients but that the combination of antiretroviral drug therapy, a good diet, and a safe low-dose multivitamin/mineral supplement may improve outcomes more than does pharmacologic therapy alone.

In another review, Webb and Villamor (*101*) concluded that vitamins C and E and the carotenoids, either individually or when combined in supplements, can influence various measures of immunity, including lymphocyte proliferation and delayed-type hypersensitivity responses. They also suggest that there is good evidence that multivitamins that include the B-vitamin group have beneficial effects on immunity and clinical outcomes in HIV-infected patients.

The NIH sponsored a conference entitled "Multivitamin/Mineral Supplements and Chronic Disease Prevention" that occurred during May 15–17, 2006. The conclusion of a comprehensive report on the conference proceedings was that there is sufficient uncertainty in the data on this subject that precludes a recommendation either in favor of or against routine multivitamin/mineral supplementation (*102*).

The long-anticipated publication of the second edition of the monograph "Food, Nutrition, and Physical Activity and the Prevention of Cancer: A Global Perspective" occurred in November, 2007

(*103*). Among the conclusions of this authoritative and comprehensive report are that there is "convincing" evidence that high-dose β-carotene supplements are a cause of lung cancer, that calcium "probably" protects against colorectal cancer, and that selenium in high doses "probably" reduces the risk of prostate cancer. Thus, supplements may have adverse or beneficial effects on the prevention of various cancers. However, the report concludes that "a general recommendation to consume supplements for cancer prevention might have unexpected adverse effects" in the general population in whom the balance of risks and benefits cannot be predicted with confidence.

The first two reviews described above focused on the outcomes of immunity and infection, primarily HIV-1 infection, whereas the latter two focus on cancers and other chronic diseases. Thus, it is not surprising that these reviews reach different conclusions about the use of supplements for disease prevention or management.

22.6. RESEARCH NEEDED ON MICRONUTRIENTS AND IMMUNITY IN OLDER PEOPLE

Several cross-sectional studies that assess relationships between micronutrient nutrition and immunity have been done in the past 20 years (*49,50*), as previously discussed. In general, significant associations between serum micronutrient concentrations or use of micronutrient supplements and various measures of immunity were found in some studies but not others (*49–52*). However, these studies compared micronutrient supplement users with nonusers but did not evaluate the use of specific supplements, and it is likely that some individual or combinations of micronutrients can improve immunity and others cannot.

The clinical trials of micronutrient supplementation and immunity done to date have usually involved healthy older subjects consuming their usual diets. In the case of some single nutrient studies, subjects lived in metabolic units and consumed standardized meals that contained about the RDA of all essential micronutrients. It is possible that the improvements in immunity found in some studies are due to correction of underlying deficiencies. However, it is also likely that micronutrient supplements enhance immunity even in the absence of underlying deficiencies, at least based on current concepts of "deficiency." This should not be surprising, since optimal immune function was not a factor in the establishment of the current Dietary Reference Intakes or in the definition of laboratory normal ranges for circulating micronutrient concentrations. In fact, daily intakes that optimize immunity may differ from both the current Dietary Reference Intakes and intakes that may prevent chronic diseases. For example, the current DRI/RDA for vitamin E (15 mg α-tocopherol equivalents for adult females and males) is substantially lower than amounts that optimize immune functions or have been associated with a reduced risk of cardiovascular heart disease (*64,65,104,105*). Similarly, the current RDA for vitamin C is adequate to prevent the development of scorbutic lesions but appears to be less than the intake that could optimize immunity or provide other health benefits (*4,106*). Recommendations for an optimal intake of any micronutrient will need to balance the impact of that nutrient on various health outcomes as well as consider possible adverse effects of relatively high doses.

Future studies that focus on clinical outcomes and have considerable statistical power are especially needed. An example is the investigation of Graat et al. (*107*). They conducted a randomized, double-blind, placebo-controlled trial in 652 Dutch men and women older than age 60 who were given either a placebo, a multivitamin/mineral supplement, 200 mg of vitamin E as α-tocopherol, or both supplements. The multivitamin/mineral supplement included 24 essential micronutrients and one "possibly essential" trace element silicon. The primary outcome measures were the incidence and severity of acute respiratory tract infections. The mean duration of participation was 441 days, and the percentage of subjects compliant with the protocol was 84%. The incidence of acute respiratory tract infections did not differ significantly among treatment groups. Surprisingly, infection severity, measured as duration of infection, restriction of activities, number of symptoms, and the presence of fever, was actually increased significantly ($p = 0.03$–0.009) in the groups ingesting vitamin E supplements. This study focused on a clinical outcome but did not include laboratory evaluation of immunity. Thus, immune system assays that might explain the study results were not available. Only 0.2% of study subjects had low plasma α-tocopherol concentrations at enrollment, and this may have precluded a beneficial effect of vitamin E supplements. The adverse effects on infection severity may be due to the long duration of high-dose supplementation in a cohort with normal plasma α-tocopherol concentrations at enrollment and are consistent with previously cited studies on β-carotene (*59,60*) that demonstrate adverse effects after long-term high-dose supplementation. These studies suggest caution in the long-term use of high-dose single micronutrients.

There is considerable evidence that patterns of physical activity and exercise can influence immunity both acutely and chronically, but very few studies have addressed interactions among physical activity, immunity, and micronutrient nutrition.

It should be emphasized that the potential of micronutrient supplements to improve immunity or exert other beneficial effects must be considered in relation to their consumption from food. This is especially true for low-to-moderate-dose supplements, for which the intake from food and supplements may be similar. Clearly, supplement use should be encouraged in conjunction with a sound diet that emphasizes fruits, vegetables, whole grains, and other sources of micronutrients and limits the intake of saturated fats. However, it is likely that beneficial intakes of some nutrients such as vitamin E may not be possible in the absence of supplement use.

The promising but variable results of studies done to date suggest continued research on nutrition and immunity in older people. Such efforts should include the following:

1. A focus on long-term, placebo-controlled, double-blind clinical trials and prospective epidemiological studies that have sufficient statistical power.
2. Study of interactions among physical activity/exercise patterns, immunity, and nutrition.
3. Evaluation of effects of nutrition on both humoral (e.g., antibody responses to vaccination) and cellular (e.g., DTH responses) immunity using clinically relevant assays and on clinical outcomes, e.g., infectious disease incidence, duration, and severity.
4. Evaluation of dietary modification alone or in combination with low doses of supplemental micronutrients. Studies of older people consuming their usual diets are also needed.
5. Long-term studies that address the persistence of the effects of micronutrients on immunity both during and after micronutrient supplementation.
6. Use of appropriate inclusion and exclusion criteria in the identification of subjects for study.
7. Study of both single micronutrients and multivitamin/minerals, with a focus on the antioxidant micronutrients and other widely used single or multiple micronutrient supplements.
8. Identification of host-specific factors (e.g., gender, age range) that influence micronutrient/immunity interactions and the basis for these effects.
9. Identification of the molecular mechanisms and genes that determine the effects of micronutrients on immunity. This will become increasingly important as new genes that influence aging are identified.

About 100 million Americans (approximately 40% of the population) take multivitamin/mineral supplements, either alone or in combination with higher doses of the antioxidant vitamins (*78,79*). Well-designed studies that assess the health impacts of this practice are urgently needed and should include evaluation of effects on the immune system.

22.7. RECOMMENDATIONS

Of course older adults and their health-care providers want to know not only what they should do to improve immunity, but also which particular combination of micronutrients is able to reduce their risk of cardiovascular heart disease, cancers, and other major diseases of the elderly. Two types of recommendations can be made, those directed to the manufacturers of micronutrient supplements and those directed to individual patients and their health-care providers.

Considerations that can guide the formulation and use of micronutrient supplements targeted to older people include the following: (1) consider all effects of the micronutrients, not just their effects on immunity; (2) adverse effects caused by micronutrient supplements can be anticipated in some individuals and formulations should be designed to minimize these effects; (3) nutrient interactions should be considered, e.g., zinc/copper; (4) dose should be a key factor; more is not necessarily better and may be more risky; (5) the anticipated duration of supplementation should be considered; higher doses may be more appropriate for short-term use; (6) micronutrient supplements are not a substitute for a good diet and regular exercise, but rather are a complementary measure; (7) nutrient supplements should include only those substances for which there are adequate convergent data that document essentiality or substantial potential health benefits.

Physicians and other health-care providers should advise their patients to eat diets low in saturated fat and high in fruits and vegetables. This can ensure consumption of significant quantities of the micronutrients and other phytochemicals that can favorably affect immunity. In addition, older subjects, especially those with poor diets, should be encouraged to take a low-dose multivitamin/mineral supplement. Taking high supplemental doses of other micronutrients that can adversely affect immunity, for example, zinc, should be persuasively discouraged. High doses of supplemental β-carotene are not recommended for smokers because of their association with the development of lung cancer and are unwise for other people, as concluded by the U.S. Preventive Services Task Force (*108*). The favorable effects of regular exercise on immunity should also be mentioned to patients. Most of this advice (low-fat diet, high intake of fruits and vegetables, regular exercise, and supplemental vitamins) may not only promote optimal immunity but is also likely to reduce the risk of cardiovascular heart disease and some cancers.

REFERENCES

1. Bogden JD, Louria DB. Micronutrients and immunity in older people. In: Bendich A, Deckelbaum RJ, eds. Preventive Nutrition: The Comprehensive Guide for Health Professionals, 2nd edition. Totowa, NJ: Humana Press, 2001:307–327.
2. Warner HR, Butler RN, Sprott RL, Schneider EL. Modern biological theories of aging. New York: Raven Press, 1987.
3. Bendich A. Antioxidant micronutrients and immune responses. Ann NY Acad Sci 1990;587:168–180.
4. Lesourd B, Mazari L. Nutrition and immunity in the elderly. Proc Nutr Soc 1999;58:685–695.
5. Ben-Yehuda A, Weksler ME. Immune senescence: mechanisms and clinical implications. Cancer Invest 1992;10:525–531.
6. Makinodan T. Patterns of age-related immunologic changes. Nutr Rev 1995;53:S27–S34.
7. Effros RB, Walford RL. Infection and immunity in relation to aging. In: Goidl EA, ed. Aging and the immune response. New York: Marcel-Dekker, 1987:45–65.
8. Bogden JD, Oleske JM, Munves EM, et al. Zinc and immunocompetence in the elderly: baseline data on zinc nutriture and immunity in unsupplemented subjects. Am J Clin Nutr 1987;45:101–109.
9. Kuvibidilia S, Yu L, Ode D, Warrier RP. The immune response in protein-energy malnutrition and single nutrient deficiencies. In: Klurfeld DM, ed. Nutrition and immunology. New York: Plenum Press, 1993:121–155.
10. Abbas AK, Lichtman AH, Pober JS. Cellular and molecular immunology. Philadelphia: WB Saunders, 1994:166–186.
11. Aronson M. Involution of the thymus revisited: Immunological trade-offs as an adaptation to aging. Mech Ageing Dev 1993;72:49–55.
12. Siskind GW. Aging and the immune system. In: Warner HR, Butler RN, Sprott RL, Schneider EL, eds. Modern biological theories of aging. New York: Raven Press, 1987:235–242.

13. Makinodan T, Kay MB. Age influence on the immune system. Adv Immunol 1980;29:287–330.
14. Weksler ME. The senescence of the immune system. Hospital Pract 1981:53–64.
15. McMurray DN. Cell-mediated immunity in nutritional deficiency. Prog Food Nutr Sci. 1984;8:193–228.
16. Schwab R, Weksler ME. Cell biology of the impaired proliferation of T cells from elderly humans. In: Goidl EA, ed. Aging and the immune response. New York: Marcel Dekker, 1987:67–80.
17. James SJ, Makinodan T. Nutritional intervention during immunologic aging: past and present. In: Armbrecht HJ, Prendergast JM, Coe RM, eds. Nutritional intervention in the aging process. New York: Springer-Verlag, 1984:209–227.
18. Meakins JL, Pietsch JB, Bubenick O, et al. Delayed hypersensitivity: indicator of acquired failure of host defenses in sepsis and trauma. Ann Surg 1977;186:241–250.
19. Christou NV, Tellado-Rodriguez J, Chartrand L, et al. Estimating mortality risk in preoperative patients using immunologic, nutritional, and acute-phase response variables. Ann Surg 1989;210:69–77.
20. Wayne SJ, Rhyne RL, Garry PJ, Goodwin JS. Cell-mediated immunity as a predictor of morbidity and mortality in subjects over 60. J Gerontol 1990;45:M45–48.
21. Roberts-Thomson IC, Whittingham S, Youngchaiyud U, McKay IR. Aging, immune response, and mortality. Lancet 1974;2:368–370.
22. Hicks MJ, Jones JF, Thies AC, Weigle KA, Minnich LL. Age-related changes in mitogen-induced lymphocyte function from birth to old age. Am J Clin Pathol 1983;80:159–163.
23. Murasko DM, Nelson BJ, Silver R, Matour D, Kaye D. Immunologic response in an elderly population with a mean age of 85. Am J Med 1986;81:612–618.
24. Markewitz A, Faist E, Lang S, et al. Successful restoration of cell-mediated immune response after cardiopulmonary bypass by immunomodulation. J Thorac Cardiovas Surg 1993;105:15–24.
25. Higa K, Noda B, Manabe H, Sato S, Dan K. T-lymphocyte subsets in otherwise healthy patients with herpes zoster and relationships to the duration of acute herpetic pain. Pain 1992;51:111–118.
26. Stites DP. Clinical laboratory methods for detection of cellular immunity. In: Stites DP, Terr AI, eds. Basic and clinical immunology, 7th edition. Norwalk, CT: Appleton & Lange 1991:263–283.
27. Giorgi JV. Lymphocyte subset measurements: significance in clinical medicine. In: Rose NR, Friedman H, Fahey JL, eds. Manual of clinical laboratory immunology, 3rd edition. Washington, DC: American Society for Microbiology, 1986:236–246.
28. Sen P, Middleton JR, Perez G, et al. Host defense abnormalities and infection in older persons. Infect Med 1994;11:364–370.
29. Perskin MH, Cornstein BN. Age-related changes in neutrophil structure and function. Mech Ageing Dev 1992;64:303–313.
30. Nagel JE, Han K, Coon PJ, Adler WH, Bender BS. Age differences in phagocytosis by polymorphonuclear leukocytes measured by flow cytometry. J Leukoc Biol 1986;39:399–407.
31. Shoham-Kesary H, Gershon H. Impaired reactivity to inflammatory stimuli of neutrophils from elderly donors. Aging Immunol Infect Dis 1992;3:169–183.
32. Corberand J, Ngyen F, Laharrague P, et al. Polymorphonuclear functions and aging in humans. J Am Geriatr Soc 1981;29:391–397.
33. Makinodan T, Lubinski J, Fong TC. Cellular, biochemical, and molecular basis of T-cell senescence. Arch Pathol Lab Med 1987;111:910–914.
34. Rubin LA, Nelson DL. The soluble interleukin-2 receptor: biology, function, and clinical application. Ann Intern Med 1990;113:619–627.
35. Manoussakis MN, Papadopoulos GK, Drosos AA, Moutsopoulos HM. Soluble interleukin-2 receptor molecules in the serum of patients with autoimmune diseases. Clin Immunol Immunopathol 1989;50:321–332.
36. Lahat N, Shtiller R, Zlotnick AY, Merin G. Early IL-2/sIL-2R surge following surgery leads to temporary immune refractoriness. Clin Exp Immunol 1993;92:482–486.
37. Bogden JD, Kemp FW, Liberatore BL, et al. Serum interleukin-2 receptor concentrations, physical activity, and micronutrient nutrition in older people. J Cell Biochem 1993;17B:86.
38. Sansoni P, Brianti V, Fagnoni F. NK cell activity and T-lymphocyte proliferation in healthy centenarians. Ann NY Acad Sci 1992;663:505–507.
39. Mariotti S, Sansoni P, Barbesino G, et al. Thyroid and other organ-specific autoantibodies in healthy centenarians. Lancet 1992;339:1506–1515
40. Scrimshaw NS, SanGiovanni JP. Synergism of nutrition, infection, and immunity: an overview. Am J Clin Nutr 1997;66:464S–477S.
41. Fraker P. Nutritional immunology: methodological considerations. J Nutr Immunol 1994;2:87–92.

42. Chandra RK. Nutrition and immunity. Contemp Nutr 1986;11:1–4.
43. Chandra RK. Immunodeficiency in undernutrition and overnutrition. Nutr Rev 1981;39:225–231.
44. McMurray DN, Loomis SA, Casazza LJ, Rey H, Miranda R. Development of impaired cell-mediated immunity in mild and moderate malnutrition. Am J Clin Nutr 1981;34:68–77.
45. Dowd PS, Heatley RV. The influence of undernutrition on immunity. Clin Sci 1984;66:241–248.
46. Beisel WR. Single nutrients and immunity. Am J Clin Nutr 1982;35:417–468.
47. Beisel WR. Nutrition and infection. In: Linder MC, ed. Nutritional biochemistry and metabolism. New York: Elsevier, 1985:369–394.
48. Bogden JD, Kemp FW, Han S, et al. Status of selected nutrients and progression of human immunodeficiency virus type 1 infection. Am J Clin Nutr 2000;72:809–815.
49. Goodwin JS, Garry PJ. Relationships between megadose vitamin supplementation and immunological function in a healthy elderly population. Clin Exp Immunol 1983;51:647–653.
50. Goodwin JS, Garry PJ. Lack of correlation between indices of nutritional status and immunologic function in elderly humans. J Gerontol 1988;43:M46–M49.
51. Kawakami K, Kadota J, Iida K, et al. Reduced immune function and malnutrition in the elderly. Tohoku J Exp Med 1999;187:157–171.
52. Kemp FW, DeCandia J, Li W, et al. Relationships between immunity and dietary and serum antioxidants, B vitamins, and homocysteine in elderly men and women. Nutr Res 2002;22:45–53.
53. Jacob RA, Kelley DS, Pianalto FS, et al. Immunocompetence and oxidant defense during ascorbate depletion of healthy men. Am J Clin Nutr 1991;54:1302S–1309S.
54. Fuller CJ, Faulkner H, Bendich A, Parker RS, Roe DA. Effect of beta-carotene supplementation on photosuppression of delayed-type hypersensitivity in normal young men. Am J Clin Nutr 1992;56:684–690.
55. Herraiz L, Rahman A, Paker R, Roe D. The role of beta-carotene supplementation in prevention of photosuppression of cellular immunity in elderly men. FASEB J 1994;8:A423.
56. Herraiz LA, Hsieh WC, Parker RS, et al. Effect of UV exposure and β-carotene supplementation on delayed-type hypersensitivity response in healthy older men. J Am Coll Nutr 1998;17:617–624.
57. Watson RR, Prabhala RH, Plezia PM, Alberts DS. Effect of beta-carotene on lymphocyte subpopulations in elderly humans: evidence for a dose–response relationship. Am J Clin Nutr 1991;53:90–94.
58. Santos MS, Meydani SN, Leka L, et al. Natural killer cell activity in elderly men is enhanced by β-carotene supplementation. Am J Clin Nutr 1996;64:772–777.
59. Omenn GS, Goodman GE, Thornquist MD, et al. Risk factors for lung cancer and for intervention effects in CARET, the Beta-Carotene and Retinol Efficacy Trial. J Natl Cancer Inst 1996;88:1550–1559.
60. Albanes D, Heinonen OP, Taylor PR, et al. Alpha-tocopherol and beta-carotene supplements and lung cancer incidence in the alpha-tocopherol, beta-carotene cancer prevention study: effects of baseline characteristics and study compliance. J Natl Cancer Inst 1996;88:1560–1570.
61. Talbott MC, Miller LT, Kerkvliet NI. Pyridoxine supplementation: effect on lymphocyte responses in elderly persons. Am J Clin Nutr 1987;46:659–664.
62. Meydani SN, Ribaya-Mercado JD, Russell RN, et al. Vitamin B-6 deficiency impairs interleukin 2 production and lymphocyte proliferation in elderly adults. Am J Clin Nutr 1991;53:1275–1280.
63. Rall LC, Meydani SN. Vitamin B6 and immune competence. Nutr Rev 1993;51:217–225.
64. Meydani SN, Meydani M, Blumberg JB, et al. Vitamin E supplementation and the in vivo immune response in healthy elderly subjects. J Am Med Assoc 1997;277:1380–1386.
65. Pallast EG, Schouten EG, de Waart FG, et al. Effect of 50- and 100- mg vitamin E supplements on cellular immune function in noninstitutionalized elderly persons. Am J Clin Nutr 1999;69:1273–1281.
66. Oleske JM, Westphal ML, Shore S, et al. Correction with zinc therapy of depressed cellular immunity in acrodermatitis enteropathica. Am J Dis Child 1979;133:915–918.
67. Fraker PJ, Gershwin ME, Good RA, Prasad A. Interrelationships between zinc and immune function. Fed Proc 1986;45:1474–1479.
68. Bogden JD, Oleske JM, Lavenhar MA, et al. Zinc and immunocompetence in elderly people: effects of zinc supplementation for 3 months. Am J Clin Nutr 1988;48:655–663.
69. Chandra RK, Hambreaus L, Puri S, Au B, Kutty KM. Immune responses of healthy volunteers given supplements of zinc or selenium. FASEB J 1993;7:A723.
70. Bogden, JD. Influence of zinc on immunity in the elderly. J Nutr Health Aging 2003;7:129–135.
71. Doherty CP, Kashein MA, Shakur MS, et al. Zinc and rehabilitation from severe protein-energy malnutrition: higher-dose regimens are associated with increased mortality. Am J Clin Nutr 1998;68:742–748.

72. Rahman MJ, Sarker P, Roy SK, Ahmad SM, Christi J, Azim T, Mathan M, Sack D, Andresson J, Raqib R. Effects of zinc supplementation as adjunct therapy on the systemic immune responses in shigellosis. Am J Clin Nutr 2005;81:495–502.
73. Martineau AR, Wilkinson RJ, Wilkinson KA, Newton SM, Kampmann B, Hall BM, Packe GE, Davidson RN, Eldridge SM, Maunsell ZJ, Rainbow SJ, Berry JL, Griifiths CJ. A single dose of vitamin D enhances immunity to mycobacteria. Am J Respir Crit Care Med 2007;176:208–213.
74. Bogden JD, Oleske JM, Lavenhar MA, et al. Effects of one year of supplementation with zinc and other micronutrients on cellular immunity in the elderly. J Am Coll Nutr 1990;9:214–225.
75. Penn ND, Purkins L. Kelleher J, et al. The effect of dietary supplementation with vitamins A, C, and E on cell-mediated immune function in elderly long-stay patients: a randomized controlled trial. Age Ageing 1991;20:169–174.
76. Chavance M, Herbeth B, Lemoine A, Zhu BP. Does multivitamin supplementation prevent infections in healthy elderly subjects? A controlled trial. Int J Vitam Nutr Res 1993;63:11–16.
77. Bogden JD, Bendich A, Kemp FW, et al. Daily micronutrient supplements enhance delayed-hypersensitivity skin test responses in older people. Am J Clin Nutr 1994;60:437–447.
78. Kniker WT, Anderson CT, McBryde JL, Roumiantzeff M, Lesourd B. Multitest CMI for standardized measurement of delayed cutaneous hypersensitivity and cell-mediated immunity. Normal values and proposed scoring system for healthy adults in the USA. Ann Allergy 1984;52:75–82.
79. Girodon F, Galan P, Monget AL, et al. Impact of trace elements and vitamin supplementation on immunity and infections in institutionalized elderly patients. Arch Intern Med 1999;159:748–754.
80. Winkler P, de Vrese M, Laue CH, Schrezenmeir J. Effect of a dietary supplement containing probiotic bacteria plus vitamins and minerals on common cold infections and cellular immune parameters. Int J Clin Pharmacol Ther 2005;43:318–326.
81. Park YK, Kim I, Yetley EA. Characteristics of vitamin and mineral supplement products in the United States. Amer J Clin Nutr 1991;54:750–759.
82. Block G, Cox C, Madans J, et al. Vitamin supplement use, by demographic characteristics. Am J Epidemiol 1988;127:297–309.
83. Cooper EL. Stress, immunity, and aging. New York: Marcel-Dekker, 1984.
84. Solomon GF. Emotions, immunity, and disease. In: Copper EL, ed. Stress, immunity and aging. New York: Marcel Dekker, 1984:1–10.
85. Watson RR, Eisinger M. Exercise and Disease. Boca Raton: CRC Press, 1992:71–178.
86. Keast D, Cameron K, Morton AR. Exercise and the immune response. Sports Med 1988;5:248–267.
87. Fry RW, Morton AR, Keast D. Overtraining in athletes. Sports Med 1991;12:32–65.
88. Nieman DC, Johanssen LM, Lee JW, Arabatzis K. Infectious episodes in runners before and after the Los Angeles marathon. J Sports Med Phys Fitness 1990;30:316–328.
89. Peters EM, Bateman ED. Ultramarathon running and upper respiratory tract infections. SAfr Med J 1983;64: 582–584.
90. Barry-Billings M, Blomstrand E, McAndrew N, Newsholme EA. A communication link between skeletal muscle, brain, and cells of the immune system. Int J Sports Med 1990;11:S122–S128.
91. Rubenoff R, Rall LC. Humoral mediation of changing body composition during aging and chronic inflammation. Nutr Rev 1993;51:1–11.
92. Meydani S. Dietary modulation of cytokine production and biologic functions. Nutr Rev 1990;48:361–368.
93. Nieman DC. Exercise immunology: future directions for research related to athletics, nutrition, and the elderly. Int J Sports Med. 2000;21(Suppl 1):S61–S68.
94. Paw MJ, deJong N, Pallast EG, et al. Immunity in frail elderly: a randomized controlled trial of exercise and enriched foods. Med Sci Sports Exerc. 2000;32:2005–2011.
95. Stallone DD. The influence of obesity and its treatment on the immune system. Nutr Rev 1994;52:37–50.
96. Spear-Hartley A, Sherman AR. Food restriction and the immune system. J Nutr Immunol 1994;3:27–50.
97. Eaton SB, Eaton SB III, Konner MJ, Shostak M. An evolutionary perspective enhances understanding of human nutritional requirements. J Nutr 1996;126:1732–1740.
98. Olshansky SJ, Carnes BA, Grahn D. Confronting the boundaries of human longevity. Am Sci 1998;86:52–61.
99. Goodwin JS. Decreased immunity and increased morbidity in the elderly. Nutr Rev 1995;53:S41–S46.
100. Bogden JD, Oleske JM. The essential trace minerals, immunity, and progression of HIV-1 I infection. Nutr Res 2007;27:69–77.
101. Webb AL, Villamor E. Update: effects of antioxidant and non-antioxidant vitamin supplementation on immune function. Nutr Rev 2007;65:181–217.

102. Agency for Healthcare Research and Quality. Multivitamin/Mineral Supplements and Prevention of Chronic Disease. US Dept of Health and Human Services. Rockville, MD, May, 2006.
103. World Cancer Research Fund, American Institute for Cancer Research. Food, Nutrition, Physical Activity, and the Prevention of Cancer. American Institute for Cancer Research, Washington DC, 2007.
104. Rimm EB, Stampfer MJ, Ascherio A, et al. Vitamin E consumption and the risk of coronary heart disease in men. N Engl J Med 1993;328:1450–1456.
105. Stampfer MJ, Hennekens CB, Manson JE, et al. Vitamin E consumption and the risk of coronary disease in women. N Engl J Med 1993;328:1444–1449
106. Bendich A, Langseth L. Health effects of vitamin C supplementation: a review. J Am Coll Nutr 1995;14:124–136.
107. Graat JM, Sohouten EG, Kok FJ. Effect of daily vitamin E and multivitamin–mineral supplementation on acute respiratory tract infections in elderly persons. JAMA 2002;288:715–772.
108. U.S. Preventive Services Task Force. Routine vitamin supplementation to prevent cancer and cardiovascular disease: recommendations and rationale. Ann Intern Med 2003;139:51–55.

23 Micronutrients: Immunological and Infection Effects on Nutritional Status and Impact on Health in Developing Countries

Ian Darnton-Hill and Faruk Ahmed

Key Points

- Deficiencies of micronutrients (vitamins, minerals and trace elements) are common—up to a third of people in low-income countries are affected.
- Women and children, especially those living in poverty, are those most at risk because of increased metabolic demands of growth, pregnancy and lactation and repeated infections.
- A vicious cycle of undernutrition leads to reduced immunity that increases disease risk and then the disease itself causes further undernutrition and so on.
- Immune systems are impacted by micronutrient deficiencies:
  - vitamin A deficiency impairs innate, cell-mediated and humoral antibody responses but probably not viral infection
  - zinc deficiency affects both innate and cell-mediated immunity but effects of supplementation on antibody production in humans are less clear than in animals
  - iron deficiency and overload impair both innate and cell-mediated immunity, with no effect on humoral antibody production
  - vitamin C deficiency in humans impairs leukocyte functions and decreases overall NK cell activity and lymphocyte proliferation
  - vitamins B6, B12, folate and E deficiencies impair Th1 cytokine-mediated immune response through insufficient production of pro-inflammatory cytokines, shifting to an anti-inflammatory Th2 cell-mediated immune response, thus increasing the risk of extracellular infections
  - supplementation with micronutrients generally reverses these impaired immune responses
- Micronutrient deficiencies can also be addressed by dietary improvement (if available and accessible), and by fortification. It is important to also address other interventions such as immunization, water and sanitation, breastfeeding and the reduction of social inequities.

Key Words: Cell-mediated immunity; developing countries; humoral immunity; immune system; innate immunity; iron; micronutrients; public health interventions; public health nutrition; vitamin A; women and children; zinc

23.1. INTRODUCTION

Just under 10 million children under 5 years of age continue to die unnecessarily in countries of the developing world *(1,2)*. Undernutrition is the direct cause of at least a third of these deaths *(3)*. Although there has been a reduction in the numbers and proportion dying, and progress toward the child survival millennium development goal (MDG4) is positive in many countries, clearly this is an unacceptable figure—especially when compared with what is possible from the single digit figures of

A. Bendich, R.J. Deckelbaum (eds.), *Preventive Nutrition*, Nutrition and Health, DOI 10.1007/978-1-60327-542-2_23,

more affluent countries. Progress toward the goals for maternal health (MDG5) and the elimination of poverty and hunger and undernutrition (MDG1) is doing even less well. Undernutrition contributes to over 3.5 million child deaths *(3)* and micronutrient deficiencies (vitamin A and zinc deficiencies) have been estimated to account for 1 million of these deaths per year or 9% of global childhood burden of disease (under 5 years), while iron deficiency is a risk factor for maternal mortality, responsible for 115,000 deaths per year, or 20% of global maternal deaths *(3)*.

Although undernutrition among children under the age of 5 years has declined from 32% in 1990 to 27% in 2006, the latest estimate shows that approximately 149 million children under the age of 5 continue to suffer from undernutrition *(2)*. Some 51 developing countries (nearly half of those for which data are available) are not making sufficient progress toward the "hunger target" of MDG1, as measured by the prevalence of underweight children under 5 years, and 18 of these countries are making no progress at all or are going backwards *(4)*. The recent unprecedented rise in international food prices has exacerbated existing factors such as inadequate health systems, the HIV/AIDS epidemic and poor human resources capacity and has led to increased food insecurity and even civil unrest. They are also likely to be a factor in increasing existing levels of undernutrition, including micronutrient malnutrition, with figures from FAO, the World Bank and the World Food Programme of increased numbers likely to be "hungry" (as defined by FAO) rising by another 75–113 million on top of the already estimated 850 million; FAO now estimates nearly 1 billion people are hungry and at risk of food security *(5)*.

The importance of micronutrient deficiencies on immunity and infection is a critical piece in the global effort to address the MDGs and global health. Vitamin A, zinc and iron will be discussed primarily, along with the other relevant vitamins and minerals such as folate and vitamin B12 which are known to have an impact on immune function and status. These micronutrients will only be discussed in terms of their immune functions in humans and the public health implications of deficiencies, as there are many other sources of information on structure, dietary sources, bioavailability, clinical manifestations, pathophysiology and the epidemiology of micronutrient deficiency. Recent reviews that do that include the following: for vitamin A *(6–8)*, all to one degree or another building on the classics of Bauerfeind *(9)* and Sommer and West *(10)*; iron and the nutritional anemias are the subject of a book edited by Kraemer and Zimmermann *(11)* and Ramakrishnan and Semba *(12)* in the excellent volume edited by Semba and Bloem *(13)* on "Nutrition and Health in Developing Countries" that again builds on earlier work by DeMaeyer et al. *(14)* and others *(15)*; public health aspects of zinc, e.g. Hotz and Brown *(16)* and one currently in press as the second IZiNCG Technical document *(17)*, Shrimpton and Shankar *(18)*; and folic acid and vitamin B12 *(19)*; as well as more general issues related to micronutrient malnutrition such as the WHO/FAO publications on vitamin and trace element requirements *(20)* and fortification with micronutrients *(21)*; and in the nutrition series by the Nutrition Society, including "Public Health Nutrition" edited by Gibney, Margetts, Kearney and Arab *(22)*.

Even just 25 years ago, it was generally assumed that, as far as the human was concerned, each vitamin served one particular main function, e.g. vitamin C preserved connective tissue; vitamin D, bone; vitamin B1, the nervous system; nicotinic acid, the skin; folic acid and vitamin B12, the blood; and, vitamin A, the eye *(23)*. This is clearly not the case and many vitamins, and minerals, are involved in many physiological actions, often, directly or indirectly, impacting on the immune system one way or another. This is more apparent with vitamin A, zinc and iron and may be vitamins C and E (and generalized undernutrition) although the mechanisms are still not always clear. In the seminal book from the WHO *(24)* 40 years ago, it was noted that the formation of specific antibodies is inhibited by many severe nutrient deficiencies, including protein, tryptophan, vitamins A and D, ascorbic acid, thiamin, riboflavin, niacin, pyridoxine, pantothenic acid, folate and vitamin B12.

This chapter examines the cycle of deficiencies of micronutrients, leading to impairment of the immune system and leading to infectious diseases which in turn lead to poor nutrition; as well as the public health implications (including for achieving the millennium development goals (MDGs)); public health interventions in the control and prevention of micronutrient deficiencies; and some conclusions. Although there is a considerable literature on micronutrients and growth, that will not be considered here, although obviously of considerable public health importance. Although much of the investigative work on micronutrients and immunity has been carried out in animal studies, it has been found that the human often works somewhat differently and so the emphasis has been on the sparser literature on humans *(24)* and on public health aspects. Much of that research has been done with relatively affluent elderly subjects, e.g. Hamer *(25)*. Solomons and Bermúdez *(26)* remind us that the elderly in developing countries are also at risk of nutrition-related and age-related decline in immune processes and hence at increased risk of infectious disease. The inflammatory response becomes dysregulated, with excess production of inflammatory cytokines such as interleukin 6 (IL-6) which advances with age. The variation in immune protective function is one factor in differential risk of cancer in the elderly, and an individual's risk to autoimmune diseases also increases *(26)*. However, the emphasis here is on young children and women in countries with developing and transitional economies with generally higher risks of micronutrient deficiencies, reduced immunity and infectious diseases.

23.2. UNDERNUTRITION AND DISEASE: A VICIOUS CYCLE

As was noted 40 years ago, interactions between malnutrition and infection contribute directly to the health of individuals and communities, and particularly so lower in socio-economic groups and less economically developed areas and countries *(24)*. They also noted that infections and immunity can be synergistic in two ways: infections are likely to have more serious consequences among persons with clinical or subclinical undernutrition, including micronutrient deficiencies, and infectious diseases have the capacity to turn borderline nutritional deficiencies into severe clinical manifestations of undernutrition, e.g. vitamin A deficiency into full-blown xerophthalmia *(8)*. One of the issues to be discussed, because of the public health aspects, is this concept of synergism and antagonism (i.e. where aspects of undernutrition appear to limit infectious disease). The authors conclude, however, that in humans, "interactions between malnutrition and infection are regularly synergistic" *(24)*; a view more recently also concluded by Caulfield et al. *(27)* and Prentice *(28)*.

Nutritionally induced determinants of synergism (between nutrition and infection) may include *(24)*

(a) reduced capacity of the host to form specific antibodies
(b) decrease in phagocytic activity of microphages and macrophages
(c) interference with production of non-specific protective substances
(d) reduced non-specific resistance to bacterial toxins
(e) alterations in tissue integrity
(f) diminished inflammatory response and alterations in wound healing and collagen formation
(g) effects originating in alterations of intestinal flora
(h) variations in endocrine activity

In response to infection, both innate and then acquired host defenses are brought into play, and both processes involve activation and propagation of immune cells and synthesis of an array of molecules requiring DNA replication, RNA expression, and protein synthesis and secretion, all consuming anabolic energy. Mediators of inflammation further increase the catabolic response *(29)*.

The nutritional status of the host critically determines the outcome of infection and includes deficiencies in single nutrients such as micronutrients, fatty acids and amino acids, with general protein–energy malnutrition greatly increasing susceptibility to infection, particularly in low-income countries and particularly in children. Ultimately, productivity and well-being are affected at the community level which perpetuates what has been called the "alarming spiral of malnutrition, infection, disease and poverty" *(29)*.

As will be seen below, many of the micronutrients have the potential to have an association with impaired immune responses. Conversely, infectious disease adversely influences the nutritional state in several indirect ways. Loss of appetite and intolerance for food result in metabolic effects. Cultural factors can lead to substitution of less nutritious diets on assumption of therapeutic effect and sometimes as purgatives, and antibiotics and some other drugs can also reduce appetite or digestion or absorption of specific nutrients *(24)*. An increased loss of body nitrogen is characteristic of all infectious disease. Classical nutritional deficiencies precipitated by infection with borderline nutrient status include keratomalacia and lack of vitamin A, scurvy and ascorbic acid, beriberi and thiamin, pellagra and insufficient niacin, macrocytic anemia due to folate or vitamin B12 deficiency, and microcytic anemia resulting from a shortage of iron. Among resource-poor societies the premature death of a mother and the lower income-generating capacity of iron-deficient and anemic workers translate into greater rates of disease and overall undernutrition *(30)*. Women and girls are often discriminated against in terms of nutrition and health, especially well-described in south Asia, including in intrahousehold distribution of micronutrient-rich foods *(31)*. Disease can also affect the ability of populations to grow and harvest food if widespread enough, e.g. endemic malaria, onchocerciasis and more recently HIV/AIDS. Consequently the cycle unless broken can lead to poor nutrition leading to impaired immune systems leading to increased incidence of infectious diseases which in turn leads to deteriorating nutritional status and so on (Fig. 23.1).

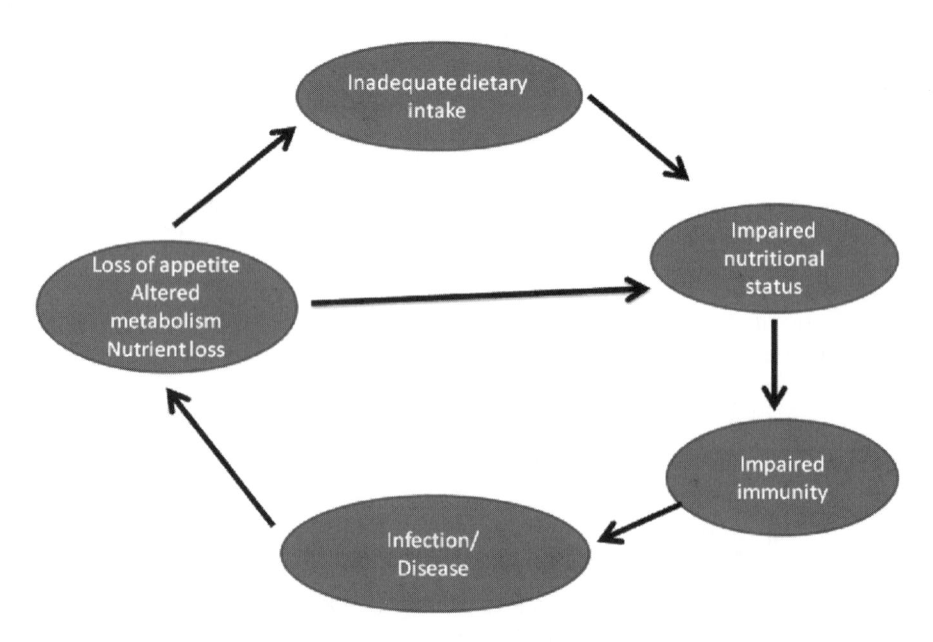

Fig. 23.1. Undernutrition/disease cycle.

23.3. MICRONUTRIENTS, IMMUNITY AND INFECTIOUS DISEASE

As noted, undernutrition can interfere with any body mechanism that interposes a barrier to the multiplication or progress of infectious agents *(24)* and that formation of specific antibodies is inhibited by many severe nutrient deficiencies, including protein (e.g. especially in kwashiorkor), tryptophan, vitamins A and D, ascorbic acid, thiamin, riboflavin, niacin, pyridoxine, pantothenic acid, folate and vitamin B12. Severe protein depletion and folate deficiency are particularly important in reducing response and activity of phagocytes, both microphages and macrophages. The integrity of skin, mucous membranes and other tissues is important in preventing entrance of infection. Such changes associated with nutritional deficiencies include (i) alterations in intercellular substances; (ii) reduction or absence of secretion of mucus; (iii) increased permeability of intestinal and other mucosal surfaces; (iv) accumulation of cellular debris and mucus to produce a favorable culture medium; (v) keratization and metaplasia of epithelial surfaces; (vi) loss of ciliated epithelium of the respiratory tract; (vii) nutritional edema, with increased fluid in the tissues; (viii) reduced fibroplastic response; and (ix) interference with normal tissue replacement and repair *(24)*. Loss of tissue integrity from deficiencies in vitamin A and ascorbic acid is regularly associated with reduced resistance. Bacterial penetration of intestinal mucosa is known to be enhanced by riboflavin and thiamin deficiencies. Laboratory animals with deficiencies of vitamins A, C and the B complex are unusually susceptible to bacterial toxins, regardless of whether antibody production is affected, but it is not clear how much this is an effect in humans.

23.3.1. Effect of Deficiencies on Immunological Status

As the early classic on nutrition and infection pointed out "throughout history, man has recognized an association between famine and pestilence" *(24)*. As noted, the direct interactions fall into two basic patterns: undernutrition generally alters resistance of the host to infection and infectious disease exaggerates existing undernutrition. All infectious diseases have direct adverse metabolic effects that, among other things, influence the amount and kind of food consumed and nutrients absorbed. In some studies with laboratory animals involving viral, protozoal or helminthic infections of systemic origin, but rarely with other infectious agents, specific vitamin and mineral deficiencies have been found to inhibit the activity of the agent more than they interfere with host resistance. In general, it has not been conclusively shown, in either natural or experimental nutritional deficiency, that a deficiency can be antagonistic to any infection in humans *(24)*. Infectious disease nearly always makes co-existing undernutrition worse while the consequences of infection are more likely to be serious in a malnourished host than a well-nourished one *(24)*, although the possible importance of this continues to be debated, especially in relation to iron supplementation in areas endemic for malaria *(28)*.

The immune system can be broadly categorized into two groups: the innate immune system and the acquired immune system. The innate immune system, the first line of defense, is naturally present and it is not influenced by previous contact with infectious agents. It includes epithelial barriers, the complement system, circulating phagocytes (neutrophils and macrophages) and other cytotoxic cells (natural killer (NK) cells). Innate immunity is regulated by two types of cytokines: pro-inflammatory cytokines such as interleukin (IL)-1, IL-6, IL-12 and tumor necrosis factor (TNF)-α, and anti-inflammatory cytokines such as IL-10, produced by neutrophils and macrophages.

On the other hand, the acquired immune system is antigen specific, where antibodies are produced by the B lymphocytes, known as humoral immunity and cell-mediated immunity which depends on the T lymphocytes system *(32)*. Acquired immunity involves the identification of an antigen by antibody or T-cell receptor on CD4+ T-helper (Th) cells or CD8+ effector T cells. The antigen presenting cells carry the antigen to regional lymph nodes, where naïve Th cells are exposed to the antigen, and proliferate and mature to form memory T cells. Memory T cells then follow either of two pathways, Th1 or Th2 memory cells. In response to an intracellular pathogen, Th1 memory cells produce interferon (IFN)-γ and IL-2, which in turn stimulate a response by cytotoxic T lymphocytes (CTLs), activation of macrophages, response of delayed-type hypersensitivity (DTH) and provide limited help to stimulate B-cell development and antibody production. Whereas Th2 memory cells in response to a pathogen produce IL-4, IL-5 and IL-10, which stimulate B cells to produce antibodies, osinophil and mast cell development and deactivation of macrophages.

It has long been accepted, in general, that malnutrition, in particular undernutrition, impairs immune function and increases the risk and severity of diseases. There are also instances where the immunomodulatory effect is independent of any nutritional value, e.g. canthaxanthine, which does not have any provitamin A activity, has been shown to have the same ability to enhance immune responses as β-carotene (at least in rats) *(33)*. In this section, an attempt has been made to elucidate the effect of various micronutrient deficiencies on immune function and the possible mechanism involved.

23.3.1.1. Vitamin A

Vitamin A deficiency is the contributing cause of over a million premature deaths each year in children globally *(34)*, as well as the commonest cause of childhood blindness, and probably a factor in several cancers *(8)*. Xerophthalmia was a recognized public health problem in much of Europe until early last century. The public health significance of vitamin A deficiency has been redefined beyond xerophthalmia in the last 15 years to include its impact on the deaths from infectious diseases in developing countries where vitamin A deficiency is frequently endemic. There has been tremendous progress in reducing the prevalence of the most severe manifestations of the disease (xerophthalmia and blindness) which are on the decline in all regions of the world *(35)*. Subclinical vitamin A deficiency, resulting from a chronic, dietary insufficiency of vitamin A, either preformed or from precursor carotenoids, and its impact on immunity and childhood infectious disease, however, is still a problem of considerable public health significance *(6,8,36,37)*. A deficiency state may arise with prolonged inadequate intake, often coupled with the high, normal demands imposed by rapid growth during childhood, pregnancy or lactation, or by excessive utilization and loss during infection *(10)*. The relative frequent occurrence in women during pregnancy and the possible consequences of that have only recently been widely recognized, which much increase the magnitude of the problem including possibly an impact on maternal mortality (at least in deficient populations such as Nepal, although apparently not in less severely deficient populations in Bangladesh) *(36,38–40)*.

Vitamin A was formerly mostly known for its role in xerophthalmia and night blindness, at least since the time of the Pharaohs were reigning in Egypt and cases of xerophthalmia have been described since those times and especially throughout the 18th and 19th centuries *(41)*. A series of scientists, Hopkins, along with McCollum and Davies and Osborne and Mendel found that animals fed only fats, protein, starch and inorganic fats not only failed to grow normally but became also more susceptible to infection and frequently died of overwhelming sepsis *(41)*. Bloch (also cited in *(41)*), studying the growth and development of children in a Danish orphanage, found that when they were given butterfat and whole milk they were less susceptible to infections of the urinary and respiratory tracts and middle ear (and less likely to develop xerophthalmia). By 1928, Green and Mellanby had declared vitamin A as an “anti-infective factor” *(10,23)*. Ellison administered daily vitamin A and reduced by half the case-fatality rates due to measles. By 1930s, it was accepted that besides the ophthalmologic manifestations of vitamin A deficient, there was also reduced resistance to some microbial infections *(41)*. As Sommer notes, at this time “further investigations on (and advocacy for) the administrations of vitamin A to treat and prevent infections virtually stopped” probably due to poor research methods giving apparently contradictory results and the advent of antibiotics. The improving vitamin A status in most of Europe and the USA also meant there was less incentive to study such anti-infective aspects *(41)*.

With Wolbach and Howe's classic description in 1923 of widespread metaplasia and keratinization of epithelial linings of the respiratory and genitourinary tracts and glandular ducts in vitamin A-depleted animals, loss of the "barrier function" of epithelial linings became one plausible explanation for the associated decreased resistance to infection (historical sources cited in *(8)*). While animal experimentation continued, clinical studies in humans from the 1920s through the 1940s continued to reveal associations between vitamin A deficiency or xerophthalmia and infectious diseases *(42)*. The inverse relationship between febrile illness and plasma vitamin A concentration, now understood as part of the acute phase response to infection, and the potential therapeutic efficacy of vitamin A in reducing childhood measles fatality, puerperal fever in women, and other clinically relevant conditions have also been described since then *(8,42)*. The regulatory roles of vitamin A in maintaining epithelial cell differentiation and function and immune competence have provided biologic plausibility to its importance in decreasing severity and mortality of infectious diseases *(10,43–45)*.

23.3.1.1.1. Vitamin A Deficiency and Immune Function. Vitamin A is one of the most extensively studied nutrients in relation to immune function. Vitamin A deficiency impairs the immune system thereby resulting in major morbidity and mortality. There are excellent reviews in the available literature on the mechanisms by which vitamin A and its metabolites influence the immune function at various levels *(42,46,47)*. More recently a review of the results from the supplementation of vitamin A in human studies has also been published *(48)*.

Innate Immunity. Vitamin A deficiency is associated with impaired innate immunity. Animal studies have shown that vitamin A deficiency is significantly associated with altered mucosal epithelial barriers in the conjunctiva of the eye *(10,49)*, respiratory *(50)*, gastrointestinal *(51)* and genitourinary tracts *(52)*. Vitamin A deficiency can result in a loss of microvilli, mucus-producing goblet cells and mucin in the small intestine *(53–55)*. Mucins are glycoproteins, secreted into the lumen, found on cell surfaces and serve as a first line of defence. Changes that occur due to vitamin A deficiency include squamous metaplasia of the conjunctiva and cornea, loss of goblet cells and abnormal keratinization of the epithelium *(10,56)*. In humans, using the lactose/mannitol urinary excretion test as an indicator of gut integrity in vitamin A supplementation trials in children suffering from severe infections, a rapid increase in intestinal integrity was shown *(57)*. A few studies have failed to show a consistent effect on the mucosal anti-infective or inflammatory markers in milk, saliva or general fluid *(48)*.

Animal studies suggest that vitamin A deficiency may lead to an increased total number of macrophages *(58)*. In addition to increased numbers of macrophages, vitamin A deficiency leads to increased IL-12 produced by macrophages, with IL-12 promoting the development of Th1 cells, which produce interferon (IFN)-γ. Increased IFN-γ leads to increased macrophage activation *(47)*. Although data from human studies are limited, clinical trials suggest that vitamin A supplementation may diminish the production of pro-inflammatory cytokines (TNF-α and IL-6) by macrophages, but only in response to infections *(48)*. Vitamin A supplementation was found to be associated with increased production of the anti-inflammatory cytokines IL-10 *(59)*. All these data suggest that vitamin A deficiency can lead to increased inflammation mediated by cytokines from macrophages, while impairing the ability of macrophages to ingest and kill bacteria.

Natural killer (NK) cells are one of the components of innate immunity which work by killing virus-infected cells as well as tumor cells. Studies on animals have shown that vitamin A deficiency impairs both the NK cell number and its lytic activity *(60,61)*. In a clinical trial among HIV-infected children in South Africa, vitamin A supplementation showed increased number of cells with the CD56 receptor expressed by the NK cells *(62)*. Vitamin A deficiency also impairs normal neutrophil development, which can lower the capacity of phagocytosis to kill bacteria *(63)*. However, the evidence on the association of vitamin A and neutrophil function in humans are limited *(48)*.

Acquired Immunity. Cell-mediated immunity can also be affected by vitamin A deficiency. Studies on animals have shown that vitamin A deficiency is associated with reduced weight of the thymus *(55)* and decreased lymphocyte proliferation in response to mitogens *(58,64)*. In murine T cells, all-trans retinoic acid has been shown to stimulate the expression of retinoic acid receptor-α and increased antigen-specific T-cell proliferation *(65)*.

Vitamin A supplementation to infants has been shown to significantly increase total lymphocyte count *(66)*, especially the CD4 subpopulation *(67)*. Similar findings have also been observed in HIV-infected children *(62)*, while when vitamin A was supplemented in HIV-infected women, no significant effect on CD4 T-cell counts was observed *(68,69)*. Human study findings on Th1-mediated response are equivocal. One study showed increased DTH response in infants following high-dose vitamin A supplementation *(70)*, whereas another study found no difference by treatment groups in the proportion of children with DTH response in a non-placebo-controlled trial of intramuscular vitamin A *(71)*. Further, in a study among children with measles, vitamin A supplementation apparently diminished the proportion of children with DTH response *(72)*.

While in animal models, vitamin A deficiency has been shown to reduce IL-2 production *(73)*, a Th1-mediated response, high-dose vitamin A *(74)* and retinoic acid *(75)* treatment in vitro showed increased IL-2 production or IL-2 receptor expression. IL-2 plays an important role in proliferation and activation of T and B lymphocytes and NK cells. In chickens, vitamin A deficiency impaired CTL function during Newcastle disease virus infection *(76)*. Studies on vitamin A-deficient children showed a depressed ex vivo production of IFN-γ, a Th1 pro-inflammatory cytokine *(77)*. Another study among non-HIV-infected infants showed no effect of vitamin A supplementation on serum concentration of IL-2 *(66)*.

Animal studies have shown that vitamin A deficiency impairs the production of Th2 cell cytokines such as IL-4, IL-5 and IL-10 *(73,78)*. Several human studies indicate that vitamin A can regulate the production of IL-10 from Th2 cells: vitamin A deficiency impairs secretion of IL-10, while supplementation of vitamin A increases the IL-10 section in vitamin A-deficient subjects *(59)*. The cytokine IL-10 plays a role in the inhibition of the synthesis of pro-inflammatory Th-1-type cytokines, such as IFN-γ and IL-2, in both T and NK cells. In vitro lymphocyte stimulation to various mitogens was higher in vitamin A-deficient rats, with higher IFN-γ and IL-2 production, indicating that vitamin A deficiency increased Th1 responses *(79)*. The results from animal studies suggest that modulation of the balance between Th1 and Th2 responses by retinoids may be influenced by the type of pathogens *(46)*. Results from human studies that examined the effect of vitamin A on either Th1 or Th2 responses also suggest that the immunological mechanisms through which vitamin A exert an effect are pathogen specific *(48)*.

The growth and activation of B lymphocytes require retinol *(80)*. The growth of B lymphocytes is also known to be mediated by the metabolites of retinol *(81)*. B lymphocytes are responsible for the production of immunoglobulins (antibodies). All-trans retinoic acid was found to be more active than retinyl acetate, retinaldehyde or retinol in restoring IgG responses in a murine model *(82)*. Vitamin A deficiency typically impairs antibody response to T-cell-dependent antigens *(44,58,79)* and in some T-cell-independent antigens *(83)*. Studies with a vitamin A-deficient animal model have shown impaired serum IgG1 antibody response to purified protein antigens *(45,58)*, impaired serum IgG1 and IgE responses to the intestinal helminth *Trichinella spiralis (73)* as well as the intestinal IgA response to cholera toxin *(79)*. Most animal studies showed no impairment of serum antibody response to viral infection in vitamin A deficiency *(61)*. The evidence for an effect of vitamin A supplementation on T-cell-dependent antibody response in humans is equivocal. Administration of a large dose of vitamin A in children, aged 1–6 years, did not result in any significant effect on the antibody response against tetanus toxoid *(71)*. Another study compared the effect of different doses of vitamin A supplementation on the antibody responses against both tetanus and diphtheria toxoids in children, 1–6 years, and also found no effect *(84)*. In contrast, others have shown significantly higher antibody response against tetanus toxoid following vitamin A supplementation in tetanus-naïve 3- to 6-year-old children *(33,85)*. The effect of vitamin A in infants on the antibody response against diphtheria toxoid was found positive, while there was no effect with tetanus toxoid *(86)*. It may be concluded that vitamin A supplements can increase the antibody response to tetanus toxoid particularly in vitamin A-deficient children who have not been exposed to tetanus. Effects of vitamin A supplements on antibody response against measles infection or measles immunization were found to be either positive *(66,87)* or negative *(88)* or no change *(72,89)*. The serum antibody response to polio vaccine showed no effect by vitamin A supplementation when given at routine immunization time *(42,90)*. In a study when vitamin A was administered to both mother and children, a significantly higher proportion of children had protective titers against type 1 poliovirus than in the placebo group *(91)*. The differential effect of vitamin A supplementation observed could be attributed to doses of vaccines, time of supplementation or baseline vitamin A status of the population studied *(48,85)*.

Vitamin A deficiency and supplementation clearly play a large part in the interaction of immunity, infection and public health. There are obvious implications for public health interventions, e.g. ensuring vitamin A supplementation twice yearly in vulnerable populations, and at the time of immunization programs, as well as in treatment, e.g. vitamin A supplementation in measles patients.

23.3.1.1.2. Public Health Implications. It is now well accepted, based on all but two of eight major epidemiological trials, that all-cause mortality among children 6 months to 5 years of age has reduced death by, on average, about a quarter when supplementation with vitamin A capsules takes place as recommended by the WHO *(10)*. A national cross-sectional survey, a large, population-based, prospective study, and several hospital-based clinical studies of xerophthalmia among Indonesian children by Sommer and colleagues in the late 1970s built on earlier work and demonstrated aspects of causation, progression, risk factors and health consequences of childhood xerophthalmia and vitamin A deficiency in low-income countries. Reports from this work, in the early 1980s, showed that non-blinding, mild xerophthalmia (night blindness and Bitot's spots) was associated with markedly increased risks of preschool child mortality *(92)*. Presumably vitamin A supplementation increased resistance to the severity of infection (measles and diarrheal diseases) by reducing the functional degree of vitamin A deficiency.

In contrast to evidence relating vitamin A deficiency to respiratory tract compromise and infection, *(10)* vitamin A supplementation has not had a consistent effect in reducing the incidence, severity or mortality of acute lower respiratory infection in children, and vitamin A supplementation of infants under 6 months of age has generally not shown a survival benefit in early infancy. National programs of varying effectiveness have now been launched in over 70 countries and UNICEF has estimated that over half a billion vitamin A capsules are distributed every year, preventing around 350,000 deaths annually *(93)*. However, although repeatedly emphasized as both effective and cost-effective public health intervention, mostly recently in *The Lancet* nutrition series *(3,94,95)*, it is still not known precisely how vitamin A increases resistance to infection, "although there is ample clinical and laboratory evidence that it does" *(4)*. A recent study from China, which somewhat surprisingly still has a problem of vitamin A deficiency in cities, looked at the effect of vitamin A supplementation on both serum vitamin A status and immune function and was able to show an impact on both *(96)*.

This considerable public health effect can be partly explained by an ability of vitamin A to lower case fatality from measles by almost half, as observed in field trials and hospital-based measles trials *(10)*, mortality from severe diarrhea and dysentery, by approximately 40% *(41)* and, based on morbidity findings from a recent supplementation trial, possibly falciparum malaria *(27)*. Vitamin A deficiency and infection interact within a "vicious cycle" *(24)*, whereby one exacerbates and increases vulnerability to the other. The bi-directional relationship complicates frequent cross-sectional evidence of depressed plasma retinol levels with diarrhea, acute respiratory infections, measles, malaria, HIV/AIDS and other infectious illnesses *(8)*. Combining mortality effects with data on the prevalence of vitamin A deficiency, it has been estimated that 1.3–2.5 million early childhood deaths each year can be attributed to underlying vitamin A deficiency *(10)*.

A strong, dose-risk gradient exists between maternal serum retinol and vertical transmission of HIV and cervico-vaginal shedding of HIV DNA *(97)*, suggesting that maternal vitamin A deficiency may affect pregnancy outcomes in HIV+ populations. To date, however, vitamin A supplementation of HIV+ pregnant women populations has shown little effect on outcomes such as low birth weight or perinatal mortality *(69)*, or in interrupting transmission of HIV from mother to infant *(97,98)*. This recent evidence of relatively high prevalences in women of the developing world, and the health impact of this, has encouraged many in international public health to examine new paradigms in the prevention and control of vitamin A deficiency in women *(38)* (Fig. 23.2). Recent evidence has suggested that maybe vitamin A may have a negative or no impact in HIV+ women. A recent study with multi-micronutrients not containing vitamin A, compared with those containing vitamin A actually appeared to neutralize the impact in delaying progression to AIDS seen with the non-vitamin A-containing multimicronutrients *(99)*. This is further discussed in the section on multimicronutrients and prevention but the excellent chapter on HIV infection for a more complete update is found in a companion volume *(97)*.

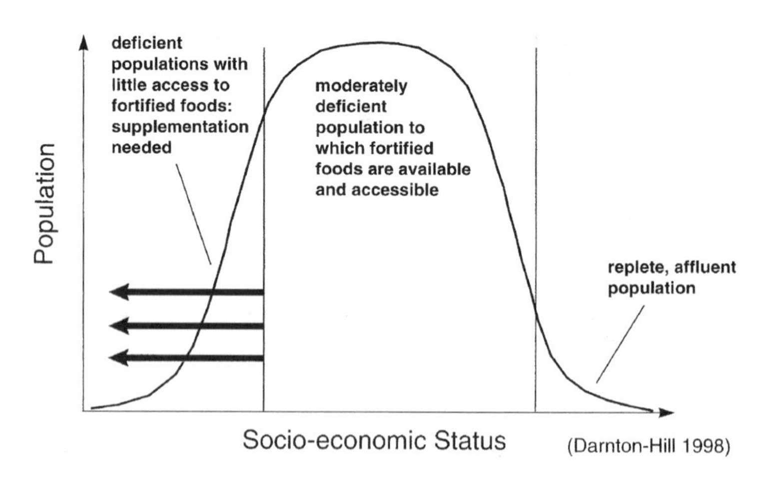

Fig. 23.2. Paradigm for increasing micronutrient intakes in populations by socio-economic status (Darnton-Hill).

23.3.1.2. Zinc

Low zinc intakes reduce resistance to infection through an effect on immune function. Zinc supplementation reduces the morbidity and mortality of common childhood diseases, including diarrhea, lower respiratory tract infection and probably malaria *(18)*. Zinc was used topically as calamine lotion as far back as 1500 BC by the Egyptians. Its current name probably originates from an early German word meaning 'tooth-like, pointed or jagged' (presumably referring to the needle-like metallic zinc crystals). Zinc mines near Udaipur in the Indian State of Rajasthan were active during 400 BC and there are references to medicinal uses in *Charaka Samhita* (300 BC). Pure zinc was not isolated in China until the 17th century although the smelting and extraction of impure forms were being undertaken around AD 1200 in India. There is a record, however, of the metallurgist Andreas Libavius receiving in 1597 from Asia a quantity of pure zinc metal, unknown in the West before then, although several different Englishmen and Germans probably isolated zinc independently in the late first half of 18th century.

The identification and quantification of zinc deficiency is bedevilled by the lack of a suitable diagnostic test. Consequently, it has been clearly demonstrated only by intervention trials showing an impact on infectious diseases, more frequently diarrhea, and less clearly respiratory tract infection. Nevertheless, there is a consensus now that zinc is likely to be a problem in many countries with high child mortality rates *(16,17,100)*. The impact on growth and maturation has of course been long recognized where the deficiency was more florid e.g. in parts of the Middle East *(101)*; the clinical manifestations of zinc deficiency, corrected by supplementation, include those on impaired growth and failure to thrive, immune effects and delayed sexual maturation. It has also been observed as an inborn error of zinc metabolism, acrodermatitis enteropathica, in patients fed incomplete parenteral solutions, in patients with Crohn's disease and occasionally in infants. Zinc-responsive night blindness has been observed in alcoholism and Crohn's disease *(101)*.

23.3.1.2.1. Zinc Deficiency and Immunity. It has long been established that zinc plays an important role in the immune system. To date there are good number of reviews in the available literature that have been published on the effect of zinc deficiency and immune function and the possible mechanisms *(102–105)*. The effects of zinc are multifaceted, from the physical barrier of the skin to gene regulation within lymphocytes *(103)*. Even with mild zinc deficiency, multiple aspects of the immune system are impaired *(102,106–108)*. Studies in animals and humans have shown that zinc deficiency is associated with impairment of both non-specific and specific immune responses *(103)*. Zinc is essential even in the earliest stages of an immune response *(109)*.

Innate Immunity. Zinc deficiency can damage epithelial linings of the gastrointestinal and pulmonary tracts *(108,110)*, and also damage epidermal cells, resulting in the skin lesions of acrodermatitis enteropathica, a genetic disorder of zinc malabsorption *(111)*. In both human and animal studies, NK cell activity was found to be depressed *(106,112)*, although one study did report increased NK cell activity *(113)*. Treatment of human peripheral blood NK cells with exogenous zinc has been found to stimulate production of IFN-γ *(114)*. Rajagopalan et al. *(115)* have suggested that zinc is required for killer cell inhibitory receptor on NK cells and so zinc deficiency results in the inhibition of the killing activity.

Zinc deficiency impairs chemotactic responses of neutrophils, while absolute numbers of neutrophils are not affected *(111,116,117)*. In both animal and humans, the chemotactic response of monocytes is also impaired and can be rapidly restored by the in vitro addition of zinc *(111,116,118,119)*. Macrophage phagocytosis in zinc-deficient animals (both mice and rats) has been found to be reduced *(120,121)*, enhanced *(118,122)* or remain unchanged *(123)*. High concentrations of zinc in vitro inhibit macrophage activity *(124)*. Information regarding the effects of zinc on macrophage function in humans is limited and more studies are needed to confirm the role of zinc on macrophage phagocytosis.

Acquired Immunity. Zinc plays an important role in cell-mediated immunity. In their review article, Shankar and Prasad *(103)* summarized the available data from both animal and human studies showing thymic atrophy, a reduction in the size of thymus, due to zinc deficiency. Thymus is the main organ for T-cell development, and thus confirms that zinc plays role in the early stage of T-cell maturation. Zinc deficiency also causes depleted numbers of T cells in the spleen, lymph nodes and peripheral blood in animals *(106,125)* and in the blood and peripheral lymphoid tissues in humans *(103)*. Studies with zinc supplementation were found to reverse these conditions *(112,125–127)*.

Delayed hypersensitivity response and cytotoxic activity of T lymphocytes are impaired in zinc deficiency and have also been found to be reversed by zinc supplementation *(106,128–130)*. Zinc supplementation in malnourished children also restored their delayed hypersensitivity responses *(131,132)*. Most in vitro studies have shown that zinc is required for T lymphocytes from animals and humans to proliferate in response to mitogen stimulation, such as with phytohemagglutinin (PHA) and concanavalin A *(103,133–136)*. Besides maintaining the proliferation, there is a role of zinc in lymphocyte homeostasis by suppression of cell death by apoptosis *(137)*. Several studies have reported that apoptosis of T lymphocytes induced by in vitro treatment of toxins and other agents can be prevented by adding high doses of zinc salts *(138,139)*. The thymic atropy seen in zinc deficiency, mentioned above, is accompanied by apoptosis of lymphocytes *(140)*.

Thymulin, a thymus-specific hormone, binds to the high affinity receptor on T cells and induces several T-cell makers for the differentiation of immature T cells and promotes T-cell functions, such as allogenic cytotoxicity, suppressor functions and IL-2 productions *(141–143)*. Thymulin also regulates the cytokine release by peripheral mature T cells *(144)* and induces the proliferation of CD8+ T cells which function as cytotoxic cells able to recognize and kill pathogens *(145)*. Thymulin requires zinc for its biological activity to be expressed *(141,142)*. In both animals and humans, the activity of thymulin in serum was found to be significantly impaired in zinc deficiency but was able to be corrected by both in vivo and in vitro zinc supplementation *(146,147)*. Further, animal study findings suggest that lymphopenia and thymic atrophy may be mediated, in part, by chronically elevated levels of glucocorticoid hormones produced as zinc deficiency advances *(148)*. However, the contribution of glucocorticoids to the effects of zinc deficiency in human lymphocytes is yet to be confirmed *(103)*.

A T-cell subpopulation study showed a significant decrease in the ratio of CD4+ to CD8+ during zinc deficiency that was corrected by zinc supplementation *(149)*. It has been suggested that zinc is required for regeneration of new CD4+ T cells *(105)*. Studies on experimental human models have also shown a decreased proportion of CD73+ in the CD8+ subset of T lymphocytes in zinc deficiency *(150)*. The CD73 molecule on cytotoxic T lymphocytes is required for antigen recognition, proliferation and cytolysis *(150)*. Zinc deficiency is also known to affect the production of a variety of cytokines, such as IL-1, IL-2, IL-4, IFN-γ *(103,151)*. The addition of zinc to human peripheral blood mononuclear cells was found to induce the release of IL-1, IL-6, TNF-α and IFN-γ *(152,153)*. Studies in the experimental human model and in patients with sickle cell disease suggest that the impaired cell-mediated immunity of zinc deficiency is caused by the imbalance between Th1 and Th2 cell functions *(149,150,154)*. While there was a decrease in the production of IFN-γ and IL-2 (Th1 response), the production of IL-4, IL-6 and IL-10 (Th2 response) was not affected during zinc deficiency *(149,150,154)*. The cytokine IFN-γ is known to downregulate a Th-2 clone, and IL-2 is involved in the differentiation of thymocytes and peripheral T lymphocytes, thus providing a possible mechanism by which zinc deficiency affects cell-mediated immunity.

Zinc deficiency is known to affect the development of B lymphocytes in the bone marrow, and as result, fewer B lymphocytes in the spleen *(103,140)*, as well as impairing B lymphocyte functions *(155)*. While some studies showed depressed B lymphocyte proliferation in response to mitogen in zinc-deficient animals *(156)*, others have shown no effect *(157)*. The later study was based on plaque-forming cell (PFC) assays measuring in vitro antibody production where data were expressed as PFC per viable splenocyte rather than PFC per whole spleen.

Animal studies have shown impaired B lymphocyte antibody responses in zinc deficiency *(125,148)*. Study findings revealed that antibody responses to T-dependent antigen sheep red blood cells *(125,158)* are more affected by zinc deficiency than is T-independent antigen dextran *(155,159)*. Further, studies on zinc-deficient mice reported a reduced antibody recall response to T-cell-dependent and T-cell-independent antigens for which they had previously been immunized *(148,155,160)*. These findings suggest that zinc deficiency can compromise B lymphocyte development, both primary and secondary antibody responses. Although the effect of zinc supplementation on antibody production has been explored in a number of vaccination trials in humans, the findings have been highly contradictory *(153)*.

23.3.1.2.2. Public Health Implications. Supplementation under experimental conditions has shown reduced poor growth rates and possibly an association with an increase in energy and protein intake. More recent field trials in Bangladesh, India, Pakistan and sites in Africa have shown conflicting results in growth, reduction in disease, severity of disease and varied results depending on the disease. Nevertheless, results of two recent meta-analyses of such studies seem to indicate a definite role for zinc supplementation (probably with other micronutrients), especially in growth and diarrheal disease *(16)*. Consequently, zinc deficiency and its impact on reducing child mortality have come much more to the fore, with sufficient evidence that the accepted treatment of diarrhea is now oral rehydration therapy with a 2-week course of zinc supplementation *(161)*. There is considerable debate now on whether this also applies to respiratory disease and the public health prevention of zinc deficiency, as opposed to therapeutic use.

Part of this is because of the difficulty of increasing intakes of zinc through dietary methods, especially poor diets low in animal-source foods *(162)*. Large variations in zinc content can be found between otherwise nutritionally similar food sources but tend to be high in meat, cheese, lentils and cereals. These tend to be components of more expensive diets with cereals being the major source of energy and zinc in large parts of the world. As the zinc is mainly located in the outer layer of the grain, a low extraction rate means that the majority of the contents of zinc and other minerals are removed, although this should also reduce the phytates that affect bioavailability. Use of zinc-rich galvanized cooking pots and canning may also contribute zinc in the diet. Unrefined cereal-based diets present the largest risk for low zinc absorption *(101)*. Iron content of the diet may also be a factor, as may be enhancers. Contributing factors to poor zinc intakes may be geophagia and large zinc losses due to intestinal parasitic infections *(101)*. Lower zinc intakes than in western-type diets have been described in Brazil around the Amazon which is a fish-based diet, and where signs of zinc deficiency were also observed *(163)*. Similarly low intakes have been described in other parts of the developing world including south Asia and Papua New Guinea. Nevertheless, Gibson and others have demonstrated the theoretical possibility and the feasibility in West Africa of increasing the bioavailability of micronutrients in plant-based diets *(164)*.

23.3.1.3. IRON DEFICIENCY, IRON DEFICIENCY ANEMIA AND OTHER NUTRITIONAL ANEMIAS

Conservative estimates indicate that 1,500 million people are anemic worldwide, with perhaps over 90% of these in the developing world, mainly south Asia and Africa *(14)*. All the figures suggest that over half of all women in developing countries are anemic. Iron deficiency, the main cause of anemia, is a major contributor to low birth weight, prematurity and maternal mortality *(11,165)*. Iron deficiency anemia (IDA) is even more prevalent in infants and young preschoolers, and while there are only very recently global data on prevalence of IDA in infants and children, in some sample populations prevalence reaches 70% or more *(166)*. Nutritional anemia, largely because of iron deficiency, remains the major nutritional problem facing the poorer nations, although even in more affluent countries, it remains a significant problem in certain, usually disadvantaged, groups. The recent WHO estimates give prevalence data for pre-school-age children and non-pregnant and pregnant women, according to information available and that were included according to pre-specified criteria *(166)*. For infants and young children, the range is from 3.4% in North America to nearly two-thirds (65.4%) in Africa.

In women, the range is 7.6% in North America to 44.7% in Africa (non-pregnant), and for pregnant women, 4.7–55%. However, individual studies have identified far higher prevalences for infants and women, especially in south Asia that show, e.g. 84.9% of pregnant women anemic (<110 g/L) with 13.1% having severe anemia (<70 g/L). In India, adolescent girls had levels of 90.1% with 7.1% having severe anemia in the 16 districts of India surveyed *(167)*.

23.3.1.3.1. Iron Deficiency and Immunity. In both humans and animals, the effects of iron deficiency on immune response have been studied extensively. As a result the relationships between iron and immune functions have been the subject of several reviews *(168–170)*. Overall, there is little evidence that shows any effects of iron deficiency, especially in humans, on B-cell-mediated immunity and antibody production. On the other hand, specific defects in several components of both innate immunity and cell-mediated immunity have been well documented. Overall though, from a public health perspective there has been limited appreciation of a role. In a recent review in *The Lancet*, besides a mention of increased susceptibility to upper respiratory infections associated with iron deficiency anemia *(171)*, the only mention was that "the effect of iron status on immune function and cognition in infants and children needs to be clarified" *(172)*.

Innate Immunity. In both human and animal studies, several components of non-specific immunity have been found to be impaired by iron deficiency. NK cell activity was found to be depressed in iron deficiency *(173,174)* presumably because the NK cell needs iron for its differentiation and proliferation. Macrophage phagocytosis in general appeared to be unaffected by iron deficiency, while bacteriocidal activity of these macrophages has been reported to be impaired *(173–175)*. The nitroblue tetrazolium (NBT) test, which reflects the capacity for generating an oxidative burst, was found to be abnormal in iron-deficient individuals and was reversed by iron treatment *(176)*. On the contrary, no abnormality in the NBT dye test was observed by others *(177,178)*. In iron deficiency, neutrophils have reduced activity of myeloperoxidase, which is involved in the killing process of pathogens *(173,179,180)*.

Acquired Immunity. Iron plays an important role in cell-mediated immunity *(168)*. In most but not all studies, the number of T cells was found to be reduced with thymic atrophy during iron deficiency *(169,173,179,181,182)*. In addition to a reduced number of T cells, more reports than not on iron deficiency show an impairment of lymphocyte blastogenesis and mitogenesis in response to a number of different mitogens *(177,179,180,183–185)*. This change is largely correctable with iron repletion *(180,184)*. Further, studies on iron-deficient patients have also reported either an absent or diminished DTH response, compared with control subjects, to a variety of antigens such as *Candida*, mumps, diphtheria, trichophyton and streptokinase–streptodornase *(177,181,183)*. Following iron supplementation, the impairment of the DTH responses, including tuberculin reactivity, were found to be reversed *(177,179,186)*.

Cytokines are important mediators of cell-mediated immunity. Although little is known about the effects of clinical iron deficiency on cytokines, it has been reported that the in vitro production of IL-2 by lymphocytes of iron-deficient children is impaired *(187,188)*. The IL-2 is the principal T lymphocyte growth factor which promotes the proliferation of T lymphocytes *(189)*.

Studies have also shown a reduced in vitro production of IFN-γ by spleen cells from iron-deficient mice *(190)*. In a study in hospitalized children with iron deficiency, they were found to have lower percentage of lymphocytes producing IFN-γ in vivo (spontaneously), while they had a higher percentage of lymphocytes producing IFN-γ with in vitro stimulation *(191)*. This study also showed a higher percentage of CD8+ cells producing IL-6 and TNF-α *(191)*. The IFN-γ is a potent macrophage activating lymphokine and an important mediator of the DTH response and cellular cytotoxicity *(192)*. Cellular iron availability also modulates the differentiation and proliferation of Th-cell subsets, with Th1 cells being more sensitive than Th2 cells to iron deficiency *(193)*. Further, the ratio of CD+ to CD8+ T lymphocytes in blood was found to be reduced in iron deficiency, whereas the number of cells remained unchanged *(194)*. Humoral immunity appears to be normal in iron-deficient individuals. In iron-deficient patients, the serum IgG, IgA and IgM concentrations were either normal or elevated *(177,181,195)*. Antibody production in response to specific immunization with most antigens was found to be well preserved in iron-deficient humans *(175,180)*.

Thus, based on the existing literature, it can be concluded that iron deficiency impairs both innate (reduces bactericidal macrophage activity and NK cell activity) and cell-mediated immunities (reduces T-cell proliferation and DTH response, decreases the ratio of CD4+ to CD8+ cells). It also impairs a variety of cytokines (IFN-γ, TNF-α, IL-2), and small decrease in IL-10, and suppresses Th-1 cells response with a small decrease in Th2 response. On the other hand, there is very little evidence to show that iron deficiency affects B-cell-mediated immunity and antibody production.

On the other hand, iron overload also affects various components of the immune system. In both animals and human patients with iron overload, they showed reduced responsiveness to contact sensitivity but normal DTH responses *(190,196,197)*. Several relatively recent reviews have summarized the effects of iron overload on immune system *(170,198,199)*. Iron overload, as seen in hereditary hemochromatosis patients, enhances suppressor T-cell (CD8) numbers and activity, decreases the proliferative capacity, numbers and activity of helper T cells (CD4) with increases in CD8/CD4 ratios, impairs the generation of cytotoxic T cells, and alters immunoglobulin secretion and increased levels of the cytokines IL-4, IL-6 and IL-10 *(198,199)*. Thus iron overload may result in increased susceptibility to infection by impairing Th1 cytokine-mediated response through diminished activity of regulatory cytokines (IFN-γ, IL-2 and IL-12), and by increasing Th2 response, by impairing the killing of intracellular pathogens by macrophages. These findings, of the effects of both iron deficiency and iron overload, emphasize the need for balanced iron status for proper immune functions.

23.3.1.3.2. Public Health Implications. The overt physical manifestations of iron deficiency include the generic symptoms of anemia, which are tiredness, lassitude and general feelings of lack of energy. While neuromaturational delays and reduced productivity, physical activity and work performance are the most important clinical features, reduced immunocompetence, thermoregulatory function and energy metabolism are also consequences *(200,201)*.

The recently published review volume on nutritional anemia gives considerably expanded information from a largely public health perspective *(11)*. Iron deficiency anemia has recently been re-recognized as an important cause of cognitive deficit in this age group *(202)*, including in the very recent and potentially influential *The Lancet* series on early child development *(203)*. Iron deficiency also has a profound effect on productivity and hence has economic implications for countries in which it is a significant public health problem *(204,205)* with physical work capacity being reduced even in moderate anemia *(206,207)*.

The greater understanding of factors in the control and prevention of nutritional anemias for public health interventions are still being evaluated but are likely to be important. For example, if populations have high levels of infection, then they will also have high levels of hepcidin—this may then block the uptake of iron that is in the diet from fortification and supplementation *(208)*; however, hepcidin has not been linked to an effect on dietary haem uptake, thus lending support for promotion of animal-source foods in poor diets *(209)*. Therefore it is becoming more apparent that treatment strategies encompass all the health concerns of a population—nutritional anemia can only be completely addressed if other diseases are concurrently treated.

23.3.1.4. OTHER MICRONUTRIENTS OF CURRENT PUBLIC HEALTH SIGNIFICANCE

Deficiencies of vitamins A, C and D, the B group of vitamins especially B6, B12, riboflavin and folate have all been associated with poorer health outcomes, although the pathways are not all clearly established and likely do not all work through impaired immunity *(24,202)*.

The framework used here of micronutrients only of current public health interest is used to justify, in an already very brief background, why micronutrients such as niacin, thiamin, vitamin B12 and selenium are not addressed more explicitly. Historically and even now in certain geographic areas, these are or have been of important public health interest. Niacin was widespread in many maize-consuming areas, including the south of the USA in early last century but is considerably less often seen these days, not least because of widespread fortification of flour with niacin among other B vitamins and iron *(210,211)*. Thiamin deficiency as beriberi was widespread throughout rice-eating populations *(20)* and is again less often seen. However its (thiamin deficiency) contribution to Wernicke–Korsakoff's syndrome in alcoholics means it is being addressed in a public health manner (in this case fortification of flour) in Australia *(212)*. Likewise selenium is being added to the water supply in Finland, although a specific problem has not been identified in humans, though it is likely to be a contributor to Keshan disease near the China/Russia border and countries such as New Zealand have low levels of it in much of their soil *(213)*. Again, the report of the FAO/WHO meeting in Bangkok in the late 1990s gives useful information on other micronutrients *(20)*. Many of the B vitamins are continuing to be of interest as fortificant pre-mixes being added to flour (both wheat and maize) include B vitamins along with iron and sometimes other fortificants such as zinc and vitamin A *(21,214)*, including more recently in Africa *(215)*.

Folate (or its most common supplemental form of folic acid) has come to prominence recently as the flour in the USA, and other countries, is now fortified with folic acid. Folate is required for DNA synthesis and so its deficiency is clinically expressed in tissues with high rates of cell turnover. The principal sign is megaloblastic anemia. However, its current public health importance is as a cause of anemia, a cause of neurological tube defects and a possible role in cardiovascular disease. The recent fortification with folic acid in several countries *(214)* was to prevent neural tube defects and it would appear to be having some success. The public health importance of folate deficiency in immune response is unclear but unlikely to play a major role, although it is associated with impaired cell-mediated immunity.

The public health importance of iodine deficiency is that it is the most common cause of preventable intellectual impairment in the world. It is also important in terms of women's reproductive outcomes and probably infant mortality. The fact that infants born to mothers who are iodine deficient are likely to suffer intellectual impairment, even when there may be no clinical manifestations of cretinism—the most extreme manifestation—makes this extremely important both in community terms and for national economic development. An estimate of iodine deficiency, or those suffering from iodine deficiency disorders (IDD) as assessed by goiter prevalence, was estimated globally at around 740 million in 1998 *(35)*. More reliably (when measured with urinary iodine), the global prevalence is around 35.2% (or 1,988,700,000 people) *(216)*. It is unlikely to have an impact on immune status and so will not be further considered in this chapter.

Scrimshaw, Taylor and Gordon concluded that vitamin A is regularly synergistic with infection; vitamin D deficiency commonly fails to show evidence of an interaction but synergism has been demonstrated; deficiencies of the vitamin B complex and some individual B vitamins behave variably, sometimes showing synergism and at other times antagonism, depending on species, the agent and host; vitamin C deficiencies are usually synergistic, but antagonism has been demonstrated; and finally, lack of minerals may result in either synergism or antagonism, depending on agent, host and species *(24)*.

23.3.1.4.1. Other Micronutrient Deficiencies and Immunity. Besides vitamin A, iron and zinc, in public health terms, there are other micronutrients that are important through having either low or deficient status in certain population groups, such as pregnant and/or lactating women, and elderly. In this section, the roles of these micronutrients on immune response are briefly discussed. Mocchegiani et al. *(217)* have identified, in addition, riboflavin, vitamins B6 and B12, calcium, and depending on local variations in inadequate diets of poorer populations, β-carotene, folate and vitamin C. Similarly, Pekarek et al. *(218)* note that single micronutrient deficiencies do not occur in isolation and that overlapping deficiencies affect more than 50% of children and women in many developing countries.

Folate deficiency has been associated with reduced cell-mediated immunity by reducing the proportion of circulating T lymphocytes and their proliferation in response to mitogen activation *(219,220)*. Folate deficiency has also been demonstrated to be associated with increased ratio of CD4+ to CD8+ T lymphocytes due to decreased CD8+ T lymphocytes proliferation, and which was reversible by in vitro addition of folate *(221)*. It has been suggested that the reduction in CD8+ cell replication in folate deficiency may be related to the finding of an increased carcinogenesis due to reduced cytotoxic activity *(221)*. Studies among post-menopausal women aged 50–70 years with diets low in folate showed an increased NK cell activity following folate supplementation *(222)*. All these findings indicate that folate deficiency is associated with impaired Th1 response.

Studies on vitamin B12 deficiency and immune response are limited. In patients with vitamin B12 deficiency (with pernicious anemia or post-gastrectomy megaloblastic anemia) a significant decrease was found in the number of lymphocytes and CD+ T cells and a reduction in the proportion of CD4+ T cells. Further, there was an abnormally high CD4+/CD8+ ratio and reduced NK cell activity *(223)*. Following treatment with methylcobalamin, CD8+ T cells were restored and NK cells activity improved *(223)*. In an elderly population with low serum vitamin B12 concentrations, a reduction in antibody response to pneumococcal polysaccharide vaccine was observed suggesting an impaired synthesis of specific immunoglobulins *(224)*. Vitamin B6 (pyridoxine) deficiency in humans has been found to be associated with reduced lymphocyte maturation, growth and proliferation, impaired NK activity, decrease in pro-inflammatory cytokines IL-1-β, IL-2, IL-2 receptors, and a decreased antibody response of DTH *(225–228)*. Thus vitamin B6 is associated with suppressed Th1 response and increased Th2 response, which is reversed following repletion of the vitamin *(229,230)*.

In animal models, vitamin C (ascorbic acid) deficiency has been associated with decreased neutrophil function and impaired delayed cutaneous hypersensitivity *(231,232)*, decreased T-cell proliferation and abnormal complement concentrations *(233)*. In humans, vitamin C deficiency was associated with decreased DTH response to several antigens, which could be reversed by high-dose supplementation *(234)*. Administration of vitamin C in humans also resulted in improvement of anti-microbiocidal and NK cell activities *(235)*. Supplementation of vitamin C has been found to enhance neutrophil chemotaxis in adult healthy volunteers *(236)*, and an increase in the proliferative response of T lymphocytes to PHA and concanavalin A in the elderly *(237)*. Thus vitamin C deficiency in humans can impair leukocyte functions, and decrease overall NK cell activity and lymphocyte proliferation.

Vitamin E is a fat-soluble vitamin important for normal function of the immune system. In the few rare cases of vitamin E deficiency in humans, impaired T-cell function and DTH response were observed *(238,239)*. Supplementation of vitamin E in healthy adults showed a significantly increased T-cell proliferation in response to PHA, an improved CD4+/CD8+ ratio and decreased parameters of oxidative stress *(240)*. In general, the elderly are at a greater risk for lower vitamin E intake. A recent review by Meydani et al. *(241)* presented a comprehensive coverage of the role of vitamin E and immunity in humans, especially in the elderly. Vitamin E supplementation above currently recommended levels has been shown to improve immune functions in the aged including delayed-type hypersensitivity skin response, increased mitogen-stimulated lymphocyte proliferation and increased production of IL-2, enhanced NK cell cytotoxic activity, and increased phagocytic activity by macrophages *(241–243)*. Antibody production in response to vaccination was shown to be significantly associated with the nutritional status of vitamin E, which was mediated through increased production of IL-2, leading to enhanced proliferation of T cells *(244)*. Thus higher vitamin E intake is associated with enhanced Th1 response and decreased Th2 response. Besides its protective role as an antioxidant, the possible mechanism for the improved immune function due to vitamin E supplementation is because of the reduced production of the T-cell suppressive factors, such as PGE_2 by macrophages *(241)*.

Selenium is an essential micronutrient that, through its incorporation into selenoproteins, plays a key role in maintaining health and an insufficiency predisposes to diseases associated with oxidative stress and encourages immune function and resistance to some viral infections *(245)*. Animal models and human studies with supplementation have been shown to enhance immune competence and resistance to viral infections. However, while the influence of selenium on immune responses is generally to enhance them, it may not always be beneficial, e.g. on antiparasitic responses or allergic asthma, which suggests the levels of selenium may affect different types of immunity *(246)*. While micronutrients can influence the ability of the host to respond to a viral infection *(247)*, the virus itself may also respond to the nutritional status of the host *(248)*. For example, a deficiency of selenium influences the expression of mRNA for the chemokine monocyte chemo-attractant protein-1, which may contribute to the development of myocarditis in the selenium-deficient host *(248)*.

Vitamin D is known to be essential to immune function *(25,249)*. Relatively little is known about vitamin D status in equatorial populations but a recent study in Tanzania showed that hypovitaminosis D is common among pulmonary tuberculosis patients (and is not explained by the acute phase response) *(249)*.

23.4. IMPACT ON INFECTIOUS DISEASES

The evidence for the impact of undernutrition on immune status is stronger than in micronutrients per se. The causal line between undernutrition, including micronutrient deficiencies to reduced immunity and then to increased incidence and/or severity of diseases leading further to a cycle of poor intakes leading to poor nutrition and so on, can continue until death or resolution. However, the direct evidence linking micronutrient deficiencies to increased disease is sometimes less clear. As noted, the actual mechanisms that lead to an inadequacy of vitamin A and subsequent diseases are somewhat unclear, especially with respiratory track infection *(8)*. Some of the stronger evidence comes from the role of micronutrient deficiencies in ageing and infectious disease. Of course, ageing is itself associated with impaired regulation of the immune system contributing to a higher incidence of morbidity and mortality from infectious, autoimmune and neoplastic diseases *(25)*. Subtle subclinical deficiencies of micronutrients such as zinc, selenium and vitamin E and inadequate macronutrient intake contribute to the decline in immune functions in the elderly *(25)*.

However a study of elderly Ecuadorians did not observe an association between any of the individual micronutrients they looked at that were often deficient in this population (vitamins C, D, B6, B12 and folic acid, as well as zinc and iron) with a history of recent pneumonia, influenza-like illness, cold or diarrhea *(25)*. However, the presence of some type of micronutrient deficiency and a history of recent infection were significantly associated. This does not imply causality but vitamin C, zinc and IFN-γ were associated, as were zinc, iron and IL-2 *(25)*.

What are the public health impacts of addressing compromised immune function through improving micronutrient status? The first section showed that the public health impact of micronutrient deficiencies extends far beyond their impact on infectious diseases; therefore it is important, when addressing micronutrient deficiencies, to go beyond a medical model. Micronutrient deficiencies affect both intellectual development and potential and have been graphically demonstrated, as in the recent *Lancet* nutrition series in early 2008 and elsewhere, include also individual earning capacity and the economic development of whole countries *(205,250)*. Similarly the correction of the deficiencies is far more than the supplementation of tablets and supplements, especially in terms of sustainability.

The challenges of getting expensive foods, the ones that are usually higher in iron and zinc and preformed vitamin A, into the diets of the poor are discussed below. Consequently both the problem and the solution need to be perceived broadly. The problem is bigger than just impaired immunity although the shame that this is, with still nearly 10 million children under 5 years of age dying each year in poorer communities and countries, is one of wider development and reduction of inequities. This must include broad solutions, not least the improvement of women's status and education and other opportunities for female children, adolescents and women *(251)*.

Consequently the following section is about the prevention, control and treatment of micronutrient deficiencies rather than about treating diseases or infectious disease control, or even other measures, such as improved hygiene and sanitation measures that would be expected to have an impact on immune status. Another whole area that is not addressed is the increased immunity over time of children, and adults, continuously exposed to disease and the likely compromising of this defense in affluent populations where exposure is usually delayed and reduced. In this context, a study with pregnant women in Indonesia showed that when they were supplemented with zinc or β-carotene (along with the routine iron and folic acid), the mothers having zinc in pregnancy had a better ability to produce interleukin 6, and those receiving β-carotene, produced less IFN-γ, independently of nutritional status or birthweight *(252)*. So the authors suggest that giving mothers antenatal nutrition might even have the unintended consequence of an increase in the incidence of allergy and atopy in their offspring. A further aspect is the possibility that vaccinations and immunization outcomes may be compromised if the child is inadequately nourished, e.g. vitamin A and triple antigen, as briefly discussed earlier.

23.5. PREVENTION, CONTROL AND TREATMENT INTERVENTIONS TO IMPROVE MICRONUTRIENT STATUS

The recognition of the magnitude of the prevalence and impact of micronutrient deficiencies, and the knowledge of the possibility of doing something about them on a large scale, has resulted in a series of international goals. A meeting in Ottawa in 1991 reviewed and recommended ways to reach these goals *(253)*. These built on experience gained over previous decades (since the early 1960s in the case of iron-fortified cereal flour and iodized salt). As more experience has been gained *(34,254,255)*, and funding increased, these have been continuously refined, and expanded. This section briefly examines the currently most commonly used interventions. The prevention and control of micronutrient deficiencies has become a higher global priority over the last couple of decades, but the extent of the programs and the level of funding are vastly under-resourced.

A suggested categorization of such interventions is seen in Table 23.1 *(255)* and is broadly given as follows:

(i) food-based approaches, including dietary diversification, nutrition education and fortification of staple and value-added foods;
(ii) supplementation with vitamin A capsules, iron-folic acid tablets and iodized oil with increasing interest in a multimicronutrient supplements and weekly low-dose supplements;

Table 23.1
Different Public Health Approaches to Modifying Micronutrient Intake Used in the Prevention and Control of Micronutrient Malnutrition

Food based
Dietary diversification
Home gardening
Nutrition education
Development of high micronutrient content varieties of staple foods (also called "biofortification")
Fortification
Staples, e.g. flour, noodles
Fats and oils, e.g. margarine, edible oils
Condiments, e.g. salt, sugar, soy sauce, fish sauce
Complementary foods for infants 6 months and older
Home-based fortification, e.g. "sprinkles"
Beverages, e.g. fortified juices, condensed milk and other dairy products

Supplementation
National distribution to all preschool children
National immunization days
Through health system centres, including MCH programs, and routine treatment
Outreach, e.g. with E.P.I. and other programs
Post-partum supplementation
"Life cycle" distribution to adolescents and young women through schools and factories
Home-based supplementation, e.g. "foodlets"

Public health measures
Improved antenatal and obstetric care
Immunization
Appropriate prevention and control of diseases such as diarrhea, respiratory tract infections, malaria
Promotion of exclusive breastfeeding
Appropriate complementary feeding
Water and sanitation measures
Appropriate birth spacing

Global equity corrections, poverty reduction and socio-political change
Increased availability and accessibility of micronutrient-rich foods
Improved health systems
Improved status and education of women

(iii) public health interventions such as immunization, adding vitamin A supplementation to other programs such as National Immunization Days and Child Health Days, promotion of breast feeding, and treatment of infectious diseases; and, change in the possibilities that are available to people by modification of the political, socio-economic and physical environment. As with so much of public health, those most vulnerable are those who are poorest.

The important point about these different approaches is that they are complementary, and should be started in concert, as they may have different time frames and differing feasibility depending on local circumstances. Behavior change to improve the intake of micronutrients is an essential part of whatever method is being used: through communications, social and political facilitation, social marketing and nutrition education. The overall strategy is to reduce the size of the most vulnerable group (to the left of the curve in Fig. 23.1) by improving the coverage of the middle group by fortification, dietary diversification and reduction of the disease burden *(256)*. The most at-risk group is likely to continue to need supplementation for many years to come. The factors listed in Table 23.1 have all been shown, to a greater or lesser degree, to have an evidence-based impact on micronutrient deficiencies prevention and control programs.

In the following section, the prevention, control and treatment of micronutrient deficiencies are described. Improving immune response is not directly addressed as it is presumed to be a function of improved micronutrient status where that is the cause of the impaired immune function and increased risk of infectious disease.

23.5.1. Food-Based Approaches and Fortification

23.5.1.1. Dietary and Horticultural Interventions

With the exception of iodine in certain ecological settings, micronutrients are found abundantly in many plant foods and animal products. However, many poor families do not have enough to eat, e.g. in Bangladesh where nearly half of all women are categorized as underweight *(257)*. But it is even more the quality of the diet, as diets characterized by poverty are not likely to include many micronutrient-rich foods which are in any case generally more expensive and often less accessible, and so diets are likely to be low in vitamins and minerals, as well as energy *(258)*. This low accessibility to food sources is aggravated by the usually low bioavailability of micronutrients in the diets eaten by poor families, and it is poor dietary quality, rather than quantity, that is considered to be the key determinant of impaired micronutrient status *(259)*. In the current environment of high food prices critically affecting the accessibility of the poor to all foods, especially micronutrient-rich foods and the shift to cereal staples such as rice, it has been demonstrated that these changes in household food expenditure patterns have a negative impact on the clinical vitamin A status of women of child-bearing age *(260)*.

Food-based approaches have been categorized as follows: (i) increasing small-scale production of micronutrient-rich foods by community fruit and vegetable gardening, school gardening and/or small animal, poultry or fish production; (ii) increasing community production of micronutrient-rich foods, such as horticultural products, oil seeds, palm oil, beverages and natural nutrient supplements; (iii) maintaining micronutrient levels in commonly eaten foods with food storage and preservation techniques, improving food safety, and better food preparation; (iv) plant breeding to increase micronutrient levels, including through genetic engineering; and (v) community strategies to increase consumption of micronutrient-rich foods *(209,261)*.

Improving dietary diversification through increasing variety and frequency of micronutrient-rich food sources through nutrition education and horticultural approaches has been shown to be effective in many settings. Measuring effectiveness should use indicators of outcomes that go beyond increased serum levels of micronutrients, to clinical outcomes (reduction in night blindness) and to social outcomes such as women's empowerment *(262,263)*. Food preparation interventions to achieve dietary diversification can include nutrition education concerning available foods and their more effective utilization; horticultural approaches such as home gardens; and improved methods of food preparation, preservation and cooking that better conserve the micronutrient content. There is also increased interest in the genetic manipulation and breeding of staples and other foods to increase micronutrient content ("biofortification") *(264–266)* (*see* Chapter 3).

While home gardening is a traditional family food production system widely practised in many developing countries in the world *(261,267)*, anecdotal experience suggests home gardening (as an intervention method for improving nutrition) has been generally successful at the pilot or local phase, but often not been scaled up successfully. Recent experience in Bangladesh has demonstrated a successful example where it has *(267)*, and some of the lessons learned are being tried, with apparent good acceptance in Cambodia, Nepal *(257)*, and parts of Africa such as Ethiopia *(268)*. An evaluation has shown that food gardening programs also strengthened the capacity of local NGOs as a contribution toward sustainability of improvements in the community *(269)*. They have also been found to increase income and empowerment of women and that can result in increased intake of micronutrient-rich foods such as eggs and meat as well as other foods such as oil and improved caring practices *(257,263,270)*. Where home gardening is traditionally practised, using such an approach to increase micronutrient intake is more likely to be successful. In Indonesia, ownership of a home garden appears to indicate long-term vitamin A intake from plant foods, which explains its relationship with vitamin A status *(271)*. In the Bangladesh national survey, young children who had not received a vitamin A supplement were half as likely to be night blind if the family had a home garden *(270)*.

Biofortification, also a food-based approach, uses traditional plant-breeding methods such as identifying plants that have cereal seeds naturally high in zinc or iron, or low in phytates, and then breeding for these, and more recently transgenic methods *(264,272)*. Effectiveness of the resultant grains to raise micronutrient status in humans has been shown in one study to date of a successful feeding trial in the Philippines (using Catholic nuns to ensure adequate control conditions) *(272)*. The use of genetic engineering is expanding the possibilities, and a relatively recent alliance among IRRI (the International Rice Research Institute) and CIMMYT (the International Maize and Wheat Improvement Center) will increase both efforts and coordination of research efforts on rice, wheat and maize aimed at "improving the lives of poor farmers" *(264,265,273)*. Poor farmers are not a group that has much benefited from transgenic food research up to this point, which has mainly benefited horticulture for western markets, despite much of the rhetoric *(266)*. Probably the best known micronutrient example of this research approach, at least in terms of micronutrients, is the "golden rice" where four different genes from the daffodil (*Narcissus pseudonarcissus*) and two from a bacterium (*Erwinia uredovora*) have been introduced to allow a non-biologically active precursor of beta-carotene, to proceed to the next three biological steps to become beta-carotene *(265,274)*. However, it is not anticipated nutrigenetics will be a significant source of micronutrients in population terms within the next decade *(266,275)*.

23.5.1.2. Fortification

Probably the most cost-effective food-based approach to improving nutrient availability and accessibility is fortification, with the proviso that the fortified foods must reach those who most need them. Not infrequently, those most at risk are outside established market systems that provide many of the "value-added foods" most likely to be fortified. It has also not been shown that the fortification of staples will be able to provide adequate micronutrient content in most of the complementary foods given to young children (due to the small volumes involved), and so commercially processed and fortified foods will generally be necessary where available and accessible. There also does appear to be increasing evidence that some animal sources in the diet are necessary for adequate micronutrient status *(209,269,276)*. Nevertheless, for the majority of many populations, fortification of foods with micronutrients has been shown to be a technologically, programmatically and economically effective method of increasing micronutrient intakes in populations *(256)*. Food fortification is likely to have played a significant role in current nutritional health and well-being of populations in industrialized countries *(211)*. Starting in the 20th century, fortification was used to target specific health conditions: goiter with iodized salt; rickets with vitamin D fortified milk; beriberi, pellagra, and anemia with B vitamins and iron enriched cereals; and more recently in the USA and other western countries, but also lately Pacific Island Nations and South Africa, risk of pregnancy affected by neural tube defects by adding folic acid to fortified flour and cereals.

A relative lack of appropriate centrally processed food vehicles, less developed commercial markets and relatively low consumer awareness and demand has meant that nearly 50 years have passed since its recognized successful impact in industrialized countries *(211)*. However, fortification is now increasingly seen as a viable option for the less developed and industrializing countries to increase micronutrient intakes *(34,277)*. As many of the previous constraints to widespread accessibility are minimized and with an increasingly global market, there is a great deal of current investment in fortification as an approach to the prevention and control of micronutrient malnutrition in poorer countries *(21,277)*. Fortification is but one arm of a micronutrient deficiency prevention and control strategy, but by becoming commercially viable, can reduce the size of the at-risk population needing other measures such as supplementation (Fig. 23.2). Where the costs are passed onto the consumer, and the food industry routinely fortifies, sustainability is potentially high *(278)*.

A single micronutrient addition to an appropriate food vehicle is increasingly an uncommon approach in food fortification programs, except iodine in salt and vitamin A in sugar. Even with iodine there is now considerable work in double fortification of salt with iodine and iron *(34)* and even triple fortification with vitamin A as well *(279)*. As Huffman et al. *(280)* and others have described, women in developing countries are often consuming diets of poor bioavailability and limited micronutrient content, leading to concurrent deficiencies of iron, vitamin A, zinc, folic acid, B6, B12 and occasionally other vitamins and minerals *(12,281,282)*. Such deficiencies have important consequences for women's own health, pregnancy outcomes and their breast-fed children's health and nutritional status *(280)*, and increasingly it seems on the birthweights of their children *(282,283)*. Mason et al. *(35)* have estimated that nearly a quarter of children have multiple deficiencies. Consequently, it is now generally recommended that fortification be with a mixture of micronutrients, often in a pre-prepared fortificant mix of iron, folic acid and other B vitamins *(21)*.

Supplement-type home fortification, e.g. "Sprinkles", are microencapsulated micronutrients, including usually ferrous fumarate, which are available in a single-dose sachet, and can be sprinkled onto complementary and weaning foods and other foods. In a randomized, controlled trial in Ghana, they were found to be as efficacious as iron drops in the treatment of anemia *(284)* and have extensive efficacy experience such as those carried out in Bangladesh, Benin, Bolivia, Canadian First Nations and Inuit areas, China, Haiti, India, Nicaragua, Pakistan, Sri Lanka and Vietnam *(285)*. Although there was initially some concern about the levels of iron being given, these have now been reduced and there seems no doubt about their efficacy. Cure rates from anemia have ranged from 55 to 90% in children in the studies conducted *(284,285)*. While the effectiveness applications need further demonstration, their use is already gaining considerable experience in the post-tsunami disaster areas in south Asia *(286)* and non-emergency settings such as Mongolia *(284)*.

In the more affluent industrialized countries, micronutrient deficiencies have been, and continue to be, addressed by food fortification, as well as by overall economic growth and general improvements in health, sanitation and nutrition that have contributed to the prevention and control of these deficiencies. These same aspects must be addressed in any prevention and control programs in non-industrialized countries. Fortification, supplementation, other food-based approaches and complementary public health measures are all necessary. This will only be done by partnerships with government, industry and the consumer. There is a need to assess more widely the impact of interventions, not least for advocacy. Ultimately the success, impact and sustainability of food fortification, like other interventions, rest with educating the consumer, developing consumer demand and demonstrating impact.

23.5.2. Supplementation

Supplementation has often been characterized as a short-term approach, criticized as an example of medicalization of a public health intervention, and presumed to have difficulty with likely sustainability, especially when supplements are supplied by foreign donors. Nevertheless, supplementation with iron and folic acid has been the method of choice to address anemia in pregnant women despite little evidence of its effectiveness and likely limited impact *(35,207,287)*, although efficacy has been repeatedly shown *(202,288)*. Vitamin A supplementation has now been in place for over 30 years in countries such as Bangladesh and so hardly merits being seen as short term, and many would argue that the need will be there for many years yet *(8,10,289)*.

Zinc supplementation is now seen as the recommended treatment for diarrhea in children in developing countries *(161)*, and while not used, at least as yet, in prevention, does reduce the risk of recurrent attacks of diarrhea for some months after treatment *(94,290–292)*. Over the past 4 years, since the release of the UNICEF–WHO Joint Statement on Clinical Management of Acute Diarrhea with new ORS and zinc, 54 countries have changed national child health policies to include zinc for treatment of diarrhea. The experience of zinc treatment for diarrhea has been instructive in terms of demand being generated before supply was assured. However the actual roll out of zinc at country level has been slow for a number of reasons, including the need for changes to national policy and treatment guidelines, as well as adequate supply of zinc supplements.

Iron and folic acid supplementation has been the traditional approach for preventing and treating iron deficiency, particularly during pregnancy *(165,293–295)* but logistics are an issue, is relatively expensive and coverage is often poor *(207)*. Compliance is usually blamed but it is likely that distribution and logistical problems are every bit as important *(296)*. The efficacy of intermittent dosages, once or twice a week, has been demonstrated, suggesting that this may be a possibility for prevention, although not to treat anemia in pregnancy *(297)*. However it does appear appropriate to recommend a dosage regimen of one or two times per week before pregnancy, e.g. to adolescents and young women in schools and factories *(298)*. It is presumed this approach would encourage compliance and reduce side effects and would certainly reduce costs *(287,288,299)* but not logistic requirements, a recognized constraint in many settings, although work in four Asian countries has shown promise with a social marketing approach *(299)*. With iron supplementation, gains in productivity and take-home pay have been shown to increase 10–30% *(205)*. Consequently there are important reasons, in addition to the already compelling health, cognitive development and reproduction consequences, to accelerate programs to prevent and control iron deficiency anemia.

Nevertheless, there is also increasing consensus that new approaches to scaling-up supplementation coverage are required *(202,207)*. Anthelminthics treatment improved the hemoglobin and serum ferritin concentrations of Tanzanian schoolchildren *(300)*, growth, appetite and anemia *(301)* with similar positive synergies in other settings *(302)*. As the strategy for improving micronutrient status moves more toward integrated approaches as a way of helping to improving child survival *(303,304)*, reducing micronutrient deficiencies will be increasingly seen as an approach to increasing child survival and development.

Interactions are also a potential issue in other multiple micronutrient intervention settings. Because immunologic function, particularly cell-mediated immunity declines with age, this probably contributes to the increased incidence of infectious diseases in the elderly *(26,305)*. Up until now, much of the work on multimicronutrient supplementation has been in the relatively affluent elderly in western societies, with a considerable amount of self-medication *(32)*. High *(305)* has concluded that "multivitamin/mineral supplements or specific micronutrients such as zinc and vitamin E maybe of value. . .oversupplementation may be harmful". Nevertheless women in North America, with generally micronutrient replete diets, are recommended to take multiple micronutrient supplements during pregnancy. On the other hand, many women in less affluent economies survive on diets of poor quality and micronutrient deficiencies are common in developing countries *(281)*. As these authors note, the ability of the newborn to maintain health, withstand disease, grow and develop normally is influenced by the gestational nutritional experience, and replacing likely deficient micronutrients would presumably correct these deficiencies. However it is not known how such micronutrients interact in depleted/infected populations, or against habitually compromised diets *(281)*. A forthcoming review of studies in twelve countries that have examined multiple micronutrients that were given antenatally in developing country settings will be published soon but the public health implication appears to be a greater attention must be paid to undernourished women in these societies, starting from before the birth of their children (and probably even before the birth of the women themselves), with special attention during adolescence and during pregnancy *(282)*. This is in line with a recent study in Ghana demonstrating, yet again, an improvement in birthweights with multiple micronutrient supplementation and improved antenatal care and which concluded that early antenatal care is crucial to a favorable outcome in such populations *(306)*.

As noted above, and as Huffman et al. *(280)* and more recently Ramakrishnan and Huffman *(12)* and others *(35,209)* have shown, women of reproductive age and very young children in developing countries are often consuming diets of poor bioavailability and limited micronutrient content, leading to concurrent deficiencies of iron, vitamin A, zinc, folic acid, B6, B12 and occasionally other vitamins and minerals. Children in poor countries, and even disadvantaged children in more affluent countries, are not infrequently suffering from more than one micronutrient deficiency also (35,298). Mason et al. *(35)* have estimated that nearly a quarter of children in developing countries have multiple deficiencies.

The use of antenatal multiple micronutrient supplements in the developing world has not yet been recommended, as the preventive role is being more rigorously evaluated than it ever was in the industrialized world *(281,282)*. Nevertheless, their widespread use and their endorsement in the WHO and WHO/UNICEF recommendations (in specific circumstances) mean this intervention remains a major potential approach to the prevention and control of micronutrient deficiencies, and several expert consultations on their use as a public health intervention are likely moving their adoption ahead in developing countries *(282,283,307)*. A meeting to review nutrition as a preventive strategy against adverse pregnancy outcomes concluded that effective interventions with micronutrients are likely to be required at an earlier stage than happens in public health programs at the present time, certainly before mid-pregnancy, and for some interventions, probably during the pre-conceptual period *(308)*. There is already an existing WHO/WFP/UNICEF joint statement on preventing and controlling micronutrient deficiencies in populations affected by emergencies *(309)*, which includes both women and young children, but not for prevention in non-emergency settings.

As also discussed, recommendations on the reduction of the incidence of low birthweight neonates by giving antenatal multiple micronutrients are also unclear at present *(281,282)* despite the largest study, well carried out by Shankar et al. in Lombok, Indonesia, which showed significantly positive impacts on birthweight and neonatal mortality *(310)*, unlike the suggestive, but non-significant findings from Nepal quoted in Nestel and Jackson *(281)*. The SUMMIT study *(310)* also had a very strong component of community support and health center strengthening which no doubt contributed to the positive results. UNICEF has provisionally recommended that a combined vitamin–mineral supplement for pregnant women, similar in composition to those freely available in the more industrialized world, is an option that needs to be considered for developing countries *(311)*. As also noted above, considerable research has been going on looking at various outcomes such as incidence of low birth weight, maternal hemoglobin, and perinatal mortality *(307)*, progression to AIDS in women with HIV *(99,312)*, morbidity and micronutrient status in children *(313)* and both efficacy and effectiveness of intervention trials of multimicronutrient supplements for antenatal care *(282)*.

23.6. CONTROL, PREVENTION AND TREATMENT

Once an infection is established, a host can do one of three things to minimize the agent's impact on its health. The immune system of the host can directly attack the growing pathogen population to contain or eliminate it (resistance); or it can attempt to minimize the harm caused by a given number of pathogens by increasing tissue repair or by detoxifying pathogen by-products (tolerance); or some combination of both *(314)*. Traditionally, the authors suggest immunologists, microbiologists and parasitologists have focussed on the ability to limit parasite numbers or on the overall ability to maintain health irrespective of parasite burden (resistance plus tolerance), with less emphasis on just tolerance although there are good examples of that such as the presence of α-thalassemia and the resulting reduced life-threatening episodes of malaria *(314)*.

With micronutrient deficiencies, the line between prevention, treatment and control is often blurred except in serious deficiency, e.g. xerophthalmia, anemia <7 g/dL and so on. Children with any stage of xerophthalmia should be treated with vitamin A according to the WHO treatment guidelines, as should pregnant women with such life-threatening levels of anemia. But, in a population with say 50% prevalence of a particular deficiency and the resultant clinical outcomes, such as may occur with anemia in pregnant women in resource-poor settings, is giving iron a treatment or prevention *(315)*? And why give folic acid where the main purpose in such environments is to prevent neural tube defects but is routinely given too late to do this (and only has a limited effect on anemia in most circumstances). In the majority of the programs, it is also clear that to address impaired immunity, integrated programs that also address food security, care, the health services, community measures and the water, sanitation and hygiene environment will all be necessary.

As there are standard guidances for treatment of the clinical outcomes of micronutrient deficiencies and these are usually well tried and efficacious, they will not be discussed further. The WHO is the technical agency of the United Nations system that takes a normative role in developing these and coordinating available research information on an evidence base and the consensus of experts (www.who.org). The prevention of micronutrient deficiencies, although undoubtedly efficacious remains challenging in terms of effectiveness, especially in hard-to-reach populations.

For example, reports from research in Pemba in Zanzibar, Tanzania, have raised again the question of the safety of iron supplementation to children with adequate stores and especially in malaria-endemic areas *(290)* although the need to treat anemia has consistently been reinforced by guidance from the WHO and UNICEF *(30)*. Earlier Mebrahtu et al. *(295)* had demonstrated that low-dose daily iron supplementation for 12 months did not increase the prevalence of malarial infection or density of parasites in their study population in Zanzibar. Caulfield et al. calculated from a meta-analysis performed for the WHO that zinc deficiency in children 0–4 years old may be responsible for approximately 20% of malaria clinical attacks and 193,000 malaria deaths each year *(27)*. While undernutrition has been demonstrated to exacerbate, palliate or have little effect on malaria outcomes, the review concluded that "although the association is complex and requires additional research, improved nutritional status lessens the severity of malaria episodes and results in fewer deaths due to malaria. Deficiencies in vitamin A, zinc, iron, folate, as well as other micronutrients are responsible for a substantial portion of malaria morbidity and mortality" *(27)*. They also concluded that undernutrition exacerbates diarrhea and respiratory infections. Prentice among others has questioned the controversy, but the uncertainty has certainly impacted on iron programs in eastern and southern African countries, despite the same Pemba study, in a large substudy, showing better mortality outcomes if children were treated with antimalarials and given iron; similar to findings from West Africa (Burkina Faso) where in a malaria-endemic area, in combination with malaria management, multiple micronutrient supplementation was more efficacious than the iron supplement and iron/zinc supplements for reducing anemia

(316). On the other hand, from a study in Zambia, iron, but not multimicronutrients, appears to reduce reinfection with *A. lumbricoides (317)*.

Similarly, vitamin A treatment and prevention in HIV-endemic populations is being questioned, along with iron and zinc but supplementation with a multimicronutrient supplement of one 1 RDA (except iron which is half an RDA) continues to be recommended, and much current work is being devoted to extending guidance. There is a bi-directional relationship between HIV infection and nutritional status with undernutrition affecting HIV transmission and disease progression while HIV infection can lead to undernutrition, including micronutrient deficiencies *(97)*. Further complicating factor are the relationship between HIV and other common infectious diseases including malaria, filiariasus, schistosomiasis and intestinal paritoses, and the importance of the immune system-destroying role of the HIV virus *(97)*.

The role of nutritional factors in HIV progression and transmission is a function of vitamins and disease progression; pregnancy outcomes and child health; vitamins and horizontal transmission of HIV; minerals (mainly selenium and zinc and perhaps iron); breastfeeding and protein–energy malnutrition. The reverse impact of HIV infection on nutritional status needs to consider impact on child growth and on micronutrient status which can be impaired through malabsorption (especially vitamins B12, A and E), decreased dietary impact secondary to anorexia and increased utilization from oxidative stress *(97)*. The prevalence of micronutrient deficiency during HIV infection varies widely according to study population and stage of disease. Multiple micronutrient supplementation is a low-cost intervention that has been effective in delaying disease progression and decreasing mortality from AIDS *(97)* and needs more research urgently, especially on appropriate micronutrient levels and the most appropriate formulation(s), along with potential roles of selenium and zinc on HIV progression and transmission.

23.6.1. Improving the Immunological Status and Resistance to Disease Through Related Public Health Interventions

It is clear that while important, improving the nutritional status, and micronutrient adequacy, will alone not be enough to improve immunological status of an individual. Most children at risk of increased risk of undernutrition and early child death live in unhealthy and deprived environments. The World Bank recently increased the definition of people living in poverty as those with under US$ 1.25 a day, and so in retrospect, there were in 2005, now 1.4 billion people defined as living in poverty (a quarter of the developing world)—although an improvement on the 1.9 billion in 1981. It is hard for most people not living in such conditions to have any idea of what this means in terms of inadequate food security, impossibly unhygienic conditions and increased risk of maternal and child death. Over 80% of all children stunted live in just 20 countries *(95)*, and 90% of the global burden of under 5 mortality live in 36 countries *(1)*. Consequently there is a need to address other issues as well, quite apart from the need to reduce inequities both within and between countries. Also there will need to be a scaling-up of water and sanitation measures, infectious disease prevention and treatment, improved measures to improve household food security and nutrition security and a reduction of parasite infections, as well as social measures. *The Lancet* has recently reported the estimate that a tenth of the global disease burden could be addressed by properly tackling water and sanitation issues *(318)*.

For maximum impact other public health interventions are essential. These integrated interventions include among other locally appropriate actions, control of infectious diseases, expansion of measles and other childhood immunization interventions, deworming for intestinal parasites (hookworms), malaria control, promotion of breast feeding, and proper health care such as oral rehydration therapy, all of which have an impact on micronutrient status *(93,302,319,320)*, and hence, in many cases, immune status. Breastfeeding is identified in the Bellagio child survival reports as the most important intervention providing 13% of the total impact on potential young lives saved *(1)*. A recent study in rural Ghana showed epidemiological evidence of a causal association between early breastfeeding and reduced infection-specific neonatal mortality *(321)*. However, the issue of HIV-infected mothers, and the possibility of insufficient vitamin A and iron in the milk, especially in premature neonates, will need continued work on the best guidance in this minority of cases. Vitamin A and iron supplementation of pregnant Indonesian women benefited the vitamin A status of their infants, but still, the authors concluded in that study, the infants may need vitamin A supplementation or increased dietary intake after 6 months *(322)*. Human milk ascorbic acid levels can be doubled or tripled by increased intake of ascorbic acid in women with low human milk ascorbic acid content with the impact far more evident in African women compared with European women *(323)*.

IFPRI (the International Food and Nutrition Policy Research Institute) has identified the four main factors contributing to infant and child undernutrition: food accessibility and availability; mother's education; women's status relative to men in the society; and health and sanitation environment *(251)*. Bendich *(324)* has demonstrated the roles of nutrients in optimizing women's health and immune function, as well as other roles of micronutrients in women's health *(247)*. While critical, the actual interventions are outside the scope of this chapter but there is increased recognition that parallel, vertical programs are no longer enough (although the funding community finds them easier to manage); but as it is the same families and communities that need investment in all these areas, and usually the same inadequate health systems trying to support them, and a lack of nutrition capacity, there are increasing attempts to recognize and implement these realities on the ground.

Reducing poverty is not always explicitly addressed as an intervention in the control and prevention of micronutrient deficiencies, but if it were to be possible, it would have the greatest impact of all. It was explicitly mentioned in UNICEF State of the World's Children Report on Nutrition *(325)*, and elsewhere, but rarely proposed as a serious intervention to address micronutrient malnutrition. It has been noted before that Europe had a significant vitamin A deficiency problem in the 19th century that was resolved without specific programs *(326)*. Likewise the relative prevalences of anemia between rich and poor countries make it clear how much iron deficiency and other nutritional anemias are problems of poor diets and poverty. Work in Indonesia and Bangladesh have demonstrated that as food prices rise, the quality of diets, i.e. micronutrient content declines *(260,327–329)*. A recent study has shown that women are at greater risk of clinical vitamin A deficiency where families have to spend more on rice and less on animal and plant foods *(260)*. Similarly, in Indonesia at the time of the financial crisis, micronutrients were the first things affected in the diets, and at the same time mothers tended to sacrifice their diet quality for their children *(328)*. These economic and development arguments for intervening in health and nutrition have an increasingly strong underpinning *(330,331)* and have been further driven by the "WHO Commission on Macroeconomics and Health" and the report of the "Copenhagen Consensus", among others. The widespread adoption of the millennium development goals further reflects this shift in geo-political consensus *(332)*. It remains to be seen if the adequate political and economic changes will be adopted and the agenda adequately funded. There is now more agreement than ever before that the social and political causes of ill-health and poor nutrition among other issues—"a defining moral ecology . . . around . . . the issues of equity and social justice in health. . .", need urgently to be addressed, including by changes in the social and political frameworks *(333)* and in such well-received reviews as the WHO Commission on Social Determinants of Health *(334)*.

Among the major remaining constraints, as the recent re-analysis of the child survival approach has reminded the international health community, are among other things, poor health systems and inadequate resources *(1,303)*. They also demonstrate convincingly that it is not that cost-effective interventions for both child survival and development and young child undernutrition are not known about, but that they are not being implemented on a sufficient scale *(95,303,320)*. Largely based on this series, but drawing on years of often poorly documented experience, international agencies, with national governments, are directing their efforts more toward an integrated approach: reduction of the diseases of childhood, reduction of the vaccine-preventable diseases, neonatal causes of death, undernutrition, water and sanitation-related disease and national policy making to strengthen health systems and coordinate funding through the various political and policy mechanisms such as the PRSPs (Poverty Reduction Strategic Plans). The consensus of the group explicitly recognizes the importance of vitamin A, iron and zinc in increasing child survival *(1,94)*. An interesting current example is "REACH: ending child hunger and undernutrition", a partnership of four UN agencies (WFP, UNICEF, FAO and WHO) and national governments and other NGO and private sector partners which is promoting an integrated approach to address maternal and young child undernutrition *(335)*.

23.7. CONCLUSION AND RECOMMENDATION

There is widespread concern that the "health-related" MDGs will not be reached in many countries. The most recent update on global progress by UNDP acknowledged there is a long way to go, and "that if some trends persist, some of the goals will be very difficult to reach" *(335)*. It has been argued that virtually all the MDGs will need scaling-up of micronutrient programs (as well as the wider scourge of undernutrition) for the MDGs to be realized *(95,255)*. The millennium development goals (MDGs) are now at the midterm of their target period, as 2015 is the date scheduled by the United Nations Organization (UN) for their attainment. Currently attempts are being made to review the current situation of the MDGs worldwide and to analyze the barriers which are preventing them from being attained, as well as to assess a number of the indicators evaluated. Although there have been improvements in some of the goals on a global level, the research carried out to date reveals barriers to the attainment of the MDGs, the insufficient weight of the developing countries in the economic and political decision-making processes and an incoherence between the economic policies and the social and health policies. Sub-Saharan Africa constitutes the most disadvantaged region, and will not, with current progress, attain the majority of the MDGs *(336)*.

At the same time, there need to be long-term investments in empowering women, in education and in reducing inequities. These interventions are most effective in the life cycle period from the beginning of pregnancy to 2 years of age, with a further window of opportunity during adolescence. Approximately 80% of all undernourished children live in just 20 countries, with the largest number of these children living in south Asia. Intensified nutrition action in these countries would greatly contribute to the achievement of all the health-related millennium development goals, as well as contributing to Goal 2 on education. The prevention and control of micronutrient malnutrition can be expected to contribute to the achievement of the MDGs *(255)*.

Micronutrient deficiencies prevention and control, given that it has been estimated that eliminating micronutrient malnutrition would save 18% of the global burden of disease, largely through mechanisms discussed in this chapter *(331)*, will need to be scaled-up to be a more important part of overall public health approaches in resource-poor settings. Noting that potential investments appear under-resourced, Behrman and colleagues have also noted the high rates of benefit-to-cost ratios and that the "gains appear to be particularly large for reducing micronutrient deficiencies in populations in which prevalences are high" *(337)*. The portion of the global burden of disease (mortality and morbidity, 1990 figures) in developing countries that would be removed by eliminating malnutrition is estimated by Mason et al. as 32% *(331)*. This includes the effects of malnutrition on the most vulnerable groups' burden of mortality and morbidity from infectious diseases only. This is therefore a conservative figure, but nonetheless much higher than previous estimates, mainly due to the calculations including micronutrient malnutrition *(331)*. Seen in relation to the overall disease burden (all population groups, all causes, all developing countries), eliminating micronutrient malnutrition (in children plus anemia in reproductive age women) would save 18% of the global burden of disease, with eliminating child underweight an additional 15%. This clearly has important implications for programming and funding, but is currently not being adequately communicated to policy makers, planners and donors.

Micronutrient deficiencies are a recognized problem in as many as a third of people living in developing and transitional economies. Their impact on immune status and disease incidence, growth, development and survival is now well appreciated although some of the underlying mechanisms are still not clear. What is often less appreciated is the impact of micronutrient deficiencies on immunological status. As has been described, there is good evidence that vitamin A, iron and zinc and other vitamin, mineral and trace element deficiencies all impact on the incidence and prevalence of infectious diseases through an impairment of the immune system resulting from these same deficiencies.

At the same time, there are many environmental, societal and cultural reasons, all aggravated by poverty, that are increasing the risk of contacting diseases for these same populations. As in undernutrition in general, there is a vicious cycle that is set up with micronutrient deficiencies further reducing resistance to infection with increased disease and further reduced appetite and absorption which then re-enforces the underlying micronutrient deficiency. Because of the broader environmental determinants of increased disease through undernutrition and impaired immunity, the interventions need to be broad ranging across health, nutrition, environment, water and sanitation and reduction of social inequities.

The efficacy of micronutrient prevention and control, and treatment, is well established. The big challenge is effective national or sub-national intervention programs being adequately scaled up. Transition from vertical to more integrated programs and getting donor support for these are a current major challenge. As is frequently quoted, the cost-effectiveness of most micronutrient interventions continues to need advocacy to policy makers; overall, it has been estimated that for "less than 0.3% of their GDP, nutrient-deficient countries could rid themselves of these entirely preventable diseases, which now cost them more than 5% of the GDP in lost lives, disability and productivity" *(330)*. Given the comparative success of many of the micronutrient deficiency prevention and control programs in many parts of the world, because of the known interventions, the challenge is now to scale up such programs to a national level to achieve results in the survival and development of children and women, through improving immunity and reducing micronutrient deficiencies in integrated, community-based programs supported by adequate resources at district, national and international levels.

REFERENCES

1. Black RE, Morris SS, Bryce J. Where and why are 10 million children dying every year? Lancet 2003;361:2226–34.
2. UNICEF. The State of the World's Children. Maternal and newborn health. Oxford & New York:Oxford University Press. 2008.
3. Black RE, Allen L, Bhutta AZ, Caulfield LE, de Onis M, Ezzati M, Mathers C, Rivera J. Maternal and child undernutrition: global and regional exposures and health consequences. Lancet 2008;371:243–60.
4. UNICEF. Progress for children. No.4 A report card on nutrition. New York:UNICEF. 2006.
5. FAO. The state of food security in the World 2008. Rome:The Food & Agricultural Organization of the United Nations System. 2008:1–56
6. Ahmed F, Darnton-Hill I. Vitamin A. In: Gibney MJ, Margetts B, Kearney JM, Arab L (eds.). Public health nutrition. The Nutrition Society Textbook Series. Oxford:Blackwell Sciences Ltd. 2004;chp11:192–215.
7. Solomons NW. Vitamin A. In: Bowman BA, Russell RM (eds.). Present knowledge in nutrition. 9th ed. Washington, DC:ILSI Press. 2006;chp12:157–83.

8. West KW, Darnton-Hill I. Vitamin A deficiency. In: Semba RD, Bloem MW (eds.). Nutrition and health in developing countries. 2nd ed. Preventive Nutrition Series. Humana Press:Totowa,NJ (Springer Science and Business Media Publishing LLC). 2008;chp13:377–433.
9. Bauerfeind JC (ed.). Vitamin A deficiency and its control. London:Academic Press. 1986.
10. Sommer A, West KP, Olson JA, Ross AC. Vitamin A deficiency: health, survival, and vision. New York:Oxford University Press. 1996.
11. Kraemer K, Zimmerman MB (eds.). Nutritional anaemia. Basel:Sight & Life/DSM. 2007.
12. Ramakrishnan U, Huffman SL. Multiple micronutrient malnutrition. What can be done? In: Semba RD, Bloem MW (eds.). Nutrition and health in developing countries. 2nd ed. Humana Press:Totowa, NJ (Springer Science and Business Media LLC). 2008;chp18:531–76.
13. Semba RD, Bloem MW (eds.). Nutrition and health in developing countries. 2nd ed. Humana Press:Totowa, NJ (Springer Science and Business Media LLC). 2008.
14. DeMaeyer EM, Adiels-Tegman M. The prevalence of anaemia in the world. World Health Stat Q 1985;38:302–16.
15. Yip R. Iron supplementation: country level experiences and lessons learned. J Nutr 2002;132(4Suppl):859S–61S.
16. Hotz C, Brown K. Assessment of the risk of zinc deficiency in populations and options for its control. Food Nutr Bull 2004;25(IZiNCG suppl):1–114.
17. Brown KH, Hess SY. International Zinc Nutrition Consultative Group (IZiNCG) Document #2. Systematic reviews of zinc intervention strategies. Food Nutr Bull. In press.
18. Shrimpton R, Shankar AH. Zinc deficiency. In: Semba RD, Bloem MW (eds.). Nutrition and health in developing countries. 2nd ed. Humana Press:Totowa, NJ (Springer Science and Business Media LLC). 2008;chp15:455–78.
19. De Benoist B, Allen LH, Rosenberg IH. Folate and vitamin B12 deficiencies: proceedings of a WHO Technical Consultation held 18–20 October, 2005. in Geneva, Switzerland. Food Nutr Bull 2008;29:2: 243pp.
20. WHO/FAO. Vitamin and mineral requirements in human nutrition. 2nd ed. Report of a Joint FAO/WHO Expert Consultation, Bangkok, Thailand 1998. Geneva:World Health Organization/Rome:Food & Agriculture Organization of the United Nations. 2004. ftp://ftp.fao.org/es/esn/nutrition/Vitrni/vitrni.html (accessed 24 December 2008).
21. WHO/FAO (Allen L, de Benoist B, Dary O, Hurrell R (eds.). Guidelines on food fortification with micronutrients for the control of micronutrient malnutrition. Geneva:World Health Organization. 2006.
22. Gibney MJ, Margetts BM, Kearney JM, Arab L (eds.). Public health nutrition. The Nutrition Society Textbook Series. Oxford:Blackwell Publishing for the Nutrition Society. 2004.
23. McLaren DS. The antiinfective vitamin arises once more. Nutrition 2000;16:1110–1.
24. Scrimshaw NS, Taylor CE, Gordon JE. Interactions of nutrition and infection. Geneva:World Health Organization. WHO Monogr Ser No.57. 1968:1–329.
25. Hamer DH, Sempértegui F, Estrella B, Tucker KL, Rodríguez A, Egas J, Dallal GE, Selhub J, Griffiths JK, Meydani SN. Micronutrient deficiencies are associated with impaired immune response and higher burden of respiratory infections in elderly Ecudorians. J Nutr 2009;139:113–9.
26. Solomons NW, Bermúdez OI. Nutrition in the elderly in developing countries. In: Semba RD, Bloem MW (eds.). Nutrition and health in developing countries. 2nd ed. Humana Press:Totowa, NJ (Springer Science and Business Media LLC). 2008;chp19:577–99.
27. Caulfield LE, Richard SA, Black RE. Undernutrition as an underlying cause of malaria morbidity and mortality in children less than five years old. Am J Trop Med Hyg 2004;71:S55–S63.
28. Prentice AM. Iron metabolism, malaria, and other infections: what is all the fuss about? J Nutr 2008;138:2537–41.
29. Schaible UE, Kaufman SHE. Malnutrition and infection: complex mechanisms and global impacts. PLoS Med2007;4:e15:7 pp. www.plosmedicine.org.
30. WHO/UNICEF/UNU. Iron deficiency anaemia: assessment, prevention, and control. A guide for programme managers. WHO/NHD/01.3. Geneva:World Health Organization. 2001.
31. Webb P, Nishida C, Darnton-Hill I. Age and gender as factors in the distribution of global micronutrient deficiencies. Nutr Rev 2007;65:233–45.
32. Chandra RK. Nutrition and immune system: an introduction. Am J Clin Nutr 1997:66:460S–3S.
33. Basu TK, Dickerson JW. Vitamins in human health and disease. Vitamin A. Wallingford:CAB International. 1996;chps11,12:148–92.
34. MI/UNICEF. Vitamin and mineral deficiency. A global report. Ottawa: Micronutrient Initiative. 2003. www.micronutrient.org (accessed 1 January, 2009).
35. Mason J, Deitchler M, Soekirman, Martorell R (eds.). Successful micronutrient programs. Special issue. Food Nutr Bull. 2004;25:(1):3–88.

36. West KP Jr. Extent of vitamin A deficiency among preschool children and women of reproductive age. J Nutr. 2002;132:2857S–66S.
37. WHO. Global prevalence of vitamin A deficiency. Geneva:World Health Organization. 2004a. http://www.sightandlife.org/booksAll/BooksHTML/Book05.html (accessed 21 December 2008).
38. Bloem MW, de Pee S, Darnton-Hill I. New issues in developing effective approaches for the prevention and control of vitamin A deficiency. Food Nutr Bull 1998;19:137–48.
39. West KP, Katz J, Khatry SK, Katz J, LeClerq SC, Pradhan EK, Shrestha SR, Connor PB, Dali SM, Christian P, Pokhrel RP, Sommer A. Double blind, cluster randomized trial of low dose supplementation with vitamin A or beta-carotene on mortality related to pregnancy in Nepal. Br Med J 1999;318:570–5.
40. Christian P. Micronutrients and reproductive health issues: an international perspective. J Nutr 2003;133:1969S–73S.
41. Sommer A. Vitamin A deficiency and clinical disease: an historical overview. J Nutr 2008;138:1835–9.
42. Semba RD. Vitamin A and immunity to viral, bacterial and protozoan infections. Proc Nutr Soc 1999;58:719–27.
43. Beaton GH, Martorell R, Aronson KJ, Edmonston B, McCabe G, Ross AC, Harvey B. Effectiveness of vitamin A supplementation in the control of young child morbidity and mortality in developing countries. ACC/SCN State of the Art Series Nutrition Policy Discussion Paper, No. 13. Geneva: Administrative Committee on Coordination—Sub-Committee on Nutrition (ACC/SCN). 1993.
44. Semba RD, Muhilal, Scott Al, et al. Effect of vitamin A supplementation on IgG subclass responses to tetanus toxoid in children. Clin Diagn Lab Immunol 1994;1:172–5.
45. Ross AC. Vitamin A status: relationship to immunity and antibody response. Proc Soc Exp Biol Med 1992;200:303–20.
46. Semba RD. The role of vitamin A and related retinoids in immune function. Nutr Rev 1998;56:S38–48.
47. Stephensen, B. Vitamin A, infection and immune function. Annu Rev Nutr 2001;21:167–92.
48. Villamor E, Fawzi WW. Effects of vitamin A supplementation on immune responses and correlation with clinical outcomes. Clin Microbiol Rev 2005;18:446–64.
49. Hatchel, DL and Sommer, A. Detection of ocular surface abnormalities in experimental vitamin A deficiency. Arch Ophthalmol 1984;102:1389–93.
50. Wong YC, Buck RC. An electron microscopic study of metaplasia of the rat tracheal epithelium in vitamin A deficiency. Lab Invest 1971;24:55–66.
51. Warden RA, Strazzari MJ, Dunkley PR, O'Laughlin, EV. Vitamin A deficient rats have only mild changes in jejunul structure and function. J Nutr 1996;126:1817–26.
52. Molly CJ, Laskin JD. Effect of retinoid deficiency on keratin expression in mouse bladder. Exp Mol Pathol 1988;49:128–40.
53. DeLuca L, Little EP, Wolf G. Vitamin A and protein synthesis by rat intestinal mucosa. J Biol Chem 1969;244:701–8.
54. Rojanapo W, Lamb AJ, Olson JA. The prevalence, metabolism and migration of goblet cells in rat intestine following the induction of rapid, synchronous vitamin A deficiency. J Nutr 1980;110:178–88.
55. Ahmed F, Jones DB, Jackson AA. The interaction of vitamin A deficiency and rotavirus infection in the mouse. Br J Nutr 1990;63:363–73.
56. Tseng SCG, Hatchell D, Tierney N, et al. Expression of specific keratin markers by rabbit corneal, conjunctival, and esophageal epithelia during vitamin A deficiency. J Cell Biol. 1984;99:2279–86.
57. Thurnham DL, Northrop-Clewes, CA, McCullough FS, Das BS and Lunn PG. Innate immunity, gut integrity and vitamin A in Gaambian and Indian infants. J Infect Dis 2000;182:S23–8.
58. Smith SM, Hayes CE. Contrasting impairments in IgM and IgG responses of vitamin A deficient mice. Proc Natl Acad Sci USA 1987;84:5878–82.
59. Aukrust P, Muller F, Ueland T, et al. Decreased vitamin A levels in common variable immunodeficiency: vitamin A supplementation in vivo enhances immunoglobulin production and downregulates inflammatory response. Eur J Clin Invest 2000;30:252–9.
60. Zhao Z, Murasko DM, Ross AC. The role of vitamin A in natural killer cell cytotoxicity, number and activation in the rat. Nat Immun 1994;13:29–41.
61. Ross A, Stephensen C. Vitamin A and retinoids in antiviral responses. FASEB J. 1996;10:979–985.
62. Hussey G, Hughes J, Potgieter S, Kossew G, Burgess J, Beatty D, Keraan M and Carelse E. Vitamin A status and supplementation and its effect on immunity in children with AIDS. Report of the XVII International Vitamin A Consultative Group Meeting, Guatemala City, Guatemala. International Life Sciences Institute, Washington D.C. 1996:p-81.
63. Twining SS, Schulte DP, Wilson PM, Fish BL and Moulder JE. Vitamin A deficiency alters rat neutrophil function. J Nutr 1997;127:558–65.
64. Ahmed F, Jones DB, Jackson AA. Effect of vitamin A deficiency on the immune response to epizootic diarrhoea of infant (EDIM) rotavirus infection in mice. Br J Nutr 1991;65:475–85.

65. Friedman A, Halevy O, Schrift M, et al. Retinoic acid promotes proliferation and induces expression of retinoic acid receptor-α gene in murine T lymphocytes. Cell Immunol 1993;152:240–8.
66. Coutsoudis A, Kiepiela P, Covadia H, et al. Vitamin A supplementation enhances specific IgG antibody levels and total lymphocyte numbers while improving morbidity in measles. Pediatr Infect Dis J 1992;11:203–9.
67. Semba RD, Muhilal, Ward BJ, et al. Abnormal T-cell subset proportions in vitamin A deficient children. Lancet 1993;341:5–8.
68. Humphrey JH, Quinn T, Fine D, et al. Short term effects of large0dose vitamin a supplementation on viral load and immune response in HIV-infected women. J Acquir Immune Def Syndr Hum Retrovirol 1999;20:44–51.
69. Fawzi WW, Msamanga GI, Spiegelman D, et al. Randomized trial of effects of vitamin A supplements on pregnancy outcomes and T-cell counts in HIVp-infected women in Tanzania. Lancet 1998;351:1477–82.
70. Rahman MM, Mahalanabis D, Alvarez J, et al; Effect of early vitamin A supplementation on cell mediated immunity in infants younger than 6 months. Am J Clin Nutr 1997;65:144–8.
71. Brown KH, Rajan MM, Chacraborty J, Aziz KM. Failure of a large dose of vitamin A to enhance the antibody response to tetanus toxoid in children. Am J Clin Nutr 1980;33:212–217.
72. Rosales FJ, Kjolhede C. A single 210-μmol oral dose of retinol does not enhance the immune response in children with measles. J Nutr 1994;124:1604–14.
73. Carman JA, Pond L, Nashold F, et al. Immunity to *Trichinella spiralis* infection in vitamin A mice. J Exp Med 1992;175:111–20.
74. Collizzi V, Malkovosky M. Augmentation of interleukin-2 production and delayed hypersensitivity in mice infected with *Mycobacterium bovis* and fed a diet supplemented with vitamin A acetate. Infect Immun 1985;48: 581–3.
75. Ballow M, Xiang S, Greenberg SJ, et al. Retinoic acid-induced modulation of IL-2 mRNA production and IL-2 receptor expression on T cells. Int Arch Allergy Immunol 1997;113:167–9.
76. Sijtsma S, Rombout J, West C, et al. Vitamin A deficiency impairs cytotoxic T lymphocyte activity in Newcastle disease virus-infected chickens. Vet Immunol Immunopathol 1990;26:191–201.
77. Wieringa FT, Dijkhuizen MA, West JCE, et al. Reduced production of immunoregulatory cytokines in vitamin A and zinc-deficient Indonesian children. Eur J Clin Nutr 2004;58:1498–504.
78. Cantorna MT, Nashold FE, Hayes CE. In vitamin A deficiency multiple mechanisms establish a regulatory T helper cell imbalance with excess Th1 and insufficient Th2 function. J Immunol 1994;152:1515–22.
79. Wiedermann U, Hanson LA, Kahu H, et al. Aberrant T-cell function in vitro and impaired T-cell dependent antibody response in vivo vitamin A deficient rats. Immunology 1993a;80:581–6.
80. Blomhoff HK, Smeland EB, Erikstein B, et al. Vitamin A is a key regulator for cell growth, cytokine production, and differentiation in normal B cells. J Biol Chem 1992;267:23988–92.
81. Buck J, Derguini F, Levi E, et al. Intracellular signaling by 14-hydroxy-4,14-retro retinol. Science 1991;254:1654–5.
82. Chun TY, Carman JA, Hayes CE. Retinoid repletion of vitamin A-deficient mice restores IgG responses. J Nutr 1992;122:1062–1069.
83. Pasatiempo AMG, Bowman TA, Taylor CE, et al. Vitamin A depletion and repletion: effects on antibody response to the capsular polysaccharide of Streptococcal pneumoniae, type III. Am.J Clin Nutr 1989;49:501–10.
84. Bhaskaram P, Jyothi SA, Rao KV, et al. Effects of subclinical vitamin A deficiency and administration of vitamin A as a single large dose on immune function in children. Nutr Res 1989;9:1017–25.
85. Semba RD, Muhilal, Scott AL, et al. Depressed immune response to tetanus in children with vitamin A deficiency. J Nutr 1992;122:101–7.
86. Rahman MM, Mahalanabis D, Hossain S, et al. Simultaneous vitamin A administration at routine immunization contact enhances antibody response to diphtheria vaccine in infants younger than six months. J Nutr 1999;129:2192–5.
87. Benn CS, Aaby P, Bale C, et al. Randomized trial of effect of vitamin A supplementation on antibody response to measles vaccine in Guinea-Bissau, West-Africa. Lancet 1997;350:101–105.
88. Semba RD, Munasir Z, Beeler J, et al. Reduced seroconversion to measles in infants given vitamin a with measles vaccination. Lancet 1995;345:1330–2.
89. Benn CS, Whittle H, Aaby P, et al. Vitamin A and measles vaccination. Lancet. 1995;346:503–4.
90. Rahman MM, Alvarez JO, Mahalanabis D, et al. Effect of vitamin a administration on response to oral polio vaccination. Nutr Res 1998;18:1125–33.
91. Bahl R, Bhandari N, Kant S, et al. Effect of vitamin A administered at expanded program on immunization contacts on antibody response to oral polio vaccine. Eur J Clin Nutr 2002;56:321–5.
92. Sommer A, Hussaini G, Tarwotjo I, Susanto D. Increased mortality in children with mild vitamin A deficiency. Lancet 1983;2:585–8.

93. Dalmiya N, Palmer A, Darnton-Hill I. Sustaining vitamin A supplementation requires a new vision. Lancet 2006;doi:10.1016/S0140-6736(06)69336–7:12–4.
94. Bhutta ZA, Ahmed T, Balck RE, Cousens S, Dewey K, Giugliani E, Haider BA, Kirkwood B, Morris SS, Sachdev HP, Shekar M, for the Maternal and Child Undernutrition Study Group. Maternal and child undernutrition 3. What works? Interventions for maternal and child undernutrition and survival. Lancet 2008;371;DOI:10.1016/S0140-6736(07)61693-6 (published online 17 January, 2008).
95. Bryce J, Coitinho D, Darnton-Hill I, Pelletier D, Pinstrup-Andersen P for the Maternal and Child Undernutrition Study Group. Maternal and child undernutrition 4. effective action at national level. Lancet 2008;371: 510–26.
96. Lin J, Song F, Yao P, Yang X, Li N, Sun S, Lei L, Liu L. Effect of vitamin A supplementation on immune function of well-nourished children suffering from vitamin A deficiency in China. Eur J Clin Nutr 2008;62:1412–8.
97. Villamor E, Manji K, Fawzi WW. Human immunodeficiency virus infection. In: Semba RD, Bloem MW (eds.). Nutrition and health in developing countries, Preventive Nutrition Series. Totowa, NJ: Humana Press (Springer Science and Business Media LLC). 2008;chp11:307–39.
98. Coutsoudis A, Pillay L, Spooner E, Kuhn L, Coovadia HM. Randomized trial testing the effect of vitamin A supplementation on pregnancy outcomes and early mother-to-child HIV-1 transmission in Durban, South Africa. South African Vitamin A Study Group. AIDS 1999;13:1517–24.
99. Fawzi WW, Msamanga GI, Spiegelman D, Wei R, Kapiga S, Villamor E, Mwakagile D, Mugusi F, Hertzmark E, Essex M, Hunter DJ. A randomized trial of multivitamin supplements and HIV disease progression and mortality. New Engl J Med 2004;351:23–32.
100. Zinc Investigators' Collaborative Group (Bhutta ZA, Black RE, Brown KH, Meeks-Gardner J, Gore S, Hidayat A, Khatun F, Martorell R, Ninh NX, Penny ME, Rosado JL, Roy SK, Ruel M, Sazawal S, Shankar A). Prevention of diarrhea and pneumonia by zinc supplementation in children in developing countries: Pooled analysis of randomized controlled trials. J Pediatr 1999;135:689–97.
101. Hallberg L, Sandstrom B, Aggett PF. Iron, zinc and other trace elements. In: Garrow JS, James WPT (eds.) Human nutrition and dietetics. 9th ed. Singapore:Churchill Livingstone/Longman. 1993;12:174–207.
102. Zalewski PD. Zinc and immunity: implication for growth, survival and function of lymphoid cells. J Nutr Immunol 1996;4:39–80.
103. Shankar AH, Prasad AS. Zinc and immune function: the biological basis of altered resistance to infection. Am J Clin Nutr 1998;68(suppl.):447S–63S.
104. Rink L, Kirchner H. Zinc-altered immune function and cytokine production. J Nutr 2000;130:1407S–11S.
105. Prasad AS. Zinc: Mechanisms of host defense. J Nutr 2007;137:1345–49.
106. Fernandes G, Nair M, Onoe K, et al. Impairment of cell mediated immune functions by dietary Zn deficiency in mice. Proc Natl Acad Sci USA 1979;76:457–461.
107. Good RA. Nutrition and immunity. J Clin Immunol 1981;1:3–11.
108. Walsh CT, Sandstead HH, Prasad AS, et al. Zinc health effects and research priorities for the 1990's. Environ Health Perspect 1994;102:5–46.
109. Chavakis T, May AE, Preissner KT, et al. Molecular mechanisms of zinc deficient leukocyte adhesion involving the urokinase receptor and B2-integrins. Blood 1999;93:2976–83.
110. Solomons NW. Zinc and Copper. In Shills ME, Young VR (eds.). Modern nutrition in health and disease. 7th ed. Philadelphia:Lea and Febiger. 1988:238–62.
111. Hambidge KM, Walravens PA, Neldner KH. The role of zinc in the pathogenesis and treatment of acrodermatitis enteropathica. In: Brewer GJ, Prasad AS (eds.). Zinc metabolism: current aspects in health and disease. New York:Alan R Liss. 1977:329–340.
112. Allen JI, Perri RT, McClain CJ et al. Alterations in human natural killer cell activity and monocyte cytotoxicity induced by zinc deficiency. J Lab Clin Med 1983;102:577–89.
113. Chandra RK, Au B. Single nutrient deficiency and cell mediated immune responses. I. Zinc. Am J Clin Nutr 1980;33:736–38.
114. Kirchner H, Salas M. Stimulation of lymphocytes with zinc ions. Meth Enzymol 1987;150:112–7.
115. Rajagopalan S, Winter CC, Wagtmann N, et al. The Ig-related killer cell inhibitory receptor binds zinc and requires zinc for recognition of HLA-C on target cells. J Immunol 1995;155:4143–6.
116. Weston WL, Huff JC, Humbert JR, et al. Zinc correction of defective chemotaxis in acrodermatitis enteropathica. Arch Dermatol 1977;13:422–5.
117. Briggs WA, Pedersen M, Mahajan S, et al. Lymphocyte and granulocyte function in zinc treated and zinc deficient hemodialysis patients. Kidney Int 1982;21:827–832.

118. Wirth JJ, Fraker PJ, Kierszenbaum F. Changes in the level of marker expression by mononuclear phagocytes in zinc deficient mice. J Nutr 1984;114:1826–33.
119. Cook-Mills JM, Wirth JJ, Fraker PJ. Possible roles for zinc in destruction of *Trypanosoma cruzi* by toxic oxygen metabolites produced by mononuclear phagocytes. Adv Exp Med Biol 1990;262:111–21.
120. Salvin SB, Horicker BL, Pan LX, et al. The effect of dietary zinc and prothymosin a on cellular immune responses of RF/J mice. Clin Immunol Immunopathol 1987;43:281–8.
121. Singh KP, Zaidi SI, Raisuddin S, et al. Effect of zinc on immune functions and host resistance against infection and tomor challenge. Immunopharmacol Immnuotoxicol 1992;14:813–40.
122. Ercan MT, Boor NM. Phagocytosis by macrophages in zinc deficient rats. Int J Rad Appl Instrum 1991;18:765–8.
123. Humphrey PA, Ashraf M, Lee CM. Hepatic cells' mitotic and peritoneal macrophage phagocyte activities during *Trypanosoma musculi* infection in zinc deficient mice. J Nutr Med Assoc 1997;89:259–67.
124. Chvapil M, Stankova L, Bernard DS, et al. Effect of zinc on peritoneal macrophages in vitro. Infect Immunol 1977;16:367–73.
125. Fraker PJ, DePasquale-Jardieu R, Zwickl CM, et al. Regeneration of T-cell helper function in zinc-deficient adult mice. Proc Natl Acad Sci USA 1978;75:5660–4.
126. Allen JI, Kay NE, McClain CJ. Severe zinc deficiency in humans: association with a reversible T-lymphocyte dysfunction. Ann Intern Med 1981; 95:154–157.
127. Mocchegiani E, Santarelli L, Muzzioli M, et al. Reversibility of the thymic involution and of age-related peripheral immune dysfunction by zinc supplementation in old mice. Int J Immunopharmacol 1995;17:703–718.
128. Pekarek RS, Sandstead HH, Jacob RA et al. Abnormal cellular immune responses during acquired zinc deficiency. Am J Clin Nutr 1979;32:1466–1471.
129. Kramer TR, Udomkesmalee E, Dhanamitta S, et al. Lymphocyte responsiveness of children supplemented with vitamin A and zinc. Am J Clin Nutr 1993;58:566–70.
130. Sazawal S, Jalla S, Mazumder S, et al. Effect of zinc supplementation on cell mediated immunity and lymphocyte subsets in preschool children. Indian Pediatr 1997;34:589–597.
131. Golden MHN, Jackson AA, GoldenBE. Effect of zinc thymus of recently malnourished children. Lancet 1977;2: 1057–9.
132. Sempertegui F, Estrella B, Correa E, et al. Effects of short-term zinc supplementation on cellular immunity, respiratory symptoms and growth of malnourished Equadorian children. Eur J Clin Nutr 1996;50:42–46.
133. Phillips JL, Azari P. Zinc transferrin. Enhancement of nucleic acid synthesis in phytohemagglutinin-stimulated human lymphocytes. Cell Immunol 1974;10:31–7.
134. Chvapil M. Effect of zinc on cells and biomembranes. Med Clin North Am 1976;60:799–812.
135. Dowd PS, Kelleher J, Guillou PJ. T-lymphocyte subsets and interleukin-2 production in zinc deficient rats. Br J Nutr 1986;55:59–69.
136. Malave I, Rodriguez J, Araujo Z. Effect of zinc on the proliferative response of human lymphocytes: mechanisms of its mitogenic action. Immunopharmacology 1990;20:1–10.
137. Sundermann FW. The influence of zinc on apoptosis. Ann Clin Lab Sci 1995;25:134–42.
138. Waring P, Egan M, Braithwaite A, et al. Apoptosis induced in macrophages and T blasts by the micotoxin sporodesmin and protection by Zn^{2+} salts. Int J Immunopharmacol 1990;12:445–457.
139. Zalewski PD, Forbes IJ. Intracellular zinc and the regulation of apoptosis. In: Lavin M, Waters D (eds.). Programmed cell death: the cellular and molecular biology of apoptosis. Melbourne: Harwood Academic Publishers. 1993: 73–86.
140. Fraker PJ, King LE, Garvey BA, et al. The immunopathology of zinc deficiency in humans and rodents: a possible role for programmed cell death. In: Klurfeld DM (ed.). Human nutrition: a comprehensive treatise. New York: Plenum Press. 1993:267-83.
141. Pleau JM, Fuentes V, Morgat JL, et al. Specific receptor for the serum thymic factor (FTS) in lymphoblastoid cultured cell lines. Proc Natl Acad Sci USA 1980;77:2861–5.
142. Saha AR, Hadden EM, Hadden JW. Zinc induces thymulin secretion from human thymic epithelial cells in vitro and augments splenocyte and thymocyte response in vivo. Int J Immunopharmacol 1995;17: 729–33.
143. Prasad AS. Zinc in human health: Effect of zinc on immune cells. Mol Med 2008;14(5–6):353–7.
144. Coto JA, Hadden EM, Sauro M, et al. Interleukin 1 regulates secretion of zinc-thymulin by human thymic epithelial cells and its action on T-lymphocyte proliferation and nuclear protein kinase C. Proc Natl Acad Sci USA 1992;89:7752–6.
145. Prasad AS. Effects of zinc deficiency on immune functions. J Trace Elem Exp Med. 2000;13:1–30.

146. Dardenne M, Pleau JM, Nabbara B, et al. Contribution of zinc and other metals to the biological activity of serum thymic factor. Proc Natl Acad Sci USA 1982;79:5370–3.
147. Prasad AS, Meftah S, Abdullah J, et al. Serum thymulin in human zinc deficiency. J Clin Invest 1988;82:1202–10.
148. DePasquale-Jardieu P, Fraker PJ. The role of corticosterone in the loss in immune function in the zinc deficient A/J mouse. J Nutr 1979;109:1847–55.
149. Beck FWJ, Prasad AS, Kaplan J, et al. Changes in cytokine production and T-cell subpopulations in experimentally induced zinc deficiency in humans. Am J Physiol 1997a;272:E1002–7.
150. Beck FWJ, Kaplan J, Fine N, et al. Decreased expression of CD73 (ecto-5'-nucleaotidase) in the CD8+ subset is associated with zinc deficiency in human patients. J Lab Clin Med 1997b;130:147–156.
151. Flynn A, Loftus MA, Finke JH. Production of interleukin-1 and interleukin-2 in allogeneic mixed lymphocyte cultures under copper, magnesium and zinc deficient conditions. Nutr Res 1984;4:673–9.
152. Driessen C, Hirv K, Rink L, et al. Induction of cytokines by zinc ions in human peripheral blood mononuclear cells and separated monocytes. Lymphokine Cytokine Res 1994;13:15–20.
153. Rink L, Gabriel P. Zinc and immune system. Proc Nutr Soc 2000;59:541–52.
154. Prasad AS, Beck FWJ, Grabowski SM, et al. Zinc deficiency: changes in cytokine production and T-cell subpopulations in patients with head and neck cancer in non-cancer subjects. Proc Assoc Am Physicians 1997;109: 68–77.
155. Fraker PJ, Gershwin ME, Good RA, Prasad A. Interrelationships between zinc and immune function. Fed Proc 1986;45:1474–9.
156. Gross RL, Osdin N, Fong L, et al. depressed immunological function in zinc deprived rats as measured by mitogen response of spleen, thymus and peripheral blood. Am J Clin Nutr 1979;32:1260–6.
157. Cook-Mills JM, Fraker PJ. Functional capacity of the residual lymphocytes in zinc deficient mice. Br J Nutr 1993;69:835–48.
158. Fraker PJ, Haas SM, Luecke RW. Effect of zinc deficiency on the immune response of young adult A/J mice. J Nutr 1977;107:1889–95.
159. Fraker PJ, Hidlebrandt K, Luecke RW. Alteration of antibody-mediated responses of suckling mice to T-cell dependent and independent antigens by maternal marginal zinc deficiency: restoration of responsivity by nutritional repletion. J Nutr 1984;114:170–9.
160. Fraker PJ, Jardieu P, Cook J. Zinc deficiency and immune function. Arch Dermatol Res 1987;123:1699–701.
161. WHO/UNICEF. Joint statement on the clinical management of acute diarrhoea. Geneva:World Health Organization. 2004b.
162. Hambidge KM, Krebs NF. Zinc deficiency: a special challenge. J Nutr 2007;137:1101–5.
163. Shrimpton R. Food consumption and dietary adequacy according to income in 1200 families, Manaus, Amazonas, Brazil. Arch Latinoam Nutr 1984;34:615–29.
164. Gibson RS, Anderson VP. A review of interventions based on dietary diversification/modification strategies with the potential to enhance intakes of total and absorbable zinc. Food Nutr Bull 2009. In press.
165. DeMaeyer EM, Dallman P, Gurney JM, Hallberg L, Sood SK, Srikantia SG. Preventing and controlling iron deficiency anaemia through Primary Health Care. Geneva: World Health Organization. 1989.
166. McLean E, Egli I, Cogswell M, de Benoist B, Wojdyla D. Worldwide prevalence of anemia in pre-school aged children, pregnant women and non-pregnant women of reproductive age. In: Kraemer K, Zimmermann M (eds.). Nutritional anaemia. Basel:Sight & Life Press. 2007;1:1–12.
167. Toteja GS, Singh P, Dhillon BS, Saxena BN, Ahmed FU, Singh RP, Prakash B, Vijayaraghavan K, Singh Y, Rauf A, Sarma UC, Gandhi S, Behl L, Mukherjee K, Swami SS, Meru V, Chandra P, Chandrawati MU, Mohan U. Prevalence of anemia among pregnant women and adolescent girls in 16 districts of India. Food Nutr Bull 2006;27: 311–5.
168. Dallman PR. Iron deficiency and the immune response. Am J Clin Nutr 1987;46:329–34.
169. Hershko C. Iron, infection, and immune function. Proc Nutr Soc 1993;52:165–74.
170. Weiss G. Iron and immunity: a double edged sword. Eur J Clin Invest 2002;32 (Suppl):70–8.
171. De Silva A, Atukorala S, Weerasinghe I, Ahluwahlia N. Iron supplementation improves iron status and reduces morbidity in children with or without upper respiratory tract infections: a randomized, controlled study in Colombo, Sri Lanka. Am J Clin Nutr 2003;77:234–41.
172. Zimmermann MB, Hurrell RF. Nutritional iron deficiency. Lancet 2007;370:511–20.
173. Dhur A, Galan P, Hercberg S. Iron status, immune capacity and resistance to infections. Comp Biochem Physiol A Mol Integr Physiol 1989;94:11–19.
174. Brock JH, Mulero V. Cellular and molecular aspects of iron and immune function. Proc Nutr Soc 2000;59:537–540.

175. Hallquist NA, McNeil LK, Lockwood JF, et al. Maternal-iron deficiency effects on peritoneal macrophage and peritoneal natural-killer-cell cytotoxicity in rat pups. Am J Clin Nutr 199255: 741–746.
176. Chandra RK. Reduced bacteriocidal capacity of polymorphs in iron deficiency. Arch Dis Child 1973;48:864–6.
177. Macdougall LG, Anderson R, MacNab GM, et al. The immune response in iron deficient children: impaired cellular defense mechanisms with altered humoral components. J Pediatr 1975;6:833–43.
178. Yetgin S, Altay C, Ciliv G, et al. Myeloperoxidase activity and bacteriocidal function of PMN in iron deficiency. Acta Haematol (Basal) 1979;61:10–4.
179. Farthing MJ. Iron and immunity. Acta Pediatr Scand Suppl 1989;361:44–52.
180. Spear AT, Sherman AR. Iron deficiency alters DMBA-induced tumor burden and natural killer cell cytotoxicity rats. J Nutr 1992;122:46–55.
181. Chandra RK. Impaired immunocompetence associated with iron deficiency. J Pediatr 1975;86:899–902.
182. Oppenheimer SJ. Iron and its relation to immunity and infectious disease. J Nutr 2001;131:616S–35S.
183. Joynson DHM, Walker M, Jacobs A, et al. Defect of cell-mediated immunity in patient with iron deficiency anaemia. Lancet 1972;II:1058–9.
184. Kuvibidila SR, Kitchens D, Baliga BS. In vivo and in vitro iron deficiency reduces protein kinase C activity and translocation in murine splenic and purified T cells. J Cell Biochem 1999;74:468–78.
185. Canonne-Hergaux F, Gruenheid S, Govoni G, et al. The NRAMP 1 protein and its role in resistance to infection and macrophage function. Proc Assoc Am Physicians 1999;111:283–9.
186. Moraes-de-Souza H, Kerbauy J, Yamamoto M, et al. Depressed cell-mediated immunity in iron-deficiency anemia due to chronic loss of blood. Braz J Med Biol Res 1984;17(2):143–50.
187. Galan P, Thibault H, Preziosi P, et al. Interleukin-2 production in iron deficient children. Biol Trace Elem Res 1992;32:421–6.
188. Thibault H, Galan P, Selz F, et al. The immune response in iron deficient young children: effect on iron supplementation on cell-mediated immunity. Eur J Pediatr 1993;152:120–4.
189. Smith KA. Interleukin 2. Annu Rev Immunol 1984;2:319–333.
190. Omara FO, Blakley BR. The effects of iron deficiency and iron overload on cell mediated immunity in the mouse. Br J Nutr 1994;72:899–909.
191. Jason J, Archibald LK, Nwanyanwu OC, et al. The effects of iron deficiency on lymphocyte cytokine production and activation: preservation of hepatic iron but not at all cost. Clin Exp Immunol 2001;126:466–73.
192. Flesch I, Kaufmann SHE. Mycobacteria growth inhibition of interferon-γ activated bone marrow macrophages and different susceptibility among strains of *mycobacterium tuberculosis*. J Immunol 1987;138: 4408–13.
193. Thorson JA, Smith KM, Gomez F, et al. Role of iron in T-cell activation: Th-1 clones differ from Th-2 clones in their sensitivity to inhibition for DNA synthesis caused by IgG mAbs against the transferrin receptor and the iron chelator desferrioxamine. Cell Immunol 1991;124:126–37.
194. Weiss G. Iron. In: Hughes DA, Darlington LG, Bendich A (eds.). Diet and human immune function. Totowa: Humana Press. 2004:203–15.
195. Bagchi K, Mohanram M, Reddy V. Humoral immune response in children with iron-deficiency anaemia. Br Med J 1980;280:1249–51.
196. Van Asbeck BS, Vergrugh HA, Van Oast BA, et al. *Listeria monocytogenes*meningitis and decreased phagocytosis associated with iron overload. Br Med J 1982;284:542–4.
197. Dwyer J, Wood C, McNamara J, et al. Abnormalities in immune system of children with beta thalassemia major. Clin Exp Immunol 1987;68:621–9.
198. Walker EM, Walker SM. Effects of iron overload on the immune system. Ann Clin Lab Sci 2000;30:354–65.
199. Wintergerst ES, Maggini S, Horning DH. Contribution of selected vitamins and trace elements to immune function. Ann Nutr Metab 2007;51:301–23.
200. Beard JL. Iron biology in immune function, muscle metabolism and neuronal functioning. J Nutr 2001;131:568–80S.
201. Lozoff B, Jimenez E, Smith JB. Double burden of iron deficiency in infancy and low socio-economic status: a longitudinal analysis of cognitive test scores to age 19 years. Arch Pediatr Adolesc Med 2006;160: 1108–13.
202. Nestel P, Davidsson L. Anemia, iron deficiency, and iron deficiency anemia. INACG publication. ILSI:Washington, DC. 2002.
203. Walker SP, Wachs T, Meeks Gardiner J, Lozoff B, Wasserman GA, Pollitt E, Carter JA, and the International Child Development Steering Group. Child development: risk factors for adverse outcomes in developing countries. Lancet 2007;369:245–57.

204. McGuire J, Galloway R. Enriching lives:overcoming vitamin and mineral deficiencies in developing countries. Washington, DC:World Bank. 1994.
205. Alderman H, Horton S. The economics of addressing nutritional anaemia. In: Kraemer K, Zimmermann M (eds.). Nutritional anaemia. Basel:Sight & Life Press. 2007;chp3:19–35.
206. Scholz BD, Gross R, Schultink W, Sastroamidjojo S. Anaemia is associated with reduced productivity of women workers even in less-physically-strenuous tasks. Br J Nutr 1997;77:47–57.
207. Darnton-Hill I, Paragas N, Cavalli-Sforza LT. Global perspectives: accelerating progress on preventing and controlling nutritional anaemia. In: Kraemer K, Zimmerman MB (eds.). Nutritional anaemia. Sight & Life/DSM,Basel. 2007:chp21:359–81.
208. Hurrell RF, Egli I. Optimizing the bioavailability of iron compounds for food fortification. In: Kraemer K, Zimmerman MB (eds.). Nutritional Anemia. Basel:Sight & Life Press. 2007;chp7:77–98.
209. Neumann CG, Murphy SP, Gewa C, Grillenberger M, Bwibo NO. Meat supplementation improves growth, cognitive, and behavioral outcomes in Kenyan children. J Nutr 2007;137:1119–23.
210. Halsted CH. Water-soluble vitamins. In: Garrow JS, James WPT (eds.). Human Nutrition and Dietetics. 9th ed. Singapore:Churchill Livingstone/Longman. 1993;14:239–63.
211. Bishai D, Nalubola R. The history of food fortification in the United States: its relevance for current fortification efforts in developing countries. Econ Dev Cult Change 2002;51:37–53.
212. Darnton-Hill I, Truswell AS. Thiamin status of a sample of homeless men in Sydney, Australia. Med J Aust 1990; 52:5–9.
213. Thomson CD. Selenium and iodine. In: Mann JI, Truswell AS (eds.). Essentials of Human Nutrition. 3rd ed. Oxford:Oxford University Press. 2007;chp10:145–51.
214. FFI. Flour Fortification Initiative; a public-private-civic investment in each nation. 2005. http://www.sph.emory.edu/wheatflour/index.php (accessed 25 March 2007).
215. MI. The Micronutrient Initiative. New solutions for hidden hunger. 2007 http://www.micronutrient.org (accessed 6 March 2007).
216. WHO. Iodine status Worldwide. WHO Global Database on Iodine Deficiency. Report by de Benoist B, Andersson M, Egli I, Takkouche B, Allen H. Geneva: World Health Organization. 2004.
217. Allen LH. Micronutrients. Health and nutrition emerging and reemerging issues in Developing Countries. IFPRI 2020 Focus No.05. Washington,DC:International Food Policy Research Institute 2001;brief10:1-4 http://www.ifpri.org/2020/focus/focus05/focus05_10.asp (accessed 7 November 2008).
218. Zimmermann MB. Global control of micronutrient deficiencies: divided they stand, united theyfall. Wageningen University Inaugural professorial presentation monograph. 2007.
219. Gross RL, Reid JVO, Newberne PM, et al. Depressed cell-mediated immunity in megaloblastic anemia due to folic acid deficiency. Am J Clin Nutr 1975;28:225–32.
220. Dhur A, Galan P, Herchberg S. Folate status and the immune system. Prog Food Nutr Sci 1991;15:43–60.
221. Courtemanche C, Elson-Schwab I, Mashiyuama ST, et al. Folate deficiency inhibits the proliferation of primary human CD8+ T lymphocytes in vitro. J Immunol 2004;173:3186–9.
222. Troen AM, Mitchell B, Sorensen B, et al. Unmetabolized folic acid in plasma is associated with reduced natural killer cell cytotoxicity among postmenopausal women. J Nutr 2006;136:189–94.
223. Tamura J, Kubota K, Murakami H, et al. Immunomodulation by vitamin B12: augmentation of CD8+ T lymphocytes and natural killer (NK) cell activity in vitamin B12 deficient patients by methyl-B12 treatment. Clin Exp Immunol 1999;116:28–32.
224. Fata FT, Herzlich B, Schiffman G, et al. Impaired antibody responses to pneumococcal polysaccharide in elderly patients with low serum vitamin B12 levels. Ann Intern Med 1996;124:299–304.
225. Chandra RK, Sudhakaran I. Regulation of immune responses by vitamin B6. Ann NY Acad Sci 1990;585:404–23.
226. Ockhuizen T, Spanhaak S, Mares N, et al. Short term effect of marginal vitamin B deficiencies on immune parameters in healthy young volunteers. Nutr Res 1990;10:483–92.
227. Rall LC, Meydani SN. Vitamin B_6 and immune competence. Nutr Rev 1993; 51:217–25.
228. Trakatellis A, Dimitriadou A, Trakatellis M. Pyridoxine deficiency: new approaches in immunosuppression and chemotherapy. Postgrad Med J 1997;73:617–622.
229. Miller LT, Kerkvliet NT. Effect of vitamin B_6 on immune competence in the elderly. Ann NY Acad Sci 1990;587: 49–54.
230. Long KZ, Santos JL. Vitamins and the regulation of the immune response. Pediatr Infect Dis 1999;18:283–90.
231. Zweiman B, Schoenwetter WF, Hildreth EA. The effect of the scobutic state on tuberculin hypersensitivity in the guinea pig. I. Passive transfer of tuberculin hypersensitivity. J Immunol 1966;96:296–300.

232. Ganguly R, Durieux MF, Waldman RH. Macrophage function in vitamin C deficient guinea pigs. Am J Clin Nutr 1976;29:762–5.
233. Johnston CS, Kolb WP, Haskell BE. The effect of vitamin C nutriture on complement component clq concentrations in guinea pig plasma. J Nutr 1987;117:764–8.
234. Jacob RA, Kelley DS, Pianalto FS. Immunocompetence and oxidant defense during ascorbate depletion of healthy men. Am J Clin Nutr 1991;54:1302S–9S.
235. Johnston CS. Complement component clq unaltered by ascorbic supplementation in healthy men and women. J Nutr Biochem 1991;2:499–501.
236. Anderson R, Oosthuizen R, Maritz R. The effects of increasing weekly doses of ascorbate on certain cellular and humoral immune functions in normal volunteers. Am J Clin Nutr 1980;33:71–6.
237. Kennes B, Dumont I, Brohee D, et al. Effect of cell-mediated immunity in older people. Gerontology 1983;29:305–10.
238. Kowdley KV, Mason JB, Meydani SN, et al. Vitamin E deficiency and impaired cellular immunity related to intestinal fat absorption. Gastroenterology 1992;102:2139–42.
239. Ghalaut VS, Ghalaut PS, Kharb S, et al. Vitamin E in intestinal fat malabsorption. Ann Nutr Metab 1995;39: 296–301.
240. Lee CYJ, Wan JMF. Vitamin E supplementation improves cell-mediated immunity and oxidative stress of Asian men and women. J Nutr 2000;130:2932–7.
241. Meydani SN, Han SN, Wu D. Vitamin E and immune response in the aged: molecular mechanism and clinical implications. Immunol Rev 2005;205:269–84.
242. Meydani SN, Barklund MP, Liu S, et al. Vitamin E supplementation enhances cell-mediated immunity in healthy elderly subjects. Am J Clin Nutr 1990;53:557–63.
243. Meydani SN, Meydani M, Blumberg JB. Vitamin E supplementation and in vivo immune response in healthy elderly subjects. A randomized controlled trial. JAMA 1997;277:1380–6.
244. Hara M, Tanaka K, Hirota Y. Immune response to influenza vaccine in healthy adults and the elderly: association with nutritional status. Vaccine 2005;23:1457–63.
245. Gill H, Walker G. Selenium, immune function and resistance to viral infections. Nutr Diet 2008;65:S41–7.
246. Hoffmann PR, Berry MJ. The influence of selenium on immune responses. Mol Nutr Food Res 2008;52:1273–80.
247. Bendich A, Chandra RK. Micronutrients and immune functions. Ann NY Acad Sci 1990;587:168–80.
248. Beck MA, Matthews CC. Micronutrients and host resistance to viral infection. Proc Nutr Soc 2000;59:581–5.
249. Friis H, Range N, Pedersen ML, Mølgaard C, Changalucha J, Krarup H, Magnussen P, Søborg C, Andersen ÅB. Hypovitaminosis D is common among pulmonary tuberculosis patients in Tanzania but is not explained by the acute phase response. J Nutr 2008;138:2474–80.
250. Victora CG, Adair L, Fall C, Hallal PC, Martorell R, Richter L, Sachdev HS, for the Maternal and Child Undernutrition Study Group. Maternal and child undernutrition 2. Maternal and child undernutrition: consequences for adult health and human capital. Lancet 2008;371:DOI:10.1016/S0140-6736(07)61692-4 (published online 17 January 2008).
251. Smith L, Haddad L. Overcoming child malnutrition in developing countries: past achievements and future choices. Washington, DC:International Food Policy Research Institute. IFPRI 2020 Brief No.64. 2000.4pp. http://www.ifpri.org/2020/briefs/number64.htm (accessed 7 February 2006).
252. Wieringa FT, Dijkhuizen MA, van der Meer JW. Maternal micronutrient supplementation and child survival. Lancet 2008;371:1751–2.
253. MI. Ending Hidden Hunger. 1991. Micronutrient Initiative:Ottawa. www.micronutrient.org (accessed 4 March 2003).
254. IOM (Institute of Medicine). Prevention of micronutrient deficiencies: tools for policy makers and public health workers. Summary and key elements. In: Howson CP, Kennedy ET, Horwitz A (eds.). Committee on Micronutrient Deficiencies, Board of International Health, Food and Nutrition, Board of the Institute of Medicine, National Academy of Sciences. Washington, DC: National Academy Press. 1998.
255. Darnton-Hill I. The Global Micronutrient Goals: lessons learned, analysis and the way forward. PhD dissertation. Available from University of Tasmania. 2008.
256. Darnton-Hill I. Overview: rationale and elements of a successful food fortification program. Food Nutr Bull 1998b;19:92–100.
257. HKI. Home gardening in Hilly and Tarai areas in Nepal: impact on food production and consumption. Nepal: HKI Nutr Bull 2001;1:1–4.
258. Calloway DH. Human nutrition: food and micronutrient relationships. Agricultural strategies for micronutrients. IFPRI Working paper 1. Washington,DC: International Food Policy Research Institute. 1995.
259. Allen LH, Ahluwahlia N. Improving iron status through diet: The application of knowledge concerning dietary iron availability in human populations. OMNI/USAID. Arlington,VA: JSI. 1997.

260. Campbell AA, Thorne-Lyman A, Sun K, de Pee S, Kraemer K, Moench-Pfanner R, Sari M, Akhter N, Bloem MW, Semba RD. Indonesian women of childbearing age are at greater risk of clinical vitamin A deficiency in families that spend more on rice and less on fruits/vegetables and animal-based foods. Nutr Res 2009;29:75–81.
261. Talukder A, Bloem MW. Homegardening activities in Bangladesh. Dhaka:Helen Keller International. 1992:1–43.
262. Darnton-Hill I, Webb P, Harvey PWJ, Hunt JM, Dalmiya N, Chopra M, Ball MJ, Bloem MW, de Benoist B. Micronutrient deficiencies and gender: social and economic costs. Am J Clin Nutr 2005;819(suppl): 1198S–205S.
263. Bushamuka VN, de Pee S, Talukder A, Keiss L, Panagides D, Taher A, Bloem MW. Impact of a homestead gardening program on household food security and empowerment of women in Bangladesh. Food Nutr Bull 2005;26: 17–25.
264. Bouis HE, Lineback D, Scheeman B. Bio-technology–derived nutritious foods for Developing Countries: needs, opportunities, and barriers. Food Nutr Bull 2002;23:342–83.
265. Lonnerdal B. Genetically modified plants for improved trace element nutrition. J Nutr 2003;133:1490S–3S.
266. Darnton-Hill I, Margetts BM, Deckelbaum R. Public health nutrition and genetics: implications for nutrition policy and promotion. Proc Nutr Soc 2004;63:173–85.
267. Talukder A, Kiess L, Huq N, de Pee S, Darnton-Hill I, Bloem MW. Increasing the production and consumption of vitamin A-rich fruits and vegetables: lessons learned in taking the Bangladesh homestead gardening programme to a national scale. Food Nutr Bull 2000;21:165–72.
268. Meskel Balcha H. Experience of World Vision Ethiopia Micronutrient program in promoting production of vitamin A-rich foods. In: Abstracts of a Workshop on long-term food-based approach towards eliminating vitamin A deficiency in Africa. MRC (South Africa)/UNU/IUNS. November 2000:25.
269. HKI. Integration of animal husbandry into home gardening programs to increase vitamin A intake from foods: Bangladesh, Cambodia and Nepal. Singapore:Helen Keller International-Asia-Pacific. 2003:1–4.
270. Kiess L, Bloem MW, de Pee E, et al. Bangladesh: Xerophthalmia free. The result of an effective vitamin A capsule program and homestead gardening (abstract) In: APHA 126th Annual Meeting Report. Washington, DC. November, 1998.
271. de Pee S, Bloem MW, Kiess L. Evaluating food-based programmes for their reduction of vitamin A deficiency and its consequences. Food Nutr Bull 2000;21:232–8.
272. HarvestPlus. Breeding crops for better nutrition. http://www.harvestplus.org (accessed 15 December 2008).
273. Crop Biotech Update. IRRI and CIMMYT to work on 4 research priorities. Global Knowledge Center on Crop Biotechnology, International Service for the Acquisition of Agri-Tech Applications SEAsiaCenter (ISAAA), AgBiotechNet. January, 2005:1–2.
274. Greger Jl. Biotechnology: mobilizing dietitians to be a resource. J Am Diet Assoc 2000;100:1306–7.
275. Nestle M. Genetically engineered 'golden' rice unlikely to overcome vitamin A deficiency. J Am Diet Assoc 2001;101:289–90.
276. Informal Working Group on Feeding of Nonbreastfed Children. Conclusions of an informal meeting on infant and young child feeding organized by the World Health Organization. Geneva:World Health Organization. Food Nutr Bull 2004;25:403–6.
277. GAIN. Global Alliance for Improved Nutrition. Geneva:GAIN http://www.gainhealth.org/gain/ch/ (accessed 12 October 2008).
278. Darnton-Hill I, Nalubola R. Food fortification as a public health strategy to meet micronutrient needs—successes and failures. Proc Nutr Soc 2002;61:231–41.
279. Zimmermann MB, Wegmüller R, Zeder C, Chaouki N, Rohner F, Torresani T, Hurrell R. Triple fortification of salt with microencapsulated iodine, iron and vitamin A: a randomized, double-blind trial. XXII IVACG Meeting, Lima, Peru: vitamin A and the common agenda for micronutrients. 2004b:Abstract M62.
280. Huffman SL, Baker J, Shumann J, Zehner ER. The case for promoting multiple vitamin/mineral supplements for women of reproductive age in Developing countries. Linkages Project, Academy for Educational Development:Washington, DC. 1999.
281. Nestel PS, Jackson AA. The impact of maternal micronutrient supplementation on early neonatal morbidity. Arch Dis Child 2008;93:647–9.
282. Shrimpton R, Huffman SL, Zehner ER, Darnton-Hill I, Dalmiya N. Multiple micronutrient supplementation during pregnancy in Developing Countries: policy and programme implications of the results of the meta-analysis. Food Nutr Bull. In press.
283. Fawzi WW, Msamanga GI, Urassa W, Hertzmark E, Petraro P, Willet W, Spiegelman D. Vitamins and perinatal outcomes among HIV-negative women in Tanzania. New Engl J Med. 2007;356:1423–31.

284. Zlotkin S, Tondeur, MC. Successful approaches-Sprinkles. In: Kraemer K, Zimmerman MB (eds.). Nutritional anaemia. Basel: Sight & Life Press. 2007;chp17:269–84.
285. Supplefer. Sprinkles for humanitarian aid interventions. Available from: www.MicronutrientSprinkles.org and www.supplefer.org (accessed 1 January 2008).
286. de Pee S, Moench-Pfanner R, Martini E, Zlotkin S, Darnton-Hill I, Bloem M. Home-fortification in emergency response and transition programming: experiences in Aceh and Nias, Indonesia. Food Nutr Bull 2007;28: 189–97.
287. Viteri FE. Supplementation for the control of iron deficiency in populations at risk. Nutr Rev 1997;55:195–209.
288. Schultink W, Merzenich M, Gross R, Shrimpton R, Dillon D. Effects of iron-zinc supplementation on the iron, zinc and vitamin A status of anaemic pre-school children in Indonesia. Food Nutr Bull 1997;18:311–7.
289. Sommer A, Davidson FR, Ramakrishnan U, Darnton-Hill I. 25 years of progress in controlling vitamin A deficiency: looking to the future. Proc XX IVACG Meeting. (Guest editors). J Nutr 2002:132(9S):2845S–990S.
290. Sazawal S, Black RE, Ramsan M, Chwaya HM, Stoltzfus RJ, Dutta A, Dhungra U, Kabole I, Deb S, Othman MK, Kabole F. Effects of routine prophylactic supplementation with iron and folic acid on admission to hospital and mortality in preschool children in a high malaria transmission setting: community-based, randomized, placebo-controlled trial. Lancet 2006;367:133–43.
291. Shrimpton R, Gross R, Darnton-Hill I, Young M. Zinc deficiency: what are the most appropriate interventions? Br Med J 2005;330:347–9.
292. Larson CP, Roy SK, Khan AI, Rahman AS, Quadri F. Zinc treatment to under-five children: applications to improve child survival and reduce burden of disease. J Health Popul Nutr 2008;3:356–65.
293. INACG/WHO/UNICEF (Stoltzfus R, Dreyfus M). Guidelines for the use of iron supplements to prevent and treat iron deficiency anemia.Washington,DC:ILSI Press. 1998.
294. Stoltzfus RJ. Defining iron-deficiency anemia in public health terms: a time for reflection. J Nutr 2001;131;(2S-II):565S–7S.
295. Mebrahtu T, Stoltzfus RJ, Chwaya HM, Jape JK, Savioli L, Montressor A, Albonico M, Tielscg JM. Low-dose daily iron supplementation for 12 months does not increase the prevalence of malarial infection or density of parasites in young Zanzibari children. J Nutr 2004;134:3037–41.
296. Galloway R, Dusch E, Elder L, Achadi, Grajeda R, Hurtado E, Favin M, Kanani S, Marsaban J, Meda N, Mona Moore K, Morison L, Raina N, Rajaratnam J, Rodriquez J, Stephen C. Women's perceptions of iron deficiency and anemia prevention and control in eight developing countries. Soc Sci Med 2002;55:529–44.
297. Beaton G, McCabe G. Efficacy of intermittent iron supplementation in the control of iron deficiency anaemia in developing countries. An analysis of experience. Final report to the Micronutrient Initiative. Ottawa:MI. 1998.
298. Gross R, de Romana GL, Tomaro J. A life-cycle approach to multi-micronutrients supplementation: rationale and programme concept. Food Nutr Bull 2000;21:270–4.
299. Cavalli-Sforza LT, Berger J, Smitasiri S, Viteri F. Summary: weekly iron-folic acid supplementation of women of reproductive age: impact overview, lessons learned, expansion plans and considerations towards achievement of the Millennium Development Goals. Nutr Rev 2005;63(II):S152–8.
300. Bhargava A, Jukes M, Lambo J, Kihamia CM, Lorri W, Nokes C, Drake L, Bundy D. Anthelmintic treatment improves the hemoglobin and serum ferritin concentration of Tanzanian schoolchildren. Food Nutr Bull 2003;24:332–42.
301. Stoltzfus RJ, Chway HM, Montresor A, Tielsch JM, Jape JK, Albonico M, Savioli L. Low dose daily iron supplementation improves iron status and appetite but not anemia, whereas quarterly anthelminthic treatment improves growth, appetite and anemia in Zanzibari preschool children. J Nutr 2004;134:348–56.
302. Hall A. Micronutrient supplements for children after deworming. Lancet. 2007;7:297–302.
303. Bryce J, el Arifeen S, Pariyo G, Lanata CF, Gwatkin D, Habicht J-P, and the multi-country evaluation of IMCI study group. Reducing child mortality: can public health deliver? Lancet 2003;362:159–64.
304. Dalmiya N. Results from the Child Health Day assessments. Preliminary findings: Ethiopia, Tanzania and Uganda. Presented at the Global Immunization Meeting at the UN, New York. 13–15 February 2007.
305. High KP. Micronutrient supplementation and immune function in the elderly. Clin Infect Dis 1999;28:717–22.
306. Tayie FAK, Lartey A. Antenatal care and pregnancy outcome in Gahna, the importance of women's education. African J Food Agric Nutr Dev 2008;8:291–303.
307. UNICEF/UNU/WHO Study team. Multiple micronutrient supplementation during pregnancy (MMSDP): a review of progress in efficacy trials. Report of a Meeting. Bangkok, Thailand organized by the Centre for International Child Health, Institute of Child Health, University College, London and UNICEF, Bangkok. Unpublished. 2004.
308. Jackson AA, Bhutta ZA, Lumbiganon P. Nutrition as a preventive strategy against adverse pregnancy outcomes. Introduction. J Nutr 2003;133:1589S–91S.

309. WHO/WFP/UNICEF. Joint statement on preventing and controlling micronutrient deficiencies in populations affected by an emergency. Geneva:World Health Organization. 2005.
310. SUMMIT (The Supplementation with Multiple Micronutrients Intervention Trial) Study Group. Effect of maternal multiple micronutrient supplementation on fetal loss and infant death in Indonesia: a double-blind cluster-randomised trial. Lancet 2008;371:215–27.
311. UNICEF/WHO/UNU. Composition of a multi-micronutrient supplement to be used in pilot programmes among pregnant women in developing Countries. New York:UNICEF. 1999.
312. Tomkins A. Improving maternal nutrition and pregnancy outcome- The role of micronutrient supplements with special reference to populations that are endemic for HIV. Presented at the UNICEF Regional Nutrition Network Meeting (Eastern & Southern Africa Region). Entebbe, Uganda. 2004:1–12.
313. Shrimpton R, Allen L (eds.). International Research on Infants Supplementation (IRIS): randomized controlled trials of micronutrient supplementation during infancy. J Nutr 2005;135(3S):628S–30S.
314. Read AF, Graham AL, Råberg L. Animal defenses against infectious agents: is damage control more important than pathogen control? PLoS Biol 2008;6(12):2638–41:e1000004.doi:10.1371/journal.pbio.1000004.
315. Brabin B, Prinsen-Geerligs P, Verhoeff F, Kazembe P. Reducing childhood mortality in poor countries. Anaemia prevention for reduction of mortality in mothers and children. Trans R Soc Trop Med Hyg 2003;97:36–8.
316. Ouédraogo HZ, Dramaix-Wilmet M, Zeba A, Hennart P, Donnen P. Effect of iron or multiple micronutrient supplements on the prevalence of anaemia among anaemic young children of a malaria-endemic area: a randomized double-blind trial. Trop Med Int Health 2008;13:1–10.
317. Nchito M, Geissler PW, Mubila L, Friis H, Olsen A. The effect of iron and multi-micronutrient supplementation on *Ascaris lumbricoides* reinfection among Zambian children. Trans R Soc Trop Med Hyg 2008:doi:10.1016/j.trstmh.2008.08.005.
318. Editorial. How to prevent a tenth of the global disease burden. Lancet 2008;371:2145.
319. Gillespie S, Mason J. Controlling of vitamin A deficiency. United Nations: ACC/SCN state of the art series: Nutrition policy discussion paper no 14, 1994.
320. Bellagio Study Group on Child Survival. Knowledge into action for child survival. Lancet. 2003;362:323–7.
321. Edmond KM, Kirkwood BR, Amenga-Etengo S, Owusus-Agyei S, Hurt LS. Effect of early infant feeding practices on infection-specific neonatal mortality: an investigation of the causal links with observational data from rural Ghana. Am J Clin Nutr 2007;86:1126–31.
322. Schmidt MK, Muslimatun S, West CE, Schultink W, Hautvast JGAJ. Vitamin A and iron supplementation of pregnant Indonesian women benefits the vitamin A status of their infants, but still may need vitamin A supplementation or increased dietary intake. Br J Nutr 2001;86:607–15.
323. Daneel-Otterbech S, Davidsson L, Hurrell R. Ascorbic acid supplementation and regular consumption of fresh orange juice increase the ascorbic acid content of human milk: studies in European and African lactating women. Am J Clin Nutr 2005;81:1088–93.
324. Bendich A. Micronutrients in women's health and immune function. Nutrition 2001;17:858–67.
325. UNICEF. The State of the World's Children. Nutrition. Oxford & New York:Oxford University Press. 1998.
326. Oomen HAPC. Vitamin A deficiency. In: Beaton GH, Bengoa JM (eds.). Nutrition in preventive medicine. WHO Monogr Ser No. 62. Geneva:World Health Organization. 1976.
327. Torlesse H, Kiess L, Bloem MW. Association of household rice expenditure with child nutritional status indicates a role for macroeconomic food policy in combating malnutrition. J Nutr 2003;133(5):1320–5.
328. Bloem MW, de Pee S, Darnton-Hill I. Micronutrient deficiencies and maternal thinness: First chain in the sequence of nutritional and health events in economic crises. In: Bendich A, Deckelbaum RJ (eds.). Preventive Nutrition: The comprehensive guide for health professionals. 3rd ed. Totowa, NJ: Humana Press. 2005:chp27:689–710.
329. Klotz C, de Pee S, Thorne-Lyman A, Kraemer K, Bloem M. Nutrition in the perfect storm: widespread health consequence of high food prices. Sight Life 2008;2:6–13.
330. World Bank. Investing in Health. World development Report 1993. New York:Oxford University Press. 1993.
331. Mason JB, Musgrove P, Habicht J-P. At least one-third of poor countries' disease burden is due to malnutrition. Disease Control Priorities Project, Fogarty International Center,NIH:Bethesda,MD. 2003;DCPP Working paper no.1:19pp.
332. UN. The Millennium Development Goals. Report 2008. New York:UNDP. 2008 (www.undp.org/publications/MDG_Report_2008_En.pdf (accessed 14 November 2008).
333. Kickbush I. Baying at the moon: addressing the politics of global health, A review of Global Health Watch 2. Zed Books 2008. Lancet 2008;372:1623–4.
334. WHO. Closing the gap in a generation: Health equity through action on the social determinants of health. Geneva:World Health Organization 2008.

335. REACH. Ending child hunger and undernutrition. www.reach-partnership.org (accessed 16 January 2009).
336. UNDP. Millennium Development Gaols. Global progress: are we on track to meet the MDGs by 2015? http://www.undp.org/MDG/basics_ontrack.shtml (accessed 14 November 2008).
337. Behrman JR, Rozenweig MR. The returns to increasing body weight. Penn Institute for Economic Research. Philadelphia: University of Pennsylvania. 2001;PIER Working paper 01:052:41 pp.

24 HIV and Nutrition

Kevin A. Sztam and Murugi Ndirangu

Key Points

- The interaction of HIV and nutrition is incompletely understood.
- Immune manifestations of protein-energy malnutrition overlap with immune manifestations of HIV.
- Protein-energy malnutrition and micronutrient deficiencies are prevalent in areas where HIV is also prevalent.
- As antiretroviral therapy becomes more accessible and life expectancy for those infected with HIV increases, there will be increased need for research involving nutrition and metabolism in conjunction with medical treatment in developing settings.
- The impact of micronutrient supplementation in different settings is beginning to be elucidated.
- Generalized recommendations of macronutrient supplementation in HIV cannot be made given very scant data.
- Breastfeeding is the recommended mode of feeding where formula feeding is not acceptable, feasible, affordable, sustainable, and safe.

Key Words: HIV; protein-energy malnutrition; resting energy expenditure; Breastfeeding

24.1. INTRODUCTION

Human immunodeficiency virus (HIV) and nutritional status maintain a critical and important relationship that is incompletely understood. Consequences of chronic viral infections may have organ-specific effects as well as general effects on health and function. HIV infection is a nutritionally progressive disorder with major metabolic changes in nutrient utilization as the balance of viral replication, immune response, and inflammation changes over time. HIV treatment has its own specific metabolic effects which have been discussed elsewhere *(1–3)*. Worldwide, nutrition and HIV are important determinants of clinical outcomes including survival, with special importance in areas of food insecurity like sub-Saharan Africa. States of normal and abnormal nutrition may affect the progression of HIV and response to treatment in different ways. Poor nutrition in the patient infected with HIV not only has implications for the individual, but in conjunction with progressive immune dysfunction also has significant implications for community and global health.

24.2. HIV BACKGROUND AND STATISTICS

24.2.1. Global Burden, Current Trends

An estimated 32.9 million people were living with HIV worldwide in 2007 with about 2.7 million of those being new infections. Africa continued to bear the burden of the epidemic with an estimated 1.9 million new infections in 2007, bringing to 22 million the number of people living with HIV. Two-thirds of all infected individuals live in this region, and three-quarters of all AIDS deaths

A. Bendich, R.J. Deckelbaum (eds.), *Preventive Nutrition*, Nutrition and Health, DOI 10.1007/978-1-60327-542-2_24,

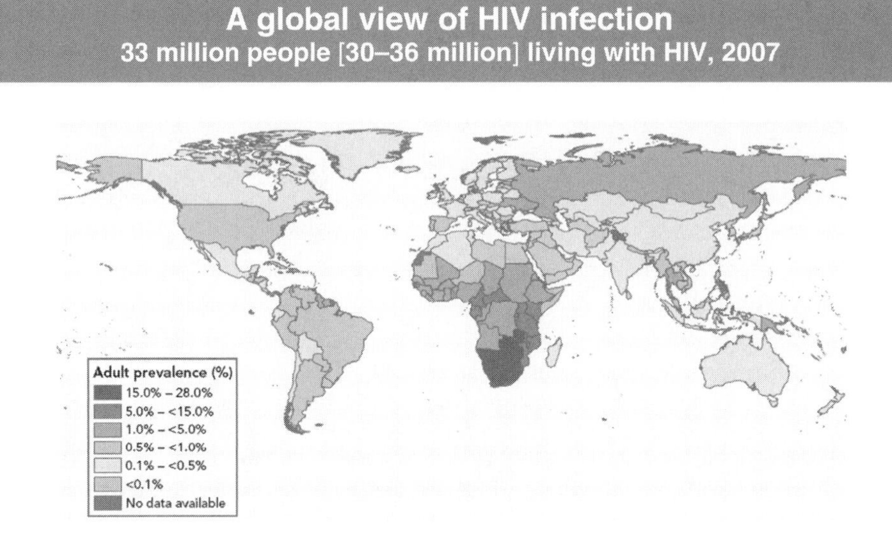

Fig. 24.1 A global view of HIV infection in 2007: 33 million people [range: *30–36* million] living with HIV (UNAIDS, 2008). Reproduced by permission of UNAIDS (www.unaids.org).

in 2007 occurred there *(4)* (Figs. 24.1 and 24.2). The magnitude of the epidemic varies across Africa, from HIV prevalence rates of below 2% in some West and Central African countries, to above 5% in countries such as Cameroon and Tanzania, to about 15% overall in Southern African countries of Botswana and Zimbabwe. The primary mode of HIV transmission is heterosexual contact.

Due to the effects of this HIV pandemic, many countries in Southern Africa have experienced a decline in life expectancy in contrast to historical gains since the 1950s in other parts of Africa and the world (Fig. 24.3).

Though life expectancy has decreased dramatically in southern Africa in the past two decades, increased access to treatment is likely to increase both life expectancy and HIV prevalence in affected areas until transmission is sufficiently interrupted. Countries in sub-Saharan Africa have experienced an increase in prevalence from 2001 to 2007 (from 5.0 to 5.7% in all age groups), the same period in which treatment became available in many locations. AIDS deaths in adults and children declined somewhat during this period *(4)*. These data clearly show that absolute number of individuals infected as well as prevalence has increased and will likely continue to do so. Clinical care issues related to treatment and chronic disease, including the interaction of nutritional status and HIV, will grow in importance.

24.2.2. Chronicity of HIV and Epidemiology Effects of Long-Term Treatment, Increasing Survival

Although the availability of antiretroviral therapy (ART) regimens in most African countries still lags behind that in western countries, medications are increasingly available due to multiple initiatives *(5–8)*. One program alone, President's Emergency Plan For AIDS Relief (PEPFAR), provides over 2 million infected people worldwide with ART *(9)*.

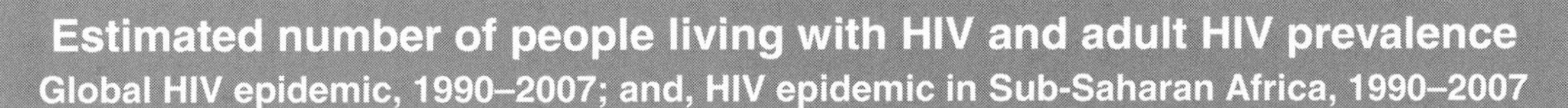

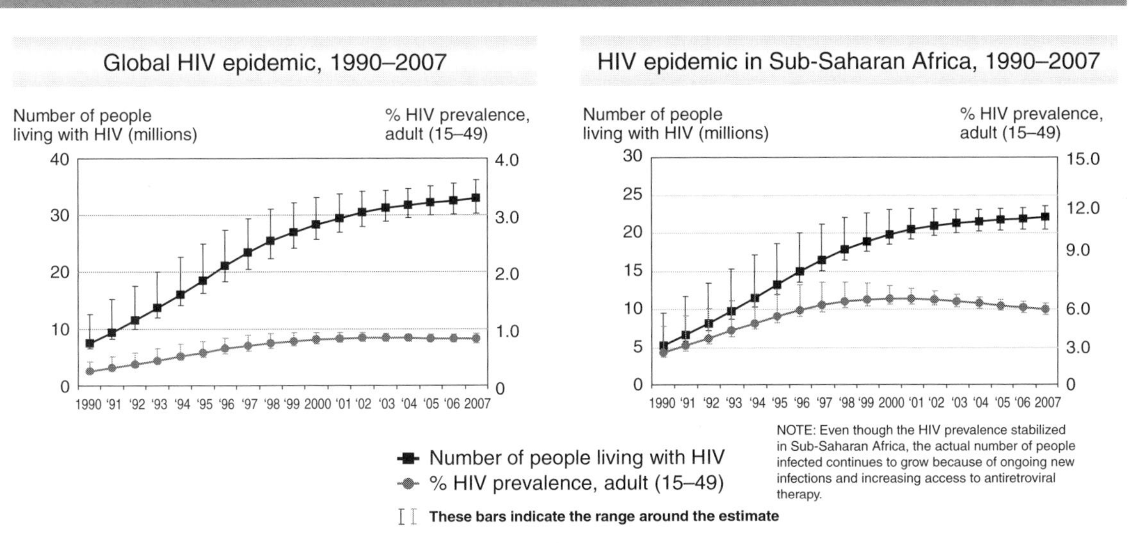

Fig. 24.2 Estimated number of people living with HIV and adult HIV prevalence. Global HIV epidemic, 1990–2007; and HIV epidemic in sub-Saharan Africa, 1990–2007 (UNAIDS 2008). Reproduced by permission of UNAIDS (www.unaids.org).

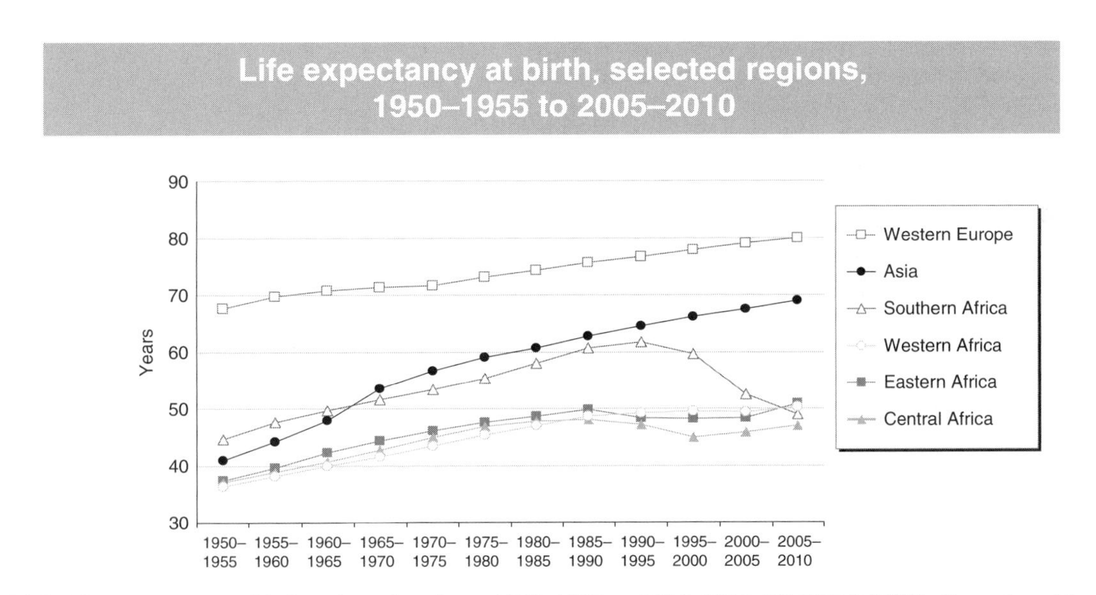

Fig. 24.3 Life expectancy at birth, selected regions, 1950–1955 to 2005–2010 (UNAIDS 2008). Reproduced by permission of UNAIDS (www.unaids.org).

As a result of increased global availability of medications, many HIV epidemics have been moderated regionally. ART interrupts the cycle of infection by controlling viral loads, increasing T-cell numbers and improving function and decreasing opportunistic infections. Treatment with ART also dramatically increases life expectancy *(9)*. Viral loads are linked to transmission rates in a dose–response relationship, with higher viral load increasing infection rates *(10)*. Other risk factors for HIV transmission include the presence of sexually transmitted diseases with ulceration, number of sexual partners, and possibly genetic HLA type I alleles *(11,12)*. Other interventions such as education, condoms *(13)*, and circumcision *(14)* help decrease incidence; however, ART itself still has the greatest potential to decrease new infections *(15)*.

With significant advances in treatment modalities and survival, and increased life expectancy in the patient treated with ART, the approach to HIV has changed to a model of chronic disease rather than acute immune dysfunction and clinical deterioration *(16)*. In this model of chronicity, support of the immune system apart from ART has become critical, and the interplay of nutritional status and HIV infection with or without medication has gained focus. Adequate nutritional status can help maximize functionality at all stages of disease, minimize side effect of therapies, and could have additional roles in improving outcomes of patients in developing settings both before and during ART *(17,18)*.

24.3. EPIDEMIOLOGY OF MALNUTRITION

24.3.1. Geographic Coexistence of Malnutrition and HIV

The HIV epidemic disproportionately affects countries already burdened by acute and chronic malnutrition. High levels of malnutrition are prevalent in many parts of Africa, for example, linked to food insecurity (Fig 24.4). Protein-energy malnutrition and micronutrient deficiencies are common as a result of poverty, environmental determinants, and economic instability.

Areas of traditional food insecurity are also areas affected by HIV. The HIV epidemic which has taxed social, economic, and cultural systems also affects food security *(19)*. Effects of the HIV epi-

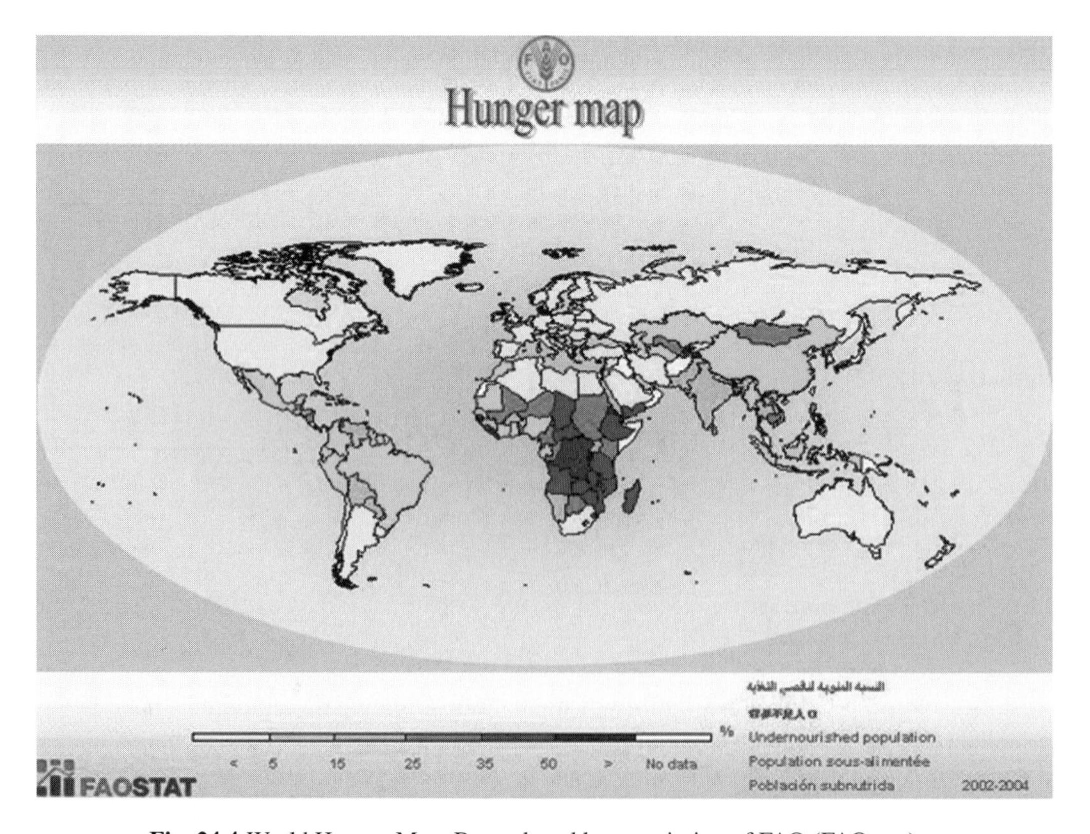

Fig. 24.4 World Hunger Map. Reproduced by permission of FAO (FAO.org).

demic on families and communities change food access and availability leading to decreased intake of micronutrients and macronutrients. Decreased intake by the individual leads to a cycle of nutrient deficiency causing associated pathologies and immunodeficiency, increased risk of infections, increased metabolic demand and losses which ultimately results in decreased intake thus exacerbating pre-existing malnutrition *(20)*.

In areas like Africa and Southeast Asia affected by high burdens of both HIV and malnutrition, multiple factors may affect food availability, access, and utilization. Food security is linked to war and displacement, productivity, geography, proximity to arable land, storage techniques, location of central markets and pricing *(21,22)*. Food availability is affected by seasonality and cyclical effects of rains or drought *(23)* and affects dietary diversity. Food access is determined by socioeconomic status. Notably in Sub-Saharan Africa, over 30% of individuals in general have insufficient food intake *(24)*. Utilization of food involves choices of foods to consume, food preparation, food additives, and sources of water.

Limited-resource areas such as parts of sub-Saharan Africa and Southeast Asia tend to have less dietary diversity than more affluent areas *(25)*. Maize and rice, two global staples, are relatively inexpensive compared to other sources of energy and are thus often the majority of calories consumed in developing areas. While price may be a prime determinant of dietary choices, food availability and cultural influences shape what an individual considers healthy or desirable food *(26)*. Disproportionate use or choice of these foods may cause dietary inadequacy, even when animal and other source proteins may be available.

For an individual patient infected with HIV, weight loss may be related to HIV infection itself or to co-infections in the immunocompromised host. In geographic areas with baseline malnutrition such as sub-Saharan Africa, the cause of weight loss may be due to food insecurity. Due to an increased frequency of concomitant illness, HIV infection contributes to malaise, inactivity, muscle loss, and deconditioning. This can reduce the ability to perform physical work. This is especially important in resource-limited countries where large portions of the population depend directly on small-scale labor-intense farming for their sustenance and livelihood. The impact on communities of decreased productivity due to death and illness is significant as HIV primarily affects those of reproductive age, also the most economically productive portion of the population *(4,27,28)*. If individuals are too ill to engage in food production, they are likely to be food insecure *(28)*. In turn, lack of adequate food intake further escalates the cycle of untreated infections and malnutrition and can have community-wide effects. AIDS-affected households experience a decline in income as costs related to managing illness and the wider the epidemic rise *(29)*. With a decline in income, the ability of a household to obtain sufficient healthy foods and adequate food quantities also declines.

The 57th World Health Assembly resolved to encourage countries and agencies such as the World Health Organization to integrate nutrition into the comprehensive response to HIV/AIDS as part of the scale-up of treatment and care *(30)*. Nutrition support programs are cited as a critical tool to help HIV-infected individuals cope with the effects of the infection and antiretroviral medications where there is food insecurity. Clearly the expansion of antiretroviral treatment in resource-limited settings holds promise of making a substantial impact on morbidity and mortality. With the availability of ART in areas of high prevalence of malnutrition, investigation must focus upon the interaction of treatments such as medications and nutritional status.

The aim of HIV programs to address malnutrition in areas of food insecurity simultaneously to treating HIV is ambitious. In contrast to supplementation as adjunctive treatment, food replacement is required to counter food insecurity as patients do not have access to or availability of local food resources. This strategy differs somewhat from targeted supplementation designed to meet time-specific or patient-specific nutritional goals. Indeed some HIV treatment programs function in areas with a high prevalence of malnutrition. Not attending to malnutrition leads inevitably to poor clinical outcomes for the individual patient. Addressing food insecurity as the cause of this malnutrition through HIV treatment programs may be insufficient to improve long-term outcomes for patients at the community level; however, the dual goals of clinical care and improving food access and food intake are intertwined. In advocacy for access to HIV care, broader population-based approaches, including efforts at economic stability, support of livelihoods, and the development of income generating activities will likely play a role. A more comprehensive approach will be needed in areas where both HIV and malnutrition are prevalent.

24.4. HIV AND NUTRITIONAL EFFECTS

24.4.1. The Cycle of Malnutrition and Infection

Nutritional status is closely associated both with the survival of HIV-infected persons and with disease progression *(31,32)*. Helper T-lymphocyte number and cellular immunity are affected by HIV, and there are secondary effects on humoral (antibody) immunity. Changes in immune function increase the risk of a variety of co-infections including those termed "opportunistic" because of this decreased host immune response *(33–35)*. Each co-infection has its own pathology and organ-specific manifestations, which in turn may affect nutritional status by increasing metabolic demand, causing nutrient losses, and reducing dietary intake *(36,37)*. In this setting weight loss can be rapid. Loss of both fat mass and fat-free mass is a common manifestation of advanced HIV infection, where decreased metabolic reserves and decreased intake together lead to additional infections by directly impairing the immune system. This interplay completes the continual cycle of malnutrition and infection in HIV (Fig. 24.5).

Weight loss is more than a symptom of HIV infection. Rather loss of body weight reflects an imbalance of nutritional inputs and metabolic demands and is prognostic. Many HIV-infected persons become nutritionally deficient in association with HIV disease and wasting syndrome is a common complication preceding death. The issue of underlying malnutrition is very important in the context of HIV infection as a body mass index less than 18.0 kg/m^2 is strongly predictive of death (adjusted hazard ratio 2.5, 95% CI 2.0–3.0) *(38)*. Weight loss in patients with HIV infection predicts clinical out-

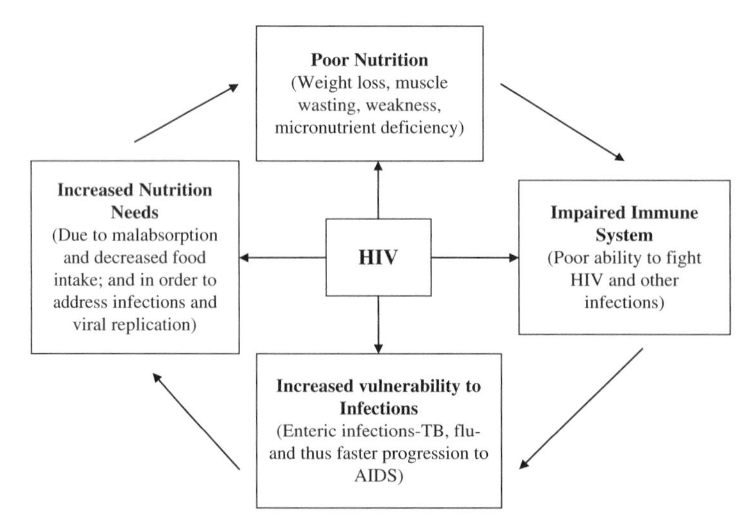

Fig. 24.5 The cycle of malnutrition and infection in the context of HIV/AIDS. Reproduced by permission of Sarah Naikoba et al., *Nutrition and HIV/AIDS a Training Manual*. (Kampala, Uganda: Regional Center for Quality of Health Care, the FANTA (Food and Nutrition Technical Assistance). Project and the LINKAGES (Project), 2006.

comes with or without ART. In HIV-infected adults, a baseline body mass index of less than 18.5 kg/m^2 at the initiation of antiretroviral therapy is significantly associated with mortality in univariate analysis [hazard ratio 2.9 (95% CI *1*.04–*8*.01)]. In fact weight loss during the first 4 weeks of antiretroviral therapy is also associated with death ($p = 0.009$) *(39)*, and this effect persists even when patients have access to ART *(40,41)*. Higher body mass index is protective, and associated with higher CD4+ T-lymphocyte count responses *(42)*. Thus, especially at the beginning of therapy, general nutrition status as measured by BMI is prognostic.

24.4.2. Nutritional Changes and Metabolic Demands in HIV

HIV and nutritional status interact in complicated ways which are not fully understood. Resting energy expenditure (REE) has been examined in multiple studies. Most studies demonstrate an increase in REE of approximately 10% in the asymptomatic patient, and up to 34% in the patient with co-infections *(43,44)*. Energy expenditure was found to be higher in HIV-infected women compared to HIV negative women *(45)*, even when adjusted for lean body mass. In the setting of co-infections, resting energy expenditure increased and energy intake decreased. Despite these changes in REE, intake remains the prime determinant in weight loss. In fact total energy expenditure may decrease due to activity levels, however, intake is still often less than energy expenditure in HIV-infected patients with rapid weight loss *(46)*.

The demands of maintaining viral replication and supporting the inflammatory immune response appear to require significant metabolic resources in both the asymptomatic state and with co-infection *(47)*. Reflecting demands against growth, disease activity in children as measured by viral load is inversely related to energy intake, growth velocity, and fat-free mass *(48)*. Energy destined for deposition into muscle, bone, fat, and other tissues of a child is directed toward the maintenance of infection and inflammation.

A redirection of metabolic resources to HIV infection may cause wasting. According to the WHO wasting is defined as unexplained weight loss of 10% of total weight *(49)*. Wasting patterns include loss of lean body mass and muscle in men, and initial fat loss in women followed by loss of muscle *(50)*. There is a preferential loss of lean body mass, particularly in men *(51)*. While malabsorption, inflammation of HIV infection, metabolic demand of co-infections, and food insecurity all contribute to weight loss, decreased intake appears to be the largest single contributor *(52)*. Decreased intake may be related to gastrointestinal symptoms, thrush, esophagitis with pain, and disease-related anorexia from effects on eating mechanics, perception of hunger, poverty, or psychosocial problems *(53,54)*. A comprehensive nutritional approach is warranted in all HIV-infected patients as weight loss "remains a major problem and may be the best predictor of mortality in [people living with HIV/AIDS]" *(18)*.

Weight loss may be reflective of overall nutritional status, but protein and fat metabolism are specifically affected in HIV infection. Patients infected with HIV have increased rates of protein turnover *(55,56)* and increased losses through the gastrointestinal tract *(57)*. If the goal protein intake is not met, inflammation from viral infection persists, and physical activity during active disease is limited, the result is muscle wasting. HIV infection may also cause specific patterns of fat loss (lipoatrophy) and redistribution (lipodystrophy) which are associated with metabolic abnormalities including dyslipidemia. These changes may be augmented by ART, especially regimens containing protease inhibitors. Chronic dyslipidemia may increase atherosclerosis and in turn cardiovascular risk in both adults and children *(58,59)*.

Aside from specific changes in metabolism, there are generalized anthropometric changes that can occur in HIV infection. The sum of the effects of losses, changes in intake, and increased metabolic demand is often protein-energy malnutrition (PEM). PEM is a state of persistent nutritional deficiency, but during which the body still actively compensates to maintain lean body mass and visceral organ function. Further malnutrition causes uncoupling from this compensated state and rapid wasting may result. Nutritional recovery then becomes less likely *(18)*. The cause of wasting is especially important in management decisions for HIV-infected individuals. If weight loss is due to the metabolic effects of HIV and co-infections, nutritional supplementation must be accompanied by treatment of infections and the initiation of ART in the appropriate patient. ART is critical in restoring helper T-cell numbers and function which offers a window for recovery and improves innate protection from further infectious insult. If the wasting is due to poor intake, however, and the patient has been screened and treated for opportunistic infections and tuberculosis, primary treatment may not include antiretroviral medications. PEM itself may decrease T-cell numbers, therefore, in certain patients low CD4 T-cell counts may be partly attributable to PEM, not solely HIV. Additionally ART can complicate medical nutritional issues such as refeeding syndrome, exacerbate anemia, and cause an immune reconstitution syndrome associated with tuberculosis *(60)* and other infections such as cryptococcus, cytomegalovirus, JC virus, hepatitis B and C, and varicella.

In contrast, a case of malnutrition may be related to poor intake from lack of food access, where replacement feeding is indicated. In practice, patients in developing settings with CD4 counts that qualify them for ART are often stabilized with enteral supplements as the first treatment. In the time before ART initiation, repletion of some nutritional stores may lessen the likelihood of metabolic complications and promote anabolism. The reversal of the catabolic state can also result in improved immune function. According to the WHO, target caloric intakes should be 10% greater than normal in asymptomatic HIV infection, and up to 30% in patients with opportunistic infections to meet the increased nutritional needs, and protein intake should increase by 10%, but should not be more than 15% of total caloric intake to provide sufficient non-protein calories to avoid protein catabolism *(61)*. In cases of extremely low CD4 counts and malnutrition where mortality is high, however, ART is usually indicated despite the potential risks of reconstitution syndromes and drug toxicity during simultaneous treatment for viremia and malnutrition.

24.4.3. Immunologic Effects of Nutrient Deficiencies

While clinicians may be concerned about a variety of effects of malnutrition, undernourished patients infected with HIV have a considerable immunologic burden. Helper T-cells, or CD4+ T lymphocytes, are a primary infective target for HIV, where the virus causes cell death. As numbers of CD4+ cells decline with natural progression of HIV, cell-mediated immunity is greatly affected. Given the central signaling role played by CD4+ T-helper cells, other immune mechanisms such as innate and humoral immunity may also be affected. This leads to a repetitive cycle of infection, inflammation, and increased metabolic demand.

In a patient with immunodeficiency related to decreased number and function of T-helper lymphocytes, malnutrition plays a synergistic role. Protein-energy malnutrition or undernutrition (PEM) is the state of energy deficiency due to chronic insufficiency of macronutrients. PEM has long been recognized as a primary cause of immunodeficiency and notably has manifestations similar to those seen in progressive HIV infection. While PEM can be accompanied by other micro- and macronutrient deficiencies, PEM itself is a cause of T-cell dysregulation and may affect other immune processes (Table 24.1) Both the quantity and the function of T cells are affected in PEM, and thymic maturation of nascent T cells is decreased due to thymic atrophy *(59)*. T-cell dysfunction resulting from PEM coupled with decreased T-cell number and function from HIV is a potent combination. This synergy of immune deficiencies is likely to play a role in the early mortality seen at ART initiation for those with

Table 24.1
Immunologic Effects of Specific Micronutrient Deficiencies

System or cell type	*Defect*	*Associated nutrient deficiency*
Lymphocytes	Decreased transferrin receptor metabolism	*Selenium, Vitamin E*
	Decreased skin test reactivity	*Protein-energy Malnutrition (PEM)*
T cells	Thymic involution	*PEM, zinc, magnesium*
	Decreased thymic export of T cells	*PEM*
	Decreased cytotoxicity	*Zinc, iron, PEM*
	Decreased Th1 cytokine production	*Zinc, selenium, iron, PEM*
	Decreased proliferative response	*Zinc, selenium, vitamin A, PEM*
	Decreased CD4/CD8 ratio	*βcarotene, vitamin A, PEM*
	Decreased Th2 cytokine response	*Vitamin A*
B cells	Decreased T-cell-dependent B-cell response	*Zinc, PEM*
	Decreased antibody response to immunization	*Vitamin A, Zinc, PEM*
	Decreased human milk IgA and secretory IgA on mucosal surfaces	*PEM*
	Decreased marrow B-cell precursors	*PEM, Zinc*
Monocytes	Increased Prostaglandin E_2 production	*Vitamin E*
Innate Immunity	Decreased leukocyte phagocytosis	*Vitamin C, copper, PEM*
Natural Killer Cells	Decreased cytotoxicity and numbers	*Zinc, selenium, vitamins A/D, PEM*
Hematopoiesis	Lymphopenia	*Vitamin A*
	Neutropenia	*Copper*
	Anemia	*Vitamin E, riboflavin, pyridoxine, B12, folate, iron, copper, PEM*

Adapted from Cunningham-Rundles, S., McNeeley, D.F., Ananworanich, J. (2004) In: Steihm ER, Ochs HD, Winkelstein JA, editors. Secondary Immunodeficiencies Immunologic Disorders in Infants and Children, 5th ed. Philadelphia, PA: Elsevier Saunders, 766.

lower BMIs. In addition, both immune function and malnutrition seen in HIV and superinfections like measles may deprive the immune system of needed building blocks for appropriate reaction to new antigens and impair B-cell memory response to antigens such as vaccines *(59)*.

Even in the setting of low CD4+ counts that might otherwise qualify a patient to receive ART, the catabolic state of malnutrition in HIV may first need to be reversed with direct nutritional therapy. Before starting ART, some hospitals in developing settings treat malnourished patients with micronutrient and macronutrient supplements such as porridge mixtures and F75, a food mixture designed specifically for the initial treatment phase of severe malnutrition in children *(62)*. Delaying ART allows treatment of PEM and micronutrient deficiencies, permits diagnosis of refeeding syndrome in the severely malnourished patient, and ultimately promotes anabolism. Treatment of malnutrition can itself support lymphocyte production. Addressing metabolic reserve and aiding in the production of blood cells prior to ART initiation also helps to minimize medication-related adverse effects including reconstitution syndrome and anemia. Specifically diagnosed deficiencies of micronutrients should be treated appropriately. Common deficiencies in malnourished HIV-infected patients mirror those seen in malnourished non-HIV-infected patients, such as fat soluble vitamins A, C, D, E, K, and B vitamins such as riboflavin, thiamine, and pyridoxine. Most anemias are best treated with transfusion in the acutely ill patient and with enteral iron or micronutrients if non-urgent. Medications for worms and malaria also play a role. Knowledge of the regional prevalence of nutritional deficiencies and treating these deficiencies, especially in the case of PEM, are both critical in optimizing therapy for HIV-infected patients.

24.4.4. Superinfections, HIV, and Malnutrition

One consequence of HIV infection is the destruction of CD4+ T lymphocytes, important as helper cells to conduct cell signaling in the normal immune response. Decimation of T-helper lymphocytes by HIV causes increased susceptibility to fungal, parasitic, and other infections. Individual infections have specific and additive effects on nutritional status. In resource-limited settings, baseline malnutrition contributes to the development of infections like tuberculosis, pneumonia, and diarrhea and impairs recovery. Multiple infections are important to consider in the nutritional assessment of the patient infected with HIV.

Tuberculosis (TB) is a leading cause of death among HIV-infected people causing roughly 200,000 annual deaths among people living with HIV, most in Africa *(63)*. People who are infected with both HIV and TB are likely to be at a greater risk for malnutrition compared to those infected with HIV alone *(64,65)*. The percentage of men and women with a BMI less than 18.5 kg/m^2 was higher in those subjects co-infected with HIV and TB than individuals with only HIV infection, and individuals who were HIV uninfected *(65)*. Co-infection is a common occurrence in developing settings.

Tuberculosis is a frequent co-infection with HIV and has significant effects itself upon nutritional status. Chronic, untreated TB infection causes weight loss, increased inflammation and has organ-specific manifestations. Co-infections such as TB have significant systemic effects, thus nutritional status may only improve when these infections are properly diagnosed and treated. Screening is important, but not always comprehensive. Even in one U.S. cohort, approximately 54% of newly diagnosed individuals had a tuberculin skin test following HIV diagnosis *(66)*. There is a need for improved diagnosis of co-infections, and a need for further investigation to evaluate the possible outcomes of nutritional interventions for patients with co-infection of HIV and tuberculosis once diagnosed *(65)*. Superinfections, like tuberculosis and others, may themselves be immunosuppressive, complicating diagnosis and treatment decisions. Disease manifestations are often apparent because symptoms are caused by the immune response, rather than by the pathogen itself. In the setting of low CD4+ T-lymphocyte counts, many infections go undiagnosed as typical symptoms may not be apparent to a clinician. Symptoms of TB may be unmasked with treatment, however, causing clinical worsening of symptoms. Similarly, treatment of HIV virus with ART may be revealing for a variety of pre-existing infections, including TB. In a patient with PEM, the outcome from this inflammatory condition, a reconstitution syndrome, can be disastrous as there is little reserve to support the proliferation of new cells. For this reason, a high degree of suspicion of infection and aggressive screening must be employed in the undernourished patient. This includes assessment of nutritional status and consideration of nutritional therapy when either ART or TB medications are initiated.

In addition to TB, other superinfections may have significant nutritional effects. Co-infections may cause bone marrow suppression, weight loss, malabsorption and diarrheal disease, and chronic inflammation. Bone marrow suppression due to viral infections, tuberculosis, and other infections may have non-specific effects on immune response and their treatment can modestly improve HIV viremia *(67,68)*. Co-infections may increase nutrient losses *(20)* and affect intake via nausea, vomiting, pain, and anorexia. Other superinfections include fungal and bacterial infections of the gastrointestinal tract. Mucositis, esophagitis, and gastroenteritis can cause pain in the mouth, chest, and abdomen; nausea and vomiting; bleeding; malabsorption; and may increase the frequency of other infections of the gastrointestinal tract after causing mucosal inflammation and damage. Protein-energy malnutrition in otherwise healthy individuals can make a patient susceptible to *Pneumocystis jirovecii* (formerly *P. carinii*) pneumonia, a superinfection common in HIV *(63)*. This infection may manifest as coughing interrupting feeding and increased work of breathing requiring additional energy expenditure.

Parasites such as cryptosporidia and giardia may cause chronic diarrhea and malabsorption. Co-infections also affect host immunity and can result in higher viral loads. Levels of viremia correlate with HIV transmission risk, and treatment of various concurrent infections including viruses, mycobacteria, and parasites not only may improve nutritional status via decreases in inflammation, but may decrease viral load via effects on immune cell production *(67,68)*. Micronutrient deficiencies, via effects on natural killer and T cells and B cells, can exacerbate the immunodeficiency of HIV and increase risk of complications during the treatment and recovery phase of HIV and increase risk of concomitant infections. To maximize nutritional status, interrupting the cycle of new infections is critical to prevent ongoing nutrient losses and to moderate metabolic demands.

24.4.5. Women, Children, and Breastfeeding

Women make up about half of people infected with HIV globally. In some resource-limited settings such as Africa, however, women are disproportionably affected by the epidemic. There were almost 12 million HIV-infected women in Africa in 2007, representing about 60% of all adult infections in this region (Fig. 24.6) Not only do women suffer the metabolic effects of HIV infection and its treatment, but there is an interaction between gender discrimination and HIV disease itself. Violence and discrimination in accessing education, property, prevention programs, and care and treatment exacerbates the risk of HIV/AIDS for women and girls *(69,70)*.

Women are the usual caretakers for HIV-infected children, who are disadvantaged by prevalent malnutrition, a large burden of infectious disease including diarrhea and pneumonia, and the lack of treatment resources especially in Africa. There were about 1.8 million HIV-infected children in Africa in 2007. This represented 90% of all infected children under 15 worldwide *(4)*. An estimated 1,500 children were infected with HIV daily, most of which occurred during pregnancy, birth, or breast

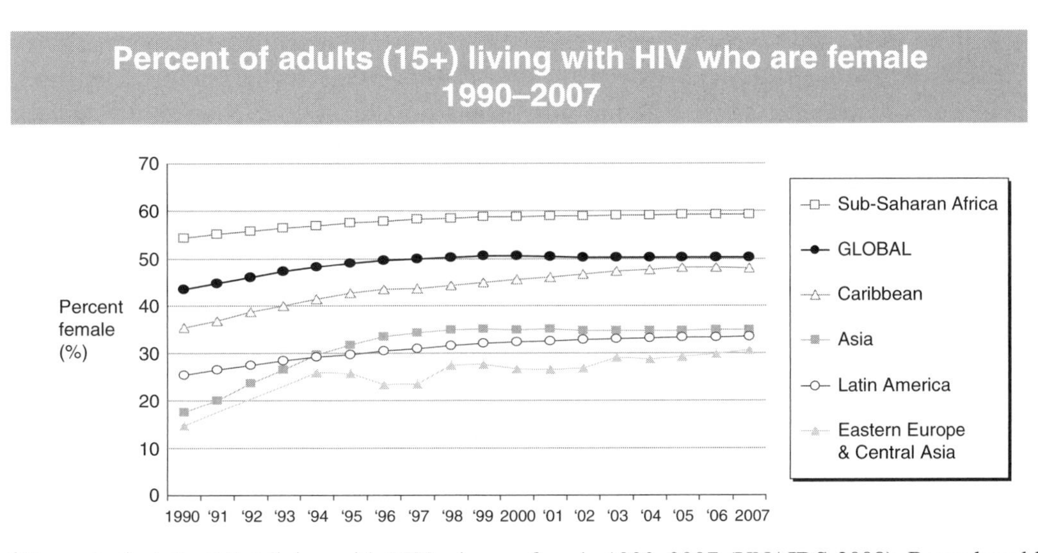

Fig. 24.6 Percent of adults (15+) living with HIV who are female 1990–2007 (UNAIDS 2008). Reproduced by permission of UNAIDS (www.unaids.org).

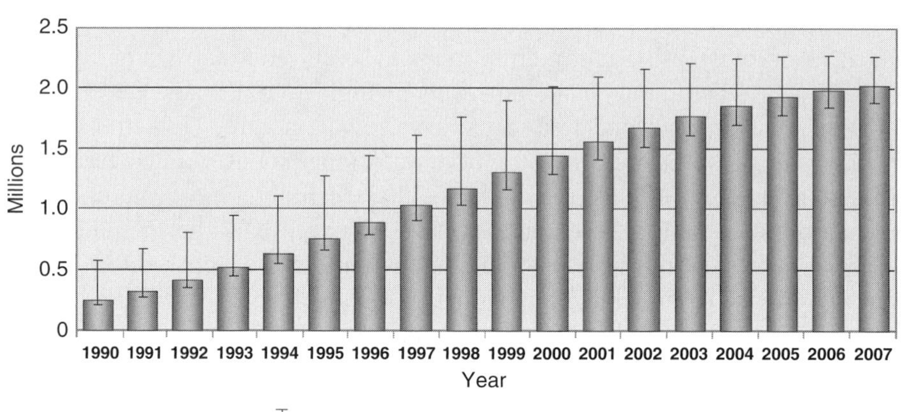

Fig. 24.7 Children living with HIV globally, 1990–2007. Reproduced by permission of UNAIDS (www.unaids.org).

feeding *(71)*. The effect on families and communities is enormous, with approximately 11.6 million children orphaned by 2007 *(4)* (Fig. 24.7).

The impact of HIV on child growth, morbidity, and mortality is severe. HIV-infected children have a fourfold higher risk of death compared with those uninfected. However, preventable conditions including contaminated water supplies, child undernutrition, and anemia contributed significantly to infant and child mortality independent of HIV infection *(72)*. These conditions are present in many areas at baseline, even without HIV. However, with prevalent HIV and the lack of trained pediatric personnel and appropriate medications, the impact is stunning. In the absence of antiretroviral therapy, HIV-infected children suffer significant stunting, undernutrition, and wasting compared to uninfected children. In a study conducted in the Congo, HIV-infected children born to seropositive mothers were significantly shorter and weighed less than both uninfected children born to seropositive mothers and uninfected children born to seronegative mothers *(73)*. Prevention of new pediatric HIV infections through treatment of mothers, provision of medications to those children already infected, and support of children and orphans affected by HIV has progressed significantly in industrialized countries, but access to care and treatment is still lacking in developing countries, especially in Africa *(74)*. Treatment has a direct positive effect on child survival and growth. To improve early mortality outcomes the WHO has adopted new recommendations advising the initiation of ART in children at diagnosis, rather than delaying ART for the potential benefits of decreased resistance, delayed metabolic consequences, and burden to caretakers. These expanded recommendations will affect the nutritional status of infants infected with HIV worldwide and decreased the nutritional consequences of wasting in children *(75)*. The long-term nutritional impact of widespread pediatric treatment has yet to be studied.

One risk factor for pediatric HIV infection is exposure to virus in breastmilk. The World Health Organization recommends early initiation of exclusive breastfeeding for all newborns; exclusive breastfeeding up to 6 months of age; safe complementary feeding from 6 months of age; good maternal nutrition; vitamin A supplementation; and counseling and support for optimizing nutrition; and infant and young child feeding *(71)*. With ART, mother-to-child-transmission (MTCT) of HIV is approximately 5% prenatally and up to 45% after 2 years of breastfeeding (Table 24.2). The WHO recommends exclusive replacement feeding if it is acceptable, feasible, affordable, sustainable, and safe

Table 24.2
Estimated Risk and Timing of Mother-to-Child Transmission of HIV in the Absence of Interventions

Timing	*Transmission rate (%)*
During pregnancy	5–10
During labor and delivery	10–15
During breastfeeding	5–20
Overall without breastfeeding	15–25
Overall with breastfeeding to 6 months	20–35
Overall with breastfeeding to 18–24 months	30–45

Reproduced by permission of WHO (www.WHO.org)

(AFASS). If not, then exclusive breastfeeding is recommended during the first months of the baby's life *(76)*. Exclusive breastfeeding to 3 months or longer was associated with a significantly lower risk of infection (hazard ratio 0·52 [95% CI 0.28–0.98]) and never having breastfed carried a similar risk of infection to mixed feeding (0.85 *(95)*.

However, the risk of morbidity and mortality from replacement feeding in resource-limited settings is substantial. Contaminated water supplies, unsanitary preparation of formula, and overdilution exposes children to morbidity and mortality from diarrhea, vomiting, and malnutrition. Successful replacement feeding has been found to depend on certain criteria, including the availability of piped water, electricity, gas or paraffin for cooking fuel, and early disclosure of HIV status *(77)*. Infants of South African mothers who met the three household criteria and chose to formula feed had the best outcome in terms of HIV-free survival. The risk of formula feeding to infants in sub-Saharan Africa has been demonstrated. Thior et al. compared the efficacy and safety of breastfeeding plus infant zidovudine prophylaxis for 6 months to formula feeding from birth plus 1 month of infant zidovudine for reducing postnatal transmission of HIV *(78)*. Breastfeeding with zidovudine prophylaxis was not as effective as formula feeding in preventing postnatal HIV transmission, but was associated with a lower mortality rate at 7 months. Both strategies had comparable HIV-free survival at 18 months. Other work is examining nevirapine prophylaxis for the breastfeeding infant.

The WHO also recommends that all HIV-infected mothers receive counseling on infant feeding including information on the general benefits of various infant feeding options and guidance on selecting the option best suited to their situation *(79)*. However, this recommendation is not always practical because there is very little communication with health-care providers as to how to translate this into practice *(80)*. Also HIV-infected women in many societies face challenges when making infant feeding choices due to stigma against formula feeding. Even when mothers are provided with infant feeding counseling to enable them make an informed choice, the intent and the actual practice may be different *(81)*. Social expectations by family and the community may force a mother to breastfeed to avoid identification as a person infected with HIV. Social concerns, including reputation and stigma, traditional advice and authority, compete with the medical concerns and risks making it difficult for women to adhere to the feeding options be they exclusive breastfeeding or replacement feeding *(81–83)*.

There may also be a relationship between mode of infant feeding and spouse's awareness of HIV status. Mothers who had disclosed their HIV status to their spouses were more likely not to breastfeed than mothers who had not disclosed their status. Mothers who had a known HIV-infected spouse, disclosed their own HIV status to family/friends, and had a previous child death were significantly more likely to plan to use exclusive replacement feeding *(84)*. However, Omwega et al. found that maternal MTCT knowledge influenced the choice of alternative infant feeding options but not breastfeeding practices *(85)*. Women with high MTCT knowledge tended to be more receptive and considered feeding alternatives. Clearly cultural influences on breastfeeding practices are strong and must continue to be considered when designing interventions to limit HIV incidence, but education may have a direct impact on infection rates.

Despite cultural and community pressures for breastfeeding, or perhaps because of them, mixed feeding with breastmilk plus other foods is common in many places. Mixed feeding may be most dangerous as it exposes infants to HIV as well as numerous pathogens that can disrupt the protective gastrointestinal mucosal layer permitting easier access to the bloodstream and immune cells. Although exclusive breastfeeding is recommended, mixed feeding is very prevalent in southern Africa and other areas *(82)*. The method of infant feeding counseling is vital to the success of preventing MTCT. While prevention programs for MTCT may contribute to mother's knowledge and understanding of feeding practices, there are many instances where the choice of exclusive breastfeeding is still being undermined by the impression that HIV transmission through breastmilk is a certainty rather than a risk *(82)*. There are still many gaps in the evidence, particularly around the pathogenesis of HIV transmission through breastfeeding *(86)*. In addition to the risk of HIV, health workers sometimes have some difficulty convincing HIV-infected mothers that breastmilk is sufficient as infant food *(82)*. Therefore accepting some risk through breastfeeding is difficult to justify. Adejuyigbe et al. reported that HIV-infected mothers in southwestern Nigeria counseled on infant feeding according to the WHO recommendations preferred exclusive breastfeeding and that mixed feeding was more common among mothers who planned to use exclusive replacement feeding *(87)*. Independent predictors of adherence to exclusive breastfeeding include knowledge of exclusive breastfeeding as a method of preventing mother-to-child transmission of HIV, getting support from the family, especially the male partner, and the socioeconomic status of the mother *(83,84)*. Indeed counseling on infant feeding may be more helpful for less educated mothers *(83)*. The choice to initiate at least some breastfeeding was more common among women with less education, who lived in homes without electricity, and who had no water source in their home or on their property *(88)*. Feeding counselors need to be sensitized to client issues and must be aware of the context in which clients make choices *(82)*.

Nutritional counseling on breastfeeding is critical to promoting benefits of breastmilk even in the context of HIV infection. Counselors need to be adequately trained and prepared for the task. Stress and frustration may be prevalent among nurses who perform counseling *(81)*. However, breastfeeding counseling, even more than traditional nursing, requires time and a fundamental knowledge of the sociocultural environments within which health and illness are addressed. The authors recommended that to improve counseling, pre-service and in-service training is fundamental. Furthermore, culturally appropriate counseling tools can be developed as a way to improve the standardization and routine of infant feeding counseling.

HIV-infected mothers in resource-limited settings need support to make realistic and feasible infant feeding choices. To promote and support optimal feeding practices, it is important to understand the barriers to exclusive breastfeeding and the reasons that mothers mix feed. Promotion of exclusive breastfeeding for at least 6 months, followed by a nutritious complementary diet, and early weaning to an animal milk formula may be the most appropriate option for the poor in countries with high levels of MTCT *(77)*. Food-based dietary guidelines should complement and strengthen policies to prevent MTCT *(89)*. Infant feeding counseling should therefore be culturally appropriate and tailored to individual situations *(81,82)*. To encourage women to adhere to good infant feeding practices, involvement of their partners, family members as well as the community for support should be encouraged *(84)*. Methods to reduce transmission and optimal breastfeeding choices will continue to change as new and increased therapeutic options for babies and mothers emerge.

When available, HAART dramatically reduces mother-to-child transmission of HIV. One of the current approaches to reduce the risk of mother-to-child transmission of HIV is antiretroviral prophylaxis during pregnancy *(90)*. The risk of mother-to-child transmission of HIV is reduced with either standard use of nevirapine alone or in combination with zidovudine administered to infants of women tested at delivery *(91)*. Using nevirapine in PMTCT is cost-effective and should be included as part of a strategy to reduce the incidence of infant infections *(92,93)*.

HAART should be delivered as part of a comprehensive strategy to prevent-mother-to-child-transmission of HIV. Some comprehensive and supportive strategies that have proven successful in the prevention of MTCT in resource-limited settings include routine antenatal voluntary counseling and testing; maternal HAART during the last trimester of pregnancy, at labor, and for up to 6 months following delivery with a goal of minimizing maternal viral load in plasma and breast milk; PCR testing of infants of seropositive mothers at 6 weeks of age; combinations of 3–6 months of exclusive breastfeeding; perinatal administration of antiretroviral therapy (ART) such as nevirapine; and provision of affordable and safe infant replacement feeds *(77,94)*.

However, there may be barriers that hindered the effective implementation of these interventions, including low acceptance of HIV testing because of fear of stigmatization by spouse, family, or community, poor compliance with complex drug regimens, the high cost of antiretroviral drugs, inadequately resourced health-care systems, and unavailability or poor acceptance of safe breast-milk alternatives *(77)*.The Drug Resource Enhancement against AIDS and Malnutrition (DREAM) initiative in Mozambique, Malawi, and Tanzania has achieved an overall reduction of HIV-1 transmission rates to levels very similar to those reported in high income countries reflecting the effectiveness of a comprehensive approach. The program provides supplementary formula and water filters for use during the first 6 months after delivery and HAART *(95)*. Infectious diarrhea with or without HIV continues to be a major killer of children worldwide. Indeed, a large impact may be made upon HIV transmission by employing interventions that are outside of the normal treatment modalities for HIV itself.

24.5. MEDICAL TREATMENT OF HIV AND NUTRITIONAL EFFECTS

24.5.1. Antiretroviral Therapy

Medications aimed at interruption of HIV function and replication are collectively termed *antiretroviral therapy* (ART). ART in current use aims to interrupt viral replication or activity by interfering with reverse transcription needed for turning RNA into stable DNA or by inhibiting an enzyme that cleaves viral pre-proteins in the cytosol into functional viral proteins. ART has multiple nutritional effects, some being specific to drugs or drug classes (*see* Chapter 4).

ART is itself a nutritional therapy, associated with an increase in weight and lean body mass *(96)* and has positive effects on the growth of children *(97)*.These changes may be greater for those who begin ART with more profound immunodeficiency at baseline *(98)*. Reversing inflammation and immunodeficiency is key in promoting anabolism and weight gain. Although nutritional status often improves with ART alone, diagnosis and treatment of co-infections and baseline nutritional status clearly play roles, and must be addressed simultaneously for optimal outcomes.

Several notable nutritional effects have been observed particular to a medication or to a class of medications. Zidovudine (AZT) can suppress bone marrow causing anemia and may cause nausea and abdominal pain. Protease inhibitors are associated with dyslipidemias and insulin resistance *(99)*. This effect may be potentiated by ritonavir, which inhibits cytochrome P450 complex and prolongs the serum half-life of protease inhibitors. HIV itself is associated with redistribution of body fat stores and atrophy in a variety of patterns, termed lipodystrophy. Some patients have a pattern of lipoatrophy of peripheral fat accompanied by accumulation of central and visceral fat. These patients may be predisposed to metabolic abnormalities during treatment including dyslipidemia most often associated with the use of protease inhibitors. Cardiovascular risk as measured by 10 year estimates of coronary heart disease were significantly higher for HIV-infected patients with lipodystrophy from the Framingham Study *(100)*. Abacavir, a nucleoside reverse transcriptase inhibitor, may be associated with increased risk of myocardial infarction *(101,102)*. Several medications can even exacerbate renal or liver disease *(103)*.

Side effects of medications may also interfere with intake, a major cause of wasting. Lactic acidosis, a potential side effect of nucleoside analogs, can cause gastrointestinal symptoms like nausea, vomiting, and right upper quadrant abdominal pain all leading to weight loss. Pancreatitis can occur with didanosine and stavudine, both nucleoside analogs. Nausea is a well-recognized effect of multiple protease inhibitors and has been associated with newer medication classes such as CCR5 and integrase inhibitors. Effective management of nausea and timing of meals may help minimize the effect of these symptoms on nutrient intake.

24.5.2. Interventions for Nutrient Deficiencies in Patients Infected with HIV

24.5.2.1. Nutritional Counseling

Given that intake is a prime determinant of weight and nutritional status for people living with HIV, a comprehensive approach to improving intake includes nutritional counseling, treatment of HIV disease, treatment of co-infections, and increased food access where needed. Nutritional counseling is important to ensure clients have a well-balanced diet, and country-specific counseling materials have been developed as part of national HIV care programs *(104)*. The value of nutrition counseling was identified early and can improve nutritional status of HIV-infected individuals *(105,106)*.

Nutritional counseling, with or without oral drink supplements, was shown to be a feasible intervention at low cost in HIV-infected patients with recent weight loss *(107)*. Fifty-three percent of South African HIV-infected patients who were counseled gained weight compared to 21% of matched controls without counseling *(108)*. Weight loss occurred in 27% of counseled and 43% of controls. Patients with the lowest CD4 count had the best response to the nutritional counseling. Nutritional counseling appeared to help prevent the adverse effects of GI tract or infection, especially in patients with CD4 counts less than 200 cells per microliter. The authors suggested that in areas with little resources nutrition education and dietary counseling are a simple yet effective means of stabilizing or increasing body weight in HIV-infected patients. Intake is a prime determinant of weight loss and gain. Counseling may improve intake in developing areas as well, thus food availability is not the only determinant of intake. This has obvious implications for the design of macronutrient supplementation programs on a large scale.

Nutritional counseling by itself may not always improve the nutritional status of persons infected with HIV. Without ART, body weight, serum cholesterol levels, and CD4 counts may decrease despite dietary counseling and continued maintenance of energy intake for 6 months *(109)*. Oral nutritional supplements for a 3-month period resulted in weight gain in HIV-infected patients, increasing the fat mass while nutritional counseling alone did not result in such an increase *(110)*. Because weight loss may progress despite dietary counseling, those who fail counseling should be identified early. Interventions should be considered to increased energy intake by both counseling and supplementation when indicated *(109)*.

24.5.2.2. Micronutrients

Several nutrition studies in the context of HIV have focused on the effects of micronutrient supplementation. Results generally have been mixed. A review was performed on 15 randomized controlled studies investigating the use of micronutrient supplements to reduce the risk of mortality and morbidity in children and adults with confirmed HIV-1 or HIV-2 infection in a hospital, outpatient clinic, or home-care setting *(111)*. There was no conclusive evidence that micronutrient supplementation effectively reduced or increased mortality and morbidity in HIV-infected adults.

There is more evidence in children to support certain micronutrient supplementation. Periodic vitamin A supplementation in children reduced all-cause mortality and improved growth in one trial and reduced diarrhea-associated morbidity in another *(111)*. Whether this effect is mediated through impact on HIV or is a reflection of improvements in immunity via another mechanism is unknown. Villamor et al. found that vitamin A supplementation given every 4 months to children less than 5 years of age improved growth in infants infected with HIV and malaria and decreased risk of stunting from diarrhea *(112)*. A large positive effect on linear growth was evident among infants with HIV infection, while weight gain was favored among children with malaria infection at baseline. Vitamin A could constitute a low-cost, effective intervention to decrease growth retardation in settings where infectious diseases are highly prevalent and antiviral therapies are not readily available. In both trials population sizes were small.

A prospective, double-blinded, placebo-controlled trial examined the immunologic, metabolic, and clinical effects of broad spectrum (33 ingredients) micronutrient supplementation in HIV-infected patients taking highly active antiretroviral therapy (HAART). The patients that received the micronutrients for 12 weeks had a significantly improved CD4 cell count *(113)*. Higher dietary intakes of vitamin A and D were positively correlated with CD4 count *(114)*. Thus there is stronger evidence of the beneficial effects of vitamin A supplementation in children *(115,116)*. Despite improved outcomes with supplementation, a South African group found that coverage of children in South Africa during campaigns was sub-optimal. Only 24% of the mothers knew why their child had been given vitamin A and only 11% of the mothers heard about this particular vitamin A supplementation program before becoming involved. The health-care providers questioned in this study also noted training deficiencies and difficulties in program implementation *(78)*. Regardless of outcomes of micronutrient supplementation at the community level, access to ART remains the prime intervention to restore immune function, improve nutritional status, and decrease morbidities and mortality for children infected with HIV.

Wiysonge et al. assessed the effect of antenatal vitamin A supplementation on the risk of MTCT of HIV as well as infant and maternal mortality and morbidity *(117)*. The authors found that evidence does not support the use of vitamin A supplementation of HIV-infected pregnant women to reduce mother-to-child transmission of HIV. They also noted that currently available data do not exclude the possibility that vitamin A supplementation could be beneficial or harmful; there was need for an appropriately powered randomized controlled trial to assess the additive effect of the intervention on the risk of MTCT of HIV in antiretroviral-treated women.

While there are some data concerning supplementation with vitamin A, many micronutrients have yet to be evaluated in the same fashion. Ultimately many HIV practitioners advocate using the U.S. dietary reference intake as the nutritional goal for supplementation and for counseling of the ideal diet *(11)*. Individual patients with severe malnutrition and specific deficiencies should be treated appropriately. Complicated interactions exist between infection and nutritional status making general recommendations of supplementation for all HIV-infected patients challenging. For the best clinical care, more data are needed from the most common clinical settings, namely developing countries.

24.5.2.3. Macronutrients

Given that intake plays a central role in weight loss, and intake correlates with CD4 counts *(118)* increasing intake would seem to be an appropriate treatment. The role of increasing calorie and protein intake in all HIV-infected patients is controversial, however, as supplementation may only increase weight but not fat-free mass. Treatment of muscle wasting is a goal of nutritional therapy in HIV because regaining fat-free mass is a surrogate for overcoming increased protein turnover and improving the capability to perform physical work. Supplements for those with nutritional depletion, additional calories, and protein are indicated and recommended *(61,107)*. However, caloric supplements in patients with a stable weight do not affect further changes in body weight *(120)*, but may be helpful where there is documented weight loss *(121)*.

Several studies have been conducted in developed countries evaluating the effects of macronutrient supplementation with and without nutritional counseling. The results of these few studies have been mixed, with some studies indicating positive change in body weight, macronutrient intake, and immunological factors. Other studies do not support these findings. In a meta-analysis of eight randomized controlled trials, a review from in the Cochrane Library found that nutritional supplements in conjunction with nutritional counseling improved energy and protein intake but not body weight *(122)*. Macronutrient supplementation with or without nutritional counseling was found to significantly improve energy and protein intake compared with no nutritional supplementation or placebo. However, there were no effects on body weight (8 trials; $n = 423$), fat mass (6 trials; $n = 305$), fat-free mass (5 trials; $n = 311$), or CD4 counts (6 trials; $n = 271$).

Oral nutrition supplementation with whole-protein-based and peptide-based formulas in another study caused a significant (10–15%) net increase in dietary energy and protein intake but no changes in body weight, body composition, or fat-free mass *(123)*. In contrast nutritional supplementation with an enterotropic peptide-based formula enriched with *n*–3 fatty acids for 3 months resulted in a significant increase in weight, fat mass, CD4 count, but not total body water and fat-free mass *(124)*. Body weight was also maintained or increased with a high energy, high protein nutrition supplement in conjunction with nutrition counseling among patients with HIV in the early stages of HIV without secondary infections. However, most of the patients who developed a secondary infection lost weight despite nutrition supplementation and counseling *(125)*. Six months of oral supplementation with arginine and *n*–3 fatty acids did not change CD4 and CD8 lymphocytes counts or total energy intake. However, there was an increase in body weight and fat mass *(126)*. Normal intake amounts may have been supplanted by the supplement itself, resulting in no change in net intake. Two randomized, double-blind controlled trials examined the effects of glutamine on AIDS-related wasting. In one trial, the subjects received glutamine with select antioxidants over a 12-week period, while the other trial prescribed glutamine, beta-hydroxymethylbutyrate, and arginine for 8 weeks. Both trials found an increase in body mass in their respective trial groups when compared to those subjects on the placebo *(127,128)*. While supplementation may address decreased intake, the addition of calories may itself increase energy expenditure and protein turnover in the fed state and may have implications for the type of macronutrient supplement and for patient selection *(129,130)*. This hypermetabolic effect may be mediated by antiretroviral treatment with protease inhibitors *(131)*. Confounders in all of these studies include baseline nutritional status, co-infections, control of viremia, and adherence to medical therapy for which each was only sometimes controlled in analysis.

Despite mixed results in studies from developed settings, the WHO recommends a 10% increase in energy intake for asymptomatic adults, a 30% increase in symptomatic adults, and up to a 100% increase in symptomatic children *(64)*. The WHO does not offer recommendations on protein intake, rather general recommendations about dietary diversity. More research is required to understand the effects of macronutrient supplementation on HIV and AIDS.

Programs designed to treat malnutrition have considered the use of modified nutrient-rich foods or food combinations to address specific nutritional needs. Table salt with iodine is one important example. Whole foods with supplemental vitamins such as A, E, as well as iron have been bred or engineered into common staples such as rice, sorghum, millet, or cassava *(132–134)*. There are other groups using unmodified but non-native foods. Recently there have been efforts to introduce such foods to communities to whom the foods may be foreign *(135)*. Acceptance of such foods requires experience of taste and knowledge of preparation and the ability to instruct patients on their use. These are tasks that may not easily fall under the rubric of publicly funded HIV treatment programs. However, increasing exposure to new nutritious foods may become easier as farming techniques change over time, and seeds and growing techniques for new crops become more available with support of governments, organizations, and increased global trade.

Given increased protein turnover in HIV *(55,56,136,137)* accessing sufficient animal protein in diets dominated by carbohydrates can be a challenge. In addition, geographic access to protein sources may be limited by terrain, climate, violence, and disease itself. Certain communities and food scientists are working on new ways to access animal protein, by growing and using local dried fish, rabbits, or other animal source proteins *(138,111)*. Natural plants and insects can be used when food access is limited or as a cultural practice *(139,140)*. Coping mechanisms for food insecurity overlap with the search for optimal methods to ensure adequate intake in HIV. New methods and foods will be needed to address both issues.

The need for food-as-medicine programs in food insecure and resource-limited settings is urgent. Several governments, international agencies, and community groups have responded in part to this need by establishing food and macronutrient supplementation programs in Africa and elsewhere. These programs are increasingly present in the context of HIV care and treatment programs. Existing programs range from prevention of malnutrition to counseling and palliative nutritional care for HIV-infected individuals *(141)*. In the context of HIV treatment programs, the aim of supplementation must be considered. Not only are metabolic demands increased in the HIV-infected patient, but in the developing settings decreased intake from HIV disease and co-infections is exacerbated by food insecurity. Large-scale efforts to address food security may be helpful and ultimately needed, but targeting the patients for nutritional therapy via individual or family support may be a manageable goal for HIV treatment programs operating in resource-limited settings. While there may be a clear indication for treatment of malnutrition for an individual patient, and while the wide scale integration of nutritional support may be viewed by many as good policy, data are only slowly increasing the evidence-base to support benefit in the clinical setting *(142)*. Clearly additional evidence is needed to support the compulsory adoption of supplementation programs by ministries of health and to provide the best nutritional therapies for patients.

Multiple questions exist concerning how best to treat patients infected with HIV. Questions remain as to the optimal choice of macronutrients, how much supplement to provide based on resting energy expenditure or other parameter, targeting the supplement to the patient himself or the whole family, duration of therapy for maximal and lasting effect, and lastly, the timing of the intervention in the course of infection. As a result, work to date has been heterogeneous in approach.

A group in Malawi examined patients who received an individual ration of corn-soy blend and vegetable oil, with primary household earner also receiving maize and beans *(143)*. There were no differences in weight gain or CD4 counts at 6 or 12 months, though there was a significant increase medication possession. Patel et al. examined rate of weight gain and percentage of children reaching weight-for-height greater than 90% ideal who received RUTF or CSB, finding that patients with less than 85% weight-for-height but greater than 80% of the international standard, and who consumed RUTF, had improved growth *(144)*.

Community- and home-based therapies are increasingly employed to improve outcomes for children and to alleviate burden on health-care facilities. Caretakers may then stay near the patient, in order to provide more accessible care, instead of leaving a patient at a health facility while the caretaker works or cares for others in the household. This home-based strategy has been used in HIV care to address clinical nutrition problems. Ninety-three HIV-infected children from Malawi with wasting were stabilized in the hospital and then allocated to full calories via RUTF, supplemental calories via RUTF, or full calories via maize soy flour. Seventy-five percent of children receiving full calories of RUTF reached 100% of weight-for-height, while 53% of those receiving maize/soy flour reached the same goal. Lower initial weight-for-height *Z* score was associated with a poor outcome *(145)*. One hundred and six HIV-infected children who received a 4-week supply of amino acid-based formula had higher hemoglobin and weight compared to a separate group who received soy/skim-milk preparation. Weight was significantly greater in the group receiving amino acid-based formula compared to control. Mortality, diarrhea frequency, and recovery scores did not differ between groups. It was unclear from this study if the groups received the same calories, and analysis by HIV serostatus was incomplete *(146)*. Food supplementation did not improve survival but effects on nutritional status trended to significance in a descriptive longitudinal study in Malawi examining the introduction of food supplementation to HIV patients in various stages of disease *(147)*. Communities are often resourceful when faced with issues such as food insecurity. One type of local support for HIV-infected individuals are community-based programs that provide a variety of services including food supplementation, dietary advice, counseling, and palliative care *(148)*. These approaches will likely be important in future interventions to address malnutrition.

There is additional unpublished data from USAID/FANTA collaboration and others in Kenya and elsewhere *(149)*. Preliminary review of these data shows that there may be some effect of supplementation at 3 months on BMI for females with weight loss on ART. Interestingly this effect was not seen in males. Additionally there were more effects of supplementation for those whose BMI and CD4 counts were higher at initiation of supplementation.

Another type of support is via HIV programs such as multiple PEPFAR-supported programs that function within the governance of national ministries of health throughout the world. Certain programs like the World Food Programme, the U.S. Agency for International Development, United Nations Children's Fund, and Food and Nutrition Technical Assistance project, among others, have coordinated to offer some nutritional supplementation via HIV care and treatment. A more unique example of nutritional support within the context of HIV care is the Academic Model for Prevention and Treatment of HIV/AIDS (AMPATH) based in western Kenya. Patients are recruited for supplementation based on criteria including food security, socioeconomic status, and anthropometrics, in the same setting in which HIV care is provided. Despite questions of sustainability AMPATH administers its own agriculture activities growing food and purchasing milk and other items for program participants *(142)*. Though some data exist showing improved ART adherence with supplementation *(135,142)*, without comparative data on outcomes it is difficult to generalize recommendations on various care models.

There is a small but growing literature concerning macronutrient supplementation for HIV-infected patients. This includes alternative means of nutrient delivery, such as ready-to-use therapeutic foods (RUTF), fortified blend foods (FBF), and fortified spreadable paste. RUTF is most commonly a peanut (groundnut)-based paste (or butter), vegetable oil, fortified with dried cow's milk, micronutrients, and sugar for taste. Its most common application is in home-based treatment of childhood malnutrition (community therapeutic feeding), as children accept the taste, dosage can be titrated to weight, and this form of RUTF does not require a cold chain and resist bacterial growth. Some observational data from Malawi and other locations have examined used of RUTF for children in program settings *(150–154)*. The most beneficial intervention for patients infected with HIV is still undetermined.

Adding whey and skimmed milk powder in FBF is used to improve protein quality in vulnerable patients and has renewed interest in HIV, including in the setting of diarrhea, a major cause of nutrient loss in developing settings *(155–157)*. Fortified blended foods are an evolution of common supplementation methods used by the World Food Program, where culturally acceptable maize- and rice-based products are mixed with other calorie, protein, and micronutrient sources to achieve a more balanced macronutrient supplement. FBF aims to treat deficiencies and to maintain normal physiology in food insecure areas; however, it is not available everywhere. The use of combinations of RUTF and FBF as part of medical treatment in conjunction with pre-ART counseling and with ART is emerging. Current studies employing combination strategy are targeting those with BMI under 18.5 or under 20 *(158)*.

As new food technologies and products emerge, evaluations of each in the setting of HIV treatment will be required. Adjunctive fortification therapies for protein and fat are now being postulated for use in developed settings. Re-evaluation of the best measures to determine the success of interventions is also needed. A range of anthropometric and clinical measurements may be needed to identify patients that might benefit from interventions, in addition to BMI. While BMI is a surrogate for physiologic homeostasis including the immunologic ability to handle infection and inflammation, an absolute cutoff for BMI that indicates clinical risk for an individual patient is not clear. A cutoff BMI of 18.5 is somewhat arbitrary and in fact is not the same weight-for-height percentile in each sex. These rough tools need refinement. Indeed, the clinical profile of the patient most likely to benefit from supplementation remains unanswered. Multiple strategies and new ideas will be required to identify appropriate patients, to address nutrient deficiencies in HIV, and to optimize combinations of macronutrients that improve clinical outcomes for all patients infected with HIV.

24.5.2.4. Other Therapies

To promote intake and anabolism, other therapies have been examined. Prolonged combination therapy with either the appetite-stimulant megestrol acetate, the anabolic steroid oxandrolone, or dietary advice may reverse weight loss associated with HIV and improve physical health *(159)*. Oxandrolone may have a role in repletion of body cell mass, for which measurements of lean body mass is a surrogate *(160)*. Testosterone, anabolic steroids, and human growth hormones can help increase lean body mass in individuals with HIV/AIDS *(161)*. Additional work investigating mechanisms of wasting in HIV infection has recently involved the cytokine inhibition (thalidomide) and adipokines such as leptin in the pathway of satiety *(162)*. More work is needed to understand the role of medical and nutritional treatment in various stages of HIV disease.

24.6. RECOMMENDATIONS

HIV-infected patients should begin nutritional care at the diagnosis of the disease to assess baseline nutritional status and provide dietary counseling *(125)*. In considering food-as-medicine in resource-limited settings, there is need for more comprehensive programs not simply to address individual deficiencies when diagnosed but also to address malnutrition and food insecurity in communities. Answers will involve more than medical personnel. Given the importance of the nutritional status of communities for health and productivity, it is surprising that there has been little specific research on the use of macronutrients in HIV, particularly in resource-limited settings. This is clearly the future application of knowledge in HIV, whether treating malnutrition or understanding metabolic changes during ART treatment. Basic questions about macronutrient supplementation remain including: What are the nutritional characteristics of patients who would most benefit? What type(s) of foods are optimal? How long should patients receive support? and When in the course of HIV disease or treatment is support most beneficial? This will require substantial mental and economic capital and require training of researchers and the integration of research into implementation programs. While there are substantial challenges to performing clinical research in general in resource-poor settings and nutritional research in particular, findings from such research could provide information for new guidelines in the management of individual patients and for national policy. Perhaps because of the difficulties and insufficient data, nutritional interventions and specific guidance are not recognized as a component of national HIV treatment guidelines until recently *(104,163)*. Providing supplemental food for HIV-infected patients remains expensive and logistically difficult and may contribute to lack of evidence and macronutrient policy *(164)*. Even specific micronutrients like vitamin A, though more widely researched, have weak evidence to support their use in guidelines *(165–168)*. Many believe a diet that approaches the RDA for most nutrients should provide for normal physiologic function *(169–171)*, but the interplay of HIV disease, its complications, and nutritional requirements is still being examined.

Addressing stigma through education and community involvement is likely to improve access to all types of HIV services, including nutrition. As noted above, lack of information seems to be a common barrier of access to treatment, either with nutritional supplements or with medication. A review article examining the barriers of access to antiretroviral treatments (ART) in developing countries *(172)* found that the most commonly citied barrier was lack of information. This study also found that lack of information took the form of a stigma associated with HIV AIDS, which ultimately discouraged people from seeking ART. Education and reduction of stigma will continue to be important in the success of nutrition programs from breastfeeding counseling to supplementation campaigns.

Communities should also be involved when making policies aimed at improving participation in programs. Investigators may be able to involve communities more directly in the design and implementation of research and in policy development *(173,174)*. There are benefits to HIV-infected individuals being involved in policy developments and may improve outcomes. The resource of the community should not be overlooked in addressing the dual epidemics of HIV and malnutrition.

A comprehensive approach to managing HIV and AIDS is needed for success of programs targeting prevention, care, and treatment. Brazil employs a comprehensive policy with universal access to HIV treatment *(175)*. This access is publicly funded to all those in need and provides access to 397 accredited hospitals, 79 day-care hospitals, 58 home-care centers and 422 outpatient facilities, 82 lymphocyte phenotyping, 71 viral load laboratories, 18 genotyping centers, and access to 17 drug pharmaceutical compounds for diagnosis, prevention, and treatment. These policies demonstrate how a developing country can develop systems of equity, independent of factors such as race, gender, or financial means which will likely be of increasing importance in future program design *(176)*.

In the United States, the CARE Act provides for a comprehensive care model in HIV care and treatment, ensuring access for vulnerable populations *(177)*. This is echoed in developing settings where advocacy for comprehensive care models continues to gain momentum. The expansion of HIV care and treatment programs has engendered new discussion about the development and strengthening of primary health-care systems worldwide, and as nutrition is critical to the health status of communities, addressing nutrition via these systems will continue to be possible and necessary.

Nutritional consequences of HIV infection can be significant. While some advances have been made in understanding nutrition in the context of this chronic infection and its treatment, our knowledge is nascent and will continue to evolve as more attention is given to optimizing medical therapies and improving clinical outcomes for those affected by this global epidemic.

REFERENCES

1. Kotler, D.P. (2008) HIV and antiretroviral therapy: lipid abnormalities and associated cardiovascular risk in HIV-infected patients. *J Acquir Immune Defic Syndr* **49**, S79–85.

2. Tebas, P. (2008) Insulin resistance and diabetes mellitus associated with antiretroviral use in HIV-infected patients: pathogenesis, prevention, and treatment options. *J Acquir Immune Defic Syndr* **49**, S86–92.
3. Triant, V.A., Lee, H., Hadigan, C., and Grinspoon, S.K. (2007) Increased acute myocardial infarction rates and cardiovascular risk factors among patients with human immunodeficiency virus disease. *J Clin Endocrinol Metab* **92**, 2506–12.
4. Joint United Nations Programme on HIV/AIDS (2008) Report on the Global AIDS Epidemic 2008. (Accessed at http://www.unaids.org/en/KnowledgeCentre/HIVData/GlobalReport/2008)
5. Fourth Annual Report to Congress on PEPFAR (2008) President's Emergency Plan for AIDS Relief. (Accessed at http://www.pepfar.gov/documents/organization/100029.pdf)
6. World Health Organization (2008) Towards Universal Access, Scaling up Priority HIV/AIDS Interventions in the Health Sector. Progress Report 2008. (Accessed at www.who.int/hiv/pub/towardsuniversalaccessreport2008.pdf)
7. Joint United Nations Programme on HIV/AIDS (UNAIDS) (2007) Towards universal access: assessment by the Joint United Nations Programme on HIV/AIDS on scaling up HIV prevention, treatment, care and support. (Accessed at http://www.unaids.org/en/PolicyAndPractice/TowardsUniversalAccess/default.asp)
8. United Nations Children's Fund (UNICEF) (2006) Technical Consultation of the Global Partners—Forum on Children Affected by HIV and AIDS: Universal Access to Prevention, Treatment and Care. (Accessed at http://www.unicef.org/aids/files/TechConsultChildren06.pdf)
9. Antiretroviral Therapy Cohort Collaboration (2008) Life expectancy of individuals on combination antiretroviral therapy in high-income countries: a collaborative analysis of 14 cohort studies. *Lancet* **372**, 293–9.
10. Quinn, T.C., Wawer, M.J., Sewankambo, N., Serwadda, D., Li, C., Wabwire-Mangen, F., Meehan, M.O., Lutalo, T., Gray, R.H. (2000) Viral load and heterosexual transmission of human immunodeficiency virus type 1. Rakai Project Study Group. *N Engl J Med* **342**, 921–9.
11. Gray, R.H., Wawer, M.J., Brookmeyer, R., Sewankambo, N.K., Serwadda, D., Wabwire-Mangen, F., Lutalo, T., Li, X., vanCott, T., Quinn, T.C., Rakai Project Team (2001) Probability of HIV-1 transmission per coital act in monogamous, heterosexual, HIV-1-discordant couples in Rakai, Uganda. *Lancet* **357**, 1149–53.
12. Buchbinder, S.P., Vittinghoff, E., Heagerty, P.J., Celum, C.L., Seage, G.R. 3rd, Judson, F.N., McKirnan, D., Mayer, K.H., Koblin, B.A. (2005) Sexual risk, nitrite inhalant use, and lack of circumcision associated with HIV seroconversion in men who have sex with men in the United States. *J Acquir Immune Defic Syndr* **39**, 82–9.
13. Garnett, G.P., White, P.J., Ward, H. (2008) Fewer partners or more condoms? Modelling the effectiveness of STI prevention interventions. *Sex Transm Infect* **84** Suppl 2, ii4–11.
14. Gray, R.H., Kigozi, G., Serwadda, D., Makumbi, F., Watya, S., Nalugoda, F., Kiwanuka, N., Moulton, L.H., Chaudhary, M.A., Chen, M.Z., Sewankambo, N.K., Wabwire-Mangen, F., Bacon, M.C., Williams, C.F., Opendi, P., Reynolds, S.J., Laeyendecker, O., Quinn, T.C., Wawer, M.J. (2007) Male circumcision for HIV prevention in men in Rakai, Uganda: a randomised trial. *Lancet* **369**, 657–66.
15. Garnett, G.P., Baggaley, R.F. (2008) Treating our way out of the HIV pandemic: could we, would we, should we? *Lancet* **373** (9657): 9–11.
16. Quinn, T.C. (2008) HIV epidemiology and the effects of antiviral therapy on long-term consequences. *AIDS* **22**, S7–12
17. Food and Nutrition Technical Assistance Project (2004) HIV/AIDS: A Guide for Nutritional Care and Support. 2nd Edition. Academy for Educational Development, Washington DC.
18. Raiten, D.J., Grinspoon, S., Arpadi, S. (2005). Nutritional considerations in the use of ART in resource-limited settings. Consultation on Nutrition and HIV/AIDS in Africa: Evidence, lessons and recommendations for action. Department of Nutrition for Health and Development, World Health Organization.
19. Kadiyala, S., Gillespie, S. (2004) Rethinking food aid to fight AIDS. *Food Nutr Bull* **25**, 271–82.
20. Friis, H. (2002) Micronutrients and HIV Infection. CRC Press LLC, Boca Raton Florida USA.
21. Bukusuba, J., Kikafunda, J.K., Whitehead R.G. (2007) Food security status in households of people living with HIV/AIDS (PLWHA) in a Ugandan urban setting. *Br J Nutr* **98**, 211–7.
22. Olupot-Olupot, P., Katawera, A., Cooper, C., Small, W., Anema, A., Mills, E. (2008) Adherence to antiretroviral therapy among a conflict-affected population in Northeastern Uganda: a qualitative study. *AIDS* **22**, 1882–4.
23. Haile, M. (2005) Weather patterns, food security and humanitarian response in sub-Saharan Africa. *Philosophical Trans R Soc London B: Bio Sci* **360**, 2169–82.
24. United Nations (2006) The Millennium Development Goals Report (Accessed at http://mdgs.un.org/unsd/mdg/Resources/Static/Products/Progress2006/MDGReport2006.pdf)
25. Ag Bendech, M., Chauliac, M., Malvy, D. (1996) Variability of home dietary habits of families living in Bamako (Mali) according to their socioeconomic status. *Sante* **5**, 285–97.
26. Shepherd, R (1999). Social determinants of food choice. *Proc Nutr Soc* **4**, 807–12.
27. UNAIDS. (2004) Report on the Global AIDS Epidemic, 2004. UNAIDS, Geneva.

28. Joffe, M. (2007) Health, livelihoods, and nutrition in low-income rural systems. *Food Nutr Bull* **2**, S227–36.
29. Gillespie, S., Haddad, L., Jackson, R. (2001) HIV/AIDS, Food and nutrition security: Impacts and actions. In Nutrition and HIV/AIDS. Nutrition Policy Paper no. 20. UN Subcommittee on Nutrition, Geneva.
30. World Health Assembly (2004) Scaling up treatment and care within a coordinated and comprehensive response to HIV/AIDS. Fifty-Seventh World Health Assembly, WHA 57.14. (Accessed at http://www.who.int/gb/ebwha/pdffiles/WHA57/A57R14-en.pdf).
31. Macallan, D.C. (1999) Wasting in HIV infection and AIDS. *J Nutr* **129**, 238S–42S.
32. Semba, R.D., Tang, A.M. (1999) Micronutrients and the pathogenesis of human immunodeficiency virus infection. *Br J Nutr* **81**, 181–9.
33. Waibale, P., Bowlin, S.J., Mortimer, E.A., Whalen, C. (1999) The effect of human immunodeficiency virus-1 infection and stunting on measles immunoglobulin-G levels in children vaccinated against measles in Uganda. *Int J Epidemiol* **28**, 341–6.
34. Niyongabo, T., Henzel D., Idi, M., et al. (1999) Tuberculosis, human immunodeficiency virus infection, and malnutrition in Burundi. *Nutrition* **14**, 289–93.
35. Taniguchi, K., Rikimaru, T., Yartley, J.E, et al. (1999) Immunological background in children with persistent diarrhea in Ghana. *Pediatr Int* **41**, 162–7.
36. Babamento, G., Kotler, D.P. (1997) Malnutrition in HIV infection. *Gastroenterol Clin North Am* **26**, 393–415.
37. Piwoz, E.G., Preble, E.A. (2000) HIV/AIDS and Nutrition: A review of literature and recommendations for nutritional care and support in Sub-Saharan Africa. SARA: US Agency for International Development.
38. van der Sande, M.A., Schim van der Loeff, M.F., Aveika, A.A., Sabally, S., Togun, T., Sarge-Njie, R., Alabi, A.S., Jaye, A., Corrah, T., Whittle, H.C. (2004) Body mass index at time of HIV diagnosis: a strong and independent predictor of survival. *J Acquir Immune Defic Syndr* **37**, 1288–94.
39. Jerene, D. (2006) Predictors of early death in a cohort of Ethiopian patients treated with HAART. *BMC Infect Dis* **6**, 136.
40. Wanke, C.A., Silva, M., Knox, T.A., Forrester, J., Speigelman, D., Gorbach, S.L. (2000) Weight loss and wasting remain common complications in individuals infected with human immunodeficiency virus in the era of highly active antiretroviral therapy. *Clin Infect Dis* **3**, 803–5.
41. Mangili, A., Murman, D.H., Zampini, A.M., Wanke, C.A. (2006) Nutrition and HIV infection: review of weight loss and wasting in the era of highly active antiretroviral therapy from the nutrition for healthy living cohort. *Clin Infect Dis* **42**, 836–42.
42. Jones, C.Y., Hogan, J.W., Snyder, B., Klein, R.S., Rompalo, A., Schuman P., Carpenter C.C., HIV Epidemiology Research Study Group. (2003) Overweight and human immunodeficiency virus (HIV) progression in women: associations HIV disease progression and changes in body mass index in women in the HIV epidemiology research study cohort. *Clin Inf Dis* **37**, S69–80.
43. Shevitz, A.H., Knox, T.A., Spiegelman, D., Roubenoff, R., Gorbach, S.L., Skolnik, P.R. (1999) Elevated resting energy expenditure among HIV-seropositive persons receiving highly active antiretroviral therapy. *AIDS* **13**, 1351–7.
44. Melchior, J.C., Raguin, G., Boulier, A., Bouvet, E., Rigaud, D., Matheron, S., Casalino, E., Vilde, J.L., Vachon, F., Coulaud, J.P., Apfelbaum, M. (1993) Resting energy expenditure in human immunodeficiency virus-infected patients: comparison between patients with and without secondary infections. *Am J Clin Nutr* **57**, 614–9.
45. Grinspoon, S, Corcoran, C, Miller, K, Wang, E, Hubbard, J, Schoenfeld, D, Anderson, E, Basgoz, N, Klibanski, A. (1998) Determinants of increased energy expenditure in HIV-infected women. *Am J Clin Nutr* **68**, 720–5.
46. Macallan, D.C., Noble, C., Baldwin, C., Jebb, S.A., Prentice, A.M., Coward, W.A., Sawyer, M.B., McManus, T.J., Griffin, G.E. (1995) Energy expenditure and wasting in human immunodeficiency virus infection. *N Engl J Med* **333**, 83–8.
47. Zulu, I., Hassan, G., Njobvu, L., Dhaliwal, W., Sianongo, S., Kelly, P. (2008) Cytokine activation is predictive of mortality in Zambian patients with AIDS-related diarrhoea. *BMC Infect Dis* **8**, 156.
48. Arpadi, S.M., Cuff, P.A., Kotler, D.P., Wang, J., Bamji, M., Lange, M., Pierson, R.N., Matthews, D.E (2000). Growth velocity, fat-free mass and energy intake are inversely related to viral load in HIV-infected children. *J Nutr* **130**, 2498–502.
49. World Health Organization (2005) Interim Who Clinical Staging Of HIV/AIDS and HIV/AIDS Case Definitions For Surveillance. Geneva. (Accessed at http://www.who.int/hiv/pub/guidelines/clinicalstaging.pdf)
50. Grinspoon, S., Corcoran, C., Miller, K., Biller, B.M., Askari, H., Wang, E., Hubbard, J., Anderson, E.J., Basgoz, N., Heller, H.M., Klibanski, A. (1997) Body composition and endocrine function in women with acquired immunodeficiency syndrome wasting. *J Clin Endocrinol Metab* **82**, 1332–7. Erratum in: *J Clin Endocrinol Metab* (1997) **82**, 3360.

51. Corcoran, C., Grinspoon, S. (1999) Treatments for wasting in patients with the acquired immunodeficiency syndrome. *N Engl J Med* **340**, 1740–50.
52. Heijligenberg, R., Romijn, J.A., Westerterp, K.R., Jonkers, C.F., Prins, J.M., Sauerwein, H.P. (1997) Total energy expenditure in human immunodeficiency virus-infected men and healthy controls. *Metabolism* **46**, 1324–6.
53. Kotler, D.P. (1998) Nutritional management of patients with AIDS-related anorexia. *Semin Gastrointest Dis* **9**, 189–99.
54. Isaac, R., Jacobson, D., Wanke, C., Hendricks, K., Knox, T.A., Wilson, I.B. (2008) Declines in dietary macronutrient intake in persons with HIV infection who develop depression. *Public Health Nutr* **11**, 124–31.
55. Macallan, D.C., McNurlan, M.A., Milne, E., Calder, A.G., Garlick, P.J., Griffin, G.E. (1995) Whole-body protein turnover from leucine kinetics and the response to nutrition in human immunodeficiency virus infection. *Am J Clin Nutr* **61**, 818–26.
56. Crenn, P., Rakotoanbinina, B., Raynaud, J.J., Thuillier, F., Messing, B., Melchior, J.C. (2004) Hyperphagia contributes to the normal body composition and protein-energy balance in HIV-infected asymptomatic men. *J Nutr* **134**, 2301–6.
57. Kotler, D.P. (2005) HIV infection and the gastrointestinal tract. *AIDS* **19**, 107–17.
58. Navas, E., Martín-Dávila, P., Moreno, L., Pintado, V., Casado, J.L., Fortún, J., Pérez-Elías, M.J., Gomez-Mampaso, E., Moreno, S. (2002) Paradoxical reactions of tuberculosis in patients with the acquired immunodeficiency syndrome who are treated with highly active antiretroviral therapy. *Arch Intern Med* **162**, 97–9.
59. World Health Organization. (2003) Nutrient requirements for people living with HIV/AIDS. Geneva, Switzerland. (Accessed at http://www.who.int/nutrition/publications/Contentnutrientrequirements.pdf)
60. Hadigan, C. (2008) In: Nutrition in Pediatrics, 4th Edition. Duggan C, Watkins J, Walker A. (eds.). BC Decker Inc Hamilton, Ontario.
61. Titanji, K, De Milito, A, Cagigi, A, Thorstensson, R, Grützmeier, S, Atlas, A, Hejdeman, B, Kroon, FP, Lopalco, L, Nilsson, A, Chiodi, F. (2006) Loss of memory B cells impairs maintenance of long-term serologic memory during HIV-1 infection *Blood* **108**, 1580–7.
62. World Health Organization. (1999) WHO Management of Severe Malnutrition: A manual for Physicians and other Senior Health Workers WHO Geneva Switzerland. (Accessed at http://whqlibdoc.who.int/hq/1999/a57361.pdf)
63. van Lettow, M., Fawzi, W.W., Semba, R.D. (2003) Triple trouble: the role of malnutrition in tuberculosis and human immunodeficiency virus co-infection. *Nutr Rev* **61**, 81–90.
64. Swaminathan, S., Padmapriyadarsini, C., Sukumar, B., Iliayas, S., Kumar, S.R., Triveni, C., Gomathy, P., Thomas, B., Mathew, M., Narayanan, P.R. (2008) Nutritional status of persons with HIV infection, persons with HIV infection and tuberculosis, and HIV-negative individuals from southern India. *Clin Infect Dis* **46**, 946–9.
65. Lee, L.M., Lobato, M.N., Buskin, S.E., Morse, A., Costa, O.S. (2006) Low adherence to guidelines for preventing TB among persons with newly diagnosed HIV infection, United States. *Int J Tuberc Lung Dis* **10**, 209–14.
66. Modjarrad, K., Chamot, E., Vermund, S.H. (2008) Impact of small reductions in plasma HIV RNA levels on the risk of heterosexual transmission and disease progression. *AIDS* **22**, 2179–85.
67. Wolday, D., Mayaan, S., Mariam, Z.G., Berhe, N., Seboxa, T., Britton, S., Galai, N., Landay, A., Bentwich, Z. (2002) Treatment of intestinal worms is associated with decreased HIV plasma viral load. *J Acquir Immune Defic Syndr* **31**, 56–62.
68. Hughes, W.T., Price, R.A., Sisko, F., Havron, W.S., Kafatos, A.G., Schonland, M., Smythe, P.M. (1974) Protein-Calorie malnutrition: a host determinant for Pneumocystis carinii infection. *Am J Dis Child* **128**, 44–52.
69. Joint United Nations Programme on HIV/AIDS (UNAIDS) (2006) Report on the global HIV/AIDS epidemic. Geneva, Switzerland. (Accessed at http://www.unaids.org/en/KnowledgeCentre/HIVData/GlobalReport/2006/default.asp)
70. United Nations Children's Fund (UNICEF) (2004) Facing the Future Together Report of the United Nations Secretary-General's Task Force on Women, Girls and HIV/AIDS in Southern Africa. (Accessed at http://womenandaids.unaids.org/regional/docs/Report20of20SGs20Task20Force.pdf)
71. World Health Organization (2008) Scale Up Of HIV-Related Prevention, Diagnosis, Care And Treatment For Infants And Children: A Programming Framework. (Accessed at http://www.who.int/hiv/paediatric/Paedsprogrammingframework2008.pdf)
72. Villamor, E., Misegades, L., Fataki, M.R., Mbise, R.L., Fawzi, W.W. (2005) Child mortality in relation to HIV infection, nutritional status, and socio-economic background. Internat Jo Epidemiol **34**, 61–8.
73. Bailey, R.C., Kamenga, M.C., Nsuami, M.J., Nieburg, P., St Louis, M.E. (1999) Growth of children according to maternal and child HIV, immunological and disease characteristics: a prospective cohort study in Kinshasa, Democratic Republic of Congo. Internat Jo Epidemiol **28**, 532–40.
74. Orne-Gliemann, J., Becquet, R., Ekouevi, D.K., Leroy, V., Perez, F., Dabis, F. (2008)Children and HIV/AIDS: from research to policy and action in resource-limited settings. *AIDS* **22**, 797–805.

75. World Health Organization (2008) Technical Reference Group Pediatric HIV CARE ART Guideline Review Report. Geneva, Switzerland. (Accessed at http://www.who.int/hiv/pub/meetingreports/artmeetingapril2008/en/index.html
76. Coutsoudis, A., Pillay, K., Spooner, E., Kuhn, L., Coovadia, H.M. (1999) Influence of infant-feeding patterns on early mother-to-child transmission of HIV-1 in Durban, South Africa: a prospective cohort study. South African Vitamin A Study Group. *Lancet* **354**, 471–6.
77. Doherty, T., Chopra, M., Jackson, D., Goga, A., Colvin, M, Persson, L.A. (2007) Effectiveness of the WHO/UNICEF guidelines on infant feeding for HIV-positive women: results from a prospective cohort study in South Africa. *AIDS* **21**, 1791–7.
78. Thior, I., Lockman, S., Smeaton, L.M., Shapiro, R.L., Wester, C., Heymann, S.J., Gilbert, P.B., Stevens, L., Peter, T., Kim, S., van Widenfelt, E., Moffat, C., Ndase, P., Arimi, P., Kebaabetswe, P., Mazonde, P., Makhema, J., McIntosh, K., Novitsky, V., Lee, T.H., Marlink, R., Lagakos, S., Essex, M., Mashi Study Team. (2006) Breastfeeding Plus Infant Zidovudine Prophylaxis for 6 Months versus Formula Feeding Plus Infant Zidovudine for 1 Month to Reduce Mother-to-Child HIV Transmission in Botswana. *JAMA* **296**, 794–805.
79. Vallenas, C., Henderson, P., Kankasa, C., Bahl, R. (2006). HIV and Infant Feeding Guidelines and Tools: Toward Global Utilization. (Accessed at http://www.globalhealth.org/images/pdf/conf2006/presentations/a4vallenas.pdf)
80. Leshabari, S.C., Blystad, A., de Paoli, M., Moland, K.M. (2007) HIV and infant feeding counselling: challenges faced by nurse-counsellors in northern Tanzania. *Hum Resour Health* **5**, 18.
81. Buskens, I., Jaffe, A., Mkhatshwa, H. (2007). 0nfant feeding practices: Realities and mindsets of mothers in southern Africa. *AIDS Care* October; **19**(9), 1101–1109
82. Matovu, A., Kirunda, B., Rugamba-Kabagambe, G., Tumwesigye, N.M., Nuwaha, F. (2008) Factors influencing adherence to exclusive breast feeding among HIV positive mothers in Kabarole district, Uganda, *East Afr Med J* **85**, 162–70.
83. Bii, S.C., Otieno-Nyunya, B., Siika, A., Rotich, J.K. (2008) Infant feeding practices among HIV-infected women receiving prevention of mother-to-child transmission services at Kitale District Hospital, Kenya. *East Afr Med J* **85**, 156–61.
84. Omwega, A.M., Oguta, T.J., Sehmi, J.K. (2006) Maternal knowledge on mother-to-child transmission of HIV and breastmilk alternatives for HIV positive mothers in Homa Bay District Hospital, Kenya. *East Afr Med J* **83**, 610–8.
85. Coovadia, H.M., Bland, R.M. (2007) Preserving breastfeeding practice through the HIV pandemic. *Trop Med IntHealth* **12**, 1116–33.
86. Adejuyigbe, E., Orji, E., Onayade, A., Makinde, N., Anyabolu, H. (2008) Infant feeding intentions and practices of HIV-positive mothers in southwestern Nigeria. *J Hum Lact* **24**, 303–10.
87. Matovu, A., Kirunda, B., Rugamba-Kabagambe, G., Tumwesigye, N.M., Nuwaha, F. (2008) Factors influencing adherence to exclusive breast feeding among HIV positive mothers in Kabarole district, Uganda. *East Afr Med J* **85**, 162–70.
88. Ogundele, M.O., Coulter, J.B.S. (2003) HIV transmission through breastfeeding: problems and prevention. *Ann of Trop Paed* **23**, 91–106.
89. Hendricks, M., Beardsley, J., Bourne, L., Mzamo, B., Golden, B. (2007) Are opportunities for vitamin A supplementation being utilised at primary health-care clinics in the Western Cape Province of South Africa? *Public Health Nutr* **10**, 1082–8.
90. Suksomboon, N., Poolsup, N., Ket-aim, S. (2007) Systematic review of the efficacy of antiretroviral therapies for reducing the risk of mother-to-child transmission of HIV infection. *J Clin Pharm Ther* **32**, 293–311.
91. Taha, T.E., Kumwenda, N., Hoover, D.R., Fiscus, S.A., Kafulafula, G., Nkhoma, C., Nour, S., Chen, S., Liomba, G., Miotti, P.G., Broadhead, R.L. (2004) Nevirapine and Zidovudine at Birth to Reduce Perinatal Transmission of HIV in an African Setting. *JAMA* **292**, 202–9.
92. Stringer, E.M., Sinkala, M., Stringer, J.S., Mzyece, E., Makuka, I., Goldenberg, R.L., Kwape, P., Chilufya, M., Vermund, S.H. (2003) Prevention of mother-to-child transmission of HIV in Africa: successes and challenges in scaling-up a nevirapine-based program in Lusaka, Zambia. *AIDS* **17**, 1377–82.
93. Fowler, M.G., Lampe, M.A., Jamieson, D.J., Kourtis, A.P., Rogers, M.F. (2007) Reducing the risk of mother-to-child human immunodeficiency virus transmission: past successes, current progress and challenges, and future directions. *Am J Obstet Gynecol* **197**, S3–9.
94. Palombi, L., Marazzib, M.C., Voetberg, A., Magid, N.A. (2007) Treatment acceleration program and the experience of the DREAM program in prevention of mother-to-child transmission of HIV. *AIDS* **21**, S65–71.
95. Carbonnel, F., Maslo, C., Beaugerie, L., Carrat, F., Wirbel, E., Aussel, C., Gobert, J.G., Girard, P.M., Gendre, J.P., Cosnes, J., Rozenbaum, W. (1998) Effect of indinavir on HIV-related wasting. *AIDS* **12**, 1777–84.
96. Verweel, G., van Rossum, A.M., Hartwig, N.G., Wolfs, T.F., Scherpbier, H.J., de Groot, R. (2002) Treatment with highly active antiretroviral therapy in human immunodeficiency virus type 1-infected children is associated with a sustained effect on growth. *Pediatrics* **109**, E25.

97. Shikuma, C.M., Zackin, R., Sattler, F., Mildvan, D., Nyangweso, P., Alston, B., Evans, S., Mulligan, K., AIDS Clinical Trial Group 892 Team. (2004) Changes in weight and lean body mass during highly active antiretroviral therapy. *Clin Infect Dis* **39**, 1223–30.
98. Grinspoon, S., Carr A. (2005) Cardiovascular risk and body-fat abnormalities in HIV-infected adults. *N Engl J Med* **352**, 48–62.
99. Hadigan, C., Meigs, J.B., Wilson, P.W., D'Agostino, R.B., Davis, B., Basgoz, N., Sax, P.E., Grinspoon, S. (2003) Prediction of coronary heart disease risk in HIV-infected patients with fat redistribution. *Clin Infect Dis* **36**, 909–16.
100. Sabin, C.A., Worm, S.W., Weber, R., Reiss, P., El-Sadr, W., Dabis, F., De Wit, S., Law, M., DArminio Monforte, A., Friis-Møller, N., Kirk, O., Pradier, C., Weller, I., Phillips, A.N., Lundgren, J.D. (2008) Use of nucleoside reverse transcriptase inhibitors and risk of myocardial infarction in HIV-infected patients enrolled in the D:A:D study: a multi-cohort collaboration. *Lancet* **371**, 1417–26. Erratum in: Lancet (2008) **372**, 292.
101. DAD Study Group. (2008) Use of nucleoside reverse transcriptase inhibitors and risk of myocardial infarction in HIV-infected patients. Strategies for Management of Anti-Retroviral Therapy/INSIGHT. *AIDS* **22**, F17–24.
102. Bartlett, J.G. (2008) Considerations prior to initiating antiretroviral therapy. www.utdol.com, (Accessed at www.utdolcom 13 December 2008)
103. FANTA (2008). FANTA-supported Materials on HIV, Food and Nutrition (Accessed at http://www.fantaproject.org/downloads/pdfs/FANTAHIVpublist.pdf)
104. Schwenk, A., Bürger, B., Ollenschläger, G., Stützer, H., Wessel, D., Diehl, V., Schrappe, M. (1994) Evaluation of nutritional counselling in HIV-associated malnutrition. *Clin Nutr* **13**, 212–20.
105. Berneis, K., Battegay, M., Bassetti, S., Nuesch, R., Leisibach, A., Bilz, S., Keller, U. (2000) Nutritional supplements combined with dietary counselling diminish whole body protein catabolism in HIV-infected patients. *Eur J Clin Invest* **30**, 87–94.
106. Schwenk, A., Steuck, H., Kremer, G. (1999) Oral supplements as adjunctive treatment to nutritional counseling in malnourished HIV-infected patients: randomized controlled trial. *Clin Nutr* **18**, 371–4.
107. van Niekerk, C., Smego, R.A., Sanne, I. (2000) Effect of nutritional education and dietary counseling on body weight in HIV-seropositive South Africans not receiving antiretroviral therapy. *J Hum Nutr Diet* **13**, 407–12.
108. Chlebowski, R.T., Grosvenor, M., Lillington, L., Sayre, J., Beall, G. (1995) Dietary intake and counseling, weight maintenance, and the course of HIV infection. *J Am Diet Assoc* **94**, 428–32.
109. de Luis Roman, D.A., Bachiller, P. et al. (2001) Nutritional treatment for acquired immunodeficiency virus infection using an enterotropic peptide-based formula enriched with n-3 fatty acids: a randomized prospective trial. *Eur J Clin Nutr* **55**, 1048–52
110. Irlam, J.H., Visser, M.E., Rollins, N., Siegfried, N. (2005) Micronutrient supplementation in children and adults with HIV infection. *Cochrane Database Syst Rev* **4**, CD003650.
111. Villamor, E., Mbise R., Spiegelman, D., Ndossi G., Fawzi W.W. (2002) Vitamin A supplements ameliorate the adverse effect of HIV-1, malaria, and diarrheal infections on child growth. *Pediatrics* **109**, E6.
112. Kaiser, J.D., Campa, A.M., Ondercin, J.P., Leoung, G.S., Pless, R.F., Baum, M.K. (2006) Micronutrient supplementation increases CD4 count in HIV-infected individuals on highly active antiretroviral therapy: a prospective, double-blinded, placebo-controlled trial. *J Acquir Immune Defic Syndr* **42**, 523–8.
113. de Luis, D.A., Bachiller, P., Aller, R., de Luis, J., Izaola, O., Terroba, M.C., Cuellar, L., González Sagrado, M. (2002) Relation among micronutrient intakes with CD4 count in HIV-infected patients. *Nutr Hosp* **17**, 285–9.
114. Mehta, S., Fawzi, W. (2007) Effects of vitamins, including vitamin A, on HIV/AIDS patients. *Vitam Horm* **75**, 355–83.
115. Wiysonge, C.S.U., Shey, M., Kongnyuy, E.J., Sterne, J.A.C., Brocklehurst, P. (2005) Vitamin A supplementation for reducing the risk of mother to-child transmission of HIV infection. *Cochrane Database Syst Rev* **4**, CD003651.
116. Institute of Medicine of the National Academies. (2008) Tables of DRI Values. (Accessed 24 Dec 08 www.iom.edu/Object.File/Master/54/391/Summary Listing.pdf)
117. Woods, M.N., Spiegelman, D., Knox, T.A., Forrester, J.E., Connors, J.L., Skinner, S.C., Silva, M., Kim, J.H., Gorbach, S.L. (2002) Nutrient intake and body weight in a large HIV cohort that includes women and minorities. *J Am Diet Assoc* **102**, 203–11.
118. Gibert, C.L., Wheeler, D.A., Collins, G., Madans, M., Muurahainen, N., Raghavan, S.S., Bartsch, G. (1999) Randomized, controlled trial of caloric supplements in HIV infection. Terry Beirn Community Programs for Clinical Research on AIDS. *J Acquir Immune Defic Syndr* **22**, 253–9.
119. Selberg, O., Süttmann, U., Melzer, A., Deicher, H., Müller, M.J., Henkel, E., McMillan, D.C. (1995) Effect of increased protein intake and nutritional status on whole-body protein metabolism of AIDS patients with weight loss. *Metabolism* **44**, 1159–65.

120. Mahlungulu, S., Grobler L.A., Visser M.E., Volmink J. (2008) Nutritional Interventions for reducing morbidity and mortality in people with HIV. *Cochrane Database Syst Rev* **4**, CD004536.
121. Hoh, R., Pelfini, A., Neese, R.A., Chan, M., Cello, J.P., Cope, F.O., Abbruzese, B.C., Richards, E.W., Courtney, K., Hellerstein, M.K. (1998) De novo lipogenesis predicts short-term body-composition response by bioelectrical impedance analysis to oral nutritional supplements in HIV-associated wasting. *Am J Clin Nutr* **68**, 154–63.
122. de Luis Roman, D.A., Bachiller, P., Izaola, O., Romero, E., Martin, J., Arranz, M., Eiros Bouza, J.M., Aller, R. (2001) Nutritional treatment for acquired immunodeficiency virus infection using an enterotropic peptide-based formula enriched with n-3 fatty acids: a randomized prospective trial. *Eur J Clin Nutr* **55**, 1048–52.
123. Stack, J.A., Bell, S.J., Burke, P.A., Forse, R.A. (1996) High-energy, high-protein, oral, liquid, nutrition supplementation in patients with HIV infection: effect on weight status in relation to incidence of secondary infection. *J Am Diet Assoc* **96**, 337–41.
124. Pichard, C., Sudre, P., Karsegard, V., Yerly, S., Slosman, D.O., Delley, V., Perrin, L., Hirschel, B. (1998) A randomized double-blind controlled study of 6 months of oral nutritional supplementation with arginine and omega-3 fatty acids in HIV-infected patients. *AIDS* **12**, 53–63.
125. Clark, R.H., Feleke, G., Din, M., Yasmin, T., Singh, G., Khan, F.A., Rathmacher, J.A. (2000) Nutritional treatment for acquired immunodeficiency virus-associated wasting using ß-hydroxy ß-methlbutyrate, glutamine, and arginine: a randomized, double-blind, placebo-controlled study. *J Parenter Enteral Nutr* **24**, 133–9.
126. Shabert, J.K., Winslow, C., Lacey, J.M., Wilmore, D.W. (1999) Glutamine-antioxidant supplementation increases body cell mass in AIDS patients with weight loss: a randomized, double-blind controlled trial. *Nutrition* **15**, 860–4.
127. Kosmiski, L.A., Bessesen, D.H., Stotz, S.A., Koeppe, J.R., Horton, T.J. (2007) Short-term overfeeding increases resting energy expenditure in patients with HIV lipodystrophy. *Am J Clin Nutr* **86**, 1009–15.
128. Paton, N.I., Ng, Y.M., Chee, C.B., Persaud, C., Jackson, A.A. (2003) Effects of tuberculosis and HIV infection on whole-body protein metabolism during feeding, measured by the [15 N]glycine method. *Am J Clin Nutr* **78**, 319–25.
129. Prod′homme, M., Rochon, C., Balage, M., Laurichesse, H, Tauveron, I., Champredon, C., Thieblot, P., Beytout, J., Grizard, J. (2006) Whole body leucine flux in HIV-infected patients treated with or without protease inhibitors. *Am J Physiol Endocrinol Metab* **290**, E685–93.
130. Ye, X., Al-Babili, S., Klöti, A., Zhang, J., Lucca, P., Beyer, P., Potrykus, I. (2000) Engineering the provitamin A (beta-carotene) biosynthetic pathway into (carotenoid-free) rice endosperm. *Science* **287**, 303–5.
131. O'Kennedy, M.M., Burger, J.T., Botha, F.C. (2006) Harnessing sorghum and millet biotechnology for food and health. *J Cereal Sci* **44**, 224–35.
132. Lucca, P., Hurrell, R., Potrykus, I. (2002) Fighting iron deficiency anemia with iron-rich rice. *J Am Coll Nutr* **21**, 184S–90S.
133. Mamlin, J., Kimaiyo, S., Lewis, S., Tadayo, H., Jerop, F.K., Gichunge, C., Petersen, T., Yih, Y., Braitstein, P., Einterz, R. (2008) Integrating Nutrition Support for Food-Insecure Patients and their Dependents Into an HIV Care and Treatment Program in Western Kenya. *Am J Public Health* 2009; 99(2): 215–21.
134. Hardin, D.S., LeBlanc, A., Young, D., Johnson, P. (1999) Increased leucine turnover and insulin resistance in men with advanced HIV infection. *J Invest Med* **47**, 405–13.
135. Yarasheski, K.E., Zachwieja, J.J., Gischler, J., Crowley, J., Horgan, M.M., Powderly, W.G. (1998) Increased plasma gln and Leu Ra and inappropriately low muscle protein synthesis rate in AIDS wasting. *Am J Physiol***275**, E577–83.
136. Mills, E.W., Seetharaman, K., Maretzki, A.N. (2007) A nutribusiness strategy for processing and marketing animal-source foods for children. *J Nutr* **137**, 1115–8.
137. Neumann, C.G. (2007) Symposium: food-based approaches to combating micronutrient deficiencies in children of developing countries. Background. *J Nutr* **137**, 1091–2
138. Glew, R.S., Vanderjagt, D. (2006) Coping strategies and nutritional health in rural Niger: recommendations for consumption of wild plant foods in the Sahel. *Int J Food Sci Nutr* **57**, 314–24.
139. Ji, K.M., Zhan, Z.K., Chen, J.J., Liu, Z.G. (2008) Anaphylactic shock caused by silkworm pupa consumption in China. *Allergy* **63**, 1407–8.
140. Abdale, F., Kraak, V. (1995) Community-based nutrition support for people living with HIV and AIDS: A technical assistance manual. New York: God's Love We Deliver, Inc.
141. Byron, E., Gillespie, S., Nangami, M. (2008) Integrating nutrition security with treatment of people living with HIV: lessons from Kenya. *Food Nutr Bull* **29**, 87–97.
142. Cantrell, R.A., Sinkala, M., Megazinni, K., Lawson-Marriott, S., Washington, S., Chi, B.H., Tambatamba-Chapula, B., Levy, J., Stringer, E.M., Mulenga, L., Stringer, J.S. (2008) A pilot study of food supplementation to improve adherence to antiretroviral therapy among food-insecure adults in Lusaka, Zambia. *J Acquir Immune Defic Syndr* **49**, 190–5.
143. Patel, M.P., Sandige, H.L., Ndekha, M.J., Briend, A., Ashorn, P., Manary, M.J. (2005) Supplemental feeding with ready-to-use therapeutic food in Malawian children at risk of malnutrition. *Health Popul Nutr* **23**, 351–7.

144. Ndekha, M.J., Manary, M.J., Ashorn, P., Briend, A. (2005) Home-based therapy with ready-to-use therapeutic food is of benefit to malnourished, HIV-infected Malawian children. *Acta Pediatrica* **94**, 222–5.
145. Amadi, B., Mwiya, M., Chomba, E., Thomson, M., Chintu, C., Kelly, P., Walker-Smith, J. (2005) Improved nutritional recovery on an elemental diet in Zambian children with persistent diarrhoea and malnutrition. *J Trop Pediatr* **51**, 5–10.
146. Bowie, C, Kalilani, L, Marsh, R, Misiri, H, Cleary, P, Bowie, C. (2005) An assessment of food supplementation to chronically sick patients receiving home based care in Bangwe, Malawi: a descriptive study. *Nutr J* **21**, 12.
147. Food and Agriculture Organization. (2008) Household food security and community nutrition. (Accessed at http://www.fao.org/ag/agn/nutrition/householden.stm)
148. Tony Castleman, US Agency for International Development, Personal communication, September 2008.
149. Lin, C.A., Manary, M.J., Maleta, K., Briend, A., Ashorn, P. (2008) An energy-dense complementary food is associated with a modest increase in weight gain when compared with a fortified porridge in Malawian children aged 6-18 months. *J Nutr* **138**, 593–8.
150. Linneman, Z., Matilsky, D., Ndekha, M., Maleta, K., Manary, M.J. (2007) A large-scale operational study of home-based therapy with ready-to-use therapeutic food in childhood malnutrition in Malawi. *Matern Child Nutr* **3**, 206–15.
151. Ciliberto, M.A., Manary, M.J., Ndekha, M.J., Briend, A., Ashorn, P. (2006) Home-based therapy for oedematous malnutrition with ready-to-use therapeutic food. *Acta Pediatr* **95**, 1012–5.
152. [No authors listed]. (2007) Ready-to-use therapeutic foods for malnutrition. *Lancet* **369**, 164.
153. Sadler, K., Myatt, M., Feleke, T., Collins, S. (2007) A comparison of the programme coverage of two therapeutic feeding interventions implemented in neighbouring districts of Malawi. *Public Health Nutr* **10**, 907–13.
154. Bounous, G., Baruchel, S., Falutz, J., Gold, P. (1993) Whey proteins as a food supplement in HIV-seropositive individuals. *Clin Invest Med* **16**, 204–9.
155. Hoppe, C., Andersen, G., Jacobsen, S., Mølgaard, C., Friis, H., Sangild, P., Michaelsen, K.F. (2008) The Use of Whey or Skimmed Milk Powder in Fortified Blended Foods for Vulnerable Groups. *J Nutr* **138**, 145S–61S.
156. Tinmouth, J., Kandel, G., Tomlinson, G., Walmsley, S., Steinhart, A.H., Glazier, R. (2006) The effect of dairy product ingestion on human immunodeficiency virus-related diarrhea in a sample of predominantly gay men: a randomized, controlled, double-blind, crossover trial. *Arch Intern Med* **166**, 1178–83.
157. Tony Castleman, US Agency for International Development, Personal communication, September 2008.
158. Mwamburi, D.M., Gerrior, J., Wilson, I.B., Chang, H., Scully, E., Saboori, S., Miller, L., Forfia, J., Albrecht, M., Wanke, C.A. (2004) Combination megestrol acetate, oxandrolone, and dietary advice restores weight in human immunodeficiency virus. *Nutr Clin Practice* **19**, 395–402.
159. Grunfeld, C., Kotler, D.P., Dobs, A., Glesby, M., Bhasin, S. (2006) Oxandrolone in the treatment of HIV-associated weight loss in men: a randomized, double-blind, placebo-controlled study. *J Acquir Immune Defic Syndr* **41**, 304–14.
160. Winson, S.K. (2001) Management of HIV-associated diarrhea and wasting. *J Assoc Nurses AIDS Care* **12**, 55–62.
161. Sweeney, L.L., Brennan, A.M., Mantzoros, C.S. (2007) The role of adipokines in relation to HIV lipodystrophy. *AIDS* **21**, 895–904.
162. Chopra, M., Darton-Hill, I. (2006) Responding to the crisis in sub-Saharan Africa: the role of nutrition. *Pub Health Nut* **9**, 544–50.
163. Brewster, D.R. (2006) Review of malnutrition treatment evidence in children. *J Pediatr Child Health* **42**, 568–74.
164. Semba, R.D., Graham, N.M.H., Caiaffa, W.T., Margolick, J.B., Clement, L., Vlahov, D. (1993) Increased mortality associated with vitamin A deficiency during HIV type 1 infection. *Arch Intern Med* **153**, 2149–54.
165. van den Broek, N.R., White, S.A., Flowers, C., Cook, J.D., Letsky, E.A., Tanumihardjo, S.A., Mhango, C., Molyneux, M., Neilson, J.P. (2006) Randomised trial of vitamin A supplementation in pregnant women in rural Malawi found to be anaemic on screening by HemoCue. *BJOG* **113**, 569–76.
166. Humphrey, J.H., Iliff, P.J., Marinda, E.T., Mutasa, K., Moulton, L.H., Chidawanyika, H., Ward, B.J., Nathoo, K.J., Malaba, L.C., Zijenah, L.S., Zvandasara, P., Ntozini, R., Mzengeza, F., Mahomva, A.I., Ruff, A.J., Mbizvo, M.T., Zunguza, C.D., ZVITAMBO Study Group. (2006) Effects of a single large dose of vitamin A, given during the postpartum period to HIV-positive women and their infants, on child HIV infection, infection-free survival, and mortality. *J Infect Dis* **193**, 860–71.
167. Fawzi, W. (2006) The Benefits and Concerns Related to Vitamin A Supplementation. *J Infect Dis* **193**, 756–9.
168. Rosales, F.J. (2002) Vitamin a supplementation of vitamin a deficient measles patients lowers the risk of measles-related pneumonia in Zambian children. *J Nutr* **132**, 3700–3.
169. Long, K.Z., Montoya, Y., Hertzmark, E., Santos, J.I., Rosado, J.L. (2006) A double-blind, randomized, clinical trial of the effect of vitamin A and zinc supplementation on diarrheal disease and respiratory tract infections in children in Mexico City, Mexico. *Am J Clin Nutr* **83**, 693–700.

170. Mahalanabis, D. (2004) Randomized, double-blind, placebo-controlled clinical trial of the efficacy of treatment with zinc or vitamin A in infants and young children with severe acute lower respiratory infection. *Am J Clin Nutr* **79**, 430–6.
171. Posse, M., Meheus, F., van Asten, H., van der Ven, A., Baltussen, R. (2008) Barriers to access to antiretroviral treatment in developing countries: a review. *Trop Med Int Health* **13**, 904–13.
172. Corneli, A.L., Piwoz, E.G., Bentley, M.E., Moses, A., Nkhoma, J.R., Tohill, B.C., Adair, L., Mtimuni, B., Ahmed, Y., Duerr, A., Kazembe, P., van der Horst, C., UNC Project BAN Study Team. (2007) Involving communities in the design of clinical trial protocols: the BAN Study in Lilongwe, Malawi. *Contemp Clin Trials* **28**, 59–67.
173. Maxwell, C., Aggleton, P., Warwick, I. (2008) Involving HIV-positive people in policy and service development: recent experiences in England. *AIDS Care* **20**, 72–9.
174. Greco, D.B., Simao, M. (2007) Brazilian policy of universal access to AIDS treatment: sustainability challenges and perspectives. *AIDS* **21**, S37–45.
175. Szekeres, G. (2008) The next 5 years of global HIV/AIDS policy: critical gaps and strategies for effective responses. *AIDS* **22**, S9–17.
176. McInnes, K., Landon, B.E., Malitz, F.E., Wilson, I.B., Marsden, P.V., Fleishman, J.A., Gustafson, D.H., Cleary, P.D. (2004) Differences in patient and clinic characteristics at CARE Act funded versus non-CARE Act funded HIV clinics. *AIDS Care* **16**, 851–7.
177. Ooms, G. (2008) Shifting paradigms: how the fight for "universal access to AIDS treatment and prevention" supports achieving "comprehensive primary health care for all". *Global Health* **4**, 11.

VI

Optimal Pregnancy/Infancy Outcomes

25 Folic Acid/Folic Acid-Containing Multivitamins and Primary Prevention of Birth Defects and Preterm Birth

Andrew E. Czeizel

Key Points

- Neural-tube defects are preventable by periconceptional folic acid or folic acid-containing multivitamin supplementation.
- The incidence of some other structural birth defects, i.e. congenital abnormalities of the heart, urinary tract, and limbs can also be reduced by folic acid-containing multivitamin use during the periconceptional period.
- All women of childbearing age who are capable of becoming pregnant should consume folic and/or a folic acid-containing multivitamin daily during the periconceptional period.
- The primary prevention of neural-tube defects and some other congenital abnormalities by periconceptional folic acid/multivitamin supplementation is much better than the so-called secondary prevention, i.e. the termination of pregnancy after the prenatal diagnosis of severe fetal defects.
- The rate of preterm birth can be reduced by the high dose of folic acid in the third trimester of pregnancy.
- Periconceptional care—beyond other benefits—is optimal for the introduction of periconceptional folic acid/multivitamin supplementation.
- Food (flour) fortification is the most practical means of supplementation of folic acid and other vitamins for all women.
- Proper preparation for conception is the earliest and most effective method for the prevention of birth defects.

Key Words: Congenital abnormalities; preterm birth; embryopathy; neural tube; periconceptional care; teratology; (structural birth defects)

25.1. INTRODUCTION

The deficiency or overdosage of certain nutrients may have a role in the origin of birth defects. First, in 1932, Hale *(1)* demonstrated that a vitamin A-free diet during early pregnancy of sows resulted in offspring without eyeballs, oral clefts, accessory ears, malposition of the kidney and defects of hind legs. Hale's conclusion was "the condition is illustrative of the marked effect that a deficiency may have in the disturbance of the internal factors that control the mechanism of development". Further development of experimental teratology became possible when small rodents were introduced for this purpose. Joseph Warkany (1902–1992), one of the founders of teratology, recognized the importance of purified diets and used these to test various vitamin deficiencies for their teratogenic effects. Warkany *(2,3)* found that maternal dietary deficiency can induce structural birth defects, i.e. *congenital abnormalities* (CAs). Marjorie M. Nelson *(4)* introduced the use of antimetabolites which made possible conversion of long-term nutritional experiments into short-term chemical testing. First, antimetabolites of folic acid were used and folate deficiency was proven to be highly teratogenic in pregnant rats *(5–7)*, producing multiple CAs, neural-tube defects, orofacial clefts, and other CAs. Later the teratogenic effect

A. Bendich, R.J. Deckelbaum (eds.), *Preventive Nutrition*, Nutrition and Health, DOI 10.1007/978-1-60327-542-2_25,

of 4-aminopteroylglutamic acid (aminopterin), a folic acid antagonist, was confirmed in man *(8–10)* as well. Recently the human teratogenic effect of some other folic acid antagonist drugs, e.g. trimethoprim has been shown *(11,12)*.

In 1964 Hibbard *(13)* reported a higher rate of CAs (3%) in the infants of folate-deficient mothers than in controls (1.6%) and later Hibbard and Smithells *(14)* showed a relationship between human embryopathy and a deficiency of folate metabolism. Smithells et al. *(15)* demonstrated the role of vitamin deficiencies in the origin of *neural-tube defects (NTD)*.

Anencephaly and spina bifida (aperta or cystica) are the major classes of NTD followed by occipital encephalocele and craniorachischisis due to the multi-site closure defects of the neural tube in humans *(16)*. However, it is necessary to differentiate the so-called isolated or non-syndromic and multiple-syndromic NTD *(17)*. In multiple NTD cases, NTD associates with other CAs and are caused by chromosomal aberrations (e.g. trisomy 18), gene mutations (e.g. Meckel–Gruber syndrome) or teratogenic factors (such as valproic acid or diabetes mellitus). The origin of isolated NTD (92% of all cases) can be explained by gene–environmental interaction. The most obvious proof for the genetic (polygenic) background of isolated NTD is the fact that recurrence in first-degree relatives is 10 times higher than their occurrence *(17)*. However, the occurrence of NTD is modified by environmental factors as well, e.g. there is a very wide range (0.5–12 per 1,000) of NTD incidences in different populations, rapid secular changes and seasonal variation of births with NTD were observed *(18)*. An important indicator for the great sensitivity of NTD due to environmental factors is its socioeconomic status dependence (a low risk in the highest class to an above-average risk in the lowest class) which was found in several populations) *(18)*. The estimated annual number of cases affected with NTD throughout the world is about 400,000 *(18)*. Our chapter reviews the prevention of *isolated NTD* cases.

25.2. INTERVENTION STUDIES FOR REDUCTION OF NEURAL-TUBE DEFECTS

R. W. Smithells (1924–2002) was the first who hypothesized that among triggering environmental factors in the origin of NTD, undernutrition could be the common and major denominator. His group therefore tested the effect of diet supplemented with folic acid and some other vitamins in the first intervention study *(19)*. In general, the closure of the neural tube occurs between postconceptional days 15 and 28 (i.e. between gestational days 29 and 42 calculated from the first day of the last menstrual period) in humans, thus the so-called critical (the most sensitive/vulnerable) period of anencephaly and spina bifida is between postconceptional days 21–26 and 23–28, respectively. The correction of undernutrition needs some time, therefore dietary supplementation should commence at least 28 days prior to conception and continue to the date of the second missed menstrual period, explaining the new term *periconceptional vitamin supplementation*. The major practical problem is that the start of the closure of neural tube overlaps with the first missed menstrual period, thus women, who do not plan and prepare their conception, are unaware of their pregnancy at the "critical period" of NTD. Periconceptional vitamin supplementation therefore needs the deliberate preparation of conception, i.e. family planning.

The final results of Smithells' study were published separately for the Yorkshire region of the UK *(20)* and Northern Ireland *(21)*, and they found *91 and 83% reduction in NTD recurrence* (Table 25.1), respectively. However, their results were not accepted by some experts due to possible selection bias because in general more educated people are willing to take part in intervention trials and these women have a lower risk for NTD. Two ethical committees refused to give permission for the original protocol of a randomized controlled trial of the Smithells' study, thus the control group was

Table 25.1
The Data from Intervention Studies for the Reduction of NTD Recurrences

Method	*Country*	*Supplement*	*With supplement No.*	*%*	*Without supplement No.*	*%*	*Risk reduction (%)*
Non-randomized	Yorkshire *(20)*		1/187	0.5	18/320	5.6	91
	Northern Ireland *(21)*	Multivitamin[a]	4/511	0.8	17/353	4.8	83
Randomized	Multicenter MRC *(22)*	Folic acid (4 mg)	2/298	0.7	13/300	4.3	84
		Folic acid + other vitamins	4/295	1.4	13/300	4.3	67
		Together	6/593	1.0	21/602[b]	4.3	71
		Other vitamins	8/302	2.6	13/300	3.5	40

[a]Included 0.36 mg folic acid.
[b]Including cases supplemented with "other" vitamins.

made up of women who had had one or more previous infants with NTD and were already pregnant when referred to the study centers or women who declined to take part in the intervention portion of the trial.

Thus in the early 1980s, the Medical Research Council (MRC) in the UK *(22)* decided to organize a multicenter (43% of participants came from Hungary) randomized double-blind controlled trial (RCT). There were four supplementation groups (Table 25.1); one of them was 4 mg folic acid because previously Laurence et al. *(23)* used this dose in their study. The MRC Vitamin Study found that a *pharmacological dose (4 mg) of folic acid supplementation alone can reduce NTD recurrence significantly by 71%* (1.0 vs. 4.3%; RR with 95% CI was 0.29, 0.12–0.71). There was a 40% reduction in the recurrent NTD after the use of other vitamins (2.6 vs. 4.3%) but this did not reach the level of significance.

Based on the results of the MRC Vitamin Study, the U. S. Centers for Disease Control (CDC) *(24)* in 1991 recommended daily supplementation of diet with 4 mg of folic acid under medical supervision in the periconceptional period for women at high risk (i.e. who had one or more previous offspring with NTD) for the reduction of NTD recurrence.

There were two major questions after the knowledge of the study design of the above two "recurrence" trials. The first question was "Does folic acid or folic acid-containing multivitamin supplementation also reduce the risk of *first* occurrence of NTD?" About 95% of women who deliver infants with NTD have no previous NTD pregnancies, thus the prevention of first occurrence of this CA group would be a real public health success. The second question was "Would a dose of folic acid less than 4 mg reduce the risk of first occurrence of NTD?" The pharmacological dose (e.g. 4 mg) of folic acid may have some adverse effects, thus it cannot be recommended for healthy pregnant women without previous NTD offspring and/or without medical supervision. The Hungarian RCT, which used 800 μg of folic acid attempted to provide data to answer the second question.

The *Hungarian RCT* was performed in the coordinating center of the Hungarian Periconceptional Service, Budapest. The goal was to test the preventive effect of a *physiological dose* (0.8 mg) of *folic acid* as one component of a multivitamin (Elevit pronatal®) used during the periconceptional period for the *first* occurrence of NTD. This micronutrient combination ("multivitamin") contained 12 vitamins including folic acid (0.8 mg), vitamin B12 (4.0 μg), B6 (2.6 mg), B2 (1.8 mg), C (100.0 mg), four minerals and three trace elements. The use of a true placebo was not allowed by the ethical committee thus a placebo-like trace element combination (including the three trace element components: copper, manganese and zinc of Elevit pronatal®) was used. The Hungarian RCT was launched on 1 February 1984 and the recruitment was closed on 30 April 1991. Pregnancy outcomes particularly informative offspring (live-born infants, stillborn fetuses and elective termination of pregnancies after prenatal diagnosis of malformed fetuses) were evaluated until the end of April 1992.

As can be seen in Table 25.2, no NTD case was found among informative offspring in the multivitamin group, while six informative offspring with NTD occurred in the placebo-like trace element, i.e. no-multivitamin group ($p = 0.01$). Thus, the Hungarian RCT demonstrated that a multivitamin containing 0.8 mg of folic acid prevented *about 90% of the first occurrences of NTD (25).*

Based on the Hungarian RCT and some observational studies, the CDC in September 1992 recommended that "all women of childbearing age who are capable of becoming pregnant should consume 0.4 mg of folic acid per day for the purpose of reducing their risk of having a pregnancy affected with spina bifida or other NTD" *(26)* and this recommendation was subsequently followed by several countries.

For ethical reasons, the Hungarian RCT could not be continued, thus a *cohort controlled trial (CCT) (27)* was designed to collect more data. Supplemented women were recruited via the coordinating center in Budapest, and all the 31 countryside centers of the Hungarian Periconceptional Service between 1 May 1993 and 30 April 1996. The examination of cases with CA was performed until April 30, 1999 due to the one year follow-up of infants. These centers supplied all participants with the same folic acid (0.8 mg) containing multivitamin (Elevit Pronatal®) during the

Table 25.2
First Occurrence of Neural Tube Defects (NTDs) in Informative Offspring of Women Taking Periconceptional Folic Acid-Containing Multivitamin (Micronutrient) Supplementation and no (Multivitamin or Folic Acid) Supplementation in the two Hungarian Intervention Studies

Intervention trials	*Supplement*		*No supplement*
Randomized controlled trial			
Number of offspring	2,471		2,391
Expected/observed number of NTD	6.9/0		6.7/6[a]
RR (with 95% CI)		0.07 (0.04–0.13)	
Cohort controlled trial			
Number of offspring	3,056		3,056
Expected/observed number of NTD	8.5/1[b]		8.5/9[c]
OR (with 95% CI)		0.11 (0.01–0.91)	
Pooled data			
Number of offspring	5,527		5,447
Expected/observed number of NTD	15.4/1		15.2/15
OR (with 95% CI)		0.08 (0.01–0.47)	

[a]Anencephaly 2, anencephaly + spina bifida (lumbal, thoracolumbal) 2, Spina bifida (thoracolumbal, lubosacral) 2
[b]Anencephaly 1
[c]Anencephaly 1, Spina bifida 8, (thoracolumbal 1, lumbal 4, lumbosacral 3)

periconceptional period. The supplemented cohort was cared for until the 14th week of gestation. Women in the unsupplemented cohort were recruited at the 14th week of pregnancy from the standard regional prenatal care clinics. They were matched to each pregnant woman of the supplemented cohort on the basis of age, socioeconomic status and place of residence and they were not supplemented with folic acid or folic acid-containing multivitamins before conception and in the first trimester of their pregnancy. CA was evaluated in the informative offspring of both cohorts, as in the Hungarian RCT, based on well-defined diagnostic criteria of CA during three time windows (during pregnancy, at birth and during the first postnatal year).

The protective effect of the folic acid-containing multivitamin for the *reduction of NTD* was confirmed because we found one offspring in the supplemented and nine offspring in the unsupplemented cohort (Table 25.2).

The efficacy of 0.4 mg, i.e. again a physiological dose of folic acid for the prevention of first occurrence of NTD was also examined in the *Chinese intervention study (28)*. This study was part of a public health campaign conducted from 1993 to 1995 and the intake of 0.4 mg folic acid daily reduced the risk of NTD about 79% in areas with high rates of NTD (6.5 per 1,000), while this reduction was 41% in areas with low rates (0.8 per 1,000).

During this same time many observational studies were also published regarding the prevention of NTD associated with higher than average intakes of folate/multivitamins containing folic acid in the international literature *(29)*. This review concentrates on intervention trials summarized above.

25.3. POSSIBLE MECHANISMS OF NTD PREVENTION BY FOLIC ACID/MULTIVITAMINS

The available epidemiological and biochemical findings suggest that the origin of NTD is not primarily the lack of sufficient folate in the diet, but arises from genetically determined changes in the uptake, metabolism, or both in maternal and, particularly, fetal cells *(30)*. Thus, a *gene–environmental interaction* between vitamin dependency (e.g. an inborn error of folate metabolism) and nutrition (such as dietary deficiency) may have a causal role in the origin of NTD. Supplementation with folic acid alone or folic acid-containing multivitamins may cause an increase in folate metabolite concentrations in tissue fluids and it may overcome the failure of the local folate metabolite supply due to genetic causes.

Obviously some genes and several environmental factors contribute to the origin of NTD. Here only one—the most established—hyperhomocysteinemia *(31,32)* is discussed, because the primary prevention of NTD by folic acid or folic acid-containing multivitamins is partly connected with the reduction of hyperhomocysteinemia.

The causes of hyperhomocysteinemia need a longer explanation. When meat, fish or plant proteins are digested, amino acids, among others, methionine, are released. There is a multi-step conversion of methionine to homocysteine. Homocysteine is a toxic metabolite; therefore humans neutralize it normally as soon as possible. On the one hand, homocysteine can be metabolized via the transsulfuration pathway to form cystathionine caused by the condensation with serine and catalyzed by cystathionine β-synthase. This enzyme requires pyridoxal 5'-phosphate, the biologically active form of vitamin B6 as a cofactor. The final product of this metabolic pathway is cysteine. On the other hand, remethylation of homocysteine to methionine is catalyzed by methionine synthase. This enzyme requires vitamin B12 as a cofactor and 5-methyl-tetrahydrofolate as the methyl donor. The latter may help to explain the importance of *folate–folic acid deficiency* in the origin of NTD.

This vitamin was discovered by Lucy Wills *(33,34)* in 1931, and she stressed the similarity of this "curative agent" with vitamin B12, as a "twin" vitamin. The generic term *folate* describes the many different naturally occurring polyglutamate forms of the vitamin, whereas *folic acid* is the synthesized monoglutamate form, chemically designated as pteroylglutamic acid. The different names of two forms of the same vitamin is disturbing and these are confused frequently, thus their joint name as vitamin B11 has been used in some countries, e.g. the Netherlands, while their name is vitamin B9 in France. Folate is required for cell division and cell maintenance, because it acts as a co-enzyme in the transfer and processing of one-carbon units and plays an important role in nucleotide (thymidine) synthesis which is essential for the *de novo* construction or repair of DNA. In addition folate is a key factor in the "site-specific" methylation of the cytosine base in DNA, which regulates gene expression. Here only the third major function of folate, the remethylation of plasma homocysteine to methionine will be summarized.

Humans cannot synthesize folate, thus it is an essential nutrient. The major dietary sources of folates are fresh and frozen green leafy vegetables, citrus fruits and juices, liver, wheat bread and legumes. Food folates are converted to monoglutamates by conjugase enzymes in the upper part of the small intestine. Monoglutamate folic acid, however, can be absorbed directly. After the active and passive absorption of monoglutamates, these are converted to dihydrofolate and then to tetrahydrofolate (THF) by reductase enzymes. THF is the parent compound of all biological active folates. The most important cause of hyperhomocysteinmia and/or lack of methionine is the polymorphism of *methylene-THF-reductase (MTHFR) gene (35)*. A 677C→T mutation has been identified in the *MTHFR*-gene and it results in a thermolabile variant (Ala225Val) of *MTHFR enzyme* with 30–50% activity of the CT (heterozygote) and with very low (less than 20%) activity of TT (homozygote) forms *(36)*. These enzyme variants cannot effectively catalyze the pathway of 5,10-methylene-THF to 5-methyl-THF, i.e. the methyl donor for methionine-synthase. The MTHFR-gene polymorphism is common in different populations *(37)*.

There are several studies that provide evidence for the confirmation of hyperhomocysteinemia-related NTD.

a) The thermolabile variants of MTHFR enzyme, thus the polymorphism of *MTHFR-gene* is an obvious risk factor for NTD *(38,39)*. The Hungarian population has a moderate total (birth + fetal) prevalence of NTD (3/1,000) *(40)*, while the frequency of TT and CT genotypes of MTHFR-gene occurs in 11.1 and 45.2% of people, respectively *(37)*. These very high frequencies need some explanation. The highly significant excess of CT heterozygosity in male first-degree relatives of patients with NTD may demonstrate CT heterozygote advantage in the context of their genetic background *(41)*. However, other common mutations of *MTHFR*-gene were identified (e.g. A1298C) and these gene polymorphisms may also have a role in folate metabolism and have an interaction in the origin of NTD *(42)*. The point is that there is an inverse relation between plasma homocysteine and folate levels in the humans. However, the plasma/serum folate level reflects the recent intake of folate/folic acid, and the concentration of folates in red blood cells represents the body storage over the last 7 weeks.
b) Hyperhomocysteinemia and/or lack of methionine can induce NTD in animal experiments *(43,44)*.
c) Mothers who gave birth to a child with NTD have higher blood *(45)* and amniotic fluid *(46)* levels of homocysteine.
d) Low maternal folate status, particularly in the first trimester was a strong risk factor for NTD ($r = 0.99$) in the MRC Vitamin Study *(29)*.
e) Folic acid-containing multivitamins and folic acid alone are able to reduce the hyperhomocysteinemia, and consequently the hyperhomocysteinemia-related NTD as it was shown earlier in this chapter.

However, the polygenic system in the origin of NTD includes other folate-independent factors as well. For example inositol deficiency leads to NTD in mice and prevents NTD in curly tail mice, in addition, the mouse mothers of NTD offspring had lower inositol concentrations than controls *(47)*. The latter may explain the lack of NTD prevention by folic acid or folic acid-multivitamins in some cases.

25.4. HISTORICAL ASPECT OF NTD

As a medical doctor I am proud to declare that the history of NTD shows that we can modify our destiny by the help of science. Before the 1960s nearly all victims of NTD had fatal outcomes. In the 1960s physicians introduced very early complex surgical and medical management tools and lives were saved in the majority of spina bifida cases (of course, anencephaly is lethal). In the 1970s, the selective criteria of surgical intervention were introduced to reduce the production of multiple handicapped children. In the 1980s prenatal screening based on maternal serum alpha-fetoprotein (MS-AFP) and/or ultrasonography was introduced. In the low-risk populations the detection rate (i.e. sensitivity) of MS-AFP testing for open spine bifida varies from 72 to 91% and the specificity from 96.2 to 98.7% *(48)*. This prenatal screening resulted in a significant drop in the birth of NTD fetuses (previously it was called "secondary prevention", i.e. the prevention of the *birth* of babies with NTD). However, it increased the number of pregnancy terminations with consequent psychological and somatic complications.

Finally, since the 1990s, we have had a chance to reduce the maldevelopment of the neural tube due to the intentional modification (supplementation) of the diet in the periconceptional period of life. Now the periconceptional folic acid or folic acid-containing multivitamin supplementation as a primary preventive method of NTD offers an appropriate alternative with the same efficacy of the so-called secondary prevention.

Obviously primary prevention is much better than the termination of pregnancy after the prenatal diagnosis of fetal defect from a moral, medical (comparing the risk of pregnancy termination and folic acid/multivitamin supplementations) and financial (periconceptional folic acid/multivitamin supplementation is much less expensive than the combined method of prenatal diagnosis, mainly followed by pregnancy termination) aspect.

25.5. OTHER EXPERIENCES OF THE HUNGARIAN RANDOMIZED CONTROLLED TRIAL

The great majority of the women in the Hungarian RCT were healthy and not malnourished *(49)*, therefore this RCT was appropriate to study several *other effects* of periconceptional folic acid-containing multivitamin supplementation.

During the preconceptional multivitamin supplementation the female menstrual cycle became more regular, i.e. the variance was lower *(50)*. Thus, multivitamin supplementation may have a beneficial effect for women with irregular menstrual cycles.

There was no difference in the sexual activity (measured by the rate of weekly sexual intercourse) of couples between the multivitamin and the no-multivitamin groups in the preconceptional period *(51)*. However, only women were supplemented and sexual activity is often determined by males.

A 7% higher rate of conceptions occurred in women within 1 year who were treated with the multivitamin preconceptionally compared with those who were not supplemented. The time taken to become pregnant was slightly but significantly shorter in the multivitamin group (3.8 ± 3.2 vs. 4.0 ± 3.3 months in the no-multivitamin group) *(52)*.

A significantly lower rate of severe (treated) morning sickness, i.e. nausea and vomiting in early pregnancy occurred after periconceptional multivitamin supplementation (3.0 vs. 6.6% in the no-multivitamin group) *(53)*.

There was no difference in maternal weight gain between the multivitamin and no-multivitamin groups before and during early pregnancy *(54)*. In Hungary this possible side effect caused the major concern among females.

All other possible side effects were monitored continuously. The number of female participants in the Coordinating Center of the Hungarian Periconceptional Service was 14,540 until January 1999, and patients with pernicious anemia have not been recorded among these reproductive aged women. Of 14,540 female participants, 66 (0.45%) were epileptic and 60 wanted to use multivitamin supplementation. There was no case with multivitamin-related side effects of epilepsy during the periconceptional period *(55)*. However, a 22-year-old epileptic woman was treated continuously with carbamazepine, who stopped with our multivitamin intake in the 12th gestational week but used another folic acid (1 mg) containing multivitamin from the 20th week, and she had repeated status epilepticus after this new supplementation parallel with the manifestation of symptoms of systemic lupus erythematosus. Her pregnancy ended in the delivery of stillborn fetus on the 39th gestational week *(55)*. Autoimmune diseases in epileptic pregnant women could damage the blood–brain barrier and the pharmacological dose (≥1 mg) of folic acid may trigger a cluster of seizures. Of 14,540 female participants, 4 (0.03%) had severe allergic exanthema, and of these 4, three discontinued the use of multivitamin (all of them had a history of drug-induced allergic diseases). Among all other possible side effects, constipation (1.8 vs. 0.8%) and diarrhea (1.4 vs. 0.4%) were reported somewhat more often after multivitamin supplementation in the preconceptional period *(54)*.

The rate of multiple births, namely twins was about 40% higher after periconceptional multivitamin supplementation *(56)*. The higher rate of twin conceptions could not be explained by maternal factors (age, parity) or by a higher rate of infertility drug use *(57)*, and at that time in vitro fertilization was not used in Hungary. This increase in twins was confirmed later in the USA *(58)* and Sweden *(59)*, but not in China *(60)*. However, the prevalence of twins is lowest in the Asian races (including Chinese population) and Chinese people have an extremely high occurrence of TT genotype of MTHFR-gene (55%) *(61)*, and this genotype associates with a very low dizygotic twin rate *(62)*. These conflicting data stimulated us to check the large database from the Hungarian Case–Control Surveillance of Congenital Abnormalities (HCCSCA), 1980–1996, including 38,151 controls without CA *(63)*. There was a somewhat higher prevalence of multiple pregnancies both after periconceptional high doses (mainly 6 mg) of folic acid and folic acid (0.1–1.0 mg) containing multivitamin supplementation *(64)*. A systematic review of the recent literature, July 1994 to July 2006 resulted in the following conclusion: "Overall. . .there is possible evidence for a relationship between periconceptional folic acid intake and increased twinning" *(65)*. However, the minor increase in the rate of twins (2.8% instead of the usual 2.0% in the Hungarian newborn population) is not a real medical problem because at present the early diagnosis of multiple pregnancies due to ultrasound scanning is followed by a special prenatal care which can result in a decrease of the previously higher perinatal mortality.

There was no significant difference in pregnancy outcomes of singletons including four types of fetal deaths: (i) chemical pregnancies (positive pregnancy test without clinical symptoms of pregnancy later), (ii) ectopic pregnancies, (iii) miscarriages (including the so-called missed abortions or blighted ova) and (iv) late fetal deaths (stillbirths) and livebirths between the multivitamin and no-multivitamin groups *(66)*. The rate of all kinds of fetal deaths together was somewhat higher in the multivitamin group than in no-multivitamin group, in addition, the sex ratio showed a slight girl excess in the multivitamin group while the well-known 51% boy predominance was seen in the no-multivitamin group. Thus, a small change in the pattern of prenatal selection cannot be excluded. This finding was confirmed in the USA *(67)* but not in China *(68)*. In my opinion the slightly higher rate of fetal death cannot be explained by terathanasia *(69)* but—if it is not caused by chance—by the higher proportion of multiple conceptions in the multivitamin group *(70)* because twin conceptions associate with some excess of fetal death. In addition, low maternal plasma folate levels were associated with an increased risk of early spontaneous abortion *(71)* and recurrent pregnancy loss *(72)* in other studies.

There was no difference in gestational age at birth and birthweight in the original cohort, in addition the rate of preterm birth and low birthweight newborns of live-born babies of mothers with or without supplementation of the multivitamin in the periconceptional period *(66)*. The findings from subsequent studies discussed below suggest that multivitamins and high dose of felic acid after the first trimester may reduce the risk of preterm birth and reduced birthweight.

Postnatal somatic (body weight, body length, head circumference) and mental (measured by three tests) development until 1 *(73)* and 6 years *(74)* of age did not show any significant difference between the multivitamin and no-multivitamin groups. Thus, the previously found higher rate of worrying, fussiness and fearfulness in girls born to mothers who previously had NTD infants and were supplemented in the next pregnancy by a multivitamin during the periconceptional period *(75)* was not confirmed. However, it is necessary to stress that our pregnant women used a folic acid-containing multivitamin until the 12th gestational week, and a major part of brain development is in the second, and mainly in the third trimester of pregnancy. The measurement of mental development of children born to mothers with folic acid or multivitamin supplementation until the end of pregnancy would be worth studying.

Last but not least, there was a significant reduction in *the total rate of informative offspring with CA* after periconceptional multivitamin supplementation. The rate of major CAs was 20.6 per 1,000 in the multivitamin and 40.6 per 1,000 in the no-multivitamin group (RR: 0.53, 95% CI: 0.35–0.70) *(76)*. After the exclusion of 6 NTD cases, the difference in the rate of major CA between the two study-groups remained very highly significant ($p < 0.0001$). In conclusion, periconceptional multivitamin supplementation reduced not only the occurrence of NTD but also the rate of other major CAs *(77)*.

The final data set of the Hungarian RCT indicated a significant reduction beyond NTD in two other CA groups: *CAs of the urinary tract* and *cardiovascular CAs (78)* (Table 25. 3). The difference was most obvious in the obstructive CAs of the urinary tract and conotruncal CAs (3 vs. 10, RR: 0.29, 95% CI: 0.09–0.97) including ventricular sepal defect of cardiovascular CAs. There was also some reduction in the prevalence at birth of congenital limb deficiencies, congenital pyloric stenosis, and Down's syndrome, but these did not reach a level of significance between the multivitamin and no-multivitamin groups (Table 25.3).

There was no difference in the rate of unidentified multiple CAs between multivitamin and no-multivitamin supplemented groups (Table 25.3).

25.6. THE POSSIBLE PREVENTION OF OTHER CONGENITAL ABNORMALITIES BEYOND NTD WITH FOLIC ACID-CONTAINING MULTIVITAMINS

The major objective the Hungarian CCT was to confirm or reject the preventive effect of periconceptional folic acid-containing multivitamin supplementation for the reduction of urinary tract and cardiovascular CAs, limb deficiencies, congenital pyloric stenosis and Down's syndrome. In addition, we wanted to collect more data regarding orofacial cleft.

However, the comparison of maternal morbidity and previous pregnancy outcomes between the two cohorts showed differences in the Hungarian CCT. The cohort of supplemented pregnant women was recruited from the Hungarian Periconceptional Service and they had a higher rate of morbidity (e.g. diabetes mellitus and epilepsy) and previous unsuccessful pregnancy outcomes (fetal death and CA, including NTD) compared to the cohort of unsupplemented pregnant women recruited at the prenatal care clinics. This is possibly due to the good reputation of the Hungarian Periconceptional Service to prepare for optimal conception and to provide care during early pregnancy which might attract women

Table 25.3

The Efficacy of Periconceptional Folic Acid-Containing Multivitamin (Micronutrient) Supplementation in the Primary Prevention of Some Major CA Groups

	Hungarian intervention trials					
	RCT		*CCT*		*Pooled data*	
CA groups	*No supplement ($n = 2,391$)*	*Supplement ($n = 2,471$)*	*No supplement ($n = 3,056$)*	*Supplement ($n = 3,056$)*	*No supplement ($n = 5,447$)*	*Supplement ($n = 5,527$)*
Urinary tract's CAs						
Renal a/dysgenesis	3	0	0	2	3	2
Cystic kidney	1	1	0	2	1	3
Obstructive CAs						
Pelvi cureteric junction	4	0	13	2	17	2
Other locations	1	1	6	8	7	9
Total	9	2	19	14	28	16
OR (95% CI)	0.21 (0.05–0.95)		0.71 (0.33–1.50)		0.56 (0.30–1.04)	
Cardiovascular CAs						
Conotruncal						
Ventricular septal defect	8	2	19	5	27	7
Others	2	1	1	3	3	4
Subtotal	10	3	20	8	30	11
Others	10	7	30	23	40	30
Total	20	10	50	31	70	41
OR (95% CI)	0.42 (0.19–0.98)		0.60 (0.38–0.96)		0.57 (0.39–0.85)	
Congenital limb deficiencies						
Terminal transverse	2	1	3	1	5	2
Others	3	0	0	0	3	0
Total	5	1	3	1	8	2
OR (95% CI)	0.19 (0.03–1.18)		0.33 (0.01–3.71)		0.25 (0.05–1.16)	

Congenital pyloric stenosis	8	2	2	0	10	2
OR (95% CI)	0.24 (0.05–1.14)		0.00 (0.00–26.8)		0.20 (0.04–0.90)	
Anal/rectal atresia/stenosis	1	0	4	1	5	1
OR (95% CI)	—		0.31 (0.02–2.52)		0.20 (0.02–1.69)	
Orofacial clefts						
Cleft lip ±palate	3	4	2	3	5	7
Cleft palate	2	0	1	1	3	1
Total	5	4	3	4	8	8
OR (95% CI)	0.77 (0.22–2.69)		1.63 (0.31–2.88)		0.99 (0.37–2.63)	
Down's syndrome	5	2	8	8	13	10
OR (95% CI)	0.39 (0.07–1.99)		1.00 (0.33–1.73)		0.76 (0.33–1.73)	
Unidentified multiple CAs	5	6	15	12	20	18
OR (95% CI)	1.16 (0.35–3.81)		0.79 (0.40–1.48)		0.89 (0.47–1.68)	

with previous pregnancy problems. Thus at the comparison of two cohorts in the Hungarian CCT, it is necessary to know that the major part of supplemented women had a high risk for adverse pregnancy outcomes while most unsupplemented women can be considered at low risk.

All CA groups were differentiated into isolated (non-syndromic) and syndromic (multiple) categories in informative offspring. First the data of isolated CA category will be presented.

There are two frequent types of *orofacial clefts* (OFC): cleft lip ± palate (about 1 per 1,000) and cleft palate (about 0.5 per 1,000). Both types of OFC are caused by polygene–environmental interaction. Tolarova reported a protective effect of a multivitamin and folic acid (10 mg) during the periconceptional period for the recurrence of cleft lips in 1982 *(79)*. However, the reduction of cleft lip ± palate (and cleft palate) was not confirmed after the use of the multivitamin containing a low dose (0.8 mg) of folic acid either in the Hungarian RCT or in the CCT (Table 25.3). However, a significant reduction was seen in the prevalence at birth of OFC in the data set of the HCCSCA, 1990–1996 *(63)* after the use of a high dose (in general 6 mg) of folic acid alone *(80)*, because Hungarian obstetricians recommended in general 6 mg of folic acid for all women in the early phase of pregnancies after the first visit in the prenatal care clinics. Thus, a dose-dependent preventive effect of folic acid for OFC cannot be excluded. The observational studies resulted in controversial results as the findings of Shaw et al. *(81)* and of Hayes et al. *(82)*. Other observational studies *(83)* showed that the efficacy of this primary preventive method depends—beyond the dose—on genetic background of the population, socioeconomic status and lifestyle, particularly diet of mothers *(84)*.

Cardiovascular CAs include a heterogeneous manifestation and origin of different CA groups with different critical periods, such as ventricular septal defect (within this CA there are four different forms), atrial septal defect type II, patent ductus arteriosus, conotruncal defects (common truncus, transposition of great vessels, tetralogy of Fallot, etc.), CAs of left ventricular outflow tract (hypoplastic left heart syndrome, coarctation of aorta, congenital stenosis or insufficiency of aortic valve, other CAs of aorta), right-sided obstructive CAs (CAs of pulmonary valve and artery, tricuspid atresia/stenosis, Ebstein's anomaly) and others. The Hungarian RCT showed a significant reduction in the prevalence at birth of cardiovascular CAs. The diagnostic criteria and method of "blind" cardiologic examination were similar in the two study-groups of RCT and CCT. About 3,056 informative offspring pairs were evaluated in the cohort of supplemented and unsupplemented matched mothers in Hungarian CCT. The occurrence of *cardiovascular CAs* (31 vs. 50) was significantly reduced, explained mainly by the lower occurrence of *ventricular septal defect* (5 vs. 19) (OR: 0.26; 95% CI: 0.09–0.72) in the supplemented cohort (Table 25.3). The protective effect of folic acid-containing multivitamins for cardiovascular CAs, mainly conotruncal defects including ventricular septal defects were also found in two U. S. studies *(85,86)*. Their and our results suggest that periconceptional use of multivitamins was associated with about a 40% reduction in risk for cardiovascular CAs. Other studies showed an association of higher plasma homocysteine level and MTHFR-gene polymorphism in mothers with a higher risk for cardiovascular CA in their offspring *(87,88)*. Three other survey studies in the USA *(89–91)* did not find the association of periconceptional multivitamin use and lowered cardiovascular CA risk. The protective effect of early postconceptional supplementation of a pharmacological dose (6 mg) of folic acid for cardiovascular CAs was also seen in the data set of the HCCSCA *(92)*. Of interest, cardiovascular CAs were induced by pteroylglutamic acid deficiency during gestation in rat fetuses *(93,94)*.

The *CAs of the urinary tract* include again very heterogeneous manifestations and origins of CA from renal a/dysgenesis and different types of cystic kidney diseases through the different localization of obstructive CAs as well as many other CAs. The rate of the urinary tract CAs showed a significant reduction in the supplemented group of the Hungarian RCT but it was not confirmed in the Hungarian CCT (Table 25.3), because there was no significant difference in children with urinary tract's CAs between supplemented and unsupplemented cohort (14 vs. 19 cases). However, the *stenosis of pelvicureteric junction* (2 vs. 13) showed a significant difference (OR: 0.19; 95% CI: 0.04–0.86) within the group of obstructive CAs (10 vs. 19). Thus, the occurrence of obstructive CAs of the urinary tract, more exactly the stenosis of pelvicureteric junction, was reduced in newborn infants born to mothers with multivitamin supplementation in both the RCT and the CCT (Table 25.3). In the 1950s Monie et al. *(95,96)* were able to produce CAs of the urinary tract in rat embryos by folic acid deficiency. In human studies Li et al. *(97)* also found a significant association between reduction in the rate of urinary tract's CAs and multivitamin use in the first trimester of pregnancy.

The group of *limb deficiencies* comprises a very heterogeneous group of CAs with several origins such as amputation (terminal transverse and amniogenic), longitudinal (radial-tibial, ulnar-fibula, split hand and foot, i.e. axial) and intercalary (phocomelia and proximal, in general femoral head deficiency) types. There was again only a decreasing trend in the supplemented cohort (1 vs. 3) of the Hungarian CCT, but it is worth mentioning that all cases had the unimelic terminal transverse type (Table 25.3). One U. S. study showed a significant reduction of congenital limb deficiencies after multivitamin supplementation *(98)*, while two others found a reduction (RR: 0.50 and 0.64) but due to the too wide confidence intervals, the differences were not significant *(89,90)*. The teratogenic effect of folic acid deficiency due to folic acid antagonists was associated with, among other CAs, limb deficiencies in human embryos *(8–10)*.

Congenital pyloric stenosis, due to the hypertrophy of the pyloric muscular sphincter at the end of the gastric canal, may be diagnosed some weeks after birth. This CA was diagnosed in two infants of the unsupplemented cohort while congenital pyloric stenosis did not occur in the supplemented cohort in the Hungarian CCT (Table 25.3). The combined rate of pyloric stenosis in the Hungarian RCT and CCT showed a significant reduction after periconceptional multivitamin supplementation. This finding was not confirmed in a U. S. study *(90)*.

In the China–U. S. Collaborative Project for Neural Tube Defect Prevention *(28)*, a somewhat lower occurrence of *rectal atresia/stenosis* was found after periconceptional folic acid supplementation *(99)*. The Hungarian RCT and CCT showed a similar trend (1 vs. 5) (Table 25.3).

Botto et al. *(100)* found a lower rate of omphalocele in newborn infants born to mothers with periconceptional multivitamin supplementation (RR: 0.4 with 95% CI: 0.2–1.0). There were 1:1 and 3:1 infants with omphalocele in the multivitamin and no-multivitamin groups of the Hungarian RCT and CCT, respectively.

There was no difference in the rate of *Down's syndrome* between the supplemented and unsupplemented cohorts in the Hungarian CCT (Table 25.3). Recent publications showed an association between polymorphism in genes involved in folate metabolism or NTD and maternal risk for Down's syndrome *(101–104)*. Thus we wanted to check this possible association in the HCCSCA *(63)*, and the high dose (mainly 6 mg) of folic acid with iron indicated some preventive effect for Down's syndrome *(105)*.

We did not find any difference in the rate of cases with *multiple CAs* in the RCT and CCT after multivitamin supplementation during the periconceptional period *(106)* (Table 25.3). However, Shaw et al. *(107)*and Yuskin et al. *(108)* reported a higher occurrence of multiple CAs after periconceptional intake of multivitamin supplements in their birth defects registries. Thus, we decided to evaluate the data set of the HCCSCA *(63)* from this aspect, but our data showed that periconceptional folic acid/multivitamin supplementation did not reduce, but also did not increase the occurrence of cases with multiple CAs *(109)*.

Thus, the results of the Hungarian CCT were more-or-less consistent with the findings of the previous RCT showing that periconceptional multivitamin supplementation provides a protective effect, beyond NTD, for some cardiovascular CAs, mainly ventricular septal defects and obstructive CAs of the urinary tract, particularly stenosis of pelvicureteric junction. These preventive effects were demonstrated despite the fact that the supplemented pregnant women were a cohort at high risk due to higher maternal morbidity and more unsuccessful previous pregnancy events.

The Hungarian findings regarding the preventive effect of periconceptional multivitamin supplementation for some other CAs beyond NTD was confirmed in several studies *(110)*. It is worth stressing that the total reduction in CAs without NTD cases was 19.93/1,000 in the Hungarian RCT, which exceeded the total prevalence of NTD (2.78/1,000) by 7.2-fold in Hungary *(40)*. This reduction resulted mainly from the partial prevention of cardiovascular CAs and the higher prevalence at birth of the cardiovascular CA-group that occur 3.5 times more often than the total prevalence of NTD. In addition the so-called secondary prevention of cardiovascular CAs is more limited than that of NTD. Thus, the prevention of other CAs by this new primary preventive method has a similar significant public health importance as the prevention of NTD. This new phase of malnutrition teratology therefore provided scientific evidence of the primary prevention of some other CAs beyond NTD by periconceptional multivitamin or folic acid supplementation.

However, it is strange that while the data concerning the prevention of NTD were accepted with enthusiasm in the international scientific community in the early 1990s and prompted new consensus *(111)* and recommendations for their practical implementation *(24,26)*, the new data regarding the prevention of other CAs have been received with reservation and without any further recommendations. On the one hand, it is understandable because different CAs have different origins, thus it is difficult to believe that such a simple vitamin supplementation can reduce these different CAs. It is necessary therefore to clearly identify the mechanism of periconceptional folic acid/folic acid-containing multivitamin use in the reduction of specific CAs. On the other hand, the possible prevention of CAs beyond NTD was confirmed in several studies and recent publications confirmed the role of folate deficiency due to antifolate drugs both in the origin of NTD *(112)* and of cardiovascular and urinary tract CAs *(11,12,113)*.

The data of our RCT and CCT, in addition the data set of the HCCSCA *(63)*, are not appropriate to study the long-term effects of folic acid and multivitamin supplementation. However, Bunin et al. *(114)* found a relation between maternal diet and subsequent reduction of primitive neuroectodermal brain tumors in young children, while the study of Thompson et al. *(115)* suggested that folic acid supplementation in pregnancy reduced the risk for common acute lymphoblastic leukemia in children (*see* Chapter 5).

On the other hand the data set of the HCCSCA *(63)* showed that the folic acid/multivitamin supplementation can reduce and/or protect from the teratogenic effect of high fever, e.g. caused by influenza *(116)*, thus we were able to confirm the previous findings of Botto et al. *(117)*. This protective effect was found in some other high fever-related maternal diseases during pregnancy as well *(118)*.

At present, 20–25% of infant mortality in industrialized countries is caused by CAs, and CAs are among the leading causes of death with a high number of life years lost and impaired life *(119)*. Another important feature of CAs is that they represent a defect condition; therefore it is difficult to achieve a complete recovery. Thus, prevention is considered the only optimal solution in the medical care of cases affected with CA.

In conclusion, several studies indicated the efficacy of folic acid-containing multivitamin supplementation for the *prevention of other CAs* beyond NTD as well and it is extremely important from the public health aspect because NTD represent only 3–12% of all major CAs. Folic acid or folic acid-containing multivitamin supplementation offers a breakthrough in the primary prevention of NTD and some other CAs because it can reduce about one-third of CAs *(120)* and provide a better alternative than secondary prevention, i.e. the termination of pregnancy after the diagnosis of severe fetal defect.

25.7. PRETERM BIRTH

The rate of preterm birth is very high in Hungary *(121)* and in the USA *(122)*, thus a major factor in the origin of infant mortality and different handicaps. Recently, the role of folate deficiency and/or hyperhomocysteinemia has been shown in the origin of preterm birth and low birthweight, i.e. intrauterine fetal growth retardation, thus I summarize these findings and show the results of our recent Hungarian study.

The deficiency of folate due to inadequate diet interferes with the growth of the conceptus *(123)*. Three intervention trials have shown some increase in birthweight after the use of prenatal folic acid-containing multivitamins whilst three others have failed to show an effect *(124)*. Some other studies showed some association between the low maternal folate intake or blood level and intrauterine growth retardation *(125–129)*. In addition, the birthweights of newborns exposed to high levels of maternal homocysteine were approximately 200 g below the birthweight of infants born to mothers with lower homocysteine levels *(130)* (*see* Chapter 26).

Mothers who delivered prematurely generally have lower dietary folate intake, for example, low folate intake (<240 μg/day) was associated with a greater than threefold increase in the risk of preterm birth and low birthweight newborns *(131)*. Lower folate intake (<500 μg/day) and low levels of serum or red blood cell folate at 24–29 gestational weeks was associated with an approximately twofold increased risk of preterm delivery in North Carolina *(132)*. In addition, in Norway, women with high homocysteine levels were more likely to have had a past reproductive history including preterm delivery and low birthweight newborns *(133)*. In China, the high preconceptional level of maternal homocysteine was associated with a fourfold increased risk of preterm delivery *(134)*.

In agreement with the above findings, folic acid-containing multivitamin supplementation in the periconceptional period *(135)* or during the first and second trimesters of pregnancy *(136)* resulted in some reduction of preterm births. A small reduction was found in the rate of preterm births and low birthweight newborns among California infants after compulsory food fortification with folic acid *(137)*. The association of short interpregnancy intervals with unfavorable pregnancy outcomes, among others higher rate of preterm birth, was also explained by the role of folate depletion during pregnancy and postpartum period *(138)*. Recently, the unpublished results of a U. S. study showed that at least 1 year folic acid supplementation was linked to a 70% decrease in very early preterm deliveries (20–28 gestational weeks) and up to a 50% reduction in early preterm deliveries of 28–32 gestational weeks *(139)*.

Thus, we decided to check whether folic acid supplementation is an appropriate method for the reduction of preterm rate or not in the data set of the HCCSCA *(63)*. About 38,151 control newborns without CA were evaluated. (Cases with CA were excluded from this analysis, because CA may have a more drastic effect for gestational age and birthweight than vitamin supplements.) The main results of this study are presented in Table 25.4

The mean birthweight exceeded the Hungarian population figure by 30, 75 and 70 grams in groups of high dose (3–9 mg, mainly 6 mg) of folic acid alone, folic acid (0.1–1.0, mainly 0.8 mg)-containing multivitamins ("multivitamin"), and multivitamin and folic acid supplementation during pregnancy, respectively. The intensive fetal growth occurs during the second, and particularly in the third trimester of pregnancy, thus mean birthweight was evaluated separately after the third trimester vitamin supplementations as well. There was no obvious difference in mean birthweight after the third trimester supplementation in the groups of folic acid alone (39 g), multivitamin (72 g) and multivitamin + folic acid (79 g).

The gestational age at delivery was 0.3, 0.2 and 0.5 week longer after folic acid alone, multivitamin and multivitamin + folic acid supplements during pregnancy than in the reference sample. Pregnant women with the third trimester supplementation showed a longer mean gestational age (39.7 ± 1.8 week) only after folic acid alone supplementation. The longer mean gestational age at

Table 25.4
Mean Birth Weight and Gestational Age at Delivery, in Addition Rate of Low Birthweight and Preterm Birth of Newborn Infants Born to Mothers with Three Kinds of Vitamin Supplementations During Pregnancy and in the Third Trimester (Including Supplementation only in III, in Addition During II–III and I–III trimesters) and of Newborn Infants Born to Mothers Without Folic Acid/Multivitamin Supplementation as Reference

Variables	*Folic acid alone*	*Multivitamins*	*Multivitamin + folic acid*	*Without FA/MV*
No. of pregnant women	19,334	694	1,441	16,308
Birth weight (g)				
Mean ± S. D.	3,287 ± 508	3,332 ± 536	3,327 ± 520	3,257 ± 513
Adjusted[a]				
P	= **0.0002**	= **0.0009**	0.11	Reference
In third trimester				
No. of pregnant women	12,485	589	1,130	–
Mean ± S. D.	3,296 ± 503	3,329 ± 522	3,336 ± 523	–
Low birthweight				
No.; %	1,058; 5.5	37; 5.3	74; 5.1	979; 6,0
Adjusted[a] OR 95% CI	0.95 0.87–1.04	0.88 0.63–1.23	0.90 0.71–1.15	Reference
In third trimester	12,485	589	1,130	–
No.; %	659; 5.3	28; 4.8	60; 5.3	–
Crude OR 95% CI	**0.87 0.79-0.97**	0.78 0.53-1.15	0.88 0.67-1.15	Reference
Gestational age (wk)				
Mean ± S. D.	39.5 ± 2.0	39.4 ± 1.8	39.7 ± 1.8	39.2 ± 2.1
Adjusted[b]				
P	***<0.0001***	0.08	***<0.0001***	
In the third trimester				
Mean ± S. D.	39.7 ± 1,8	39.3 ± 1.8	39.7 ± 1.8	–
Preterm birth				
No.; %	1,529; 7.9	44; 6.3	73; 5.1	1,825; 11.2
Adjusted[b] OR 95% CI	***0.70 0.65–0.75***	***0.62 0.45–0.85***	***0.45 0.35–0.57***	Reference
Adjusted[c] OR 95% CI	***0.69 0.64–0.74***	***0.54 0.40–0.74***	***0.42 0.33–0.54***	Reference
Adjusted OR 95% CI°	***0.73 0,65–0.81***	*0.82 0.50–1.36*	***0.39 0.23–0.66***	
In third trimester	12,485	589	1,130	–
No.; %	668; 5.3	41; 7.0	57; 5.0	–
Crude OR 95% CI	***0.45 0.41–0.49***	***0.59 0.43–0.82***	***0.42 0.32–0.52***	Reference
Crude OR 95% CI°	**0.44 0.38–0.51**	0.80 0.48–1.34	**0.28 0.14–0.54**	

[a]adjusted for gestational age, maternal age and socioeconomic status, birth order.
[b]adjusted for maternal age and socioeconomic status, birth order.
[c]adjusted for maternal age and socioeconomic status, birth order, plus threatened preterm birth/cervical incompetence, vulvovaginitis/bacterial vaginosis and placental disorders.
°Singletons born to primiparous mothers with only medically recorded vitamin supplements.
Bold numbers show significant associations.

delivery explains a certain part of larger mean birthweight in the supplemented groups, particularly in the group of multivitamin + folic acid.

The rate of low birthweight newborns was lower in all vitamin supplemented groups than in our reference value, but this difference was significant only after folic acid alone supplementation in the third trimester.

The rate of preterm births was significantly lower in all vitamin supplemented groups compared with our reference value. A further significant reduction was found in the rate of preterm births only after folic acid alone supplementation in the third trimester.

Thus, the fetal growth promoting effect of different vitamin supplements was limited (30–75 g), and it can be explained mainly by the general effect (i.e. not third trimester specific) of multivitamins and there was no dose-effect relation with folic acid. However, we found an obvious preterm birth reducing effect (30–55%) of these vitamin supplements, particularly of high dose folic acid alone in the third trimester.

Folate demand is increased during pregnancy *(140)*, and without adequate dietary intake of folate or folic acid supplementation, concentrations of folate in maternal serum, plasma and red blood cells decrease from the fifth month of pregnancy onwards *(141)*. Thus, the moderate fetal growth promoting effect of folic acid-containing multivitamin supplementation seems to be reasonable. However, it is more difficult to explain the preterm preventive effect of a high dose of folic acid.

Our first hypothesis for the explanation of preterm reduction due to folic acid supplementation was based on the well-known fact that pregnant women who used vitamin supplements were likely different than those that did not use these supplements. Thus, the preterm birth reducing effect of folic acid supplement may be associated with a better preconceptional and/or antenatal care and/or in general the better lifestyle and more health consciousness behaviors of these pregnant women. This hypothesis was supported by a somewhat lower socioeconomic status and somewhat higher rate of smokers in the group of unsupplemented pregnant women. In addition, medical doctors prescribe more frequently these vitamin supplements for women with pregnancy complications. In fact, these risk factors occurred more frequently in pregnant women with vitamin supplements. However, we considered these factors among confounders in the calculations of adjusted OR and the previous association between folic acid supplementation and lower rate of preterm births was confirmed.

Our present hypothesis is that the reduced maternal folate status is associated with elevated homocysteine related placental vasculopathy *(133,142–145)*. This placental vasculopathy may be a causal factor in the origin of preterm birth which can be neutralized with high dose folic acid supplementation during pregnancy, particularly in the third trimester. Johnson et al. *(146)* showed an interaction between a pregnant woman's dietary folate intake and the presence or absence of a deletion allele in a folate-metabolizing gene that codes for the production of dihydrofolate-reductase. The presence of the deletion allele increased the risk of preterm delivery threefold, but this increase in the risk for preterm birth was fivefold if folate intake was low (<400 μg/day). Maternal nutritional deficiency, i.e. lower vitamins and mineral consumption, may also be associated with decreased blood flow and increased maternal infections *(147)*. Maternal undernutrition is prevalent in low- and middle-income countries, resulting in substantial increases in mortality and overall disease burden, partly due to higher rate of preterm birth *(148)*.

Thus, our findings may have an argument for the use of higher doses of folic acid during pregnancy *(149)*. Of 136 million total births per year worldwide, more than 4 million newborns die within the first days or weeks of life. The major causes of their deaths are preterm birth, birth asphyxia and infections *(150)*. Thus, the reduction of preterm births by folic acid supplementation would be very important but these findings need confirmation.

25.8. RECOMMENDATIONS FOR NUTRITIONAL INTERVENTIONS TO REDUCE NTD AND OTHER CONGENITAL ABNORMALITIES

At present the folic acid supplementation is a real public health challenge *(151)*. There are three possibilities to provide appropriate folic acid and other B vitamin intakes for women of childbearing age who are capable of becoming pregnant.

25.8.1. Consumption of Folate-Rich and Other Vitamin-Rich Diets

The preconceptional period is an appropriate time to change the dietary habits and to improve the lifestyle of prospective parents, particularly mothers due to their good compliance as they want to do their best to have a healthy baby. Thus it is an important task to advise all women to have a folate-rich and other vitamin-rich diet from the preconceptional time onwards.

There is evidence that appropriate nutritional status of pregnant women can promote the postnatal, including adult health, of offspring. The origin of many common complex diseases (hypertension, coronary heart disease, diabetes mellitus, and obesity) may be related to the quality of fetal and infant life *(152)*. Appropriately preparing parents for pregnancy can also help to educate their children from the earliest time (i.e. birth) of their life about a healthy diet and lifestyle and there is a good chance that these habits will be fixed for their later life.

McPartlin et al. *(140)* suggested that the optimal daily intake of folate/folic acid in the pre- and postconceptional period is about 660 μg per day. Thus, the recommended intake of folate/folic acid advised to a woman of childbearing age, who is sexually active, is 700 μg *(153)*. This quantity is much higher than the recent RDA, because there is an increased requirement for folate during pregnancy due to (a) decreased absorption, (b) accelerated breakdown of folate to p-aminobenzoylglutamate and its acetylated derivative p-acetamidobenzoylglutamate, (c) increased urinary loss and (d) fetal transfer. The calculated total fetal and placental THF content is 800 μg/100 g at term *(154)*, thus fetal blood has a higher THF level than maternal blood and is indicative of active placental transfer *(155)*.

The usual daily intake of folate is about 180–200 μg/d mg in Hungary *(49)* and this consumption is not significantly higher in other countries *(156)*. Thus, it is difficult to imagine about a 3.5-fold increase in folate intake every day in anticipation of conception, which would require the consumption of 500 g of raw spinach, 900 g boiled spinach or 900 g of raw broccoli *(157)*, i.e. about 15 servings of broccoli on each day. Furthermore, some part of dietary folate is lost through cooking and processing. Finally, an extreme increase in the consumption of extra folate from natural food is relatively ineffective at increasing folate status *(158)*.

Three other B vitamins, B12, B6 and B2, have important roles in folate and/or homocysteine metabolism as we have shown previously. In Hungary the daily intake of vitamin B12 is adequate to prevent deficiency, but the consumption of vitamin B2 and particularly vitamin B6 is lower than the RDA.

In conclusion, a diet rich in vitamins, particularly folate, is important in the prevention of NTD and other CAs, but cannot alone completely neutralize the genetic predisposition for these CAs.

25.8.2. Periconceptional Supplementation

Good evidence is available to advise all women capable of becoming pregnant to have periconceptional (i. e. at least 1 month before and until 3 months after conception) folic acid or folic acid-containing multivitamin supplementation to reduce the occurrence of NTD and some other major CAs.

The absorption of folic acid in the gastrointestinal tract is quick and easy. Thus, it would be a simple and useful approach; however, about 50% of pregnancies are unplanned in the USA, Hungary and many other industrialized countries. If women have unplanned pregnancies and are not using a supplement routinely, they cannot take advantage of this new primary preventive method during the preconceptional period.

There are two public health tasks to help prospective pregnant women increase their use of periconceptional folic acid or multivitamins. The first is a strong and widespread *educational campaign*, to suggest the start of the use of folic acid or multivitamins immediately after the discontinuation of oral contraceptive pills or other contraceptive methods when couples decide to have a baby. However, unfortunately these campaigns have had only a limited success in all countries *(159)*.

The second important task is to establish a network of pre-or *periconceptional care* within the primary heath care *(160)*. The Hungarian Periconceptional Service (HPS) was launched in 1984 *(161)* and the Hungarian RCT and CCT—presented previously—were based on the HPS. We prefer to use the term periconceptional instead of preconceptional because prenatal care usually begins in about the 8–12th week of pregnancy, and thus the most sensitive and vulnerable early period of fetal development from the third postconceptional week until the eighth week is omitted from medical health service, thus embryos are uncared for, in general, unprotected.

The HPS begins 3 months before pregnancy and continues for 3 months in the postconceptional period including information-counseling, examinations and interventions performed by qualified nurses *(162)*. The HPS consists of three steps (Table 25.5):

I) *Check-up of reproductive health* is a "preconceptional screening" for risk identification and assessment, and referral of couples or persons at high risk to appropriate secondary care.
II) *The 3-month preparation for conception.* The preparation for conception is an appropriate period to stop smoking, alcohol drinking, and use of narcotic and unnecessary drugs with hazards to germ cells and later the fetus, in addition an optimal time for the launch of periconceptional folic acid or multivitamin supplementation.
III) *The better protection of early pregnancy.* Females are asked to visit the HPC immediately after the first missed menstrual period to confirm the conception by a sensitive pregnancy test and ultrasound scanning; in addition, they are asked to continue the intake of folic acid or multivitamins, etc. and are referred to the prenatal care clinics with the discharge summary of the HPS.

Our Hungarian experiences have shown that periconceptional care is feasible and economical, in addition to providing an appropriate opportunity for nutritional interventions as well *(163)*.

However, there are at least three further questions which need discussion.

The first question is connected with the choice of supplement: *folic acid alone or folic acid-containing multivitamins (164)*. The use of multivitamins containing folic acid and other B vitamins in the Hungarian RCT *(25)* and CCT *(27)* and in the study of Smithells et al. *(19–21)* showed a higher efficacy (about 90%) in the reduction of NTD than the MRC Vitamin Study *(22)* using a high dose of folic acid alone (about 70%) and Chinese–U. S. study using a low dose of folic acid (41–79%) *(28)* for the prevention of NTD. The usual important argument against the use of other vitamins beyond folic acid is that the supplementation group of "other vitamins" in the MRC Vitamin Study *(22)* did not result in a significant reduction in recurrent NTD. However, it is worth mentioning that there was a 40% reduction in the recurrent NTD near to the level of significance after the supplementation of other vitamins without folic acid.

Another argument is that folic acid-containing multivitamins seem to be effective in the reduction of cardiovascular, urinary tract and limb reduction CAs, but there are very limited data concerning a similar preventive effect of folic acid alone for the reduction of these CAs.

Finally, hyperhomocysteinemia plays a role in the origin of at least of some part of NTD and vitamins B12, B2 and B6 are important cofactors in the folate-homocysteine metabolism. The methyl group of 5-methyl-THF is used by methionine-synthase to recycle homocysteine back to methionine,

Table 25.5
Three Steps and Different Items of the Hungarian Periconceptional Service

1) Check-up of reproductive health
a) Family history of females and males, and obstetric history of females.
b) Case history and available medical records of females, e.g. epilepsy, diabetes, etc.
c) Vaginal and cervical smear screening of sexually transmitted infections/disorders.
d) Sperm analysis for the detection of subfertility and pyosperm.
e) Psychosexual exploration of couples.
f) Blood examination for the revealing of rubella seronegative women and women without previous varicella (they are vaccinated), HIV positivity, in addition carrier screening for cystic fibrosis and recently for predictive gene diagnostic tests.

2) The 3-month preparation for conception
a) Protection of germ cells: avoidance of smoking, alcohol, narcotic and unnecessary drugs.
b) Discontinuation of contraceptive pills and IUD (condoms are provided).
c) Occupational history of females (exemption of women at high risk).
d) Menstrual history and measurement of basal body temperature for detection of hormonal dysfunction (and treatment, if necessary).
e) Start of preconceptional multivitamin supplementation.
f) Suggestion to check dental status.
g) Guidelines for physical exercise.
h) Guidelines for healthy diet.

3) The better protection of early pregnancy
a) Necessary further examinations and/or treatments in the disorders of couples detected at the check-up examination and women with hormonal dysfunction.
b) To achieve conception on the optimal day (prior to the day of ovulation).
c) Early pregnancy confirmation by pregnancy test and ultrasound scanning.
d) Postconceptional multivitamin supplementation.
e) Avoidance of teratogenic and other risks.
f) Referral of pregnant women to prenatal care clinics.

and methionine-synthase is a *vitamin B12*-dependent enzyme. Vitamin B12 is important for DNA and RNA biosynthesis as well. These functions may explain why vitamin B12 deficiency is an independent risk factor in the origin of NTD *(165,166)*. Vitamin B12 completely abolished the embryotoxicity of L-homocysteine in rat embryos, which was shown to be mediated by catalysis of the spontaneous oxidation of L-homocysteine to the less-toxic L-homocysteine *(44)*. Thus, a hypothesis was developed that L-homocysteine embryotoxicity is explained by the inhibition of transmethylation reactions by increased embryonic 5-adenosylhomocysteine level.

There is no vitamin B12 deficiency in Hungary and industrialized countries; however, the major concern about higher doses of folic acid is its so-called masking effect in patients with pernicious anemia. Thus, the combined use of folic acid and vitamin B12 may prevent this possible adverse effect but only 1–3% of an oral dose of vitamin B12 can be absorbed from the gastrointestinal tract of patients with pernicious anemia, therefore a much higher dose (e.g. 10 or 25 μg) of vitamin B12 is needed for this purpose *(167)* instead of the usual low doses (in general 4 μg).

The conversion of homocysteine to cystathione by cystathione synthase requires pyridoxine, i.e. *vitamin B6*-dependent enzyme. A disturbance in the above processes results in decreased homocysteine remethylation causing hyperhomocysteinemia and the shortage of methionine *(168);* therefore, cells are not able to methylate important compounds such as proteins, lipids and myelin.

MTHFR is a key factor in the neutralization of homocysteine and is a *vitamin B2* (riboflavin)-dependent enzyme. The function of thermolabile MTHFR is impaired due to lack of riboflavin *(169).*

In addition, vitamin C has an important role in preventing the oxidation of THF, thus helping to keep the folate metabolic pool complete *(170).* Folate conjugase is a zinc metalloenzyme and folate polyglutamates are not well absorbed by the intestine in zinc-deficient humans *(171).*

On the other hand, folic acid alone is much less expensive than folic acid-containing multivitamins. However, some countries (e.g. Turkey) reimburse the major part of the cost of multivitamins or the full cost of folic acid (e.g. Italy) used during the periconceptional period. In addition, as far as I know there is no-multivitamin product containing "only" folic acid, vitamin B12, B6 and B2 in the market for the prevention of NTD and some other CAs.

The second question concerns the *optimal dose of folic acid.* At present there is no unequivocal evidence and universal consensus regarding the optimal dose of folic acid. The U. S. National Academy of Sciences *(172)*—followed the recommendations of the CDC *(26)*— suggested that all women who are capable of becoming pregnant should consume 400 μg folic acid/day to prevent NTD. However, in 1992 there was no scientific evidence for this recommendation because the low dose (360 μg) of folic acid in the study of Smithells et al. *(19–21)* was a component of multivitamin product. Later, the Chinese–U. S. study *(28)* confirmed the practical basis of this suggestion because this dose reduced significantly the first occurrence of NTD. The U. S. –Chinese study had another important message: the efficacy of prevention by folic acid depends on the incidence of NTD. The periconceptional multivitamin supplementation containing 800 μg of folic acid resulted in a very effective (about 90%) reduction in the first occurrence of NTD in the Hungarian intervention trials (Table 25.3). At present, most experts agree that 700 μg seems to be the necessary dose of folic acid during pregnancy *(153)*, and the study of Daly et al. *(173)* showed that there was no obvious greater reduction of NTD after the use of higher doses of folic acid. Thus, the recent general recommendation is that all pre- and pregnant women need 700 μg of folate-folic acid daily, partly by the consumption of 200–400 μg of folate from diet, and partly by supplementation with 300–400 μg of folic acid/day.

However, Wald et al. *(149)* calculated a dose–effect relation of folic acid in the reduction of hyperhomocysteinemia and the preventable part of NTD, and on the basis of their estimation suggested daily 5.0 mg (5,000 μg) of folic acid for all potentially pregnant women.

On the other hand, the upper tolerable dose of folate-folic acid for healthy persons including pregnant women is 1 mg by the U. S. Institute of Medicine *(174)* and the European Commission Scientific Committee on Food *(175).* Thus, it is worth differentiating the physiological dose of folic acid (less than 1 mg daily) for preventive purposes in healthy persons and the pharmacological dose of folic acid (more than 1 mg daily) for the treatment of patients. At present we cannot exclude possible adverse events after the use of high doses of folic acid *(176)*, though we did not find any serious pathological effects in four pregnant women who attempted suicide with very high doses of folic acid *(177).*

The major concern is the masking effect of high doses of folic acid in patients with pernicious anemia *(178).* However, it is necessary to differentiate two types of vitamin B12 deficiency. The classical vitamin B12 deficiency is caused by the lack of intrinsic factor, and these patients have a very limited absorption of dietary vitamin B12. The common new vitamin B12 deficiency without anemia and/or neuropathy is caused by the low production of stomach acid that is needed to unbind food-bound vitamin B12; they have enough intrinsic factor and can absorb dietary vitamin B12 well. Persons with new vitamin B12 deficiency have no masking effect of folic acid and are treatable with dietary vitamin B12 *(179).*

The optimal dose of folic acid for the *reduction of recurrent NTD risk* is the third question. There is an international consensus due to the results of the MRC Vitamin Study *(22)* that 4 mg of folic acid is the best for this preventive effort. However, the study of Smithells et al. *(19–21)* also clearly demonstrated the high efficacy (about 90%) reduction in the recurrent NTD by a multivitamin containing 0.36 mg of folic acid. The results of the MRC Vitamin Study showed about a 70% reduction in recurrent NTD after the use of 4 mg folic acid. In the data set of the Hungarian CCT, 34 informative offspring had previous sibs or parents with NTD without recurrence in the supplemented cohort with a multivitamin containing 0.8 mg folic acid *(27)*. In addition, there were 4 prospective parents with spina bifida and 111 parents with one or two previous offspring with NTD in the data set of our genetic counselling clinic until the end of 2007, we recommended that they take the multivitamin containing 0.8 mg folic acid used by us. Of their 115 off spring, only one had recurrent NTD (i.e. 0.9% instead of expected 3.5%).

25.8.3. Food Fortification

Food fortification seems to be the most practical means of supplementation with folic acid and other vitamins for women with unplanned pregnancies. This public health initiative is comparable to the prevention of goiter by the addition of iodine to salt.

In Ireland, vitamin B12 (in 1981) and folic acid (in 1987) were added to breakfast cereals, which constitute a significant part of the diet. This fortification has probably contributed to the significant fall in the prevalence at birth of NTDs in the 1980s (from 4.7 to 1.3/1,000), because only a small part of this decline could be attributed to termination of affected pregnancies *(153)*.

In February 1996, the U. S. Department of Health and Human Services *(180)* ordered food fortification with folic acid of all cereal grain products at a level of 0.14 mg/100 g beginning January 1998. This adds only about 100 μg of folic acid to the average daily diet of women of reproductive age, nevertheless the mean plasma folate concentrations increased from 4.6 to 10.0 μg/mL and the prevalence of low folate levels (<3 μg/mL) decreased from 20.0 to 1.7%. In addition, the mean total homocysteine concentration decreased from 10.1 to 9.4 μmol/L and the prevalence of high homocysteine concentrations (>13 μmol/L) decreased from 18.7 to 9.8% *(181)*. At the same time, there was a 26% reduction in the total (birth + fetal) prevalence of NTD *(182,183)*. The estimated benefit–cost ratio of U. S. folic acid fortification is 40:1 *(184)*, thus the estimated economic benefit in U. S. dollars is 312–425 million annually.

Canada also introduced a mandatory flour fortification with folic acid (0.15 mg/100 of white flour) in September 1998 *(185)* and a 42% reduction was found in the total prevalence of NTD *(186)*. Chile has also a mandatory flour fortification project using a higher dose of folic acid (0.22 mg/100 g flour) which resulted in 40% reduction in the rate of NTD *(187,188)*. Recently Costa-Rica has also introduced the food fortification with folic acid *(189)*.

In Hungary, three vitamins: folic acid (0.16 mg), vitamin B12 (0.8 μg) and B6 (0.864 mg) have been added to 100 g bread flour from the third quarter of 1998 *(190)*. The intake of folic acid, vitamin B12 and B6 is about 0.2, 0.01 and 1.08 mg, respectively, from 200 g of bread. In the second step in food fortification, flour was fortified with the above 3 B vitamins and marketed in 2003 *(191)*. However, both bread and flour fortification programs have been voluntary without real success due to the higher price of these products.

In the United Kingdom, 0.28 mg folic acid/100 g flour was recommended *(192)* but the mandatory flour fortification was postponed. The authorities of Ireland decided to start mandatory flour fortification: all breads manufactured or marketed, with the exception of minor bread products, are fortified with folic acid at a level 0.12 mg/100 of bread as consumed *(153)*. Their hope was to reduce the incidence of NTD-affected pregnancies by about 24%, to eradicate folate deficiency and related anemia and to achieve a possible modest reduction in cardiovascular disease in adults on the basis of previous meta-analysis *(193)*. An improvement in stroke mortality was observed in Canada and the USA from 1990 to 2002, partly due to folic acid flour fortification *(194)*. The data in the Yang et al. paper suggest that mandatory fortification of flour with folic acid prevented 10 times more deaths from stroke than prevented cases of NTD in the USA and Canada. The debated U-shaped risk relation between the dose of folic acid and the incidence of certain cancers, mainly colorectal, is not discussed here *(195)*. However, the study of Fenech *(196)* suggested that the consumption of 0.7 mg of folic acid resulted in a blood concentration which prevented new mutations caused by environmental factors in human cells.

The mandatory flour fortification would be especially important for the large proportion of women with lower levels of education and income who, in general, have difficulties buying more expensive foods rich in folate and other vitamins and who have unplanned pregnancies. The wider use of this public health action can prevent a large proportion of the 400,000 offspring affected with NTD in the world.

25.9. CONCLUSION AND RECOMMENDATIONS

A considerable number of structural birth defects, i.e. congenital abnormalities, particularly neural-tube defects, are preventable due to the recent introduction of folic acid/multivitamin supplementation during the periconceptional period in pregnant women. Obviously, the folate-rich diet is important but cannot solve the problem of neural-tube defects and some other congenital abnormalities. The periconceptional multivitamin/folic acid supplementation is an optimal option for educated women who are preparing for their conception. Finally, the food fortification with folic acid seems to be the most effective method to achieve this goal in the whole population of reproductive age, especially for women with lower socioeconomic and/or educational status. Thus, G. P. Oakley *(197)* is right:"Inertia on folic acid fortification equals public health malpractice."

NOTE

The total data set of the flungarian RCT is available *(198)*.

REFERENCES

1. Hale, F. (1932) Pigs born without eyeballs. *J Hered* **24**, 105–9.
2. Warkany, J. (1971) Congenital malformations induced by maternal dietary deficiency: Experiments and their interpretation. *Harvey Lec*, 1952–1953 **18**, 89–102.
3. Warkany, J. (1971) *Congenital Malformations. Notes and Comments*. Year Book Medical Publications, Chicago, IL.
4. Nelson, M.M. (1955) Mammalian fetal development and antimetabolities. In: Rhoads EP, ed. *Antimetabolites and Cancer*. American Association for the Advancement of Science Monograph, Washington D.C.
5. Evans, H.M., Nelson, M.M., Asling, C.V. (1951) Multiple congenital abnormalities resulting from acute folic acid deficiency during gestation. *Science* **114**, 479.
6. Nelson, M.M., Asling, C.W., Evans, H.M. (1952) Production of multiple congenital abnormalities in young by maternal pteroylglutamic acid deficiency during gestation. *J Nutr* **48**, 61–79.
7. Nelson, M.M., Wright, H.V., Asling, C.W., Evans, H.M. (1955) Multiple congenital abnormalities resulting from transitory deficiency of pteroylgutamic acid during gestation in the rat. *J Nutr* **56**, 349–69.
8. Thiersch, J.B. (1952) Therapeutic abortions with a folic acid antagonist, 4-aminopteroylglutamic acid (4-amino PGA) administered by the oral route. *Am J Obstet Gynecol* **63**, 1298–304.
9. Meltzer, H.J. (1956) Congenital anomalies due to attempted abortion with 4-aminopteroglutamic acid. *J Am Med Assoc* **161**, 1253.

10. Warkany, J., Beaudry, P.H., Hornstein, S. (1959) Attempted abortion with 4-aminopteroylglutamic acid (aminopterin): malformations of the child. *Am J Dis Child* **97**, 274–81.
11. Hernandez-Diaz, S., Werler, M.M., Walker, A.M., Mitchell, A.A. (2000) Folic acid antagonists during pregnancy and risk of birth defects. *N Engl J Med* **343**, 1608–14.
12. Czeizel, A.E., Rockenbauer, M., Sørensen, H.A.T., Olsen, J. (2001) Teratogenic risk of trimethoprim-sulfonamides. A population-based case-control study. *Reprod Toxicol* **15**, 637–46.
13. Hibbard, B.M. (1964) The role of folic acid in pregnancy with particular reference to anaemia, abruption and abortion. *J Obstet Gynecol* **71**, 529–42.
14. Hibbard, E.D., Smithells, R.W. (1965) Folic acid metabolism and human embryopathy. *Lancet* **1**, 1254.
15. Smithells, R.W., Sheppard, S., Schorah, C.J. (1976) Vitamin deficiencies and neural tube defects. *Arch Dis Child* **51**, 944–9.
16. Van Allen, M.I., Kalousek, D.K., Chernoff, G.F., et al. (1993) Evidence for multi-site closure of the neural tube in humans. *Am J Med Genet* **47**, 723–43.
17. Czeizel, A.E., Tusnády, G. (1984) *Aetiological Studies of Isolated Common Congenital Abnormalities in Hungary*. Akadémiai Kiadó, Budapest.
18. Elwood, J.M., Little, J., Elwood, J.H. (1992) *Epidemiology and Control of Neural Tube Defects*. Oxford University Press, Oxford.
19. Smithells, R.W., Sheppard, S., Schorah, C.J., et al. (1980) Possible prevention of neural tube defects by periconceptional vitamin supplementation. *Lancet* **1**, 339–40.
20. Smithells, R.W., Sheppard, S., Wild, J., Schorah, C.J. (1989) Prevention of neural tube defect recurrences in Yorkshire: final report. *Lancet* **2**, 498–9.
21. Nevin, N.C., Seller, M.J. (1990) Prevention of neural tube defect recurrences. *Lancet* **1**, 178–9.
22. MRC Vitamin Study Research Group. (1991) Prevention of neural tube defects: results of the Medical Research Council vitamin study. *Lancet* **338**, 131–7.
23. Laurence, K.M., James, N., Miller, M.H., et al. (1981) Double-blind randomised controlled trial of folate treatment before conception to prevent recurrence of neural-tube defects. *Br Med J* **282**, 1509–11.
24. CDC. (1991) Use of folic acid for prevention of spina bifida and other neural tube defects. *J Am Med Assoc* **266**, 1191–2.
25. Czeizel, A.E., Dudás, I. (1992) Prevention of the first occurrence of neural-tube defects by periconceptional vitamin supplementation. *N Engl J Med* **327**, 1832–5.
26. CDC. (1992) Recommendations for the use of folic acid to reduce the number of cases of spina bifida and other neural tube defects. *MMWR* **41**, 1233–8.
27. Czeizel, A.E., Dobó, M., Vargha, P. (2004) Hungarian two-cohort controlled study of periconceptional multivitamin supplementation to prevent certain congenital abnormalities. *Birth Defects Res, Part A*. **70**, 853–61.
28. Berry, R.J., Li, Z., Erickson, J.D., et al. (1999) Prevention of neural-tube defects with folic acid in China. China-US Collaborative Project for Neural Tube Defect Prevention. *N Engl J Med* **341**, 1485–90.
29. Wald, N.J. (1994) Folic acid and neural tube defects: the current evidence and implications for prevention. In: Bock G, Marsh J, eds. *Neural Tube Defects. CIBA Foundation Symposium 181*. John Wiley and sons, Chichester, pp. 192–211.
30. Yates, R.W., Ferguson-Smith, M.A., Shenkin, A., et al. (1987) Is disordered folate metabolism the basis for genetic predisposition to neural tube defects? *Clin Genet* **31**, 279–87.
31. Bolander-Gouaille, C. (2002) *Focus on Homocysteine and the Vitamins involved in its metabolism*, 2nd edn. Springer Verlag, France, Paris.
32. Massaro, E.J., Rogersm J.M. (eds.) (2002) Folate and Human Development. Humana Press, Totowa, New Jersey.
33. Wills, L. (1931) Treatment of "pernicious anaemia" of pregnancy and "tropical anaemia" with special reference to yeast extract as a curative agent. *Br Med J* **I**, 1059–64.
34. Hoffbrand, A.V. (2001) The history of folic acid. *Br J Haematol* **113**, 579–89.
35. Goyette, D., Summer, J.S., Milos, R. (1994) Human methylenetetrahydrofolate reductase: isolation of cDNA, mapping and mutation identification. *Nat Genet* **7**, 195–200.
36. Frosst, P., Blom, H.J., Milos, R. (1995) A candidate genetic risk-factor for vascular disease: a common mutation in methylenetetrahydrofolate reductase. *Nat Genet* **10**, 111–3.
37. Wilcken, B., Bamforth, F., Li, Z., et al. (2003) Geographical and ethnic variation of the 677C/T allele of 5, 10 methylenetetrahydrofolate reductase (MTHFR): findings from over 7000 newborns from 16 areas world wide. *J Med Genet* **40**, 619–25.
38. Van der Put, N.M., Steegers-Theunissen, R.P.M., Frosst, P., et al. (1995) Mutated methylentetrahydrofolate reductase as a risk factor for spina bifida. *Lancet* **346**, 1070–1.

39. Ou, C.Y., Stevenson, R.F., Brown, V.K., et al. (1996) 5, 10 methylenetetrahydrofolate reductase genetic polymorphism as a risk factor for neural tube defects. *Am J Hum Genet* **63**, 610–4.
40. Czeizel, A.E., Révész, C. (1970) Major malformations of the central nervous system in Hungary. *Br J Prev Soc Med* **24**, 205–22.
41. Weitkamp, L.R., Tackels, D.C., Hunter, A.G.W., et al. (1998) Heterozygote advantage of the MTHFR gene in patients with neural-tube defects and their relatives. *Lancet* **351**, 1554–5.
42. van der Put, N.M.J., Gabreels, F., Stevens, E.M.B., et al. (1998) A second common mutation in the methylenetetrahydrofolate reductase gene: An additional risk factor for neural-tube defect? *Am J Hum Genet* **62**, 1044–6.
43. Coelho, C.N.D., Klein, N.W. (1990) Methionine and neural-tube closure in cultured rat embryos: morphological and biochemical analysis. *Teratology* **42**, 437–51.
44. Vanaerts, L.A.G.J.M., Blom, H.J., Deabreu, R., et al. (1994) Prevention of neural tube defects by and toxicity of L-homocysteine in cultured postimplantation rat embryos. *Teratology* **50**, 348–60.
45. Steegers-Theunissen, R.P.M., Boers, G.H.J., Trijbels, F.J.M., Eskes, T.K.A.B. (1991) Neural-tube defects and derangement of homocysteine metabolism. *N Engl J Med* **24**, 199–200.
46. Steegers-Theunissen, R.P.M., Boers, G.H.J., Blom, H.J., et al. (1995) Neural tube defects and elevated homocysteine levels in amniotic fluid. *Am J Obstet Gynecol* **172**, 1436–41.
47. Greene, N.D., Copp, A.J. (2005) Mouse models of neural tube defects: investigating preventive mechanism. *Am J Med Genet C Semin Med Genet* **135**, 31–41.
48. Canadian Task Force on the Periodic Health Examination. (1994) Periodic health examination, 1994 update. 3. Primary and secondary prevention of neural tube defects. *Can Med Assoc J* **151**, 21–8.
49. Czeizel, A.E., Susánszky, E. (1994) Diet intake and vitamin supplement use of Hungarian women during the preconceptional period. *Int J Vitam Nutr Res* **64**, 300–5.
50. Dudás, I., Rockenbauer, M., Czeizel, A.E. (1995) The effect of preconceptional multivitamin supplementation on the menstrual cycle. *Arch Gynecol Obstet* **256**, 115–23.
51. Czeizel, A.E., Rockenbauer, M., Susánszky, E. (1996) No change in sexual activity during periconceptional multivitamin supplementation. *Brit J Obst Gynecol* **103**, 569–73.
52. Czeizel, A.E., Métneki, J., Dudás, I. (1996) The effect of preconceptional multivitamin suppelementation on fertility. *Int J Vitam Nutr Res* **66**, 55–8.
53. Czeizel, A.E., Dudás, I., Fritz, G., et al. (1992) The effect of periconceptional multivitamin-mineral supplementation on vertigo, nausea and vomiting in the first trimester of pregnancy. *Arch Gynecol Obstet* **251**, 181–5.
54. Czeizel, A.E. (1993) Randomized, controlled trial of the effect of periconceptional multivitamin supplementation on pregnancy outcome. In: Wharton BA, ed. *Maternal-Child Issues in Nutrition Wyeth-Ayerst Nutritional Seminar Series.* Excerpta Medica, Princeton, NJ, pp. 13–24.
55. Erős, E., Géher, P., Gömör, B., Czeizel, A.E. (1998) Epileptogenic activity of folic acid after drug induces SLE. (Folic acid and epilepsy). *Eur J Obstet Gynecol Reprod Biol* **80**, 75–8.
56. Czeizel, A.E., Métneki, J., Dudás, I. (1994) Higher rate of multiple births after periconceptional multivitamin supplementation. *N Engl J Med* **330**, 1687–8.
57. Czeizel, A.E., Métneki, J., Dudás, I. (1994) The higher rate of multiple births after periconceptional multivitamin supplementation: An analysis of causes. *Acta Genet Med Gemellol* **43**, 175–84.
58. Werler, M.M., Cragan, J.D., Wasserman, C.R., et al. (1997) Multivitamin supplementation and multiple births. *Am J Med Genet* **71**, 93–6.
59. Erickson, A., Källen, B., Aberg, A. (2001) Use of multivitamins and folic acid in early pregnany and multiple births in Sweden. *Twin Res* **4**, 63–6.
60. Zhu, L., Gindler, J., Wang, H., et al. (2003) Folic acid supplementation during early pregnancy and likelihood of multiple births: a population-based cohort study. *Lancet* **361**, 380–4.
61. Botto, L.D., Yang, G. (2000) 5,10,methylenetetrahydrofolate reductase gene variants and congenital abnormalities. A HuGE review. *Am J Epidemiol* **15**, 862–77.
62. Hasbargen, U., Lohse, P., Thaler, C.J. (2000) The number of dichorionic twin pregnancies is reduced by the common MTHFR 677CT mutation. *Hum Reprod* **15**, 2059–662.
63. Czeizel, A.E., Rockenbauer, M., Siffel, Cs., Varga, E. (2001) Description and mission evaluation of the Hungarian Case-Control Surveillance of Congenital Abnormalities, 1980–1996. *Teratology* **63**, 176–85.
64. Czeizel, A.E., Vargha, P. (2004) Periconceptional folic acid/multivitamin supplementation and twins. *Am J Obst Gynecol* **191**, 790–4.
65. Muggli, E.E., Halliday, J.H. (2007) Folic acid and risk of twinning: a systematic review of the recent literature, July 1994 to July 2006. *MJA* **186**, 243–8.

66. Czeizel, A.E., Dudás, I., Métneki, J. (1994) Pregnancy outcomes in a randomized controlled trial of periconceptional multivitamin supplementation. Final report. *Arch Gynecol Obstet* **255**, 131–9.
67. Windham, G.C., Shaw, G.M., Todoroff, K., Svan, S.H. (2000) Miscarriage and use of multivitamins or folic acid. *Am J Med Genet* **90**, 261–2.
68. Gindler, J., Li, Z., Berry, R.J., et al. (2001) Folic acid supplements during pregnancy and risk of miscarriage. *Lancet* **358**, 796–800.
69. Hook, E.B., Czeizel, A.E. (1997) Can terathanasia explain the protective effect of folic acid supplementation on birth defects? *Lancet* **350**, 513–5.
70. Czeizel, A.E. (2001) Miscarriage and use of multivitamins or folic acid. *Am J Med Genet* **104**, 179.
71. George, L., Mills, J.L., Johansson, A.L.V., et al. (2002) Plasma folate levels and risk of spontaneous abortion. *J Am Med Assoc* **288**, 1867–73.
72. Nelen, W.L., Blom, H.J., Steegers, E.A. et al. (2000) Homocysteine and folate levels as risk factor for recurrent early pregnancy loss. *Obstet Gynecol* **99**, 519–24.
73. Czeizel, A.E., Dobó, M. (1994) Postnatal somatic and mental development after periconceptional multivitamin supplementation. *Arch Dis Child* **70**, 229–33.
74. Dobó, M., Czeizel, A.E. (1998) Longterm somatic and mental development of children after periconceptional multivitamin supplementation. *Eur J Pediatr* **157**, 719–23.
75. Holmes-Siedle, M., Dennis, J., Lindenbaum, R.H., Galliard, A. (1992) Long-term effects of periconceptional multivitamin supplementation for prevention of neural tube defects: a seven to 10 year follow up. *Arch Dis Child* **67**, 1436–41.
76. Czeizel, A.E. (1993) Prevention of congenital abnormalities by periconceptional multivitamin supplementation. *Br Med J* **306**, 1645–8.
77. Czeizel, A.E. (1998) Periconceptional folic acid-containing multivitamin supplementation. *Eur J Obstet Gynec Reprod Biol* **75**, 151–61.
78. Czeizel, A.E. (1996) Reduction of urinary tract and cardiovascular defects by periconceptional multivitamin supplementation. *Am J Med Genet* **62**, 179–83.
79. Tolarova, M. (1982) Periconceptional supplementation with vitamins and folic acid to prevent recurrence of cleft lip. *Lancet* **2**, 217.
80. Czeizel, A.E., Tímár, L., Sárközi, A. (1999) Dose-dependent effect of folic acid on the prevention of orofacial clefts. *Pediatrics* **104**, e66.
81. Shaw, G.M., Lammer, E.J., Wasserman, C.R., et al. (1995) Risks of orofacial clefts in children born to women using multivitamins containing folic acid periconceptionally. *Lancet* **345**, 393–6.
82. Hayes, C., Werler, M.M., Willett, W.C., Mitchell, A.A. (1996) Case-control study of periconceptional folic acid supplementation and oral clefts. *Am J Epidemiol* **143**, 1229–34.
83. Czeizel, A.E. (2002) Prevention of oral clefts through the use of folic acid and multivitamin supplements: evidence and gap. In: Wyszynski, D.F. ed. *Cleft Lip and Palate. From Origin to Treatment*. Oxford University Press, New York, pp. 443–57.
84. Van Rooij, I.A.L.M., Vermeij-Keers, C., Kluijtmans, L.A.J., et al. (2003) Does the interaction between maternal folate intake and the methylenetetrahydrofolate reductase polymorphism affect the risk of cleft lip with or without cleft palate? *Am J Epidemiol* **157**, 583–91.
85. Botto, L.D., Khoury, M.J., Mulinare, J., Erickson, J.D. (1996) Periconceptional multivitamin use and the occurrence of conotruncal heart defects. Results from a population-based case-control study. *Pediatrics* **98**, 911–7.
86. Botto, L.D., Mulinare, J., Erickson, J.D. (2000) Occurrence of congenital heart defects in relation to maternal multivitamin use. *Am J Epidemiol* **151**, 878–84.
87. Kapusa, L., Haagmans, M.L.M., Steegers, E.A.P., et al. (1999) Congenital heart defects and derangement of homocysteine metabolism. *J Pediatr* **135**, 773–4.
88. van Beynum, I.M., Kapusta, L., den Heijer, M., et al (2006) Maternal MTHFR 677C-T is a risk factor for congenital heart defects: effect modification by periconceptional folate supplementation? *Eur Heart J* **27**, 981–7.
89. Shaw, G.W., O'Malley, C.D., Wasserman, C.R., et al. (1995) Maternal periconceptional use of multivitamin and reduced risk for conotruncal heart defects and limb deficiencies among offspring. *Am J Med Genet* **59**, 536–45.
90. Werler, M.M., Hayes, C., Louik, C. (1999) Multivitamin use and risk of birth defects. *Am J Epidemiol* **150**, 675–82.
91. Scanlon, K.S., Ferencz, C., Loffredo, C.A., et al. (1998) Preconceptional folate intake and malformations of the cardiac outflow tract. Baltimore-Washington Infant Study Group. *Epidemiology* **9**, 95–8.
92. Czeizel, A.E., Tóth, M., Rockenbauer, M. (1996) A case-control analysis of folic acid supplementation during pregnancy. *Teratology* **53**, 345–51.

93. Baird, C.D., Nelson, M.M., Monie, I.W., Evans, H.M. (1954) Congenital cardiovascular anomalies induced by pteroylglutamic acid deficiency during gestation in the rat. *Circ Rev* **2**, 544–8.
94. Monie, I.W., Nelson, M.M. (1963) Abnormalities of pulmonary and other vessels in rat fetuses from maternal pteroylglutamic acid deficiency. *Anat Rec* **147**, 397–401.
95. Monie, I.W., Nelson, M.M., Evans, H.M. (1954) Abnormalities of the urinary system of rat embryos resulting from maternal pteroylglutamic acid deficiency. *Anat Rec***120**, 119–36.
96. Monie, I.W., Nelson, M.M., Evans, H.M. (1957) Abnormalities of the urinary system of rat embryos resulting from transitory deficiency of pteroylglutamic acid during gestation in the rat. *Anat Rec* **127**, 711–24.
97. Li, D.K., Daling, J.R., Mueller, B.A., et al. (1995) Periconceptional multivitamin use in relation to the risk of congenital urinary tract anomalies. *Epidemiology* **6**, 212–8.
98. Yang, Q., Khoury, M.J., Olney, R.S., et al. (1997) Does periconceptional multivitamin use reduce the risk for limb deficiency in offspring? *Epidemiology* **8**, 157–61
99. Myers, M.F., Li, S., Correa-Villasenon, A., et al. (2001) Folic acid supplementation and risk for imperforate anus in China. *Am J Epidemiol* **154**, 1051–6.
100. Botto, L.D., Mulinarem J., Erickson, J.D. (2002) Occurrence of omphalocele in relation to maternal multivitamin use: a population-based study. *Pediatrics* **109**, 904–8.
101. James, S.J., Pogribna, J., Pogribny, I.P., et al. (1999) Abnormal folate metabolism and mutation in the methylenetetrahydrofolate-reductase gene may be maternal risk factors for Down syndrome. *Am J Clin Nutr* **70**, 495–501.
102. Hobbs, C.A., Sherman, S.L., Yi, P., et al. (2000) Polymorphisms in genes involved in folate metabolism as maternal risk factors for Down syndrome. *Am J Hum Genet* **67**, 623–30.
103. Barkai, G., Arbuzova, S., Berkenstadt, M., et al. (2003) Frequency of Down's syndrome and neural-tube defects in the same family. *Lancet* **361**, 1331–5.
104. Gueant, J.L., Gueant-Rodriguez, R.M., Anello, G., et al (2003) Genetic determinants of folate and vitamin B12 metabolism: a common pathway in neural tube defects and Down syndrome? *Clin Chem Lab Med* **41**, 1473–7.
105. Czeizel, A.E., Puhó, E. (2005) Maternal use of nutritional supplements during the first month of pregnancy and a reduced risk of Down's syndrome. A case-control study. *Nutrition* **21**, 698–704.
106. Czeizel, A.E., Medveczki, E. (2003) No difference in the occurrence of multimalformed offspring after periconceptional multivitamin supplementation. *Obstet Gynecol* **102**, 1255–61.
107. Shaw, G.M., Croen, L.A., Todoroff, K., Tolarova, M.M. (2000) Periconceptional intake of vitamin supplement and risk of multiple congenital anomalies. *Am J Med Genet* **93**, 188–93.
108. Yuskin, N., Honein, M.A., Moore, C.A. (2005) Reported multivitamin supplementation and the occurrence of multiple congenital anomalies. *Am J Med Genet Part A* **136A**, 1–7.
109. Czeizel, A.E., Puho, H.E., Bánhidy, F. (2006) No association between periconceptional multivitamin supplementation and risk of multiple congenital abnormalities. A population-based case-control study. *Am J Med Genet Part A* **140A**, 2469–77.
110. Botto, L.D., Olney, R.S., Erickson, J.D. (2004) Vitamin supplements and the risk for congenital anomalies other than neural tube defects. *Am J Med Genet Part C* **125C**, 12–21.
111. Lumley, J., Watson, L., Watson, M., et al. (2001) Periconceptional supplementation with folate and/or multivitamins for preventing neural tube defects. *Cochrane Database Syst Rev***3**, CD00156.
112. Hernandez-Diaz, S., Werler, M.M., Walker, A.M., Mitchell, A.A. (2001) Neural-tube defects in relation to use of folic acid antagonists during pregnancy. *Am J Epidemiol* **153**, 961–8.
113. Czeizel, A.E., Puho, E. (2004) A possible association between congenital abnormalities and the use of different sulfonamides during pregnancy. *Congenit Anom (Kyoto)* **44**, 79–86.
114. Bunin, G.R., Kuijten, R.R., Buckley, J.D., et al. (1993) Relation between maternal diet and subsequent primitive neuroectodermal brain tumours in young children. *N Engl J Med* **329**, 536–41.
115. Thompson, J.R., Gerald, P.F., Willoughby, L.N., Armstrong, B.K. (2001) Maternal folate supplementation in pregnancy and protection against acute lymphoblastic leukaemia in childhood: a case-control study. *Lancet* **358**, 1935–40.
116. Ács, N., Bánhidy, F., Puho, E., Czeizel, A.E. (2005) Maternal influenza during pregnancy and risk of congenital abnormalities on offspring. *Birth Defects Res Part A* **73**, 989–96.
117. Botto, L.D., Erickson, J.D., Mulinare, J., et al. (2002) Maternal fever, multivitamin use, and selected birth defects: evidence of interaction? *Epidemiology* **13**, 485–8.
118. Czeizel, A.E., Ács, N., Bánhidy, F., et al. (2007) Primary prevention of congenital abnormalities due to high fever related maternal diseases by antifever therapy and folic acid supplementation. *Curr Woman's Health Rev* **3**, 1–12.

119. Czeizel, A.E., Sankaranarayanan, K. (1984) The load of genetic and partially genetic disorders in man. I. Congenital anomalies: estimates of detriment in terms of years of life lost and years of impaired life. *Mutat Res* **128**, 499–503.
120. Tarusco, D. (ed.) (2004) Folic Acid: From Research to Public Health Practice. Rapporti ISTISAN 04/26, Roma.
121. Bjerkedahl, T., Czeizel, A.E.R., Hosmer, D.W. (1989) Birth weight of single livebirths and weight specific early neonatal mortality in Hungary and Norway. *Paediatr Perinat Epidemiol* **3**, 129–40.
122. Goldenberg, R.L., Culhane, J.F., Iams, J.D., Romero, R. (2008) Epidemiology of preterm birth. *Lancet* **371**, 75–84.
123. Institute of Medicine, (1990) *Subcommittee on Nutritional Status and Weight Gain during Pregnancy: Nutrition during Pregnancy*. National Academy Press, Washington, DC.
124. Spencer, N. (2003) Social and environmental determinants of birth weight. In: *Weighing the Evidence —How is Birthweight Determined*? Radcliffe Medical Press, Oxford, pp. 87–121.
125. Ek, J. (1982) Plasma and red cell folate in mothers and infants in normal pregnancies. Relation to birth weight. *Acta Obstet Gynecol Scand* **61**, 17–20.
126. Goldenberg, R.L., Tamura, T., Cliver, S.P., et al. (1992) Serum folate and fetal growth retardation: a matter of compliance? *Obstet Gynecol* **79**, 719–22.
127. Neggers, Y.H., Goldenberg, R.L., Tamura, T., et al. (1997) The relationship between maternal dietary intake and infant birth weight. *Acta Obstet Gynecol Scand* **165**(Suppl), 71–5.
128. Scholl, T.O., Johnson, W.G. (2000) Folic acid influence on the outcome of pregnancy. *Am J Clin Nutr* **71**, 1285–303S.
129. Relton, C.L., Pearce, M.S., Parker, L. (2005) The influence of erythrocyte folate and serum vitamin B12 status on birth weight. *Br J Nutr* **93**, 593–9.
130. Murphy, M.M., Scott, J.M., Arija, V., et al. (2004) Maternal homocysteine before conception and throughout pregnancy predicts fetal homocysteine and birth weight. *Clin Chem* **50**, 1406–12.
131. Scholl, T.O., Hediger, M.L., Schall, J.I., et al. (1996) Dietary and serum folate: their influence on the outcome of pregnancy. *Am J Clin Nutr* **63**, 520–5.
132. Siega-Riz, A.M., Savitz, S.A., Zeisel, S.H., et al. (2001) Second trimester folate status and preterm birth. *Am J Obstet Gynecol* **191**, 851–7.
133. Vollset, S.E., Refsumm H., Irgens, L.M., et al. (2000) Plasma total homocysteine, pregnancy complications, and adverse pregnancy outcomes: The Hordaland Homocysteine Study. *Am J Clin Nutr* **71**, 962–3
134. Ronnenberg, A.G., Goldman, M.B., Chen, D., et al. (2002) Preconception homocysteiner and B vitamin status and birth outcomes in Chinese women. *Am J Clin Nutr* **76**, 1385–91.
135. Shaw, G.M., Liberman, R.F., Todoroff, K., Wasserman, C.R. (1997) Low birth weight, preterm delivery, and periconceptional vitamin use. *J Pediatr* **130**, 1013–4.
136. Scholl, T.O., Hediger, M.L., Bendich, A., et al. (1997) Use of multivitamin/mineral prenatal supplements: influence on the outcome of pregnancy. *Am J Epidemiol* **146**, 134–41.
137. Shaw, G.M., Carmincheael, S.L., Nelson, V., et al. (2004) Occurrence of low birthweight and preterm delivery among California infants before and after compulsory food fortification with folic acid. *Publ Health Rep* **119**, 170–3.
138. Smits, L.J.M., Essed, G.G.M. (2001) Short pregnancy intervals and unfavorable pregnancy outcomes: Role of folate depletion. *Lancet* **358**, 2074–7.
139. Bukowski, R., Malone F.D. (2009) Preconceptional folate supplementation and risk of spontaneous preferm birth; a cohort study. PLoSMid 6(5). e1000061
140. McPartlin, J., Halligan, A., Scott, J.M., Darling, M., Weir, D.G. (1993) Accelerated folate breakdown in pregnancy. *Lancet* **341**, 148–9.
141. Cikot, R.J.L.M., Steegers-Theunissen, R.P.M., Thomas, C.M.G., et al. (2001) Longitudinal vitamin and homocysteine levels in normal pregnancy. *Br J Nutr* **85**, 49–58.
142. van der Molen, E.F., Verbruggen, B., Nokalova, I., et al. (2000) Hyperhomocysteinemia and other thrombotic risk factors in women with placental vasculopathy. *Br J Obstet Gynecol* **107**, 785–91.
143. Medina, M.A., Urdiales, J.L., Amores-Sanchez, M.I. (2001) Roles of homocysteine in cell metabolism, old and new function. *Eur J Biochem* **268**, 3871–82.
144. Ferguson, S.E., Smith, G.N., Walker, M.C. (2001) Maternal plasma homocysteine levels in women with preterm premature rupture of membranes. *Med Hypotheses* **56**, 85–90.
145. Nilsen, R.M., Vollset, S.E., Svein, A., et al. (2008) Folic acid and multivitamin supplementation use and risk of placental abruption: A population-based registry study. *Am J Epidemiol* **10**, 1093/aje/kwm373.
146. Johnson, W.G., Scholl, T.O., Spychala, J.R., et al. (2005) Common dihydrofolate reductase 19-base deletion allele: a novel risk factors for preterm delivery. *Am J Clin Nutr* **81**, 664–8.
147. Goldenberg, R.L. (2003) The plausibility of micronutrient deficiency in relationship to perinatal infection. *J Nutr* **133**, 1645–8S

148. Black, R.E., Allen, L.H., Bhutta, Z.A., et al. (2008) Maternal and child under nutrition: global and regional exposures and health consequences. *Lancet* **371**, 243–60.
149. Wald, N.J., Law, M.R., Morris, J.K., Wald, D.S. (2001) Quantifying the effect of folic acid. *Lancet* **358**, 2069–73.
150. WHO Reg. 1. (2205) Make every mother and child count. http://www.who.int/whr/2005/download/en
151. Eichholzer, M., Tönz, O., Zimmermann, R. (2006) Folic acid: a public-health challenge. *Lancet* **367**, 1352–61.
152. Barker, D.J.P. (ed.) (1992) *Fetal and Infant Origins of Adult Disease*. British Medical Journal Publication, London, England.
153. Food Safety Authority of Ireland (2006) Report of the Committee on Folic Acid Food Fortification, Dublin.
154. Iyengar, L., Apte, S.V. (1972) Nutrient stores in human foetal livers. *Br J Nutr* **27**, 313–7.
155. Strelling, M.K. (1976) Transfer of folate to the fetus. *Dev Med Child Neurol* **28**, 533–5.
156. Department of Health (2000) Committee on Medical Aspects of Food and Nutrition Policy (COMA): Folic Acid and the Prevention of Disease, London.
157. Bower, C., Wald, N.J. (1995) Vitamin B12 deficiency and the fortification of food with folic acid. *Eur J Clin Nutr* **49**, 87–93.
158. Cuskelly, G.J., McNulty, H., Scott, J.M. (1996) Effect of increasing dietary folate on red-cell folate: implications for prevention of neural tube defects. *Lancet* **347**, 657–9.
159. Botto, L.D., Lisi, A., Robert-Gnansia, E., et al. (2005) International retrospective cohort study of neural tube defects in relation to folic acid recommendations: are the recommendations working? *Br Med J* **330**, 571.
160. Healthy People 2000. (2000) National health population and disease prevention objectives. US Department of Health and Human Service. Public Health Service. DHHS Publ No. 91-502.13.
161. Czeizel, A.E., Dobó, M., Dudás, I., et al. (1998) The Hungarian periconceptional service as a model for community genetics. *Community Genet* **1**, 252–9.
162. Czeizel, A.E. (1999) Ten years of experience in the periconceptional care. *Eur J Obstet Gynecol Reprod Biol* **89**, 43–9.
163. Czeizel, A.E., Gasztonyi, Z., Kuliev, A. (2005) Periconceptional clinics: A medical healthcare infrastructure of new genetics. *Fetal Diagn Ther* **20**, 518–45.
164. Czeizel, A.E. (2004) The primary prevention of birth defects: Multivitamin or folic acid? *Int J Med Sci* **1**, 50–61.
165. Kirke, P.N., Molloy, A.M., Daly, L. E, et al. (1993) Maternal plasma folate and vitamin B12 are independent risk factors for neural tube defects. *Q J Med* **86**, 703–8.
166. Mills, J.L., McPartlin, J.M., Kirke, P.N., et al. (1995) Homocysteine metabolism in pregnancies complicated by neural-tube defects. *Lancet* **345**, 149–51.
167. Herbert, V., Bigaouette, J. (1997) Call for endorsement of petition to the Food and Drug Administration to always add vitamin B12 to any folate fortification or supplement. *Am J Clin Nutr* **65**, 572–3.
168. van der Griend, R. (1999) Combination of low-dose folic acid and pyridoxine for treatment of hyperhomocysteinemia in patients with premature arterial disease and their relatives. *Atherosclerosis* **143**, 177–83.
169. McNulty, H., McKinely, Wilson, B., et al. (2002) Impaired functioning of thermolabile methylenetetrahydrofolate reductase is dependent on riboflavin status: Implications for riboflavin requirements. *Am J Clin Nutr* **76**, 436–41.
170. Stokes, P.L. (1975) Folate metabolism in scurry. *Am J Clin Nutr* **28**, 126–9.
171. Tamura, T., Shane, B., Baer, M.T., et al. (1978) Absorption of mono- and polyglutamyl folates in zinc-depleted man. *Am J Clin Nutr* **31**, 1984–7.
172. US National Academy of Sciences. (1988) *Dietary Reference Intakes: Folate, Other B Vitamins and Choline*. National Academy Press, Washington DC.
173. Daly, S., Mills, J.L., Molloy, A.M., et al. (1997) Minimum effective dose of folic acid for food fortification to prevent neural-tube defects. *Lancet* **350**, 1666–9.
174. Institute of Medicine. (1998) *Dietary Reference Intakes: A Risk Assessment Model for Establishing Upper Intake Levels for Nutrients*. National Academy Press, Washington DC.
175. European Commission Scientific Committee on Food. (2000) Tolerable Upper Limit Levels for Vitamins and Minerals, Brussels.
176. Butterworth, C.E., Tamura, T. (1989) Folic acid and toxicity: a brief review. *Am J Clin Nutr* **50**, 353–8.
177. Czeizel, A.E., Tomcsik, M. (1999) Acute toxicity of folic acid in pregnant women. *Teratology* **60**, 3–4.
178. Lindenbaum, J., Healton, E.B., Savage, D.G., et al. (1988) Neuropsychiatric disorders caused by cobalamin deficiency in the absence of anaemia or macrocytosis. *N Engl J Med* **318**, 1720–8.
179. Oakley, G.P., Jr, (1997) Let's increase4 folic acid fortification and include vitamin B-12. *Am J Clin Nutr* **6**, 1889–90.
180. US Department of Health and Human Services. (1996) Food standards: amendment of standards of identity for enriched grain products to require addition of folic acid. *Fed Regist* **61**, 8781–7.

181. Jacques, P.F., Selhub, J., Bostom, A.G., et al. (1999) The effect of folic acid fortification on plasma folate and total homocysteine concentrations. *N Engl J Med* **340**, 1449–54.
182. Honein, M.A., Paulozzi, L.J., Methens, T.J., et al. (2001) Impact of folic acid fortification of the US Food Supply on the occurrence of neural tube defects. *J Am Med Assoc* **285**, 2981–6.
183. Williams, L.J., Mai, C., Edmonds, L.D., et al. (2002) Prevalence of spina bifida and anencephaly during the transition to mandatory folic acid fortification in the United States. *Teratology* **66**, 33–9.
184. Grosse, S.D., Waltzman, N.J., Romano, P.S., Mulinare, J. (2005) Reevaluating the benefits of folic acid fortification in the United States: economic analysis, regulation, and public health. *Am J Pub Health* **95**, 1917–22.
185. Vermeulen, M.J., Boss, S.C., Cole, D.E.C. (2002) Increased red cell folate concentrations in women of reproductive age after Canadian folic acid food fortification. *Epidemiology* **13**, 238–40.
186. De Wals, P., Tairou, F., Van Allen, M., et al. (2007) Reduction in neural-tube defects after folic acid fortification in Canada. *N Engl J Med* **357**, 135–42.
187. Freire, W.B., Hertrampf, E., Cortes, F. (2000) Effect of folic acid fortification in Chile: preliminary results. *Eur J Pediatr Surg* **10**, 42–3.
188. Hertrampf, E., Cortes, F. (2004) Folic acid fortification of wheat flour in Chile. *Nutr Rev* **62**, S44–8.
189. Chen, L.T., Rivery, M.A. (2004) The Costa-Rica Experience: Reduction of NTDS following food fortification programs. *Nutr Rev* **62**, S40–3.
190. Czeizel, A.E., Merhala, Z. (1998) Bread fortification with folic acid, vitamin B12 and vitamin B6 in Hungary. *Lancet* **352**, 1225.
191. Czeizel, A.E., Kökény, M. (2002) Bread fortification with folic acid in Hungary. *Br Med J* **325**, 391.
192. Department of Health. (2002) *Scientific Review of the Welfare Food Scheme. Report on Health and Social Subjects 51*. TSO, London.
193. Wald, D.S., Law, M., Morris, J.K. (2002) Homocysteine and cardiovascular disease: evidence on causality from e meta-analysis. *Br Med J* **325**, 1202–6.
194. Yang, Q., Botto, L.D., Erickson, J.D., et al. (2006) Improvement in stroke mortality in Canada and the United States, 1990 to 2002. *Circulation* **113**, 1335–43.
195. Bresalier, R.S. (2008) Chemoprevention of colorectal cancer? Why all the confusion. *Curr Opin Gastroenterol* **24**, 48–50.
196. Fenech, M. (2001) The role of folic acid and vitamin B 12 in genomic stability of human cells. *Mutat Res* **475**, 57–67.
197. Oakley, G.P. (2002) Inertia on folic acid fortification: Public health malpractice. *Teratology* **66**, 44–54.
198. Czeizel, A.E. (2004) Randomized Controlled Trial of Multivitamin Supplementation on Birth Defects and Pregnancy Outcomes, 1984–1994. Complementary and Alternative Medicine Data Archive, Data Set 16 October 2004. Sociometric Corporation, National Institute of Health.

26 Maternal Nutrition and Preterm Delivery

Theresa O. Scholl

Key Points

The rate of preterm delivery in the United States has increased by approximately 20% for all births in the past 15 years. Research, on the influence of maternal nutritional status and diet, has suggested that the following may be associated with this increasing risk:

- Low weight before and inadequate weight gain during pregnancy.
- Fasting and long intervals (>12 hours) between meals.
- Maternal anemia and iron deficiency anemia early in pregnancy.
- High hemoglobin, hematocrit or ferritin in the third trimester.
- Low intake of iron, folate, omega three fatty acids, calcium and zinc.
- High maternal levels of homocysteine.
- Failure to use prenatal vitamin/mineral supplements.
- High levels of triglycerides, cholesterol and free fatty acids.

Key Words: Folate; iron; lipids; multivitamins; nutrition; pregnancy; preterm delivery; zinc

26.1. DEFINITION AND IMPORTANCE OF PRETERM DELIVERY

There has been a substantial increase in preterm births (<37 completed weeks gestation) in the United States during the past few decades *(1)*. Between 1990 and 2005 the rate increased by 13.4% for singletons (9.7% in 1990, 11.0% in 2005) and by 19.8% for all births, singleton and plural (10.6% in 1990, 12.7% in 2005).

Preterm delivery is an important public health problem. Disorders related to short gestation and low birth weight are the leading cause of neonatal mortality (<28 days) and rank second only to congenital defects as the leading cause of mortality during the first year of life (infant mortality) *(2)*. In addition to their increased risk of death, infants delivered preterm are at greater immediate risk of serious complications (e.g. respiratory distress syndrome, bronchopulmonary dysplasia, intraventricular hemorrhage, and necrotizing enterocolitis). While neonatal intensive care and the use of corticosteroids and surfactants have increased survivorship, in the longer term (1–3 years of age), such children remain at increased risk (e.g. seizures, cerebral palsy and mental retardation) and experience persistent learning and behavioral deficits *(3–6)*. In later life children with low weight at birth (<2500 g), approximately 60% of whom were born preterm experience more social and medical disability *(7)*, are at increased risk of chronic diseases including cardiovascular disease and type 2 diabetes *(8)*.

There is no way to prevent preterm delivery that is universally or consistently effective. The few interventions that at times reduce risk are not applicable to most women or all populations. For example while treatment with antibiotics eradicates bacterial vaginosis, a potent risk factor, antibiotics do not consistently reduce and in some cases increase the risk of preterm delivery *(9–11)*. While the injection

A. Bendich, R.J. Deckelbaum (eds.), *Preventive Nutrition*, Nutrition and Health, DOI 10.1007/978-1-60327-542-2_26,

of 17 alpha-hydroxyprogesterone caproate has been shown to reduce risk, its use is currently limited to those women with a prior history of delivering before term *(9)*. Thus, it is important to prevent preterm delivery and to do that it is necessary to understand its etiology. This review focuses on the influence that maternal nutrition and maternal nutritional status have on risk of preterm delivery.

26.2. RISK FACTORS ASSOCIATED WITH PRETERM DELIVERY

One problem hampering the identification of risk factors is the heterogeneous nature of preterm delivery. The most common cause is a spontaneous preterm delivery from preterm labor or preterm premature rupture of the fetal membranes (PROM). In addition, there are numerous complications like preeclampsia, placental abruption, or fetal growth restriction that lead to an indicated preterm birth *(12)*. Each of the proximate causes may have separate risk factors. Consequently, only a fraction of the factors and exposures that give rise to preterm have been identified. These include a prior history of preterm delivery, by far the strongest risk factor – but also being poor, African American, a cigarette smoker, at the extremes of maternal age, having inflammation or an infection. Of the risk factors that are recognized, several suggest that maternal diet and nutritional status (underweight, inadequate weight gain during pregnancy, anemia) are important *(13)*.

26.3. MATERNAL WEIGHT AND WEIGHT GAIN

Most research on maternal nutritional status has focused on the relationship between pregravid weight, total weight gain, and birth weight, and the body of evidence on this topic has been extensively reviewed by the Institute of Medicine (IOM) *(14)*. Studies have been virtually unanimous in showing a positive relationship between pregravid weight or body mass index (BMI), gestational weight gain, and birth weight *(14–15)*. Pregravid weight or BMI and weight gain appear to have independent and additive effects on birth weight. The average magnitude of this effect in women with a normal weight-for-height (BMI 19.8–<26.0) is approximately 20 g of birth weight for every 1 kg of total gain, and pregravid weight-for-height is a strong effect modifier of birth weight *(16)*.

In several studies, maternal underweight has been associated with an increased risk of preterm delivery *(17–19)*. There is even a greater likelihood of delivering preterm when a low weight gain (<0.12 kg/week in the second and third trimester) accompanies pregravid underweight (BMI <19.8); in one study, this amounted to a threefold increase in risk of delivering moderately preterm (32–<37 weeks) and nearly a tenfold increased risk for delivering very preterm (20–<32 weeks) *(19)*. Although the finding of an association between low pregravid maternal weight/BMI and preterm delivery is fairly consistent, the extent to which it represents a size bias in the estimation of gestation from fundal height or by ultrasound or the extent to which gestational weight gain and diet during pregnancy can overcome the deficit are not well known.

The pattern of weight gain and rate of gain also appear to be significant in that both early and later weight gain have independent effects on outcome *(20–21)*. In pregnant adolescents from Camden, an early inadequate weight gain increased the risk of small-for-gestational age (SGA) births *(20)*. Preterm delivery was increased with inadequate weight gain rates late in pregnancy but unrelated to early weight gain, even when the total pregnancy weight gain never fell below the targets set in clinical standards. Other studies of adult gravidas have also shown that low rates of weight gain, often in the latter half of pregnancy, are associated with preterm delivery *(19,22–24)*.

High or excessive rates of gain (~0.75–0.80 kg/week) and maternal obesity *(17)* also carry an increased risk of preterm delivery. In the case of excessive gain, it has been speculated that the association is a function of late edema from preeclampsia or hypertension *(17,19)*. While obese women also experience increased preterm delivery, this may reflect an excess of indicated deliveries associated with obesity-related complications like gestational diabetes and preeclampsia *(17)*.

Carmichael and Abrams *(24)* reviewed and summarized the extant literature (13 published studies) on gestational weight gain and preterm delivery. They found that 11 of these studies reported a significant association between maternal weight gain and preterm delivery principally that risk increased when gestational weight gain was inadequate. Gestational weight gain in later pregnancy was consistently associated with increased preterm delivery while inadequate gain in early pregnancy was not.

26.4. DIET AND GESTATIONAL WEIGHT GAIN

One reason that maternal weight gain during pregnancy and gestation duration are related may be the maternal diet. Although this association appears to be a reasonable one, a link between energy intake and gestational gain has not often been described.

The first report of a positive relationship between diet and weight gain was made by Thomson *(25)*, who found a correlation of 0.30 between intake and weight gain in Scottish primigravidae eating “to appetite.” Among Camden gravidae, a significantly lower energy intake (approximately 150–300 kcal/day less) was associated with an inadequate gestational gain *(26,27)*. When women from upstate New York were asked a series of behavioral questions about changes in food intake during pregnancy *(28)*, the consumption of “a little or a lot less food” during pregnancy was associated with a greater than twofold increase in weight gain below IOM guidelines. Conversely, eating “much more food” related to a twofold increased risk of an excessive weight gain. By comparison, a reduction in physical activity was associated with a 70% increase in the risk of an excessive weight gain.

26.5. DIET AND PRETERM DELIVERY

The relationship between poor diet and inadequate gestational weight gain and the observation that a low pregravid weight or BMI and low rate of weight gain were each associated with preterm delivery suggests that a poor maternal diet could be a factor. In the African country of Gambia, the food supply fluctuates between dry and rainy seasons. Pregnancies conceived in the months when food is scarce and women are at their lowest weights were shorter than pregnancies conceived when food is more plentiful *(29)*. During the Dutch Famine of 1944–1945, third trimester exposure to intense famine shortened gestation by about 4 days while exposure during the first trimester was associated with a clear excess of preterm birth. Famine-related amenorrhea which would have made gestational dating insecure for most women could have led to underestimated risk *(30)*. The famine is best known for its effect on fetal growth and maternal weight.

Fasting during pregnancy may have effects analogous to famine and food shortage. In sheep a short interval of food deprivation around the time of conception—from 2 months before to the first month after conception—increased the risk of preterm birth. The nutritional restriction was brief, maternal weight reduced by only 15%; restriction was followed by ad libitum feeding for the remainder of gestation *(31)*. Since the nutritional demands of the fetus early in gestation are modest, it was unclear why preterm delivery had been triggered or what triggered it. The authors speculated that the shortage of an essential nutrient(s) was more likely than a lack of calories *(32)*. However, an under-expanded plasma volume may be another underlying cause.

Later in pregnancy a fast of 24–48 hours elicits labor and/or delivery. This occurs across species: ewes *(33–34)*, mares *(33–35)*, and primates *(36)*. Maternal fasting reduced circulating glucose levels, increased free fatty acid production, raised fetal cortisol, and increased prostaglandin synthesis *(36)*. Fasting may elicit a similar response in humans: abstaining from food and water during the Yom Kippur fast increased the likelihood of labor onset in women who are close to term *(37)*.

One study suggested that women in preterm labor were more likely to have detectable ketone bodies than controls *(38)*. While this might mean that women eat less early in labor, it is also possible that many of the pregnant women are in "accelerated starvation" and metabolize fat stores *(39)*. A pattern of infrequent eating and snacking throughout the day also may increase risk *(40)*. Likewise, periods of extended fasting—13 hours or more during the second and third trimesters—have been associated with increased levels of corticotropin releasing hormone (CRH) *(41)*. Higher CRH concentrations correlate with shorter gestation duration and an increased preterm delivery risk *(42)*.

Thus, famine and fasting, perhaps in concert with low levels of maternal glucose and other circulating nutrients, may influence CRH production and increase risk of preterm delivery. Poorly nourished animals have a plasma volume that is not fully expanded *(43)*, and this may explain why hydration can temporarily arrest preterm labor in some women. Consequently, hypovolemia may be another reason why pregnancies are not supported until term.

26.6. IRON

Iron is an essential element in the production of hemoglobin for the transport of oxygen to tissues and in the synthesis of enzymes that are required to use oxygen for the production of cellular energy *(14)*. Supplementation with iron is generally recommended during pregnancy to meet the energy needs of both mother and rapidly growing fetus. Anemia (low hemoglobin levels) and iron deficiency anemia (IDA) sometimes serve as indicators of overall poor maternal nutritional status during pregnancy. When overall dietary intake is inadequate, anemia is one of the most obvious symptoms. However, not all anemia are nutritional in origin – some arise from infection or from chronic disease.

When detected early in pregnancy, IDA is associated with a lower energy and iron intake, an inadequate gestational weight gain over the whole of pregnancy, as well as with a greater than twofold increase in the risk of preterm delivery *(44–45)*. Maternal anemia, when diagnosed before mid-pregnancy, is also associated with an increased risk of preterm birth *(46–48)*. During the third trimester, anemia may be a good prognostic sign reflecting expansion of the maternal plasma volume. Thus during the third trimester, anemia is often associated with decreased as opposed to increased risk of preterm birth *(49)*.

Scanlon and colleagues used data from Pregnancy Nutritional Surveillance to examine the relationship between maternal anemia and preterm delivery in 173,031 low-income gravidas *(50)*. Preterm delivery was increased for anemic women and women with low hemoglobin during the first or second trimester. For women with moderate to severe anemia, risk was approximately doubled, for the others risk of preterm delivery was increased between 10 and 40%. During the third trimester, the association reversed—anemia and low hemoglobin were each associated with a decreased risk of preterm birth, and there was little relationship between high maternal hemoglobin and preterm delivery *(50)*. The association of maternal anemia in early pregnancy with preterm delivery and other poor outcomes has been confirmed in recent studies from the developing world *(48,51)*.

The increased risk of preterm delivery may be specific to iron-deficiency anemia, and not anemia from causes other than iron deficiency. Scholl et al. *(44)* reported data on 755 pregnant women receiving initial antenatal care at 16.7 +/–5.4 weeks gestation in Camden, New Jersey. Serum ferritin (<12.0 μg/L) was used to characterize iron deficiency anemia *(14)*. While anemia based upon low hemoglobin was high (27.9%) at the initial antenatal visit, prevalence of iron deficiency anemia (anemia with serum ferritin concentrations <12.0 μg/L) was low, amounting to 3.5%. After controlling for confounding variables, the risk for preterm delivery increased more than twofold for women with IDA in early pregnancy, while anemia from other causes were not associated with any increased risk for preterm delivery. Consistent data were obtained from Papua, New Guinea. Severe anemia (<80 g/L) in early pregnancy, which was attributed to iron deficiency, increased risk of low birth weight approximately sixfold in primiparae. Risk was not increased when IDA was diagnosed at delivery *(51)*.

Thus, it seems reasonable to presume that some but not all of what seems to be anemia or IDA is caused by the expansion of the maternal plasma volume. At present this state is poorly differentiated from anemia or iron-deficiency anemia late in pregnancy but is easier to distinguish during the first or second trimester.

Later in pregnancy, high levels of hemoglobin or hematocrit are associated with an increase in preterm delivery. High hemoglobin suggests failure of the plasma-volume to expand and often correlates clinically with increased risk of preeclampsia. A number of studies *(e.g. 52–53)* have documented such a "U-shaped" relationship between low and high maternal hemoglobin or hematocrit and preterm delivery (Table 26.1).

A high concentration of the iron storage protein ferritin, during the third trimester of pregnancy, has been strongly associated with an increased risk for preterm delivery *(54–55)*. In Camden *(55)* high ferritin levels (>41.5 ng/L) during the third trimester, likely stemming from the failure of ferritin to decline as plasma volume expanded, increased risk of very preterm delivery more than eightfold. High maternal ferritin has been associated with IDA early in pregnancy *(55)* and with indicators for maternal infection including clinical chorioamnionitis *(55)* and infant sepsis *(56)*.

Despite the long-standing association between maternal anemia and preterm delivery, it is uncertain if adverse pregnancy outcomes can be prevented by supplementing pregnant women with iron. There have been two clinical trials in the United States, both of which suggested that gestation duration may be lengthened or preterm delivery rates reduced with iron supplementation by mid-pregnancy.

Iron-replete women ($N = 275$) were enrolled by 20 weeks gestation and randomly assigned to placebo or iron (30 mg/day as ferrous sulphate) until week 28 gestation *(57)*. Supplemented women had significantly longer gestations (+0.6 week) but risk of preterm delivery, based solely from the mother's recall of her LMP, was not altered. Supplemented women had infants with higher birth weight and bore fewer low birth weight (<2500 g) or preterm and low birth weight infants (<37 weeks and <2500 g). Prophylactic supplementation did not improve maternal iron status or reduce anemia prevalence. However, the proportion of women with absent iron stores or IDA was reduced among the iron supplemented.

Another trial in iron replete North Carolina women *(58)* randomly assigned them to prenatal supplements with and without iron (ferrous sulphate 30 mg), before week 20. Infant birth weight was increased, and risk of preterm delivery reduced by 85% ($p = 0.05$) among the iron supplemented. However, apart from the frequency of iron depletion (ferritin <20 μg/L) which was marginally lower ($p = 0.08$) among the iron supplemented women, most measures of iron status did not differ between groups *(58)*.

Not all trials have given these results. Non-anemic Australian women, enrolled before 20 weeks gestation, showed no improvement in birth weight or gestation when they received supplemental iron (ferrous sulphate 20 mg) *(59)*. But fewer women assigned to iron had later iron deficiency (35 vs 58%) or iron deficiency anemia (3 vs 11%) *(59)*.

Table 26.1
Maternal Nutrition and Preterm Delivery, Study Characteristics

Nutrient	*References*	*Study design, size, and setting*	*Age, year*	*SES*	*Race/ethnicity*	*Findings*
Iron/anemia	Klebanoff et al.*(47)*	Case–control study of 1706 (725 preterm, 981 term) deliveries from Kaiser Permanente Births Defects Study, Hct levels measured throughout gestation.	—	—	14% Asian 28% Black 21% Mexican 37% White	Hct values fell during early second trimester and began to increase again at 31–33 week. Moderate relationship between second trimester anemia (Hct <10th percentile for ethnicity and gestation) and preterm delivery.
	Zhou et al. *(48)*	Observational study of 829 women from Shanghai. Population homogeneous for race, parity, prenatal care, smoking. Hgb sampled at 4–8 weeks, 16–20 and 28–32 weeks gestation.	27.7 ± 4.7 years.	—	100% Chinese	Inverse U-shaped relationship between maternal Hgb at 4–8 weeks and preterm delivery and low birth weight (LBW). Odds of preterm delivery increased 2.5-fold for women with Hgb >130 g/L or between 90 and 99 g/L. Odds increased 3.5-fold for entry Hgb between 60 and 89 g/L.
	Steer et al. *(49)*	Retrospective analysis of over 150,000 women from the North West Thames region of London, lowest recorded Hgb during pregnancy.	—	Mixed	73% White 14% Indo-Pakistani 5% Black 8% Other	Increased risk of preterm delivery (<37 week completed) with Hgb levels ≤85 g/L (AOR = 1.62, CI 95%:1.35–2.18) and >115 g/L.
	Steer et al. *(49)*	Retrospective analysis of over 150,000 women from the North West Thames region of London, lowest recorded Hgb during pregnancy.	—	Mixed	73% White 14% Indo-Pakistani 5% Black 8% Other	Increased risk of preterm delivery (<37 week completed) with Hgb levels ≤85 g/L (AOR = 1.62, CI 95%:1.35–2.18) and >115 g/L.

Scanlon et al. *(50)*	Retrospective analysis of data from 173,631 pregnant WIC participants.	70% 20–35 years.	Low	55% White 20% Black 19% Hispanic 6% Other	Risk of preterm delivery increased with low Hgb in first and second trimester. SGA not associated with anemia. Odds of preterm delivery with first trimester severe anemia increased 1.7 ×
Garn et al. *(52)*	Observational study of over 50,000 women followed in the National Collaborative Perinatal Project (NCPP), lowest pregnancy values of hemoglobin (Hgb).	—	Mixed	Whites and Blacks	Maternal Hgb and Hct levels had "U-shaped" relationship to preterm delivery (≤37 week), with the risk being higher at Hgb <100 g/L and >120 g/L. and hematocrit (Hct). The increased risk with Hgb >120 indicated a failure of plasma-volume expansion.
Murphy et al. *(53)*	Observational study of nearly 55,000 women in the Cardiff Births Survey, Hgb levels ascertained at entry to prenatal care.	—	Mixed	Welsh	Preterm delivery (<37 completed weeks) showed a "U-shaped" relationship to Hgb levels. At <13 week, risk was increased ~50% with Hgb <104 g/L; at week 13–19 risk not increased; at week 20–24, risk was increased for low (104 g/L) and high (>145 g/L) Hgb > 50%.

(Continued)

Table 26.1
(Continued)

Nutrient	*References*	*Study design, size, and setting*	*Age, year*	*SES*	*Race/ethnicity*	*Findings*
	Cogswell et al. *(57)*	Randomized placebo controlled trial of WIC participants, not anemic at entry (<20 weeks). Participants supplemented until week 28, 30 mg iron as ferrous sulphate or placebo. At week 28 controls with low ferritin received iron (30 or 60 mg/d).	24.3 ± 5 years.	Low	Low 56% White 25% Black 16% Hispanic 3% Other	Iron supplementation did not reduce anemia or increase ferritin or other iron status measures. Iron supplementation increased birth weight (+206 g) and decreased risk of LBW and preterm LBW but not preterm delivery.
	Siega-Riz et al. *(58)*	Randomized placebo controlled trial of 429 WIC eligible, not anemic at entry (<20 weeks). Supplemented to weeks 26–29, multivitamins w/wo 30 mg iron as ferrous sulphate.	24.3 ± 5 years.	Low	36% White 62% Black 2% Other	Iron supplementation did not reduce anemia or increase ferritin or other iron status measures. Iron supplementation increased birth weight (+108 g) and decreased preterm delivery (7.5% vs 13.9%).
	Makrides et al. *(59)*	Randomized placebo controlled trial, Australian participants not anemic at entry (<20 weeks). 459 participants supplemented until delivery, 20 mg iron as ferrous sulphate or placebo.	24.3 ± 5	Low	95% White 2.3% Aboriginal 2% Hispanic 0.7% Other	Iron supplementation reduced anemia, IDA at delivery, 6 months postpartum. Iron supplementation did not increase birth weight or improve other pregnancy outcomes.

Iron/iron-deficiency anemia	Scholl et al. *(44,45)*	Observational study of iron-deficiency anemia (IDA, anemia with serum ferritin <12 μg/L) among 779 Camden, NJ women.	18.4 ± 3.7	78% Medi-caid	66% Black	IDA (12.5% of all anemia) was associated with lower energy and iron intakes early in pregnancy. The risk or preterm delivery increased >2 × with IDA, but was not with anemia from other causes.
Iron/ferritin	Goldenberg et al. *(54)*	Observational study of serum ferritin (19, 26, and 36 weeks) and preterm delivery in 580 gravidas who participated in a randomized study of zinc supplementation.	—	100% Low	100% Black	Ferritin in highest quartile at week 28 associated with twofold increase in odds of preterm delivery (<37 weeks) and infant LBW, threefold increase in odds of very preterm delivery (≤32 weeks) and a fourfold increase in birth weight ≤1,500 g.
	Scholl *(55)*	Observational study of serum ferritin (15 and 28 weeks) in 1,162 gravidas from the Camden Study.	12–29 years.	100% Low	33.3% Hispanic 57.3% Black 9.4% White	At week 28 high concentrations of ferritin (>90th percentile) from failure of ferritin to fall with gestation, associated with ninefold increase in very preterm delivery, fourfold increase in preterm delivery and fivefold increase in LBW ($p < 0.01$ for each).
	Goldenberg et al. *(56)*	Observational study of 223 gravidas with PROM before 32 weeks. Data from randomized antibiotic trials at five clinical centers.	24.8 ± 5.9 years.	—	American	Plasma ferritin increased between admission for PROM and delivery. Ferritin levels significantly higher (+36 μg/L) in women whose infants developed sepsis.

(Continued)

Table 26.1
(Continued)

Nutrient	*References*	*Study design, size, and setting*	*Age, year*	*SES*	*Race/ethnicity*	*Findings*
Zinc	Scholl et al. *(26)*	Observational study of dietary zinc intake in 818 Camden, NJ pregnant women.	19.0 ± 4.1 years.	Low	60% Black	Low dietary zinc intake associated with twice greater preterm delivery (<37 week) and infant LBW, 3–4 times greater very early preterm delivery (<33 week) ($p < 0.05$ for each). Risk or preterm delivery with PROM increased 3.5 times with low zinc ($p < 0.05$).
	Neggers et al. *(64)*	Observational study of 476 Alabama women.	21.8 ± 4.9 (12–42) years.	Low	76.5% Black	Low plasma zinc at entry to prenatal care (7.0–12.2 μmol/L) increased risk of infant LBW 8 × compared with women in the highest quartile (15.9–25.4 μmol/L, $p < 0.05$). Plasma zinc was linearly related to gestation duration (0.17 week/μmol/L zinc, $p < 0.05$).
	Sikorski et al. *(66)*	Observational study of zinc status in 70 women with term deliveries.	26 ± 4.4 years.	—	Polish	Zinc index compiled from zinc assayed in blood, scalp and pubic hair, and colostrum. Zinc index in patients with PROM was lower (4.33 ± 1.18) compared with patients without PROM (5.97 ± 1.39, $p < 0.05$). Zinc index was inversely correlated with parity ($r = -0.61$).
	Mahommed et al. *(67)*	Cochrane review of randomized trials of zinc during pregnancy (>9,000 subjects enrolled in 17 trials) for meta-analysis.	—	Mixed	—	The summary odds ratio for preterm delivery was 0.86 (95% CI 0.76–0.98) and statistically significant. No reductions in LBW, SGA, or other outcomes to support the finding.

	Cherry et al. *(68)*	Randomized, controlled trial of zinc supplementation among pregnant New Orleans teenagers, assigned to placebo ($N = 556$) or 30 mg zinc supplement ($N = 581$).	17.6 (13.5–19.6) years.	Low	95% Black	With stratification by maternal weight at delivery, zinc supplementation was related to a lower risk of preterm delivery (38 week) among normal weight women (18.3 vs 28%, $p < 0.05$). Zinc supplemented low weight multiparas had increased gestation duration (+1.2 week, $p < 0.008$). There was a tendency ($p < 0.15$) for weight gain rates to be greater among zinc supplemented women (0.52 kg/week) compared with placebo (0.42 kg/week).
	Goldenberg et al. *(69)*	Randomized, controlled trial of zinc supplementation among pregnant Alabama women with plasma zinc below median at entry to prenatal care, assigned to multivitamin tablet with additional 25 mg zinc ($N = 294$) or placebo ($N = 286$).	23.5 ± 5.5 years.	Low	100% Blacks	Birth weight increased with zinc supplementation by +126 g ($p < 0.05$) and gestation duration by 0.5 week ($p = 0.06$). Women with pregravid BMI <26 had a birth weight increase +248 g ($p < 0.005$) and an increase in gestation of +0.7 week ($p = 0.08$).
Folic acid/ folate	Scholl et al. *(61)*	Observational study of folate from diet and supplements and serum folate in Camden, NJ ($N = 832$).	18.8 years.	Medicaid 80%	67% Black	Women with low folate intake ($\leq 240\ \mu g/d$) had three times greater preterm delivery and infant LBW ($p < 0.05$). Risk of preterm delivery without PROM increased 3 times ($p < 0.05$). Odds of preterm delivery increased 1.5% per unit decrease in serum folate ($p < 0.05$).

(Continued)

Table 26.1
(Continued)

Nutrient	References	Study design, size, and setting	Age, year	SES	Race/ethnicity	Findings
	Shaw et al. *(80)*	Observational study before vs after folate food fortification in 5.9 million California births, 1990–2000.		Mixed		Unadjusted analyses showed no change in % preterm or LBW. Adjusted data (age, parity, ethnicity, education, year, fortification period). Showed reductions in risk of preterm (–4%), LBW (–6%) VLBW (–9.8%).
	Siega-Riz et al. *(82)*	Observational study of folate from diet and supplements and serum, RBC folate in North Carolina 1995–2000.		Mixed	40% Black 54% White	Women with low folate intake ($\leq$500 μg/d) in second trimester 1.8 $\times$ more preterm deliveries <37 weeks. Odds ratios for preterm <37 weeks 1.8 with low serum folate (<16.3 ng/mL and low RBC folate (<626.6 ng/mL).
	Blot et al. *(87)*	Nonrandomized, double-blind study of iron and ascorbic acid vs iron, ascorbic acid, and folate (350 μg/d) at 6 mo. Gesta in 200 women.	27.5 $\pm$ 4.5 years	34% Low	French	Gestation duration increased by 0.8 week with folate supplementation, birth weight increased by +158 g, birth length by +1.7 cm ($p < 0.05$).
	Tchernia et al. *(88)*	Observational study of serum folate in 100 women using iron or iron and vitamin C.	—	—	French	Serum folate lower by 1.1 μg/L in women delivering $\leq$39 week ($p < 0.01$).
		Observational study of red cell folate in 100 high-risk women.	—	—	French	Gestation duration reduced by 0.8 week in women with RBC folate $\leq$200 μg/L ($p < 0.025$).

		Supplementation trial (open) of iron vs iron + folate (350 μg/d) in 108 women.	—	—	French	Gestation duration increased by 0.8 week among iron and folate supplemented women ($p < 0.001$).
Folic acid	Fleming et al. *(84)*	Randomized controlled trial of folic acid (0.5 mg or 5 mg/day) with and without iron plus a placebo group in 146 gravidas.	—	—	Western Australia	Small sample sizes in each of the five trial arms ($N \sim 20$). No difference in gestation or birth weight was detected among groups.
	Fletcher et al. *(85)*	Double blind, randomized study of folic acid plus iron (5 mg/d) or iron in 643 gravidas.	27.3 ± 5.7 years	—	English	Gestation duration (39.7 weeks (London) iron vs 39.7 folic acid + iron) and proportion less than 38 weeks (8.7% vs 8.7%) were not different.
	Giles et al. *(86)*	Randomized study of folic acid (5 mg/d) or iron supplementation in 692 gravidas. Patients stratified by gestation at entry to care: <10 weeks, 10–20 weeks, 20–30 weeks, >30 weeks.	25.0 ± 5.2 years.	—	Australian (Melbourne)	Folic acid group contained more primigravidas than placebo group. No difference between groups in gestation duration or birth weight.
Folic acid/ homocysteine	Rajkovic et al. *(90)*	Observational study of plasma homocysteine in 20 nulliparae with preeclampsia and 20 controls.	20 ± 4 years.	—	45% Black 55% White	Preeclamptic gravidae had higher homocysteine (8.66 vs 4.99 g mol/L), and delivered significantly earlier than controls (35 ± 4 weeks vs 40 ± 1 week) ($p < 0.05$ for each).

(Continued)

Table 26.1
(Continued)

Nutrient	*References*	*Study design, size, and setting*	*Age, year*	*SES*	*Race/ethnicity*	*Findings*
	Vollset et al. *(91)*	Retrospective cohort analysis of 5,883 women linking current homocysteine to past pregnancy outcomes.	40–42 years.	—	Norwegian	Current homocysteine (highest quartile) associated with past history of adverse pregnancy outcome including preeclampsia, preterm delivery, LBW, stillbirth, neural tube defects, and club foot.
	Ronnenberg et al. *(92)*	Case–control study of pre-pregnancy homocysteine, B vitamins (B_6, B_{12}, folate) and pregnancy outcome in 405 women from Anquin, China.	24–27 years.	—	100% Chinese	Risk of preterm delivery (<37 weeks) increased 3.6 × for women with elevated homocysteine (≥ 12.4 μ md/L). Preterm delivery, 50–60% lower for women with non-deficient levels of B_{12} and B_6. Serum folate not associated with preterm delivery.
	Malinow et al. *(94)*	Observational study of circulating serum homocysteine and serum folate in 35 healthy nulliparous gravidas at delivery.	24.2 ± 5.9 years.		—	American Maternal homocysteine correlated inversely ($p < 0.05$) with gestation duration ($r = -.42$) serum folate correlated positively ($p < 0.05$) with gestation duration ($r = .23$). Maternal folate and homocysteine correlated ($r = .54$) with each other ($p < 0.05$). Maternal serum B_{12} was not related to gestation duration ($r = .08$, $p = 0.67$).
Calcium	Belizán et al. *(99)*	Randomized, controlled trial of calcium supplementation among 1,194 gravidas in Rosario, Argentina.	23.7 ± 5.5 years.	Mixed	Argentinian	Percent with preterm delivery (<37 week) among 579 calcium-supplemented women 6.3 vs 6.8% for 588 women receiving placebo.

López-Jaramillo et al. *(102,103)*	Randomized, controlled trial of calcium supplementation among 56 nulliparous clinic patients with overall low calcium intake at risk for pregnancy-induced hypertension in Quito, Ecuador.	19.4 ± 1.8 years.	Low income	Ecuadorian	Gestation duration longer for 22 supplemented women (2 g calcium/d) vs placebo group ($N = 34$). Gestation 39.2 ± 1.2 week for calcium supplemented, 37.4 ± 2.3 week for placebo (p <0.01).
Bucher et al. *(104)*	Medline and Embase searches for randomized trials of calcium and preeclampsia during pregnancy (2,459 subjects enrolled in 14 trials) for meta-analysis.	—	Mixed	—	The summary odds ratio for preterm delivery was 0.69 (95% CI 0.48–1.01) and not statistically significant. All the trials which included preterm delivery showed a reduced risk with calcium supplementation, during pregnancy only one was statistically significant.
Levine et al. *(105)*	Randomized trial of calcium supplementation (500 mg/d) or placebo in 4,589 healthy nulliparas at 13–21 weeks gestation. Eligible women underwent test of compliance before enrollment in trial.	21 ± 4 years.	Mixed	35% White 17% Hispanic 47% Black 2% Other	Calcium supplemented gravidas delivered at 38.9 ± 2.5 weeks vs 38.9 ± 2.4 weeks for placebo (NS). Preterm delivery (<37 weeks) and very preterm delivery (<34 weeks) occurred in 10.8% and 4.2% of supplemented vs 10.0% and 3.7% of controls (NS). LBW (<2500 g) was 8.8% in supplemented, 9.6% in controls (NS).
Crowther et al. *(106)*	Randomized trial of calcium supplementation (1.8 g/d) or placebo in 456 nulliparous with singleton pregnancies.	25 ± 5 years.	Mixed	Australian	Treatment with calcium reduced risk of preeclampsia and preterm birth with more than twofold reduction in risk for each.

(*Continued*)

Table 26.1
(Continued)

Nutrient	*References*	*Study design, size, and setting*	*Age, year*	*SES*	*Race/ethnicity*	*Findings*
	Villar and Repke *(107)*	Randomized, controlled trial of calcium supplementation, 190 teenagers in Baltimore, MD.	≤17 years.	Low-income clinic patients	93.7% Black	Gestation duration longer for 94 supplemented teenagers (2 g calcium/d) vs placebo group ($N = 95$). Gestation 39.2 ± 2.7 week for calcium supplements, 37.9 ± 3.9 week for placebo ($p < 0.01$). Lower percentage of preterm delivery (<37 week) among supplemented (7.4%) vs placebo (21.1%, $p < 0.01$).
	Villar et al. *(108)*	Randomized trial of calcium supplements (1.5 g/d) vs placebo in 8,325 nulliparae <20 weeks at entry from populations w low calcium intake (<600 mg/d).	22.6 ± 4.4 years	Mixed		Calcium supplemented gravidae <20 years had fewer preterm delivery <37 weeks (OR = 0.82) and very preterm deliveries <34 weeks (OR = 0.64). No effect on preeclampsia but reduced risk of eclampsia (OR = 0.68) and severe gestational hypertension (OR = 0.71, severe preeclamptic complications (OR = 0.76) and severe maternal morbidity/mortality (OR = 0.80) with calcium for women, all ages.
Fish oils (*n*–3 fatty acids)	Olsen et al. *(109)*	Randomized, controlled trial of supplementation with fish oil, 533 women assigned 2/1/1 to fish oil, olive oil, or no oil supplementation, Aarhus, Denmark.	29.4 ± 4.4 years.	—	Danish	Mean gestation duration longer ($p < 0.006$) for the 266 women receiving fish oil (283.3 ± 11.1 d) compared with olive oil (279.4 ± 13.1 d) or no oil groups (281.7 ± 11.6 d).

	Makrides et al. *(114)*	Cochrane review of randomized trials of fish oil/prostaglandin precursor supplements during pregnancy (>2,700 subjects enrolled in six trials) for meta-analysis.	—	Mixed	—	No decrease in preterm delivery but very preterm delivery was reduced (0.69, (95% CI 0.49–0.99) and gestation duration increased by 2.5 days ($p < 0.05$). No reductions in LBW, SGA or other outcomes.
Multivitamin/ Minerals	Scholl et al. *(117)*	Observational study of 1,430 gravidas from the Camden Study who entered care during trimesters 1 and 2. Prenatal supplement users compared to gravidas entering care in same time frame who did not use supplements.	12–29 years.	Low	58.9% Black	Prenatal supplement use associated with two- to fourfold reduction in odds of preterm delivery and very preterm delivery ($p < 0.001$). Significant reductions in odds of very LBW (six-—to sevenfold) and infant LBW (twofold) ($p < 0.001$). Prenatal supplement use corroborated by assay of circulating micronutrients at 15 and 28 weeks ($p < 0.05$).
	Wu et al. *(118)*	Livebirths from 1988 NMIH Survey ($N \sim 9{,}000$) analyzed to examine effect of regular of multivitamin/mineral supplements and smoking. Gravidas regularly using supplements (3+times/week) compared to those who used them less often.	25.5 ± 5.9 years.	Mixed	American	No difference between women who regularly used multivitamins and those using them less in preterm delivery, LBW, or SGA.

(Continued)

Table 26.1
(Continued)

Nutrient	*References*	*Study design, size, and setting*	*Age, year*	*SES*	*Race/ethnicity*	*Findings*
	Catov et al. *(119)*	Observational study of 1,832 gravidas and periconceptional multivitamin use within 6 months of first prenatal visit (9.9 +/–3.9 weeks).	25 ± 5.5 years (mean)	—	35% Black	Periconceptional prenatal supplement use associated with threefold reduction in odds of preterm delivery (<34 weeks). All women and 85% reduction in SGA (<5th percentile) non-obese women. No change in preterm delivery 34–<37 weeks or SGA 5th—<10th percentiles.
	Fawzi et al. *(120)*	Randomized study of 1,075 HIV-1 infected gravidas from Tanzania randomly assigned to placebo, vitamin A, or multivitamins in a 2 × 2 factorial design.	—	Low	Tanzanian	Women receiving multivitamins had significantly decreased ($p < 0.01$) infant LBW by (44%), very preterm birth (39%) and fetal growth restriction (43%) CD4, CD8, and CD3 counts were significantly increased ($p < 0.05$). Vitamin A had no effect on outcomes.
	Fawzi et al. *(121)*	Randomized study of 8,468 HIV negative gravidae from Tanzania randomly assigned to multivitamins or placebo all received iron (60 mg) plus folate (0.25 mg).	—	Low	Tanzanian	Women receiving multivitamins had significantly decreased ($p < 0.01$) infant LBW by (20%) and SGA by 27%. Gestation duration increased by +0.2 week ($p = 0.02$) but preterm birth unchanged.

	Christian et al. *(122)*	Cluster randomized controlled trial of 426 rural communities in SE Nepal. Communities randomized to five regimes of supplementation vs vitamin A control.	55% 20–29 years.	Low	Nepalese	None of the supplements reduced risk of preterm delivery. In comparison to vitamin A control, multiple micronutrients increased birth weight (+64 g) and reduced LBW by 14%. Folic acid plus iron also increased birth weight (+37 g, NS) and reduced LBW by 16%.
	Ramakrishnan et al. *(123)*	Randomized trial of multivitamin supplements vs iron.	23 ± 5 years.	Low	Mexican	Multivitamin and iron supplements did not yield differences in birth weight, gestation duration, LBW, IUGR, or ponderal index.
Dyslipidemia	Catov *(127)*	Case–control study of periconceptional multivitamin use within 6 months of first prenatal visit (9.9 +/–3.9 weeks).		Mixed		Periconceptional prenatal supplement use associated with threefold reduction in odds of preterm delivery (<34 weeks) for all women and 85% reduction in SGA (<5th percentile) for non-obese women. No change in preterm delivery 34–<37 weeks or SGA 5th–<10th percentiles.
	Chen and Scholl *(128)*	Observational study of fasting free fatty acids at week 30 in Camden ($N = 523$).	22.5 ± 5 years.		53% Hispanic 34% Black	Women with high FFAs (>432 uM) had 3 × increased risk for preterm delivery <37 weeks.
	Khoury et al. *(129)*	Randomized trial ($N = 290$) low risk gravidae, nonsmokers with no prior pregnancy complications assigned to usual (control) or cholesterol lowering diet (intervention).	21–38 years		100% White	Diet-lowered total and LDL cholesterol, cord lipids NS. Preterm delivery <37 weeks reduced significantly from 7.4% (controls) vs 0.7% (intervention).

Some burning questions remain about the aforementioned iron trials. It is unclear how an iron supplement would give rise to improved birth weight and extend gestation duration without also improving maternal anemia or iron deficiency or why the results of the Australian trial differ from the two trials done in the United States. Mechanisms whereby maternal anemia and iron deficiency might influence the outcome of pregnancy include chronic hypoxia that initiates a stress response, increased oxidative stress that damages the maternal–fetal unit, and reduced immune function with increased maternal infection *(60)*.

26.7. MICRONUTRIENTS

During pregnancy, low intakes of two micronutrients, zinc *(26)* and folate *(61)*, are associated with lower energy intake, iron-deficiency anemia at entry to care, an inadequate gestational gain during pregnancy, as well as an increased risk of preterm delivery. Zinc is an element involved directly either as a metalloenzyme in the production of enzymes that include DNA and RNA polymerase or as a catalyst in the synthesis of other enzymes *(62)*. Folic acid functions as a coenzyme in the transfer of single-carbon atoms to intermediates in the synthesis of amino acids and nucleic acids *(63)*. While many other nutrients in addition to these two would be limited in a marginal maternal diet, inadequate intake of zinc, folate, or both potentially leads to impaired cell division and alterations in protein synthesis. Such alterations are most notable and have the greatest potential to do harm during times of rapid tissue growth, such as pregnancy *(14)*.

26.7.1. Zinc

Observational studies of circulating levels of zinc or dietary zinc intake *(26,64–66)* have often suggested that higher intakes or greater circulating concentrations of zinc are associated with improved pregnancy outcomes including preterm birth (Table 26.1). Clinical trials have yielded equivocal results often focusing on entire groups of low-income women where the mean zinc intake is below the RDA for pregnancy, an approach that selects a population, as opposed to individuals, at risk. A recent review of 17 clinical trials involving approximately 9,000 pregnant women showed a small but significant 14% reduction in risk of preterm delivery when zinc supplemented women were compared to women on placebo *(67)*. There was a caveat – no ancillary support for this result was evidenced – that is, there was no accompanying increase in infant birth weight or improvement in other neonatal outcomes, no dimunition in maternal complications such as preeclampsia or preterm premature rupture of membranes (PPROM) among the zinc-supplemented women.

In two trials, effects of zinc were conditional on maternal weight *(68–69)* with a lower rate of preterm delivery evidenced in zinc-supplemented women who were not overweight (Table 26.1). Cherry et al. *(68)* reported that low serum zinc concentrations were more common in underweight and multiparous women. The frequency of preterm delivery was reduced in the zinc-treated normal-weight women; zinc treatment of underweight multiparous women also was associated with a gestational age increase of nearly 3 weeks.

The trial conducted by Goldenberg and colleagues *(69)* recruited women with plasma zinc levels below the median and randomly assigned them to zinc or placebo. When stratified by body mass index, zinc supplementation increased gestation duration of approximately half a week ($p = 0.06$) and increased in birth weight along with lengthening the duration of gestation. Women with a BMI <26 benefited most from zinc treatment with a 248 g increase in infant birth weight and a 0.7 cm larger infant head circumference. Thus, consistent with prior results, effects were increased for women with lower pregravid body mass index (Table 26.1).

In contrast, two recent trials from the developing world (Peru, Bangladesh), where one might suppose that zinc deficiency would be prevalent, were negative *(70–71)*. It is possible that if gravidas from the developing world have multiple nutritional deficiencies, then these may reduce the bioavailability of zinc or otherwise limit fetal growth.

Low plasma zinc has a number of potential causes. In addition to its effect on protein synthesis, zinc also has an antiseptic action. In theory, a low zinc intake could be associated with an increased risk of infection during pregnancy leading to fragile fetal membranes (and PROM). A genetic polymorphism in matrix metalloproteinase-9, one of the genes that codes for zinc-dependent enzymes, has recently been associated with an increased risk of PROM *(72)*. Conversely, a low plasma zinc level could be an acute phase response to a stressor, such as maternal infection *(62)*. Keen *(73)* cautioned on the importance of distinguishing a secondary zinc deficiency arising as an acute phase response to infection or inflammation from a primary (dietary) deficiency in zinc.

26.7.2. Folate

During gestation marginal maternal folate nutrition has the potential to impair cellular growth and replication in the fetus and/or the placenta which, in turn, could increase the risk of preterm delivery and infant LBW *(14)*. Such adverse outcomes are more common in poor women than those who are better off. Pregnant women living under circumstances where preterm delivery is prevalent have been reported to consume diets with a lower density of vitamins and minerals, including folate and to limit consumption of folate-containing dietary supplements *(74–76)*.

In the United States, it is recommended that during pregnancy women consume 600 mcg folic acid per day, which includes 400 mcg of synthetic folic acid from supplements or fortified cereals, in order to reduce risk of neural tube defects; less than one-third of childbearing-age women do so *(77)*. Also, in the United States, fortification of flour and cereals with folic acid (1998–present) resulted in a 19% decline in infants with neural tube defects, increased serum and red cell folate and a decline in homocysteine levels *(78–79)*. Rates of preterm delivery, very low birth weight, and low birth weight also declined between 4 and 10% among California live born infants, when rates before and after fortification of the food supply with folate were compared *(80)*.

The influence of dietary and circulating folate on preterm delivery and infant low birth weight was studied in Camden, one of the poorest cities in the continental United States *(61)*. Low intakes of folate from diet and supplements and low circulating levels of folate were associated with many maternal characteristics reflecting poorer maternal nutritional status as well as with increased risks of preterm delivery and infant low birth. There was an interaction between a deletion allele for a folate-metabolizing gene coding for dihydrofolate reductase (DHFR) *(81)* and folate intake. DHFR is an enzyme that converts dietary folate to the reduced folate forms used by cells. Presence of the DHFR deletion allele and a low folate intake increased risk of low birth weight eightfold and risk of preterm birth fivefold for Camden women. Even after mandatory fortification of the US food supply, a similar result was observed in North Carolina. Low maternal folate status in the second trimester, assessed from both dietary and circulating measures, was associated with approximately a twofold increased risk of preterm delivery *(82)*. Among black but not white women, an interaction between low folate intake and a variant of a different folate metabolizing gene, cystolic serine hydroxymethyltransferase, was evidenced that resulted in a small increase in risk *(83)*.

Some randomized studies of routine folic acid supplementation *(84–88)* have suggested increased maternal hemoglobin, greater gestational weight gain, as well as an increase in mean gestation duration (Table 25.1). Further study of folate on risk of preterm delivery and LBW is obviously important *(89)*.

Many of the trials have demonstrated a beneficial effect when a folic acid supplement was used in combination with iron during pregnancy.

26.7.3. Homocysteine

Hyperhomocysteinemia, a metabolic effect that may be the result of folate deficiency, can occur when dietary folate intake is low or the requirement is increased. For instance, genetic factors and the interaction between genes and the environment may increase the metabolic requirement for folate and the risk of spontaneous abortion, preterm delivery, and other untoward gestational events. This may include many women with indicated preterm delivery from serious complications, where higher levels of homocysteine have been reported *(90)*.

Homocysteine levels from more than 5,000 Norwegian women were linked to past data in birth registries *(91)*. Higher maternal homocysteine correlated with older age at delivery, higher cholesterol, less multivitamin usage, lifestyle factors (cigarette smoking and high coffee consumption) and a past reproductive history that included preeclampsia and preterm delivery. In China, preconceptional levels of homocysteine were associated with nearly a substantially increased risk of preterm delivery as were low levels of the other B vitamins (B6, B12) whereas folate was not. *(92)*. Other studies have reported that higher maternal homocysteine measured before *(93)* or during pregnancy is associated with significantly lower infant birth weight *(93–94)* and shorter gestation duration *(94)*.

In summary, while observational studies of folate and pregnancy suggest a potential benefit of folic acid supplementation during pregnancy, a decrease in serious complications and an improvement in birth weight and gestation, randomized trials indicate that routine folic acid supplementation is not uniformly beneficial. Some who are at risk from common genetic polymorphisms that alter folate metabolism or because of environmental factors associated with poor diet would seem to benefit the most. Homocysteine may prove to be a more sensitive indicator of risk than diet, serum, or red cell folate.

26.8. OTHER NUTRIENTS

26.8.1. Calcium

Another element that has received attention for its possible association with preterm delivery is a macromineral, calcium. During pregnancy, there is an increased physiologic demand for calcium. A full-term infant accretes about 30 g of calcium, primarily in the third trimester when the fetal skeleton is actively ossifying, and to meet these needs there is enhanced absorption of calcium from the maternal gut *(95–96)*. Diets low in calcium both in general and especially during pregnancy have been associated with an increased blood pressure via heightened smooth muscle reactivity that may lead to preterm delivery *(97)*. Calcium supplementation trials during pregnancy have been shown to lower blood pressure levels *(98–99)*.

Maternal growth and pregnancy often coincide and approximately half of all pregnant teenagers continue to grow while pregnant *(100)*. Bone ultrasound measures of the os calcis during pregnancy showed greater loss for growing teenage gravidas (–5.5%) than for mature women (–1.9%) *(101)*. Thus, in the case of an adolescent pregnancy, calcium may be limited by maternal diet, but simultaneously driven by the need to retain enough calcium to mineralize two skeletons.

Two calcium supplementation trials *(102–103)* among high-risk women with very low intakes in Quito, Ecuador, and teenagers in Baltimore *(107)* showed promising results in decreasing the incidence of preterm delivery (Table 26.1). On the other hand, a large calcium supplementation trial of over 1,000 adult women from Argentina showed the expected decrease in the incidence of pregnancy induced hypertension (PIH), but no effect on preterm delivery *(104)*. A meta-analysis *(105)* of 14 randomized controlled trials of calcium supplementation involving several thousand gravidas showed significant reductions in systolic and diastolic blood pressure and preeclampsia with the administration of calcium salts. However, the analysis yielded no effect of calcium intake on preterm delivery or fetal growth restriction (Table 26.1).

A large randomized double-blind placebo controlled trial of more than 5,000 low-risk nulliparous women were supplemented with 2,000 mg/day elemental calcium or placebo before mid-pregnancy, and this was continued until they delivered *(106)*. Calcium did not reduce risk of preeclampsia and had no effect on obstetrical outcomes including preterm delivery or infant birth weight. However, it should be borne in mind that the women recruited into this trial had calcium intakes that were atypically high, averaging 1,100 mg/day before supplementation. A trial of calcium supplementation in Australian gravidae ($N = 456$) with apparently adequate dietary calcium intakes (median 1,100–1,200 g/day) and other characteristics very similar to those in the US study *(106)* reported a positive effect *(107)*. There was a greater than twofold reduction in the risk of preeclampsia and a reduction in the risk of preterm delivery from 10 (placebo) to 4.4% (calcium supplemented) with reductions in admissions for threatened preterm labor ($p = 0.03$) and preterm premature rupture of membranes ($0 = .08$).

More recently a WHO trial of supplemental calcium (1.5 g calcium/day) or placebo in gravidae with low calcium intake (<600 mg/day) showed no reduction in risk of preeclampsia *(108)*. However, there was a significant effect of calcium on preterm delivery in women <20 years where odds of preterm delivery were reduced by 22% (10.6 vs 12.8%) and very preterm delivery (<32 weeks gestation) by 56% (2.4 vs 3.8%). In women of all ages, the reduction was small (6.9 vs 7.2% preterm) and not statistically significant. Thus, it seems that the capacity of supplemental calcium to decrease risk may be confined largely to high-risk populations either where there is a dietary restriction of calcium intake or where, as in the case of adolescents, there is an increased demand for calcium to meet the needs of the growing fetus and the mother herself.

26.8.2. *n*-3 Fatty Acids

Consumption of marine foods rich in *n*–3 fatty acids has been associated with longer gestation duration. Ecologic data from the Faroe Islands, where 50% of the diet is derived from marine sources, show that island women bear infants with substantially higher birth weights (about 200 g) and longer gestations than Danes *(109)*. This difference in gestation has been hypothesized to arise with high consumption of *n*–3 fatty acids via a reduction in inflammatory cytokines which induce cyclooxygenase2 and phospholipase A2 and increase prostaglandin F_2 and E_2 or from a lowering of the thromboxane–prostacyclin ratio and the ensuing reduction in myometrial activity.

The clinical trial conducted by the People's League of Health *(110)* on 5,022 London women between 1938 and 1939 suggested that supplementing the usual diet of British women with minerals, vitamins, and halibut oil extended gestation. Fewer infants were born before the 40th week of gestation to the supplemented women (20% among supplemented women were born before week 40 vs 24% among the unsupplemented), although there was no difference in mean birth weight.

A trial *(111)* of 533 Danish women who received fish oil, olive oil, or no supplement by week 30 gestation showed that women taking the fish-oil supplement had longer average gestations (4 days) and bore infants with higher average weights (+107 g) which was mostly attributable to the change in gestation. The effect of fish oil was strongest for women with low fish consumption at entry and amounted to an increase of 7.4 days in this group. There was little effect of the supplement on women with high consumption at entry (–1.6 days) (Table 26.1). Thus, the hypothesized effect of *n*-3 fatty acids on gestation appears to have a threshold, beyond which there is no effect. Furthermore, the effect appeared to be specific to the initiation of idiopathic preterm labor rather than PROM.

Recently, a multicenter trial of high-risk women enrolled gravidae with a prior history of preterm delivery, fetal growth restriction or preeclampsia. Women were randomized to either fish oil or olive oil. Fish oil reduced the recurrence risk of preterm delivery approximately twofold but did not alter the recurrence of the other poor pregnancy outcomes *(112)*. A recent prospective and observational study of seafood intake in early pregnancy bolstered this relationship: risk of preterm delivery was increased nearly fourfold among women with little or no intake of fish by 16 weeks' gestation *(113)*.

A Cochrane review ($N = 2793$) reported that pregnant women in fish oil supplement groups showed a small but significant lengthening of gestation (+2.6 days) compared to controls. This was not supported by a reduction in risk of preterm delivery (<37 completed weeks), although there was a 45% reduction in the odds of giving birth <34 completed weeks′ gestation *(114)*.

Thus, although it has been suggested that compliance has been a problem in many fish oil trials *(115)* evidence from meta-analysis *(114)* suggests only a small overall effect of fish oil supplements.

26.8.3. Multivitamin-Mineral Supplements

In the United States, there is limited supplement use by reproductive-age women; overall about one-quarter of women (26% white, 15.5% black) report regularly taking vitamin or mineral supplements *(116)*. In one study only 16% of low-income Massachusetts gravidae took vitamins before pregnancy *(76)*. This varied by ethnicity, with white gravidae (23%) reporting use about twice as frequently as blacks (11%) or Hispanics (10%). During pregnancy 9% of whites, 20% of blacks, and 13% of Hispanics used prenatal vitamins erratically (1–3 times per week) or not at all. Reasons for non-compliance included: maternal confidence that diet was good, an unstable home life, and side effects attributed by the women to the supplement.

In Camden, 17% of low-income minority gravidae reported using supplements before they got pregnant *(117)*. Periconceptional use was more likely to occur among women with a history of an adverse pregnancy outcome, principally spontaneous abortion in past pregnancies, and was associated with increased spotting and bleeding during the current pregnancy. Thus, low-income women appeared to use supplements when they previously had or currently anticipated a problem with their pregnancies. Other predictors of the failure to use vitamin/mineral supplements before or during include being black, unmarried, under the age of 20 with less than a high school education, being multiparous and having a late entry to prenatal care *(117–118)*.

Information on the effects of prenatal multivitamin/mineral supplements on pregnancy outcome is limited. Camden women who use such supplements during pregnancy have reduced risks of preterm delivery (twofold lower than controls) and very preterm delivery (fourfold reduction in risk with first trimester use, twofold reduction with second trimester use). Prenatal supplement use was corroborated by assays of circulating micronutrients *(117)*.

A recent prospective study of 1,823 women reported that periconceptional multivitamin use (at least once a week during the preceeding 6 months) was associated with a threefold reduced risk of preterm delivery before 34 completed weeks gestation. There was, however, no reduction in risk for deliveries between 34 and <37 weeks gestation *(119)*. The National Maternal and Infant Health Survey *(118)* showed no effect of multivitamin use on pregnancy outcomes including preterm delivery. Then again the survey was geared toward regular multivitamin use (3 days/week or more) and did not differentiate women who used vitamins less often from those who did not.

Fawzi and colleagues *(120)* examined the effect of supplementing more than 1,000 HIV positive women from Tanzania with multivitamins (most at 2 RDAs), vitamin A, or placebo in a double blind-placebo controlled trial. Multivitamin supplementation decreased the risk of very preterm delivery (approximately twofold reduction) along with decreasing risk of infant low birth weight and fetal growth restriction. A second trial *(121)* was undertaken to demonstrate effectiveness among the HIV negative ($N = 8468$). Results suggested that while supplements decreased the risk of low birth weight and births that were small for gestation, there was no reduction in risk for preterm delivery <37 weeks (16.9 multivitamin vs 16.7 % placebo), or <34 weeks (4.9% multivitamin, 5.6% placebo). On the other hand, gestation duration was significantly, albeit slightly, lengthened (+0.2 week) among the multivitamin supplemented *(121)*. Maternal anemia was reduced by supplementation in both HIV positive and HIV negative women.

However, two contemporary clinical trials of multivitamin supplementation, one in Nepal and the other in Mexico, found no effect of multivitamins (1 RDA) on preterm delivery. In Nepal, a quasi-experimental study assigned pregnant women by community to vitamin A plus the following: folic acid, folic acid + iron, folic acid + iron + zinc, or multiple micronutrients *(122)*. The control group received vitamin A alone. The multivitamin formulation increased birth weight by 64 g and decreased low birth weight by 16%, both changes were statistically significant. The folic acid + iron regimen also decreased low birth weight by 16% but the increase in infant birth weight (37 g) was not significant. Folic acid without iron had no effect on pregnancy outcome. None of the regimens decreased risk of preterm delivery.

A second randomized study was conducted in Mexico *(123)*. Women were randomly assigned to receive iron with or without multivitamins. In comparison to iron alone, multivitamins did not alter birth weight, decrease risk of low birth weight or of preterm delivery. In Guinea Bissau, a randomized trial compared women taking 1 or 2 RDAs of 15 micronutrients to controls on iron + folate. Data showed increased birth weight in women supplemented with 2 RDAs (+88 g) and reduced the odds of a birth below 3,000 g by 58%. Effects on preterm delivery and gestation duration were not reported *(124)*. Thus, despite the modest number of studies to date, data at least from the developing world, are beginning to suggest that it may be important to use micronutrient supplements that exceed 1 RDA to detect effects on pregnancy outcome.

26.8.3.1. Dyslipidemia

Women who deliver preterm have an increased risk of cardiovascular mortality and morbidity years after the event *(125–126)*. One study linked birth and death data from Norwegian women *(125)*. Women with a prior history of preterm delivery had an eightfold increased risk of cardiovascular death when a history of preeclampsia also was recorded and a threefold increased risk when preeclampsia was not. In Pittsburgh, elderly women, aged 70–79 years, with a self-reported history of delivering a preterm and low birth weight infant had an approximately threefold increased risk of cardiovascular morbidity in later life compared with women who reported a term delivery *(126)*.

Consequently, during pregnancy women who deliver preterm may have increased cardiovascular risk factors. Recently Catov *(127)* reported that elevations in maternal plasma triglyceride and cholesterol levels by 15 weeks gestation were associated with an increased risk (2–2.8-fold) of spontaneous preterm delivery. Triglycerides must be hydrolyzed to fatty acids to cross the placenta. In Camden we found that women who had elevated fasting free fatty acids showed a greater than 3.5-fold increased risk of preterm delivery. The associations persisted in women who had spontaneous preterm delivery and after excluding women with gestational diabetes mellitus and/or preeclampsia. Additional stratified analyses showed that the association was independent of pregravid maternal obesity *(128)*. There has been one small-scale clinical trial. A diet to reduced cholesterol levels in pregnant women ($N =$ 290) showed that women randomized to this intervention had lower levels of total and LDL cholesterol as well as a reduced risk of preterm delivery compared to controls (0.7 vs 7.4%) *(129)*. Cardiovascular disease is linked to diet particularly to saturated fat which raises cholesterol and increases risk of heart disease from atherosclerosis. The link between pregnancy dyslipidemia and the maternal diet remains to be explored.

26.9. RECOMMENDATIONS

While the relationship between nutrition and growth, both prenatal and postnatal, is well established, maternal nutrition has emerged as a factor that may increase risk of preterm delivery. Few observational and experimental studies have examined the influence of maternal diet and nutritional status on gestation duration and preterm birth; thus, for greater confidence the preliminary positive findings require replication.

In the developing world, the interaction between nutrition and infection is a well-known cause of childhood malnutrition, morbidity from infectious disease and death. Similarly, infection/inflammation is an important risk factor for preterm delivery. It is plausible that women who develop infection/inflammation during pregnancy have an impaired immune response associated with subclinical micronutrient deficits, but there is virtually no research in this important area from either the developed or the developing world *(130)*. Likewise, it is possible that inflammation may alter the mother's nutritional requirements and lead to low circulating levels of some micronutrients and increased levels of others *(131)*. Dyslipidemia also correlates with C-reactive protein (CRP)—a marker for inflammation *(126)*. Again, virtually no research has been done on this topic in pregnant women. In animals, fasting during early or later pregnancy increases risk of preterm delivery. In humans, data from the Dutch famine *(30)* suggested that early starvation increased risk as do fasting and long intervals between meals *(40–41);* while the mechanism(s) is not known maternal nutrition is clearly implicated.

Populations at high-risk of preterm birth appear to have a poor quality diet but it is unclear if the maternal diet is a risk factor or a risk marker. For example, among minority groups from the inner cities of the United States, the consumption of fresh fruits and vegetables and whole grains is virtually nonexistent *(132)*. Likewise, diets of lower social class women from London failed to meet "...basic maternal needs for a range of nutrients characteristic of whole grain, vegetable and fruit and dairy produce" *(74)*. An alternative nutrient source, vitamin and mineral supplements, is used more frequently by the middle-classes than the poor *(76)*. Thus, it is prudent to eat a healthful diet that includes fish and to complement it with folic acid-containing multivitamin/mineral supplements before and during pregnancy. And, since a poor diet rarely occurs in isolation, it is also important to seek prenatal care and to maintain a lifestyle free of cigarettes, alcohol, and drugs.

REFERENCES

1. Martin J.A., Hamilton B.E., Sutton P.D., Ventura S.J. Menacker F., Kirmeyer S., Munson M.L. (2007) Births: Final Data for 2005. *Natl Vital Stat Rep* **56**, 1–5.
2. Anderson R.N., and Smith B. (2005) Leading Causes of Neonatal and Postneonatal Deaths — United States, 2002. *Natl Vital Stat Rep* **54**, 966. Available at http://www.cdc.gov/nchs/deaths.htm
3. Aylward G., Pfeiffer S., Wright A., and Velhurst S. (1989) Outcome studies of low birth weight infants published in the last decade: a meta analysis. *J Pediatr* **115**, 515–8.
4. Saigal S., Szatmari P., Rosenbaum P., et al. (1990) Intellectual and functional status at school entry of children who weighed 1000 grams or less at birth: a regional perspective of births in the 1980's. *J Pediatr* **116**, 409–16.
5. Astbury J., Orgill A., Bajuk B., et al. (1983) Determinants of developmental performance of very low birth weight survivors at one and two years of age. *Dev Med Child Neurol* **25**, 709–11.
6. Blackman J., Lindgren S., Hein H., et al. (1987) Long term surveillance of high risk children. *Am J Dis Child* **141**, 1293–8.

7. Moster D., Lie R.T., and Markestad T. (2008) Long-term medical and social consequences of preterm birth. *N Eng J Med* **359**, 262–73.
8. Barker D.J.P., Gluckman P.D., Godfrey K.M., et al. (1993) Fetal nutrition and cardiovascular disease. *Lancet* **341**, 938–41.
9. Spong C.Y. (2007) Prediction and prevention of recurrent spontaneous preterm birth. *Obstet Gynecol* **110**, 405–15.
10. Iams J.D., Romero R, Culhane J.F., and Goldenburg R.L. (2008) Primary, secondary, and tertiary interventions to reduce the morbidity and mortality of preterm birth, *Lancet* **371**, 164–75.
11. Simhan H.N., and Caritis S.N. (2007) Prevention of preterm delivery. *N Eng J Med* **357**, 477–87.
12. Savitz D.A., Blackmore C.A., and Thorp J.M. (1991) Epidemiologic characteristics of preterm delivery: etiologic heterogeneity. *Am J Obstet Gynecol* **164**, 467–71.
13. Berkowitz G.S., and Papiernik E. (1993) Epidemiology of preterm birth. *Epidemiol Rev* **15**, 414–43.
14. Institute of Medicine, National Academy of Sciences. (1990) *Nutrition during Pregnancy*. Washington, DC: National Academy Press.
15. Johnston E.M. (1991) Weight changes during pregnancy and the postpartum period. *Prog Food Nutr Sci* **15**, 117–57.
16. Abrams B.F., and Laros R.K. (1986) Prepregnancy weight, weight gain, and birth weight. *Am J Obstet Gynecol* **154**, 503–9.
17. Nohr E.A., Bech B.H., Vaeth M., Rasmussen K.M., Henriksen T.B., and Olsen J. (2007) Obesity, gestational weight gain and preterm birth: a study within the Danish National Birth Cohort. *Pediatr Perinat Epidemiol* **21**, 5–14.
18. Kramer M.S., Coates A.L., Michoud M.C., Dagenais S., Hamilton E.F., and Papageorgiou A. (1995) Maternal anthropometry and idiopathic preterm labor. *Obstet Gynecol* **86**, 744–8.
19. Dietz P.M., Callaghan W.M., Cogswell M.E., Morrow B., Ferre C., and Schieve L.A. (2006) Combined effects of prepregnancy body mass index and weight gain during pregnancy on the risk of preterm delivery. *Epidemiol* **17**, 170–7.
20. Hediger M.L., Scholl T.O., Belsky D.H., Ances I.G., and Salmon R.W. (1989) Patterns of weight gain in adolescent pregnancy: effects on birth weight and preterm delivery. *Obstet Gynecol* **74**, 6–12.
21. Hediger M.L., Scholl T.O., and Salmon R.W. (1989) Early weight gain in pregnant adolescents and fetal outcome. *Am J Hum Biol* **1**, 665–72.
22. Abrams B., Newman V., Key T., and Parker J. (1989) Maternal weight gain and preterm delivery. *Obstet Gynecol* **74**, 577–83.
23. Hickey C.A., Cliver S.P., McNeal S.F., et al. (1995) Prenatal weight gain patterns and spontaneous preterm birth among nonobese black and white women.*Obstet Gynecol* **85**, 909–14.
24. Carmichael S., and Abrams B. (1997) A critical review of the relationship between gestational weight gain and preterm delivery. *Obstet Gynecol* **89**, 865–73.
25. Thomson A.M. (1959) Diet in relation to the course and outcome of pregnancy. *Br J Nutr* **13**, 509–23.
26. Scholl T.O., Hediger M.L., Schall J.I., Fischer R.L., and Khoo C.S. (1993) Low zinc intake during pregnancy: its association with preterm and very preterm delivery. *Am J Epidemiol* **137**, 1115–24.
27. Scholl T.O., Hediger M.L., Khoo C.S., et al. (1991) Maternal weight gain, diet and infant birth weight: Correlations during adolescent pregnancy. *J Clin Epidemiol* **44**, 423–8.
28. Olson C.M., and Strawderman M.S. (2003) Modifiable behavioral factors in a biopsychosocial model predict inadequate and excessive gestational weight gain. *J Am Diet Assoc* **103**, 48–54.
29. Rayco-Solon P., Fulford A.J., and Prentice A.M. (2005) Maternal preconceptional weight and gestational length. *Am J Obstet Gynecol* **192**, 1133–6.
30. Stein Z., Susser M., Saenger G., et al. (1975) *Famine and Human Development: The Dutch Hunger Winter of 1944/1945*. New York, NY: Oxford University Press.
31. Bloomfield F.H.., Oliver M.H., Hawkins P., Campbell M., Phillips D.J., Gluckman P.D., Challis J.R.G., and Harding J.E. (2003) A periconceptional nutritional origin for noninfectious preterm birth. *Science* **300**, 606–7.
32. Miller G. (2003) Hungry Ewes. *Science* **300**, 50–2.
33. Fowden A.L., Ralph M.M., and Silver M. (1994) Nutritional regulation of uteroplacental prostaglandin production and metabolism in pregnant ewes and mares during late gestation. *Exp Clin Endocrinol* **102**, 212–21.
34. Fowden A.L., and Silver M. (1983) The effect of the nutritional state on uterine prostaglandin F metabolite concentrations in the pregnant ewe during late gestation. *Q J Exp Physiol* **68**, 337–49.
35. Silver M., and Fowden A.L. (1982) Uterine prostaglandin F metabolite production in relation to glucose availability in late pregnancy and a possible influence of diet on time of delivery in the mare. *J Reprod Fertil* **32**(suppl), 511–9.
36. Binienda Z., Massmann A., Mitchell M.D., Gleed R.D., Figueroa J.P., and Nathanielsz P.W. (1989) Effect of food withdrawal on arterial blood glucose and plasma 13,14-dihydro-15 keto-prostaglandin $F_{2\alpha}$ concentrations and noctur-

nal myometrial electromyographic activity in the pregnant rhesus monkey in the last third of gestation: a model for preterm labor. *Am J Obstet Gynecol* **160**, 746–50.
37. Kaplan M., Eidelman A.I., and Aboulafia Y. (1983) Fasting and the precipitation of labor. *JAMA* **250**, 1317–8.
38. Frentzen B.H., Johnson J.W.C., and Simpson S. (1987) Nutrition and Hydration: relationship to preterm myometrial contractility. *Obstet Gynecol* **70**, 887–91.
39. Metzger B., Vileisis R.A., Ravnikac V., and Freinkel N. (1982) Accelerated starvation and the skipped breakfast in late normal pregnancy. *Lancet* **319**, 588–92.
40. Siega-Riz A.M., Herrmann T.S., Savitz D.A., and Thorp J.M. (2001) Frequent of eating during pregnancy and its effect on preterm delivery. *Am J Epidemiol* **153**, 647–52.
41. Herrmann T.S., Siega-Riz A.M., Hobel C.J., Aurora C., and Dunkel-Schetter C. (2001) Prolonged periods without food intake during pregnancy increase risk for elevated maternal corticotropin-releasing hormone concentrations *Am J Obstet Gynecol* **185**, 403–12.
42. Smith R., Mesiano S., and McGrath S. (2002) Hormone trajectories leading to human birth. *Regul Pept* **108**, 159–64.
43. Rosso P., and Streeter M.R. (1979). Effects of food or protein restriction on plasma volume expansion in pregnant rats. *J Nutr* **109**, 1887–92.
44. Scholl T.O., Hediger M.L., Fischer R.L., and Shearer J.W. (1992) Anemia vs iron deficiency: increased risk of preterm delivery in a prospective study. *Am J Clin Nutr* **55**, 985–8.
45. Scholl T.O., and Hediger M.L. (1994) Anemia and iron-deficiency anemia: compilation of data on pregnancy outcome. *Am J Clin Nutr* **59**(suppl), 492–501S.
46. Klebanoff M.A., Shiono P.H., Berendes H.W., and Rhoads G.G. (1988) Facts and artifacts about anemia and preterm delivery. *JAMA* **262**, 511–5.
47. Klebanoff M.A., Shiono P.H., Selby J.V., et al (1991) Anemia and spontaneous preterm birth. *Am J Obstet Gynecol* **164**, 59–63.
48. Zhou L.M., Yang W.W., Hua J.Z., Deng C.Q., Tao X., and Stolzfus R.J. (1998) Relation of hemoglobin measured at different times in pregnancy to preterm birth and low birth weight in Shanghai, China. *Am J Epidemiol* **148**, 998–1006.
49. Steer P., Alam A., Wadsworth J., and Welch A. (1995) Relation between maternal haemoglobin concentration and birth weight in different ethnic groups. *Br Med J* **310**, 489–91.
50. Scanlon K.S., Yip R., Schieve L.A., et al (2000) High and Low hemoglobin levels during pregnancy: differential risks for preterm birth and small for gestational age. *Obstet Gynecol* **96**, 741–8.
51. Brabin B., Ginny M., Supau J., et al. (1990) Consequences of maternal anaemia on outcome of pregnancy in malaria endemic area in Paoua, New Guinea. *Ann Trop Med Parisatol* **84**,11–24.
52. Garn S.M., Ridella S.A., Petzold A.S., and Falkner F. (1981) Maternal hematologic levels and pregnancy outcomes *Semin Perinatol* **5**, 155–162.
53. Murphy J.F., O'Riordan J., Newcombe R.G., et al. (1986) Relation of haemoglobin levels in first and second trimesters of pregnancy. *Lancet* **1**, 992–5.
54. Goldenberg R.L., Tamura T., DuBard M., Johnston K.E., Copper R.L., and Neggers Y. (1996) Plasma ferritin and pregnancy outcome. *Am J Obstet Gynecol* **175**, 1356–9.
55. Scholl T.O. (1998) High third-trimester ferritin concentration: associations with very preterm delivery, infection, and maternal nutritional status. *Obstet Gynecol* **92**, 161–5 and **93**, 156–7.
56. Goldenberg R.L., Mercer B.M., Miodovnik M., Thurnau G.R., Meis P.J., Moawad A., Paul R.H., Bottoms S.F., Das A., Roberts J.M., McNellis D., and Tamura T. (1998) Plasma ferritin, premature rupture of membranes, and pregnancy outcome. *Am J Obstet Gynecol* **179**, 1599–604.
57. Cogswell M.E., Parvanta I., Ackes L., Yip R., and Brittenham G.M. (2003) Iron supplementation during pregnancy, anemia and birth weight: a randomised controlled trial. *Am J Clin Nutr* **78**, 773–81.
58. Siega-Riz A.M., Hartzema A.G., Turnbull C, et al. (2006) the effects of prophylactic iron given in prenatal supplements on iron status and birth outcomes: a randomized controlled trial. *Am J Obstet Gynecol* **194**, 512–9.
59. Makrides M., Crowther C.A., Gibson R.A., Gibson R.S., and Skaeff C.M. (2003) Efficacy and tolerability of low-dose iron supplements during pregnancy: A randomized controlled trial. *Am J Clin Nutr* **78**, 145–53.
60. Allen L.H. (2001) Biological mechanisms that might underlie iron′s effects on fetal growth and preterm birth. *J Nutr* **131**, 581–9S.
61. Scholl T.O., Hediger M.L., Schall J.I., Khoo C.S., and Fischer R.L. (1996) Dietary and serum folate: their influence on the outcome of pregnancy. *Am J Clin Nutr* **63**, 1–6.
62. Pilch S.M., and Senti F.M., eds. (1984) Assessment of the zinc nutritional status of the U.S. population based on data collected in the Second National Health and Nutrition Examination Survey, 1976—Bethesda, MD: Life Sciences Research Office, Federation of American Societies for Experimental Biology.

63. Pilch S.M., and Senti F.M., eds. (1984) Assessment of the folate nutritional status of the U.S. population based on data collected in the Second National Health and Nutrition Examination Survey, 1976—Bethesda, MD: Life Sciences Research Office, Federation of American Societies for Experimental Biology.
64. Neggers Y.H., Cutter G.R., Acton R.T., et al. (1990) A positive association between maternal serum zinc concentration and birth weight. *Am J Clin Nutr* **51**, 678–84.
65. Kirksey A., Wachs T., Yunis F., Srinath U., Rahmanifar A., McCabe G., Galal O., Harrison G., and Jerome N. (1994) Relationship of maternal zinc nutriture to pregnancy outcome and infant development in an Egyptian village. *Am J Clin Nutr* **60**, 782–92.
66. Sikorski R., Juszkiewicz T., Paszkowski T. (1990) Zinc status in women with premature rupture of membranes at term. *Obstet Gynecol* **76**, 675–7.
67. Mahomed K., Bhutta Z., and Middleton P. (2007) Zinc supplementation for improving pregnancy and infant outcome. *Cochrane Database Sys Rev* **2**, CD000230.
68. Cherry F.F., Sanstead H.H., Rojas P., et al. (1989) Adolescent pregnancy: association among body weight, zinc nutriture and pregnancy outcome. *Am J Clin Nutr* **50**, 945–54.
69. Goldenberg R.L., Tamura T., Neggers Y., et al. (1995) The effect of zinc supplementation on pregnancy outcome. *JAMA* **274**, 463–8.
70. Osendarp S.J.M., van Raaiji J.M.A., Wahed M.A., et al. (2000) A randomized, placebo controlled trial of the effect of zinc supplementation during pregnancy on pregnancy outcome in Bangladeshi urban poor. *Am J Clin Nutr* **71**, 114–9.
71. Caufield L.E., Zavaleta N., Figueroa A., et al. (1999) Maternal zinc supplementation does not affect size at birth or pregnancy outcome. *J Nutr* **129**, 1563–8.
72. Ferrand P.E., Parry S., Sammel M., Macones G.A., Kuivaniemi H., Romero R., and Strauss J.F.(2007) A polymorphism in the matrix metalloproteinase-9 promoter is assocated with increased risk of preterm premature rupture of membranes in African Americans. *Mol Hum Reprod* **8**, 494–501.
73. Keen C.L., Clegg M.S., Hanna L.A., Lanoue L., Rogers J.M., Daston G.P., Oteiza P., and Uriu-Adams J.Y. (2003) The plausibility of micronutrient deficiencies being a significant contributing factor to the occurrence of pregnancy complications. *J Nutr* **133**, 1597S–605S.
74. Wynn S.W., Wynn A.M., Doyle W., and Crawford M.A. (1994) The association of maternal social class with maternal diet and the dimensions of babies in a population of London women. *Nutr Health* **9**, 303–15.
75. Subar A.F., and Block G. (1990) Use of vitamin and mineral supplements: demographics and amounts of nutrients consumed (The 1987 Health Interview Survey). *Am J Epidemiol* **133**, 1091–101.
76. Suitor C.W., and Gardner J.D. (1990) Supplement use among culturally diverse group of low-income pregnant women *J Am Diet Assoc* **90**, 268–71.
77. Centers for Disease Control (1999) Knowledge and use of folic acid by women of childbearing age. *MMWR* **48**, 325–7.
78. Honein M.A., Paulozzi L.J., Mathews T.J., et al. (2001) Impact of folic acid fortification on the US food supply on the occurrence of neural tube defects. *JAMA* **285**, 2981–6.
79. Neuhouser M.L., Beresford S.A.A. (2001) Folic acid: are current fortification levels adequate? *Nutrition* **17**, 868–72.
80. Shaw G.M., Carmichael S.L., Nelson V., Selvin S., and Schaffer D.M. (2004) Occurrence of low birthweight and preterm delivery among California infants before and after compulsory food fortification with folic acid. *Public Health Rep* **119**, 170–3.
81. Johnson W.G., Scholl T.O., Spychala J.R., et al. (2005) Commoin dihydrofolate reductase 19-base deletion allel: a novel risk factor for preterm delivery. *Am J Clin Nutr* **81**, 664–8.
82. Siega-Riz A.M., Savitz D.A., Zeisel S.H., Thorp J.M., and Herring A. (2004) Second trimester folate status and preterm birth. *Am J Obstet Gynecol* **191**, 1851–7.
83. Engel S.M., Olshan A.F., Siega-Riz A.M., Savitz D.A., and Chanock S.J. (2006) Polymorphisms in folate metabolizing genes and risk for spontaneous preterm and small-for-gestational age birth. *Am J Obstet Gynecol* **195**, 1231.e1–1231.e11.
84. Fleming A.F., Martin J.D., Hahnel J.R., and Westlake A.J. (1974) Effects of iron and folic acid antenatal supplements on maternal haematology and fetal wellbeing. *Med J Aust* **2**, 429–36.
85. Fletcher J., Gurr A., Fellingham F.R., Prankerd T.A.J., Brant H.A., and Menzies D.N. (1971) The value of folic acid supplements in pregnancy. *J Obstet Gynaecol Br Commonw* **78**, 781–5.
86. Giles P.F.H., Harcourt A.G., and Whiteside M.G. (1971) The effect of prescribing folic acid during pregnancy on birth-weight and duration of pregnancy. *Med J Aust* **2**, 17–21.
87. Blot I., Papiernik E., Kaltwasser J.P., et al. (1981) Influence of routine administration of folic acid and iron during pregnancy. *Gynecol Obstet Invest* **12**, 294–304.

88. Tcherina G., Blot I., Rey A., et al. (1982) Maternal folate status, birthweight and gestational age. *Dev Pharmacol Therap* **4**(suppl 1), 58–65.
89. Mohammed K. (1993) Routine folate supplementation in pregnancy. In: Enkin MW, Keirse MJCN, Renfreq MJ, Nielson JP (eds), *Pregnancy and Childbirth Module "Cochrane Database of Systematic Reviews"* Review No. 03158.
90. Rajkovic A., Catalano P.M., and Malinow M.R. (1997) Elevated homocyst(e)ine levels with preeclampsia. *Obstet Gynecol* **90**, 168–71.
91. Vollset S.E., Refsum H., Emblem B.M., et al. (2000) Plasma total homocysteine, pregnancy complications and adverse pregnancy outcomes: The Hordaland homocysteine study. *Am J Clin Nutr* **71**, 962–8.
92. Ronnenberg A.G., Goldman M.B., Chen D., Aitken I.W., Willett W.C., Selhub J., and Xu X. (2002) Preconception homocysteine and B vitamin status and birth outcomes in Chinese women. *Am J Clin Nutr* **76**, 1385–91.
93. Murphy MM, Scott JM, Arija V, et al. (2004) Maternal homocysteine before conception and throughout pregnancy predicts fetal Homocysteine and birth weight. *Clin Chem* **50**, 1406–12.
94. Malinow M.R., Rajkovic A., Duell P.B., Hess D.L., and Upson B.M. (1998) The relationship between maternal and neonatal umbilical cord plasma homocyst(e)ine suggests a potential role for maternal momocyst(e)ine in fetal metabolism. *Am J Obstet Gynecol* **178**, 228–33.
95. Repke JT. (1991) Calcium magnesium, and zinc supplementation and perinatal outcome. *Clin Obstet Gynecol* **34**, 262–7.
96. NIH Consensus Development Panel on Optimal Calcium Intake (1994) Optimal calcium intake. *JAMA* **272**, 1942–8.
97. Zhang J., Villar J., Sun W., et al. (2007) Blood pressure dynamics during pregnancy and spontaneous preterm birth. *Am J Obstet Gynecol* **197**, 162.e1–e6.
98. Repke J.T., and Villar J. (1991) Pregnancy-induced hypertension and low birth weight: the role of calcium. *Am J Clin Nutr* **54**, 237–41S.
99. Belizán J.M., Villar J., Conzalez L., et al. (1991) Calcium supplementation to prevent hypertensive disorders of pregnancy. *N Eng J Med* **325**, 1399–405.
100. Scholl T.O., Hediger M.L., and Schall J.I. (1997) Maternal growth and fetal growth: pregnancy course and outcome in the Camden Study. *Ann NY Acad Sci* **817**, 281–91.
101. Sowers M.F., Scholl T.O., and Harris L. (2000) Bone loss in adolescent and adult pregnant women. *Obstet Gynecol* **96**, 189–93.
102. López-Jaramillo P., Narváez M., Weigel R.M., Yépez R. (1989) Calcium supplementation reduces the risk of pregnancy-induced hypertension in an Andes population. *Br J Obstet Gynaecol* **96**, 648–55.
103. López-Jaramillo P., Narváez M., Felix C., and López A. (1990) Dietary calcium supplementation and prevention of pregnancy hypertension. *Lancet* **335**, 293.
104. Bucher, H.C., Guyatt, G.H., Cook, R.J., Hatala, R., Cook, D.J., Lang, J.D., and Hunt, D. (1996) Effect of calcium supplementation on pregnancy-induced hypertension and preeclampsia. *JAMA* **275**, 1113–7.
105. Levine, R.J., Hauth, J.C., Curet, L.B., Sibai, B.M., Catalano, P.M., Morris, C.D., Der Simonian R., Esterlitz, J.R., Raymond, E.G., Bild, D.E., Clemens, J.D., and Cutler, J.A. (1997) Trial of calcium to prevent preeclampsia. *N Eng J Med* **337**, 69–76.
106. Crowther C.A., Hiller J.E., Pridmore B., Bryce R., Duggan P., Hague W.M., and Robinson J.S. (1999) Calcium supplementation in nulliparous women for the prevention of pregnancy-induced hypertension, preeclampsia and preterm birth: an Australian randomized trial FRACOG and the ACT Study Group. *Aust NZ J Obstet Gynecol* **39**, 12–8.
107. Villar J., and Repke J.T. (1990) Calcium supplementation during pregnancy may reduce preterm delivery in high-risk populations. *Am J Obstet Gynecol* **163**, 1124–31.
108. Villar J., Abdel-Aleem H., Merialdi M., et al (2006) World health organization randomized trial of calcium supplementation among low calcium intake pregnant women. *Am J Obstet Gynecol* **194**, 639–49.
109. Olsen S.F., Hansen H.S., Sorensen T., et al. (1986) Intake of marine fat, rish in (n-3) polyunsaturated fatty acids, may increase birthweight by prolonging gestation. *Lancet* **2**, 367–9.
110. People's League of Health. (1942) Nutrition of expectant and nursing mothers. *Lancet* **2**, 10–2.
111. Olsen S.F., Sorensen J.D., Secher N.J., et al. (1992) Randomized controlled trial of effect of fish-oil supplementation on pregnancy duration. *Lancet* **339**, 1003–7.
112. Olsen S.F., Secher N.J., Tabor A., et al. (2000) Randomized clinical trials of fish oil supplementation in high-risk pregnancies. *Br J Obstet Gynecol* **107**, 382–95.
113. Olsen S.F., and Secher N.J. (2002) Low consumption of seafood in early pregnancy as a risk factor for preterm delivery: Prospective cohort study. *BMJ* **234**, 1–5.

114. Makrides M., Duley L., Olsen S.F. (2008) Marine oil, and other prostaglandin precursor, supplementation for pregnancy uncomplicated by preeclampsia or intrauterine growth restriction (review). *Cochrane Database Syst Rev* **3**, CD003402.pub2.
115. Secher N.J. (2007) Does fish oil prevent preterm birth? *J Perinat Med* **35**, S25–7.
116. Block G., Cox C., Madans J., Schreiber G.B., Licitra L., and Melia N. (1988) Vitamin supplement use, by demographic characteristics. *Am J Epidemiol* **127**, 297–309.
117. Scholl T.O., Hediger M.L., Bendich A., Schall J.I., Smith W.K., and Krueger P.M. (1997) Use of multivitamin/mineral prenatal supplements: influence on the outcome of pregnancy. *Am J Epidemol* **146**, 134–41.
118. Wu T., Buck G., Mendola P. (1998) Can regular multivitamin/mineral supplementation modify the relation between maternal smoking and select adverse birth outcomes. *Ann Epidemiol* **8**, 179–83.
119. Catov J.M., Bodnar L.M., Ness R.B., Markovic N., and Roberts J.M. (2007) Association of preconceptional multivitamin use and risk of preterm or small-for-gestational-age births. *Am J Epidemiol* **166**, 296–303.
120. Fawzi W.W., Msamanga G.I., Spiegelman D., Urassa E.J.N., MaGrath N., Mwakagile D., Antelman G., Mbise R., Herrera G., Kapiga S., Willett W., and Hunter D.J. (1998) Randomized trial of effects of vitamin supplements on pregnancy outcomes and T cell counts in HIV-1-infected women in Tanzania. *Lancet* **351**, 1477–82.
121. Fawzi W.W., Msamanga G.I., Urassa E.J.N., Hertzmark E., Petraro P., Willett W., and Spiegelman D. (2007) Vitamins and Perinatal outcomes among HIV-Negative women in Tanzania. *N Eng J Med* **356**, 1423–31.
122. Christian P., Khatry S.K., Katz J., Pradhan E.K., LeClerq S.C., Shrestha S.R., Adhikari R.K., Sommer A., West Jr K.P. (2003) Effects of alternative maternal micronutrient supplements on low birth weight in rural Nepal: double blind randomised community trial. *BMJ* **326**, 1–6.
123. Ramakrishnan U., Gonzalez-Cossio T., Neufeld L.M., Rivera J., and Martorell R. (2003) Multiple micronutrient supplementation during pregnancy does not lead to greater infant birth size than does iron-only supplementation: a randomized controlled trial in a semirural community in Mexico. *Am J Clin Nutr* **77**, 720–5.
124. Kaestel P., Michaelsen K.F., Aaby P., and Friis H. (2005) Effects of prenatal multimicronutrient supplements on birth weight and Perinatal mortality: a randomized, controlled trial in Guinea-Bissau. *Eur J Clin Nutr* **59**, 1081–9.
125. Irgens H.U., Reisaeter L., Irgens M., and Lie R.T. (2001) Long term mortality of mothers and fathers after preeclampsia: population based cohort study. *BMJ* **323**, 1213–7.
126. Catov J.M., Newman A.B., Roberts J.M., et al (2007) Preterm delivery and later maternal cardiovascular disease risk. *Epidemiology* **18**, 733–9.
127. Catov J.M., Bodnar L.M., Kip K.E., Hubel C., Ness R.B., Harger G., and Roberts J.M. (2007) Early pregnancy lipid concentrations and spontaneous preterm birth. *Am J Obstet Gynecol* **197**, 610.e1–e7.
128. Chen X., and Scholl T.O. (2008) Association of elevated free fatty acids during late pregnancy with preterm delivery—Camden study. *Obstet Gynecol*, 112:297–303.
129. Khoury J., Henriksen T., Christophersen B., and Tonstad S. (2005) Effect of a cholesterol-lowering diet on maternal, cord, and neonatal lipids, and pregnancy outcome: a randomized clinical trial. *Am J Obstet Gynecol* **193**, 1292–301.
130. Goldenberg R.L. (2003) The plausibility of micronutrient deficiency in relationship to perinatal infection. *J Nutr* **133**, 1645S–8S.
131. Tomkins A. (2003) Assessing micronutrient status in the presence of inflammation. *J Nutr* **133**, 1649S–55S.
132. Rogers M.A., Simon D.G., and Zucker L.B. (1995) Indicators of poor dietary habits in a high risk population. *J Am Coll Nutr* **14**, 159–64.

27 Linking Prenatal Nutrition to Adult Mental Health

Kristin Harper, Ezra Susser, David St. Clair, and Lin He

Key points

- Studies on Dutch and Chinese cohorts in the Netherlands have found an association between prenatal exposure to famine and risk of developing schizophrenia.
- The mechanisms responsible for this increased risk are unknown at present, but could include elevated mutation rate, altered epigenetic regulation, and/or gene–environment interactions.
- As examples of possible contributing nutritional factors, the effects of selenium, zinc, and folate deficiencies upon brain development, mutation rate, and methylation, as well as their interaction with human polymorphisms, are discussed.
- The interaction between micronutrients in the body is complex and must be considered in any study examining the association between a given deficiency and risk of schizophrenia.
- Understanding the role of micronutrient deficiencies in the development of schizophrenia may help elucidate risk factors for other mental disorders.

Key Words: Fetal malnutrition; famine; micronutrients; schizophrenia

27.1. INTRODUCTION

In most settings, malnutrition is strongly associated with poverty, and even in the exceptional circumstances in which an entire population is affected by malnutrition, the poor tend to be the most affected *(1,2)*. An extensive body of work has examined the effects of prenatal and childhood malnutrition on child development and health (*see* Delisle *(3)* for a review). Also, researchers have related specific micronutrients to the risk of specific neurodevelopmental disorders in children. For example, folate supplements reduce the risk of neural tube defects, and low iodine intake can cause cretinism *(4–6)* (*see* chapter 25 and 26).

The impact of prenatal malnutrition may not, however, be limited to health and mental health effects that are evident at birth or in childhood. Recent work has provided support for the view that there are also latent effects which become evident in adult life *(7,8)*. This work suggests that we need to extend our perspective to encompass health throughout the life course *(9)*. By limiting our purview to reproductive and child outcomes, we limit our understanding of optimal prenatal and childhood nutrition and of the potential benefits of improved nutrition. The potential effects on adult health should also be considered and cannot be inferred from the health effects detectable in early life.

An extensive literature has now established that early life experiences can have an impact on adult health *(10)*. For example, low birthweight is related to increased risk of cardiovascular disease. Sometimes this work is interpreted loosely as demonstrating that prenatal maternal nutrition influences

A. Bendich, R.J. Deckelbaum (eds.), *Preventive Nutrition*, Nutrition and Health, DOI 10.1007/978-1-60327-542-2_27,

offspring's adult health. However, birthweight is probably not a good index of maternal nutrition during pregnancy. Studies based on the Dutch famine, described below, suggest that, to the extent that birthweight does reflect maternal nutrition, it reflects mainly nutritional intake during the last trimester of gestation *(2,11)*.

There have, in fact, been very few studies with direct measures of both maternal prenatal nutrition and adult health in offspring. The central studies in this field have been follow-ups of the birth cohorts exposed and unexposed to prenatal famine during the Dutch Hunger Winter of 1944–1945. These studies find some evidence of effects of prenatal famine on obesity and insulin resistance, but as yet, the effects are modest and are not consistent across studies *(12–16)*.

Despite the evidence noted earlier that prenatal nutrition influences neurodevelopment, even fewer studies have examined the relation of maternal nutrition to adult mental health. Again, the results derive mainly from studies of the Dutch Hunger Winter. The strongest positive result links early gestational exposure to famine with an increased risk of schizophrenia *(17–20)*. There have been reports of correlations between exposure to the famine and other psychiatric disorders. Males exposed to the Hunger Winter during the first trimester had a significantly greater likelihood of being treated for addiction as adults *(21)*. Additional studies have suggested that exposure to famine during other periods of gestation may influence risk of developing other psychiatric disorders. For example, males exposed during either the first or second trimester were more likely to develop antisocial personality disorder as young adults *(22)*. Also, second and third trimester exposure increased the odds of being hospitalized for major affective disorders *(23,24)*, though no association was found between prenatal exposure and a depressive symptom score in a follow-up of a cohort in midlife *(25)*. Nor has an effect of prenatal famine exposure on IQ, as measured at 18 years in males at military induction, been found *(26)*. The link between exposure to famine and schizophrenia was recently replicated, when a second series of studies based on the Chinese famine of 1959–1961 produced congruent findings *(27,28)*. At this point, therefore, the strongest result linking prenatal starvation with the later development of adult mental disorders is the increased risk of schizophrenia demonstrated in both the Dutch Hunger Winter and Chinese famine studies.

In light of the state of the evidence, therefore, we focus this chapter primarily on the evidence for a latent effect of prenatal malnutrition on schizophrenia. By "latent" effect, we mean that the mental disorder is not detectable in childhood, though we do not exclude the possibility of subtle manifestations in childhood. Establishing one clear example of a latent effect on a major adult mental disorder opens the door to considering a range of other potential effects. We discuss some of the mechanisms through which prenatal nutritional deficiencies could influence the risk of developing schizophrenia, and we highlight three specific micronutrients whose intake would be expected to fall during famine and which could work through the mechanisms discussed.

27.2. PRENATAL FAMINE AND SCHIZOPHRENIA

Schizophrenia is a disorder characterized by psychotic symptoms such as delusions and hallucinations, as well as deficits in other domains, such as motivation and affect. As the disorder is often associated with long-term disability, it ranks among the top ten causes of disability in the WHO classification *(29)*. Currently most investigators consider schizophrenia to be a neurodevelopmental disorder, in the sense that it has some early origins in brain development, but the disorder is not diagnosed until adolescence or adulthood. Typically, the full syndrome required for diagnosis emerges between age 16 and 29 years. In many cases, however, certain signs or symptoms are apparent long before the full syndrome and in other cases the syndrome does not emerge until midlife or even late-life.

The hypothesis that prenatal nutrition may be related to schizophrenia dates back to the mid-20th century *(30)*. It was not tested, however, until the end of the 20th century *(18)*. The first test was based around the historical circumstance of the Dutch Hunger Winter of 1944–1945 *(2,31)*. This famine was precipitated by a Nazi blockade of the occupied region of Holland in October 1944 and exacerbated by the severe winter which soon followed. The food shortage was most severe in the occupied cities of western Holland; in the rural areas, there was more access to supplementary food.

Three remarkable features of this famine made it possible to study its effects on schizophrenia in adulthood. First, the food rations distributed to the population were documented. Although individuals found ways to supplement the official ration, the caloric content of the ration was strongly correlated with intake. When rations fell, food intake fell. Second, the peak period of famine was of short duration. The famine ended abruptly in May 1945, when the Allied troops liberated western Holland. In the last months of the famine, the ration fell to extremely low levels, and supplementary food was increasingly difficult to find. In addition, the population was nutritionally depleted. Thus the period of most severe starvation was approximately from February 1945 until liberation in early May. The increased severity in these last months is reflected in data on mortality, fertility, and birthweight *(2)*. Third, information could be obtained on schizophrenia admissions in adulthood for the individuals born in the famine cities of western Holland before, during, and after the famine. The Dutch national psychiatric registry recorded hospital admissions for specific diagnoses from 1970 onward. Taken together, these features made it possible to define birth cohorts exposed to famine at specific periods of gestation and to test whether the exposure was linked to an increased risk of schizophrenia.

Although the result for schizophrenia emerged through a series of studies *(17)*, we summarize here the main findings in a single figure. In Fig. 27.1, we define the birth cohort of October 16–December 31, 1945 as severely exposed to famine in early gestation. Based on their birth dates, we can infer that the vast majority of this birth cohort was conceived or in early gestation during the peak of the famine. This inference is supported by the excess of neural tube defects and other congenital anomalies of the central nervous system in this birth cohort (Fig. 27.1). It is also supported by the drop in the birth rate, which reached a nadir in September 1945 and remained low until the end of 1945 (Fig. 27.1); the decline in fertility correlated strongly with rations around the time of conception in these Dutch birth cohorts *(2)*. Fig. 27.1 also illustrates that this exposed birth cohort had a sharply increased risk of schizophrenia in adulthood, a result based on the national psychiatric registry data. Finally, the figure illustrates that among males there was an excess of Schizoid Personality Disorder diagnosed at age 18 in the same exposed birth cohort *(20,32)*. This result derived from military induction data: all males born 1944–1946 were subject to a military draft at age 18 years, and the induction examination included a psychiatric assessment. Current diagnostic practice would probably classify the individuals with Schizoid Personality as having a "Schizophrenia Spectrum" personality disorder; familial aggregation studies and other evidence suggest that "Schizophrenia Spectrum" personality disorders are etiologically related to schizophrenia *(33)*.

In sum, the results from the Dutch famine studies suggest that in the birth cohort conceived or in early gestation at the height of the famine, there was a sharp increase in neurodevelopmental disorders at birth, in adolescence, and in adulthood. It appears then that the same exposure led to different neurodevelopmental disorders at different points in the life course. These disorders may also share other features. For instance, neural tube defects and schizophrenia have both been associated with a genetic variant in the folate pathway, MTHR677TT *(34)*, and as noted above, schizophrenia and schizophrenia spectrum personality disorders share a genetic diathesis. At this point, however, the possibility of an interrelationship among all these three disorders is intriguing but not established.

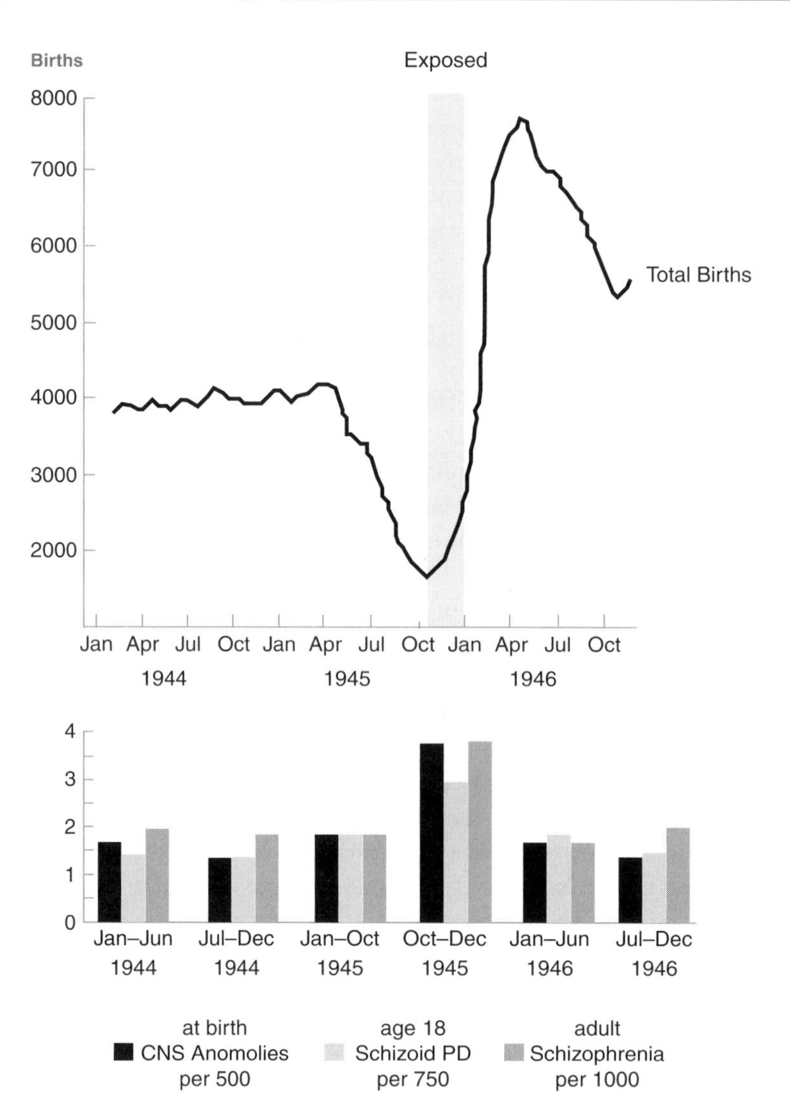

Fig. 27.1. Dutch famine birth cohorts of October 16–December 31, 1945. Reproduced from Susser et al. *(137)*, with permission from the New York Academy of Sciences.

Although these data offered fairly compelling evidence for a link between prenatal malnutrition and the risk of schizophrenia in offspring, the number of exposed cases was modest, and a single study is rarely sufficient basis for a causal inference. Also, other interpretations were plausible. For instance, during periods of famine people often resort to food substitutes—such as tulip bulbs in the Dutch Hunger Winter—that might be toxic to the developing brain. Although starvation would have led to the ingestion of these food substitutes, a causal pathway mediated by ingestion of toxic food substitutes would have different implications for the pathophysiology and ultimately for preventive interventions.

It proved difficult, however, to find another historical circumstance in which this result could be tested. Famines are common, but the documentation of psychiatric outcomes in a defined population for decades after a famine is rare indeed. It was nearly a decade before the finding was replicated, and this study was done by an independent group *(27)*. The study was done in the Wuhu region of

Anhui Province, China and based on the devastating famine which afflicted China following the Great Leap Forward. In Wuhu, the peak of this famine was in 1959 and 1960. The key data available in Wuhu were the number of births in each year and the number of people born in these years who subsequently received outpatient or inpatient treatment for schizophrenia. The authors were able to demarcate a district that was served by the same psychiatric hospital over the main decades of risk for schizophrenia in the relevant birth cohorts. Also, the population of this district was remarkably stable over these decades, in part because of tradition and in part because changes in district of residence were controlled.

The Chinese famine was long lasting, and the Wuhu data on birth rates were available for years rather than months or weeks. Following the result for the Dutch study, however, it could be hypothesized that the schizophrenia risk should peak in those birth years in which the birth rate dropped. (The Dutch famine results in Fig. 27.1 indicate that the schizophrenia risk peaked shortly after the nadir in the monthly birth rate.) This is exactly what was observed (Fig. 27.2). Although the measure of exposure was less precise, the numbers were much larger, making these two studies quite complementary.

Two studies with concordant results substantially strengthened the hypothesis that prenatal malnutrition had latent effects on adult schizophrenia. The lead investigators of the two studies (ES, DS, and LH) now joined together in the completion of a third study that could further solidify (or undermine) the hypothesis. This was conducted in the region of Liuzhou, in southern China, using essentially the same design as the Wuhu study. Again the results were concordant; that is, the risk of schizophrenia peaked in the annual birth cohorts in which the birth rate dropped *(28)*.

Some special features of this third study added strength to the hypothesis. It was based on a very large number of people, even larger than the Wuhu study. It was conducted in a region of China that differed in customs, ethnic diversity, and historical famine experience from the Wuhu region. Finally,

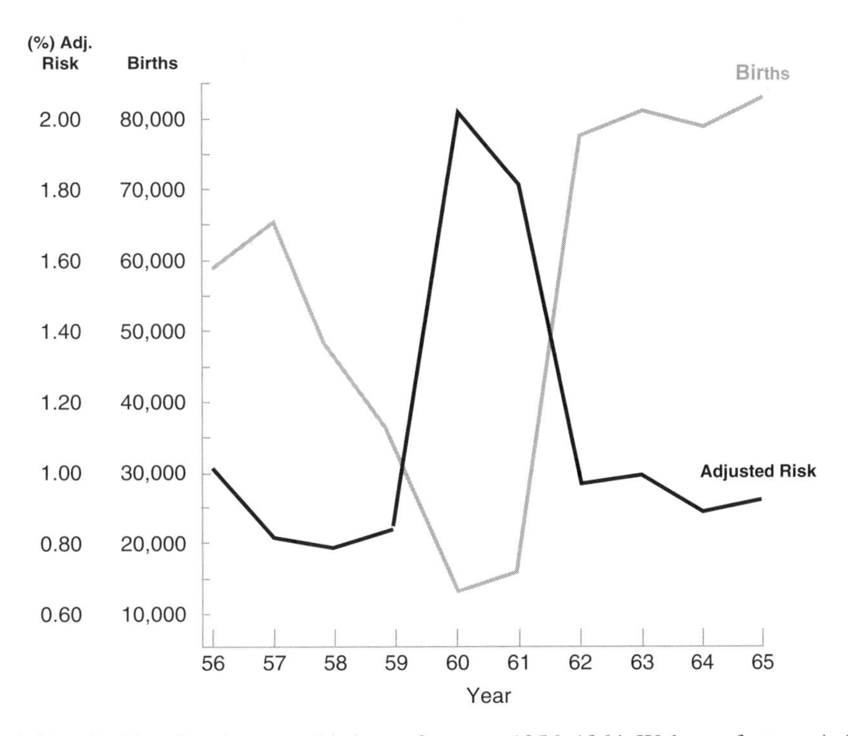

Fig. 27.2. Adjusted risk of schizophrenia versus birth rate for years 1956–1964. Wuhu prefecture, Anhui. Reproduced from Susser et al. *(137)*, with permission from the New York Academy of Sciences.

and most important, it permitted us to clearly differentiate the impact in urban and rural areas. This was important because, in contrast to Holland, the famine in China affected mainly the rural areas. Urban residents received rations and generally suffered little or no starvation, while the rural population suffered starvation on a massive scale. Therefore, an increased risk of schizophrenia due to prenatal malnutrition should be evident in the rural but not the urban area. This is what we observed *(28)*.

27.3. MECHANISM

The two Chinese studies, together with the earlier Dutch study, make a very compelling case that prenatal exposure to famine increases risk of schizophrenia and possibly other forms of major mental illness in later life. The specific risk factors and mechanisms involved are unclear, however. Various biological pathways by which prenatal nutritional adversity may produce increased risk of schizophrenia have been enumerated elsewhere *(35–39)*.

Not all the plausible pathways are related to nutrition per se. For instance, one intriguing speculation is that the effects of famine are mediated through the hypothalamic–pituitary–adrenal (HPA) axis. Some investigators have argued that exposure to high levels of glucocorticoids, induced by a stressful prenatal environment, might "program" the HPA axis for life or disturb brain development *(40–42)*. Famine might be thought of as a stressor, since fetal undernutrition is associated with greater transplacental transfer of glucocorticoids and reduced HPA axis function *(43)*. A recent study found that mothers who experienced the death or major illness of a close relative during the first trimester of pregnancy were more likely to give birth to children that would later develop schizophrenia *(44)*. Is it possible that exposure to nutritional stress in early gestation might have similar effects to those caused by bereavement?

In this chapter, however, we focus on micronutrient deficiencies as potential causes. We first describe three distinct—but not mutually exclusive—pathways by which prenatal micronutrient deficiencies might affect the risk of schizophrenia. Each of these pathways represents a form of interplay between genes and environment. Thus, we examine possible roles for (1) mutation rate; (2) epigenetic regulation; and (3) gene–environment interactions. Then we consider three specific micronutrients—selenium, zinc, and folate—to demonstrate the plausibility of these pathways.

Nutritional deficiency could indirectly affect brain development by interfering with DNA stability. A number of micronutrients play key roles in DNA synthesis, methylation, and repair. Antioxidants protect against oxidative damage, a major cause of DNA damage, and trace metals are integral parts of many of the proteins involved in DNA transcription and repair. For these reasons, the rate of *de novo* mutations may increase under nutritional deprivation, and this could result in genetic changes that give rise to schizophrenia in later life. Recently, researchers have demonstrated that individuals with schizophrenia are more likely to harbor novel micro-insertions and deletions in their genomes than controls *(45–49)*. The analysis of affected genes suggests that those from signaling networks involved in neurodevelopment are disproportionately represented. A role for *de novo* mutations in generating schizophrenia is compatible with other evidence *(50)*. Advancing paternal age is a major source of new mutations in humans, and the relative risk of schizophrenia rises with age of the father *(51)*. Also, schizophrenia has been detected in almost all populations that have been studied, which suggests that the disease has persisted over a long period in human populations despite its association with decreased reproductive success. Recurrent *de novo* mutations transmitted over a limited number of generations could explain this persistence.

Epigenetic regulation of genes critical for brain development may also be important in the development of schizophrenia *(52,53)*. It is known that imprinted genes play important roles in the growth of the fetus and placenta and also in the development of the brain, affecting cognition and behavior long after infancy is over *(54)*. For this reason, errors in imprinting could increase the risk of schizophrenia, by changing the expression of genes either involved in neuropsychiatric pathology specifically or related to fetal growth in general. Some researchers have posited that the mode of inheritance and development of schizophrenia are compatible with epigenetic changes in gene regulation. Also, because epigenetic errors might occur more frequently with advancing age, the result that schizophrenia is associated with increased paternal age is consistent with an epigenetic mechanism, as well as a DNA mutation one *(55)*. In animal models of schizophrenia, the methyl donor L-methionine has been shown to play an important role in epigenetic regulation *(53,56)*. To our knowledge, the only published study on humans is again from the Dutch Hunger Winter: exposure to famine in early gestation was associated with altered methylation of insulin-like growth factor 2 (IGF2) in midlife *(57)*.

Finally, gene–environment interactions may play a role in the development of schizophrenia. For example, the results of a study of Finnish children born to schizophrenic mothers and subsequently adopted suggest that these children's genetic background may make them especially susceptible to their household environment, with a greater risk of exhibiting thought disorders when placed with a family that communicates poorly and a lower risk when placed with a family that communicates well *(58)*. Might similar patterns with regard to prenatal nutritional environment be important? In some cases, variation in genes involved in micronutrient-dependent pathways could result in changes to the *in utero* environment that are conducive to the development of schizophrenia—but only in the presence of a deficiency of the relevant nutrient.

In the remainder of this chapter, we will describe what is known about the effects of three specific micronutrient deficiencies, selenium, zinc, and folate, in regard to the mechanisms just discussed. Although famine would reduce the intake of many different micronutrients, the functions of these particular micronutrients make them especially promising for future study.

27.3.1. Selenium

It has been proposed previously that selenium deficiency influences the risk of developing schizophrenia *(59)*. Ecologic studies have linked selenium deficient soil with high rates of schizophrenia. An analysis of U.S. state and county medical hospital records from 1965 found that of 219 environmental variables, low selenium levels in fodder crops had the strongest positive correlation with high schizophrenia prevalence *(60)*. A subsequent analysis of prevalence data drawn from nine U.S. schizophrenia surveys conducted between 1880 and 1963 reported a significant correlation between low selenium soil and high schizophrenia rates at the state level *(61)*. We cannot infer from these data that prenatal or postnatal selenium is causally related to schizophrenia. They do, however, strengthen the rationale for studies explicitly focused on the relationship between selenium deficiency during prenatal development and the later development of schizophrenia.

Selenium is thought to play an important role in brain development, but exactly what this role is has yet to be elucidated. When animals are deprived of sufficient levels of selenium, supply to the brain is prioritized *(62)*, although in severe deficiency the activity of important selenium-containing enzymes falls *(63)*. That the brain receives such high priority suggests that selenium plays a particularly important role in the function of this organ *(64)*, a hypothesis supported by recent findings. It has been shown that selenite is an essential component of the medium used to culture central neurons *(65)* and that selenoprotein P (SePP) promotes the survival of cultured neurons *(66)*. Selenium has also been shown to protect against neuronal damage and death in response to free radicals *(67)*. Finally, mRNA levels of brain-derived neurotrophic factor fall in pups born to dams fed a selenium deficient diet, as does the production of selenoenyzymes required for the expression of thyroid enzymes essential for normal brain development at certain stages *(68)*.

The role of selenium in guarding against oxidative stress is better characterized, for selenoenzymes catalyze a number of reactions known to neutralize free radicals *(65)*. Selenium is an integral component of the enzyme glutathione peroxidase (GPX), which protects against reactive oxygen species. As might be expected, this enzyme's activity is responsive to available selenium, falling with serum levels *(69)*. Selenoprotein R and thioredoxin, another selenoprotein, also play a protective role by reducing the products of oxidative stress *(65)*. Since oxidative stress causes increased DNA damage, selenium deficiency could result in a higher mutation rate, increasing the likelihood that the sequence of a schizophrenia-related genetic region will be altered.

There is also good evidence that selenium levels influence the methylation of genes, and thus affect epigenetic regulation. Selenium reverses the hypermethylation of genes, such as those coding for tumor suppressors, in human prostate cancer cells, reactivating them by modifying methylation patterns *(70)*. In contrast, young rats fed a selenium-rich diet have hypermethylated liver and colon DNA compared to those fed selenium-deficient diets *(71)*, while young rats with a selenium-adequate diet have hypermethylated colon DNA only *(72)*. These results suggest that the effect of selenium deficiency upon methylation may depend on the type of cells being examined. Further, selenium deficiency apparently has different effects upon methylation pathways across species such as mice and rats *(73)*. Thus, while the data indicate an important role for selenium in epigenetic regulation, the nature of this role appears to be quite complex.

Finally, gene interactions with selenium deficiency appear to influence the development of a number of human diseases. Polymorphisms in the selenoenzyme GPX-1 that change the enzyme's activity *(74)* have been found at higher prevalence in individuals with Keshan disease *(75)*. In addition, Keshan disease is linked to low selenium levels. It appears that the interaction between these genetic polymorphisms and selenium deficiency is multiplicative in determining the risk of the disease. A similar situation may exist for lung cancer. Specific GPX-1 polymorphisms are more common in lung cancer cells *(76)*, and supplementation with selenium appears to significantly reduce the incidence of this type of cancer in individuals with low baseline levels of the element *(77)*. If low levels of selenium contribute to schizophrenia, then environmental interactions with such polymorphisms in the prenatal environment may be important.

27.3.2. Zinc

The hypothesis that prenatal zinc deficiency contributes to schizophrenia is not a new one *(78,79)*. Zinc deficiency is important in the neurodevelopment of humans, and insufficient zinc levels early in pregnancy can lead to serious deformities, such as anencephaly. In a randomized controlled trial, pregnant women given zinc supplements upon beginning prenatal care had babies with improved neurobehavioral development, as measured by fetal heart rate and movement patterns *(80)*. Studies in animal models are consistent with findings in humans and provide a better picture of the mechanisms at work. Embryonic rat neural cells are especially susceptible to zinc deficiency, dying at a higher rate than other types of cells *(81)*. Not surprisingly, exposure of fetal rats to zinc deficiency results in reduced fetal growth and number of brain cells *(82)*. There is also evidence that maternal zinc deficiency in mice suppresses the development of neural stem cells, which could lead to future neuroanatomical and behavioral abnormalities *(83)*. In addition to physical changes found in animal models, a number of behavioral sequelae have been noted which persist into adulthood even after the provision of a normal diet *(84–86)*.

Zinc also plays an important role in guarding against DNA damage by reactive oxygen species. Rats fed a low zinc diet *(87)* and cells grown in zinc-deficient culture *(88)* experience increased oxidative stress. In addition to playing an important role in DNA transcription factor function, many of the proteins involved in base and nucleotide excision repair are either zinc finger or zinc-associated proteins. Zinc deficiency alters the expression of genes *(66)* and the conformation of proteins *(89)* needed to respond adequately to DNA damage. For this reason, dietary deficiency can lead to single and double-strand breaks in DNA *(90,91)*, as well as oxidative modifications to it *(89)*.

Zinc also plays an important role in methylation, which is central to epigenetic regulation. It is a cofactor for DNA methyltransferase *(92)*, and zinc-finger domains are found in methyl-DNA-binding proteins *(93)*. Proteins featuring zinc-finger domains establish and maintain methylation patterns *(94)* and are associated with epigenetic reprograming events *(95)*. Thus, zinc is essential for proper functioning at many levels in the process of epigenetic regulation, and inadequate amounts of this metal have the potential to alter methylation patterns.

Though gene–environmental zinc interactions may be important, little is known about this subject in humans. Mutations in various genes result in the inability to secrete zinc in milk *(96)* and in the disorder acrodermatitis enteropathica *(97)*. Little is known, however, about potential interactions between the genetic changes responsible for these human diseases and normally encountered variation in zinc intake. In contrast, in animal models, it is known that loss of function in the ZIP1 and 3 proteins, which mediate cellular zinc uptake, results in mice that are extremely sensitive to the teratogenic effects of zinc deficiency during pregnancy *(98)*. While loss of these genes does not manifest in any unusual clinical signs when dietary intake of zinc is normal, an interaction with low zinc intake exists. Similarly, mice with nonfunctioning metallothionein-I and II (MTI and II) genes, which encode zinc-binding proteins, are very sensitive to the teratogenic effects of zinc deficiency *(99)*, while mice that overexpress MTI are more resistant to these effects than controls *(100)*. It is not clear what sort of polymorphisms in zinc homeostasis proteins are present in humans, but if functional genetic differences are present, an interaction with zinc levels that affects brain development is a possibility.

27.3.3. Folate

We have long been intrigued by the notion that folate deficiency might link prenatal famine to schizophrenia *(101)*. An excess of neural tube defects was found in the same Dutch cohorts in which we found an excess of schizophrenia, and neural tube defects are known to be related to periconceptional folate intake. As noted earlier, a polymorphism in an important enzyme in the folate pathway, MTHFR, has been associated with both neural tube defects *(102)* and with schizophrenia *(34)*. Furthermore, there is strong evidence that a folate deficiency could activate the pathways discussed here.

It is well established that the folate metabolic pathway is central to DNA synthesis and repair, and we have previously hypothesized that mutagenic effects of folate deficiency might explain the Dutch and Chinese famine results *(37)*. Both *in vitro* and *in vivo* studies suggest that folate deficiency induces genomic instability *(103–105)*. Also, reduced levels of folate in seminal fluid have been associated with genomic instability in sperm *(106,107)*. This raises the intriguing possibility that the effect might originate in the father's germ cells rather than *in utero*.

The folate pathway is also essential for the methylation of DNA. This is one of the most important means by which epigenetic effects are established. In some of the earliest animal studies that documented epigenetic effects in the prenatal period, folate and other carbon donors such as methionine and betaine were used extensively *(108–110)*. It is probably fair to say that, at this time, folate level is the best established source of epigenetic variation *in utero*.

Finally, there is clearly potential for gene–environment interaction that disturbs brain development. With respect to neural tube defects, it is widely believed (though in our view not yet proven) that periconceptional folate supplements are most important for women who are homozygotic for the MTHFR677TT polymorphism, as noted earlier. It is theorized that these women (and/or their offspring) require high folate levels to overcome their slower metabolic rate at certain points in the folate pathway, such as the conversion of homocysteine to less toxic metabolites *(102)*. Also noted earlier, this MTHFR polymorphism has been associated with schizophrenia in many studies *(34,111)*, although this association cannot be considered definitive. There are many other genetic variants of enzymes in the folate pathway that might lead to susceptibility in the presence of folate deficiency, though these have not yet been as well studied.

27.3.4. Complexities

The relationship between the nutrients described is complex. For example, there is evidence that high levels of folate interfere with absorption of zinc in the gut *(112–114)*. And because selenium and folate influence one-carbon metabolism differently, results suggest that selenium deprivation ameliorates some of the effects of folate deficiency *(115)*. Similarly, selenium and iodine levels interact. These elements are both required for the production of thyroid hormones essential for brain development, and in areas with dual selenium and iodine deficiency, adverse neurodevelopmental outcomes such as cretinism appear to be higher *(116)*. However, high levels of selenium coupled with low levels of iodine actually exacerbate hypothyroidism *(117)*. Zinc deficiency, in the presence of iodine and selenium deficiencies singly or together, adds even more complexity to this picture *(118)*. Because micronutrient deficiencies are often found in multiples, and many micronutrients play overlapping but different roles in neurodevelopment, taking into account the interactions between them is a challenging necessity.

27.4. MICRONUTRIENTS, INFECTIONS, AND SCHIZOPHRENIA

Many studies have implicated exposure to infection *in utero* in the development of schizophrenia *(119–123)*. Because so many types of prenatal infection have been linked to increased rates of schizophrenia, a hypothesis that alterations in cytokine production in the fetal brain are responsible has been proposed *(124)*. It has been demonstrated that micronutrient deficiencies, including those discussed here, alter the immune response. This may enhance the risk to the fetus associated with prenatal infection exposure. Gestational zinc deficiency causes immune dysfunction in the mother and also has adverse effects on the developing immune system of the baby *(125)*. Because zinc is necessary in order for immunoglobulins to cross the placenta, perinatal zinc deficiency may reduce the availability of maternal antibodies to the fetus *(126)*. Similarly, selenium deficiency has been found to alter cytokine and chemokine expression in response to influenza infection. For example, in an adult mouse model, there is an increase in proinflammatory cytokines and chemokines and a decrease in anti-inflammatory ones *(127)*. This leads to more severe lung pathology; it remains to be seen whether similar results will be found in adult or fetal humans. In humans, marginal or deficient adults who received selenium supplements were able to more rapidly clear administered, attenuated viruses than those who did not receive supplements *(128,129)*. Thus, maternal micronutrient deficiencies may result in increased susceptibility to infection and/or increased morbidity in the fetus.

On the other hand, maternal infections can result in the sequestration of important nutrients, such as folate, selenium, and zinc, as part of the acute phase response *(130–132)*. Even temporary drops in the availability of these nutrients may result in damage to the fetus, if they occur during important developmental windows. In this way, infections may initiate or exacerbate micronutrient deficiencies in pregnant women, with effects on fetal neurodevelopment.

Because micronutrient deficiencies can affect the susceptibility to infection of a mother and her fetus, as well as the extent of the pathology resulting from infection, and because infections can alter micronutrient levels available to the developing fetus, future studies on micronutrient deficiencies may complement prior results linking prenatal exposure to infection and later development of schizophrenia.

27.5. CONCLUSION

The example of schizophrenia demonstrates that prenatal malnutrition is associated with latent effects on adult mental disorders. The challenge now is to understand the causal pathways that account for these latent effects. Until we do, we cannot be entirely sure that the relationship is causal and cannot use these findings to tailor interventions to prevent schizophrenia or other diseases. This is the focus of our current work, and the mechanisms and micronutrients described in this chapter point the way toward some areas for future study that may prove useful. The rapid advent of new genomic technologies will enable us to test some of the hypothesized pathways among the large numbers of persons who were exposed to the Chinese famine. Also, in the past 20 years, large randomized trials and quasi-experimental interventions have been conducted with early prenatal nutritional supplements, and in coming years the follow-up of the offspring from these trials will permit testing of other hypothesized pathways. Finally, we have already made use of archived biological specimens from pregnancy/birth cohorts established in the 1950s and 1960s to investigate associations between prenatal micronutrients and schizophrenia *(133,134)*. A new wave of pregnancy/birth cohorts recently initiated are much larger and their biological specimens are far richer. By 2030, we can expect cohorts such as the Norwegian Moba *(135)* and the US National Child Study *(136)* to answer some of the questions posed in this chapter about prenatal nutrition and schizophrenia.

So far, most research on prenatal nutritional deficiency and adult mental health has focused on schizophrenia. We have yet to determine, however, whether this exposure also influences the risk of developing other mental disorders over the life course. Understanding the role of a given micronutrient deficiency in one disorder, schizophrenia, may shed light on risk factors for others.

27.6. RECOMMENDATION

At this point, the study of prenatal nutritional deficiency and its latent effects on adult mental health is in its infancy. Therefore, more study on this subject is needed before specific recommendations can be made. In the meantime, providing pregnant women with access to good nutrition has long been known to improve the general health of infants and may also contribute to their life-long mental health.

27.7. ACKNOWLEDGEMENTS

This chapter is an expanded version of a previous article *(137)*. The authors thank Kim Fader for her help and the Robert Wood Johnson Health & Society Scholars Program for its financial support.

REFERENCES

1. Becker, J. (1998) *Hungry ghosts: Mao's secret famine*. New York: Henry Holt.
2. Stein, Z.A., Susser, M., Saenger, G., and Marolla, F. (1975) *Famine and human development: the Dutch hunger winter of 1944–1945*. New York: Oxford University Press.
3. Delisle, H.F. (2008) Poverty: the double burden of malnutrition in mothers and the intergenerational impact. *Ann NY Acad Sci* **1136**, 172–184.
4. Pharoah, P.O., Buttfield, I.H., and Hetzel, B.S. (1971) Neurological damage to the fetus resulting from severe iodine deficiency during pregnancy. *Lancet* **1(7694)**, 308–310.
5. Hetzel, B.S. (2000) Iodine and neuropsychological development. *J Nutr* **130(2S Suppl)**, 493S–495S.
6. MRC Vitamin Study Research Group. (1991) Prevention of neural tube defects: results of the Medical Research Council Vitamin Study. *Lancet* **338(8760)**, 131–137.
7. Barker, D.J. (1998) *Mothers, Babies and Health in Later Life*. Edinburgh, Scotland: Churchill Livingstone.
8. Brown, A.S. and Susser, E.S. (2008) Prenatal nutritional deficiency and risk of adult schizophrenia. *Schizophr Bull* **34(6)**, 1054–1063.

9. Bresnahan, M. and Susser, E. (2007) Belated concerns and latent effects: the example of schizophrenia. *Epidemiology* **18(5)**, 583–584.
10. Kuh, D. and Ben-Shlomo, Y. (2004) *A life course approach to chronic disease epidemiology*. New York: Oxford University Press.
11. Lumey, L.H. (1992) Decreased birthweights in infants after maternal *in utero*exposure to the Dutch famine of 1944–1945. *Paediatr Perinat Epidemiol* **6(2)**, 240–253.
12. Huang, J.S., Lee, T.A., and Lu, M.C. (2007) Prenatal programming of childhood overweight and obesity. *Matern Child Health J* **11(5)**, 461–473.
13. Kyle, U.G. and Pichard, C. (2006) The Dutch Famine of 1944–1945: a pathophysiological model of long-term consequences of wasting disease. *Curr Opin Clin Nutr Metab Care* **9(4)**, 388–394.
14. Painter, R.C., Roseboom, T.J., and Bleker, O.P. (2005) Prenatal exposure to the Dutch famine and disease in later life: an overview. *Reprod Toxicol* **20(3)**, 345–352.
15. Roseboom, T., de RS, and Painter, R. (2006) The Dutch famine and its long-term consequences for adult health. *Early Hum Dev* **82(8)**, 485–491.
16. Stein, A.D., Kahn, H.S., Rundle, A., Zybert, P.A., van der Pal-de Bruin, and Lumey, L.H. (2007) Anthropometric measures in middle age after exposure to famine during gestation: evidence from the Dutch famine. *Am J Clin Nutr* **85(3)**, 869–876.
17. Susser E., Hoek H.W., and Brown, A. (1998) Neurodevelopmental disorders after prenatal famine: The story of the Dutch Famine Study. *Am J Epidemiol* **147(3)**, 213–216.
18. Susser E., Neugebauer, R., Hoek, H.W. et al. (1996) Schizophrenia after prenatal famine. Further evidence. *Arch Gen Psychiatry* **53(1)**, 25–31.
19. Hulshoff Pol, H.E., Hoek, H.W., Susser E. et al. (2000) Prenatal exposure to famine and brain morphology in schizophrenia. *Am J Psychiatry* **157(7)**, 1170–1172.
20. Hoek, H.W., Brown A.S., and Susser E. (1998) The Dutch famine and schizophrenia spectrum disorders. *Soc Psychiatry Psychiatr Epidemiol* **33(8)**, 373–379.
21. Franzek, E., Sprangers, N., Janssens, A.C.J.W., van Duijn, C.M., and van de Wetering, B.J.M. (2008) Prenatal exposure to the 1944–5 Dutch "hunger winter" and addiction later in life. *Addiction* **103**, 433–438.
22. Neugebauer, R., Hoek, H.W., and Susser, E. (1999) Prenatal exposure to wartime famine and development of antisocial personality disorder in early adulthood. *JAMA* **282(5)**, 455–462.
23. Brown, A., Susser, E.S., Lin, S.P., Neugebauer, R., and Gorman, J.M. (1995) Increased risk of affective disorders in males after second trimester prenatal exposure to the Dutch Hunger Winter of 1944–55. *Br J Psychiatry* **166**, 601–606.
24. Brown, A., van Os, J., Driessens, C., Hoek, H.W., and Susser, E.S. (2000) Further evidence of relation between prenatal famine and major affective disorder. *Am J Psychiatry* **157**, 190–5.
25. Stein, A.D., Pierik, F.H., Verrips, G.H.W., Susser, E.S., and Lumey, L.H. (in press). Self-reported quality of life and depressive symptoms in adults after gestational exposure to the Dutch Famine of 1944–1945. *Epidemiology*.
26. Stein, Z., Susser, M., Saenger, G., and Marolla, F. (1972) Nutrition and mental performance. *Science* **178(62)**, 708–713.
27. St.Clair, D., Xu, M., Wang, P. et al. (2005) Rates of adult schizophrenia following prenatal exposure to the Chinese famine of 1959–1961. *JAMA* **294(5)**, 557–562.
28. Xu, M.Q., Sun, W.S., Liu, B.X. et al. (2009). Prenatal malnutrition and adult schizophrenia: further evidence from the 1959-61 Chinese famine. *Schizophr Bull* **35(3)**, 568–576.
29. World Health Organization (2001). *The World Health Report 2001: Mental Health: New Understanding, New Hope*. Geneva: Switzerland: World Health Organization.
30. Pasamanick, B., Rogers, M.E., and Lilienfeld, A.M. (1956) Pregnancy experience and the development of behavior disorders in children. *Am J Psychiatry* **112(8)**, 613–618.
31. Lumey, L., Stein, A.D., Kahn, H.S. et al. (2007) Cohort Profile: the Dutch Hunger Winter Families Study. *Int J Epidemiol* **36(6)**, 1196–1204.
32. Hoek, H.W., Susser, E., Buck, K.A., Lumey, L.H., Lin, S.P., and Gorman, J.M. (1996) Schizoid personality disorder after prenatal exposure to famine. *Am J Psychiatry* **153(12)**, 1637–1639.
33. Owen, M.J., Craddock, N., and Jablensky, A. (2007) The genetic deconstruction of psychosis. *Schizophr Bull* **33(4)**, 905–911.
34. Gilbody, S., Lewis, S., and Lightfoot, T. (2007) Methylenetetrahydrofolate reductase (MTHFR) genetic polymorphisms and psychiatric disorders: a HuGE review. *Am J Epidemiol* **165(1)**, 1–13.
35. Neugebauer, R. (2005) Accumulating evidence for prenatal nutritional origins of mental disorders. *JAMA* **294(5)**, 621–623.

36. Picker, J.D. and Coyle, J.T. (2005) Do maternal folate and homocysteine levels play a role in neurodevelopmental processes that increase risk for schizophrenia? *Harv Rev Psychiatry* **13(4)**, 197–205.
37. McClellan, J.M., Susser, E., and King, M.C. (2006) Maternal famine, *de novo*mutations, and schizophrenia. *JAMA* **296(5)**, 582–584.
38. Brown, A.S. and Susser, E. (2008). Prenatal nutritional deficiency and risk of adult schizophrenia. *Schizophr Bull* **34(6)**, 1054–1063.
39. Susser, E. and Opler, M. (2006) Prenatal events that influence schizophrenia. In: Sharma T, Harvey PD, eds. *The Early Course of Schizophrenia*. New York: Oxford University Press.
40. Weinstock, M., Matlina, E., Maor, G.I., Rosen, H., and McEwen, B.S. (1992) Prenatal stress selectively alters the reactivity of the hypothalamic-pituitary adrenal system in the female rat. *Brain Res* **595(2)**, 195–200.
41. Seckl, J. (2001) Glucocorticoid programming of the fetus; adult phenotypes and molecular mechanisms. *Mol Cell Endrocrinol* **185**, 61–71.
42. Welberg, L.A., Seckl, J.R., and Holmes, M.C. (2001). Prenatal glucocorticoid programming of brain corticosteroid receptors and corticotrophin-releasing hormone: possible implications for behaviour. *Neuroscience* **104(1)**, 71–79.
43. Vieau, D., Sebaai, N., Léohardt, M., Dutriez-Casteloot, I., Molendi-Coste, L., Laborie, C. et al. (2007) HPA axis programming by maternal undernutrition in the male rate offspring. *Psychoneuroendocrinology* **32**, S16–20.
44. Khashan, A., Abel, K.M., McNamee, R. et al. (2008) Higher risk of offspring schizophrenia following antenatal maternal exposure to severe adverse life events. *Arch Gen Psychiatry* **65(2)**, 146–52.
45. Walsh, T., McClellan, J.M., McCarthy, S.E. et al. (2008) Rare structural variants disrupt multiple genes in neurodevelopmental pathways in schizophrenia. *Science* **320(5875)**, 539–543.
46. Xu, B., Roos, J.L., Levy, S., van Rensburg, E.J., Gogos, J.A., and Karayiorgou, M. (2008) Strong association of *de novo*copy number mutations with sporadic schizophrenia.*Nat Genet* **40(7)**, 880–885.
47. International Schizophrenia Consortium. (2008) Rare chromosomal deletions and duplications increase risk of schizophrenia. *Nature* **455(7210)**, 237–241.
48. Stefansson, H., Rujescu, D., Cichon, S. et al. (2008) Large recurrent microdeletions associated with schizophrenia. *Nature* **455(7210)**, 232–236.
49. Vrijenhoek, T., Buizer-Voskamp, J.E., van dS, I. et al. (2008) Recurrent CNVs disrupt three candidate genes in schizophrenia patients. *Am J Hum Genet* **83(4)**, 504–510.
50. McClellan, J.M., Susser, E., and King, M.C. (2007) Schizophrenia: a common disease caused by multiple rare alleles. *Br J Psychiatry* **190**, 194–199.
51. Malaspina, D., Harlap, S., Fennig, S. et al. (2001) Advancing paternal age and the risk of schizophrenia. *Arch Gen Psychiatry* **58(4)**, 361–367.
52. Petronis, A. (2004) The origin of schizophrenia: genetic thesis, epigenetic antithesis, and resolving synthesis. *Biol Psychiatry* **55(10)**, 965–970.
53. Sharma, R.P. (2005) Schizophrenia, epigenetics and ligand-activated nuclear receptors: a framework for chromatin therapeutics. *Schizophr Res* **72(2–3)**, 79–90.
54. Malaspina, D. (2001) Paternal factors and schizophrenia risk: *de novo*mutations and imprinting. *Schizophr Bull* **27(3)**, 379–393.
55. Perrin, M., Brown, A.S., and Malaspina, D. (2007) Aberrant epigenetic regulation could explain the relationship of paternal age to schizophrenia. *Schizophr Bull* **33(6)**, 1270–3.
56. Dong, E., Agis-Balboa, R.C., Simonini, M.V., Grayson, D.R., Costa, E., and Guidotti, A. (2005) Reelin and glutamic acid decarboyxlase$_{67}$promoter remodeling in an epigenetic methionine-induced mouse model of schizophrenia. *PNAS* **102(35)**, 12578–12583.
57. Heijmans, B.T., Tobi, E.W., Stein, A.D. et al. (2008 epub) Persistent epigenetic differences associated with prenatal exposure to famine in humans. *PNAS* **105 (44)**, 17046–17049.
58. Wahlberg, K., Wynne, L.C., Oja, H. et al. (1997) Gene–environment interaction in vulnerability to schizophrenia: findings from the Finnish adoptive family study of schizophrenia. *Am J* Psychiatry **154**, 355–62.
59. Brown, J. and Foster, H.D. (1996) Schizophrenia: an update of the selenium deficiency hypothesis. *J Orthomol Med* **11(4)**, 211–22.
60. Foster, H. (1988) The geography of schizophrenia: possible links with selenium and calcium deficiencies, inadequate exposure to sunlight and industrialization. *J Orthomol Med* **3(3)**, 135–40.
61. Brown, J. (1994) Role of selenium and other trace elements in the geography of schizophrenia. *Schizophr Bull* **20(2)**, 387–98.
62. Behne, D., Hilmert, H., and Scheid, S. (1988) Evidence for specific selenium target tissues and new biologically important selenoproteins. *Biochimica et Biophysica Acta* **966(1)**, 12–21.

63. Castaño, A., Cano, J., and Machado, A. (1993) Low selenium diet affects monamine turnover differentially in substantia nigra and striatum. *J Neurochem* **61(4)**, 1302–7.
64. Benton, D. (2002) Selenium intake, mood, and other aspects of psychological functioning. *Nutr Neurosci* **5(6)**, 363–374.
65. Schweizer, U. and Schomburg, L. (2006) Selenium, selenoproteins and brain function. In: Hatfield D, Berry MJ, Gladyshev VN, eds. *Selenium: its molecular biology and role*. New York: Springer.
66. Yan, J. and Barrett, J.N. (1998) Purification from bovine serum of a survival-promoting factor for cultured neurons and its identification as Selenoprotein-P. *J Neurosci* **18(21)**, 8682–8691.
67. Savaskan, N., Bräuer, A.J., Kühbacher, M. et al. (2003) Selenium deficiency increases susceptibility to glutamate-induced excitotoxicity. *FASEB J* **17(1)**, 112–114.
68. Mitchell, J., Nicol, F., Beckett, G.J., and Arthur, J.R. (1998) Selenoprotein expression and brain development in preweanling selenium- and iodine-deficient rats. *J Mol Endocrinol* **20**, 203–210.
69. Zimmerman, C., Winnefeld, K., Streck, S., Roskos, M., and Haberl, R.L. (2004) Antioxidant status in acute stroke patients and patients at stroke risk. *Eur Neurol* **51**, 157–61.
70. Xiang, N., Zhao, R., Song, G., and Zhong, W. (2008) Selenite reactivates silenced genes by modifying DNA methylation and histones in prostate cancer cells. *Carcinogenesis* **29(11)**, 2175–2181.
71. Davis, C., Uthus, E.O., and Finley, J.W. (2000) Dietary selenium and arsenic affect DNA methylation in vitro in Caco-2 cells and in vivo in rat liver and colon. *J Nutr* **130**, 2903–2909.
72. Uthus, E., Ross, S.A., and Davis, C.D. (2006) Differential effects of dietary selenium (Se) and folate metabolism in liver and colon of rats. *Biol Trace Elem Res* **109(3)**, 201–214
73. Uthus, E. and Ross, S.A. (2007) Dietary selenium affects homocysteine metabolism differently in Fisher-344 rats and CD-1 mice. *J Nutr* **137**, 1132–1136.
74. Hu, Y. and Diamond, A.M. (2003) Role of glutathione peroxidase 1 in breast cancer: loss of heterozygosity and allelic differences in the response to selenium. *Cancer Res* **63**, 3347–3351.
75. Lei, C., Niu, X., Wei, J., Zhu, J., and Zhu, Y. (2009) Interaction of glutathione peroxidase-1 and selenium in endemic dilated cardiomyopathy. *Clinica Chimica Acta* **399 (1–2)**, 102–108.
76. Hu, Y., Benya, R.V., Carroll, R.E., and Diamond, A.M. (2005) Allelic loss of the gene for the GPX-1 selenium containing protein is a common event in cancer. *J Nutr* **135**, S3021–S3024.
77. Reid, M., Duffield-Lillico, A.J., Garland, L., Turnbull, B.W., Clark, L.C., and Marshall, J.R. (2002) Selenium supplementation and lung cancer incidence: an update of the nutritional prevention of cancer trial. *Cancer Epidemiol Biomarkers Prev* **11**, 1285–1291.
78. Andrews, R. (1990) Unification of the findings in schizophrenia by reference to the effects of gestational zinc deficiency. *Med Hypotheses* **31**, 141–153.
79. Andrews, R. (1992) An update of the zinc deficiency theory of schizophrenia. Identification of the sex determining system as the site of action of reproductive zinc deficiency. *Med Hypotheses* **38**, 284–91.
80. Merialdi, M., Caulfield, L.E., Zavaleta, N., Figueroa, A., and DiPietro, JA. (1999) Adding zinc to prenatal iron and folate tablets improves fetal neurobehavioral development. *AmJ Obstet Gynecol* **180(2)**, 483–490.
81. Harding, A., Dreosti, I.E., and Tulsi, R.S. (1988) Zinc deficiency in the 11 day rat embryo: a scanning and transmission electron microscope study. *Life Sci* **42**, 889–896.
82. McKenzie, J., Fosmire, G.J., and Sandstead, H.H. (1975) Zinc deficiency during the latter third of pregnancy: effects on fetal rat brain, liver, and placenta. *J Nutr* **105(11)**, 1466–1475.
83. Wang, F., Bian, W., Kong, L.W., Zhao, F.J., and Guo, J.N.H. (2001) Maternal zinc deficiency impairs brain nesting expression in prenatal and postnatal mice. *Cell Res* **11(2)**, 135–141.
84. Halas, E. and Hanlon, M. (1975) Intrauterine nutrition and aggression. *Nature* **257**, 221.
85. Halas, E. and Sandstead, H.H. (1975) Some effects of prenatal zinc deficiency on behavior of the adult rat. *Pediatr Res* **9(2)**, 94–7.
86. Halas E., Hunt, C.D., and Eberhardt, M.J. (1986) Learning and memory disabilities in young adult rats from mildly zinc deficient dams. *Physiol Behav* **37**, 451–458.
87. Bruno, R., Song, Y., Leonard, S.W. et al. (2007) Dietary zinc restriction in rats alters antioxidant status and increases plasma F2 isoprostanes. *J Nutr Biochem* **18**, 509–518.
88. Ho, E. and Ames, B.N. (2002) Low intracellular zinc induces oxidative DNA damage, disrupts p53, NFkB, and AP1 DNA binding, and affects DNA repair in a rat glioma cell line. *PNAS* **99(26)**, 16770–16775.
89. Ho, E. (2004) Zinc deficiency, DNA damage and cancer risk. *J Nutr Biochem* **15**, 572–578.
90. Castro, C., Kaspin, L.C., Chen, S.S., and Nolker, S.G. (1992) Zinc deficiency increases the frequency of single-strand DNA breaks in rat liver. *Nutr Res* **12**, 721–736.

91. Olin, K., Shigenaga, M.K., Ames, B.N. et al. (1993) Maternal dietary zinc influences DNA strand break and 8-hydroxy-2′-doxygunaosine levels in infant rhesus monkey liver. *Proc Soc Exp Biol Med* **203**, 461–466.
92. Bestor, T. (1992) Activation of mammalian DNA methyltransferase by cleavage of a ZN binding regulatory domain. *EMBO J* **11(7)**, 2611–2617.
93. Salozhin, S., Prokhorchuck, E.B., and Georgiev, G.P. (2005) Methylation of DNA—one of the major epigenetic markers. *Biochemistry* **70(5)**, 525–532.
94. Ohlsson, R., Renkawitz, R., and Lobanenkov, V. (2001) CTCF is a uniquely versatile transcription regulator linked to epigenetics and disease. *Trends Genet* **17(9)**, 520–527.
95. Loukinov, D., Pugacheva, E., Vatolin, S. et al. (2002) BORIS, a novel male germ-line-specific protein associated with epigenetic reprogramming events, shares the same 11-zinc-finger domain with CTCF, the insulator protein involved in reading imprinting marks in the soma. *PNAS* **99(10)**, 6806–6811.
96. Chowanadisai, W., Lönnerdal, B., and Kelleher, S.L. (2006) Identification of a mutation in SLC30A2 (ZnT-2) in women with low milk zinc concentration that results in transient neonatal zinc deficiency. *J Biol Chem* **281(51)**, 39699–39707.
97. Wang, K., Zhou, B., Kuo, Y.M., Zemansky, J., and Gischier, J. (2002) A novel member of a zinc transporter family is defective in acrodermatitis enteropathica. *Am J Hum Genet* **71**, 66–73.
98. Dufner-Beattie, J., Huang, Z.L., Geiser, J., Xu, W., and Andrews, G.K. (2006) Mouse ZIP1 and ZIP3 genes together are essential for adaptation to dietary zinc deficiency during pregnancy. *Genesis* **44**, 239–51.
99. Andrews, G. and Geiser, J. (1999) Expression of the mouse metallothionein-I and II genes provides a reproductive advantage during maternal dietary zinc deficiency. *J Nutr* **129**,1643–1648.
100. Dalton, T., Fu, K., Palmiter, R.D., and Andrews, G.K. (1996) Transgenic mice that overexpress matallotionein-I resist dietary zinc deficiency. *J Nutr* **126**, 825–833.
101. Susser, E., Brown, A.S., and Gorman, J.M. (1999) *Prenatal exposures in schizophrenia*. Arlington: American Psychiatric Publishing.
102. van der Put, N., van Straaten, H.W.M., Trijbels, F.J.M., and Blom, H.J. (2001) Folate, homocysteine and neural tube defects: an overview. *Exp Biol Med* **226**, 243–270.
103. Fenech, M. (2001) The role of folic acid and Vitamin B12 in genomic stability of human cells. *Mutat Res* **475(1–2)**, 57–67.
104. Teo, T. and Fenech, M. (2008 epub) The interactive effect of alcohol and folic acid on genome stability in human WIL2-NS cells measured using the cytokinesis-block micronucleus cytome assay. *Mutat Res* **657 (1)**, 32–38.
105. Bagnyukova, T.V., Powell, C.L., Pavliv, O., Tryndyak, V.P., and Pogribny, I.P. (2008) Induction of oxidative stress and DNA damage in rat brain by a folate/methyl-deficient diet. *Brain Res* **1237**, 44–51.
106. Young, S., Eskenazi, B., Marchetti, F.M., Block, G., and Wyrobek, A.J. (2008) The association of folate, zinc and antioxidant intake with sperm aneuploidy in healthy non-smoking men. *Hum Reprod* **23(5)**, 1014–1022.
107. Boxmeer, J., Smit, M., Utomo, E. et al. (in press). Low folate in seminal plasma in associated with increased sperm DNA damage. *Fertil Steril.*
108. Wolff, G., Kodell, R.L., Moore, S.R., and Cooney, C.A. (1998) Maternal epigenetics and methyl supplements affect *agouti*gene expression in A*vy/a*mice. *FASEB J* **12**, 949–957.
109. Cooney, C., Dave, A.A., and Wolff, G.L. (2002) Maternal methyl supplements in mice affect epigenetic variation and DNA methylation of offspring. *J Nutr* **132**, S2393–2400.
110. Waterland, R.A. and Jirtle, R.L. (2003) Transposable elements: targets for early nutritional effects on epigenetic gene regulation. *Mol Cell Biol* **23(15)**, 5293–5300.
111. Lewis, S., Zammit, S., Gunnell, D., and Davey-Smith, G. (2005) Meta-analysis of the MTHFR C677T polymorphism and schizophrenia risk. *Am J Med Genet Part B: Neuropsychiatr Genet* **135B(1)**, 2–4.
112. Milne, D., Canfield, W.K., Mahalko, J.R., and Sandstead, H.H. (1984) Effect of oral folic acid supplements on zinc, copper, and iron absorption and excretion. *Am J Clin Nutr* **39**, 535–539.
113. Ghishan, F., Said, H.M., Wilson, P.C., Murrell, J.E., and Greene, H.L. (1986) Intestinal transport of zinc and folic acid: a mutual inhibitory effect. *Am J Clin Nutr* **43**, 258–262.
114. Keizer, S., Gibson, R.S., and O'Connor, D.L. (1995) Postpartum folic acid supplementation of adolescents: impact on maternal folate and zinc status and milk composition. *Am J Clin Nutr* **62**, 377–384.
115. Davis, C., and Uthus, E.O. (2003) Dietary folate and selenium affect dimethylhydrazine-induced aberrant crypt formation, global DNA methylation and one-carbon metabolism in rats. *J Nutr* **133**, 2907–2914.
116. Vanderpas, J., Contempré, B., Duale, N.L. et al. (1990) Iodine and selenium deficiency associated with cretinism in northern Zaire. *Am J Clin Nutr* **52**, 1087–1093.
117. Vanderpas, J., Contempré, B., Duale, N.L. et al. (1993) Selenium deficiency mitigates hypothyroxinemia in iodine-deficient subjects. *Am J Clin Nutr* **57(S2)**, S271–S275.

118. Ruz, M., Codoceo, J., Galgani, J. et al. (1998) Single and multiple selenium-zinc-iodine deficiencies affect rat thyroid metabolism and ultrastructure. *J Nutr* **129(1)**, 174–180.
119. Brown, A.S., Begg, M.D., Gravenstein, S. et al. (2004) Serologic evidence of prenatal influenza in the etiology of schizophrenia. *Arch Gen Psychiatry* **61(8)**, 774–780.
120. Brown, A.S., Begg, M.D., Gravenstein, S. et al. (2005) Serologic evidence of prenatal influenza in the etiology of schizophrenia. *Obstet Gynecol Surv* **60(2)**, 77–78.
121. Brown, A. (2006) Prenatal infection as a risk factor for schizophrenia. *Schizophr Bull* **32(2)**, 200–2.
122. Mortensen, P., Nørgaard-Pedersen, B., Waltoft, B.L., Sørensen, T.L., Hougaard, D., and Yolken, R.H. (2007) Early infections of *Toxoplasma gondii*and the later development of schizophrenia. *Schizophr Bull* **33(3)**, 741–744.
123. Sørensen, H., Mortensen, E.L., Reinisch, J.M., and Mednick, S.A. (2009). Association between prenatal exposure to bacterial infection and risk of schizophrenia. *Schizophr Bull* **35 (3)** 631–637.
124. Meyer, U., Feldon, J., and Yee, B.K. (2008). A review of the fetal brain cytokine imbalance hypothesis of schizophrenia. *Schizophr Bull* DOI: 10.1093/Schbul/sbno22.epub.
125. Wellinghausen, N. (2001) Immunobiology of gestational zinc deficiency. *Br J Nutr* **85(S2)**, S81–S86.
126. Caulfield, L., Zavaleta, N., Shankar, A.H., and Merialdi, M. (1998) Potential contribution of maternal zinc supplementation during pregnancy to maternal and child survival. *Am J Clin Nutr* **68**, 499S–508S.
127. Beck, M., Nelson, H.K., Shi, Q. et al. (2001) Selenium deficiency increases the pathology of an influenza virus infection. *FASEB J* **15**, 1481–1483.
128. Food Standards Agency (2003). Report N05012: Functional markers of selenium in man. [Accessed 1019/2008 from http://www.foodstandards.gov.uk/science/research/researchinfo/nutritionresearch/optimalnutrition/n05programme/n05listbio/n05012/.]
129. Broome, C., McArdle, F., Kyle, J.A. et al. (2004) An increase in selenium intake improves immune function and poliovirus handling in adults with marginal selenium status. *Am J Clin Nutr* **80**, 154–162.
130. Brown, K. (1998) Effect of infections on plasma zinc concentration and implications for zinc status assessment in low-income countries. *Am J Clin Nutr* **S68**, S425–S429.
131. Tomkins, A. (2003) Assessing micronutrient status in the presence of inflammation. *J Nutr* **133**, S1649–S1655.
132. Duggan, C., MacLeod, W.B., Krebs, N.F. et al. (2005) Plasma zinc concentrations are depressed during the acute phase response in children with falciparum malaria. *J Nutr* **135**, 802–807.
133. Brown, A.S., Bottiglieri, T., Schaefer, C.A. et al. (2007) Elevated prenatal homocysteine levels as a risk factor for schizophrenia. *Arch Gen Psychiatry* **64(1)**, 31–39.
134. Insel, B., Schaefer, C.A., McKeague, I.W., Susser, E.S., and Brown, A.S. (2008) Maternal iron deficiency and the risk of schizophrenia in offspring. *Arch Gen Psychiatry* **65(10)**, 1136–1144.
135. Magnus, P., Irgens, L.M., Haug, K. et al. (2006) Cohort profile: the Norwegian mother and child study. *Int J Epidemiol* **35(5)**, 1145–1150.
136. Branum, A., Collman, G.W., Correa, A. et al. (2003) The National Children's study of environmental effects on child development. *Environ Health Perspect* **111(4)**, 642–646.
137. Susser, E., St Clair, D., and He, L. (2008) Latent Effects of Prenatal Malnutrition on Adult Health: The Example of Schizophrenia. In: Kaler SG, Rennert OM, eds. *Reducing the Impact of Poverty on Health and Human Development: Scientific Approaches.*Boston, MA: Blackwell Publishing on behalf of the New York Academy of Sciences, pp. 185–192.

VII

Nutrition Transitions Around the World

28 Nutritional Habits and Obesity in Latin America: An Analysis of the Region

Jaime Rozowski, Oscar Castillo, Yéssica Liberona, and Manuel Moreno

Key Points

- The dietary pattern in Latin America has become "westernized," moving away from traditional habits.
- Socioeconomic level is a major determinant of food intake in the region.
- Recent decades have witnessed a dramatic increase in the prevalence of obesity in most countries.
- Most worrisome is the increase in the prevalence of obesity in children.
- Programs with a clear concentration in education and change in lifestyle are showing some effectiveness.
- To be successful, programs have to be directed to the general population providing a safe environment to increase physical activity.
- Objectives and goals of a program to prevent obesity are outlined.

Key Words: Obesity; nutritional habits; Latin America; physical activity

28.1. INTRODUCTION

Latin American (LA) countries, although located in the same region, show a significant heterogeneity in their ethnicity, socioeconomic level, and health. The region holds 555 million inhabitants and is the most urban region in the developing world *(1)*. It shows great inequalities in socioeconomic level (SEL), with more than 1/3 of the population living in poverty. All this makes it extremely hard to discuss average values of whatever indicator is used. For instance, although the infant mortality rate as an average in Latin American countries is 19.7 per thousand born alive, the range observed in these countries is between 3.7/1000 and 129/1000. Likewise, life expectancy at birth, though showing an average of 70 years, ranges from 56.8 to 79 years, a 22-year difference *(2)*.

Urbanization has been a major issue in LA during the last decades, reaching currently more than 70% of the population. Of them, almost 80% live in cities with more than 700,000 inhabitants. Undeniably, while urbanization in industrialized countries usually brings along general progress in technology and social development, in developing countries the benefits of urban growth usually show only in the higher socioeconomic levels. In the lower socioeconomic groups, urbanization generally leads to a relative decrease in income which, from the nutritional point of view, produces a shift toward consumption of high-calorie foods with a reduced energy density, with a trend to incorporate into their diets foods that are more commonly consumed in industrialized nations. In fact, this trend is observed in the majority of the countries in the region in what could be called the "westernization of eating habits," characterized by an increased consumption of the so-called fast foods coupled with a tendency to decrease physical activity. This phenomenon is observed across all socioeconomic levels.

A. Bendich, R.J. Deckelbaum (eds.), *Preventive Nutrition*, Nutrition and Health, DOI 10.1007/978-1-60327-542-2_28,

The incorporation of these new food habits trends has led the LA region as a whole and every country in particular to change morbidity and mortality profiles which show an increase in the prevalence of chronic diseases above that of infectious diseases, the usual killers in the past. Currently, chronic diseases are the main cause of death in most of the LA countries *(1)*.

An example is the case of Chile, where the infant mortality rate in 1955 was approximately 120/1,000 live births, mainly as a result of malnutrition and infectious diseases. At that time, cancer and cardiovascular diseases accounted for 25.8% of all deaths in the country.Now the infant mortality rate is around 7.3/1000 live births, with cancer and cardiovascular diseases accounting for more than 50% of all deaths. At the same time, the mortality rates attributed to diarrhea and gastroenteritis of presumed infectious origin decreased from 20/1000 to 0.035/1000 live births *(3)*.

A similar pattern has been seen in other countries of the region. Fig. 28.1 shows the infant mortality rates from 1960 to 2007 in five countries that represent a range of socioeconomic level. Although all of these countries show a reduction in this indicator, this decline has been variable (with Chile showing the largest relative decrease in almost five decades). The reduced infant mortality and the general improvement of health care are reflected in the proportion of individuals older than 65 years of age as shown in Fig. 28.2, which depicts data in this population group since 1960. The proportion of the older-than-65-year group has increased 66% in Brazil and 50% in Chile and Uruguay in comparison with the year 1960 *(4)*. In countries with slower development, such as Bolivia and Guatemala, the increase in this segment of the population has been minimal. The aging of the population, when coupled with a decline in early deaths and an increase in life expectancy, necessarily forecasts an increase in the prevalence of the chronic diseases of the aged.

The increase in adult population in most countries in the LA region causes a growing demand on health services, not only to take care of the nutritional needs of the mother and child (malnutrition and micronutrient deficiency) but also to provide for the treatment of nontransmissible chronic diseases (NTCD) and their associated risk factors. Thus, it seems urgent to decrease the nutrition-related risk factors involved in NTCD, such as overweight and high intake of saturated fats; one mechanism is through health promotion via healthy lifestyle programs.

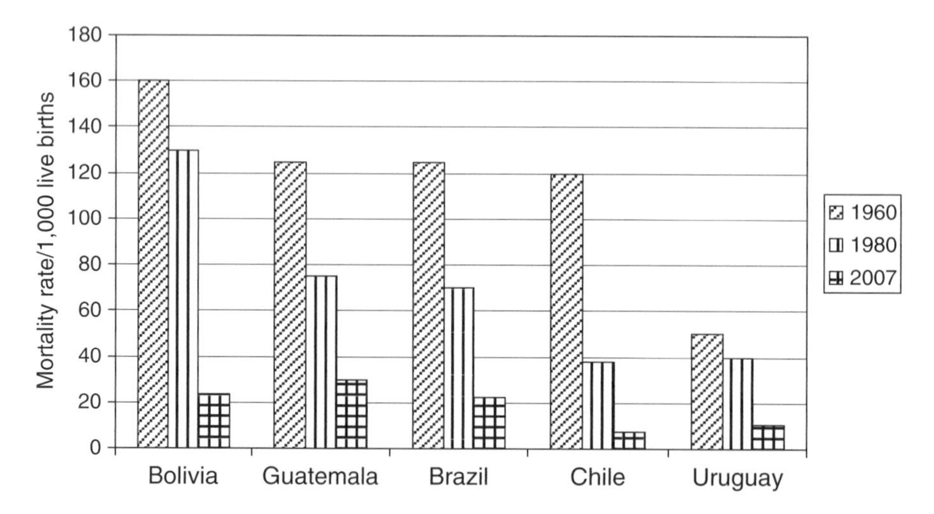

Fig. 28.1. Infant mortality rate in selected countries in Latin America, 1960, 1980, and 2007. Adapted from Ref. *(1)*.

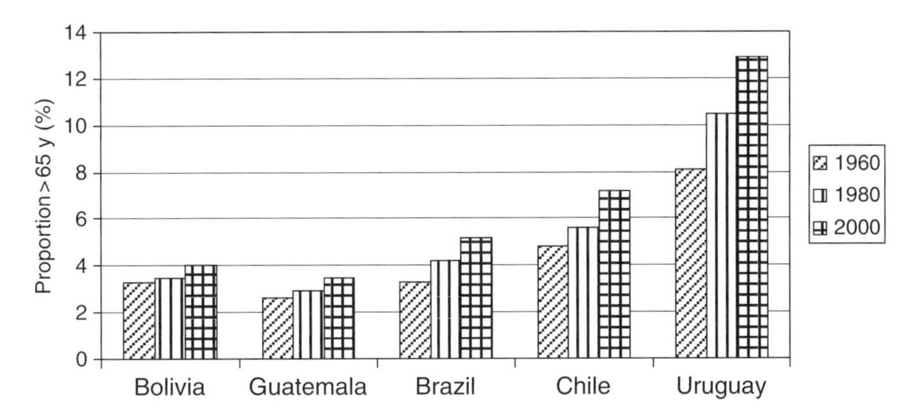

Fig. 28.2. Proportion of people older that 65 years of age in selected countries in Latin America, 1960, 1980, and 2000. Adapted from Ref. *(4)*.

28.2. SOCIOECONOMIC CHANGES

A thorough discussion of socioeconomic changes in the region is beyond the scope of this chapter. However, it is important to discuss some aspects that are relevant from the nutritional point of view. The gross national product (GNP) of the different countries in Latin America has shown substantial increases in the last decades. Nevertheless, this increase has not always been accompanied by an improvement in the health situation. For instance, Brazil, which in 1990 had one of the largest GNP per capita in Latin America, still had an infant mortality rate of 57.5/1000 live births that year, much higher than those of countries with lower GNP such as Chile and Uruguay *(5)*. This situation still continues today, as it can be seen in Fig. 28.3, where the relationship between per capita income and infant mortality is shown, pointing out the inequality of health services in these countries *(6)*.

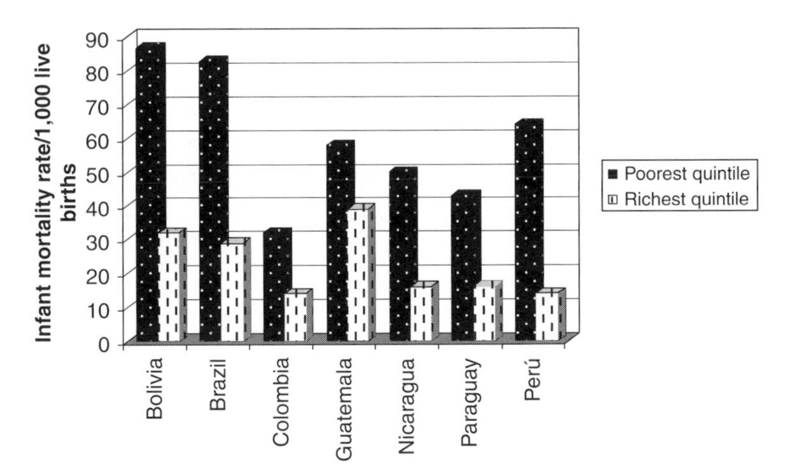

Fig. 28.3. Infant mortality rate according to income in selected countries in Latin America, 2006. Adapted from Ref. *(6)*.

The improvement in GNP and in the general well-being of many countries has produced an influx of foreign fast-food enterprises with a wide range of food items at very low cost and, therefore, at the reach of people from all socioeconomic levels. Also involved in the development of obesity has been the increase in portion sizes, which have become progressively larger in the last decades. Recently Rolls et al. showed that larger food portions led to greater energy intake in volunteers, independently of body mass index *(7)*.

28.3. FOOD HABITS

A thorough analysis of food habits in Latin America is hampered by the lack of national food consumption surveys. Except for a few countries, most information available is from small surveys that are often not representative of the population. In addition, most of these surveys are published in local journals that are not always widely available.

There is a great diversity of national nutritional habits in Latin America. In the Southern cone countries (Chile, Argentina, and Uruguay), wheat represents the main component of the national diets, with the addition of meat and dairy products in Argentina and Uruguay. Corn is the main staple in Mexico and most of Central America. The rest of the countries in that area have a rather diverse diet, combining the three main cereals (wheat, corn, and rice). In the Andean countries (Peru, Bolivia, and Ecuador) and Paraguay, potatoes represent an important contributor to their diets. In many countries, sugar is an important contributor of calories, representing about 10% of caloric intake as an average. However, in some groups, it can provide as much as 20% of dietary calories *(8)*.

Socioeconomic level is a major determinant of food intake. The structure of national diets has shown a tendency to change in most of the countries when income per capita has increased *(8)*. Lipid calories have increased because of a higher supply of fat (butter, margarine, and different types of oils), with a high proportion of them from animal fat. Calories from complex carbohydrates have shown a tendency to decrease in time while the ones from sugar tend to increase. Protein calories tend to remain stable or to increase slowly but with a rapid rise in the intake of proteins from animal origin. One example of this is shown in Fig. 28.4, which shows the changes in food availability in Chile between 1970 and 2003 according to the FAO Food Balance Sheets *(9)*. The increase in availability is observed mainly in fats and sugar, accompanied by a decrease availability of healthier foods like legumes. Red meat availability has also increased, showing that as time passes the Chilean diet gets further from a healthy pattern. In addition, increases in body mass index (BMI) parallel increases in income and dietary fat content.

Following we compare the availability of different nutrients in countries of LA. We have chosen to use food availability data provided by FAO balance sheets rather than compare nutrition surveys that are often based on different methods (24-hour recall, food, registry, etc.), usually done in small populations which are not always reliable.

28.3.1. Energy

There is a great variation in the amount of calories available in the different countries from the region. Most countries show a historical increase in calorie availability, particularly in the case of Peru and Ecuador, which showed an increase in almost 400 calories between 1981 and 2003 (Fig. 28.5). Other countries underwent a decrease in calorie availability, particularly the case of Cuba and Venezuela. The highest energy availability is seen in Argentina and Mexico.

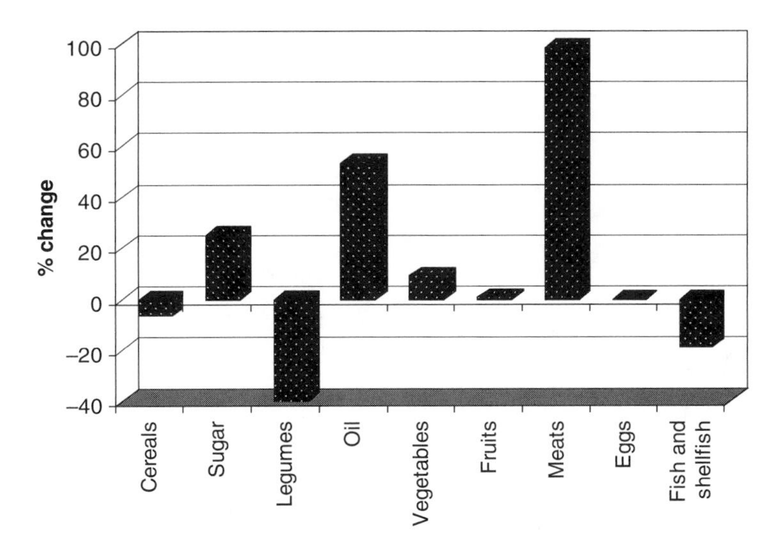

Fig. 28.4. Percent changes in food availability in Chile, 1987–2003. Adapted from Ref. *(9)*.

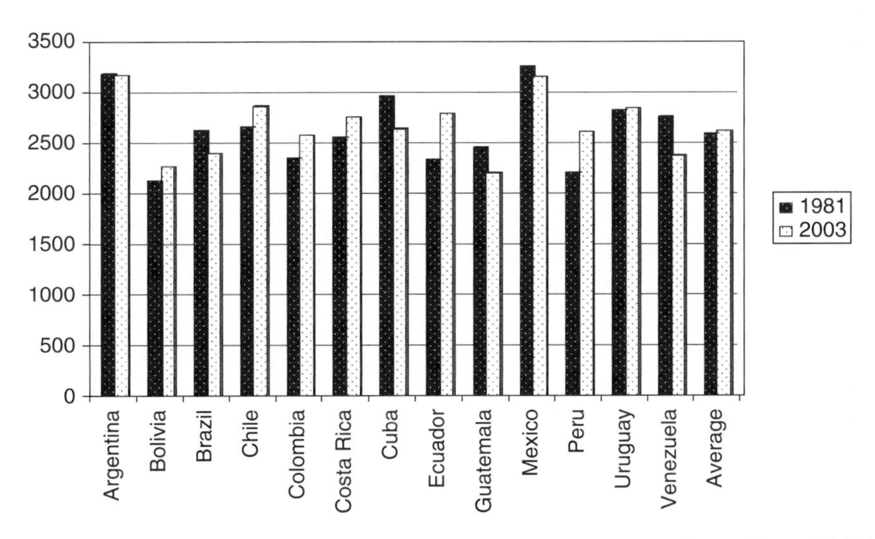

Fig. 28.5. Changes in the availability of energy in selected countries in Latin America, 1981 and 2003. Adapted from Ref. *(9)*.

28.3.2. Fat

There is a clear trend in the region to increase the consumption of fats, with many countries showing an increase in availability in the last 20 years (Fig. 28.6). Nevertheless, only two countries, Argentina and Uruguay, consume a proportion of calories from fats that are considered elevated (>30%). Cuba is the only country that has shown a substantial decrease in the availability of fat (25%), while the largest increase has been seen in Chile (24%). The increase in fat consumption is a hallmark of the westernization of the diet. This has been confirmed in several countries *(10–12)* where the pattern of fat consumption has been shown to be correlated with an increase in blood lipids *(13)*. In terms of the type of fats consumed, there has been a shift from fats of vegetable origin to animal fats, most certainly due to the increased consumption of meat and prepared foods.

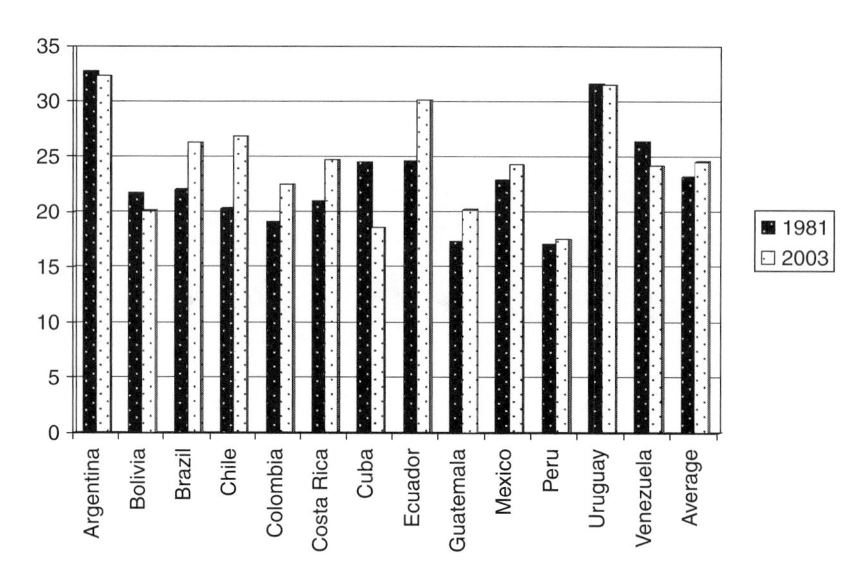

Fig. 28.6. Changes in fat availability in selected countries in Latin America, 1981 and 2003. Adapted from Ref. *(9)*.

28.3.3. Carbohydrates

In general, most countries from the region show a decrease in carbohydrate consumption, with the exception of Cuba, and, more importantly, they also show a shift from complex carbohydrates to refined ones, mainly sugar. In general, several studies show that poorer populations consume more carbohydrates than the better-to-do ones *(14)*.

28.3.4. Proteins

Protein availability in the region has remained constant in the last 20 years, with a tendency to consume relatively more protein of animal origin.

The changes in consumption, as in the case of carbohydrates, are due to a transition from having traditional dishes (usually based mainly on cereals) to a more convenient and faster pattern of eating. As mentioned earlier, historically, cereals have been the main staple in Latin America followed by vegetables, fruits, and roots. However, this is changing to a diet that is more refined and has more sugar, fats, animal proteins, and salt.

28.3.5. Salt

Consumption of salt has increased markedly in the last decade, most probably due to a higher consumption of convenience foods. An increased salt intake has a direct correlation with the appearance of high blood pressure, and many institutions like WHO have recommended a decrease in salt intake. Many substitutes of sodium have appeared in the markets which replace 50% of the sodium for potassium, but their use is still not widespread.

28.3.6. Food Consumption According to Socioeconomic Level (SEL)

It is obvious in all countries that there are differences in consumption according to SEL. A notable example is the case of Chile. The last national nutrition survey (which took place in 1974) showed that while only 5% of the population in the highest income quintile was consuming less than the

Table 28.1
Apparent Daily Consumption of Calories and Proteins per Person, Chile 1988 and 1998

	Calories		*Proteins (g)*		*Lipids (g)*	
Quintile of Income	*1988*	*1998*	*1988*	*1998*	*1988*	*1998*
I	1640	1961	44.5	56.3	42.2	56.7
II	1617	2083	43.3	60.2	43.3	62.8
II	1734	2143	47.4	61.0	47.5	64.8
IV	1988	2324	55.5	67.8	57.3	73.0
V	2200	2513	65.0	73.9	72.1	82.3
All	1869	2335	51.9	74.2	52.0	78.2

Source: Adapted from Ref. *(19)*
Based on food expenses in Greater Santiago (n = 5076 households for 1988 and 8445 households for 1998).

requirement of protein (0.75 g/kg body weight), 51% of people in the poorest quintile consumed less protein than the requirements *(15)*. Crovetto et al. estimated consumption based on the amount of money spent on food, showing a remarkable increase in the 1997 survey compared with 1988. Table 28.1 shows an analysis of food consumption based on the 1988 and 1997 surveys *(16,17)*. Macronutrient intake increased in all income levels in that period of time, the increase being more pronounced in lipids (50% on the average) than in the other macronutrients. The average increase in calories was more pronounced in poorer groups (quintile I–III). The same was observed for proteins and lipids, which is consistent with the prevalence of obesity in those groups (*see* below).

The direct, positive relationship between income and food intake has also been observed in smaller studies in Chile. In general, the proportion of calories from fat has increased steadily and correlates directly with socioeconomic level, reaching almost 29% in the wealthier groups. Qualitatively, the principal foods consumed by the rich and the poor in Chile are essentially the same, but the wealthier 60% of the population consumes a diet quantitatively different from the poorer 40%. Over time and socioeconomic sectors, the dominant contributors to total calories in the diet have been bread and other cereals. Although bread consumption in Chile has decreased about 25% in the last two decades, it was estimated to be at a level of 90 kg/person/year in 1987 and even higher in rural areas (140 kg/person/year). In comparison, developed countries consume an average of 50 kg/person. Besides being a traditional staple in the Chilean diet, bread is an inexpensive item widely used by the poorer groups to satiate hunger.

The consumption of specific items in the low-income populations can be determined as much by their availability as it is by price. Fig. 28.7 shows the relationship between price and calorie content of different foods in Chile. The cheapest calorie comes from sugar and the most expensive is lettuce. Income is a clear determinant of the amount of calories, and, in addition the choices of food availability in poor areas tend to be more limited than in the high-income neighborhoods, where the presence of large competing supermarkets allows for a much wider availability of different types of foods.

28.4. PREVALENCE OF OBESITY

Obesity can be described as a nutritional disease with serious consequences. Its prevalence increases with age and though the increase in weight with age has been accepted as inevitable, data show that

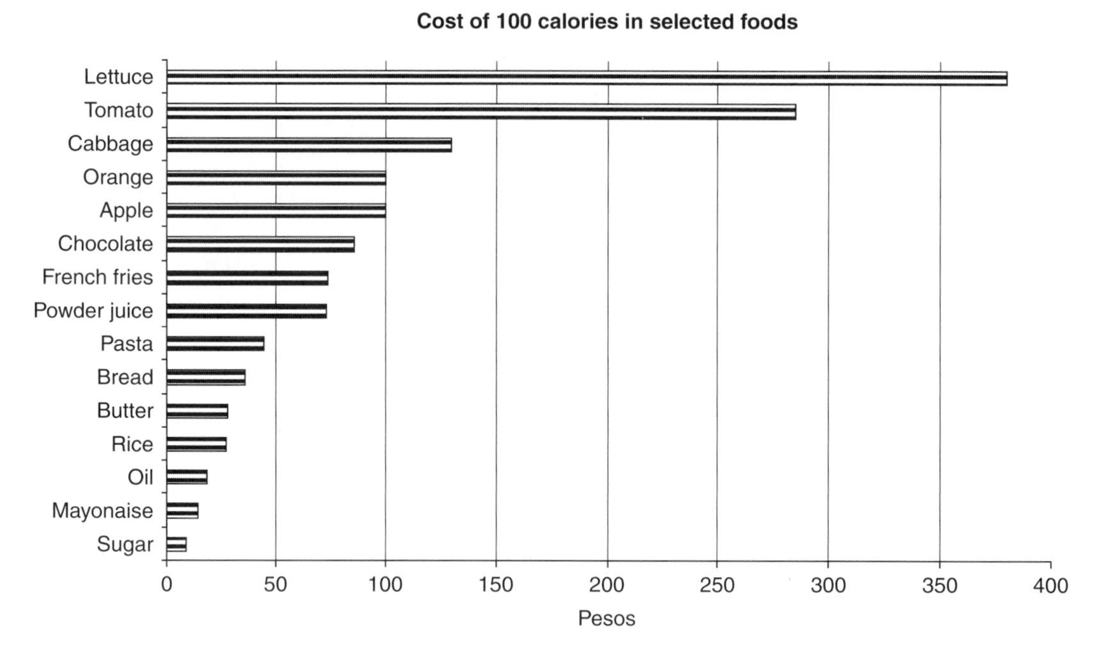

Fig. 28.7. Cost of 100 calories in selected foods in Chile. $1 = 600 pesos.

this weight gain augments the risk of mortality in women *(18)*. The treatment of obesity is difficult and usually unsatisfactory, therefore making its prevention very important *(19,20)* .

In the last decades, overnutrition has become an important public health problem, both in developed and developing countries. WHO has recently estimated that approximately 1.6 billion people in the world show overweight and at least 400 million of them are obese, and it is expected that by 2015 2.3 billion people will be overweight and 700 million will be obese, making this disease a true epidemic *(21)*. In addition, obesity is a major risk factor for mortality, particularly from cardiovascular disease *(19,22)* .

Recent decades have witnessed a dramatic increase in the prevalence of obesity not only in developed countries but also in developing ones, including those in Latin America. Table 28.2 summarizes the prevalence of obesity observed in several selected countries of the region *(23)*.

Table 28.2
Prevalence (%) of Obesity Among Adults in Seven Latin American Cities

	Men	*Women*
Buenos Aires (Argentina)	23.1	16.8
Bogota (Colombia)	12.8	22
Santiago (Chile)	23.6	29.4
Quito (Ecuador)	10.3	22.4
Mexico City (Mexico)	31.7	30.4
Lima (Peru)	21.1	23.4
Barquisimeto (Venezuela)	23.5	26.1

Source: Adapted from Ref. *(23)*

A study in 3305 individuals 35–64 years old from Mexico City showed that at baseline and 7 years later (with no intervention), the prevalence of overweight (BMI $\geq$ 25 kg/m^2) was 77.8% and 79.98%, and the prevalence for obesity (BMI $\geq$ 30 kg/m^2) was 26.61% and 34.7% respectively *(24)*. The authors stress the need for intervention to stop the "obesity epidemics." In the same country, in the city of Morelos, obesity prevalence in children and adolescents (11–24 years old) was 21.2% *(25)*.

Recent data from the 2006 Mexican National Survey of Health and Nutrition showed that compared to the national survey performed in 1999, the average waist circumference increased from 81 to 91 cm in women 20–49 years old, reflected by an increase in obesity prevalence from 25 to 32%. In adult men, obesity prevalence increased from 19 to 24%.

Brazil, the largest country in the region, has shown a steady decrease in malnutrition and an increase in the prevalence of obesity. A national survey was carried out in 1989 using a representative sample of 14,455 households and compared with a previous survey of 55,000 households performed in 1974–1975. An increase in prevalence was observed in both overweight and obesity in the period of time between the two surveys. Women were 2.5 times as likely as men to be obese, but the highest relative increase was seen in men, where prevalence almost doubled in 15 years. In men the prevalence was much higher in urban areas than in rural areas and showed a direct relationship with income. In both sexes, the more developed regions in the country were associated with a higher prevalence of overweight/obesity *(26)*.

While obesity prevalence in Brazil showed an increase in obesity in lower socioeconomic groups, a decrease in the prevalence of obesity in well-to-do women has been observed. As a matter of fact, this seems to be the only favorable change observed in the region at present *(26)*. In their analysis, the authors showed that in women obesity is associated with income and education (negatively). Men from the same region only showed a positive association with income. In the more developed regions, obesity prevalence was associated negatively only with education in both sexes while in men it was positively associated with income. Similarly, Jacoby et al. in urban Peru showed that in women, education was negatively associated with overweight and obesity *(27)*.

A serious problem that we face now is the remarkable prevalence of obesity in young children. The reasons for this are diverse. Nowadays children from developed countries and from late developing countries perform much less physical activity, and tend to eat or drink high-fat, high-carbohydrate foods. A study in Brazil showed that in children 6–9 years old and 10–18 years old, the prevalence of obesity increased, from 1974 to 1997, from 4.9 to 17.4% and 3.7 to 6%, respectively, and from 4.9 to 18.4% in urban children *(28)*.

In Mexico, the National Nutrition Survey of 2006 showed a prevalence of overweight plus obesity in children 5–11 years old has increased from 17.1 to 25.7% in males and from 20.1 to 27% in females, respectively, in a period of 7 years. The prevalence of obesity in this age group was 9.4% in men and 8.7% in women *(29)*. The same survey showed that only 35% of adolescents were physically active.

In Chile, the increase of obesity in preschool children has been notoriously rapid, rising from 5.8 in 1994 to 7.4% in 2006 *(3)* and the obesity in children entering school rising from 6 in 1987 to 19.4% in 2006 *(18,30)*. De Onis et al. published a review of obesity in preschool children in different countries in the world, including Latin America (Table 28.3). In LA, the highest prevalence was observed in Argentina, Chile, Peru, and Bolivia while the lowest prevalence was observed in Haiti. In general there is a correlation of overweight in preschool children and the purchasing power parity (PPP) in countries in LA. The countries with higher PPP tend to have a higher prevalence of overweight in preschool children. However there are some instances where it does not correlate. For instance, Bolivia, with some of the lowest PPP ($3.810) in the region, shows a prevalence of obesity of 9.3%, while Argentina, with one of the highest PPP ($11.760), shows a prevalence only slightly higher (9.9%). When we evaluate the prevalence of overweight and obesity related to socioeconomic status, we find

Table 28.3
Prevalence of Obesity (%) in Preschool Children from Selected Latin American Countries (w/h > 95%)

Country	*Obesity (%)*	*GDP/capita ($)*
Argentine	7.3	12.377
Bolivia	6.5	2.424
Brazil	4.9	7.625
Chile	7.0	9.417
Colombia	2.6	6.248
Costa Rica	6.2	8.650
Dominican Rep.	2.8	6.033
El Salvador	2.2	4.497
Guatemala	4.0	3.821
Haiti	2.8	n.a.
Mexico	3.7	9.023
Nicaragua	2.8	2.366
Panama	3.7	6.000
Paraguay	3.9	4.426
Peru	6.4	4.799
Uruguay	6.2	9.035
Venezuela	3.0	5.794

(De Onis, World Bank)
n.a.: Not available

a similar pattern than that in developed countries, showing poorer people with a higher prevalence of overweight and obesity than well-to-do individuals. This is accompanied by a deficient diet and a lack of physical activity. This was recently confirmed by a study of Chilean girls aged 8–13 years, which found a lower prevalence of overweight and obesity in the middle-high socioeconomic level, presenting a healthier food consumption characterized by a higher intake of dairy products, and a reduced frequency of bread, sweet and salty snacks, and sugar drinks consumption *(32)*.

What is the cause of such a high obesity in preschool children? Undeniably, there has been a historical decrease in physical activity in children of all ages. Although it is hard to establish the reasons for this decrease, two factors seem to be most relevant. First, the increased amount of time spent in activities that do not require any effort, like TV watching or computer use, in detriment of physical activity. In the Mexican survey, it was found that 51% of adolescents spend more than 12 hours watching television. Of these, more than 50% of them watch TV more than 21 hours per week *(30)*. In Chile, Olivares et al. found that 22.3% of a sample of 1,701 schoolchildren of low socioeconomic status watched TV more than 3 hours per day during the week and more than double of them (47%) during the weekend *(33)*.

Unfortunately, food is the most advertised product category in children's television, with most of the products directed to sweetened products and fast-food restaurants *(34)*. Considering the time spent in front of the TV, it is particularly hard to teach children proper eating habits.

The other factor responsible for the decrease in physical activity is safety concern of parents regarding their children playing on the street, although attraction for sedentary activities appear to be more important. Fortunately, programs are being developed in different countries promoting exercise, and data are showing they are being successful (*see* next section).

A study even showed that there was a positive significant correlation between consumption of sugar-sweetened drinks and obesity in children *(35)*. These results have been questioned by other authors *(36)*.

The consumption of fast foods has been often blamed for the increase in the prevalence of obesity in children from LA countries. However, it is hard to blame the food industry for the increase in obesity. Rather, it is the whole environment (including lack of exercise, lack of nutrition education and consumption of high-calorie food) that is conductive to this condition. Another important concern is that obesity during childhood markedly increases the probabilities to be obese in adult life. Obese boys and girls at 1 year of age have a 20 and 21% chance, respectively, of being obese at age 35. However, if young men and women are obese at age 18, the probability increases to 78 and 66%, respectively *(37,38)* .

An added component that we have to consider when we discuss obesity is the effect of early malnutrition and stunting on the appearance of obesity at later years *(39)*. The studies of Barker strongly suggest that early undernutrition results in increased adiposity in later life. If this is so, we will be faced in the future with the consequences of malnutrition observed in the last decades in Latin America children which will impose an unbearable burden to the health system when they reach adult life.

28.4.1. Prevalence of Conditions Related to Obesity

28.4.1.1. Diabetes

Obesity is a disease which not only represents a heavy financial burden by itself but is also a risk factor for other diseases like diabetes, lipid disorders, hypertension. This increases the risk of contracting cardiovascular disease, kidney disease, amputation of extremities, and blindness. However a problem that is prevalent when trying to compare studies in different countries is the difference in diagnosis criteria used.

There is a limited amount of information on the prevalence of diabetes in Latin America, coming from relatively small studies. In a survey done in a representative sample of people older than 20 years old in the south of Chile (a combination of rural and urban population), using internationally accepted diagnosis criteria, showed that the overall prevalence of diabetes was 5.4%. The prevalence by age was 1.88, 10.75, and 11.3% for 20–44 years, 45–64 years, and above 65 years old groups, respectively. Probably the most concerning characteristic of this study was that 45.2% of those who had diabetes did not know about it, stressing the importance of screening the adult population for the detection of this disease. The prevalence was slightly higher in the urban areas (5.8%) than in the rural areas (4.5%) *(40)*.

In Mexico, the average country prevalence of diabetes diagnosed in the 2006 survey was 7.3% in women and 6.5% in men. However, prevalence in the 50–59-year-old group doubled the national average and increased even further (21.3% in women and 16.8% in men) in the 60–69-year-old group. It is interesting to note that prevalence in Mexican-Americans, that is, Mexican adults living in the United States, was substantially higher (22.2%) than the values observed in those Mexicans living in Mexico. This could be due to the fact that those Mexicans living in the United States are further along in the nutritional transition, suggesting that conditions in Mexico will worsen in the future as the country moves along in this transition *(1)*.

In Chile, Mella et al. in 1979 found a diabetes prevalence of 5.3% in 1,100 adult individuals in Santiago *(41)*. The diagnosis was done according to the World Health Organization DATA Group *(42)*. Of the identified diabetics, 43.7% did not know their condition, similar to the data described above. The prevalence was similar in men and women, although in most age groups men had a higher prevalence, and there was no difference according to socioeconomic level.

The 2003 Health Survey in Chile showed a prevalence of 4.8 in men and 3.8% in women. Prevalence increased with age reaching 15.2% in the age group older than 65 years *(43)*.

Interestingly, Santos et al. showed that Aymara natives from Chile, although they have a relatively higher BMI than the non-native population, show less prevalence of diabetes *(44)*, suggesting that other factors like food, physical activity, and genetics may play an important protective role.

Our own study in Chilean pregnant women showed a prevalence of gestational diabetes of 14.3% of the women, 8.3% of them in the second trimester and 6% in the third trimester of pregnancy (Huidobro, A, Parodi, CG, Prentice AM, Fulford T, Rozowski J. Incidence of gestational diabetes: Cohort study in Chilean women. Submitted for publication).

28.4.1.2. Lipid Disorders

Cardiovascular disease is currently responsible for 31% of all deaths in Latin America. Although there is a lack of representative surveys of lipid disorders in LA, an interesting case is the one of Chile. Since 1969, cardiovascular diseases constitute the first cause of death in Chile, currently representing 30% of all deaths. In 1961–1963, Arteaga et al. studied 467 blue-collar workers 15–64 years old *(45)* and found average serum cholesterol level of 188.4 mg/dL. The same group in the early 1970 s compared 100 subjects with coronary heart disease (CHD) with 123 apparently healthy individuals and found average levels of serum cholesterol of 210 and 278 mg/dl for the healthy and the CHD individuals, respectively *(46)*. Triglycerides levels were 52.9 and 92.3 mg/dL, respectively.

In 1983 Chamorro et al. studied 828 professional males (middle- to high-income level) without history of CHD *(47)*. They observed a marked increase in cholesterol and triglyceride levels with age showing values that seemed to be higher than those of previous studies *(48)*. In the population studied, 52% of the subjects older than 41 years of age had serum cholesterol levels above 220 mg/dL.

An interesting finding was reported in 1988 by Albala et al. *(49)* who compared obese women (height/weight index > 120%) of high and low socioeconomic level (SEL). Serum cholesterol levels were much higher in women from high SEL compared with those in low SEL (227 and 179 mg/dL, respectively). Triglyceride levels were also significantly higher in the high SEL group (150 mg/dL) compared with the low SEL group (107 mg/dL). Calorie intake was similar in both groups. The authors attribute the low cholesterol levels in the obese women of low SEL to the level of fat in the diet coming mainly from lard in this group as opposed to dairy products in the diet of women of high SEL. One could also speculate that an unbalanced diet based on bread and cereals for some reason is protective against alterations in blood lipids.

In the 2003 Chilean Health Survey, it was found that 35.1% of men and 35.6% of women above age 17 had serum cholesterol levels above 200 mg/dL *(43)*. In Mexico, the 2006 National Survey found that 22.7% of adult men and 28.8% of adult women had cholesterol levels above 200 mg/dL *(29)*.

The social and economic changes that are taking place in Latin America have a strong influence on the dietary habits of its people. This has been correlated with an increase in the incidence of chronic diseases. The prevalence of obesity in many countries of the region is reaching levels seen in industrialized countries. Nevertheless, the limited resources make it difficult to neither prevent nor treat obesity. The key elements that need to be addressed are nutrition education and the promotion of physical activity. The need to counteract the changes that are taking place in lifestyle and nutrition should be of outmost importance to governments, which should aim to develop a nutrition literacy and a consciousness in the population toward making appropriate food choices and increase physical activity. Only these will be able to maximize health benefits and diminish health hazards in the population.

Given the variation in conditions in different countries of LA, it is difficult to generalize the impact of this disease. As economic development progresses in each country, sooner or later all will be faced with a range of problems in adults that will seriously burden the health system. Coherent prevention programs must be implemented to reduce the impact of diseases of "nutritional excess" on the health system and to improve the quality of life of the population as it ages *(50)*. An additional concern is that the chronic, prolonged illness of an adult has an effect on the health and well-being of other members of the household *(51)*.

28.4.1.2.1. How Do We Tackle the Obesity Problem? People that achieve economic progress increase their food-purchasing power, which is corresponded by an increased availability of food by industry. Although the role of food companies in health has long been debated, the reality is that their function is viewed by their shareholders to make a profit for them. Only public education will be able to change eating trends, and most surely food companies will respond to them. The influence of foreign food styles and imports in Latin American countries has been increasing steadily and certainly will keep this trend.

Fast-food restaurants have shown a striking increase in some Latin American countries, like Chile, Argentina, Brazil, and Uruguay, where most of these traditionally American fast-food restaurants have had explosive growth. As an example, a well-known international fast-food chain increased the number of outlets in Chile from 1 in 1990 to 70 in 2006. This trend is seen all over the region *(52)*.

These types of restaurants are attractive for the population because of the low cost, easy accessibility, and good taste. Unfortunately their foods are high in calories, saturated fat, and salt. They are particularly attractive for the low-income population because they provide a clean place where they can eat a tasty meal, within their financial means. Aware of their success, many middle-income local food service stores have turned into fast-food outlets with mass produced items. As a result, items like hamburgers and French fries are more frequently consumed compared to years ago, when their intake was only sporadic.

In the future, consumption of prepared foods, either from large supermarkets or from fast-food outlets, is bound to increase, unless it is counteracted by improving nutrition education of the general population. It is expected that industry will respond to public demand, if it exists, to develop foods that have a low fat content, are rich in complex carbohydrates, and are low in salt.

We have to consider special care when we propose programs to prevent obesity. In the United States, billions of dollars have been spent and nevertheless obesity prevalence keeps increasing *(53)*. We are fighting with not only environmental components such as food availability at a low price but also biological components, since the signal to eat is stronger than the signal to not eat, brought about by millions of years of adaptation to a hunting and gathering lifestyle *(20)*.

28.4.1.2.2. Current Programs Intended to Prevent Obesity. There are some activities in the region implemented to promote physical activity and improve nutrition education. One of them is the AGITA Sao Paulo experience in Brazil *(54)*. This is an intervention program directed to promote physical activity in a community-wide intervention designed to increase knowledge on the benefits and levels of physical exercise. The message is that the majority of the population should accumulate at least 30 minutes of physical activity during most days of the week. Activities are encouraged in three settings: home, transport, and leisure time. More than 160 groups are involved, and the Brazilian government decided to make it a nationwide program (Agita Brazil). The impact of the program has reached other countries of the region.

An evaluation of this program showed that the Agita message reached 55.7% of the population, and among these, 23.1% knew the main message. Recall of Agita and knowledge of its purpose were well distributed among different socioeconomic levels, being known by 67% of the most educated. The prevalence of people reaching the recommendation was 54.8% (men 48.7%, women 61%); risk of being sedentary was smaller among those who knew the Agita message (7.1%) compared with those who did not know (13.1%).

Based upon this Agita Sao Paulo experience, it appears that a multilevel, community-wide intervention to promote physical activity may obtain good results as a prevention strategy.

The Brazilian government has also addressed the prevention of nutrition-related noncommunicable diseases by means of innovative legislative and regulatory actions, mass communications, and capacity building, creating a comprehensive approach for addressing poor dietary and activity patterns in Brazil related to obesity. Some of these measures are new nutrition-related initiatives in the labeling area, shifts in the types of food purchased for the school food program, use of mass media to communicate components of the food guidelines, establishment of a smart shopping initiative, and training of teachers and health workers. This has represented an effort which took several years to get underway *(55,56)*.

In Chile, there is currently underway a 4-year intervention program in 4,800 children from basic school in the communities of Maipú and Puente Alto in Santiago, Chile. This intervention is based on the training of teachers to improve nutrition education in class and optimization of the physical education class. Our baseline results indicated that more than half the children studied were obese or overweight before the intervention. In addition, more than 50% of the children were not able to touch their toes in a seating position (data not shown).

28.4.1.2.3. Obesity Program for Prediabetic Subjects. The Chilean Ministry of Health and selected obesity specialized centers implemented an interdisciplinary pilot program for overweight adults at risk of diabetes. The objective of the pilot program was to demonstrate the potential to decrease the risk of type 2 diabetes (T2D) and cardiovascular risk factors (CVRF), which could be eventually extended to the public health system.

The program was focused on the promotion of healthy ways of life in adults who were overweight or obese and therefore at risk of diabetes. Beneficiaries within the public system of health were detected during adult preventive examinations of health or from other health-related activities. The patients were recruited between December 2004 and June 2005. The investigators defined a group, or basket, of services, which were financed by FONASA (National Health Fund) at nearly US$ 160 per patient, including the visits of medical staff and dietitians, group and individual workshops, as well as laboratory tests. All medications for diabetes or obesity were excluded. A 6-month protocol for intervention was defined to cover the different activities.

The patients were between 18 and 45 years old, with a body mass index of 25–38 kg/m^2 and fasting blood glucose in the range of 100–125 mg/dL or with direct family member with T2D. During the first 4 months of the program, the subjects were scheduled for 3 physician visits, 4 dietitian consultations, 14 physical activity sessions, and 4 group workshops (2 with a psychologist or therapist, 1 with the physician, and 1 with the dietitian).

Fasting blood samples were drawn at the beginning of the program and then again in the fourth month for the determination of blood glucose, insulin, and lipid levels. The homeostasis model assessment (HOMA) index was calculated to assess insulin resistance. In this program, 276 patients were recruited of which 160 (141 women), completed the whole period. In this subgroup at the start and the end of the intervention, respectively (all these comparisons were significant with a p value <0.05):

- prevalence of BMI $\geq$ 30 kg/m^{2}was reduced from 69 to 52% of subjects;
- prevalence of systolic blood pressure $\geq$ 140 mm Hg was diminished from 24 to 6%;
- prevalence of diastolic blood pressure $\geq$ 90 mm Hg reduced from 28 to 9%;
- blood glucose $\geq$ 100 mg/dL was reduced from 61 to 19%;
- plasma insulin $\geq$ 12.5 lU/mL reduced from 49 to 34%;
- HOMA $\geq$ 2.5 reduced from 63 to 42%.

The results of this program showed that the intervention significantly decreased median blood glucose, insulin, and HOMA index and increased HDL cholesterol. Intervention also significantly decreased the prevalence of obesity from 68.8 to 51.9%, abdominal obesity, and systolic and diastolic blood pressure.

The analysis of intervention studies with lifestyle change for overweight prediabetic patients has proved to be successful, mostly if it includes any cognitive behavioral intervention, such as the ones used in this program *(57,58)*. Even though weight loss after 4 months was only ~4% of the initial body weight, in parallel, a significant reduction was obtained in the main biochemical tests, such as the fasting blood sugar and insulin, with a low cost considering the complexity and duration of the intervention.

Before the implementation of this intervention program, FONASA did not support any such multidisciplinary program for the integrated attention of patients with overweight or obesity. The results of this program have stimulated the national health authorities to extend this program to different primary care centers in the country.

An important lesson is drawn from this program. It was relatively cheap (US$ 160/person) to carry out, much less than other more sophisticated programs that usually are not that effective. The cost and benefits of the program stimulated the health authorities to expand it, indicating that an inexpensive approach can be effective in the management of the disease.

In 2002, the FAO/WHO Expert Consultation on Diet, Nutrition and the Prevention of Chronic Diseases met in Geneva to examine the science base of the relationship between diet and physical activity patterns and the major chronic diseases related to nutrition *(59)*. Although this meeting provided a wealth of evidence on the problem of obesity at all ages, conclusions from it were already known in terms that imbalance between declining energy expenditure due to physical inactivity and high energy in the diet is the main determinant of the obesity epidemic.

The strategies developed in this report were based on the following:

1. proper feeding for infants and young children including the promotion of exclusive breastfeeding, avoiding the use of added sugars and starches when feeding formula, instructing mothers to accept their child's ability to regulate energy intake rather than feeding until the plate is empty, and assuring the appropriate micronutrient intake needed to promote optimal linear growth;
2. for children and adolescents, promoting an active lifestyle, limiting television viewing, promoting the intake of fruits and vegetables; restricting the intake of energy-dense, micronutrient-poor foods (e.g., packaged snacks) and restricting the intake of sugar-sweetened soft drinks;
3. modifying the environment to enhance physical activity in schools and communities, creating more opportunities for family interaction (e.g., eating family meals), limiting the exposure of young children to heavy marketing practices of energy dense, micronutrient-poor foods, and providing the necessary information and skills to make healthy food choices;
4. in developing countries, like in Latin America, giving special attention to avoid the overfeeding population groups with stunted growth. Nutrition programs designed to control or prevent undernutrition need to assess stature in combination with weight to prevent providing excess energy to children of low weight-for-age but normal weight-for-height;
5. in countries in economic transition, maintaining the healthy components of traditional diets (e.g., high intake of vegetables and fruits).

These strategies are well known but in Latin America they are often relegated to a second place due to more pressing problems that have a political priority, like access to health services, poverty, housing, etc. However, probably the prevention of obesity is extremely profitable from the point of view of future savings in health care. Obviously the two most important aspects are nutrition education and the promotion of physical activity. There are two main impediments to provide education. First, schools are reluctant to change the curriculum since they are always evaluated based on performance in areas like mathematics, science, and history. Second, it is very common to suspend the physical education class to either catch up in other subjects or hold other types of events. In addition, a very relevant barrier for the promotion of physical activity is security. If there is no safety in the areas dedicated to exercise like parks and open fields, people don't want to risk being outside and definitely will not allow their children to leave the safety of home.

28.4.1.2.4. Is There a Basic Concept to Prevent Obesity that can be Applied in General? Below we outline activities directed to the prevention of obesity concentrating in the two fundamental aspects: nutrition education and the promotion of exercise.

Because of the limited resources available, an effective program of obesity prevention should be focused and have as a general goal the creation of a level of awareness that will guide individuals in the food choices they make in daily living. It should involve all levels of education (including elementary schools, both public and private) as well as local health facilities to promote exercise involving community groups. The program should also stimulate and welcome the participation of the food industry to improve labeling (regulations on labeling are very basic in many of the LA countries) and to diminish the use of food items known to have an undesirable effect on health.

The program has to be part of a coherent health and nutrition policy with clear targets for promoting the concept of healthy nutrition . The general framework of a prevention program can be defined by the following components:

1. To develop a general awareness of the problem
2. To identify in the public those groups at risk
3. To remove or add those factors that affect the development of the disease (for instance, to instruct people to reduce the consumption of saturated fats [remove] and to promote physical activity [add])
4. To treat those individuals in whom the disease is already present.

The goals of a particular program, once the diagnosis has been made, must be clearly defined and attainable, taking into account the different socioeconomic groups and the realities of daily living. This is an important aspect since the implementation of many programs that have been described as effective are much too expensive for the health system to apply them at the national or local level.

The following describes aspects taken from programs for nutrition education and for the promotion of exercise. Although the discussion of all programs is far beyond the scope of this chapter, seemingly successful programs are currently in operation in some countries (i.e., Agita Brazil). Many programs have been used to implement actions. Several programs and recommendations were used to formulate the goals and specific objectives outlined below. These include the North Karelia Program in Finland *(60–62)*, the MRFIT program in the United States *(63,64)*, the recommendations of the National Academy of Sciences *(65)*, the 2002 FAO/WHO Expert Consultation on Diet, Nutrition and the Prevention of Chronic Diseases *(59,66)*. The original goals and objectives were adapted from the Model Standards of Healthy Communities *2000 (67)*, although the objective years have been adapted to the reality of the region.

28.5. NUTRITION EDUCATION RECOMMENDATIONS

The nutrition literacy of Latin Americans is low. An education program aimed at the general population can create an awareness directed to the proper food choices. The vehicles for the program should include

1. posters disseminated in public places like health clinics and hospital waiting rooms, government offices, schools, banks, pharmacies, public transportation, and other places where people have the opportunity to read announcements;

2. Nutrition education leaflets with a specific message for distribution through health services, voluntary organizations, and schools;
3. Articles in newspapers;
4. Radio announcements of short duration (1–2 minutes) dedicated to a particular subject (e.g., high-caloric foods, physical activity);
5. School programs to provide nutrition instruction and to transfer the message to the family (e.g., in the North Karelia project, school children gave their fathers a Father's Day card containing an health message);
6. Television "shorts" with specific messages and development of specific television programs.

Although the cost of some of these components can be very high, it can be diminished by enlisting governmental and nongovernmental organizations (e.g., government-owned or university-owned radio and television stations).

The goals of this program should be

1. to increase awareness of the relationship between diet and heart disease, high blood pressure, diabetes, and certain types of cancer;
2. to increase awareness of the ideal range of body weight and provide information on sound weight control strategies;
3. to provide information and education on healthy food choices at the point of purchase, at home, the workplace, and school;
4. to provide information on selection and preparation of healthy diets;
5. to stimulate the food industry to make available to the public a wide choice of low-fat, low-sodium processed foods and to develop clear labels for these products. This aspect also involves the establishment of government policy on food labeling.

The specific objectives for the year 2015 should be

1. that the proportion of the population that can identify the principal dietary factors related to chronic diseases should be at least 70%;
2. that 75% of the nation's schools should provide sound nutrition education as part of the curriculum in basic and middle cycles;
3. that 60% of the adult population should be able to identify foods that are low in sodium, low in saturated fats, high in calories, and good sources of fiber;
4. that 75% of adults should understand the concept of losing weight by reducing caloric intake or by increasing physical activity, or both;
5. that 75% of adults should be able to understand food labels to make nutritious food selections;
6. to maintain fat consumption in the diet to 30 % or less of calories and saturated fat intake to 10% or less of calories. Although a substantial proportion of the population seems to be at these levels already, there is a trend in some groups to increase fat consumption;
7. to decrease salt consumption so that at least 50% of home food preparers do not add salt during preparation and at least 80% of people do not add salt at the table before tasting;
8. that 60% of adults have a basic knowledge of the food groups and make an effort to increase consumption of complex carbohydrates and fiber-containing foods.

28.5.1. The Promotion of Exercise

A program to stimulate physical activity can be implemented as part of the nutrition education outlines above. In addition, certain measures need to be more specific. The vehicles for this program are similar to those indicated above.

The goals of the exercise-promotion program should be

1. to increase awareness that physical inactivity results in an increase of developing obesity and its disease correlates and of the beneficial effect of exercise on the prevention of cardiovascular disease, hypertension, and diabetes;
2. to increase awareness of the ideal range of body weight and provide information on sound weight-control strategies (same as goal #2 of nutrition education);
3. to disseminate information on appropriate exercise regimens to reduce the risk of cardiovascular disease in adults.

The specific objectives of this program for the year 2015 should be

1. to reduce the prevalence of obesity in adults age 20 or more to 20% in adult females and 10% in adult males;
2. to increase the proportion of people 15 years and older that engage in moderate exercise to 60% or more;
3. to increase to 50% the proportion of overweight people that are engaged in some form of physical activity combined with appropriate dietary practices to achieve ideal body weight;
4. to increase to 60% the proportion of people 20 years and above who are aware that regular exercise reduces the risk of heart disease, helps maintain appropriate body weight, retards the development of osteoporosis, and enhances self-esteem.

The objectives and goals outlined above were established taking Chile as a model but obviously have to be adapted to different countries. The particular situations and the current prevalence will command the establishment of specific goals. However, we have to look LA as continuum where some countries precede others in the nutritional and epidemiological transition.

REFERENCES

1. Pan American Health Organization. http://www.paho.org/spanish/dd/ais/coredata.htm. accessed October 15, 2008.
2. Pan American Health Organization. Health in the Americas. Edition 2002. Pan American Health Organization, Washington, DC.
3. Basic indicators of health, Ministry of Health, Chile. http://deis.minsal.cl/deis/indicadores/indi2007.pdf . accessed September 15, 2008.
4. CELADE (Economic Commission of Latin America and the Caribean). Anuario Estadístico, 2002.
5. The World Bank. World Tables 1992. Johns Hopkins University, Baltimore, MD, 1992.
6. World Bank. World development indicators 2008. The World Bank, Washington D.C. 2008.
7. Rolls BJ, Morris EL, Roe SR. Portion size of food affects energy intake in normal weight and overweight men and women. *Am J Clin Nutr* 2002;76:1207–1213.
8. Bermudez O, Tucker K. Trends in dietary patterns in Latin American populations.Cad Saude Publica2003;19(suppl 1):S87–S99.
9. Food and Agriculture Organization (FAO) www.fao.org/faostat. accessed August 3, 2008.
10. Sichieri R, Coitinho DC, Leao MM, et al. High temporal, geographic and income variation in body mass index among adults in Brazil. *Am J Public Health* 1994;84:793–798.
11. Aguirre-Arenas J, Escobar-Pérez M, Chávez-Villasana A. Evaluación de los patrones alimentarias y la nutrición en cuatro comunidades rurales. Salud pública Mex 1998;40:398–407.
12. Castillo O, Rozowski J. Tendencias en el consumo de grasas. Rev Child Nutr 2000;27(suppl.1):105–112.
13. Fornes N, Martins I, Hernan M, Velásquez-Meléndez G, Ascherio A. Frequency of food consumption and lipoprotein serum levels in the population of an urban area, Brazil. Rev Saude Publica 2000;34:380–387.

14. Gamboa E, López N, Vera L, Prada G. Displaced and local children's alimentary patterns and nutritional state in Piedescuesta, Colombia. Rev Salud Pública 2007;9:129–139.
15. Ministry of Health. Nutrition Survey of the Chilean population (ECEN). July 1974–June 1975. Santiago, Chile 1976.
16. Crovetto MM. Cambios en la estructura alimentaria y consumo aparente de nutrientes de los hogares del gran Santiago 1988–1997. Rev Chil Nutr 2002;1:24–32.
17. Albala C, Vio F, Kain J, Uauy R. Nutrition transition in Chile: determinants and consequences. Public Health Nutr 2002;5:123–128.
18. Manson JA, Willet WC, Stampfer Mi, et al. Body weight and mortality among women. *N Eng J Med* 1995;333: 677–685.
19. Pi-Sunyer X. A clinical view of the obesity problem. *Science* 2003;299:859–860.
20. Taylor R, McAuley K, Barbezat W, Farmer V, et al. Two-years follow-up of an obesity prevention initiative in children: the APPLE project. *Am J Clin Nutr* 2008;88:1371–1377.
21. World Health Organization. www.who.int/mediacentre/factsheets. accessed May 10, 2008.
22. Bjorge T, Engeland A, Tverdal A, Smith GD. Body mass index in adolescence in relation to cause-specific mortality: a follow-up of 230, 000 Norwegian adolescents. *Am J Epidemiol* 2008;168:30–37.
23. Schargrodsky H, Hernández-Hernández R, Marcet B. CARMELA: Assesment of cardiovascular risk in seven latin American cities. *Am J Med* 2008;121:58–65.
24. Gonzales-Villalpando C, Rivera-Martinez D, Cisneros, Castólo M, Gonzales-Villalpando ME, Simon J, Williams K, Haffner S, Stern M. Seven-year incidence and the progression of obesity. Characterization of body fat pattern evolution in low-income Mexico City urban population. *Arch Med Res* 2003;34:348–353.
25. Lazcano-Ponce EC, Hernandez B, Cruz-Valdez A, Allen B, Diaz R, Hernandez C, Anaya R, Hernandez-Avila M. Chronic disease risk factors among healthy adolescents attending public schools in Morelos, Mexico. *Arch Med Res* 2003;34:222–236.
26. Monteiro CA, Conde WL, Popkin BM. Independent effects of income and education on the risk of obesity in Brazilian adult population. *J Nutr* 2001;131:881–886.
27. Jacoby E, Goldstein J, Lopez A, Nunez E, Lopez T. Social class, family, and life-style factors associated with obesity among adults in Peruvian cities. *Prev Med* 2003;37:396–405.
28. Wang Y, Monteiro C, Popkin B. Trends of obesity and underweight in older children and adolescents in the United States, Brazil, China, and Russia. *Am J Clin Nutr* 2002;75:971–977.
29. Olaiz-Fernández G, Rivera-Dommarco J, Shamah-Levy T, Rojas R, Villapando-Hernández S, Hernández-Ávila M, Sepúlveda-Amor J. Encuesta Nacional de Salud y Nutrición 2006. Cuernavaca, México: Instituto Nacional de Salud Pública, 2006.
30. Situación de obesidad 1 básico por región (Prevalence of obesiy in 1st grade by region). http://www.sistemas.junaeb.cl/estadosnutricionales2007/tablaobesidadregion. accessed December 06, 2007.
31. De Onis M, Blossner M. Prevalence of trends of overweight among preschool children in developing countries. *Am J Clin Nutr* 2000;72:1032–1039.
32. Olivares S, Bustos N, Lera L, Zelada M. Estado nutricional, consumo de alimentos y actividad física en escolares mujeres de diferente nivel socioeconómico de Santiago de Chile. *Rev Med Chil* 2007;135:71–78.
33. Olivares S, Kain J, lera L, Pizarro F, Moron C. Nutritional Status, Food consumption and physical activity among Chilean school children: a descriptive study. *Eur J Clin Nutr* 2004;58:1278–1285.
34. Olivares S, Albala C, Garcia F, Jofre I. Television publicity and food preferences of school age children of the Metropolitan Region. *Rev Med Chile* 1999;127:791–799.
35. Ludwig DS, Petersen KE, Gortmaker SL. Relation between consumption of sugar-sweetened drinks and childhood obesity: a prospective, observational analysis. *Lancet* 2001;357:505–508.
36. Wolff E, Dansinger M. Soft drinks and weight gain: how strong is the link? *Medscape J Med* 2008;10:189. Published online August 12, 2008 .
37. Guo SS, Roche AF, Cameron Chumlea W, Gardner JD, Siervogel MR. The predictive value of childhood body mass index values for overweight at age 35 y. *Am J Clin Nutr* 1994:59:810–819.
38. Martorell R, Stein AD, Schroeder DG. Early nutrition and later adiposity. *J Nutr* 2001;131:874S–880S.
39. Barker DJP. The effects of nutrition of the fetus and neonate on cardiovascular disease in later life. *Proc Nutr Soc* 1992;51:135;144.
40. Baechler R, Mujica V, Aqueveque S, Lola I, Soto A. Prevalencia de diabetes mellitus en la VII región de Chile. *Rev Med Chile* 2002;130:1257–1264.
41. Mella I, García de los Rios M, Parker M, et al. Prevalencia de la diabetes en el Gran Santiago, Chile. *Rev Méd Chil* 1981;109:869–875.

42. WHO Study Group. Expert Committee of Diabetes Mellitus. Technical Repon Series Nr. 646, 1980.
43. Encuesta Nacional de Salud 2003. Ministry of Health, Chile 2004.
44. Santos JL, Perez-Bravo F, Carrasco E, Calvillan L, Albala C. Low prevalence of type 2 diabetes despite a high average body mass index in the Aymara from Chile. *Nutrition* 2001;17:304–309.
45. Arteaga A, Valiente S, Taucher E. Relación entre colesterol y acidos grasos de la dieta. *Rev Med Chile* 1963;91:888–894.
46. Arteaga A, Soto S, Valdivieso J, et al. Estudio nutricional y metabólico en la enfermedad coronaria. *Rev Med Chile* 1973;101:519–523.
47. Chamorro O, Costa E, Valenzuela O, et al. Riesgo cardiovascular en dos poblaciones laborales chilenas. *Rev Med Chil* 1980;108:697–699.
48. Chamorro G, Arteaga A, Casanegra P, et al. Factores de riesgo de enfermedad cardiovascular aterioesclerótica y prueba de esfuerzo en hombres de nivel profesional en Santiago. *Rey Méd Chile* 1983;111:1009–1017.
49. Albala C, Villarroel P, Olivares S, et al. Mujeres obesas de alto y bajo nivel socioeconómico: Composición de la dieta y niveles séricos de lipoproteínas. *Rey Méd Chile* 1989;117:3–9.
50. Kane RL, Radosevich DM, Vaupel JW.Compression of morbidity: issues and irrelevancies. In:Improving the Health of Older People.Kane RL, Evans JO, Macfayden, eds.New York:Oxford University Press on behalf of The World Health Organization, 1990;30–49.
51. Over M, Ellis RP, Huber J, Solon O. The consequences of adult health. ln:Health of Adults in the Developing World. Feachem RGA, Kjellstrom T, Murray CJL, Over M, Phillips MA, eds.Washington, DC:The World Bank, 1991.
52. Latin Business Chronicle. www.latinbusinesschronicle.com . accessed August 29, 2008.
53. Flegal K, Carroll M, Orden C, Johnson C. Prevalence and trends in obesity among US adults, 1999–2000. *JAMA* 2002;288:1728–1732.
54. Matsudo V, Matsudo S, Andrade D, et al. Promotion of physical activity in a developing country: the Agita Sao Paulo experience. *Pub Health Nutr* 2002;5:253–261.
55. Coitinho D, Monteiro C, Popkin B. What Brazil is doing to promote healthy diets and active lifestyles. *Pub Health Nutr* 2002;5:263–267.
56. Matsudo S, Matsudo V. Coalitions and networks: facilitating global physical activity promotion. *Promot Educ* 2006;13:133–138.
57. Norris SL, Zhang X, Avenell A, et al. Long-term non pharmacological weight loss interventions for adults with prediabetes. *Cochrane Database Syst. Rev.* 2005; Jan 25; (1): CD004096.
58. Shaw K, O'Rourke P, Del Mar C, Kenardy J. Psychological interventions for overweight or obesity. *Cochrane Database Syst Rev* 2005; Apr 18; (2): CD003818.
59. WHO. Technical report series 916. Diet, nutrition and the prevention of chronic diseases. Geneva, 2003.
60. Kottke TE, Nissinen A, Puska P, et al. Message dissemination for a community-based cardiovascular disease prevention programme (the North Karelia Project). *Scand J Prim Health Care* 1984;2:99–104.
61. Pietinen P, Vartiainen E, Korhonen HJ, et al. Nutrition as a component in community control of cardiovascular disease (the North Karelia Projecfl). *Am Clin Nutr* 1989;49:1017–1024.
62. Papadakis S, Moroz I. Population-level interventions for coronary Herat disease prevention: what have we learned since the North Karelia Project?. *Curr. Opin. Cardiol.* 2008;23:452–461.
63. Benfari RC. (for the MRFIT). The Multiple Risk Factor Intervention Trial (MRFIT). III. The Model for intervention. *Prev Med* 1981;10:426–442.
64. Domanski M, Mitchell G, Pfeffer M, Neaton J, Norman J, Svendsen K, Grima R, Cohen J, Stamler J, MRFIT Research Group. Pulse pressure and cardiovascular disease-related mortality: follow-up study of the Multiple Risk Factor Intervention Trial (MRFIT). *JAMA* 2002;287:2677–2683.
65. Committee on Diet and Health, National Research Council. Diet and Health: Implications for Reducing Chronic Disease Risk. Washington, DC: National Academy Press, 1989.
66. Department of Health and Human Services (USDHHS). Healthy People 2000. National Health Promotion and Disease Prevention Objectives. Washington, DC: Public Health Service, 1990.
67. American Public Health Association. Healthy Communities 2000: Model Standards. APHA, Washington, DC, 1991.

29 Effects of Western Diet on Risk Factors of Chronic Disease in Asia

Koji Takemoto, Supawadee Likitmaskul, and Kaichi Kida

Key Points

- Increasing adoption of the "Western diet" is associated with an increased prevalence of childhood obesity, dyslipidemia, and type 2 diabetes in Asia.
- Chronic disease risk factors, such as hypercholesterolemia and type 2 diabetes, are now as high or higher in Asian children as compared to US or European children.
- Low-fat and well-balanced Asian diets are recommended for Asian children.
- School programs can provide effective lifestyle interventions against risks for chronic disease.

Key Words: Adults; children; hypercholesterolemia; Japan; Korea; lifestyle-related chronic diseases; metabolic syndrome; nutritional recommendations; obesity; salt; risk factors; Thailand; type 2 diabetes

29.1. INTRODUCTION

Infectious diseases and malnutrition were the major health problems in most Asian countries before World War II and still are today in some parts of Asia. The patterns of diseases have changed dramatically in many Asian countries after World War II, in parallel with rapid economic development. The latter has been accompanied by Westernization of lifestyles. From 1970 to 2005, the gross domestic product increased by 22 times in Japan, 90 times in Korea, and 25 times in Thailand, and lifestyles relating to housing, clothing, eating habits, and physical activities were Westernized in these countries. The Westernization of diet is demonstrated as an increase of protein and fat consumption at meals and the prevalence of American-style fast foods, such as hamburgers and fried chicken.

In Japan, Korea, and Thailand, the three leading causes of death today are neoplasms, heart disease, and cerebrovascular disease, similar to Europe and North America. The cost of cardiovascular diseases (CVDs) and diabetes, which are closely related with lifestyle, was US $65 billion in Japan in 2005, accounting for 8.3% of the national budget. These chronic diseases, which are often the result of lifestyle—particularly the diet—are not only a medical but also a growing economic problem in Asia similar to Western countries. Therefore, more attention is required to prevent risk factors for these lifestyle-related chronic diseases through nutritional intervention.

A. Bendich, R.J. Deckelbaum (eds.), *Preventive Nutrition*, Nutrition and Health, DOI 10.1007/978-1-60327-542-2_29,

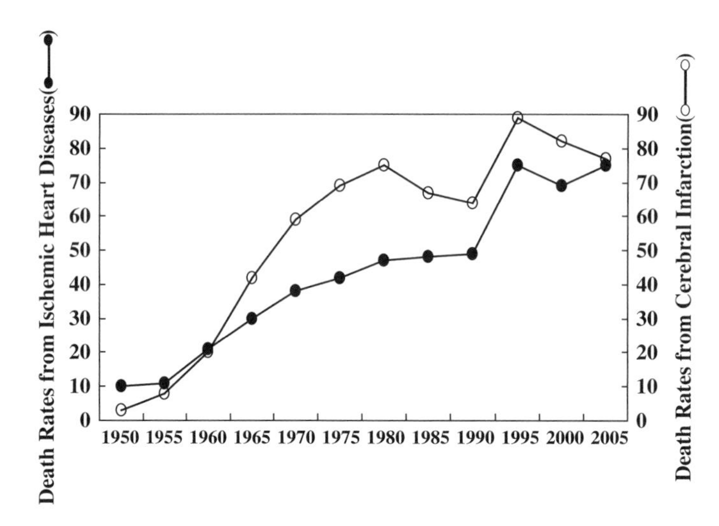

Fig. 29.1. Yearly change in death rates from ischemic heart diseases and cerebral infarction in Japan.

This chapter reviews the relationship between Westernization of diets and lifestyle-related chronic diseases and their risk factors in Asian countries. Nutritional recommendations are described regarding prevention of risk factors for chronic diseases that often result from lifestyle choices.

29.2. EPIDEMIOLOGY OF CHRONIC DISEASES AND THEIR RISK FACTORS

29.2.1. Chronic Diseases

Cardiovascular disease has been increasing linearly each year and has become one of the major causes of death in Japan, Korea, and Thailand *(1)*. Although the death rate from ischemic heart diseases (IHD) has increased severalfold in Japan and Korea in the last 20–50 years, it is still substantially lower than in Europe and North America (9.9/100,000 in 1950 to 75.0/100,000 in 2005 in Japan *(2)* and 2.2/100,000 in 1983 to 12.5/100,000 in 1992 in Korea; *see* Fig. 29.1 and Ref. *(3)*). In Thailand, the death rate from heart disease increased from 40.3 per 100,000 people in 1987 to 56 per 100,000 people in 1992, a period of only 5 years. The prevalence of IHD in one serially studied Thai community increased from 7 per 1,000 in 1976 to 17 per 1,000 in 1983 *(4)*.

Cerebrovascular diseases also are emerging as a leading cause of death in Japan and other Asian countries (*1*; Fig. 29.1). Although the death rate from total cerebrovascular diseases in Japan has gradually decreased during the past 30 years, the death rate from cerebral infarction and its contribution to total cerebrovascular diseases have greatly increased over 50 years (from 4/100,000 in 1950 to 77/100,000 in 2005; *see* Fig. 29.1 and Ref. *(2)*). In Korea, the death rate from cerebrovascular diseases increased from 65.4 per 100,000 in 1983 to 80.4 per 100,000 in 1992 *(3)*. In Thailand, the death rate from hypertension and cerebrovascular diseases increased from 12.8 per 100,000 in 1987 to 16.9 per 100,000 in 1992, and prevalence of hemiplegia as a result of cerebrovascular diseases in one town increased from 1.5 per 1,000 in 1976 to 6.6 per 1,000 in 1983 *(4)*.

Type 2 diabetes is caused mostly by insulin resistance (IR) associated with obesity and it is steeply increasing in many parts of the world, particularly in Asian countries (Fig. 29.2 and Ref. *(5)*), in parallel with an increase of obesity resulting from Westernization of lifestyle regarding eating habits and physical activities. The total number of patients with type 2 diabetes worldwide was estimated to be 133 million in 1995 and is anticipated to be 300 million in 2025, although the rise in prevalence will occur more rapidly in Asia than in Europe and North America *(6)*. In Japan, the number of patients

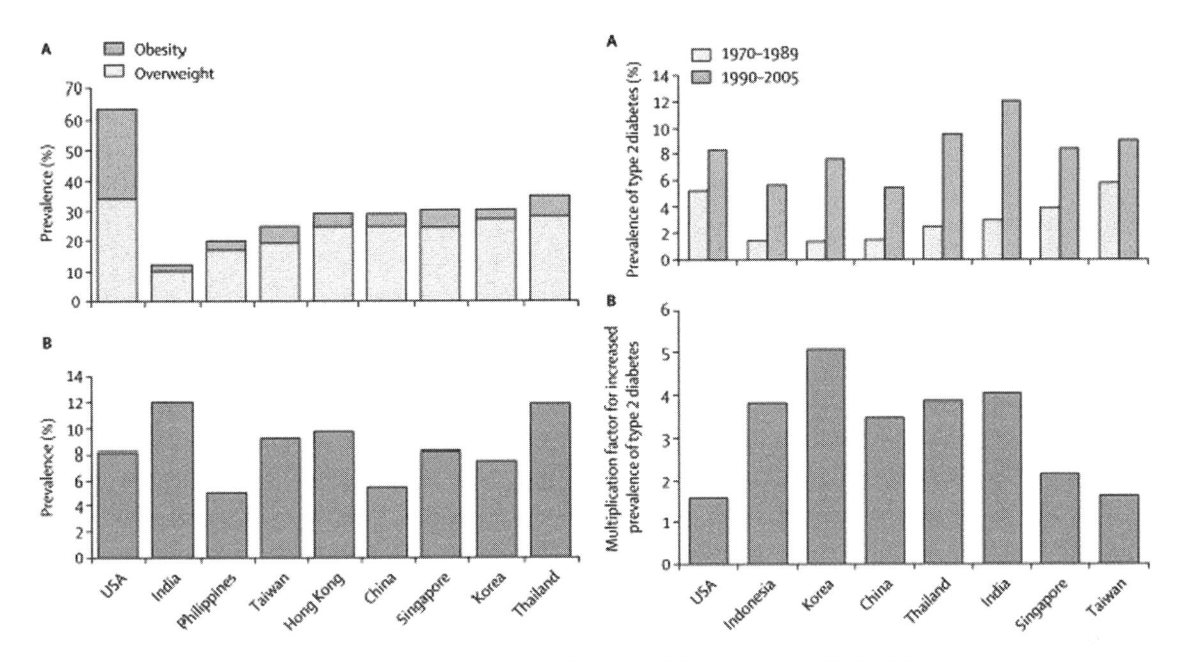

Fig. 29.2. Epidemic obesity and type 2 diabetes in Asia (Ref. *(5)*).

with type 2 diabetes whose HbA1c was 6.1% or higher or who were on treatment for type 2 diabetes increased from 6.9 million in 1997 to 7.4 million in 2002 *(7)*. Type 2 diabetes (the major type of adulthood diabetes) was believed not to develop in childhood; however, it is emerging rapidly in children in some ethnic groups, including Asian populations, Native Canadians, Native Americans, African-Americans, and Hispanics, who are perhaps genetically more insulin resistant than Caucasians *(8–15)*.

In Japan, all school children aged 6–15 years are screened for diabetes every year with a urine test according to the Law of School Health; data obtained from the diabetes screening program for school children have revealed a steep linear increase of type 2 diabetes in school children in parallel with an increase of obesity of this age group in Japan (*see* Fig. 29.3 and Table 29.1; and refs. *(8,16,17)*). Childhood type 2 diabetes has increased by 2.5 times over the past 25 years, and its incidence is now significantly higher than that of childhood type 1 diabetes in Japan ($5–7/10^5$ vs $1.5/10^5$ per year, respectively) *(8,17–20)*. Similarly, in Thailand, childhood type 2 diabetes (newly diagnosed in Siriraji Hospital, the biggest hospital in Bangkok) increased from 5.9 cases per year between 1987 and 1996 to 15.9 cases per year between 1997 and 1999 *(21)*. In Singapore, childhood type 2 diabetes (newly diagnosed in the second biggest hospital) was five times greater over the past 3 years *(22)*.

In New Zealand, type 2 diabetes was reported to account for 35% of diabetes diagnosed before the age of 30 years among Northland Maori, whereas in previous studies in the 1980s, the majority of young New Zealand Maori patients had type 1 diabetes; this change indicates an increasing rate of type 2 diabetes in young New Zealand Maori *(23)*. Similarly, in a 5-year follow-up study with Australian Aborigines, it was reported that the prevalence of impaired glucose tolerance and type 2 diabetes was 1.4 and 1.4%, respectively, at a mean age of 13.3 years and 8.1 and 2.7%, respectively, at a mean age of 18.5 years after 5 years of follow-up *(24)*. Therefore, it is evident that rapid Westernization of lifestyles, particularly eating habits that increase fat intake, has magnified IR and caused an epidemic of type 2 diabetes in children as well as in adults among these ethnic groups. Type 2 diabetes should be recognized as one of the major health problems of children in these ethnic groups, because complications such as retinopathy and nephropathy develop in childhood type 2 diabetes as fast as or even faster than in childhood type 1 diabetes *(25–27)*.

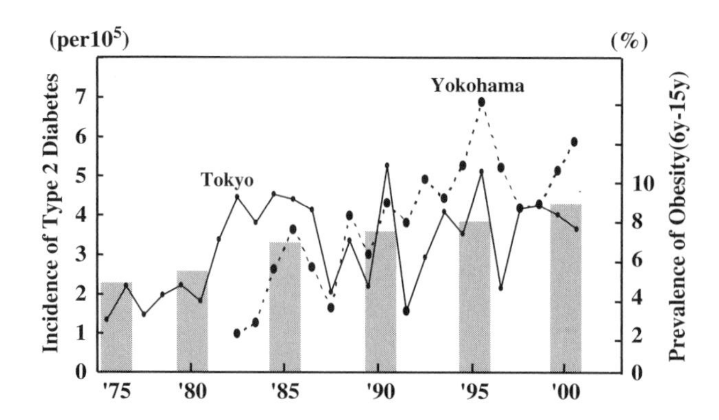

Fig. 29.3. Incidence of type 2 diabetes in Japanese school children.

29.2.2. Risk Factors

Both hypercholesterolemia and hypertension are major risk factors for IHD and cerebrovascular diseases among Japanese people, similar to Caucasian populations *(28,29)*. Nationwide surveys demonstrated that serum total cholesterol levels increased by 10–35 mg/dL between 1960 and 2003 in both adults and children in Japan (Table 29.2 and Fig. 29.4; and refs. *(30,31)*). Currently, the average total cholesterol levels of Japanese adults are still lower than those of American adults, but those of Japanese children are now similar or even higher than those of American children *(32)*. All school children aged 9–12 years (approx. 10,000 school children) have been screened for serum cholesterol, blood pressure, and obesity in the Japanese city of Matsuyama every year since 1989. Their serum total cholesterol levels have increased by 10 mg/dL over 15 years. The prevalence of hypercholesterolemia (>220 mg/dL) is 6.7% (343 of 5,085) in school children aged 9–10 years and 3.9% (174 of 4,414) in those aged 12–13 years in Matsuyama; their average total cholesterol levels are 176 ± 28 mg/dL in boys ($n = 2{,}619$) and 177 ± 27 mg/dL in girls ($n = 2{,}466$) at age 9–10 years and 166 ± 27 mg/dL in boys ($n = 2{,}309$) and 171 ± 27 mg/dL in girls ($n = 2{,}105$) at age 12–13 years *(8)*.

The prevalence of hypertensive disease (hypertension and diseases from hypertension) in Japan increased from 8.0 per 1,000 people in 1965 to 15.6 per 1,000 in 1975 and 30.7 per 1,000 in 1985 *(33)*. In 2003, the prevalence of hypertension among adults aged 30 years or older was 37.3 and 30.7% in men and women, respectively *(31)*. Among school children, the prevalence of hypertension was 19 per 1,000 in boys and 13 per 1,000 in girls aged 9–10 years and 18 per 1,000 in boys and 13 per 1,000

Table 29.1
Distribution of Children with Type 2 Diabetes by Overweight

Overweight (%)	*Boys*	*Girls*	*Total*
Number	106	126	232
<20	9 (8.5%)	29 (23.0%)	38 (16.4%)
20–39	28 (26.4%)	51 (40.5%)	79 (34.0%)
40–59	30 (28.3%)	27 (21.4%)	57 (24.6%)
>60	39 (36.8%)	19 (15.1%)	58 (25.0%)

Source: Ref. *(17)*.
Obesity = % overweight > 20%

Table 29.2
Yearly Changes in Serum Total Cholesterol Levels in Japan

Age (years)	*1960 (31)*	*1970 (31)*	*1980 (31)*	*1990 (31)*	*2003 (31)*
Males					
30–39	167 ($n = 676$)	188 ± 33 ($n = 239$)	192 ± 38 ($n = 897$)	196 ± 35 ($n = 620$)	201 ± 36 ($n = 263$)
40–49	175 ($n = 1043$)	194 ± 45 ($n = 1043$)	197 ± 38 ($n = 1533$)	204 ± 37 ($n = 788$)	205 ± 35 ($n = 270$)
50–59	175 ($n = 878$)	181 ± 45 ($n = 623$)	199 ± 39 ($n = 1081$)	200 ± 37 ($n = 758$)	208 ± 35 ($n = 401$)
60–69	179 ($n = 428$)	169 ± 43 ($n = 334$)	193 ± 40 ($n = 594$)	197 ± 38 ($n = 674$)	201 ± 34 ($n = 513$)
Females					
30–39	187 ($n = 396$)	184 ± 42 ($n = 184$)	178 ± 34 ($n = 591$)	186 ± 32 ($n = 992$)	188 ± 30 ($n = 463$)
40–49	192 ($n = 325$)	179 ± 36 ($n = 337$)	191 ± 39 ($n = 705$)	200 ± 36 ($n = 1124$)	200 ± 35 ($n = 476$)
50–59	207 ($n = 216$)	194 ± 44 ($n = 142$)	211 ± 41 ($n = 735$)	218 ± 37 ($n = 995$)	220 ± 34 ($n = 662$)
60–69	194 ($n = 104$)	189 ± 44 ($n = 33$)	213 ± 39 ($n = 597$)	223 ± 38 ($n = 870$)	218 ± 35 ($n = 690$)

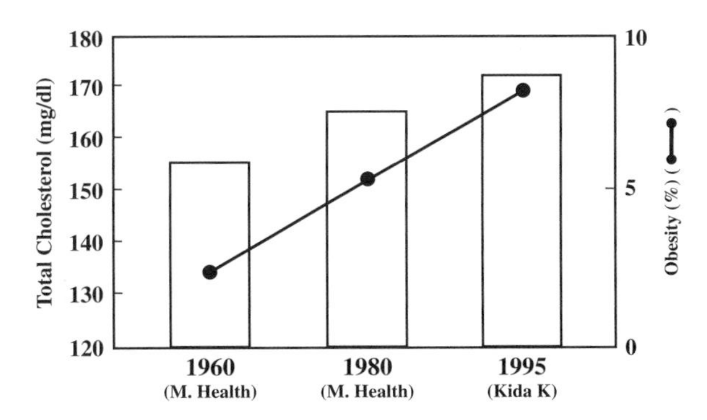

Fig. 29.4. Change in serum cholesterol levels in Japanese children.

in girls aged 12–13 years *(8)*. It is well established that obesity is associated with hyperlipidemia and hypertension among Japanese *(34,35)*. The average body mass index (BMI; weight/height^2) among Japanese adults has increased during the past 40 years, and estimated prevalence of adulthood obesity (BMI > 30) in Japan was 2.1 and 3.1% in men and women, respectively, in 1994 and 2.8 and 3.5% in men and women, respectively, in 2003, which is still lower than that in Western countries *(31,36)*. Nevertheless, the prevalence of obesity (20% overweight or more) among Japanese school children has increased remarkably (threefold) during the past 30 years (*see* Fig. 29.5 and Ref. *(37)*).

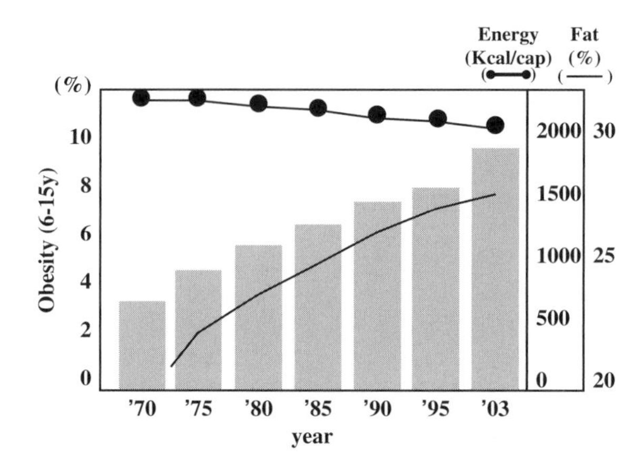

Fig. 29.5. Nutrition and childhood obesity in Japan.

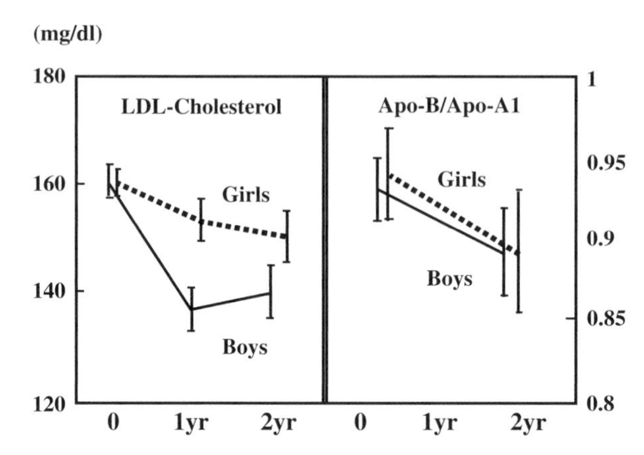

Fig. 29.6. Effects of intervention on LDL-cholesterol levels and ratio of apoB/apoA1. Children with 140 mg/dL or more of serum LDL cholesterol and those with an apoB/apoA1 ratio of 0.8 or higher were given nutritional education by school dietitians and follow-up for 2 years.

School children with hypercholesterolemia or obesity in Matsuyama are provided with health education, including diet and exercise recommendations, by school dietitians and school nurses. Significant improvement of serum total cholesterol or obesity is achieved in more than 50% of children after these interventions (*(8)*; *see* Figs. 29.6 and 29.7, Table 29.3). This suggests that school-based programs for motivation and education play an important role in reducing risk factors for lifestyle-related chronic diseases that occur later in life.

In Korea, the average total cholesterol levels of children are comparable with those of Japanese or American children, and the prevalence of hypercholesterolemia (>200 mg/dL) among school children is reported to be 5–23% *(38)*. The prevalence of obesity has increased by 70% in boys and by 35% in girls over only 4 years, from 1984 to 1988; these rates are now 12.4% among boys and 11.6% among girls in Seoul and Cheju *(39,40)*. In Thailand in 1985, a survey of coronary risk factors and nutritional conditions was performed among 3,495 workers and reported that the prevalence of hypercholesterolemia (>200 mg/dL), hypertension (>141/91), and obesity (BMI > 25) was 71.3, 9.6, and 25.5%, respectively, in men and 65.4, 4.3, and 21.4%, respectively, in women. In 1991, another survey was taken among 519 hospital staff workers, who were expected to be more health conscious. This

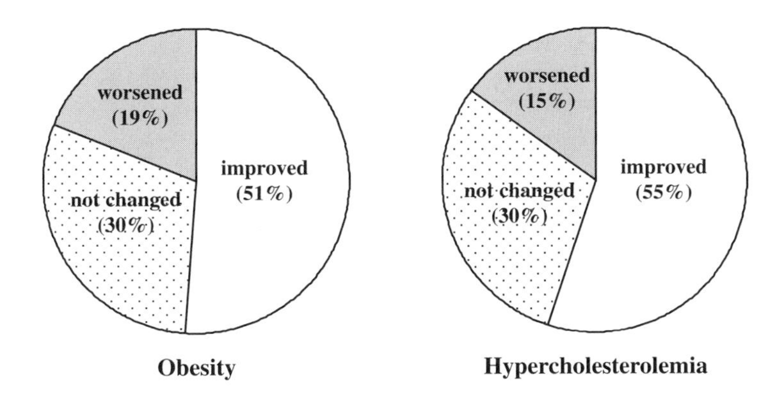

Fig. 29.7. Effects of school program on obesity and hypercholesterolemia in school children (2 years follow-up).

Table 29.3
Effect of Intervention of Obesity Among School Children in Matsuyama (Longitudinal Study)

	Percent overweight	
	9–10 years	*12–13 years*
Intervention group		
Boys ($n = 339$)	33.0 ± 0.6	30.1 ± 0.9a
Girls ($n = 261$)	31.0 ± 0.7	23.4 ± 1.0b
Nonintervention group		
Boys ($n = 82$)	31.1 ± 1.3	33.9 ± 1.6
Girls ($n = 111$)	31.2 ± 1.0	30.0 ± 1.6

[a]$p < 0.05$, mean ± SE; [b]$p < 0.001$, mean ± SE.

study demonstrated that the prevalence of hypercholesterolemia, hypertension, and obesity was 33.4, 4.6, and 18.2%, respectively, in men and 40.2, 1.8, and 27.0%, respectively, in women – lower than those in the former survey but still high *(4,41)*. Therefore, risk factors for lifestyle-related chronic diseases, including hypercholesterolemia, hypertension, and obesity, have increased over only a few decades in Asian countries and are now compatible with or even exceed those in Western countries.

29.2.3. Metabolic Syndrome

Visceral obesity and insulin resistance lead to type 2 diabetes, dyslipidemia, hypertension, and the development of atherosclerosis, in what is labeled the "metabolic syndrome." In Japan the diagnostic criterion for metabolic syndrome was published for adults in 2004 and for children and adolescents in 2007 (Table 29.4). Additionally it was confirmed that males and females who fulfilled the criterion for metabolic syndrome in 2005, classified by age and sex (where data are available), were 0.9 and 0% in 20–29 years, 9.0 and 2.2% in 30–39 years, 13.3 and 3.1% in 40–49 years, 23.0 and 6.0% in 50–59 years, 29.3 and 15.1% in 60–69 years, 29.7 and 19.3% in >70 years, respectively. There were especially higher rates in males *(42)*.

The homeostasis model assessment of IR(HOMA-R) is a conventional index for assessment of IR that is calculated by a formula of fasting blood glucose (mg/dL) × fasting serum insulin (μunit/mL)/450. The IR assessed by HOMA-R is highly correlated with BMI (0.77; $p < 0.01$) or

Table 29.4
Metabolic Syndrome in Children and Adolescents in Japan (Ministry of Health Welfare and Labor)

	Risk factor	
1	Waist circumference	>80 cm (Males: >85 cm, Females: >90 cm)
2	Triglycerides	>120 mg/dL (>140 mg/dL) and/or
	HDL-C	<40 mg/dL (<40 mg/dL)
3	Blood pressure	Systolic >125 mmHg (>130 mmHg) and/or diastolic >70 mmHg (85 mmHg)
4	Fasting blood glucose	>100 mg/dL (>110 mg/dL)

HDL-C: High-density lipoprotein cholesterol; values in parenthesis: adult.

percent body fat measured by the bioimpedance method ($r = 0.88$; $p < 0.01$) in children (unpublished data,) whichsuggests that IR develops according to progression of visceral obesity in children, similar to IR development in adults. It was recently discovered that the IR associated with visceral obesity is regulated by the adipocytokines secreted by the fat cells in the adipose tissues including tumor necrosis factor α and resistin, which induces IR, and adiponectin and leptin, which suppresses IR. Obesity (as expressed in BMI or percent body fat) is associated with decreasing levels of serum adiponectin ($r = -0.31$; $p < 0.01$ or $r = -0.32$; $p < 0.01$) and leptin resistance, shown by the increased levels of serum leptin ($r = 0.81$; $p < 0.01$ or $r = 0.88$; $p < 0.01$) in children ($n = 109$) (unpublished data).

29.3. NUTRITIONAL INTAKES

In Japan, the total energy intake per day has not changed or gradually decreased for the past 40 years (2,098, 2,184, 2,119, and 1,920 kcal in 1950, 1965, 1980, and 2003, respectively; *see* Fig. 29.7; ref *(31)*). On the other hand, the percent energy intake from fat has increased 3.3 times over the past 40 years (7.7, 14.8, 23.6, and 26.3% in 1950, 1965, 1980, and 2003, respectively). Intake of fat from animal, plant, and ash origin has increased by 4.2, 1.9, and only 1.7 times, respectively, over 35 years, and the ratio of fat from animal, plant, and fish is now 4:5:1 in Japan (*see* Fig. 29.8; Ref. *(31)*). The percent energy intake from saturated fatty acids (S), monounsaturated fatty acids (M), and

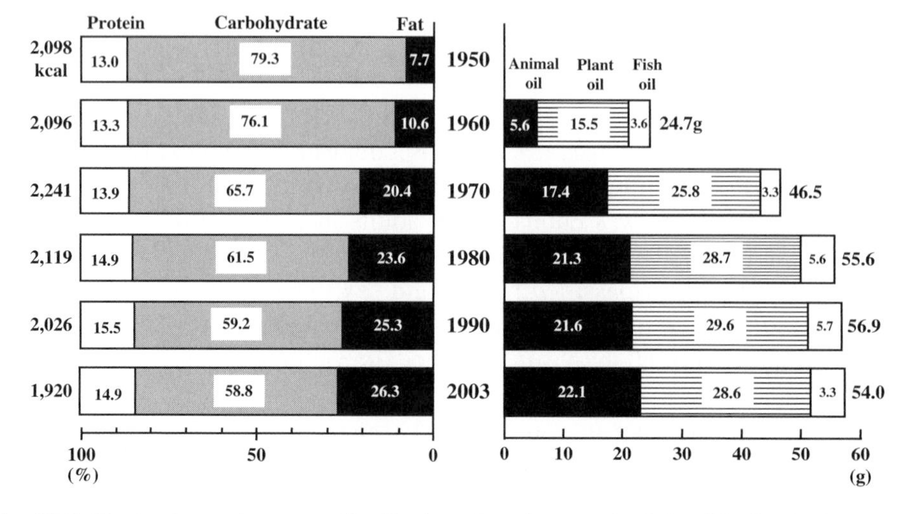

Fig. 29.8. Yearly change in energy distribution of nutrients and origin of fat intake in Japan.

Table 29.5
Yearly Change in Intake of Nutrients Per Capita Per Day in Korea

Nutrients	*1971*	*1976*	*1981*	*1986*	*1991*	*1992*
Energy (kcal)	2,072	1,926	2,052	1,930	1,930	1,875
Carbohydrate (g)	422	380	394	343	325	313
Energy (%)	81.5	78.9	76.8	71.0	67.4	66.8
Fat (g)	13.1	20.0	21.8	28.1	35.6	34.5
Energy (%)	5.7	9.3	9.6	13.1	16.6	16.6
Protein (g)	67.0	60.4	67.2	74.2	73.0	74.2
Animal origin (%)	11.6	20.2	32.2	41.2	42.7	46.6

polyunsaturated fatty acids (PUFA) in Japan was 2, 3, and 4%, respectively, in 1955; however, these percentages rose to 8, 9, and 9%, respectively, in 1985,which indicates that fat intake, particularly fat of animal origin, greatly increased over 30 years *(43)*. The ratio of *n*–6 PUFAs/*n*–3 PUFAs increased from 2.8 in 1955 to 3.8 in 1985 *(43)* and 4.1 in 1990 *(44)*. The increase in the ratio of *n*–6 PUFAs/*n*–3 PUFAs might be the result of a relative decrease in intake of fat of fish origin and an increase in intake of plant oils containing a large amount of linoleic acid. In fact, the energy intake from linoleic acid increased 2.5 times, from 2.6% in 1955 to 6.4% in 1985 in Japan *(43, 45)*. The protein intake per capita per day did not change in the past 30 years in Japan, but protein of animal origin increased 1.7 times and protein of plant origin decreased by 34% during this period *(31)*. The change in the quantity and quality of nutrients might reflect Westernization of diets of Japanese people. Furthermore, occasions to eat meals outside at restaurants or even at American-style fast food bars, where fat-rich foods such as hamburgers and fried chicken are served, have increased remarkably in Japan during the past 30 years (11.3, 16.9, and 18.5% in 1965, 1980, and 2001, respectively) *(31)*. Therefore, the eating habits of Japanese have been Westernized in quantity, quality, and manner.

A similar pattern is evident in Korea *(46)* (Table 29.5). Eating habits have been greatly Westernized during the past 20 years. The total energy intake per capita per day has slightly decreased, whereas the percent energy from fat has increased threefold (5.7, 9.6, and 16.6% in 1971, 1981, and 1992, respectively) *(46)*. The ratio of fat from animal, plant, and ash was 19:69:12 in 1971 and 25:62:13 in 1990. The ratio of saturated to monounsaturated to polyunsaturated fatty acids (S:M:P) was reported to be 2:2.8:1 and that of *n*–6/*n*–3 was 4.9 in 1990 *(46)*. The protein intake per capita per day was increased only 1.1-fold, whereas protein of animal origin increased 4.4 times from 1981 to 1992 *(46)*. The expenditure on fast foods and instant foods increased from 7.7% in 1973 to 12.9% in 1983, and occasions to eat meals outside the home also increased from 2.8% in 1977 to 8.9% in 1986 *(46)*. In Thailand, Westernization of diet can be seen in increased fat intake by people in Bangkok. Fat currently accounts for 30% of total energy intake *(46)*.

29.4. ATHEROSCLEROSIS AND NUTRITION

It is well established that the origin of dietary fat and its composition of fatty acids are related to the risk of CVD. The epidemiological study of Eskimos in Greenland and many other studies demonstrated that fish oils or *n*–3 PUFAs could play a role in preventing CVDs *(47–51)*. Epidemiological studies of Japanese subjects showed similar results. The death rates from IHD and cerebral infarction were 2.6 and 9.2 times lower, respectively, in the town of Higashi-Izu, where people eat more fish than in Tokyo *(52)*. More recently, a large-scale epidemiological investigation was performed to determine

Table 29.6
Cohort Study on Eating of Fish and Sex- and Age-Adjusted Death Rate (Relative Risk) from Chronic Disease in Japan, 1966–1982

	Eating of fish					
Case of death	*Every day*	*Sometimes*	*Rarely*	*Never*	x^2	*p*
Total death	1.0	1.07	1.12	1.32	9.13	<0.0001
Cerebrovascular diseases	1.0	1.08	1.10	1.10	4.54	<0.0001
Heart diseases	1.0	1.09	1.13	1.24	3.92	<0.0001
Hypertension	1.0	1.55	1.89	1.79	4.14	<0.0001
Liver cirrhosis	1.0	1.21	1.30	1.74	3.77	<0.0001
Stomach cancer	1.0	1.04	1.04	1.44	2.14	<0.05
Liver cancer	1.0	1.03	1.16	2.62	2.11	<0.05
Uterus cervix cancer	1.0	1.28	1.71	2.37	4.14	<0.0001
Observed 1 person year	1,412,740	2,186,368	203,945	28,943		

the effect of eating fish on the risk of chronic diseases among adults in Japan. From 1966 to 1982, the 55,523 deaths out of a cohort of 265,118 were analyzed regarding their intake of fish. The study revealed that the total death rate and the death rate from cerebrovascular disease, heart disease, hypertension, and cancer were significantly lower in people who ate fish every day compared to those who did so sometimes, rarely, or never (Table 29.6; Ref. *(53)*).

The degree of development of atherosclerosis in living subjects (measured by pulse wave velocity of the aorta) was 7.0 ± 1.1 m/s in a fishermen's village, where 90% of inhabitants ate fish every day and death from IHD was low. The pulse wave velocity was significantly slower (less sclerotic) than the value of 7.7 ± 1.3 m/s in a farmer's village, where only 11% of inhabitants ate fish every day and death from IHD was high *(54)*. Furthermore, renarrowing of coronary arteries after percutaneous transluminal coronary angioplasty for IHD was significantly decreased, from 37% in controls ($n = 43$) to 20% in subjects who were given 1.6 g/day of eicosapentenoic acid (EPA) orally for 3–4 months ($n = 30$) *(55)*, as confirmed by a large-scale, double-blind investigation *(56)*. A multicenter study in Japan demonstrated that EPA was effective in arteriosclerosis obliterans (diameters of skin ulcers were 7.4 ± 1.87 mm in controls ($n = 25$) vs 6.0 ± 1.59 mm in subjects treated with 1.8 g/day of EPA for 6 weeks ($n = 18$)) *(57)*. These data indicate that oils of fish origin or *n*–3 PUFAs could be beneficial in preventing atherosclerosis and related chronic diseases.

Another characteristic of the Japanese diet is a high intake of salt (NaCl) which likely plays an essential role in hypertension as an important risk factor for atherosclerosis. The current NaCl intake per capita per day in Japan is 12.2 g/day (i.e., 6.2 g/1,000 kcal/day) and has not changed in the past 30 years *(31)*. About 50% of total NaCl is from soy sauce, soy bean paste, and salted vegetables *(31)*. People in the eastern part of Japan consume 1.3 times more salt than those in the Kinki district, including Osaka and Kyoto *(31)*. Public education in a community for 8 weeks successfully reduced NaCl intake of the residents by 3.8 g/day and lowered the systolic and diastolic blood pressure by 4.7 and 3.9 mmHg, respectively *(58)*.

29.5. NUTRITIONAL RECOMMENDATIONS

Taking the rapid increase in lifestyle-related chronic disease and Westernization of diet into consideration, much attention should be given to reduction of these chronic diseases by establishing a nutritionally appropriate lifestyle in Japan. The most recent recommendations appear in the sixth

Table 29.7
Nutritional Recommendations in Japan, Korea, and Thailand

	Japan (59)	*Japan Matsuyama*	*Korea (64)*	*Thailand (65)*
Fat intake	25–30 (1–17 years)	25–30 (1–5 years)	20	>30
Total energy (%)	20–25 (18 years)	20–25 (>6 years)		
Fat intake from animal:plant:fish	4:5:1	3:4:2		
Fat intake from animal meat (g)		1.0–1.5		
S:M:P	3:4:3	1:1.5:1.5	1:1–1.5:1	1:1:1
n–6 polyunsaturated fatty acid				
n–3 polyunsaturated fatty acid	4	3		

edition of *Recommended Dietary Allowance for Japanese* issued by the Ministry of Health and Welfare of Japan in 1999 (Table 29.7; Ref. *(59)*).

The recommended energy intake (E) (kcal/day) was calculated by the following formula:

$$E(\text{kcal/day}) = \text{standard basal metabolic rate for age (kcal/kg/day)} \times \text{body weight (kg)} \times \text{intensity of life (1.3–1.9)}$$

The recommended energy intake from fat is 45% of total energy intake for infants under age 6 months, 30–40% for infants between 6 and 12 months, 25–30% for young people aged 1–17 years, and 20–25% for adults aged 18 years and older. The recommended ratio of S:M:P among fatty acids is 3:4:3, and the recommended ratio of *n*–6 PUFAs/*n*–3 PUFAs is 4.0, according to the results of analyses of the present nutritional conditions of Japanese. The recommendation of S:M:P is in contrast to the previous recommendation, which was 1:1.5:1 *(60)*. The recommended intake of NaCl is 10 g/day or less and should be reduced to 7–8 g/day, if possible. There are a few arguments regarding these recommendations. There are no apparent reasons to elevate the upper limit of fat intake to 30% of the total energy for young people from 25% for adults. In fact, hypercholesterolemia is found in 8.2–9.6% of school children (Kida K, Matsuyama study, 2003). Furthermore, pathological studies in Japan as well as in the United States revealed that atherosclerosis occurred with high frequency even in young people and was related to blood cholesterol levels *(61–63)*. Therefore, our study group of Matsuyama recommends that fat intake should not exceed 25% of the total energy in school children and adolescents. Even infants, with the exception of babies and young infants, should not be exposed to fat-rich diets, because the period of infancy is critical toward setting eating habits in later life (Table 29.7). Although no definite consensus has been established regarding the ideal composition of fat and fatty acids, it is obvious from animal experiments and epidemiological studies that fat of animal origin should be reduced and the ratio of *n*–6 PUFAs/*n*–3 PUFAs should be lowered as much as possible. It might be feasible for Japanese to take more fat of fish origin and less of animal origin while also reducing linoleic acid (*n*–6 PUFA) intake by avoiding oil or food fortified with linoleic acid; thus, the ratio of *n*–6 PUFAs/*n*–3 PUFAs may be lowered to the levels of 2.8, which was the ratio in 1955. Accordingly, our group recommends that ratios of S:M:P among fatty acids and of *n*–6 PUFAs/*n*–3 PUFAs be 1:1.5:1.5 and 3.0–2.0, respectively (Table 29.7). Regarding salt intake, the recommended intake of salt (10.0 g/day) is still high compared with those in many other parts of the world. We recommend salt intake to be 7–8 g/day (3.5–4.0 g/1,000 kcal) for adults and children above age 6 years.

In Korea, the recommended fat intake is 20% of total energy intake and recommended ratio of S:M:P and *n*–6 PUFAs/*n*–3 PUFAs are 1.0:1.0–1.5:1.0 and 4–10, respectively *(64)*. In Thailand, it is recommended that the fat intake should not exceed 30% of total energy intake, with equal distribution of S, M, and PUFAs among fatty acids *(65)*.

REFERENCES

1. WHO: World Health Statistics Annual, 2008.
2. Statistics and Information Department, Minister's Secretariat, Ministry of Health and Welfare of Japan: Vital Statistics of Japan (2005), 2007.
3. National Statistics Office, Republic of Korea: Annual report on Case of Death Statistics, 1994.
4. Leelagul P, Tanphaichitr V. Current status on diet-related chronic diseases in Thailand. *Intern Med* 1995; 11:28–33.
5. Kun-Ho Y, Jin-Hee L, Ji-Won K, Jae HC, Yoon-Hee C, Seung-Hyun K, Paul Z, Ho-Young S. Epidemic obesity and type 2 diabetes in Asia. *Lancet* 2006 November 11; 368(9548):1681–1688.
6. King H, Aubert RE, Herman WH. Global burden of diabetes, 1995–2025, nemerial estimates, and projection. *Diabetes Care* 1998; 21:1414–1431.
7. Ministry of Health, Welfare and Labor of Japan: Rapid Report on the Survey of Diabetes (2002), 2003.
8. Kida K. Type 2 Diabetes in Children and Adolescents in Asia. In: Silink M, Kida K, Rosenblom A, eds. *Type 2 Diabetes in Childhood and Adolescence*. Martin Duniz, London, 2003, pp. 51–61.
9. Rosenbloom AL, Young RS, Joe JR, Winter WE. Emerging epidemic of type 2 diabetes in youth. *Diabetes Care* 1999; 22:345–354.
10. Fagot-Campagna A, Pettitt DJ, Engelgau MM, et al. Type 2 diabetes among North American children and adolescents: epidemiologic review and a public health perspective. *J Pediartr* 2000; 136:664–672.
11. Dabelea D, Hanson RL, Bennett PH, Roumain J, Knowler WC, Pettitt DJ. Increasing prevalence of typediabetes in American Indian children. *Diabetologia* 1998; 41:904–910.
12. Balanchard JF, Dean H, Anderson K, Wajda A, Ludwig S., Depew N. Incidence and prevalence of diabetes in children aged 0–14 years in Manitoba, Canada, 1085–1993. *Diabetes Care* 1997; 20:512–515.
13. Neufeld ND, Chen Y-DI, Raffel LJ, Vadheim CM, Landon C. Early presentation of type 2 diabetes in Mexican-American youth. *Diabetes Care* 1998; 21:80–86
14. Pinhas-Hamiel O, Dolan LM, Daniels SR, Standiford D, Khoury PR, Zeitler P. Increased incidence of non-insulin-dependent diabetes among adolescents. *J Pediatr* 1996; 128:608–615.
15. Scott CR, Smith JM, Cradock MM, Pihoker C. Characteristics of youth-onset noninsulin-dependent diabetes mellitus at diagnosis. *Pediatrics* 1997; 100:84–91.
16. KidaK. Japanese school programs combat type 2 diabetes. *Diabetes Voice,(International Diabetes Federation)* 2003; 48:51–54.
17. Urakami T, Harada K, Kubota S, Owada M, Nitadori Y, Kitagawa T. Annual incidence and clinical characteristics of type 3 diabetes in children as detected by urine glucose screening in the Tokyo Metropolitan Area. *Diabetes Care* 2005; 28:1876–1881.
18. Kitagawa T, Owada M, Urakami T, Ymauchi K. Increased incidence of non-insulin dependent diabetes mellitus among Japanese school children correlates with an increased intake of animal protein and fat. *Clin Pediatr* 1998; 37:111–116.
19. Kikuchi N, Shiga K, Tokuhiro E. Epidemiology of childhood onset NIDDM. *Clin Endocrinol(Tokyo)* 1997; 45:823–827 (in Japanese).
20. Kida K, Mimura G, Ito T, Murakami K, Ashkenzi I, Laron Z, Data committee of Childhood Diabetes of the Japan Diabetes Society (JDS). Incidence of type 1 diabetes mellitus in children aged 0–14 in Japan, 1986–1990, including an analysis for seasonality of onset and month of birth: JDS study. *Diabet Med* 2000; 17:59–63.
21. likitmaskul S, Tuchinda C, Punnakanta L, Kiattisakthavee P, Chaichanwattanakul K, Angsusingha K. An increase of type 2 diabetes in Thai children and adolescents in Thailand. *J Pediatr Endocrinol Metab* 2000; 13(suppl.4):1210.
22. Lee WRW, Yap KPF, Loke KY, Hamidah K, Chia YY, Ang S. Characterisitics of childhood onset type 2 diabetes in Asia and Singapore. *J Pediatr Endocrinol Metab* 2000; 13(suppl.4):1209.
23. McGrath NM, Parker GN, Dawson P. Early presentation of type 2 diabetes mellitus in young New Zealand Maori. *Diab Res Clin Prac* 1999; 43:205–209.
24. Braun B, Zimmermann MB, Kretcher N, Spargo RM, Smith RM, Gracey M. Risk factors for diabetes and cardiovascular disease in young Australian aborigines. *Diabetes Care* 1996; 19:472–479.
25. Takahashi H, Sato Y, Mastui M, Urakami T, Ocular Fundus Findings and systemic factors by type of childhood diabetes. *J Jpn Soc Ophthalmol* 1995; 46:695–699 (in Japanese).

26. Yokoyama H, Okudaira M, Otani T, et al. Existence of early-onset NIDDM Japanese demonstrating severe diabetic complications. *Diabetes Care* 1997; 20:844–847.
27. Yokoyama H, Okuraira M, Otani T, et al. Higher incidence of diabetic nephropathy in type 2 diabetes than in type1 diabetes in early-onset diabetes in Japan. *Kidney Int* 2000; 58:302–311.
28. Tarui S. Distribution of phenotypes of hyperlipidemia and relationship between serum lipid levels and development of cardiovascular complications in Japan. Report of Study Group on Primary Hyperlipidemia, Ministry of Health and Welfare of Japan, 1987, pp. 17–26.
29. Ueda K, Omae T, Hsuo Y, et al. Progress and outcome of elderly hypertensives in a Japanese community: results from a long-term prospective study. *J Hyperten* 1988; 6:991–997.
30. Ministry of Health and Welfare of Japan: Report of the 5th National Survey of Cardiovascular Diseases (2000), 2003.
31. Division of Health Promotion and Nutrition, Ministry of Health and Welfare of Japan: Report of National Nutrition Survey (2003), 2004.
32. Couch SC, Cross AT, Kida K, et al. Rapid westernization of children's blood cholesterol in 3 countries: evidence for nutrient-gene interaction. *Am J Clin Nutr* 2000;72:1266s–1272s.
33. Health and Welfare Statistics Association of Japan: Trends Public Health, 1991.
34. Tarui S. Study on etiology and pathophysiology of obesity in adults and children. Report of Sogo kenkyu, Ministry of Education and Culture of Japan, 1991.
35. Tokunaga K, Fujiok S, Matsuzawa Y. Ideal body weight estimated from body mass index with lowest morbidity. *Int J Obes* 1991; 15:1.
36. Division of Health Promotion and Nutrition, Ministry of Health and welfare of Japan: Report of Epidemiological studies on Obesity (1994), 1995.
37. Division of Statistics, Ministry of Education and Science of Japan: Report on Statistics of school health (2003), 2006.
38. Yang MK. Childhood hyperlipidemia in Korea—a review. *J Korean Pediatr Soc* 1993; 36:1049–1058.
39. Cho KB, Park SB, Park SC, Lee DH, Lee SJ. The prevalence and trend of obesity in children and adolescents. *J Korean Pediatr Soc* 1995; 38:500–528.
40. Ham BH, Kin DH, Park YK, Lee JH, Kin HS. Incidence and complications of obesity in pubescent school children. *J Korean Pediatr Soc* 1995; 38:500–528.
41. Tnphaichitr V, Leelahagul Role of nutrition on healthy lifestyle. *Intern Med (Bangkok)* 1995; 11:31–40.
42. Division of Health Promotion and Nutrition, Ministry of Health and welfare of Japan: *JHealth Welf Stat* 2007; 54(16).
43. Sakai K, Ishikawa A, Okuyama H. Yearly change in quality and quantity of fat intake in Japan. *Aburakagaku (Oil Chem)* 1990; 39:196–201.
44. Hirahara F. Yearly change in quality and quantity of dietary fat of Japanese. *Fat Nutr (Toyama,Japan)* 1995; 4:73–82.
45. Lands WE, Hamazaki T, Yamazaki K, et al. Changing dietary patterns. *Am J Clin Nutr* 1990; 51:991–993.
46. Ministry of Health and Social Affairs: Republic of Korea (1992) National Nutrition Survey Report, 1994.
47. BangHO, Dyeberg J, Sinclair HM. The composition of the Eskimo food in northwestern Greenland. *Am J. Clin Nutr* 1980; 33:3657–3661.
48. Kromhout D, Bosschieter EB, Coulander CDL. The inverse relation between ash consumption and 20-years mortality from coronary heart disease. *N Engl J Med* 1985; 3(12):1205–1209.
49. Burr ML, Fehily AM, Gilbert JF, et al. Effects of changes in fat, fish and fiber intakes on death and myocardial infarction: diet and reinfarction trial (DART). *Lancet* 1989; ii: 757–761.
50. Dolecek TA, Grandits G. Dietary polyunsaturated fatty acids and mortality in the multiple risk factor intervention trial (MRFrr). *World Rev Nutr Diet* 1991; 66:205–216.
51. de Lorgeril M, Renaud S, Mamele N, et al. Mediterranean alpha-linolenic acid-rich diet in secondary prevention of coronary heart disease. *Lancet* 1994; 343:1451–1459.
52. Yasugi T. Serum lipid level and incidence of ischemic arteriosclerotic diseases in Japanese population through the past 30 years with special reference on serum lipids and eicosapentaenoic acid levels. In: Fidge NH, Nestel PJ, eds. *Atherosclerosis*. Elservier, Amsterdam, 1986, pp. 55–59.
53. Hirayama T. Fish consumption and health—the cause—specific death rate in reference to frequency of eating of fish. *Med Chugai(Tokyo)* 1992; 46:157–162.
54. Hamazaki T, Urakaze M, Sawazaki S, Yamazaki K, Taki H, Yano S. Comparison of pulse wave velocity of the aorta between inhabitants of fishing and farming village in Japan. *Atherosclerosis (Tokyo)* 1990; 18:416.
55. Yamaguchi T, Ishiki T, Nakamura M, Saeki F. Effect of fish oil on restenosis after PTCA. *Atherosclerosis* 1988; 73: 157–160.

56. Bairati I, Roy L, Meyer F. Double-blind, randomized, controlled trial of fish oil supplements in prevention of recurrence of stenosis after coronary angioplasty. *Circulation* 1992; 85:950–956.
57. Sakurai K, Tanabe T, Mishima Y, et al. Clinical evaluation of MND-21 on chronic arterial occlusion-double-blind study in comparison with ticlopidine. *Myakkangaku(Angiol)(Tokyo)* 1988; 28:597JO4.
58. Tanaka H, Date C, Yamaguchi M. Salt intake and hypertension. *Igaku-no-Ayumi(Tokyo)* 1994; 169:533–536.
59. Division of Health Promotion and Nutrition, Ministry of Health and Welfare of Japan: Recommended Dietary Allowance for the Japanese, 6th Rev., 1999.
60. Division of Health Promotion and Nutrition, Ministry of Health and Welfare of Japan :Recommended Dietary Allowance for the Japanese, 5th Rev.,1994.
61. Sakurai I, Miyakawa K, Komatsu A, Sawada T. Atherosclerosis in Japanese youth with reference to differences between each artery. *Ann NY Acad Sci* 1990; 598:410–417.
62. Tnaka K, Masuda J, Imamura T, et al. A nation-wide study of atherosclerosis in infants, children and young adults in Japan. *Atherosclerosis* 1988; 72:143–156.
63. Newman WP, Wattigney W, Berenson GS. Autopsy studies in the United States Children and Adolescents; Relationship of Risk Factors to Atherosclerosis . *Ann NY Acad Sci* 1991; 623:16–25.
64. The Korean Nutrition Society: Recommended Dietary Allowance for Koreans, 6th Rev., 1995.
65. Department of Health, Ministry of Public Health of Thailand: Recommended Daily Dietary Allowance and Dietary Guideline for Healthy Thais, 2nd ed., 1989.

30 Goals for Preventive Nutrition in Developing Countries

Osman M. Galal and Gail G. Harrison

Key Points

- Low- and middle-income countries face complex problems of persistent undernutrition and the emergence of nutrition-related chronic diseases of adulthood.
- The persistence, and worsening in some areas, of childhood and maternal malnutrition are urgent situations that call for international action as well as national priority.
- Obesity and related chronic diseases are rising rapidly on a global basis, fueled by urbanization, aging populations, technological development, and globalizing food supplies.
- Several countries are making major investments and infrastructure changes in attempts to break the cycles of poverty and malnutrition; these await definitive evaluation but are potentially very important.
- Articulating specific goals for nutritional improvement is essential for progress.

Key Words: Developing countries; malnutrition; nutrition transition

30.1. INTRODUCTION

Low- and middle-income countries increasingly face extremely complex problems in the arena of public health nutrition. Persistent endemic malnutrition affecting primarily children and women of reproductive age contributes hugely to global burdens of disease and disability. Recent estimates are that one-third of child deaths worldwide and 11% of the global disease burden are due to largely preventable maternal and child undernutrition *(1)*. The burden is decidedly nonrandom, with 80% of the world's undernourished children living in just 20 countries, concentrated in sub-Saharan Africa, Asia, and parts of the Middle East *(2)*. Increasing numbers of complex humanitarian emergencies emerging from natural disasters and from political instability impose severe food insecurity and risk of malnutrition on displaced and disrupted populations, requiring governments and other agencies to find ways to manage. And the burdens anticipated from global climate change will be dominated by increased food insecurity, threatened agricultural production, and population movements as an anticipated 50 million "environmental refugees" will depart lands made uninhabitable by drought, flooding, or rising sea levels *(3)*. All of this occurs embedded in a global "nutrition transition" characterized by aging population structures, rapid urbanization in low-income countries, the emergence of obesity and associated diseases as public health problems almost everywhere, and an increasingly globalized and expensive food supply *(4,5)*.

The Millennium Development Goals (MDGs) adopted by the United Nations at the turn of this century and meant to be achieved by the year 2015 form a framework on which many if not most low- and middle-income countries base their public health and welfare planning and evaluation *(6)*. The

A. Bendich, R.J. Deckelbaum (eds.), *Preventive Nutrition*, Nutrition and Health, DOI 10.1007/978-1-60327-542-2_30,

eight MDGs (Fig. 30.1), which are associated with 21 quantifiable targets measured by 60 indicators, include explicit and implicit attention to improving nutrition. While only the first goal mentions hunger directly, it will be impossible to approach several of the other goals without significant improvements in nutrition in most countries.

1. Eliminate extreme poverty and hunger.
2. Achieve universal primary education.
3. Promote gender equality and empower women.
4. Reduce child mortality.
5. Improve maternal health.
6. Combat HIV/AIDS, malaria, and other diseases.
7. Ensure environmental sustainability.
8. Develop a Global Partnership for Development.

Fig. 30.1. The Millennium Development Goals. (From: http://www.undp.org/mdg/basics.shtml (accessed November 5, 2008)

The last several years have seen a significant expansion in the evidence and analyses available on which to base planning and action to improve nutrition *(7,8)* as well as a number of large-scale investments by national governments. However, progress is very uneven and requires significant enlargements of national and international commitments over sustained periods of time if the avoidable loss of human capital attributable to malnutrition is to be decreased significantly.

30.2. THE HEALTH TRANSITION

Figure 30.2 shows the general relationships among the processes termed the "health transition," and which include both the "demographic transition" marked by declines in mortality and fertility, and the "epidemiologic transition" in which as the population ages there is an emergence of adult chronic disease and a change in the major causes of mortality.

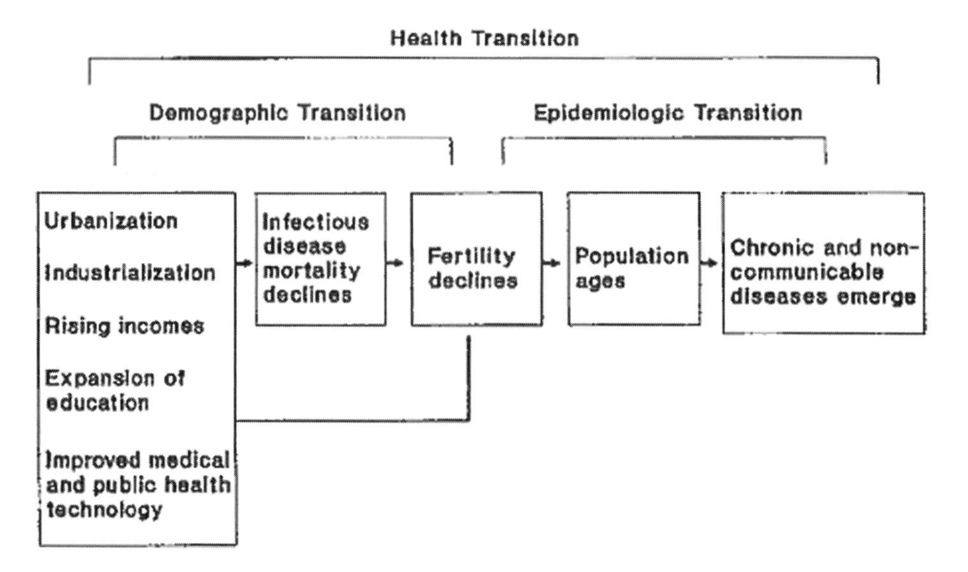

Fig. 30.2. Relationship among demographic, epidemiologic, and health transitions.

In the original formulation by Omran more than 35 years ago (9), there were three proposed eras through which countries passed at different stages of socioeconomic development: the era of "pestilence and famine," in which life expectancy was low and the major causes of death were those associated with malnutrition, infection, and reproduction; the era of "receding pandemics" where life expectancy rises to 30–50 years, morbidity is still dominated by nutritional and infectious causes, and major mortality fluctuations are less common; and finally the era of "degenerative and man-made diseases," with life expectancy over 50 years and major causes of death including cardiovascular diseases, cancer, diabetes, and other chronic ailments. Acknowledging the heterogeneity of social and economic development among human societies, Omran suggested that there were at least three models of the transition: the traditional or Western model, an accelerated model (typified by Japan), and a delayed or contemporary model. The latter describes the incomplete transition that characterizes most developing countries today, with obesity and undernutrition co-existing in the same populations. Recent analyses have shown that up to 15% of households in some nations are dual burden households, in which there are both underweight and overweight persons (10). Because of the prevalence of this double burden of disease in most developing nations, both pre-transition and post-transition problems must be dealt with simultaneously.

Figure 30.3, from Jamison and Mosley (11), shows one widely utilized summary by age group of the "unfinished agenda" of health problems associated with underdevelopment and those "neglected and emerging problems" which are increasingly coming to dominate the health agenda. The relative importance of the "emerging problems" on the health agenda is a function not only of socioeconomic development but also of demographic shifts, lifestyle, and health services. As populations age adult problems become more important as causes of morbidity, mortality, disability, and health-care costs. The elderly are becoming more numerous and more visible almost everywhere. More than half (59%) of the world's elderly population—>65 years—now live in developing countries (12).

Important Health Problems in Developing Countries by Age Group[a]

Age group	*Problems on the unfinished agenda*	*Neglected and emerging problems*
Young children (0–4 yr)	Diarrheal disease Acute respiratory infection Measles, tetanus, polio Micronutrient deficiencies Malaria	Injury
School-age children (5–14 yr)	Schistosomiasis Geohelminth infections	Adolescent pregnancy Disability
Young adults (15–44 yr)	Maternal mortality Malaria Excess fertility	Tuberculosis Injury AIDS STDs Mental illness
Middle-aged (45–64 yr)		Cardiovascular disease Cancers COPD
Elderly (65+ yr)		Disability Depression

Adapted from ref. 11.

Fig. 30.3. Important health problems in developing countries by age group.

30.3. THE NUTRITION TRANSITION

Food consumption patterns, nutritional status, and diet-related morbidity and mortality show characteristic shifts with socioeconomic development, urbanization, education, and industrialization; these shifts have been most thoroughly articulated by Popkin (5), who has postulated that the pace of dietary change seems to have recently accelerated to varying degrees in different regions of the world. The transition seems to converge on a rapid shift, when economic resources allow, to a dietary pattern characterized by higher intakes of fat, animal products, sugar and refined foods, and lower intakes of fiber. The associated changes in nutritional status, often attributed to dietary changes because of their parallel in time but also affected by changes in other risk factors such as sanitation and physical activity patterns, include declines in general undernutrition and micronutrient deficiencies and increased prevalence of obesity and the chronic diseases for which obesity is a risk factor (cardiovascular disease, type II diabetes, and some cancers).

The pace of dietary change and emergence of obesity as a global public health problem in many developing countries have been very rapid; one estimate is that the annual rate of increase in obesity in developing countries is four to five times greater than that of the United States and European countries *(13)*. The rapid dietary change in the developing world has been fueled largely by urbanization. It is estimated that 40% of the population of the developing world's people lived in cities in the year 2000, and 50% will be urban dwellers by 2015 *(14)*. In 1990, seven of the world's "megacities" (population more than 10 million) were in developing countries; six more were added in the decade 1990–2000 and there are expected to be seven more added by 2025 *(15)*. Almost all of these 26 megacities (except only New York, Tokyo, Los Angeles, and Osaka-Kobe) will be in low- or middle-income countries.

The urban environment is often more conducive to good nutritional status in children, but urbanization also brings disadvantages. Declines in breastfeeding rates and duration, sedentary lifestyles, and household food insecurity linked to the cash economy for very poor urban families, bring new nutritional risks which residents of the world's large cities face in unprecedented numbers. The urbanization of large populations typically results in diets with larger amounts of fat, animal products, sugar, and processed foods than the diets of those living in rural areas. In addition, technical innovations have increased the number of service sector jobs in urban areas and reduced the amount of energy expenditure required for work. As a result, the amount of energy expended by individuals on a daily basis has declined, contributing to increasing prevalence of obesity *(16,17)*.

Whether early undernutrition might predispose to later adult obesity and associated conditions is a question with potentially important implications. Possible relationships between undernutrition early in life and later risk of obesity and related chronic disease have received attention largely due to the body of work published by Barker and colleagues from England *(18,19)*. Briefly, these investigators observed that in the United Kingdom (as in the United States) obesity and related morbidity and mortality risks are more prevalent among the lower socioeconomic portion of the population, as are the risk of low birth weight and poor post-natal growth. In a series of correlational studies, they showed a relationship between low birth weight and risk in adult middle age of high blood pressure, non-insulin-dependent diabetes, and other risk factors for coronary heart disease. They hypothesized several possible mechanisms by which early undernutrition, either in utero or in early infancy, may "program" or predispose the individual for increased risk of adult chronic disease; these include modification of gene expression by the nutrient environment, modification of the endocrine milieu through permanent influences on endocrine structures or responses, reduced cell number and altered organ structure, and/or selection of clones of cells resulting in differences in proportions of different cell types.

Since the publication of the work of Barker et al., several studies have shown that nutritional trauma during critical developmental periods early in life is related to obesity and chronic disease *(20)*. Studies in the United States *(21)* and Norway *(22)* have demonstrated an inverse association between height and coronary heart disease, as well as other chronic diseases in adulthood. If ultimately borne out, the hypothesis will have major implications for paradigms of planning in public health nutrition, since one of the most powerful weapons against chronic diseases of middle age in adults will be the improvement of nutritional status of mothers and infants.

30.4. THE VARIETY OF NUTRITIONAL PATTERNS AND COMPLEXITY OF THE CHALLENGES

The variation among low- and middle-income countries with regard to the distribution and prevalence of various nutrition problems and risks is at least as great as those between higher-income and less developed countries. The major determinants of the prevalence of various forms of malnutrition as well as other health problems are extremely variable. Some of the specific challenges facing public health nutrition planners in different economic contexts are highlighted below.

30.4.1. The Lowest Income Countries: The Unfinished Agenda Dominates

It has been estimated that over the next decade perhaps 30–40 countries will remain in the lowest income bracket, in which health problems are dominated by infectious diseases of childhood, undernutrition, and high fertility *(11)*. The major problems continue to be diarrheal disease, acute respiratory illnesses in young children, measles, tetanus, malaria, micronutrient deficiencies, parasitic diseases, and reproductive mortality. In these predominantly sub-Saharan and South Asian countries, the focus of nutrition goals must continue to be on the establishment of basic primary health-care infrastructure, breastfeeding, maternal nutrition, immunization against vaccine-preventable diseases, control of diarrheal disease, and family planning. A year 2000 analysis of trends in child undernutrition in the 15-year period between 1970 and 1995 *(23)* showed a 15.5% reduction in rates of child underweight, attributable primarily to improvement in formal education, in national food availability, and in basic sanitation and health services. However, the problems of undernutrition remain large and persistent. Every year 13 million children are born undernourished, 112 million are underweight, and 178 million children <5 years of age are stunted; 55 million are wasted (thin) and of these 19 million suffer from severe acute malnutrition, defined as a weight-for-height 3 standard deviations below the reference median *(24)*. The consequences for adult health and human capital are major *(25)*. The malnutrition situation in sub-Saharan Africa has actually been worsening in recent years, and the prospects for improving child nutrition and health in South Asia are bleak unless maternal undernutrition is effectively addressed.

There are a number of proven effective interventions to reduce maternal and child undernutrition, ranging from micronutrient supplementation, breastfeeding promotion, health promotion/education for effective complementary feeding of older infants and young children, through sanitation and infectious disease control interventions *(7,8)*. Major investment in these approaches and serious attempts to reach the most vulnerable groups in society have been mostly lacking or misdirected toward activities that have not been shown effective (such as stand-alone growth monitoring) *(26)*. Making nutrition a priority and drawing on available evidence and experience to reduce malnutrition in low-income countries need to be both national and international priorities as investments in the future.

30.4.2. Middle-Income Countries: Social Polarization of Priorities

In a larger group of somewhat higher-income countries, predominantly in East Asia, Latin America, and the Middle East, the infectious disease, malnutrition, and reproductive health burdens remain substantial at the same time as rapid urbanization, industrialization, and economic development have led to the emergence of new health problems including injuries, occupational injuries, and preventable chronic diseases. Diet-related chronic disease emerges as a major priority quickly, since it affects first the higher income, urban, educated segment of the population. Progress and experience are decidedly mixed for these countries, many of whom struggle with increasing fragile economies. There are, however, many examples of significant initiatives in public nutrition education (some are reviewed by Doak in Ref. *(27)*). And several countries have made very bold and major investments and changes in infrastructure, whose impact is mostly yet to be seen. All illustrate that significant impacts on nutrition will require relatively large and sustained, long-term commitments.

A major experiment in improving health and nutrition is playing out in several large, middle-income countries in Latin America, with the development of and investment in conditional cash transfer (CCT) programs. The nature of CCT programs is that they have twin goals of immediate poverty reduction through cash transfers to poor families and long-term poverty reduction through investment in childrens' health and education. Receipt of the cash transfers requires parents (and sometimes other household members) to take advantage of preventive health services, participate in health and nutrition education and in some cases utilize nutritional supplements for malnourished children, keep older children in school, and otherwise invest in their future development. Several national programs are in place, including in Mexico, Brazil, Nicaragua, Honduras, Colombia, Chile, and Jamaica *(28)*. These programs are expensive investments toward trying to break the "cycle of poverty" through significant development of human capital in one generation. They have naturally attracted much attention and evaluation *(29,30)*. The earliest of these programs was Mexico's *Oportunidades* (originally *Progresa*) program, begun in 1997 and designed to facilitate sound evaluation. The program was implemented first in randomly selected eligible communities, with delayed introduction in the control communities. The delay was not as long as the 2 years originally planned, but was long enough to permit evaluation. Compared to control communities, in the first year of the program significant positive impact was seen on child height (most in the youngest and most malnourished children), anemia rates in children, childhood illness, use of preventive medical services, dietary quality, and function/mobility among older adults *(31)*. A recent *(32)* attempt to quantify the cash transfer portion of the program concluded that a doubling of cash transfers (measured as twice the length of time in the program) was associated with better height-for-age in children, lower prevalence of overweight, and better indices of child motor and cognitive development.

Another example of impressive national commitment and infrastructure development in nutrition and food security can be seen in Brazil, another very large middle-income country. Like Mexico, Brazil clearly suffers from both the nutrition-related chronic diseases associated with a nutrition transition and from significant problems of undernutrition and food insecurity in some population segments. Following a national presidential election some years ago in which the successful candidate ran on a platform of "zero hunger," a number of targeted activities were undertaken to reduce food insecurity and hunger in addition to institution of a CCT program. These have included making school feeding programs available everywhere in the country, a "workers" nutrition program, and a vitamin A deficiency reduction program. Published evaluations of these programs thus far are process oriented and available only in local literature *(33)*; impact evaluations await more experience and research. At the same time Brazil has accomplished several ambitious activities designed to provide some protection against nutrition-related chronic diseases; these include legislation and regulation, information and communication to the public, and capacity building at the level of school teachers and health workers *(34)*. Additionally civil society organizations have contributed, notably the *Agita São Paulo*program, a community-based health promotion program emphasizing physical activity developed by the Physical Fitness Research Laboratory from Sao Caetano do Sul (CELAFISCS) that grew into *Agita Brazil* and eventually was adopted by the World Health Organization as *Agita El Mundo (35)*.

That the effects of inflation and price rises in the absence of explicit planning and investment in nutrition have adverse effects can be seen in the example of Egypt. Egypt is a large lower middle-income country where economic investment in improving nutrition has been minimal and social/economic inequalities in the population have been increasing; Demographic and Health Survey data from 2001, 2003, and 2005 show steady deterioration in the nutritional status of children under 3 years of age (measured by stunting) *(36)*. These alarming rates (almost 25% of 18–23-month-old children are stunted) are accompanied by extremely high rates of iron deficiency and iron deficiency anemia in children and women (45% among children 16–23 months and 35% in adolescent girls). Among the 20 countries that account for the large majority of malnourished children in the world, only two are in the Middle East—Egypt and Yemen, the latter of which has vastly lower per capita income *(2)*. There are efforts in Egypt currently to propose targeted, effective programs to counteract this trend, but the overall trend illustrates clearly that left alone, malnutrition is not a problem that will fix itself even in a country where the per capita food supply is quite adequate. At the same time, Egypt suffers one of the highest adult obesity rates in the world along with the predictable and expensive co-morbidities and health-care costs as well as significant government expenditures on largely untargeted consumer subsidies of several basic food commodities *(37)*.

30.4.3. Higher-Income Countries: The Challenge of Equity

In high-income countries diet-related chronic diseases dominate the public health nutrition agenda. Problems of undernutrition and access to food affect mainly the lower income segments of the population and generally have to fight for space on the political agenda and for resource allocation. Further, at least in the United States and the United Kingdom, obesity and its attendant problems also are most prevalent among low socioeconomic status segments of the population. So-called "diseases of affluence" are actually diseases of poverty in some of the richer countries. Policy makers often find it difficult to conceptualize the simultaneous occurrence of undernutrition and obesity, with the result that improvement of access to food and dietary quality often get a back seat to other priorities.

The nations which comprise Eastern Europe and the former Soviet Union represent rather special cases; these have been categorized as "industrialized, non-market economies" *(38)*. Having long since passed through demographic, epidemiologic, and nutrition transitions, they have been recently undergoing an epidemic of excess mortality among relatively young adults from preventable causes. These include trauma, poisoning, respiratory diseases (often secondary to smoking), and complications of pregnancy and child birth *(39)*. Dietary and nutritional problems include high alcohol consumption, very high costs of food, limited dietary variety in terms of fruits and vegetables, micronutrient deficiencies in some groups, including iodine deficiency disease, and undernutrition particularly in the elderly *(40)*. Recognition of the substantial inequities in nutritional well-being and food security within present-day Europe characterizes a recent Food and Nutrition Action Plan for the region *(41)*.

30.5. CHALLENGES OF PROTECTING THE NUTRITIONALLY VULNERABLE IN THE COURSE OF ECONOMIC ADJUSTMENT PROGRAMS, RISING FOOD PRICES, AND CLIMATE CHANGE

Current and imminent economic and environmental changes pose particular challenges in protecting against deterioration in nutritional status, to say nothing of making strides toward improving it. For the last two decades, economic adjustment or "restructuring" programs in a number of low-income countries have resulted in, among other things, curtailment of untargeted or minimally targeted consumer subsidy programs. Without specific protections, such adjustments have the greatest impact on the poorest consumers and most nutritionally vulnerable families. Articulation of specific goals and strategies to protect nutrition in the process of these adjustment programs is essential *(42)*. Processes of climate change well underway are projected to reduce agricultural yields in some of the poorest areas and countries with pronounced regional effects *(43,44)* as well as to greatly increase migration from agricultural lands made untenable by drought, flood, or sea level rise thus further stimulating the growth of cities and the challenges posed for food security for poor urban populations. And most rapidly and recently, unprecedented rises in the prices of basic foodstuffs globally are creating needs for careful planning and reasoned action. Global rice prices tripled between January and May 2008, and overall grain prices increased by 60% between 2006 and 2008 *(45)*. The rapid rise in food prices is linked to a number of causes—increased energy costs, demands for grain for biofuel production, increasing urban migration, adverse weather in some areas, and rapid economic growth particularly in India and China that has shifted food consumption patterns and increased demand particularly for animal-source products *(46)*.

30.6. GOALS FOR PREVENTIVE NUTRITION

While nutritional improvements have generally paralleled economic development, there is far from a one-to-one or deterministic relationship between the two. Experience to date indicates that medical and public health interventions, per capita incomes, and behavior and lifestyle variables must all be taken into consideration. A few countries and regions have managed to improve nutritional status dramatically without major increases in per capita incomes, while others which have experienced large increases in income have not made concomitant strides in nutritional improvements.

Some time ago Caldwell *(47)* analyzed data on health and income, identifying 11 countries (plus the state of Kerala in India) with health achievements over and above what would be expected by per capita incomes and another 11 countries with the poorest performance. The areas which did the best in achieving favorable health indicators were (in descending order) Kerala, Sri Lanka, China, Burma (Myunmar), Jamaica, India, Zaire, Tanzania, Kenya, Costa Rica, Ghana, and Thailand. They averaged a life expectancy of 61 years and infant mortality of 64/1000 live births, and a per capita income average of $501. The countries which did worst at that time in this analysis were (in order from the poorest performance) Oman, Saudi Arabia, Iran, Libya, Algeria, Iraq, Yemen AR, Morocco, Ivory Coast, Senegal, and Sierra Leone, with average per capita incomes of $4462, life expectancy 51 years, and infant mortality rates 124/1000 live births. The analysis showed that the strongest relationships with improved health status at low cost were variables which reflected the education of women of reproductive age, the practice of family planning, and the education of men. Medical care variables were less strongly related to health indices, and income showed even weaker relationships.

The foregoing analysis strongly supports the idea that societies which have an egalitarian tradition which includes relative independence for women and an emphasis on the value of education have been able to take advantage of advances in medical and public health knowledge and information to improve their situations. Those which have severe differentials between male and female educational opportunities, and for which other cultural and behavioral characteristics mitigate against reduced fertility, have done poorly with regard to health even with increased incomes.

Beyond these social and structural priorities however, the explicit articulation of goals for protection and improvement of nutritional status is important. This process has the effect of continuous reinforcement of the importance of nutrition to policy makers and program administrators. It also provides the basis for monitoring of progress. The incorporation of specific and measurable nutrition and dietary goals into national health policy plans has characterized some of the most successful national experiences, including those of Norway *(48)*, Costa Rica *(49)*, Cuba *(50)*, and others.

The Millennium Development Goals only partly address the nutrition agenda, and in no specific detail. The impact of major investments such as those described in some Latin American countries awaits the test of time, but it is clear that if ignored, nutritional status will deteriorate in constrained economic situations.

While the simultaneous existence of problems of malnutrition and adult nutrition-related chronic disease is now generally acknowledged, the informed articulation of realistic goals for overall nutritional improvement in developing countries is far behind where it needs to be. As Popkin *(51)* has pointed out, a failure to acknowledge the role of nutrition in coronary heart disease, cancer, and other chronic diseases will lead to the domination of medical interventions in the allocation of resources. The experience of the industrialized countries to date argues that primary prevention strategies associated with lifestyle changes are the most effective strategies with regard to these post-transition health problems. The challenge will be to articulate goals for preventive nutrition clearly and to avoid the polarization which results from a conceptual framework which pits undernutrition and children and the poor against emerging problems associated with diets dense in energy and fat and with sedentary lifestyles. If the hypotheses of Barker et al. articulated more than a decade ago prove to be correct, then the best weapon against chronic diseases of adulthood will prove to be protection and improvement of the nutritional status of mothers and children.

30.7. RECOMMENDATIONS

Our recommendations for policy makers and planners in developing countries are several: (1) develop explicit goals for reduction of malnutrition, undernutrition, and food insecurity in vulnerable population groups; (2) allocate resources and direct policies to assure the achievement of these goals even in the face of overall economic stress; (3) embrace primary prevention strategies first for damming the flood of diet-related chronic diseases of adulthood; and (4) develop and maintain nutrition surveillance systems that will allow the tracking of nutrition problems in specific circumstances, in order to enable prompt response.

ACKNOWLEDGEMENT

The authors acknowledge with thanks the technical assistance of Eanas Aboobakar and Susan Silah.

REFERENCES

1. Black, R.E., Allen, L.H., Bhutta, Z.A., Caulfield, L.E., de Onis, M., Ezzati, M., Mathers, C., Rivera, J. for the Maternal and Child Undernutrition Study Group (2008). Maternal and child undernutrition: global and regional exposures and health consequences. *Lancet* **371**: 243–260.
2. Bryce, J., Coitinho, D., Darnton-Hill, I., Pelletier, D., Pinstrup-Andersen, P., for the Maternal and Child Undernutrition Study Group (2008). Maternal and child undernutrition: effective action at the national level. *Lancet* **371**: 510–526.
3. Intergovernmental Panel on Climate Change. *Climate Change 2007: Synthesis Report* (2007). Cambridge, U.K.: Cambridge University Press.
4. Harrison, G.G., Hamide, S. (2009) Overweight, obesity and related diseases. In: Pond, W., Nichols, B. and Brown, D. (eds.), *Adequate Food for All in the 21st Century*. NY: Taylor and Francis.
5. Popkin, B.M. (2002). An overview on the nutrition transition and its health implications: The Bellagio meeting. *Public Health Nutr* **5** (1A): 93–103.
6. United Nations (2007). *The Millennium Goals Report, 2007*. NY: United Nations.
7. Allen, L., Gillespie, S. (2001). *What Works? A Review of the Efficacy and Effectiveness of Nutrition Interventions*. SCN Nutrition Policy Paper No. 19. SCN in collaboration with the Asian Development Bank, Manila. Available on-line at: http://www.unsystem.org/scn/Publications/NPP/npp19.pdf
8. Bhutta, Z.A., Ahmed, T., Black, R.E., Cousens, S., Dewey, K., Glugliani, E., Haider, B.A., Kirkwood, B, Morris, S.S., Sachdev, H.P.S., Shekar, M., for the Maeternal and Child Undernutrition Study Group (2008). What works? Interventions for maternal and child undernutrition and survival. *Lancet* **371**: 417–440.

9. Omran, A. (1971). The epidemiologic transition: a theory of the epidemiology of population change. *Milbank Mem Fund Q* **49**: 509–538.
10. Doak, C.M., Adair, L.S., Bentley, M., Monteiro, C., Popkin, B.M. (2005). The dual burden household and the nutrition transition paradox. *Int J Obes* **29**: 129–136.
11. Jamison, D.T., Mosley, W.H. (1991). Disease control priorities in developing countries: health policy responses to epidemiologic change. *Am J Public Health* **81**: 15–22.
12. Kinsella, K. (2001). *An Aging World: 2001.* International Population Reports, US Department of Health and Human services and US Department of Commerce, Washington DC.
13. Popkin, B., Gordon-Larsen, P. (2004). The nutrition transition—worldwide obesity dynamics and their determinants. *Int J Obes* **28**: S2–S9.
14. Solomons, N.W., Gross, R. (1995). Urban nutrition in developing countries. *Nutr Rev* **53**: 90–95.
15. United Nations Human Settlements Program (UN-HABITAT), (2008). *State of the World's Cities 2008/2009: Harmonious Cities.* London: Earthscan Publishing.
16. Popkin, B., Bisgrove E. (1988). Urbanization and nutrition in low-income countries. *Food Nutr Bull* **10**: 3–23.
17. Popkin, B.M. (2006). Technology, transport, globalization and the nutrition food policy. *Food Policy* **31**: 554–569.
18. Barker, D.J.P. (1994). *Mothers, Babies and Disease in Later Life.* London: BMJ Publishing Group.
19. Barker, D.J.P. (1989). Rise and fall of western diseases. *Nature* **338**: 371–372.
20. Cameron N., Demerath, E.W. (2002). Critical periods in human growth and their relationship to diseases of aging. *Yearb Phys Anthropol* **45**: 159–184.
21. Palmer, J.R., Rosenberg, L., Shapiro, S. (1990). Stature and the risk of myocardial infarction in women. *Am J Epidemiol* **132**: 27–32.
22. Waaler, H.T. (1984). Height, weight and mortality. The Norwegian experience. *Acta Med Scand* (Suppl.) **679**: 2–50.
23. Smith L.C., Haddad, L. (2000) *Explaining Child Malnutrition in Developing Countries: A Cross-Country Analysis.* Research Report #111. Washington DC: International Food Policy Research Institute.
24. Anonymous (2008). Executive Summary: The Lancet's series on maternal and child undernutrition. *Lancet* **371**: 1–5.
25. Vitora, C.G., Adair, L., Fall, C., Hallal, P.C., Martorell, R., Richter, L., Sachdev, H.S., for the Maternal and Child Undernutrition Study Group (2008). Maternal and child undernutrition: consequence for adult health and human capital. *Lancet***371**: 340–357.
26. Morris, S.S., Cogill, B., Uauy, R. for the Maternal and Child Undernutrition Study Group (2008). Effective international action against undernutrition: why has it proven so difficult and what can be done to accelerate progress? *Lancet* **371**: 608–621.
27. Doak, C. (2002). Large-scale interventions and programmes addressing nutrition-related chronic diseases and obesity: examples from 14 countries. *Public Health Nutr* **5**: 275–277.
28. de Janvry, A., Sadoulet, E. (2006). Making conditional cash transfers for improving uptake of health interventions in low- and middle-income countries: a systematic review. *JAMA* **298**: 1900–1910.
29. Rawlings, L.B., Ribio, G.M. (2005). Evaluating the impact of conditional cash transfer programs. *World Bank Res Obs* **20**: 29–55.
30. Das, J., Do, Q-T., Ozler, B. (2005). Reassessing conditional cash transfer programs. *World Bank Res Obs* **20**: 57–80.
31. Rivera, J.A., Sotres-Alvarez, D., Habicht, M.P., Shamah, T., Villalpando, S. (2004). Impact of the Mexican program for education, health and nutrition *(Progresa)* on rates of growth and anemia in infants and young children: a randomized effectiveness study. *JAMA* **291**: 2563–2570.
32. Fernald, L.C.H., Gertler, P.J., Neufeld, L.M. (2008). Role of cash in conditional cash transfer programmes for child health, growth and development: an analysis of Mexico's *Oportunidades. Lancet* **371**: 828–837.
33. Santos, L.M., Santos S.M., Santana, L.A., Henrique, F.c., Mazza, R.P., Santos, L.A., Santos, L.S. (2007). Evaluation of food security and anti-hunger public policies in Brazil, 1995–2002. *Cad Saude Publica* **23**: 2681–2693 (in Portuguese).
34. Coitinho, D., Monteiro, C.A., Popkin, B.M. (2002). What is Brazil doing to promote healthy diets and active lifestyles? *Public Health Nutr* **5**: 263–267.
35. Matsudo, V., Matsudo, S., Andrade, D., Arauojo, T., Andrade, E., de Oliveira, L.c., Braggion, G. (2002). Promotion of physical activity in a developing country; the *Agita São Paulo* experience. *Public Health Nutr* **5**: 253–261.
36. El-Zanaty, F., Way, A., and ORC Macro. *Egypt 2005 DHS Final Report* (English). Calverton, MD: Macro International. http://www.measuredhs.com/pubs (accessed November 23, 2008).
37. Galal, O.M. (2002). The nutrition transition in Egypt: obesity, undernutrition and the food consumption context. *Public Health Nutr* **5**: 141–148.
38. Kinsella, K.G. (1992). Changes in life expectancy 1900–1990. *Am J Clin Nutr* **55**: 1196S–1202S.

39. Barr, D.A., Field, M.G. (1996). The current state of health care in the former Soviet Union: Implications for health care policy and reform. *Am J Public Health* **86**: 307–312.
40. Tulchinsky, t.H., Varavikova, E.A. (1996). Addressing the epidemiologic transition in the former Soviet Union: Strategies for health system and public health reform in Russia. *Am J Public Health* **86**: 313–320.
41. Robertson, A. (2000). WHO tackles food inequalities: Europe's first comprehensive food and nutrition action plan debate, November 1999, Malta. *Public Health Nutr* **3**: 99–101.
42. Selowsky, M. (1991). Protecting nutrition status in adjustment programs. *Food Nutr Bull* **13**: 293–302.
43. Brown, M.E., Funk, C.C. (2008). Food security under climate change. *Science* **319**: 580–581.
44. Lobell, D.B., Burke, M.B., Tebaldi, C., Mastrandrea, M.D., Falcon, W.P., Naylor, R.L. (2008). Prioritizing climate change adaptation needs for food security in 2030. *Science* **319**: 607–610.
45. World Bank (2008). *G8 Hokkaido-Tokyo Summit: Double Jeoparrdy: Responding to High Food and Fuel Prices*. Washington DC: World Bank.
46. Von Braun, J. (2007). The world food situation: new driving forces and required actions. *Food Policy Rep***18**: 1–27.
47. Caldwell, J.C. (1986). Routes to low mortality in poor countries. *Popul Dev Rev* **12**: 171–220.
48. Norum, K.R., Johanson, L., Botten, G., Gjomeboe, G.E., Oshaug, A. (1997). Nutrition and food policy in Norway: Effects on reduction of coronary heart disease. *Nutr Rev* **55**: S32–S39.
49. Mohs, E., Scrimshaw, N.S. (1995). Health policies and strategies. In: Monuz, C. (ed.), *The Nutrition and Health Transition of Democratic Costa Rica*. Boston: International Foundation for Developing Countries.
50. Amador, M., Pena, M. (1991). Nutrition and health issues in Cuba: strategies for a developing countries. *Food Nutr Bull* **13**: 311–317.
51. Popkin, B. (1999). Urbanization, lifestyle changes and the nutrition transition. *World Dev* **27**: 1905.

31 Preventive Nutrition and the Food Industry: Perspectives on History, Present, and Future Directions

Derek Yach, Zoë Feldman, Dondeena Bradley, and Robert Brown

Key Points

- Despite overall progress in health status, major nutrition challenges remain at both ends of the spectrum—that of feast and that of famine.
- The emergence of new business models that tackle social problems while remaining profitable offers promise that the long-term nutrition needs of people can be met.
- Urbanization and industrialization in the late 19th century in Europe and the United States led to what was to become the modern food industry—characterized by high safety standards and the production of foods affordable and permanently available to most people.
- The food industry has advanced scientific knowledge across a range of major food-related challenges: adding or restoring nutrients to foods through fortification; reducing levels of salt and saturated fats; and developing foods with specific health benefits.
- New private–public interactions hold promise that partnerships will emerge to tackle obesity and undernutrition. The May 2008 pledge by the CEOs of eight leading food companies to support WHO's Strategy on Diet, Physical Activity, and Health being one example.

Key Words: Business; corporate social responsibility; food industry; public health; public–private partnerships; undernutrition

31.1. INTRODUCTION

31.1.1. Nutrition, Health and Development Crisis

Over the last 50 years, there have been unprecedented gains for human health around the world *(72)*. Life expectancy has soared in many developing countries and is now converging with trends in developed countries. A large part of this progress can be attributed to improved nutrition. However, in two parts of the world, there has been significant regression. First, in sub-Saharan Africa, life expectancy has declined for over a decade, driven directly by the HIV/AIDS epidemic and indirectly by the continued and worsening food crisis. Additionally, in several countries previously part of the Soviet Union, including Russia, life expectancy among men has declined, a process driven by unhealthy diets and considerable alcohol and tobacco consumption *(1)*.

Despite overall progress, current major nutrition challenges remain at both ends of the spectrum—that of feast and that of famine. Globally, there are 1 billion overweight or obese people, 1 billion who are hungry, and about 2 billion who are micronutrient deficient. Nearly 4 million women and children

A. Bendich, R.J. Deckelbaum (eds.), *Preventive Nutrition*, Nutrition and Health, DOI 10.1007/978-1-60327-542-2_31,

under 5 years of age die each year because of factors related to undernutrition, and many more millions of adults die prematurely due to unhealthy diets *(2,3)*. The intimate links between stunting in early childhood and obesity in late childhood are creating risks for diabetes and cardiovascular disease at younger ages in developing countries around the globe.

Increased food prices are having severe impacts on the nutritional status of populations *(71)*. The World Bank estimates that doubling or more of food prices between 2006 and 2008 pushed 100 million people into poverty worldwide *(4)*. Families respond to increased prices by eating a smaller variety of foods, often of lower nutritional quality, reducing portion sizes, eating wild foods and seed stock (in rural areas), seeking credit or borrowing, begging and, in time, starving. These responses compromise already extremely vulnerable populations, especially those where stunting and micronutrient deficiencies are common. Permanent declines in physical and mental growth leading to economic and broad societal impacts inevitably follow such crises.

Further, rising food prices have already led to civil strife in countries on all continents and to growing distrust of food companies in some countries due to their perceived inaction *(5)*. The causes for increased food prices have been well described by the World Bank, many academics, and non-government organizations (NGOs). Lester Brown, writing a decade ago for the Worldwatch Institute, predicted almost exactly what is unfolding today *(6)*. Causes of the food price increases include rising oil prices, surging demand for grain—largely for the production of meat, and especially in China—and greater use of biofuel. The current food system has evolved over decades in a largely unplanned manner and without consideration for the complexity and implications of linkages between health, nutrition, agricultural, economic, trade, and security issues.

The underlying causes for the nutrition crisis also include several nutrition-specific factors; specifically, decades of neglect with regard to nutritional science (especially in emerging markets), a failure of governance with respect to the major players involved in nutrition, a weak response by government donors and foundations to invest in basic nutrition (in contrast to growing support for humanitarian aspects of food aid), and a reluctance to develop private–public partnerships *(7)*. A tiny fraction of total nutrition science output comes from emerging markets—and virtually none from the poorest countries of the world. Well over 80% of all scientific output in the top medical and nutrition journals address overweight and nutrition, with about 15% addressing micronutrients and 5% focusing on stunting and hunger *(7)*.

31.1.2. Emerging Engagement with Business

The emergence of new business models that tackle social problems while remaining profitable offers promise that the long-term nutrition needs of people can be met. Muhammad Yunus, founder of Grameen Bank and 2007 Nobel Prize recipient, recently called for the development of "social business entrepreneurs" and backed this call by working with multinational food corporation Danone to develop new ways of addressing the nutrition needs of poor families in Bangladesh *(8)*. In 2006, Groupe Danone signed an agreement with Grameen Group in a joint social business venture to provide nutrition to low-income populations. Two years later, Bill Gates, founder of Microsoft, urged that a new form of "creative capitalism" was needed *(9)*. PepsiCo CEO Indra Nooyi, at the helm of a $40 billion company, defines a "good business" as one that addresses financial performance while simultaneously addressing health and environmental needs. She calls this "performance with purpose" *(10)*.

There is a compelling role for the private sector, and specifically for the food industry, to continue to play in addressing the global nutrition crisis. This is done through a uniquely acquired understanding of, and adaptation to, consumer needs, combined with technological capabilities and business acumen. Corporations have experience building the infrastructure to deliver products and solutions where they are needed, as well as in maximizing benefits related to cost, scale, distribution, and agriculture. The public sector brings particular strength in identifying global and national priorities based on broad health needs; abilities and the mandate to development norms and standards to guide how products should be developed and marketed. Together, the public and private sectors must collaborate to co-create the strategic plan for action. Governments recently recognized this when they endorsed the Global Strategy on Diet and Physical Activity at the May 2008 World Health Assembly *(7)*. Only if both parties work together to write the action plan will the unique capabilities of each be maximized to ensure developments in health on a global scale.

Many leading food and beverage companies are already contributing to improved global nutrition. Here, a few examples:

PepsiCo has consistently been a leader in industry. In the 1950s and 1960s, many new products were introduced that focused on health and wellness, from Tropicana's first not-from-concentrate orange juice in 1954 to Diet Pepsi with zero calories in 1964. Beginning around 1996, PepsiCo began to develop internal nutrition criteria to guide product development, marketing, and labeling. In recent years, these efforts have expanded to a global scale. PepsiCo now has in place a large and diverse research and development team with capability to support product transformation aimed at limiting specific nutrients that may contribute to the prevalence of chronic disease, adding needed nutrients to address particular needs of a given population, and developing new products to address needs of consumers across the broad economic spectrum. Some examples from around the globe include oat-based snacks through Gamesa-Quaker in Mexico, expansion of baked snacks in Brazil, 70% total fat reduction in Baked Lays and Baked Walkers in the United Kingdom, and SoBe Life with PureVia—a zero-calorie, all-natural sweetener derived from the stevia plant—in Latin America *(10)*.

Since 2005, global food and beverage giant Unilever has been conducting an assessment of its 22,000+ item portfolio, a process that has led Unilever to plan on improving the nutritional quality of more than half the portfolio *(11)*. The Nutrition Enhancement Programme has led to a reduction of sugar, salt, and fat in many Unilever products. Additionally, smaller portion sizes, micronutrient fortification, and public–private partnerships with organizations such as UNICEF, The World Heart Federation, The United Nations World Food Programme, and GAIN have been central to Unilever's governing philosophies for the past decade. Sustainable development has also been a cornerstone of Unilever's "best business practices," including investing in sustainable agriculture and developing sustainable packaging. The work of these two companies is exemplary of only a few examples of potential collaborative efforts between the public and private sectors.

31.1.3. Building Trust and Creating Partnerships Between Public and Private Sectors

Creating trust and building relationships is at the heart of good business. Industry, government, non-government organizations, and academia each provide unique expertise. Working together to harness knowledge from all angles can provide the highest possible rate of success. Distrust regarding both intentions and capabilities present roadblocks that must be addressed. Partnerships built on a foundation of trust, shared values, and common aspirations seem to be the most productive way to reach realistic, measurable goals. In fact, as Rosabeth Kanter Moss of the *Harvard Business Review (12)* notes,

> When giants transform themselves from impersonal machines into human communities, they gain the ability to transform the world around them in very positive ways. . .Values turn out to be the key ingredient in the most vibrant and successful multinational (corporations). I refer . . . to the serious nurturing of values in hearts and minds.

Partnerships that have the greatest chance of succeeding will be those that are formed with a strong understanding of the practical challenges to be overcome in the communities they hope to help. Business practices, cultural nuances, and the presence or the absence of distribution systems and infrastructure affect the daily reality of bringing products and solutions to individuals. Even with excellent global policy, translation into the language of local needs is necessary. Through dialogue regarding capabilities and shared values, public–private partnerships have the potential to effectively deal with these issues *(13)*.

31.2. HISTORY OF THE FOOD INDUSTRY

The sheer magnitude of the world's 10 leading food and beverage companies is unparalleled in any other industry (Tables 31.1 and 31.2). The top five food consumer products companies include Nestle, Unilever, PepsiCo, Kraft Foods, and Groupe Danone, which together exceeded $28 billion USD in profits in 2008. On the beverage side, Coca-Cola & Coca-Cola Enterprises, Inbev, Heineken Holding, and SABMiller dominate the global market, profiting over $12 billion USD that same year. Each day, these 10 companies reach billions of consumers in every country around the globe

Table 31.1
Fortune Magazine's **Largest Food Consumer Products Companies Worldwide, 2008**

Rank	*Company*	*Global 500 Rank*	*Revenues, in Billions $USD*	*Profits, in Billions $USD*
1	Nestle	57	89.6	8.9
2	Unilever	122	55.0	5.3
3	PepsiCo	184	39.4	5.7
4	Kraft Foods	195	37.2	2.6
5	Groupe Danone	423	20.1	5.7

Table 31.2
Fortune Magazine's **Largest Beverage Companies Worldwide, 2008**

Rank	*Company*	*Global 500 Rank*	*Revenues, in Billions $USD*	*Profits, in Billions $USD*
1	Coca-Cola	275	28.9	5.9
2	Coca-Cola Enterprises	403	20.9	.711
3	Inbev	431	19.8	3.0
4	Heineken Holding	491	17.2	.553
5	SABMiller	493	17.1	2.0

Table 31.3
Significant Food-Related Technologies by Inventor and Year

Technology	*Inventor*	*Year*
Canning	Nicholas Appert	1810
Dried instant milk "powder"	TS Grimwade	1855
Commercial refrigeration	Alexander Twining	1856
Pasteurization	Louis Pasteur	1862
Decaffeination	Unknown	1900
Freeze-drying	Unknown	1940s
UHT processing	Unknown	1960s

through their diverse range of products. Together, their capabilities for global nourishment are tremendous.

31.2.1. Historical Review of the Development of the Food Industry

Urbanization and industrialization in the late 19th century in Europe and the United States led what was to become the modern food industry. Prior to this time, virtually everyone either grew or raised all of their food. People lived near where food was grown and tended to mainly eat seasonal produce. The pre-industrialized era was characterized by famines, food safety outbreaks, and high levels of food insecurity. However, as urban populations increased, city dwellers began to use their newfound wages to buy food rather than be personally responsible for growing it. Food companies met the needs of burgeoning city dwellers and were able to convert perishable locally grown and raised foods into products that could survive shipping and storage, without compromising either taste or basic nutrient composition. The development of new transport and food-processing technologies facilitated these profound changes.

Prior to train transportation, and later with the development of automobiles, food distribution methods kept locally grown food within a limited area. This contributed to the lack of variety and ability to move plentiful food to places where shortages were occurring. In the second half of the 1800s, there was a boom in modern technology, including thermal food processing, factory production techniques, transportation, packaging, and refrigeration, stimulating the creation and development of what is now known as the modern food industry.

31.2.2. Early Industry Development

The principal processing breakthrough that led to spreading foods around the world to new consumers was a heat-processing technique known as pasteurization, developed by Louis Pasteur in 1862 that destroyed pathogenic microorganisms and led to the development of canned, jarred, and bottled goods. By canning locally grown fruits, vegetables, milk, and meats, consumers could have foods on a regular and annual basis, and people discovered foods never before seen, or even imagined. Regardless of geography, people were able to balance their diets with a variety of foods that were safe and nutritionally adequate.

New technologies led to better ways of drying foods, refrigeration, vacuum packing, and the making of cans, bottles, and other food packaging materials. Safe and shelf-stable foods became widely available. As food industry innovation grew, so did the need for governmental oversight. There had been local dairy laws around the country to ensure safety of the milk supply and other local regulations, but nothing on a national scale to protect consumers from "unscrupulous" manufacturers. Unsanitary and inhumane conditions in slaughterhouses were the topic of Upton Sinclair's 1906 expose *The Jungle*, which aroused the United States Congress to pass both the Federal Meat Inspection Act and the Pure Food and Drug Act that same year. The latter led to the establishment of a branch of the US Department of Agriculture that eventually became the independent US Food and Drug Administration (FDA). The authority of the FDA was to investigate the safety of food additives and ensure truthful packaging claims; for example, Carnation's "Sterilized Cream" was required to have nomenclature more true to its origin, and was renamed Evaporated Milk in 1906 *(14)*. However, the new law ultimately had little enforcement power and, therefore, was not sufficiently effective. It was not until 1938 that Congress finally passed a strong Food, Drug, and Cosmetic Act. It has remained in effect to present time, with amendments added over the years to keep the law current with science and consumer demands; for example, the Food Additive Amendment in 1958 and the Nutrition Labeling and Education Act of 1990.

Successful food companies had to modernize and broaden their systems of food production, processing, and distribution. In 1925, the average US homemaker prepared all her meals at home. By 1965, 75–90% of all foods used were factory-processed to some degree. The food industry's impact on the way of life of people in developed countries became a very powerful force. The industry has responded to the need of urban populations for safe and reliable sources of food, to increasing demands on time for food growing, preparation, and cooking.

In Paul Roberts' *The End of Food*, the author notes that the new uniform, rationalized, intensified food system in the United States was extremely powerful, allowing farmers to produce four times their output since the beginning of the 20th century *(15)*. In this sense, industry—from agriculture and petrochemical to food—had responded to a growing need of consumers for safe and plentiful food. The food industry slowly became responsible for a relatively recent trend in food production and consumption: "convenience foods." Born of the technological boom in the early-to-mid part of the 20th century, "convenience foods" include those food products that are developed in response to a consumer demand to outsource meals, from their own kitchens to those of the largest multinational corporations. There has been a consistent decline in the consumers' ability to prepare or even understand their own food, notes Roberts, which is why the food industry has been put to task to dramatically simplify what was once a very complex process—cooking.

Refrigerated rail cars, using lake ice and thick insulation, began moving fresh meats and produce, as well as beer, across the country in the late 1880s. However, it was the development of manufactured ice in the early 20th century that boosted the ability of rail lines to ship refrigerated foods. As a direct result of these modern refrigeration techniques, the first supermarket chain, Piggly Wiggly, opened in Memphis, Tennessee, in 1916. Canned food-processing plants, operated locally by growers and cooperatives, continued to evolve throughout the country.

Bottled soft drinks were first produced in 1891, with Coca-Cola's John Pemberton unintentionally introducing what is commonly thought to be the world's first carbonated soft drink in Georgia in 1886. (In actuality, that title belongs to Vernors Ginger Ale and Hires Root Beer, both of which debuted in 1866 (and are now owned by Dr. Pepper Snapple Group). Pemberton initially developed Coca-Cola as a cocawine in 1885, but, the following year, Fulton County passed prohibition legislation, forcing Pemberton to develop a non-alcoholic version of his cocawine. Pemberton claimed that this non-alcoholic beverage could cure illness. Pepsi-cola followed soon after, developed by pharmacist Caleb Bradham in North Carolina in the 1890s, and officially trademarked in 1903. Royal Crown Cola was developed by another southern pharmacist, Claude Hatcher, in Georgia in 1905. And, finally, Dr. Pepper, first formulated by German pharmacist Charles Alderton in 1885 in Texas, was introduced to the world at the 1904 Louisiana Purchase Exposition.

31.2.3. A Maturing Industry

The nutritional status of Americans was being questioned as a result of the poor nutritional status of young men enlisting for service during World War II. The immediate motivation for the calling of the National Nutrition Conference for Defense in 1941 was the shocking discovery made by army officials that out of 1,000,000 men examined under the Selective Service Act, 400,000 had been found unfit for general military service. Brigadier General Lewis B. Hershey, deputy director of Selective Service, in speaking of these rejections to the Conference, said, "Probably one-third of these are suffering from disabilities directly or indirectly connected with nutrition." An outcome of this conference was the recommendation for flour and bread enrichment using the existing standards developed by the FDA, later known as the National Fortification Act. This is one prime example of how industry engaged in supporting a national crisis related to the nutritional status of an entire population.

Industry has also responded to population health crises overseas. Nearly 30 years before World War II, the British, Welsh, and Scottish governments developed the Education (Provision of Meals) Acts in their respective countries after the disclosure from recruiting army officials that over half of the young men eligible to fight in the Boer War were unable to do so due to malnutrition and underdevelopment *(16)*.

Following the World War II, when there was extensive rationing of foods on the home front and the food industry focused on the wartime effort (producing portable foods for field soldiers and easily transported stable products), innovation soon picked up again. The post-war affluence in America blossomed during the 1950s allowing for suburban homes to be abundant with refrigerators and other nascent electrical appliances such as mixers, blenders, toasters, can openers, and electric knives. During the 1960s and 1970s, with the threat of food shortages and depression-era poverty, environmentalists and advocates alike began pointing out the possible downsides of the nation's food supply. Pesticide safety, the presence of food additives, sugar and salt intakes, obesity and heart disease links to the food supply, and concerns regarding "artificial" foods prompted the food industry to once again respond to consumers' needs. "Diet" soft drinks and other artificially sweetened products were introduced, and changes were made to the types of saturated fats used in foods. In addition, salt levels were steadily reduced over time. Often the prevailing public health concern has led to unexpected health problems.

During this time, there were also major concerns regarding high levels of saturated fat available in the diet. Significantly influential research around heart health was published at this time in the academic literature, including the Framingham Heart Study, which began in 1948 and remains the largest study of cardiovascular disease (CVD) risk factors to date. At that time, little was known about the causes of both heart disease and stroke; thus, the objectives of the study were to identify common characteristics that contribute to CVD by following its development over a long period of time with a large group of participants. Since its inception, Framingham has led to the development and publication of hundreds of studies integral for better understanding not only CVD risk factors, but related co-morbid disorders and other long-term health effects. In addition, studies in Finland 30 years later came to similar conclusions as research in the United States; a 1994 article in *British Medical Journal* used these cross-sectional population surveys to assess risk-factor levels and observed changes in mortality statistics. Researchers concluded that, of nearly 30,000 men and women from the initial population surveys assessed decades later, observed changes in risk factors predicted a decline in mortality from ischaemic heart disease. Changes in the three risk factors—serum cholesterol concentration, blood pressure levels, and smoking—suggested that decline in mortality from ischaemic heart disease shows a clear association between risk factors and disease *(17)*. These findings are consistent with much of the modern published literature that has closely followed the Framingham Heart Study.

As a result of these longitudinal cross-sectional studies, concerns arose about the high levels of saturated fats available in the typical diet. The scientific literature led food companies to consider developing products with little to no levels of saturated fat. Among the first to produce such products were the major global dairies, introducing low-fat and skimmed milk products to their portfolios. The industry response is a cautionary one: complete information is rarely available to guide development of food products able to meet multiple nutrition needs. The dairy industry played a significant role in reducing population levels of saturated fats in the diet. This must have contributed to the continued declines in rates of ishaemic heart disease.

31.3. RESEARCH IN INDUSTRY

31.3.1. Background

The food industry has helped to advance scientific knowledge across a range of major food-related challenges. Here the focus is only on selected nutritional innovations. It should be recognized though that high levels of research and innovation were required to develop and the modern ability to feed billions with scaleable production abilities. Safe and convenient foods did not merely arise, but rather required innovation in packaging to assure safety and long shelf lives; innovation in mass standardized production; innovation in distribution and in marketing. Most major food companies are, in addition, deeply involved in the full range of agricultural research aimed at developing stable, healthy, and ecologically sustainable plant species.

31.3.2. The Food Industry's Role in the Evolution of Infant Formula

Many advances in infant nutrition have been made and infant formulas currently on the market are recognized as secondary alternatives to breast milk. The World Health Organization recommendation of promoting 6 months' worth of exclusive breastfeeding remains poorly implemented around the world. If it were, the positive impact on undernutrition and on the development of several chronic diseases would be profound *(18)*.

The quest for alternatives began early in the 19th century. The first chemical comparison of cow's milk to human milk was published by Johann Franz Simon in 1838. His research showed that human milk had a higher carbohydrate and lower protein composition as compared with cow's milk *(19)*. To increase the carbohydrate content of cow's milk, Meade Johnson Company released a milk additive called Dextri-Maltose in 1912. In 1919, Henry John Gerstenberger developed a new infant formula with a fat blend based on animal and vegetable fats that more closely resembled the fat composition of human milk. This new formula became the basis for the SMA (Simulated Milk Adapted) Nutrition company. SMA was also the first company to include cod liver oil in their infant formula for prevention of rickets. The practice of adding cod liver oil to infant formulas was quickly adapted by Nestle and others *(20)*. Nestle, Wyeth Laboratories, Mead Johnson, and Abbott Laboratories have continued their efforts to develop a human milk mimetic by sponsoring research in infants and examining the impact of fatty acid composition on visual acuity, growth and development, protein composition on growth and infant health, and iron absorption and status on growth in infants *(18–26)*. The scientific development of better approaches to infant feeding has not to date been matched by simultaneous investment in global promotion of breastfeeding. This remains a challenge.

31.3.3. Development of Heart-Healthy Products

The WHO predicts that coronary heart disease (CHD) will be the primary cause of global mortality by 2030 *(27,28)*. Estimates indicate that worldwide deaths from CHD will increase from 17.1 million in 2004 to 23.4 million in 2030 *(29)*. Many factors contribute to the development of CVD, including elevated total and LDL cholesterol, depressed HDL cholesterol, and elevated triglycerides. Diet and lifestyle are also important factors to consider, and recommendations for prevention of CHD have been documented in the literature. The food industry has long recognized this link between diet and disease risk and has successfully sponsored research to substantiate a heart health claim on the relationship between soluble fiber from certain foods and risk of CHD.

The first petition for a heart health claim was submitted in 1995 by the Quaker Oats Company, which sponsored several of the 37 studies used to substantiate the claim *(30–35)*. The FDA concluded that the totality of publicly available scientific evidence supported the relationship between consumption of whole oat products and reduced risk of CHD, and, further, that the type of soluble fiber found in oats, beta-glucan, was primarily responsible for the hypocholesterolemic effect *(36,37)*. Other companies, such as Kellogg Company *(38,39)* and Procter and Gamble *(40,41)*, have sponsored research on the impact of psyllium fiber on cholesterol, while Cargill *(42,43)* and Sapporo Breweries *(44)* sponsored research on barley fiber and cholesterol reduction *(41–45)*. As of August 2008, food companies, along with academia and government, have sponsored research that has expanded the soluble fiber heart health claim to include not only oats and oat products, but psyllium fiber and barley fiber products as well *(36,38)*.

The FDA has authorized roughly 17 heart health claims including but not limited to fiber, phytosterols and phytostanols, and unsaturated fatty acids. Health claims provide food manufacturers a means to communicate with consumers about maintaining healthy dietary practices. In addition, the American Heart Association (AHA) Heart Check program allows food manufacturers to bear the AHA symbol on foods that meet certain nutrient criteria. Health claims provide guidance and incentive to food manufacturers to develop healthier food options for consumers. Another example of a health claim comes from Finland, where researchers have discovered that plant sterols can stabilize cholesterol levels *(44)*. Omega-3 fatty acids have also been repeatedly shown to have decrease levels of triglycerides and decrease risk of arrhythmias.

In 2006, Frito-Lay, part of the diverse global PepsiCo portfolio, announced that the company was committed to reducing saturated fat in its leading potato chip brands, Lay's and Ruffles, by 50% by switching to mid-oleic sunflower oil, which is low in saturated fat—a commitment which has been carried out successfully. In addition to reducing levels of saturated fat, sunflower oil also increases mono-and-polyunsaturated fats *(45)*, which are shown to decrease risk of cardiovascular disease. That same year in the United Kingdom, the Food Standards Agency published voluntary salt reduction targets for food manufacturers and retailers in the hope that reducing salt would decrease risk of developing high blood pressure. The reduction targets applied to products in 85 different food categories. Significant progress in reducing levels of saturated fat and salt in food products has been made in the last several years, largely in response to public health calls for such reductions.

31.3.4. Sports Nutrition

Gatorade, currently an integral part of the PepsiCo portfolio, is an isotonic blend of water, sodium, and sugar designed to optimize hydration *(46)*. It was developed at the University of Florida in 1965 to help the football players avoid dehydration during practices and games in the extreme heat and drew inspiration from work in Bangladesh on ways of preventing diarrhea during cholera epidemics. Oral rehydration solution and Gatorade are based on the same scientific insight that sugar aids absorption of an electrolyte solution. Gatorade-supported research has been instrumental in proving the benefits of hydration with carbohydrate and electrolyte beverages on sports performance *(47–50)* and the well-being of athletes *(51,52)*.

31.3.5. Emerging Research on Cocoa

Mars, Inc. is a global leader in cocoa and chocolate science. Mars was one of the first companies to describe the health benefits of cocoa flavanols, the development of improved cocoa varieties through understanding of cocoa genetics, and working through private and public partnerships to create new ecological and socioeconomic opportunities for the millions of farmers throughout the tropics who depend on cocoa crops for their livelihood. Early research on cocoa focused primarily on its effects as an antioxidant, however, as scientists became better able to measure the cocoa and it components after consumption it became apparent that most of the antioxidant activity is lost through digestion. Since these initial studies, research has demonstrated positive effects on cardiovascular disease, specifically blood vessel relaxation, and research is beginning to emerge for inflammation and diabetes *(53–55)*.

By establishing key partnerships with public and private organizations, Mars has been able to work collaboratively to find solutions for many of the pest and disease problems that negatively impact worldwide cocoa crops. For example, Mars, in partnership with the US Department of Agriculture-Agricultural Research Service (USDA-ARS) is exploring opportunities to help control invasive pests, like the cocoa pod borer in Asia, through the use of naturally produced, biological control attractants called pheromones. It is through partnerships with academia, industry, government, and public organizations that Mars has established a scientific leadership position by continuing to invest in clinical and agricultural science.

31.3.6. Product and Industry Approaches to Reduce Obesity

By the 1980s, the food industry realized that foods promoting a healthier diet were a growing market, and that consumers were searching for products that were both healthy and convenient. Low-fat foods started the trend, promising "guilt-free" snacking to those trying to watch their weight. The National Dairy Council began the "got milk?" campaign in 1993 to explain the health benefits of milk and calcium, and in recent years have used these advertisements to promote the weight management benefits of three dairy servings a day. Dr. Robert Atkins sparked the low-carbohydrate diet craze of the 1990s with his book, *Dr. Atkins' New Diet Revolution*, which led several food companies to flood the market with a variety of "low carb" foods. This even gave way to Atkins-branded products such as high protein bars, shakes, and meals. Food companies began to respond to this craze by starting to implement portion control on a wide scale.

31.3.7. Food Safety

Legislative bodies promulgate laws that regulate food safety, and regulatory agencies enforce compliance with these laws and regulations. However, primary responsibility for producing safe and wholesome food lies with the manufacturer of the product. The history of food safety legislation is replete with examples of corporate misconduct followed by public outrage and finally legislative action. In the European Union (EU), previously the laws and extent of laws on food safety varied greatly between the different Member States, although in recent years laws have been harmonized. The common laws apply to food produced in and imported into any of the EU Member States. In 2002, the EU Food Law, Regulation 178/2002, established common principles and responsibilities for food safety, consolidating an integrated "farm to fork" approach. It places the primary legal responsibility on the food business operator for assuring food safety at all stages of production, processing, and distribution. The food business operator is obliged to have systems in place to trace all suppliers of substances used in their food products, to label products to permit traceability, and to initiate procedures to act if any food safety non-compliance is detected. EU regulatory standards for compliance are developed based on extensive consultation with stakeholder groups, including industry as well as taking into account existing international laws.

Global food standards and guidelines are agreed under the committees of the Codex Alimentarius Commission established in 1963 by the Food and Agriculture Organization (FAO) and the World Health Organization (WHO). The purpose is to protect the health of consumers, to ensure fair practices in the food trade, and to promote coordination of food standards by international governmental and non-governmental organizations. Codex provides key globally agreed principles on food safety.

In order to produce a safe and wholesome product, a manufacturer must control at a minimum of three things: (1) the raw materials and ingredients that make up the product, (2) the process by which the product is produced, and (3) the environment in which the product is produced. Many manufacturers use Hazard Analysis and Critical Control Points (HACCP) as a template for assessing and controlling these threats. HACCP plans are focused on identifying and controlling potential hazards that are "upstream" of the finished product. HACCP plans generally embody seven principal steps:

1. Hazard analysis: Identify potential hazards that may make a food unsafe or unfit.
2. Identify critical control points (CCP): Points in the process in which hazards identified in one can be prevented or controlled.
3. Establish critical limits at each CCP: Establish critical limits, i.e., minimum or maximum values for physical, chemical, and biological hazards.
4. Monitoring: Establish a monitoring protocol at each CCP.
5. Corrective actions: Establish steps to be taken when a critical limit is exceeded.
6. Records: Maintain documentation demonstrating compliance with HACCP plan.
7. Validation: Ensures that manufacturing plants are correctly implementing the HACCP plan.

Generally each manufacturing process, and sometimes each production line, will have its own HACCP plan. These plans are critical to ensure a safe food environment.

31.3.8. Micronutrient Fortification

Adding or restoring nutrients to foods has been a major success story in countries where fortification has been practiced. Selective nutrient addition in the food supply has reversed population-wide nutrient-deficiency diseases, replaced important nutrients lost in processing, prevented undernourishment in at-risk populations, provided targeted functional foods for individuals with unique needs, and helped supplement the diets for individuals who seek "optimal" nutrition. Internationally, fortification is a cost-effective way to reach populations in need by adding nutrients to staple foods in the diet.

One of the earliest successful fortification programs was addition of iodine to salt. Iodine fortification was initiated in the United States and some areas of Switzerland in the early 1920s to prevent goiter, cretinism, and other symptoms of severe iodine deficiency. The further one lived from the sea, and therefore the less seafood consumed, the greater was the risk of iodine deficiency. Thus, land-locked Switzerland and the entire middle of the United States had iodine-deficient diets, and fortification of a staple food-like salt was highly successful in providing this nutrient to those in need *(56)*.

Iodine deficiency is still prevalent in many areas of the world that are geographically distant from the oceans, such as India and China. Since fortification worked well in the United States, some nations have begun adding iodine to salt as a way to reach needy populations. Wherever iodized salt is distributed, the deficiency is mostly eradicated, but acceptance is slow in coming to many locations. Iodine deficiency is the leading cause of preventable mental retardation, and approximately 38 million newborns are still at risk of iodine deficiency.

In the early 1930s, vitamin D was first added to cow's milk to aid in absorption of calcium and phosphorus, preventing development of rickets. This disease was common in many parts of the world where sunlight was scarce, since vitamin D is not found in many foods, but primarily comes from exposure to sunlight on the skin. In 1938, voluntary enrichment of flours and breads was initiated in the United States to prevent the development of deficiency diseases in the general population. It had been known that processing grains removed a significant amount of the naturally occurring nutrients, and the science of nutrition had progressed sufficiently to demonstrate that these nutrients were critically low for many population groups. Enrichments included thiamin to prevent beriberi; niacin to prevent pellagra; riboflavin which is essential for proper functioning of vitamin B_6 and niacin; and iron to prevent iron-deficiency anemia.

Mandatory requirements for flour fortification became effective in 1943 after the US government published its first Recommended Dietary Allowances. These requirements were the outcome of a National Nutritional Conference for Defense, convened by President Roosevelt in 1941, to explore why so many military recruits were in poor health. Although bread was not part of the mandatory program, the industry voluntarily fortified bread such that by 1942, 75% of the breads in the United States were fortified.

In the subsequent 60 years, fortification has mainly been used for voluntary addition of nutrients in an increasingly wide variety of foods with a vast array of nutrients. Other than the standard of identity for enriched flours and breads changing in 1998 to require the addition of folic acid to prevent birth defects, most of the regulated fortification history in the United States involves the FDA and USDA wanting to restrict overfortification of the food supply. Internationally, there are numerous success stories of fortification eradicating or reducing public health problems due to nutrient deficiencies. Unfortunately, fortification programs in developing countries began only in the past two decades. Many more people could be helped worldwide if fortification were adopted in more nations, accepted at the local level, and supported by industry in developing countries.

Vitamin A fortification of staple foods, such as margarine and sugar, has made significant strides in eradicating blindness and promoting growth in children in developing countries. It is estimated that childhood mortality could be decreased by 23% by improving vitamin A status of children under 5 years. Currently, supplementation is the most effective means for providing vitamin A, and interest in strategies to use food fortification and diet diversification has increased as a long-term solution.

Iron fortification of wheat flour has effectively combated anemia, which is a serious condition stunting millions of children and women worldwide. However, anemia is so widespread that fortification and supplementation has only reached the tip of the iceberg for those in need. Nearly 50 countries worldwide now fortify flour, but food science research must continue to explore new forms of iron and add to a wider variety of foods to increase exposure of this vital mineral to needy populations. Iron fortification is challenging because it is difficult to find stable forms of iron to use for any particular food, since iron is chemically reactive, and can degrade or cause the fortified food to be less desirable. This requires researchers to continuously test how iron in its various forms is affected by food processing while maintaining the quality of various food products.

Zinc intake is low in the diets of Middle East countries where grains are the dietary staple. Supplement studies show that growth is stunted until zinc is adequate, and in the few locations where grains are enriched with zinc, improvement has been seen. However, like iron, zinc can be unstable, and the lack of awareness of zinc deficiency has prevented it from gaining ground in national fortification programs.

The Food and Agriculture Organization of the United Nations has outlined strategies and goals for fortification for all nations. Included are the following:

1. There should be appropriate fortification of foods used in food-aid programs, with donors being required to provide relevant nutritional information particularly through adequate labeling.
2. Levels of fortification should be evaluated and adjusted according to bioavailability of the nutrient(s) in the diets of target populations.
3. It is important to evaluate the potential of local food industries to become involved in the production of high-quality fortified food products, including those destined for use in food-aid programs, in areas where problems of micronutrient deficiency are likely to occur.
4. The impact of food fortification on the nutritional status of target populations should be monitored so that appropriate corrective action can take place as required or successful approaches can be reapplied in other areas. *(56)*

One of the challenges of a successful fortification program includes the product formulation. It is important to choose the right food vehicle for fortification that will be widely consumed by the target population. Added vitamins and minerals can be sensitive to many factors, such as oxygen, light, pH, and moisture. It is important to consider the package as well, since it can help control exposure of the product to some of these environmental factors. For example, riboflavin is light sensitive, and vitamin C can degrade with oxygen exposure, especially when transition metals are available. Transition metals, like iron and zinc, can be challenging to add to foods since they readily undergo oxidation–reduction reactions depending on the food matrix, causing color changes and off-flavors to develop. Once a stable product has been developed, it is also important to keep in mind the expected shelf life and storage temperatures of the product, as both can have an effect on the stability of the product.

Enrichment of flours and breads was essentially the only food fortification until the 1960s. A few standards of identity in the 1950s did require addition of nutrients, such as vitamin A to margarines and the B vitamins and iron to additional cereal grains. However, formulated meal replacements were introduced in the early 1960s that included nearly all the required nutrients, such as Mead Johnson's Nutrament and Carnation Instant Breakfast. Their advertising with a focus on "complete and balanced" nutritional content led to media reports in the late 1960s that breakfast cereals, a staple of children's diets, were lacking in nutrient density. Such products quickly became fortified and the general public responded enthusiastically. Additionally, the White House Conference on Food, Nutrition, and Health in 1969 issued as one of its recommendations that more food should be fortified to reduce population-level malnutrition.

The FDA, however, became concerned that companies would use fortification as a marketing tool, and overfortification of the food supply would lead to nutrient imbalances. The agency's hands were significantly tied in what they could do because, in 1961, they lost a court case in which they attempted to halt the sale of fortified sugar (which had 19 added nutrients and made the claim that it was more nutritious than other sugars). The court said that the FDA had no authority under the law to prohibit the sale of a food "simply because it is not in sympathy with its use." The FDA was told it could not prohibit food fortification unless it could be shown to be unsafe. Thus, the FDA approached their concerns from the standpoint of avoiding nutrient imbalances.

In addition to tackling overfortification, the FDA was intent on limiting dietary supplements for the same reason—nutrient imbalance by the consuming public. However, in 1976, Congress stepped in and amended the Food, Drug, and Cosmetic Act so that the FDA could not set maximum potencies of supplements except for safety reasons. When the FDA then attempted in 1977 to set a maximum for vitamins A and D for safety reasons, the courts denied their efforts. This left the FDA with little clout but to issue its 1980 guidelines (21 CFR Section 104.20) for appropriate food fortification, but as previously noted, they remain non-binding.

USDA-regulated products—those containing meat and poultry—are also subject to the FDA's fortification policy in Section 104.20 because the agency's Food Safety, and Inspection Service (FSIS) adopted the policy in 1980. Interestingly, such products adhere closely to the policy's principles because FSIS has prior-label-approval authority, and the agency reviewers can enforce the policy as they review labels before they are marketed. The net result is that very few products under USDA jurisdiction are fortified.

The overwhelming trend for the last 15 years has been to fortify an increasing number of foods. Fortunately, much of the fortification has followed the FDA principles to add nutrients to correct a nutrient deficit in the diet, such as adding calcium, magnesium, zinc, folic acid, vitamin B_6, and vitamin D, or a limiting amino acid to improve protein quality. Although the nutrient deficits are not full deficiencies, dietary improvement to achieve "optimal" nutrition is desired by the public and is increasingly being shown scientifically to improve overall health outcomes *(57)*.

As a sharp contrast, there is an increasing trend that concerns the FDA: fortification for marketing reasons and a "nutrient horsepower race" that has the potential to cause intake imbalances. In some respects, we have the equivalent of the fortified sugar from 1961, under the pretense that the fortified food is better if it has more nutrients than the competitor. Candy, crackers, desserts, soft drinks, and water are just some of the foods that are currently fortified, and levels are sometimes as high as at 100% of the Daily Value per serving *(58,59)*.

The last 5 years have seen tremendous growth in public–private partnerships to address micronutrient fortification needs in developing countries. Iron, vitamin A, zinc, and iodine deficiencies are still common in developing countries. GAIN, the Global Alliance for Improved Nutrition, initially funded jointly by the Bill and Melinda Gates Foundation and the Norwegian government, focuses on reducing the prevalence of micronutrient deficiencies among target groups by increasing regular consumption of fortified foods. The organization is a self-described "alliance of governments, international organizations, the private sector and civil society" *(60)*. The GAIN Business Alliance, chaired by multinational food company Unilever, is a global network dedicated to finding market-based solutions to address malnutrition. Additionally, the Micronutrient Initiative, which seeks to provide vulnerable populations with proper vitamins and minerals needed to thrive, works in a similar fashion as GAIN. MI works in partnership with governments, the private sector, UN agencies, and civil society organizations to prevent hunger. The scope of both of these organizations is impressive, with GAIN already reaching 650 million people with fortified products.

31.4. INDUSTRY AND PUBLIC HEALTH

31.4.1. Chronic Disease

WHO's latest estimates indicate that chronic non-communicable diseases already dominate as major causes of death and disease globally *(27,28)*. The nutritional and activity underpinnings of the global burden of disease have been well described. Table 31.4 shows the top 15 risks globally *(7)* and how the food industry has become involved in each area to increase population health.

WHO responded to the chronic disease concerns by initiating a series of major reports and consultations with the food industry, non-governmental organizations (NGOs), and governments beginning in 2002. In 2003, the WHO released Technical Report #916, entitled "Diet, Nutrition and the Prevention of Chronic Diseases" *(61)*. The Consultation's experts looked at diet within the context of macroeconomic implications of public health recommendations on agriculture, and the global supply and

Table 31.4
Top 15 chronic NCD risks globally and industry response

Risk	*Industry Response*
Blood pressure	Reduce levels of sodium; increase levels of potassium
Tobacco	NR
Cholesterol	Convert to heart-healthy oils; remove trans fats
Underweight	Begin implementation of public–private partnerships to combat global under- and malnutrition
Unsafe sex	NR
Fruit and vegetable intake	More aggressive marketing and better availability of fruit and vegetables, juices and products with nutrient profiles of fresh equivalents
High BMI	Reduce energy density of products, reduce portion sizes, support programs to increase physical activity in populations
Physical inactivity	Support programs whose agenda is to increase physical activity in population
Alcohol	NR
Unsafe water, sanitation, and hygiene	Implement public–private partnerships with "safe water" organizations
Indoor smoke from solid fuels	NR
Iron deficiency	Support fortifying food products with iron in populations in need
Urban air pollution	NR
Zinc deficiency	Support fortifying food products with zinc in populations in need
Vitamin A deficiency	Support fortifying products with Vitamin A in populations in need
Unsafe health-care injections	NR

NR – Not related to the food industry

demand for foodstuffs, both fresh and processed. In many developing countries, food policies remain focused only on undernutrition and are not addressing the prevention of chronic disease. The primary purpose of the consultation was to examine and develop recommendations for diet and nutrition in the prevention of chronic diseases. In order to achieve the best results in preventing chronic diseases, the strategies and policies that are applied must fully recognize the essential role of diet, nutrition, and physical activity. The major outcome of the report was a science-based set of guidelines for diet and physical activity. At a macro level, the report had many implications for industry which were summarized in the Action Plan for Chronic Diseases adopted by governments in May 2008. The key recommendations for industry were as follows:

1. Limit the levels of saturated fats, trans-fatty acids, free sugars, and salt in existing products.
2. Continue to develop and provide affordable, healthy, and nutritious choices to consumers.
3. Practice responsible marketing that supports the strategy, particularly with regard to the promotion and marketing of foods high in saturated fats, trans-fatty acids, free sugars, or salt especially to children.
4. Issue simple, clear, and consistent food labels and evidence-based health claims.
5. Provide information on food composition to national authorities.
6. Assist in developing and implementing physical activity programs.

In response to the WHO policy actions, the CEOs from eight major multinational food and non-alcoholic beverage companies (Kellogg Company, Kraft Foods Inc., Mars Incorporated, Nestlé S.A., PepsiCo Inc., General Mills, The Coca-Cola Company, and Unilever) signed a Global Commitment to

Action in support of implementing the World Health Organization's Strategy on Diet, Physical Activity, and Health. This unprecedented step by industry was communicated to the WHO Director-General and has led to ongoing interactions aimed at finding ways of solving chronic disease challenges related to prevention. Through this coalition, five key global commitments will be carried out over the next 5 years:

1. **Innovate** product composition and availability to provide healthier product options that address both excess and deficient consumption of specific nutrients and calories. Trans fats and salt were explicitly highlighted.
2. **Provide** clear nutrition information to consumers, including consumers in regions where nutrition information is not required.
3. **Globalize** individual company and regional measures to ensure responsible marketing and advertising of foods and non-alcoholic beverages to children, bringing increasing proportions of the industry into the fold.
4. **Target** individual company communications and forge public–private partnerships to cultivate awareness and adoption of healthier lifestyles worldwide.
5. **Commit** time, expertise, and resources to support public–private partnerships to accomplish the objectives of the WHO Strategy *(7)*.

Progress is underway in relation to these five commitments. Most leading food companies are increasingly investing in ways to transform their core portfolios with a focus on reducing sugar, salt, saturated fats, and removing trans-fatty acids. In the United States and Europe, there has been progress in accepting the importance of simple fact-based statements on the front-of-pack of food and beverage products with regard to major nutrients and calories per serving. This will increasingly go global. Leading food companies agreed in December 2008 to restrict marketing of all products not meeting nutritionally acceptable criteria to children under 12 years old in all markets and supporting many initiatives to increase levels of physical activity.

These industry actions will have a broad-based impact on chronic diseases *(61,62)*. More focused approaches are being developed to address obesity and the concomitant increase in type 2 diabetes. According to one study, the effect of longevity on obesity is actually a decline in life expectancy in the United States in the 21st century *(63)*. According to Yach et al. *(64)*, "overweight and obesity have become to diabetes what tobacco is to lung cancer. " Worldwide, there are 155 million obese school-aged children, and approximately 22 million obese children under the age of 5. The global economic burden of obesity and diabetes as co-morbid disorders is tremendous, particularly in the United States, where, in a 5-year span, the medical-related cost of diabetes is more than doubled, from \$44 to \$92 billion *(64)*. Outside the United States, the economic consequences of obesity and diabetes range from 0.2 to 3.8% of the GDP.

In the United States, several industry initiatives are under way. An example of a one initiative that is demonstrating significant impact on population health involves a partnership between several private multinational corporations, the American Heart Association and the William J. Clinton Foundation, through the Alliance for a Healthier Generation. The Alliance includes several food companies that pledged to only serve products in schools that meet specific nutritional guidelines for fat, saturated fat, sugar, and sodium. The Alliance School Beverage Guidelines of 2006 significantly limit portion size and the number of calories in beverages available to students during the school day. In a recent progress assessment, calories from beverages were shown to have decreased by 58% since 2004, and shipments of full-calorie soft drinks have also decreased by nearly two-thirds, showcasing the industry's commitment to shifting the school beverage landscape.

Since 2007, 13 of the United States' largest food and beverage companies, including McDonald's and Unilever, have become a member of the Children's Food and Beverage Advertising Initiative, through which they commit to the following:

- Include healthier dietary choices that incorporate one company's food or beverage products.
- Not engage in food and beverage product placement in editorial and entertainment content.
- Reduce the use of third-party licensed characters in advertising that does not meet the initiative's product or messaging criteria.
- Limit products shown in interactive games to healthy dietary choices or incorporate healthy lifestyle messages in the games.
- Not advertise to children in elementary schools.

For food company actions at national or global levels to reach scale and have impact, governments and other players, including NGOs and academics, will need to take complementary actions. For example, governments need to develop incentives for food companies to increase research and development in the same way they have favored pharmaceutical research to date. Governments need to step up their involvement in all aspects of promoting active living including ensuring that school children have regular physical education. Governments and WHO need to develop enabling norms that apply to all companies so that multinational, medium and small food and beverage companies will all operate on a level playing field.

There is a need for positive examples of how to reduce or prevent obesity in large community settings. Several food companies are investing in developing community-based models to reduce obesity. For example, the PepsiCo Foundation, in collaboration with the Centers for Disease Control and Prevention and the Friedman School of Nutrition Science and Policy at Tufts University, co-funded Shape Up Somerville, a 3-year, effective set of interventions designed to help prevent obesity in high-risk, elementary school children using environmental changes in the community, in schools and at home that affect behavior *(65)*. Citywide policy changes were made that directly affected the behaviors of children and their families. Internationally, the PepsiCo Foundation has started to fund four major community-based interventions in Mexico, India, China, and the United Kingdom.

31.4.2. Undernutrition

In May 2008, Gordon Brown, with United Nations Development Programme, launched an initiative aimed at drawing on business's core capabilities to contribute to the attainment of the Millennium Development Goals *(66)*. Several food companies responded to this call to action and have committed to use their distribution systems to get food aid to remote areas; to develop new nutritious and affordable food products for the poorest communities; to lever their agricultural research to develop plans with higher yields; and to invest in nutrition science of benefit to the public and private sectors.

PepsiCo has committed to support Millennium Development Goal #1—to eradicate extreme poverty and hunger—by developing a nutritional product for malnourished children in India, South Africa, and Nigeria, using the company's extensive research and development teams and distribution networks.

31.4.3. Desired Actions by Business to Address the Long-Term Food and Nutrition Crises

Businesses can have greater impact acting collectively rather than individually. In May 2008, CEOs of eight major food companies pledged in a letter to the WHO Director-General to develop and market fortified nutritious products to the poorest communities. This is in addition to broader commitments that CEOs made to support WHO implement the action plan of the Global Strategy on Diet and Physical Activity *(61)*. The companies are gearing up to develop specific steps that will demonstrate their on-the-ground progress. Food, retail, food service, chemical, and pharmaceutical companies have expertise, distribution systems, and access to customer insights that, if well harnessed, could leapfrog progress in addressing the global food and nutrition crises. The PepsiCo Foundation supports the World Food Programme to strengthen their distribution systems, and for Save the Children to meet basic nutrition needs of families in India and Bangladesh.

Business could increasingly address the entire range of the agricultural investment climate, including access to microcredit for small farmers, research on better seeds, training, provision of water saving irrigation systems, and long-term purchase guarantees. PepsiCos work with potato farmers in Peru, where the company supports the local Peruvian potato industry; citrus farmers in Indian Punjab; corn farmers in rural Mexico; and oats farmers worldwide revolve around these principles. Business could also support local sourcing and use of indigenous foods *(7)*.

In an environment of soaring prices, businesses need to be hyperefficient and reduce waste along their supply chains and reduce fuel costs by bringing production closer to consumers. This is a particular problem in Africa, where up to 40% of fresh produce is lost through poor supply chain management. Furthermore, the potential of using nutritious components of current waste streams for affordable nutrition is being explored. Retail chains can work with governments and food companies to develop a balanced food basket of local staples priced to be affordable to the poor *(7)*.

All the evidence suggests that Margaret Chan, the Director-General of WHO, is correct in her recent statement that

> food choices are highly sensitive to price. The first items to drop out of the diet are usually the healthy foods. . .fatty processed foods or low-energy nutrient staples are often the cheapest way to fill hungry stomachs.

There is broad consensus about the need for nutrition interventions to give priority to young women and children under 2 or 3 years of age if the long-term effects of stunting on growth and intellectual development are to be prevented. Business could, and in some cases is, supporting this through the development and marketing of products for women and children that address key nutrient needs. Business, through joint efforts with governments, could support truly effective social marketing campaigns for breastfeeding—the most cost-effective and poorly marketed nutritional intervention.

Business could support programs to fortify staples and develop a wider range of ready-to-eat therapeutic foods. One in seven people—or 854 million worldwide—go hungry each night, and 25,000 die each day from causes related to hunger or malnutrition *(67,68)*. The United Nations World Food Programme (UN-WFP), the largest humanitarian organization with the mandate to combat global hunger, is transforming food aid to create a new paradigm for affordable and accessible nutrition for the world's poor. UN-WFP is currently aiding in the development of ready-to-use therapeutic foods, or RUTFs, and ready-to-use supplementary foods, or RUSFs, in order to treat severe and acute malnutrition in women and children. At the June 2008 meeting of the High-Level Task Force of the United Nations on the Global Food Crisis, the UN Global Compact office has completed a report (Food Sustainability 2008) highlighting the need for certain private sector actions. These include a need to openly address impediments to food company involvement in addressing the complimentary food needs of children worldwide to a far greater extent *(69)*.

The demand for meat is a response to a lack of protein in the diet of emerging market populations. Global meat consumption is about 100 g/person/day with there being a 10-fold variation between high- and low-consumption countries *(70)*. If continued demand for protein is met with meat, the consequences for the environment, human health, and vulnerable populations will be dire. Business and academia need to lead through their research and development teams to develop ways of stimulating increased consumption of less energy/grain intensive protein sources from plants, fish, and in vitro meat cultures. As Ricardo Uauy and his colleagues stressed in the recent *Lancet* review, a goal of achieving a more equitably distributed global consumption of 90 g/person/day by the 2030s is possible if work started in earnest today *(6)*. At that level of consumption, populations' needs for animal protein and iron would be easily met.

Business has an important advocacy role to play on issues that affect agricultural productivity. Business also needs to be part of the dialogue at country level about how to create an enabling environment for investing in agriculture and local food production. While business can do lots more, its combined impact will be minimal if a range of essential government actions and policies are not addressed. Governments need to create innovative and complementary opportunities that include incentives for businesses including: setting clear nutritional guidelines for fortification and for ready-to eat products; offering agreements to endorse approved products and support their distribution to clinics and schools; eliminating duties on imported vitamins and other micronutrients; and providing tax and other incentives for industry to invest with donors in essential nutrition and agricultural research.

While extensive global effort has been targeted to Millennium Development Goal #1 (eradicating extreme poverty and hunger), the specific roles of private and public sector partnerships in supporting this has been missing. In those countries where progress on the MDG 1 goals is waning, PepsiCo is developing a more orchestrated approach to its core business capabilities in consumer fundamentals, health policy, nutrition, product and food production research and development, distribution systems, and marketing.

One specific effort is to develop affordable and nutritious products for mothers and families at the base of the economic pyramid in India, Nigeria, and South Africa. The level of iron and zinc deficiencies and stunting in these populations is extremely high; health professionals agree that such deficiencies lead to a multitude of lifelong physical and intellectual problems that undermine national competitiveness. PepsiCo is partnering with both governments and leading academic researchers in these three countries, as well as with key non-governmental organizations, such as the Global Alliance for Improved Nutrition. Consistent with our sustainability agenda, we are giving a great deal of attention to crafting sustainable business models to offer affordable, nutritious foods, and beverages to consumers in these areas *(7)*.

31.4.4. Investing in Nutrition Science

31.4.4.1. Recommendations

During the past decade, rapid expansion in a number of relevant scientific fields has helped to clarify the role of diet in preventing and controlling morbidity and premature mortality resulting from noncommunicable diseases (NCDs). Some of the specific dietary components that increase the probability of occurrence of these diseases in individuals, and interventions to modify their impact, have also been identified. Furthermore, rapid changes in diets and lifestyles that have occurred with industrialization, urbanization, economic development, and market globalization have accelerated over the past decade. This is having a significant impact on the health and nutritional status of populations, particularly in developing countries and in countries in transition. Changes in the world food economy are reflected in shifting dietary patterns; because of these changes in dietary and lifestyle patterns, chronic NCDs are becoming increasingly significant causes of disability and premature death in both developing and newly developed countries.

The need for investment in 21st century nutrition science is most obvious in emerging economies of Asia and Africa. Both are beset by dual burdens of under- and overnutrition crises. The human capacity to address these needs is weak and, in the case of India, has weakened since the 1970s. Evidence of this weak capacity is clear when studying the focus and quantity of nutrition research output from

Table 31.5
Journal Publications in At-Risk Nutrition Areas

Country	*Obesity/Overweight (83%) N = 1423*	*Micronutrition (13%) N = 220*	*Undernutrition (4%) N = 73*
United States (%)	48.1	64.1	41.1
China (%)	2.2	0.9	4.1
India (%)	0.4	4.1	1.4
Mexico (%)	0.6	3.2	2.7

researchers in emerging economies through the prism of publications in leading medical and nutrition journals. For example, Table 31.5 shows the number of full-length publications in leading nutrition journals in four specific countries, 2006–2007 *(7)*:

The overwhelming majority of scientific journals focused on obesity and overweight in their publications for 2006–07; only 13% focused on micronutrition, and a mere 4% on undernutrition *(7)*. Low levels of research in nutrition science retards societal progress in the field and could erode long-term productivity. Further, lack of emphasis on public sector nutrition science creates serious obstacles for corporate innovation for two reasons: first, the scientists we need are not available locally and, second, the attitudes of leaders in nutrition remain fixed in a 20th century mode. This latter obstacle creates road blocks for food industry innovation. Currently, governments in developed countries provide a wide range of incentives to the pharmaceutical industry to develop medicated solutions to nutritional problems. These include tax incentives for pharmaceutical companies to invest in new drug development, and extensive funding of public research entities like the US National Institutes of Health and the UK MRC which tend to support medicated versus food/activity-based solutions to problems like obesity or even under nutrition. Equal effort should be given to the development of more sustainable agricultural and food-based incentives to shift the level of research in the food sector from its current levels of about 0.5–1.15% of revenues closer to the pharmaceutical industry levels of about 15–20% of revenues. Greater public support for basic and applied research aimed at food-based solutions could transform the ability of countries to more effectively address the nutrition needs of populations.

31.5. CONCLUSIONS AND RECOMMENDATIONS

> In a world filled with complex health problems, WHO cannot solve them alone. Governments cannot solve them alone. NGOs, the private sector and foundations cannot solve them alone. Only through new and innovative partnerships can we make a difference.
>
> *Dr. Gro Harlem Brundtland, Director-General, WHO, 13 May 2002, WHA, Geneva*

> Our goal must be a world in which good health is a pillar of individual well-being, national progress, and international stability and peace. This cannot be achieved without partnerships involving governments, international organizations, the business community, and civil society.
>
> *Kofi Annan, former Secretary-General of the United Nations*

Multinational food and beverage corporations have a vested interest in helping to improve population health. The worlds' 10 largest global food and beverage companies, seven of which are US based, include Nestle SA, Cargill, Kraft Foods, Unilever, Tyson Foods, PepsiCo, General Mills, Groupe

Danone, Kellogg Company, and ConAgra Foods. Together, revenues gross in the multiple trillions of dollars on an annual basis. It is paramount to the success of impacting the global nutrition crisis that industry creates a platform of trust between the public and private sectors through acknowledgment of our common values and shared goals, especially those that transcend cultural, regional, or socio-economic differences. It is the responsibility of the private sector to develop innovative strategies that utilize and share business know-how with partners in the government, academic, research, and non-governmental arenas. The same comprehensive performance with purpose approach that is employed to reach corporate goals must be applied to the urgent purpose of alleviating the global nutrition crisis. It is the industry's in-depth knowledge of the individual consumer that puts industry in a unique position to solve problems in practical ways, to reach individuals in communities, bringing the promise of global nutrition policy into the daily lives of those who need it most *(12)*.

There is much that still needs to be done. Obesity remains the only major public health problem for which there is no example of a sustained decline in any large population anywhere, outside of a war, famine, or recession. That alone suggests there is a need for a new and innovative plan of action to enable more sustainable partnerships that would draw upon all of our unique capabilities and expertise. In the June 2008, Pacific Health Summit meetings on nutrition, Dr. Margaret Chan (WHO D-G), Tachi Yamada (Gates Foundation), Ann Veneman (UNICEF), and Sir William Castell (Wellcome Trust) echoed recent calls by NIH Director Dr E. Zahouni for private–public partnerships to address global nutrition challenges. PepsiCo is committed and already engaged in such partnerships and we are confident that they will yield outcomes of benefit to health. Sustained progress requires that a firmer basis is built for public–private collaboration. Only through new and innovative partnerships will come great opportunity and great change.

ACKNOWLEDGEMENTS

We thank many colleagues at PepsiCo for critical inputs. Substantive ideas were provided by Danielle Dalheim, RD; James Holden, MS; Anne Kurilich, PhD; Renee Mellican, PhD; Heather Nelson-Cortes, PhD; Kari Ryan, PhD, RD; Steve Saunders, PhD; and Kimberly White, PhD.

REFERENCES

1. McKee, M. (1999). Alcohol in Russia. *Alcohol and Alcoholism*, 34, 6, 824–29.
2. Black, R. (2008). Maternal and child undernutrition: global and regional exposures and health consequences. *The Lancet*, 371, 9608, 243–60.
3. Wilk, B. & Bar-Or, O. (1996). Effect of drink flavor and NaCl on voluntary drinking and hydration in boys exercising in the heat. *Journal of Applied Physiology*, 80, 1112–7.
4. Nakamichi, T. (2008). "G8 leaders blast food-export bans," *Wall Street Journal*, 8 July.
5. Brown, L.R., Flavin, C., French, H.F., Starke, L., Abramovitz, J.N., Dunn, S., Gardner, G., Mattoon, A.T., McGinn, A.P. & O'Meara, M. *State of the World 1999*. Worldwatch Institute, W.W. Norton and Company, London, 1999.
6. *Lancet* Nutrition Series (2008). http://www.thelancet.com/collections/series/undernutrition. Accessed 19 October 2008.
7. Yach, D. (2008). Personal correspondence.
8. Yunus, M. http://muhammadyunus. org/content/view/56/83/lang,en/. Accessed 19 October 2008.
9. Weber, T.J. (2008). "Gates wants creative capitalism," *Wall Street Journal*, 24 January.
10. PepsiCo Annual Report (2007). http://www.pepsico.com/AnnualReports/07/index.html; Accessed 19 October 2008.
11. Unilever, www.unilever.com, Accessed 28 November 2008.
12. Moss, R.K. (2008). Transforming giants: what kind of company makes it its business to make the world a better place? *Harvard Business Review*.
13. Bradley, D. (2008). Beyond product: the private sector drive to perform with the purpose of alleviating global undernutrition. *Global Forum for Health Research*, 5, 171–73.
14. Weaver, J. (1974) *Carnation, The First 75 Years*. Carnation Company, 1–15.

15. Roberts, P. (2008). *The End of Food*. USA: Houghton Mittlin Harcowt.
16. Young, I. (2002). Is healthy eating all about nutrition? British Nutrition Foundation. *Nutrition Bulletin*, 27, 7–12.
17. Vartiainen, E., Puska, P., Pekkanen, J. & Toumilehto, J. (1994). Changes in risk factors explain changes in mortality from ischaemic heart disease from Finland. *British Medical Journal*, 309, 23–27 (2 July 1994)
18. Butte, N.F., Lopez-Alarcon, M.G. & Garza, C. (2002). Nutrient adequacy of exclusive breastfeeding for the term infant during the first six months of life. *World Health Organization Report*, Geneva.
19. Spaulding, M. *Nurturing Yesterday's Child: A portrayal of the Drake Collection of Paediatric History*. Philadelphia, BC Decker, 1991.
20. Apple, R.D. *Mothers and Medicine: A social History of Infant Feeding*. Madison, Wisconsin, University of Wisconsin Press, 1987.
21. Berseth, C.L., Van Aerde, J.E., Gross, S., Stolz, S.I., Harris, C.L. & Hansen, J.W. (2004). Growth, efficacy, and safety of feeding an iron-fortified human milk fortifier. *Pediatrics*, 114, 6, e699–706.
22. Birch, E.E., Hoffman, D.R., Uauy, R., Birch, D.G. & Prestidge, C. (1998). Visual acuity and the essentiality of docosohexaenoic acid and arachidonic acid in the diet of term infants. *Pediatric Research*, 44, 2, 201–9.
23. Mace, K., Steenhout, P., Klassen, P. & Donnet, A. (2006). Protein quality and quantity in cow's milk-based formula for healthy term infants: past, present and future. *Nestle Nutrition Workshop Ser Pediatric Program*, 58, 189–203.
24. Makrides, M., Neumann, M.A., Jeffery, B., Lein, E.L. & Gibson, R.A. (2000). A randomized trial of different ratios of linoleic to a-linolenic acid in the diet of term infants: effects on visual function and growth. *American Journal of Clinical Nutrition*, 71, 120–9.
25. Morley, R., Abbott, R., Fairweather-Tait, S., MacFadyen, U., Stephenson, T. & Lucas, A. (1999). Iron fortified follow on formula from 9 to 18 months improves iron status but not development or growth: a randomised trial. *Archives of Disease in Childhood*, 81, 3, 247–52.
26. O'Connor, D.L., Hall, R., Adamkin, D. & Auestad N. (2001). Growth and development in preterm infants fed long-chain polyunsaturated fatty acids: a prospective, randomized controlled trial. *Pediatrics*, 108, 2, 359–71.
27. Vanderhoof, J., Gross, S., Hegyi, T. & Clandinin, T. (1999). Evaluation of a Long-Chain Polyunsaturated Fatty Acid Supplemented Formula on Growth, Tolerance, and Plasma Lipids in Preterm Infants up to 48 Weeks Postconceptional Age. *Journal of Pediatric Gastroenterology Nutrition*, 29, 3, 318–26.
28. World Health Organization (2008). "World Health Report 2008." www.who.int/en . Accessed 18 November 2008.
29. World Health Organization 2008. World Health Statistics. www.who.int/whosis/whostat/ENWHS08Full.pdf. Accessed 18 November 2008.
30. Lichtenstein, A.H., Appel, L.J., Brands, M. & Carnethon, M. (2006). Diet and lifestyle recommendations revision 2006: a scientific statement from the American Heart Association Nutrition Committee. *Circulation*, 114, 1, 82–96.
31. Anderson, J.W., Spencer, D.B., Hamilton, C.C., Smith, S.F., Tietyen, J., Bryant, C.A. & Oeltgen, P. (1990). Oat-bran cereal lowers serum total and LDL cholesterol in hypercholesterolemic men. *American Journal of Clinical Nutrition*, 52, 3, 495–9.
32. Braaten, J.T., Wood, P.J., Scott, F.W., Wolynetz, M.S., Lowe, M.K., Bradley-White, P. & Collinsn M.W. (1994). Oat beta-glucan reduces blood cholesterol concentration in hypercholesterolemic subjects. *European Journal of Clinical Nutrition* 48, 7, 465–74.
33. Davidson, M.H., Dugan, L.D., Burns, J.H., Sugimoto, D., Story, K. & Drennan, K. (1996). A psyllium-enriched cereal for the treatment of hypercholesterolemia in children: a controlled, double-blind, crossover study. *American Journal of Clinical Nutrition*, 63, 1, 6–102.
34. He, J., Klag, M.J., Whelton, P.K., Mo, J.P., Chen, J.Y., Qian, M.C., Mo, P.S. & He, G.Q. (1995). Oats and buckwheat intakes and cardiovascular disease risk factors in an ethnic minority of China. *American Journal of Clinical Nutrition*, 61, 2, 366–72.
35. Paul, G.L., Ink, S.L. & Geiger, C.J. (1999). The Quaker Oats Health claim: A case study. *Journal of Nutraceutical functions & Medical Foods*. 1, 4, 5–32.
36. Van Horn, L., Moag-Stahlberg, A., Liu, K.A., Ballew, C., Ruth, K., Hughes, R. & Stamler, J. (1991). Effects on serum lipids of adding instant oats to usual American diets. *American Journal of Public Health*, 81, 2, 183–8.
37. Food and Drug Administration. (1997). Food labeling: health claims; soluble fiber from certain foods and risk of coronary heart disease. Final rule. *Federal Register*, January 23, 62, 15, 3583–601.
38. Wolever, T.M., Jenkins, D.J., Mueller, S., Boctor, D.L., Ransom, T.P., Patten, R., Chao, E.S., McMillan, K. & Vulgoni, D. (1994). Method of administration influences the serum cholesterol-lowering effect of psyllium. *American Journal of Clinical Nutrition*, 59, 5, 1055–9.
39. Anderson, J.W., Davidson, M.H., Blonde, L., Brown, W.V., Howard, W.J., Ginsberg, H., Allgood, L.D. & Weingand, K.W. (2000). Long-term cholesterol-lowering effects of psyllium as an adjunct to diet therapy in the treatment of hypercholesterolemia. *American Journal of Clinical Nutrition*, 71, 6, 1433–8.

40. Sprecher, D.L., Harris, B.V., Goldberg, A.C., Anderson, E.C., Bayuk, L.M., Russell, B.S., Crone, D.S., Quinn, C., Bateman, J., Kuzmak, B.R. & Allgood, L.D. (1993). Efficacy of psyllium in reducing serum cholesterol levels in hypercholesterolemic patients on high- or low-fat diets. *Annals of Internal Medicine*, 119, 7, 545–54.
41. Delaney, B., Nicolosi, R.J., Wilson, T.A., Carlson, T., Frazer, S., Zheng, G.H., Hess, R., Ostergren, K., Haworth, J. & Knutson, N. (2003). Beta-glucan fractions from barley and oats are similarly antiatherogenic in hypercholesterolemic Syrian golden hamsters. *Journal of Nutrition* 133, 2, 468–75.
42. Li, J., Kaneko, T., Qin, L.Q., Wang, J. & Wang, Y. (2003). Effects of barley intake on glucose tolerance, lipid metabolism, and bowel function in women. *Nutrition*, 19(11–12), 926–9.
43. Food and Drug Administration. Food labeling: health claims; soluble fiber from certain foods and risk of coronary heart disease. (2008). Final rule. *Federal Register*, May 1, 73, 85, 23947–53.
44. Keenan, J.M., Goulson, M., Shamliyan, T., Knutson, N., Kolberg, L. & Curry, L. (2007). The effects of concentrated barley beta-glucan on blood lipids in a population of hypercholesterolaemic men and women. *British Journal of Nutrition*, 97, 6, 1162–8.
45. Shimizu, C., Kihara, M., Aoe, S., Araki, S., Ito, K., Hayashi, K., Watari, J., Sakata, Y. & Ikegami, S. (2008). Effect of high beta-glucan barley on serum cholesterol concentrations and visceral fat area in Japanese men—a randomized, double-blinded, placebo-controlled trial. *Plant Foods for Human Nutrition*, 63, 1, 21–5.
46. Tuomilehto, J., Tikkanen, M.J., Hogstrom, P., Keinanen-Kiukaannieme, S., Piironen, V., Toivo, J., Salonen, J.T., Nyyssonen, K., Stenman, U.H., Alfthan, H. & Karppanen, H. (2008). Safety assessment of common foods enriched with natural nonesterified plant sterols. *European Journal of Clinical Nutrition*, 60, 633–42.
47. Binkoski, A.E., Kris-Etherton, P.M., Wilson, T.A., Mountain, M.L. & Nicolosi, R.J. (2005). Balance of unsaturated fatty acids is important to cholesterol lowering diet: Comparison of mid-oleic sunflower oil and oil on CVD risk factors. *Journal of the American Dietetic Association*, 107, 1080–86.
48. Cade, R., Spooner, G., Schlein, E., Pickering, M. & Dean, R. (1972). Effect of fluid, electrolyte, and glucose replacement during exercise on performance, body temperature, rate of sweat loss, and compositional changes of extracellular fluid. *Journal of Sports Medical and Physical Fitness*, 12, 150–6.
49. Davis, J.M., Jackson, D., Broadwell, M, Queary, J. & Lambert, C. (1997). Carbohydrate drinks delay fatigue during intermittent, high-intensity cycling in active men and women. *International Journal of Sports and Nutrition*, 7, 261–73.
50. Dougherty, K.A., Baker, L.B., Chow, M. & Kenney, W.L. (2006) Two percent dehydration impairs and six percent carbohydrate drink improves boys basketball skills. *Medicine Science and Sports Exercise*, 38(9), 1650–8.
51. Dueck, C.A., Matt, K.S., Manor, M.M. & Skinner J.S. (1996) Treatment of athletic amenorrhea with a diet and training intervention program. *International Journal of Sport and Nutrition* 6, 24–40.
52. Utter, A.C., Kang, J., Nieman, D.C., Vinci, D.M., McAnulty, S.R. & Dumke, C.L. (2003). Ratings of perceived exertion throughout an ultramarathon during carbohydrate ingestion. *Perceptual and Motors Skills*, 97, 175–84.
53. Coyle, E.F. & Montain, S.J. (1992) Benefits of fluid replacement with carbohydrate during exercise. *Medicine Science and Sports Exercise* 24, S324–30.
54. Stofan, J.R., Zachwieja, J.J., Horswill, C.A., Murray, R., Anderson, S.A. & Eichner, E.R. (2005) Sweat and sodium losses in NCAA football players: a precursor to heat cramps? *International Journal of Sport Nutrition ad Exercise Science*, 15, 641–52.
55. Balzer, J. (2008). Sustained benefits in vascular function through flavanol-containing cocoa in medicated diabetic patients a double-masked, randomized, controlled trial. *Journal of the American College of Cardiology*, 51, 22, 2141–9.
56. Hooper, L. (2008). Flavonoids, flavonoid-rich foods, and cardiovascular risk: a meta-analysis of randomized controlled trials. *American Journal of Clinical Nutrition*, 88, 1, 38–50.
57. Selmi, C. (2006). The anti-inflammatory properties of cocoa flavanols. *Journal of Cardiovascular Pharmacology*, 47, S163–71.
58. Allen, L., Benoist, B.D., Dary, O. & Hurrell, R. (2006) Guidelines on Food Fortification with Micronutrients World Health Organization and Food and Agriculture Organization of the United Nations.
59. Food Fortification: Technology and Quality Control. (1997). Report of an FAO Technical Meeting, Rome, Italy, 20–23 November, 1995. Conclusions and Recommendations. Food and Agriculture Organization.
60. Berner, L.A., Clydesdale, F.M. & Douglass, J.S. (2001) Fortification contributed greatly to vitamin and mineral intakes in the United States, 1989–1991. *Journal of Nutrition*, 131, 2177–83.
61. Hegenbart, S. (1997) "It's a gas," Food product design. http://www.foodproductdesign.com/articles/1997/05/its-a-gas.aspx
62. GAIN; www.gainhealth.org. Accessed 22 October 2008.
63. World Health Organization (2005). "Diet, Nutrition and the Prevention of Chronic Diseases," WHO Technical Report Series 916, Geneva.

64. Bekefi, T. (2006). "Business as a partner in tackling micronutrient deficiency: Lessons in multisector partnership." Corporate Social Responsibility Initiative, Harvard University, Boston MA.
65. Olshansky, S.J., Passaro, D.J., Hershow, R.C., Layden, J., Carnes, B.A., Brody, J., Hayflick, L. Butler, R.N., Allison, D.B. & Ludwug, D.S. (2005). A potential decline in life expectancy in the United States in the 21st century. *Obstetrical and Gynecological Survey*¸ 60, 7, 450–52.
66. Yach, D., Stuckler, D. & Brownell, K.D. (2006). Epidemiologic and economic consequences of the global epidemics of obesity and diabetes. *Nature*, 12, 62–6.
67. Economos, C.D., Hyatt, R.R., Goldberg, J.P., Must, A., Naumova, E.N., Collins, J.J. & Nelson, M.E. (2007). A community intervention reduces BMI z-score in children: Shape Up Somerville first year results. *Obesity*, 15, 1325–36.
68. Brown, G. (2008). http://www.dfid.gov.uk/mdg/call-to-action.asp. Accessed 20 October 2008.
69. World Bank (2008). "Food price crisis imperils 100 million in poor countries, Zoellick says," Development Committee Press Briefing, Spring Meetings 2008, 12 April, Washington DC USA.
70. Sheeran, J. (2008). Innovating against hunger and undernutrition. *Global Forum for Health Research*, 5, 174–6
71. World Health Organization (2008). "Action Plan—World Health Assembly," 24 May 2008, http://www.who.int/mediacentre/news/releases/2008/wha02/en/index.html. Accessed 22 October 2008.
72. McMichael, A.J., Powles, J.W., Butler, C.D. & Uauy, R. (2007). Food, livestock production, energy, climate change and health. *The Lancet*, 370, 1253–63.

32 The Role of Preventive Nutrition in Clinical Practice

A. Julian Munoz, Jamy D. Ard, and Douglas C. Heimburger

Key Points

- The impact of nutrition on altering risk for chronic disease is generally slow and only gradually evident after lengthy exposure to a given dietary pattern.
- Changing detrimental eating patterns to healthier patterns can have a significant impact on preventable disease risk factors for the individual and the population at large.
- From the nutrition perspective much of disease prevention is about making small and important changes in the diets of as many people as possible.
- Dietary Guidelines attempt to provide evidence-based recommendations for diet and physical activity that may lead to higher quality of life and improve longevity.
- Fruits, vegetables, and whole grains comprise the basis of a healthful diet; reduced-fat dairy products are also key components.
- Adherence to the Dietary Guidelines is low among the US population and a significant effort is needed to improve this problem.
- The seriousness of the current epidemic of nutrition-related disorders demands both community-level interventions as well as clinical preventive services aimed to impact patients' personal health practices.
- Assessment of individual disease risk and dietary intake is the starting point for tailoring dietary advice based on recommended guidelines.
- Office visits offer an opportunity to provide nutrition assessment and counseling and impact morbidity and mortality. Brief clinician counseling can be effective in inducing positive lifestyle changes.
- Evidence supports the use of formal systems to screen for risky dietary patterns and sedentary lifestyle in clinical settings, increasing the likelihood of prioritizing nutrition counseling.
- A patient-centered collaborative approach is the basis for comprehensive and complex behavior change, where the physician first assesses patients' readiness for behavior change, helps address barriers to change, and negotiates goals for lifestyle changes.

Key Words: Behavior; cancer; counseling; diabetes; dietary guidelines; exercise; obesity; prevention

32.1. INTRODUCTION

Nutrition plays a central role in health by virtue of the simple fact that everyone must eat as a matter of survival. Complexity arises when the choices of foods and accompanying nutrients lead to either health benefits or detrimental effects. Nutritional factors contribute significantly to the burden of preventable illnesses and premature deaths, as detailed in previous chapters in this volume. Indeed, nutritional factors are associated with 4 of the 10 major causes of death: coronary heart disease (CHD), cancer, stroke, and type 2 diabetes *(1)*. It is estimated that about 16 and 9% of mortality from any cause

A. Bendich, R.J. Deckelbaum (eds.), *Preventive Nutrition*, Nutrition and Health, DOI 10.1007/978-1-60327-542-2_32,

in US men and women, respectively, could be eliminated by the adoption of desirable dietary behaviors *(2)*. The impact of nutrition on altering risk for chronic disease is generally slow and only gradually evident after lengthy exposure to a given dietary pattern. However, because we all must eat, changing detrimental eating patterns to healthier patterns can have a significant impact on preventable disease risk factors for the individual and the population at large *(3)*.

In the United States, many of the preventable deaths from problems such as heart disease and cancer occur in individuals with moderate risk. If a relatively small reduction in risk for a disease occurred in this group of persons with moderate risk, it would produce major benefits for the population at large. It has been estimated that between 300,000 and 800,000 deaths per year could be prevented in the United States, if Americans followed national dietary recommendations *(4)*. These benefits occur because of sheer volume: there are large numbers of people with moderate risk, whereas those with the highest risk make up a much smaller proportion of the population. Even making large reductions in risk for this smaller, high-risk group is less powerful than small risk reductions on the larger group. For example, modifying diets to reduce coronary heart disease (CHD) in the general population is thought to be worthwhile, because most deaths occur not among those at high risk due to high serum cholesterol levels, but in people who have only moderate elevations in serum cholesterol (i.e., 200–240 mg/dL). Therefore, from the nutrition perspective much of disease prevention is about making small and important changes in the diets of as many people as possible.

A lower level of impact at the individual level for population-targeted nutrition interventions is to be expected because of the numerous individual level factors that can modify the nutrition–disease risk relationship. Genetics, environment, and physical activity are all examples of factors that can modify the effect of nutrition on disease risk at the individual level. Although genetic factors can affect individual susceptibility, they appear to account for only a small part of the observed variation in disease incidence among populations, as exemplified by the tendency of immigrants to acquire the disease rates of their adoptive countries. More attention is being given to the impact of the local environment and its ability to influence dietary patterns through enhancing availability of various food sources and types. A major challenge for each individual is to consume a total energy intake that is matched to energy expenditure. Energy expenditure via leisure time physical activity is an important component of overall energy balance. As health-care providers, it is increasingly important to consider these factors as we attempt to supplement dietary recommendations for the general population with more sophisticated, individually based dietary interventions in the attempt to achieve optimal disease risk reduction.

32.2. TRENDS IN DIET AND DISEASE

At the turn of the 20th century, the leading causes of death were infectious diseases and curing them would have reduced death rates. Today, most of the leading causes of death in Western countries are strongly influenced by lifestyle, and medical resources are mainly invested in treating diseases associated with specific lifestyles. Heart disease, cancer, and stroke account for two-thirds of all deaths in the United States. One-third will die from CHD before age 65, and many others will be disabled by these illnesses and their complications.

Changes in eating patterns parallel these disease trends. In lieu of the high-fiber, low-fat foods once consumed as the basis of our diet, refined starches, sweets, saturated fats, and salt comprise a major share of today's typical American diet *(5)*. Table 32.1 lists 8 of the top 15 causes of death in the United States that are strongly influenced by nutrition. As the table shows, five are strongly linked with dietary habits and three are associated with alcohol abuse. The table also details the many dietary

Table 32.1
Dietary Influences on the Major Causes of Death and Morbidity in the United States

Cause of Death or Morbidity	*Factors Associated with Decreased Risk*	*Factors Associated with Increased Risk*
Death		
Heart diseases	Intake of complex carbohydrates, particular fatty acids (e.g., monounsaturated, polyunsaturated, and ω-3 fatty acids from fish), soluble fiber, polyphenols, soy proteins, antioxidants (vitamins E, C; β-carotene, selenium), folic acid, moderate alcohol	Intake of saturated fat, cholesterol; excess calories, sodium; abdominal distribution of body fat
Cancer	Intake of fruits and vegetables (for β-carotene, vitamins A, C, D, and E, folic acid, calcium, selenium, phytochemicals), fiber	Intake of excess calories, fat, alcohol, red meat, salt- and nitrite-preserved meats, possibly grilled meats; abdominal distribution of body fat
Cerebrovascular diseases	Intake of potassium, calcium, ω-3 fatty acids	Sodium, alcohol consumption (as with hypertension)
Accidents		Excess alcohol consumption
Diabetes mellitus	Fiber intake	Intake of excess calories, fat, alcohol; abdominal distribution of body fat
Suicide		Excess alcohol consumption
Chronic liver disease and cirrhosis		Excess alcohol consumption
Hypertension and hypertensive renal disease	Intake of fruits and vegetables, potassium, calcium, magnesium, ω-3 fatty acids	Intake of sodium, alcohol, excess calories, total and saturated fat; abdominal distribution of body fat
Morbidity		
Obesity		Intake of excess calories and fat
Osteoporosis	Intake of calcium, vitamin D, vitamin K	Intake of excess of vitamin A, sodium, protein
Diverticular disease, constipation	Fiber intake	
Neural tube defects	Folic acid intake	

contributions to obesity, atherosclerosis, osteoporosis, diverticular disease, and neural tube defects, which cause significant morbidity and indirect mortality.

In the past, the goal and measure of success of our health-care system has been to increase life expectancy, regardless of well-being or quality of life. However, reducing morbidity—that is, improving quality of life and maximizing the period of good health—may be the more important outcome. The Dietary Guidelines for Americans 2005, a joint publication of the US Department of Human Health Services (USDHHS) and the Department of Agriculture (USDA) published every 5 years since 1980 *(6)*, attempt to provide evidence-based recommendations for diet and physical activity that, if adopted, may lead to achievement of a higher quality of life and optimum health. The intent of the Dietary Guidelines is to summarize knowledge and provide authoritative advice regarding individual nutrients and food choices and translate these into recommendations, which serve as the basis for federal food and nutrition education programs, including Healthy People 2010 *(7)*. Attempts to assist in implementing the recommendations are exemplified by the USDA's www.MyPyramid.gov and the DASH Eating Plan *(8)*.

32.3. CURRENT DIETARY HABITS IN THE UNITED STATES

Public awareness of dietary intake has increased substantially in recent years, likely fueled by a constant stream of media coverage related to the increased prevalence of obesity in the United States and our growing obsessions with fad diets. This has also been associated with changes in nutrition labeling practices and promotion of the USDA/DHHS MyPyramid, a pictorial depiction of how American diets should be structured (*see* Fig. 32.1). Certainly, food manufacturers and producers are aware of the increased health consciousness of the society and have launched promotional efforts to tout the beneficial effects of their food products.

In spite of increased awareness, large dietary surveys of the population reveal multiple areas for improvement in current dietary patterns. Surveys suggest that people in higher educational and income brackets have been more responsive to public health recommendations. Intake of total fat has declined to 34 and 32% of calories for men and women, respectively *(9)*. Fruit and vegetable intake is less than optimal, with a major portion of vegetable intake obtained from white potatoes, typically fried, and a significant portion of fruit intake obtained from juices and other beverages. In addition, fiber intake is less than recommended, partially because of less than optimal fruit and vegetable intake. The low fiber intake is also a result of higher-than-recommended consumption of refined grain products such as white bread, pasta, and rice. These products displace the higher fiber alternatives. Added sugar consumption is also on the rise as sweetened beverages and desserts are more common in schools, work settings, and at home. The intake of milk and dairy products has declined over time as well, particularly among children and adolescents, being replaced primarily by carbonated beverages and fruit-flavored drinks *(10)*. Another trend involves increased frequency of meals consumed outside the home. As time pressures increase and commercially prepared foods are more available, palatable, and affordable, the proportion of the population consuming meals away from home has grown dramatically. The odds of eating out at least one or more times per week were 40% higher in 1999–2000 relative to 1987 *(9)*. The reported number of commercially prepared meals consumed per week is positively related to energy intake; persons who eat out more frequently consume higher amounts of total energy. In addition, the portion sizes that are available in restaurants have increased, particularly for many entrée and dessert items.

The challenge for the panel convened to develop the newest national dietary guidelines was to move the population from this current dietary pattern to an overall eating pattern that is associated with lower disease risk and promotes optimal health.

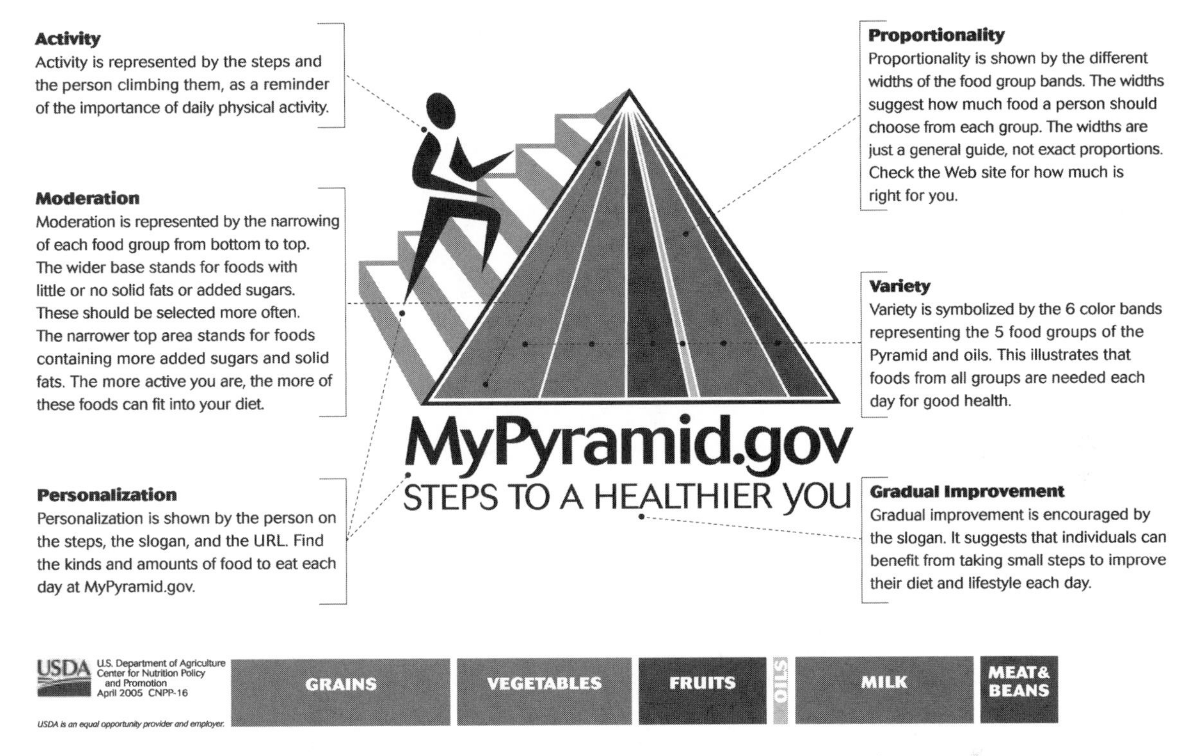

Fig. 32.1. USDA/DHHS Food Guide Pyramid (www.MyPyramid.gov).

32.4. EATING FOR OPTIMUM HEALTH

There have been at least seven reports of dietary guidelines for Americans published in recent years and many more reports of national dietary guidelines outside of the United States. It is noteworthy that there is close agreement on the general recommendations made in these reports, which enhances their credibility. The most widely publicized guidelines are those embodied in the USDA/DHHS MyPyramid (*see* Fig. 32.1), which pictorially prioritizes food groups in terms of their importance to health. In 2005, the Pyramid underwent a significant change to reflect new scientific evidence related to optimal eating patterns. Many of the major messages from the new guidelines are consistent with previous versions; however, because many Americans experienced difficulty in applying previous guidelines to their daily routine while receiving apparently conflicting information on a regular basis about the types of nutrients to consume, significant effort has been invested in increasing the personalization of the Pyramid and translating the latest scientific findings into dietary guidance. Another major change was to include physical activity recommendations as a key part of the Pyramid and Guidelines. The complete key recommendations for the general population from the 2005 Dietary Guidelines are shown in the box.

32.4.1. Dietary Guidelines

The nine major messages of the Dietary Guidelines are summarized below.

Dietary Guidelines for Americans 2005: Key Recommendations for the General Population *(6)*

Adequate Nutrients Within Calorie Needs

- Consume a variety of nutrient-dense foods and beverages within and among the basic food groups while choosing foods that limit the intake of saturated and trans fats, cholesterol, added sugars, salt, and alcohol.
- Meet recommended intakes within energy needs by adopting a balanced eating pattern, such as the US Department of Agriculture (USDA) Food Guide or the Dietary Approaches to Stop Hypertension (DASH) Eating Plan.

Weight Management

- To maintain body weight in a healthy range, balance calories from foods and beverages with calories expended.
- To prevent gradual weight gain over time, make small decreases in food and beverage calories and increase physical activity.

Physical Activity

- Engage in regular physical activity and reduce sedentary activities to promote health, psychological well-being, and a healthy body weight.
  - To reduce the risk of chronic disease in adulthood: Engage in at least 30 minutes of moderate-intensity physical activity, above usual activity, at work or home on most days of the week.
  - For most people, greater health benefits can be obtained by engaging in physical activity of more vigorous intensity or longer duration.
  - To help manage body weight and prevent gradual, unhealthy body weight gain in adulthood: Engage in approximately 60 minutes of moderate- to vigorous-intensity activity on most days of the week while not exceeding caloric intake requirements.
  - To sustain weight loss in adulthood: Participate in at least 60–90 minutes of daily moderate-intensity physical activity while not exceeding caloric intake requirements. Some people may need to consult with a health-care provider before participating in this level of activity.
- Achieve physical fitness by including cardiovascular conditioning, stretching exercises for flexibility, and resistance exercises or calisthenics for muscle strength and endurance.

Food Groups to Encourage

- Consume a sufficient amount of fruits and vegetables while staying within energy needs. Two cups of fruit and 2½ cups of vegetables per day are recommended for a reference 2,000-calorie intake, with higher or lower amounts depending on the calorie level.
- Choose a variety of fruits and vegetables each day. In particular, select from all five vegetable subgroups (dark green, orange, legumes, starchy vegetables, and other vegetables) several times a week.
- Consume three or more ounce-equivalents of whole-grain products per day, with the rest of the recommended grains coming from enriched or whole-grain products. In general, at least half the grains should come from whole grains.
- Consume three cups per day of fat-free or low-fat milk or equivalent milk products.

Fats

- Consume less than 10% of calories from saturated fatty acids and less than 300 mg/day of cholesterol, and keep trans-fatty acid consumption as low as possible.
- Keep total fat intake between 20 and 35% of calories, with most fats coming from sources of polyunsaturated and monounsaturated fatty acids, such as fish, nuts, and vegetable oils.
- When selecting and preparing meat, poultry, dry beans, and milk or milk products, make choices that are lean, low fat, or fat-free.
- Limit intake of fats and oils high in saturated and/or *trans*-fatty acids, and choose products low in such fats and oils.

Carbohydrates

- Choose fiber-rich fruits, vegetables, and whole grains often.
- Choose and prepare foods and beverages with little added sugars or caloric sweeteners, such as amounts suggested by the USDA Food Guide and the DASH Eating Plan.
- Reduce the incidence of dental caries by practicing good oral hygiene and consuming sugar- and starch-containing foods and beverages less frequently.

Sodium and Potassium

- Consume less than 2,300 mg (approximately one teaspoon of salt) of sodium per day.
- Choose and prepare foods with little salt. At the same time, consume potassium-rich foods, such as fruits and vegetables.

Alcoholic Beverages

- Those who choose to drink alcoholic beverages should do so sensibly and in moderation—defined as the consumption of up to one drink per day for women and up to two drinks per day for men.
- Alcoholic beverages should not be consumed by some individuals, including those who cannot restrict their alcohol intake, women of childbearing age who may become pregnant, pregnant and lactating women, children and adolescents, individuals taking medications that can interact with alcohol, and those with specific medical conditions.
- Alcoholic beverages should be avoided by individuals engaging in activities that require attention, skill, or coordination, such as driving or operating machinery.

Food Safety

- To avoid microbial foodborne illness:
    - Clean hands, food contact surfaces, and fruits and vegetables. Meat and poultry should not be washed or rinsed.
    - Separate raw, cooked, and ready-to-eat foods while shopping, preparing, or storing foods.
    - Cook foods to a safe temperature to kill microorganisms.
    - Chill (refrigerate) perishable food promptly and defrost foods properly.
    - Avoid raw (unpasteurized) milk or any products made from unpasteurized milk, raw or partially cooked eggs or foods containing raw eggs, raw or undercooked meat and poultry, unpasteurized juices, and raw sprouts.

32.4.1.1. Consume a Variety of Foods within and Among the Basic Food Groups while Staying Within Energy Needs

A basic premise of the Dietary Guidelines is that nutrient needs should be met primarily through consuming foods. Foods provide an array of nutrients (as well as phytochemicals, antioxidants) and other compounds that may have beneficial effects on health. In some cases, fortified foods may be useful sources of one or more nutrients that otherwise might be consumed in less-than-recommended amounts. Supplements may be useful when they fill a specific identified nutrient gap that cannot or is not otherwise being met by the individual's intake of food. Nutrient supplements cannot replace a healthful diet. Individuals who are already consuming the recommended amount of a nutrient in food will not achieve any additional health benefit if they also take the nutrient as a supplement. In fact, in some cases, supplements and fortified foods may cause intakes to exceed the safe levels of nutrients. Another important premise of the Dietary Guidelines is that foods should be prepared and handled in such a way that reduces risk of foodborne illness.

Consuming a variety of foods provides the opportunity to achieve recommended nutrient intakes. As previously mentioned, the goal of achieving recommended nutrient intakes is to prevent chronic disease, a change in focus from the earliest guidelines that focused on preventing deficiencies of various nutrients. The basic food groups include fruits; vegetables; grains; milk, yogurt, and cheese; and meat, poultry, fish, dry beans, eggs, and nuts. The Dietary Guidelines Advisory Committee specifically noted, however, that additional efforts are still warranted to promote higher dietary intakes of vitamin E, calcium, magnesium, potassium, and fiber by children and adults and to higher intakes of vitamins A and C by adults. The probability of nutritional adequacy for these nutrients was less than 60% for adult men and women, and the mean intake for potassium and fiber was less than adequate in all age groups.

While achieving adequate nutrition using a variety of foods is the main thrust of this first recommendation, the Guidelines remind us that particular attention should be paid to maintaining an energy-balanced state by matching calorie intake to energy expenditure. Consuming a large variety of foods has been associated with higher energy intake *(11)*. To manage calorie intake, we are advised to limit intake of foods that have low nutritive value, such as added sugars, solid fats, and alcoholic beverages. Foods that are high in these components are typically energy dense and have few essential nutrients that play a role in disease prevention. In addition, the strategy of substituting nutrient-rich foods for those that are nutrient-poor and calorically dense improves the nutrient profile and reduces caloric intake.

For several population subgroups, the Dietary Guidelines delineate a few special nutrient recommendations. First, adolescent females and women of childbearing age are encouraged to increase dietary intake of iron-rich foods and maintain a folic acid intake of 400 μg per day. The recommendations for increased iron and folic acid intake for this group are targeted at reducing the incidence of iron deficiency and neural tube defects, respectively. Secondly, a substantial proportion of the population over the age of 50 has a reduced capacity to absorb vitamin B_{12} in its naturally occurring food-bound form. However, the crystalline form of vitamin B_{12} can be readily absorbed, and it is recommended that individuals over the age of 50 meet their recommended daily allowance of vitamin B_{12} by consuming foods fortified with B_{12} or by taking vitamin B_{12} supplements. Finally, the Committee recommends that individuals at high risk for low levels of vitamin D, including the elderly, persons with dark skin, and persons exposed to little UVB radiation, obtain extra vitamin D from vitamin D-fortified foods and/or supplements that contain vitamin D.

32.4.1.2. Control Calorie Intake to Manage Body Weight

The management of body weight appears early in the set of major recommendations as a result of the increasing prevalence of obesity in the United States. Significant attention has been devoted to altering the proportions of carbohydrates, fats, and proteins in the diet as a way to control weight. However, the emphasis of this second recommendation is that total energy intake, from any combination of energy sources, is a primary determinant of body weight. To maintain one's body weight, energy intake must be matched with energy expenditure. To lose weight, an energy deficit must be created by either reducing caloric intake or increasing energy expenditure with physical activity, or a combination of the two.

Because a majority of the population is considered overweight (body mass index [BMI] $\geq 25\,kg/m^2$) and nearly one-third is obese (BMI $\geq 30\,kg/m^2$), much of the emphasis for this recommendation is directed toward preventing further weight gain and promoting weight loss. An essential key to limiting weight gain or losing weight via calorie restriction is to reduce one's intake of added sugars, solid fats, and alcohol. As noted in the first recommendation, these items are calorically dense and provide few essential nutrients. Another key factor is to consume foods that are low in energy density, such as fruits and vegetables. Limiting portion sizes, particularly of foods that are high in energy density, is another recommended strategy for managing body weight.

32.4.1.3. Be Physically Active Every Day

The enhanced emphasis on physical activity and its relationship to diet and disease prevention and health promotion is evident by the inclusion of a figure to represent activity on MyPyramid (steps on the side of the pyramid and a person climbing them—*see* Fig. 32.1). In general, the guidelines call for moderate physical activity for at least 30 minutes per day on most days of the week to promote fitness and reduce the risk of chronic health conditions such as obesity, hypertension, diabetes, and CHD. Moderate-intensity physical activity, which expends 3–5 metabolic equivalents or METs (1 MET involves consumption of 3.5 mL oxygen/kg/min), is achieved by walking at a brisk pace of 3–4 mph, bicycling on level ground, light swimming, gardening, or mowing a lawn. The same benefits may be achieved by participating in vigorous physical activity 20 minutes per day on 3 days of the week. Vigorous physical activity expends six or more METs and is achieved by jogging, running, aerobic dancing, competitive sports, heavy yard or construction work, brisk swimming, or fast bicycling. It should be emphasized, particularly to patients with differing levels of physical fitness, that the primary determinant of the impact of physical activity on health outcomes is the total volume of physical activity rather than its intensity alone. Therefore, most individuals can achieve significant health benefits using moderate physical activity, as long as its duration is sufficient.

Additional outcomes of interest include prevention of unhealthy weight gain and avoidance of weight regain following weight loss. Based on a report from the Institute of Medicine in 2002 cited in the Dietary Guidelines, most adults need up to 60 minutes of moderate to vigorous physical activity on most days to prevent unhealthy weight gain *(12)*. For those who previously lost weight and are trying to avoid weight regain, the recommendation is to participate in 60–90 minutes of moderate physical activity daily.

32.4.1.4. Increase Daily Intake of Fruits and Vegetables, Whole Grains, and Nonfat or Low-Fat Milk and Milk Products

For many Americans, the intake of the food groups in this set of recommendations is well below recommended levels. For this reason, the Guidelines group them together. The other rationale for grouping them together is that they form the basis of a dietary pattern that has been shown to have multiple positive effects on disease risk factors, called Dietary Approaches to Stop Hypertension (DASH). The guidelines also review evidence for the roles of each of the food groups in disease prevention and treatment.

As noted previously, fruit and vegetable intake is less than optimal in the United States, resulting in lower than desirable levels of fiber, vitamins, minerals, and phytonutrient intakes. The range of intake recommended for fruits and vegetables on a daily basis is 2½–6½ cups, depending on an individual's energy needs. The goal for persons who require 2,000 calories per day to maintain their weight is to consume 4½ cups of fruits and vegetables daily. Daily consumption should be from a variety of fruits and vegetables ranging from citrus and other fruits to dark green leafy and bright orange vegetables. Scientific evidence demonstrates that increased consumption of fruits and vegetables is important for reducing risk for cardiovascular disease, cancer, and diabetes. Diets high in fruits and vegetables have also been associated with maintenance of body weight, successful weight loss, and long-term maintenance of weight loss *(13)*.

The emphasis on whole grain intake is related to two factors: (1) whole grains are an important source of energy and 14 nutrients including fiber; (2) inclusion of whole grains in the diet will likely displace refined grains that are less nutrient dense. Similar to the impact of fruits and vegetables, diets high in whole grains reduce the risk of cardiovascular disease and type 2 diabetes. Whole grain intake is also associated with successful weight control. The Guidelines recommend daily intake of at least three 1-ounce equivalents of whole grain foods, preferably in the place of refined grains.

Nonfat and low-fat milk and other milk products are key sources of calcium, magnesium, potassium, and vitamin D. The milk food group is key for achieving and maintaining peak bone mass and preventing osteoporosis. Milk and other dairy products have also been associated with lower levels of insulin resistance in several population studies. There is growing interest in the role of milk for regulating body weight and enhancing weight loss. It was the Committee's opinion that the current evidence did not support a definitive effect of the milk group on weight loss. However, studies do show that consumption of the milk group is not associated with weight gain. The recommended level of intake is three cups of milk or equivalent milk products per day, preferably nonfat or low-fat products, such as skim milk and yogurt.

32.4.1.5. Choose Fats Wisely for Good Health

The Guidelines provide specific instruction regarding types of dietary fat and their relationships to CHD. While low fat intake (less than 20% of calories) is often produces weight reduction because of low total energy intake, the combination of low fat and high carbohydrate intake, particularly from refined grains sources, can elevate serum triglycerides and depress high-density lipoprotein (HDL) cholesterol levels. On the other hand, high fat intakes of more than 35% of calories are associated with obesity and CHD because of high energy intake and increased saturated fat intake, respectively. Therefore, total fat intake between 20 and 35% of calories is recommended for all Americans of 18 years and older. For children aged 2 and 3 years, the lower limit of fat intake is 30%, while for children age 4–18 years, the lower limit is 25% of calories.

The primary objective for this recommendation is to reduce the risk of CHD, which involves lowering low-density lipoprotein (LDL) cholesterol. This can be achieved by keeping dietary intake of saturated fat below 10% of calories, trans fat below 1% of calories, and cholesterol below 300 mg per day, by limiting intake of animal fat, partially hydrogenated vegetable oils, and eggs and organ meats, respectively. In addition to these recommendations, consuming monounsaturated fatty acids (MUFA) and polyunsaturated fatty acids (PUFA) can be beneficial for CHD risk. Diets that substitute MUFA for saturated fat reduce LDL cholesterol. Fatty fishes such as salmon, tuna, and lake trout are high in PUFA known as *n*–3 fatty acids, and intake of two servings per week of fatty fish is associated with decreased risk of sudden cardiac death. More details on dietary fats and CHD risk are provided in Chapter 20.

32.4.1.6. Choose Carbohydrates Wisely for Good Health

While fat intake has garnered much of the public's attention since the early 1980s, in recent years significant attention has been shifted toward carbohydrates. This has produced a variety of misconceptions regarding the effects of carbohydrates, resulting in a large number of people trying to avoid carbohydrates altogether. In most instances, restricting carbohydrates leads to lower energy intake, simply because of decreased food intake; however, this dietary pattern is not sustainable for most individuals. Carbohydrates, when chosen properly, are an important part of a healthful diet and serve as a key energy source for the body.

Because maintaining fiber intake is important for promoting healthy laxation and reducing the risk of type 2 diabetes and CHD, the Guidelines recommend that carbohydrate choices be fiber-rich. The recommended amount of dietary fiber (14 g per 1,000 calories) can be achieved within the acceptable carbohydrate range of 45–65% of calories per day by choosing foods from the fruit, vegetable, and grain groups that are high in fiber while avoiding excessive intake of refined grains and added sugars (i.e., sugars and syrups that are added to foods during processing or preparation or at the table). Whole fruits are preferred over fruit juice. Added sugars may increase the palatability of food but they add little nutrition other than energy. Prospective studies in multiple populations have shown that added sugar intake, particularly in sweetened beverages, is associated with weight gain over time. The objective of reducing added sugars is to lower total energy intake, enabling individuals to meet their nutrient requirements by using more nutrient-dense foods and not exceeding the calorie level required to maintain their weight.

32.4.1.7. Choose and Prepare Foods with Little Salt

The median daily sodium intakes for US men and women ages 31–50 are 4,300 mg and 2,900 mg, respectively, well above the Committee's recommendation of 2,300 mg per day. The Guidelines recommend this as the upper limit of sodium intake for the general population primarily because of the direct relationship between sodium intake and blood pressure. The use of the upper limit is intended to protect special populations that may be more sensitive to the hypertensive effects of sodium (blacks, middle-aged and older individuals, and persons with hypertension, diabetes, or kidney disease) but may be unaware of their sensitivity. Lowering sodium intake can be difficult in a food environment where sodium is ubiquitous in the processing of food. In addition to limiting added salt in food preparation and at the table, useful strategies include limiting the use of prepared sauces, broths and soups, salty snack foods, and canned foods. The preference for high sodium intake is a learned behavior that individuals can modify relatively easily to increase acceptance of lower sodium foods.

32.4.1.8. If You Drink Alcoholic Beverages, Do So in Moderation

In 1999–2001, 60% of the US adult population reported drinking alcoholic beverages, with 95% having moderate consumption, or less than one to two drinks per day. Several studies have shown that moderate alcohol intake is associated with a lower risk of CHD (*see* Chapter 20). However, it is not recommended that nondrinkers start consuming alcohol to lower their risk of CHD because excessive drinking poses significant health risks that offset any potential advantages. For young adults, this is particularly true because alcohol intake has not been associated with any health benefits in this age group; on the contrary, it is associated with increased risk of traumatic injury and death. It should be clear that abstention is an important option for all individuals, particularly for those who have difficulty restricting their drinking to moderate levels, individuals with specific medical conditions, women who may become pregnant or who are pregnant, and breast-feeding women.

32.4.1.9. Keep Food Safe to Eat

Foodborne illnesses are responsible for a number of hospitalizations and deaths each year in the United States. *Salmonella*, *Listeria*, and *Toxoplasma* are responsible for more than 75% of the 5,000 annual deaths related to foodborne diseases. These illnesses can be prevented with the use of proper food safety techniques. Basic food safety recommendations include (1) cleaning hands, contact surfaces, and fruits and vegetables; (2) separating raw, cooked, and ready-to-eat foods during food preparation and storage; (3) cooking foods to a safe temperature; (4) refrigerating perishable foods properly; and (5) avoiding higher risk foods, particularly for the very young, pregnant women, the elderly, and immunocompromised persons.

32.4.2. Summary of the Dietary Guidelines

Experimental data and lessons derived from populations with low rates of chronic diseases such as atherosclerosis, obesity, diabetes, and cancer point strongly to several common dietary factors that should form the foundation for our daily eating patterns. As compiled in the Dietary Guidelines for Americans, fruits, vegetables, and whole grains comprise the basis of a healthful diet; reduced-fat milk and milk products are also key components. Maintaining adequate intakes of these food groups will ensure high levels of key nutrients such as fiber, vitamins C, A, and E, potassium, magnesium, and calcium. Other foods should be added with the goal of limiting chronic disease risk (primary prevention) or treating known risk factors and disease (secondary prevention). Therefore, choices for dietary fats should emphasize MUFA and PUFA while limiting intake of saturated and *trans*-fatty acids and cholesterol. Limiting added sugars helps to avoid weight gain from excessive calories and reduces risk for type 2 diabetes. Matching energy expenditure to energy intake is a key factor in modifying disease risk and should be the final consideration in the dietary pattern. Maintaining a physically active lifestyle, at work and at leisure, has an independent health benefit in addition to assisting with weight management.

Currently, adherence to the Dietary Guidelines is low among the US population and a significant effort is needed to improve this problem. A growing body of evidence demonstrates that following a diet that complies with the Dietary Guidelines may reduce the risk of chronic diseases and reduce mortality among individuals aged 45 years and older *(14)*. Figure 32.2 depicts data from US Department of Agriculture (USDA) illustrating the degree of change in the overall dietary pattern of Americans needed to be consistent with a food pattern encouraged by the Dietary Guidelines.

It is helpful to understand the relationship between the Dietary Guidelines for Americans and others that are recommended by various national health organizations. Summaries of the key features of dietary guidelines promulgated by national organizations for preventing and/or managing major chronic diseases are shown in Table 32.2. There is significant convergence among the various guidelines and no significant conflicts or discrepancies. Each disease indication, however, may require spe-

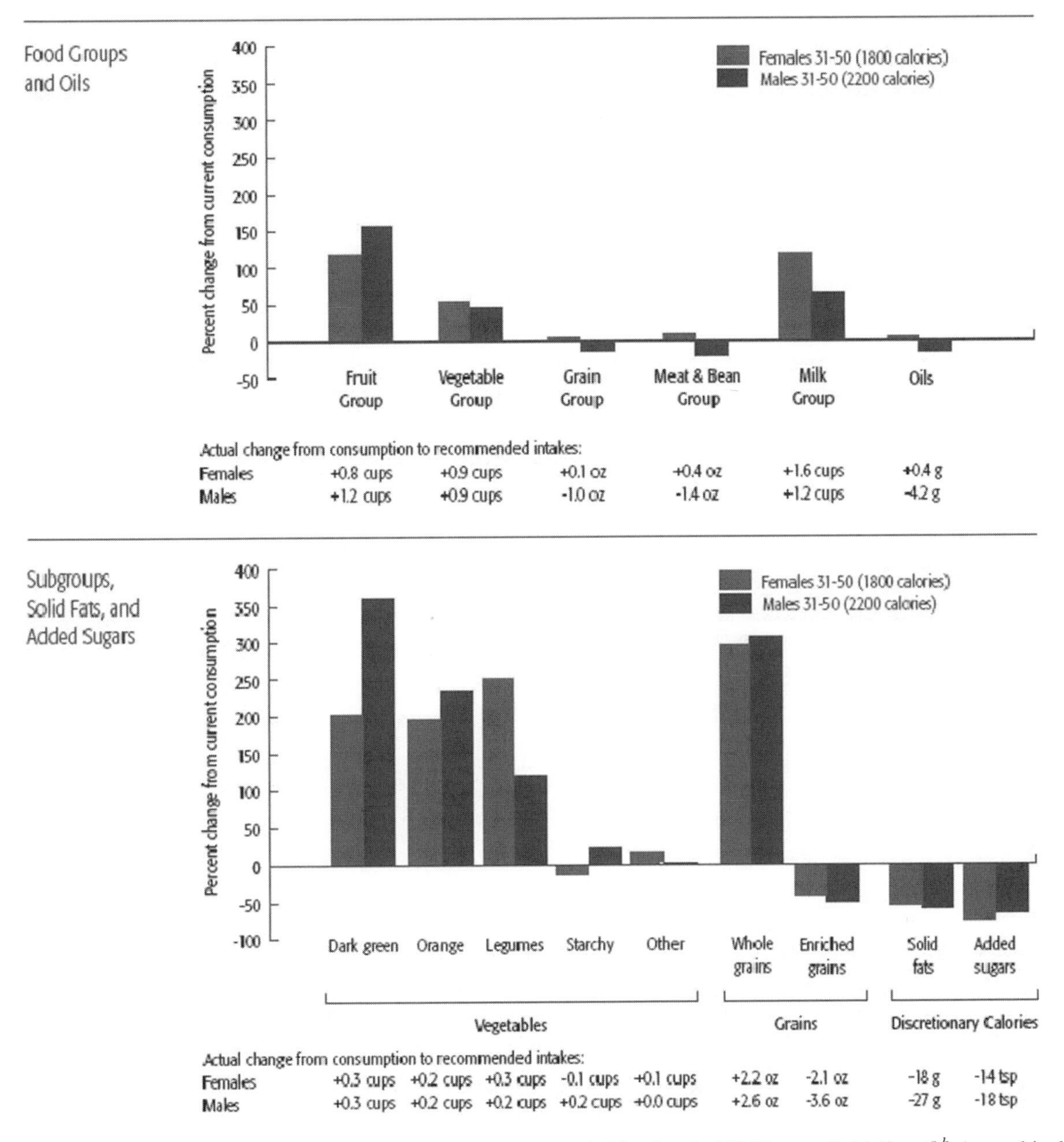

Fig. 32.2. Variance of current US consumption from recommended intakes in US Dietary Guidelines [a,b] A graphical depiction of the degree of change in average daily food consumption by Americans that would be needed to be consistent with the food patterns encouraged by the *Dietary Guidelines* for Americans. The zero line represents average consumption levels from each food group or subgroup by females 31–50 years of age and males 31–50 years of age. Bars above the zero line represent recommended increases in food group consumption, while bars below the line represent recommended decreases. From http://www.health.gov/dietaryguidelines/dga2005/document/html/chapter1.htm [a] USDA Food Guide in comparison to National Health and Nutrition Examination Survey 2001–2002 consumption data. [b] Increases in amounts of some food groups are offset by decreases in amounts of solid fats (i.e., saturated and trans fats) and added sugars so that total calorie intake is at the recommended level.

Table 32.2

Dietary Guidelines Promulgated by National Organizations[a]

	Organization				
	United States Department of Agriculture and Department of Health and Human Services Dietary Guidelines for Americans 2005	National Cholesterol Education Panel ATP III Therapeutic Lifestyle Change (TLC) Diet (2002)	National High Blood Pressure Education Program/Joint National Committee 7 Dietary Approaches to Stop Hypertension (DASH, 2006)	American Diabetes Association (2004)	American Cancer Society (2006)
Indication or objective	General health promotion and disease prevention	Elevated cholesterol/heart disease prevention	Pre-hypertension and hypertension	Diabetes	Cancer prevention
Nutrient/food group					
Total energy	Adequate energy intake to maintain a healthy weight		Reduce energy intake to lose weight if overweight	Reduced energy intake and modest weight loss can improve glycemia and insulin resistance	Balance calorie intake with physical activity. Choose foods and drinks in amounts that help achieve and maintain a healthy weight.
Fruits/vegetables	Two cups of fruit and 2.5 cups of vegetables per 2000 calories per day		8–10 servings per day		$\geq$5 servings per day
Lean meats, poultry, and fish			Six servings or less per day		Limit intake of processed and red meats
Low-fat dairy	2–3 servings per day		2–3 servings per day		

Grains	Three or more ounce equivalents of whole grain products		6–8 servings of whole grains and whole grain products		Choose whole grains over processed (refined) grains
Fat	20–35% of daily energy intake	25–35% of daily energy intake	27% of daily energy intake		
Saturated fats	Less than 10% of daily energy intake	Less than 7% of daily energy intake	6% of daily energy intake	Less than 10% of daily energy intake; those with LDL cholesterol ≥100 may benefit from lowering saturated fat to <7%.	
Polyunsaturated fats	Up to 10% of daily energy intake	Up to 10% of daily energy intake		Up to 10% of daily energy intake	
Monounsaturated fats	10–20% of daily energy intake	Up to 20% of daily energy intake		15–20% of daily energy intake; combination of MUFA and carbohydrates should equal 60–70% of total energy intake	
Trans fats	Less than 1% of daily energy intake	Intake should be kept low		Intake should be minimized	
Cholesterol	Less than 300 mg per day	Less than 200 mg per day	150 mg per day	Less than 300 mg per day; those with LDL cholesterol ≥100 may benefit from lowering cholesterol intake to 200 mg per day	

(*Continued*)

Table 32.2
(Continued)

	Organization				
Carbohydrates	45–65% of daily energy intake	50–60% of daily energy intake	55% of daily energy intake	Total amount of carbohydrate is more important than the source or type	
Sugar	Limit added sugars			Sucrose and sucrose-containing foods do not need to be restricted	
Protein		Approximately 15% of daily energy intake	18% of daily energy intake	10–20% if renal function is normal	
Alcohol	Up to two drinks per day for men and up to one drink per day for women; persons in special circumstances (e.g., pregnancy, history of alcoholism) should abstain		Less than two drinks per day for men and less than one drink per day for women	If individuals choose to drink, limit to two drinks per day for men or one drink per day for women	Drink no more than one drink per day for women or two per day for men
Sodium	Up to 2,300 mg per day		2,300 mg per day	Less than 2,400 mg per day	
Potassium			4,700 mg per day		
Calcium			1,250 mg per day	1,000–1,500 mg per day	
Magnesium			500 mg per day		

[a] Recommendations are generally based on approximately 2,000 calories daily energy intake. For further detail, refer to the websites listed in this chapter.

cific attention to particular aspects of the respective diet to target isolated risk factors or metabolic pathways.

32.5. APPLICATION

It is the responsibility of health-care providers to take population-based science and apply it to a variety of individuals with unique characteristics and circumstances. Nowhere is this more true than in nutrition. Assessment of individual disease risk and dietary intake is the starting point for tailoring dietary advice based on recommended guidelines. Because much of nutrition is related to behavior (food preparation, food choices, physical activity, etc.), simply understanding and imparting the facts of nutrition may lead to less-than-optimal implementation. The keys to reaching individuals at moderate risk for chronic disease are related to understanding their perspectives on disease risk and providing clear, consistent health messages that are supported by the guidelines.

32.5.1. The Role of Health-Care Providers in Reaching Healthy People 2010 Nutritional Goals

Addressing the challenge of improving the health of our communities is a shared responsibility that requires the active participation and leadership of federal, state, and local governments, policy-makers, professionals, business executives, educators, community leaders, and the public itself. The seriousness of the current epidemic of nutrition-related disorders demands both community-level interventions as well as clinical preventive services aimed to impact patients' personal health practices.

Health-care providers are a primary source of support to reach the goals of increasing quality and years of healthy life and eliminate health disparities. Health-care providers can serve as effective public policy advocates and further catalyze intervention efforts in the family, community and in the media, improving awareness of resources and referral services that can assist with the management of nutrition-related problems. Most Americans have a primary care physician whom they see on average at least once every 2 years *(15)*. These office visits offer an opportunity to provide nutrition assessment and counseling and impact morbidity and mortality. Primary care providers are expected to be key players in nutrition counseling, especially given data suggesting that patients consider physicians to be one of the most credible sources of nutrition information *(16)*. Healthy People 2010, a national endeavor of the US Department of Health and Human Services designed to identify the most significant preventable threats to public health and to establish and monitor national goals aimed to reduce the burden of highly prevalent chronic diseases *(7)*, requests physicians and health-care providers to put preventive nutrition into their practices through screening and counseling. The nutrition goals of Healthy People 2010 aim to increase the proportion of population that make healthy food choices to 75% of adults and children, eating at least two servings of fruit and at least three daily servings of vegetables a day. Healthy People 2010 obesity goal is to reduce the number of obese adults to less than 15% and increase the proportion of physician office visits made by patients with a diagnosis of cardiovascular disease, diabetes, or hyperlipidemia that include counseling or education related to diet and nutrition.

However, evidence indicates that nutrition-related risk assessment and counseling interventions are delivered less frequently than other preventive interventions such as cancer screenings, with less than half of primary care physicians routinely provide dietary counseling *(17)* and with time spent in nutrition counseling in primary care averaging 1 minute *(18)*. Perhaps as a result of these types of practical challenges, the goals of Healthy People 2010 are unlikely to be realized. As previously noted, during the time frame that this initiative has been implemented, the United States has experienced a dramatic increase in the prevalence of obesity among all age, gender, and ethnic groups.

Although most patients look to their physicians for lifestyle recommendations and physicians report a positive view about the importance of nutrition, surveys describe several barriers to effective preventive nutrition counseling by primary care physicians including lack of time, uncertainty of the effectiveness of nutrition counseling and patient compliance, inadequate skills in providing counseling, lack of financial incentives, and lack of a systematic, organized approach within the practice *(19,20)*. With the current epidemic of nutrition-related diseases, the need for nutrition education and practical tools for health-care providers is urgent.

Although time is an important limitation in the primary care setting, evidence demonstrates that brief clinician counseling is effective in inducing positive lifestyle changes such as smoking cessation, reducing alcohol consumption, and increasing physical activity *(21,22)*. Although several studies have shown efficacy of dietary counseling to significantly reduce dietary fat and cholesterol intake and increase fruit and vegetable consumption *(23)*, the effectiveness of this intervention when applied to low-risk populations in the primary care setting has not been fully demonstrated. Clinician counseling seems to be more effective when tailored to individual risk factors, needs, preferences, and abilities of each patient identified through a systematic approach. Considering the time constraints of primary care practice, these processes need to be brief and part of an organized office system with referrals of complex cases to qualified nutrition professionals. Delivering preventive care requires a team approach and it is important for the entire office staff to participate in delivering these services. Counseling to promote a healthy diet may involve the receptionist who can provide information that will reinforce educational messages that the patient will hear during the visit, the nurse who while weighing the patient can emphasize nutritional information, and the clinician who can discuss diet-related risk factors for particular conditions such as heart disease or diabetes.

Traditionally, physicians have counseled patients to change habits by sharing facts about illnesses or using their expert power. However, research shows that these simple means of persuasion are not effective for promoting lasting behavioral changes, and better counseling methods are needed. The most effective approaches start by assessing the degree of motivation and readiness to implement changes. Most patients are initially ambivalent about changing long-standing lifestyle behaviors; they fear that it will be difficult, uncomfortable, or depriving. In addition to motivation, physicians need to assess patient support, stressful life events, psychiatric status, time availability and constraints, and appropriateness of goals and expectations to help increase the likelihood of long-term lifestyle change. Determining readiness for behavior change is essential for success because attempts to initiate changes when patients are not ready often leads to frustration and may hamper future efforts. Current thinking is to use a patient-centered collaborative approach for comprehensive and complex behavior change. In this approach, the physician first assesses patients' readiness for behavior change, and then helps them address barriers to change. Goals are developed only when patients are ready and have thought about the benefits and challenges of the lifestyle change.

32.5.2. Assessing Patients' Readiness to Make Behavioral Changes

Several methods for assessing patients' readiness to change have been described. They include anchoring patient's interest for change in a numerical scale, e.g., 0–10, and measuring confidence in reaching proposed goals with the same numerical scale. Another efficient method for assessing patient readiness is to use direct and targeted questioning that can engage patients in self-reflection that may enhance readiness for change. Assessing patients' readiness for change and helping them increase it are interactive processes. Periodic assessments of patients' locations and movement on the readiness continuum should guide counseling. Several useful behavior-change models can be used to increase readiness for change. Among these, the Health Belief Model, the Social Learning Theory, and the Transtheoretical or Stages of Change Model are the most widely used *(24)*. Elements of each model may be appropriate for counseling individual patients, depending on the care provider's personal counseling style and patients' characteristics.

The Health Belief Model conceives that health behavior is a function of people's perceptions regarding their vulnerability to illness and their perceptions of the effectiveness of treatments. In this model, behavior change is determined by whether people perceive themselves to be susceptible to a particular health problem; believe the problem is serious and that prevention is effective and not overly costly in regard to money, effort, or pain; and are exposed to a cue to take health action.

The Social Learning Theory holds that patients must believe that they have the needed skills to change behavior (called self-efficacy) before they will take action. An important component of skill development comes through modeling; physicians can provide needed guidance in this area. Modeling is most effective when it addresses prior attempts to change behavior, the strategies that were and were not successful, and provides ideas to help patients succeed. The Transtheoretical or Stages of Change Model proposes that at any specific time, patients are in one of five discreet stages of change: precontemplation, contemplation, preparation, action, and maintenance. Patients move among the stages in the process of change and may repeat stages several times before they achieve lasting change. Within the Transtheoretical Model, the health professional's tasks include both assessing patients' stages of change and using behavioral counseling strategies to help them advance from one stage to the next.

Other important characteristics for successful counseling include empathic communication, which refers to making the patient feel that you understand their situations, perspectives, and feelings; and establishing a patient–provider partnership to work together in developing a treatment plan.

32.6. NECESSARY CONDITIONS FOR PRODUCTIVE BEHAVIORAL COUNSELING

1. **Knowledge.** For an unhealthy habit to change, the burden of being informed lies both on the clinician and the patient. Clinicians must not only know about the mechanisms of, treatments for, and supportive community resources for addressing problems related to poor lifestyle habits, they must also be facile with practical solutions to common barriers patients face when attempting changes. Listening to patients may be the source of understanding the common barriers to and facilitators of change. Patients' knowledge deficits need to be addressed if they are to implement and maintain behavior change, but focused teaching by clinic staff keeps the learning manageable at each visit. If patients are expected to change using clinically proven approaches, they need to be given a sufficiently detailed plan.
2. **Motivation.** Just informing a patient about health risks, while necessary, is usually not sufficient for a decision to change. The clinician's goal is to try to influence the patient's choices before serious medical consequences ensue. Table 32.3 summarizes a strategy for promoting the patient's decision to change.

   "Motivational reflection" is a technique used to promote the patient's decision to change. The strategy is to help the patient identify—likely over several visits—all the personally important present and future consequences of changing or continuing the habit. To enact this strategy successfully, the patient must feel that the clinician genuinely cares about them, has a sincere interest in understanding their perspective and is "on their side". Table 32.4 shows how this motivational reflection process can be initiated and followed up.

   The clinician should also monitor the patient's progress and provide encouragement, praise, and accountability at follow-up visits. Using a patient log of adherence to the plan and follow-up visits or telephone calls from office staff to discuss the patient's problems and progress indicate both that the clinician is interested in the patient's success and holds the patient accountable to following the plan.

Table 32.3
Addressing the Patient's Motivation

Task	*Method*	*Purpose*
1. Determine the patient's readiness to change	Ask "Have you thought about losing weight in the last month?"	If the patient is not thinking about changing, he/she is not ready for a solution. Go to 2
2. Facilitate the patient's understanding of personally important consequences of changing and not changing. Use the motivational reflection technique	Ask, "Can you think how losing weight might improve your appearance, finances, relationships, ability to work and do fun stuff?"	You are helping the patient identify the things in his/her life important enough to cause a decision to change
3. Give a strong, clear message that you believe the patient needs to change	Say, "For all the reasons you just told me, as your doctor I encourage you strongly to change."	In addition to the intrinsic motivation developed in 2, you are adding your social influence to encourage a change

3. **Addressing Barriers to Change.** Even motivated, informed patients encounter barriers to starting or continuing plans to change their dietary habits and activity patterns. When this occurs, two common barriers need to be explored. First, there may be one or more powerful antecedent conditions that provoke the unhealthy behaviors. For example, patients often eat, drink, and smoke to self-medicate negative feelings like stress, boredom, anxiety, and depression. If the patient has no alternative in dealing with those feelings, the unhealthy behavior will continue until a plan for coping differently is developed and practiced.

The second barrier is that some physical, social, financial, emotional, environmental, or time barrier prevents or decreases the patient's ability to carry out the plan. For example, a patient commits to a modest walking program but does not start because the weather was too hot. Making a point to ask about and assist patients in developing their own solutions for these barriers should increase their chances of successfully making changes and preventing relapses.

32.6.1. Responding to Patients' Resistance

Recognizing the patient's ambivalence about behavior change is critical to minimizing resistance. When the physician initiates a discussion of an unhealthy lifestyle behavior, the patient may respond with verbal and/or nonverbal resistance to discussing the issue. The clinician's response to the patient's resistance influences whether the rest of the encounter moves the patient closer to deciding to change. Resistance manifests itself in a variety of behaviors, e.g., patient silence, eyes shifting down or away, "yes, but" excuses, or insincere and nonspecific promises quickly made to do something about the behavior. If the clinician ignores these resistance signals and tells the patient what he/she should do, a

Table 32.4
Promoting the Decision to Change

More Helpful	• **Asking the patient to identify positive consequences** important to the patient that would occur if the adverse behavior is changed, specifically related to work, family and leisure activities, finances, appearance, health, religion • **Asking the patient to identify important negative consequences if the adverse behavior continues** • **Asking the patient what good things would result** if the first consequence occurred; repeating with each consequence
Less Helpful	• **Assuming the patient will not change and being silent** • **Telling the patient** the reasons he/she has to change • **Trying to frighten the patient into changing** by informing only about the worst outcomes

common outcome is that the patient is even less interested in discussing the topic at future visits, and the clinician may give up the power struggle by becoming silent about the issue.

To respond more effectively to a patient's resistance, it is useful to understand its sources. Patients' resistance is often the result of guilt about the behavior, perceived lack of autonomy, a sense of hopelessness about the ability to change the behavior, and/or the perception that the physician will preach, nag, or exhort the patient to do something he/she is not ready to do. Resistance is reduced when the patient feels in control of the decision making. Table 32.5 lists helpful responses to patients' resistance.

32.7. THE 5A'S BRIEF BEHAVIORAL COUNSELING PROTOCOL

The 5A's protocol is a useful approach that addresses enabling conditions for change and is summarized in Table 32.6.

32.7.1. Addressing Non-Adherence to Recommendations

Even the most compliant patients will at times fail to adhere to recommendations. Although this may be discouraging to both the practitioner and the patient, a number of strategies are available to help reduce mutual frustration. Not blaming the patient is a major step. Although the physician may feel inclined to label the patient as undisciplined, non-compliant, or difficult for failing to adhere to recommendations, this can lead to more frustration and further damage the patient–physician partnership. Instead, health-care professionals should work with their patients to determine the causes for non-adherence. After causes have been identified, problem-solving skills can be applied to develop effective solutions. For example, it is possible that the patients did not understand the importance of adhering to recommendations, that the recommendations were unrealistic, or that there were unanticipated barriers. In each case, working together with the patient to develop strategies for adherence or establish more realistic goals is indicated. Continued encouragement can help patients initiate and adhere to recommendations.

Table 32.5
Responses to Patient Resistance

More Helpful	• **Acknowledge** it by telling the patient you sense his/her reluctance to discuss the topic • **Reassure** by telling the patient you will not nag, preach, judge, or tell him/her what to do • **Ask permission** to discuss the topic, after acknowledging and reassuring. If the answer is "no," do not proceed • **If a tentative "yes" is given, proceed** and avoid preaching, judging, or telling the patient what to do • **Keep the patient in control of the decision making** by asking permission and offering options • **Exit** the conversation at any time if resistance increases
Less Helpful	• **Not noticing**, ignoring, or exerting pressure to overcome the patient's resistance • **Telling the patient** what habit has to be changed • **Giving specific instructions** about how to change a harmful behavior without assessing the behavior or soliciting the patient's ideas

32.7.2. Effectiveness of Systems for Delivering Clinical Preventive Services

Several studies provide evidence that formal systems for delivering preventive services, including screening and counseling, increase their likelihood of delivery in the clinical setting *(25)*. A system refers to a process integrating staff roles and tools for the routine delivery of preventive care, which defines individual responsibilities, flow of activities, and monitoring. The preventive service delivery system usually includes the use of established evidence-based preventive care protocols, pre-auditing charts, reminder notices, flow sheets, educational materials, and recordkeeping for patients, among others *(26,27)*. In one study, the use of such a system increased the documentation of cholesterol screening from 70 to 84%, smoking assessment from 56 to 80%, up-to-date Pap smears from 70 to 81%, yearly mammograms from 30 to 48%, and documentation of pneumococcal vaccination from 22 to 48% *(25)*.

The Physician-Based Assessment and Counseling for Exercise (PACE) program is another example of a system that was implemented to improve the rate and quality of counseling for physical activity in the primary care setting. In a controlled trial conducted in 17 physician practices, sedentary patients

Table 32.6
The 5A's Brief Behavioral Counseling Protocol for Diet and Physical Activity

UAB *Nutrition Academic Award** NAA	*A Guide to Help Patients Succeed in Lifestyle Changes*
5A's Step	*Counseling Script*
1. Address issue	Mr/Ms. _________, I would like to discuss your _________ with you today. Would that be ok?
2. Assess and motivate	*Recent attempts* Have you tried to in the past 3 months? **If "yes," ask:** What did you do? What problems did you have? What did you do that was successful? Why did you stop? *Readiness to change* Have you thought about trying to _________ in the last month? **If "no," do motivation and exit.** *Motivation to change* Mr/Ms _________, what activity now, or in the future, would you miss if you were to become ill? What is the worst thing about missing that? What other activities would you hate to miss? What is the worst thing about missing them? What good things (relationships, work, leisure, and other activities) might happen in your life if you change your _________?
3. Advise	As your physician, it is important that you _________ for all the reasons you just told me and for other important medical reasons.
4. Assist	What high-fat, fried, sugary foods could you eat less of tomorrow? Ask only if "Yes" to Readiness to Change question. **Negotiate goal(s) and specifics—how much, how often, what size, etc.**, to achieve goal. **Educate briefly** about recommended behavior change (e.g., diet, exercise, smoking cessation) or health risks. **Request** patient to keep **a log** for review at next session. **Assess barriers:** What could keep you from starting this program and continuing until our next visit?
5. Arrange follow-up	I will see you in _________ weeks to see how you are progressing.

who received 3–5 minutes of counseling about physical activity plus a telephone call 2 weeks later were significantly more likely to increase physical activity than those who were not counseled *(28)*.

32.7.3. How to Develop a Preventive Nutrition System

The key function of a preventive nutrition care protocol is to efficiently identify individuals who most need to change nutrition behaviors and the frequency with which interventions should be applied. The US Preventive Services Task Force (USPSTF) found fair evidence that brief, low- to medium-intensity behavioral dietary counseling in the primary care setting can produce small-to-medium changes in average daily intake of core components of an overall healthy diet (especially saturated fat and fruits and vegetables) in unselected patients. However, it considered the strength of evidence limited because it relies on self-reported dietary outcomes, limited use of measures corroborating reported changes, limited follow-up data, and possible selection bias in study participants *(29)*. The USPSTF concluded that the evidence is insufficient to recommend for or against routine behavioral counseling to promote a healthy diet in unselected patients in primary care settings. In contrast, the USPSTF did recommend intensive behavioral dietary counseling for adult patients with known risk factors for cardiovascular and diet-related chronic diseases. It found good evidence that medium- to high-intensity counseling interventions can produce medium-to-large changes in average daily intake of core components of a healthy diet (including saturated fat, fiber, fruits, and vegetables) among adult patients at increased risk for diet-related chronic diseases. The USPSTF concluded that such counseling is likely to improve important health outcomes and that benefits outweigh potential harms. Therefore, screening for diet-related disorders and triaging patients based on their risk seems to be a reasonable approach to prioritize nutrition counseling and intervention.

32.7.4. Tools for Screening and Counseling

A critical initial step in establishing a preventive system is to define the tools to be used for screening and counseling. Valid assessment methods are needed to identify patients who could benefit most from interventions and to evaluate intervention success. Nutritional screening is fundamental for preventive services. The ideal assessment tool needs to be simple, inexpensive, rapidly administered, accurate, and precise. Researchers have developed several tools to measure diet quality attempting to characterize the relationship between diet and health outcomes. These measures of diet quality, or diet quality indices, are based on national guidelines such as the Dietary Guidelines for Americans and the Food Guide Pyramid. However, applying these measures consumes significant time partly because they use relatively complex weighting and calculations, making them difficult to implement in the clinical setting. The Diet Quality Index (DQI) is based on weighting of selected nutrient and food intake recommendations of the Food and Nutrition Board *(30)*. The Healthy Eating Index (HEI) was developed by the US Department of Agriculture to measure adherence to the Dietary Guidelines for Americans and the Food Guide Pyramid *(31)*. It was modified by McCullough et al. to target food choices and macronutrient sources associated with reduced chronic disease risk in clinical and epidemiologic investigations; this index is called Alternate Healthy Eating Index (AHEI) *(32)*. Other diet quality indices have been tailored to specific countries or designed to evaluate prevention efforts for specific diseases and/or subpopulations *(33)*. Specific macronutrient screening tools useful in clinical environments are also available. The Dietary Fat Screener is a brief self-administered screening tool including 17 questions that takes 5 minutes to complete, designed to rank individuals with regard to their usual fat intake. Results can be analyzed using prediction equations to estimate intakes of total and saturated fat and cholesterol and percent of calories from fat. The Fruit/Vegetable/Fiber Screener is a self-administered brief screening tool that includes seven questions about fruit and vegetable intake and three questions about foods high in fiber. It takes about 5 minutes to complete and ranks individuals' usual intakes of fruits and vegetables. These last two tools are commercially available at http://nutritionquest.com/products/questionnairesscreeners.htm

The National Cancer Institute (NCI) Risk Factor Monitoring and Methods Branch (RFMMB) has developed several short instruments that assess intakes of fruits and vegetables, percent of energy intake from fat and/or fiber. These tools have been evaluated in cross-sectional general population studies and are being evaluated currently for use in self-selected samples in intervention research. They have also been used in large population studies such as the National Health Information Survey (NHIS) and the California Health Information Survey (CHIS). They are available at http://riskfactor.cancer.gov/diet/screeners/fruitveg/instrument.html.

Most of these methods are too complicated and time-consuming for routine clinical use, so there is a need for fast, simple, and valid tools for physical activity and food assessment designed for primary care settings *(34)*. Table 32.7 shows some proposed tools.

Among the tools referenced in Table 32.7, the Starting the Conversation (STC) questionnaire is a validated seven-item scale that assesses food patterns, nutrient and fat intake, and provides tailored strategies for behavioral counseling *(35)*. The Summary of Diabetes Self-Care Activities (SDSCA), another validated assessment tool, includes four dietary items that address the most commonly targeted dietary recommendations: lowering fat, increasing fiber, fruit, and vegetable intake. The SDSCA has been found to be sensitive to change in intervention studies with diabetes patients. The questions seem applicable to non-diabetic adults but a full validation in the non-diabetic population is needed *(36,37)*.

The PACE+ project developed a two-item assessment of fruit and vegetable intake for adolescents. These items are consistent with Healthy People 2010 goals and have been validated in a diverse population of middle school-aged adolescents *(38)*. A one-page self-administered food screener that includes the top sources of fat, fruits, and vegetables in the American diet and can be completed and scored in less than 5 minutes has been validated recently against a 100-item food frequency questionnaire by Block et al. Its first 15 item-session is designed to capture dietary fat intake and the second 7 item-session captures fruit, vegetable, fiber, and micronutrient intake *(39)*. MyPyramid Tracker, part of www.MyPyramid.gov, is an online dietary and physical activity assessment tool that provides information on diet quality and physical activity status, and links to nutrient and physical activity information. The food calories/energy balance feature calculates energy balance by subtracting physical activity energy expenditure from energy intake.

32.7.5. Auditing Preventive Nutrition Systems

Another step in implementing a system involves establishing short-term or intermediate goals that can be objectively evaluated. It is helpful to conduct pre- and post-intervention chart audits to determine how well the practice is meeting the standards set. Starting small and adding more services is a prudent approach.

In addition, simple tasks like placing needed preventive services on a reminder note placed on the fronts of patients' charts have been shown to improve the delivery of screening tests, counseling, and immunizations *(40)*. The day before a patient's visit, a staff member reviews the patient's chart and preventive care flow sheet, and records on a "reminder note" the screening or counseling activity needed. Preventive care flow sheets help office staff and clinicians monitor and document counseling provided to patients. These flow sheets can also serve as reminders of which preventive services the practice's setting recommends.

A more universal tool to apply preventive care and recommended by the Clinician's Handbook of Preventive Services of the Agency for Healthcare Research and Quality is the Health Risk Profiles (HRPs, http://www.ahrq.gov/ppip/manual/appendc.htm). These are tools that help office staff and clinicians identify whether age, gender, or health-related behaviors put the patient at risk for cardiovascular disease, diabetes, certain types of cancer, certain infectious diseases, and other conditions. The identification of risk factors guides the clinician in deciding which screenings, immunizations, patient education, and follow-up to provide each patient. Office staff can complete the HRP annually as part of a comprehensive health assessment for all adult patients and more frequently for children. Clinicians can use the HRP to identify and track patients' health risks and behaviors and provide appropriate immunizations, screenings, counseling, or referral for the identified risks and behaviors. After completing the HRP, the preventive care flow sheet should be initiated to document patient education, counseling and referrals, the results of screening tests or examinations, and immunizations given.

Table 32.7
Rapid Preventive Medicine Screening Tools Suitable for Clinical Practice

Physical activity	Rapid Assessment of Physical Activity (RAPA)[37] http://depts.washington.edu/hprc/publications/rapa.htm
	Physical Activity Scale for the Elderly (PASE) http://www.neriscience.com/web/default.asp
	PACE project assessment http://www.sandiegochi.com/pace_written_materials.html
	Summary of Diabetes Self-Care Activities http://care.diabetesjournals.org/cgi/content/abstract/23/7/943
	National Health Interview Survey (NHIS) ftp://ftp.cdc.gov/pub/Health_Statistics/NCHS/Survey_Questionnaires
	Behavioral Risk Factor Surveillance System—BRFSS http://www.cdc.gov/brfss/questionnaires/pdf-ques/2008brfss.pdf
	Guidelines for Adolescent Prevention Services (GAPS) http://www.ama-assn.org/ama/upload/mm/39/gapsmono.pdf
Risky drinking	Five Shot questionnaire http://www.alcoholconcern.org.uk
	Alcohol Use Disorders Identification Test (AUDIT) http://whqlibdoc.who.int/hq/2001/WHO_MSD_MSB01.6a.pdf
	NHIS ftp://ftp.cdc.gov/pub/Health_Statistics/NCHS/Survey_Questionnaires
	BRFSS http://www.cdc.gov/brfss/questionnaires/pdf-ques/2008brfss.pdf
	GAPS http://www.ama-assn.org/ama/upload/mm/39/gapsmono.pdf
Eating patterns	Food diaries or logs
	24-hour dietary recall
	Summary of Diabetes Self-Care Activities http://care.diabetesjournals.org/cgi/content/abstract/23/7/943
	PACE+ http://www.sandiegochi.com/pace_written_materials.html
	Block Brief 2000 Questionnaire http://nutritionquest.com/products/questionnaires_screeners.htm
	"Rate Your Plate" http://www.diabetes.org/all-about-diabetes/chan_eng/i3/i3p4.htm
	NHIS ftp://ftp.cdc.gov/pub/Health_Statistics/NCHS/Survey_Questionnaires
	BRFSS http://www.cdc.gov/brfss/questionnaires/pdf-ques/2008brfss.pdf
	GAPS http://www.ama-assn.org/ama/upload/mm/39/gapsmono.pdf
	Starting the Conversation (STC)[35]http://www.ncpreventionpartners.org

HRPs can be modified to fit specific patient populations' needs. The preventive services listed on the flow sheets should be based on the risks identified by the HRPs. Any risks found on the HRP must be addressed in the appropriate area of the flow sheet.

It is advantageous to use creativity in designing educational materials keeping in mind that patients' education begins when they enter the clinical setting. Using colorful posters and lively messages is an effective way of reinforcing information that patients receive when interacting with office staff. It is also important to change the message formats periodically.

32.8. RECOMMENDATIONS

Since the impact of nutrition on altering risk for chronic disease is generally slow and only gradually evident after lengthy exposure to a given dietary pattern, it is critical to understand that, from the nutrition perspective, much of disease prevention is about making small and important changes in the diets of as many people as possible. Currently, adherence to the Dietary Guidelines is low among the US population and we recommend that a significant effort be undertaken to address this problem. The seriousness of the current epidemic of nutrition-related disorders demands both community-level interventions and clinical preventive services aimed to impact patients' personal health practices. Significantly, evidence supports the use of formal systems to screen for risky dietary patterns and sedentary lifestyle in clinical settings. We support the prioritization of nutrition counseling as a patient-centered collaborative approach. Such collaboration can form the basis for comprehensive and complex behavior change, where the physician first assesses patients' readiness for behavior change, helps address barriers to change, and negotiates goals for lifestyle changes.

ACKNOWLEDGEMENTS

This chapter was adapted, with permission, from Heimburger DC, Ard JD (eds). *Handbook of Clinical Nutrition*. 4th ed. Philadelphia: Mosby Elsevier, 2006. Portions of the material on behavioral counseling were written with support from Nutrition Academic Award grant 1 KO7 HL 03934 from the National Heart, Lung, and Blood Institute, National Institutes of Health, Frank A. Franklin, MD, PhD, Principal Investigator.

GENERAL WEBSITES

American Cancer Society, www.cancer.org
American College of Sports Medicine, http://www.acsm.org/
American Diabetes Association, http://www.diabetes.org
American Heart Association, http://www.americanheart.org
National Cholesterol Education Program, http://www.nhlbi.nih.gov/about/ncep/index.htm
National High Blood Pressure Education Program, http://www.nhlbi.nih.gov/guidelines/hypertension/
Nutrition Data.com, http://www.nutritiondata.com
US DHSS/USDA Dietary Guidelines for Americans 2005, http://www.healthierus.gov/dietaryguidelines/ (will be revised in 2010)
US DHSS/USDA MyPyramid, www.MyPyramid.gov

REFERENCES

1. Jemal A, Siegel R, Ward E, et al. Cancer Statistics 2008. *CA Cancer J Clin* 2008; 58:71–96.
2. Kant AK, Graubard BI, Schatzkin A. Dietary patterns predict mortality in a national cohort: The national health interview surveys, 1987 and 1992. J Nutr 2004; 134:1793–9
3. Frazao, E. The high costs of poor eating patterns in the United States. In: Frazao, E., ed. America's Eating Habits: Changes and Consequences. Washington, DC: U.S. Department of Agriculture (USDA), Economic Research Service (ERS), AIB-750, 1999.
4. McGinnis JM, Foege WH. Actual causes of death in the United States. JAMA 1993; 270:2207–12
5. Wells HF, Jean CB. Dietary Assessment of Major Trends in U.S. Food Consumption, 1970–2005, Economic Information Bulletin No. 33. Economic Research Service, U.S. Dept. of Agriculture. March 2008.
6. U.S. Department of Health and Human Services and U.S. Department of Agriculture. *Dietary Guidelines for Americans, 2005*. 6th Edition, Washington, DC: U.S. Government Printing Office, January 2005. http://www.healthierus.gov/dietaryguidelines/
7. U.S. Department of Health and Human Services. Healthy People 2010. 2nd ed. With Understanding and Improving Health and Objectives for Improving Health. 2 vols. Washington, DC: U.S. Government Printing Office, November 2000. http://www.healthypeople.gov/Publications/
8. Appel LJ, Moore TJ, Obarganek E, et al. A clinical trial of the effects of dietary patterns on blood pressure. DASH Collaborative Research Group. *N Engl J Med* 1997; 336:117–24.
9. Kant AK, Graubard BI. Eating out in America, 1987–2000: Trends and nutritional correlates. *Prev Med* 2004; 38:243–9.
10. Cavadini C, Siega-Riz AM, et al. US adolescent food intake trends from 1965 to 1996. *Arch Dis Child* 2000; 83:18–24.
11. McCrory MA, Fuss PJ, McCallum JE, Yao M, Vinken AG, Hays NP, Roberts SB. Dietary variety within food groups: association with energy intake and body fatness in men and women. *Am J Clin Nutr* 1999; 69:440–7.
12. Food and Nutrition Board, Institute of Medicine: Dietary Reference Intakes for energy, carbohydrate, fiber, fat, fatty acids, cholesterol, protein, and amino acids (macronutrients), Washington, DC: Institute of Medicine, National Academy Press, 2002. (www.nap.edu/books/0309085373/html/index.html).
13. Rolls BJ, Ello-Martin JA, et al: What can intervention studies tell us about the relationship between fruit and vegetable consumption and weight management? *Nutr Rev* 2004; 2:1–17.
14. Kant AK, Graubard BI, Schatzkin A. Dietary patterns predict mortality in a national cohort: The national health interview surveys, 1987 and 1992. *J Nutr* 2004; 134:1793–9.
15. Cherry, D.K. National Health Statistics Report Number 3, "National Ambulatory Medical Care Survey: 2006 Summary," August 6, 2008.
16. Kushner R. Barriers to providing nutrition counseling by physicians: a survey of primary care practitioners. *Prev Med* 1995; 24:546–52.
17. Cornuz J, Ghali WA, Di Carlantonio D, Pecoud A, Paccaud F. Physicians' attitudes towards prevention: importance of intervention-specific barriers and physicians' health habits. *Fam Pract* 2000; 17:535–40.
18. Eaton CB, Goodwin MA, Stange KC. Direct observation of nutrition counseling in community family practice. *Am J Prev Med* 2002; 23:174–9.
19. Ockene IS, Hebert JR, Ockene JK, et al. Effect of physician-delivered nutrition counseling training and an office-support program on saturated fat intake, weight, and serum lipid measurements in a hyperlipidemic population: Worcester Area Trial for Counseling Hyperlipidemia (WATCH). *Arch Intern Med.* 1999;159:725–31.
20. Kushner R: Barriers to providing nutrition counseling by physicians: a survey of primary care practitioners. *Prev Med* 1995; 24:546–52.
21. Whitlock EP, Orleans CT, Pender N, Allan J. Evaluating primary care behavioral counseling interventions: an evidence-based approach. *Am J Prev Med* 2002; 22:267–84.
22. Absetz P, Valve R, Oldenburg B, Heinonen H, Nissinen A, Fogelholm M, Ilvesmäki V, Talja M, Uutela A. Type 2 diabetes prevention in the "real world": one-year results of the GOAL Implementation Trial. *Diabetes Care* 2007; 30:2465–70.
23. Riebe D, Greene GW, Ruggiero L, Stillwell KM, Blissmer B, Nigg CR, Caldwell M. Evaluation of a healthy-lifestyle approach to weight management. *Prev Med* 2003; 36:45–54.
24. Glasgow RE, Goldstein MG, Ockene JK, Pronk NP. Translating what we have learned into practice. Principles and hypotheses for interventions addressing multiple behaviors in primary care. *Am J Prev Med* 2004; 27(2 Suppl):88–101.
25. Gottlieb NH, Huang PP, Blozis SA, Murphy, Smith M. The impact of *Put Prevention into Practice* on selected clinical preventive services in five Texas sites. *Am J Prev Med* 2001; 21:35–40.
26. Schwartz RP, Hamre R, Dietz WH, Wasserman RC, Slora EJ, Myers EF, Sullivan S, Rockett H, Thoma KA, Dumitru G, Resnicow KA. Office-based motivational interviewing to prevent childhood obesity: a feasibility study. *Arch Pediatr Adolesc Med* 2007; 161:495–501.

27. Werch CE, Bian H, Moore MJ, Ames S, DiClemente CC, Weiler RM. Brief multiple behavior interventions in a college student health care clinic. *J Adolesc Health* 2007; 41:577–85.
28. Calfas KJ, Long BJ, Sallis JF, et al. A controlled trial of physician counseling to promote the adoption of physical activity. *Prev Med* 1996; 25:225–33.
29. Guide to Clinical Preventive Services, 2007. AHRQ Publication No. 07-05100. Agency for Healthcare Research and Quality, Rockville, MD. http://www.ahrq.gov/clinic/pocketgd07.
30. Patterson RE, Haines PS, Popkin BM. Diet quality index: capturing a multidimensional behavior. *J Am Diet Assoc* 1994; 94:57–64.
31. Kennedy ET, Ohls J, Carlson S, Fleming K. The Healthy Eating Index: design and applications. *J Am Diet Assoc* 1995; 95:1103–8.
32. McCullough ML, Feskanich D, Stampfer MJ, Giovannucci EL, Rimm EB, Hu FB, Spiegelman D, Hunter DJ, Colditz GA, Willett WC. Diet quality and major chronic disease risk in men and women: moving toward improved dietary guidance. *Am J Clin Nutr* 2002; 76:1261–71.
33. Fransen HP, Ocké MC. Indices of diet quality. *Curr Opin Clin Nutr Metab Care* 2008; 11:559–65.
34. Glasgow RE, Ory MG, Klesges LM, Cifuentes M, Fernald DH, Green LA. Practical and relevant self-report measures of patient health behaviors for primary care research. *Ann Fam Med* 2005; 3:73–81.
35. Keyserling TC, Samuel Hodge CD, Jilcott SB, Johnston LF, Garcia BA, Gizlice Z, Gross MD, Saviñon CE, Bangdiwala SI, Will JC, Farris RP, Trost S, Ammerman AS. Randomized trial of a clinic-based, community-supported, lifestyle intervention to improve physical activity and diet: the North Carolina enhanced WISEWOMAN project. *Prev Med* 2008; 46:499–510.
36. Ammerman A, Haines P, DeVellis R, Strogatz D, Keyserling T, Simpson R, Siscovick D. A brief dietary assessment to guide cholesterol reduction in low-income individuals: design and validation. *J Am Diet Assoc* 1991; 91:1385–90.
37. Topolski TD, LoGerfo J, Patrick DL, Williams B, Walwick J, Patrick MB. The Rapid Assessment of Physical Activity (RAPA) among older adults. *Prev Chronic Dis* 2006; 3(4):A118.
38. Svilaas A, Ström EC, Svilaas T, Borgejordet A, Thoresen M, Ose L. Reproducibility and validity of a short food questionnaire for the assessment of dietary habits. *Nutr Metab Cardiovasc Dis* 2002; 12:60–70.
39. Block G., Gillespie C., Rosenbaum E. H., Jenson C. A rapid food screener to assess fat and fruit and vegetable intake. *Am J Prev Med* 2000; 18:284–8.
40. Lannon CM, Flower K, Duncan P, Moore KS, Stuart J, Bassewitz J. The Bright Futures Training Intervention Project: implementing systems to support preventive and developmental services in practice. *Pediatrics* 2008; 122:163–71.

Appendix A: Table of Contents from Volumes 1 to 3 and Primary and Secondary Preventive Nutrition

Foreword, by Charles H. Hennekens and JoAnn E. Manson v

Preface vii

Cotributors xvii

PART I: PUBLIC HEALTH IMPLICATIONS OF PREVENTIVE NUTRITION

A CANCER PREVENTION

1 Public Health Implications of Preventive Nutrition 1
Jeffrey B. Blumberg

2 Diet and Childhood Cancer: *Preliminary Evidence* 17
Greta R. Bunin and Joan M. Cary

3 Prevention of Upper Gastrointestinal Tract Cancers 33
Elizabeth T. H. Fontham

4 Diet and Nutrition in the Etiology and Primary Prevention of Colon Cancer 57
Roberd M. Bostick

5 Nutrition and Breast Cancer 97
Geoffrey R. Howe

6 Preventive Nutrition and Lung Cancer 109
George W. Comstock and Kathy J. Helzlsouer

7 Nonnutritive Components in Foods as Modifiers of the Cancer Process 135
John A. Milner

B CARDIOVASCULAR DISEASE

8 Dietary Fat and Coronary Heart Disease 153
K. C. Hayes

9 Antioxidant Vitamins and Cardiovascular Disease 171
Julie E. Buring and J. Michael Gaziano

10 Iron and Heart Disease: *A Review of the Epidemiologic Data* 181
Christopher T. Sempos, Richard F. Gillum, and Anne Condon Looker

A. Bendich, R.J. Deckelbaum (eds.), *Preventive Nutrition*, Nutrition and Health, DOI 10.1007/978-1-60327-542-2,

11 Homocysteine, Folic Acid, and Cardiovascular Disease Risk 193
Shirley A. A. Beresford and Carol J. Boushey

12 Omega-3 Fatty Acids from Fish: *Primary and Secondary Prevention of Cardiovascular Disease* 225
William E. Connor and Sonja L. Connor

PART II PREVENTION OF MAJOR DISABILITIES

A CATARACTS AND MACULAR DEGENERATION

13 The Relationship Between Nutritional Factors and Age-Related Macular Degeneration 245
Shirley Hung and Johanna M. Seddon

14 Antioxidant Status and Risk for Cataract 267
Allen Taylor and Paul F. Jacques

B OSTEOPOROSIS

15 Osteoporosis: *Vitamins, Minerals, and Other Micronutrients* 285
Robert P. Heaney

C IMMUNE FUNCTION

16 Antioxidant Nutrients and Prevention of Oxidant-Mediated, Smoking-Related Diseases 303
Ronald Anderson

17 Micronutrients and Immunity in Older People 317
John D. Bogden and Donald B. Louria

18 Impact of Vitamin A on Immunity and Infection in Developing Countries 337
Richard D. Semba

PART III OPTIMAL BIRTH OUTCOMES

19 Folic Acid-Containing Multivitamins and Primary Prevention of Birth Defects 351
Andrew E. Czeizel

20 Nutritional Prevention of DNA Damage to Sperm and Consequent Risk Reduction in Birth Defects and Cancer in Offspring 373
Alan A. Woodall and Bruce N. Ames

21 Maternal Nutrition and Preterm Delivery 387
Theresa O. School and Mary L. Hediger

22 Dietary Polyunsaturated Fatty Acids for Optimal Neurodevelopment: *Recommendations for Perinatal Nutrition* 405
Ricardo Uauy-Dagach, Patricia Mena, and Patricio Peirano

PART IV BENEFITS OF PREVENTIVE NUTRITION IN THE UNITED STATES AND EUROPE

23 Potential Benefits of Preventive Nutrition Strategies: *Lessons for the United States* 423
Walter C. Willett

24 Nutrient Addition to Foods: *The Public Health Impact in Countries with Rapidly Westernizing Diets* 441
Paul A. Lachance

25 Nutrition and Food Policy in Norway: *Effects on Reduction of Coronary Heart Disease* 455
Kaare R. Norum, Lars Johansson, Grete Botten, Gunn-Elin Aa. Bjørneboe, and Arne Oshaug

26 Prevention of Pediatric Obesity: *Examining the Issues and Forecasting Research Directions* 471
Myles S. Faith, Angelo Pietrobelli, David B. Allison, and Steven B. Heymsfield

PART V IMPLICATIONS OF PREVENTIVE NUTRITION FOR THE FAR EAST, SOUTH AMERICA, AND DEVELOPING AREAS

27 Effect of Westernization of Nutritional Habits on Obesity in Latin America: *Recommendations for the Region* 487
S. Jaime Rozowski and Manuel Moreno

28 Prevention of Malnutrition in Chile 505
Fernando Mönckeberg

29 Effects of Western Diet on Risk Factors of Chronic Diseases in Asia 523
Kaichi Kida, Takuo Ito, Sei Won Yang, and Vichai Tanphaichitr

30 Goals for Preventive Nutrition in Developing Countries 535
Osman M. Galal and Gail G. Harrison

PART VI NUTRITION-RELATED RESOURCES

A Books Related to Preventive Nutrition 549
B Nutrition-Related Journals and Newsletters 551
Index 559

Table of Contents of Second Edition

Foreword ... vii
Preface ... ix
Cotributors ... xv

Part I Cancer Prevention

1 Diet and Childhood Cancer: *Preliminary Evidence* ... 3
Greta R. Bunin and Joan M. Cary

2 Prevention of Upper Gastrointestinal Tract Cancers ... 21
Elizabeth T. H. Fontham

3 Diet and Nutrition in the Etiology and Primary Prevention of Colon Cancer ... 47
Roberd M. Bostick

4 Preventive Nutrition and Lung Cancer ... 97
George W. Comstock and Kathy J. Helzlsouer

5 Nonnutritive Components in Foods as Modifiers of the Cancer Process ... 131
John A. Milner

Part II CARDIOVASCULAR DISEASE

6 Dietary Fat and Coronary Heart Disease ... 157
K. C. Hayes

7 Iron and Heart Disease: *A Review of the Epidemiologic Data* ... 175
Christopher T. Sempos, Richard F. Gillum, and Anne C. Looker

8 Homocysteine, Folic Acid, and Cardiovascular Disease Risk ... 191
Shirley A. A. Beresford and Carol J. Boushey

9 n-3 Fatty Acids from Fish and Plants: *Primary and Secondary Prevention of Cardiovascular Disease* ... 221
William E. Connor and Sonja L. Connor

Part III PREVENTION OF MAJOR DISABILITIES: IMPROVEMENT IN HEALTH OUTCOMES

10 The Relationship Between Nutritional Factors and Age-Related Macular Degeneration ... 247
Shirley Hung and Johanna Seddon

11 Osteoporosis: *Minerals, Vitamins, and Other Micronutrients*...271
Robert P. Heaney

12 Antioxidant Nutrients and Prevention of Oxidant-Mediated Diseases...293
Ronald Anderson

13 Micronutrients and Immunity in Older People...307
John D. Bogden and Donald B. Louria

14 Impact of Vitamin A on Immunity and Infection in Developing Countries...329
Richard D. Semba

PART IV OPTIMAL PREGNANCY/INFANCY OUTCOMES

15 Folic Acid-Containing Multivitamins and Primary Prevention of Birth Defects...349
Andrew E. Czeizel

16 DNA Damage to Sperm from Micronutrient Deficiency May Increase the Risk of Birth Defects and Cancer in Offspring...373
Craig A. Mayr, Alan A. Woodall, and Bruce N. Ames

17 Maternal Nutrition and Preterm Delivery...387
Theresa O. Scholl

18 Dietary Polyunsaturated Fatty Acids for Optimal Neurodevelopment: *Recommendations for Perinatal Nutrition*...415
Ricardo Uauy, Patricia Mena, and Patricio Peirano

PART V GLOBAL PREVENTIVE NUTRITION STRATEGIES

19 Effects of Western Diet on Risk Factors of Chronic Diseases in Asia...435
Kaichi Kida, Takuo Ito, Sei Won Yang, and Vichai Tanphaichitr

20 Potential Benefits of Preventive Nutrition Strategies: *Lessons for the United States*...447
Walter C. Willett

21 Goals for Preventive Nutrition in Developing Countries...465
Osman M. Galal and Gail G. Harrison

PART VI NUTRITION-RELATED RESOURCES

Books Related to Preventive Nutrition...479
Websites of Interest...483
Index...489

Table of Contents of 3rd Edition

Series Editor's Introduction vii

Foreword xi

Preface xiii

Cotributors xxi

Value-Added ebook/PDA xxv

PART I PREVENTIVE NUTRITION OVERVIEW

1 Preventive Nutrition: *A Historic Perspective and Future Economic Outlook* 3
Nancy D. Ernst and J. Michael McGinnis

PART II CANCER PREVENTION

2 Prevention of Cancers of the Esophagus and Stomach 25
Elizabeth T. H. Fontham and L. Joseph Su

3 Non-Nutritive Components in Foods as Modifiers of the Cancer Process 55
Keith W. Singletary, Steven J. T. Jackson, and John A. Milner

4 Dietary Supplements and Cancer Risk: *Epidemiological Research and Recommendations* 89
Marian L. Neuhouser, Ruth E. Patterson, Alan R. Kristal, and Emily White

5 Soy Consumption and Cancer Prevention: *A Critical Review* 123
Jin-Rong Zhou and John W. Erdman, Jr.

6 Tomato, Lycopene, and Prostate Cancer 157
Jessica K. Campbell and John W. Erdman, Jr.

PART III CARDIOVASCULAR DISEASE PREVENTION

7 Iron and Heart Disease: *A Review of the Epidemiological Data.* 173
Christopher T. Sempos, Richard F. Gillum, and Anne C. Looker

8 Homocysteine, Folic Acid, and Cardiovascular Disease Risk 191
Shirley A. A. Beresford and Arno G. Motulsky

9 n-3 Fatty Acids from Fish and Plants: *Primary and Secondary Prevention of Cardiovascular Disease* 221
William E. Connor and Sonja L. Connor

10 Antioxidant Vitamin Supplementation and Cardiovascular Disease 245
Howard N. Hodis, Wendy J. Mack, and Alex Sevanian

11 Health Effects of Trans Fatty Acids 279
Ronald P. Mensink and Susanne H. F. Vermunt

PART IV DIABETES AND OBESITY

12 Obesity and Insulin Resistance in Childhood and Adolescence 293
Erik Bergström and Olle Hernell

13 Prevention of Pediatric Obesity: *Examining the Issues and Forecasting Research Directions* 321
Myles S. Faith, Christina J. Calamaro, Angelo Pietrobelli, Meredith S. Dolan, David B. Allison, and Steven B. Heymsfield

14 Can Childhood Obesity Be Prevented?: *Preschool Nutrition and Obesity*.................. 345
Christine L. Williams

15 Obesity and Chronic Disease: *Impact of Weight Reduction* 383
Henry I. Frier and Harry L. Greene

PART V BONE DISEASES

16 Osteoarthritis: *Role of Nutrition and Dietary Supplement Interventions*.................... 405
Timothy E. McAlindon

17 Calcium Requirements During Treatment of Osteoporosis in Women....................... 425
Richard Eastell

18 Osteoporosis: *Protein, Minerals, Vitamins, and Other Micronutrients* 433
Robert P. Heaney

PART VI: PREVENTION OF MAJOR DISABILITIES: *Improvement in Health Outcomes*

19 Antioxidant Status and Risk for Cataract 463
Mark Siegal, Chung-Jung Chiu, and Allen Taylor

20 Antioxidant Nutrients and Prevention of Oxidant-Mediated Diseases....................... 505
Ronald Anderson

21 Nutritional Supplements and Upper Respiratory Tract Illnesses in Young Children in the United States 521
Linda A. Linday

22 Micronutrients and Immunity in Older People... 551
John D. Bogden and Donald B. Louria

23 Vitamin A and the Prevention of Morbidity, Mortality, and Blindness......................573
Richard D. Semba

Part VII Optimal Pregnancy/Infancy Outcomes

24 Folic Acid-Containing Multivitamins and Primary Prevention of Birth Defects.............603
Andrew E. Czeizel

25 Maternal Nutrition and Preterm Delivery..629
Theresa O. Scholl

26 Dietary Polyunsaturated Fatty Acids for Optimal Neurodevelopment: *Recommendations for Perinatal Nutrition*...665
Ricardo Uauy, Patricia Mena, Adolfo Llanos, and Patricio Peirano

27 Micronutrient Deficiencies and Maternal Thinness: *First Chain in the Sequence of Nutritional and Health Events in Economic Crises*..................................689
Martin W. Bloem, Saskia de Pee, and Ian Darnton-Hill

Part VIII: Preventive Nutrition: Global Perspectives

28 Potential Benefits of Preventive Nutrition Strategies: *Lessons for the United States*.........713
Walter C. Willett

29 Nutrition and Food Policy in Norway: *Effects on Reductionof Coronary Heart Disease*.....735
Kaare R. Norum, Gunn-Elin A. Bjørneboe, Arne Oshaug, Grete Botten, and Lars Johansson

30 Prevention of Malnutrition in Chile..753
Fernando Mönckeberg Barros

31 Effect of Westernization of Nutritional Habits on Obesity Prevalence in Latin America: *Analysis and Recommendations*...................................771
Jaime Rozowski, Oscar Castillo, and Manuel Moreno

32 Effects of Western Diet on Risk Factors of Chronic Diseases in Asia......................791
Kaichi Kida, Koji Takemoto, Sei Won Yang, and Supawadee Likitmaskul

Part IX: Critical Issues for the 21st Century

33 Alcohol: *The Balancing Act*..807
William E. M. Lands

34 Influence of Medication on Nutritional Status...833
Joseph I. Boullata

35 Health Claims for Foods and Dietary Supplements: *Current and Emerging Policies*.........869
Annette Dickinson

36 Teaching Preventive Nutrition in Medical Schools 889
Martin Kohlmeier and Steven H. Zeisel

37 Preventive Nutrition Throughout the Life Cycle: *A Cost-Effective Approach to Improved Health*.. 901
Adrianne Bendich and Richard J. Deckelbaum

Appendix A: Related Readings.. 923
Appendix B: Web Sites of Interest.. 927
Index.. 931

Appendix B: Nutrition and Health Series

1. Fluid and Electrolyte Disorders in Pediatrics, edited by Leonard G. Feld and Frederick Kaskel, 2010
2. Iron Deficiency and Overload: From Basic Biology to Clinical Practice, edited by Shlomo Yehuda and David Mostofsky, 2009
3. Handbook of Drug-Nutrient Interactions, Second Edition, edited by Dr. Joseph Boullata and Dr. Vincent Armenti, 2009
4. Handbook of Clinical Nutrition and Aging, Second Edition, edited by Dr. Connie Bales and Dr. Christine Ritchie, 2009
5. Probiotics in Pediatric Medicine, edited by Sonia Michail and Philip M. Sherman, 2009
6. Handbook of Nutrition and Pregnancy, edited by Carol J. Lammi-Keefe, Sarah Collins Couch and Elliot H. Philipson, 2008
7. Nutrition and Health in Developing Countries, Second Edition, edited by Richard D. Semba and Martin W. Bloem, 2008
8. Nutrition and Rheumatic Disease, edited by Laura A. Coleman, 2008
9. Nutrition in Kidney Disease, edited by Laura D. Byham-Gray, Jerrilynn D. Burrowes, and Glenn M. Chertow, 2008
10. Handbook of Nutrition and Ophthalmology, edited by Richard D. Semba, 2007
11. Adipose Tissue and Adipokines in Health and Disease, edited by Giamila Fantuzzi and Theodore Mazzone, 2007
12. Nutritional Health: Strategies for Disease Prevention, Second Edition, edited by Norman J. Temple, Ted Wilson, and David R. Jacobs, Jr., 2006
13. Nutrients, Stress, and Medical Disorders, edited by Shlomo Yehuda and David I. Mostofsky, 2006
14. Calcium in Human Health, edited by Connie M. Weaver and Robert P. Heaney, 2006
15. Preventive Nutrition: The Comprehensive Guide for Health Professionals, Third Edition, edited by Adrianne Bendich and Richard J. Deckelbaum, 2005
16. The Management of Eating Disorders and Obesity, Second Edition, edited by David J. Goldstein, 2005
17. Nutrition and Oral Medicine, edited by Riva Touger-Decker, David A. Sirois, and Connie C. Mobley, 2005
18. IGF and Nutrition in Health and Disease, edited by M. Sue Houston, Jeffrey M. P. Holly, and Eva L. Feldman, 2005
19. Epilepsy and the Ketogenic Diet, edited by Carl E. Stafstrom and Jong M. Rho, 2004
20. Handbook of Drug–Nutrient Interactions, edited by Joseph I. Boullata and Vincent T. Armenti, 2004
21. Nutrition and Bone Health, edited by Michael F. Holick and Bess Dawson-Hughes, 2004
22. Diet and Human Immune Function, edited by David A. Hughes, L. Gail Darlington, and Adrianne Bendich, 2004
23. Beverages in Nutrition and Health, edited by Ted Wilson and Norman J. Temple, 2004
24. Handbook of Clinical Nutrition and Aging, edited by Connie Watkins Bales and Christine Seel Ritchie, 2004
25. Fatty Acids: Physiological and Behavioral Functions, edited by David I. Mostofsky, Shlomo Yehuda, and Norman Salem, Jr., 2001

26. Nutrition and Health in Developing Countries, edited by Richard D. Semba and Martin W. Bloem, 2001
27. Preventive Nutrition: The Comprehensive Guide for Health Professionals, Second Edition, edited by Adrianne Bendich and Richard J. Deckelbaum, 2001
28. Nutritional Health: Strategies for Disease Prevention, edited by Ted Wilson and Norman J. Temple, 2001
29. Clinical Nutrition of the Essential Trace Elements and Minerals: The Guide for Health Professionals, edited by John D. Bogden and Leslie M. Klevay, 2000
30. Primary and Secondary Preventive Nutrition, edited by Adrianne Bendich and Richard J. Deckelbaum, 2000
31. The Management of Eating Disorders and Obesity, edited by David J. Goldstein, 1999
32. Vitamin D: Physiology, Molecular Biology, and Clinical Applications, edited by Michael F. Holick, 1999
33. Preventive Nutrition: The Comprehensive Guide for Health Professionals, edited by Adrianne Bendich and Richard J. Deckelbaum, 1997

Appendix C: Books Related to Preventive Nutrition

1. American Academy of Pediatrics Staff. 1998. *Pediatric Nutrition Handbook, 4th edition.* Elk Grove Village, IL: American Academy of Pediatrics.
2. Bales, C.W., Ritchie, C.S. 2003. *Handbook of Clinical Nutrition and Aging.* Totowa, NJ: Humana Press.
3. Barker, D. J. P., Bergmann, R. L., Ogra, P. L. (eds.) 2008. *The Window of Opportunity: Pre-pregnancy to 24 Months of Age.* Unionville, CT: Karger Publishers, Inc.
4. Bendich, A.B., Deckelbaum, R.J. 2001. *Preventive Nutrition: The Comprehensive Guide for Health Professionals, 2nd Edition.* Totowa, NJ: Humana Press.
5. Bendich, A.B., Deckelbaum, R.J. 2000. *Primary and Secondary Preventive Nutrition.* Totowa, NJ: Humana Press.
6. Bendich, A.B., Deckelbaum, R.J. 2005. Pre*ventive Nutrition: The Comprehensive Guide for Health Professionals, 3rd Edition.* Totowa, NJ: Humana Press.
7. Bendich, A.B., Deckelbaum, R.J. (eds.) 1997. *Preventive Nutrition: The Comprehensive Guide for Health Professionals.* Totowa, NJ: Humana Press.
8. Berdanier, C.D. 1998. *CRC Desk Reference for Nutrition.* Boca Raton, FL: CRC Press.
9. Berdanier, C.D., Failla, M.L. 1998. *Advanced Nutrition: Micronutrients.* Boca Raton, FL: CRC Press.
10. Blaylock, R.L. 2002. *Health and Nutrition Secrets That Can Save Your Life.* Health Press (NM).
11. Bogden, J.D., Klevay, L.M. 2000. *Clinical Nutrition of the Essential Trace Elements and Minerals: The Guide for Health Professionals (Nutrition and Health).* Totowa, NJ: Humana Press.
12. Boullata, J., Armenti, V.T. 2004. *Handbook of Drug–Nutrient Interactions.* Totowa, NJ: Humana Press.
13. Bray, G.A., Ryan, D.H. (eds.) 1999. *Nutrition, Genetics, and Obesity,* Pennington Center Nutrition Series Volume 9. Baton Rouge, LA: Louisiana State University Press.
14. Brody, T. 1998. *Nutritional Biochemistry, 2nd edition.* San Diego, CA: Academic Press.
15. Burckhardt, P., Dawson-Hughes, B., Heaney, R.P. July 9, 2004. *Nutritional Aspects of Osteoporosis.* Academic Press, 2nd Edition.
16. Chernoff, R. 1999. *Geriatric Nutrition: The Health Professional's Handbook, 2nd Edition.* Gaithersburgh, MD: Aspen Publishers, Inc.
17. Cho, S.S., Prosky, L., Dreher, M. (eds.). 1999. *Complex Carbohydrates in Foods.* New York: Marcel Dekker, Inc.
18. Chow, C.K. (ed.) 1999. *Fatty Acids in Foods and Their Health Implications, 2nd Edition.* New York: Marcel Dekker, Inc.
19. Clifford, A.J., Muller, H-G. (eds.) 1998. *Mathematical Modeling in Experimental Nutritional.* New York: Plenum.
20. Combs, G.F. Jr., 1998. *Vitamins: Fundamental Aspects in Nutrition and Health, 2nd edition.* San Diego, CA: Academic Press.
21. Coulston, A.M., Rock, C.L., Monsen, E.R. (eds.) 2001. *Nutrition in the Prevention and Treatment of Disease.* Academic Press, 1st Edition.
22. Driskell, J. 1999. *Sports Nutrition.* Boca Raton, FL: CRC Press.
23. Duyff, R.L. 2002. *American Dietetic Association Complete Food and Nutrition Guide.* Wiley, 2nd Edition.

24. Eitenmiller, R.R., Landen, W.O., Jr. (eds.) 1999. *Vitamin Analysis for the Health and Food Sciences.* Boca Ranton, FL: CRC Press.
25. El-Khoury, A.E. (ed.) 1999. *Methods for Investigation of Amino Acids and Protein Metabolism.* Boca Raton, FL: CRC Press LLC.
26. Escott-Stump, S. 1998. *Nutrition and Diagnosis-Related Care, 4th Edition.* Hagerstown, MD: Lippincott Williams & Wilkins.
27. Frost, G., Dornhorst, A., Moses, R. (eds.) 2003. *Nutritional Management of Diabetes Mellitus (Practical Diabetes).* John Wiley & Sons.
28. Gershwin, M., Keen, C.L., German, J.B. (eds.) 1999. *Nutrition and Immunology: Principles and Practice.* Totowa, NJ: Humana Press.
29. Gibson, G.R., Roberfroid, M.B. 1999. *Colonic Microbiota, Nutrition and Health.* Boston, MA: Kluwer Academic.
30. Goldstein, D.J. (ed) 1999. *The Management of Eating Disorders and Obesity.* Totowa, NJ: Humana Press.
31. Goldstein, D.J. (ed), 2005. M*anagement of Eating Disorders and Obesity*, *2nd Edition.* Totowa, NJ: Humana Press.
32. Goldstein, M.C., Goldstein, M.A. 2002. *Controversies in Food and Nutrition.* Greenwood Press.
33. Holick, M.F., Dawson-Hughes, B. (ed), 2004. *Nutrition and Bone Health.* Totowa, NJ: Humana Press.
34. Holick, M.F. (ed), 1998. *Vitamin D: Physiology, Molecular Biology, and Clinical Applications.* Totowa, NJ: Humana Press.
35. Houston, S.M., Holly, J.M., Feldman, E.L. (ed), 2004. *IGF and Nutrition in Health and Disease (Nutrition and Health Ser).* Totowa, NJ: Humana Press.
36. Huang, Y. Sinclair, A. (ed), 1998. *Lipids in Infant Nutrition.* Champaign, IL: AOCS Press.
37. Hughes, D.A., Darlington, G.L., Bendich, A. (ed),. 2003. *Diet and Human Immune Function.* Totowa, NJ: Humana Press.
38. Kessler, D.B., Dawson, P. (eds.) 1999. *Failure to Thrive and Pediatric Undernutrition: A Transdisciplinary Approach.* Baltimore, MD: Paul H. Brookes Publishing.
39. Lammi-Keefe, C. J., Couch, S.C., Philipson, E. H. (eds.) 2008. *Handbook of Nutrition and Pregnancy.* New York, NY: Humana Press/Springer.
40. Mann, J., Truswell, A.S., Truswell, S. (eds.) 1998. *Essentials of Human Nutrition.* New York: Oxford University Press.
41. Mann, J., Truswell, S. 2002. *Essentials of Human Nutrition.* Oxford University Press, 1st Edition.
42. Mcardle, W.D., Katch, F.I., Katch, V.L. 1999. *Sports and Exercise Nutrition.* Philadelphia, PA: Williams & Wilkins.
43. McCormick, D.B. (ed). 1998. *Annual Review of Nutrition.* Palo Alto, CA: Annual Reviews Inc.
44. Miller, G.D., Jarvis, J.K., McBean, L.D. (ed), 1999. *Handbook of Dairy Foods and Nutrition, 2nd Edition.* Boca Raton, FL: CRC Press.
45. Miller, T.L., Gorbach, S.L. (eds.) 1999. *Nutrition Aspects of HIV Infection.* New York: Oxford University Press.
46. Morrison, G., Hark, L. 1999. *Medical Nutrition and Disease, 2nd Edition.* Malden, MA: Blackwell Science, Inc.
47. Morrow, K.D. 2004. *The Color of Nutrition: Create natural balance for our health, wellbeing, and optimal weight.* IUniverse, Inc.
48. Mostofsky, D.I., Yehuda, S., Salem, N. Jr. (ed), 2001. *Fatty Acids: Physiological and Behavioral Functions.* Totowa, NJ: Humana Press.
49. Nelson, M.E., Knipe, J. 2002. *Strong Women Eat Well: Nutritional Strategies for a Healthy Body and Mind.* Perigee Books.
50. Nieman, D.C., Pedersen, B.K. (eds.) 2000. *Nutrition and Exercise Immunology.* Boca Raton, FL: CRC Press.

51. Oberleas, D., Harland, B.F., Bobilya D.J. 1999. *Minerals: Nutrition and Metabolism.* New York: Vantage Press.
52. O'Dell, B.L., Sunde, R.A. 1997. *Handbook of Nutritionally Essential Mineral Elements, Vol. 2.* New York: Marcel Dekker Inc.
53. Papas, A. (ed). 1998. *Antioxidant Status, Diet, Nutrition, and Health.* Boca Raton, FL: CRC Press.
54. Peckenpaugh, N.J., Poleman, C.M. 1999. *Nutrition Essentials and Diet Therapy, 8th Edition.* Philadelphia, PA: WB Saunders Company.
55. Pence, B.C., Dunn, D.M. 1998. *Nutrition and Women's Cancers.* Boca Raton, FL: CRC Press.
56. Pennington, J.A.T., Bowes, A.D., Church, H. 1998. *Bowes and Church's Food Values of Portions Commonly Used, 17th Edition.* Philadelphia, PA: Lippincott.
57. Rugg-Gunn, A.J., Nunn, J.H. 1999. *Nutrition, Diet, and Oral Health.* Oxford [England]: Oxford University Press.
58. Sadler, M.J., Strain, J.J., Caballero, B. (eds.) 1998. *Encyclopedia of Human Nutrition.* San Diego, CA: Academic Press Inc.
59. Semba, R.D., Bloem, M.W. (ed), 2001. *Nutrition and Health in Developing Countries.* Totowa, NJ: Humana Press.
60. Shils, M., Olson, J.A., Shike, M. Ross, A.C. (eds.). 1999. *Modern Nutrition in Health and Disease, 9th edition.* Baltimore, MD: Williams & Wilkins.
61. Simopoulos, A.P. (ed.) 1998. *The Return of ω3 Fatty Acids into the Food Supply. I. Land-Based Animal Food Products and Their Health Effects.* Basel, Switzerland: Karger, S., AG.
62. Spallholz, J.E., Mallory Boylan, L., Driskell, J.A. 1999. *Nutrition: Chemistry and Biology, 2nd Edition.* Boca Raton, FL: CRC Press LLC.
63. Stafstrom, C.E., Rho, J.M. (ed), 2004. *Epilepsy and the Ketogenic Diet.* Totowa, NJ: Humana Press.
64. Stipanuk, M. (ed) 1999. *Biochemical and Physiological Aspects of Human Nutrition.* Philadelphia, PA: Saunders, W.B.
65. Taylor, A. (ed.) 1999. *Nutritional and Environmental Influences on the Eye.* Boca Raton, FL: CRC Press LLC.
66. Tarnopolsky, M. (ed.) 1999. *Gender Differences in Metabolism: Practical and Nutritional Implications.* Boca Raton, FL: CRC Press.
67. Touger-Decker, R., Sirois, D.A., Mobley, C.C. (ed), 2004. *Nutrition and Oral Medicine.* Totowa, NJ: Humana Press.
68. Veith, W.J. (ed.) 1999. *Diet and Health, 2nd Edition.* Boca Raton, FL: CRC Press.
69. Watson, R.R. (ed.) 1998. *Nutrients and Foods in AIDS.* Boca Raton, FL: CRC Press.
70. Watson, R.R., Preedy, V.R. 2003. *Nutrition and Heart Disease: Causation and Prevention.* CRC Press.
71. Willett, W. 1998. *Nutritional Epidemiology, 2nd Edition.* New York: Oxford University Press.
72. Wilson, T., Temple, N.J. (ed), 2003. *Beverages in Nutrition and Health.* Totowa, NJ: Humana Press.
73. Wilson, T., Temple, N.J. (ed), 2001. *Nutritional Health: Strategies for Disease Prevention.* Totowa, NJ: Humana Press.
74. Wiseman, G. 2002. *Nutrition and Health.* T&F STM.
75. Wolinsky, I. (ed.) 1998. *Nutrition in Exercise and Sport, 3rd Edition.* Boca Raton, FL: CRC Press.

Appendix D: Web Sites of Interest

http://nutrition.org

The American Society for Nutrition (ASN) is a non-profit organization dedicated to bringing together the world's top researchers, clinical nutritionists, and industry to advance our knowledge and application of nutrition for the sake of humans and animals. Our focus ranges from the most critical details of research and application to the broadest applications in society, in the United States and around the world.

http://cdc.gov

CDC is globally recognized for conducting research and investigations and for its action-oriented approach. CDC applies research and findings to improve people's daily lives and responds to health emergencies—something that distinguishes CDC from its peer agencies. CDC works with states and other partners to provide a system of health surveillance to monitor and prevent disease outbreaks (including bioterrorism), implement disease prevention strategies, and maintain national health statistics. CDC also guards against international disease transmission, with personnel stationed in more than 25 foreign countries. CDC is now focusing on achieving the four overarching Health Protection Goals to become a more performance-based agency focusing on healthy people, healthy places, preparedness, and global health. CDC is one of the 13 major operating components of the Department of Health and Human Services (HHS).

http://ods.od.nih.gov

The NIH Office of Dietary Supplements, established in 1995, provides educational materials and tools and research opportunities for health professionals and consumers including information about dietary supplement ingredients and safety.

http://www.springer.com/series/7659

Nutrition and Health Book Series information on the Springer web site provides information about all volumes published in the series, book reviews, and book-ordering instructions.

http://www.ifst.org

IFST (Institute of Food Science & Technology) is based in the United Kingdom, with members throughout the world, with the purpose of serving the public interest in the application of science and technology for food safety and nutrition as well as furthering the profession of food science and technology. Eligibility for membership can be found at the IFST home page, an index and a search engine are available.

http://www.nysaes.cornell.edu/cifs/start.html

The Cornell Institute of Food Science at Cornell University home page provides information on graduate and undergraduate courses as well as research and extension programs. Links to related sites and newsgroups can be found.

http://www.blonz.com

Created by Ed Blonz, Ph.D., "The Blonz Guide" focuses on the fields of nutrition, foods, food science and health supplying links, and search engines to find quality sources, news, publication, and entertainment sites.

http://www.hnrc.tufts.edu

The Jean Mayer United States Department of Agriculture (USDA) Human Nutrition Research Center on Aging (HNRC) at Tufts University. This research center is one of six mission-oriented centers aimed at studying the relationship between human nutrition and health, operated by Tufts University under the USDA. Research programs; seminar and conference information; publications; nutrition, aging, medical, and science resources; and related links are available.

http://www.fao.org

The Food and Agriculture Organization (FAO) is the largest autonomous agency within the United Nations, founded "with a mandate to raise levels of nutrition and standards of living, to improve agricultural productivity, and to better the condition of rural population," emphasizing sustainable agriculture and rural development.

http://www.eatright.org

The American Dietetic Association is the largest group of food and nutrition professionals in the United States, members are primarily registered dietitians (RDs) and dietetic technicians, registered (DTRs). Programs and services include promoting nutrition information for the public; sponsoring national events, media and marketing programs, and publications (*The American Dietetic Association*); and lobbying for federal legislation. Also available through the web site are member services, nutrition resources, news, classifieds, and government affairs. Assistance in finding a dietitian, marketplace news, and links to related sites can also be found.

http://www.foodsciencecentral.com

The International Food Information Service (IFIS) is a leading information, product and service provider for professionals in food science, food technology, and nutrition. IFIS publishing offers a wide range of scientific databases, including FSTA—Food Science and Technology Abstracts. IFIS GmbH offers research, educational training, and seminars.

http://www.ift.org

The Institute of Food Technologists (IFT) is a membership organization advancing the science and technology of food through the sharing of information; publications include *Food Technology* and *Journal of Food Science*; events include the Annual Meeting and Food Expo. Members may choose to join a specialized division of expertise (there are 23 divisions); IFT student associations and committees are also available for membership.

http://www.osteo.org

The National Institutes of Health Osteoporosis and Related Bone Diseases—National Resource Center (NIH ORBD-NRC) mission is to "provide patients, health professionals, and the public with an important link to resources and information on metabolic bone diseases, including osteoporosis, Paget's disease of the bone, osteogenesis imperfecta, and hyperparathyroidism. The center is operated by the National Osteoporosis Foundation, in collaboration with the Paget Foundation and the Osteogenesis Imperfecta Foundation."

http://www.ag.uiuc.edu/~food-lab/nat

The Nutrition Analysis Tool (NAT) is a free web-based program designed to be used by anyone to analyze the nutrient content of food intake. Links to an "Energy Calculator" and "Soy Food Finder" are also available. NAT is funded by C-FAR at the University of Illinois.

http://www.calciuminfo.com

This is an online information source created, copyrighted, and maintained by GlaxoSmithKline Consumer Healthcare Research and Development. The nutritional and physiological role of calcium is presented in formats designed for healthcare professionals, consumers, and kids. References and related links, educational games for kids, calcium tutorials, and a calcium calculator are easily accessible.

http://vm.cfsan.fda.gov

The Center for Food Safety and Applied Nutrition (CFSAN) is one of five product-oriented centers implementing the FDA's mission to regulate domestic and imported food as well as cosmetics. An overview of CFSAN activities can be found along with useful sources for researching various topics such as food biotechnology and seafood safety. Special interest areas, for example, advice for consumers, women's health, and links to other agencies are also available.

http://www.bcm.tmc.edu/cnrc

The Children's Nutrition Research Center (CNRC) at Baylor College of Medicine is one of six USDA/ARS human nutrition research centers in the nation assisting healthcare professionals and policy advisors to make appropriate dietary recommendations. CNRC focuses on the nutrition needs of children, from conception through adolescence, and of pregnant and nursing women. Consumer news, seminars, events, and media information are some of the sections available from this home page.

http://www.usda.gov

The United States Department of Agriculture (USDA) provides a broad scope of service to the nation's farmers and ranchers. In addition, the USDA ensures open markets for agricultural products, food safety, environmental protection, conservation of forests and rural land, and the research of human nutrition. Affiliated agencies, services, and programs are accessible through this web site.

http://www.nalusda.gov

The National Agriculture Library (NAL), a primary resource for agriculture information, is one of four national libraries in the United States and a component of the Agriculture Research Service of the US Department of Agriculture. Access to NAL's institutions and resources is available through this site.

http://www.fns.usda.gov/fns

The Food and Nutrition Service (FNS) administers the US Department of Agriculture's (USDA) 15 food assistance programs for children and needy families with the mission to reduce hunger and food insecurity. Details of nutrition assistance programs and related links can be found.

http://www.agnic.org

The Agriculture Network Information Center (AgNIC), established through the alliance of the National Agriculture Library (NAL) and other organizations, provides public access to agriculture-related resources.

http://www.who.int/nut/welcome.htm

The World Health Organization (WHO) has regarded nutrition to be of fundamental importance for overall health and sustainable development. The global priority of nutritional issues, activities, mandates, resources, and research are presented in detail.

http://www.clinicaltrials.gov/ct

ClinicalTrials.gov is a registry of federally and privately supported clinical trials conducted in the United States and around the world. ClinicalTrials.gov gives you information about a trial's purpose, who may participate, locations, and phone numbers for more details. This information should be used in conjunction with advice from health-care professionals.

http://www.faseb.org

A multi-society, interdisciplinary, scientific community that sponsors meeting featuring plenary and award lectures, symposia, oral and poster sessions, career services, and exhibits of scientific equipment, supplies, and publications.

Subject Index

Note: Locators followed by *t* and *f* refer to tables and figures respectively.

A

Abacavir, 626
Acceptable, feasible, affordable, sustainable, and safe (AFASS), 622–623
Acceptable macronutrient distribution ranges (AMDR), 27
Acetylcholine, 473
Achlorhydria, 490
Acid-suppression therapies, 477
  See also Achlorhydria
Acquired immunity, 574–575, 577–579, 580–581
Acrodermatitis enteropathica, 577, 713
Adenocarcinoma, 152–155
  Barrett's esophagus and medications, 152
  Helicobacter pylori, 154–155
  obesity and diet, 153–154
  tobacco and alcohol, 152–153
"Adequate Intakes" of the Food and Nutrition Board, 450
Adipokines, 29, 375, 419–429, 631
  source/activities, 422*t*
Adipokines, nutrition, obesity, 419–429
  adipokines/cytokines
    adiponectin, 424–425
    adipsin, 427
    IL-6, 427–428
    leptin, 422–424
    resistin, 426
    TNF-α, 428
    visfatin, 426–427
  adipose tissue as endocrine organ, 419–420
  nutrition, adipokines, cytokines, 428–429
  obesity, 420–422
Adiponectin (APN), 424–425
  APN receptors, 424
  binding effects, 425
  cardiovascular diseases, role in, 425
  functions, 425
  growth factors, inhibition, 425
AdipoR1 *vs.* AdipoR2 receptors, 424
Adipose tissue
  BAT, 419
  endocrine organ, 419–420
  IR, effects, 420
  in lean/obese individuals, 421*f*
  obesity, impact on, 420–421
  WAT, 419
Adipsin, 427
Adult chronic diseases, 37
  age-related eye diseases, 39
  cardiovascular disease and cancer, 37–39
  osteoporosis, 40–42
Adult mental health, 705–715
African leafy vegetables (ALV), 73
AGA, *see* American Gastroenterological Association (AGA)
Age-adjusted coronary mortality, decline in, 3
Age-related bone loss, 451
Age-related cataract (ARC), 501, 502
Age-related macular degeneration (AMD), 39, 526
  clinical trials on, 530*t*–533*t*
Age-related maculopathy (ARM), 501–538
Aging, 545
Aging and immunity
  B-cell functions, decline in, 549
  immunity with aging, general changes
    hypersensitivity skin test responses, 546
  immunity with aging, specific changes, 546*t*
    involution of thymus, 546–547
    T-lymphocyte functions, 547–549
  infectious disease morbidity, 549
  neutrophil activity, influence on, 549
  vaccination, effects, 549
Agita El Mundo, 762
Agita São Paulo program, 762
Agouti-related peptide (AgRP), 423
Agriculture–Nutrition Advantage Conceptual Framework, 73*f*
Agrodiversity interventions, 51–75
  approaches and endpoints, 67–68
  capital, types of, 54*t*
  and capital inputs, 68
  and education, 68–70
  gender considerations, 71–72
  and household income generation, 70
  improved human well being, 67
  micronutrient bias in literature, 70–71
  nutritional and health indicators, improvements in, 69*f*
  nutritional improvement by capital inputs, 69*f*
  nutritional indicators, improvements in, 68*f*

Agrodiversity interventions (*cont.*)
results, 55–61
study quality, 67
AgRP, *see* Agouti-related peptide (AgRP)
Alcohol, harmful effects, 13
17 alpha-hydroxyprogesterone caproate, 674
Alpha-linolenic acid (ALA), 7
Alternate Healthy Eating Index (AHEI), 816
AMD, *see* Age-related macular degeneration (AMD)
American Gastroenterological Association (AGA), 473
Anergy, 548
Antenatal multiple micronutrient supplements, 591
Antiarrhythmic actions, *N*-3 fatty acids
animal studies, 250–251
clinical trials, 252–253
population studies, 251–252
"Anti-infective factor," *see* Vitamin A
Antioxidant/B-vitamins/atherosclerosis, 285–316
B-vitamins, inflammation, and atherosclerosis, 288
completed randomized controlled trials, 299–300
arterial imaging trials, 310–312
with CVD as primary outcome, 305–310
with CVD as secondary outcome, 300–305
nutri-genomic trials, 312–315
observational studies
blood B-vitamin and tHCY concentrations, study, 297–299
ecological studies, 288
epidemiological studies, 288–294
serum antioxidant concentrations, study, 294–297
ongoing randomized controlled trials, 315
reactive oxygen species, antioxidants, and atherosclerosis, 286–287
Antioxidants, 199–203, 285–316
energy balance, 202–203
lycopene and tomato-based products, 199–200
nutritional, 501–537
selenium, 200–201
vitamin E, 201–202
Antiretroviral therapy (ART), 558, 612, 614
anabolism/weight gain, 625
complications, 618
HIV treatment, 625–626
MTCT of HIV, 622
nutritional effects, 625
side effects, 626
Anti-TNF-α therapy, 428
Antrum, 472
APN, *see* Adiponectin (APN)
ARC, *see* Age-related cataract (ARC)
AREDS cohort, 526, 527
AREDS II study, 527
ARM, *see* Age-related maculopathy (ARM)
ART, *see* Antiretroviral therapy (ART)
"Artificial" foods, 775
Ascorbic acid, *see* Vitamin C
Atherosclerosis, 285–316, 617
B-vitamins, inflammation and, 288
and fish oil, 254
and nutrition, 751–752
eating fish/sex-/age-adjusted death rate, study in Japan, 752*t*
hypertension, risk factor, 752
"Available" calcium, 456, 457*f*

B

Baltimore Longitudinal Study of Aging (BLSA), 529
"Barker hypothesis," 31
Barrett's esophagus, 477
and medications, adenocarcinoma, 152
"Barrier function" of epithelial linings, 573
Basic fibroblast growth factor (bFGF), 425
BAT, *see* Brown adipose tissue (BAT)
B cells, 549
Beaver Dam Eye Study, 521
Behavioral counseling, conditions
barriers to change, 812
knowledge, 811
motivation, 811
addressing patient's motivation, 812*t*
screening and counseling, tools, 818*t*
dietary fat screener, 816
fruit/vegetable/fiber screener, 816
SDSCA, 817
Bending set point, 449
Beriberi, 570, 582, 780
See also Thiamin
BFGF, *see* Basic fibroblast growth factor (bFGF)
Biofortification, 587, 588
Birth defects, 31–32, 643–672
deficiency or over dosage, role in, 643
Blindness, causes, 501, 502
BLSA, *see* Baltimore Longitudinal Study of Aging (BLSA)
Blue Mountain Eye Study (BMES), 526
Body mass index (BMI), 10, 12, 29, 30, 164, 182, 185, 186, 341, 345, 359*t*, 360*t*, 361*t*, 362, 373, 375, 377, 378, 379, 383, 397, 401, 410, 423, 428, 617, 619, 620, 630, 631, 674, 675, 683*t*, 692, 726, 731, 734, 736, 747, 748, 749, 750, 783*t*, 801
Body weight
appetite and, 85
drug-induced changes to, 86*t*
and risks, 12–13
Bone accumulation, 448, 449
Bone health, nutrition effects, 443–444
Bone marrow suppression, 620
Bone mass/density
bending set point, 449
influential factors, 444
Bone matrix, 444

Bone remodeling, 449
Bone strength, 444
Brain tumors, 128–129
child's diet, 134–135
and maternal consumption of cured meats during pregnancy, 131*t*
maternal diet, 130–134
N-nitroso hypothesis, 129–130
Breast cancer, 176–177, 339
animal fat *vs.* vegetable fat, 8
increased risk, fat intake, 7
low-fat diet and, 352
Breastfeeding, 621
AFASS, replacement feeding, 622–623
HIV, reduced risk, 623
HIV-infected mothers, role, 624
mixed feeding, risk of, 624
zidovudine prophylaxis *vs.* formula feeding, 623
Brown adipose tissue (BAT), 419
Business, 770
B-vitamins, 285–316, 325–332

C
Calcitriol, 460
Calcium, 11, 694–695
acquired bone mass conservation, secondary prevention, 449
estrogen-replete women, study, 449
premenopausal women, case, 449
adolescent pregnancy, case, 694
blood pressure levels, impact, 694
bone mass acquisition, primary prevention, 447–449
calcium intake-retention, relationship, 448*f*
and cancer risk, 232–233
double-blind placebo controlled trial, 695
low dietary intake, effects, 694
menopause, 449–451
menopausal bone loss, cause, 450
postmenopausal bone loss, 451*f*
WHO criteria, 450
nutrient–nutrient interactions, 453–456
See also Nutrient–nutrient interactions, calcium
pre-eclampsia and supplementation of, 35–36
requirement for calcium, ascertaining, 445–447
human life stages, retention curves, 446*f*
threshold behavior, 445*f*
in women, estimation, 447*t*
reserves, 445
senescence, 451–453
age-related bone loss, 451
intention-to-treat analysis, 452–453
"per protocol" analysis, 453
sources, 456–457
available calcium per serving for variety of foods, 457*f*
iron absorption, interference with, 457
supplementation trials, results, 694–695
Calcium citrate malate (CCM), 457
Calreticulin, 425
Cancer, 127–141, 145–167, 175–189, 195–210
dietary fat and, 7–8
excess body fat as cause, 8
risk, 219–238
Candida, 580
Canning, 579, 773
Carbohydrate, dietary, 501–537
and cortical cataract, 526*f*
Cardiometabolic disease, 371–385
differential effects of lifestyle changes on, 376
spectrum of progression of, 375*f*
See also Diabetes (type 2) and cardiometabolic disease
Cardiovascular birth defects, 31, 32, 33
Cardiovascular CAs, 654
Cardiovascular disease (CVD), 25, 37–39, 433, 744
maternal diet during pregnancy, 33
Cardiovascular effects of *trans* fatty acids, 6, 273–282
coronary heart disease, effects of risk factors
endothelial cell function, 280
hemostasis, 279–280
LDL oxidation, 279
low-grade systemic inflammation, 280
epidemiological studies, 280–281
metabolism
conversion, 276–277
digestion, absorption, and incorporation into blood lipids, 275–276
oxidation, 277
tissue levels, 276
serum lipids and lipoproteins, effects
mechanism, 279
serum total, LDL, and HDL cholesterol, 277–278
trans fatty acids in foods, 274–275
Carnitine
absorption, 96
distribution, 97
metabolism, 100
α-carotene, 535
and cataract extraction, 521
and cortical cataract, 520
and mixed cataract, 521
and nuclear cataract, 520
and PSC cataract, 521
β-carotene, 535, 552
and cataract extraction, 520
and cortical cataract, 519
high-dose supplements, effects, 559
and mixed cataract, 520
and nuclear cataract, 519
and PSC cataract, 520

Carotenoids
α-carotene, *see* α-carotene
β-carotene, *see* β-carotene
cryptoxanthin, 536
lutein/zeaxanthin, 517–519, 534
lycopene, 536
retinol/vitamin A, 521–522, 535–536
riboflavin, 522
total carotenoids, 517, 534
CAs of urinary tract, 654–655
Cataracts, 10, 13, 39, 42, 43, 501–538
CCK2, *see* Cholecystokinin-2 (CCK2)
Cell-mediated immunity, 571, 574, 577, 578, 580, 582, 590
Centrum®, 510, 511
Cerebrovascular diseases, 744
hemiplegia, 744
Japan, major cause of death in, 744
Cervical cancer, 178–179
Charaka Samhita, 577
CHD, *see* Coronary heart disease (CHD)
Chemical pregnancies, 650
Chemokines, 425
Childhood cancer, 127–141
Childhood, preventive nutrition in, 36–37
Cholecystokinin-2 (CCK2), 473
Chronic acid-suppressive therapy, 471
Chronic disease(s)
cardiovascular disease, 744
cellular/biochemical mechanisms, 26
cerebrovascular diseases, 744
childhood type 2 diabetes, 745, 746*f*
chronic NCD risks globally/industry response, 783
IHD, 744
industry/public health, 782–785
Action Plan, recommendations, 783
future global commitments, 784
risk reduction and economic cost, 26–27
Type 2 diabetes, 744
Chronic disease in Asia, effects of western diet, 743–754
atherosclerosis and nutrition, 751–752
epidemiology/risk factors
chronic diseases, 744
epidemic obesity/type 2 diabetes in Asia, 745*f*
metabolic syndrome, 749–750
IHD/cerebral infarction, yearly change in death rates, 744*f*
nutritional intakes, 750–751
yearly change per capita per day in Korea, 751*t*
nutritional recommendations, 752–754
in Japan, Korea, and Thailand, 753*t*
risk factors, 746–749
children with type 2 diabetes by overweight, distribution, 746*t*
cholesterol levels, yearly changes in Japan, 747*t*
nutrition/childhood obesity in Japan, 748*f*
obesity in school children, effect of intervention in Matsuyama, 749*t*
Chronic lower respiratory tract disease (CLRT), 24
Chronic NCD risks globally/industry response, 783*t*
CIMMYT, 588
Circadian rhythm, 428
Clinical Trial of Nutritional Supplements and Age-Related Cataract (CTNS), 502, 510
Cognitive decline, vitamin B in prevention of, 325–332
Cohort controlled trial (CCT), 646
Co-infection, 620
Colorectal cancer, 559
risk of
red meat consumption, 8
reduced, folic acid, 9
Conditional cash transfer (CCT), 762
programs, goals, 762
Congenital abnormalities (CA), 643
congenital pyloric stenosis, 651
limb deficiencies, 655
non-syndromic/syndromic, 654
rectal atresia/stenosis, 655
urinary tract/cardiovascular, 651
Congenital pyloric stenosis, 651
"Convenience foods," 774
Copper deficiency, 463
Corn, 726
Coronary Artery Risk Development in Young Adults (CARDIA), 39
Coronary heart disease (CHD), 777
decline in, 4
effects of *trans* fatty acids on other risk factors for
endothelial cell function, 280
hemostasis, 279–280
LDL oxidation, 279
low-grade systemic inflammation, 280
HDL and, 5
prevention, *see* *N*-3 fatty acids from fish and plants
saturated fat intake, 6
trans-fatty acid intake, 6–7
Corpus, 472
Corticosteroids, 464
Corticotropin releasing hormone (CRH), 676
Cow's/human milk, 776
C-reactive protein (CRP), 420
CRH, *see* Corticotropin releasing hormone (CRH)
Crohn's disease, 577
Cryptosporidia/giardia, causes, 621
Cryptoxanthin, 536
CTNS, *see* Clinical Trial of Nutritional Supplements and Age-Related Cataract (CTNS)
Cysteine, 647
Cystolic serine hydroxymethyltransferase, 693
Cytokines, 580

D
Daffodil *(Narcissus pseudonarcissus),* 588
Delayed hypersensitivity skin test (DTH), 547
"Demographic transition," 758
Dextri-Maltose, 776
DHFR, *see* Dihydrofolate reductase (DHFR)
Diabetes, 4, 389–412, 733–734
behavior change intervention studies, 409–410
body weight and functional status in older adults with, 397
burden of, 390
clinical issues impacting older adults with and at risk for
dietary habits of older adults with diabetes, 401
hypoglycemia in older adults, 402
hypoglycemia unawareness and treatment of hypoglycemia, 402–403
self-management behaviors, 403–404
self-monitoring and dietary treatment of hypoglycemia, 402
cultural issues and nutrition in, 408–409
diagnosis and classification of, 391
typologies in older adults, 391–392
dietary recommendations, 394–396
balancing diet and medication, 396
diet-specific intervention in older adults with, 410
goals of treatment, 390–391
lifestyle change and physical limitations, 404
medication and nutritional management, 392
medication use and glycemic control, 392–394
overweight as cause of, 8
prevalence in older adults, 390
psychosocial and behavioral issues
cognitive dysfunction, 406–407
depression, 404–405
personal beliefs and nutrition, 407–408
social support, 405–406
and quality of life in older adults, 409
weight loss intervention in older adults with, 411
See also Diabetes (type 2) and cardiometabolic disease; Type 2 diabetes (T2D)
Diabetes (type 2) and cardiometabolic disease
diabetes progression
insulin resistance, obesity, and metabolic syndrome, 374–375
prediabetes and abnormalities of carbohydrate metabolism, 372–374
nutritional strategies for treatment of prediabetic states and prevention of type 2 diabetes, 376
alterations in composition of dietary carbohydrate, 382–383
alterations in composition of dietary fat, 380–382
genetic bases of type 2 diabetes, 383–384
low-fat *versus* high-fat diets, 378–379
pharmacological agents that affect nutrient availability, 380
weight loss, 376–378
prediabetes, and metabolic syndrome, 371–372
prevention, and treatment of, 371–385
Diet, 433–438 443–464
Diet and childhood cancer, 127–141
brain tumors, 128–129
child's diet, 134–135
and maternal consumption of cured meats during pregnancy, 131*t*
maternal diet, 130–134
N-nitroso hypothesis, 129–130
leukemia, 135
child's diet, 139–140
and diet, 132*t*–133*t*
maternal diet, 135–139
and maternal use of multivitamins and other supplements during pregnancy, 137*t*–138*t*
other cancers, 141
vitamin K, 140
Diet and nutrition, role in prostate cancer, 198
antioxidants, 199–203
energy balance, 202–203
lycopene and tomato-based products, 199–200
selenium, 200–201
vitamin E, 201–202
cruciferous vegetables, 208–209
dairy products, calcium, and vitamin D, 207–208
fat intake and fatty acids, 203–206
cruciferous vegetables, 208
dairy products, calcium, and vitamin D, 207–208
fat intake, 203
fatty acids, 204–206
fish intake, 204
meat, 204
soy/phytoestrogens, 206–207
zinc, 208–209
soy/phytoestrogens, 206–207
Dietary adequacy, 65
Dietary and horticultural interventions, 587–588
Dietary Approaches to Stop Hypertension (DASH) study, 30, 39
Dietary carbohydrate, 382, 395, 501–538
Dietary fat, 5–7
and body fatness, 8–9
and cancer, 7–8
Dietary Fat Screener, 816
Dietary guidelines for optimum health, 798–804, 805*f*
alcoholic beverages intake, 803–804
control of calorie intake, 801
daily intake of fruits/vegetables/low-fat milk, 801–802
foodborne illnesses, prevention, 804
low carbohydrate intake, effects, 803
low fat intake, 802–803
median daily sodium intakes, 803
by National Organizations, 806*t*–808*t*

Dietary guidelines for optimum health (*cont.*)
physical activity, importance, 801
Dietary or nutritional recommendations, 14–15
Dietary patterns, 525
USDA Food Guide Pyramid, use of, 525
Dietary Reference Intakes (DRI), 559
Dietary supplements and cancer risk, 37, 43, 219–238
definition of dietary supplements in US, 220
discussion, 234
importance of time-integrated measure of supplement use, 235
measurement error in assessment of supplement use, 234–235
supplement use as marker for cancer-related behavior, 235–236
effects of micronutrients in multivitamins cannot be isolated, 224
hypothesized mechanisms of effect, 220–222
issues of study design as related to research on vitamin supplements, 224
case–control studies, 225
cohort studies, 225
randomized controlled trials (RCT), 224
micronutrient intakes from foods and supplements differ markedly, 222–224
nutrient intake among supplement users, 223*t*
prevalence of supplement use in US, 222
vitamin supplements and cancer risk
calcium, 232–233
folic acid, 233–234
methods, 225
multivitamins, 229–230
randomized, controlled trials of β-carotene, 226*t*–227*t*
randomized, controlled trials of selenium, 228–229
randomized, controlled trials of vitamin E, 229
vitamin C, 231
vitamin E, 231–232
Diet–heart hypothesis, 5
Diet–hormone receptor status interaction, 184–185
Diet/nutrients and reduced disease risk, guidelines, 26–27
Diet/obesity/lipids, 433–438
barriers to action, 434
culture/politics, problems
strategies for solutions, 436–438
Diet, osteoporosis, and fracture prevention, 23, 41, 443–464
calcium
acquired bone mass conservation, secondary prevention, 449
bone mass acquisition, primary prevention, 447–449
calcium sources, 456–457
menopause, 449–451
nutrient–nutrient interactions, 453–456
requirement for calcium, ascertaining, 445–447
senescence, 451–453
magnesium, 462
nutrition/glucocorticoid-induced osteoporosis, 464
nutrition/hip fracture, 463–464
osteoporotic fracture, nutrition effects, 443–444
protein, 460–461
trace minerals, 462–463
vitamin D, 443, 458–460
vitamin K, 461–462
Diet Quality Index (DQI), 816
Dihydrofolate reductase (DHFR), 693
Diphtheria, 580
Disease prevention, micronutrients for, 42–43
Diseases, gastric acid secretion, 473–478, 474*t*–476*t*
GERD, 473–477
infections with *h. pylori* (HP), 477
ulcers, 477–478
"Diseases of affluence," 763
Down's syndrome, 651, 655
DREAM, *see* Drug Resource Enhancement against AIDS and Malnutrition (DREAM)
DRI, *see* Dietary Reference Intakes (DRI)
Drug–nutrient interaction, 79–111
classification of, 80*t*
Drug Resource Enhancement against AIDS and Malnutrition (DREAM), 625
Drugs and nutrients, interactions between, 80–81
DTH, *see* Delayed hypersensitivity skin test (DTH)
Duodenal ulcer, 473
Dutch HungerWinter, 707
Dyslipidemia, 697–698

E
Econutrition, 51–75
agrodiversity interventions, 52–75
approaches and endpoints, 67–68
and capital inputs, 68
and education, 68–70
gender considerations, 71–72
and household income generation, 70
micronutrient bias in literature, 70–71
study quality, 67
defined, 52
home gardening, 54–63
investing in agrodiversity interventions, 63–67
recommendations
questions, suggestions, future research, 72–75
Ectopic pregnancies, 650
Endometrial cancer, 178–179
Enterochromaffin-like (ECL) cells, 472
"Epidemiologic transition," 758
Epigenetic regulation, 710
methyl donor L-methionine, role in, 711
selenium deficiency, effect on, 712
Erwinia uredovora, 588
Esophagus, cancer of, 145–146

adenocarcinoma, 152–155
by alcohol and tobacco consumption, 147
squamous cell carcinoma, 146–152
Estrogen, 449, 450
non-skeletal effects, 450
EU Food Law, Regulation 178/2002, 778
Exercise-promotion program, goals, 740
Eye diseases, age-related, 39

F
Fat and *n*–3 fatty acid content of seafood and fish oils, 266*t*–267*t*
ω-3 fats, 537
Federal Meat Inspection Act, 774
Female cancers, 8, 175–189
diet–hormone receptor status interaction, 184–185
gene–diet interaction
dietary patterns, 181–182
fruits and vegetables, 184
genetic susceptibility, 179–180
isothiocyanates, 181
phytoestrogens, 182–183
polymorphisms, 180–181
regional diets, 183
incidence and mortality of, 178*t*
leading sites of, 177*t*
multi-regional and multi-ethnic incidence of female cancers
breast cancer, 176–177
cervical and endometrial cancer, 178–179
ovarian cancer, 177–178
obesity and energy balance
energy balance and physical activity, 186
hormonal regulation, 186–188
obesity and risk of female cancers, 185–186
Ferritin, 677
Fetal deaths, types, 650
Fetal malnutrition, 710
FFA, *see* Free fatty acids (FFA)
Fish oil, 696
Fish oil
in diabetic patients, 264–265
hypertension, 265
Folate, 582, 648, 713–714
deficiency, 583
dietary sources, 648
epigenetic variation *in utero,* source of, 713
gene expression, key factor, 648
MTHFR677TT polymorphism, 713
nucleotide synthesis, role in, 648
preterm delivery/maternal nutrition, 673–698
deficiency, hyperhomocysteinemia, 694
DHFR, impact on, 693
dietary/circulating folate, influence on, 693
low intakes, risk of, 693
Folate–folic acid deficiency, 647
Folic acid, 643–665
absorption, 94–95
and cancer risk, 233–234
distribution, 97
metabolism, 98–99
Folic acid/multivitamins, birth defects/preterm birth prevention, 643–665
congenital abnormalities, prevention, 651–656
efficacy of periconceptional folic acid-containing multivitamin, 652*t*–653*t*
See also Congenital abnormalities (CA)
Hungarian RCT, 649–651
intervention studies for reduction of neural-tube defects, 644–647
maternal dietary deficiency, defects, 643, 651
mechanisms of NTD prevention, 647–649
NTD, history, 649
NTD, nutritional interventions to reduce
folate-rich/vitamin-rich diets, consumption of, 660
food fortification, 664–665
periconceptional supplementation, 660–664
See also Neural tube birth defects (NTD)
preterm birth, 657–659
Food, Drug, and Cosmetic Act, 774
Food and Drug Administration (FDA), 165, 219, 299, 307, 774
Food and Nutrition Board of the Institute of Medicine, 446
Foodborne illnesses, prevention, 804
Food fortification, 31, 588–589, 664–665
Food intake, 83–84
appetite and body weight, 85
drugs that alter, 84*t*
gastrointestinal function, 87*t*–89*t*
GI tract, 85–90
constipation, 90
diarrhea, 89
nausea and vomiting, 86–89
oral cavity, 85–86
Food industry, 437, 769–789
history of, 772–776
and public health, 782–788
research in, 776–782
Food prices rise, causes, 764
Food-related technologies, 773*t*
Food safety, 778–779
techniques, 804
Food Safety, and Inspection Service (FSIS), 782
Fracture
prevention, 443–464
incidence in older adults, reduced, 11
Framingham Heart Study, 775
Framingham osteoporosis cohort, 460
Free fatty acids (FFA), 419–420
Free radical theory of aging, 545

Fruit/vegetable/fiber screener, 816
FSIS, *see* Food Safety, and Inspection Service (FSIS)
Fundus, 472

G

Gastric acid, secretions, treatments, nutritional consequences, 471–494
  excess gastric acid, pharmacological interventions
    antacids, 478
    drug–drug interactions, 486–487
    H_2RAs, 479
    PPI, 479–486
  gastric acid
    central/peripheral region, secretion, 473
    diseases of gastric acid secretion, 473–478
    regulation and secretion, 472–473
    stomach anatomy, 472*f*
  reduced gastric acid, nutritional consequences
    calcium/gastric acid, 490–491
    infectious diseases, 493–494
    iron absorption/gastric acid, 492
    vitamin B_{12}/gastric acid, 487–488
    vitamin C/gastric acid, 488–489
  stimulation, 473
Gastrin, 472, 473
Gastrin cells (G cells), 472
Gastroesophageal reflux disease (GERD), 473–477
  definition, 473
  NERD group, 474–477
  symptoms, 473
Gastrointestinal cancers, 547
Gastrointestinal tract cancers, prevention of, 145–167
  cancer of esophagus, 145–146
    adenocarcinoma, 152–155
    squamous cell carcinoma, 146–152
  cancer of stomach, 155
    anatomic subsites, 155
    histologic types, 155
    risk factors, 155–164
Gatorade, 777–778
  hydration benefits, 778
Gene–diet interaction
  dietary patterns, 181–182
  fruits and vegetables, 184
  genetic susceptibility, 179–180
  isothiocyanates, 181
  phytoestrogens, 182–183
  polymorphisms, 180–181
  regional diets, 183
Gene–environment interactions, 710
Genetic engineering, 588
"Geometrical isomers," 273
GERD, *see* Gastroesophageal reflux disease (GERD)
Gestational weight gain, 675
GI, *see* Glycemic index (GI)
GIO, *see* Glucocorticoid-induced osteoporosis (GIO)
Global Alliance for Improved Nutrition (GAIN), 782
Glucocorticoid-induced osteoporosis (GIO), 464
Glucose
  drug-induced changes to, 92*t*
  metabolism, 91
Glutathione peroxidase (GPX), 712
Glycemic index (GI), 10, 526
Glycoxidation, 526
Goblet cells, 573
"Golden rice," 588
GPX, *see* Glutathione peroxidase (GPX)
GPX-1 polymorphisms, 712

H

HACCP, *see* Hazard Analysis and Critical Control Points (HACCP)
Hazard Analysis and Critical Control Points (HACCP), 779
HB-EGF, *see* Heparin-binding epidermal growth factor (HB-EGF)
HCCSCA, *see* Hungarian Case–Control Surveillance of Congenital Abnormalities (HCCSCA)
Health economics of preventive nutrition, 23–44
  adult chronic diseases, 37
    age-related eye diseases, 39
    cardiovascular disease and cancer, 37–39
    osteoporosis, 40–42
  guidelines for diet/nutrients and reduced disease risk, 26–27
    guidelines for macronutrient intakes and implementations, 27–28
  interactions with diet and other lifestyle factors, 23
    diet, smoking, and health, 25–26
    nutrition and general health status, 24–25
  micronutrients for disease prevention, 42–43
  obestiy, 28–30
  optimizing pregnancy outcomes, 31
    birth defects, 31–32
    low birth weight and premature birth prevention, 32–35
    preventive nutrition in childhood, 36–37
    reduction in pre-eclampsia and other pregnancy associated adverse effects, 35–36
Health transition
  demographic/epidemiologic/health transitions, relationship, 758*f*
  "demographic transition," 758
  "epidemiologic transition," 758
  eras, causes of death
    degenerative and man-made diseases, 759
    pestilence and famine, 759
    receding pandemics, 759
  health problems by age group, developing countries, 759*f*
  models, 759

Healthy eating index (HEI), 27, 525, 816
and standards for scoring criteria, 28
Heart-healthy products, development, 777
Helicobacter pylori (HP), 154–155, 473, 477
diagnosis, 166
gastric carcinogenesis, cause of, 477
treatment, 166
Hemiplegia, 744
Heparin-binding epidermal growth factor (HB-EGF), 425
Hepatitis B antigens, 549
High-density lipoprotein (HDL), 5
Hip fracture, 463–464
Histamine, 472, 473
Histamine H2-Receptor Antagonists (H_2RAs), 479
HIV and nutrition, 593, 611–632
global burden/current trends, 611–612
HIV infection in 2007, global view, 612*f*
HIV and nutritional effects
malnutrition/HIV infection cycle, 616–617, 616*f*
nutrient deficiencies, immunologic effects, 618–620, 619*t*
nutritional changes/metabolic demands in HIV, 617–618
superinfections/HIV/malnutrition, 620–621
women, children, breastfeeding, 621–625
HIV chronicity/epidemiology effects, 612–614
ART treatment, 614
HIV transmission, risk factors, 613–614
life expectancy at birth, 613*f*
people with HIV/adult HIV prevalence, estimation, 613*f*
malnutrition, epidemiology of
food availability/security, hamper, 615
HIV infection, symptoms, 615
HIV treatment programs, aim, 615, 616
malnutrition/HIV, coexistence, 614–616
nutrition support programs, aid, 615
medical treatment
ART, 625–626
nutrient deficiencies in HIV patients, interventions, 626–631
HIV infection
nutritional changes/metabolic demands
malnutrition, impact on metabolism, 618
PEM, 617
REE, study, 617
weight loss/wasting, 617
symptoms
deconditioning, 615
malaise, 615
muscle loss, 615
weight loss, 615
T-cells/CD4+ T lymphocytes, target of, 618
women, children, breastfeeding
pediatric HIV infection, risk factor, 622
HLA type I alleles, 614
Home gardening, 54–63
advantages, 62–63
Homocysteine, 326, 327*f*, 647, 694
HPA axis, *see* Hypothalamic–pituitary–adrenal (HPA) axis
HPS, *see* Hungarian Periconceptional Service (HPS)
H_2RAs, *see* Histamine H2-Receptor Antagonists (H_2RAs)
Human immunodeficiency virus (HIV), 611
Humoral immunity, 571, 581
Hungarian Case–Control Surveillance of Congenital Abnormalities (HCCSCA), 650
Hungarian Periconceptional Service (HPS)
steps/items of, 661, 662*t*
Hungarian RCT, goal, 645–646
25-hydroxyvitamin D [*25* (OH)D], 458
Hypercholesterolemia, 746, 748
Hypergastrinemia, 477
Hyperhomocysteinemia, 647, 694
related to NTD, evidences, 648
Hypertension, 752
Hypoglycemia, 402
unawareness and treatment, 402–403
Hypothalamic–pituitary–adrenal (HPA) axis, 710
Hypothalamic suprachiasmatic nucleus, 428
Hypovolemia, 676

I

IDA, *see* Iron deficiency anemia (IDA)
IGF-1, 461
IHD, *see* Ischemic heart diseases (IHD)
IL-2, *see* Interleukin-2 (IL-2)
IL-6, 427–428
IL-2 receptor (IL-2R), 549
Immune system, 550
Immunity
iron deficiency and, 580–581
in older people, 545–561
See also micronutrients and immunity
Immunologic theory of aging, 545
Inducible nitric oxide synthase (iNOS), 421
"Industrial diets," 3–4
Infant formula, role of food industry, 776
cow's/human milk, comparison, 776
SMA, basis for, 776
Innate immunity, 573–574, 577, 580
Insulin resistance (IR), 420
Interferon (IFN)-γ, 573
Interleukin-2 (IL-2), 547, 549
International Plant Genetic Resources Institute (IPGRI), 73
Intrauterine growth retardation, 33–34
Iodine deficiency, 582
Iodine deficiency disorders (IDD), 582
Iodine fortification, 779
IR(HOMA-R), homeostasis model assessment, 749

Iron, 676–677
anemia/IDA, 676
hemoglobin/preterm delivery, “U-shaped” relationship, 677
deficiency and immunity, 580–581
acquired immunity, 580–581
innate immunity, 580
ferritin, increased risk of preterm delivery, 677
high hemoglobin, risk, 677
IDA, 579
iron-replete/supplemented women, trials, 677
public health implications, 581
Iron deficiency anemia (IDA), 579
Iron deficiency and immunity, 580–581
acquired immunity, 580–581
cytokines, effect on, 580
T cells activity, impaired, 580
innate immunity
NBT test, 580
neutrophil function, impaired, 580
NK cell activity, depressed, 580
Iron deficiency anemia (IDA), 676
Iron fortification, 780
Ischemic heart diseases (IHD), 744

J

Janus tyrosine kinase 2 (JAK2), 423
Japan
cerebrovascular diseases, major cause of death, 744
children with type 2 diabetes by overweight, distribution, 746*t*
cholesterol levels, yearly changes in, 747*t*
metabolic syndrome in children and adolescents, 750*t*
nutritional recommendations, 752–754
nutrition/childhood obesity in Japan, 748*f*

K

Kaplan–Meier curves, 548*f*
12-kDA polypeptide, *see* Resistin
16-kDa polypeptide, *see* Leptin
52-kDa protein, *see* Visfatin
Keratinization of epithelium, abnormal
vitamin A deficiency, 573
Keratomalacia, 570
Keshan disease, 712
Killer cells, 549, 552

L

Lactic acidosis, symptoms, 626
Lactose/mannitol urinary excretion test, 573
Lens
carotenoids found in, 516
events, 506*t*
opacity, 506*t*
Leptin
binding to receptors, effects, 423
db/db mice, 422–423
function of, 423
leptin deficiency, neuroendocrine/endocrine alterations, 423–424
mRNA expression, 423
ob/ob mice, 422
Leptos, 423
Leukemia, 135
child’s diet, 139–140
and diet, 132*t*–133*t*
maternal diet, 135–139
foods containing inhibitors of DNA topoisomerase II, 135–136
foods containing *N*-nitroso compounds, 139
other foods and nutrients, 139
vitamin and/or mineral supplements, 136–139
Leukocyte elastase, 424
Limb deficiencies, 655
α-linolenic acid content of various oils and foods, 268*t*
Lipid(s), 433–438
calories, 726
control in developing economies, barriers to, 433–438
disorders, 734–738
-soluble vitamins
absorption, 96–97
distribution, 97
metabolism, 100–102
Lipoatrophy, 617
Lipodystrophy, 626
Long-chain polyunsaturated fatty acids (LC PUFA), 37
Low birth weight, 32–33
studies of birth defect and LBW prevention, 33–35
Low-density lipoprotein (LDL), 5
saturated fats’ influence on, 6
LPR, *see* Lymphocyte proliferative responses (LPR)
Lung cancer, 552
high-dose β-carotene supplements, cause for, 559
See also GPX-1 polymorphisms
Lutein, 517–519, 534
and cataract extraction, 519
and cortical cataract, 518
and mixed cataract, 519
and nuclear cataract, 517–518, 518*f*
and PSC cataract, 518–519
Lycopene, 536
Lymphocyte proliferative responses (LPR), 548

M

Macronutrients, 628–631
absorption and distribution, 90–91
excretion, 93
electrolytes, 109–110
trace minerals, 110–111
intakes and implementations, guidelines, 27–28

metabolism, 91–92
excretion, 93
glucose, 91
lipid, 91
protein, 91–92
Major histocompatibility complex (MHC), 547
Malaria, endemic, 570
Malnutrition, 51–75
Malnutrition/HIV infection cycle, 616–617, 616*f*
Mars, Inc., 778
Maternal diet, leukemia
foods containing inhibitors of DNA topoisomerase II, 135–136
foods containing *N*-nitroso compounds, 139
other foods and nutrients, 139
vitamin and/or mineral supplements, 136–139
Maternal dietary folate deficiency, defects, 643, 651
Maternal nutrition, 673–698
deficiency, 659
and preterm delivery, 673–698
MCP-1, *see* Monocyte chemoattractant protein (MCP)-1
MDGs, *see* Millennium Development Goals (MDGs)
Meckel–Gruber syndrome, 644
Medical Research Council (MRC), 645
Medical treatment, HIV
ART, 625–626
nutrient deficiencies in HIV patients, interventions
macronutrients, 628–631
micronutrients, 627
nutritional counseling, 626
Menopausal bone loss, cause, 450
Menopause, 449–451
menopausal bone loss, cause, 450
postmenopausal bone loss, 451*f*
WHO criteria, 450
Metabolic syndrome, 389–412, 749–750
ATP III criteria for diagnosis of, 399*t*
children and adolescents, Japan, 750*t*
Metallothionein-I and II (MTI and II) genes, 713
Metaplasia, 573
Methionine, 647
Methylene-THFreductase (MTHFR) gene polymorphism, 648
MHC, *see* Major histocompatibility complex (MHC)
Microcytic anemia, 570
Micronutrients
deficiencies and immunity, 545–561, 583–584
folate deficiency, 583
vitamin B_{12} deficiency, 583
vitamin C (ascorbic acid) deficiency, 583
vitamin D, 584
vitamin E, 583
deficiency, 550
immune systems, 567
resolving, added variety to food, 65*f*
staple food form and, 66*f*
disease prevention, 42–43
fortification, 779–782
importance of, 11
over-the-counter, 554
preventive nutrition strategies, 43*t*
and reduced risk of adverse pregnancy outcomes, 42*t*
supplements, 545
high-dose, effects in HIV-1-infected patients, 558
Micronutrients, immunological/infection effects, 567–596
control, prevention and treatment, 592–595
See also Public health interventions, categorization
immunity/infectious disease
effect of deficiencies on immunological status, 571–584
infectious diseases, impact on, 584–585
undernutrition/disease cycle, 569–570, 570*f*
Micronutrients and immunity, 545
combinations of micronutrients, clinical trials, 554–556
placebo/micronutrient groups, DTH response, 555*t*
cross-sectional studies, 551–552
in older people, 545–561
limitations, 556–557
nutrition, immunity, and aging, 550–551
dietary deficiencies, impact on aging, 550
water-soluble vitamins, 550
nutrition–immunity relationships, factors, 557–558
physical/psychological stress, impact, 557
research in older people, 559–560
considerations, 561
single micronutrients, clinical trials
pyridoxine on lymphocyte responses, impact, 553
Shigella flexneri-infected children, study, 553
UV radiation-induced photosuppression of DTH, study, 552
vitamin B_6 deficiency, effects, 553
vitamin E, study, 553
zinc supplements, study, 553
See also Aging and immunity
Millennium development goals (MDGs), 567–569, 757, 758*f*
Minerals, 103
absorption, 103–107
drugs that alter macro-mineral status, 104*t*–106*t*
drugs that alter trace mineral status, 107*t*
distribution, 108–109
excretion, 109–111
Miscarriage, 650
Mixed feeding, 624
Monocyte chemoattractant protein (MCP)-1, 421
Monocytes, 549
Monoglutamate folic acid, 648
Mother-to-child-transmission (MTCT), 622
HAART, 625
infant feeding counseling, 624

Mother-to-child-transmission (MTCT) (*cont.*)
  risk of HIV, 623*t*
"Motivational reflection" technique, 811
MRC, *see* Medical Research Council (MRC)
MTCT, *see* Mother-to-child-transmission (MTCT)
MTHFR gene, methylene–THFreductase (MTHFR) gene
MTHFR polymorphism, 714
Mucin, 573
Multivitamins
  antioxidant indices
    and cataract extraction, 525
    and cortical cataract, 524
    and mixed cataract, 525
    and nuclear cataract, 523–524, 523*f*
    and PSC cataract, 524–525
  and cancer risk, 229–230
  folic acid/folic-containing, 643–665
  maternal use during pregnancy and childhood cancer, 137*t*–138*t*
  mineral supplements
    dyslipidemia, 697–698
Mumps, 580
Mutation rate, 710
Myeloperoxidase, 580
Myocardial infarction (MI), reduced risk
  fish consumption, 7
  vitamin E, 13
Myristic acid, 6

N

Natural killer (NK) cells, 573–574
Nausea, 626
Neural tube birth defects (NTD), 31, 32, 643, 644, 646*t*
  anencephaly and spina bifida, 644
  CCT, 646
  Chinese intervention study, 647
  critical period, 644
  folate-folic acid deficiency, origin of, 647
  folic acid-containing multivitamin, effects, 647
  gene–environmental interaction, role in, 647
  Hungarian RCT, goal, 645–646
  hyperhomocysteinemia, related to, 648
  interventions to reduce, recommendations
    folate-rich/vitamin-rich diets, consumption, 660
    food fortification, 664–665
    periconceptional supplementation, 661–664
  MRC vitamin study, 645
  optimal dose of folic acid, reduced risk, 664
  randomized double-blind controlled trial (RCT), 645
  Smithells' study, 644–645
N-3 fatty acids, 205–206, 249–268, 628, 688, 695–696, 803
  fish oil, reduced risk of preterm delivery, 696
  myometrial activity, reduction in, 695
N-3 fatty acids from fish and plants, 249–268
  antiarrhythmic actions
    animal studies, 250–251
    clinical trials, 252–253
    population studies, 251–252
  effects on plasma lipids and lipoproteins, 254–255
    effects of fish oil on normal subjects, 255
    fish oil and LDL turnover, 263
    fish oil effects on plasma lipids and lipoproteins, 263–264
    implications of fish oil studies in hypertriglyceridemic patients, 257–259
    mechanism of hypolipidemic effects of fish oil, 259–263
    reduction of postprandial lipemia after fatty meals, 259
    studies in hyperlipidemic patients, 255–257
  experimental atherosclerosis and fish oil, 254
  fish oil in diabetic patients, 264–265
    hypertension, 265
  thrombosis, 254
Niacin, 780
Night blindness, 572, 575
Nitroblue tetrazolium (NBT) test, 580
N-nitroso hypothesis, 129–130
Nonsteroidal anti-inflammatory drugs (NSAIDs), 486
Nontransmissible chronic diseases (NTCD), 724
NSAIDs, *see* Nonsteroidal anti-inflammatory drugs (NSAIDs)
NTCD, *see* Nontransmissible chronic diseases (NTCD)
Nurses' Health Study (NHS), 510
Nutrient–nutrient interactions, calcium
  calcium balance in normal women, variance, 454*f*
  intestinal absorption of calcium, influence
    caffeine, 454
    fiber, 454
  renal conservation of calcium, influence
    acid ash residue, 455
    aluminum, 456
    phosphorus, 455–456
    protein and sodium, 454–455
    sodium/potassium salt effects on urine calcium, 455*f*
Nutrient reserves, role, 444–445
Nutrition, metabolic syndrome, and diabetes in the senior years, 389–412
  behavior change intervention studies of self-management and lifestyle change in, 409–410
  body weight and functional status, 397
  burden of diabetes, 390
  clinical issues and risk for diabetes
    hypoglycemia, 402
    hypoglycemia unawareness and treatment of hypoglycemia, 402–403
    self-management behaviors, 403–404
    self-monitoring and dietary treatment of hypoglycemia, 402
  clinical issues and risk of diabetes

dietary habits, 401
cultural issues and nutrition in diabetes, 408–409
diabetes lifestyle change and physical limitations, 404
diagnosis and classification of diabetes, 391
typologies of diabetes, 391–392
diet-specific intervention, 410
establishing medication and nutritional management goals, 392
medication use and glycemic control, 392–394
general diabetes dietary recommendations, 394–396
balancing diet and medication, 396
goals of diabetes treatment, 390–391
imparting diabetes dietary information, 411–412
prevalence of diabetes, 390
preventive nutrition issues and diabetes risk factors and metabolic syndrome, 399–400
prevention of type 2 diabetes—targeting prediabetes, 400–401
psychosocial and behavioral issues related to self-care and dietary intake
cognitive dysfunction, 406–407
depression, 404–405
personal beliefs and nutrition, 407–408
social support, 405–406
quality of life and diabetes diet, 409
special nutrition intervention situations in diabetes preventive nutrition
enteral and parenteral nutrition, 398–399
hospitalization, 397
long-term care, 398
weight loss intervention, 411
Nutritional anemia, 579, 581
Nutritional antioxidants, dietary carbohydrate, ARM, cataract, 501–538
antioxidants/ARM, epidemiological studies, 528*t*
carotenoids and ARM, 533–536
GI/AMD, study, 527
intervention trials, 527–529
multivitamins, antioxidant indices, and ARM, 536–537
vitamin C and ARM, 529
vitamin E and ARM, 529–533
antioxidant supplements/cataract, epidemiological studies, 503*t*–504*t*
carotenoids, study, *see* Carotenoids
cataract extraction, 512
cortical cataract, 511
intervention trials/cataract, comparison of, 505*t*–510*t*
mixed cataract, 511–512
multivitamins/antioxidant indices, study, 522–525
nuclear cataract, 511
PSC, 511
vitamin C, study, *see* Vitamin C
vitamin E, study, *see* Vitamin E
dietary patterns, indices, carbohydrate intake, epidemiological studies, 525–526
Nutritional deficiencies, classical, 570
immune system, effects on, 571
iron, 579–581
vitamin A, 572–576
zinc, 576–579
Nutritional habits and obesity, Latin America, 723–740
cancer/cardiovascular diseases, major causes of death, 724
food habits
calories in food intake, diversity, 726
carbohydrates, 728
energy, 726–727, 727*f*
fats, increased consumption, 727, 728*f*
food availability, percent changes in, 727*f*
food consumption, SEL, 728–729
proteins, 728
salt, 728
SEL, important factor, 726
wheat, corn, rice/main cereals, 726
infant mortality rates, decreased, 724*f*
nutrition education recommendations
goals/objectives, 739
promotion of exercise, goals, 740
obesity prevalence, conditions, 729–733
Brazil, survey, 731
consumption of fast foods, high risk, 733
control of obesity, 735
diabetes, 733–734
lipid disorders, 734–738
mortality, risk factor, 730
obesity program for prediabetic subjects, 736–738
prevalence (%) of obesity, preschool children, 732*t*
prevalence (%) of obesity among adults, 730*t*
prevention program, framework, 738
prevent obesity, nutrition education programs to, 735–736
people older that 65 years of age, rates, 725*f*
socioeconomic changes
improvement in GNP, impact, 726
infant mortality rate according to income, 725*f*
urbanization, major issue, 723
westernization of eating habits, 723
Nutritional therapy, *see* Antiretroviral therapy (ART)
Nutrition and Vision Project (NVP), 510
Nutrition in age of polypharmacy, 79–111
clinical implications, 81–82
digestion, 90
food intake, 83–84
appetite and body weight, 85
GI tract, 85–90
interactions between drugs and nutrients, 80–81
macronutrients
absorption and distribution, 90–91

Nutrition in age of polypharmacy (*cont.*)
excretion, 93
metabolism, 91–92
mechanisms of altered nutritional status, 82–83
medication use, 79–80
minerals, 103
absorption, 103–107
distribution, 108–109
excretion, 109–111
vitamins, 93–94
absorption, 94–97
distribution, 97
excretion of vitamins, 102–103
metabolism, 98–102
Nutrition transition, 759–761
risk factors, 760
undernutrition, early/later stage effects, 760
urbanization, disadvantages, 760
NVP, *see* Nutrition and Vision Project (NVP)

O

Obesity, 4, 28–30, 419–429, 433–438
adipose tissue in lean/obese individuals, 421*f*
cause/effects, 420
childhood, 29
chronic activation of immune system, effects, 420
increased risk of birth defects, 32
in Latin America, 723–740
macrophage infiltration, 421
risk of other diseases, 29
statistics, 25–26, 28–29
status in China, 29
stomach, cancer of, 165
Obesity and energy balance
energy balance and physical activity, 186
hormonal regulation, 186–188
interplay between AMPK and mTOR, 188*f*
obesity and risk of female cancers, 185–186
Obesity reduction, product/industry approach
food safety, 778–779
EU Food Law, Regulation 178/2002, 778
HACCP plans, steps, 779
manufacturer, criteria, 779
micronutrient fortification, 779–782
FAO, goals/strategies, 781
iodine fortification, 779
iron fortification, 780
product formulation, importance, 781
vitamin A fortification, 780
vitamin D to cow's milk, addition, 780
Ob–Rb receptor, 423
Omega-3 fatty acids, 777
consumption of, 9
Onchocerciasis, 570
Oportunidades program, 762
Organizational nutritional guidelines (websites), 163
Orofacial clefts (OFC), 654
Osteoporosis, 40–42, 443–464
definition, 443
GIO, 464
syndrome
vitamin D deficiency, role, 459
Ovarian cancer, 177–178
Oxidative stress, 502, 526, 712
Oxyntic/pyloric glands, 472

P

Palmitic acid, 6
Paracellular diffusion, 458
Parathyroid hormone (PTH), 444, 447, 451
Parietal cells, 472
Pasteur, Louis, 773
Pasteurization, 773
Pathophysiology, 371–385
PBEF, *see* Pre-B-cell colony-enhancing factor (PBEF)
PCM, *see* Protein-calorie malnutrition (PCM)
PDGF, *see* Platelet-derived growth factor (PDGF)
Pellagra, 780
PepsiCo, 771
Periconceptional use of multivitamins, 32, 33
Periconceptional vitamin supplementation, 644, 660–664
HPS, steps, 661, 662*t*
increased use of, public health tasks for
educational campaign, 661
pre-or periconceptional care, 661
Phagocytosis/intracellular killing, 549
Phytohemagglutin (PHA), 553
PIH, *see* Pregnancy induced hypertension (PIH)
Placental vasculopathy, 659
Plant breeding, 587
Plant sterols, 777
Plasma lipids and lipoproteins, *N*-3 fatty acids effects on, 254–255
changes in plasma cholesterol levels, 258*f*
changes in plasma triglyceride levels, 257*f*
effects of baseline diet, 261*f*
effects of fish oil on normal subjects, 255
effects of high-carbohydrate control and fish oil diets, 262*f*
fish oil and LDL turnover, 263
fish oil effects on plasma lipids and lipoproteins, 263–264
implications of fish oil studies in hypertriglyceridemic patients, 257–259
increase in plasma triglyceride levels, 259*f*
mechanism of hypolipidemic effects of fish oil, 259–263
reduction of postprandial lipemia after fatty meals, 259
studies in hyperlipidemic patients, 255–257
Platelet-derived growth factor (PDGF), 425
Pneumocystis jirovecii, 621

Polypharmacy, 79–111
Posterior subcapsular cataract (PSC), 511
Poverty Reduction Strategic Plans (PRSPs), 595
PPROM, *see* Preterm premature rupture of membranes (PPROM)
Pre-B-cell colony-enhancing factor (PBEF), 427
Pre-eclampsia, 35
  initiation and progression of, 35
  oxidative stress, 36
Pregnancy induced hypertension (PIH), 694–695
Pregnancy outcomes, 31
  birth defects, 31–32
  food fortification, 31
  low birth weight and premature birth prevention, 32–35
  preventive nutrition in childhood, 36–37
  reduction in pre-eclampsia and other pregnancy associated adverse effects, 35–36
Premature birth prevention, 32–33
Premature rupture of the fetal membranes (PROM), 674
Prenatal nutrition, 705–715
Prenatal nutrition to adult mental health, linking, 705–715
  mechanism, 714
    complexities, 714
    epigenetic regulation of genes, 710–711
    folate, 713–714
    gene-environment interactions, 710
    selenium, 711–712
    zinc, 712–713
    See also Schizophrenia
  micronutrients, infections, and schizophrenia, 714
  prenatal famine and schizophrenia, 706–710
    Dutch famine birth cohorts, 707, 708*f*
    Dutch HungerWinter, 707
    famine, effects on schizophrenia, 707
    HPA axis, risk factor, 710
    prenatal malnutrition, latent effects, 709
    risk of schizophrenia *vs.* birth rate (1956-1964), 709*f*
    Wuhu study, 709
President's Emergency Plan For AIDS Relief (PEPFAR), 612
Preterm birth
  folate deficiency, 657
    homocysteine related placental vasculopathy, 659
  folic acid supplementation, longer mean gestational age, 657–658
  infant mortality/handicaps, 657
  mean birthweight and gestational age at delivery, 658*t*
Preterm delivery, 673–698
Preterm delivery/maternal nutrition, 32–33, 673–698, 678*t*–691*t*
  definition/importance
    infants, risk associated, 673–674
  diet/gestational weight gain, relation, 675
  diet/preterm delivery, 675–676
    CRH, preterm delivery risk, 676
    famine-related amenorrhea, risk of, 675
    maternal fasting, effects, 676
  iron, 676–677
  maternal diet during pregnancy, 33
  maternal weight/weight gain, 674–675
    maternal underweight, effects, 674
    obesity-related complications, 675
    pregravid maternal weight/BMI/preterm delivery, association, 674
    pregravid weight/BMI, 674
    rate/pattern of gain, significance, 674
    SGA births, risk, 674
  micronutrients
    folate, 693–694
    homocysteine, 694
    zinc, 692–693
  nutrients, other
    calcium, 694–695
    multivitamin-mineral supplements, dyslipidemia, 696–698
    n-3 fatty acids, 695–696
  prenatal multivitamins and reduced risk, 33
  risk factors, 674
    fetal growth restriction, 674
    placental abruption, 674
    preeclampsia, 674
    PROM, 674
Preterm premature rupture of membranes (PPROM), 692
Preventive nutrition, role in clinical practice, 793–819
  application
    assessing patients' readiness to make behavioral changes, 810–811
    health-care providers, role of, 809–810
  5A'S brief behavioral counseling protocol, 815*t*
    addressing non-adherence to recommendations, 813–814
    auditing preventive nutrition systems, 817–819
    effectiveness of systems for delivering clinical preventive services, 814–815
    PACE+ project, 817
    preventive nutrition system, developing, 816
    screening and counseling, tools, 816–817, 818*t*
  behavioral counseling, conditions
    responses to patients' resistance, 812–813, 814*t*
  dietary habits in US, 796
    USDA/DHHS Food Guide Pyramid, 796, 797*f*
  eating for optimum health
    dietary guidelines, 798–804
  trends in diet and disease, 794–796
    dietary guidelines, importance, 796
    major causes of death in US, dietary influences, 795*t*
Preventive nutrition/food industry, 769–788
  emerging business
    improved global nutrition, examples, 771
    public/private sectors, role, 771

Preventive nutrition/food industry (*cont.*)
social business entrepreneurs, development, 770
food industry, history
development of, 773
early industry development, 773–774
food-related technologies, 773*t*
largest beverage companies worldwide, 2008, 772*t*
largest food companies worldwide, 2008, 772*t*
maturing industry, 775–776
industry/public health
chronic disease, *see* Chronic disease(s)
nutrition science, investing in, 787–788
undernutrition, 785–787
nutrition, health, development crisis
food price rise, impact on health, 770
life expectancy, decline in, 769
nutrition crisis, causes, 770
undernutrition, risks of diabetes/cardiovascular disease, 770
present/future directions, 769–789
public–private sector partnerships, 771–772
research in industry
background, 776
cocoa, research on, 778
evolution of infant formula, role of food industry, 776
food safety, 778–779
heart-healthy products, development, 777
micronutrient fortification, 779–782
product/industry approach to reduce obesity, 778
sports nutrition, 777–778
Preventive nutrition in developing countries
challenges
economic adjustment programs/rising food prices/ climate change, 763–764
goals, 757–765
health transition, 758–759
demographic/epidemiologic/health transitions, relationship, 758*f*
eras, causes of death, 759
nutritional patterns/complexity/health problems
higher income countries, 763
lowest income countries, 761
middle income countries, 761–763
nutrition transition, 759–761
"Probiotic bacteria," supplements with, 556
Programmed senescence, 545
Proinflammatory cytokine, *see* IL-6
PROM, *see* Premature rupture of the fetal membranes (PROM)
Prostate cancer, 8, 12, 38, 195–210, 220, 222, 223, 236, 559, 712
age-adjusted incidence and mortality rates of, 196*f*
diet and nutrition, 198
antioxidants, 199–203
cruciferous vegetables, 208–209
dairy products, calcium, and vitamin D, 207–208
fat intake and fatty acids, 203–206
soy/phytoestrogens, 206–207
global burden of
incidence, 196–197
influence of PSA screening on studies of prostate cancer prevention, 198
migrant studies, 197–198
mortality, 197
incidence rates among men, 197*f*
role of nutrition and diet in, 195–210
Protease inhibitors, 625
Protein, 11
adequate protein intake, importance, 461
calciuric effects of, 460
intake/age-related bone loss, relativity, 460
Protein-calorie malnutrition (PCM), 81, 550
Protein calories, 726
Protein-energy malnutrition (PEM), 614, 618
definition, 618
Proton pump, *see* Triphosphatase pump
Proton Pump Inhibitor (PPI)
gastroesophageal reflux disease, medical therapy for, 480
HP infection, medical therapy for, 480–486
NSAIDs, medical therapy by, 486
PSC, *see* Posterior subcapsular cataract (PSC)
Pteroylglutamic acid, *see* Folic acid
PTH, *see* Parathyroid hormone (PTH)
Puberty, 547
Public health benefits of preventive nutrition, 3–15
alcohol, 13
body weight, 12–13
recommendations, 14–15
sources of evidence, 5
specific dietary components
calcium, vitamin D, and dairy products, 11–12
dietary fat, 5–7
dietary fat and body fatness, 8–9
dietary fat and cancer, 7–8
protein, 11
salt and processed meats, 12
starches and complex carbohydrates, 10–11
vegetables and fruits, 9–10
vitamin supplements, 13–14
Public health interventions, categorization, 585–586, 586*t*, 593
food-based approaches/fortification
biofortification, 588
dietary/horticultural interventions, 587–588
fortification, 588–589
supplementation, 590–591
Public Health Nutrition, 568
Public–private sector partnerships, 771–772
Pure Food and Drug Act, 774

R
Rectal atresia/stenosis, 651
RELM, *see* Resistin-like molecules (RELM)
Remodeling transient phenomenon, 448
Replacement feeding, 622, 623
Required nutrient density (RND), 65
Resistin, 426
  RELM, 426
Resistin-like molecules (RELM), 426
Resting energy expenditure (REE), 617
Retina, 529, 533
Retinol, *see* Vitamin A
Riboflavin, 780
  cataract, 522
Rickets, 780
Ritonavir, 625

S
Schizophrenia, 706
  advancing paternal age, source of, 710
  de novo mutations, role in, 710
  epigenetic regulation of genes, cause, 710–711
  famine, effects on, 707
  folate deficiency, prenatal famine to, 713
  gene–environment interactions, cause, 711
  micronutrient deficiencies, effects, 714
  neurodevelopmental disorder, 706
  prenatal malnutrition, latent effects on, 709
  risk of schizophrenia *vs.* birth rate (1956–1964), 709*f*
  selenium deficiency, results, 711
Scurvy, 570
Selenite, 711
Selenium, 584, 711–712
  deficiency
    epigenetic regulation, effect on, 712
    gene interactions, effect on, 712
    methylation, effect on, 712
  epigenetic regulation, 712
  GPX, component of, 712
  Keshan disease, 712
  methylation of genes, 712
  neuronal damage, protection from, 711
Selenoenyzymes, 711, 712
Selenoprotein (SePP), 711
  oxidative stress, reduction, 712
Senescence, 451–453
Serum cholesterol, 5
SGA, *see* Small-for-gestational age (SGA)
Shigella flexneri, 553
Simulated Milk Adapted (SMA), 776
Skeleton, function of, 444
Small-for-gestational age (SGA), 674
Smoking and diet, 25
Socioeconomic level (SEL), 723
Somatostatin, 472, 473
"Sprinkles," 589
Squamous cell carcinoma, 146–152
  nutrition
    biochemical studies, 150
    chemoprevention studies, 151–152
    dietary studies, 148–150
  thermal irritation, 147–148
  tobacco and alcohol consumption, 146–147
Staphylococcus aureus, 553
Stearic acid, 6
Stomach, cancer of, 155
  anatomic subsites, 155
  epidemiologic studies of diet and risk, 160–163
  histologic types, 155
  risk factors, 155–164
    epidemiologic studies of diet and, 160–163
    fruits and vegetables, 159
    H.pylori, 155–157
    micronutrients, 159
    obesity, 165
    other diet, 164
    salted, pickled, and smoked foods, 157–159
    tobacco and alcohol, 157
Stomach anatomy, 472*f*
Streptokinase–streptodornase, 580
Suboptimal dietary folic acid, 9
Sugar, 726
Superinfections, 620–621
  Pneumocystis jirovecii, 621
  See also Tuberculosis (TB)

T
T-cell activation, 424
  See also Leptin
T-cell growth factors, *see* Interleukin-2 (IL-2)
T2D, *see* Type 2 diabetes (T2D)
Testosterone, 631
Tetanus toxoid, 549
Tetrahydrofolate (THF), 648
TG, *see* Triglycerides (TG)
The International Rice Research Institute (IRRI), 588
Thiamin, 780
  deficiency
    beriberi, 582
    Wernicke–Korsakoff's syndrome, 582
Thrombosis, *N*-3 fatty acids, 254
Thymulin, 578
Thymus, 577
  involution of, 546–547
    hypothesis, 547
    positive/negative selection, 547
T-lymphocytes
  DTH responses, 547
  in elder/younger people, 547
  functions, 547–549

T-lymphocytes (*cont.*)
LPR study, 548
subsets with aging, example, 548, 549
TNF-α, 428
'Tooth-like, pointed or jagged,' *see* Zinc
Total carotenoids, 517, 534
Trans fatty acids, 6–7, 273
chemical structure of, 274
desaturation and elongation of, 277
effects of exchanging 1% of energy from saturated or *trans* monounsaturated, 278
in foods, 274–275
potential conversions of linoleic acid into its positional and geometrical isomers, 274
risk of CHD, 6–7
simplified scheme of metabolism, 275
Trans fatty acids, cardiovascular effects of, 273–282
effects of *trans* monounsaturated fatty acids on serum lipids and lipoproteins
mechanism, 279
serum total, LDL, and HDL cholesterol, 277–278
effects on other risk factors for coronary heart disease
endothelial cell function, 280
hemostasis, 279–280
LDL oxidation, 279
low-grade systemic inflammation, 280
epidemiological studies, 280–281
metabolism
conversion, 276–277
digestion, absorption, and incorporation into blood lipids, 275–276
oxidation, 277
tissue levels, 276
Trichinella spiralis, 574
Trichophyton, 580
Triglycerides (TG), 420
Triphosphatase pump, 472
Tuberculosis (TB), 620
co-infection with HIV, effects, 620
Type 2 diabetes (T2D), 371–385, 420, 744
children with, due to overweight, 746*t*
prevention of, 400–401
See also Diabetes (type 2) and cardiometabolic disease

U
Ulcers, 477–478
Undernutrition
early childhood, 36–37
effects, 571
long-term food/nutrition crises, business action to, 785–787
Undernutrition/disease cycle, 569–570, 570*f*
synergism, nutritionally induced determinants, 569
Unilever, 771
Urine calcium, 455*f*, 460
USDA-ARS, *see* US Department of Agriculture-Agricultural Research Service (USDA-ARS)
USDA/DHHS Food Guide Pyramid, 796, 797*f*
US Department of Agriculture-Agricultural Research Service (USDA-ARS), 778
US Department of Agriculture (USDA), 796
US Department of Human Health Services (USDHHS), 796
USDHHS, *see* US Department of Human Health Services (USDHHS)
US Dietary Guidelines (2000), 26

V
Vagotomy, 477
Varicella zoster, 549
Vascular dementia, B vitamins in prevention of, 325–332
Visceral adipose tissue, 421
Visceral *vs.* subcutaneous WAT, 421–422
Visceral WAT, 421
Visfatin, 426–427
Vision loss
high-dose zinc supplement, reduced risk, 529
Vitamin(s), 93–94
absorption, 94–97
lipid-soluble vitamins, 96–97
water-soluble vitamins, 94–96
distribution, 97
lipid-soluble vitamins, 97
water-soluble vitamins, 97
drugs that alter, 93*t*–94*t*
excretion of, 102–103
metabolism, 98–102
lipid-soluble vitamins, 100–102
water-soluble vitamins, 98–100
water-soluble, 550
See also individual vitamins
Vitamin A, 13–14, 535–536, 574
capsules, 575
and cataract extraction, 521–522
and cortical cataract, 521
deficiency, 572
acquired immunity, 574–575
HIV-infected women, clinical trials, 574–575
and immune function, 573–575
immune function, effects on, 573–575
innate immunity, 573
macrophages, impaired ability of, 573
maternal mortality, 572
neutrophil function, impaired, 574
night blindness, 572
squamous metaplasia, 573
xerophthalmia, 572
fortification, 780
intake, reduced risk of measles, 572
lack during pregnancy, effects, 643

maternal mortality reduction, 34
and mixed cataract, 521
and nuclear cataract, 521
and PSC cataract, 521
public health implications, 575–576
HIV+ pregnant women, outcomes, 576
increased micronutrient intakes, paradigm, 576*f*
multimicronutrients with/without vitamin A, 576
vitamin A capsules, 575–576
Vitamin B
in prevention of cognitive decline and vascular dementia, 325–332
B vitamin blood level associations with cognition and dementia, 328
B vitamin deficiency and cognitive impairment, 327
B vitamins, homocysteine, and brain function, 326
dietary B vitamins and cognition, 327–328
folate, 329
fortification with folic acid and cerebrovascular effects, 329–330
intervention trials with B vitamins, 330
vitamin B_{12}, 328–329
Vitamin B_6
metabolism, 98
Vitamin B_{12}
absorption, 95–96
deficiency, 570
"twin" vitamin, 548
Vitamin C, 97, 462
antioxidants/cataract, epidemiological studies
vitamin C/cataract extraction, 513–514
vitamin C/cortical cataract, 512
vitamin C/mixed cataract, 513
vitamin C/nuclear cataract, 512, 513*f*
vitamin C/PSC cataract, 512
and cancer risk, 231
deficiency, 583
metabolism, 100
and preeclampsia, 36
Vitamin D, 11, 443, 458–460, 584
absorption *vs.* calcium intake, 458*f*
deficiency, role in osteoporosis syndrome, 459
fracture reduction, 459
metabolism, 100–102
25 (OH)D, 458
paracellular diffusion, 458
Vitamin E, 583
antioxidants/cataract, epidemiological studies
vitamin E/cataract extraction, 516
vitamin E/cortical cataract, 515
vitamin E/mixed cataract, 516
vitamin E/nuclear cataract, 514–515
vitamin E/PSC cataract, 515
and cancer risk, 231–232
myocardial infarction (MI), reduced risk, 13
and preeclampsia, 36
Vitamin K
diet, osteoporosis, and fracture prevention, 461–462
diet and childhood cancer, 140

W

Wasting
definition, 617
food intake, cause of, 617, 626
patterns, 617
See also Weight loss
WAT, *see* White adipose tissue (WAT); White adipose tissue (WAT)
Water-soluble vitamins
absorption, 94–96
distribution, 97
metabolism, 98–100
Weight loss, 615, 616–617
Wernicke–Korsakoff's syndrome, 582
See also Thiamin
Western diet, 743–754
"Westernization" of diets, 3
Wheat, 726
White adipose tissue (WAT), 419, 420
MCP-1, 421
Women's health initiative (WHI), 6, 25, 30, 185, 222, 223, 230, 232, 234, 236, 237, 337–365, 452
clinical applications, 353–364
summary of findings, 354–364
clinical measurement procedures, 342*t*–343*t*
clinical trials (CT) and observational study (OS), 340
baseline characteristics of study sample, 341
data collection and time points, 341
progress to date, 341–344
recruitment, 340–341
diet modification study
dietary intervention, 345–346
results, 347–348
role of low-fat diet in modifying other health risks, 353
study hypotheses, 344–345
study population, 345
trial and body weight, 349–353
trial duration, data collection, and time points, 346–347
publication of trial findings, 344*t*
scientific rationale for DM trial, 338–339
breast cancer, 339
cardiovascular disease, 339–340
colorectal cancer, 339
vitamin and mineral supplement use, 223
World Hunger Map, 614*f*

X
Xerophthalmia, 572

Z
Zeaxanthin, 517–519, 534
  and cataract extraction, 519
  and cortical cataract, 518
  and mixed cataract, 519
  and nuclear cataract, 517–518, 518*f*
  and PSC cataract, 518–519
Zero hunger, 762
ZES, *see* Zollinger–Ellison syndrome (ZES)
Zidovudine (AZT), 625
Zinc
  alcoholism/Crohn's disease, 577
  calamine lotion, use as, 576
  Charaka Samhita, medicinal uses in, 577
  deficiency, effects, 712
  deficiency and immunity, 577–579
    acquired immunity, 577–579
    innate immunity, 577
  DNA damage, prevention, 712, 713
  gene–environmental interactions
    acrodermatitis enteropathica, 713
  methylation, role in, 713
  preterm delivery/maternal nutrition, 692–693
    low plasma zinc, effects, 693
    PPROM, zinc-supplemented women, 692
    zinc-treated normal-weight/underweight women, trials, 692
  public health implications
    poor zinc intakes, factors, 579
    zinc supplementation, role, 579
Zinc deficiency and immunity, 577–579
  acquired immunity
    B lymphocytes, effect on, 578
    cytokines, effect on, 578
    thymic atropy, 578
    T lymphocytes activity, impaired, 578
  innate immunity
    acrodermatitis enteropathica, 577
    macrophage activity, inhibition, 577
    neutrophil function, impaired, 577
Zollinger–Ellison syndrome (ZES), 473

About the Editor

Dr. Richard J. Deckelbaum is the Robert R. Williams Professor of Nutrition, Professor of Pediatrics, and Professor of Epidemiology at Columbia University, where he serves as Director of the Institute of Human Nutrition. He also founded the Children's Cardiovascular Health Center at Columbia University Medical Center (CPMC). In 1996, Dr. Deckelbaum initiated the first M.D. Program in International Health and Medicine as a collaboration with colleagues at Israel's Ben-Gurion University of the Negev and Columbia University. Dr. Deckelbaum's primary research interests concern human plasma lipoproteins, the metabolism of intravenous lipid emulsions, and the cellular effects of dietary fats and free fatty acids on gene expression. His work has contributed to better understanding of mechanisms whereby human lipoproteins are structurally remodeled in the plasma compartment, factors modulating receptor–lipoprotein interactions, and nutrient–gene interactions.

Author of over 300 scientific articles, reviews, and chapters, Dr. Deckelbaum has served on numerous national task forces, as well as review boards for nutrition and clinical research. He has chaired a national conference on *Preventive Nutrition: Pediatrics to Geriatrics* and served as a member of the United States Department of Agriculture and Health and Human Services Advisory Committee for Dietary Guidelines for the year 2000. In addition, he chaired the International March of Dimes Task Force on Nutrition and Optimal Human Development. A recent honor includes his appointment to the Food and Nutrition Board of the Institute of Medicine, National Academies of Science. Currently, he continues in projects related to health and science as a bridge between different populations in the Middle East, Africa, and Asia. Dr. Deckelbaum is president of GHEC, the Global Health Education Consortium, whose mission is to improve the ability of the global health workforce to meet the needs of underserved and vulnerable populations through improved education and training. Additionally, he has led other national and international initiatives directed toward applying basic nutrition science toward improving health in individuals and in populations.

About the Series Editor

Dr. Adrianne Bendich is Clinical Director, Medical Affairs at GlaxoSmithKline (GSK) Consumer Healthcare where she is responsible for leading the innovation and medical programs in support of many well-known brands including TUMS and Os-Cal. Dr. Bendich had primary responsibility for GSK's support for the Women's Health Initiative (WHI) intervention study. Prior to joining GSK, Dr. Bendich was at Roche Vitamins Inc. and was involved with the groundbreaking clinical studies showing that folic acid-containing multivitamins significantly reduced major classes of birth defects. Dr. Bendich has co-authored over 100 major clinical research studies in the area of preventive nutrition. Dr. Bendich is recognized as a leading authority on antioxidants, nutrition and immunity, and pregnancy outcomes, vitamin safety, and the cost-effectiveness of vitamin/mineral supplementation.

Dr. Bendich is the editor of nine books including *Preventive Nutrition: The Comprehensive Guide For Health Professionals* co-edited with Dr. Richard Deckelbaum, and is series editor of *Nutrition and Health* for Humana Press with 29 published volumes including *Probiotics in Pediatric Medicine* edited by Dr. Sonia Michail and Dr. Philip Sherman; *Handbook of Nutrition and Pregnancy* edited by Dr. Carol Lammi-Keefe, Dr. Sarah Couch, and Dr. Elliot Philipson; *Nutrition and Rheumatic Disease* edited by Dr. Laura Coleman; *Nutrition and Kidney Disease* edited by Dr. Laura Byham-Grey, Dr. Jerrilynn Burrowes and Dr. Glenn Chertow; *Nutrition and Health in Developing Countries* edited by Dr. Richard Semba and Dr. Martin Bloem; *Calcium in Human Health* edited by Dr. Robert Heaney and Dr. Connie Weaver and *Nutrition and Bone Health* edited by Dr. Michael Holick and Dr. Bess Dawson-Hughes.

Dr. Bendich served as associate editor for *Nutrition*, the International Journal, served on the editorial board of the *Journal of Women's Health and Gender-based Medicine*, and was a member of the Board of Directors of the American College of Nutrition.

Dr. Bendich was the recipient of the Roche Research Award, is a *Tribute to Women and Industry* awardee and was a recipient of the Burroughs Wellcome Visiting Professorship in Basic Medical Sciences, 2000–2001. In 2008, Dr. Bendich was given the Council for Responsible Nutrition (CRN) Apple Award in recognition of her many contributions to the scientific understanding of dietary supplements. Dr. Bendich holds academic appointments as adjunct professor in the Department of Preventive Medicine and Community Health at UMDNJ and has an adjunct appointment at the Institute of Nutrition, Columbia University P&S, and is an adjunct research professor, Rutgers University, Newark Campus. She is listed in *Who's Who in American Women*.